Detlev Möller

Chemistry of the Climate System

Also of Interest

Chemische Thermodynamik
Grundlagen, Übungen, Lösungen
Walter Schreiter, 2013
ISBN: 978-3-11-033106-6, e-ISBN: 978-3-11-033107-3

Chemical Energy Storage
Robert Schlögl (Ed.), 2012
ISBN: 978-3-11-026407-4, e-ISBN: 978-3-11-026632-0

Sustainable Process Engineering
Prospects and Opportunities
Andrzej Benedykt Koltuniewicz, 2014
ISBN: 978-3-11-030875-4, e-ISBN: 978-3-11-030876-1

Green
The International Journal of Sustainable Energy Conversion
and Storage
Editor-in-Chief: Schlögl, Robert
ISSN: 1869-8778

Detlev Möller

Chemistry
of the Climate System

2nd fully revised and extended edition

DE GRUYTER

Univ.-Prof. Dr. rer.nat.habil. Detlev Möller
Brandenburgische Technische Universität
Arbeitsgruppe für Luftchemie und Luftreinhaltung
Volmerstr. 13
12489 Berlin
moe@btu-lc.fta-berlin.de

This book has 186 figures and 160 tables.

ISBN 978-3-11-055399-4
e-ISBN 978-3-11-0331949-3

Library of Congress Cataloging-in-Publication Data
A CIP catalog record for this book has been applied for at the Library of Congress.

Bibliographic information published by the Deutsche Nationalbibliothek
The Deutsche Nationalbibliothek lists this publication in the Deutsche Nationalbibliografie;
detailed bibliographic data are available on the Internet at http://dnb.dnb.de.

To my parents to whom I owe everything
and to all people who love nature, life and science.

Nothing in nature makes sense except in the light of evolution

Adapted from *Theodosius Dobzhansky* (1900–1975; born in Ukraine,
emigrated to the United States in 1927). In: Nothing in biology
makes sense except in the light of evolution.
The American Biology Teacher 35 (1973) 125–129.

Nicht die Wahrheit, in deren Besitz irgend ein Mensch ist, oder zu seyn
vermeynet, sondern die aufrichtige Mühe, die er angewandt hat, hinter
die Wahrheit zu kommen, macht den Wert des Menschen. Denn nicht durch
den Besitz, sondern durch die Nachforschung der Wahrheit erweitern sich seine
Kräfte, worin allein seine immer wachsende Vollkommenheit besteht.
Der Besitz macht ruhig, träge, stolz –

It is not the truth which a man possesses, or thinks he possesses, but the sincere
endeavour which he has used to come to truth, makes the worth of a man.
For not through the possession of truth, but through the search for it,
are those powers expanded in which alone his ever-growing perfection consists.
Possession makes restful, indolent, proud –

Gotthold Ephraim Lessing (Eine Duplik, 1778)
[Lessing 1883, English version from Rolleston 1889, pp. 200/201]

Preface

Half a century ago, Christian Junge, the founding 'father' of atmospheric chemistry summarized the existing knowledge in his field of research. In 1960 Junge counted only 75 research articles dealing with atmospheric chemistry and radioactivity. Nowadays the publication rate and number of researchers in atmospheric chemistry are orders of magnitude greater. Many major advances have been made since then. For example in the early 1970s, the role of OH radicals in oxidizing gases, leading to their removal from the atmosphere, was discovered. The necessary ingredients to produce OH are ozone, water vapor, UV-B solar radiation. The catalytic role of NO in producing ozone was also recognized at the beginning of the 1970's. Until then it was generally believed that tropospheric ozone was produced in the stratosphere and transported downwards into the troposphere. The feedstocks for the creation of ozone are CO, CH_4 and many biogenic gases. Both natural and anthropogenic processes are responsible for their emissions. Although the main photochemical chain reactions are reasonably well known, their quantification needs much further research. They all are parts of the biogeochemical cycles of carbon, nitrogen and sulfur. They can also play a role in climate, as does particulate matter, which, contrary to the greenhouse gases (CO_2, CH_4), tends to cool the earth and atmosphere.

In his book, Detlev Möller gives a thorough overview of the main chemical processes that occur in the atmosphere, only a few of which have been mentioned above. The novel title of this book "chemistry of the climate system" should direct the attention of the reader to the fact that understanding atmospheric chemistry is incomplete without considering interfacing neighboring reservoirs such as the hydrosphere, the lithosphere and the biosphere.

An overview of the topics treated, is provided in the Introduction. It emphasizes that drawing strong borderlines between disciplines makes no sense and this is also valid for the various systems because they overlap and the most important processes can happen at their interfaces. Therefore, chemistry of the climate system combines atmospheric with water, soil and biological chemistry. Another general approach of this book lies in the incorporation of historical facts: despite the orders of magnitude more publications each year nowadays than in past, we should not forget that careful observations were made and serious conclusions drawn by many of our scientific ancestors.

The text has its roots in a book written in German, entitled "Luft" and published by DeGruyter in 2003. Although the text has been entirely rewritten and many chapters have been replaced, the main emphasis on regarding the atmosphere as a *multiphase system* is essentially unchanged. Moreover, by adding *interfacial chemistry*, the system is enlarged to a *multireservoir system*, encompassing the climate system. At the end, however, is the central focus is chemistry. The author avoids using the term *environmental* chemistry, emphasizing that substances having specific physical and chemical properties can modify (chemical) systems in various directions depending on the mixture and initial conditions. Specialists in physics, chemistry and biology as well its many subdisciplines need to understand what their disciplines can bring to the subject of climate system chemistry. This book is about fundamental aspects of (chemical) climate change. But this book does not attempt to review all of the research on this topic. At many sites in this book, "Further Reading" is provided referring the reader to more specialized textbooks.

The book comprises three large fundamental chapters: Chemical evolution, physico-chemical fundamentals and substances and chemical reactions. Following the Introduction, the chapter "Chemical Evolution" gives a brief history from the Big Bang to the Anthropocene, the human influenced earth system. Detlev Möller describes the historical dimension in connecting the past with future developments. Changing chemical air composition is based on three pillars: land use change, burning of fossil fuels and agricultural fertilizing, all caused by the rising global population. The resulting air chemical "episodes" (acid rain, ozone, particulate matter) are fairly well understood and end-of-pipe technologies to control air pollution have been introduced. The remaining future challenge, limiting global warming through reduced emission of CO_2, however can be achieved only by moving the anthroposphere into a solar era, as suggested in chapter 2.8.4. Hence, Möller defines global sustainable chemistry as a coupling biogeochemical cycles with anthropogenic matter cycles, almost importantly CO_2 cycling.

Chapter 4 briefly overviews the physical and chemical principles of transporting and transforming substances in natural reservoirs, again with an emphasis on multiphase and interfacial processes. The texts on chemical reactions, multiphase processes, atmospheric removal and characteristic timescales are well written and easy to read even for non-chemists. The third main chapter treats "Substances and Chemical Reactions in the Climate System" according to the elements and their compounds depending on the conditions of reactions and whether the substance exists in the gaseous or condensed phase. The structure is adapted from "classical", substance oriented textbooks in chemistry: hydrogen, oxygen, nitrogen, sulfur, phosphorous, carbon and halogens. Many excellent figures summarize chemical pathways under different natural conditions and make it easy to understand complex chemical processes.

Different appendixes provide useful information on abbreviations, quantities, units and the earth geological time-scale. The respect brought to the history of science is fulfilled by a nice biography including some not so well-known scientists.

All together, this well written and useful book fills a gap in the attempt to provide books written from the perspective of a particular discipline (chemistry) on an interdisciplinary subject (the climate system) for readers and scientists from different disciplines to learn (or teach) some chemistry outside of the laboratory retort or industrial vessel.

Paul Crutzen
Nobel prize in Chemistry 1995

Authors preface to the 2nd edition

Now, four years after publishing the first edition of "Chemistry of the Climate System", we recognize continued growth in scientific literature and particularly new insights in the field of atmospheric aerosol (fine particulate matter and especially the role of biological material) but also in HONO and OH chemistry; however, it is not my aim to consider them all in the 2nd edition, giving an actual review. My intention is rather, to provide the reader with a monographic textbook for a deeper understanding of physico-chemical processes in a very complex system to be comprised of different states of matter (gas, liquid and solid) within different reservoirs, namely the atmosphere, hydrosphere, lithosphere, and biosphere (the *climate system*). Scientific progress is large in fundamental sciences – laboratory and theoretical studies. But studying complex environmental processes by doing "outdoor" studies (field experiments) – and even extending one reservoir and/or phase – became extremely complicated and has only been carried out within the last 1–2 decades in large campaigns including many institutes and scientists. Our knowledge also increased and feedback for detailed laboratory work was the result. Better understanding of the climate system realized by deeper consideration and better coupling of the subsystems, most of all atmosphere with ocean (as an example to explain the discrepancy between the further rise of atmospheric CO_2 and stagnant air temperature). But also the biosphere has been understood to be much more complex in its feedback to atmospheric changes (the still unanswered question of where all the CO_2 remains).

Unfortunately, controlling of our climate system is much more fare comparing to its understanding four years ago. Economy largely disregards results from climate research. Visible human made catastrophes such as "Fukushima" result not in an energy change forward to solar technologies but back to coal combustion, boosting the carbon dioxide problem. Climate change is likely not a catastrophe in the traditional sense but is a slow process with winners and losers (humans, animals, and plants). I hope that the new edition will find new readers who want to learn more about our climate system. Only knowledge is the key to sustainable development and to realize that the benefits of the winners will be small compared to the fee the losers have to pay.

Needless to say that (I hope) all errors have been omitted, information (especially monitoring and emission data) is updated, many Figures are improved, some paragraphs are rewritten for better understanding and new key knowledge has been considered. Moreover, two new sections under the aspect of "chemical climatology" have been included: precipitation and cloud chemistry monitoring. The Chapter on atmospheric trace species concerning hydrochloric acid, nitrous acid and nitric acid with its salty particulate matter has been enlarged.

Last but not least, I would like to express my deep thanks to my co-workers who accompanied me over many years, decades even, in atmospheric chemistry and without whom this book would not have been possible: Dr. Karin Acker (1988–

present), Dr. Wolfgang Wieprecht (1982—present), Dr. Günther Mauersberger (1985—2006), Dr. Renate Auel (1987—2008), Gisela Hager (1987—1997), Jürgen Hofmeister (1988—present), and Dieter Kalass (1992—present). Many others, not named here, were "only" for a few years a part of my group. During the 1980s and 1990s, we had a great time in atmospheric chemistry with many highlights.

Detlev Möller
Berlin, February 2014

Prologue

10 years ago the publishing house de Gruyter asked me to write a book titled "Luft" (Air), which was published in 2003. Paul Crutzen said that it was written in the wrong language. It was written in German, my mother tongue. I percieved it as a compliment that they wanted me to reach more international readers. My first thank goes to de Gruyter, who offered me the opportunity to write *another* book on *air* in English about 3 years ago. The German book "Air" belongs to a planned series on water, air and soil (last one is not written yet) which means that there were certain restrictions and wishes by the editor. But for this new book I had absolute freedom to choose my own content; therefore I would like to express a further thank to de Gruyter. The only wish from the editor was that this book should contribute to the discussion on *climate change*.

This new book is not a (revised) translation of "Air" − I have only used a few fundamental issues such as physico-chemistry and basic air chemistry, which now is permanent knowledge of any textbook. I am a chemist; my original background is physical chemistry but since 35 years I deal with air chemistry (and air pollution studies). Chemistry is the science of matter and matter distributes and cycles through nature.

The climate system is a part and the atmosphere another part of this. The earth's climate system provides a habitable zone. I agree with scientists who argue that human beings as a part of nature have altered the "natural system" to such an extent that we are now unable to reverse the present system back to a preindustrial or even prehuman state. But the key question nowadays is how to maintain the function of our climate system so that sustainable survival of humans is possible. We can state that humans have become a global force in the chemical evolution with respects to climate change by interrupting naturally evolved biogeochemical cycles. But humans also have all the facilities to turn the "chemical revolution" into a sustainable chemical evolution.

Understanding on how the climate system works is provided by the *natural* sciences (physics, chemistry and biology). This book focuses on the chemistry of the climate system. But differentiating precisely between physics and chemistry makes absolutely no sense when trying to understand the climate system. However, without understanding biological laws − and we can probably include social laws here as well considering that the biosphere has long ago been transformed into the noosphere − we will neither understand climate change nor find solutions for *climate control*. Maybe this book will provoke people by bridging a state-of-the-art textbook knowledge with ideas or opinions beyond "pure" chemistry. Without a paradigm change within the next 2−3 decades (for example carbon dioxide cycling) we will have little chance of climate control.

My knowledge is limited; therefore specialists will find gaps and missing things in my book. A few weeks ago I visited a lecture by Professor *Norimichi Takenaka* (Japan), who studied the decomposition of ammonium nitrite in drying dew droplets with the formation of N_2 — a disproportioning and fascinating pathway for atmospheric NH_3 and NO_x. Unfortunately, I did not know about this reaction, even though is was already known in the 19th century, and it is an exact key example for interfacial chemistry, which is one focus of my book and my special interest in the last few years after studying multiphase chemistry for many years. I believe that interfaces (or heterogeneous systems in other terms) in certain natural places provide special conditions for chemical reactions and therefore result in a turnover of matter. Our knowledge about what happens in atmospheric multiphase chemistry, despite much progress within the last two decades, is still limited (I hope that some is summarized in this book). But interfacial chemistry may be more important (and almost more interesting) for controlling the climate system at the interface along the atmosphere and earth surface, including natural waters, soils, plants, microorganism and others.

Special thanks go to Professor Volker A. Mohnen, to whom I owe important scientific stimulation over the last 20 years, including the impulse to write this book.

Finally I would like to thank my coworkers who accepted that I was not in the institute a lot throughout the last two years (and who kept the business going); I wrote this book in my office at my home in Berlin (day and night, weekday and weekend: my special thank goes to my wife Ursula and my family), surrounded by hundreds of books on chemistry (but also on history, physics and biology) of the *climate system* from the last 200 years (and a few older ones). This book is the first book to be titled "Chemistry of the Climate System" to my knowledge and I hope — no, I am sure that it is not going to be the last one. It stands between "special" and "generic"; it is partly like a textbook and a monograph (and it may even appear to be an ecopamphlet). Hence, this book is recommended to anybody who likes nature and chemistry or anyone who wants to learn more about climate system chemistry.

Detlev Möller
Berlin and Cottbus, November 2010

List of principal symbols

a	albedo, radius of particles	F	*Faraday* constant
a	acceleration	f	force
a	surface of particles	f	free energy (*Helmholtz* energy)
ads	adsorption (index)	G	molar free enthalpy
ap	apparent (index)		(*Gibbs* energy)
aq	aqueous (index)	g	gravity constant
Acy	acidity	g	gaseous (index)
α	dissociation degree	g	free enthalpy (*Gibbs* energy)
α	*Bunsen* absorption coefficient	Γ	transport coefficient
α	mass accommodation coefficient	Γ_s	surface excess
		γ	surface tension
β	*Ostwald*'s solubility	γ	uptake coefficient
C_p	molar heat capacity at constant pressure	H	*Henry* coefficient
		H	molar enthalpy
C_V	molar heat capacity at constant volume	H_0	*Hammet* function
		h	*Planck*'s constant
c	concentration	h	(reference) height
c	velocity of light	*het*	heterogeneous (index)
c_N	number concentration	I	electric current
chem	chemical conversion (index)	i	specific component or particle (index)
coll	collision (index)		
D	diffusion coefficient	J	emittance
d	diameter	j	photolysis rate
d	diffusion (index), see also *diff*	K	equilibrium constant
des	desorption (index)	K_f	cryoscopic constant
diff	diffusion (index)	K_n	*Knudsen* number
dry	dry deposition (index)	K_z	turbulent vertical diffusion coefficient
E	energy		
E	radiant flux, irradiance	$\mathbf{k}$	*Boltzmann* constant
E	electrical potential	k	reaction rate constant
E_A	activation energy	k_g	mass transfer coefficient
e	molecular electric charge	κ	coefficient for absorption (a) or scattering (s)
ε	emission of light		
ε	general physical quantity	κ	*von Kármán* constant
ε	fraction (0 ... 1) of scavenged (washout) aerosol particles in air	L	radiance
		l	mean-free path
		LWC	liquid water content
eff	effective (index)	λ	wave length
F	flux	λ	scavenging coefficient
F	molar free energy (*Helmholtz* energy)	M	molar mass

M	third body	*rxn*	reactive or reaction (index)
m	mass	*RH*	relative humidity
m	molality	ϱ	density
m_m	mass of molecule	ϱ_m	mass density
μ	chemical potential	*S*	salinity
N	number (of objects or subjects)	*S*	molar entropy
		S	actinic flux
N_A	*Avogadro* constant	*S*	total particle surface per volume of air
N(r)	cumulative number size distribution	S_0	solar constant
n	mole number	*Sc*	*Schmidt* number
n_0	*Loschmidt* constant	*sol*	solution or dissolved (index)
n(r)	differential number size distribution	$\mathfrak{S}$	saturation ratio
η	number density	σ	*Stefan-Boltzmann* constant
η	dynamic viscosity	σ	absorption cross section
0	zero – reference concerns time or distance (index)	*T*	temperature
		t	time
$\ominus$	index for standard conditions	τ	residence time
ν	frequency	τ	shear stress
ν	kinematic viscosity	τ_c	characteristic time
Ω	steariant	τ_t	turnover time
P	radiant power	θ	solar zenith angle
p	pressure	θ	surface coverage degree (0 … 1)
p	area	*U*	molar inner energy
p	particulate (index)	*u*	wind speed
q	square (surface)	u_*	friction velocity
Φ	quantum yield	*V*	volume
φ	azimuth angle	V_m	molar volume
φ	fluidity	v	velocity
R	rate	v_d	dry deposition velocity
R	precipitation amount	*W*	(molar) work
R	gas constant	*wet*	wet deposition (index)
r	radius, distance	*Q*	(molar) heat
r	correlation coefficient	*x*	mixing ratio
r	resistance	*x*	displacement (horizontal)
r_a	aerodynamic resistance	*y*	displacement (horizontal)
r_b	quasi-laminar resistance	*z*	displacement (vertical)
r_c	surface resistance	*z*	collision number
rem	removal (index)	*z*	number of elementary charges
rev	reversible (index)	z_0	roughness length of the surface

Contents

1 Introduction

Out of the recognized complexity of nature arose the three basic sciences; physics, chemistry and biology. Further progress in the understanding of natural processes created numerous sub-disciplines and cross-disciplines, termed with a variety of prefixes and combinations, often creating misunderstandings unless careful definitions are used. In order to overcome disciplinary borders, a new super-science was established, *earth system science*, to study the earth as a system, with an emphasis on observing, understanding and predicting global environmental changes involving interactions between land, atmosphere, water, ice, biosphere, societies, technologies and economies. We will define the climate system to be a part of the earth system, with emphasis on the atmosphere but involving interactions between land, atmosphere, water, ice, biosphere, societies etc. Hence, "Chemistry of the Climate System" is neither simply air nor environmental chemistry. Humans − by decoupling their life cycle from natural conditions − have altered "natural" biogeochemical cycles. The Russian geochemist *Vladimir Ivanovich Vernadsky* understood by *noosphere* (called the *anthroposphere* by *Paul Crutzen*) a new dimension of the biosphere, developing under the evolutionary influence of humans on natural processes (Vernadsky 1926). This Introduction has two aims; first, to note the importance of definitions (further definitions will be given at many other parts of this book), and second, to demonstrate respect for our scientific ancestors (further historical remarks are given at many parts of this book).

Antoine Lavoisier, who revolutionized the science of chemistry in the eighteenth century and replaced the mythical "phlogiston" with the term (and concept) of oxygen, clearly understood the importance of accurate definitions. In his words: "We cannot improve the language of any science without at the same time improving the science itself; nor can we, on the other hand, improve a science without improving the language or nomenclature" (Lavoisier 1798). *Imre Lakatos* (1981) wrote: "Philosophy of science without history is empty; history of science without philosophy is blind".

1.1 Air and atmosphere – a multiphase and multi-component system

The typical dictionary definition of atmosphere is "the mixture of gases surrounding the earth and other planets" or "the whole mass of an aeriform fluid surrounding the earth". The terms air and atmosphere are widely used as synonyms. The word "air" derives from Greek ἀήρ and Latin aer or ær. The term "atmosphere", however, originated from the Greek ατμός (= vapor) and σφαιρα (= sphere), and was not regularly used before the beginning of the nineteenth century. The Dutch astronomer and mathematician *Willebrord van Roijen Snell* translated the term "*damphooghde*" into Latin "atmosphæra" in 1608. *Otto von Guericke*, who invented the air pump and worked on his famous experiments concerning the physics of the air in the 1650s, used the term "ærea sphæra" (Guericke 1672). In addition, in the nineteenth century the term "Air Ocean" was also used, in analogy to the sea.

It makes even more sense to define the atmosphere as being the reservoir (space) surrounding our (and any) planet, and air to be the mixture of substances filling the atmospheric space. With this in mind, the term *air chemistry* is more adequate than *atmospheric chemistry*. From a chemical point of view it is possible to say that *air* is the substrate with which the *atmosphere* is filled. This is in analogy to the *hydrosphere* where *water* is the substance (Chapter 2.5). Furthermore, *air* is an atmospheric suspension containing different gaseous, liquid (water droplets) and solid (dust particles) substances and therefore it provides a multiphase and multi-component chemical system. Solar radiation is the sole primary driving force in creating gradients in pressure, temperature and concentration which result in transport, phase transfer and chemical processes (Fig. 1.1).

The physical and chemical status of the atmosphere is called *climate* (see Chapter 3.2 for details). Considering that the incoming solar radiation shows no trend over several hundred years (despite periodic variations), and accepting that natural biogenic and geogenic processes vary but also do not show trends on these time scales, it is only mankind's influence on land use and emissions into the atmosphere

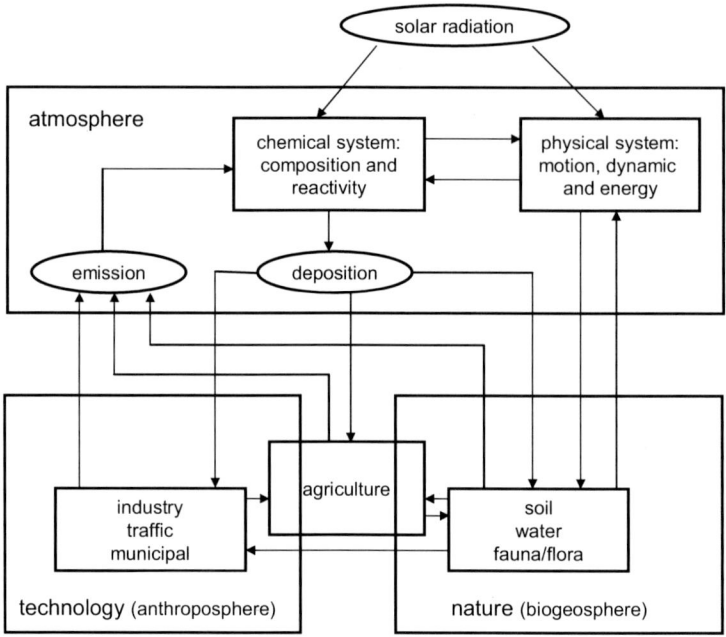

Fig. 1.1 Scheme of the physico-chemical interactions between the atmosphere, biosphere and anthroposphere (the Climate System).

that changes air chemical composition, and thus the *climate*. Human activities have an influence on natural processes (biological, such as plant growth and diversity, and physical, such as radiation budget), resulting in a cascade of consequent physical and chemical developments (feedback).

Without discussing in any detail the biological and physical processes within the climate system in this book, the chemical composition of the atmosphere and its variation in time and space, as well as its trends, is essential for an understanding of climate change. Atmospheric substances with their physical and chemical properties will have many effects in the climate system; we list the most important among them here together with impacts (there are many more impacts, parallel and synergistic effects):

- Formation of cloud condensation nuclei (CCN) and subsequent cloud droplets: hydrological cycle,
- being "greenhouse" gases (GHG): radiative interaction (warming the atmosphere),
- being ozone-depleting substances (ODS): radiative interaction (increasing UV radiation penetration into the lower troposphere),
- formation of atmospheric aerosol: radiative interaction (cooling the atmosphere),
- oxidation capacity: lifetime of pollutants,
- acidity: chemical weathering,
- toxicity: poisoning the environment (affecting life functions),
- nutrition: bioavailability of compounds essential for life.

We see that the climate system has physical and chemical components, interacting and (at least partly) determining each other. Physical quantities in the climate system show strong influences on chemical processes:

- temperature and pressure: reactions rate and (chemical and phase) equilibria,
- radiation (wavelength and intensity): photochemistry,
- motion: fluxes of matter (bringing substance together for chemical reactions).

Table 1.1 shows the present composition of our air. It shows that the concentration range from the main constituents to the trace species covers more than 10 orders of magnitude. Each component in air has a "function" in the climate system and in biogeochemical cycling (see Chapter 2).

Changes in the chemical composition of air caused by humans are termed *air pollution*. The terms *air pollution* and *pollutant* need some comments. To start with, the term *pollutant* should be used only for man-made (anthropogenic) emitted substances, despite the fact that most of them are also of natural origin. Air pollution represents a deviation from a natural chemical composition of air (providing a reference level) at a given site and period (note that climate change and variation is an ongoing natural process). Depending on the residence time of the pollutant, we can characterize the scale of pollution from local via regional to global. Air pollution nowadays is a global phenomenon because long-lived pollutants can be found to be increasing at any site of the globe. *Remote* air just means that the site is located far away from the sources of emissions and, consequently, this air has lower concentrations of short-lived (reactive) substances compared with sites close

Table 1.1 Composition of the dry remote atmosphere (global mean concentrations); note that the concentration of water vapor in air is highly variable from less than 0.5 % in saturated polar continental air to more than 3 % in saturated (i. e. 100 % humidity) tropical air (see also Fig. 2.37).

substance	formula	mixing ratio (in ppm)	comment
nitrogen	N_2	780 830[a]	constant
oxygen	O_2	209 451[a]	constant
argon	Ar	9 339[a]	constant
carbon dioxide	CO_2	380	increasing
neon	Ne	18.18	constant
helium	He	5.24	constant
methane	CH_4	1.73	increasing
krypton	Kr	1.14	constant
hydrogen	H_2	0.5	constant
dinitrogen monoxid	N_2O	0.32	increasing
carbon monoxid	CO	0.120	increasing
xenon	Xe	0.087	constant
ozone	O_3	0.03	variable
particulate matter	–	0.01[d]	variable
carbonyl sulfide	COS	0.00066	increasing
nitric acid[b]	HNO_3	≤ 0.001	variable
radicals[c]	–	< 0.00001	highly variable
hydroxyl radical	OH	0.0000003	highly variable

[a] related to the "clean" atmosphere $O_2 + N_2 + Ar + CO_2$ (= 100 %); Tohjima et al. (2005) note 207 392 ppm for O_2 against 380 ppm CO_2.
[b] and many other trace substances (NH_3, NO_x, HCl, NMVOC, H_2O_2, DMS, CFC's et. al.)
[c] e. g. HO_2, NO_3, Cl
[d] corresponds to 10−20 µg m^{-3}

to sources of pollutants. Although *polluted* air is human-influenced air, *clean* air is not synonymous with *natural* air. The natural atmosphere no longer exists; it was the chemical composition of air without man-made influences. However, this definition is also not exact because humans are part of nature. In nature situations may occur, such as volcanic eruptions, sand storms and biomass burning, where the air is being "polluted" (rendered unwholesome by contaminants) or in other words, concentrations of substances of *natural* origin are increased. Therefore, the *reference state* of natural air is a climatological figure where a mean value with its variation must be considered. The term *clean air* is also used politically in air pollution control as a target, i. e., to make our air cleaner (or less polluted) in the sense of pollutant abatement. A clean atmosphere is a political target, it represents an air chemical composition (defined in time and scale) which should permit sustainable development. The largest difficulty, however, lies in the definition of what *sustainable* means. This term comprises the whole range of categories from simple scientific questions (for example, impact threshold) to political decisions (global eco-management) and also to philosophical questions (for example, what human life needs).

Therefore, air pollution in terms of the changing chemical composition of the atmosphere must be identified through a *problem*, not simply by measured concentrations. The problem lies between "dangerous" and "acceptable" climate impact, a definition that is beyond the direct role of the scientific community despite the fact that scientists have many ideas about it (Schneider 2006).

Doubtless the air of settlements and towns was extremely polluted in the past. Heavy metals have been found in Greenland ice cores dating back to the Roman Empire; thus demonstrating that metallurgical operations of immense volume took place in that era. The modern quality of air pollution, since the Industrial Revolution of the nineteenth century, is best characterized by a continuous worldwide increase in emissions as the era of fossil fuel combustion began (see Chapter 2.8). In the last 150 years, serious air pollution problems (see Chapter 1.3.4) have been described, analyzed and solved (to an extent, by end-of-pipe technologies), such as soot, dust and smoke plagues, winter and summer smog and acid rain. The chemicals (or emitted compounds) behind these phenomena are soot, sulfur dioxide (SO_2), nitrogen oxides (NO_x) and volatile organic compounds (VOC) – all short-lived species, but because of global use of fossil fuels, the pollution problem is distributed globally. Some air pollution problems connected with long-lived compounds – now in terms of persistent compounds, such as agricultural chemicals (food chain accumulation) and halogenated organic compounds (ozone layer destruction) – have been solved by legislation banning their use. Unfortunately some long-lived emitted compounds, the so-called "greenhouse" gases, have increased significantly in global mean concentration (Table 1.2).

Ozone is not among long-lived species but globally it is a secondary product from methane oxidation (see Chapter 5.3.2.2). CH_4 and N_2O (mainly byproducts of agriculture) as well as CO_2 (a byproduct of fossil fuel combustion) are coupled with the two columns of human existence; food and energy. There is no doubt that the CO_2 problem can be solved only through sustainable technology change (Chapter 2.8.4). It should be noted that these three pollutants (CO_2, CH_4 and N_2O), even with much higher atmospheric concentrations, are harmless for life (they are in fact key substances in biogeochemical cycles, see Chapter 2.4.3), but there is also no doubt that these three substances contribute to about 90% of global warming. Beside, N_2O acts as an ozone-depleting substance in the stratosphere and CH_4 contributes to O_3 formation in the troposphere and is probably responsible

Table 1.2 Increase of some climate relevant gases in air (in ppb).

substance	1850	present (2012)	increase (in ppb)	increase (in %)
carbon dioxide CO_2	285 000	394 000[a]	107 000	38
methane CH_4	700	1 816[b]	1 116	259
dinitrogen monoxide N_2O	270	324[c]	54	20
ozone O_3	10	33[d]	23	330

[a] Mauna Loa station
[b] mean of Mace Head and Cape Grimm stations value
[c] global mean
[d] European surface near

for up to 80% of the global O_3 increase. Future air pollution control is synonymous with climate control and must be focused on CO_2, CH_4 and N_2O.

Let us now return to the atmosphere as a *multiphase system*. While gases and particles (from molecules via molecular clusters to nano- and micro-particles) are always present in the air, although with changing concentrations, condensed water (hydrometeors) is occasionally present in air, depending on the presence of so-called cloud condensation nuclei (CCN) and water vapor supersaturation at the site of fog and cloud formation. With the formation, transportation and evaporation of clouds, huge amounts of atmospheric energy are transferred. This results in changing radiation transfer and thus "makes" the weather, and, on a long-term scale, the climate. Furthermore, clouds provide an effective "chemical reactor" and transport medium and cause redistribution of trace species after evaporation. When precipitating, clouds remove trace substances from the air (we term it wet deposition) to the earth's surface. As a consequence, beside the continuous process of dry deposition, clouds may occasionally lead to large inputs of trace substances into ecosystems. The amount of condensed water in clouds and fog is very small, with around 1 g per m^{-3} air or, in the dimensionless term, liquid water content $1 \cdot 10^{-6}$ (identical with 1 ppb). Thus 99.99% or more of total atmospheric water remains in the gaseous phase of an air parcel. Hydrometeors may be solid (ice crystals in different shapes and forms) or liquid (droplets ranging from a few μm up to some tens of μm). We distinguish the phenomenon of hydrometeors into clouds, fog and precipitation (rain, drizzle, snow, hail etc.). This atmospheric water is always a chemical aqueous solution where the concentration of dissolved trace matter (related to the bulk quantity of water) is up to several orders of magnitude higher than in the gaseous state of air. This analytical fact is the simple explanation why collection and chemical analysis of hydrometeors began much earlier than gas-phase measurements.

Due to permanent motion, namely advection and turbulent diffusion, having stochastic characteristics on different time and spatial scales, it is extremely complicated to model chemistry and transport (so-called chemistry-transport models, CTM, which are also a basis for climate modeling) in space and time. At the earth-air interface, exchange of matter occurs, emission as well as deposition.

1.2 Chemistry and environmental research

In understanding the meaning of chemistry of the climate system (the *chemical climate system*), we have to discuss these terms further. It seems likely that in the past, as in the present, difficulties of language gave rise to many of the misconceptions upon which alchemical practice was based and later environmental studies were carried out. In early chemistry, it was the practice to give names to substances before their precise chemical nature had been ascertained. In modern environmental research, it seems similarly common to define disciplines before the complexity of nature has been approximately understood. In this way, many of the so-called environmental disciplines tended not to see the wood for the trees, but to become lost within the details (even though it is important to study them) without seeing how the system functions. A "classical" example may be the enlargement of our view from ecosystem to earth system research. The existing sciences to study and to understand nature were physics, chemistry and biology − no more. The system of sciences is pyramidal. To understand just the physics of the systems, chemistry and biology is not needed, but to describe the chemistry of a system we also need physics (physicochemistry) and to gain insights in biology we cannot exclude physics and chemistry (biophysics and biochemistry).

However, it makes sense to establish scientific disciplines and sub-disciplines simply to limit the area of interest (the discipline itself). Complications arise by establishing separate languages and causing misunderstandings among scientists from different disciplines. "To look for definitions, to separate physics and chemistry fundamentally is impossible because they deal with the very same task, the insight into matter," the German chemist *Jean D'Ans* (1881−1969) wrote in the preface to his *Einführung in die allgemeine und anorganische Chemie* (Berlin 1948).

In the *Encyclopedia Britannica* published in Edinburgh in 1771 (shortly before the discovery of the chemical composition of air) chemistry is defined as "to separate the different substances that enter into the composition of bodies [analytical chemistry in modern terms]; to examine each of them apart; to discover their properties and relations [physical chemistry in modern terms]; to decompose those very substances, if possible; to compare them together, and combine them with others; to reunite them again into one body, so as to reproduce the original compound with all its properties; or even to produce new compounds that never existed among the works of nature, from mixtures of other matters differently combined [synthesis chemistry in modern terms]". *Julius Adolf Stöckhardt* wrote in his textbook, *The Principles of Chemistry*: "Wherever we look upon our earth, chemical action [a better translation from the original German is "chemical processing"] is seen taking place, on the land, in the air, or in the depths of the sea" (English translation, Cambridge 1850, p. 4). Thus chemistry is a priori the science of mineral, animal and vegetable matter (cf. also Chapter 2.1.2.1).

Basically there is no need to prefix the sciences (environmental, ecological, geo-, air, hydro-, etc.). "Other" sciences such as geology and meteorology, for example, can be reduced to the natural sciences. It is worth noting the language roots of "geo" and "meteor". In Greek, γή (ge) means earth, land and ground (poetically γαία (gaia))[1]. In the composition of words, Greek γέα (gea), γης (ges), γην (gen) and γεω (geo) occur. Already before the year 600 B.C., the Greek word μετέωρος (metéoros) was in use, meaning "a thing in the air, altitude or above ground"[2]. Until the end of the eighteenth century, μετέωρος denoted all celestial phenomena; aqueous, vaporous, solid and light. *Aristotle*'s *Meteorologia* (see Chapter 1.3.1) is a book on natural philosophy (in modern terms: earth science). Hence, in a more narrow (historical) sense, geosciences (geology, geography, geochemistry, geophysics

---

[1] In Latin, the word for earth, land and ground is "tellūs" (in sense of the celestial body) and "terra" (in sense of matter).
[2] μετέωρολογια [metéorologia] means the "science of celestial, heavenly things" but also "vague talk" and "philosophical shenanigans" (Benselers Greek−German School Dictionary (*Griechisch-Deutsches Schulwörterbuch*, Leipzig and Berlin 1911).

etc.) deal with the subject of the *solid* earth (the geosphere). Geochemistry studies the composition and alterations of the solid matter of the earth; geology is the scientific study of the origin, history, and structure of the earth; geography studies the earth and its features and the distribution of life on the earth, including human life and the effects of human activity. Geography (which is an old discipline, founded as a modern science by *Humboldt*) is thus the science of the earth's surface − the physical and human landscapes, the processes that affect them, how and why they change over time, and how and why they vary spatially. In other words, geography is an interfacial science between the *solid* and the *gaseous* earth (the atmosphere); indeed some geographers consider climatology to be a sub-discipline of geography (and vice versa, some meteorologists do not include climatology in meteorology). Finally, the *liquid* earth (the hydrosphere) is the subject of hydrology[3]. It is important to note that gaseous water (vapor) in the atmosphere is not considered to be an object in hydrology, but liquid water in the atmosphere is. Logically we now can state that the science of studying the "gaseous earth" (atmosphere) is meteorology. Older (synonymous) terms for the science of the study of earth's atmosphere are *aerology* and *atmospherology*. For example, meteorology has been defined as the physics and chemistry of the atmosphere (Scharnow et al. 1965) but also reduced to just the physics of the atmosphere (Liljequist and Cehak 1984). Some definitions focus meteorology on weather processes and forecasting (which surely was the beginning of that science). A satisfactory definition is "the study dealing with the phenomena of the atmosphere, including physics, chemistry and dynamics, extending to the effects of the atmosphere on the earth's surface and the oceans". Hence meteorology is more than "only" the physics and chemistry of the atmosphere. Nowadays the term "atmospheric sciences" is also used to summarize all the sub-disciplines needed to explain atmospheric phenomena and processes. To return to chemistry, (atmospheric) chemistry is just one of the sciences to understand the (chemical) processes in the atmosphere (see below for definition of atmospheric chemistry).

What we see is that all disciplines overlap. The original focus of geology as the science of the (solid) earth, hydrology as the science of liquid water, and meteorology as the science of the atmosphere, is still valid and should not be diffused, but the modern study of earth sciences looks at the planet as a large, complex network of physical, chemical, and biological interactions (Sammonds and Thompson 2007). This is known as a systems approach to the study of the earth. The systems approach treats the earth as a combination of several subsystems, each of which can be viewed individually or in concert with the others. These subsystems are the geosphere, the atmosphere, the hydrosphere, and the biosphere.

As mentioned in the preface, this book will treat the chemistry of the *climate system* according to the elements and its compounds depending from the conditions

---

[3] In Greek ὕδωρ (idor) means originally rain water and generally water; it appears in composite words as ὕδρο (idro). This is the derivation of the prefix "hydro" (the pronunciation of Greek letter ὕ and ὑ is like "hy"). In Latin, water as a substance is "aqua", but natural waters are referred to as "unda", derived from the Greek "hy-dor" (ὕδωρ); ὕδρα (Greek) and hydra (Latin) is the many-headed water snake in Greek mythology. From Gothic "vato" and Old High German "waz-ar", are clearly seen the roots of English "water" and German "Wasser".

of reactions and whether the substance exists in the gaseous or condensed (aqueous and solid) phase. Not yet discussing how we should understand a climate system (see Chapter 3.2), we will initially set it in a chain of subsystems:

Cosmic system → solar system → earth system → climate system → sub-systems (e. g. atmosphere).

In other words, each system to be defined lies in another "mother" system, the surrounding or environment where an exchange of energy and material is realized via the interfaces. Consequently there is no fully closed system in our world.

As shown above, the climate system is a subsystem of the earth system. The chemistry of the earth system − when not considering life − is *geochemistry*. The term *geochemistry*, like many other scientific terms, has variable connotations. If geochemistry means simply the chemical study of the earth or parts of the earth, then geochemistry must be as old as chemistry itself, and dates from the attempts of Babylonian and Egyptian metal-workers and potters to understand the nature and properties of their materials (Tomkeiev 1944).

It is self-evident that biochemistry deals with chemical processes in organisms and thus the chemical interaction between organisms occurs via geochemical processes, consequently, biogeochemistry is *the* chemistry of the climate system which we have defined as that part of the earth system affecting life. Subdividing the geogenic part of the climate system into "other" systems, we have the atmosphere, hydrosphere, cryosphere, and lithosphere; thus atmospheric, aquatic and soil chemistry are the sub-disciplines of geochemistry.

Traditionally we also may subdivide the climate system into the atmosphere, lakes and rivers, the oceans, the soils, minerals, and volcanoes, consisting of gases, liquids and solids. Organisms − plants and animals − are distributed among these compartments. Studying the chemistry of a natural system is done by chemical quantitative analysis of the different matters. An uncountable number of analyses are needed to gain a three-dimensional distribution of the concentration of specific substances or, in other words, the chemical composition of the climate system. This is the static approach to obtain an averaged chemical composition over a given time period, depending on the lifetime of the substances; the climatological mean. However, concentration changes occur with time because of transportation (motion) and transformation (chemical reactions). Hence a dynamic approach is needed too. Chemical reactions can be studied − which means establishing the kinetics and mechanisms − only in the laboratory under controlled (and hence replicable) conditions in a large variety of so-called smog chambers. Many attempts to gain kinetic data from field experiments have been unsuccessful because the number of degrees of freedom in a natural system is always larger than the number of measurable parameters. *Chemical climatology* seeks to establish the four-dimensional concentration field in space and time. The timescale can show variations and trends. Variations can be characterized stochastically and periodically.

We will define air (or atmospheric) chemistry as the discipline dealing with the origin, distribution, transformation and deposition of gaseous, dissolved and solid substances in air. This chain of matter provides the atmospheric part of the biogeochemical cycles (see for example Schulze et al. 2001). A more general definition, but one that is appealing as a wonderful phrase, is given by the German air chemist

Christian Junge: "Air chemistry is defined ... as the branch of atmospheric science concerned with the constituents and chemical processes of the atmosphere ..." (Junge 1963). In other words, air chemistry is the science concerned with the origin and fate of the components in air. The origin of air constituents concerns all source and formation processes, the chemicals of air itself, but also emissions by natural and man-made processes into the atmosphere. The fate of air includes distribution (which is the main task of meteorology), chemical conversion, phase transfers and partitioning (reservoir distribution) and deposition of species. Deposition is going on via different mechanisms from gas, particulate and droplet phases to the earth's ground surface, including uptake by plants, animals and humans. Removal from the atmosphere is the input of matter to another sphere (cf. Fig. 1.1).

The chemical composition of air depends on the natural and man-made sources of the constituents (their distribution and source strength in time and space) as well the physical (e. g. radiation, temperature, humidity, wind) and chemical conditions (other trace species) which determine transportation and transformation. Thus, atmospheric chemistry is not a pure chemistry and also includes other disciplines which are important for describing the interaction between atmosphere and other surrounding reservoirs (biosphere, hydrosphere, etc.). Measurements of chemical and physical parameters in air will always contain a "geographical" component, i. e., the particularities of the locality. That is why the terms "chemical weather" and "chemical climate" have been introduced. For example, diurnal variation of the concentration of a substance may occur for different reasons. Therefore general conclusions or transfer of results to other sites should be done with care. On the other hand, it is a basic task in atmospheric chemistry not only to present local results of chemical composition and its variation in time, but also to find general relationships between pollutants and their behavior under different conditions.

The elemental chemical processes occurring in the atmospheric aqueous phase are not different from those described by "water chemistry", now often also termed *aquatic chemistry* (e. g. Stumm and Morgan 1995, Sigg and Stumm 1996). However aquatic chemistry deals with the composition of natural water and aqueous solutions (or common water, to distinguish it from atmospheric water in the droplet form). In air, the chemistry in droplets is permanent in interaction with the surrounding gas phase chemistry. Hence the term *multiphase chemistry* reflects best the processes in the air. Concerning the different phenomenology of atmospheric waters, it is obvious to speak of rain, snow, fog, cloud and dew chemistry in the sense of analyzing the chemical composition of the solution.

Analytical chemistry as a sub-discipline of chemistry has the broad mission of understanding the composition of all matter. Much of early chemistry was analytical chemistry since the questions of which elements and chemicals are present in the world around us and what is their fundamental nature are very much in the realm of analytical chemistry. Before 1800, the German term for analytical chemistry was "*Scheidekunst*" ("separation craft"); in Dutch, chemistry is still generally called "*scheikunde*". Before developing reagents to identify substances by specific reactions, simple knowledge about the features of the chemicals (odor, color, crystalline structure etc.) was used to "identify" substances. With *Lavoisier*'s modern terminology of substances (1790) and his law of the conservation of mass, chemists acquired the basis for chemical analysis (and synthesis). The German chemist *Carl*

Remigius Fresenius wrote the first textbook on analytical chemistry (1846) which is still generally valid.

The term "atmospheric chemistry" appears to have been used for the first time[4] in German by *Hans Cauer* in 1949 (Cauer 1949a). It was soon used as the label for a new discipline. The first monograph in the field of this new discipline was written by *Junge*, entitled *Air Chemistry and Radioactivity* (New York and London 1963), soon after he had published a first long chapter entitled "Atmospheric Chemistry" in a book in 1958 (Junge 1958). This clear term identifies a sub-discipline of chemistry and not meteorology or physics. The "discipline" was called "chemical meteorology" before that time. However before the 1950s, chemical meteorology was mainly looking for the relationship between condensation nuclei, its chemical composition and the formation of clouds and rain. That term is still in use as a subtitle of the journal *Tellus* (Series B: Chemical and Physical Meteorology, started in 1982 when Tellus, created in 1949, was split into Series A and B) and for a professorship at the University of Stockholm (*Henning Rodhe* held the chair from 1980 to 2008), created in 1979 by *Bert Bolin*. Another term, "chemical climatology" came into use with the famous book *Air – The Beginning of a Chemical Climatology* (London 1872) by *Robert Angus Smith*. At that time knowledge of chemical processes in air was still rather limited, hence the book's focus on the description of concentrations of air species in time and space in analogy to physical parameters ("meteorological elements" such as temperature, pressure, wind etc). Consequently meteorology has often been defined as the physics and chemistry of the atmosphere. Despite some ideas on chemical processes in air (see the next Chapter) at the beginning of the nineteenth century, only the formation of ozone under strong UV light had been roughly understood by the 1930s. Atmospheric chemistry as a scientific discipline using laboratory, field and (later) modeling studies was vigorously developed after identification of the Los Angeles smog at the beginning of the 1950s.

It was not before the 1970s that scientists recognized the occurrence of global changes. It is worth mentioning that already in the Roman epoch, 2000 years ago, mining and smelting of ores led to hemispheric distribution of heavy metals which have been detected in Greenland ice cores. The rapid reduction of European forest from more than 80% cover 2000 years ago to about 25% at present has surely had a significant influence on air chemistry (due to emissions, deposition and climatic changes), but is not something that can be quantified. Local pollution was order(s) of magnitude higher in medieval cities than it is today, and consequently "urban air chemistry" must have been quite different. Efforts at pollution control and legislative measures have been applied since that time. Emission of chemically reactive

---

[4] I found that the term "*Chemie der Atmosphäre*" (chemistry of the atmosphere) was first used by *Anton Baumann* (chemist at *Forstliche Versuchsanstalt*, forest experimental station of University Munich) in 1892 (and later) in the periodical "*Jahresberichte über die Fortschritte auf dem Gesamtgebiete der Agrikultur-Chemie*" (Annual Reviews on Progress in the Whole Field of Agricultural Chemistry), where the reports were subdivided into atmosphere, water, and soil; "atmosphere" divided further in "chemistry of the atmosphere" and "physics of the atmosphere". In 1886, the heading of that report in *Jahresberichte* was, for the first time, termed "*Chemie der Atmosphäre und der atmosphärischen Niederschläge*" (Chemistry of the atmosphere and atmospheric precipitation) by the rapporteur *Richard Hornberger* (professor of meteorology at Forest Academy in Münden, Germany).

substances (SO_2, NO, VOC) on a scale from local to regional began with the exponential increase of fossil fuel use (coal after 1870 and oil after 1950).

Presently, after the successful introduction of end-of-pipe technologies for the reduction of these substances, we are observing a drastic decrease of many trace species in the air and subsequent changing of "chemistry" (in terms of acidity, oxidation potential, deposition pattern) that often is not "linear" with the emission change. For example, the number of days with excessive ozone concentration in Germany fell practically to zero at the end of the 1990s, but a further increase of annual means of ozone concentration is still observed at many sites. This suggests that different processes determine ozone formation and distribution, since the abatement strategy reduced only the short-lived so-called ozone precursors. It is worth noting here that many chemical features such as acidity and redox potential are quantified from budgets (acid − bases, oxidant − reductants). Hence the same difference can be given by concentrations at a low level and vice versa. However, low level or high level concentrations of individual substances may lead to other effects. The matter (in terms of physics and chemistry) of the climate system is extremely complex.

With the focus of many disciplines in science and engineering on our environment, many new sub-disciplines have arisen, such as biometeorology, bioclimatology, environmental chemistry, ecological chemistry, atmospheric environmental research, environmental meteorology, environmental physics and so on. Atmospheric chemistry is again subdivided into sub-sub-disciplines; tropospheric chemistry, stratospheric chemistry, cloud chemistry, precipitation chemistry, particle chemistry, polar chemistry, marine chemistry and so on. It is not the aim here to validate the sense or non-sense of such sub-disciplines. Basic research is progressing with individuals and in small groups on limited topics and hence with growing understanding (learning) and specialization results in establishing "scientific fields". There is no other way to proceed in fundamental science. However, the complexity of the climate system (and even just of the atmosphere) calls for an inter-disciplinary approach and we have learnt that, especially at the "interfaces" between biological, physical, chemical and geological systems, crucial and key processes occur in determining the function of the whole system. As a consequence, a chemist without understanding of the fundamental physical processes in the atmosphere (and vice versa) can only work on discrete research topics and will not be able to describe the climate system.

1.3 A historical perspective of air, water and chemistry

Humans have always dealt with and been fascinated by the properties of our atmosphere. In ancient times, the motivation to observe the atmosphere was clearly the driving force which increased the understanding of nature. Atmospheric (weather) observations were closely associated with astronomy, and everything above the earth's surface was named "heaven" or "ether". The weather phenomena − fog, mist and clouds, precipitation (rain, snow, and hail) and dew − have been described since Antiquity. A phenomenological understanding of the physical (but not the chemical) processes associated with hydrometeors was complete only by the end of the nineteenth century. Today the physics and chemistry in the aerosol-cloud-precipitation chain are relatively well understood – also with relation to climate. However, it seems that because of the huge complexity a mathematical description of the processes (i.e., the parameterization of the chemistry and also for climate modeling) is still under construction.

1.3.1 From Antiquity to the Renaissance: Before the discovery of the composition of air

Before the sixth century B.C., air was identified as emptiness. Greek natural philosophers assigned air and water beside earth and fire to the four elements (in Latin, *materia prima*, primary matter). *Thales of Miletus* (624−546 B.C.) was the first person who is known to have tried to answer the question of how the universe could possibly be conceived as made not simply "by gods and daemons". He defined water as a primary matter and regarded the earth as a disc within the endless sea. *Pythagoras* (about 540−500 B.C.) was probably the first to suggest that the earth was a sphere, but without explanation (only based on esthetic considerations). *Parmenides of Elea* (about 540−480 B.C.), however, explained the spheroid earth due to his observations of ships floating on the sea; he was a scholar of *Xenophanes* from Kolophon (about 570−480 B.C.), the founder of Eleatic philosophy. *Xenophanes* in turn was a student of *Anaximander* from Miletus (about 611−546 B.C.). With *Anaximander*, a student of *Thales*, and *Anaximenes* (from Miletus, about 585−528 B.C.), the cycle of pre-Socratic philosophers is closed. *Anaximenes* assumed – in contrast to *Thales* – that air is a primary element (root or primordial matter) that can change its form according to density: diluted into fire, it may condense to wind and, by further condensation, into water and finally into soil and rocks. This was very likely the first "poetic" description of the idea that all material on earth is subject to cycling, where "dilution" and "condensation" are the driving processes. *Empedocles of Acragas* (495−435 B.C.) introduced the four elements; earth, water, air and fire. The list was then extended by *Aristotle* (384−322 B.C.) by a fifth one, the *æther* (explaining the heavenly, in Greek αιθέϱας). Thus, the first to describe a number of weather phenomena and the water cycle was *Aristotle* in his *Meteorologica* (Aristoteles 1829, Aristotle 1923). From his *Meteorologica* we know that *Aristotle* believed that weather phenomena were caused by mutual interaction of the four elements (fire, air, water, earth), and the four prime contraries: hot, cold, dry and moist. He contributed many accurate explanations of atmospheric phenomena. The description of the water cycle (reasons for rain), as presented above, could have been taken from a modern textbook. *Theophrastus* (about 372−287 B.C.), the successor of *Aristotle* in the Peripatetic school and a native of *Eresus in Lesbos*, compiled a book on weather forecasting, called the "Book of Signs". His work consisted of ways to predict the weather by observing various weather-related indicators, such as a halo around the moon, the appearance of which is often followed by rain. *Archimedes of Syracuse* in Sicily (287−212 B.C.) indirectly contributed with his buoyancy principle to the design of the hot-air balloon, an invention which added much to our knowledge of the vertical structure of the atmosphere in the nineteenth and the beginning twentieth century, and to the basis for theoretical investigation of the buoyant rise of cumulus clouds.

The Romans were not interested in the continuation of Greek doctrines, however, they preserved the Greek learning. After the fall of the Roman Empire, around the fifth century, the occident forgot this ancient scientific heritage and replaced it with one single doctrine, that of the Bible. In the Middle Ages religious belief prevailed, with the view that all "heavenly" things were governed by God (which, after all, was the belief of peoples all over the world and which led to the idea of the existence of special gods for many atmospheric phenomena). Medieval monks began to observe the weather and take records, out of personal interest. In the Middle Ages, any meteorological (i. e., weather) observation was linked to astrology. The idea that the motion of the stars and planets influenced all processes on earth and in the atmosphere inhibited progress in the natural sciences. In the orient, however, *Aristotle*'s doctrine remained vital, and from there it was reintroduced to Europe in the twelfth century, probably via Sicily, where famous alchemistic laboratories were established. It is very likely that the first physical treatment of rainwater was performed by the great Arab scientist *Abd al-Rahman al-Khazini* who worked in Merv (formerly in Persia, now Turkmenistan) between 1215 and 1230 and was a student of (*Abū 'r-Raihān Muḥammad ibn Aḥmad*) *al-Bīrūnī* (937−1048) who first introduced the weighing of stones and liquids to determine their specific weight (Durant 1950, Hall 1973). *Al-Khazini* is known for his book, *Kitab Mizan al-Hikma* (The book of the Balance of Wisdom), completed in 1121, which has remained a central piece of Muslim physics ever since. *Al-Khazini* was the first to propose the hypothesis that the gravity of bodies varies depending on their distance from the center of the earth and he defined the specific weight of numerous substances including that of rainwater defining it to be exactly 1.0 g per cm^{-3} (Szabadváry 1966).

Between the Greek philosophers, who recognized the atmosphere only by visual observations and reflection, generalizing it in philosophical terms, and the first instrumental observations, there is a gap of almost 1500 years. Agricultural development and the interest in understanding plant growth (i. e., the beginning of commercial interests) initiated chemical research in Europe in the seventeenth century. Chemistry, first established as a scientific discipline around 1650 by *Robert Boyle*, had been a non-scientific discipline (alchemy) until then (Boyle 1680). Alchemy never employed a systematic approach and because of its "secrets" no public communication existed which would have been essential for scientific progress. In contrast, physics, established as a scientific discipline even earlier, made progress, especially with regard to mechanics, thanks to the improved manufacturing of instruments in the sixteenth century. Astronomers, observing the object of their discipline through the atmosphere, also began to discover the earth's atmosphere. There are two personalities to whom deep respect must be paid for initiating the scientific revolution in both the physical and chemical understanding of atmospheric water; *Isaac Newton*, who founded the principles of classical mechanics in his *Philosophiæ Naturalis Principia Mathematica* (1687), and, one hundred years later, *Lavoisier*, with his revolutionary treatment of chemistry (1789), which made it possible to develop tools to analyze matter; this is why he is called "the father of modern chemistry" (Lavoisier 1789).

We should not forget that solely the estimation of volume and mass was the fundament of the basic understanding of chemical reactions and physical principles after *Boyle*. While instruments to determine mass (resp. weight) and volume had been known for thousands of years, the new instruments (thermometer, barometer) to supply scientists with the necessary data to test the physical laws were only available from the time of *Galileo Galilei*. Around the year 1600, *Galileo* established an apparatus to determine the weight of the air and invented a crude thermometer. Independently of *Galilei*, the thermometer was invented in Holland by *Cornelius Jacobszoon Drebbel* (1572−1633) and was first used in 1612 by the physician *Santorio* (Hellmann 1920). The Italian mathematician and physicist *Torricelli*, a student of *Galilei*, produced a vacuum for the first time and discovered the principle of the barometer in 1643. *Torricelli* also proposed an experiment to show that atmospheric pressure determines the level of a liquid (he used mercury). *Torricelli's* student *Vincenzo Viviani* finally conducted this experiment successfully and *Blaise Pascal*, a contemporary French scientist, carried out very careful measurements of the air pressure at Puy de Dôme near Clermont in France. He noticed the decrease of pressure with altitude and concluded that there must be a vacuum at high altitudes. In 1667 *Robert Hook* (1635−1703), an assistant of *Boyle's*, invented an anemometer for measuring wind speed. In 1714, *Fahrenheit*, a German glassblower and physicist, born in Danzig (modern Gdansk in Poland) and later working in Holland, worked on the boiling and freezing of water, and from this work he developed a temperature scale. *Horace-Bénédict de Saussure*, a Swiss geologist and meteorologist, invented the hair hygrometer for measuring relative humidity in 1780. According to Umlauft (1891), Grand Duke *Ferdinand II* of Toscana (who reigned 1621−1670) invented the first hygrometer (*Torricelli* was his court mathematician). *Benedetto Castelli*, a friend of *Galilei's*, used the first rain gauge in 1639 to measure rainfall.

Recall that air and water had been regarded as "elements" convertible into each other since *Aristotle*. The statement by *René Descartes*, the French philosopher, mathematician, scientist and writer, that water vapor is not (atmospheric) air, is remarkable as this was 15 years before the introduction of the term "gas" by *van Helmont*. The gaseous substances that were observed in alchemical experiments were named fumes, vapors and airs. Atmospheric air (called common air) was still regarded as a uniform chemical substance. The meaning of different terms in different languages (e. g. French, English and German) has been changing over time; the words were used in slightly different senses by various scientists. There was obviously a need for a new word to name and distinguish the laboratory airs (i. e., gaseous substances) from atmospheric (common) air. The new word was proposed by *van Helmont* in his posthumously published book, *Ortus medicinae i. e. initia physicae inaudita* (Amsterdam 1652, p. 86): "hunc spiritum, incognitum hactenus, novo nomine Gas voco" (I call this entity, unknown hitherto, by the new name of Gas), and "ideo paradoxi licentia, in nominis egestate, halitum illum Gas vocavi, non longe a Chao veterum secretum" (Licence of a Paradox, for want of a name, I have called that vapour, Gas, being not far severed from the Chaos of the Aunfients) (p. 59; English translation by J. Chandler, 1662).

Adelung (1796, p. 425) quoted *Helmont* in wishing: "… dass unsere Naturkundige ein schicklicheres Wort, welches nicht so sehr das Gepräge der Alchymie an sich hätte, ausfündig machten" (… that our natural scientists find a proper word which does not have so much the aura of alchemy). *Adelung* believed that *Helmont* had derived the word "gas" from the Dutch Geest (ghost). Carbon dioxide, hitherto called spiritus sylvestris (wild spirit), was renamed by Helmont as "gas sylvestre". There are other ideas on the origin of the word gas, however. Paracelsus (1493–1541) denoted the "atmosphere" to be Chaos – air and chaos were synonymous for him. The Greek word χάος denotes both an empty sphere and the initiation. Emptiness is not synonymous with nothing because the Greek philosophers stated that the world is born from chaos or, in other words, the chaos is creative of life (Genz 1994). From the primordial chaos (or mysterium magnum) the four elements water, fire, earth and air are derived (by separatio). The Dutch pronunciation of "chaos" is close to "gas" when the letter "o" is omitted (Egli 1947). *Antoine-Laurent Lavoisier* wrote in his book, *Opuscules physiques et chimiques* (Paris 1774, p. 5):

> *Gas* vient du mot hollandais *Ghoast*, qui signifie *Esprit*. Les Anglais expriment la même idée par le mot *Ghost*, et les Allemands par le mot *Geist* qui se prononce *Gaistre*. Ces mots ont trop de rapport avec celui de *Gas*, pour qu'on puisse douter qu'il ne leur doive son origine.

The observation that remote water (*materia prima*) only comes from the atmosphere (atmospheric water) certainly promoted experiments to derive the philosopher's stone from it. Despite much progress in science at the beginning of the seventeenth century, the belief of convertibility between air and water, and water and soil (and vice versa) was widely accepted until the chemical composition of the air and the structure of water was discovered by *Cavendish, Scheele, Priestley, Lavoisier* and others after 1770.

1.3.2 Discovery of the composition of air and water

In the eighteenth century the interest in natural processes generally expanded. Travellers and biologists were interested in describing the climate and its relation to culture and biota, and in the late 1700s chemists began to understand the transformation between solid, liquid and gaseous matter. A fundamental interest in biological processes, such as plant growth, nutrition and respiration among others, stimulated the study of the water cycle and the gas exchange between plant and air.

John Mayow, an English chemist and physiologist, showed that air contains a gas which is a special agent for combustion and respiration and is fixed from calcified metals (i. e., carbon dioxide). It might be said that modern chemistry had its beginning with the work of *Stephen Hales*, who early in the eighteenth century began his important study of the elasticity of air. He also pointed out that various gases, or "airs" as he called them, were contained in many solid substances. His many careful measurements (published in, Vegetable Statics, or *an account of some statical experiments on the sap in* vegetables, London 1727) of the absorption of water and its transpiration to the atmosphere were the basis for the understanding that air and light are necessary for the nutrition of green plants. The careful studies of *Hales* were continued by his younger colleague, *Joseph Black*, a Scottish chemist, whose experiments concerning the weights of gases and other chemicals were the first steps in quantitative chemistry. *Black* made valuable studies on carbon dioxide, which he named fixed air, and found that candle lights do not burn in this gas, creatures cannot exist, and that it is a product of respiration.

In the second half of the eighteenth century, air was found to consist of two different constituents, maintaining respiration and combustion (O_2) and not maintaining it (N_2). The discovery of nitrogen is generally credited to *Rutherford* whose *Dissertatio Inauguralis de Aero Fixo Dicto, aut Mephitico* (On Air said to be Fixed or Mephitic) was published in Edinburgh in 1772. Seventeen years after *Black's* dissertation on "fixed air" (CO_2), just before *Priestley's*, (and *Scheele's*) discovery of "good air" (O_2), *Rutherford* conducted experiments where he removed oxygen from air by burning substances (i. e., charcoal) and afterwards carbon dioxide by absorption with lime; the rest (nitrogen) he denoted as "phlogistigated air", despite the fact that it was not flammable. *Priestley* wrote in 1771 about the goodness of air (air quality in modern terms) and noted that injured or depleted air can be restored by green plants. In 1772 *Priestley* started his studies on air using mercury for locking gases. After a break in 1776 he systematically began to investigate different kinds of "air": nitrous (salpetric) air (NO_x), acid (muriatic) air (HCl), and alkaline air (NH_3). He stated that these "kinds of air" are not simple modifications of ordinary (atmospheric) air. He published his observations in a book titled *Observations on Different Kinds of Air* (1772); Priestley (1775). By heating red mercury oxide, he produced dephlostigated air (O_2) in 1774.

The facts on air composition were expressed most clearly by *Scheele*, in his booklet *Abhandlung von der Luft und dem Feuer* (Treatise on Air and Fire), which was published in 1777 (Scheele 1777). From laboratory scripts it is now known that *Scheele* discovered oxygen − dephlogisticated air − before *Priestley* and by similar methods: heating silver carbonate, red mercury oxide, salpeter and magnesium nitrate[5]. *Scheele* also discovered chlorine (Cl_2), and he named the ingredients of air as *"Feuerluft"* (O_2) and *"verdorbene Luft"* (N_2). *Scheele* found evidence that one unit of oxygen produces one volume of carbon dioxide and defined that

$$Feuerluft\ (O_2)\ =\ Phlogiston + fixe\ Luft\ (CO_2)$$

This is incorrect, it should be written (in the old terms): carbon = *fixe Luft* (fixed air) + *phlogiston*, i. e., when carbon is burning, it is transformed into carbonic acid (CO_2) while releasing "phlogiston".

These days it is difficult to understand what "phlogiston" meant to eighteenth-century scientists. The phlogiston theory, founded by *Johann Becher* and developed further by *Georg Ernst Stahl* − both of them German chemists − was to some extent derived from the old belief that there was a fire element and that all combus-

---

[5] It is not generally accepted that the priority in discovering oxygen must be attributed to *Scheele*, because *Priestley* published (direct) his results before *Scheele*. However, there is strong evidence in support of *Scheele* from the so-called "Braunbuch" (brown book), a bound collection of *Scheele's* laboratory scripts, prepared by *Berzelius* in 1829, 40 years after the early death of *Scheele*. There is also a letter from *Scheele* written to Lavoisier in 1774, explaining the experiment for isolating oxygen, as well as (and this is the most important evidence for the priority) the published correspondence of *Torbern Bergmann* (to whom the much younger *Scheele* reported on his experiments) between 1765 and 1775. In his *Abhandlungen von der Luft und dem Feuer*, *Scheele* describes no less than 10 different methods for preparing oxygen (*Feuerluft*).

Table 1.3 Historic data on air composition (in vol-%).

substance	Lavoisier[e] (1778)	Cavendish[e] (1783)	Benedict (1912)	Krogh[b] (1919)	Cadle and Johnstone (1956)	NASA[d] (2008)
N_2	79.19[a]	79.16[a]	79.016[a]	79.0215[a]	79.02	79.014[a, c]
O_2	20.81	20.84	20.954	20.9485	20.95	20.950
CO_2	–	–	0.031[b]	0.030	0.03	0.0380

[a] within these figures novel gases are
[b] corrected to 0.30 by Krogh due to 0.002% CO formation while oxygen absorbing by potassium pyrogallate
[c] Ar amounts 0.9340%
[d] Earth Fact Sheet reference
[e] after Humboldt (1799)

tible bodies contained a common principle (element), phlogiston (which in Greek means "flammable" or "inflammable"), which is released in the process of combustion. Substances rich in phlogiston, such as wood, burn almost completely; metals, which are low in phlogiston, burn less well. The phlogiston theory created great confusion and essentially embedded the understanding of the chemistry of phase-transfer processes and solid-gas reactions. Chemists spent much of the eighteenth century evaluating *Stahl*'s theory before it was finally proved to be false by *Lavoisier*. *Lavoisier* founded his theory on combustion on the discovery of the chemical composition of air. *Priestley*, reported in Paris in 1774 of his discovery (O_2) and said that he had no name for this gas. *Lavoisier* repeated the experiments of *Priestley* and dealt especially with the question of calcining caustic substances (metal oxides) as well as their reduction by charcoal. In *Reflexions sur le Phlogistique* (1783), *Lavoisier* showed the phlogiston theory to be inconsistent with observation. He believed that this element (oxygen, denoted as dephlogiston) is an immanent part of acids and this gave him the name *oxygéne* (from Greek οξνς – acid). He also named the other element, called by Scheele "*verdorbene Luft*" (bad air, and by *Priestley* "phlogistigated air") "*azote*" (this was nitrogen). *Lavoisier* was the first person who quantified the composition of air, in 1778 (Table 1.3).

Finally, in 1787, *Lavoisier* together with the French chemists, *de Morveau, Berthollet* and *de Fourcroy* established in Paris a new chemical nomenclature, that has remained valid until today. *Lavoisier* wrote in 1789 the *Traité élémentaire de Chimie* (Elementary Treatise of Chemistry), the first modern textbook on chemistry, and presented a unified view of new theories of chemistry, containing a clear statement of the law of conservation of mass, and denied the existence of phlogiston. In addition, it contained a list of elements, or substances that could not be broken down further, which included oxygen, nitrogen, hydrogen, phosphorus, mercury, zinc, and sulfur.

Another remarkable scientist was *Cavendish* who did not publish his results on air studies until 1783 (*Cavendish* remained a supporter of the phlogiston theory until his death). Already in 1772 he privately told *Priestley* about his experiments with "mephistic air" (nitrogen); thus it seems likely that *Cavendish* already knew before *Rutherford* about "inflammable air" (N_2). In 1781 he realized that water is produced in a reaction of hydrogen ("flammable air") with oxygen ("vital air") and soon he noted that there are also acidic substances not containing any oxygen. In 1781 he sampled atmospheric air at different sites and analyzed it gravimetrically after sorption of water-soluble gases (CO_2, NH_3 and water vapor) (see Table 1.3). *Cavendish* had already tested to find whether airy nitrogen is a uniform matter and found that there is a small residue (noble gases). He did not conclude, however, that these remains are an element (argon).

The debate about who actually discovered the chemical composition of water (H_2O) was called the "water controversy" in the nineteenth century. With respect to the discovery of the chemical composition of water, three scientists must be regarded as candidates (Kopp 1869): *Cavendish*, who was probably the first (in 1781) to carry out experiments to form water by combining phlogiston and dephlogisticated air (O_2), also called good, pure, vital, fire air (in German: *gute Luft*, *Dephlogiston*, *Lebensluft*, *reine Luft*, *Feuerluft* etc.); *Watt*, who formulated the composition of water in 1783 in a similar way to *Cavendish*; and finally *Lavoisier*, who, in 1783, made the first public announcement that water consisted of inflammable air (H_2) and dephlogisticated air (O_2).

Hydrogen (H_2) was probably already known to *Paracelsus* and *Helmont* (without using the name) in the sixteenth century but was often confused with other combustible gases. Hydrogen was produced by the treatment of metals with acids, but any "flammable air" was called "sulfurous". *Stahl* maintained that phlogiston is exhausted by metals and combines with the acid to form a flammable substance. *Cavendish* (1766), however, was able to show that the flammable air produced by the dissolution of iron in sulfuric acid and of zinc in hydrochloric (muriatic) acid was phlogiston itself and did not contain anything of acid. Today we know that this gas is hydrogen. However, at that time, other flammable airs (produced, for example, when organic matter is decomposed: CO, PH_3) were hardly distinguished. *Cavendish* was the first to study this flammable air (H_2) in different mixtures with common air to investigate its explosion (1766). Priestley (1775) found that this flammable air (H_2) exploded much more vehemently when brought together with the newly discovered pure dephlogisticated air (O_2) than with common air. *Cavendish* observed that after the explosion the inside of the glass vessel became dewy ("… that common air deposits its moisture by phlogistication"). In explosions in which *Cavendish* (1784) used electric sparks he found "… liquor in the globe …; it consisted of water united to a small quantity of nitrous acid" (Cavendish 1893). This statement is most remarkable to me; it forms the first evidence of HNO_3 formation under atmospheric conditions by lightning. *Blagden*, *Cavendish*'s assistant from 1782 to 1789, reported to *Lavoisier* about *Cavendish*'s experiments in 1781 and, together with *Laplace*, the great French mathematician, he repeated *Cavendish*'s experiments. He was able to invert the experiment, i. e., he decomposed water (by directing water vapor over a red-hot iron wire) into hydrogen and oxygen. *Lavoisier* (1790) estimated the composition of 100 g of water as 85 g oxygen and 15 g inflammable gas (hydrogen), which is relatively close to the correct quantities: 89 + 11.

When reading these old papers with our present scientific knowledge it is often difficult, if not impossible, to understand what the scientists meant by different terms; confusion also results from attributing the same term to different substances (we may only conclude that in those days such distinctions were not always possible): phlogisticated air for both N_2 and H_2, acid air for both CO_2 and O_2. Kopp (1869) accepted that phlogiston was actually hydrogen.

Gay-Lussac and *Humboldt* carried out air analysis from different sites and validated the ratio 21/79 for oxygen/nitrogen as a constant. In 1804, *Gay-Lussac* made several daring ascents of over 7000 m above sea level in hydrogen-filled balloons – a feat not equalled for another fifty years – that allowed him to investigate other aspects of gases. Not only did he gather magnetic measurements at various altitudes, but he also measured pressure, temperature and humidity, and took samples of air, which he later analyzed chemically. *Robert Bunsen* showed in 1846 that the oxygen content in air varies slightly between 20.84 and 20.95 % (measurement error was 0.03 %).

Lord Rayleigh was the first who observed (between 1882 and 1892) that oxygen and other gases produced from different sources always showed the same density but not airy nitrogen (Rayleigh et al. 1896). While "airy nitrogen" had a density of 1.2572 g L^{-1}, nitrogen from decomposition of organic substances showed a density of 1.2505 g L^{-1}. The difference of 7 mg was already far away from measurement errors. In his address on the occasion of receiving the Nobel Prize (1904) *Rayleigh* explained how he made his discovery, showing the (from today's point of view) simple but accurate experiments and conclusions:

> The subject of the densities of gases has engaged a large part of my attention for over 20 years. ... Turning my attention to nitrogen, I made a series of determinations ... Air bubbled through liquid ammonia is passed through a tube containing copper at a red heat where the oxygen of the air is consumed by the hydrogen of the ammonia, the excess of the ammonia being subsequently removed with sulfuric acid. ... Having obtained a series of concordant observations on gas thus prepared I was at first disposed to consider the work on nitrogen as finished. ... Afterwards, however ... I fell back upon the more orthodox procedure according to which, ammonia being dispensed with, air passes directly over red hot copper. Again a good agreement with itself resulted, but to my surprise and disgust the densities of the two methods differed by a thousandth part – a difference small in itself but entirely beyond experimental errors. ... It is a good rule in experimental work to seek to magnify a discrepancy when it first appears rather than to follow the natural instinct to trying to get quit of it. What was the difference between the two kinds of nitrogen? The one was wholly derived from air; the other partially, to the extent of about one-fifth part, from ammonia. The most promising course for magnifying the discrepancy appeared to be the substitution of oxygen for air in the ammonia method so that all the nitrogen should in that case be derived from ammonia. Success was at once attained, the nitrogen from the ammonia being now 1/200 part lighter than that from air. ... Among the explanations which suggested themselves is the presence of a gas heavier than nitrogen in air ... .

This new gas was identified by *Ramsay* in 1894 who made spectroscopic studies, identified this gas as an element and named it argon (Ar), derived from Greek αργόν = slack (*Ramsay* was awarded the Nobel Prize together with *Rayleigh* in 1904). While investigating for the presence of argon in a uranium-bearing mineral, he instead discovered helium, which since 1868 had been known to exist, but only in the Sun. This second discovery led him to suggest the existence of a new group of elements in the periodic table. *Ramsay* and his co-workers quickly (1898) isolated neon (Ne), krypton (Kr), and xenon (Xe) from the earth's atmosphere (Rayleigh and Lord 1901, Ramsay 1907).

1.3.3 Discovery of trace substances in air

Nitrogen (N_2), oxygen (O_2), water vapor (H_2O), carbon dioxide (CO_2) and rare gases are the permanent main gases in air. Only water shows large variation in its concentration and CO_2 is steadily increasing due to fossil fuel burning. Already in the first half of the nineteenth century other gaseous substances had been supposed and later detected in air. Due to the fact that the concentration of almost all trace gases are orders of magnitude smaller than those of the main gases (Table 1.1), it was only with the development of analytical techniques in the late nineteenth century that they were proved to be present in air.

Ammonia (*Scheele*, identified nitrogen in "alkaline air" (NH_3); the formula was established in 1785 by *Berthollet*) was found in air by Scheele in 1786 by observing that a precipitate originated on the cork a bottle containing hydrochloric acid, identified as salt ammonia (NH_4Cl) and it was later confirmed by *Théodore de Saussure* in the early 1800s. Still in 1900, it was stated that ammonia never exists freely (i.e., in gaseous form) in air but only in compounds with carbonate and others (Blücher 1900).

Other atmospheric trace gases were known from the experiments by *Priestley*, around 1774 (HCl, NO_x, HNO_3, SO_2) but not yet identified in air. *Cavendish* (1785) and *Priestley* (1788) described the HNO_3 formation in moisture air under the influence of electric discharges. Only around 1900 were all these gases (NH_3, HNO_3, and HNO_2) directly identified in the atmosphere. In the nineteenth century, the terms nitric acid, sulfurous acid etc. were used in the same sense for dissolved species (nitrate, sulfite) as well as anhydrites (e.g. SO_2). Nitric acid (HNO_3) was "known" as a result of thunderstorms and (without knowing the details) life processes. Atmospheric H_2S was known from mineral springs and rotting organic material. "Hydrocarbon" (not yet specified as methane, CH_4) was known from marshes and swamps (called swamp gas) and many natural gas sources (from which it was already sometimes used as fuel). This gas was feared by coal miners, who called it "firedamp" because it caused dangerous explosions. Natural sources of phosphurated hydrogen (phosphine PH_3) have been identified as sewage sludge, swamps and human flatus. In the early nineteenth century phosphine (which is spontaneously inflammable) was also known from cemeteries where it sometimes burned with blue flames.

Ozone, the first atmospheric trace species, was discovered by *Schönbein* while conducting electrolysis experiments with water in 1841 (he never identified the constitution of ozone). *Van Marum*, subjecting oxygen to electrical discharges in 1785, noted "the odor of electrical matter" and the accelerated oxidation of mercury. Thus, *van Marum* reported the odor of ozone but he failed to identify it as a unique form of oxygen. Its chemical composition (only consistent from oxygen), however, was proposed many years later by *Thomas Andrews* in 1856. The formula O_3 was proposed by *William Odling* in 1861. Ozone, as a natural component of air, was found in 1866 (Andrews 1867), despite the fact that the so-called *Schönbein* paper (ozonometry) had already been used in England in 1848 for atmospheric "monitoring".

Hydrogen peroxide (H_2O_2) was discovered by *Thénard* in 1818 while treating barium peroxide with sulfuric acid (Thénard 1819). He called it *l'eau oxygénée* (oxygenated water). *William Prout* first proposed its presence in the atmosphere and he called it deutoxide of hydrogen. This term was introduced by the Scottish chemist *Thomas Thomson*. It was subsequently termed "peroxide of hydrogen". In his famous book, *Chemistry, Meteorology and the Function of Digestion* (1834), *Prout* wrote:

> ... a combination of water and oxygen is a frequent, if not a constant, ingredient in the atmosphere. This ingredient, which we suppose to be a vapour, and analogous to (we do not say identical with) the deutoxide of hydrogen, may be imagined to act as a foreign body, and thus to be the cause of numerous atmospheric phenomena, which at present are very little understood ... The oxygen and vapour in this combination are so feebly associated ... (Prout 1834, p. 569).

Prout provided other observations on: "... the bleaching qualities of dew, and of the air itself; as to the large proportion of oxygen sometimes contained in snow water and in rain water..." (p. 570), which we can deduce from current knowledge on the presence of atmospheric oxidants such as H_2O_2 and photolytic-induced formation of radicals (OH, O_2^-) in surface water, for example, dew.

The first evidence of H_2O_2 in rain during a thunderstorm was provided by *Georg Meissner* in 1862 (Meissner 1863). *Schönbein* (1869) confirmed this observation and *Heinrich Struve* detected it in snow (Struve 1869). *Struve* even proposed in 1870 that H_2O_2 is produced during all burning processes in air (Struve 1871). The German chemist *Emil Schöne*, however, was the first scientist to study atmospheric H_2O_2 in detail in rain, snow and air near Moscow (Petrowsko-Rasumowskaja, an agricultural research station) in the 1870s (Schöne 1874, 1878, 1893, 1894).

It is notable that *Prout's* ideas were established before the discovery of ozone by *Schönbein* in 1839 (Schönbein 1844). The study of H_2O_2 in air was closely connected with studying the chemistry of O_3 in the nineteenth century (Engler 1879; Rubin 2001). It is remarkable that the existence of H_2O_2 in air (as gas, as well as dissolved in hydrometeors) was definitely established before 1880 but the existence of O_3 was still being discussed around 1880. Definite proof of the existence of O_3 in the atmosphere was not provided until the first spectrometric measurements at the end of the nineteenth century (Möller 2004).

In 1766 *Cavendish* separated hydrogen from other gases and showed that it burned to water. In connection with *Lavoisier's* discovery of the role of airy oxygen (1777) it became clear that water is a chemical compound. Only in 1900 was *Armand Gautier* the first to proclaim the presence of hydrogen in atmospheric air. This was verified in 1902 by *Rayleigh's* spectroscopic studies in air.

It is important to note that all the trace species mentioned and discovered or assumed to be in air were believed to be natural or, in other words, substances with a (at that time still unknown) special function in nature. The assimilation of gases and the uptake of nitrogen dissolved in water by plants and the decomposition of dead biomass as source of gases led to a first understanding of matter cycles by early agricultural chemists (e. g. Knop 1868).

1.3.4 Dust and acid rain: Air pollution

With the intense industrial development in the middle of the nineteenth century, air pollution as a new atmospheric aspect became the object of interest of researchers; more precisely, the impacts of air pollutant (forest decline, human health, corrosion) were the first foci of research. Already in the late nineteenth century, some impacts could be related to individual air pollutants (cause-receptor relationship, e. g. Stöckhardt 1850, 1871). The techniques available to measure trace species, however, were still very limited. In spite of the fact that quantitative relationships were missing, legislation concerning air pollution was introduced in the nineteenth century. Nevertheless, air pollution remained a local problem until the 1960s. It was only then, with concern over acid rain (despite the fact that it had already been described in England in 1852 by *Robert Angus Smith*), that the first regional environmental problem appeared in Europe. In the 1980s global problems were first recognized in relation to climate change due to the global change of the air's chemical composition. Localized catastrophic environmental events, like the smog events in Los Angeles (1944) and London (1952), helped to initiate atmospheric chemistry as a new discipline from the beginning of the 1950s.

It is remarkable that the mixing of air (as it was still regarded as a uniform body) and water with pollutants (accurately referred to as "foreign bodies" in the old terminology) has been known since *Aristotle*. The role of precipitation in cleaning the environment is wonderfully described by *John Evelyn*, who wrote the first book on air pollution (Evelyn 1661: 8 ff.):

> … in Clouds of Smoake and Sulphur, so full of Stink and Darkness … It is this horrid Smoake which obscures our Churches, and makes our Palaces look old, which fouls our Clothes, and corrupts the Waters, so as the very Rain, and refreshing Dew which fall in the several Seasons, precipitate this impure vapour, which, with its black and tenacious quality, spots and contaminates whatever is exposed to it. … poysoning the *Aer* with so dark and thick a Fog, as I have been hardly able to pass through it, for the extraordinary stench and *halitus* it send forth;… *Arsenical* vapour, as well as *Sulphur*, breathing sometimes from this intemperate use of *Sea-Coale* … our *London* Fires, there results a great quantity of volatile Salts, which being very sharp and dissipated by the Smoake, doth infect the *Aer*, and so incorporate with it, that the very Bodies of those corrosive particles …

Evelyn's remarks volatile salts and their corrosive effects after distribution in the air may form the first evidence for gaseous (and, consequently, dissolved) HCl in the urban air (ibid., p. 28). His expression that the "… traveler … sooner smells than sees the city …" (Evelyn 1661: 19) gives us an idea of the level of air pollution. The terms "smoake" and "clouds" in *Evelyn*'s booklet (only once does he use the term "fog") surely mean what we now call *smog*, an artificial expression coined by *des*

Voeux in his paper "Fog and Smoke" for a meeting of the Public Health Congress in London in 1905.

For centuries, until the end of the twentieth century, when the air pollution problems associated with the combustion of fossil fuels seem to have been solved (the problem of climate change due to carbon dioxide remains unsolved, however), sulfur dioxide and soot (smoke) were the key air pollutants. Coal has been used in cities on a large scale since the beginning of the Middle Ages; and this "coal era" has not yet ended. Remarkably, the term "smog" is not used in *Marsh's* book *Smoke* (Marsh 1947). Concerning the "relationship between fog and smoke", *Marsh* wrote that fog was a natural phenomenon, whereas smoke, passing through fog, could not dissipate as it could in non-foggy weather because of the absence of air currents, and the "clean natural fog gradually becomes more and more impregnated with smoke" (this is not correct, because smoke particles act as condensation nuclei, and thus "clean natural fog" could not possibly have appeared in cities like London in those years). Smoke and fog as contemporaneous phenomena were scientifically described by *Julius Cohen* who had studied chemistry in Munich. He wrote:

> Town fog is mist made white by Nature and painted any tint from yellow to black by her children; born of the air of particles of pure and transparent water, it is contaminated by man with every imaginable abomination. That is town fog. (Cohen 1895, p. 369)

Cohen conducted laboratory experiments and concluded: "The more dust particles there are, the thicker the fog" (Cohen 1895, p. 371). Carbonic acid (CO_2) and sulfurous acid (SO_2) were observed to increase rapidly during fog, and, "… although I have no determinations of soot to record, the fact that it increases also is sufficiently evident," he wrote. With these terms the acid anhydrides CO_2 and SO_2 were mentioned in the literature of the nineteenth century and not the acids H_2CO_3 and H_2SO_3 (sometimes also named gaseous carbonic acid). Fog water particles become coated with a film of sooty oil. Consequently, fog persists longer than under clean conditions. *Francis Russell* used the expression "smoky fog" and wrote that "town fogs contain an excess of chlorides and sulfates, and about double the normal, or more, of organic matter and ammonia salts" (Russell 1895, p. 234). During the last fortnight of February 1891, the weight of the fog deposit in Kew (just outside London) was 0.84 g m^{-2} which contained 42.5% carbon, 4.8% hydrocarbons, 4% sulfuric acid, 0.8% hydrochloric acid, 1.1% ammonia, and 41.5% mineral matter (Russell 1895).

Dust (in the past often called "solid bodies" and nowadays "particulate matter" but in a more scientific sense "atmospheric aerosol particles") has been observed since ancient times, and with the beginning of the nineteenth century some chemical species (iodine, phosphorus), microorganisms and plant remains were considered as its source. In the 1850s *Louis Pasteur* sampled air at Arbois (France) to investigate the hypothesis of so-called "spontaneous" generation. He found many different "germs" in collected dust which were able to germinate with different substrates. *Pasteur* also studied air from different sites (rural and urban) and different altitudes and found a decrease of air-borne germs with height (up to 2000 m a.s.l. at Mt. Montauvert); Pasteur (1862). Before *Pasteur's* findings, it was believed that air contains "miasmas", foul smelling gases, transferring diseases. Scientific understanding of dust in the atmosphere began with *Thomas Graham*'s definition of a colloid in 1861. The first direct observations of fine dust particles dispersed in air were made in 1870 by *John Tyndall* (Tyndall 1870) and in 1880 by *John Aitken* (Aitken 1880). *Aitken* also found that the presence of fine particles is necessary for the formation of rain. *Lord Rayleigh* had shown that these dust particles, by their scattering action on the small waves of light at the violet end of the spectrum are the cause of blue sky. Besides fine dust particles, soot and coarse particles have been known. At the end of the nineteenth century it was known that the coarse fraction was soil dust and organic matter (with the latter constituting up to one-third of the total). *Gaston Tissandier* first stated in 1879 that dust is partly of cosmic origin (Tissander 1879). The generic term "aerosol" was first introduced into literature by the German meteorologist *August Schmauß* (Schmauss 1920, Schmauss and Wigand 1929).

Atmospheric waters were first studied alchemically, by distillation of rainwater in the eighteenth century (Möller 2008). The first semi-quantitative chemical analysis of rain and snow was conducted by *Andreas Sigismund Marggraf* who collected and analyzed rainwater and snow water for purely analytical interests between 1749 and 1751 in Berlin; he also analyzed different natural (potable) waters to check their quality (Marggraf 1753, 1786). He furthermore found ammonium, because of its odor, after repeated distillation of rainwater. In total, he found in rainwater (by identifying the crystals); nitrate (salpeter), calcium, sodium and chloride (common salt), and organic substrate ("sticky and oily brown remains"). He assumed that its origin was from salty and earthy components. He also found the rainwater to be rotting. In snow, he found more hydrochloric acid than nitric acid, and vice versa in rain.

More detailed analyses of rainwater at the beginning of the nineteenth century were performed for the same reasons and in combination with the application of newly developed methods in analytical chemistry. Systematic studies of deposition (precipitation chemistry) only began with *Justus von Liebig*, known as the "father of the fertilizer industry", who discovered that plants assimilate (chemically fixed) nitrogen dissolved in rain. Since then, agricultural interests have formed an important base for rainwater chemistry monitoring. Air pollution in urban areas but also damage to forests (in Germany) stimulated several studies at the end of the nineteenth century. Deposition studies (bulk sampling) due to the smoke problem started after 1910. The aim to understand matter cycles, first between local and regional scales, initiated precipitation chemistry in the 1930s which led to systematic research from the 1950s.

We can learn from history that all kinds of persons were interested in the subject from a philosophical perspective and/or with respect to the application of techniques (engineering) but always motivated by the specific problems (e. g. pollution) of their era. We also hold deep respect for our scientific ancestors for their brilliant conclusions, based on scientific experiments with very simple techniques and limited quantitative measurements. The great interest in historical data from the era before fossil fuel combustion lies in determining background concentrations, in other words, the natural reference concentrations for assessing the human-influenced changes in chemical air composition. The endeavor remains to learn from previous studies to ask the appropriate open questions and draw the right conclusions for further studies.

Today, an uncountable number of sites of air chemistry study exist, often only active for short periods with sometimes barely more than a dozen samples that are collected and analyzed for whatever purpose. The history of atmospheric chemistry studies, at least since the systematic monitoring in the second half of the nineteenth century – which is certainly unknown to most modern air chemists – not only encourages respect for our scientific ancestors but may definitely help to avoid many scientifically meaningless studies of the kind that have appeared over the last few decades.

2 Chemical evolution

The term evolution[1] was used first in the field of biology at the end of the nineteenth century. In the context of biology, evolution is simply the genetic change in populations of organisms over successive generations. Evolution is widely understood as a process that results in greater quality or complexity (a process in which something passes by degrees to a different stage, especially a more advanced or mature stage). However, depending on the situation, the complexity of organisms can increase, decrease, or stay the same, and all three of these trends have been observed in biological evolution. Nowadays, the word has a number of different meanings in different fields. *Geological evolution* is the scientific study of the earth, including its composition, structure, physical properties, and history; in other terms: the earth change over time or the process of how the earth has changed over time. The term *chemical evolution* is not well-defined and is used in different senses.

Chemical evolution is not simply the change and transformation of chemical elements, molecules and compounds as is often asserted – that is the nature of chemistry itself. It is essentially the process by which increasingly complex elements, molecules and compounds develop from the simpler chemical elements that were created in the Big Bang. The chemical history of the universe began with the generation of simple chemicals in the Big Bang. Depending on the size and density of the star, the fusion reactions can end with the formation of carbon or they can continue to form all the elements up to iron.

The origin of life is a necessary precursor for biological evolution, but understanding that evolution occurred once organisms appeared and investigating how this happens does not depend on understanding exactly how life began. The current scientific consensus is that the complex biochemistry that makes up life came from simpler chemical reactions, but it is unclear how this occurred. Not much is certain about the earliest developments in life, the structure of the first living things, or the identity and nature of any last universal common ancestor or ancestral gene pool. Consequently, there is no scientific consensus on how life began, but proposals include self-replicating molecules such as RNA, and the assembly of simple cells. Astronomers have recently discovered the existence of complex organic molecules in space. Small organic molecules were found to have evolved into complex aromatic molecules over a period of several thousand years. Chemical evolution is an exciting topic of study because it yields insight into the processes which lead to the generation of the chemical materials essential for the development of life. If the chemical evolution of organic molecules is a universal process, life is unlikely to be a uniquely terrestrial phenomenon and is instead likely to be found wherever the essential chemical ingredients occur.

---

[1] From Greek ἐξελίγμός and ἐξελίσσω (Latin evolutio and evolvere), to evolve (develop, generate, process, originate, educe).

In colloquial contexts, evolution usually refers to development over a long time scale, and the question is not important whether evolution tends toward more complexity. Many definitions tend to postulate or assume that complexity expresses a condition of numerous elements in a system and numerous forms of relationships among the elements. At the same time, what is complex and what is simple is relative and changes with time.

Natura non facit saltus (Latin for "nature does not make jumps/leaps") has been a principle of natural philosophy at least since *Aristotle*'s time, used as an axiom by *Gottfried Wilhelm Leibniz, Isaac Newton* and *Charles Darwin* as well as others. A modern understanding of evolution includes continuing development, but also leaps (catastrophes, see Chapter 2.2.2.6). This is referred to as "transformation of quantity into quality" (*dialectic leap*) and may characterize the current discussion on the impacts of climate change. However, it is hard to envisage a physical situation in which a quantifiable parameter can increase indefinitely without a critical condition occurring.

Physical processes – starting with the Big Bang – created the first atoms (which form chemical elements) and physical conditions permanently affecting the subsequent chemical and biological evolution. Compared with the Big Bang as the beginning of physical evolution, the creation of molecules and life can be referred to as the starting point of a chemical and biological evolution, respectively. Life became a geological force with oxygenic photosynthesis (Chapter 2.2.2.4 and 2.4.4) and created an interactive feedback with chemical and physical evolution. After forming the geosphere and the first atmosphere in the sense of a potentially habitable system, and later the biosphere with the modern atmosphere, a habitable climate system evolved. But life created a further dimension, *human intelligence*, which becomes another geological force (human evolution – today approaching a critical condition which we call *crisis*). Human intelligence disengaged humankind from the rigorous necessities of nature and provided unlimited scope for reproduction (at least in the past). Man in all his activities and social organizations is part of, and cannot stand in opposition to or be a detached or external observer of, nature. However, the new dimension (or quality) of human intelligence as a result of biological evolution – without some global ecomanagement – could change the climate system in a direction not providing the internal principle of self-preservation. Mankind converts the biosphere into a noosphere (Chapter 2.4.1). Chemical evolution is now interloped with human evolution. Changing fluxes and concentrations of chemicals in bio- (or rather noo-) geochemical cycles with a subsequent changing climate system seems to be the creation of a human-chemical evolution.

Under the evolution of the earth and the climate system, we will simply understand the historical development from earliest times until the present. Theories for how the atmosphere and ocean formed must begin with an idea of how the earth itself originated (Kasting 1993). An understanding of our atmosphere and the climate system is incomplete without going into the past. "The farther backward you can look the farther forward you can see" (*Winston Churchill*). "Evolution is God's, or Nature's method of creation. Creation is not an event that happened in 4004 B.C.; it is a process that begun some 10 billion years ago and is still under way" (*Theodosius Grygorovych Dobzhansky*). "Progress is not an objective fact of nature and cannot therefore be used to justify a normative ethic" (Ruse 1999, p. 221).

2.1 The pre-biological period

2.1.1 Origin of elements, molecules and the earth

Our galaxy is probably 13.80 ± 0.06 Gyr old (Bennett et al. 2013) and was formed by the hot Big Bang, assuming that the whole mass of the galaxy was concentrated in a primordial core. Based on the principles of physics, it is assumed that density and temperature were about 10^{94} g cm^{-3} and 10^{32} K, respectively (Tolstikhin and Kramers 2008). The initial products of the Big Bang were neutrons which, when released from dense confinement (quarks), began to decay into protons and electrons: $n^o = e^- + p^+$. As the half-life for this reaction is 12.8 minutes, we can assume that soon after the Big Bang, half of all the matter in the universe was protons and half electrons. Temperatures and pressures were still high and nuclear reactions possibly led to the production of helium via the interaction of neutrons and protons (remember that the proton already represents hydrogen), see Fig. 2.1. Recall that it is the number of protons in the nucleus that defines an element, not the number of protons plus neutrons (which determines its weight). Elements with different numbers of neutrons are termed isotopes, and different elements with the same number of neutrons plus protons (nucleons) are termed isobars.

Hydrogen and helium produced in the Big Bang served as the "feed stock" from which all heavier elements were later created in stars. The fusion of protons to form helium is the major source of energy in the solar system. This proceeds at a very slow and uniform rate, with the lifetime of the proton before it is fused to deuterium of about 10 Gyr (note that the proton lifetime concerning its decay is $> 10^{30}$ yr). From He to Fe, the binding energy per nucleon increases with atomic number, and fusions are usually exothermic and provide an energy source. Beyond Fe, the binding energy decreases and exothermic reactions do not occur; elements are formed through scavenging of fast neutrons until ^{209}Bi. Heavier elements only are produced in shock waves of supernova explosions.

The most abundant elements (Fig. 2.2) up to Fe are multiples of ^{4}He (^{12}C, ^{16}O, ^{24}Mg, ^{28}Si, ^{32}S, etc.). During the red giant phase of stellar evolution, free neutrons are generated which can interact with all nuclei and build up all the heavy elements up to Bi; all nuclides with the atomic number ≥ 84 are radioactive. Recently (2003) it has been found that even ^{209}Bi decays, but extremely slowly ($\tau_{1/2} = 1.9 \cdot 10^{19}$ yr). The build-up of elements of every known stable isotope depends on different conditions of density and temperature. Thus, the production process required cycles of star formation, element formation in stellar cores, and ejection of matter to produce a gas enriched with heavy elements from which new generations of stars could form. The synthesis of material and subsequent mixing of dust and gas between stars produced the solar mix of elements in the proportions that are called "cosmic abundance" (Fig. 2.2 and cf. Table 2.13).

In addition to stable elements, radioactive elements are also produced in stars. There are four natural radioactive decay series; ^{232}Th ($\tau_{1/2} = 1.405 \cdot 10^{10}$ yr), ^{238}U ($\tau_{1/2} = 4.468 \cdot 10^9$ yr), ^{227}Ac ($\tau_{1/2} = 21.77$ yr), and ^{237}Np ($\tau_{1/2} = 2.14 \cdot 10^6$ yr). Np is extremely rare on earth, however. The initial elements of the Th and Ac series are ^{244}Pu ($\tau_{1/2} = 1.4 \cdot 10^{11}$ yr) and ^{239}Pu ($\tau_{1/2} = 2.411 \cdot 10^4$ yr), respectively. Natural Pu is also extremely rare. Thus the unstable but long-lived isotopes

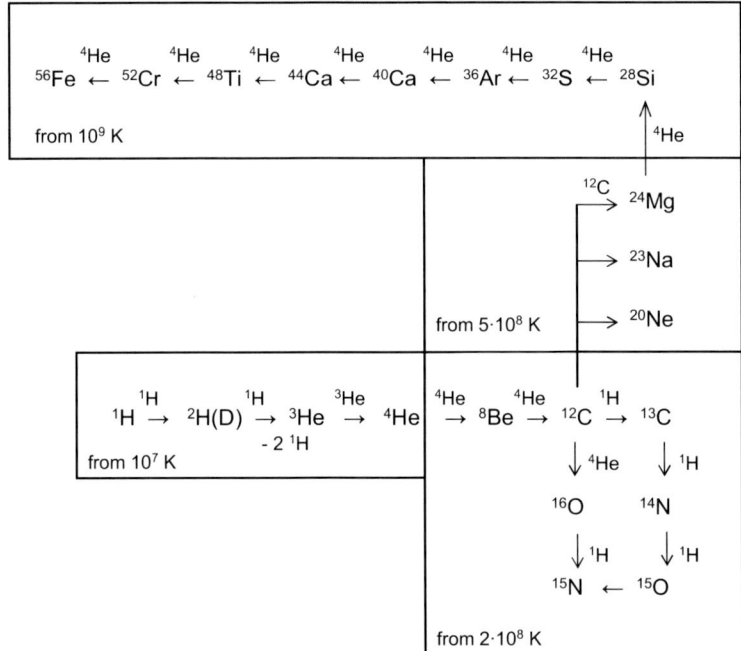

Fig. 2.1 Scheme of thermonuclear formation of chemical elements (fusions reaction in stars).

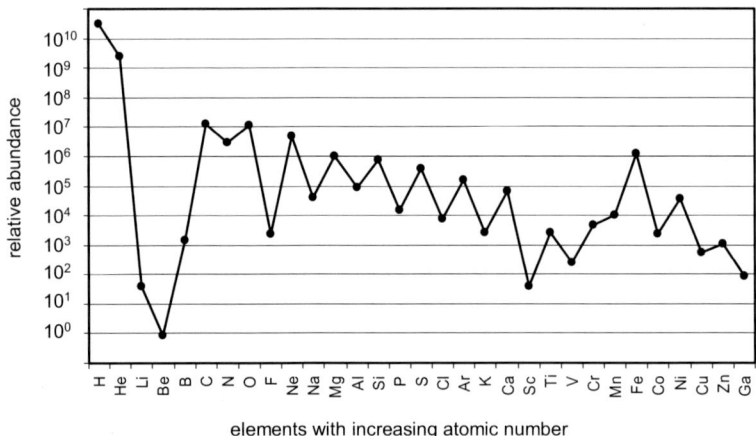

Fig. 2.2 Abundance of chemical elements in space.

(^{232}Th, ^{235}U, ^{236}U) make up the internal heat source that drives volcanic activity and processes related to internal convection in the terrestrial planets; these three nuclides (and ^{244}Pu) are used for determination of the age of the earth, meteorites and other celestials. The four decay series naturally produce 16 radioactive elements with atomic numbers < 82 (Rb, K, Cd, In, Te, La, Nd, Sm, Gd, Lu, Hf, Ta, Re, Os, Pt, Pb) which all have extreme long lifetimes of from 10^{10} to 10^{24} years. Note also that many unstable nuclides can be produced artificially.

The Milky Way, sometimes called simply "the Galaxy", is the galaxy in which our solar system is located. It is extremely difficult to define when the Milky Way formed, but the age of the oldest star in the Galaxy yet discovered, is estimated to be about 13.7 Gyr, nearly as old as the Universe itself. A few days before passing this manuscript to the publisher, the discovery of the oldest star ever observed was published (Keller et al. 2014). SMSS J031300.36-670839.3 (abbreviated to SM0313), located 6000 light-years from Earth, is 13.7 Gyr old, besting the current record-holder HD 140283 (13.6 Gyr), discovered in 2013 (Bond et al. 2013).

The region between the stars in a galaxy, termed the interstellar medium, has very low densities, but is filled with gas, dust, magnetic fields, and charged particles. Approximately 99% of the mass of the interstellar medium is in the form of gas (where denser regions are termed interstellar clouds) with the remainder primarily in the form of dust. The total mass of the gas and dust in the interstellar medium is about 15% of the total mass of visible matter in the Milky Way. The exact nature and origin of interstellar dust grains is unknown, but they are presumably ejected from stars. One likely source is from red giant stars late in their lives. Interstellar dust grains are typically a fraction of a micron across, irregularly shaped, and composed of carbon and/or silicates.

The formation of molecules is impossible in stars because of the high temperature, but in the interstellar medium (T between 10 and 20 K) chemical reactions are possible which have the potential to create molecules. There are essential differences between laboratory and interstellar chemistry, namely the much larger time scale available in interstellar space.

The role of the interstellar dust in molecular growth is important because the dust particles provide a surface (heterogeneous chemistry) where reactions may occur under much higher density (collision probability). Radiation can break down the surface molecules and produce a wider variety of molecules. The study of interstellar chemistry began in the late 1930s with the observation of molecular absorption spectra in distant stars within the Galaxy. Now called "Large Molecule Heimat" (LHM). The species CH, CH^+, and CN have electronic spectra in an accessible wavelength region where the earth's atmosphere is still transparent. The character of these observable interstellar clouds is low density and essentially atomic with a small diatomic molecular component. Our discovery that the universe is highly molecular is quite recent. At present, nearly 200 molecular species are listed. Among others, the following non-organic molecules and radicals, which will later be highlighted in atmospheric chemistry, have been detected interstellar: CO, CO_2, HCO, HCN, OCS, H_2S, SO_2, N_2, NH_2, NO, HNO, N_2O, O_2, O_3, OH, HO_2, H_2O, H_2O_2. Du et al. (2011) demonstrate that H_2O_2 – an important intermediate in the formation of water – can be produced on the interstellar dust grains and released into the gas phase through non-thermal desorption via surface exothermic reactions; in the early phase, the gas-phase abundance of H_2O_2 can be much higher than the current detected value. The molecular abundances do not follow the cosmic abundance of the elements. The rich variety of observed species includes ions and free radicals. In particular, of the observed species with six or more atoms (presently 60 species)[2], all contain carbon. Thus, the chemistry of positively identified polyatomic species observed in the gas phase is carbon chemistry (Klemperer 2006). The first identification of organic polymers in interstellar grains was made by Wickramasinghe (1974) and the first association of a biopolymer with interstellar dust was made by Hoyle and Wickramasinghe (1977).

The heterogeneity of interstellar and circumstellar regions gives rise to a variety of chemistries. Most of the molecular material in our galaxy and elsewhere occurs in *giant molecular clouds* (GMCs). It appears that these are the simplest regions whose bulk is not penetrated by optical radiation from either the galactic radiation field or that from nearby stars. In these regions, temperatures are 10–20 K and the molecular processes, not being at thermodynamic equilibrium, require energy input to initiate. However, high-energy cosmic rays penetrate and produce volume ionization. The chemistry is initiated by the primary ionization of H_2 and He, which constitute > 99% of the cosmic material in molecular clouds, providing primarily H_2^+ and He^+. H_2^+ is very rapidly converted to H_3^+ by reaction with H_2. Initially, the presence of nonpolar H_3^+ was surmised from observations of rotational transitions of the very abundant highly polar ion HCO^+, produced by proton transfer from H_3^+ to CO. Because the abundance of a collision complex will scale with the abundances of the collision partners, their collision frequency, and the binding energy of the complex, it appears that the species most likely to attract are an ion and H_2. The most abundant ion in dense molecular clouds is HCO^+. Thus, the species of interest initially is H_2—HCO^+ (Herbst and Klemperer 1973). Within these giant molecular clouds, the greatest concentration and diversity of molecules are found in pockets, known as *hot cores*, near certain recently formed luminous stars. Hot cores are very compact (fractions of a light year), warm (a few hundred K, compared to 10 or 20 K for the general interstellar gas), and dense (more than 10^6 hydrogen molecules per cm^3) condensations, with remarkably rich millimeter-wave emission-line spectra.

From the abundance of "reactive" volatile elements in space in the order H—O—C—N, it appears that the simplest molecules derived (apart from H_2, O_2, and N_2) are bonds between the following elements (the bonding energy in kJ mol^{-1} is given in parenthesis); H—C (416), H—O (464), H—N (391), C—O (360) and N—O (181). Because of the excess of hydrogen in space, the hydrides (OH_2, CH_4, NH_3, SH_2) should have the highest molecular abundance among the compounds derived from such elements; furthermore other simple gaseous molecules are CO, CO_2, and HCN. Correspondingly, the simplest non-gaseous stable molecules are ammonium and nitrate but also hydrocarbons. Because of the H excess, highly oxidized compounds (e. g. NO_x and nitrates) are unlikely. Moreover gaseous NO_x molecules are much more unstable compared with the other listed compounds; most of the oxygen is bonded in H_2O, CO_x, FeO and SiO_2.

Fig. 2.3 shows schematically the possible reactions, established by modeling as well as kinetic and thermodynamic considerations. All these reactions are sufficient to produce and destroy polyatomic species such as H_2O, HCN, NH_3, and HCHO.

[2] When the first edition was published (2010), only 49 species were known.

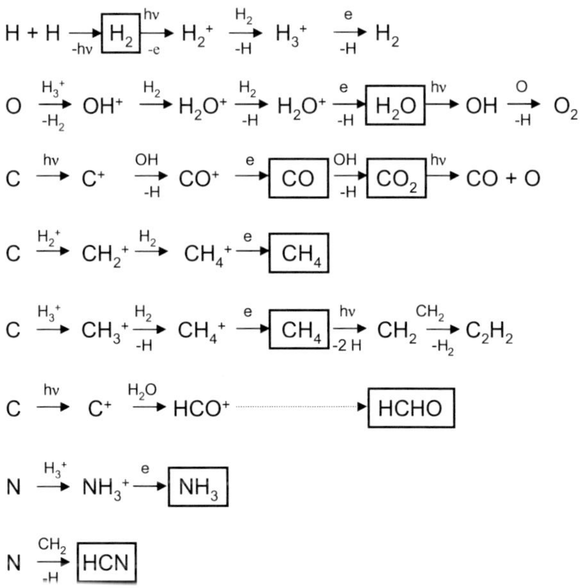

$$H + H \xrightarrow[-h\nu]{} \boxed{H_2} \xrightarrow[-e]{h\nu} H_2^+ \xrightarrow[-H]{H_2} H_3^+ \xrightarrow[-H]{e} H_2$$

$$O \xrightarrow[-H_2]{H_3^+} OH^+ \xrightarrow{H_2} H_2O^+ \xrightarrow[-H]{H_2} H_2O^+ \xrightarrow[-H]{e} \boxed{H_2O} \xrightarrow{h\nu} OH \xrightarrow[-H]{O} O_2$$

$$C \xrightarrow{h\nu} C^+ \xrightarrow[-H]{OH} CO^+ \xrightarrow{e} \boxed{CO} \xrightarrow[-H]{OH} \boxed{CO_2} \xrightarrow{h\nu} CO + O$$

$$C \xrightarrow{H_2^+} CH_2^+ \xrightarrow{H_2} CH_4^+ \xrightarrow{e} \boxed{CH_4}$$

$$C \xrightarrow{H_3^+} CH_3^+ \xrightarrow[-H]{H_2} CH_4^+ \xrightarrow{e} \boxed{CH_4} \xrightarrow[-2H]{h\nu} CH_2 \xrightarrow[-H_2]{CH_2} C_2H_2$$

$$C \xrightarrow{h\nu} C^+ \xrightarrow{H_2O} HCO^+ \dashrightarrow \boxed{HCHO}$$

$$N \xrightarrow{H_3^+} NH_3^+ \xrightarrow{e} \boxed{NH_3}$$

$$N \xrightarrow[-H]{CH_2} \boxed{HCN}$$

Fig. 2.3 Interstellar formation of molecules.

Overall, the original nebula is likely to have been composed of about 98 % gases (H, He, and noble gases), 1.5 % ice (H_2O, NH_3, and CH_4), and 0.5 % solid materials (Schlesinger 1997 and literature therein).

According to conventional astrophysical theory, our solar system (the sun and its planetary system) was formed from a cloud of gas and dust that coalesced under the force of gravitational attraction approximately 5 Gyr ago. This matter was formed from a collapsed supernova core, a neutron star with radiant energy and protons in the solar wind. The formation of planetesimals and planets was accompanied in many cases by high temperatures and violent conditions, and most interstellar dust particles were destroyed. However, the class of meteorites known as carbonaceous chondrites contains small particles with unusual isotopic ratios which indicate that they did not form in the solar nebula, but rather must have been formed in a region with an anomalous composition (e. g. as outflow from an evolved star) long before the formation of the solar system. Therefore these particles must have been part of the interstellar grain population prior to the formation of the solar nebula. Other debris from the supernova remains as gases and particulate matter, termed *solar nebula*. This system cooled, particles rose by condensation growth and the sun grew by gravitational settlement about 4.6 Gyr ago. Cooling and subsequent condensation occurred with distance from the protosun, resulting in an enlargement of heavier elements (e. g. Fe) at the inner circle, corresponding to the condensation temperature of matter (Table 2.1). Mercury formed closest to the sun, mostly from iron and other materials in solar nebula that condense at high temperatures (above 1400 K). It also shows the highest density (5.4 g cm^{-3}) of all

Table 2.1 Temperature-dependent condensation of compounds and formation of minerals, after Bérczi and Lukács (2001).

T (in K)	elements, compounds, reactions	mineral
1600	CaO, Al_2O_3, REE oxides[a]	oxides (e.g. perovskite)
1300	Fe, Ni alloy metals	Fe—Ni
1200	$MgO + SiO_2 \rightarrow Mg\,SiO_3$	enstatite (pyroxene)
1000	alkali oxides + Al_2O_3 + SiO_2	feldspar
1200–490	$Fe + O \rightarrow FeO$; $FeO + MgSiO_3$	olivine
680	$H_2S + Fe \rightarrow FeS$	troilite
550	Ca minerals + H_2O	tremolite
425	olivine + H_2O	serpentine
175	ice-H_2O crystallizes	water-ice
150	gaseous NH_3 + ice-$H_2O \rightarrow [NH_3 \cdot H_2O]$	ammonia-hydrate
120	gaseous CH_4 + ice-$H_2O \rightarrow [CH_4 \cdot 7\,H_2O]$	methane-hydrate
65	CH_4, Ar crystallize	methane and argon ice

[a] REE = rare earth element

earth-like planets, and in contrast, Jupiter contains more hydrogen and helium, with an average density of only 1.25 g cm^{-3}.

The earth, like the other solid planetary bodies, formed by the accretion of large solid objects in a short time between 10 and 100 Myr (Fig. 2.4); the postaccreationary period is dated from ~4.5 Gyr ago (Kasting and Catling 2003). Earlier theories (Walker 1977, Holland 1984) suggest that the earth was formed largely in the form of small grains, but interspersed with occasional major pieces. The largest particles (proto-planets) developed a gravitational field and attracted further material to add to its growth. We assume that all this primary material was cold (10 K) at first. The energy of the collisions between the larger microplanets, as well as interior radioactive and gravitational heating, generated a huge amount of heat, and the earth and other planets would have been initially molten. The molten materials were also inhomogeneously distributed over the protoplanet. The moon formed rather late in this process, about 45 Myr after the inner planets began to form. The current theory is that a Mars-sized planetoid, sometimes named Theia, collided with the earth at this time. As astronomical collisions go, this was a mere cosmic fender-bender. The bodies, both molten, merged fairly smoothly, adding about 10% to the earth's mass. The moon formed from the minimal orbiting debris (about 0.01 earth masses) resulting from this low-speed crash (Wood and Halliday 2005, Wood et al. 2006). During the formation of the earth by the accumulation of cold solids, very little gaseous material was incorporated. Evidence of this comes from the extremely low level of the non-radiogenic noble gases in the atmosphere of the earth. Among those, only helium could have escaped into space, and only xenon could have been significantly removed by absorption into rocks. Neon, argon, krypton would have been maintained as an atmospheric components. Most of the helium found on earth is ^{4}He, the result of the radioactive decay of uranium and thorium; primordial helium is ^{3}He.

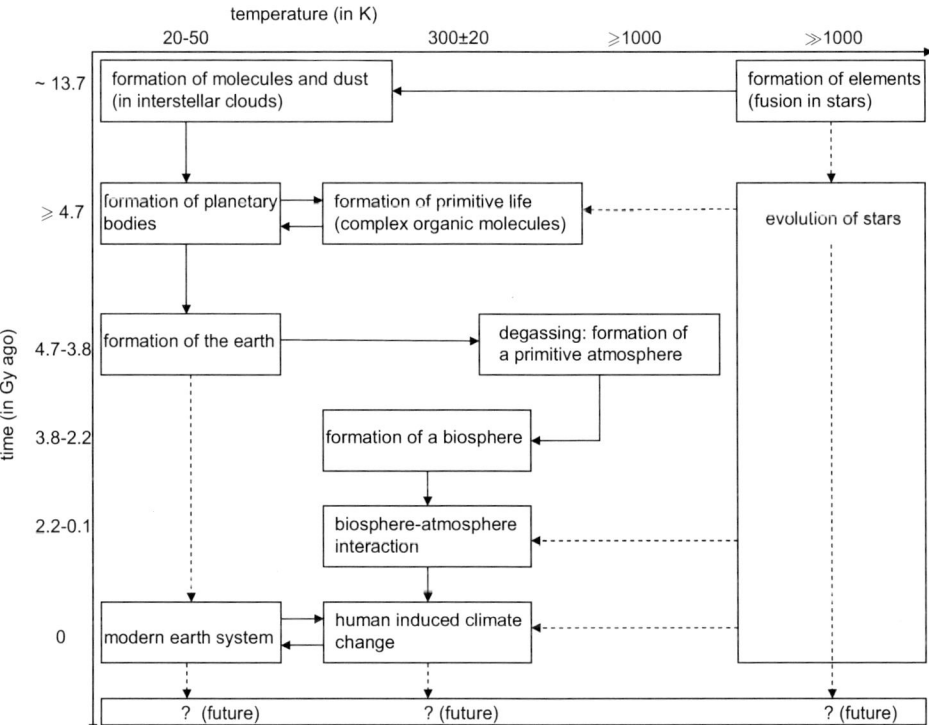

Fig. 2.4 Evolution of the Solar System.

The heavier molten iron sank to become the core, while materials of lower density (particularly the silicates) made their way to the surface. The lightest of all became the crust as a sort of "scum" on the surface. This crust melted and reformed numerous times, because it was continuously broken up by gigantic magma currents that erupted from the depths of the planet and tore the thin crust. The dissipation of heat into space began the cooling of our planet. In the magma ocean, blocks began to appear, formed from high-melting-point minerals sinking again into the heart of the earth. Approximately 500 million years after the birth of the earth, this incandescent landscape began to cool down. When the temperature fell below 1000 °C, the regions of lower temperatures consolidated, become more stable, and initiated the assembly of the future crust. Only with the further cooling of the planet, might those fragments become numerous and large enough to form a first, thin, solid cover, a true primitive crust. This primordial crust might have developed as a warm expanse of rocks (some hundreds of degrees Celsius), interrupted by numerous large breaks, from which enormous quantities of magma continued to erupt. The composition of the crust began to change by a sort of distillation. Disrupted by highly energetic convective movements, the thin lithospheric covering would have been fragmented into numerous small plates in continuous mutual movement, separated and deformed by bands of intense volcanism. During this continuous remelting of the "protocrust", heavier rock gradually sank deeper into the mantle, leaving behind a lighter magma richer in silicates. Thus, around the basalts appeared andesites: fine granular volcanic rocks, whose name derives from the Andes, where several volcanoes are known to form rocks of this type. Gradually, a granitic crust emerged.

At this early point in the history of the solar system, there was a relatively short period (50 Myr or so) of intense meteoric bombardment (termed the *late heavy bombardment* LHB) which would have continually opened new holes in the crust, immediately filled by magma. The scars left by this intense meteoric bombardment have been almost totally erased on the earth by subsequent reworking of the crust. The evidence for the LHB is quite strong, however. It comes mostly from lunar astronomy (big craters formed significantly later than the lunar maria large, which are dark, basaltic plains on earth's Moon, formed by ancient volcanic eruptions.) and the lunar rocks recovered from space exploration. The implication is that the post-Hadean granitic crust was not the product of gradual distillation, but of catastrophic reworking after the protocrust was destroyed by the LHB (Koeberl 2006). As Ryder (2003) pointed out, our best information on the LHB comes from the moon. Since much of the moon's original crust remains, why would we think that the crust of the earth was destroyed? Some even argue that life evolved before the LHB and survived to tell the tale (Russell and Arndt 2005).

Harkins (1917) showed that 99 % of the material in ordinary meteorites consists of seven, even-numbered elements − iron (Fe), oxygen (O), nickel (Ni), silicon (Si), magnesium (Mg), sulfur (S) and calcium (Ca). However, Payne (1925) and Russell (1929) showed in the 1920s that the solar atmosphere is mostly hydrogen (H) and helium (He). Today there are theories (e. g. Manual et al. 1998) that the sun's chemical composition (at least in the core) is similar to that of meteorites, i. e., Fe is the most abundant element.

Today, the more external part of the crust or lithosphere constitutes the superficial covering of the earth. Two kinds of crust are easily distinguished by composition, thickness and consistency; continental crust and oceanic crust. Continental crust has a thickness that, in mountain chains, may reach 40 kilometers. It is composed mainly of metamorphic rock and igneous blocks enriched with potassium, uranium, thorium and silicon. This forms the diffuse granitic bedrock of 45 % of the land surface of the earth. The oceanic crust has a more modest thickness, in the order of 5−6 kilometers, and is made up of basaltic blocks composed of silicates enriched with aluminium, iron and manganese. It is continuously renewed along mid-ocean ridges (cf. Table 2.2).

At all over the entire history of the earth the sum of all (non-radioactive) elements is constant despite exhausting of hydrogen and helium into space. The loss of hydrogen to space (and later its deep burial in hydrocarbons) is the reason for the changing redox[3] state from a low oxygen to a more oxidized environment. This happens with photolysis of water (2.1) and hydrides such as CH_4, NH_3 and H_2S (2.2) and later with marine photosynthesis (2.3)

$$H_2O \xrightarrow{h\nu} O + H_2(\uparrow) \tag{2.1}$$

---

[3] Shorthand for reduction-oxidation reaction.

Table 2.2 Geographic quantities of the atmosphere, ocean and continents; after Holland (1984) and Weast (1980).

mass of the earth	$6.0 \cdot 10^{27}$ g (density 5.52 g cm^{-3})
mass of the atmosphere	$5.2 \cdot 10^{21}$ g
mass of the troposphere (up to 11 km)	$4.0 \cdot 10^{21}$ g
volume of the earth	$1.08 \cdot 10^{21}$ m^3
volume of the troposphere (up to 11 km)	$5.75 \cdot 10^{18}$ m^3
volume of world's ocean	$1.37 \cdot 10^{18}$ m^3 (density 1.036 g cm^{-3})
area of northern hemispheric ocean	$1.54 \cdot 10^{14}$ m^2
area of southern hemispheric ocean	$2.10 \cdot 10^{14}$ m^{2b}
area of continents northern hemisphere	$1.03 \cdot 10^{14}$ m^2
area of continents southern hemisphere	$0.46 \cdot 10^{14}$ m^2
depth of the crust[c]	35 km (locally varies between 5 and 70 km)
mass of the crust[c]	$4.9 \cdot 10^{25}$ g[d]
depth of the upper mantle[c]	60 km (locally varies between 5 and 200 km)
mass of the upper mantle[c]	$4.3 \cdot 10^{25}$ g[e]
depth of the mantle	2890 km
mass of the lower mantle[f]	$3.4 \cdot 10^{27}$ g[f]
thickness of the earth's atmosphere[a]	1000 km

[a] It is not a definite number. The reason that there is no definite number is because there is no set boundary where the atmosphere ends.
[b] total ocean area $3.62 \cdot 10^{14}$ m^2 after Schlesinger (1997)
[c] the lithosphere comprises the crust and the upper mantle
[d] assuming 35 km depth and 2.7 g cm^3 density
[e] assuming 60 km depth and 3.3 g cm^3 density
[f] between 60 and 2890 km; density about 6.0 g cm^{-3}

$$XH_n \xrightarrow{h\nu} X + \frac{n}{2} H_2 (\uparrow) \xrightarrow{mO} XO_m \text{ (oxide and anhydride)}$$
$$\xrightarrow{H_2O} H_2XO_{m+1} \text{ (oxoacid)} \tag{2.2}$$

$$n\,CO_2 + n\,H_2O \xrightarrow{h\nu} (CH_2O)_n (\downarrow) + O_2 \tag{2.3}$$

With an increasing state of oxidation, a rise of acidity also occurs, and the two combine until achieving an equilibrium state in geochemical evolution. Oxidation/ reduction and the acidity potential are interlinked where organisms create a biogeochemical evolution by separating oxidative and reductive processes among different living species. Table 2.3 summarizes the most important chemical relationships between such components. It is remarkable that only C, N and S compounds are gaseous and/or dissolved in water in all redox states, which makes these compounds globally distributable and exchangeable among different reservoirs to provide global cycles (see Chapter 2.4.2). The other minor elements listed in Table 2.3 provide important compounds for life and the geogenic (abiotic) environment but are much less volatile or almost immobile (Si, P). Some oxygenated halogens are unstable. Chemical evolution alone can change the distribution of the elements among different molecules and reservoirs creating a heterogeneous world (Fig. 2.4).

Table 2.3 Substances in reduced state (hydrides) and in oxidized form (oxides) as well as the corresponding oxoacids. If not mentioned, the species are gaseous under standard conditions; aq − exists only dissolved in water.

hydrogen excess (reduced state)	oxygen excess (oxidized state)	oxoacids
H_2	OH, HO_2	H_2O (liquid)
OH_2 (H_2O, liquid)	O_2, O_3	H_2O_2
CH_4	CO, CO_2	H_2CO_3 (*aq*)
NH_3	NO, NO_2, NO_3, N_2O_5	HNO_2, HNO_3
SH_2 (H_2S)	SO_2, SO_3	H_2SO_3 (*aq*), H_2SO_4 (*aq*)
SiH_4	SiO, SiO_2 (solid, unsoluble)	$Si(OH)_2$, H_4SiO_4 (solid, unsoluble)
PH_3	P_2O_5 (solid)	H_3PO_4 (solid)
AsH_3	As_2O_3 (solid)	$As(OH)_3$ (*aq*)
HBr	BrO, Br_2O, Br_2O_3 (solid)	$HOBr$, $HBrO_3$ (*aq*)
HCl	ClO, Cl_2O, ClO_2, Cl_2O_6 (liquid)	$HOCl$ (*aq*), $HClO_2$ (*aq*), $HClO_3$ (*aq*)
HJ	JO, J_2O_5 (solid)	HJO_3 (solid)

2.1.2 Origin of organic bonded carbon

Space consists of 98 % hydrogen (3/4) and helium (1/4); of the remaining 2 %, three-quarters is composed of just two elements, namely oxygen (2/3) and carbon (1/3). Based on the molar ratios a formula for the "space molecule" would be about $H_{2600}O_3C_2$. It is logical that, concerning compounds, water (H_2O) and hydrocarbons (C_xH_y) are the most abundant molecules in space. Carbon (among other elements in the four main groups of the periodical system; silicon, germanium, tin and lead) offers comparable affinities to electropositive and electronegative elements (to the left and right of the periodical system). In other words, it has a symmetric or even harmonic position in the word of matter. Therefore, the number of stable molecules (compounds with all other elements within the periodical system) must be larger than of any other element out of these four main groups. The question may arise why "organic matter" is composed of carbon and not of silicon. The answer is simple and twofold: (a) silicon is too heavy to create organisms that would be mobile against earth's gravity, and (b) only carbon can provide gaseous molecules in the most electronegative state (CH_4) and the most electropositive state (CO_2). Nevertheless, the non-organic world, the "fundament" of life is composed from solid SiO_2. The diversity of silicon compounds is no less than that of carbon. However most of the silanes, the counterpart of hydrocarbons (in other words, hydrosilicon), are unstable and the variety of molecules with unsaturated bonding is much less then for carbon. The ability to create element chains, rings and networks (in two and three dimensions) is analoguous for carbon and silicon and explains why these elements are the basic matter of materials, more soft for organic and more rigid for inorganic.

2.1.2.1 What is organic chemistry?

Before we discuss the source of organic carbon, first we have to address what "organic" means in chemistry because in the colloquial language it is equated with "life." The term derives from ὄργανον (Greek) and *organum* (Latin) which means tool, device and voice and was first used in the seventeenth century in French with various derived forms and enlarged senses (*organe, organiser, organisme, organisation, organique*). "Organic" compounds were known for centuries but were not distinguished from "inorganic" matter. A systematic classification of chemistry was done into mineral, vegetable and animal according to its origin by the French chemist *Nicolas Lémery* who wrote *Cours de chymie* (1675, cited after Kopp 1831). According to Walden (1941), the first use of the term "organic chemistry" is now attributed to the Swedish chemist *Jöns Jacob Berzelius* who termed it "*organisk kemie*" in a book published in 1806. After *Lavoisier's* revolutionary book, *Traité élémentaire de chimie* (1789), *Berzelius* wrote the first textbook *Lärbok i kemien* (1817−1830) in six volumes. It was soon published in French (1829) still with the now traditional title, *Traité de chimie minerale, vegetale et animale* in 8 volumes with the subtitle "Chimie organique" (2ème partie − 3 volumes). In Germany the first handbook, subtitled "organic compounds", is the third volume of *Handbuch der theoretischen Chemie* by *Leopold Gmelin* (1819), later rearranged into separate volumes of inorganic and organic chemistry (from 1848). It was believed since that time that organic matter (also termed "organized") could not be synthesized from its elements and that a special force, the *vital force*, is needed for its production. However, organic molecules can be produced by processes not involving life. *Friedrich Wöhler* destroyed the theory of vital force by the synthesis of urea in 1828; an event generally seen as the turning point.

First *Lavoisier* found systematically that vegetable matter is composed from C, H and O and that in animal matter additionally N and P are present. *Liebig* ("Organic chemistry in its application to agriculture and physiology", 1840) defined the task of organic chemistry as follows:

> The object of organic chemistry is to discover the chemical conditions which are essential to the life and perfect development of animals and vegetables, and, generally, to investigate all those processes of organic nature which are due to the operation of chemical laws. (The first phrase in Part I: "Of the chemical processes in the nutrition of vegetables").

This definition is still very much focused on the original idea that all "organic" is definitely equal to "life", despite the fact that it was written by *Liebig* after *Wöhler's* rejection of the "vital force" in association with organic compounds. Gmelin (1848) wrote that carbon is the only element never missing and hence it is the only essential constituent in an organic compound. There has been no change in the definition since that time: the lexical database WordNet (Princeton University) defines organic chemistry as:

> … the chemistry of compounds containing carbon (originally defined as the chemistry of substances produced by living organisms but now extended to substances synthesized artificially).

Table 2.4 Different simple inorganic carbon compounds.

oxides and oxoacids	CO	carbon monoxide
	CO_2	carbon dioxide
	H_2CO_3	carbonic acid
	$H_2C_2O_6$	oxalic acid
sulfides	CS_2	carbon disulfide
	COS	carbonyl sulfide
halogens[a]	CCl_4	carbon tetrachloride
	C_2Cl_6	dicarbon hexachloride
	C_2Cl_4	dicarbon tetrachloride
	C_2Cl_2	dicarbon dichloride
	$COCl_2$	carbonyl chloride
nitriles	HCN	hydrocyanic acid
rhodanides	HSCN	rhodanic acid
carbides	Ca_2C	calcium carbide[b]

[a] here given as example for chlorine, but exchangeable by F, Br, and I
[b] as an example of many metallic carbides (cf. Eq. 2.8–2.10); unstable against water

According to this definition it appears, however, that (for example) carbon dioxide is an organic compound. It has been convenient to distinguish between inorganic and organic carbon compounds to explain the cycles between the biosphere and the atmosphere. On the other hand, when we state that there is a prime chemistry considering a limited number of elements and an unlimited (or at least immense) number of molecules in defined numeric relationship between the elements (including carbon), there is no need to separate chemistry into inorganic and organic chemistry. In contrast to the countless number of "organic" carbon compounds, the number of "inorganic" carbon compounds is rather small (of course, we count here only simple compounds found in nature), Table 2.4.

It seems that a useful condition to specify an "organic" compound is the annex to the given definition that there must be at least one C—H bonding in the molecule.

2.1.2.2 Origin of carbon

Now let us look at the origin of organic compounds on earth. There is evidence that most of the earth's volatiles may have been supplied by a "late heavy bombardment" (LHB) of comets and carbonaceous meteorites, scattered into the inner Solar system following the formation of the giant planets. How much in the way of intact organic molecules of potential prebiotic interest survived delivery to the earth has become an increasingly debated topic over the last several years. The principal source for such intact organics was probably the accretion of interplanetary dust particles of cometary origin. Holland et al. (2009) have shown based on the discovery of primordial Kr in samples derived from earth's mantle[4] that the noble gases

---

[4] Noble gas isotopes are key tracers of both the origin of volatiles found within planets and the processes that control their eventual distribution between planetary interiors and atmospheres.

in earth's atmosphere and oceans are dominantly derived from later volatile capture rather than impact degassing or outgassing of the solid earth during its main accretionary stage.

Before life appeared on earth, there were two possible sources of organic molecules on the early earth:

- Terrestrial origins – organic synthesis driven by impact shocks or by other energy sources (such as ultraviolet light or electrical discharges),
- extraterrestrial origins – delivery by objects (e. g. carbonaceous chondrites) or the gravitational attraction of organic molecules or primitive life-forms from space.

An earlier hypothesis suggested that the primitive (or better termed, primary) atmosphere consisted of NH_3 and CH_4. This idea was supported by the finding of both species being in some meteorites and the belief that the solar nebula also contains a small amount of ammonia and methane (Schlesinger 1997). The existence of a NH_3—CH_4 atmosphere was believed to be a precondition for the origin of life. The well-known *Miller-Urey* experiment in 1953 showed that under UV radiation organic molecules can be formed in such an atmosphere (Miller 1953, Miller and Urey 1959). However, the intensive UV radiation at the earth's beginning would have destroyed NH_3 and CH_4 soon and no processes are known to chemically form both species in air. Furthermore, it became evident that it is difficult to synthesize prebiotic compounds in a non-reducing atmosphere (Stribling and Miller 1987). Whether the mixture of gases used in the *Miller-Urey* experiment truly reflects the atmospheric content of the early earth is a controversial topic. Other less reducing gases produce a lower yield and variety. It was once thought that appreciable amounts of molecular oxygen were present in the prebiotic atmosphere, which would have essentially prevented the formation of organic molecules; however, the current scientific consensus is that such was not the case. Hence organic matter must be available on the early earth (a) as a precondition for creating life, and (b) to explain either CH_4 and/or CO_2 in the first atmosphere.

There is no longer any doubt that organic compounds are relatively common in space, especially in the outer solar system where volatiles are not evaporated by solar heating (cf. Fig. 2.3). For example, over 100 carbon-bearing radicals and ions and organic compounds have been identified in dense interstellar gas and dust clouds which have temperatures in the range $10-100\,°K$ and within the solar system, primarily through spectroscopy at the highest radio frequencies. At least 80 organic compounds are known to occur in carbonaceous meteorites (Zinner et al. 1995). A few years after this work, more than 100 molecular species, the bulk of them organic, have been securely identified (Irvine 1998). There is considerable evidence for significantly heavier organic molecules, particularly polycyclic aromatics, although precise identification of individual species has not yet been obtained.

In addition, complex organic compounds can be synthesized on very short time scales (10^3 years) in the low-density circum-stellar environment during the last stages of stellar evolution (Kwok 2004). Estimates of these sources suggest that the LHB in the early atmosphere between 3.9 and 4.5 Gyr ago made available quantities of organics (estimated in the order of 10^{12} g C per year) comparable to those produced by other energy sources (Chyba and Sagan 1992). It is also supposed that a rain of material from comets could have brought significant quantities of such complex organic molecules to earth. The low temperature kinetics in interstellar clouds leads to very large isotopic fractionation, particularly for hydrogen, and this signature is present in organic components preserved in carbonaceous chondritic meteorites. Outer-belt asteroids are the probable parent bodies of the carbonaceous chondrites, which may contain as much as 5% organic material, and they contain it mainly in unoxidized form, a substantial fraction in the form of solid, heavy hydrocarbons (Briggs and Mamikunian 1963, Botta and Bada 2002).

Carbonaceous chondrites are the second most abundant subset of chondritic meteorites and are thought to be some of the most primitive examples of solid materials from our early solar system because they contain (Klemperer 2006):

- Ca—Al-rich inclusions which have been dated and found to be the oldest solids that formed in our Solar system; in fact this is where the age of our solar system comes from,
- pre-solar grains (crystals which formed from other stars before being incorporated in our solar system) and
- have bulk compositions most similar to that of the solar photosphere (since the sun makes up > 90% of the mass of the Solar system, it is thought that the first materials would also have this composition).

The quantitative information on carbonaceous chondrites is difficult to evaluate. They are much more friable than most other meteorites, and therefore survive the fall through the atmosphere less often than the others. Many groups of carbonaceous chondrites show abundant evidence for hydrothermal alteration, alteration from contact with an aqueous fluid and/or heat. Carbonaceous chondrites are also destroyed by erosion on the ground much more rapidly. Presumably that is the reason why in the older literature reports on such kinds of meteorites are missing. On the other site, analytical methods (GC-MS) to study such material were not developed before the 1960s. Carbonaceous chondrites constitute about 3% of all meteorites from cometary debris that hits the earth. They are encrusted by outer layers of dark material, thought to be a tar-like substance composed of complex organic material formed from simple carbon compounds after reactions initiated mostly by irradiation by ultraviolet light. Therefore, surface photochemical reactions may provide a "chemical evolution" in contrast to the first earth atmosphere where traces of oxygen and the UV light limit the homogeneous formation of larger organic molecules.

The analysis of organic compounds in carbonaceous meteorites provides information about chemical evolution in an extraterrestrial environment and the possible compounds that could have been present on the earth before and during the origin of life. Probably the first study with detailed analysis of organic compounds in meteorites was carried out by Kaplan et al. (1963), who examined seven carbonaceous chondrites (Orgueil, Cold Bokkeveld, Mighei, Murray, Felix, Lancé and Warrenton), five non-carbonaceous chondrites (Richardton, Bjurböle, Karoonda, Abee and Hvittis) and one achondrite (Norton County). Both aliphatic and aromatic compounds were detected; in addition carbonyl groups appeared to be present as well as unsaturated groups of the vinyl or allylic type. Amino acids and sugars were encountered in all the meteorites studied. Seventeen amino acids were detected;

serine, glycine, alanine, and the leucines. Glutamic acid, asparatic acid and threonine were found to be the most abundant. The absence of rotation, the type and distribution pattern of amino compounds in chondrules and matrix, the lack of pigments, fatty acids and presumably nucleic acids in addition to other biochemical criteria, suggest that the organic material has been synthesized by chemical rather than known biochemical processes.

Great care has been taken to assure that the interior of the Murchison meteorite that fell to earth in September 1969 near Murchison, Australia, has not been contaminated by terrestrial life. The Murchison meteorite contains 92 amino acids, only 19 of which are found naturally on earth. After a reanalysis of the Murchison meteorite, Martins et al. (2008) advanced proposals that life's raw materials were delivered to the early earth (and other planetary bodies) by exogenous sources, including carbonaceous meteorites. These authors found a large variety of the key component classes in terrestrial biochemistry, including amino acids, sugar-related compounds, carboxylic acids and nucleobases, as indigenous components in the Murchison meteorite. Some investigators have found microorganisms in them. Others have found other life forms that they call *nannobacteria*. They look like fossilized bacteria, as small as 50–200 nm. This is far smaller than biologists believe bacteria could be. Discussion is continuing that nannobacterial structures in sedimentary rocks are probably by-products of bacterial degradation of organic matter and not evidence for minute life forms called nannobacteria (Schieber and Arnott 2003). Beside Murchison, the Murray (Kentucky, USA) and Orgueil (France, 1864) carbonaceous meteorites provide different organic acids (Sephton 2002, Martins et al. 2008).

A meteorite that landed in Monahans, Texas, in 1998 was cut open and water was found in it; it was the first time that scientists have detected water in a meteorite, an essential ingredient for life of primordial origin (Zolensky et al. 1999). The fluids are dominantly sodium and potassium chloride brines, but they also contain divalent cations such as iron, magnesium, or calcium. The Monahans belongs to a class of meteorites known as ordinary chondrites, which astronomers have believed are fragments of asteroids that contain little or no water. One explanation for the water in this meteorite is that its parent asteroid acquired it after the rock formed. A water-rich, icy projectile, such as a comet, could have plowed into the newborn asteroid and spilled some of its water. Alternatively, the water might have been incorporated into the asteroid as it coalesced (Kargel 1992, Schmitt et al. 2007).

In January 2000 the Tagish Lake meteorite fell on a frozen lake in Canada and may provide the most pristine material of its kind (Pizzarello et al. 2001). Analysis has shown this carbonaceous chondrite to contain a suite of soluble organic compounds (~ 100 ppm) which includes mono- and dicarbonyl acids, dicarboximides, pyridine carboxylic acids, a sulfonic acid, and both aliphatic and aromatic hydrocarbons, in small concentrations (< 0.1 ppm) amino acids, amines and amides also have been found, see Table 2.5. The insoluble carbon (total 2.4%) exhibits exclusively aromatic character, deuterium enrichment, and fullerenes containing "planetary" helium and argon. These findings provide insight into an outcome of early solar chemical evolution which differs from any seen so far in meteorites. A continuous influx of meteoritic organic material would have enriched the prebiotic organic inventory necessary for life to assemble on the early earth.

Table 2.5 Water soluble organic compounds (WSOC)[a] in the Tagish Lake meteorite, fell in January 2000 in Canada, after Pizzarello et al. (2001) – the first pristine carbonaceous chondrite found on earth.

substance class	concentration (ppm)	number of identified species
aliphatic hydrocarbons	5	12
aromatic hydrocarbons	≥ 1	13
dicarboxylic acids	17.5	18
carboxylic acids	40	7
pyridine carboxylic acids	7.5	7
dicarboximides	5.5	4
sulfonic acids	≥ 20	1
amino acids	< 0.1	4
amines	< 0.1	3
amides	< 0.1	1

[a] formula of insoluble fraction: $C_{100}H_{46}N_{10}O_{15}$

From a large collection of micrometeorites extracted from Antarctic old blue ice in the size range of 50 to 100 μm (of which carbonaceous micrometeorites represent 80% of the samples and contain 2% carbon), Barbier et al. (1998) concluded that they might have brought more carbon to the surface of the primitive earth than that involved in the present surficial biomass. One assumes that the original carbon on earth from chondrites was bitumeous and tar (40% of the reduced carbon Tagish Lake meteorite are carboxylic acids). It is supposed that between 100 and 300 km depth below the earth surface we have a patchwork in which the carbonaceous chondrite material comprises 20% on average (Gold 1999). With the carbonaceous chondrite type of material as the prime source of the surface carbon, the question arises as to the fate of this material under heat and pressure, and in the conditions it would encounter as buoyancy forces drove some of it toward the surface. The detailed mix of molecules will depend on pressure and temperature, and on the carbon-hydrogen ratio present.

What would be the fate of such a mix? Would it all be oxidized with oxygen from the rocks, as some chemical equilibrium calculations have suggested? Evidently not, for we have clear evidence that unoxidized carbon exists at depths between 150 km and 300 km in the form of diamonds. We know diamonds come from there, because it is only in this depth range that the pressures would be adequate for their formation. Diamonds are known to have high-pressure inclusions that contain CH_4 and heavier hydrocarbons, as well as CO_2 and nitrogen (Melton and Giardini 1974). The presence of at least centimeter-sized pieces of very pure carbon implies that carbon-bearing fluids exist there, and that they must be able to move through pore-spaces at that depth, so that a dissociation process may deposit the pure carbon selectively; a process akin to mineralization processes as we know them at shallower levels. The fluid responsible cannot be CO_2, since this has a higher dissociation temperature than the hydrocarbons that co-exist in the diamonds; it must therefore have been a hydrocarbon that laid down the diamonds (CH_4). Assuming for the carbonaceous matter the formula $C_nH_mN_xO_y$, the following products could be produced under thermal dissociation. The decomposition may be oxidative (2.5) or reductive (2.6 and 2.7) where oxygen and hydrogen are produced even from the carbon substrate (2.4):

$$C_nH_mN_xO_y \xrightarrow{T, p} CH_4, C, H_2, CO, CO_2, N_2, H_2O, O_2 \qquad (2.4)$$

$$C_nH_mN_xO_y \xrightarrow{O_2, T, p} CO_2 + H_2O \qquad (2.5)$$

$$C_nH_mN_xO_y \xrightarrow{H_2O, T, p} CH_4 + O_2 + H_2 \qquad (2.6)$$

$$C_nH_mN_xO_y \xrightarrow{H_2, T, p} CH_4 + H_2O \qquad (2.7)$$

Other atoms that may also be present, such as oxygen and nitrogen, will form a variety of complex molecules with the carbon and hydrogen. Thus it is easy to understand that reduced carbon in the form of CH_4, as well as in oxidized form (CO_2) and H_2O, will be produced. At sufficient depth, methane will behave chemically as a liquid, and it will dissolve the heavier hydrocarbons that may be present, and therefore greatly reduce the viscosity of the entire fluid. The continuing upward stream would acquire more and more of such unchangeable molecules, and the final product that may be caught in the reservoirs we tap for oil and gas, is the end product of this process.

The dominant fraction of carbon on earth (see Table 2.12) is termed *kerogen*; a mixture of organic chemical compounds that make up a portion of the organic matter in sedimentary rocks. Terrestrial kerogen is almost all collected in sedimentary layers, which lie near the earth's surface, showing H/C ~ 0.5 (different kerogen types are distinguished according to H/C from < 0.5 to > 1.25). It is insoluble in normal organic solvents because of the huge molecular weight (upwards of 1000 Daltons). The soluble portion is known as *bitumen* (petroleum belongs chemically to bitumen, as the liquid form). When heated to the right temperatures in the earth's crust, some types of kerogen release crude oil or natural gas, collectively known as hydrocarbons (*fossil fuels*). When such kerogens are present in high concentration in rocks such as shale, and have not been heated to a sufficient temperature to release their hydrocarbons, they may form oil-shale deposits.

The classical theory assumes that kerogen is produced from humic substances similar to the formation of humic and fulvic acids from microbial degradation of dead biomass (algae and higher plants but also bacteria) in soils and aquatic sediments (Killops and Killops 2005). The main process is (poly-)condensation (release of water but also CH_4 and CO_2) under formation of polymeric carbon structures. The final process would be carbonization with formation of graphite. A mean formula of kerogen is such as $C_{1000}H_{500-1800}O_{25-300}N_{10-35}S_{5-30}$ (Killops and Killops 2005).

Kerogen materials have been detected in interstellar clouds and dust around stars (Papoular 2001). Kerogens and coals in evolved stages are very similar. The main difference is that coals are found in the form of bulk rocks and kerogens in dispersed form (sand-like). The idea that kerogens have been transported to earth by carbonaceous chondrites is supported by the finding of kerogen-like material, mainly as solvent unextractable macromolecular matter, analogous to terrestrial kerogen or poorly crystalline graphite in different meteorites (Nakamura 2005).

Some of those hydrous carbonaceous chondrites show evidence for heating during aging and the kerogen-like amorphous carbonaceous materials lose their labile fractions, and become more and more graphitized.

Finally we have to ask where the carbonaceous matter found in such meteorites was produced. The only explanation is inorganic synthesis from the elements and simple compounds: $n(C + H_2) = (CH_2)_n$ and $n(CO + H_2) = (CH_2O)_n$ in the solar nebula and condensed onto particles. The chemical formula $(C_{100}H_{46}O_{15} N_{10})$ of the insoluble fraction found in the Tagish Lake meteorite (Table 2.5) indicates an enrichment of elemental carbon over hydrogen and oxygen and suggests a chemical aging according to reaction (2.4). The summarized chemical composition can be roughly simplified into $(C_7H_5NO)_n$ and further transformed into building blocks such as $(4C + HCN + CH_2O + CH_2)_n$ representing a mixture from elemental carbon (C), nucleic acids (HCN), polyalcohols (CH_2O) and aliphatic compounds (CH_2). The formula $(C_6H_4O)_n$ – neglecting now the HCN building block – can be retransformed by adding $4 H_2O$ (which was probably lost during the chemical aging) into $(C_6H_{12}O_6)_n$ – sugar – which is exactly a multiple of the building block $(CH_2O)_n$; here, CH_2O (formaldehyde) is also geochemist's shorthand for more complex forms of organic matter. Hence the formation of black carbon matter in the chondrites is a carbonization similar to the formation of coal from biomass with loss of water (dehydrogenation). It is remarkable that the formula of the carbonaceous material (a) is not very different from that of terrestrial kerogen (it contains more chemically bonded H_2O and much less N) and (b) also explains the huge excess of inorganic carbon (C or oxidized into CO_2) compared with "biomass" carbon (C_{ORG}) by aging/carbonization.

2.1.3 Origin of nitrogen

Nitrogen, such as carbon, is not one of the more abundant elements on earth – its mean mass fraction is about only 0.002%; thus oxygen is 23000 times and carbon 10 times more abundant (Table 2.13). In space, however, nitrogen is the fourth most abundant element (when not considering hydrogen and the noble gases); the ratios of O/N and C/N amount to about 10 and 5 respectively. The molecular nitrogen (N_2) now present in earth's atmosphere is considered to have remained here since the planet was first formed, 4.6 Gyr ago. Variation of the mass of O_2, CO_2 and H_2O, but not that of N_2, must have led to variation in total atmospheric pressure with time. The residence time of N_2 in the atmosphere, relative to exchange with and storage in crustal rocks, is estimated to be about 1 Gyr (Berner 2006).

As a likely constituent of many volatile-rich solid bodies in the outer solar system (e.g., Saturn's moons), ammonia is expected to combine with water to form ammonia hydrates, most likely ammonia dihydrate $NH_3 \cdot 2 H_2O$ (Lunine et al. 1988). The dihydrate of ammonia was isolated for the first time by Bertie and Shehata (1984). As shown in Fig. 2.3, ammonia is relatively easily produced in interstellar regions. Cosmochemical arguments suggest that all these materials will be dominated by water, with variable amounts of materials plausibly including ammonia, methane clathrate hydrates, salts, and other materials (Tobie et al. 2005). In the outer regions of the solar system, in the planets Uranus and Neptune and in 100 billion comets, over 100 earth masses of ices and organics exist. The comets spend most of their lives deep frozen at temperatures below 50 K (Schmidt et al. 2005). The formation of ice and stone asteroids finds a natural explanation – they are some kind of remains of the building materials of the solar system. The Tunguska phenomenon in 1908 (which has attracted some 120 hypotheses) has been attributed to an icy ammonia-water asteroid (Plekharov 2008, Baillie 2008).

The idea that icy chunks fall down to the earth is now widely accepted in the astronomical community. It lies in the nature of such collisions that no residue remains for collection and chemical analysis. It is reported in the literature that sometimes after the observed mysterious fall of celestials to earth – and without any residue being found – an intensive foul smell appeared, variously described as "sulfurous", "metallic" and/or "ammoniacial" (Bérczi and Lukács 1997). Beech (2006) argues, considering the survival time in dependence on the fall velocity, that it is unlikely that large ice chunks that have fallen down to the ground (those that have been proved) are extraterrestrial, and their origin must lie somewhere within the earth's atmosphere (for example, extraordinary hail formation). Hence it is even more unlikely that icy meteorites fell on an early earth with a hot surface and atmosphere. It remains highly speculative to assume that icy NH_3—H_2O-bodies from the outer part of our solar system coalesced with planetesimal bodies forming protoplanets and thus coming down to earth while forming it by the accretion of such solid (and cold) bodies. If so, soon afterwards all volatile material evaporated.

Asteroids (and comets), however, can be mixed by coalescence with stony and icy bodies. If large enough, they may transport volatilizable material into earth's atmosphere (the Tunguska event is probably associated with an ammonia spike in the Greenland ice core record). The stony body can fall to the earth's surface but evaporates (in the Tunguska event there is speculation of cosmic carbon input; on the other hand, nitrate increase as would be expected from NO formation by heating the atmosphere (see Chapters 2.6.4.4 and 5.4.1) has not been found (Rasmussen et al. 1984, 1999, 2008).

There is evidence for hydrates on Europa, the sixth moon of the planet Jupiter, of a likely origin from the cosmos (Prieto-Ballesteros et al. 2005). Several substances besides water ice have been detected on the surface of Europa by spectroscopic sensors, including CO_2, SO_2, and H_2S. These substances might occur as pure crystalline ices, as vitreous mixtures, or as clathrate hydrate phases, depending on the system conditions and the history of the material. Clathrate hydrates are crystalline compounds in which an expanded water-ice lattice forms cages that contain gas molecules. The molecular gases that may constitute clathrate hydrates may have two possible ultimate origins: they might be primordial condensates from the interstellar medium, solar nebula, or Jovian sub-nebula, or they might be secondary products generated as a consequence of the geological evolution and complex chemical processing of the satellite. Primordial ices and volatile-bearing compounds would be difficult to preserve in pristine form on Europa without further processing because of its active geological history. But dissociated volatiles derived from differentiation of a chondritic rock or cometary precursor may have produced secondary clathrates that may be present now.

In the discussion above there are indications that ammonia is a common substance in space, distributed from grains (interstellar dust) over asteroids to satellites.

Hence it is logical to assume that during the formation of the planet (icy) ammonia (and water and likely other substances) was carried to earth. The survival of ammonia would be extremely limited, first because the planetary bodies were soon molten after aggregation and will lose all volatiles (see Chapter 2.2.1) and form a primitive atmosphere. Secondly, most icy matter falling to the earth's surface would afterwards evaporate (by frictional heating) in that atmosphere, but it would input the volatile matter, e. g., ammonia and other substances (water, methane), which easily explains the "first" atmosphere (cf. Table 2.7).

Ammonium is present as a trace constituent in all granites, with an average concentration of 45 ppm (NH_4^+), equivalent to 35 ppm of elemental N (Schlesinger 1997, Hall 1999). It shows wide variations related to petrography and region, but the only significant correlation between ammonium and other geochemical parameters is that it is most abundant in peraluminous granites and least abundant in peralkaline granites. These variations can be related to (a) the amount of sedimentary material in the magmatic source region, and (b) redox conditions in the source region. The ammonium content of granitic magmas can also be modified by fractionation or contamination. Hydrothermal alteration has a major effect on the ammonium content of granitic rocks, and variation due to this cause may exceed the magmatic variation. Most hydrothermally altered granites are enriched in ammonium as a result of the transfer of ammonium from sedimentary "country rock" (existing adjacent rock) by the hydrothermal fluids.

Ammonium muscovite ($NH_4Al_2AlSi_3O_{10}(OH)_2$) and ammonium phlogopite ($NH_4Mg_3AlSi_3O_{10}(OH)_2$) have been synthesized hydrothermally at gas pressures of 2000 bar and temperatures between 550 °C and 730 °C. Both micas are stable only in environments of high ammonia fugacity. Ammonia or nitrogen, or both, are released by thermal decomposition, cation exchange, or oxidation. The ammonia/nitrogen ratio in the gas depends primarily on the hydrogen fugacity and the temperature of the environment. Calculations show that, even in a predifferentiated earth, nitrogen may have predominated. The total amount of nitrogen present on the surface of the earth could be accounted for by the decomposition of a layer of ammonium muscovite 170 meters thick (Eugster and Munoz 1966).

Sublimates are often found, especially of chloride, from potassium, ammonium, sodium and others close to volcanoes. Today there are no indications that NH_3 is found in volcanoes. However it is not unlikely that a large share of sublimates have undergone extensive recycling, so that they should not be considered juvenile degassing products. There are no grounds to believe that all salmiac (NH_4Cl) was destroyed in an early state of the earth and that the HCl (and Cl_2) in present volcanoes is a product from other mineral chlorides and N_2 from gaseous deposits, even due to recycling of sediments. There is agreement that these substances are secondary, derived from rainwater but also vegetation deposition into the crater. However despite the fact that nowadays NH_3 is not (more) detected in volcanic gases it is not proof that NH_3 was not degassed in the early earth.

Another explanation for the origin of NH_3 considers the possibility that nitrides (N^{3-}), formed deep within the earth, are the initial compounds; it is known that nitrides (e. g., Fe_5N_3 and Mg_3N_2) form NH_3 in contact with water. This was first proposed by Gautier (1903) to explain nitrogen in volcanic exhalations (after oxidation of the NH_3). *Julius Stoklasa*, a well-known Czech agricultural chemist, found at the Vesuvius eruption in April 1906 the largest amount, (30 ppm) of NH_3 extracted from an olivine bomb; organic contamination, in the samples of lava examined, was impossible (Stoklasa 1906). Later, respected researchers also noted (but without citing the early work of *Stoklasa*) the possibility of NH_3 formation from nitrides in the early earth (Oparin 1924, 1938).

The volcanic nitrogen found was attributed in the older literature to mixing with air (which is not likely because of missing argon) and decomposition of organic nitrogen from deposits into the crater (Bischof 1863). Today it is believed that all ammonia/ammonium (Sutton et al. 2008) on earth is the result of biological activity, i. e., by assimilation of atmospheric N_2 and subsequent nitrification into NH_3.

2.2 Evolution of the atmosphere

We call the present atmosphere of the earth "secondary", but the discussion is still open whether there was a primordial atmosphere and what the composition of the primary one was. The so-called primitive (or primordial) atmosphere could have been formed by the remaining gas phase (solar wind) from solar nebula (H_2, He). However, it is unlikely that the earth had such a primitive atmosphere. There are several assumptions (e. g. the collision of earth with another celestial body, by which we assume today the formation of the moon) according to which this primitive atmosphere must have quickly eroded to space. Hydrogen from such an atmosphere would also outgas quickly to space. For the further evolution of the earth's atmosphere the possible existence of a first H_2-He atmosphere plays no role. The key question for all gases forming an atmosphere is of its origin. As discussed in Chapter 2.1.1, we are forced to conclude that the acquisition of gases, or substances that would be gaseous at the pressures and temperatures that prevailed in the region of the formation of the earth was limited to the small value implied by the low noble gas values. Assuming that gaseous material, except noble gases, was absent or of less importance in the mass budget of the initial earth, all gases believed to have been present in the primordial or primitive atmosphere must be a result of volatilization of materials from the inner part of the earth.

2.2.1 Degassing of the earth: The formation of the atmosphere

Assuming that very little gaseous material (e. g. CO_2, CH_4, NH_3) was incorporated into the primary earth aggregate, it is likely that corresponding solid substances from which, under the current conditions (heat and pressure), gases could evolve were available in the crust or the inner earth. Most scientists assume that earth's atmosphere, about 4.5 billion years ago, consisted mainly of CO_2 under high pressure (~ 250 bars) and temperature ($> 300 \,°C$), with N_2 and H_2O (and a little HCl) being minor important species. Those volatile elements and compounds were degassed from the inner earth (see also Walker 1977, Holland 1978, Brimblecombe 1996, Wayne 1996, Warneck 2000).

2.2.1.1 Volcanic gases

Walker (1977), who proposed an initially hot earth with strong tectonic activities, volcanisms and cross-mantle interchange, indicated that the rate of degassing must have declined over time. Degassing introduced water and carbon dioxide to the primitive atmosphere, he wrote. But where did the huge amounts of CO_2 and H_2O – oxidized species – come from? Volcanic eruptions and continuous degassing from earth are significant sources of trace substances in today's atmosphere (see also Chapter 2.6.4.3) but are also considered to be the origin of the first prebiological atmosphere (cf. Fig. 2.4).

There are about 500 active volcanoes on the earth. Of these about 3 % erupt each year and of that number about 10 % have sufficient explosive power to transport gases and particles to the stratosphere (Brasseur et al. 1999). Magmatic gases released from volcanoes today contain water vapor and carbon dioxide as the main components, with smaller contributions of SO_2/H_2S, HCl, HF, CO, H_2, and N_2, but also traces of organic compounds, and volatile metal chlorides and SiF_4.

Conclusions about the atmospheric composition of the early earth have been drawn from the gas exhalations of the so-called Hawaiian volcanoes type (Table 2.6). Oxygen is almost entirely missing from volcanic exhalations. Where O_2 is detected in volcanic exhalations (sometimes in those from fumaroles) it is assumed that there was a mixture with atmospheric air. The fact that oxygen has not been observed in volcanic gases should not be taken into account for the volcanic activities of the early earth. Modern volcanic gases are believed to be more oxidized than those at an early time in the earth's formation. The upper mantle was probably highly reduced in the immediate aftermath of the impact that formed the moon. Care is also required concerning the composition of volcanic exhalations at the

Table 2.6 Mean composition of volcanic exhalations (in vol-%).

substance	Hawaiian volcanoes			mean from different Hawaiian volcanoes[b]
	Walker (1977)		Day and Shepherd (1913)[a]	
H_2O	79.31	–	–	18–97
CO_2	11.61	57	55	1–50
SO_2	6.48	32	3	<1–30
N_2	1.29	6	30	<1–38
H_2	0.58	3	8	<1–4
CO	0.37	2	4	≪1–4
S_2	0.24	–	–	<1–8
HCl/Cl_2	0.05	–	–	≪1–3
H_2S	–	–	–	≪1–1
Cl_2	–	–	–	≪1–1
Ar	–	–	–	≪1–0.5
sum	100	100	100	100

[a] Kilauea eruption 1912
[b] Clarke (1920), Saukov (1953)

earth's beginning, i. e., the present composition of volcanic exhalations may not absolutely represent the former one due to recycling of rocky materials through volcanoes.

At high pressures deep beneath the earth, volcanic gases are dissolved in molten rock (magma) and these are released into the atmosphere during eruptions. As magma rises toward the surface, where the pressure is lower, gases held in the melted material begin to form tiny bubbles. The increasing volume taken up by gas bubbles makes the magma less dense than the surrounding rock, which may allow the magma to continue its upward journey. Closer to the surface, the bubbles increase in number and size so that the gas volume may exceed the melt volume in the magma and create a magma foam. The rapidly expanding gas bubbles of the foam can lead to explosive eruptions in which the melt is fragmented into pieces of volcanic rock, known as *tephra*. Volcanic gases undergo a tremendous increase in volume when magma rises to the earth's surface and erupts. Such enormous expansion of volcanic gases, primarily water, is the main driving force of explosive eruptions. If the molten rock is not fragmented by explosive activity, a lava flow will be generated.

The composition of volcanic exhalations differs from volcano to volcano. There are different ways of classifying volcanoes, but concerning the gases emitted, the classification by composition of material erupted (lava) is presented here:

- If the erupted magma contains a high percentage ($> 63\%$) of silica, the lava is called felsic (feldspar and silica). The most common felsic rock is granite.
- If the erupted magma contains $52-63\%$ silica, the lava is of intermediate composition. These volcanoes generally only occur above subduction zones.
- If the erupted magma contains $< 52\%$ and $> 45\%$ silica, the lava is called mafic or basaltic.
- Some erupted magmas contain $< 45\%$ silica and produce ultramafic lava. These are very rare.

As shown in the next Chapter 2.2.1.2, the rock types will determine the composition of the degassed matter. Gautier (1906) first stated this in summing up his views on the chemistry of volcanism due to fissuring and subsidence in the crust of the earth, whereby rocks are lowered into the heated region and gases are then developed under the enormous pressures and in immense quantities.

Volcanism from the crust to the earth's surface is the driving force in recycling rocky material today. *Subduction* is the process in which one tectonic plate is pushed downward beneath another plate into the underlying mantle when plates move towards each other. The plate that is denser will slide under the thicker, less dense plate. Faulting (the process in which rocks break and move or are displaced along the fractures) occurs in the process. The subducted plate usually moves in jerks, resulting in earthquakes. The area where the subduction occurs is the subduction zone. Magma is produced by the melting plate. It rises through fractures in the crust and reaches the surface to form volcanoes.

Hofmann and White (1982) suggested that oceanic crust recycled into the mantle during subduction could be the source of plume volcanism. The oceanic crust sinks into the deeper mantle and accumulates at some level of density compensation, possibly the core-mantle boundary. The accumulated layer locally reaches thicknesses exceeding 100 km. This model has proved to be very successful and is now widely accepted by the scientific community. The oceanic material recycled into the mantle is a combination of oceanic basalts from mid-ocean ridges, seamounts and ocean islands as well as sedimentary material deposited on the ocean floor. Moreover, a large amount of seawater (including dissolved matter) flows into the magma. In this way atmospheric gases (oxygen, nitrogen and noble gases) dissolved in seawater can also go through subduction zones into the mantle (Holland and Ballentine 2006).

One has to draw the conclusion that modern volcanism provides a mixture of recycled atmospheric and surface material with primordial rocky gas evolution. As a rough but plausible assumption, we can state that in an early earth, without free oxygen, the volcanic gases could consist more of reduced gases, for example, H_2S over SO_2, CO over CO_2 and even CH_4. At that time, when carbon recycling was not yet possible and hence carbon sediments did not occur, the question arises of the origin of carbon (CH_4, CO and CO_2) (see Chapter 2.1.2). The fractions of H_2 and CO found in volcanic exhalations agree approximately with those expected from the thermo-chemical equilibrium with their precursors H_2O and CO_2, assuming oxygen pressures of 10 mPa or less (Warneck 2000).

It is remarkable that during the Kilauea eruption (Hawaii) in 1912, Day and Shepherd (1913) did not find HCl and Ar; which proves that there was no admixture of atmospheric air and the gases were truly magmatic. The authors also proved for the first time the abundance of water in the molten lava. It is not unlikely that this type of eruption (known as "hot-spot") shows similarities with the eruption of early magma.

2.2.1.2 Gases occluded and produced from rocks

Just seven elements (Si, Al, Fe, Ca, Na, Mg, K) in oxidized form comprise 97% of the earth's crust (Table 2.13); it is notable that silica contributes 53% of the total. With the exception of oxygen (which amounts to 46% of all crust elements), none of such elements is in a volatile form (the only exception is SiH_4). In space, carbon, nitrogen and sulfur amount to 33% of the total abundance of material, but in the earth's crust they only constitute 0.057% (0.02%, 0.002%, and 0.035%, respectively). This fact of the depletion of C, N and S by about two orders of magnitude in the earth's crust shows that these elements are in partitioning among different composites.

The very high abundance of carbon in space (Table 2.13) − taking the O/C ratio to be 0.5 in space and 2300 in the earth's crust − suggests that interstellar matter and/or other planetesimal bodies and planets are rich in carbon relative to earth-like planets. As discussed above, carbon is among the interstellar gases, in form of hydrocarbon, deposited on carbonaceous chondrites but it is also found in the form of carbides in meteorites and on earth. Elemental carbon in the form of graphite can react with many elements (especially alkali and alkaline earth metals) to form graphitic mixtures and carbides. Carbides, produced at high temperatures (> 1000 °C) from carbon and the metal or its oxide, and have been also found in nature (they are produced synthetically for industrial purpose). In the literature, only the occurrence of the extremely rare silicon carbide SiC (of which there are five different compounds) and iron carbides (FeC_3) is reported. SiC has been found in diamonds (Leung et al. 1990). Analysis of SiC grains found in the Murchison carbonaceous chondrite meteorite has revealed anomalous isotopic ratios of carbon and silicon, indicating an origin from outside the solar system. SiC is even older than our solar system, having wandered through the Milky Way for billions of years as stardust that was generated in the atmospheres of carbon-rich red giant stars and from supernova remnants. The gravitational coalescence of our solar system trapped micron-size silicon carbide grains in the meteorites that were forming from the accretion of the debris in clouds of interstellar gas (Hoppe et al. 1994).

FeC_3, also known as *cohenite*, particularly when found mixed with nickel and cobalt carbides in meteorites (Hutchison 2007), was first identified by *Ernst Weinschenk*, a German pioneer of microscopy and petrography in Munich. A general formula of *cohenite* is $(FeNiCo)_3C$ suggesting that this mineral also comprises the earth's core. In molten iron (above 1100 °C) carbon can be dissolved up to 4.3%, about double the carbon content in *cohenite* (2.2%). The carbon solubility increases with temperature and when the solution slowly cools down, carbon in excess of 4.3% separates as graphite. We have no information on carbon in the earth's core but we can speculate on it and assume that some was also mixed within other upper layers. A carbon content of 1.4−2.3% has been found in native iron (Clarke 1920).

We also may speculate that unstable carbides existed on the early earth and converted according to the following reactions (M = Al, Be, Ca, Na, K, Li and others):

$$M_4C \xrightarrow{H_2O} 4\,M^+ + CH_4 \tag{2.8}$$

$$M_2C_2 \xrightarrow{H_2O} 2\,M^{2+} + C_2H_2 \tag{2.9}$$

$$M_4C_3 \xrightarrow{H_2O} 4\,M^{3+} + C_3H_4 \tag{2.10}$$

Similarly nitrogen can form with metal nitrides, in the form of salts of ammonia (N^{3-}) with alkali and earth alkaline metals (Na, Li, Ca, Mg etc.), as covalent compounds (with Si, P, S and others) as well as metallic carbides (with Cr, Co, Mn, U etc.), but all of them have been hydrolyzed over time in the formation of ammonia (see also next Chapter 2.2.1.3):

$$M_3N \xrightarrow{H_2O} 3\,M^+ + NH_3 \quad \text{and} \quad M_nN_m \xrightarrow{H_2O} n\,M^{\frac{3m}{n}+} + m\,NH_3; \tag{2.11}$$

M_3N_2 is formed from Be, Mg, Sr, Ba and Ca. Under high temperature and absence of water such nitrides can produce so-called subnitrides with the formation of nitrogen gas:

$$2\,M_3N_2 \xrightarrow{T} 3\,M_2N + \tfrac{1}{2}N_2 \tag{2.12}$$

Another possibility − which we will argue later is relatively unlikely − is the assumption that in the earth's crust ammonium chloride (NH_4Cl) was present, which quickly and completely dissociates into NH_3 and HCl with increasing temperature. The "charm" of this hypothesis lies in the explanation of the atmospheric HCl source (and Cl_2, which is formed easily via HCl photolysis). However thermal hydrolysis of chlorides which may be primordial can also explain degassing of HCl; $FeCl_2$ has been detected in meteorites.

$$FeCl_2 + H_2O \rightarrow FeO + 2\,HCl \qquad (2.13)$$

$$KCl + H_2O \rightarrow KOH + HCl \qquad (2.14)$$

Today chlorides only exist dissolved in the oceans and in marine sediments – to be discussed below as the recycling process. The reactions (2.13) and (2.14) are remarkable through a reservoir separation of alkalinity into the crust and acidity into the atmosphere.

It has long been known that nearly if not quite all rocks, upon heating to redness, give off large quantities of gas, a fact which was noted by *Priestley*, as early as 1781 (Clarke 1920). Already one hundred years ago distinguished scientists studied the gases occluded and produced from rocks while heating them. Among those researchers, the French chemist *Armand Gautier* and the American geologist *Thomas Chamberlain* made the most elaborate research upon the gases extractable from rocks. A summary of their findings can be found in Chamberlain (1916) and Clarke (1920). For the purpose of this book, it is not important to cite all such data and to discuss the variations. Here we present the main conclusions to support the ideas of the origin of volcanic gases (see Chapter 2.2.1.1) and the formation of the primitive earth atmosphere by degassing of the earth's crust (see Chapter 2.2.1). Gautier (1906) has shown that a large number of reactions are possible, starting with just water, carbon dioxide, and the solid constituents of lava; the nitrogen *Gautier* attributes to the presence of nitrides in rocks (see above), but he also assumes the existence of metallic carbides and that hydrocarbons may be generated (see Eq. 2.8–2.10).

The rocks give off on average 5–10 times their own volume of gases (excluding H_2O vapor). Before heating, the rocks were dried in the experiments to remove hygroscopic moisture. The produced steam (H_2O) however was dominant and exceeds by a factor of 4–5 all other gases from the rocks. The evolved gases from granite and basalt are (% of volume in parenthesis):

$$H_2\ (10\text{–}90\%),\ CO_2\ (15\text{–}75\%),\ CO\ (2\text{–}20\%),\ CH_4\ (1\text{–}10\%),\ N_2\ (1\text{–}5\%),$$

and in traces ($\ll 1\%$) HCl, SiF_4 and H_2S have been detected. Taking mean values, about 0.2% of the rocks' mass is volatile (corresponding to 200 ppm hydrogen, 600 ppm carbon and 200 ppm nitrogen). Water, however, is the dominant volatile compound in rocks belong traces of sulfur, nitrogen and halogens. The mean amount of water liberated from heated rocks (using the values cited by *Clarke*) corresponds to about 20 000 ppm or 2%, much more than is nowadays found in rocks as "free" water (see below), suggesting that most of the water vapor produced while heating the rocks originates from OH-bonded water due to silicate condensation, as discussed below. But in melt inclusions from past volcanic eruptions (650 BP Mt. Pelée and 1902 Santa Maria) much more H_2O with about 5% (2–7) have been found (Villemant et al. 2003); the rocks chemical composition corresponds to *Clarke* values (in parenthesis content in %): SiO_2 (73.5), Al_2O_3 (14.5), Na_2O (4.5), Fe_2O_3 (2.5), CaO (2), K_2O (2), MgO (0.5), TiO_2 (0.3), and MnO (0.1).

Most minerals of earth's upper mantle contain small amounts of hydrogen, structurally bound as hydroxyl (OH). The OH concentration in each mineral species is variable, in some cases reflecting the geological environment of mineral formation. Of the major mantle minerals, pyroxenes are the most hydrous, typically containing $\sim 200-500$ ppm H_2O, and they probably dominate the water budget and hydrogen geochemistry of mantle rocks that do not contain a hydrous phase. Garnets and olivines commonly contain $\sim 1-50$ ppm H_2O. Nominally anhydrous minerals constitute a significant reservoir for mantle hydrogen, possibly accommodating all the water in the depleted mantle and providing a possible mechanism to recycle water from earth's surface into the deep mantle.

The most important ion in rocks is the silicate SiO_4^{4-} (in analogy to sulfate SO_4^{2-}) from the weak orthosilicic acid $H_4SiO_4 = Si(OH)_4$. It is in equilibrium with silicon dioxide (SiO_2) by condensation (and liberation of H_2O) via metasilicic acid (H_2SiO_3):

$$H_4SiO_4 \ (= Si(OH)_4) \rightleftarrows H_2SiO_3 \ (= OSi(OH)_2) + H_2O \rightleftarrows SiO_2 + H_2O$$

SiO_2 finally is the anhydride of the acids; its solubility in water is about 0.12 g L^{-1} (and increases strong with temperature). Orthosilicic acid condenses to amorphic and/or polymeric SiO_2.

$$n\,H_4SiO_4 \rightarrow (SiO_2)_n + 2n\,H_2O \tag{2.15}$$

Thus, water can be "stored" in silicates and liberated by heating of hydrated silicates. This group of metamorphic rocks includes serpentine $(Mg,Fe)_3Si_2O_5(OH)_4$ and tremolite $Ca_2Mg_5Si_8O_{22}(OH)_2$. Serpentine, a basic orthosilicate, is a very common secondary mineral, resulting from a hot water alteration of magnesium silicates (mostly peridotite), present in magma, a process termed *serpentinization*:

$$3\,MgCaSi_2O_6 + 3\,CO_2 + 2\,H_2O \rightarrow Mg_3SiO_5(OH)_4 + 2\,CaCO_3 + 4\,SiO_2 \tag{2.16}$$

The water content in serpentines lies between 5 and 20% (Clarke 1920). The decomposition of serpentine may be written as follows:

$$Mg_3SiO_5(OH)_4 \rightarrow Mg_2SiO_4 + MgSiO_3 + 2\,H_2O \tag{2.17}$$

Hence water is stored in silicate and liberated according to the conditions of temperature and free water. Water-containing silicates of chlorite and serpentine groups were also found in meteorites.

There is also evidence from deep drilling that free water occurs in rocks. Today, the deepest hole ever created by humankind lies on the Kola Peninsula in Russia. It is 12 262 m deep and is termed the *Kola Superdeep Borehole*. At this depth, rock was found to be saturated in water, which filled the cracks. Because free water should not be found at those depths, scientists theorize that the water is comprised of hydrogen and oxygen atoms which were squeezed out of the surrounding rocks due to the incredible pressure. The water was then prevented from rising to the surface because of the layer of impermeable rocks above it. The last of the cores to be plucked from the borehole were dated to be about 2.7 billion years old. Alternatively, the water could simply have been deposited from carbonaceous chondrite material. It is assumed that water also came with the collision of some comets during the LHB period (Kastling and Catling 2003). Furthermore, water was available in much larger amounts in the crust of the early earth because the liquid water nowadays forming the hydrosphere was then included in the former crust, and additionally water was abiotically produced by chemical dissociation of hydrocarbons from carbonaceous chondrite material.

The following reactions rapidly changed the mantle redox state to a more oxidized level and would explain the outgassing of H_2, H_2S and SO_2.

$$FeS + 3 H_2O \rightarrow FeO + SO_2 + 3 H_2 \tag{2.18}$$

$$FeS + H_2O \rightarrow FeO + H_2S \tag{2.19}$$

$$3 FeO + H_2O \rightarrow Fe_3O_4 + H_2 \tag{2.20}$$

Summarizing the gases possibly liberated from rocks discussed above by chemical reactions these are H_2O, H_2, H_2S, CH_4 (and other simple hydrocarbons), HCl, NH_3, and N_2 – all in reduced state. The formation of oxidized carbon (CO/CO_2), however, is hard to explain in carbonate-free rocks; it is generally accepted among the scientific community that no carbonates existed in the early earth (or even in negligible amounts) and that they are later produced from atmospheric CO_2 with CaO and through silicate weathering (cf. Chapter 2.5.5). On the other hand, degassed CO_2 from deeper layers could react with CaO deposits to produce carbonate at thermally more stable sites. As suggested above, CH_4 could be produced in the early earth from carbides which can be easily oxidized – but in the presence of free oxygen.

Today, we can safely discount the presence of carbides in the earth's crust (although there is no evidence for the absence of carbides at an early stage). Hence, there must have been carbon in other compounds in the crust which can be oxidized to CO_2 or reduced to CH_4 depending on the reaction conditions. The small mean *Clarke*-amount of carbon in rocks (0.02%), which probably increases with depth, and could be very inhomogeneously distributed, is the only source of carbon gases. With the assumption (see previous Chapter 2.2.1.1) that all rocky carbon exists as hydrocarbons, it is easily to explain that under pressure, heat and hydrogen CH_4 is produced. It has been shown that the coexistence of water vapor, hydrogen, and both oxides of carbon, is possible depending upon varying conditions of temperature and concentration (Clarke 1920); the source of CO or CO_2 however could not be answered in *Clarke*'s time. Oxygenated hydrocarbons (Eq. 2.4) also produce oxygen under high temperature and pressure CO and CO_2, and when water is added (Eq. 2.5). This O_2 is used subsequently (we assume overall reducing conditions), for example, to oxidize hydrocarbon with increasing yield of CO and CO_2. Another initial production of free oxygen in the earth's mantle can be also explained by the thermal decomposition of metal oxides, transported to hotter regions; for example, FeO, giving oxygen and metallic iron; the heavy iron moving toward the earth's core, leaving the oxygen to escape.

The free oxygen, however, could have oxidized the reduced carbon existing in heavy hydrocarbons into carbon dioxide and water:

$$C_6H_2O + 6 O_2 \rightarrow 6 CO_2 + H_2O$$

or generally

$$(CH_2O)_n + n O_2 \rightarrow n CO_2 + n H_2O.$$

The destruction of hydrocarbons under pressure and higher temperatures produced CH_4 as well as elemental C (in oxygen-poor conditions) and CO_2 as well as H_2O

(in oxygen-rich conditions) as a continuous process over geological epochs. Under oxygen-free conditions the product from thermal dissociation (cf. Eq. 2.4) is $C + CO_2 + H_2$. Hydrogen can also be produced via reactions (2.4) and (2.6) and transform deep carbon into CH_4 and H_2O (Eq. 2.4). Reaction (2.6) can invert under the conditions deep within the earth. In other words, the process shown below in Equation (2.21) represents an inorganic formation of hydrocarbons ("fossil fuels"). Although the biogenic theory for petroleum was first proposed by *Georg Agricola* in the sixteenth century, various abiogenic hypotheses were proposed in the nineteenth century, most notably by *Alexander von Humboldt*, the Russian chemist *Dimitri Mendeleev* (Mendeleev 1877) and the French chemist *Marcellin Berthelot*, and renewed in the 1950s (e. g. Kudryavtsev 1959) and later by *Thomas Gold* (Gold 1999), see also Chapter 2.3.6.

$$CO_2 \, (+ \, H_2O) \underset{O_2}{\overset{H_2}{\rightleftharpoons}} CO \, (+H_2O) \underset{O_2}{\overset{H_2}{\rightleftharpoons}} C_nH_m \, (+ \, H_2O) \qquad (2.21)$$

Under pressure and high temperature ($> 900\,°C$), equilibriums are established between CO, CO_2, H_2O, H_2 and CH_4:

$$2\,CO + 2\,H_2O \rightleftarrows 2\,CO_2 + 2\,H_2 \qquad (2.22)$$

$$4\,CO + 2\,H_2 \rightleftarrows 2\,H_2O + CO_2 + 3\,C \qquad (2.23)$$

$$2\,CO + 2\,H_2 \rightleftarrows CO_2 + CH_4 \qquad (2.24)$$

$$CO_2 + H_2 \rightleftarrows CO + H_2O \qquad (2.25)$$

The degassing of CO_2 from primordial carbonates is not improbable. However, carbonates are very rare in meteorites. Hence this was probably not a dominant source of CO_2 in the early earth and may have become dominant in the volcanic source due to carbonate subduction.

2.2.1.3 The pre-biological primitive atmosphere

We have seen above that

- meteorites or their parent asteroids as well as comets ferry water, carbon (including organics) and nitrogen to earth;
- reactions under high pressure and temperature provide volatile substances.

Hence all gases compiling and cycling through the atmosphere (compounds of nitrogen, carbon and sulfur, and water) originally volatilized from the crust in the degassing period to create a first "primitive" atmosphere (H_2O, NH_3, CH_4, CO_2, HCl, H_2S, SO_2). This was characterized by intensive atmospheric photochemical processes forming a secondary atmosphere (Table 2.7) that was slightly oxidized ($NH_3 \rightarrow N_2$ and $CH_4 \rightarrow CO_2$). It is assumed that water soon condensed creating the first, hot oceans.

Despite the difference between the atmosphere of the earth today and those of the neighboring planets, Venus and Mars (Table 2.8), one can assume a similar atmosphere at the beginning because the debris forming these planets was similar. Afterwards only the distance to the sun, and hence the radiation budget with differ-

Table 2.7 Evolution of the earth's atmosphere.

atmosphere	time ago (Gyr)	composition	origin	fate
primordial[a]	4.5−4.6	H_2, He	solar nebula	erosion to space
primitive (first)	~ 4.5	NH_3 (?), CH_4 (?), CO_2, H_2O	degassing	photolysis
secondary	4.5−4.0	N_2, CO_2, H_2O,	degassing	washout
intermediate (third)	4.0−2.3	N_2, CO_2	secondary phase	remaining
present (fourth)	2.3−0.5	N_2, O_2	photosynthesis	biosphere-atmosphere equilibrium

[a] speculative

Table 2.8 Composition of the atmospheres of the inner terrestrial planets (in ppm when not given in %), after Brimblecombe (1996).

substance	Venus	Earth	Mars
carbon dioxide	96 %	0.03 %	95 %
nitrogen	3.5 %	77 %	2.7 %
oxygen	< 0.001 %	21 %	0.13 %
water vapor	< 0.5 %	0.01 %	0.03 %
helium	10	5.24	< 100
argon	70	9340	16 000
argon-36	35	31	5
neon	5−13	18	2.5
krypton	0.5	1,1	0.3
xenon	< 0.04	0.08	0.08
carbon monoxide	50	0.1	700
sulfur dioxide	150	0.01−0,1	
hydrogen chloride	1	0.001	
D/H ratio[a]	0.022	0.00015	0.0009

[a] deuterion/hydrogen

ent resulting surface temperatures, led to the different fates of the atmospheres. In the atmosphere of Venus, carbon dioxide and water maintained the greenhouse effect (no cooling and formation of liquid water). With sufficiently high temperatures (750 K at surface), the water vapor could rise high enough in the atmosphere for the water molecules to be broken up by ultraviolet radiation from the sun. The freed hydrogen could then escape from the atmosphere, leaving the oxygen only in the form of carbon and sulfur oxides. In contrast, the atmosphere of Mars is too cold (between 130 and 250 K) to maintain liquid water. The atmosphere of Mars is relatively thin, and the atmospheric pressure on the surface varies but with average of just 0.6 kPa it is smaller than on earth by a factor of 170 times. There is clear evidence for water bodies on Mars in the past. Probably, after freezing, the water was lost by sublimation and with a similar photochemical fate as on Venus.

The primitive earth long remained covered in darkness, wrapped in dense burning clouds into which water vapor poured continuously from volcanic emissions. When temperatures finally cooled sufficiently, the clouds began to melt into rain. At first, falling on incandescent rock, the rain evaporated, but the evaporation gradually cooled the crust until the water could accumulate in the depressed regions of the earth's surface, forming the first oceans. It is assumed that the atmospheric pressure was a few bars and temperature about 85 °C (Kasting 1993). On the primordial continents, the first river networks were created, and they transported detritus torn from elevated regions and deposited it on the bottom of the primordial seas. The metamorphism and remelting of the products of the erosion ultimately produced magma and lava increasingly rich in silicates, and therefore of different composition from the mantle and the primitive crust.

It has been argued (Morse and MacKenzie 1998, Kasting and Howard 2006) that liquid water was present on parts of the earth's surface as early as 4.4 Gyr ago, i.e., within 200 million years after its formation. Water can remain as liquid up to its critical temperature (374 °C for pure water, or approximately 400 °C for water with modern ocean salinity). This is because the surface pressure exerted by a fully vaporized ocean (about 270 bars) is comparable to the critical pressure of seawater (285−300 bars). Simultaneously with the first rain, however, all soluble gases (CO_2, NH_3, HCl, Cl_2, H_2S, SO_2) were washed out to a different extent through large differences in their solubility. Relative to the *Henry* constant of CO_2 (taken to be 1), SO_2 (400), HCl ($6 \cdot 10^7$) and NH_3 ($1.8 \cdot 10^3$) are very soluble and will be scavenged quantitatively from the gas phase, whereas Cl_2 (2.7) has a similar solubility to CO_2, and H_2S is much less soluble (0.05) and will therefore remain in the atmospheric gas phase. Modern volcanic emissions show a ratio of SO_2—S/H_2S of about 10. We can speculate that in early volcanoes the percentage of reduced gases was higher and possibly SO_2 was of minor importance in the sulfur budget.

It is a matter of speculation, without having information on the atmospheric concentrations, to assess the rainwater content. Assuming a much higher ammonia concentration than that for HCl, it is not unlikely (in contrast to the common assumption that seawater became acidic due to dissolved CO_2) that in the very early earth rain and seawater were slightly alkaline. HCl is neutralized quantitatively by NH_3 into NH_4Cl ($NH_4^+ + Cl^-$). Possible excess NH_3 reacts with dissolved CO_2 and provides a buffer medium:

$$NH_3 + CO_2 + H_2O \rightleftarrows NH_4^+ + HCO_3^- \tag{2.26}$$

$$NH_3 + H^+ \rightleftarrows NH_4^+ \tag{2.27}$$

$$NH_4^+ + OH^- \rightleftarrows NH_3 + H_2O \tag{2.28}$$

$$HCO_3^- + H^+ \rightleftarrows H_2CO_3 \rightleftarrows CO_2 + H_2O \tag{2.29}$$

$$HCO_3^- + OH^- \rightleftarrows CO_3^{2-} + H_2O \tag{2.30}$$

Adopting the much higher CO_2 partial pressure in the Archean, the rainwater (when only taking into account CO_2) would have had a temperature of 70 °C and a pH of 3.7 (Kasting and Howard 2006). This should have produced incredibly intensive weathering; but that is not what is observed in the paleoweathering record.

Table 2.9 Mean composition of world's ocean (in ppm; H_2O in %) with 3.5 % salinity, after Turekian (1968), Weiss (1970), Millero (2006), Pilson (1998).

substance	concentration	species	concentration	dissolved gases	concentration
H_2O	96.5 %	F	13.0	CO_2	~ 80–90
Cl^-	19 353	Sr	7.94	N_2	~ 12.5
Na^+	10 781	Si	2.9	O_2	< 7
SO_4^{2-}	2 712	Rb	0.12	Ar	~ 0.45
Mg^{2+}	1 284	P	0.09	NH_3	< 0.06[b]
Ca^{2+}	411.9	I	0.064	H_2O_2	< 0.003
K^+	399	Ba	0.021		
HCO_3^-	126	Mo	0.01		
Br^-	67.3	Ni	0.0066		
BO_3^-	25.7	Fe	0.0034		
N^a	~ 15	U	0.0033		

[a] in form of highly variable concentrations of NO_3^- (< 0.7 ppm) NO_2^- (< 0.02 ppb) and NH_4^+ (< 10 ppb) as well as dissolved N_2 (12 ppm)
[b] Clarke (1920) cites much higher values (0.16–1.22, 0.14–0.34, 0.4)

Taking into account atmospheric HCl scavenging, the pH would be even lower. The amount of chloride in today's oceans ($2.6 \cdot 10^{21}$ g) would result in a seawater pH of 1.2 when it originates from atmospheric HCl scavenging in a short time. On the other hand, retransferring the total Cl seawater amount into air, the atmospheric concentration would correspond to an incredible 470 g m^{-3}. Therefore one has to assume that all the HCl from volcanic emissions over the entire history of the earth has been deposited in the oceans. We do not know whether the chloride content of the oceans increases or whether there is equilibrium, i. e., Cl cycling from seawater into magma and recycling by volcanic eruption into the atmosphere with subsequent deposition. Volcanic emissions of HCl have been estimated in a wide range (see Chapter 2.4.3.3) from 0.4 to 170 Tg yr^{-1}. Distributing the seawater Cl over the entire history of the earth results in a flux of 0.6 Tg yr^{-1}, representing a lower estimate of present volcanic HCl emissions. Hence it is likely that chloride recycles through subduction and volcanic release such as other seawater components (e. g. carbonate).

The continuous HCl deposition (even when taking into account a much higher level of volcanic activity in the early earth) led to acidification (remember that today's ocean is slightly alkaline) and a nearly balanced equivalence between Na^+ and Cl^-. Hence it is supported that acid-buffering or even alkaline compounds would have been scavenged simultaneously, such as NH_3.

The parts of the earth's crust becoming the ocean bottom were likely to be highly alkaline because of NaO and MgO and, of much less importance, CaO (according to abundance of the cations in seawater, Table 2.9). Large amounts of soluble oxides led to dissolved Na^+, Mg^{2+} and Ca^{2+} and OH^-, which converts bicarbonate into less soluble carbonates (Eq. 2.30) as well as ammonium back to NH_3 (Eq. 2.27) with subsequent degassing from the ocean. This is simply the explanation for the chemical composition of the seawater (Table 2.9).

Nitrogen (N_2), the principal constituent of the earth's atmosphere today, is believed to be produced from ammonia photolysis in the pre-biological atmosphere:

$$2\,NH_3 + h\nu \rightarrow N_2 + 3\,H_2\uparrow. \tag{2.31}$$

When we adopt the idea that water was carried by icy ammonia hydrate bodies to the earth not only at the very beginning of the earth's formation around 4.6 Gy ago but also during the LHB ~ 4 Gyr ago − when the oceans had already been recycled by the hot surface together with evaporation of dissolved species − there was competition between NH_3 photolysis, an irreversible transformation process into N_2 (no abiotic process is known on earth that produces NH_3 and CH_4 under natural conditions), and NH_3 scavenging by rain. It also remains open to speculation how much of the ammonia was probably produced from nitrides.

Contrary to the air depletion by scavenging, the air was enriched relatively with insoluble N_2 and less soluble compounds such as CO_2 and H_2S. As described later there is a continuous flux of CO_2 through the oceans to the sediments converted as carbonate. Due to the low oxygen level, H_2S remains in the atmosphere for the first half of the earth's history. Small amounts of SO_2 from volcanic exhalations may have been in the air and in seawater after wet deposition. It is likely that reduced matter (e. g. S-IV, Fe^{2+}) still existed in seawater because of the continued absence of oxygenic photosynthesis by cyanobacteria (see below).

We assume that, just before beginning of life (about 4 Gyr ago; Mojzsis et al. 1996), the air consisted of N_2 (10^5 Pa), CO_2 (10^4 Pa) and H_2 (10^2 Pa) and small amounts of O_2 (10^{-8} Pa). These small amounts of oxygen remained due to inorganic photochemical processes and could be up to 1% in the upper atmosphere. The CO_2 level, 600 times greater than the present concentration (and other greenhouse gases), was probably needed to compensate for a predicted reduction in solar luminosity of about 30% (Kasting 1993). Further trace gases are assumed to come from different geogenic sources: NO (lightning), HCl (sea salt and volcanoes), CO (volcanoes), HCHO (photochemistry). Because O_2 was not yet produced by water dissociation via photosynthesis, the earth's surface was a strong oxygen sink through oxidation of reduced metals (e. g. Fe and U). The atmosphere of a planet with a hydrological cycle will keep atmospheric H_2 sufficiently high due to rainout of oxidized species (H_2O_2, HO_2, OH etc.) onto a reduced surface which also limits the oxygen and ozone level (Segura et al. 2007). Table 2.7 shows a simplified scheme of the evolution of the atmosphere.

We now discuss the chemical fate of the other degassed compounds, such as CH_4, CO_2, and H_2O in the first atmosphere (cf. Table 2.7). Methane photolysis occurs at α-Lyman wavelength (121.6 nm) into H, H_2, CH, CH_2 and CH_3 and, in the atmosphere of the Titan, it is thought to promote the propagation of hydrocarbon chemistry (Wilson and Atreya 2000). A small presence of O_2 and OH radicals is a result of H_2O photolysis. Assuming CH_4 and H_2O to represent first the degassing products, the formation of CO_2 is an irreversible subsequent step.

$$CH_4 + 2\,O_2\,(+\,h\nu) \rightarrow CO_2 + 2\,H_2O \text{ (via } CH_x) \tag{2.32}$$

CO_2 cannot be reduced under lower atmospheric conditions; it will scavenge while forming carbonic acid.

$$2\,H_2O + h\nu \rightarrow 2\,H_2\uparrow + O_2 \text{ (via H and OH)} \tag{2.33}$$

The water photolysis under low O_2 pressure always led to a loss of hydrogen into space (2.33). The diffusion rate of the H_2 (or H after it has been broken down by photolysis) through the homopause and exobase is limited. The definition of the homopause (80−90 km altitude) is the point at which the molecular and eddy diffusion coefficients are equal or, in other words, the critical level below which an atmosphere is well-mixed. The exobase ($\sim$ 550 km) is the height at which the atmosphere becomes collisionless; above that height the mean free path of the molecules exceeds the local scale height (RT/g).

The lifetime of CH_4 with respect to oxidation by OH radicals is around 10 years in today's atmosphere. In the early atmosphere OH levels would be so low that only the photolysis remains as a sink, resulting in a residence time of CH_4 in the order of 10^5 years (Kasting and Siefert 2002). It is assumed that methane remained until formation of the oxygenic environment in air (2.2−2.7 Gyr ago) at relatively high concentration (ppm level) to maintain a warming potential (a "greenhouse effect"). Therefore, CH_4 must be produced from the crust at rates compensating its atmospheric oxidation. It is unlikely that H_2 was available at high concentrations for hydrogenations, e. g.

$$CO_2 + 4\,H_2 \rightarrow CH_4 + 2\,H_2O. \tag{2.34}$$

CO_2 photolysis occurring in the upper atmosphere produces O_2 which can fix hydrogen back into water. The presence of oxygen in even small concentrations (10^{-8}), however, would inhibit the formation of more complex organic molecules due to CO formation (and subsequent CO_2) as proposed for the origin of life (the *Miller-Urey* hypothesis; see Chapter 2.2.2.1).

The photolysis of carbon dioxide (beside that of water) also provides small amounts of oxygen (and related radicals). Hence, the presence of CO_2 and H_2O as well as UV radiation can also produce simple hydrocarbons. As mentioned, strong UV radiation limited the synthesis of more complex molecules and radical reactions led back to the radiative relatively stable molecules $CO_2 + H_2O$.

$$CO_2 + H_2O + h\nu \rightarrow H\,(\uparrow) + O_x + H_xO_y + CO + C_xH_y\ (?)$$
$$\rightarrow CO_2 + H_2O. \tag{2.35}$$

The following main elementary reactions are behind the above scheme (Eq. 2.35, cf. also 2.33):

$$CO_2 + h\nu \rightarrow CO + O \tag{2.35a}$$

$$H_2O + h\nu \rightarrow OH + H\uparrow \tag{2.35b}$$

$$OH + O \rightarrow O_2 + H\uparrow \tag{2.35c}$$

$$O + O_2 \rightarrow O_3 \tag{2.35d}$$

$$O_3 + h\nu \rightarrow O + O_2 \tag{2.35e}$$

Only with the occurrence of photosynthesis about 2.7 Gyr ago, was O_2 available in a stepwise excess (compared to the low photolytic production in air), but first it was consumed by oxidizing Fe^{2+} and other reduced compounds. Only after reaching a redox equilibrium did the seawater become saturated with O_2 and oxygen may have escaped to the atmosphere. This certainly had quite an impact on further evolution.

Small amounts of oxygen abiotically produced in the atmosphere had been toler-ated for the first 2.5 Gyr. Besides free oxygen in the lower atmosphere, oxygen was deposited due to oxidation of reduced materials on the crustal surface of the conti-nents.

2.2.2 Biosphere-atmosphere interaction

2.2.2.1 Origin of life

If water was as common in the solar system as is implied by the facts presented in Chapter 2.1, then that would suggest that there were many environments in the solar system where the conditions were right for the development of life. What is life? For instance, *Lynn Margulis* (quoted by Horgan 1997) has stated that to pro-ceed "... from a bacterium to people is less of a step than to go from a mixture of amino acids to that bacterium". At the beginning of the seventeenth century, the ultimate origin of life was considered to be primarily a theological issue. However, it was thought possible that small creatures, such as maggots and even mice, could arise from non-living material by *spontaneous generation*, a theory first propounded by *Aristotle*. Today, there is no doubt that bacterial life is created, exists and sur-vives in space (Maurette 2006). But, what is life? Where did we come from? These two fundamental questions remain (still) unanswered in science. The existence of humans (and all animals) depends on free oxygen in the atmosphere and this com-pound is almost completely produced from oceanic cyanobacteria. Hence, the ori-gin of life lies in the darkness of the evolution of molecules in structured systems (a chemical plant we call a cell) to provide work-sharing synthesis via non-equilibrium electron transfer processes (in other terms, redox processes; see Chapter 2.2.2.3). Cells represent a dissipative structure whose organization and stability is provided by irreversible processes running far from equilibrium. Falkowski and Godfrey (2008) states that the question posed above reflects our ignorance of basic chemistry of the electron transfers that bring the ensemble of molecules in cells to "life". In the next two Chapters we will discuss briefly how chemical evolution is triggered by biological evolution.

Today we know that even in environments normally characterized as not sup-porting life a number of highly specialized animals have been detected. Species of tube worms, bivalves, gastropods and crustaceans are capable of surviving in com-plete darkness, under extreme pressures and at water temperatures that range from 10 °C to 400 °C. Microbial life exists in the pore spaces of the rocks down to depths of 6.7 km in bore holes. In the Kola Deep Borehole, 24 distinct species of plankton microfossils were found and they were discovered to have carbon and nitrogen coverings rather than the typical limestone or silica. Despite the harsh environment of heat and pressure, the microscopic remains were remarkably intact. At that depth researchers had estimated that they would encounter rocks at 100 °C, but the actual temperature was about 180 °C – much higher than anticipated. At that level of heat and pressure, the rocks began to act more like a plastic than a solid (Kozlovsky 1987). These organisms survive by eating bacteria that use hydrogen sulfide as their primary energy source. Recent investigations of microbial *extremophiles* from deep marine sediments, crustal rocks and polar ice sheets have invalidated many long held paradigms and established that the biosphere is far more extensive than was previously recognized.

The possibility of the exogenous origin of life was considered by scientists as early as *Helmholtz*, and *Arrhenius* termed it *panspermia*, an idea with ancient roots. Today, the *panspermia* hypothesis has finally achieved some measure of scientific respectability. Although it remains the orthodox view that life evolved in situ on this planet and, possibly, many others, there is mounting evidence of at least some extraterrestrial input to the formative stages of planet-based biology. Here is a summary of the relevant facts (Hoover 2006, Russel and Hall 2006, Wickramasinghe 2004):

- Discovery of increasingly complicated organic molecules between the stars,
- evidence for liquid water on comets,
- evidence for microfossils in carbonaceous meteorites,
- existence of habitable zones.

One of the fundamental requirements for life as we know it is the presence of liquid water on (or below) a planet's surface (Kasting et al. 1993). Life began very early in earth's history, perhaps before 4 Gyr ago, and achieved remarkable levels of metabolic sophistication before the end of the Archean, around 2.5 Gyr. Wherever life developed, the conditions can be characterized as follows

- liquid water at about 40 °C,
- dissolved nutrients (ammonium, carbonate, sulfide),
- hydrogen and basic organic molecules,
- protection against hard radiation,
- inorganic substrate for fixing.

Conditions like those have been assumed to exist in "deep environments" such as the sea-floor near "black smokers", in the interior of comets but also in deep rocks in the earth. A black smoker, or sea vent, is a type of hydrothermal vent found on the ocean floor. They are formed in fields hundreds of meters wide when super-heated water (~ 400 °C) from below earth's crust comes through the ocean floor. This water is rich in dissolved minerals from the crust, most notably sulfides. New and unusual species are constantly being discovered in the neighborhood of black smokers. A species of phototrophic bacterium has been found living near a black smoker off the coast of Mexico at a depth of 2500 m using the faint light from the black smoker to power its metabolism. This is the first organism discovered in nature to use a light other than sunlight for photosynthesis (Beatty et al. 2005).

However, a homogeneous mixture such as aqueous solutions (ocean) or gases such as the atmosphere (*Miller-Urey* experiment) providing all necessary educts can only synthesize molecules which much less complex than found in organisms; a heterogeneous and very likely interfacial surrounding is essential. There were two fundamental problems: first, to explain how the giant polymers those are essential to life, especially proteins and nucleic acids, were synthesized under natural conditions from their sub-units and, second, to understand the origin of cells. *Cell Theory* is one of the foundations of modern biology. Its major tenets are:

- All living things are composed of one or more cells;
- the chemical reactions of living cells take place within cells;
- all cells originate from pre-existing cells; and
- cells contain hereditary information, which is passed from one generation to another.

The debate is ongoing about how cell membranes and hereditary material (DNA and RNA) first evolved. Membranes are essential to separate the inner parts of the cell from the outer environment and being a selectively permeable barrier for certain chemicals. Both DNA and RNA are needed for a cell to be able to replicate and/or reproduce. Most organisms use DNA (deoxyribonucleic acid). DNA is a stable macromolecule consisting (usually) of two strands running in opposite directions. These strands twist around one another in the form of a double helix and are built up from components known as nucleotides. Biologists believe that RNA (ribonucleic acid) evolved on earth before DNA.

DNA could maintain its structure in a vacuum, perhaps almost indefinitely, in the very low temperatures of space. Freeze drying in a vacuum (as exists in space) would ensure that free water in the cell diffuses out. The ability of bacteria to remain viable after exposure to high vacuum and extreme cold suggests the nuclei of comets are ideal sites to search for potentially viable microbes. Comets are formed from interstellar gases and grains, containing interstellar bacteria and organic molecules. Radiogenic heating by nuclides such as ^{26}Al maintains a warm liquid interior for nearly one million years, and this is enough for bacterial replication. A typical doubling time for bacteria would be 2−3 h. A continued cascade of doubling with unlimited access to nutrients would lead to a culture that enveloped the interior of a 10 km radius comet in less than a week. When such a comet refreezes the bacterial cells will become frozen. Comets eject organic particles − most of the bacteria will decompose under the harsh irradiative conditions and provide huge amounts of complex organic molecules, detected in space and now believed to be a result of living biomass production in a (primitive) habitable zone in space. To support the idea of life transportation in space (Wickramasinghe 2004) the logical scheme is as follows: the dust in interstellar clouds must always contain the minutest fraction of viable bacteria (less than one in 10^{21}) and retain viability until formation of a new star system from interstellar matter; comets condense in the cooler outer periphery as a prelude to planet formation. Each such comets incorporate billions of viable bacteria, and these bacteria again begin to replicate in the warm interior regions of the comets.

The standard theory for the origin of life on the earth is based on the "soup" theory (*Oparin-Haldane* theory), summarized as follows:

1. The early earth had a chemically reducing atmosphere.
2. This atmosphere, exposed to energy in various forms, produced simple organic compounds ("monomers").
3. These compounds accumulated in a "soup".
4. By further transformation, more complex organic polymers and ultimately life developed in the soup.

The Russian biochemist *Oparin* wrote in 1924:

> There is no fundamental difference between a living organism and lifeless matter. The complex combination of manifestations and properties so characteristic of life must have arisen in the process of the evolution of matter.

Oparin suggested that the organic compounds could have undergone a series of reactions leading to more and more complex molecules. He proposed that the molecules formed colloid aggregates, or "coacervates", in an aqueous environment. The coacervates were able to absorb and assimilate organic compounds from the environment in a way reminiscent of metabolism. They would have taken part in evolutionary processes, eventually leading to the first life forms. *Oparin* postulated that the infant earth had possessed a strongly reducing atmosphere, containing methane, ammonia, hydrogen, and water vapor. In his opinion, these were the raw materials for the evolution of life:

> At first there were the simple solutions of organic substances, the behavior of which was governed by the properties of their component atoms and the arrangement of those atoms in the molecular structure. But gradually, as the result of growth and increased complexity of the molecules, new properties have come into being and a new colloidal-chemical order was imposed on the more simple organic chemical relations. These newer properties were determined by the spatial arrangement and mutual relationship of the molecules ... In this process biological orderliness already comes into prominence. Competition, speed of growth, struggle for existence and, finally, natural selection determined such a form of material organization which is characteristic of living things of the present time.

Oparin proposed that the "spontaneous generation of life" that had been attacked by *Louis Pasteur*, did in fact occur once, but was impossible now due to the fact that conditions found in the early earth had changed and the presence of living organisms would immediately consume any spontaneously-generated organism.

Independently, *John Haldane* speculated in 1928 about the chemical origin of life on earth (Haldane 1928, 1954). *Haldane* coined the term "prebiotic soup", and this became a powerful symbol of the *Oparin-Haldane* view of the origin of life. The sea became a "hot dilute soup" containing large populations of organic monomers and polymers. *Haldane* envisaged that groups of monomers and polymers acquired lipid membranes, and that further developments eventually led to the first living cells.

The Hadean environment (4.6−3.8 Gyr) was highly hazardous to life (Sleep et al. 1989). Frequent collisions with large objects, up to 500 kilometers in diameter, would have been sufficient to vaporize the ocean within a few months of impact, with hot steam mixed with rock vapor leading to high altitude clouds completely covering the planet. After a few months the height of these clouds would have begun to decrease but the cloud base would still have been elevated for about the next 1000 years. After that, it would have begun to rain at low altitude. For another 2000 years rains would slowly have drawn down the height of the clouds, returning the oceans to their original depth only 3000 years after the impact event. The Late Heavy Bombardment, potentially caused by the movements in position of the gaseous giant planets, that pockmarked the moon and other inner planets (Mercury, Mars, and presumably earth and Venus) between 3.8 and 4.1 billion years ago, would likely have sterilized the planet had life evolved by that time (Ryder

2001). However it is likely that these comets carried large amounts of the "raw materials" of life (water, nitrogen organics) to earth, forming a *deep hot biosphere*.

With the assumption of primordial complex organic molecules, life could also arise deep in the earth − protected against collisions and atmospheric phenomena. The "soup" needed for the formation of life or development from more simple extraterrestrial bacteria within the carbonaceous chondrites was available: H_2O, NH_3, and organics. Evidence of the early appearance of life comes from the Isua supercrustal belt in Western Greenland and from similar formations in the nearby Akilia Islands; isotopic fingerprints are preserved in the sediments, and Mojzis et al. (1996) have used it to suggest that life already existed on the planet by 3.85 billion years ago. Lazcano and Miller (1994) suggest that the rapidity of the evolution of life is dictated by the rate of recirculating water through mid-ocean submarine vents. Complete recirculation takes 10 million years, thus any organic compounds produced by then would be altered or destroyed by temperatures exceeding 300 °C. They estimate that the development of a 100 kilobase (1 kb: unit of length for DNA fragments equal to 1000 nucleotides) genome of a DNA/protein primitive heterotroph into a 7000 gene filamentous cyanobacterium would have required only 7 million years.

For the further evolution of the earth's atmosphere, the final answer of the question of where life originated is not so important. Today's atmosphere is a result of the evolution of the earth's biosphere and is developed under special physical and chemical conditions which have changed over time. Organic compounds were synthesized from the elements, in space *and* on earth. Conditions for the development of self-organizing organic matter (what we call life) were manifold and may not be specific to the earth alone.

2.2.2.2 The rise of oxygen and ozone: Biogeochemical evolution

Knowledge of the atmospheric composition and surface temperature throughout the Archean (2.5−4.0 Gyr) is essential for understanding the origin and early evolution of life on earth. Two factors should have inevitably affected this environment; reduced solar luminosity and reduced levels of oxygen. The global geothermal heat flow was substantially higher during earth's first billion years (Turcotte 1980), and the vigorous geothermal outgassing probably dispersed reduced chemical species throughout sunlit aquatic environments. Perhaps the substantial decline in thermal activity between 4 and 3 Gyr created opportunities for oxygenic photosynthesis to develop.

It is now believed that life appeared very early on earth, 3.8 Gyr ago or earlier. Photosynthetic microbial communities have left a relatively robust fossil record, in part because their productivity was particularly high on stable submerged continental platforms and margins, and thus contributed to sediments with excellent potential for long-term preservation (Marais 2000). The cyanobacterial microfossil record is robust throughout the Proterozoic (around 2.5 to 0.5 Gyr). The record of organic biomarkers − molecules that are highly diagnostic for their parent organisms − is consistent with the microfossil record. For example, only cyanobacteria are known to synthesize 2-methyl bacteriohopanepolyols, which are transformed in sediments to 2-methylhopanes. The latter have now been identified in rocks as old as 2.5 to 2.7

Gyr (Brocks et al. 1999). The discovery of sterane biomarkers in 2.7 Gyr sediments demonstrates not only the existence of eukaryotic organisms, but also that free oxygen was available for sterol biosynthesis. The extremely low $^{13}C/^{12}C$ values in 2.8 Gyr old kerogens have been attributed to methanotrophic bacteria, which require both oxygen and methane. The substantial deposition rates of ferric iron in massive banded iron sediment formations before 2.5 Gyr are clearly consistent with an abundant biological source of free oxygen. Indeed, vast sedimentary deposits of organic carbon, reduced sulfide, ferric iron, and sulfate on continental platforms and along coastal margins are among the most prominent and enduring legacies of billions of years of oxygenic photosynthetic activity.

The dominant scientific view is that the early atmosphere had 0.1% oxygen or less (Copley 2001). Assuming an O_2 level of 10^{-8} of the present level or less before 4 Gyr due to photochemical steady-states, with the evolution of biological life it is believed that there was a concentration increase on 10^{-5}. The oxygen levels in the Archean probably remain low: less than 10^{-5} the present atmospheric level in the upper atmosphere and 10^{-12} near the surface (Kasting et al. 1979). Much later ($\sim$ 2.2–2.4 Gyr ago) significant levels of oxygen arose in the atmosphere establishing the present (fourth) atmosphere (Towe et al. 2002).

However, there appears to be a dilemma. As mentioned above, the presence of oxygen in the early atmosphere would destroy organic molecules (via radical pathways, see Chapter 5.3.2). On the other hand, the presence of oxygen is the precondition for production of ozone (O_3) which protects the earth from UV radiation. Without this layer, organic molecules would break down and life would soon be eliminated. With increasing oxygen levels in the atmosphere the ozone concentration rose – and as we have learned from photochemical modeling – faster than that of O_2. O_3 and O_2 are linked within a photo-stationary equilibrium (see Chapter 5.3.1). With increasing oxygen (and subsequent O_3), the absorption of UV(B) became more complete. Before oxygen levels in the atmosphere were significant, a water column of about 10 m was sufficient to protect the layers below against UV. Only with reduced UV were aquatic organisms able to live near the surface and finally they were able to live on dry land and cover the continents. Thus it is necessary to state that neither missing nor present O_2 prevents colonization of the land but the presence of hard UV radiation. When the rise of atmospheric oxygen levels due to photosynthetic processes is simulated, it is found that the amount of ozone present is not sufficient to provide an effective shield against solar UV radiation until the oxygen mixing ratio reaches about 0.1 times that of the present, which corresponds to a time before the Silurian and the spread of life on land. Between 2.2 and 2.4 Gyr ago a huge and rapid rise in atmospheric oxygen levels from less than 0.0001% to at least 0.03% is assumed (Rye and Holland 1998, Kasting 2001, 2006), now often called the "Great Oxidation Event" (Bendall et al. 2008).

This picture is consistent with the marine $\delta^{13}C$ record, which indicates that a large amount of oxygen was added to the atmosphere between 2.22 and 2.06 Gyr (Holland 1984). The CO_2 concentration was about 2% (60 times more than the present). The CO_2 concentration began to decrease with the accumulation of biomass produced via photosynthesis, which does not return all the carbon and hydrogen contained in plant debris to return to the atmosphere as CO_2 and H_2O (the form in which it was taken up by the plants) because the organic carbon is buried in marine sediments, leaving excess oxygen behind in the atmosphere – this excess

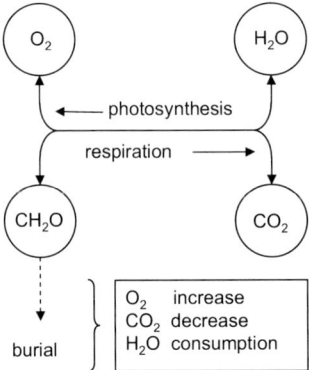

Fig. 2.5 Schematic CO_2—O_2 linkage: photosynthesis, respiration and organic carbon burial.

Table 2.10 Composition of the prebiotic earth atmosphere; after Kasting (1993), Kasting and Catling (2003).

substance	concentration (in ppm)	change with height
N_2	800 000	constant
CO_2	200 000	constant
H_2O	8 000	decrease
H_2	1 000	constant
CO	70	increase
CH_4	0.5	decrease
O_2	0.000001	increase

oxygen would otherwise be used up as the organism decay. Thus, for every carbon atom laid down as biological debris, approximately two oxygen atoms (O_2) would be liberated, cf. Fig. 2.5.

Our information on oxygen levels in the paleoatmosphere (Table 2.10) comes from studying paleosols, that is, a soil that formed on a landscape in the past at the interface of the atmosphere and the lithosphere. Chemical changes caused by weathering are driven by the acids and oxidants in soil water. Atmospheric CO_2 and O_2 dissolved in rainwater contribute significantly to the acid and oxidant budgets of soil waters and therefore drive some of the chemical changes wrought by weathering. The relationship between oxidant availability and iron mobility during weathering is the primary tool for estimating oxygen levels. In most soils H_2CO_3 is the most important weathering acid. If oxygen is supplied much more rapidly to a weathering horizon than carbon dioxide is consumed, then essentially Fe^{2+} in the weathering horizon will be oxidized to Fe^{3+} and retained as a component of ferric oxides and oxohydroxides (Rye and Holland 1998). Other evidence for the rise in atmospheric oxygen was provided by the sulfur isotope ratios in rocks older than 2.3 Gyr (Kasting 2001).

It is remarkable that the timing of the initial O_2 rise, the so-called Great Oxidation Event, is now relatively well established, but the question of what triggered it remains debated. Researchers agree that O_2 was produced initially by cyanobacteria, the only prokaryotic organisms capable of oxygenic photosynthesis; they can live anaerobically and aerobically. But cyanobacteria are thought to have emerged by 2.7 Gyr ago, on the basis of evidence from organic biomarkers in well-preserved sedimentary rocks (Brocks et al. 1999); some researchers now assume that oxygenic photosynthesis developed much earlier. The hypothesis that oxygenic photosynthesis evolved well before the atmosphere became permanently oxygenated seems well supported (Bendall et al. 2008). To sustain low oxygen levels despite near-modern rates of oxygen production from $\sim 2.7-2.5$ Gyr ago thus requires that oxygen sinks must have been much larger than they are now.

It has been proposed (Kasting 2001) that the appearance of atmospheric oxygen 200−400 Myr later was due to compensating the net photosynthetic production rate of oxygen with the flux of reduced gases (H_2 and CH_4) from volcanic outgassing and so-called serpentinization (ferrous iron released from basalt is partially oxidized to form magnetite) on the sea floor, whereas a sudden rise of O_2 occurred after changing the redox state. However, such reduced gases were continuously emitted by abiogenic processes into the atmosphere and photolytically destroyed, whereas hydrogen is escaping into space and oxygen is consumed for the production of CO_2 from CH_4. As mentioned above, another reduced gas was accumulated in the atmosphere from the very beginning, namely H_2S. In the modern atmosphere, H_2S is oxidized (by OH radicals) to SO_2 and further to SO_3/H_2SO_4. In the ancient atmosphere, H_2S would be photolyzed to H (which is escaping) and sulfur, which form S_{2n} molecules ($n = 1 \ldots 4$) surviving and accumulating in air. With the rise of atmospheric oxygen, therefore, the reduced sulfur pool must be oxidized first quantitatively before the biogenic oxygen production led to rising atmospheric levels. We can calculate how much oxygen and how much time was needed for the sulfur oxidation. By using the value of $\sim 1 \cdot 10^{12}$ g yr^{-1} for the H_2S—S emission (Berresheim and Jaeschke 1983), 2.7 Gyr ago about $(2-10) \cdot 10^{21}$ g S was accumulated (the upper level relates to a higher volcanic activity by a factor of 5) corresponding to $(5-25) \cdot 10^{21}$ g oxygen to build up SO_2 and sulfuric acid. By comparison with the O_2 production by photosynthesis being $6 \cdot 10^{13}$ g (Holland 2006), a transition time between 100 and 500 Myr was needed to establish the redox equilibrium and hence a rise of oxygen in the atmosphere. The amount of sulfur oxidized in this episode is by a factor of 2−10 more than the total dissolved sulfur (as sulfate) in modern oceans. Hence large sulfate sediments have been formed. Another theory (Holland 2002, Kump and Barley 2007) to explain why the Great Oxidation Event was so much later than the first occurrence of oxygen by cyanobacteria is that the rise of atmospheric oxygen occurred because the predominant sink for oxygen in the Archaean era − enhanced submarine volcanism − was abruptly and permanently diminished during the Archaean-Proterozoic transition. Subaerial volcanism only became widespread after a major tectonic episode of continental stabilization at the beginning of the Proterozoic. Submarine volcanoes are more reducing than subaerial volcanoes, so a shift from predominantly submarine to a mix of subaerial and submarine volcanism more similar to that observed today would have reduced the overall sink for oxygen and led to the rise of atmospheric oxygen. This change led to the oxygenation of the atmosphere and to a large increase in the sulfate concentration of seawater.

The accumulation of O_2 in the atmosphere led to the biological innovation of aerobic respiration, which harnesses a more powerful metabolic energy source. The toxic O_2 and the oxygen containing radicals also caused different biological problems, now termed *oxidative stress*. The organisms answered this stress by developing mechanisms to protect themselves against oxidants (*antioxidants*). The organisms in existence at around 2 Gyr ago had two ways: first go back to anaerobic regions and live without oxygen, or secondly to live in tolerance with oxygen. Choosing the second, evolution created with the respiration by heterotrophic organisms (biotic back reaction 2.21 and 2.43) a unique, biogenic-controlled equilibrium between atmosphere and biosphere, between reducing and oxidizing regions of the earth.

Because aerobic metabolism generates 18 times more energy (ATP) per metabolic input (hexose sugar) than does anaerobic metabolism, the engine of life became supercharged. This sequence of evolutionary steps enabled the emergence of complex, multicellular, energy-efficient, eukaryotic organisms (Dismukies et al. 2001). When our biosphere developed photosynthesis, it developed an energy resource orders of magnitude larger than that available from oxidation-reduction reactions associated with weathering and hydrothermal activity. The onset of oxygenic photosynthesis most probably increased global organic productivity by at least two to three orders of magnitude.

Since there are no findings that the ocean has ever been frozen, the atmospheric "greenhouse" effect should have been much higher. Although increased CO_2 abundance can provide sufficient warming, constraints on atmospheric CO_2 from ancient soils suggest that an additional greenhouse gas (CH_4) should have been present in significant amounts to explain the high temperature in the Archean atmosphere. In recent years, the idea that methane (CH_4) maintained in relatively high concentrations (100−1000 ppm) by ancient biota (methanogenic bacteria) have been proposed. Assuming a biogenic CH_4 source comparable to that of today and adopting a much longer residence time (10^5 years) it would have generated over 1000 ppm (Kastling and Siefert 2002). This is enough to have had a major warming effect on the climate and avoided a freezing of the oceans. Kastling and Siefert (2002) discuss that the factor that limited the CH_4 abundance was probably the production of an organic haze which created an "anti-greenhouse effect", which would have lowered surface temperatures. Thus, microorganisms have probably determined the basic composition of earth's atmosphere since the origin of life. During the first half of earth's history, this may have resulted in a planet that looked much like Saturn's moon Titan (an orange-tinted haze). During the latter half of earth's history, microorganisms created the breathable, oxygen-rich air and clear blue skies that we enjoy today. Atmospheric evolution on an inhabited planet is determined largely by its microbial populations.

The rise of oxygen corresponds precisely with earth's first well-documented glaciations, suggesting that decreasing CH_4 cooled the atmosphere. It is controversial whether the earth's ocean was frozen or not. The existence of a "Snowball earth" around 600−750 Myr ago remains controversial. In this episode (lasting a few million years) the earth's surface may have frozen over entirely to a depth of a 1 km or more. A tropical distribution of the continents is, perhaps counter-intuitively, necessary to allow the initiation of a snowball earth. The tropics are subject to more rainfall, which leads to increased river discharge and erosion (cf. Eq. 2.37) of silicate rocks thus reducing atmospheric CO_2. A serious argument against a snowball effect is the extreme and unlikely high level (factor up to 500 comparing top present) of CO_2 necessary to melt a global ice cover provided only from volcanic eruptions after the initiation of a snowball earth (Pierrehumbert 2004). Moreover neither glaciations nor melting are fast processes and the presence of ice on the continents and pack ice on the oceans would inhibit both silicate weathering and photosynthesis, which are the two major sinks for CO_2. Hence, this would be a rather unstable situation with the potential for fluctuating rapidly between the cooling and warming states. On the other hand, post-glaciations periods are characterized by a considerable increase of organism size and complexity. This development of multicellular organisms may have been the result of increased evolutionary pressures resulting from multiple icehouse-hothouse cycles; in this sense, Snowball earth episodes may have "pump primed" evolution. Alternatively, fluctuating nutrient levels and rising oxygen may have played a part. Newer data shows evidence from the magnetic field fossilized in sedimentary rocks that, more than 600 million years ago, ice occupied tropical latitudes. However, sedimentary rocks deposited during these cold intervals indicate that dynamic glaciers and ice streams continued to deliver large amounts of sediment to open oceans throughout the glacial cycle. The sedimentary evidence therefore indicates that, despite the severity of glaciations, some oceans must have remained ice-free. Significant areas of open ocean have important implications for the survival and diversification of life and for the workings of the global carbon cycle (Allen and Etienne 2008).

Oxygen probably continuously increased to about 2 % with beginning of the Cambrian (600 Myr ago) (Kasting and Donahue 1980). This O_2 level would absorb 100 % of solar light with wavelength < 250 nm and 89 % of the wavelength < 302 nm (today 97 % of the wavelength < 302 nm is absorbed). The water column necessary for protection reduced at this time to about 1 meter and it is assumed that just after this time (0.5 Gyr ago) an erratic biological development on land begun. At that time, at the end of the Ordovician and the beginning of the Silurian, the land was desolate and empty. It cannot be excluded that bacteria, lichen and algae already covered some parts of the land. The evolution from algae to land plants must have been a lengthy process. Then, after a short time, land plant photosynthesis increased O_2 in the atmosphere and we assume that with the beginning of the Silurian (400 Myr ago) the ozone layer was sufficient to protect all life (Fig. 2.6).

A second step in the rise of O_2 up to the present 21 % was in connection with the colonization of the land. From the Early Devonian (380 Myr ago), the evolution of flora gains momentum. The appearance of the first true trees is dated to about 370 Myr ago. According to the scientific consensus, the first verified land animal was a one-centimeter myriapod which appeared 428 million years ago. The earliest land animals probably lived in oxygen-poor shallow pools. The land at that time would have been much more nutrient-rich than the water, as plants colonized the land before animals and left their decaying plant matter everywhere. About 180 million years ago the first mammals began to develop on land along with primitive birds. It took about 20 million years for animals to develop the art of breathing air and so to live on land. During the early Jurassic (warm tropical greenhouse conditions worldwide), then, evolution seems to have polarized: on the one hand, there were the ruling land animals, the great dinosaurs (for the next 135 million years), which filled the ecological roles now taken up by medium-sized and large mammals; on the other hand the first mammals had appeared (Benton and Harper 1997).

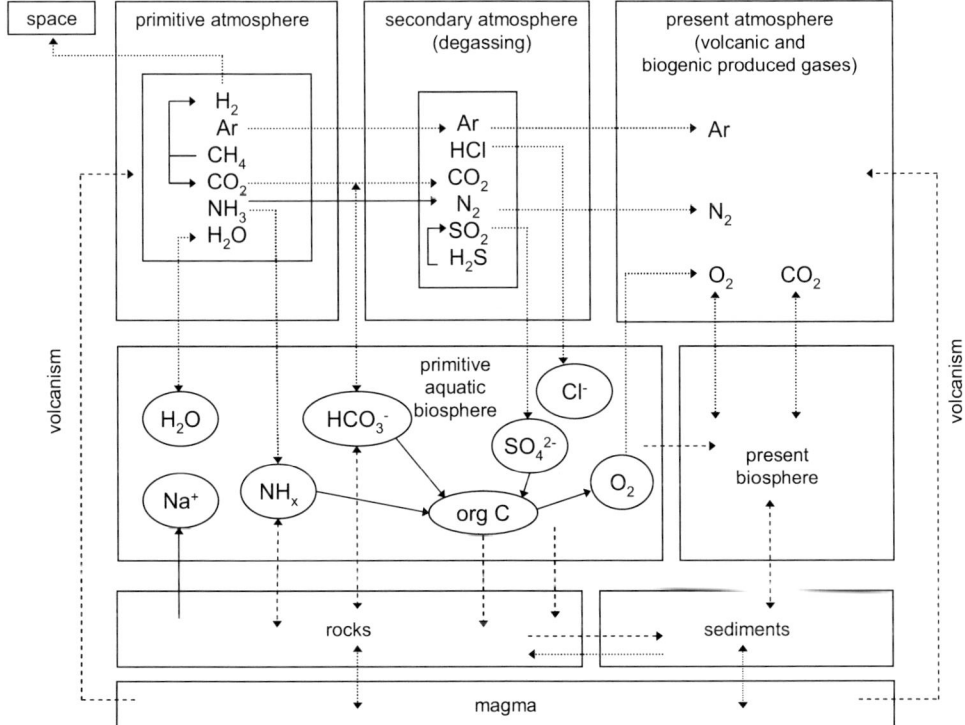

Fig. 2.6 Scheme of the evolution of the atmosphere, hydrosphere and biosphere.

In the Cretaceous (150−70 Myr) the main drivers may have been changes in ocean chemistry, probably associated with the very active mid-ocean ridges of the time. Ocean crust was forming at a record rate, with all the usual geological belching and stretching that accompanies such events. Water temperatures became exceedingly, perhaps excessively, hot. Methane and carbon dioxide levels may have increased, and perhaps spiked at far above Early Mesozoic levels. It was the first appearance of the flowering plants, the angiosperms. By the end of the Cretaceous, a number of modern plant forms had evolved. During the Paleocene the climate worldwide was warm and tropical, much as it had been for most of the preceding Mesozoic. The Neocene saw a drastic cooling in the world's climate, possibly caused by the Himalayan uplift (Tibetan plateau) that was generated by the Indian subcontinent ramming into the rest of Asia (which is still going on now). During the Pleistocene, the continuing cooling of the climate resulted in an ice age, or rather a series of ice ages with interspersed warm periods.

On land, carbon can be buried from litter and stepwise accumulation in soils but only at very low rates because of the presence of oxygen and thus mineralization was favored. Hence only biomass under more or less anaerobic conditions (in lakes, marshes and the sea) can be deposited to the bottom and form sediments. Microorganisms, however, may facilitate the oxidation of sedimentary organic matter to inorganic carbon when sedimentary rocks are exposed by erosion. Thus, microorganisms may play a more active role in the biogeochemical carbon cycle than previously recognized, with profound implications for control on the abundance of oxygen and carbon dioxide in earth's atmosphere over geological time (Petsch et al. 2001). Only a small fraction ($\sim 0.1\%$) of the organic matter synthesized in the oceans is buried as sediments and thus responsible for most of our atmospheric O_2. On land most photosynthesis is carried out by higher plants, not by microorganisms; but terrestrial photosynthesis has little effect on atmospheric O_2 because it is nearly balanced by the reverse processes of respiration and decay.

An atmosphere having the present concentration of oxygen is hard to explain: only 15% O_2 would be too low to allow the present life, and an O_2 level of 25% would burn (self-oxidation) buried biomass. Accepting that biological life (it remains a hypothesis) is causatively related with the changing air composition (Fig. 2.5), we also have to take into account the feedback mechanisms, principally suggested by the *Gaia* hypothesis (Crutzen 2002). It is not important for our understanding to believe that the continental life was a result of the protecting ozone layer or whether the beginning land plants first created the protecting ozone layer via O_2 production. What remains important is the idea that there exists a close relationship between biota and air. Evolution of one reservoir is the history of the evolution of the other one. Holland (2006) divided the last 3.85 Gyr of earth's history into five stages:

1. During stage 1 (3.85−2.45 Gyr ago) the atmosphere was largely or entirely anoxic, as were the oceans, with the possible exception of oxygen oases in the shallow oceans.
2. During stage 2 (2.45−1.85 Gyr) atmospheric oxygen levels rose to values estimated to have been between 0.02 and 0.04 atm. The shallow oceans became mildly oxygenated, while the deep oceans continued to be anoxic.
3. During stage 3 (1.85−0.85 Gyr) atmospheric oxygen levels did not change significantly. Most of the surface oceans were mildly oxygenated, as were the deep oceans.
4. Stage 4 (0.85−0.54 Gyr) saw a rise in atmospheric oxygen to values not much less than 0.2 atm. The shallow oceans followed suit, but the deep oceans were anoxic, at least during the intense Neoproterozoic ice ages.
5. Atmospheric oxygen levels during stage 5 (0.54 Gyr-present) probably rose to a maximum value of ~ 0.3 atm during the Carboniferous before returning to the present value (0.21 atm). The shallow oceans were oxygenated, while the oxygenation of the deep oceans fluctuated considerably, perhaps on rather geologically short time-scales.

2.2.2.3 Photosynthesis: Non-equilibrium redox processes

The basic principles of photosynthesis are presented in this Chapter 2.2.2.3 and the next one 2.2.2.4. Here we will discuss the chemical evolution of the *assimilation* process (see for definition what photosynthesis means the next Chapter 2.2.2.4). Let us understand as assimilation generally the conversion of nutrients into the fluid or solid substance of the body of an organism, by the processes of digestion and absorption. It is not the aim here to discuss biological chemistry (biochemistry), but the *pathway* of inorganic molecules (CO_2, H_2O and O_2), which are the "fundaments" of our climate and therefore climate system, through the organism. It is often said that our biosphere is far from redox equilibrium, or in other words, without photosynthesis, atmospheric oxygen would soon disappear (see also the discussion on this topic in Chapter 2.2.2.5). Establishing a redox equilibrium requires that all redox couples (oxidants and reductants) in a natural system (such as waters, the atmosphere or within an organism) must be in equilibrium or, in other words, the rates of oxidation are equal to the rates of reduction − the net flux of electrons is zero. This is not the case in real systems because of different time scales between chemical kinetics (single reaction rates), transport rates and microbial catalysis. Moreover, many reactions are irreversible in a subsystem (for example, sulfate production in the atmosphere) and the products must transfer into another system (in that example, in soil having microbial anaerobic properties) for closing a cycle in the sense of dynamic but not thermodynamic equilibrium (further reading: Archer and Barber 2004).

Without life on earth, probably most of the geochemical redox potentials would become in equilibrium due to tectonic mixing of all redox couples over the entire earth history. Consequently, the role of green plants is the unique "transfer" of photons (solar radiation) into electrons (and its transfer onto carrier molecules) creating electrochemical gradients and promoting synthesis and degradation. That is the role of photosynthesis; the oxidizing of water (into oxygen) and reducing of carbon dioxide (into hydrocarbons) in two separate processes. However, reduction of water represents a redox couple: positive-charged hydrogen (H^+) is reduced to "neutral" H and negative-charged oxygen (OH^-) is oxidized to "neutral" O.

The following Equations (2.36) to (2.44) do not represent *elemental* chemical reactions but gross turn-over mechanisms (termed in biology *metabolization*). The (bio-)chemical processes consist of many steps (reactions chains) and include organic catalysts (enzymes), complex biomolecules being carriers of reducing (H) and oxidizing (O) properties as well as structured reactors with specific functions (hierarchic cell organs), transport channels and organic membranes (being the separating plates between different "reactors").

As discussed in Chapters 2.1.2 and 2.2.2.1, organisms consist of organic bonded carbon. In biological evolution, the first primitive organisms must have based their development and growth on already existent organic compounds by re-synthesis. We know that the first forms of life must have existed under anaerobic conditions. *Fermentation* is the process of deriving energy from the oxidation of organic compounds, a very inefficient process, the bacteria produce ethanol and carbon dioxide from fructose (and other organic material), but many other products (e. g. acids) and carbon dioxide may have been produced:

$$C_6H_{12}O_6 \rightarrow 2\,CH_3CH_2OH + 2\,CO_2 \tag{2.36}$$

An important success was achieved by the first autotrophic forms of life (methanogens and acetogens) which transfer carbon from its oxidized (inorganic) form (CO_2) to the reduced (organic) forms that results in bacterial growth (in contrast, heterotrophic organisms can use carbon only from living or dead biomass: higher plants, animals, mushrooms, most bacteria). This process is termed *anoxygenic photosynthesis*:

$$2\,CO_2 + 4\,H_2 \rightarrow CH_3COOH + 2\,H_2O, \tag{2.37}$$

$$CO_2 + 4\,H_2 \rightarrow CH_4 + 2\,H_2O. \tag{2.38}$$

Serpentinization, arc volcanism and ridge-axis volcanism provided hydrogen (Tian et al. 2005), where the geochemical processes may involve primordial hydrocarbon and water destruction. The next step in biological evolution, the *oxygenic photosynthesis*, sharply increased the productivity of the biosphere. Today the first photosynthetic prokaryotes range from cyanobacterial and algal plankton to large kelp. Such organisms have used H_2O as electron donator:

$$2\,H_2O + h\nu \rightarrow 4\,H^+ + 4\,e^- + O_2\uparrow. \tag{2.39}$$

Generally, the process of photosynthesis is written as (cf. Eq. 2.3)

$$CO_2 + H_2O + h\nu \rightarrow CH_2O + O_2\uparrow \tag{2.40}$$

where CH_2O is a synonym for organic matter (a building block of sugar $C_6H_{12}O_6$). The creation of a photosynthetic apparatus capable of splitting water into O_2, protons, and electrons was the pivotal innovation in the evolution of life on earth. For the first time photosynthesis had an unlimited source of electrons and protons by using water as the reductant. By freeing photosynthesis from the availability of volatile-reduced chemical substances (such as H_2S, CH_4 and H_2), the global production of organic carbon could be enormously increased and new environments opened for photosynthesis to occur. There has been great progress in understanding the process of photosynthesis (see Chapter 2.2.2.4). The biological chemistry of oxygen is still a mystery, however (Falkowski and Godfrey 2008). Equation (2.40) does not represent a chemical reaction but only a gross turn-over. The two steps, H_2O splitting and CO_2 reduction, are chemically and biologically separated (within different cell compartments):

$$2\,H_2O \rightarrow 4\,H\downarrow + O_2\uparrow \tag{2.41}$$

$$CO_2 + 4\,H \rightarrow CH_2O + H_2O \tag{2.42}$$

The "cycle" is closed by respiration, the process of liberation of chemical energy in the oxidation of organic compounds:

$$CH_2O + O_2 \rightarrow CO_2\uparrow + H_2O \tag{2.43}$$

It is remarkable that in this way a stoichiometric ratio of 1:1 between fixed carbon and released oxygen is established. Therefore, a net oxygen production is only possible when the rate of reaction (2.43) is smaller than that of reaction (2.42), or in other words, the organic matter produced must be buried and protected against oxidation.

This was the first closed biogeochemical cycle. Before "inventing" photosynthesis, when fermentation was the only process in transforming organic matter into biomass by primitive life (Eq. 2.36), no cycle was provided. Organic matter, derived either from hydrocarbons found in carbonaceous chondrite material or (as we consider to be less probable) synthesized photochemicals in air, is finally transformed via Equation (2.43) into CO_2.

We assume in the modern world that photosynthesis is balanced with respiration (see also Chapter 2.3.5.3). The reaction (2.42) is of "*Fischer-Tropsch*" type synthesis. This is also known from inorganic nature under conditions deep within the earth (Eq. 2.5 and 2.21) and in the upper atmosphere (Eq. 2.35), but under extreme conditions outside the climate system or in other words where no life exists. We can also write the formation and degradation of "biomass" in the overall reaction:

$$CO_2 + H_2O + NH_3 + H_2S \underset{\text{oxidation}}{\overset{\text{reduction}}{\rightleftharpoons}} C_xH_yN_nS_mO_o + O_2 \qquad (2.44)$$

Water oxidation in oxygenic photosynthesis is a concerted four-electron process. Thus we can ask was there a transitional multielectron donor before water was adopted as the universal reductant for oxygenic photosynthesis? Dismukes et al. (2001) argue that bicarbonate is thermodynamically a better electron donor than water for O_2 evolution:

$$HCO_3^- \rightleftharpoons OH^- + CO_2 \ (K = 2.8 \cdot 10^{-7}). \qquad (2.45)$$

This implies that the Archean period with its high concentration of dissolved bicarbonate had available a stronger reductant than water for the first inefficient attempts at evolution of an oxygenic reaction center. Dismukes et al. (2001) hypothesize that bicarbonate (in more detail: Mn-bicarbonate oligomers), not water, was the transitional electron donor that facilitated the evolution of the bacterial photosynthetic precursor of cyanobacteria. The post-Archean period was brought on by the enormous reduction of atmospheric CO_2. Although it is unclear how this transition occurred, it would have required the evolution of a stronger inorganic catalyst and a stronger photooxidant to split water efficiently.

The dissociation of water (into hydrogen and oxygen) however is only possible under natural conditions in the upper stratosphere (see Chapter 5.3.7). The most appropriate process under technical conditions, created artificially is electrolysis. Water, however, is not non-reactive and is dissociated (protolyzed) into ions ($H^+ + OH^-$) in solution. In the lower atmosphere, H_2O is decomposed by $O^1(D)$ into the most reactive OH radicals. At the end of myriad reactions, OH is turned back to H_2O. There is no known way to obtain free hydrogen for reducing the nutrients important to life such as CO_x, SO_x and NO_x outside non-living systems.

Basically, the water-splitting process in higher plants and some bacteria, where coloring matter (such as chlorophyll) is able to absorb photons and transfer them (similar to a photovoltaic cell) into electrons, works very similarly to an electrolytic cell (Fig. 2.7) with a cathodic (electron donor) and an anodic site (electron acceptor). Chlorophyll (like other chromophoric substances) consists of several conjugated π-electron systems containing electrons easily excitable by light absorption. For most compounds that absorb light, the excited electrons simply return into the ground energy level while transforming the energy into heat. However, if a suitable electron acceptor is nearby, the excited electron can move from the initial molecule to the acceptor. This process results in the formation of a positive charge on the initial molecule (due to the loss of an electron) and a negative charge on the acceptor and is, hence, referred to as *photoinduced charge separation*:

$$M + h\nu \rightleftharpoons M^+ + e^- \qquad (2.46)$$

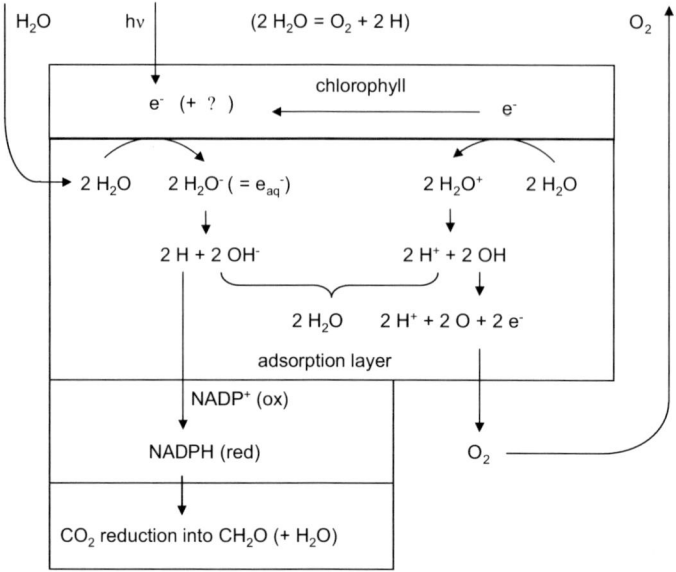

Fig. 2.7 Basic chemistry of the water splitting process.

The site where the separational change occurs is called the *reaction centre*. The first step on the negative side is the formation of an *aquated electron* (H_2O^-), cf. Fig. 2.7:

$$H_2O + e^- \rightarrow H_2O^- \;(\equiv e_{aq}^-). \tag{2.47}$$

The energy provided by the donator (or in other terms the electrode potential) must be equivalent to the reaction enthalpy of (2.47). Consequently, the total system "proton (wavelength) − chromophor (photon-electron-transfer) − electron (excited state or potential)" is a result of a coupled (quantum-)chemical and biological evolution.

The aquated electron undergoes several reactions (see Chapter 5.3.4). It can be transported via water molecules and special cellular substances (*electron transfer*) to other molecules (respectively reactive centers) to achieve reductions (2.48a), but it can also decay into atomic hydrogen and the hydroxyl ion (2.48b).

$$H_2O^- + R \rightarrow H_2O + R^- \tag{2.48a}$$

$$H_2O^- \rightarrow H + OH^- \tag{2.48b}$$

Reaction (2.48b) is synonymous with $H^+ + e^- \rightarrow H$ (in the acidity balance the OH^- from dissociated water remains). At the electropositive site of the photosystem (M^+) occurs

$$H_2O \rightarrow H_2O^+ + e^-, \tag{2.49}$$

and again, the potential difference must be equivalent with the energy for the dissociation of an electron from a water molecule. H_2O^+ splits into $H^+ + OH$ (see Fig. 2.7). It is important to consider (similarly to electrode processes) that all species are in an electrical double layer and first undergo surface processes before diffusion into the solution occurs. The highly reactive OH radical (see also Chapter 5.3.2) will be further oxidized to oxygen:

$$OH \rightarrow O + H^+ + e^-. \tag{2.50}$$

The freed electrons "react" back according to Equation (2.46), $M^+ + e^- \rightarrow M$, closing the electron transfer chain. Subsequent to reaction (2.50) is the evolution of molecular oxygen and its release to the air. Biological evolution created the reaction system in such way as to avoid oxygen diffusing to reducing (electronegative) sites; see later the discussion on oxidative stress.

As a consequence of the hydrogen formation, the process represents a *hydrogenation* and reaction (2.48a) as well as (2.48b) resulting in the formation of hydrogen-richer molecules (RH) via $H + R \rightarrow RH$ or $R^- + H^+ \rightarrow RH$. The proton (or hydrogenium ion H_3O^+) and the aquated electron are both *the* chemical fundamental species. The water molecule (H_2O) is the supporter and, because of the special structure of water (see Chapter 2.5.1.1), both species are rapidly transferred in aqueous solution when there are gradients in the electrical potential or acidity, respectively. In both cases, changing these natural system properties (for example, by human influences due to acidification and oxidative stress) will shift or even interrupt the natural (bio-)chemical cycles.

The free hydrogen is used for carbon reduction according to Equation (2.42). The process is complex (called the *Calvin* cycle) and the starting point is the transfer of H onto $NADP^+$ (nicotinamide adenine dinucleotide phosphate: $C_{21}H_{29}N_7O_{17}P_3)^5$, which is the oxidized form of NADPH (which is the reduced form of $NADP^+$):

$$NADP^+ + H \rightleftarrows NADPH \tag{2.51a}$$

or, considering the reactive center:

$$(2.51b)$$

The so-called photosystems I and II are connected by an electron transport chain and produce NADPH for use in the *Calvin* cycle (cf. Fig. 2.7). Similarly to reaction (2.51), very few functional groups act as redox couples, for example:

$$(2.52)$$

---

[5] The IUPAC name is: [(2R,3R,4S,5S)-2-(6-aminopurin-9-yl)-5-[[[[(2S,3S,4R,5R)-5-(5-carbamoyl-pyridin-1-yl) -3,4-dihydroxy-oxolan-2-yl] methoxy-oxido-phosphoryl] oxy-hydroxy-phos-phoryl]-oxymethyl]-4-hydroxy-oxolan-3-yl]oxyphosphonic acid.

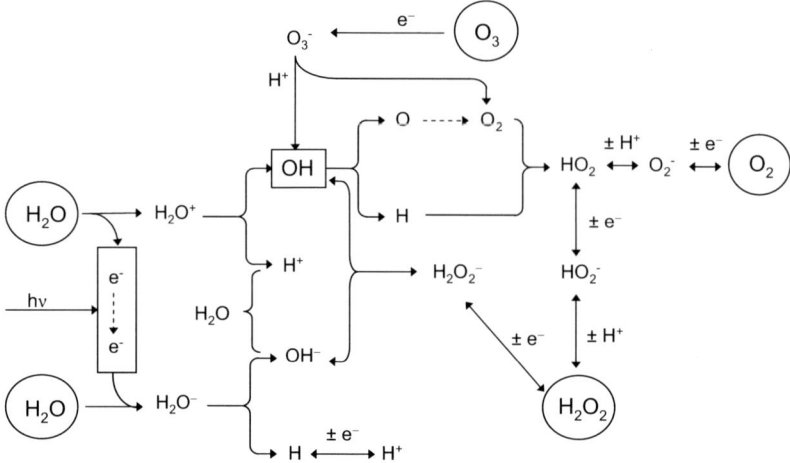

Fig. 2.8 Photosynthesis and oxidative stress.

Without here touching any biological structure, it is remarkable that the base chemical reaction in the reduction of carbon dioxide (see 2.34, 2.38 and 2.42) is the inverse reaction of CO oxidation under atmospheric conditions ($CO + OH \rightarrow CO_2 + H$):

$$CO_2 + H \rightarrow CO + OH \tag{2.53}$$

OH is deactivated in this reducing medium (or, in other words, electronegative center) by adding H into H_2O, or in terms of electron transfer, by adding an electron into OH^-; see also the remarks above on the fate of OH in an oxidizing medium. Subsequent to reaction (2.53), the formyl radical HCO, a building block of sugars, is formed:

$$CO + H \rightarrow HCO \ (H{-}\overset{\bullet}{C}{=}O). \tag{2.54}$$

Similarly, an electron transfer onto CO_2 led to formic acid:

$$CO_2 \xrightarrow{e^-} CO_2^- \xrightarrow{H^+} HCO_2 \xrightarrow{e^-} HC(O)O^- \rightleftharpoons \xrightarrow{H^+} HCOOH \tag{2.55}$$

Fig. 2.8 shows schematically the basic (elemental) chemical reactions in the biological water splitting process and formation of oxygen, but also several competitive pathways. Normally, there is water input and oxygen output, whereas hydrogen (in terms of reducing power) remains in the system; the redox potentials are well balanced in their ability to maintain the biological productivity in the sense of plant growth. A key species in the oxygen evolution is the OH radical, which has the main feature of abstracting hydrogen from organic molecules (to oxidize them and itself to turn into the water molecule), i. e., to adverse the biological function in building up organic molecules. Any change of the biochemical system to a more electropositive state or oxidant offer (increase in oxygen, ozone, and hydrogen peroxide) will turn the reaction chain to the inverse of the photosynthesis; finally back to water according to the following reactions:

$$H + OH^- \rightarrow H_2O + e^- \tag{2.56}$$

$$H + O_2 \rightarrow HO_2 \rightleftarrows O_2^- + H^+ \tag{2.57}$$

$$O_2 + e^- \rightarrow O_2^- \xrightarrow{H^+} HO_2 \xrightarrow{e^-} HO_2^- \xrightarrow{H^+} H_2O_2 \xrightarrow{e^-} OH + OH^- \tag{2.58}$$

$$O_3 + e^- \rightarrow O_3^- \xrightarrow{H^+} OH + O_2 \tag{2.59}$$

It is therefore assumed that under increased oxic conditions – termed *oxidative stress* – not only will the *Calvin* cycle be interrupted due to reactions such as (2.56) and (2.57), but also the potential concentration of OH increases as a result of the reactions (2.57) to (2.59). All these findings emphasize a central role of H_2O_2 in radical chains during biochemical processes; its ubiquitous presence in living cells is proved (see Möller 2009a). This suggests that H_2O_2 may not be only a harmful by-product of many normal metabolic processes but can also have the function via the Fenton reaction (OH formation via electron transfer onto H_2O_2, see reaction 2.59, right-hand site) of attacking unwanted particles such as viruses and bacteria. The enzymatic deactivation of H_2O_2 prevents damage[6] by its transformation into oxygen:

$$H_2O_2 \xrightarrow{-H^+} HO_2^- \xrightarrow{-e^-} HO_2 \xrightarrow{-H^+} O_2^- \xrightarrow{-e^-} O_2 \tag{2.60}$$

The fundamental role of H_2O_2 in environmental chemistry lies in the simple redox behavior between water and oxygen depending on available electron donors or acceptors:

$$H_2O \xleftarrow{\text{reduced}} H_2O_2 \xrightarrow{\text{oxidized}} O_2$$

2.2.2.4 A short history of understanding the process of photosynthesis

As we have seen, today's biosphere-atmosphere or, more simply, the plant-air interaction, is due to oxygenic photosynthesis. It is worth undertaking a short review of how our understanding of plant growth rose (Table 2.11). The main aim of this chapter is to touch upon the history of understanding the gas exchange between plant (biosphere) and air (atmosphere). As mentioned, oxygenic photosynthesis created *the* force for driving biogeochemical cycles.

Plant growth has been both a curiosity and a source of much documentation and experimentation since the beginnings of agriculture. The "substance" of plants has been explored and debated since the time of the early Greek philosophers (Korcak 1992). Before the discovery of the chemical composition of air and the proof that the organic matter of plants is solely by uptake from atmospheric CO_2, humus was considered to be the sole and direct source of carbon for plants. The

---

[6] By *catalase*, which has one of the highest turnover numbers of all enzymes; one molecule of catalase can convert millions of molecules of hydrogen peroxide to water and oxygen per second.

Table 2.11 Milestones in discovery the photosynthesis.

year	discoverer	work
1577	Paracelsus	rain as nutrient
1648	Jan van Helmont	"willow tree experiment"
1674	John Mayow	life and non-life supporting airs (gases)
1699	John Woodward	water as essential compound for plants
1738	Stephen Hales	air and light are necessary
1754	Joseph Black	(re-)discovery of carbon dioxide
1779	Jan IngenHousz	carbon in plants is of atmospheric origin
1799	Jean Senebier	plants consume CO_2 and release O_2
1804	Théodore de Saussure	binding of H_2O and release of O_2
1838	Jean Boussingault	air as source of nitrogen
1840	Justus von Liebig	inorganic nutrition theory
1842	Julius von Mayer	plants transform light into chemical energy
1862	Julius von Sachs	chlorophyll is involved in "photosynthesis"
1872	Eduard Pflüger	respiration is located on the cellular level
1877	Schloesing and Muntz	nitrification is a biological process
1894	Theodore Engelmann	microscopic observation of active cells
1905	Blackman and Matthaei	T dependence of photosynthesis
1925	Otto Warburg	dark and light reaction
1937	Robert Hill	H_2O oxidation and CO_2 fixation are separate processes
1940	Ruben and Kamen	C^{14} experiments
1945	Melvin Calvin	sunlight acts on the chlorophyll (not on CO_2)

roots of "humus theory" lie in the ancient transmutation between the four "elements". *Johann Wallerius*, known for the first monograph on mineralogy (1750), wrote in 1761: "Plants derive no growth from any mineral earths ... The substances that promote plant growth must be identical or analogous with substances pre-existing in the plant, or capable of being transmuted and combined into a nature that belongs to plants" (Browne 1943, Wallerius 1750, 1770). The replacement of the humus theory by the theory of mineral nutrition of plants lasted the full nineteenth century. The first researcher to doubt the validity of the humus theory was *Carl Sprengel*. In 1826, he published an article in which the humus theory was refuted, and in 1828 another, extended journal article on soil chemistry and mineral nutrition of plants which contained in essence the "Law of the Minimum" (Sprengel 1826, 1828). *Sprengel*'s doctrines are presented again in the books published by *Liebig* in 1840 and 1855.

Paracelsus gained fame as the first scientist to lecture in German, instead of the traditional Latin, allowing for understanding and involvement in science among lay persons, and recognition of the importance of experimentation in chemistry (Browne 1943). More germane to plant nutrition, he initiated a new concept of plant nutrition, which was not aligned to the *Aristotelian* four elements. In his *Aurora thesaurusque philosophorum* (Paracelsus 1577) (cited after Browne 1943) *Paracelsus* stated:

... So also every vegetable of the earth must give nutriment to the three things of which they consist. If they fail to do that the *prima condita* perish and die in their three species. These nutriments are earth and Rain, that is the Liquor, each of the three parts of which nourishes its own kind – sulfur for sulfur, mercury for mercury, and salt for salt, for Nature contains these, one with the others.

In modern terms, the "three principles" mean organic constituents (sulfur), water (mercury), and mineral matter (salt). *Jan van Helmont* wrote in his posthumously published book *Ortus medicinae* (1648); cited[7] after Hoff (1964):

> For I took an earthen Vessel, in which I put 200 pounds of earth that had been dried in a Furnace, which I moystened with Rain-water, and I implanted therein the Trunk or Stem of a Willow Tree, weighing five pounds; and at length, five years being finished, the Tree sprung from thence, did weigh 169 pounds, and about three ounces ... At length, I again dried the earth of the Vessel, and there were found the same 200 pounds, wanting about two ounces. Therefore 164 pounds of Wood, Barks, and Roots, arose out of water onely.

This often cited (and different interpreted) "willow tree experiment" would be just the first in a long sequence of experiments and explanations of how plants grow and where the matter is coming from (Krikorian and Steward 1968). *Helmont* also found that the gas bubbling in a brewery during fermentation is the same as that obtained by burning charcoal, as both gases turned limewater milky – a test which is used even today to identify carbon dioxide; he called it *spiritus sylvestre* (wild gas). *John Mayow* concluded (Mayow 1674) from his experiments on respiration of animals that there is a constituent of the air that is absolutely necessary for life (O_2), and another not supporting life. One hundred years later carbon dioxide was re-discovered by *Joseph Black*; he named it "fixed air" (Black 1754); it was renamed carbonic acid by *Lavoisier* in 1781. *Black* showed that it is present in the atmosphere and that it combines with other chemicals to form compounds.

In 1699, the classic water culture experiment of *John Woodward* dispelled the concept of water as the sole substance of plants (Russell 1926). *Woodward* noted better growth of spearmint in water containing garden soil than when it was grown in rain water, or impure water. However, it was not until the composition of air from different gases became known (in the 1770s, see Chapter 1.3.2) that their significance for plant nutrition was studied. It can be said that *Stephen Hales* was the first to understand that air and light are necessary for the nutrition of green plants when making his measurements (*Vegetable Statics*, London 1727) of the absorption of water and its transpiration to the atmosphere. In 1771 *Priestley*, observed in his work on the purification of air by plants that green plants give off oxygen and thus improve the air.

Jan IngenHousz, who was in close contact with *Priestley* and *Lavoisier*, in the summer 1779 performed experiments on plants and published his findings in *Exper*

---

[7] John Baptista van Helmont. Oriatrike: Or, Physick Refined. The Common Errors therein Refuted, And the whole Art Reformed & Rectified. Being A New Rise and Progress of Philosophy and Medicine, for the Destruction of Diseases and Prolongation of Life. trans. John Chandler (London, Lodowick Loyd, 1662), p. 109. For the original Latin, see: Ortes medicinae, Id est, initia physica inaudita. Progresses medicinae novess, in morborum ultionem, ad vitam longam (Amsterdam: Ludovicus Elsevir, 1648); reprinted Brussels: Culture et Civilisation, 1966

iments upon Vegetables (London 1779)[8]. He clearly recognized both the meaning of light and the fact that all the carbon contained in plants is of atmospheric origin. Without doubt he was the first to describe and understand the essentials of the process of *photosynthesis*, but he did not know it as such (Magiels 2001). *Ingen-Housz* rightly concluded that the green parts of plants in the light of the sun produce oxygen, which is beneficial for other forms of life, and that they produce CO_2 when in the dark.

Jean Senebier discovered in 1799 that the regeneration of air is based on the use of "fixed air" (CO_2) or − in modern terms − he demonstrated that green plants consume CO_2 and release O_2 under the influence of light. In 1893, *Charles Barnes* proposed that the biological process for "synthesis of complex carbon compounds out of carbonic acid, in the presence of chlorophyll, under the influence of light" should be designated as either "photosyntax" or "photosynthesis" (Gest 2002). Before this year, the process had been termed *assimilation*.

In 1804 *Théodore de Saussure* discovered that the increase in weight of plants cannot solely be caused by the uptake of carbon and minerals, but is based on the binding of the water components, too, and accompanied by the release of oxygen during photosynthesis (Saussure 1804). In 1838, *Boussingault* conducted an elegant series of experiments and showed that legumes had higher nitrogen levels than cereals and, based on some crop rotation studies over five years, concluded that the atmosphere was the source of this nitrogen (it could have been particulate matter, nitrogen gas or ammonia, he did not specify which). In 1848, *Justus von Liebig* (erroneously) argued, with no new evidence, that ammonia in air was the source of plant nitrogen and that this supply limited growth. This hypothesis was accepted because of his great reputation. Already in 1840 *Liebig* had published a book entitled *Die organische Chemie in ihrer Anwendung auf Agricultur und Physiologie* (Organic chemistry in its application to agriculture and physiology); Liebig (1840). *Liebig*'s theories on the atmospheric source of ammonia-nitrogen for plant growth led *John Lawes* in 1843 to establish the now famous Broadbalk Field wheat experiment at Rothamsted (Hertfordshire, UK; Russell 1926) and soon initiated many studies on the chemical composition of rainwater.

In 1862−64, *Julius von Sachs* could finally prove that chlorophyll is involved in photosynthesis. He worked out the overall equation of photosynthesis:

$$6\,CO_2 + 6\,H_2O + \text{solar energy} \rightarrow C_6H_{12}O_6 + O_2.$$

These results are in accord with the first law of thermodynamics, whose discoverer *Julius von Mayer* postulated in 1842 that plants take up energy in the form of light and that they transform it into another, chemical state of energy (Mayer 1845). In 1872 *Eduard Pflüger* defined respiration as a process located on the cellular level (Pflüger 1872).

---

[8] John Ingen-Housz. *Experiments upon vegetables, discovering* Their great Power of purifying the Common Air in the Sun-shine, And of Injuring it in the Shade and at Night: To which is joined, A new Method of examining the accurate Degree of Salubrity of the Atmosphere. London, P. Elmsly (1779), 68 + 302 + 15 pp. "This work is part of the results of above 500 experiments, all which were made in less than three month, having begun them in June, and finished them in the beginning of September, working from morning till night" Ingen-Housz writes on p. xlii in the preface (the book was printed in December 1779).

In the 1870s, the French biologists *Jean Schloesing* and *Achille Muntz* discovered that nitrification is a biological process (Schloesing and Muntz 1877). Although years before (1859), *Louis Pasteur* had theorized that the process was biological, he was never able to prove it. The phenomenon of nitrification, i. e., the formation of nitrites and nitrates from ammonia and its compounds in the soil, was formerly held to be a purely chemical process.

At the end of the nineteenth century, with a vast development of new techniques, a period characterized by a step-wise understanding of the mechanisms of the photosynthesis begun. In 1894 *Engelmann* constructed a gadget out of a modified microscope condenser that allowed him to expose parts of photosynthetically active cells (of the green alga *Spirogyra*) to a thin ray of light. His aim was to discover which components of the cell functioned as light receptors. A few years later (1902–1905) the first experiments to determine the temperature dependence of photosynthesis were performed by *Blackman* and *Matthaei*, and these showed that carbon dioxide fixation is based on normal, temperature-dependent biochemical reactions (Matthaei 1902, Blackman and Matthaei 1905). In 1925 *Otto Warburg* put the results of *Blackman* down to the existence of two classes of photosynthetic reactions: the light and the dark reaction.

Robert Hill demonstrated in 1937 that the oxidation of H_2O to O_2 and CO_2 fixation into carbohydrates are separate processes (Hill 1937). He observed O_2 evolution by chloroplast suspensions when artificial electron acceptors, other than CO_2, are used. This reaction, which *Hill* called "the chloroplast reaction", later became known as the "*Hill* reaction". This, and the postulate of *Warburg* that the fixation of carbon dioxide is energy consuming but independent of light, was confirmed by *Ruben, Kamen* and his coworkers in 1939–1941 after the isotope technique had found its way to biochemistry (Ruben 1939, Ruben et al. 1941). In 1945 *Melvin Calvin* commenced research to determine the pathway by which CO_2 becomes fixed into carbohydrate.

Calvin and his co-workers showed that sunlight acts on the chlorophyll in a plant to fuel the manufacture of organic compounds, rather than on carbon dioxide, as was previously believed (now called the *Calvin* cycle). In the period from 1950 until the present there have been numerous investigations to identify the detailed mechanisms, intermediates and controlling systems of photosynthesis. Gest (1993) suggested the following general definition for "photosynthesis":

Photosynthesis is a series of processes in which electromagnetic energy is converted to chemical energy used for biosynthesis of organic cell materials; a photosynthetic organism is one in which a major fraction of the energy required for cellular syntheses is supplied by light. Molecular oxygen and CO_2 are not included in the "common denominator definition" of photosynthesis because photosynthetic bacteria do not produce O_2, and CO_2 is not necessarily a required carbon source.

2.2.2.5 The carbon and oxygen pools and global cycling

Carbon and oxygen were formed during the fusion process from helium within the first steps after Big Bang (cf. Fig. 2.3):

$$^4He + {}^4He \rightarrow {}^8Be, \quad {}^8Be + {}^4He \rightarrow {}^{12}C \quad and \quad {}^{12}C + {}^4He \rightarrow {}^{16}O.$$

It is likely that most of the oxygen (and other volatile elements) in the particulate matter of the solar nebula was chemically bonded with metals forming oxides (Farquhar and Johnson 2008).

The carbon we have on the surface or in the sediment of the earth (note that this excludes possible carbon stocks in rocks and deep in the earth) is estimated to be around 80 % in the form of carbonate rocks, and 20 % in unoxidized form, frequently referred to as organic carbon (possible carbon stocks deep in the earth are not counted), cf. Table 2.12. If the vast amounts of present carbonate sediments ($5000 \cdot 10^{19}$ g C) had originally been exhaled in reduced forms (CH_4 or CO) it would have required oxygen from the atmosphere for oxidation. This huge amount of required oxygen, however, is not balanced with the O_2 produced via photosynthesis, it is two to three times greater (see discussion later). Thus, the earth's present carbonate reservoir was probably initially exhaled from the earth as CO_2 which was oxidized from reduced carbon derived from chondrite-type debris. There is no argument against the hypothesis that this is a more or less continuous process over the entire history of the earth, explaining the deposition of carbonate sediments. In a process of continuous recycling, the proportion of ^{13}C would continuously increase in the atmosphere, and hence the younger carbonates should be isotopically heavier then the old ones; this is not the case. Marine carbonates of all ages back to the Archean show the same narrow range of the carbon isotopic ratio

Table 2.12 Reservoir distribution (in 10^{19} g element, despite for the water molecule); after Schlesinger (1997), if not other noted.

reservoir	C	C[h]	C[k]	O	H$_2$O	S[g]
atmosphere	0.075	0.0766	–	119	1.7	negl.
ocean	3.8/0.07[a]	3.8–4.0/–[a]	–/0.06[a]	12500[b]	14000	128
land plants	0.06	0.054–0.061	0.095	negl.	negl.	negl.
soils, organic	0.15	0.15–0.16	0.16	? (negl.)	? (negl.)	negl.
fossil fuels	0.7	0.4		negl.	negl.	0.001
sediments	~ 5000/1500[m]	6600–10 000[j]	6000/1500[m]	4745[c]	1500	247[i]
clathrates[d]	1.1			–	–	–
rocks[n]	3200–9300[e]			1200[f]	~ 2 000 000	?

[a] carbonate / dissolved organic carbon (DOC)
[b] in water molecules
[c] held in Fe_2O_3 and evaporitic $CaSO_4$
[d] methane hydrates, after Kvenvolden and Lorenson (2001)
[e] estimated by using mean element abundance (Clarke 1920) and assuming a mass of crust being $4.9 \cdot 10^{25}$ g (Table 2.2), sediments likely are included (the chemical form is not specified)
[f] held in silicates
[g] after Möller (1983)
[h] after Pidwirny (2008)
[i] held in in $CaSO_4$
[j] not specified into carbonate and organic C
[k] after Pédro (2007)
[m] carbonate / buried organic (kerogene)
[n] in mantle: $5 \cdot 10^{23}$ g (Pédro 2007)

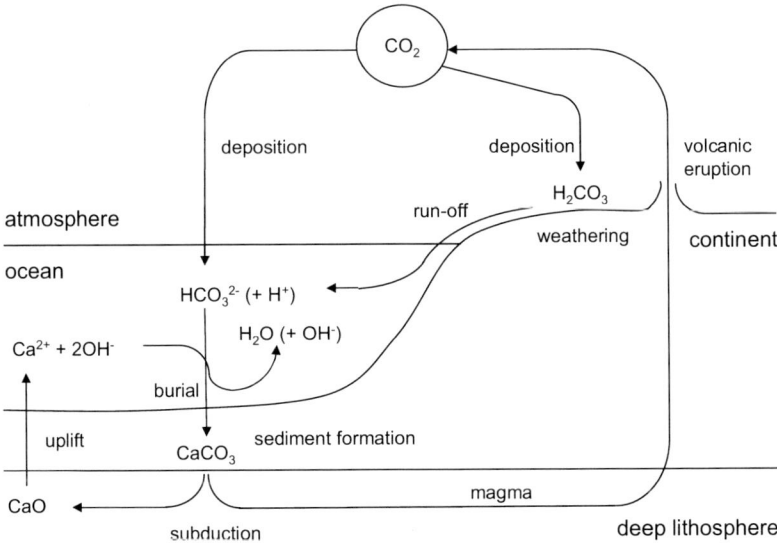

Fig. 2.9 Scheme of the inorganic carbon cycle (inorganic carbon burial).

(Schidlowski et al. 1975). What it shows is a reasonably continuous process of laying down carbonate rocks according to Equation (2.61); no epoch having enormously more per unit time, nor enormously less (cf. Fig. 2.9).

$$CaO + CO_2 \rightarrow CaCO_3 \qquad (2.61)$$

Most, but not all, of this carbonate has been an oceanic deposit, deriving the necessary CO_2 from the atmospheric-oceanic CO_2 store. The amount that is at present in this store is, however, only a very small fraction of the amount required to lay down the carbonates present in the geological record. The atmospheric-oceanic reservoir at present holds only about 0.05 % of the earth's carbon (cf. Table 2.12).

The carbonate sediments also support the hypothesis that CO_2 was a product of the earth's inner processes (not only atmospheric CH_4 oxidation), thermal dissociation of primordial carbonates − which is not very likely due to its spare finding in meteorites − or other carbon species, already formed within the solar nebula. Transforming back the CO_2 stored in inorganic and organic sediments to the atmosphere one would have 60 to 80 bars CO_2 in the first atmosphere. Another 140 bars is provided by the water from the ocean and the sediments transformed back into the atmosphere.

The amount of carbon estimated in the rocks of the lithosphere (Table 2.12) corresponds to about 3 % carbon content. This is very close to the assumption by Gold (1999) of carbonaceous chondrite material which may comprises 20 % of the material in the depth range between 100 and 300 km. In this material, carbon amounts to 5 %. Hence this layer would provide some $3 \cdot 10^{24}$ g C from which only 2 % have been released to provide the surface carbon. A supply of hydrocarbons at depth may thus provide CO_2 in three different ways. One is through volcanic path-

Table 2.13 Chemical abundance of elements and oxides in mass %; after Ronov and Yaroshevsky (1969), Morgan and Anders (1980), Carmichael (1989), Lide (2005). According to conventions, the oceans are included in the earth crust.

space		meteorites[b]		earth		earth core		earth mantle		earth crust[c]	
H	[a]	O	52.80	Fe	32.1	Fe	88.8	O	44.8	O	46.1
He	[a]	Si	15.37	O	30.1	Ni	5.8	Mg	22.8	Si	28.2
O	49.3	Mg	13.23	Si	15.1	S	4.5	Si	21.5	Al	8.23
C	24.6	Fe	11.92	Mg	13.9			Fe	5.8	Fe	5.63
Ne	6.4	S	2.10	S	2.9			Ca	2.3	Ca	4.15
Fe	5.4	Al	1.15	Ni	1.8			Al	1.2	Na	2.36
N	5.9	Ca	0.90	Ca	1.5			Na	0.3	Mg	2.33
Si	3.4	Ni	0.65	Al	1.4					K	2.09
Mg	3.0	Na	0.62								
S	2.5	Cr	0.19								
		K	0.13								
sum 100		sum 97.91		sum 98.8		sum 99.1		sum 99.7		sum 99.65	

[a] H and He represent 98 % of total element mass (H 76.5 % and He 23.5 %); the other elements are set to 100 % as sum
[b] after Fersman (1923): 0.07 % C, 0.05 % Cl, $3 \cdot 10^{-4}$ % N
[c] after Clarke (1920): 0.03−0.19 % C, 0.06 % Cl, 0.12 % P, 0.08 % S (N and Br negl.)

ways and oxidation with oxygen supplied by the magma. Another is by the ascent of hydrocarbons through solid rocks and oxidation at shallow levels, most likely by bacterial action, with subsequent escape of CO_2 to the atmosphere. A third way will be the escape of methane and other hydrocarbons into the atmosphere, where, in the presence of atmospheric O_2 they would oxidize to CO_2.

Oxygen is the most abundant element on earth and also in the space, excluding hydrogen and helium (Table 2.13). Our present atmosphere contains $119 \cdot 10^{19}$ g molecular oxygen (Table 2.12) which represents only 0.006 % of total oxygen on the earth, which is almost completely fixed in oxides (cf. Table 2.12), very likely of primordial origin. The total amount of volatile oxygen consumed over the earth's history has been estimated by Warneck (2000) to be $3100 \cdot 10^{19}$ g. As already noted, this value must be equivalent to the buried organic carbon; the amount of free oxygen in the atmosphere we can neglect in this budget. The values given in Table 2.12 for organic carbon sediments ($1500 \cdot 10^{19}$ g) correspond to $4000 \cdot 10^{19}$ g oxygen. In this value are included:

- H_2S oxidation into sulfate,
- FeO oxidation into Fe_2O_3,
- CO oxidation into CO_2 (from volcanic exhalations),
- H_2 oxidation into H_2O.

The residence time of O_2, based on the rapid exchange of carbon between the biosphere and CO_2 in the ocean-atmosphere system, amounts 5000 years (Warneck 2000), corresponding to 24 000 Tg oxygen yr^{-1} due to photosynthesis and respiration. The exchange rate of oxygen due to weathering and net emission from buried carbon is only 300 Tg yr^{-1} (corresponds to 6 Myr turnover time). Due to the low solubility of O_2 in seawater, the atmosphere acts as a buffer and reservoir for free O_2.

Modeling the long-term carbon cycle shows strong relationships between the key parameters, atmospheric oxygen and carbon dioxide (inverse correlated) as well as carbon burial, which is correlated with oxygen (Berner 2003). The (calculated) range of the amounts is:

- Atmospheric O_2 concentration: 13–30 %,
- atmospheric CO_2 concentration (factor to pre-industrial level): 1–25,
- C burial: (2.3–6.2) · 10^{18} mol Myr^{-1} (or 28–74 Tg yr^{-1}).

From the dates given by Berner (2003), only within the period of the last 410 Myr, has global organic carbon burial amounted to ~ 2200 · 10^{19} g carbon; using the low burial rate, in the period back to first oxygenic photosynthesis (2.7 Gyr), it would give a value of more than 6000 · 10^{19} g carbon burial, together with a multiple of the estimated organic carbon sediments (~ 1500 · 10^{19} g). This suggests that organic carbon is also cycling through subduction and volcanic release to the atmosphere, at an average rate of 15 Tg C yr^{-1}.

Photosynthesis (Eq. 2.40) as the only source of free oxygen, however, is balanced with mineralization of biomass (Eq. 2.43). In other words, there is no net production of oxygen when there is no burial in oxygen-depleted environments. It is logical that the burial rate is higher when the atmospheric oxygen concentration (and therefore in waters too) is low. This is not in contradiction with the general relationships shown above presented by Berner (2003) because at the early stages of the earth, most of the oxygen from photosynthesis was used to oxidize several elements from a reduced stage (mainly sulfur and iron but also hydrogen and reduced carbon exhalations from deep within the earth).

The oxygen required for iron and sulfide oxidation (4745.10^{19} g O from Table 2.12) corresponds to a carbon equivalent of about 1800 · 10^{19} g C additional to the C_{org} in sediments (1500 · 10^{19} g C). However, with imprecise knowledge of the volume of plant material buried in different epochs, with the strong possibility that deposits of organic carbon are in significant part due to upwelling hydrocarbons and not all to plant debris, and with an inexact knowledge of the ratio of oxidized to unoxidized carbon in the primary carbon supply to the surface, no firm judgment can yet be made (Gold 1999).

The loss of hydrogen (and helium) to space is one of the fundamental characteristics distinguishing the terrestrial planets from the giant planets. The earth was unable to capture hydrogen gravitationally from the solar nebula; it must have come in as a constituent of condensed material (mainly H_2O). Kasting and Catling (2003) estimated the H loss to be about 3.1 · 10^8 H atoms $cm^{-2} s^{-1}$, corresponding with an O_2 production rate behind of 6.7 · 10^{11} g yr^{-1}. Roughly half of the hydrogen that is escaping to space today comes from CH_4, almost all of which is produced biologically. Thus, the actual abiotic O_2 production rate from H_2O photolysis is only half (3 · 10^{11} g yr^{-1}). By comparison, the production rate from photosynthesis is ~ 5.8 · 10^{13} g yr^{-1} (Holland 2002). In the prebiotic period oxygen could not be accumulated in the atmosphere (Walker 1977); only the oxidation of volcanic H_2 (~ 5 · 10^{12} g yr^{-1} according to Holland, 2002) would consume 4 · 10^{13} g oxygen

yr^{-1} what is ~ 200 times greater than the abiotic O_2 production from water photolysis. Hence H_2 would have accumulated until the escape rate to space was high enough to balance the H_2 flux from volcanoes. This mixing level was assessed to be a few time of 10^{-3} (Kasting and Catling 2003).

In the period 100–300 Myr ago, large oxygen variations did occur with the maximum oxygen concentration around 30 % (280 Myr ago) and the minimum oxygen about 13 % (200 Myr ago). The periods 0–100 Myr ago and 350–450 Myr ago experienced about 20 % oxygen with only small variations. Several processes may have a stabilizing effect on the atmospheric oxygen level. The laying down of inorganic oxidized sediments (mainly sulfate) fixes oxygen. The escape of hydrogen from the earth into space, which is more dominant when the atmospheric oxygen level is low, will leave more oxygen behind. Higher oxygen levels in the atmosphere, and hence in groundwater, will diminish the areas of swamps and anoxic lakes, ponds and seas, the locations in which plant material undergoes fermentation and methanogenesis leaving reduced substances such as H_2 and CH_4. Conversely, low oxygen would favor anoxic deposition, leaving more oxygen behind. Hence the present reservoir distribution (Table 2.12) of elements/compounds is the result of the evolutionary processes described above. The biospheric carbon turn-over (and that of sulfur as discussed below) is so vast that small (never measurable) annual variations will be buffered and result in a well-balanced redox and acidity state of the climate system over geological time. It seems that the climate system itself has established a buffer system to stabilize it even in the case of (not too large) "catastrophic" effects. Consequently, disturbing the redox state of the environment, for example by human activities, can lead to changing reservoir distributions.

In the carbon cycle we have to consider long-term cycling, including rock weathering and volcanism. Over much longer time scales, atmospheric CO_2 concentrations have varied tremendously due to changes in the balance between the supply of CO_2 from volcanism and the consumption of CO_2 by rock weathering. CO_2 in the atmosphere is consumed in the weathering of rocks:

$$CO_2 \xrightarrow{H_2O} H_2CO_3 \, [H^+ + HCO_3^-]$$
$$\xrightarrow{CaCO_3} Ca\,(HCO_3)_2 \, [Ca^{2+} + 2\,HCO_3^-] \tag{2.62a}$$

This comes about by the first global reaction, first deduced by Ebelmen (1845):

$$CO_2 + (Ca, Mg)SiO_3 + (Ca, Mg)\,CO_3 + SiO_2 \tag{2.62b}$$

Carbonic acid is strong enough to dissolve silicate rocks – in small quantities, of course, and over long time scales. To illustrate this, we take an orthosilicate which is dissolved into orthosilicic acid (where SiO_2 is the anhydride) and bicarbonate:

$$CaH_2SiO_4 + (2\,H^+ + 2\,HCO_3^-) \rightarrow H_4SiO_4\,(\rightleftarrows SiO_2 + 2\,H_2O)$$
$$+ Ca\,(HCO_3)_2 \tag{2.63}$$

SiO_2 is moderately soluble in water (5–75 mg L^{-1} in river water and 4–14 mg L^{-1} in seawater). The products are then transported in river water to the oceans. There organisms such as foraminifera use calcium carbonate to make shells. Other organisms such as diatoms make their shells from silica. When these organisms die, they fall into the deepest oceans. Most of the shells redissolve but a fraction of them are buried in sediments of the seafloor. The overlaying sediments are carried down to the depths by subduction. Temperature and pressure transform the shells back to silicate minerals, in the process releasing CO_2 back to the surface of the earth through volcanoes and into the atmosphere to begin the cycle again, over a geological time scale. This inorganic (no photosynthesis) but biotic (mineral production) carbon cycle is not linked with the oxygen cycle but with water (H_2O) and acidity (H^+). Simply said, insoluble rock carbonate is transformed into more soluble bicarbonate where atmospheric CO_2 is fixed as dissolved bicarbonate. The volcanic carbon dioxide released is roughly equal to the amount removed by silicate weathering; so the two processes, which are the chemical reverse of each other, sum to roughly zero, and do not affect the level of atmospheric carbon dioxide on time scales of less than about 10^6 years. As a planet's surface becomes colder, however, atmospheric CO_2 levels should tend to rise. The reason is that removal of CO_2 by silicate weathering followed by carbonate deposition should slow down as the climate cools, and would cease almost entirely if the planet were to glaciate globally. On planets like earth that have abundant carbon (in carbonate rocks) and some mechanism, like plate tectonics, for recycling this carbon, volcanism should provide a more-or-less continuous input of CO_2 into the atmosphere.

Additionally one has to include in the budget still permanent CO_2 degassing (unknown value) from the crust (accepting the deep carbon hypothesis). There is also no doubt that degassed CH_4 is partly a product of deep rock chemistry (cf. Eq. 2.4–2.7).

Four revolutionary changes have occurred in the chemical evolution of carbon:

1. The first change (and what separates the earth's atmosphere from those of the other terrestrial planets) was the absorption of atmospheric CO_2 by rain in the early earth during the formation of the oceans with subsequent seawater carbonate equilibriums.
2. With the modern photosynthesis by cyanobacteria (~ 2.3 Gyr ago), a huge consumption (equivalent to the atmospheric oxygen increase by a factor of 300) of CO_2 into biomass occurs within a few million years.
3. A third drop in atmospheric CO_2 was associated with the O_2 increase in the late Silurian (420 Myr ago) due to the evolution of land plants.
4. Today atmospheric CO_2 is increasing at a rate that has probably never occurred over the entire history of the earth due to the release by human activity of carbon in "fossil fuels" buried over millions of years in a period of barely hundreds of years.

The amount of CO_2 removed from the atmosphere each year by oxygenic photosynthetic organisms is massive (Fig. 2.10). It is estimated that photosynthetic organisms remove about $120 \cdot 10^{15}$ g C per year. This is equivalent to $4 \cdot 10^{18}$ kJ of free energy stored in reduced carbon, which is roughly 0.1 % of the visible radiant energy incident on the earth per annum (see also Chapter 2.3.5.3). Each year the photosynthetically reduced carbon is oxidized, either by living organisms for their survival, or by combustion; the burial rate is small, approximately between 60 Tg C yr^{-1} (from data from Berner 2003) and 200 Tg C yr^{-1} (Fig. 2.10). Only the present

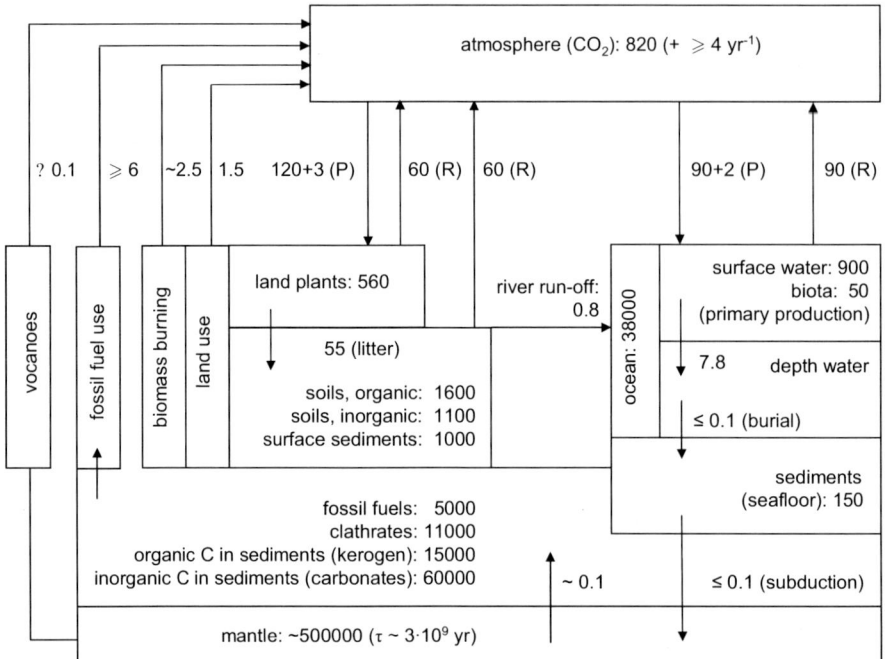

Fig. 2.10 Scheme of the carbon cycle and reservoirs; fluxes in 10^{15} g yr^{-1} and pools in 10^{15} g (data adapted from Schlesinger 1997, Houghton 2005a, Pédro 2007, Denman et al. 2007, Nieder and Benbi 2008). R respiration, P photosynthesis. River-runoff (0.8 Tg C yr^{-1}) consists from (all in Tg C yr^{-1}) 0.2 rock weathering, 0.4 soil weathering, and 0.2 hydrogen carbonate precipitation (Denman et al. 2007); The preindustrial ocean-atmosphere exchange amounts 70 Tg C yr^{-1} (20 anthropogenic additional), Denman et al. (2007) and the oceanic surface water contains 112 ± 17 Pg accumulated anthropogenic C (Sabine et al. 2004).

atmospheric oxygen requires a carbon equivalent via photosynthesis of $44 \cdot 10^{19}$ g C, two orders of magnitude more than present estimates of economically extractable fossil fuels ($0.7 \cdot 10^{19}$ g C in the mid 20th century, of which coal represents more than 90%). However, by burning all the fossil fuels that are still available in a short time, the oxygen content of the atmosphere would be reduced by only 1%. According to an estimate by Warneck (2000), the atmospheric CO$_2$ concentration would rise to about 800 ppm, twice the present level and about three times more than the preindustrial level. This is large compared with the CO$_2$ variation over the last few hundred thousand years but small concerning the time scale over epochs. The amount of combustible fossils fuels is only 40 times larger than the yearly biological turnover of carbon. The "problem", however, consists in the human time scale of a few hundred years and the vulnerable infrastructural systems of mankind.

Burning of fossil fuels amounts now to $\sim 10 \cdot 10^{15}$ g C yr^{-1}, which is a mere 10% of the terrestrial carbon uptake by photosynthesis, however it interrupts the carbon cycle due to the large residence time of CO$_2$ in the atmosphere (see Chapter 2.8.3.3). The oceans mitigate this increase by acting as a sink for atmospheric CO$_2$. It is estimated that the oceans remove about $2 \cdot 10^{15}$ g C yr^{-1} from the atmosphere. This carbon is eventually stored on the ocean floor. Although these estimates of sources and sinks are uncertain, the net global CO_2 concentration is increasing. Direct measurements show that currently each year the atmospheric carbon content is increasing by about $3 \cdot 10^{15}$ g. Over the past two hundred years, CO_2 in the atmosphere has increased from about 280 parts per million (ppm) to its current level of near to 400 ppm. Based on predicted fossil-fuel use and land management, it is estimated that the amount of CO_2 in the atmosphere will reach 700 ppm by end of this century. The consequences of this rapid change in our atmosphere are unknown. Because CO_2 acts as a "greenhouse" gas, some climate models predict that the temperature of the earth's atmosphere may increase by 2−8 °C. Such a large temperature increase would lead to significant changes in rainfall patterns. This could enhance weathering and subsequent CO_2 fixing as described above; it seems that the key question (or problem?) lies in the different time scales of the interactive processes contributing to liberation and fixing of CO_2. Little is known about the impact of such drastic atmospheric and climatic changes on plant communities and crops. Current research is directed at understanding the interaction between global climate change and photosynthetic organisms.

2.2.2.6 Life limits by catastrophic events: Mass extinctions

It appears that at any stage of chemical and biological evolution, (catastrophic) events may occur which change the conditions of the climate system in such a manner as to decimate life on earth by up to 95 %. In *Darwin*'s time (1850s−1860s), the view that catastrophes were an integral part of evolution was seen as a fall-back to ancient theories of catastrophism and biblical theories. This view has changed recently, and modern ideas of catastrophism are part of an active discussion of the *internal* and *external* causes of evolution. Internal causes are causes internal to the biota (such as competition, evolution of diseases), external causes are those external to the biota (such as volcanic eruptions, collisions with celestial bodies or nuclear war). It is logical to include among such *catastrophic* impacts first of all collisions with large bodies (comets and asteroids). Until now, the only known extinction-level meteorite impact was at Chicxulub, on Mexico's Yucatan peninsula, which is linked to the demise of the dinosaurs 65 Myr ago (Cretaceous/ Tertiary limits).

Other impacts that are also discussed concern volcanism and/or degassing of poisonous and/or climate-driven gases such as H_2S, SO_2, CO_2, and CH_4. Multiple effects and cascades of interactions are possible as they are in principle also modelled and discussed with the ongoing climate change. Warming by emission of CO_2 and CH_4 can have feedback effects on oxygen degassing in oceans which limits life but also triggers more warming by the release of CH_4 from clathrates in permafrost layers. Extreme eruptions (for example in Siberia) may directly defrost soils which consequently release CH_4. Eruptions may have caused dust clouds and acid aerosols which would have blocked out sunlight and thus disrupted photosynthesis both on land and in the upper layers of the seas, causing food chains to collapse. These eruptions may also have caused acid rain when the aerosols washed out of the atmosphere. Cooling may result in glaciation and falling sea levels. The year 1783 has been referred to in Europe as "Annus Mirabilis" given the large number of extreme climatic, volcanic and tectonic events that took place. From June 1783 to February 1784 was the second largest historical eruption of Laki (Iceland), injected some 100 Tg of SO_2 up to the lower stratospheric altitude (cf. Table 2.39) and caused exceptional dry and sulfur-rich haze, called dry fog which resulted in at least 10 000 extra death only in England and likely a multiple in France where the excess mortality was almost 40% (Courtillot 2005). There is no (geological) doubt that there will be a new such eruption at some time in the future (on timescale of a few centuries to a millennium).

Upwelling H_2S in oceans may kill all marine life and after escaping into the atmosphere, a layer of only a few meters is sufficient to kill most land plants and animals. Ozone-depleting gases (H_2S, SO_2, CH_4) can penetrate into the stratosphere (sometimes directly ejected by immense eruptions) and lead to life-damaging UV increase through ozone reduction. Therefore, a "horror scenario" of climate war against life can occur, where the weapons are poisonous gases, acid rain, UV radiation, but also no sunlight and other direct impacts by fire balls (CH_4 burning and flames) and hot lava and magma from volcanic eruptions.

The five more significant well-known mass extinctions are the extinctions at the Ordovician/Silurian ($\sim$ 445 Myr ago), late Devonian ($\sim$ 365 Myr ago), Permian/Triassic ($\sim$ 251 Myr ago), Triassic/Jurassic ($\sim$ 208 Myr ago) and Cretaceous/Tertiary limits ($\sim$ 65 Myr ago), the most significant being at the Permian/Triassic limit, and the most well-known at the Cretaceous/Tertiary limit (Benton 1986, Gore 1989, Benton 2005). Despite the Cretaceous/Tertiary extinction, the causes of the four other mass extinctions in history remained unexplained, including the largest, the end-Permian event of 250 million years ago – informally called "the Great Dying" – which killed 90% of sea life and 80% of land life.

2.3 The earth's energy sources

All chemical reactions need energy for activation (see Chapter 4.2.1) independent of whether they are exothermic or endothermic. Deep in the earth, heat and pressure are thermodynamic quantities initiating chemical reactions, as discussed in Chapters 2.1.2 and 2.2.1. In the atmosphere and at the earth's surface, however, the only available energy to promote chemical reactions is direct solar radiation. The fundaments of photochemical processes are described later (Chapter 4.2.3); in this Chapter the radiation transfer and physical processes in relation to it are briefly summarized. Beside initialization of chemical reactions, solar energy determines the earth's temperature regime (heat and subsequent pressure gradients) allowing transportation of matter in fluid systems such as the atmosphere and hydrosphere. The mean surface temperature of the earth is 288 K (varying between 222 K and 321 K); without the present atmosphere the mean surface temperature would be only 255 K. This narrow temperature regime allows phase transfer of water and the coexistence of its solid, liquid and gaseous phase, which surely is the main condition for our habitat/biosphere. Beside the distance from the planet to the sun (which determines the solar constant and hence the habitable zone in the solar system) the planetary atmospheric composition determines the radiation transfer and budget through the atmosphere. However there exist interactive correlations, as discussed above, concerning the different fates of the similar primitive atmospheres of the terrestrial planets.

2.3.1 Solar radiation

2.3.1.1 The sun and its radiation output

For each second of the solar nuclear fusion process, 700 Mt of hydrogen is converted into the heavier atom helium. Since its formation ~ 4.6 Gyr ago, the sun has used up about half of the hydrogen found in its core. The solar nuclear process also creates immense heat which causes atoms to discharge photons. Temperatures at the core are about $1.5 \cdot 10^7$ K. Each photon that is created travels about one micrometer before being absorbed by an adjacent gas molecule. This absorption then causes the heating of the neighboring atom and it re-emits another photon that again travels a short distance before being absorbed by another atom. This process then repeats itself many times over before the photon can finally be emitted to outer space at the sun's surface. For the last 20 % of the journey to the surface the energy is transported more by convection than by radiation. It takes a photon approximately 10^5 years or about 10^{25} absorptions and re-emissions to make the journey from the core to the sun's surface. The journey from the sun's surface to the earth takes about 8 minutes.

The irradiative surface of the sun, or photosphere, has an average temperature of about 5800 K. Most of the electromagnetic radiation emitted from the sun's surface lies in the visible band centered at 500 nm, although the sun also emits significant energy in the ultraviolet and infrared bands, and small amounts of energy in the radio, microwave, X-ray and gamma ray bands. The total quantity of energy emitted from the sun's surface is approximately $6.3 \cdot 10^7$ W m^{-2}. The energy emitted by the sun passes through space until it is intercepted by planets, other celestial objects, or interstellar gas and dust. The intensity of solar radiation striking these objects is determined by a physical law known as the *inverse-square law*. This law merely states that the intensity of the radiation emitted from the sun varies with the squared distance from the source. For example, the intensity of radiation from the sun is 9140 W m^{-2} at the distance of Mercury; but only 1370 ± 5 W m^{-2} at the distance of earth − a threefold increase in distance results in a ninefold decrease in intensity of radiation. This quantity is called *solar constant* I_K. It is important to note that this quantity is related to a plane perpendicular to the radiation beam. Therefore, the earth receives solar radiation only hemispherically ($I_K \pi r_{\text{Earth}}^2$), and on a global average of $I_K / 4$ (343 W m^{-2}).

The *solar cycle*, or the *solar magnetic activity cycle*, is the main source of periodic solar variation driving variations in space weather. The cycle is observed by counting the frequency and placement of *sunspots* visible on the sun. The solar cycle was discovered in 1843 by *Samuel Heinrich Schwabe*. *Rudolf Wolf* compiled and studied these and other observations, reconstructing the cycle back to 1745, eventually pushing these reconstructions to the earliest observations of sunspots by *Galileo* and contemporaries in the early seventeenth century. Starting with *Wolf*, solar astronomers have found it useful to define a standard sunspot number index, which continues to be used today. Several periodic cycles are evident, most notably the 11-year (131 ± 14 month) cycle. Every solar proton event will generate nitrogen monoxide (NO) in the upper atmosphere (see also Chapter 2.6.4.4) and consequent ozone depletion (see Chapter 5.3.7) (Jackman et al. 2000). But only major large events (those with > 30 MeV) will generate sufficient NO_y to be observable above this terrestrial background. An ultra-high resolution nitrate analysis of ice cores from Greenland shows an impulsive nitrate deposition, relatively consistent with those of the last five solar cycles (Smart et al. 2007).

2.3.1.2 Solar radiation transfer through the atmosphere

Radiation describes any process in which energy emitted by one body travels through a medium or through space, ultimately to be absorbed by another body. The radiant energy is the energy of electromagnetic waves. Sunlight (solar radiation), in the broad sense, is the total spectrum of the electromagnetic waves given off by the sun. On earth, sunlight is filtered through the atmosphere, and the solar radiation is obvious as daylight when the sun is above the horizon. The World Meteorological Organization defines sunshine as direct irradiance from the sun measured on the ground of at least 120 W m^{-2}. Direct sunlight has a *luminous efficiency* of about 93 lumens per watt of radiant flux, which includes infrared, visible, and ultra-violet light. Bright sunlight provides *luminance* of approximately 10^6 cd m^{-2} (candelas per square meter) at the earth's surface. There are many systems of units for optical radiation: Table 2.14 summarizes the radiometric, purely physical, quantities, used for description of photophysical and photochemical processes. In Chapter 4.2.3 more information is presented concerns photodissociation, whereby *spectral* and *actinic* related quantities are of interest. The *actinic radiation* (and hence related quantities such as actinic flux) is the solar radiation that can initiate photochemical reactions. The term "spectral" simply means that a quantity is measured per wavelength per interval (see Table 2.15).

Table 2.14 Radiometric quantities.

quantity	symbol	dimension
radiant energy	Q^a	J
radiant power[b]	$E = dQ/dt$	W (J s^{-1})
radiant flux[c]	$P = dE/da$	W m^{-2}
radiance[d]	$L = dE/(da \cdot \cos \Theta) = dS/d\Omega$	W m^{-2} sr^{-1}
radiant intensity[e]	$S = de/d\Omega = \int L \, d\Omega$	W sr^{-1}
radiant exitance[f]	J	W m^{-2}
irradiance[g]	$I = S \cos \Theta = \int L \cos \Theta \, d\Omega$	W m^{-2}

[a] often denotes W (dimension: energy); [b] also called (incorrectly) flux and denotes Φ (dimension: energy / time); [c] also called radiant flux density and denotes I (dimension: energy / time · square); [d] spherical radiant flux (dimension: energy / time · square · unit angle); [e] such as actinic flux (dimension: energy / time · unit angle); [f] also called emittance (light emission); [g] often denotes E

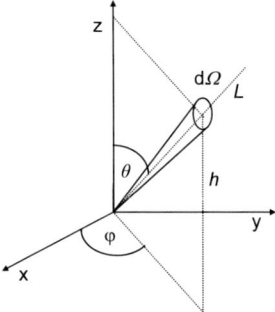

Fig. 2.11 Scheme of the geometry of irradiation; θ solar zenith angle, Ω solid angle, φ azimuth direction angle.

The solar constant I_k represents a *radiant flux* E (energy $\cdot$ area^{-1} $\cdot$ time^{-1})

$$E = \frac{dP}{da} \ [\text{W} \cdot \text{m}^{-2} \quad or \quad \text{J} \cdot \text{s}^{-1} \cdot \text{m}^{-2}] \tag{2.64}$$

a is the area/surface; the radiant power p, often confusingly denotes radiant flux, is the radiant energy per unit of time (in Js^{-1}):

$$P = \frac{dQ}{dt} = \int I(\Omega)d\Omega \ [\text{J} \cdot \text{s}^{-1}] \tag{2.65}$$

The radiant flux may be the total emitted from a source (exitance), or the total landing (irradiance) on a particular surface (da). *Irradiance E* is the total amount of radiant flux incident upon a point on a surface from all directions above the surface. *Radiant exitance J* is the total amount of radiant flux leaving a point on a surface into all directions above the surface. *Radiant intensity S* is the radiant flux radiated from a point on a light source into a unit solid angle Ω in a particular direction. The *radiant energy Q* denotes the radiation transported by solar light to a surface or emitted from the surface. It is important to distinguish between radiation transfer through a plane and a spherical surface. Hence, the spherical radiant flux, denotes *radiance L* is described (see Fig. 2.11) by

$$L = \frac{d^2P}{da \cos\theta \, d\Omega} = \frac{dE}{\cos\theta \, d\Omega} = \frac{dE}{\sin\theta \cos\theta \, d\varphi \, d\theta} \tag{2.66}$$

Ω steradiant (cf. Fig. 2.11), defined to be the solid angle with the unit sr. The total solid angle (steradiant) Ω of a sphere is given by $4\pi \cdot$ sr and therefore it follows $1 \text{ sr} = \Omega/4\pi$. With the area of a sphere ($a = 4\pi \cdot r^2$), $a = (\Omega_{\text{total}}/1 \text{ sr})r^2$ is valid. For a partial area of a sphere it follows $\Delta a = (1 \text{ sr}/1 \text{ sr})r^2 = r^2$, and thus 1 sr is the solid angle (steradiant) when the partial area is equal to the square of the radius (r^2). In Equation (2.66) the area (da) is given by (Fig. 2.11) $da = r \sin\theta \cdot r \sin\varphi$ and because $ds = r^2 d\Omega$, it follows that $d\Omega = \sin\theta \cdot d\varphi \cdot d\theta$. The factor $\cos\theta = h/L$ (θ solar zenith angle) in Equation (2.66) operates so that the radiation is always regarded as perpendicular to the direction of arrival of light along the line L (Fig. 2.11).

Table 2.15 Spectral distribution of extraterrestrial solar radiation; definition according to WMO (1986).

light range	wave length (in nm)	radiant flux density (in W m^{-2})	percentage to total flux (in %)
UV-C light	100–280	7.0	0.5
UV-B light	280–315	16.8	1.2
UV-A light	315–400	84.1	6.2
total UV light	100–400	107.9	7.9
visible light	400–760	610.9	44.7
infrared light	760– 10^6	648.2	47.4
total	100– 10^6	1367.0	100.0

The radiant flux can also be described by photons – a more usual approach for describing photochemical processes. The photon is the matter symbol of the electromagnetic wave. Since the velocity of a photon is given by the light velocity (c = 2.998 · 10^8 m s^{-1} in vacuum); the speed of light is proportional to the refractive index of the medium and is in the air only slightly less than c. We can therefore describe the energy of a photon by the quantity hv (h Planck's constant); frequency v and wavelength λ are combined by $\lambda = c \:/\: v$.

Photometric quantities (luminance and derived luminous quantities) are used to describe the light within the visible range in relation to the human eye. *Luminous flux* (Φ_v) is energy per unit time (dW/dt) that is radiated from a source over visible wavelengths. More specifically, it is energy radiated over wavelengths sensitive to the human eye, from about 400 nm to 760 nm (Table 2.15). Thus, luminous flux is a weighted average of the *radiant flux* in the visible spectrum. It is a weighted average because the human eye does not respond equally to all visible wavelengths. The *lumen* is the standard unit for the luminous flux of a light source. It is an SI-derived unit based on the candela. It can be defined as the luminous flux Φ_v emitted into unit solid angle (1 sr) by an isotropic point source (cf. Fig. 2.11) having a *luminous intensity* (I_v) of 1 cd. The unit lumen (lm) is then equal to cd · sr.

For general purposes, the energy output of the sun can be considered constant. This of course is not entirely true. Scientists have shown that the output of the sun is temporally variable (Fig. 2.12). Some researchers have also suggested that the increase in the average global temperature over the last century may have been solar in origin. This statement, however, is difficult to prove because accurate data on solar output of radiation only goes back to about 1978. The incoming solar flux (solar constant) distributes among all wave lengths (cf. Table 2.15) extraterrestrially above the earth's atmosphere perpendicular to the source with a mean distance between sun and earth of 1496 · 10^8 m. The light range sensible to human eyes lies between 400 and 760 nm (visible light); light < 400 nm is called ultraviolet (UV) and light > 760 infrared (IR).

Only about 30 % of the solar energy intercepted at the top of earth's atmosphere passes directly through to the surface (cf. Fig. 2.13). On the way through the atmosphere, direct solar radiation undergoes *scattering, absorption* and *reflection* on mol-

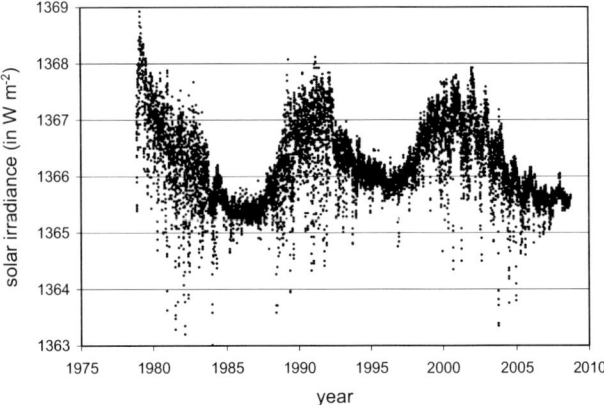

Fig. 2.12 Composite database of solar irradiance compiled from many satellite TSI data 1978–present (daily means). I acknowledge the receipt of the dataset from PMOD/WRC, Davos, Switzerland, and acknowledge unpublished data from the VIRGO Team (see e. g. Fröhlich and Lean 1998).

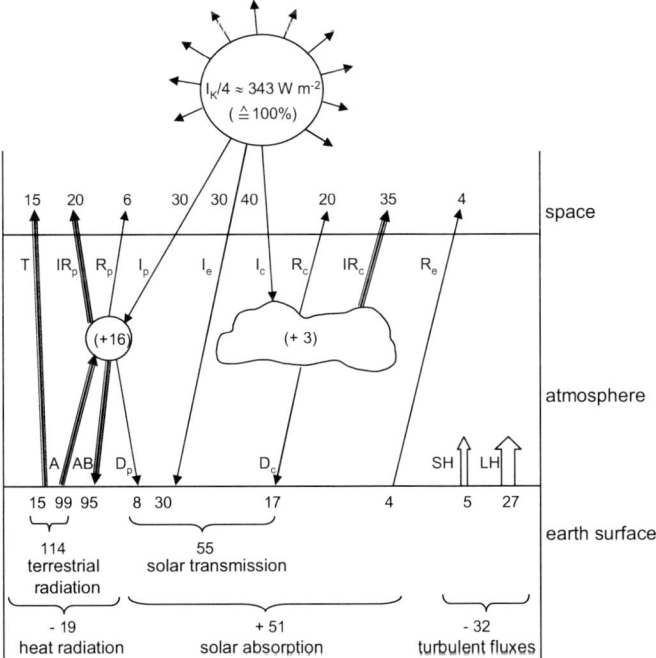

Fig. 2.13 Radiation and heat balance of the system earth-atmosphere. Percentages are given in relation to the incoming solar radiation (100 % $\cong$ 343 W m^{-2}). I interception of solar radiation by molecules/particles (p) clouds (c), and earth surface (e), D diffuse radiation by molecules/particles (p) clouds (c), R reflexion (albedo) by molecules/particles (p) clouds (c), and earth surface (e), IR infrared dissipation to space by molecules/particles (p) clouds (c), T terrestrial radiation back to space (without absorption), A absorption of terrestrial radiation by molecules, AB atmospheric back-radiation, SH and LH sensible and latent heat, resp.

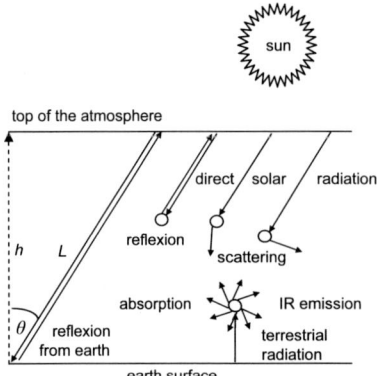

Fig. 2.14 Scheme of solar radiation transfer through the earth atmosphere; θ solar zenith angle, h height of the atmosphere, L length of the radiation beam through the atmosphere.

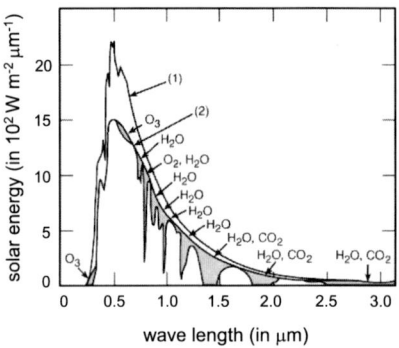

Fig. 2.15 Solar spectrum at the top of earth atmosphere (1) and at earth surface (2); absorbing gases are denoted.

ecules and suspended particles, such as dust particles and cloud droplets (Fig. 2.14). The atmosphere reflects and scatters some of the received visible radiation. Gamma rays, X-rays, and ultraviolet radiation less than 200 nm in wavelength are selectively absorbed in the upper atmosphere by oxygen and nitrogen and turned into heat energy. Most of the solar ultraviolet radiation with a range of wavelengths from 200 to 300 nm is absorbed by ozone (O_3) and oxygen found in the stratosphere. Infrared solar radiation with wavelengths greater than 700 nm is partially absorbed by carbon dioxide, ozone, and water present in the atmosphere in liquid and vapor forms (Fig. 2.15). Hence in the troposphere only radiation with wavelength > 290–300 nm can penetrate and initiate photochemical processes (see Chapter 4.2.3). As mentioned in Chapter 2.2.2.2, this was a precondition for the evolution of life on the earth's surface. Photochemical reactions only occur with radiation < 750 nm (ozone photolysis into O^3P). The three relevant ozone absorption bands (named after their discoverers, *Hartley*, *Huggins* and *Chappuis*) are encompassed by its range of 240–750 nm. Another band, called the *Wulf* band (near IR > 750 nm), is only of photophysical interest, for instance, for observing ozone from space.

The process of scattering is elastic and occurs when small particles and gas molecules diffuse part of the incoming solar radiation in random directions (Fig. 2.14) without any alteration to the wavelength of the electromagnetic energy, i.e., no energy transformation results. Hence, scattering reduces the amount of incoming radiation reaching the earth's surface. The sky is bright also from directions from where there is no direct sun light as a result of scattering; that light is called diffused solar radiation (skylight). Scattering occurs on particles when their diameter is similar to or larger than a wavelength (*Mie* scattering) and on molecules, generally much smaller than the wavelength of the light (*Rayleigh* scattering). The sky has a blue appearance in the daytime because *Rayleigh* scattering is inversely proportional to the fourth power of wavelength, which means that the shorter wavelength of blue light will scatter more than the longer wavelengths of green and red light. When the sun is near the horizon, the light passes a longer distance through the atmosphere and red light remains after scattering out of blue light.

Absorption is defined as a process in which solar radiation is retained by a substance and converted into heat energy, or in other words, into *inner energy* (rotation, vibration, and translation) but also into dissociation (photolysis, see Chapter 4.2.3). The absorbed light will finally be emitted as long-wave radiation (dissipated heat). The wave is spherical or, in other words, the emission is uniform in all directions (Fig. 2.14). Reflection is the third process of altering direct solar radiation through the atmosphere where sunlight is redirect by 180° after it strikes very large particles (cloud droplets, liquid and frozen) and the earth's surface. The angle of incidence equals the angle of reflection. This redirection causes a 100% loss of the insolation. Sunlight reaching the earth's surface unmodified by any of the above atmospheric processes is termed direct solar radiation. Roughly 30% of the sun's visible radiation (wavelengths from 400 nm to 700 nm) is reflected back to space by the atmosphere or the earth's surface (cf. Table 2.16). The reflectivity of the earth or any body is referred to as its *albedo*, defined as the ratio of light reflected to the light received from a source, expressed as a number between zero (total absorption) and one (total reflectance), see Fig. 2.13. The reflectivity or albedo of the earth's surface varies with the type of material that covers it. For example, some surface-type reflectivities are:

− Clouds, 40 to 90%,
− fresh snow, up to 95%,
− dry sand, 35 to 45%,
− broadleaf deciduous forest, 5 to 10%,
− coniferous forest, 10 to 20%,
− grass type vegetation, 15 to 25%.

Therefore, and due to global averaging ($I_K/4$), only 240 W m^{-2} reaches the earth's surface, varying with time and site. Further reading: Wendisch and Yang (2012).

Table 2.16 Earth radiation budget in % of incoming solar radiation $(100\% = 343\ \mathrm{W\ m^{-2}})$. ↓↓ input, ↑↑ output, IR infrared radiation.

where		%	type of radiation / interaction	sum	budget
top of the atmosphere	↓↓	100	solar radiation (solar constant)	100	$\Delta = \pm\,0$
	↑↑	20	albedo (clouds)		
		6	albedo (particles)		
		4	albedo (earth surface)	30	
		35	IR from clouds		
		20	IR from molecules		
		15	IR from earth surface	70	
atmosphere	↓↓	99	absorption of terrestrial radiation by molecules		$\Delta = \pm\,0$
		40	reflexion/scattering of solar radiation by clouds		
		30	reflexion/scattering of solar radiation by particles		
		32	latent and sensible heat	201	
	↑↑	20	albedo (clouds)		
		6	albedo (particles)		
		25	diffuse radiation (particles and clouds)	51	
		35	IR from clouds		
		20	IR from molecules		
		95	atmospheric back-radiation to the Earth surface	150	
earth surface	↓↓	30	direct solar radiation		$\Delta = \pm\,0$
		17	diffuse radiation (clouds)		
		8	diffuse radiation (particles)	55	
		95	atmospheric back-radiation	95	
	↑↑	4	albedo of solar radiation (transmission to space)	4	
		15	terrestrial radiation (absorption by molecules)		
		99	terrestrial radiation (transmission to space)		
		5	sensible heat		
		27	latent heat	146	

2.3.2 Absorption and emission of light

Thermal equilibrium means that a body (molecule, particle, surface) receives from its environment as much energy as it emits. The higher the temperature of the body, the higher is the radiant flux density of emitted (thermal) radiation. The percentage of absorbed radiation is called *absorptivity* (or *absorbance*) degree ε (and *emissivity*, respectively), complementary with the reflectivity $(1 - \varepsilon)$. For each temperature the emitted radiation is linear with the absorption degree to obtain the thermal equilibrium. In other words, the more a surface absorbs the more it emits. This is valid for each frequency. A black surface is defined with $\varepsilon = 1$. With this the radiant emittance (J) of all surfaces can be related to that of black bodies/surfaces (J_b):

$$J(\lambda) = \varepsilon(\lambda) J_b(\lambda) \tag{2.67}$$

This called *Kirchhoff*'s law. Hence, a body which does not absorb radiation of a given wavelength also cannot emit on this wavelength.

The emissivity of earth's atmosphere varies according to cloud cover and the concentration of gases that absorb and emit energy in the thermal infrared (i. e., wavelengths around 8 to 14 μm). These gases are often called "greenhouse" gases, from their role in the "greenhouse" effect. The main naturally-occurring "greenhouse" gases are water vapor, carbon dioxide, methane, and ozone. The human problem lies in an increase of the latter gases in the troposphere. The major constituents of the atmosphere, N_2 and O_2, do not absorb or emit in the thermal infrared.

Thus, the earth obtains energy from solar radiation and loses it (to maintain a thermal equilibrium) by reflection, emission and terrestrial radiation; the temperature of the earth's surface and atmosphere thus corresponds to the thermal budget.

2.3.2.1 Absorption (*Lambert-Beer* law)

Due to the interaction of solar radiation with molecules and particles of the atmosphere, the radiant flux decreases with the path x through the atmosphere (Fig. 2.14). *Johann Heinrich Lambert* showed in 1760[9] that the reduction of light intensity is proportional to the length of path x (or layer thickness) and the light (radiant flux) itself, $\Delta E = x \cdot E$, from which we derive the equation

$$\frac{dE}{dx} = -m' \cdot E \tag{2.68}$$

Where m' is the extinction module. The minus sign denotes that the radiation decreases. Now, expressing the path x by the solar zenith angle θ, we obtain $x = h \cdot \sec \theta$ and finally

$$E = E_0 \exp(-m \cdot \sec \theta) \tag{2.69}$$

where $m = m' \cdot h$ and E_0 the irradiation. For $\theta > 60°$ equation (2.69) must be corrected due to the curvature of the earth. *August Beer* (1848) found that the extinction of light further depends on the concentration of substances within the irradiated medium, i. e. $m = \kappa \cdot c$, where κ is the extinction coefficient (fraction of light lost due to scattering and absorption per unit distance in a participating medium). The extinction coefficient, depending on wavelength, is further separated into coefficients for absorption and scattering for molecules as well as particles: $\kappa = \kappa_{abs}^{gas} + \kappa_{abs}^{particle} + \kappa_{scat}^{gas} + \kappa_{scat}^{particle}$. By combination of *Lambert*'s and *Beer*'s relationships we obtain the *Lambert-Beer* law:

---

[9] *Photometria, sive, De mensura et gradibus luminis, colorum et umbrae.* (Photometry, or, On the Measure and Gradations of Light, Colors, and Shade), Augsburg 1760

$$E = E_0 \exp \ (-\kappa \cdot c \cdot \sec \ \theta) \qquad (2.70)$$

The amount of light absorbed by a substance, depending on the wavelength, can be calculated according to

$$E_{abs}(\lambda) = E(\lambda)\,\sigma(\lambda) \cdot c \qquad (2.71)$$

The *absorption cross section* σ, depending on wavelength and temperature and specific for each substance, characterizes the effective area of a molecule for scavenging of a photon. Equation (2.71) is used to calculate the photolysis rate (see Chapter 4.2.3).

Between 0.3 and 0.7 μm (visible range) and 8−12 μm, with the exception of ozone bands, there is virtually no absorption in the atmosphere − therefore these ranges are called *atmospheric windows*; solar and terrestrial radiation can penetrate the atmosphere unopposed. Between 1 and 8 μm H_2O (2.5−3.5 μm and 4.5−7.5 μm) and CO_2 (2.2−3.5 μm and 4−4.5 μm as well as 15−20 μm) absorb terrestrial radiation partially, and at > 15 μm nearly completely.

2.3.2.2 Emission (*Planck's* law and *Stefan-Boltzmann's* law)

Max Planck first found in 1900 that none of the previous thermodynamic and electrodynamic laws could explain the spectrum of black-body radiation and therefore proposed the *quantum theory*. This suggests that energy in a radiation field is not exchanged continuously, but only as integer multiples of the so-called *energy quantum hv* (now often called *Planck's* quantum). *Planck's* law gives the intensity radiated by a black body as a function of frequency (or wavelength), see Fig. 2.16: here written as the specific emission per area da and frequency dv or wavelength $d\lambda$ (note that in some formulas instead of $2h$ it is written as $8\pi h$ which is related to the whole spherical emission):

$$E(v, T)\,da\,dv = \frac{2hv}{c^2} \ \frac{v^2}{\exp\left(\dfrac{hv}{kT}\right) - 1} \ da\,dv \qquad (2.72a)$$

$$E(\lambda, T)\,da\,d\lambda = \frac{2hc^2}{\lambda^5} \ \frac{1}{\exp\left(\dfrac{hv}{kT}\right) - 1} \ da\,d\lambda \qquad (2.72b)$$

Note that $\lambda = c/v$ and $dv = (c/\lambda^2)\,d\lambda)$, k is the *Boltzmann* constant. A simplification was formerly known for large frequencies ($hv \gg kT$) as *Wien's* law:

$$E(v, T)\,da\,dv \approx \frac{2hv^3}{c^2} \exp\left(-\frac{hv}{kT}\right) da\,dv \qquad (2.73a)$$

$$E(\lambda, T)\,da\,d\lambda = \frac{2hc^2}{\lambda^5} \exp \ \left(-\frac{hv}{kT}\right) da\,d\lambda \qquad (2.73b)$$

The *Planck* curves for different T proceed such that those for lower T always lie below the curves for higher T − in other words, the curves never intersect. The

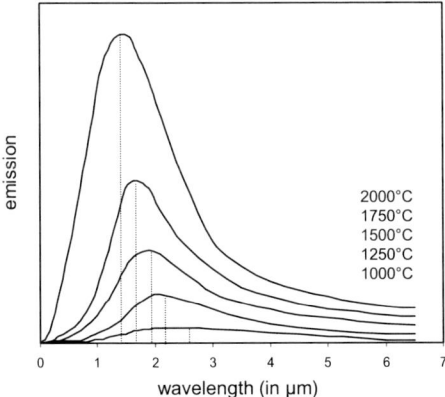

Fig. 2.16 *Planck*-curves, showing the spectral radiation distribution of black bodies for different temperatures (in K).

radiation maximum displaces with increasing T to shorter wavelength (*Wien*'s displacement law). From Equation (2.73) with $dI/dv = 0$ *Wien*'s displacement law can finally be derived (by using an approximate solution because only a numerical solution is possible) − the statement that there is an inverse relationship between the wavelength of the peak of the emission of a black body and its temperature:

$$v_{max} = \frac{2.82}{h}\mathbf{k}T = 5.88 \cdot 10^{10}\ T \quad \text{or expressed in wavelength} \quad \lambda_{max} = \frac{b}{T} \quad (2.74)$$

where b is *Wien*'s displacement constant ($b = 2.898 \cdot 10^6$ nm $\cdot$ K). The total energy emitted from a black body corresponds to the area below the *Planck* curve:

$$E_{total} = \int_0^\infty E(v, T)\ dv \quad (2.75)$$

from which the *Stefan-Boltzmann* law is derived:

$$E_{total} = \frac{2\pi^5}{15}\frac{\mathbf{k}^4}{c^2 h}T^4 = \sigma T^4 \quad (2.76)$$

$\sigma = 5.67 \cdot 10^{-8}$ W $\cdot$ m^{-2} $\cdot$ K^{-4} (*Stefan-Boltzmann* constant). It means that the total emitted black-body radiation only depends on its temperature (of the fourth power).

According to *Kirchhoff*'s law (Eq. 2.67), the emission of a body for a given T and λ is in a definite ratio to its absorptivity, where this ratio is independent from the material of the body and identical with the black-body radiation. Therefore, the emission can also be calculated for non-black bodies (with $\varepsilon(\lambda) < 1$) using *Planck*'s law. The emission for very low frequencies (large wavelength), or in other words, the body emits to the far left of the radiation maximum ($hv \ll \mathbf{k}T$), can be calculated using the so-called *Rayleigh-Jeans* approximation with $\exp(hv/\mathbf{k}T) \approx 1 + hv/\mathbf{k}T$ in application on Equation (2.72):

$$E(v, T)\, dv = \frac{2\,v^2}{c^2}\, \mathbf{k}T dv \quad \text{and} \quad E(\lambda, T)\, d\lambda = \frac{2c}{\lambda^4}\, \mathbf{k}T d\lambda \qquad (2.77)$$

The specific emission at temperature T is the same as for a black body according to *Kirchhoff*'s law at temperature T':

$$E = \beta \cdot T = \varepsilon \cdot T_{black} = \beta \cdot T', \quad \text{and it follows} \quad T = \varepsilon \cdot T' \qquad (2.78)$$

Where β summarizes $2\,v^2 \mathbf{k}/c^2$ or $2\,c\mathbf{k}/\lambda^4$ respectively. With this, the temperature of the (cold) emitting non-black body by the factor ε (absorption degree) is smaller than the temperature of the comparable black body. In analogy it follows for hot emitters ($hv \gg \mathbf{k}T$) from Equation (2.73) that the true temperature T is higher than the black temperature T':

$$\frac{1}{T} - \frac{1}{T'} = \frac{\mathbf{k}}{hv}\, \ln\, \varepsilon \qquad (2.79)$$

where for $\varepsilon = 1$ it follows $T = T'$.

2.3.3 Terrestrial radiation and radiation budget

As already mentioned, 30 % of incoming solar radiation is reflected back into space (without changing the wavelength, almost all within the visible range), mostly by clouds, to be about two-thirds (Table 2.16 and Fig. 2.13). It is therefore assumed that clouds play a dominant role in regulating the climate, i. e., the temperature of the atmosphere. Several parameters such as coverage, depth, liquid water content, and cloud droplet size are among so-called *cloud climatology*. These important quantities have to be considered in the radiation budget but are difficult to monitor as well to model. Therefore, climate models contain a large uncertainty in this relationship.

Of the remaining 70 % of solar radiation (corresponding to 240 W m^{-2}), however, less than half (30 %) actually reaches the earth's surface directly. The remaining 40 % is scattered by clouds and particles in the atmosphere, from which a part of the diffuse radiation (25 %) is also transmitted to the earth's surface. Hence the total of direct solar radiation and diffuse sky radiation received by a unit horizontal surface is called global radiation.

Naturally, the radiation budgets at the borders of the atmosphere (at the upper reaches and at the earth's surface) are balanced, i. e., ± 0 (Table 2.16 and Fig. 2.13); all percentages are climatological means and based on the incoming solar radiation at the top of the atmosphere. Insofar as the energy fluxes of irradiation and emission are equal, it results in a certain temperature according to *Planck*'s law. The temperature remains constant if the fluxes do not change. Otherwise, T may increase (flux > 150 %) or decrease (flux < 150 %). A valid equation (a = mean global albedo) is:

$$E_+ = \frac{S_0}{4} = E_- = \sigma\, T_{Earth}^4 + a\frac{S_0}{4} \quad \text{and} \quad \sigma\, T_{Earth}^4 = (1 - a)\frac{S_0}{4} \qquad (2.80)$$

With an albedo of 0% and a radiation flux of 343 W m^{-2}, an equilibrium temperature of 6 °C results, almost unrealistically. When $a = 30\%$ and subsequent 240 W m^{-2}, $T_{earth} = -18$ °C results. This is exactly the global mean temperature at about 5000 m altitude or 550 hPa. This quantity is called the *effective radiation temperature* of the earth (255 K or -18 °C). It is worth noting here that in such case no life would exist on earth. Fortunately, the global mean temperature close to the earth's surface (at 1 m altitude) is about 15 °C (280 K), i. e., 33 K higher than the effective radiation temperature. This *natural* effect is due to the absorbance of some trace gases (H_2O, CO_2, CH_4, N_2O and others, now called "greenhouse" gases) in the atmosphere which absorb infrared (IR) terrestrial radiation and emit the energy in a more IR (remember: in all directions and thus partly back to the earth's surface), called *atmospheric back-radiation*. The anthropogenic effect lies in increasing the atmospheric concentration of such "greenhouse" gases due to different activities (see Chapters 2.7 and 2.8). This effect is termed the *greenhouse effect*, but it does not correspond to the effects in a greenhouse, only to the *impact* of warming (by different physical reasons).

The shortwave radiation budget of the earth's surface amounts to 51% (see Table 2.16 and Fig. 2.13), i. e., $+0.51 \cdot I_K/4 \triangleq 175$ W m^{-2} (solar absorption). This energy is transformed into terrestrial radiation (budget -19%) and turbulent fluxes of latent and sensible heat (-32%). The latter energy in form of turbulent fluxes (110 W m^{-2}) is than transferred through cloud processing and finally dissipated in IR back to space (35% corresponding to 120 W m^{-2}). The difference of -3% (10 W m^{-2}) is balanced by the positive budget of shortwave solar radiation, i. e., $+3\%$ is the net absorbance of clouds, resulting in atmospheric heating. More net absorbance, however, results by absorption of solar radiation through molecules and particles ($+16\% \triangleq 55$ W m^{-2}). The energy budget of the two groups of interactors (molecules/particles and clouds) with radiation, both shortwave as well as longwave, is naturally zero: $\pm 72\%$ for clouds and $\pm 129\%$ for molecules/particles, respectively.

In the terrestrial radiation (budget -19%) at the earth's surface, one has to consider a split between absorption of terrestrial radiation by molecules (99%) and so-called atmospheric back-radiation (95%), while 20% is radiated back into space. The budget (-4%) must be added to the transmission of the 15% of terrestrial radiation which can pass the atmosphere without absorption (through the atmospheric windows) to space. This back-radiation is the reason that the earth's surface has a higher temperature than would be expected from the effective radiation temperature, as mentioned above.

A part of the solar transmission absorbed by the earth's surface is transferred into *heat*, a different form of energy compared with thermal radiation. The *latent heat transport* is realized through water. Water (surface water and soil humidity) evaporates (see Chapter 2.5.2) by taking up the enthalpy needed for evaporation. As a consequence, the earth's surface cools down. Through vertical turbulent transport of water vapor it finally condenses into clouds while releasing the heat back to the atmosphere. The latent heat transport goes through heated air due to a temperature gradient from the surface to higher altitudes. To balance the ascending heated air (as a consequence there is a density gradient), at another site cooler air sinks down. This turbulent heat (32%) finally − together with the 3% of shortwave radiation stored in clouds − is dissipated in far IR back to space (35%).

In contrast to the global mean values used in discussing the energy budget of earth's atmosphere, the "true" geographic budgets are different (because of the different characteristics of the earth's surface and the solar zenith angle) and depend on time of day and as well as season. It is logical to mention that such differences drive the weather on earth. Below are listed some typical annual means of global radiation:

- 168 W m^{-2} global mean,
- 130 W m^{-2} middle Europe,
- 200 W m^{-2} Spain,
- 280 W m^{-2} Sahara.

2.3.4 Geothermal energy

Geothermal energy is heat from within the earth. If we return for a moment to the formation of the earth (see Chapters 2.1.1 and 2.2.1) it is evident that enormous quantities of heat originated from the formation of the planet by accretion, due to gravitational binding energy, and heat was also continuously produced inside the earth by the slow decay of radioactive elements. Often geothermal heat is called *renewable* on the grounds that the water (steam and hot water produced inside the earth) is replenished by rainfall and the heat is continuously produced inside the earth. However, it should be emphasized, all the heat generated in the earth's core – being very large – is limited, and is finally dissipated through the atmosphere to space. Geothermal energy can sometimes find its way to the surface in the form of volcanoes and fumaroles (holes where volcanic gases are released), hot springs and geysers. Hydrothermal circulation in its most general sense is the circulation of hot water, occurring most often in the vicinity of sources of heat within the earth's crust.

Knowledge of the structure and physical states of the interior regions of the earth is derived mainly from seismic data augmented by moment-of-inertia considerations. The chemical compositions of those regions, however, are primarily obtained as implications from meteorite data. The earth's fluid core comprises 30.8 % of the mass of the earth and is thought to consist of iron and one or more light elements, such as sulfur. A small, apparently solid object resides at the center, called the inner core. It is about the size of the moon and three times its mass, and comprises 1.65 % of the mass of the earth. The composition of the inner core is inextricably connected to ideas about the energy source that powers the geomagnetic field of the earth (Hollenbach and Herndon 2001).

The temperature within the earth increases with increasing depth. Within the first 100 m the temperature is nearly constant at about 11 °C. Below the zone of circulating ground water, however, temperature almost always increases with depth. However, the rate of increase with depth (geothermal gradient g) varies considerably with both the tectonic setting and the thermal properties of the rock (10–200 K km^{-1}). In Iceland, geothermal energy, the main source of energy, is extracted from those areas with geothermal gradients ($\geq$ 40 K km^{-1}). Highly viscous or par-

Table 2.17 Estimation of the geothermal energy of inner earth regions.

region	distance (km)	mean T (°C)	density (g cm^{-3})	matter	heat (J)
crust	0– 30[a]	~ 350	~ 3.5	rocks	~ 2.0 · 10^{22}
outer mantle	30– 300	~ 2000	~ 4	rocks	~ 5.6 · 10^{28}
inner mantle	300–2890	~ 3000	~ 5	rocks	~ 2.2 · 10^{30}
outer core	2890–5150	~ 5000	~ 8	Fe—Ni	~ 1.5 · 10^{31}
inner core	6150–6371	~ 6000	~ 8.5	Fe	~ 3.6 · 10^{30}

[a] continental crust, oceanic is 5–10 km depth

tially molten rock at temperatures between 650 and 1200 °C is postulated to exist everywhere beneath the earth's surface at depths of 80 to 100 km, and the temperature at the earth's centre, 6370 km deep, is estimated to be 5650 ± 600 °C (Steinle-Neumann et al. 2001). The temperature of the oceanic lithosphere base (95 km depth) is estimated to about 1400 °C (Stein and Stein 1992, Denlinger 1992). The continental crust shows a temperature of 700 °C at 18 km and 800 °C at 32 km depth (Mechie et al. 2004).

If that rate of temperature change were constant, temperatures deep within the earth would soon reach the point where all known rocks would melt. We know, however, that the earth's mantle is solid, because of the high pressure. The temperature gradient dramatically decreases with depth for two reasons. First, radioactive heat production is concentrated within the crust of the earth, and particularly within the upper part of the crust, as concentrations of uranium, thorium, and potassium are highest there: these three elements are the main producers of radioactive heat within the earth. Second, the mechanism of thermal transport changes from conduction (within the rigid tectonic plates) to convection (in the portion of the earth's mantle). Despite its solidity, most of the earth's mantle behaves as a fluid over long time-scales, and heat is transported by advection, or material transport. Thus, the geothermal gradient within the bulk of the earth's mantle is only of the order of 0.3 K km^{-1}, and is determined by the adiabatic gradient associated with mantle material.

By using the law of heat flow (also called *Fourier*'s law) $F_Q = M \cdot g$ applied to the earth, where F_Q is the heat flux at a point on the earth's surface and M the thermal conductivity of the rocks there (for granitic rocks, $M = 3.0$ W m^{-1}K^{-1}), we can calculate that $F_Q \approx 0.03$ W m^{-2}. This estimate is corroborated by thousands of observations of heat flow in boreholes all over the world. Compared with the global mean of global radiation (168 W m^{-2}) this value is totally negligible. In contrast, a variation in surface temperature induced by climate change and the *Milankovitch* cycle can penetrate below the earth's surface and produce an oscillation in the geothermal gradient with periods varying from daily to tens of thousands of years and amplitude which decreases with depth and having a scale depth of several km. Melt water from the polar ice caps flowing along the ocean bottoms tends to maintain a constant geothermal gradient throughout the earth's surface.

Geothermal energy in total can be estimated to about $2 \cdot 10^{31}$ J, very roughly assuming mean temperatures, densities, and heat capacities (Table 2.17).

Terrestrial heat flow is about 45 TW (Pollack et al. 1993). Geophysicists believe that not more than about 10 to 11 TW comes from the core, and most geophysicists are more comfortable with a figure of about 4 or 5 TW (this figure can be derived by using the above estimate of 0.03 W m^{-2} surface heat flow). Part of the core's heat flow is thought to represent power dissipated by the geo-dynamo that is thought to produce the geomagnetic field (Gubbins et al. 1979). The surface geothermal energy is not usable due to the very small temperature difference. However, using geothermal energy shows a large prospective potential when drilling deeper than 5000 m where the efficiency already amounts to 30%. Useful accessible resource base for electricity production have been estimated between 330 and 42 000 TWh per year (Bertani 2003). Geothermal heat produced in 2000 49 TWh electricity ($\sim 0.3\%$ share to world electricity production) and 53 TWh heat for direct use.

2.3.5 Renewable energy

In physics, *energy* is a scalar physical quantity that describes the amount of *work* that can be performed by a *force* (see also Chapter 4.1). Energy is neither lost nor produced, but only transformed between different forms of energy. It appears then that the term "renewable" (in the sense of regenerative) is not from the physical sciences but is more colloquial. Energy and matter are interchangeable with each other ($E = mc^2$). The ultimate source of energy in our terrestrial system is solar energy or, going back to its origin, the fusions as described in Chapter 2.1.1. No source of energy is "endless" but it may be regarded as such with respect to human time-scales. That is, without doubt, geothermal and solar energy. In terms of the terrestrial system, geothermal energy is part of the system but solar energy is part of the solar system, hence it is extraterrestrial. With this, solar energy and from it continuously (!) regenerated forms of energy (wind, water, biomass) are forms of "renewable" energy. Additionally, tidal energy, the result of interplanetary gravitational forces and the rotation of the earth, is another form of "renewable" energy.

2.3.5.1 Wind energy

The term "wind" is another expression for a *mass flow* (of a fluid). Wind is caused by the uneven heating of the earth's surface by the sun, resulting in different air pressure between air masses. Therefore air flows from a region of higher pressure into a region having lower pressure until it has compensated for the pressure difference. The causative force is termed *pressure gradient force*. Other forces influencing the mass flow in the atmosphere (*Coriolis* force, buoyancy forces, and friction forces) have to be taken into account for describing the flow.

Fluid dynamics is based on *Newton*'s axioms. It has been convenient to describe the action of a force on a body in the following forms of *Newton*'s law (see also Chapter 4.1):

$$\left(\frac{dv}{dt}\right)_m = \frac{1}{m}f, \quad \left(\frac{dm}{dt}\right)_v = \frac{1}{v}f \quad \text{and} \quad \frac{d}{dt}(mv) = \sum_i f_i = m \cdot a \qquad (2.81)$$

where v is velocity (vector quantity), m is mass, f is force and a is acceleration (dv/dt), mv is momentum. With the latter equation, we have an equation of motion for the fluid, more precisely the air. As the force (or sum of forces) is known, the change of wind speed follows from this (we take the symbol u instead of the symbol v for velocity) over time while mass is constant. Taking into account that energy is force multiplied by distance ($E = Fdx$) and assuming constant wind velocity and neglecting other forces beside pressure gradient force f_p, it follows simply for the wind energy E_w

$$f_p = \frac{m \cdot u}{dt} \quad \text{and} \quad E_w = \frac{m \cdot u}{dt} dx = m \cdot u^2 \qquad (2.82)$$

Wind power is the conversion of wind energy into a useful form, such as electricity, using wind turbines.

2.3.5.2 Water energy

The term "water energy" denotes several forms of energy usable by humans. We can distinguish between kinetic and potential energy and heat of water, all caused from solar radiation, partly via transformation into wind energy. Waves are caused by the wind blowing over the surface of the ocean. There is tremendous energy in the ocean waves. The useful worldwide resource of wave water energy has been estimated to be greater than 2 TW. As already remarked, tidal energy is a form of "water energy" but it is not caused by solar energy.

The most widely used form of renewable energy is hydroelectricity. As shown in Fig. 2.13 and Table 2.16, 27% of solar radiation (after solar absorption) is used for global evaporation of water and therefore transformed into latent heat in the atmosphere. Only a small percentage is usable as potential energy (uplift of surface water via evaporation and precipitation into river water above sea level) estimated to about 100 TW (corresponding to 0.1% of solar absorption).

The total power of latent heat, delivered first as kinetic energy (and finally dissipated in heat again) after water vapor condensation (cloud formation) amounts to 49 000 TW globally, 25 times more than the total wind power.

2.3.5.3 Bioenergy

According to Equation (2.40), solar energy is transformed into chemical bonding energy within biomass $(CH_2O)_n$ via the photosynthesis by autotrophs (plants, blue-green algae, autotrophic bacteria); approximately 475 kJ of free energy is stored in plant biomass for every mole of CO_2 fixed during photosynthesis (compare it with the bonding energy of C—H to be 416 kJ mol^{-1}). Any analysis of biomass energy production must consider the potential efficiency of the processes involved. Although photosynthesis is fundamental to the conversion of solar radiation into stored biomass energy, its theoretically achievable efficiency is limited both by the limited wavelength range applicable to photosynthesis, and the quantum requirements of the photosynthetic process. Only light within the wavelength range of

400 to 700 nm (photosynthetical active radiation, PAR, also called daily incident radiation) can be utilized by plants, effectively allowing only 45% of total solar energy to be utilized for photosynthesis. Furthermore, fixation of one CO_2 molecule during photosynthesis necessitates a quantum requirement of ten (or more), which results in a maximum utilization of only 25% of the PAR absorbed by the photosynthetic system. On the basis of these limitations, the theoretical maximum efficiency of solar energy conversion is approximately 11%. In practice, however, the magnitude of photosynthetic efficiency observed in the field, is further decreased by factors such as poor absorption of sunlight due to its reflection, respiration requirements of photosynthesis and the need for optimal solar radiation levels. The net result is an overall photosynthetic efficiency of 3−6% of total solar radiation.

About 120 Gt of carbon is fixed from atmospheric and dissolved aquatic CO_2 per annum by terrestrial plants (Fig. 2.10). This is the *gross primary production* (GPP), the rate at which an ecosystem's producers capture and store a given amount of chemical energy as biomass in a given length of time. Some fraction of this fixed energy is used by primary producers for cellular respiration and maintenance of existing tissues. Whereas not all cells contain chloroplasts for carrying the photosynthesis, all cells contain mitochondria for oxidizing organics, i. e., the yield of free energy (*respiration*). The remaining fixed energy is referred to as *net primary production* (NPP):

$$NPP = GPP - \text{plant respiration.}$$

NPP is the primary driver of the coupled carbon and nutrient cycles, and is the primary controller of the size of carbon and organic nitrogen stores in landscapes.

Aboveground NPP ranges from 35 to 2320 g $m^{-2} a^{-1}$ (dry matter) and total NPP from 182 to 3538 g $m^{-2} a^{-1}$ (Scurlock and Olson 2002). However, quantifying primary production at a global scale is difficult because of the range of habitats on earth, and because of the impact of weather events (availability of sunlight, water) on its variability. Direct observations of NPP are not available globally, but computer models based on remote sensing and derived from local observations have been developed to represent global terrestrial NPP showing a range from 40 to 80 Gt C yr^{-1} (Cramer and Field 1999). For example, the model by Potter et al. (1993) estimates a global terrestrial net primary production of 48 Gt C yr^{-1} with a maximum light use efficiency (LUE) of 0.39 g C $m^{-2} MJ^{-1}$ absorbed photosynthetically active radiation (APAR). Over 70% of terrestrial net production takes place between 30° N and 30° S latitude. It is estimated that the total (photoautotrophic) net primary production (NPP) of the earth was 104.9 Gt C yr^{-1}. Of this, 56.4 Gt C yr^{-1} (53.8%) was the product of terrestrial organisms, while the remaining 48.5 Gt C yr^{-1} was accounted for by oceanic production (Staley and Orians 2000). For the year 1997−2004 period, Werf et al. (2006) found that on average approximately 58 Gr C yr^{-1} was fixed by plants. Current "best guess" amounts to 60 Gt C yr^{-1} (Haberl et al. 2007); see Table 2.18. Consistent data on terrestrial net primary productivity (NPP) is urgently needed to constrain model estimates of carbon fluxes and hence to refine our understanding of ecosystem responses to climate change. Recent climatic changes have enhanced plant growth in northern mid-latitudes and high latitudes. Research findings by Ramakrishna et al. (2003) indicate that global changes in climate have eased several critical climatic constraints to plant growth,

Table 2.18 Global estimates of GPP (after Gough 2012) and NPP in Gt C yr^{-1} and biomass carbon in Gt C and biome area (after Amthor et al. 1998).

biome	GPP	NPP		Plant carbon[b]	Soil carbon[b]	Area in 10^{12} m^2
		Gough (2012)	Amthor et al. (1998)			
Forest, tropical	40.8	16.0−23.1	13.7	244	123	14.8
Forest, temperate	9.9	4.6−9.1	5.0	92	90	7.5
Forest, boreal	8.3	2.6−4.6	3.2	22	135	9.0
Savannah, tropical	31.3	14.9−19.2	17.7	6	264	22.5
Grassland, temperate	8.5	3.4−7.0	4.4	9	295	12.5
Deserts	6.4	0.5−3.5	1.4	7	168	21.0
Tundra	1.6	0.5−1.0	1.0	6	121	9.5
Croplands	14.8	4.1−8.0	6.3	3	117	14.8
Other[a]	−	−	3.0	85	542	55.4
Total	121.7	48.0−69.0	59.0	486	2057	150.8

[a] Ice, lakes, peatlands, human areas, extreme deserts
[b] assuming that phytomass is 45% C

such that net primary production increased by 6% (3.4 Gt C over 18 years, 1982 to 1999) globally. The largest increase was in tropical ecosystems.

In contrast, for the oceanic NPP there seem high uncertainties, but it is agreed that ocean phytoplankton is responsible for approximately half the global NPP. Oceans (especially at high latitudes) typically represent a net sink of atmospheric carbon (cf. Fig. 2.10). These regions are dominated by diatoms which typically grow and sink faster than other phytoplankton groups, and thus can represent an important carbon transfer mechanism to the deep sea. The low latitudes, conversely, represent a source of CO_2 to the atmosphere. Satellite-based ocean chlorophyll records indicate that global ocean annual NPP has declined more than 6% since the early 1980s (Gregg et al. 2003). The reduction in primary production may represent a reduced sink of carbon here via the photosynthetic pathway.

The fate of NPP is heterotrophic respiration (R_h) by herbivores (animals) who consume 10−20% of NPP, and respiration of decomposers (microfauna, bacteria, fungi) in soils. The largest fraction of NPP is delivered to the soil as dead organic matter (litter), which is decomposed by microorganisms under release of CO_2, H_2O, nutrients and a final resistant organic product, *humus*. Hence, a *net ecosystem production* (NEP) is defined:

$$NEP = NPP - consumers\ respiration\ (R_h).$$

Another part of NPP is lost by fires (biomass burning, see Chapter 2.6.4.5), by emission of volatile organic substances (VOC) and human use (food, fuel and shelter); loss of NPP (in 10^{15} g C yr^{-1}):

- biomass burning: ~4.0 (Malingreau and Zhuang 1998)
- humans: 18.7 (Krausmann et al. 2008)
- VOC emission: 1.2 (Guenther et al. 1995)

NEP finally represents the burial carbon, a large flux at the beginning of the bio-spheric evolution but nowadays limited at about zero. Most of the NPP goes in litter, and is finally mineralized back to CO_2. Werf et al. (2006) estimated for the 1997–2004 period that approximately 95 % of NPP was returned back to the atmosphere via R_h; another 4 % (or 2.5 Gt Y yr^{-1}) was emitted by biomass burning (see Chapter 2.6.4.5) – the remainder consisted of losses from fuel wood collection and subsequent burning.

Historically, global estimates of litter production have ranged from 75 to 135 Gt C yr^{-1} (often cited as dry mass or matter, DM, not identical with carbon; Werf et al. (2006) use a dry matter carbon content of 45 %). Steady-state pools of standing litter represent global storage of around 174 Pg C (94 and 80 Pg C in non-woody and woody pools, respectively), whereas the pool of soil C in the top 0.3 m, which turns over on decadal time scales, comprises 300 Pg C. Several estimates from Matthews (1997) suggest values in the middle of this range, from 90 to 100 Gt C yr^{-1}, accounting for both above-ground and below-ground litter. Above-ground litter production may be 5–10 Gt C yr^{-1} including mainly forest, woodland, and grassland. The global litter pool estimated by Matthews (1997) from the measurement compilation is 136 Gt C. Inclusion of the remaining ecosystems may add ∼ 25 Gt C, raising the total to ∼ 160 Gt C. An additional ∼ 150 Gt C is estimated for the coarse woody detritus pool. Global mean steady-state turnover times of litter estimated from the pool and production data range from 1.4 to 3.4 years; mean turnover time from the partial forest/woodland measurement compilation is ∼ 5 years, and turnover time for coarse woody detritus is ∼ 13 years.

Considering only the terrestrial net primary production (∼ 60 Gt C yr^{-1}), the biomass energy amounts to ∼ 30 · 10^{20} J stored per year (simply calculated from C bonding), which corresponds to 76 TW, respectively ∼ 2400 EJ.

2.3.5.4 Comparison between the earth's energy sources – potential for humans

In Fig. 2.17 the flow of solar radiation as well as other energy sources for utilizing in different forms of energy is represented. It is remarkable that nearly all solar and atmospheric energy is dissipated into heat and IR radiation, finally back to space – the efficiency for usable energy is small (concerns NPP ∼ 2 · 10^{-4} % related to solar absorption). The amount of energy in fossil fuels (in estimated ressources ∼ 2.6 · 10^{23} J) is only one-twentieth of the yearly incoming solar flux, but was buried 50–100 Myr ago and over a period of millions of years.

In 1990, the global energy consumption based on fossil fuels (coal, oil and gas) was estimated to 3.2 · 10^{20} J, corresponding to a CO_2 emission of 8.0 · 10^{15} g carbon. This value (cf. Fig. 2.17) is 16 % of the annual terrestrial photosynthesis (and 7.3 % of global one). Since the year 1751, roughly 321 · 10^{15} g carbon has been released to the atmosphere from the consumption of fossil fuels and cement production (the last source amounts only to about 4 % of total CO_2 in 1995) (Marland et al. 2008). It is logical that there are two problems, first the finite nature (on human time-scales) of this source of energy, and second, the problem of CO_2 accumulation in the atmosphere (see Chapter 2.7.3.1 and 2.8.3). The fossil-fuel energy corre-

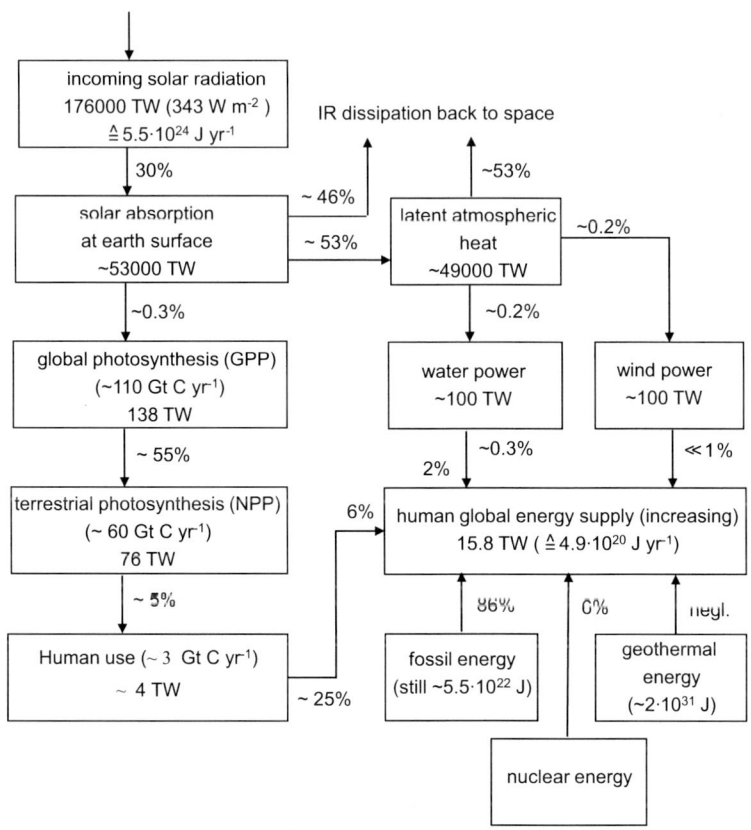

Fig. 2.17   Scheme of energy fluxes on earth. Note that world yearly energy supply often is given in TWh (138,300 in 2006; see Fig. 2.48), corresponding with TW per year (in this case for the year 2006: 15.8); to the world energy supply (not in Figure) other carrier contribute, such as biofuels (~ 10%); data related to 2006 (20 TW per year in 2011).

sponds to the energy captured by NPP within 20−30 years. It should be emphasized that the efficiency of energy capture by carbonization of biomass (fossil-fuel formation) is extremely low. The only "advantage" of fossil fuels lies in the high density of energy and its ready actual availability. With little doubt, globally, fossil fuels may satisfy the energy demands of humanity for the next 500 years but no more. With the CO_2 problem in mind, it is the challenge to solve in the near future, let's say the next 50 years, the technical realization of sustainable energy sources, which are exclusively geothermal and solar energy (further reading: Hodgson 2010).

Geothermal energy is an interesting reservoir. The energy estimated in the crust alone is in the order of all the available fossils fuels (cf. Table 2.17 and Fig. 2.17); the energy in the outer mantle would allow a human energy supply for some millions of years. It is difficult to state that there is thermal equilibrium, i.e., when extracting heat from 10 km depth then the thermal flow from deeper layers will compensate the gradient. Moreover, the technical infrastructure for deep drilling and heat exchange is extreme. Hence it seems unlikely that geothermal energy will solve the global problem of energy demand, but it will be a useful local/regional source in future, characterized by large geothermal gradients.

It appears then that only solar radiation provides continuous potential for energy. In solar-rich regions, the annual averaged solar flux amounts to more than 250 W m^{-2}. Assuming an efficiency of 10 % photon-electron conversion by photovoltaic cells (which presently is the standard for "cheap" technology), the annual energy flux amounts $0.8 \cdot 10^9$ J per square meter. To provide the present annual energy supply, an area of $\sim 5 \cdot 10^5$ km^2 is needed, about the size of France. The global agricultural area (under arable and permanent crops as defined by the FAO) amounts to $1.44 \cdot 10^7$ km^2, roughly 30 times more than the area needed for the global energy supply. It is remarkable that the energy used from that agricultural area (using the daily global mean food consumption given by the FAO in 2800 kcal per person) results only in yearly energy flux of $1.8 \cdot 10^6$ J m^{-2}. In other words, the photovoltaic energy production is at least two orders of magnitude greater per area, even when considering more than 50 % loss by conversion into transportable chemical energy.

Human use of biomass has become a major component of the global biogeochemical cycles. The use of land for biomass production (e. g. crops) is among the most important pressures on biodiversity. At the same time, biomass is indispensable for humans as food, animal feed, raw material and energy source. It seems that biomass − with the exception of a source of organic substances and food − will not play any role in the future as an energy source. An estimate by Krausman et al. (2008) shows that from approximately 18.7 Pg C of usable global biomass, 6.6 Pg is lost (destroyed during harvest) and from the remaining 12.1 Pg the following percentages represent the different uses (data based on the year 2000):

- Feed (58 %),
- food (7 %),
- raw materials (20 %),
- biofuels (10 %).

The global annual food supply is equivalent to $\sim 3 \cdot 10^{19}$ J, and will probably increase by a factor of two to three in the next 50−100 years. Then it would amount a percentage of 10 % to the global NPP ($\sim 5 \cdot 10^{20}$ J yr^{-1}).

2.3.6 Abiogenic versus biogenic formation of "fossil fuels"

It was previously noted that the net ecosystem production (NEP) nowadays is small or even negligible but it had a strong influence in the period when colonization of the continents by plants did occur. During this time, with still reduced oxygen, respiration was not yet significant and large quantities of GPP could be buried. As a result, coal, oil and natural gas have been formed and it seems unnecessary to go deeper in discussing the biogenic theory of fossil-fuel formation. None the less, it seems that there are strong reasons to believe that besides a natural process of carbonization of buried former biomass, there were abiogenic processes in the formation of oily and gaseous hydrocarbons, as already remarked in discussing gases

occluded and produced from rocks (Chapter 2.2.1.2). Now, with little doubt we
have to consider large carbon inputs to the early earth from heavy bombardment.
It is only a question of understanding the geochemical processes that occurred
under high temperature and pressure, and over a long time scale, to see how the
carbon compounds delivered to earth have been turned into petroleum (liquid
crude oil and long-chain hydrocarbon compounds). The possible formation of
methane (CH_4) needs no further comment. It is however remarkable that various
abiogenic hypotheses were proposed long before the presence of evident organic
matter from space was proved. The most notable advocates of such hypotheses
were the nineteenth-century scientists, such as *Alexander von Humboldt*, the Russian
chemist *Dmitri Mendeleev* and the French chemist *Marcellin Berthelot*. In the middle of the twentieth century, respected Soviet scientists (*Nikolai Kudryavtsev* and
his colleagues), developed theories on abiogenic formation of natural gas and oil,
but these were almost unknown to the West until end of the 1980s. Fortunately,
these hypotheses need no further consideration. The astronomer *Thomas Gold* was
the most prominent proponent of the abiogenic hypothesis in the West (Glasby
2006). Currently there is little direct research on abiogenic petroleum or experimental studies into the synthesis of abiogenic methane. However, several research areas,
mostly related to astrobiology and the deep microbial biosphere and serpentinite
reactions, continue to provide insight into the contribution of abiogenic hydrocarbons into petroleum accumulations. Similarly, such research is advancing as part
of the attempt to investigate the concept of panspermia and astrobiology, specifically using deep microbial life as an analog for life on Mars.

It is worth of noting here that the controversial discussion of biogenic versus
abiogenic makes no sense because both chemical and biological evolutionary processes are occurring, at different times and sites, but also simultaneously. Therefore
arguments such as that biomarkers are indicative of the biological origin of petroleum dissipate when we accept that the biosphere is integrated with the geosphere
and hence chemical and biological processes are overlapping.

2.3.7 The energy problem

Energy provision is based on three pillars: renewable sources (all based on solar energy either directly in the form of radiation or indirectly as wind and water energy,
as well as biomass), geochemical-stored solar energy (fossil fuels), and nuclear energy (geo heat and fusion and fission processes). A fourth source should be listed
here but seems to be of less interest in solving global problems: gravitational energy
(for example, tides). For thousands of years, humans have been using wood for
heating, cooking, and building. Disregarding local pollution (soot and smoke), it
was a perfect solution: renewable and CO_2 neutral.

For smelting and other high-temperature processes, however, the temperature
obtainable by wood burning was not high enough. Therefore, coal was probably in
use 2000 years ago. Technical inventions such as the steam machine and the dy-
namo led to the exponential growth in coal consumption and, therefore, mining by
the end of the nineteenth century. Furthermore, wood growth became smaller than
wood consumption in the eighteenth century.

Another invention, the mobile combustion engine (motor), made two other fossil fuels – gaseous and liquid hydrocarbons – the dominant energy (and material) carriers over the past 100 years, along with coal. In that time, a global infrastructure (transportation, storage, chemical industry, power plants, traffic, etc.) had been developed on the basis of fossil fuels and optimized with excellent engineering. It would be of infinite strategic value to use this base for an indefinite time and look for an alternative carbon carrier to replace fossil fuels. The ultimate source of reactive carbon compounds is carbon dioxide. We "only" have to copy the principles of biosphere-atmosphere interaction using intelligent and sustainable technical solutions.

It is remarkable that basically all sub-global air pollution problems (dust and smoke, sulfur and nitrogen pollution) in connection with fossil fuel combustion have been (or can be) solved by end-of-pipe technologies. Hence, the last and apparently insoluble problem remains CO_2 emissions and the subsequent increase in the greenhouse effect. CCS technology is another end-of-pipe approach and far from any sustainable chemistry. But it seems the only practical way to start the abatement of CO_2 emissions.

Despite the question of how to solve the climate problem (in other words, how to avoid a further increase in atmospheric CO_2) we have to find answers to where our energy will come from after the consumption of all fossil fuels. The answer is simple: through renewable sources, among which direct solar radiation is surely the dominant source. Several conceptions have been developed over the past decades and the newest one, DESERTEC[10], seems to overcome the loss of electric current over long-distance transportation. But none of these ideas solve two problems: the storage of energy and the material substitution of carbon from fossil fuels (natural biomass net production can no longer match the annual carbon consumption by humans). Obviously, chemicals can store energy and release it in suitable reactors in the form of heat for further energetic use. There is no doubt that the oxidation (combustion) of substances provides a large heat yield in most simple "reactors" (burners); no substance other than carbon provides a variety of species most suitable for combustion. Among all combustible substances, carbon has the largest abundance (see also Chapter 2.8.4.2 for carbon economy).

2.4 The biosphere and global biogeochemical cycles

Throughout the entire history of our planet, chemical, physical and biological processes have changed the composition and structure of its reservoirs. Beginning with a highly dynamic inner earth 4.6 billion years ago, geochemical and geophysical

---

[10] This conception is based on concentrating solar thermal power (CSP) plants in northern Africa and high voltage direct current (HVDC) transmission lines. HVDC transmission is more efficient than the use of hydrogen as an energy vector. Loss of power during transmission can be limited to only about 3% per 1000 km. The main reason for favouring CSP over photovoltaics (PV) is its ability to supply power on demand, 24 hours a day. PV is more expensive than CSP and needs expensive systems for storing electricity, such as pumped storage.

processes have created the fundamentals for the earth to become a habitat. With this, the formation of the hydrosphere (the oceans and a hydrological cycle, see the next section, Chapter 2.5) was the most important precondition for the evolution of living matter. Despite large changes of the chemical composition of the atmosphere, hydrosphere and lithosphere (the geospheres) over the ages, these spheres or reservoirs are well-defined concerning such essential parameters as interfaces, volume, mass and others. Let us look first at the biosphere, a not-well-characterized reservoir which is much more dynamic concerning the parameters mentioned. Living phytomass becomes, via photosynthesis, the driving geological (in truth, biological) force moving material around the system naturally. In the following sections the sulfur, nitrogen, and chlorine cycles are briefly described, although many other cycles are of interest (almost all elements contribute to biogeochemical cycles) for understanding the biosphere and the geosphere. The carbon cycle is presented (in linkage with that of oxygen) in Chapter 2.2.2.5. Some authors include the water or *hydrological cycle* in biogeochemical cycles. As noted with emphasis in Chapter 2.4.4, biological water splitting is *the* driving force of all redox processes and therefore cycles. However, because of the overwhelming dimension and importance of water and its cycle, this issue is described in a separate section (Chapter 2.5.2).

The British scientist *James Lovelock* together with *Lynn Margulis* developed the Gaia hypothesis, that the earth is a self-controlling system (*Gaia*: the earth goddess in Greek), and proposed that our present atmosphere is far from the chemical equilibrium that is assumed for other planets (Lovelock and Margulis 1974, Crutzen 2002). One expression of this is the difference in the redox potentials between biosphere (reducing medium) and atmosphere (oxidizing medium). It is believed that living organisms are responsible (to a large extent) for the chemical composition of the present atmosphere and, from the opposite point of view, the chemical composition of the atmosphere determines the biota. Remarkably, the composition of the biosphere is similar to that of the present atmosphere.

2.4.1 The biosphere and the noosphere

The *biosphere* is considered to represent the earth's crust, atmosphere, oceans, and ice caps, and the living organisms that survive within this habitat (Hover 2006). Hence, the biosphere is more than a sphere in which life exists. It is the totality of living organisms with their environment, i.e., those layers of the earth and the earth's atmosphere in which living organisms are located. Another common definition such as "the global sum of all ecosystems", however, calls for a definition of what is meant by an "ecosystem". An ecosystem is a natural unit consisting of all plants, animals and microorganisms (biotic factors) in an area functioning together with all of the non-living physical (abiotic) factors of the environment.

Some life scientists and earth scientists use *biosphere* in a more limited sense (we prefer it too because then the use of the *climate system* makes more sense, see Chapter 3.2). For example, geochemists define the biosphere as being the total sum of living organisms (the "biomass" or "biota" referred to by biologists and ecologists). Thus the three major biospheric pools are *live phytomass* (the mass of animals is negligible in a global context), *consumers* and *litter* (dead biomass). In this

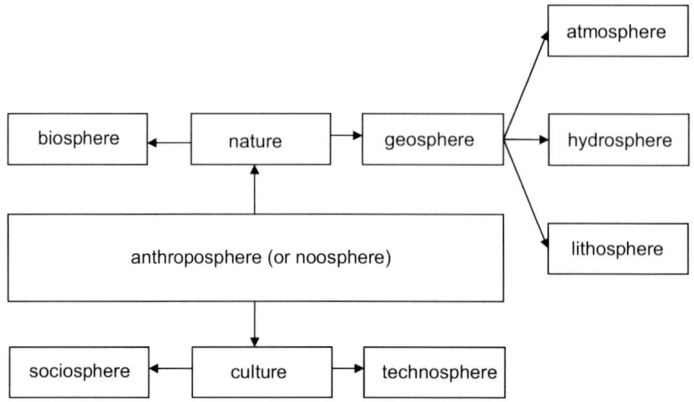

Fig. 2.18 Scheme of the noosphere, integrating natural and social (humans) sub-systems; only for the geosphere as an example, related sub-systems are shown.

sense, the *biosphere* is one of four separate components of the geochemical model, the other three being the *lithosphere, hydrosphere* and *atmosphere*. The narrow meaning used by geochemists is one of the consequences of specialization in modern science. Some might prefer the word *ecosphere*, coined in the 1960s, as all encompassing of both biological and physical components of the planet. In Chapter 3.2 we will introduce the *climate system* being the sphere affecting life, which comprises those parts of the lithosphere, hydrosphere and atmosphere where life exists (the reader may now see that this definition is close to that of the biosphere in a wider sense).

The term "biosphere" was coined in 1875 by the famous Austrian geologist *Eduard Suess*

> ... one thing seems to be foreign on this large celestial body consisting of spheres, namely, organic life. But this life is limited to a determined zone at the surface of the lithosphere. The plant, whose deep roots plunge into the soil to feed, and which at the same time rises into the air to breathe, is a good illustration of organic life in the region of interaction between the upper sphere and the lithosphere, and on the surface of continents it is possible to single out an independent biosphere. (Suess 1875; quoted after Smil 2002).

The Russian geochemist *Vladimir Ivanovich Vernadsky* was the first to propose the idea of biogeochemical cycling (having asked: What is the impact of life on geology and chemistry of the earth?). *Vernadsky*, who had met *Suess* in 1911, popularized the term biosphere in his book *The Biosphere* (first published in Russian in 1926, not translated into a full English version until more than 60 years later), hypothesizing that life is the geological force that shapes the earth (Vernadsky 1926, 1944, 1945, 1998).

The term "ecosystem" was coined in 1930 by *Roy Clapham* to denote the combined physical and biological components of an environment (Willis 1997). British ecologist *Arthur Tansley* later refined the term, describing it as: "The whole system, ... including not only the organism-complex, but also the whole complex of physical factors forming what we call the environment" (Tansley 1935). He championed the term *ecosystem* in 1935 and *ecotope* in 1939 (Cooper 1957). *Vladimir Vernadsky* defined "ecology" (originally intended as the "economy of nature") as the science of the biosphere.

In *Vernadsky*'s theory of how the earth develops, the *noosphere* is the third stage in a succession of phases of development of the planet, after the geosphere (inanimate matter) and the biosphere (biological life). He was one of the first scientists to recognize that the oxygen, nitrogen and carbon dioxide in the earth's atmosphere results from biological processes. In the 1920s, he published works arguing that living organisms could reshape the planets as surely as any physical force. The term "noosphere" was first coined by the French mathematician and philosopher, *Edouard Le Roy* (1927). *Le Roy*, building on *Vernadsky*'s ideas and on discussions with *Teilhard de Chardin* – both attended *Vernadsky*'s lectures on biogeochemistry at the Sorbonne in 1922–1923 – came up with the term "noosphere", which he introduced in his lectures at the Collège de France in 1927 (Le Roy 1927). *Vernadsky* saw the concept as a natural extension of his own ideas, predating *Le Roy*'s choice of the term (Smil 2002, p. 13).

Pierre Teilhard de Chardin was a Jesuit who introduced in 1925 the concept of noosphere, the "sphere of mind" superimposed on the biosphere or the sphere of life. Not until after the appearance of humans on the earth did the *noogenesis* begin. The term *noogenesis* comes from the Greek word νούς = mind. *Teilhard de Chardin* wrote in an essay entitled "Hominization" (1925): "And this amounts to imagining, in one way or another, above the animal biosphere a human sphere, a sphere of reflection, of conscious invention, of conscious souls (the noosphere, if you will)". He later wrote: "Man discovers that he is nothing else than evolution become conscious of itself. The consciousness of each of us is evolution looking at itself and reflecting upon itself." (Teilhard de Chardin 1961, p. 221). This conception is based on philosophical writings, however, and completely neglects *Vernadsky*'s biogeochemical approach. The divergence is perhaps best expressed as an opposition between the anthropocentric view of life (*Teilhardian* biosphere) and the biocentric view of the natural economy (*Vernadskian* biosphere).

Vernadsky first took up the term "noosphere" in 1931, as a new dimension of the biosphere under the evolutionary influence of humankind (Vernadsky 1944). He wrote: "The Noosphere is the last of many stages in the evolution of the biosphere in geological history" (Vernadsky, 1945). The biosphere became a real geological force that is changing the face of the earth, and the biosphere is changing into the noosphere. In *Vernadsky*'s interpretation (1945), the noosphere is a new evolutionary stage of the biosphere, when human reason will provide further sustainable development both of humanity and the global environment:

> In our century the biosphere has acquired an entirely new meaning; it is being revealed as a planetary phenomenon of cosmic character ... In the twentieth century, man, for the first time in the history of earth, knew and embraced the whole biosphere, completed the geographic map of the planet earth, and colonized its whole surface. Mankind became a single totality in the life on earth ... The noosphere is the last of many stages in the evolution of the biosphere in geological history (Vernadsky 1945, p. 10).

Today the term "anthroposphere" is also used (Fig. 2.18). The idea of a close inter-relation between humans and the biosphere is topical in understanding the "earth system", i. e., climate change, and is used by Schellnhuber (1999) with the terminology "global mind" and by Crutzen and Stoermer (2000) with "anthropocene" to characterize the present epoch[11].

The *Gaia hypothesis* is an ecological hypothesis proposing that the biosphere and the physical components of the earth (atmosphere, cryosphere, hydrosphere and lithosphere) are closely integrated to form a complex interacting system that maintains the climatic and biogeochemical conditions on earth in a preferred *homeostasis*, originally proposed by *James Lovelock* as the earth-feedback hypothesis (Lovelock 1979). In short, the biosphere itself becomes an organism, it acts much like an organism and appears to be internally controlled. Gaia is the "invisible hand" of biogeochemistry. *Lovelock* suggests that life processes regulate the radiation balance of the earth to keep it habitable. As a well-known example, a self-controlling cycle between marine sulfur emissions and air temperature was established by Charlson et al. (1987), initiating much useful air chemical and climate research (Fig. 2.19). Behind this relationship is the so-called *Twomey* effect (Twomey 1974), which describes how cloud condensation nuclei (CCN) from anthropogenic pollution may increase the amount of solar radiation reflected by clouds (Fig. 2.20).

While not strictly a "hypothesis" in the scientific sense, since it cannot be falsified, the Gaia hypothesis has great merit as an educational tool, regarding the concepts of biogeochemical cycling. Do biogeochemical processes indeed combine forces to favour life on earth? Critics suggest that the processes of living organisms in the past were not necessarily favourable for the long-term survival of the then existing life. For example, the increase of oxygen in the atmosphere caused by expanding photosynthesis greatly diminished opportunities for anaerobic organisms. However, new life forms evolved that took advantage of the new situation (for example, highly mobile multicellular organisms, including *Homo sapiens*, the humans). A conclusive example of how life does not necessary regulate its own environment for optimum survival conditions (what we now call "sustainable development") is the human species, originally nothing more than one population among millions of others in the biosphere. *Friedrich Engels* wrote (in 1876, first published in German in Die Neue Zeit 1895; see also Kryzhanovskii 1970):

> Labour is the source of all wealth, the political economists assert. And it really is the source − next to nature, which supplies it with the material that it converts into wealth. But it is even infinitely more than this. It is the prime basic condition for all human existence, and this to such an extent that, in a sense, we have to say that labour created man himself ... Labour begins with the making of tools.

In 1939 *Haldane* wrote in the preface to the first English edition of *Engels' Dialectic of Nature*:

> I believe that the Chapters of the book which deal with biology are the most immediately valuable to scientist's to-day ... Had Engels' method of thinking

---

[11] Zalasiewicz et al. (2008) published the first proposal for the formal adoption of the Anthropocene epoch by geologists, and this adoption is now pending.

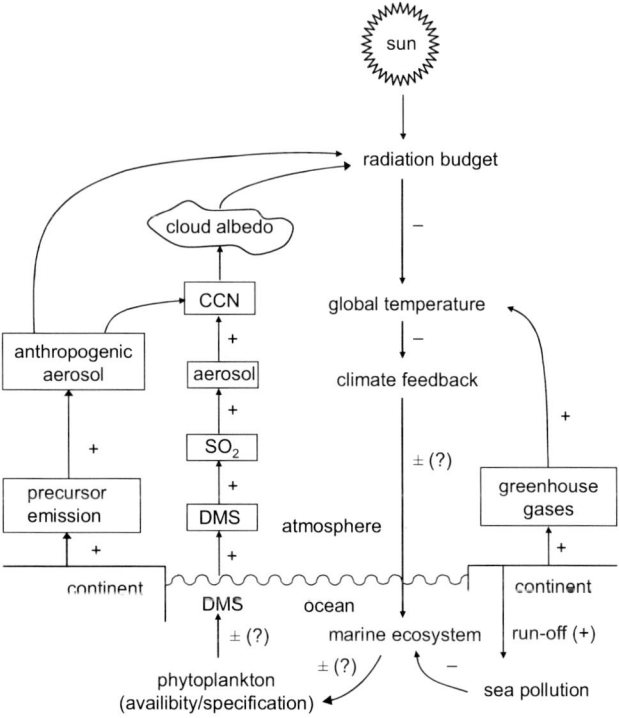

Fig. 2.19 Relationship between marine sulfur emission (as dimethyl sulfide DMS), climate and feedback, including human perturbation as an example of the Gaia hypothesis, called the CLAW hypothesis; modified after Charlson et al. (1987).

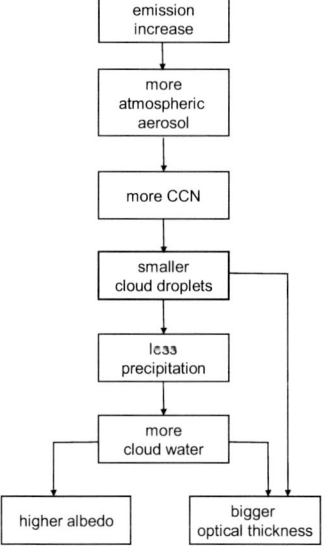

Fig. 2.20 Scheme of the so-called *Twomey*-effect.

been more familiar, the transformations of our ideas on physics which have occurred during the last thirty years would have been smoother. Had his remarks on Darwinism been generally known, I for one would have been saved a certain amount of muddled thinking.

The incarnation (*Teilhard de Chardin*'s "hominization") creates the human who used tools with more and more dimensions concerning changing nature, i. e., creating the noosphere, and unintentional *climate change*. It appears then that, at least since the Industrial Revolution, biogeochemical cycles are not merely natural, but are modified by humans, or in another word, *biogeonoochemically*. In the following description of biogeochemical cycles, therefore, the human dimension (man-made emissions, see also Chapter 2.8.1) is always included.

2.4.2 Biogeochemical cycling: The principles

Fundamentally, life on earth is composed of six major elements; namely H, C, N, O, S and P. These elements are the building blocks of all the major biological macromolecules including proteins, nucleic acids, lipids and carbohydrates (Falkowski and Godfrey 2008). The production of macromolecules requires an input of energy, which is almost exclusively derived from the Sun (see Chapter 2.3.1.1). The character of biological energy transduction is non-equilibrium redox chemistry. Besides energy, the production of macromolecules requires an input of a chemical substrate as the raw material, which has transportable (volatilable, solvable) and transformable (oxidable, reductable, dissociable) geochemical conditions. Biological evolution gave rise to specified cells and organisms, responsible for driving and maintaining global cycles of the first five elements (phosphorous does not belong to gaseous compounds in the air and thus is not of global importance). Because all redox reactions are air paired (oxidation-reduction), the resulting network is a linked chemical system of the elemental cycles (Fig. 2.21). For example, reduction of carbonate, sulfate and nitrate requires hydrogen and the oxidation of organic compounds (including carbon, sulfur and nitrogen) requires oxygen.

According to the biological cycle, which is that part of the biogeochemical cycle where biomass is produced and decomposed (Fig. 2.22), all volatile compounds occurring in the biochemical cycles can be released to their physical surroundings. These are:

- Carbon: CH_4, NMVOC (non-methane hydrocarbons), CO, CO_2
- Nitrogen: NH_3, N_2O, N_2, NO, RNH_2 (organic amines)
- Sulfur: H_2S, COS, $(CH_3)_2S$ (DMS), RSH (organic sulfides)

As seen, sulfur and nitrogen also belong in organic compounds. The sulfur and nitrogen cycle is inherent linked with the carbon cycles. As discussed before (Chapter 2.2.2.5), oxygen is closely coupled with the carbon cycle, and is chemically combined with compounds of nitrogen and sulfur (and others). Other cycles, such as those of phosphorous and of trace metals, are important for the biosphere but play a minor role in the atmosphere. With a few exceptions, most of the compounds of

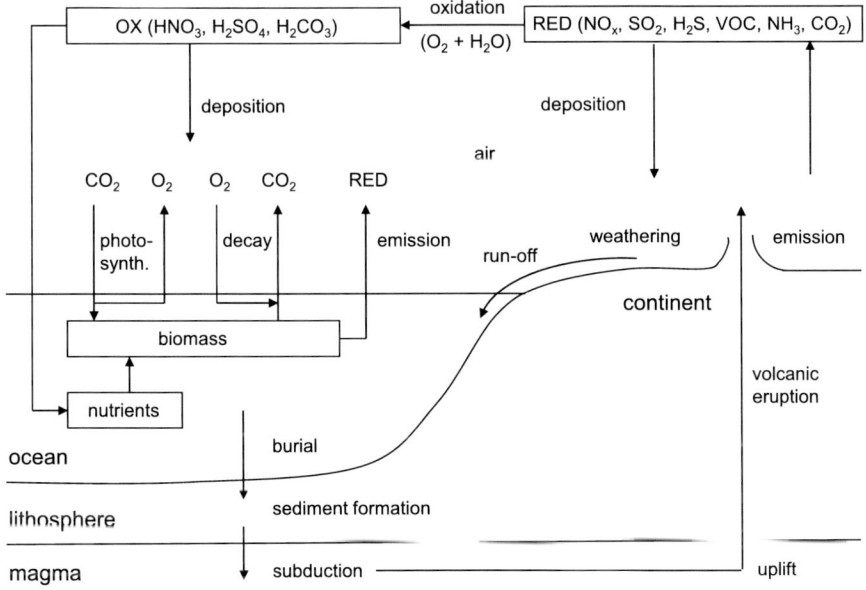

Fig. 2.21 Scheme of global cycling: The global redox process.

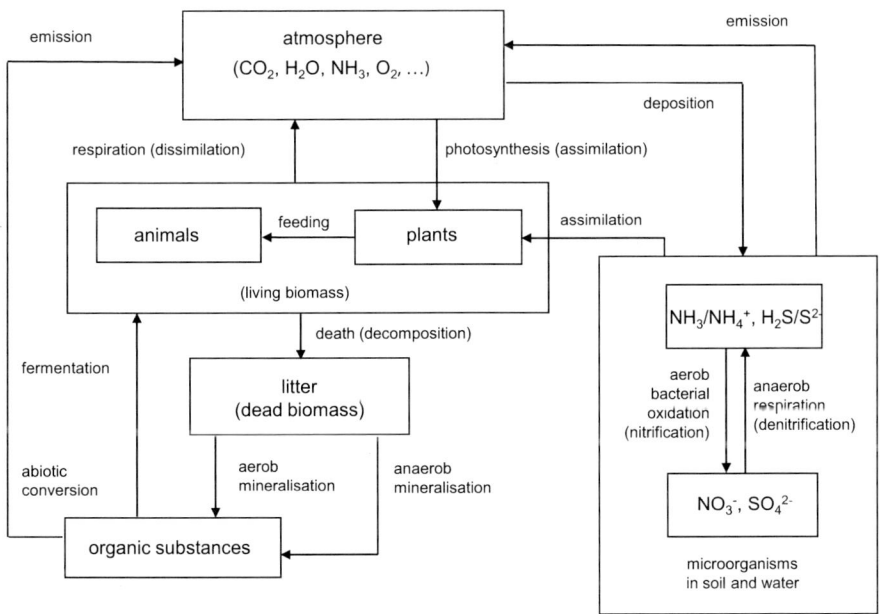

Fig. 2.22 Scheme of biosphere-atmosphere interaction.

these cycles found in the atmosphere are condensed within particulate matter. A major part of these species is within soils or dissolved in waters. Chlorine is not among the crucial elements important for life but it is the dominant part of seasalt, one of the most important natural sources, and plays an important role in atmospheric chemistry. Beside chlorine other halogens such as iodine and bromine have been found − despite their much lower significance as sources compared with that of chlorine − to be very important in multiphase chemical processes in the atmosphere and at interfaces.

The chemical composition of the biosphere and/or the climate system is the result of continuous geochemical and biological processes within natural cycles, as already mentioned in the Chapters above. As just discussed, since the appearance of humankind, a new driving force has developed; the human matter turnover as a consequence of man-made (or anthropogenic) emissions. Therefore, today's biogeochemical cycles are very clearly no longer natural but anthropogenically-modified cycles. Moreover, human-induced *climate change* already causes permanent change to natural fluxes; for example, through weathering, shifting redox and phase equilibriums. As will be discussed in Chapter 3.2, *climate* includes not only "traditional" meteorological elements such as temperature, radiation and humidity, but also acidity and oxidation capacity, to name some of the most important parameters influencing fluxes. From a physicochemical point of view, the driving forces in transport and transformation are *gradients* in pressure, temperature and concentration. Substances, once released from anthropogenic sources into the air, become an inherent part of biogeochemical cycles. It is worth noting here that different sources can provide more or less specific isotopic ratios which are very useful for source-appointment studies. Overall, concerning chemical reactions in the climate system, isotopic differences in reactions rate play no role apart from on geological time scales.

A biogeochemical cycle comprises the sum of all transport and conversion processes that an element and its compounds can undergo in nature. The substances undergoing the biogeochemical cycle pass through several reservoirs (atmosphere, hydrosphere, pedosphere, lithosphere and biosphere) where certain concentrations accumulate because of flux rates, determined by transport and reaction. The atmospheric reservoir plays a major role because of its high dynamic in transport and reaction processes and the global linkage between biosphere and atmosphere (Fig. 2.22). A global cycle may be derived from the budget of composition of the individual reservoirs, with a (quasi) steady state being considered to exist. It shows variations on different time scales and may be disturbed by catastrophic events (e. g., volcanism, collision with other celestial bodies). The largest perturbation, however, is the one that can be observed over the past hundred years, caused by humans. The scale of anthropogenic global sulfur and nitrogen fluxes is now in the same order as the natural ones.

Sulfur and nitrogen, among many other elements, maintain and propagate life. Biological systems constantly synthesize, change, and degrade organic and inorganic chemical species. For example, nitrogen and sulfur create compounds like amino acids, which serve as building blocks for proteins, enzymes and genetic materials (Fig. 2.23). The biological cycle includes the synthesis of living organic matter (plants) from inorganic compounds, which are mainly in the upper oxidized level (carbon dioxide, water, sulfate, and nitrate). During biogenic reduction to more reduced sulfur and nitrogen species, a variety of volatile S and N compounds are

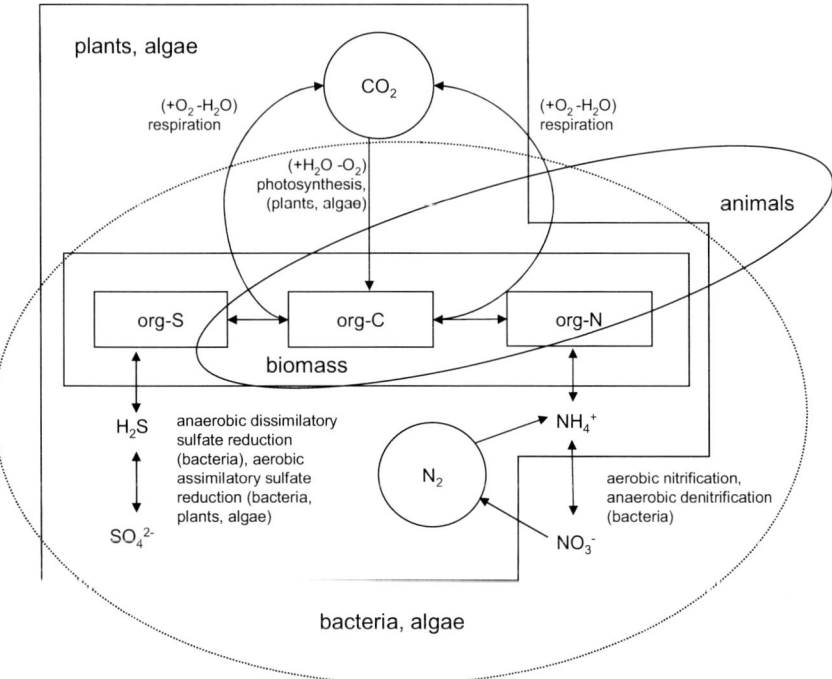

Fig. 2.23 Scheme of biological cycling: The role of different species.

formed which can enter the atmosphere, in other words they undergo a phase transfer (emission).

Biological processes which lead to this reduction can be considered as the driving force of atmospheric cycles. Emitted biogenic compounds return to ecosystems (mostly) in oxidized form. This interaction is a result of the biogeochemical evolution of the earth. Thus, the enhancement of global sulfur and nitrogen fluxes since the beginning of the Industrial Revolution about 150 years ago can lead not only to new reservoir concentrations (pools) but also to changes of the parameters of physical and chemical reservoirs, e. g., radiation, temperature, reactivities, which themselves influence internal production rates, and transfer fluxes (positive and/or negative feedback). These interactions are not yet well understood.

It is essential to know better the natural fluxes (anthropogenic fluxes are rather well estimated) and their variations, but even more important is knowledge about the biosphere-atmosphere interaction: What processes control the environmental parameters that themselves maintain life on earth? We can then ask another question: What is the threshold of quantitative change (of any parameter) leading to a new quality of life? Finally, we have to answer (or to define) what changes in air, soil and water quality humankind can even accept in future against the background of sustainable development.

With respect to the climate system (biospheric evolution), and in comparison with the man-made activities (noospheric evolution), it is logical to look to natural processes from which substances are released into the atmosphere. Fig. 2.24 shows

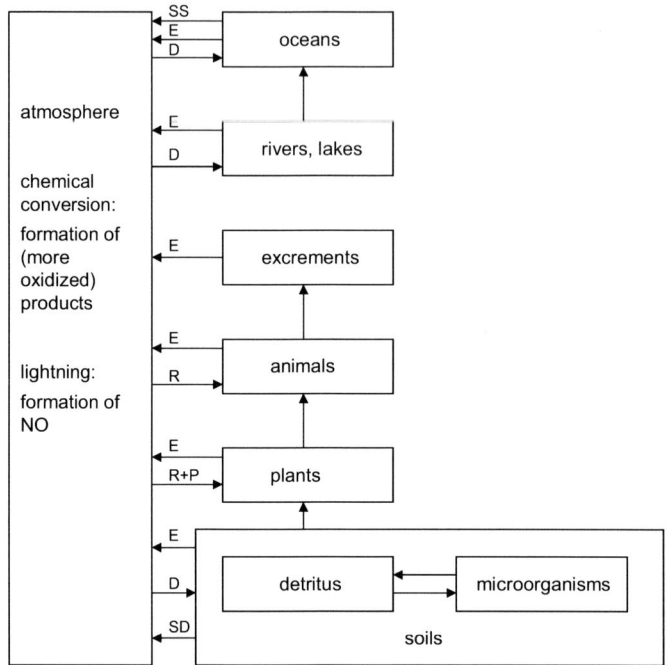

Fig. 2.24 Biosphere-atmosphere interaction: fluxes of emission and deposition. E emission of gaseous substances, SD soils dust emission, SS sea spray emission, D deposition (dry, wet and sedimentation), R respiration, P photosynthesis.

the principal interaction between the biosphere (here again considered in a more narrow definition) and the atmosphere. Uptake of gases by *assimilation* occurs in plants because of photosynthesis and respiration and in animals due to respiration. Uptake of gases by sorption onto solid and aquatic surfaces of the earth is a physico-chemical process, termed *dry deposition* (see Chapter 4.4.1), however, assimilative uptake is also said to belong to dry deposition. Other processes of deposition (wet deposition and sedimentation) do not depend on surface properties, but have to be taken into account as input flux to the biosphere. Biogenic emission (which, as mentioned above, accounts for a significant fraction of NEP) is basically a loss of matter from the ecosystem. By turning into other ecosystems via atmospheric transport and transformation, however, it could have several functions (information by pheromones, self protection, climatic regulation, and nutrient spreading).

2.4.3 Global biogeochemical cycles

The following sections describe the cycles in general, the basic processes, such as emissions, deposition and chemical reactions are presented in detail in subsequent Chapters. The aim here is to point out the most important issues in the biosphere-atmosphere interaction and the part of humans in a changing climate system.

2.4.3.1 Nitrogen

Even though the atmosphere is 78% nitrogen (N_2), most biological systems are nitrogen limited on a physiological timescale because most biota are unable to use molecular nitrogen (N_2). Two natural processes convert nonreactive N_2 to reactive N; *lightning* and *biological fixation*. Reactive nitrogen is defined as any single nitrogen species with the exception of N_2 and N_2O. It includes:

- NO_y ($NO + NO_2 + N_2O_3 + N_2O_4 + HNO_2 + HNO_3 + NO_3 + N_2O_5 + HNO_4$ + organic NO_x + particulate NO_2^- and NO_3^-),
- NH_x ($NH_3 + NH_4^+$) and
- organic bonded N (mostly NH_2^- but also SCN and other structures or functional groups with special biochemical functions);

Note that $NO_x = NO + NO_2$ and is often defined $NO_z = NO_y - NO_x$.

Atmospheric NO production via lightning is based on the same thermal equilibrium ($N_2 + O_2 \leftrightarrow 2\,NO$), which also takes place in all anthropogenic high-temperature processes (e. g., burning), see Chapter 2.7.1.3. This natural source, however, is too limited to supply the quantity of reactive nitrogen within the global biological nitrogen cycle (Fig. 2.25). Biological nitrogen *fixation* by microorganisms in soils and oceans produces organic nitrogen (as reduced N^{3-} in form of NH_2 bonding N) within the fixing organisms. This nitrogen is lost from the organisms after their death as ammonium (NH_4^+) via mineralization.

Nitrification is the biological oxidation of ammonium (NH_4^+) to nitrate (NO_3^-), with nitrite (NO_2^-) as an intermediate under aerobic conditions: ($NH_4^+ \rightarrow NO_2^- \rightarrow NO_3^-$). Under oxygen-limited conditions nitrifiers can use NO_2^- as a terminal electron acceptor to avoid accumulation of the toxic NO_2^-, whereby N_2O and NO are produced. During that process, because separate bacteria (Meiklejohn 2006) oxidize NH_4^+ into NO_2^- and NO_2^- into NO_3^-, the process can lead to the temporary accumulation of NO_2^- in soil and water. These bacteria are generally chemoautotrophic, requiring only CO_2, H_2O and O_2. The nitrifying bacteria *Nitrosomas*, which converts NH_4^+ to NO_2^-, is also able to reduce NO_2^-. This nitrite can decompose abiotically, yielding NO or NO_2, substantially favoured in acidic soils

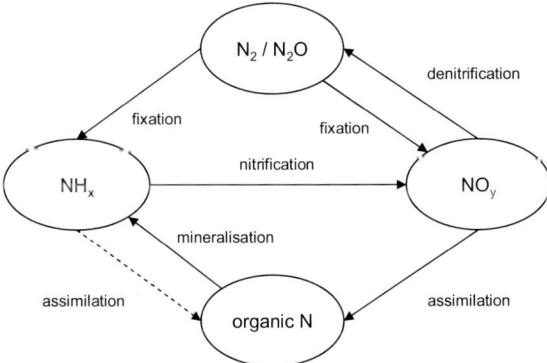

Fig. 2.25 The biological nitrogen cycle.

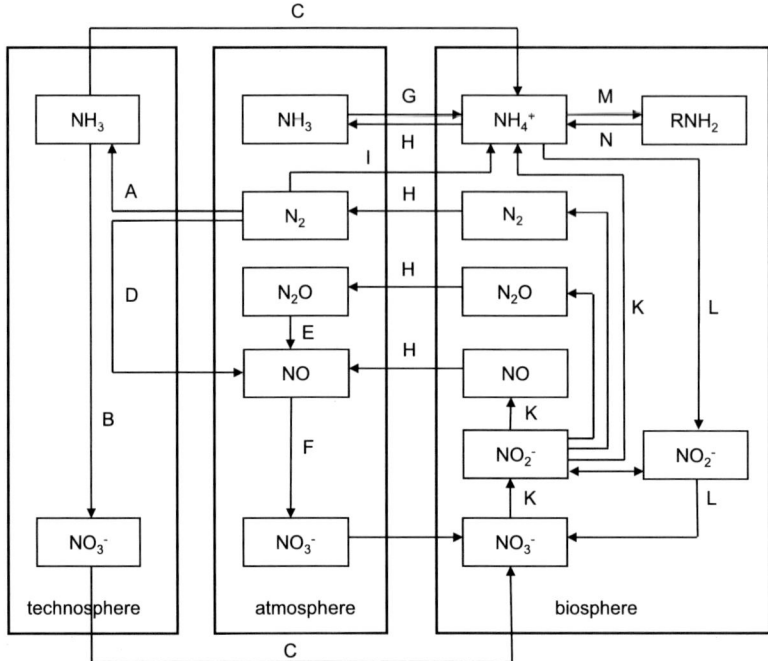

Fig. 2.26 The biogeochemical nitrogen cycle. A ammonia synthesis (man-made N fixation), B oxidation of ammonia (industrial production of nitric acid), C fertilizer application, D formation of NO due to high-temperature processes, E Oxidation of N_2O within the stratosphere, F oxidation of NO within the troposphere, G ammonia deposition and transformation into ammonium, H biogenic emission, I biogenic N fixation, K denitrification, L nitrification, M assimilation (biogenic formation of amino acids), N mineralization. RNH_2 organic bonded N (e. g. amines).

(see Fig. 2.26 and 2.27). The debate about whether NO is an intermediate or is produced by nitrifiers itself via NO_2^- reduction remains open (Firestone and Davidson 1989). The reduction of NO_2^- to NO and N_2O by nitrifiers also avoids accumulation of potentially toxic nitrite.

Denitrification is a microbial process for growth of some bacteria in soil and water that reduces nitrate (NO_3^-) or nitrite (NO_2^-) to gaseous nitrogen oxides (almost all N_2O and NO) and molecular N_2 by essentially aerobic bacteria. The general requirements for denitrification to occur are: (a) the presence of bacteria possessing the metabolic capacity; (b) the availability of suitable reductants such as organic carbon; (c) the restriction of O_2 availability; (d) the availability of N oxides. Current knowledge shows that the NO flux from the soil will depend both on physical transfer processes from the site of the denitrification to the atmosphere and on the relative rates of production and consumption of NO (Galbally 1989). Field measurements show that N_2 is the main product. NO_2 has been found as soil emission (Slemr and Seiler 1984). Colbourn et al. (1987) suggested that the NO_2 they detected was due to oxidation of NO either in the soil air or in their chamber system. However, there are abiotic processes in soils that produce NO_2 from nitrite.

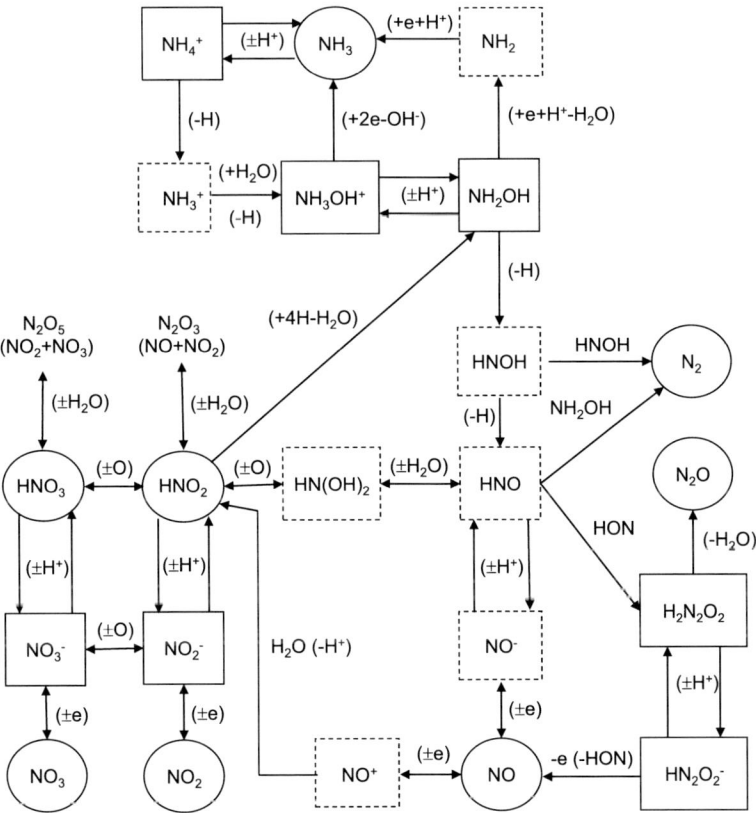

Fig. 2.27 Scheme of the biochemical, soil chemical, and atmospheric chemical nitrogen oxidation and reduction in the biosphere (nitrification, ammonification, and denitrification). Symbols for redox processes: −e: red, +e: ox, −H: ox ($\equiv$ e + H$^+$), +H: red ($\equiv$ + H$^+$ − e), −O: red ($\equiv$ + 2 H$^+$ − H$_2$O), +O: ox ($\equiv$ + H$_2$O − 2 H). Circles: gases; doted boxes: intermediates.

The accepted sequence for denitrification is demonstrated in Fig. 2.27 − the inverse process corresponds to nitrification:

$$\text{HNO}_3 \text{ (nitric acid)} \rightarrow \text{HNO}_2 \text{ (nitrous acid)} \rightarrow \text{HNO (nitroxyl)}.$$

Nitroxyl exists only as an intermediate and acts as a very weak acid (pK = 11.4) with NO$^-$ being the protolytic anion, from which NO is formed via electron transfer (and vice versa). HNO is the transfer point to NO (as just described) and to N$_2$ and N$_2$O in parallel pathways (Fig. 2.27). H$_2$N$_2$O$_2$ (hyponitrous acid), probably produced by enzymatic dimerisation of HNO, is formally the acid in the form of its anhydride N$_2$O (see for details Chapter 5.4.4.2).

Consequently, the biological nitrogen cycle is closed, beginning and ending with N$_2$: N$_2$ fixation $\rightarrow$ NH$_4^+$ *nitrification* $\rightarrow$ NO$_3^-$ *denitrification* $\rightarrow$ N$_2$ (Fig. 2.26). Each intermediate, which is produced within the biological cycle (NO, N$_2$O, NH$_3$, HNO$_2$, organic N species, see Fig. 2.27), can escape the "biosphere" (in terms of organisms, soil, water and plants) by physical exchange processes depending on many environmental factors. On the other hand, plants and organisms can take up the same species from its environment and, moreover, abiotically oxidized components (e. g., NO_2, HNO_3). The bacterial processes of denitrification and nitrification are the dominant sources of N_2O and NO in most systems. Only denitrification is recognized as a significant biological consumptive fate for N_2O and NO. The chemical decomposition of HNO_2 (or chemical denitrification), that is, the reduction of NO_2^- by chemical reductants under oxygen-limited conditions and at low pH, can also produce N_2, N_2O and NO. Chemical denitrification generally occurs when NO_2^- accumulates under oxygen-limited conditions, which may occur when nitrification rates are high, e. g., after application of NH_4^+-based mineral fertilizers or animal manure. This process may account for 15–20% of NO formation. Abiotic HNO_2 emission from soil NO_2^-, due to a simple chemical acid-base equilibrium (see Chapter 4.2.5.2), was shown by Su et al. (2011) and recently by Oswald et al. (2013) showed that ammonia-oxidizing bacteria can directly release HNO_2 in larger amounts than expected from equilibrium calculations whereas soil emission of HNO_2 and NO feature similar optimum curves. HNO_2 is produced and emitted during nitrification occurring under low soil water content (less than 40%) and may contribute up to 50% of the reactive nitrogen release from soils (Oswald et al. 2013). Emission and/or deposition (biosphere-atmosphere exchange) is therefore a complex function of biological, physical and chemical parameters describing the system.

In the process of *ammonification*, hydroxylamine (NH_2OH) is an important intermediate in two directions, denitrification direct to ammonium as well as the oxidation of ammonium to nitrate (nitrification). As long as nitrogen remains in its reduced form (NH_4^+), it remains in the local environment because of its affinity for soil absorption and its rapid uptake by biota. NH_4^+ is in equilibrium with NH_3, which can escape to the atmosphere, depending on pH, temperature, soil moisture, soil type and atmospheric NH_3 partial pressure. The equilibrium between emission and deposition (gas uptake) is called the *compensation point*; similar factors control the emission/dry deposition of NO.

In addition to being important to biological systems, reactive nitrogen also affects the chemistry of the atmosphere. At very low NO concentrations, ozone (O_3) is destroyed by reactions with radicals (especially HO_2), although at higher levels of NO (larger than 10 ppt), there is a net O_3 production (because HO_2 reacts with NO to form NO_2). The photolysis of NO_2 is the only source of photochemically produced O_3 in the troposphere. Although N_2O is not viewed as a reactive form of nitrogen in the troposphere, it adsorbs IR radiation and acts as a greenhouse gas. In the stratosphere, N_2O will be oxidized to NO_x and impacts the O_3 concentration. NH_3 is the major source of alkalinity in the atmosphere and a source of acidity in soils. A small part of atmospheric NH_3 ($\leq 5\%$) is only oxidized by OH radicals, where a main product has been estimated to be N_2O (Dentener and Crutzen 1994), thus contributing around 5% to estimated global N_2O production. Deposited NH_3/ NH_4^+ will be *nitrified* in soils and water to NO_3^-, where two moles of H^+ are formed for each mole of NH_3/NH_4^+. Thus, any change in the rate of formation of reactive nitrogen (and N_2O), its global distribution, or its accumulation rate can have a fundamental impact on many environmental processes.

Table 2.19 Global turn-over of nitrogen (in Tg yr^{-1}); data adapted from Galloway et al. (2004).

	natural	anthropogenic	total
fixation			
terrestrial	107	32	139
oceanic	86–156	0	86–156
atmospheric	5.4[a]	36[c]	41
industrial	0	100[b]	100
total	200–250	170	270–420
emission	63	50	113
deposition	–	–	200
transport to groundwater	–	–	48
oceanic denitrification	–	–	147–454
atmospheric burden (emission + atmospheric flux)[d]			154

[a] by lightning (NO formation)
[b] ammonia synthesis (NH$_3$)
[c] high-temperature NO formation
[d] should be equal to deposition

The linkage between the biosphere and the atmosphere has to be assumed to have been in equilibrium prior to the current industrial age. Nitrogen species (NO$_x$, N$_2$O, N$_2$), emitted from plants, soils and water into the atmosphere, which are biogenically produced within the redox cycle between *nitrification* and *denitrification*, will be oxidized to nitrate and return to the biosphere. This cycle has been increasingly disturbed since the beginning of the Industrial Revolution more than 100 years ago. The most important aspect in the anthropogenically-modified N cycle is the worldwide increase of nitrogen fertilizer application. Without human activities, biotic fixation provides about 100 Tg N yr^{-1} on the continents, whereas human activities have resulted in the fixation of around an additional 170 Tg N yr^{-1} by fossil-fuel combustion (~ 30 Tg N yr^{-1}), fertilizer production (~ 100 Tg N yr^{-1}), and cultivation of crops (e. g., legumes, rice; about 40 Tg N yr^{-1}); Galloway et al. 1995, 2004), Table 2.19. The oceans receive about 100 Tg N yr^{-1} (about 60 Tg N yr^{-1} by atmospheric deposition and about 40 Tg N yr^{-1} via rivers), which is incorporated into the oceanic nitrogen pool. The remaining about 230 Tg N yr^{-1} are either retained on continents, in water, soils and plants, or denitrified to N$_2$. Thus, although anthropogenic nitrogen is clearly accumulating on continents, we do not know the rates of individual processes. Galloway et al. (1995) predict the anthropogenic N fixation rate will increase by 60% (based on 1990) by the year 2020, primarily due to increased fertilizer use and fossil-fuel combustion. About two-thirds of the increase will occur in Asia, which by 2020 will account for over half the global anthropogenic N fixation.

In contrast to sulfur species, there are no differences in principle between natural and anthropogenic processes in the formation and release of reactive nitrogen species. Industrial nitrogen fixation (in separated steps: N$_2$ → NH$_3$, N$_2$ → NO$_x$, NO$_x$

→ NO$_3$) proceeds via the same oxidation levels as biotic *fixation* and *nitrification*, either on purpose in chemical industries (ammonia synthesis, nitric acid production) or unintentionally in all high-temperature processes, namely combustion, as a byproduct due to N$_2$ + O$_2$ → 2 NO.

2.4.3.2 Sulfur

Like nitrogen, sulfur is an important element in biomolecules with specific functions. In contrast to nitrogen, where the largest pool is the atmosphere (molecular N$_2$), for sulfur the largest pool is the oceans (as dissolved sulfate); both components are chemically stable. In air, carbonyl sulfide (COS) represents the major sulfur component, due to its long residence time. Similarly to nitrogen and carbon, the sulfur content in the lithosphere is small because of degassing volatile sulfur compounds in the early history of the earth. Primordial sulfides and elemental sulfur are almost all oxidized in the atmospheric turn-over (Canfield 2004). Moreover, humans have extracted them by mining from the lithosphere to such an extent that the remaining resources are negligible at the present time (Fig. 2.28). The great "role" of life again is the reduction of sulfate. Volcanism is an important source of sulfur dioxide (SO$_2$) and promotes the formation of a strong acid (H$_2$SO$_4$) which

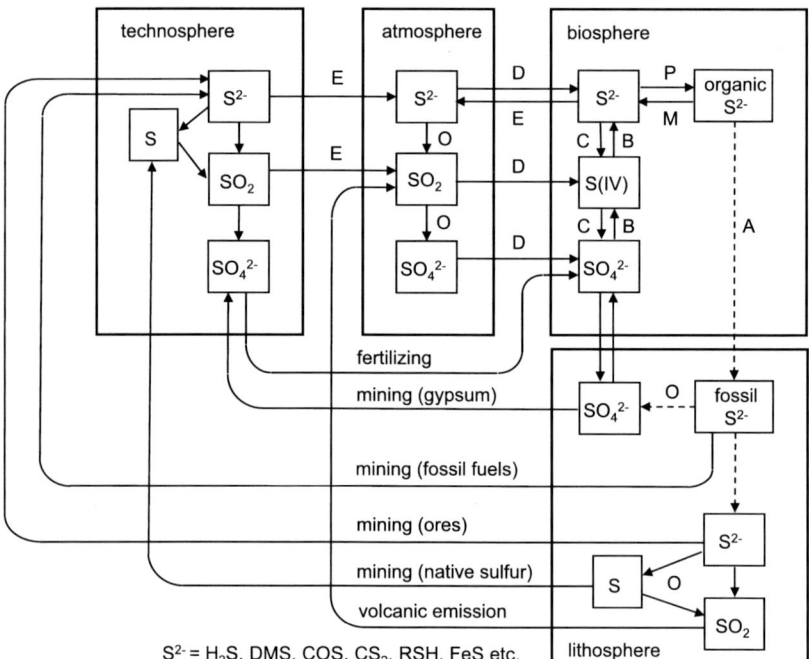

Fig. 2.28 The biogeochemical sulfur cycle. A burial (formation of sediments), B assimilation (bacterial sulfate reduction), C aerobic oxidation, D deposition, E emission, M mineralization, P plant assimilation, O oxidation.

may play an important role in weathering. Consequently, the dominant anthropo-genic SO_2 emission since the Industrial Revolution has resulted in significant acidi-fication of many parts of the world (see also Chapter 2.7.2.1).

In anaerobic environment (e. g., anoxic water basins, sediments of wetlands, lakes, coastal marine ecosystems) sulfur bacteria reduce sulfate to support respiratory metabolism, using sulfate as a terminal electron acceptor instead of molecular oxy-gen (*dissimilatory sulfate reduction*). This process is the major pathway for H_2S production globally. However, since the anaerobic environment is often not in direct contact with the atmosphere, the escape of H_2S is limited because of re-oxidation in an oxic layer. Reduced sulfur provides substrates for microbial oxidation to sul-fate from which certain bacteria can obtain energy. Such microorganisms are present in high numbers at the oxic-anoxic interface and can completely oxidize H_2S and other reduced sulfur compounds in a layer smaller than a millimeter (Andreae and Jaeschke 1992). Consequently, the large amount of H_2S which is produced in coastal areas cannot usually be transferred to the atmosphere. Within the biological sulfur cycle, immense amounts of sulfur are turned over: 550 Tg yr^{-1} in the oceanic and assimilative bacterial sulfate reduction (100−200 by land plants and 300−600 by sea algae) (see Table 2.20). COS is the most abundant tropospheric sulfur gas on earth.

Most biota, however, need sulfur to synthesize organosulfur compounds (cysteine and methionine as examples of the major sulfur amino acids). In contrast to animals, which depend on organosulfur compounds in their food to supply their sulfur requirement, other biota (bacteria, fungi, algae, plants) can obtain sulfur in the aerobic environment from sulfate reduction (*assimilatory sulfate reduction*). Most of the reduced sulfur is fixed by intracellular assimilation processes and only a minor fraction is released as volatile gaseous compounds from living organisms. However, after the death of organisms, during microbial degradation, volatile sulfur

Table 2.20 Global turn-over of sulfur. Data adapted from Andreae (1990), recalculated from mol in mass (rounded).

process	Tg S yr^{-1}
bacterial dissimilatory sulfate reduction	
costal zone	70
shelf sediments	190
depth sediments	290
assimilatory sulfate reduction	
land plants	100−200
ocean algue	300−600
anthropogenic SO_2 emission	≈ 100[a]
total biogenic gaseous sulfur emission	≈ 50
total natural sulfur emission (without sea salt)	≈ 70

[a] today, global anthropogenic SO_2 emission is considerable less (Table 2.67)

compounds may escape to the atmosphere, mainly H_2S, but also organic sulfides like CH_3SH, CH_3SCH_3 (DMS), $CH_3S_2CH_3$ (DMDS) and CS_2, COS. CS_2 and COS in air were first detected as recently as 1976; Crutzen (1976) proposed that COS was a possible source of sulfur for the stratospheric sulfate in the *Junge* layer in quiescent volcanic periods, because of its relative abundance and long lifetime compared with other sulfur trace gases.

In the open ocean waters, DMS is the predominant volatile sulfur compound, being formed by phytoplankton. The precursor of DMS is dimethylsulfoniopropionate (DMSP), which is produced within phytoplankton cells and is thought to have a number of important physiological functions. Comparisons of the results of three-dimensional models of the sulfur cycle globally (Langner and Rodhe 1991, Pham et al. 1995) with field observations suggest that the marine DMS emission figure of 16 Tg S yr^{-1} (given by Bates et al. 1992) is reasonable. This, however, contrasts with the range of 19−58 Tg DMS-S yr^{-1} for marine emission given by Andreae (1990); see also discussion on oceanic DMS emision in Chapter 2.6.3. The uncertainty factor in the DMS emission estimate of 2 to 3 is therefore an unresolved issue. This is a serious problem because of the dominant role of DMS in the natural sulfur budget (50−80 % of the total natural sulfur emission). The *CLAW hypothesis* (named after the authors of the paper; Charlson et al. 1987) highlighted the important potential role in climate regulation of DMS production in the oceans (see Fig. 2.19). Since the hypothesis was presented, it has become increasingly apparent that a complex network of biological and physicochemical processes control DMS emissions from the oceans. However, several of the crucial biological controls on DMS production remain uncertain, presenting a major hindrance in our ability to decipher whether DMS cycling could contribute to the regulation of a warming climate (Archer 2007).

What then is the fate of DMS in the marine troposphere? The gas phase oxidation pathways of DMS involve OH and NO_3 (the last mainly in NO_x-polluted air masses) as oxidants forming either dimethylsulfoxide (DMSO) or methansulfonic acid (MSA) or SO_2 itself. DMS oxidation by IO is known to be negligible (Andreae 1990); oxidation by Cl and BrO are interesting possibilities. DMS oxidation by OH is rapid (see Chapter 5.5.1), so that a residence time equal to or less than one day could be assumed, which is significantly shorter than the three days assumed by Langner and Rodhe (1991). The lower estimate of DMS residence time, τ, is consistent with an estimate via the ratio of burden to flux ($\tau = M/F$), using a DMS concentration average for the mixing layer (typically 1 km) of 20−100 ppt. Considering a DMS emission range of 15−58 Tg S yr^{-1}, the possible range of τ lies between 0.06 and 1.2 days. DMSO and MSA also have some uncertain aspects. DMSO reacts rapidly with OH, probably resulting in the formation of SO_2 and MSA (Andreae 1990). Mihalopoulos et al. (1993) estimated the annual average wet deposition of MSA to be 0.51 µeq m^{-2} d^{-1}, corresponding to a flux of 2 Tg S yr^{-1}, which is between 5 % and 10 % of the annual DMS emission. Ayers et al. (1986) found a molar ratio of 6 % of MSA to excess sulfate in the aerosol phase in a study conducted at Cape Grim Baseline Air Pollution Station (Tasmania, Australia). Both figures support the idea that most DMS is oxidized to SO_2 and probably

SO_4^{2-}. The marine S(IV) budget is negligibly enlarged as SO_2 is produced from H_2S and CS_2 at only $1-2$ Tg S yr^{-1}.

DMS is now believed to be the most probable natural sulfate precursor, and SO_2 from fossil-fuel combustion is the dominant man-made one. Could this natural DMS-derived "sulfate function" be a result of the earth's evolution, and are there feedbacks between the climate system and the sulfur cycle (Lovelock and Margulis 1974, Charlson et al. 1987)? Has the natural functioning of the atmospheric sulfur cycle been perturbed by human activities? Temperature records supporting the hypothesis that anthropogenic sulfate aerosol influences clear-sky and cloud albedo, and thus climate, have been advanced by several investigators (see, for example, Hunter et al. 1993), who have suggested that any natural role of sulfur in climate has been subsumed by anthropogenic pollution. The direct climatic effect of sulfate aerosol is due simply to reflection of sunlight back to space, while indirect climatic effects of sulfate result from aerosol influence on cloud albedo and/or extent. Sulfate aerosols contribute to cooling, either directly or indirectly through their role in cloud formation. Also, changes in the chemical composition of aerosols may either increase or decrease the number of CCN. Changes in concentration of the number of cloud droplets can affect not only the albedo but also the cloud lifetime and precipitation patterns. Precipitation is both an important aspect of climate and the ultimate sink for submicrometer particles and scavenged gaseous pollutants.

Climate change and decreasing seawater pH (ocean acidification) have widely been considered as uncoupled consequences of the anthropogenic CO_2 perturbation. Recently, experiments in seawater enclosures showed (Hopkins et al. 2011) that concentrations of DMS were markedly lower in a low-pH environment. Six et al. (2013) conclude that global DMS emissions will decrease by about $18(\pm 3)\%$ by 2100 compared with pre-industrial times as a result of the combined effects of ocean acidification and climate change. The reduced DMS emissions induce a significant additional radiative forcing. However, Arnold et al. (2013) studied that DMS production is sensitive both to acidification and temperature and that "global change" may not result in decreased DMS as suggested in earlier studies.

Sulfur, or more precisely sulfur dioxide (SO_2), is the oldest known pollutant. Without knowing the chemical species, its influence on the air quality was described several hundred years ago in European cities where it was prevalent because of coal burning (e. g., Evelyn 1661). More than 100 years ago, SO_2 was first identified as the cause of forest damage in the German Erzgebirge. However, it became of huge environmental interest only in the middle of the twentieth century after the well-known London episode in 1952, where increased concentrations of SO_2 and particulate matter in the presence of fog led to an unusually high mortality rate for the particular time of year. Subsequently, the atmospheric chemistry of SO_2 and global sulfur distribution has been studied intensively. Since then, the anthropogenic sulfur emission (and, consequently atmospheric SO_2 concentration) is continuously decreased in Europe. Today, after the introduction of measures for the desulfurization of flue gases from all power-plants, SO_2 no longer plays a role as a pollutant in Europe (see Chapter 2.7.2.1 and 2.8.1).

Because of the fact that the only natural source of SO_2 (volcanism) shows large variations and the mean annual estimate is around 10 Tg yr^{-1}, anthropogenic SO_2 emissions currently still account for around 80% of the total global flux of SO_2, and more than 90% are injected into the northern hemisphere (Langner and Rodhe 1991, Dignon 1992, Spiro et al. 1992). Nevertheless, 10% of the global anthropogenic sulfur emissions account for 50% of the sulfur budget of the southern hemisphere. Thus, even in remote areas, we must assume that the sulfur budget is markedly disturbed by human activities. Moreover, the ratio between the direct SO_2 removal flux and the flux due to its longer-lived sulfates, which cover regions up to several thousands of kilometers across, still remains an imprecise estimate. Many attempts have been made to quantify the global sulfur budget, beginning with the first estimates by Eriksson (1959, 1960 and 1963) and continuing to the most recent one by Pham et al. (1995). It is remarkable that no more updated natural sulfur emission estimates have appeared, at the time of writing, for some 15 years. While the estimate of the ratio between natural and anthropogenic sulfur emissions was a primary goal of the first established global sulfur budgets that were established (see, for example, Möller 1983, 1995a), in more recent estimates the question of sulfate aerosols and their precursors has been the main point of interest (Möller 1995b).

2.4.3.3 Chlorine

Chlorine is one of the most abundant elements on the surface of the earth. Until recently, it was widely believed that all chlorinated organic compounds were xenobiotic, that chlorine does not participate in biological processes and that it is present in the environment only as chloride. However, over the years, research has revealed that chlorine takes part in a complex biogeochemical cycle, that it is one of the major elements of soil organic matter and that the amount of naturally formed organic chlorine present in the environment can be counted in tonnes per km^2 (Öberg 1998). More than 4000 organohalogen compounds, mainly containing chlorine or bromine but a few with iodine and fluorine, are produced by living organisms or are formed during natural abiogenic processes, such as volcanoes, forest fires, and other geothermal processes. The oceans are the single largest source of biogenic organohalogens, which are biosynthesized by myriad seaweeds, sponges, corals, tunicates, bacteria, and other marine life. Terrestrial plants, fungi, lichen, bacteria, insects, some higher animals, and even humans, also account for a diverse collection of organohalogens (Gribble 2003).

The abundance of chlorine in the lithosphere seems not to be well established – in literature it is found in a range from 0.013% to 0.11%; Clarke (1920) suggested 0.045%. With the last value, we obtain a total mass of chlorine (using the mass of the lithosphere, see Table 2.2) of $2.2 \cdot 10^{22}$ g, which is comparable with the chloride dissolved in the oceans ($2.6 \cdot 10^{22}$ g). Elemental chlorine is one of the most reactive species and therefore is not found in nature with the exception of small volcanic emissions (see Chapter 2.6.4.3). In Chapter 2.2.1.2 we discussed the origin of chlorine from rock degassing in the form of HCl. Inorganic chlorine in the form of chloride (and similar to other halogens such as fluorine, bromine and iodine) is therefore stored in seawater and salt stocks from former oceans. Hence the formation of sea salt (see Chapter 2.6.4.2) is the dominant source of particulate chloride and gaseous HCl due to subsequent heterogeneous reactions (see Chapter 5.8.2).

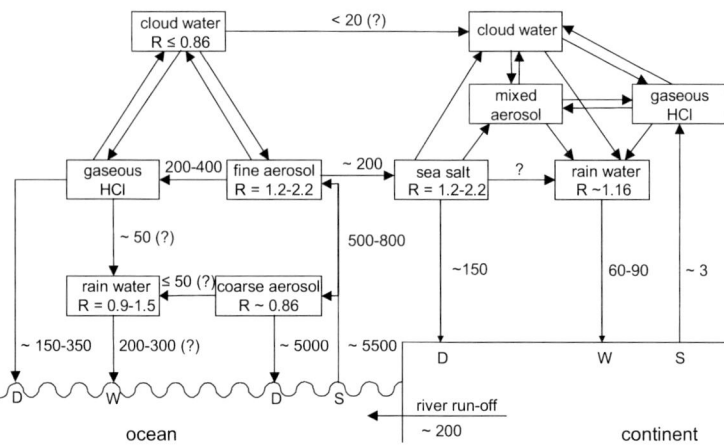

Fig. 2.29 The global inorganic chlorine cycle (data from Möller 1990, 2003). S − source, W − wet deposition, D − dry deposition, R − Na/Cl ratio.

The circulating amounts of sea salt are immense (Fig. 2.29) and provide chloride (together with sodium) to all parts of the world. Globally, large amounts of fine sea salt (i. e., not removed by sedimentation) are to be assumed to be emitted into the free troposphere and available for degassing processes between 100 and 800 Tg a^{-1} as Cl (e. g., Erickson and Duce 1988, Möller 1990). These values are based on total sea-salt chloride emission in the order of 5000 Tg yr^{-1} (see discussion in Chapter 2.6.4.2). Erickson et al. (1999) estimated (based on global modelling) the integrated annual production of sea-salt Cl^- to be substantially smaller, with 1785 Tg Cl yr^{-1} (this is the latest estimation found in the literature) and the subsequent HCl flux to be only 7.6 Tg Cl yr^{-1} (see Table 2.21). Due to its ready solubility in water, Cl^- is found in all natural waters and is finally transported back to the oceans.

Loss of chlorine (as HCl) from marine particulate matter was observed more than 50 years ago and attributed to surface reactions (acidification) by acids produced from gaseous SO_2 and its oxidation onto sea-salt particles (e. g., Junge 1954, Duce 1969, Hitchcock et al. 1980). However, the Cl deficit in atmospheric aerosol particles seems to have been known for even longer and was explained by Cauer (1949b) through HCl degassing. McInnes et al. (1994) have shown that in the marine remote boundary layer the Cl depletion from individual particles is highly variable, in the range from sub-μm to above μm particles. Polluted air masses may result into a complete loss of Cl (Roth and Okada 1998). During their study in northern Finland in summer 1996, Kerminen et al. (1998) found that the fraction of chloride lost from sea-salt aerosol increased with decreasing particle size for all samples. They also estimated the average contribution from sulfate (after subtraction of $(NH_4)_2SO_4$), nitrate, and selected organic anions (methanesulfonate, oxalate). The overall calculated loss was very close to 100 % for sub-μm particles (up to 1.8 μm), about 43−65 % for particles up to 7.5 μm and little less for particles in the range 7.5−15 μm, and seems to depend on the air transport time over continental areas and the concentration of sea-salt aerosol in the air. Cl degassing observed

Table 2.21 Emission of chlorine compounds (in Tg Cl yr^{-1}). After Lobert et al. 1999, Khalil et al. 1999, Graedel and Keene 1996, 1999, Keene et al. 1999, McCulloch et al. 1999).

substance	ocean	soils	biomass burning	fossil fuel burning	other man-made emission
CH_3Cl	0.46	0.0001	0.640	0.075	0.035
$CHCl_3$	0.32	0.0002	0.0018	–	0.062
CH_3CCl_3	–	–	0.013	–	0.572
C_2Cl_4	0.016	–	–	0.002	0.313
C_2HCl_3	0.020	–	–	0.003	0.195
CH_2Cl_2	0.16	0.0003	–	–	0.487
$CHClF_2$	–	–	–	–	0.080
total organic	1.0	0.0006	0.65	0.08	1.7
total inorganic	1785[a]	15[b]	6.3[c]	4.6[d]	2[d]

[a] sea salt including release of HCl (7.6) and ClNO$_2$ (0,06) after Erickson et al. (1999)
[b] soil dust chloride
[c] HCl and particulate chloride
[d] HCl

also from continental particulate matter is mainly believed to occur by gaseous HNO$_3$ adhering to the particles (e. g., Pakkanen 1996, Keene and Savoie 1998, Zhuang et al. 1999, Yao et al. 2003). As a consequence, the particulate matter is enriched in sodium. In Fig. 2.29, the measured ratio R_{meas} = [Na]/[Cl] is given for different chlorine reservoirs; in sea water $R_{seawater}$ = 0.86 (molar ratio; mass ratio amount 0.56). Thus, deviations to higher values indicate Cl loss. However over continents and in polluted air masses, so-called excess chloride may occur, which is mainly caused by human activity (coal combustion, waste incineration, salt industries). This excess chloride can be calculated according to Eq. (2.83); however, there are two preconditions: (a) that there are no sodium sources other than sea salt, and (b) that the local reference value of $R_{seasalt}$ is known. In older literature, instead of $R_{seasalt}$, authors used the seawater bulk value $R_{seawater}$ = 0.86.

$$[Cl^-]_{ex} = [Cl^-]_{seasalt}\left(\frac{R_{seasalt}}{R_{sample}} - 1\right) = [Cl^-]_{sample} - \frac{[Na^+]_{sample}}{R_{seasalt}} \qquad (2.83)$$

The Cl loss x (in %) in a sample can be calculated according to

$$x = 100\left(1 - \frac{0.86}{R_{seasalt}}\right) \qquad (2.84)$$

where $R_{seasalt}$ is the reference value at a given site. Möller (1990, 2003) estimated, based on experimental data for sea salt entering the Northern European continent, $R_{seasalt}$ = 1.16 which corresponds to x = 26% mean Cl loss. Taking into account that sea salt is already depleted in Cl to a large extent (50−75%) when entering the continents (and will be further depleted by acid reaction during transport over the continents), the HCl flux from acid degassing must be much larger then the above estimate by Erickson et al. (1999) (see Fig. 2.29). It is worth noting that, despite large differences in the estimates of the tropospheric inorganic chlorine budget, the freshwater-to-ocean transport of chloride agrees with the 220 Tg estimated by Graedel and Keene (1996). This run-off is due to wet deposition of fine sea-salt aerosol (often termed SSA) transported to the continents, which is to a large extent Cl depleted.

Other secondary chlorine species (atomic Cl, ClO, ClOOCl etc.) have been made responsible for Arctic ozone depletion, whereas the sources of the chlorine atoms are poorly understood (Keil and Shepson 2006). The Cl atom reacts similarly to OH (e. g. in oxidation of volatile organic compounds; Cai and Griffin 2006). However, the photolysis of HCl is too slow (even in the stratosphere) to provide atomic Cl. Thus, the only direct Cl source from HCl is due to its reaction with OH, but with a fairly low reaction rate constant (Rossi 2003). There are several chemical means of production of elemental Cl (and other halogens) from heterogeneous chemistry (see Chapter 5.8.2); in the troposphere the photolysis of chloroorganic is not very important, with a few exceptions (see Chapter 5.8.1).

Thornton et al. (2010) estimated by measuring and modelling the global Cl source from photolysis (see reaction 5.184) of nitryl chloride ($ClNO_2$) to be 8–22 Tg Cl yr^{-1} whereas $ClNO_2$ is nocturnally produced, especially in polluted areas, according to reaction 5.181. Platt et al. (2004) estimated that the global methane sink due to reaction with Cl atoms (see reaction 5.385) in the marine boundary layer could be as large as 19 Tg yr^{-1}; see Chapter 5.8.2 for the Cl formation process via bromine autocatalysis.

As seen from Table 2.21, some volatile organochlorine compounds are emitted both naturally and anthropogenically. More than 200 chlorinated gases have been identified in air (Graedel 1979, Graedel et al. 1986), and more than 1000 organic chlorine compounds have been identified in nature (Öberg 2002). As mentioned above, today more then 4000 compounds are known from natural processes (Gribble 2004). It is now an established fact that natural organohalogens are a normal part of the chlorine cycle in the environment (Khalil and Rasmussen 1999). The group of reactive gases consists of chloromethane or methyl chloride (CH_3Cl), chloroform ($CHCl_3$), phosgene ($COCl_2$), dichloromethane (CH_2Cl_2), chlorinated ethylenes (C_2HCl_3, C_2Cl_4), chlorinated ethanes (CH_4Cl_2, $C_2H_2Cl_4$), from natural and man-made sources (cf. Table 2.21). The long-lived or unreactive gases are the chlorofluorocarbons (CCl_2F_2, CCl_3F, $C_2Cl_3F_3$, $C_2Cl_2F_4$, C_2ClF_5, CCl_3F) and carbon tetrachloride (CCl_4). These are all of man-made origin and will no longer be produced because of the Montreal Protocol and its amendments (international agreements to phase out global production of compounds that can deplete the stratospheric ozone layer), see also Chapter 5.3.7. Synthesized organic chlorine compounds have been widely used as solvents, cleaning materials, pesticides, pharmaceuticals and plastics. In organic chemical synthesis, chlorinated compounds (mostly via radical attack of elemental chlorine) are used for other synthesis because Cl is easily exchangeable with other functional groups. Many compounds belong to the category of persistent organic pollutants (POPs). Because chloroorganics are lipophilic, they accumulate in the organs of animals and can have effects when they exceed a certain toxic threshold. Pesticides such as DDT (dichloro-di-phenyl-trichloroethane), which is the best known, are banned in many countries. It seems that chlorinated (or generally halogenated) organic compounds play very special roles as biomolecules. Halogenated natural products are medically valuable and include antibiotics (chlorotetracycline and vancomycin), antitumour agents (rebeccamycin and calichemycin), and human thyroid hormone (thyroxine). Halogenation is essential to the biological activity and chemical reactivity of such compounds, and often generates versatile molecular building blocks for chemists working on synthetic organic molecules (Gribbles 2004).

The organic chlorine in soil was originally suggested to be of anthropogenic origin, resulting from the atmospheric transport and deposition of man-made chlorinated compounds. However, the total atmospheric deposition of organic chlorine in remote areas can only explain a small fraction of the organic chlorine found in soil. Furthermore, it has been shown that soil constituents which originate from the period before industrialization also contain organic chlorine. Very little is known about the biogeochemical cycling (formation, mineralization, leaching etc.) of chlorinated organic matter in soil. For example, the net formation of organic chlorine in spruce forest soil is closely related to the degradation of organic matter. The ecological role of this formation is so far unknown, but recent findings suggest (Öberg 2002) that the amount of organically-bound halogens in soil increases with decreasing pH, and that production seems to be related to lignin degradation, in combination with studies which suggest that production of organochlorine is a common feature among white-rot fungi. This makes it tempting to suggest a relationship between lignin degradation and production of organohalogens. Such a relation may result from an enzymatically catalyzed formation of reactive halogen species as outlined below. Öberg (2002) enlightens four paradoxes that spring up when some persistent tacit understandings are viewed in the light of recent work as well as earlier findings in other areas. The paradoxes are that it is generally agreed that: (1) chlorinated organic compounds are xenobiotic even though more than 1000 naturally produced chlorinated compounds have been identified; (2) only a few, rather specialized, organisms are able to convert chloride to organic chlorine even though it appears the ability among organisms to transform chloride to organic chlorine is more the rule than the exception; (3) all chlorinated organic compounds are persistent and toxic even though the vast majority of naturally produced organic chlorine are neither persistent nor toxic; (4) chlorine is mainly found in its ionic form in the environment even though organic chlorine is as abundant or even more abundant than chloride in soil.

Considering the important role of chlorine (and other more reactive halogens; but Cl is only industrially used due to its cheap production from electrolysis) in organic synthesis, it is a small step to assume that the evolution of the metabolisms of organisms (especially animals) results in the use of chloride which is transformed into Cl atoms used in specific organosynthesis reactions and also provide functional molecules. Again hydrogen peroxide (cf. Chapters 2.2.2.3 and 5.8.2 and Eq. 2.60) plays a central role in oxidizing chloride in aqueous solution:

$$ \text{HCl} \xrightarrow{\text{H}_2\text{O}_2} \text{HOCl} \underset{\text{H}_2\text{O}}{\overset{\text{H}^+ + \text{Cl}^-}{\rightleftharpoons}} \text{Cl}_2 \tag{2.85} $$

The ubiquitous role of chloride (as dissolved sodium chloride) in animal and human cells and blood plasma is manifold: as just discussed, it provides Cl for organosynthesis and to control the electrolytic properties such as osmosis (a process where water molecules move through a semipermeable membrane from a dilute solution into a more concentrated solution), for nutrient and waste transport, as well as providing electrical gradients (based on conductivity) for information transfer through neurons.

2.4.4 What is the role of life in the earth's climate system?

If we return for a moment to the Gaia hypothesis, with the proposition that the biosphere itself is an organism, then it is logical to state that biogenic emissions have "sense" in making a suitable climate for the organisms and in controlling the climate system to maintain the optimum conditions for life. However, we also may state that the climate is first a result of geophysical and chemical processes, and that evolving life adapts to these conditions. We know that oxygen is definitively a result of the photosynthesis of plants, hence it is of biological origin. Without any doubt we can state that life did change the climate − with "sense" or without − and that feedback did influence the evolution of life. What is life or a living thing? Each living thing is composed of "lifeless" molecules, independent of its dimension and complexity, which are subject to the physical and chemical laws that are characteristic of inanimate bodies. A living thing (to avoid the definition of "life") has certain characteristics which are common to living matter and are not found in nonliving objects, such as:

- the capacity for self-replication (they grow and reproduce in forms identical in mass, shape and internal structure),
- the ability to extract, transform and use energy from their environment (in the form of nutrients and sunlight), and
- an organized structure, where each component unit has a specific purpose or function.

All these characteristics result in non-equilibrium with themselves and with their environment. Non-living things tend to exist in equilibrium with their surroundings. An earth-like planet, not developing life, would therefore be oxidized with aging. Only photolytic dissociation and thermal degradation would occur, depending on incoming radiation (distance from the Sun) and available thermal heat (planetary size). At a final stage, the atmosphere would be composed solely of oxides and acids. The large CO_2 content would increase the atmospheric temperature. Missing free oxygen (because it is fixed in oxides, volatile and non-volatile) in the atmosphere (and subsequent ozone) would prevent a UV-absorbing layer and therefore allow almost all photodissociations close to the planetary surface. With time, all water would disappear due to photolytic splitting into hydrogen, which escapes into space. The oxygen from water splitting cannot accumulate until all primordial reduced atoms are oxidized. Finally, free oxygen could be possible in the case of an excess over the equivalent of atoms in reduced state. Conversely, no free oxygen would occur when the reduction equivalent exceeds that of oxygen. The planet becomes irreversibly uninhabitable, especially because of absence of water. There is no doubt that this process would occur over a long time, potentially over the planetary lifetime.

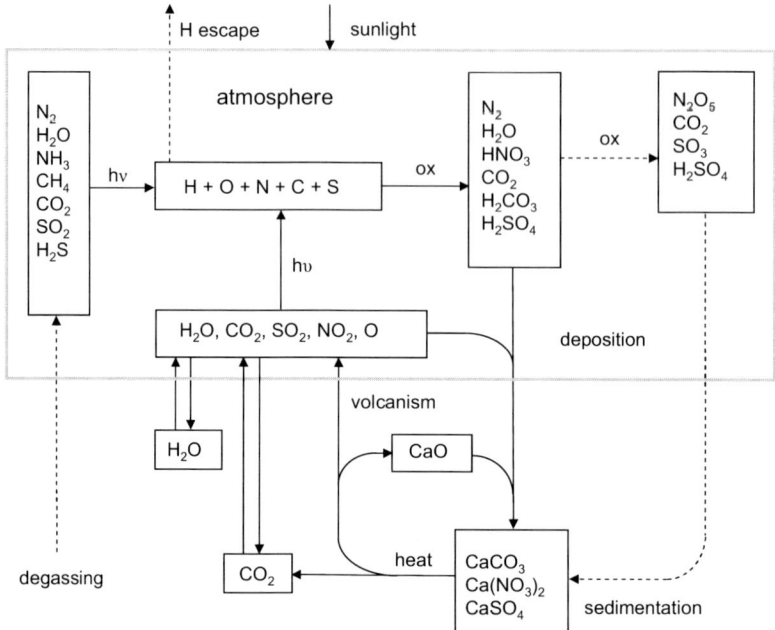

Fig. 2.30 Scheme of geochemical cycling over geological epochs.

Fig. 2.30 suggests that there is cycling, but only geochemically. Thus, the answer to the question why we observe natural abiogenic emissions (volcanism, soil dust and sea salt) is simple: these fluxes arise due to fundamental physical reasons, such as differences in pressure and instability in material.

Nonetheless, it is a fact that life developed on earth. The possible origin and conditions for its early development we have discussed before (Chapter 2.2.2.1). It is worth noting here that the evolution of life could have been interrupted or steered in other directions by several circumstances. One is always evident, the collision of the earth with another planetary body. Besides this catastrophic event, during all the stages of evolution, slow changes of physical and chemical conditions in other directions could have been caused dramatic changes. Complex systems are described by non-linear relationships. Mathematically it means that the "state equation" has several solutions for one set of initial parameters. From a given point in time and space, there is no singular predictability. That is precisely society's current problem in discussing the magnitude of the undeniable climate change that is occurring.

The non-living world tends to dissipate structures and therefore to increase entropy. However, because of its huge energy pools, the earth's internal geothermal heat and the Sun's radiation, both likely to remain over the entire expected lifetime of the earth, provide gradients to force geochemical cycling with the irreversible direction of oxidation and acidification (Fig. 2.30).

Only the living world is able to reduce entropy by creating structures, or in other words, to move the system back from equilibrium. Indeed, the central role of life

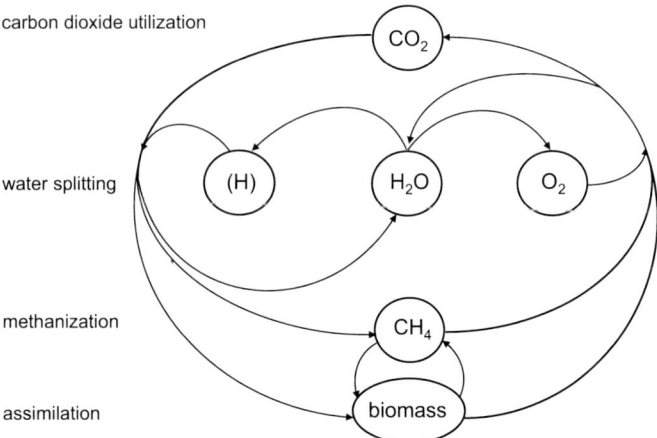

Fig. 2.31 Scheme of the water-carbon interlinked cycle.

(more exactly, autotrophic organisms) is to maintain the cycle shown in Fig. 2.31, starting with light-induced electrolytic water splitting, and subsequent parallel – but, important to note, in separated organs – reduction of CO_2 into hydrocarbons and the oxidation of them back to CO_2. Globally, this cycle represents a dynamic equilibrium. Consequently, there are established stationary concentrations in the climate system. As discussed in Chapter 2.2.2.5, it does not mean that the concentrations remain constant over time or, in other words, that there is no climate change or variation. The equilibrium may shift, characterized by different "equilibrium constants" resulting in different ratios of concentrations, namely between the "pool" species given in Fig. 2.31 (O_2, oxidized carbon, reduced carbon, and H_2O). Within short geological time intervals, only the atmospheric CO_2 concentration seems to be sensitive due to the "greenhouse effect". Reduced (CH_4) and oxidized (CO_2) carbon are inversely correlated biologically: less carbon dioxide in the atmosphere results in more reduced carbon buried in the lithosphere. The correlation between O_2 and CO_2 in air is similar; because of the huge oxygen level, small changes in carbon will not lead to measureable changes in oxygen.

The most important "invention" of life is the water-splitting process and, together with the abiotic surroundings, to create oxic (more oxidizing) and anoxic (more reducing) environments in the climate system. In other words, to separate oxygen (O) and hydrogen (H) to avoid simply the reaction back to water. Consequently, the continuous processes of oxidation and reduction of sulfur, nitrogen and carbon provide global cycling according to the following scheme:

$$\overset{\overset{\textstyle H}{\longleftarrow}}{\underset{\underset{\textstyle O}{\longrightarrow}}{\begin{array}{l} CH_4 \leftrightarrow C \leftrightarrow CO_2 \\ NH_3 \leftrightarrow N \leftrightarrow N_2O_5 \\ H_2S \ \leftrightarrow S \ \leftrightarrow SO_3 \end{array}}}$$

This shows clearly that at the earth's beginning physico-chemical conditions created a habitable zone and then the biosphere itself determined the atmospheric concentration of CO_2 and O_2. The other main atmospheric constituents, N_2 and H_2O, also play a well-defined role in biogeochemistry. Molecular nitrogen is first a gaseous "buffer" because of its chemically stability but it is also the only reservoir for biogenic (and abiogenic) nitrogen fixation (see Chapter 2.4.3.1). It is amazing to see clearly that there is *no* (!) other gaseous compound to limit oxygen (and carbon dioxide) within the concentration thresholds required to maintain a habitable earth, or in other words, not to burn biomass due to high oxygen levels or to shift the atmosphere into a hot greenhouse. The reservoir distribution of water (see Chapter 2.5.2) is mainly determined by the temperature of the climate system; the range of changes covers extremes from a frozen ocean to a steamy atmosphere.

There is an urgent need to understand the long-term response of "life" (or the biosphere in a narrow sense) to human disturbances with climatically relevant substances and possible geologically induced catastrophic events. We need to address the following questions: (a) does the biosphere (in the wider sense) have capacities for buffering and "repairing" disruptions? (b) how long are the transient times? and (c) how can humans adapt to the changes? Here is one (albeit simple) example to illustrate it. In the 1970s it was identified that the stratospheric ozone layer is disturbed by (among others) persistent halogen organic compounds (although understanding of the phenomenon came later). With the "Montreal Protocol" in 1992 it was internationally agreed to stop further use of so-called *ozone depleting substances* (ODS) from 1996. Now, after fifteen years, observations show clearly the reduction of such substances that have a relatively short lifetime. Models show that the recovery of the ozone layer back to the 1970s level is expected in 30 to 50 years. Hence, there is no further need (if the control and non-use of ODS remain assured) for action, we can just "sit out" this self-solving problem.

2.5 The hydrosphere and the global water cycle

Water is the most abundant compound in the climate system (Tables 2.9 and 2.22). It is the only substance that can exist simultaneously in the earth's climate system in all three states of matter; gaseous (vapor), liquid (natural waters) and solid (ice and snow). As intensively discussed in Chapters 2.1 and 2.2, water together with hydrocarbons is among the most abundant compounds in space and was delivered to the early earth through asteroids and other celestial bodies. The term "water" is used in two senses. First, *pure* water is a chemical substance (Chapters 2.5.1 and 5.3.4) and we call the discipline that studies it *water chemistry* (or chemistry of water) which deals with the chemical and physical properties of water as molecule, liquid and ice. Second, *natural* water(s) is a solution; the discipline that studies it is called hydrochemistry (sometimes hydrological chemistry), chemical hydrology, and aquatic chemistry. Water in nature is always a solution, mostly being a diluted system, and in equilibrium or non-equilibrium but in exchange with the surrounding medium, solids (soils, sediments, rocks, vegetation) and gases (atmospheric and soil air). The fundamentals of these physical and chemical processes will be described in more detail in Chapter 4. The aim of this section is to present the key features of water in the climate system, its properties, occurrence, distribution and cycling.

Table 2.22 Global water reservoirs (in 10^{18} g) and fluxes (in 10^{18} g yr^{-1}) on earth; adapted from A: Berner and Berner (1987), B: Berner and Berner (1996), C: Schlesinger (1997), D: Trenberth et al. (2007).

reservoirs	A	B	C	D
oceans	1370 000	1400 000	1350 000	1335 040
ice caps and glaciers	29 000	43 400	33 000	26 350
groundwater	9 500	15 300	15 300	15 300
lakes	125	125	–	178
rivers	1.7	1.7	–	
soil moisture	65	65	122	122
permafrost	–	–	–	22
atmosphere total	13	15.5	13	12.7
biosphere (plants)	0.6	2	–	–
total				
fluxes				
evaporation ocean	423	434	425	413
precipitation ocean	386	398	385	373
evaporation land	73	71	71	73
precipitation land	110	107	111	113
atmospheric transport[a]	37	36	40	40
river run-off	37	36	40	40

[a] from marine to continental atmosphere (equal to river run-off)

The unique chemical and physical properties of water mean that it plays the key role in the climate system:

- solvent (for compounds essential for life, but also pollutants, dissolved and non-dissolved),
- chemical agent (for photosynthesis in plants and photochemical oxidant production in air),
- reaction medium (aqueous-phase chemistry),
- transport medium (in the geosphere, for example, oceanic circulation, rivers, clouds, and in the biosphere in plants, animals and humans),
- energy carrier (latent heat: evaporation and condensation; potential energy in currents and falling waters),
- geological force (weathering, ice erosion, volcanic eruptions).

Natural water is an aqueous solution of gases, ions and molecules, but it also contains undissolved, suspended and/or colloidal inorganic particles of different size and chemical composition and biogenic living and/or dead matter such as cells, plants, and animals etc., sometimes termed *hydrosol*. Water occurs in the climate system in different forms:

- liquid bulk water (natural waters): in rivers, lakes, wetlands, oceans, and groundwater (held in aquifers),
- humidity (soil water): adsorbed onto soil particles,

- liquid droplet water: in clouds, fog, rain, but also as dew on surfaces,
- ice-particulate water in the atmosphere: snow, hail, grains,
- water vapor (humidity) in the atmosphere (one gaseous component among many other components of air),
- hydrates: chemically bonding water molecules onto minerals,
- clathrate hydrates: crystalline water-based solids physically resembling ice, inside which small non-polar gas molecules are trapped (existing under high pressure in the deep ocean floor),
- bulk ice: snow cover, glaciers, icebergs.

Water and the hydrosphere are practically synonymous, but not completely so. The hydrosphere is the sum total of water on earth, except for that portion in the atmosphere. The hydrosphere combines all water underground, which constitutes the vast majority of water on the planet, as well as all freshwater in streams, rivers, and lakes; saltwater in seas and oceans; and frozen water in icebergs, glaciers, and other forms of ice.

2.5.1 Water: Physical and chemical properties

Water is unusual in *all* its physical and chemical properties (Table 2.23). Its boiling point (abnormally high), its density changes (maximum density at 4 °C, not at freezing point), its heat capacity (highest of any liquid except ammonia), and the high dielectric constant as well as the measurable ionic dissociation equilibrium, for example, are not what one would expect by comparison of water with other similar substances (hydrides). All the physical and chemical properties of water make our climate system unique and have shaped the course of chemical evolution. Water is the medium in which the first cell arose, and the solvent in which most biochemical transformations take place. *Thales of Miletus*, who introduced the primacy of water (see Chapter 1.3.1), suggested that in the beginning there was only water, somehow everything was made of it and is now composed of water, though perhaps not in the typical form of water (for example, air is a gaseous form of water and rocks are a solid form of water). In the changes that occur, water is not created or destroyed − it just changes its properties (Lloyd 1970).

2.5.1.1 Water structure: Hydrogen bond

Under certain conditions, an atom of hydrogen is attracted by rather strong forces to two atoms instead of only one, so that it may be considered to be acting as a bond between them. This is called the *hydrogen bond*. This statement is from *Linus Pauling* in his book, *The Nature of the Chemical Bond* (1939). At that time, the hydrogen bond was recognized as mainly ionic in nature. The energy associated with the hydrogen bond is about 20 kJ mol^{-1}. Due to hydrogen bonding, water molecules form dimers, trimers, polymers, and clusters. The hydrogen bonds are not necessarily linear (Fig. 2.32). The ion mobility of H_3O^+ and OH^- are anomalously high: $350 \cdot 10^{-4}$ and $192 \cdot 10^{-4}$ cm^2 V^{-1} s^{-1} (25 °C) in comparison with

Table 2.23 Physical and chemical properties of water; data from Höll (2002), Franks (2000), Forsythe (2003).

property	value	dimension
molar mass	18.015268	g mol^{-1}
freezing point at 1 bar	1.00	°C
boiling point at 1 bar	100.0	°C
vapor pressure at 25 °C	3.165	kPa
latent heat of melting at 1 bar	332.5	J g^{-1}
latent heat of evaporation at 1 bar	2257	J g^{-1}
specific heat capacity of water	4187	J g^{-1} K^{-1}
specific heat capacity of ice	2108	J g K^{-1}
specific heat capacity of water vapour	1996	J g K^{-1}
critical temperature at 1 bar	647.096	K
critical pressure	220.64	bar
critical density	322	g L^{-1}
maximum density (at 3.98 °C)	1.0000	g cm^{-3}
density of water at 25 °C	0.99701	g cm^{-3}
density of ice at melting point (0 °C)	0.91672	g cm^{-3}
density of gas at boiling point (100 °C)	0.0005976	g cm^{-3}
viscosity, dynamic	0.8903	cP^a
viscosity, kinematic	0.008935	stokes^b
surface tension of water at 25 °C	72	dyn cm^{-1}
dielectric constant at 25 °C	78.39	–
Prandtl number at 25 °C	6.1	–
cryoscopic constant	1.8597	K kg mol^{-1}
O—H bond dissociation energy	492.2148	kJ mol^{-1}
bond energy, average at 0 K (H—O—H → O + 2 H)	458.9	kJ mol^{-1}
conductivity, electrolytic, at 25 °C	0.05501	µS cm^{-1}
conductivity, thermal, for water at 25 °C	0.610	W m^{-1} K^{-1}
conductivity, thermal, for at ice −20 °C	2.4	W m^{-1} K^{-1}
conductivity, thermal, for vapour at 100 °C	0.025	W m^{-1} K^{-1}
electron affinity at 25 °C	−16	kJ mol^{-1}
energy, internal (U) for water at 25 °C	1.8883	kJ mol^{-1}
enthalpy of formation, ΔH_f, at 25 °C	−285.85	kJ mol^{-1}
enthalpy ($H = U + PV$), at 25 °C	1.8909	kJ mol^{-1}
enthalpy of vaporization (liquid), at 0 °C	45.051	kJ mol^{-1}
enthalpy of sublimation (ice Ih), at 0 °C	51.059	kJ mol^{-1}
Gibbs energy of formation^c, ΔG_f, at 25 °C	−237.18	kJ mol^{-1}
surface enthalpy (surface energy) at 25 °C	0.1179	J m^{-2}
surface entropy (= $-d\gamma/dT$) at 25 °C	0.0001542	J m^{-2} K^{-1}
ionic dissociation constant, [H$^+$][OH$^-$]/[H$_2$O], 25 °C	$1.821 \cdot 10^{-16}$	mol L^{-1}
O—H bond length (liquid, ab initio)	0.991	Å
H—O—H bond angle (liquid, ab initio)	105.5	°
redox potential E_0: water oxidation^d	1.229	V
redox potential E_0: water reduction^e	−0.8277	V

^{a} centipoise (= 0.008903 g cm^{-1} s^{-1})
^{b} (= $0.8935 \cdot 10^{-6}$ m^2 s^{-1})
^{c} = chemical potential (μ)
^{d} 2 H$_2$O → O$_2$(g) + 4 H$^+$ + 4 e$^-$
^{e} 2 H$_2$O + 2 e$^-$ → H$_2$(g) + 2 OH$^-$

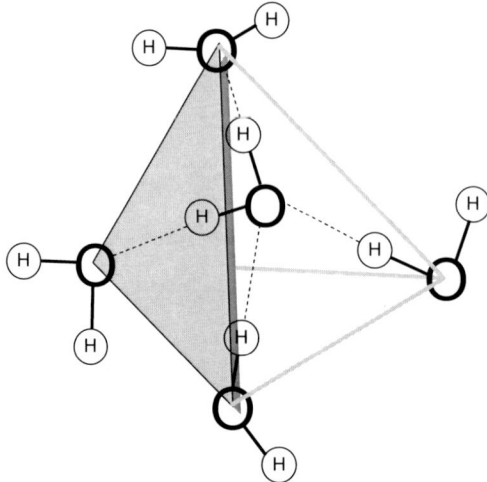

Fig. 2.32 Water structure and hydrogen bonds (dotted lines) in a tetrahedral grid; note that the structure is not in a plane.

$(50-75) \cdot 10^{-4}$ cm^2 V^{-1} s^{-1} for most other ions. Chapters 4.2.5.2 and 5.3.4 deal with the chemistry of the proton (H_3O^+) and aquated electron (H_2O^-), both fundamental species in nature (see Chapter 2.2.2.3). The mobility is a result of the special structure of liquid water where the H_2O molecules are linked in chains, stabilized by hydrogen bonding, which also give liquid water great internal cohesion (Fig. 2.32). Fig. 2.33 (in the scheme the tetrahedral structure shown in Fig. 2.32 remained but is not clearly seen) illustrates that the proton (H^+) is not really transported but only the charge − similarly to the OH^- transport − and thus explaining the high ionic mobility. The water grid (quasi-crystalline) is restructured. Therefore, the high evaporation heat and entropy, the high surface tension and relatively high viscosity are a result of the hydrogen bonding structure (Fig. 2.32). Bernal and Fowler (1933) first proposed a water structure model in the sense of a "quasi-crystalline structure". Nowadays many models of the structure of liquid water have been proposed (e. g. Rick 2004) but there is no consensus even on the number of H_2O molecules forming species ("polymers").

When water freezes, the crystalline structure is maintained (Fig. 2.32) and determined by the prevailing condition; at least nine separable ice structures exist. Normally, ice has a hexagonal structure (E_n) when cooling down liquid water; each O atom is surrounded by a regular tetrahedron of a further four O atoms. The positioning of H is very complex. The four hydrogen bonds around an oxygen atom form a tetrahedron in a fashion found in the two types of diamonds. Thus, ice, diamond, and close packing of spheres are somewhat topologically related. Water ice is unusual because its density is less than that of the liquid water with which it is in equilibrium. This is an important property for the survival of life in water. When the ice melts, a few hydrogen bridges (probably every fourth one) begin to break, the H_2O molecules close ranks, and the density consequently increases.

Many salts crystallized from aqueous solutions are not water-free but take the form of well-defined hydrates. Other solid phases contain water associated in

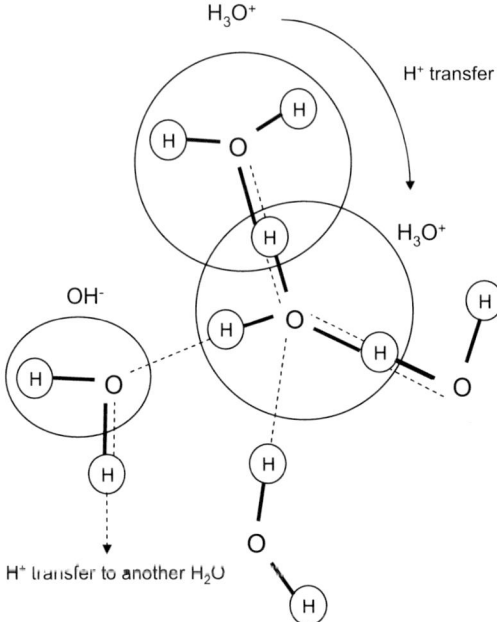

Fig. 2.33 Formation and transport of hydronium (H_3O^+) and hydroxyl (OH^-) ions in the water grid (as shown in Fig. 2.32) by charge transfer, i. e. changing hydrogen bonding (dotted line) from strong covalent to weak interaction.

changing amounts. A classic case is water coordinated onto oxoanions, for example, $CaSO_4 \cdot 2\,H_2O$. The most frequent are cation complexes with water, for example, alums like $[Al\,(OH_2)_6]^{3+}$; this is an explanation for the large water content in rocks. Framework silicates (e. g. zeolites) hold huge amounts of water within their cavities; faujasite is the mineral with the highest water content: $Na_2Ca\,[Al_2Si_4O_{12}]_2 \cdot 16\,H_2O$ (Strunz 1955). Similar to the zeolitic grids, water $(H_2O)_n$ can build up cage-like inclusion structures (clathrate hydrates) where, in a skeleton of 46 H_2O, there exist six cavities of the same size and two more, smaller ones. The guest molecules in high-pressure clathrates are Ar, Kr, CH_4, and H_2S.

Polywater (also called anomalous water), which was first described in the 1960s in the Soviet Union and controversially discussed in the 1970s, does not exist, however, and was probably a mixture of colloidal silicic acid.

2.5.1.2 Water as solvent

The polarity of water gives it important properties that the biosphere needs to function: it is a universal solvent and it adheres and is cohesive. Thus, water facilitates chemical reactions and serves as a transport medium. Even in a covalent bond, atoms may not share electrons equally. In H_2O, the unequal electron sharing creates two electric dipoles along each of the O—H bonds. The H—O—H bond angle is 104.5°, 5° less than the bond angle of a perfect tetrahedron which is 109.5°.

The structure shown in Fig. 2.32 is idealized, only in ice is it fixed (crystalline), but in liquid water at any moment, depending on the temperature, each water molecule forms hydrogen bonds with an average of 3.4 other water molecules. They are in continuous motion in the liquid state, hence hydrogen bonds are constantly and swiftly being broken and formed (Fig. 2.33). The protolytic equilibrium is described in more detail in Chapter 4.2.5.3. Hydrogen bonding is unique for water. The bonds readily form between an electronegative atom (usually oxygen, nitrogen or sulfur) and a hydrogen atom covalently bonded to another electronegative atom in the same or another molecule: $H^{\oplus}{-}O^{\ominus}{-}H^{\oplus}{-}{-}{-}O^{\ominus}{-}$ and $H^{\oplus}{-}O^{\ominus}{-}H^{\oplus}{-}{-}{-}N{-}$. However, hydrogen atoms covalently bonded to carbon atoms (which are not electronegative) do not participate in hydrogen bonding; hence hydrocarbons are insoluble in water. But organic compounds with oxygen (and nitrogen) containing functional groups (like alcohols, aldehydes, acids, ketones etc.) are water-soluble. The more oxygen groups and the less carbon atoms in a compound, the more soluble it is in water. It is the polarity *and* the hydrogen bond affinity that makes water a solvent for many chemically different substances: oxygenated and/or nitrogen-containing organic compounds (most biomolecules, which are generally charged or polar compounds), salts (electrostatic interacting solid grids), but also non-polar gases (biologically important CO_2, O_2, and N_2) and all polar gases (for example, SO_2, NH_3, HCl, HNO_3 which are important to the atmosphere). It needs no further explanation that the solubility of non-polar molecules is much less than that of polar substances. The property to interact with water is also called hydrophilicity (affinity to water: attraction) and inversely hydrophobia (non-affinity to water: repulsion).

2.5.1.3 Water properties in relation to the climate system

We would normally expect that water − when comparing H_2O with H_2S, for example − would boil at −80 °C instead of 100 °C. This large difference is due to the relatively strong hydrogen bond (weak van der Waals interaction is in the order of only 4 kJ mol^{-1}) which must be broken before separating H_2O molecules from the bulk liquid. This is expressed by the heat of vaporization (Table 2.23), which is one of the highest of all liquids (only H_2O_2 is higher). This effect produces a liquid over most of the earth's surface (between 0 °C and 100 °C). Another property following from the structure of water is the large heat capacity (see Chapter 4.1.3). Water can absorb a lot of heat for a small change of temperature. The temperatures of large natural water bodies are relatively constant and they act as thermal buffers in the climate system. Without the large heat of vaporization, soils (as seen in dry areas) would be overheated, but the solar heat is taken for water evaporation, cooling the surface. The same amount of heat is released when the vapor condenses in the atmosphere to form cloud droplets. This causes the transport of heat from one site of the earth to another (see also Fig. 2.17). Without appreciable changing of the surface temperature, around 25 % of the solar radiation (and 50 % of the solar transmission) is dissipated in the atmosphere. It needs no detailed explanation that this process of water evaporation and condensation forces the water cycle and makes our weather (and finally our climate).

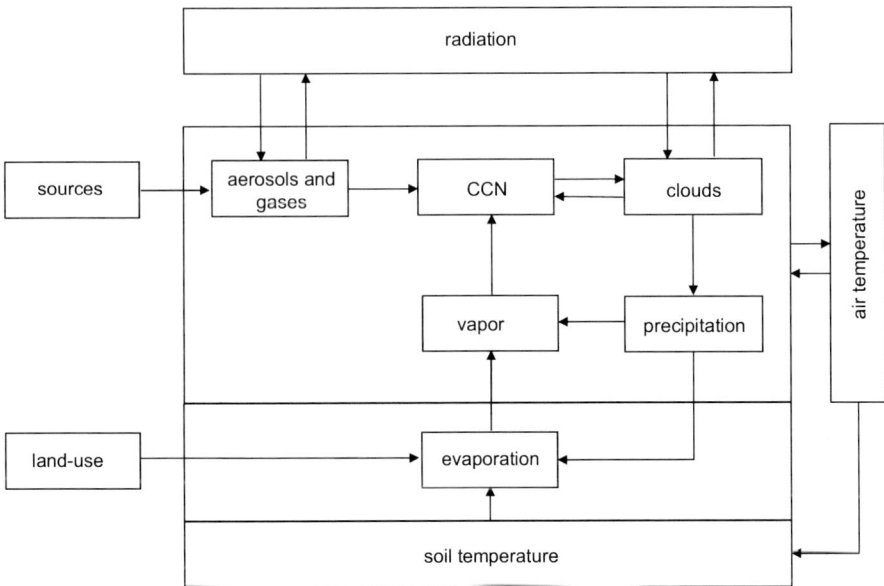

Fig. 2.34 Scheme of interaction between the water cycle and changes in the climate system.

The high surface tension of water plays a major role in the formation of drops in clouds and fog. Surface tension is related to the cohesive properties of water, or in other words, water is attracted to other water. This property results in the growth of cloud droplets into bigger drops after collision. Water can also be attracted to other materials. This is called adhesion, a property of water which leads to capillary action, an important transport process for plants (besides transpiration).

Liquid water is fairly transparent to visible light, which allows photosynthesis to a considerable depth in large water bodies (up to more than 100 m in clear water). Water vapor (H_2O molecules) in the atmosphere absorbs strongly in the infrared region, which is important to the heat budget of the earth (see Chapter 2.3.3). The resulting "greenhouse" effect of atmospheric water is, by far, the strongest determinant of the earth's surface climate. Naturally occurring greenhouse gases have a mean warming effect of about 33 °C and water causes about 36–70 % of it. The global mean surface temperature (based on long-term near surface measurements) is 15 °C (for land surfaces it is 8.5 °C and for sea surfaces it is 16.1 °C). Without water in the atmosphere, the global mean temperature would be definitively below the freezing point of water on land, which would be ice-covered and would not provide conditions for life. Large parts, at least, of the ocean (and likely globally) would also be frozen.

Furthermore, atmospheric humidity is highly variable and responds to changes in atmospheric temperature (see Fig. 2.37), thus providing the most important feedback mechanism tending to amplify global climate changes induced by other factors (Fig. 2.34). Furthermore, clouds contribute about 50 % of planetary albedo (Fig. 2.13), and absorption of terrestrial radiation by clouds is equivalent to that of all "greenhouse" gases other than water vapor. Water vapor in the upper troposphere and lower stratosphere is very critical in determining the rate at which radiative energy emitted by the atmosphere escapes to space. Being able to measure and to forecast the evolution of the spatial and temporal patterns in water vapor and clouds is a key to understanding the climate system, namely the water resources and ecosystem problems.

2.5.2 Hydrological cycle and the climate system

In ancient Greek, ἄτμις (atmis: water vapors) denotes the transfer of water (by evaporation) from the telluric form (hydrosphere) into ἀηρ (aer), the water vapor of the atmosphere and its return as precipitation to the earth, (with water) one of the two lower elements. We know from *Herodotus* that in the fifth century B.C. this theory was known and accepted and described by *Hippocrates*.

The long-term mean values for hydrological reservoirs and fluxes (as depicted in Table 2.22 and Fig. 2.35) are uncertain. Various versions have been published and tracing some of the references indicates a cascade whereby one source cites another that in turn cites another and the original value is often not very certain (Trenberth et al. 2007); many of these values in turn come from Baumgartner and Reichel (1975). Trenberth et al. (2007) used values from the Global Precipitation Climatology Project (GPCP) from 1988 to 2004, which result in an annual mean global precipitation of $(489.9 \pm 2.9) \cdot 10^3 \, km^3$, separated into a global mean ocean precipitation $(372.8 \pm 2.7) \cdot 10^3 \, km^3$ and $(112.6 \pm 1.4) \cdot 10^3 \, km^3$ over land.

The water or hydrological cycle is the continuous circulation of water throughout the earth and between its systems. At various stages, water moves through the atmosphere, the biosphere, and the geosphere, in each case performing functions

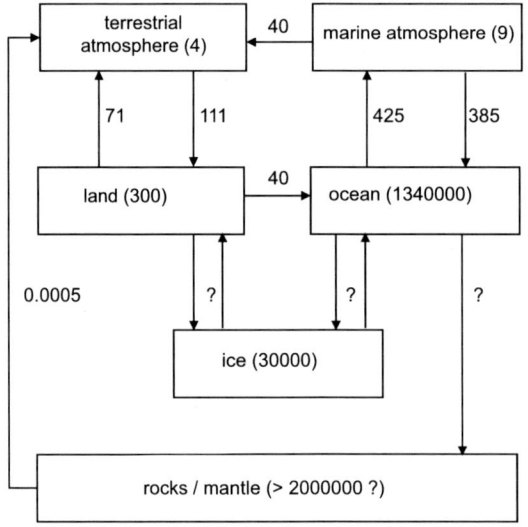

Fig. 2.35 Water reservoirs and the hydrological cycle; reservoir concentrations in 10^{18} g and fluxes in 10^{18} g yr^{-1}. Data from Table 2.22.

essential to the survival of the planet and its life-forms. Thus, over time, water evaporates from the oceans; then falls as precipitation; is absorbed by the land; and, after some period of time, makes its way back to the oceans to begin the cycle again. The total amount of water on the earth has not changed in many billions of years, though the distribution of water has changed. Compared with other biogeo-chemical cycles, water provides the largest turnover of material on earth (Fig. 2.35).

The water cycle is driven by processes that force the movement of water from one reservoir to another. Evaporation from the oceans and land is the primary source of atmospheric water vapor (Fig. 2.36). Water vapor is transported, often over long distances (which characterize the type of air masses), and eventually condenses into cloud droplets, which in turn develop into precipitation. Globally, there is as much water precipitated as is evaporated, but over land precipitation exceeds evaporation and over oceans evaporation exceeds precipitation (Fig. 2.35). The excess precipitation over land equals the flow of surface and groundwater from continents to the oceans. Flowing water also erodes, transports and deposits sediments in rivers, lakes and oceans, affecting the quality of water.

This natural cycling of water is now perturbed by human activities. Together with changing vegetation patterns due to land-use management (see Chapter 2.6.5.3),

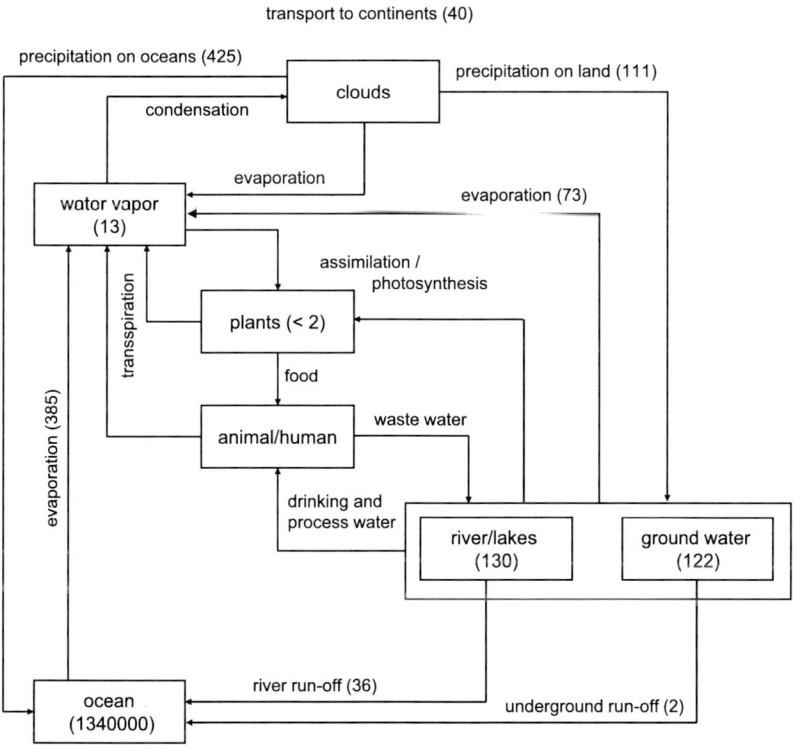

Fig. 2.36 The global water cycle; reservoir concentrations in 10^{18} g and fluxes in 10^3 km^3 yr^{-1}. Data from Table 2.22.

these factors complicate the prediction of the consequences of climate change on the global water cycle.

The global water cycle includes sub-cycles, such as the cloud processing (condensation onto CCN and evaporation with forming new CCN), the plant water cycle (assimilation and transpiration), and the human water cycle (water processing for drinking and processing as well as waste water) (see Fig. 2.36). As water cycles through the climate system, it interacts strongly with other biogeochemical cycles, notably the cycles of carbon, nitrogen, and other nutrients. These linkages directly affect water quality and the availability of potable water and industrial water supplies. It is estimated that 70–80 % of worldwide water use is for irrigation in agriculture, while 15–20 % of worldwide water use is industrial. The remainder of worldwide water use is for household purposes (drinking, bathing, cooking, sanitation, and gardening).

The already mentioned feedback between the water and carbon cycle (see also Fig. 2.31) is not only given by the essential role of water in sustaining all forms of life. Observational evidence indicates that transpiration rates of plants are high at the same time as CO_2 fixing by the plants, and hence CO_2 flux from the atmosphere to the plant canopy is large. When an environment is humid, plants grow more rapidly, draw more CO_2 from the atmosphere, and release more water to the atmosphere.

Seasonal-to-interannual variability in the global water cycle is largely determined by ocean and land processes and their impact on the atmosphere. The prediction of this variability relies on the persistence in surface conditions that tend to provide the atmosphere with consistent anomalies in fluxes over periods of weeks or months or even years. The time-scale of "memory" in the atmosphere is fairly short, but due to the atmosphere's connection to the land and ocean, each of which is characterized by a much longer memory, the climate system becomes complex in time and space; for example the atmospheric residence time of trace substances (see Chapter 4.5) is not constant but a function of time and space. Current dynamic climate models on regional and global scales demonstrate limited ability to predict precipitation on time scales beyond a few days. It is obvious that a better understanding of the variability will have important consequences for agricultural and water resource management. The El Niño and La Niña cycles are the most obvious examples of a coupled phenomenon that produces significant seasonal-to-interannual variability.

As mentioned in Chapter 2.5.1.3, water has an important influence on atmospheric circulation and plays a key role in the maintenance of the climate system as a moderator of the earth's energy cycle. Water cools its surroundings as liquid and ice are converted into water vapor. On the average, this latent cooling is balanced by the latent heat released when the vapor is condensed into clouds. Water is a very effective means of storing and transporting energy in the atmosphere. In general, latent heat is the principal source of energy that drives cyclogenesis and sustains weather systems, like the convective cells that generate tornadoes and tropical storms that evolve into hurricanes. The different types of clouds in the atmosphere are linked to the climate system by a multitude of dynamic and thermodynamic processes, including numerous feedback mechanisms. In the present-day climate, on average, clouds cool our planet, the net cloud radiative forcing at the top of the atmosphere is about -20 W m^{-2}. One of the most interesting questions concerning clouds is: how will they respond to a change in climate? A slight change in cloud amount or a shift in the vertical distribution of clouds might have a considerable impact on the energy budget of the earth. IPCC states clearly that cloud processes and related feedbacks are among the physical processes leading to large uncertainties in the prediction of future climate. The main reason for this is that many microphysical and dynamic processes controlling the life cycle and radiative properties of clouds are not adequately implemented in global climate models. The interaction of aerosols and clouds and the resulting radiative forcing (indirect and semidirect aerosol effect) is one of the major fields of active cloud research at present.

Therefore, the hydrological cycle is not only a cycle of water; it is a *cycle of energy* as well.

2.5.3 Atmospheric water

Atmospheric water includes physical water in all aggregate states, i. e. as gaseous, liquid (in droplet form) and solid (ice particles). The historic term "atmospheric waters" has the meaning of *hydrometeors* in current terminology, i. c., meteoric water. For historical reasons, *dew* has been considered to belong these "waters", as it was before Wells (1814) stated that dew is not from water drops fallen from the heaven. The *phenomena*, fog and clouds, precipitation (rain, snow, hail) as well as dew, have been well described since Antiquity. A *phenomenological* understanding of physical (but not chemical) processes associated with hydrometeors was complete only by the late nineteenth century. Today we understand the physics and chemistry in the chain aerosol-cloud-precipitation and in relation to the climate relatively well. However, it seems that because of the huge complexity a *mathematical* description (i. e., parameterization for chemistry and climate modelling) is under continuous development.

Clouds and precipitation are not only the atmospheric link in the global water cycle but also an important reservoir for chemical processing and the transportation of trace substances (see Fig. 2.38). Naturally, clouds are far above the surface (with the exception of fog) and thus not easy to study even nowadays. Precipitation (rain, snow, hail), however, has always been easy to observe by human sensors (seeing, feeling, smelling and tasting) and to collect for volume estimation and analysis. Precipitation was probably long considered a climatic precondition for survival by early humans, but, with extreme events, it could also be catastrophic for housing as well as farming. The mixing of air and water with pollutants (accurately referred to in old terminology as "foreign bodies") was known since *Aristotle*; the role of precipitation in cleansing the surroundings was wonderfully described by John Evelyn (see Chapter 1.3.4). *Aristotle* asked in his *Meteorologica*: "Since water is generated from air, and air from water, why are clouds not formed in the upper air?" He explained this as follows (Aristotle 1923):

> But when the heat which was raising it leaves it, in part dispersing to the higher region, in part quenched through rising so far into the upper air, then the vapour cools because its heat is gone and because the place is cold, and condenses again

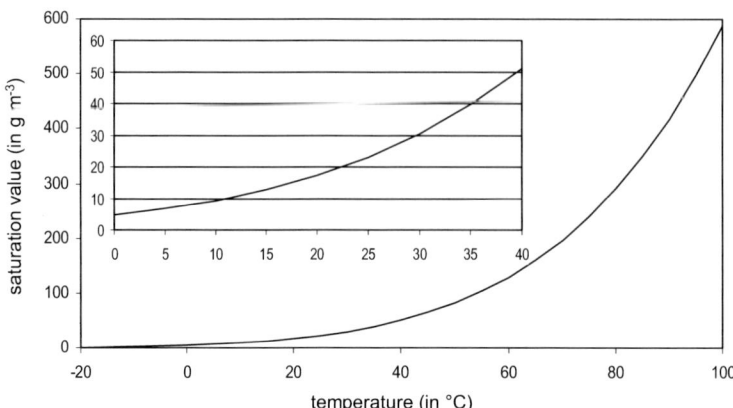

Fig. 2.37 Saturation value of the absolute humidity (in g m^{-3}) over water in dependence from the temperature. Data from Sonntag and Heinze (1981).

and turns from air into water. And after the water has formed it falls down again to the earth. The exhalation of water is vapour: air condensing into water is cloud. Mist is what is left over when a cloud condenses into water, and is therefore rather a sign of fine weather than of rain; for mist might be called a barren cloud. So we get a circular process that follows the course of the sun... From the latter [clouds] there fall three bodies condensed by cold, namely rain, snow, hail... When the water falls in small drops it is called a drizzle; when the drops are larger it is rain ... When this [vapour] cools and descends at night it is called dew and hoar-frost."

Only a very small percentage of all the water in the climate system is actually present in the atmosphere (Table 2.12). Of the atmospheric water, most is in the vapor phase (Fig. 2.37); the liquid water content (LWC) of clouds is only in the order 1 g m^{-3}, the cloud ice water content (IWC) still less, down to 0.0001 g m^{-3}. But clouds play a huge role in the climate system, whereas precipitation closes the cycle for water and also for substances dissolved in it (wet deposition). Some of the processes (droplet formation, transfer processes, deposition, and chemistry) will be described later. The aim of this chapter is to describe the phenomenology of water in the atmosphere so far as we need it for an understanding of chemical processes.

2.5.3.1 Water vapor

John Tyndall wrote the following very clear phrase (Strachan 1866, p. 123):

Aqueous water is always diffused through the atmosphere. The clearest day is not exempt from it; indeed, in the Alps, the purest skies are often the most treacherous, the blue deepening with the amount of aqueous vapour in the air. Aqueous vapour is not visible; it is not fog; it is not cloud, it is not mist of any kind. These are formed of vapour which has been condensed to water; but the true vapour is an impalpable transparent gas. It is diffused everywhere throughout the atmosphere, though in very different proportion.

The content of water vapor in air varies from nearly zero up to about 4 Vol-%, depending on temperature and saturation (Fig. 2.37). Normally the chemical composition of air is based on dry air (cf. Table 1.1). Under normal conditions (20 °C and 60 % RH), air contains around 1 % water vapor (absolute humidity).

The density of water vapor is less than that of other gaseous air constituents ($N_2 + O_2$); hence wet air at the same temperature and pressure has a lower density than dry air. Consequently, at same pressure dry air has a somewhat higher temperature (called *virtual* temperature) than wet air to obtain the same density. This follows from the ideal gas equation

$$pV = n\mathbf{R}T \quad \text{or} \quad p = \varrho \frac{\mathbf{R}T}{M}$$ (2.86)

where p − pressure of air, V − volume of air, n − number of moles, $\mathbf{R}$ universal gas constant, and T − temperature of air, ϱ − density of air (m/V); all mean values of air. There are *intensive* quantities (p, T, and ϱ) and *extensive* quantities (n, m, V, energy). Extensive quantities depend on the system size or the amount of material in the system; intensive do not. If one cubic meter of air is divided then the intensive quantities will not change but the extensive properties will. Many derived ratios from two extensive quantities (for example, density, chemical potential, specific fluxes) are intensive quantities. At this stage we will ignore the fact that air is a mixture with particulate matter in the form of droplets and solids and we will only consider the gases. It is important to note that air is not a mixture of different gases but a solution, because all gases can form solutions with each other in all ratios. The specific properties of individual gases will not change; therefore the properties of air can be calculated from those of the constituents according to the dilution rule, i. e., the property E of the solution (e. g. pressure, mass) can be calculated from the part x_i of the specific gases i, called *mixing ratio* (based on mass or volume, respectively) contributing to air:

$$E = \sum x_i E_i$$ (2.87)

We may consider p, V, and n as sums of the so-called *partial* quantities:

$$n = \sum n_i, \quad V = \sum V_i \quad \text{and} \quad p = \sum p_i$$ (2.88)

The partial volume V_i is that volume which is occupied from the amount (expressed as mass or mole) of the gas i under the same pressure as the air. The partial pressure is that pressure which would have the amount of the gas i in the same volume of air. The mole number is given by $n_i = m_i/M_i$ (m_i − mass and M_i − molar mass of the substance i) and the density by $\varrho_i = m_i/V$. Without significant error we calculated the mean molar mass of dry air only from the main constituents N_2, O_2, and Ar to be:

$$M_{dry\ air} = 0.78\ M_{N_2} + 0.21\ M_{O_2} + 0.01\ M_{Ar}$$
$$= 28.96 \text{ (more exactly 28.9644)}$$ (2.89)

The mean molar mass of wet air (x – mixing ratio of water vapor) is given by

$$M_{wet\ air} = (1 - x) \cdot 28.96 + x \cdot 18.0 \qquad (2.90)$$

Under the same conditions (pressure and density) it follows from Eq. (2.86)

$$\frac{p}{R\varrho} = \frac{T_v}{(1 - x) \cdot 28.96 + x \cdot 18} \qquad (2.91)$$

Thus with increasing humidity x the virtual temperature T_v increases with the condition of constant density and pressure (no changing quotient). In other words, at same T and p the density becomes smaller or at the same T and ϱ the pressure increases. It follows that $T/T_v = 1 - 0.378\ x$.

The water vapor in air is a result of vaporization of water from the earth's surface. We can consider liquid water to be condensed gas. At any given time, a certain number of molecules can escape the liquid from the surface to the surrounding air (we call it *evaporation*). Because of air motion (turbulent mixing and advection) there is no equilibrium, i.e., transfer of water molecules from the air back to the surface (we call it *condensation*) in the same flux as evaporation. Such equilibrium can only be reached in a closed undisturbed chamber. Hence the vapor pressure dependency shown in Fig. 2.37 is theoretical and cannot be directly applied to the atmosphere. If the equilibrium between condensed and vaporous water is reached, the pressure is called *saturation pressure* p^∞. Such conditions are important for cloud formation but are also frequently observed in the tropics. The *relative humidity RH* is the ratio of the vapor pressure at temperature T to the saturation vapor pressure at the same temperature expressed as a percentage.

$$RH = \frac{100 \cdot p_{H_2O}}{p^\infty(T)} \qquad (2.92)$$

The water vapor pressure p_{H_2O} is numerically identical with the mixing ratio x_{H_2O}. The *absolute humidity* (or water vapor concentration or density) is the mass of water vapor in volume of air at a given temperature: p_{H_2O}/V. From Eq. (2.86) it follows

$$\varrho_{H_2O} = p_{H_2O}\ \frac{18}{RT} = p^\infty(T)\ \frac{RH}{RT}\ 0.18 \qquad (2.93)$$

Note that in all the equations the saturation vapor pressure is measured above a plane that is an "endless" water surface. Another important quantity is the *dew point*, the temperature when, at a given absolute humidity, the saturation vapor pressure is reached, i.e., the relative humidity becomes 100%. Under these conditions, water vapor starts to condense – when surfaces (in air, CCN) are available.

2.5.3.2 Clouds

Clouds are the intermediate in the water cycle between vaporized water from the surface and precipitation back to the surface. Clouds also have several important functions in air chemistry; transportation, removal as well redistribution of atmos-

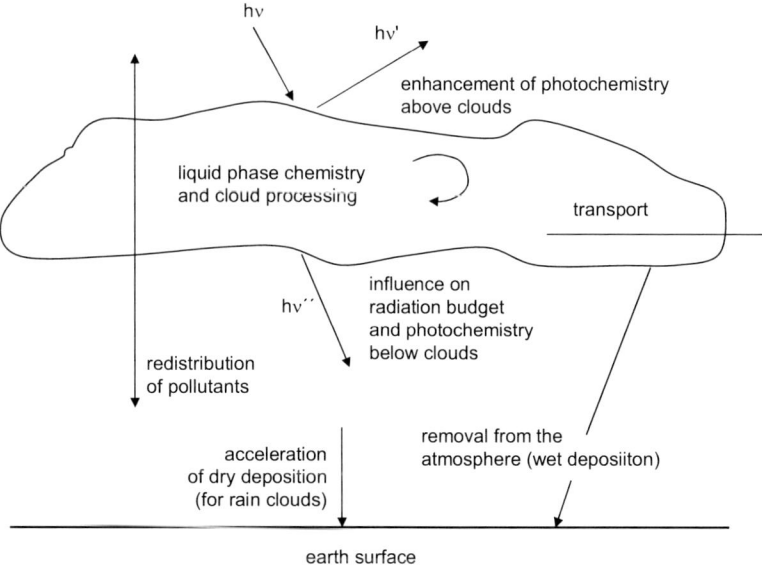

Fig. 2.38 Role of clouds in the climate system: influencing transport and mixing, removal, chemical transformation and radiation budget.

pheric compounds, and the radiation budget (Fig. 2.38). At any given time about 60% of the earth's surface is masked by cloud of one form or another (Summer 1988). Despite its relatively low spatial occupancy of the troposphere, cloud provides an aqueous-phase medium for surface and bulk chemical reactions for unique transformations. Formation of clouds and fog droplets requires two conditions, first the presence of aerosol particles acting as condensation nuclei (CCN) for water vapor condensation (see Chapter 4.3.6) and then a slight supersaturation. As demonstrated in Fig. 2.37, an air parcel with given absolute humidity can increase its relative humidity and finally become supersaturated only by cooling. Air does not, normally, decrease in temperature spontaneously. Uplift of air within the atmosphere causes a drop in temperature. There are three ways to initiate this uplift. Air, becoming hot at the earth's surface, will continue to rise (we call this process *convection*) as long as its temperature remains higher than that of the air surrounding it. Second, air may be forced to rise by barriers such as mountains. Third, air passing over rough surfaces will become *turbulent*, causing an exchange of warmer surface air with cooler air above. These three categories will clearly act at different scales and, most importantly, involve uplift at different rates. The clouds associated with these different scales and types of processes vary in their morphology (Table 2.24).

In determining their ability to nucleate clouds, the chemical composition of aerosol particles is much less important than their size, a result that will clarify aerosol effects on climate (Rosenfeld 2006). This is what is expected from theory because the radius-to-volume ratio determines the molecular transfer from the gas phase, whereas the hygroscopicity (surface characteristics of CCN) determines the uptake coefficient (Chapter 4.3.7.4). The other very important parameter for cloud forma-

Table 2.24 Cloud types (after Summer 1988 and WMO 1992).

category	type	base height (in km)	depth (km)	typical base-temperature (°C)	LWC (g m^{-3})
high	cirrus (CI)	5–13	0.6	−20 to −60	0.05
	cirricumulus (CC)	5–13	0.6	−20 to −60	
	cirrostratus (CS	5–13	0.6	−20 to −60	
medium	altocumulus (AC)	2–7	0.6	+10 to −30	0.1
	altostratus (AS)	2–7	0.6	+10 to −30	0.1
low	nimbostratus (NS)	1–3	2	+10 to −15	0.5
	stratocumulus (SC)	0.5–2		+15 to −5	
	stratus (ST)	0–0. 5	0.5	+20 to −5	0.25
vertical	cumulus (CU)	0.5–2	1	+15 to −5	1.0
	cumulonimbus (CB)	0.5–2	6	+15 to −5	1.5

tion is the CCN number, determining the cloud droplet number (cf. Fig. 2.20). As shown originally by Twomey (1991), and recently reviewed by Lohmann and Feichter (2005), the sensitivity of climate to CCN number density is nonlinear, with the effect being much stronger at low particle numbers.

The chemical composition of the CCN determines the water chemistry of the individual droplet and the ability to absorb gases. In the short period (on average about one hour) of the existence of drops of different sizes and chemical composition, the chemical composition of the interstitial gas phase and that of the droplet phase is changing. Depending on the cloud microphysics and cloud dynamics, the droplets can have one of two fates, either they evaporate or they precipitate. Only in one out of ten cases, on average, does the cloud precipitate. Subsequent evaporation and condensation again until final dissipation back to water vapor is the more frequent process, called *cloud processing*. During that process, the aerosol particles acting as nuclei and coming to residues are generally growing and becoming more water-soluble. The cloud amplifies production of CCN. Recent research suggests that clouds are able to take up water-soluble organic molecules, which are then oxidized and form SOA after evaporation of cloud droplets (Ervens et al. 2008).

For a description of cloud chemical climatology, the most important cloud parameters are (see Table 2.24):

- Radius and size distribution of droplets or ice particles,
- liquid or ice water content (if possible as vertical profile),
- base height and top height of clouds,
- chemical composition of cloud water (if possible size-resolved),
- interstitial gas-phase chemical composition of cloud,
- frequency and duration of cloud events, degree of cloud cover,
- type of cloud and weather situation.

Some results of cloud chemistry monitoring at Mt. Brocken, one of the longest time-series worldwide, will be summerized in Chapter 3.4.3.

Many attempts, although limited because of the experimental difficulties, have been made to study the chemistry in single droplets and in different size fractions. The literature is huge and it is beyond the scope of this book to review the results. Cloud modelling is one of the most important and effective means of investigating cloud processes with sophisticated representations of cloud microphysics, and it can reasonably well resolve the time evolution, structure and life cycles of a single cloud and its systems (Barth et al. 2003).

Ice particles found within polar stratospheric clouds (PSCs) and upper tropospheric cirrus clouds can dramatically impact the chemistry and climate of the earth's atmosphere. The formation of PSCs and the subsequent chemical reactions that occur on their surfaces are key components of the massive ozone hole observed each spring over Antarctica. Cirrus clouds also provide surfaces for heterogeneous reactions and significantly modify the earth's climate by changing the visible and infrared radiation fluxes. Although the role of ice particles in climate and chemistry is well recognized, the exact mechanisms of cloud formation are still unknown, and thus it is difficult to predict how anthropogenic activities will change cloud abundances in the future.

Beside the role of clouds in atmospheric chemistry and the water cycle, probably their most important impact on the climate system is through influencing solar and terrestrial radiation. Counteracting the role of "greenhouse" gases in absorbing terrestrial radiation, clouds act in scattering and reflecting incoming solar radiation. For details on cloud, aerosol and climate interactions, see Hobbs (1993), Crutzen and Ramanathan (1996), Heintzenberg and Charlson (2009), and for details on cloud (and precipitation) physics, see Summer (1988), Pruppacher and Klett (1997), Strangeways (2007).

2.5.3.3 Haze, mist and fog

The atmosphere becomes "milky" through light scattering on particles in the size range between 50 and 200 nm. WMO uses for atmospheric obscuration the following terms: Fog, Ice fog, Steam fog, Mist, Haze, Smoke, Volcanic ash, Dust, Sand, and Snow. With the exception of smoke, volcanic ash, dust and sand (all these are particles belonging to atmospheric aerosol; see also Chapter 4.3.5) the other particles belong to hydrometeors. Snow is precipitation (see the next section, Chapter 2.5.3.4) as a final fate of cloud development, whereas fog, mist and haze are particles (in that order) with decreasing water content. Fog is a cloud that is in contact with the ground. Fog is distinguished from mist only by its density, as expressed in the resulting decrease in visibility; Fog reduces visibility to less than 1 km, whereas mist reduces visibility to no less than 2 km. Before fog (or cloud) and mist is formed through heterogeneous nucleation, aerosol particles that can act as CCN must be activated, i.e., they become wetted at RH depending on the so-called deliquescence point (around 60% RH). The particles are generally secondary, produced from various organic (biogenic terpenes, NMVOC) and inorganic (SO_2, NH_3, NO_x) gaseous precursors. Haze particles are formed when water condenses on dry particles (Heintzenberg et al. 1998). Some texts differentiate between dry and wet haze, but usually (and this is recommended) haze is the term for wetted particles that will probably become CCN. Dry haze is simply dust and smoke (each particle in the atmosphere contains more or less water) in the size range below about 100 nm. A well-known phenomenon is *blue haze*: SOA from plant organic emissions. A hazy condition often occurs in the summer and affects large areas from cities to mountains. Such a haze is often caused by excessive amounts of pollutants resulting from combustion (smog and smoke).

Depending on size, the particles produce different optical effects. Continuation of this condensation leads to the formation of fog. The formation of a fog layer occurs when a moist air mass is cooled to its saturation point (dew point). This cooling can be the result of radiative processes (*radiation* fog), advection of warm air over cold surfaces (*advection* fog), evaporation of precipitation (*precipitation* or *frontal* fog), or air being adiabatically cooled while being forced up a mountain (*upslope* fog). Another type of fog is the so-called *valley* fog. This fog forms as a result of air being radiatively cooled, during the evening, on the slopes of topographical features. This air, becoming denser than the surrounding air, starts going down the slope. This results in the creation of a pool of cold air at the valley floor. If the air is cold enough to reach its dew point, fog formation occurs. Generally, it is separated between warm ($> 0\,°C$) and cold ($< 0\,°C$) fog. Globally, warm fog is dominant (about 80 % of all events). Compared with other cloud types, fog droplets are generally smaller, although great variability in size distributions of fog droplets have been observed. In some instances, larger than expected fog droplets were observed (Gerber 1991). Also in contrast with other cloud types, fogs have low liquid water content (LWC). Most fogs have LWC ranging from 0.05 to 0.5 gm^{-3}.

Fogs forming in polluted areas act as efficient mechanisms for the deposition of some chemical species on vegetation (wet deposition) (Lovett and Kinsman 1990). Aqueous-phase chemical reactions are taking place in fog droplets (e. g. Facchini et al. 1992, Bott and Carmichael 1993, Acker et al. 2007) and when these drops evaporate after settling on vegetation, they leave behind concentrated chemicals that may be detrimental to vegetation growth. Also, fog water represents an important resource for people living in arid regions (Amedie 2001). Fog-water collectors have been developed so that fogs can be used as a source of irrigation and drinking water (Klemm et al. 2012). Consequently, knowing and understanding the chemical composition of fog water is important for health and agricultural productivity issues.

2.5.3.4 Precipitation

We stated that not all clouds precipitate. Indeed, from only a very small proportion of clouds does precipitation actually reach the ground surface below. The basic problem is that cloud water droplets or ice particles are frequently too small to fall from the cloud base or to survive on the way to the ground because they evaporate. Whereas a cloud droplet is on average 8 µm in diameter, a rain drop is between 500 and 5000 µm (0.5−5 mm); this means that a small rain drop is as large in volume as 240 000 cloud drops. Assuming 240 cloud droplets cm^{-3} (cf. Table 2.25), there is only one rain drop in 1 L of air. Several microphysical processes occur in clouds depending on temperature, vertical resolution, dynamic and other parameters that result in growth of a particle (Fig. 2.39) and different precipitation forms (Table 2.26).

Table 2.25 Cloud droplet characteristics including standard deviation low-level stratiform clouds; observational data from Miles et al. (2000).

parameter		marine cloud	continental cloud
droplet numer	N (cm^{-3})	74 ± 45	288 ± 159
mean droplet diameter	D (µm)	14.2 ± 3.4	8.2 ± 3.9
effective droplet diameter	σ (µm)	5.8 ± 2.0	3.1 ± 1.2
liquid water content	LWC (g m^{-3})	0.18 ± 0.14	0.19 ± 0.21

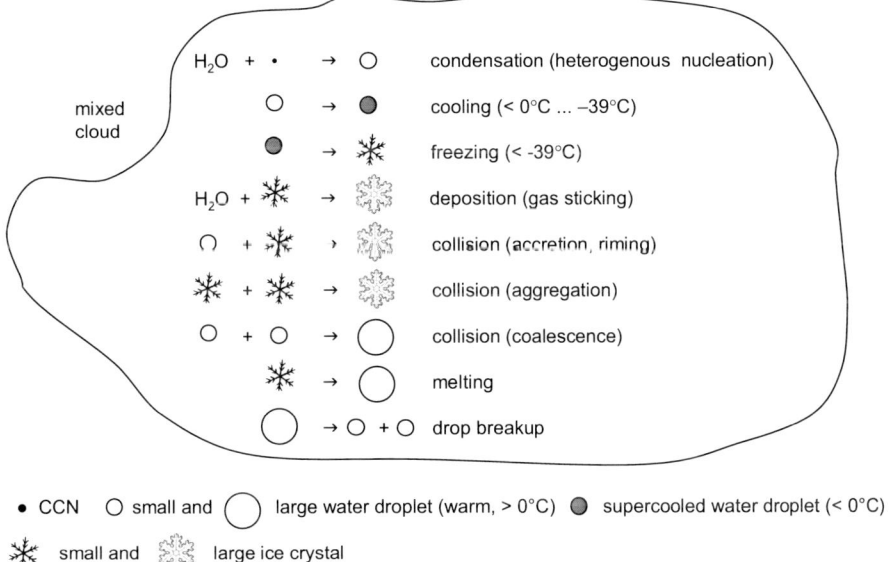

Fig. 2.39 Droplet growth and dynamic in a mixed cloud (containing droplets and ice particles).

In a pioneering work, Mason and Ludlam (1951) wrote: "While the rate of spontaneous nucleation in water vapor is inappreciable until the supersaturation reaches about 400%, the presence of foreign nuclei [we call them now CCN]) in the atmosphere allows cloud formation with supersaturations of only about 0.1%". Earlier theories of cloud-droplet growth by diffusion and by coagulation cannot explain the formation of rain drops (so called precipitation elements). The collision between particles is described by three different processes (Fig. 2.39); *coalescence* (for liquid on liquid), *aggregation* (for solid on solid) and *accretion* (for liquid on solid). The process of coalescence is thus the only process occurring in warm clouds and the process of aggregation is the only process that may occur where cloud temperature is below −39°C. This is the temperature of spontaneous freezing of supercooled drops (see also 4.3.3.4). In mixed clouds, where both supercooled droplets and ice particles co-exist, ice particles tend to grow at the expense of adjacent supercooled water droplets due to different vapor pressures (the *Bergeron-Findeisen*

Table 2.26 Precipitation forms; according to the magnitude of precipitation, rain, showers, drizzle, snow and hail is distinguished.

form	description
rain	water drops, $D > 0.5$ mm
drizzle	fine drops, $D < 0.5$ mm and close together
freezing rain	supercooled water drops, freeze on impact with surfaces
freezing drizzle	as freezing rain but $D < 0.5$ mm
snowflakes	loose aggregates of ice crystals $> -5\,°C$
snow pellets	white, opaque ice grains $D = 2-5$ mm
snow grains	small ice grains, $D < 1$ mm
ice pellets	frozen rain drops, melted and refrozen snow, $D < 5$ mm
ice prism	ice crystals in the form of needles at very low temperature
hail	balls or pieces of ice $D = 5-50$ mm, sometimes more

process). The crystals grow by vapor deposition at a rate (maximum at about $-12\,°C$) that produces individual snow crystals in some 10 to 20 minutes. The modern practice of cloud seeding (to initiate precipitation artificially) is mostly based upon the introduction of artificial ice nuclei to supply more of the ice particles. Similarly the *seeder-feeder* mechanisms work especially in orographic clouds: there are vertical layers of warm and cold clouds, where the warm (*feeder*) cloud obtains falling ice crystals from the upper cold (*seeder*) cloud to initiate particle growth. This mechanism also explains why the rain drops collected at ground level for chemical analysis contain much less substances (they are diluted) than cloud water. In warm clouds not only coalescence may cause rain formation. It is suggested that in cumulus clouds, a water-drop chain reaction may be initiated by a small number of either wet hailstones or giant salt nuclei and that in tropical clouds the presence of the latter may render the intervention of ice crystals unnecessary. The processes of rain formation are not completely understood even today.

The chemical composition of precipitation is a result of three different processes:

– *nucleation scavenging*: the chemical composition of the CCN determines the initial cloud (fog) composition (dissolved and suspended material),
– *in-cloud scavenging*: in the cloud, aerosol particles (of minor importance) and gases are taken up; chemical reactions occur within the droplet,
– *sub-cloud scavenging*: the falling rain drop (snow flakes) absorbs gases and taken up particles.

For a description of precipitation chemical climatology the most important precipitation parameters are (see Chapter 3.4.2 for long-term precipitation chemistry):

– precipitation intensity (depth per hour, in mm h^{-1}),
– precipitation rate (depth per time),
– precipitation amount (total amount of precipitated water in L),
– frequency and duration of precipitation events (including time without precipitation),
– distribution of drop sizes,
– precipitation water chemical composition (event-based),
– precipitation form (see Table 2.26).

2.5.4 Dew, frost, rime, and interception

In the last two sections we considered hydrometeors, drops and ice particles in clouds, fog, mist and precipitation. This section deals with the formation of interfacial water, either from water vapor (dew and frost) or from hydrometeors (rime and interception). These forms of atmospheric water need the contact with a surface: soils and vegetation but also artificial surfaces. Some meteorologists classify these phenomena as belonging to precipitation – we will not (see Chapter 4.4 for more details). Precipitation is physically a *sedimentation* process due to gravitational force; see Fig. 2.40 for different deposition processes.

Condensation of water vapor occurs when the *dew point* is undershot onto cooler surfaces resulting in a subsequent diffusion process in the direction of the surface (that is *dry deposition*, similar to surface removal of other molecules than H_2O, but these are ad- or absorbed onto the surface – only water can condense into liquid in the earth's climate system). When the temperature is above 0 °C, liquid drops (not film) are formed on the surface – most likely on condensation or crystallization centers – and we call it *dew*. If $T < 0$ °C, water vapor is transferred into ice on the surface (*frost*) in different sizes and forms, mostly depending on temperature and wind; on trees, filigreed ice needles and structures are formed (*hoar frost*).

Droplets in clouds, fog and mist are in continuous motion because of the dynamic processes but also due to the advective motion of air. In this way they can impact with surfaces such as trees, towers and buildings and due to the adhesive properties of water, the droplet sticks. When the temperature is higher than 0 °C,

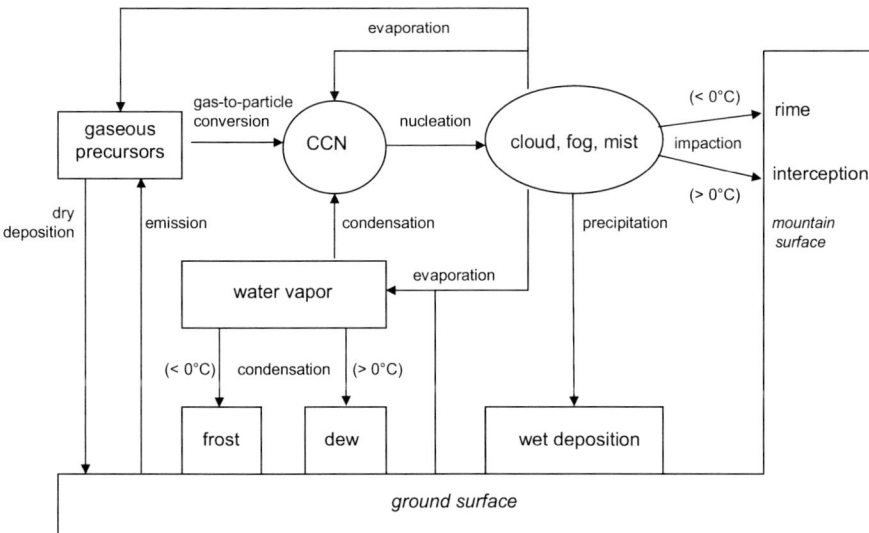

Fig. 2.40 Simplified scheme of atmospheric water forms, transfers and removal (deposition); note that different scavenging processes not completely shown.

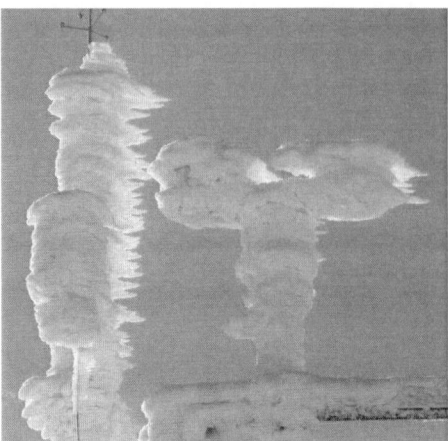

Fig. 2.41   Riming of meteorological instruments in winter 1992/1993 at the Cloud Chemistry Monitoring Station Mt. Brocken (Harz, Germany, 1142 m a.s.l.).

the impacted drops can grow by collision and finally flow down; this process[12] is called *cloud-droplet interception*. At temperatures below 0 °C the supercooled droplets will freeze on the surface to form *rime*. Riming can lead to severe problems at mountain tops (Fig. 2.41). Cloud-water (and rainfall) interception are hydrological processes of particular interest in Tropical Mountain Cloud Forests (TMCF). Studies in these systems have shown the important contributions of cloud/fog water to the hydrological balance (Gomez-Peralta et al. 2008).

   William Charles Wells was the first to explain satisfactorily the phenomenon of dew. After decisive experiments on dew, he published his book, *An Essay on Dew and several appearances connected with it*, in London in 1815. This was the first and now still accepted scientific description of dew formation, coming after a long debate. *Wells* showed that apparently all these phenomena (including hoar frost and mist) result from the effects of radiation of heat from the earth's surface during the absence of the Sun.

   The first statement in the literature that dew contains traces of atmospheric origin (and not alchemistic elements) besides water is by *Lampadius* (based on his experiment from 1796), that dew consists of pure water and some carbonic acid (2%) but more than in rain water (Lampadius 1806). *Lampadius* also wrote that dew may contain substances emitted from plants (effluvia), and it was thus used in the past as medicine; the bleaching properties of dew have been known for centuries and dew has been used for the cleansing of clothes. Textiles have long been whitened by grass bleaching (spreading the cloth upon the grass for several months), a method virtually monopolized by the Dutch from the time of the Crusades to the eighteenth century. Probably the first chemical study of dew was conducted by the

---

[12] Foresters and hydrologists name the interception of rain drops by leaves and needles of trees *rainfall interception* (or sometimes canopy-interception; however, this term is not to separate from cloud and fog interception, and should therefore not be used). Because of evaporation, interception of liquid water leads to loss of that precipitation for the drainage basin.

French chemist *Jean-Sébastien-Eugène Julia de Fontenelle* (1790–1842) who in 1819 collected 4 liters of dew in the marshes of Cercle, France, and found chloride, sodium, potassium, sulfate, calcium and carbonate (Fontenelle 1823), and mentioned: "This water was inodorous, without colour, and clean; in a short time it deposited small flakes of nitrogenous matter" (cited by Pierre 1859, p. 41, and Smith 1872, p. 241). The first quantitative estimates of ammonia in dew are known from Boussingault (1853) at Mt. Liebfrauenberg (3–50 mg L^{-1}) and the German agricultural chemists *Wolf* and *Knop* who took measurements in 1860 in Möckern near Leipzig (1–6 mg L^{-1}). While *Boussingault* collected dew using a sponge, *Wolf* and *Knop* collected dew with a glass cup from grass leaves before sunrise (in Knop 1868, annex pp. 77–78). Interest in dew formation, its physics and chemistry, rose after 1960 for several reasons: first, due to dry deposition onto wetted surfaces (Chang et al. 1967, Brimblecombe and Todd 1977); second, in the last two decades because of ecological considerations (dew as a source of moisture for plants, biological crusts, insects and small animals, e. g. Jacobs et al. 2000) and its potential use as potable water (Beysens et al. 2006, Muselli et al. 2006); and finally from an air chemical point of view, now also termed interfacial chemistry (Rubio et al. 2006, Acker et al. 2007).

2.5.5 Soil water and groundwater: Chemical weathering

Water falling to the surface of continents in any form of precipitation, upon striking the surface, undergoes major modifications both in mode of transport and chemical composition. *Infiltration* is the flow of water from the ground surface into the ground. Once infiltrated, the water becomes soil moisture or groundwater.

In the *cryosphere*, precipitated snow undergoes transformation into ice layers. Drilling of so-called ice-cores gives us insights into the chemical composition of the palaeoatmosphere. Ice cores have been taken from many locations around the world, especially in Greenland and Antarctica. In January 1998, the collaborative ice-drilling project between Russia, the United States and France at the Russian Vostok station in East Antarctica yielded the deepest ice core ever recovered, reaching a depth of 3623 m. Preliminary data indicates the Vostok ice-core record extends through four climate cycles, with ice slightly older than 400 000 years. This data set contains ice-core chemistry, timescale, isotope, and temperature data analyzed by several investigators. Dome F in Antarctica, also known as Dome Fuji, is another example of well-described ice-cores; recently a depth of 3035 m was reached and, according to a preliminary dating, it reaches back to 720 000 years ago. However, the processes of laying down snow and its conversion into ice are complicated, and air is slowly exchanged by molecular diffusion through pore spaces in firn. Thermal diffusion causes isotope fractionation in firn when there is rapid temperature variation, creating isotope differences which are captured in bubbles when ice is created at the base of firn. Firn is partially-compacted névé[13], a type of snow that has been left over from past seasons and has been recrystallized

---

[13] Young, granular type of snow which has been partially melted, refrozen and compacted. This type of snow is associated with glacier formation through the process of nivation.

Table 2.27 Typical residence times of water found in various reservoirs; after Pidwirny (2006).

reservoir	average residence time
glaciers	20 to 100 years
seasonal snow cover	2 to 6 months
soil moisture	1 to 2 months
groundwater: shallow	100 to 200 years
groundwater: deep	10 000 years
lakes	50 to 100 years
rivers	2 to 6 months

into a substance denser than névé. It is ice that is at an intermediate stage between snow and glacial ice.

In the biosphere, rain that is not lost back to the atmosphere by evaporation from the ground or from trees may pass deep underground, only to emerge at a much later date (Table 2.27) in a river or lake. Water coming into contact with rocks (and derived soils) reacts with primary minerals contained in them. The minerals dissolve to varying extents, and some of the dissolved constituents react with one another to form new, secondary minerals. Dissolution is mainly controlled by the water acidity provided from plant mineralization (humic acids), atmospheric carbonic acid and "acid rain". The overall process is called *chemical weathering* (see Chapter 2.2.2.5, Eqs. 2.62 and 2.63; Berner and Berner 1996).

2.5.6 Surface water: Rivers and lakes

Rivers and other forms of surface water actually account for a relatively small portion of the planet's water supply, but they loom large in the human imagination as the result of their impact on our lives. The first human civilizations developed along rivers in Egypt, Mesopotamia, India, and China, and today many great cities lie alongside rivers. Rivers provide us with a means of transportation and recreation, with hydroelectric power, and even − after the river water has been treated − with water for drinking and bathing. River waters typically begin with precipitation, whether in the form of rainwater or melting snow. They also are fed by groundwater exuding from bedrock to the surface. As we have already noted, however, a river is really the sum of its tributaries, and thus hydrologists speak of river systems rather than single rivers. Finally, the river discharges into an ocean, a lake, or a desert basin.

The chemical composition of river water is significantly different from that of seawater (Table 2.28). At first approximation, seawater is mainly a solution of Na^+ and Cl^- while river water is a solution of Ca^{2+} and HCO_3^-. Interestingly the ratio Na/Cl is the only one which is relatively similar for rivers and oceans, suggesting that both components are within a global cycle from seaspray through cloud transportation to continents and precipitation. Carbonate is approximately in equilibrium with atmospheric CO_2 and the concentration difference between river water

Table 2.28 The composition of average river water and seawater (in mg L^{-1}); data from Langmuir (1997).

substance	river water	seawater	seawater to river water	ratios	river water	seawater
HCO_3^-	58.6	146	2.5	Ca/HCO_3	0.26	2.8
Ca^{2+}	15.0	412	27	$Ca/SO4$	1.4	0.15
SO_4^{2-}	10.6	2707	255	Na/Cl	0.77	0.55
Cl^-	7.8	19383	2485	Na/Mg	1.46	8.66
Na^+	6.3	10764	1794	Na/K	2.22	102.5
Mg^{2+}	4.1	1243	303			
K^+	2.3	105	39			

Table 2.29 Chemical composition of different rivers (in mg L^{-1}); data from Livingstone (1963) and Holland (1978).

	HCO_3^-	SO_4^{2-}	Cl^-	NO_3^-	Mg^{2+}	Na^+	K^+	Fe	SiO_2
North America	68	20	8	1	6	9	1.4	0.16	9
South America	31	4.8	4.9	0.7	1.5	4	2	1.4	11.9
Europa	95	24	6.9	3.7	5.6	5.4	1.7	0.8	7.5
Asia	79	8.4	8.7	0.7	5.6	9.3	–	0.01	11.7
Africa	43	13.5	12.1	0.8	3.8	11	–	1.3	23.2
Australia	32	2.6	10	0.05	2.7	2.9	1.4	0.3	3.9
World	58.1	11.2	7.8	1	4.1	6.3	2.3	0.67	13.1

and seawater is determined by the pH. Ions transported by rivers are the most important source of most elements in the ocean. Rivers collect dissolved matter from precipitation, weathering and pollution. Their chemical composition shows large seasonal and interannual variations as well as great differences between the continents, reflecting the processes of the sources (Table 2.29).

2.5.7 The oceans

The chemical composition of the oceans defines the impact of the other geochemical reservoirs on the oceanic reservoir leading to the addition of elements to, or removal from, seawater (Elderfeld 2006). The ocean is the global disposal center of chemicals. In Chapter 2.2 the evolution of the ocean was discussed with the interaction of degassing from the surface and the atmosphere. Evaporation of river water will not make seawater (compare Tables 2.28 and 2.9). In addition, on comparison of the amount of material supplied to the oceans by rivers with the amount in the ocean, it is clear that most of the elements have been replaced many times. Thus chemical reactions occurring in the ocean balance the river input by sediment removal. The main input and removal fluxes for seawater ions can be roughly estimated from the averages of river composition and the global run-off value

$(36-40) \cdot 10^{18}$ g yr^{-1}. Sillén (1967) first constructed a simplified multicomponent equilibrium model based on nine components: HCl, H_2O and CO_2 which represent acid volatiles from inside the earth, and KOH, CaO, SiO_2, NaOH, MgO and Al(OH)$_3$ which corresponds to the bases of the rocks. For closing the global mineral cycle, a *reverse weathering* in the sea (Eq. 2.94) has been proposed by Mackenzie and Garrels (1966).

$$\text{clay mineral} + HCO_3^- + H_4SiO_4 + \text{cation} \rightarrow \text{cation rich silicate}$$
$$+ CO_2 + H_2O \qquad (2.94)$$

However, most of the sedimentary mass is of detrital origin. Nutrient elements (Si, P, V, N, and trace metals) are removed in the ocean as biological debris to sediments. The main sink of substances is by hydrothermal reactions (volcanic activity) at locations of seafloor spreading and circulation through the ocean crust. Volcanic activities in the oceans are extensive, and produce submarine lava flows which are unstable in seawater. The high temperature (200−400 °C) is not only important for basalt-seawater reactions but also triggers circulation. Subduction (see Chapters 2.2.1.1 and 2.6.4.3) transfers seawater and its constituents back to the magma.

Because of the large volume of seawater, almost all ions are conservative, meaning it does not change its concentration over geological periods. However, there are differences; for example, Cl$^-$ is the most conservative ion in the climate system because it takes no part in any chemical reaction[14] but Na$^+$, the counterpart is involved in rock weathering (it is a major constituent of silicate minerals), cation exchange, and possibly volcanic-seawater reactions and reverse weathering.

Seawater, compared with all other natural waters, is remarkably constant in composition. However, all elements involved in biological turnovers (oxygen and carbon being the most important) may have variations (and possibly trends), again due to changing key parameters of the climate system, of which the temperature is the most important one. Concerning the global climate, the oceans have two essential functions:

− They are the largest pool of carbon (almost in the form of HCO$_3^-$) in the climate system (see Table 2.9) and the only ultimate sink of atmospheric CO_2 (burial in sediments, Sabine et al. 2004); temperature change results in solubility change as well as changes in circulation and hence total fluxes, see Fig. 2.9.
− The global oceanic circulation is closely interlinked with the atmospheric circulation; hence oceans transport energy and material and exchange it with the atmosphere. Change in one reservoir results in change in the other, but on different time scales.

Wind-driven *circulation* occurs within the surface layer (50−300 m), creating a number of current rings, or *gyres*, that flow clockwise in the northern hemisphere and counterclockwise in the southern hemisphere. The circulation pattern is a result of complicated interaction of wind and water but also other factors are involved, such as the earth's rotation and friction. *Coastal upwelling* is a special case of surface circulation which has an important effect on the biological productivity in the

---

[14] Reactions, as discussed in Chapter 5.8.2, are negligible comparing with the huge amount of chloride.

ocean. Below a few hundred meters, the oceanic circulation is a result of gradients in density due to differences in temperature and salinity. In contrast to freshwater lakes, where the density maximum is at 4 °C (see Chapter 2.5.1), the deep ocean is vertically stratified into various water masses, with the densest water at the bottom and the lightest at the top. All density-related stratification owes its origin to conditions at the surface: temperature differences due to the radiation budget; salinity due to evaporation (increase in salinity) and precipitation (decrease in salinity); but also from ice forming (dissolved ions remain in liquid water) and melting (temperature and salinity are interlinked properties). As a result of deep stratification, deep circulation is much more horizontal than vertical. Diffuse upwelling is very slow (about 1 m yr^{-1}) and the residence times of deep water are in the order of hundreds, even thousands, of years. For further reading, see Berner and Berner (1996), Bigg (2003), Lehr and Keeley (2005), and Follows and Oguz (2007).

2.6 Sources of atmospheric constituents

For a given sub-system such as the atmosphere, the sources of compounds to be emitted into the air are crucial, similar to any chemical reactor in the sense of one boundary condition (the others are temperature, pressure, radiation, mixing, and water[15] in different phases). For the atmosphere as a three-dimensional reservoir, the spatial distribution of sources, together with the source strength (or in other words the emission flux) must be known. This topic is presented very differently in the literature. Normally, a distinction is made between natural and anthropogenic sources and emissions. This is important for the knowledge to assess the degree of man-made changes in the climate system, but for modelling the physics and chemistry of the climate system, for example, this separation is unimportant. We follow here the "classical" treatment; first the main source categories are described briefly, where the typical compounds belong to the sources and also emissions are mentioned. In the next section, Chapter 2.7, the compounds themselves are the focus with the emissions from different sources. It is not the aim of this book to provide details of sources and emissions; the reader is referred to more specialized literature. The aim is to present a comprehensive critical overview and to provide the "best known" emission values presently.

2.6.1 Source characteristics

Natural sources of atmospheric components are principally distinguished into biogenic and abiogenic (or geogenic). They may relate to biological processes in the soil (e. g., bacteria), in vegetation (plants), in the oceans (substances produced by algae) and non-biological processes in the earth's crust (volcanoes or gas/oil seep-

---

[15] Water is the atmospheric compound with the largest flux by far and the most significant for weather, climate and chemistry (see Chapter 2.5.2).

age), or in the atmosphere (lightning), but also simply in dispersion of particles by wind-surface interaction. These emissions often depend strongly on climate, soil or water characteristics, and thereby show a strong temporal variation in seasonality, diurnal cycle or both. Soils acts as source as well as sink; the direction of the flux is changing with the compensation point. Thus, soil biological activities (as found for COS by Conrad and Meuser 2000) may depend from the ambient concentration, stressing the necessity to measure fluxes as function of concentration to obtain reliable source or sink data for atmospheric budgets.

Thus, natural sources often have a distinctly different character than anthropogenic sources, which are comparably constant with respect to seasonality. In addition, the source strength as well as the spatial distribution of natural sources may differ substantially from year to year. An extreme example is volcanic emission.

Abiogenic sources can be distinguished as physical processes (dispersed soil dust and sea spray), geochemical processes (gaseous emanations, volcanic eruptions), and atmospheric chemical processes (lightning causing nitrogen oxides). Biogenic processes in wild and agricultural plants are the same in principle. The main difference lies in the input of chemicals (mainly fertilizers) onto farmland with the purpose to control and to accelerate the plant growth. Cultivated soils produce soil dust by wind erosion: is that a contribution by anthropogenic or natural sources? Nonetheless, desert dust is globally the dominant source in this category. Changes in land use over hundreds or even thousands of years (for example, deforestation) have changed the quantity and the mix of VOC emissions from soils and vegetation. Biomass burning (wildfires) is nowadays almost all attributed to human activity, whether intentional or not; the only natural cause is lightning strikes, which are of minor importance globally.

We see that it becomes more and more difficult to distinguish clearly between natural and anthropogenic emissions. This book will not draw a sharp line − it is more important to understand different sources and the controlling parameters of the emissions. With our understanding that humans are part of the biosphere and have changed it in an evolutionary way into the noosphere, we have to consider the (climate) system as a whole. The only difference between natural and man-made emissions lies in the fact that man-made emissions are unbalanced in nature or, in other words, are in an open-loop. As a somewhat extravagant example, we could assert that it does not matter how one reduces the concentration of chlorine in the upper troposphere; whether by cleansing volcanic plumes of chlorine (if there were such a facility) or by banning its use. The effects are always based on the sum of molecules in the atmosphere. However, some sources or substances used without anticipating any environmental problem, due to small volume and/or chemical virtual harmlessness, such as halogenated and resistant organic compounds, and some heavy metals, have resulted in serious problems and required counteractions.

An emission (and deposition) represents a flux (mass per time). Concerning the source characteristics, we separate emissions from

1. point sources (volcanoes, fumaroles, factories etc.),
2. specific sources (plants, animals, vehicles etc.), and
3. diffuse sources (surfaces such as soils, lakes, landscapes etc.).

The flux can be related to an area (a standard area such as ha and m^{-2}) and is then termed a specific emission. The matter state is

4. gaseous (volatile elements, molecules) or
5. solid (particles, see Chapter 4.3.5).

2.6.2 Biogenic sources

2.6.2.1 Vegetation and microorganisms (soils and waters)

For the atmosphere, the earth's surface acts as both a source and a sink for trace gases and particles. Field measurements (mostly via concentration gradients) only quantify the net flux F, i.e., emissions-deposition (Matson and Harriss 1995). The so-called *compensation point* is reached when $F = 0$. Hence, to calculate individual fluxes (emission and deposition) separately, empirical relationships and models are needed. The extrapolation of local measurements to broader ecosystems or grids concerning the spatial resolution of climate models under different meteorological conditions remains a major challenge.

Key to an understanding of the exchange between the surface and the atmosphere is the planetary boundary layer (PBL). This layer, with a variable thickness between several hundred meters and a few kilometers, is defined as that region of the lower troposphere that responds to surface forcing (frictional drag, heat transfer etc.). Temperature gradients influence vertical mixing and hence vertical transport rates. In the case of positive T gradients (increase with height), gases accumulate near the surface and therefore surface emission is reduced and even stopped depending on the equilibrium conditions. In the case of strong convection producing rapid mixing, surface trace gases can reach the upper troposphere in just a few hours. Meteorological conditions strongly influence surface-near concentrations of gases and particles.

As shown in Fig. 2.22, organisms themselves emit gases via respiration; breathing and transpiration. Plants can emit trace substances only during respiration (together with CO_2); it has been assessed that 2−3 % of their photosynthetic capacity is released in the form of diverse organic substances, though significantly more under stress conditions. As discussed in Chapter 2.4.2, the emission of trace gases by plants is always in relation to the assimilation and respiration process (Kesselmeier 2001, Kesselmeier et al. 2002).

Fig. 2.42 shows a simplified biochemistry, the organic synthesis during CO_2 reduction through photosynthesis. In a strong reducing medium, hydrogen (H) and carbon monoxide (CO) are the building blocks of organic compounds (similar to the *Fischer-Tropsch* synthesis). As seen, simple compounds in the reaction chain produced are also present in the main emitted substances: formaldehyde and acetaldehyde ($HCHO$ and CH_3CHO), methanol (CH_3OH), methane (CH_4), acetone (CH_3OCH_3). The formation of acids (formic acid and acetic acid: $HCOOH$ and CH_3COOH) needs an oxidative environment; remember, plants oxidize during respiration.

C$_1$ chemistry

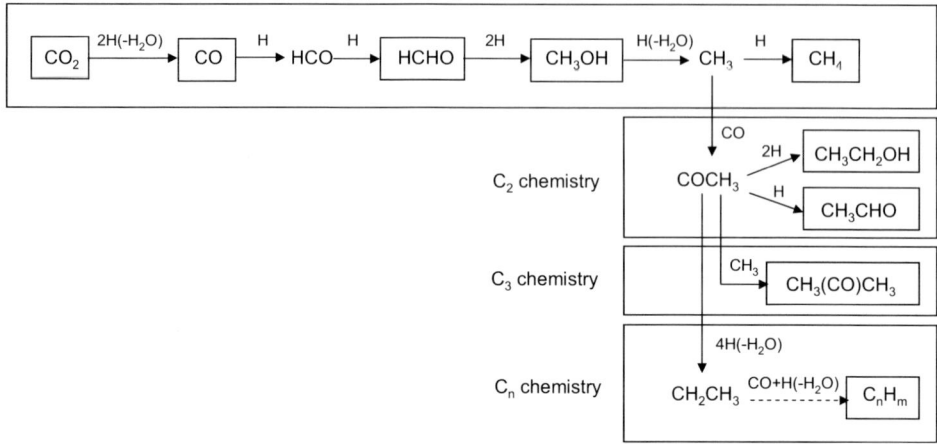

Fig. 2.42 Simplified reaction scheme of photosynthetic formation of organic compounds. In boxes are stable substances which are possibly emitted. The radical intermediates may undergo several competitive reactions, esp. HCO (formyl) and CH$_3$ (methyl) depending from the ratio between reactants (H, CO, and O$_2$). In oxidative environment, acids are formed from aldehydes.

In plants nitrate is reduced to nitrite in the cytoplasm after which the nitrite is reduced to NH$_3$ in the chloroplasts (Galbally 1989). Consequently, plants emit NO and NH$_3$. Plants also take up NH$_3$ and NO$_2$ very rapidly. This NO$_2$ is chemically produced by atmospheric oxidation of biogenic and anthropogenic emitted NO, either above the canopy, or below the canopy from soil-emitted NO. In the last case, a "closed" N cycle between soil and plant occurs, resulting in a smaller ecosystem emission. The uptake of NO$_2$ faster than that of NO through stomata can be explained by different rates of conversion into some readily metabolized forms (e. g., NO$_2^-$), perhaps by photolytically-produced solvated electrons: NO$_2$ + e_{aq}^- → NO$_2^-$. Similarly, sulfur species (almost all organic sulfides) are produced by assimilatory sulfate reduction. Plants emit DMS and COS, but also take them up (Yonemura et al. 2007).

The largest amount of emitted species are based on carbon itself (see Chapter 2.7.3 and Table 2.30). Care should be taken when using emission values denoted as "terrestrial emission" because it is probably the sum of plant (vegetation) and soil emission.

Soils containing microorganisms predominantly emit inorganic compounds (NH$_3$, NO, HNO$_2$, N$_2$, H$_2$S, COS) because of nutrient cycling. It is evident that plants take up gases from air in the assimilation process, but only during the daytime. In contrast, soils can absorb them during both day and night. Animal faeces are rich in nitrogen and hence are used in internal cycles, but by decomposition they emit many different volatile substances. Microorganisms are distributed in soils, aquatic systems and oceans. Most of the material turnover in ecosystems is due to microorganisms. All substances exchanged with the local environment are therefore found in the reservoir (soil, lake, river, and ocean), dissolved, suspended

Table 2.30 Average global annual emission of organic substances emitted from terrestrial vegetation in Tg yr^{-1}. − no value given.

substance	RETRO (2008)[b]	Lathière et al. (2005)	Guenther et al. (1995)	Jacob et al. (2002)
isoprene	539.07	460	506	−
monoterpenes	194.66	117	126	−
methanol[d]	269.72	106	−	−
CO	96.01	−	−	−
acetaldehyde	37.44	−	−	−
formaldehyde	33.76	−	−	−
acetone	30.26	42	−	33 (± 9)
propene	15.66	−	−	−
propane	−	−	−	12
i-pentane	−	−	−	5
ethane	4.12	−	−	−
ethene	3.84	−	−	−
i-butane	−	−	−	3.6
total[a]	−	752 (± 16)[c]	1150	−

[a] not the sum of the above listed compounds but all
[b] only terrestrial fluxes as computed by the MEGAN model in MOZECH. Note that the computed values simulate exactness not being true
[c] this error is unrealistic small
[d] further estimate (in Tg yr^{-1}): 100 (Galbally and Kristine 2002)

and/or adsorbed. Thus, physical equilibrium and transport processes determine the reservoir distribution and phase partitioning. Between surfaces (soils and waters) and their surrounding air, a permanent exchange exists where the sensitive equilibrium is influenced by the air concentration. The escape of NO from soils is driven by molecular diffusion within the soil pore spaces and by advective and convective air flows through the soil due to pressure and thermal forcing. Liquid water in soil pores can drastically reduce the NO escape due to much slower diffusivities. Field studies have shown that soil water forms a substantial barrier to the escape of NO from soils; however, a simultaneous increase in N$_2$O has been observed (Galbally 1989). The increased formation of N$_2$O could be explained by increasing oxygen deficiency, and therefore higher reduction capacity.

Waters (rivers and lakes) may − especially when they are polluted − provide an important source for atmospheric emissions locally, but play no role globally. Wetlands are among the most important ecosystems on earth; an excellent book on wetland chemistry is Reddy and DeLaune (2008). Wetlands occupy about 5–8% of the global land surface (Table 2.31) and have a global NPP of $(4-9) \cdot 10^{15}$ g C yr^{-1} (Mitsch and Gosselink 2007), which corresponds to about 10–20% of total land surface NPP (see Chapter 2.3.5.3). Natural wetlands contribute about 20% to the global CH$_4$ emission (115–145 Tg yr^{-1}; cf. Table 2.69). Rice paddies are anthropogenic wetlands and emit 60–80 Tg CH$_4$-C yr^{-1} (Megonigal et al. 2004, Whalen 2005). In the 1960s it was believed that huge sulfur emissions in the form of hydrogen sulfide were emitted from wetlands (Möller 1983), but after reevaluation it now appears they are negligible on a global scale.

Table 2.31 Global wetlands (from data in Mitsch and Gosselink 2007).

type	area (in 10^6 km^2)
polar/boreal	2.4 − 3.5
temperate	0.7 − 1.0
subtropical/tropical	1.9 − 4.8
rice paddies	1.3 − 1.5
total	6.8 − 12.8

2.6.2.2 Animals

It is possible to determine the emission of different compounds by specific animals similarly to plants, depending on biological parameters (age, status, health etc.) and physical parameters (temperature, radiation, humidity, acidity etc.) (see Tables 2.32 and 2.65). Animals (including humans) contribute to atmospheric emissions in different pathways; directly through breathing, transpiration and flatulence, and indirectly through conversion of excrement (see, for example, NH_3 emissions in Table 2.65). As the total biomass of animals compared with that of plants is negligible, with the exception of emissions of NH_3 and CH_4, there is no global significance in these emissions.

Methane production by termites was first reported by Cook (1932), who observed the evolution of a gas from a species of termite. Besides CH_4 termites may emit large quantities of carbon dioxide, and molecular hydrogen into the atmosphere. Termites inhabit many different ecological regions, but they are concentrated in tropical grasslands and forests. Studies indicated that methane is produced in a termite's digestive tract during the breakdown of cellulose by symbiotic microorganisms. Later studies showed large variations in the amount of CH_4 produced by different species. Estimates of the contribution to the global CH_4 from termites vary widely, from negligible up to 15%. Zimmermann et al. (1982) estimated global annual emissions calculated from laboratory measurements of 150 Tg of methane and 5000 Tg of carbon dioxide. As much as 200 Tg of molecular hydrogen may also be produced. Seiler et al. (1984) estimated the flux of CH_4 and CO_2 from termite nests into the atmosphere to be in the range 2−5 Tg only.

Fermentation of feed in the rumen of cows, goats, sheep and several other animals belonging to the class of animals called ruminants is the largest source of methane, from enteric fermentation by methanogenic bacteria and protozoa. In the European Union in the late 1990s, approximately two-thirds of emitted CH_4 come from enteric fermentation and one-third from livestock manure (Moss et al. 2000). The composition of the animal feed is a crucial factor in controlling the amounts of methane produced, but a sheep can produce about 30 liters of methane each day and a dairy cow up to about 200 liters. The reason that ruminants are so important to mankind is that much of the world's biomass is rich in fibre. Ruminants can convert this into high quality protein sources (i.e., meat and milk) for human consumption and this benefit needs to be balanced against the concomitant production of methane.

Table 2.32 Emission factors for wild animals' and human emissions (in kg per animal/ person and year).

species	assumed life weight (in kg)	CH_4	NH_3	author
deer (red deer, reindeer)	100	25	1.1	Sutton et al. (1995)
moose	350	50	2.2	Crutzen et al. (1986)
roe deer	15	4	0.2	scaled from red deer[a]
boar		1.5	1	Niethammer and Krapp (1986)
birds	0.8	–	0.12	Sutton et al. (1995)
large birds	2.4	–	0.36	Sutton et al. (1995)
humans		0.1	0.05	Crutzen et al. (1986)

[a] Use animal weights to similarly scale emissions for other species

2.6.3 The ocean as source

The ocean as a source of trace gases is between biogenic and abiotic processes; hence we consider the ocean here in a separate subchapter. Whereas the generation of sea-salt particles from the surface of seawater is a purely physical process (see Chapter 2.6.4.2), it has been identified as a considerable global source of different nitrogen, sulfur and carbon species (Table 2.33). These species are produced in the ocean by similar biological processes to those in soils, but photochemical processes in the surface water (e. g. Zhou and Mopper 1997) also provide products that can exchange with the air. It is assumed that the surface seawater is saturated or super-saturated with these gases (for example, CO and dimethyl sulfide). The principles regulating the flux between ocean and atmosphere are well understood (Liss and Slinn 1983). However, there is a lack of systematic measurements of concentrations in surface air over the oceans along with the necessary measurements to evaluate the equilibrium partial pressure in surface water.

For example, alkyl nitrates (see Chapter 5.4.5.2) have been measured (Chuck et al. 2002) in surface seawater and air from the North and South Atlantic that indicate a (natural) flux out of the oceans. Although production processes for trace gases in seawater are often poorly known or totally unknown, in the case of these alkyl nitrates, laboratory experiments indicate that they can be formed photochemically by the reaction of alkyl radical (ROO) and NO (Dahl et al. 2003). In a number of cases, we lack even basic knowledge of how the gases are formed (and destroyed) in the surface oceans and this is a real handicap in trying to predict how the fluxes may change under an altered climate (Liss 2007).

By far the most important trace gas emission from the ocean influencing our climate is dimethyl sulfide (DMS), a gas that is biogenically produced by algae in marine waters. It is formed by cleavage of the precursor molecule dimethylsulfonio-propionate (DMSP) that acts as an osmolyte and cryoprotectant for the organisms. A small fraction (probably less than 10 %) of the total DMS formed in seawater is transferred to the atmosphere across the air-sea interface. In the atmosphere, it is unstable and is converted into a variety of products some of which are particulate, making DMS a major natural provider of fine particles over the open oceans. In this way, it plays an important role in climate as a source of cloud condensation nuclei (CCN), as well as supplying the element sulfur to the terrestrial environment, where it is essential for plant growth. In the atmosphere, DMS is subject to oxidation by free radicals including OH and NO_3 to yield a number of gaseous and particulate products.

The major difficulty in estimating global sea-air fluxes of DMS is in interpolating measured concentrations of DMS in seawater to create seasonally resolved gridded composites (for example, Kettle et al. 1999). Attempts to correlate DMS with variables that can be mapped globally, e. g., chlorophyl, have not yielded reliable relationships. Using the parameterizations of Liss and Merlivat (1986), Wanninkhof (1992), and Erickson (1993), and using published wind fields and the maps of sea-surface DMS concentration of Kettle et al. (1999), Kettle and Andreae (2000) calculated the fluxes of DMS from the global oceans at approximately 15, 33, and 21 Tg S yr^{-1}, respectively. This shows that the air-sea exchange (model assumptions for the gas transfer velocity; see also Chapter 4.3.7) may be the dominant factor in controlling the emission value. Nightingale et al. (2000) reported the first estimation of the power dependence of gas transfer on molecular diffusivity in the marine environment. A gas exchange relationship that shows a dependence on wind speed intermediate between those of Liss and Merlivat (1986) and Wanninkhof (1992) is found to be optimal. Recent estimates of globally integrated DMS fluxes from ocean to atmosphere tend to the higher value given by Möller (2003) as 35 ± 20 Tg S yr^{-1}. Anderson et al. (2001) estimated 0.86 and 1.01 Tmol S yr^{-1} (about 30 Tg S yr^{-1}) and Simo and Dachs (2002) have globally quantified as 23–35 Tg S yr^{-1}, and Kloster (2006) and Kloster et al. (2006) simulate a global annual mean DMS emission of 28 Tg S yr^{-1}.

The ocean's influence on volatile organic compounds (VOCs) in the atmosphere is poorly understood. Sinha et al. (2007) show that compound-specific PAR and biological dependency should be used for estimating the influence of the global ocean on atmospheric VOC budgets. They estimated the mean fluxes in a Norwegian fjord for methanol, acetone, acetaldehyde, isoprene, and DMS as −0.26, 0.21, 0.23, 0.12, and 0.3 (all in ng m^{-2} s^{-1}), respectively. It was observed that under conditions of high biological activity and a PAR of ~ 450 μmol photons m^{-2} s^{-1}, the ocean acted as a net source of acetone. However, if either of these criteria were not fulfilled then the ocean acted as a net sink of acetone. This new insight into the biogeochemical cycling of acetone at the ocean-air interface has helped to resolve discrepancies from earlier studies, such as Jacob et al. (2002) who reported the ocean to be a net acetone source (27 Tg yr^{-1}), and Marandino et al. (2005) who reported the ocean to be a net sink of acetone (−48 Tg yr^{-1}). The ocean acted as net source of isoprene, DMS and acetaldehyde but a net sink of methanol. This is also supported by shipboard measurements made off Mace Head Research Station (Ireland) (Lewis et al. 2005), which demonstrated a strong inverse correlation of CH_3OH with DMS, with both species showing wind-speed dependence indicative of air-sea exchange. Marine concentrations of acetone and acetaldehyde are similar, at around 0.4–0.5 ppb. The statement of Singh et al. (2001), that large and widespread (most likely oceanic) sources must be present for oxygenated hydrocarbons, is therefore supported by these results. A large enrichment of acetone and also of

Table 2.33 Global oceanic trace gas emissions (in Tg element yr^{-1}).

substance	flux	author
N_2O	1.4–2.6	Khalil and Rasmussen (1992)
	4 (1.2–6.8)	Nevison et al. (1995)
	4.7	IPCC (2001)
	0.9–1.7	Rhee at al. (2009)
NH_3	8–15	cf. Table 2.63
CO	47 (5–41)	Seiler and Crutzen (1980)
	23 (9–38)	Khalil and Rasmussen (1990a)
	6 (3–14)	Bates et al. (1995)
	5 (0–14)	Khalil (1999)
	9	RETRO (2008)
CH_4	3.5 (3–4)	Lambert and Schmidt (1993)
	0.4 (0.2–0.6)	Bates et al. (1996)
	10	Watts (2000)
	0.6–1.2	Rhee at al. (2009)
isoprene	5	Guenter et al. (1995)
alkanes and	30	Bonsang et al. (1988)
alkenes	0.3–0.6	Plass-Dülmer et al. (1995)
	~1	Rudolph (1997)
	7.6	RETRO (2008)
alkanes	1 (0–2)	Williams and Koppmann (2007)
alkenes	6 (3–12)	Williams and Koppmann (2007)
acetone	10–27 ($\pm$ 50%)	Jacob et al. (2002)
CH_3Cl	0.46	Khalil et al. (1999)
$CHCl_2$	0.32	Khalil et al. (1999)
CH_2Cl_2	0.16	Khalil et al. (1999)
C_2HCl_3	0.02	Khalil et al. (1999)
C_2Cl_4	0.02	Khalil et al. (1999)
DMS	35 ($\pm$ 20)	Möller (2003) and citations therein

other carbonyl compounds (e. g. acetaldehyde) has been measured by Zhou and Mopper (1997) in seawater samples of the surface micro-layer of the ocean and a sea-air flux was suggested for the less soluble compounds (such as acetone and acetaldehyde).

Isoprene (see Chapter 2.7.3.4) is a gas reactive in the atmosphere, which is particularly important for the formation of organic particles. It is known to be emitted from the oceans (Broadgate et al. 1997), but generally not thought to be of sufficient magnitude to lead to particle formation. However, Meskhidze and Nenes (2006) have used satellite imagery to show that over a biologically productive region of the Southern Ocean, CCN number density and cloudiness are enhanced compared with the less productive surrounding area. Controversially, they attribute the additional CCN to particles formed from isoprene emitted from the biologically productive water, where it might be expected that DMS would be the more obvious precursor. This finding is of importance not only because it shows that marine trace gas emissions can have a direct effect on CCN and cloudiness and hence climate, but it also reopens the question of the role of oceanic isoprene emissions on the marine atmosphere.

Recent reassessment has shown (Rhee et al. 2009) global oceanic emission estimates of CH_4 to be ~ 10 times lower and of N_2O to be ~ 3 times lower than the "established" emission values used in the recent IPCC report (Denman et al. 2007). The reasons for the discrepancies are unknown, but three studies (Conrad and Seiler 1988, Bates et al. 1996, Rhee et al. 2009) indicate that the open ocean is supersaturated with respect to atmospheric CH_4 at a level of ~ 0.04, about an order of magnitude lower than the ΔCH_4 value of 0.3 suggested by Ehhalt (1974) and used by others (e. g. Watts 2000 and the IPCC reports 1995, 2001, 2007). The much lower N_2O estimate implies (Rhee et al. 2009) that upwelling activities and/or the amount of dissolved N_2O in upwelling subsurface waters of the Atlantic are weaker than in the other oceans.

Emissions of organohalogen (Br or I) gases (one to two carbon atoms) from the oceans can have significant effects on the oxidant chemistry of the atmosphere, and in some cases in particle formation. There is a growing database of shipboard measurements from which sea-to-air fluxes can be calculated (e. g. Chuck et al. 2005). However, in contrast to DMS, there is little understanding of how these compounds are formed, with both direct biological production and indirect photochemical formation from organic precursor being invoked, but with little knowledge of the detailed mechanisms by which these processes operate.

The "best guess" emissions (in Tg element yr^{-1}) from the oceans are summarized here (cf. also Table 2.33 with older data):

OVOC	DMS	NH_3	NMVOC	N_2O	CH_4	org. Cl
50 ± 20	30 ± 15	10 ± 5	7 ± 4	1 ± 1	1 ± 0.5	1 ± 1

2.6.4 Geogenic sources

Geogenic or abiotic/abiogenic sources are characterized by physical and/or chemical processes on earth resulting in emission of trace gases and particles into the atmosphere. A special – so-called secondary emission – source is lightning; through physico-chemical processes (electric discharges) new products are produced from the ambient molecules (N_2 and O_2). The whole atmospheric chemistry itself continuously produces new secondary, products, but we only call an emission "secondary" when there are also direct, primary sources outside the atmosphere. When considering the budgets (and resulting effects and impacts on the climate system) of compounds it is necessary to estimate secondary sources as well as the primary sources.

2.6.4.1 Soil dust

The magnitude and distribution of atmospheric dust is strongly controlled by dust emissions, which depend on the extent and type of terrestrial vegetation and land use, as well as on soil properties and meteorological conditions. Soil particles are entrained into the air by wind erosion caused by strong winds over bare ground. While large sand particles quickly fall to the ground, smaller particles (less than about 10 µm) remain suspended in the air as mineral (or soil) dust aerosol. Billions of tons of mineral dust aerosols are released each year from arid and semi-arid regions to the atmosphere (Tanaka 2009). Mineral dust particles are estimated to be the most common aerosol by mass (see Chapter 4.3.5). Estimates of its global source strength range from 1000 to 5000 Mt yr^{-1} (Duce 1995).

Table 2.34 shows estimates depending on particle size. In recent years, the soil dust burden has been studied by modeling because of its contribution to climate (negative) forcing (Jaenicke 1993, Tegen and Fung 1995, Tegen et al. 2004, Hara et al. 2006) and to air pollution and air chemistry (Andreae and Crutzen 1997, Prospero 1999, Goudie and Middleton 2006). The chemical composition of dispersed soil varies over a wide range, Table 2.35 provides the global average of composition of the earth's crust. The agreement of older and modern estimates is remarkable with the exception of estimates for sulfur and carbon. It seems that the use of this mean composition is also not valid for large-scale deposition estimates (Table 2.36);

Table 2.34 Estimates of global particular emissions (Penner et al. 2001; IPCC 2001 estimates). NH − Northern hemisphere, SH − Southern hemisphere, OM − organic matter, BC − black carbon.

type and characteristics		NH	SH	global	author
OM	biomass burning	28	26	54 (45−80)	A
	fossil fuel burning	28	0.4	28 (10−30)	B, C
	biogenic	−	−	56 (0−90)	D
BC	biomass burning	2.9	2.7	5.7 (5−9)	A
	fossil fuel burning	6.5	0.1	6.6 (6−8)	B, C
industrial dust		−	−	100 (40−130)	E, F, G
sea salt	<1 µm	23	31	54 (18−90)	IPCC (2001)
	1−16 µm	1420	1870	3290 (1000−6000)	IPCC (2001)
soil dust	<1 µm	90	17	110	IPCC (2001)
	1−2 µm	240	50	290	IPCC (2001)
	2−20 µm	1470	282	1750	IPCC (2001)
	total	1800	349	2150 (1000−3000)	IPCC (2001)

A: Scholes and Andreae (2000)
B: Penner et al. (1993)
C: Cook et al. (1999)
D: Penner et al. (1995)
E: Wolf and Hidy (1992)
F: Andreae (1995)
G: Gong et al. (1997)

Table 2.35 Average composition of known terrestrial matter (in %); − no value given (see also Table 2.13).

element	Wedepohl (1995)	Mason and Moore (1982)	Clarke (1920)
O	−	46.60	47.33
Si	28.8	27.72	27.74
Al	7.96	8.23	7.85
Fe	4.32	5.00	4.50
Ca	3.85	3.63	3.47
Na	2.36	2.83	2.46
Mg	2.20	2.09	2.24
K	2.14	2.59	2.46
Ti	0.40	0.44	0.46
H	−	0.14	0.22
P	0.076	0.105	0.12
Mn	0.072	0.095	0.08
F	−	0.0625	0.10
Ba	−	0.0425	0.08
Sr	−	0.0375	0.02
S	0.070	0.0260	0.12
C	−	0.0200	0.19

Table 2.36 Deposition of soil dust on oceans, in Tg yr^{-1} (Duce et al. 1991).

ocean	deposition	deposition of iron
Northern pacific	470	1.6
Southern pacific	39	0.14
Northern Atlantic	220	0.76
Southern Atlantic	24	0.08
Northern Indian ocean	100	0.35
total	900	3.2

from the data listed in Table 2.36, a mean Fe concentration in dust deposition amounts to 0.35%, in contrast to the mean lithosphere composition of around 4.5%. The mean sulfur content of soils varies between 0.009 and 11.5% (Ryaboshapko 1983) in contrast to the mean crust composition of between 0.03 and 0.1% (Table 2.35). There are only two estimates of the contribution of soil dust to atmospheric sulfate (in Tg S yr^{-1}): 20 (10−30) and 3−30 (Ryaboshapko 1983, Andreae 1985, 1990). A contribution of 20 Tg S yr^{-1} to continental sulfate is significant in comparison with the assessed continental emission, in the range between 3 and 7 Tg S yr^{-1}, from biogenic sources including biomass burning.

Soil-forming material such as igneous, sedimentary and metamorphic rocks determines the chemical soil composition, which varies by more than one order of magnitude (Table 2.37). It is logical that material containing volatile elements such as C, N and S but also halogens are concentrated in soils and are resuspended in

Table 2.37 Variation of chemical composition of soil-forming rocks (igneous, sandstone and limestone), in % from date from Clarke (1920).

element	concentration range
Al	0.8−16
Fe	0.5−3.1
Mg	1−8
Ca	3−43
Na	0.05−4

the air by wind. The particle size is mainly in the range 1−3 µm, but sub-µm particles are also observed.

Locally, especially in urban areas where there is traffic on the streets, resuspension of soil dust by moving vehicles contributes to about one-third of the PM_{10} levels; this fraction is mainly between 2.5 and 10 µm (see also Chapter 4.3.5).

2.6.4.2 Sea salt

Sea-salt aerosol (SSA) has multiple impacts (beside other particulate matter categories) on atmospheric properties: response to climate by optical properties (e. g., Mahowald et al. 2006); providing cloud condensation nuclei (e. g., Clarke et al. 2006); being a heterogeneous surface for multiphase chemical reactions, e. g., SO_2 oxidation (Luria and Sievering 1991); and a source for reactive chlorine (e. g., Finlayson-Pitts et al. 2003), whereas gaseous HCl is the most abundant form of chlorine. Only in the last few years has evidence been found that SSA is also occurring in the very fine fraction below 250 nm (down to 10 nm) depending on sea state (e. g., Geever et al. 2005, Clarke et al. 2006). This aerosol may be the dominant contributor to both light scattering and cloud nuclei in those regions of the marine atmosphere where wind speeds are high and/or other aerosol sources are weak (O'Dowd et al. 1997, Murphy et al. 1998, Quinn et al. 1998).

Sea-salt particles are very efficient CCN, and therefore characterization of their surface production is of major importance for aerosol indirect effects. For example, Feingold et al. (1999) showed that in concentrations of 1 particle per liter, giant salt particles are able to modify strato-cumulus drizzle production and cloud albedo significantly. A semi-empirical formulation was used by Gong et al. (1997) to relate the size-segregated surface emission rates of sea salt aerosols to the wind field and produce global monthly sea-salt fluxes for eight size intervals between 0.06 and 16 mm dry diameter. Sea-salt aerosols are generated by various physical processes, especially the bursting of entrained air bubbles during whitecap formation (Blanchard 1983, Monahan et al. 1986), resulting in a strong dependence on wind speed. Sea-salt particles cover a wide size range (about 0.05 to 10 mm diameter), and have a correspondingly wide range of atmospheric lifetimes. Thus, as for dust, it is necessary to analyze their emissions and atmospheric distribution in a size-resolved model.

Several studies have quantified the flux and concentration of sea-salt particles. Blanchard (1985) expressed that it is impossible to estimate the total sea-salt emission more accurately than between 1000 and 10 000 Tg yr^{-1}; the flux is strongly dependent on the height above sea level: for the 0−1 m level the flux is > 10 000 Tg and for the ~ 5000 m level the flux is < 1000 Tg. Jaenicke (1982) estimated the transport of sub-μm sea-salt particles into higher atmospheric levels to be 180 Tg yr^{-1}. Using detailed wind-field data, Erickson and Duce (1988) estimated much higher fluxes than *Blanchard*. They give fluxes through a plane 15 m above the sea for dry deposition (8000−22 000 Tg yr^{-1}) and wet deposition (1500−4500 Tg yr^{-1}) resulting in a total flux of 10 000−30 000 Tg yr^{-1}. These estimates have often used empirical relations for the surface concentration, assumed a relationship between dry and wet deposition, and calculated the dry deposition. Tegen et al. (1997) found the global average burden of sea salt to be 22 mg m^{-2} (with a source strength of 5900 Tg yr^{-1}) using a 3D tracer model, whereas Takemura et al. (2000) calculated the source strength to be 3321 Tg yr^{-1} and the global burden to be 11 mg m^{-2} using a general circulation model. Both studies calculate the sea-salt concentration directly from an empirical dependence on surface wind.

The corresponding sulfate figure is 300−800 Tg S yr^{-1}, from which probably a mere 10 % (50 Tg yr^{-1}) is reaching the upper PBL. The global averaged mass percentage of sulfate in seawater amounts to 7.68 %. Assuming no deviation in pure sea salt, this figure must be multiplied with the total estimate of sea-salt emission to the atmosphere. Using a sea-salt production of 1800 Tg yr^{-1}, a figure for sea-salt sulfate emission of 44 Tg S yr^{-1} (Eriksson 1960) was used for more than 20 years by all investigators of the global sulfur cycle (Cullis and Hirschler 1980). Higher values of the sea-salt sulfate contribution, between 100 and 300 Tg S yr^{-1}, are published by Ryaboshapko (1983), Möller (1983, 1984a), and Varhélyi and Gravenhorst (1983).

Petrenchuk (1980) estimates that the river runoff of sea salt is 300−400 Tg yr^{-1}. This should equal what is deposited over land (as nothing is assumed to be accumulated on land). Using yearly river data of 35.6 · 10^3 km^3 and an average chlorine concentration in rivers of 6.4−7.8 mg L^{-1}, Petrenchuk (1979) finds 230−280 Tg yr^{-1} of chlorine. This should be approximately the same as the chlorine deposited over land from sea-salt aerosols. Grini et al. (2002) model deposits of 161 Tg yr^{-1} of sea salt over land, corresponding to 88 Tg yr^{-1} of chlorine. Given the uncertainties and assumptions in making such a comparison, the estimates compare reasonably well.

Several studies in the last few years have shown that "sea salt" aerosol actually contains a substantial amount of organic matter, consisting both of insoluble material (biological debris, microbes etc.) and water soluble constituents. Organic materials and sea salt are present as internal mixtures, consistent with a production mechanism that involves fragmentation of organic-rich surface film layers during the bursting of air bubbles on the sea surface (Andreae et al. 2009 and literature therein).

2.6.4.3 Volcanism and emanation

Volcanoes emit gases, aerosols, and fine particulate matter in modes ranging from continuous to ephemeral (cf. Chapter 2.2.1.1), Table 2.38. From 1975 to 1985, an

Table 2.38 Composition of volcanic gases; − no value given.

substance	in vol-% (Symonds et al. (1988, 1994)			in mol-% (Edmonds et al. 2005)[a]	in vol-% (Textor et al. 2004)[b]	in mol-% (sevens workshop 2000)[c]
	Kilauea summit (1170 °C)	Ertá Ale[d] (1130 °C)	Momo-Tombo[e] (820 °C)			
H_2O	37.1	77.2	97.1	75−85	50−90	94−99
CO_2	48.9	11.3	1.44	0.1−3	1−40	0.3−1.4
SO_2	11.8	8.34	0.50	10−13	1−25	0.1−4.5
H_2S	0.04	0.68	0.23	0.1−0,8	1−10	0.02−1.5
HCl	0.08	0.42	2.89	0.3−0,6	1−10	0.001−1.0
HF	−	−	0.26	0.1−0,5	<0.001	0.0003−0.4
HBr	−	−	−	−	?	0−0.0003
CO	1.51	0.44	0.01	0.015−0.025	−	0−0.0018
COS	−	−	−	−	$10^{-4}-10^{-2}$	−
CS_2	−	−	−	−	$10^{-4}-10^{-2}$	−
H_2	0.49	1.39	0.70	−	−	0.0001−0.4
N_2	−	−	−	−	−	0.003−0.07
CH_4	−	−	−	−	−	0−0.0006
Ar	−	−	−	−	−	0−0.00007
NH_3	−	−	−	−	−	0−0.00003

[a] Kilauea Volcano, Hawaii (2004−2005)
[b] means, based on values from Symonds et al. (1988, 1994), Cadle (1980), Chin and Davis (1993)
[c] 39 samples from two fumaroles at Satsuma-Iwojima and one at Kuju volcano in 2000 (T: 100−800 °C)
[d] Ethiopia
[e] Nicaragua

average of 56 volcanoes erupted yearly. While some showed continuous activity, others erupted less frequently or only once, so that 158 volcanoes actually erupted over this time period. This number increases to 380 volcanoes with known eruptions in the twentieth century, 534 volcanoes with eruptions in historical times, and more than 1500 volcanoes with documented eruptions in the last 10 000 years (McClelland et al. 1989). Hence emission estimates found in literature are always averages over different long time scales. Therefore the true annual emission for a given year is almost unknown and varies considerably from year to year. In years with large volcanic eruptions the emission could be many times higher than the average. On a short time scale, volcanic emissions play no role compared with other sources of emissions − with the exception of supervolcanic events. However due to emission into the upper troposphere, and occasionally direct into the lower stratosphere, volcanoes play an important role in providing trace species in layers of the atmosphere where the residence times increase significantly (Graf et al. 1997, 1998).

Chin and Jacob (1996) found that volcanic sulfur comprises 20−40% of the sulfate in the mid-troposphere globally and 60−80% over the North Pacific. It also

Table 2.39 SO$_2$ emissions of some large volcanoes eruptions, in Tg S.

volcanoes	year of eruption	emission	author
Roza (USA)[a]	14.7 Myr ago	6210[b, e]	Thordarson and Self (1996)
Laki (iceland)	1783–1784	61[c]	Thordarson et al. (1996)
Pinatubo (Indonesia)	1991	10	Bluth et al. (1997)
Agung	1963	3.5[d]	Self and King (1996)
El Chichón (Mexico)	1982	3.5	Bluth et al. (1997)
Nyamuragira (Congo)	1994	2	Bluth et al. (1997)
St. Helens (USA)	1980	0.5	Bluth et al. (1997)

[a] The Roza Member represents a compound pahoehoe flood basalt lava flow field, with an area of 40 300 km^2 and a volume of 1300 km^3
[b] over 10 years
[c] over 8 month
[d] over 2 days
[e] additionally from lava degassing: 1410 Mt S

accounts for a significant percentage of the sulfate burden in the upper troposphere at high latitudes. These disproportionate effects are realized even though volcanic S accounts for only 7% of the total S emitted to the atmosphere in their chemical transport model, yet it accounts for 18% of the column sulfate burden globally. Thornton et al. (1996) have shown that volcanic SO$_2$ becomes an important cloud condensation nucleus source in the North Pacific. There is a large number of global estimates of volcanic SO$_2$, with a variation between 0.75 and 30 Tg S yr^{-1}; the value with the most agreement seems to be 10 ± 5 Tg S yr^{-1} (see Tables 2.39 and 2.40). Other emission values are listed in Table 2.41.

Some recent analyses suggest that up to 40% of the global tropospheric sulfate burden may be volcanic; much of it derived from open-vent volcanoes, fumaroles fields and mild eruptions (Oppenheimer et al. 2003). This is much more than previous estimates of secondary sulfate formation suggest, with about only 5% (see Table 2.49). However, the values (burden and emission) are not directly comparable due to the fact that normally volcanic emissions are injected into the free troposphere where they spend much more time then in the lower troposphere. On the other hand, it is possible that the SO$_2$ emissions from permanent degassing activities globally are significant larger than is conventionally believed.

The historical record of volcanic activity, for example, goes back only a few thousand years, yet there are innumerable apparently extinct volcanoes that are still potentially active. Volcanic gases are globally imbalanced on time-scales comparing to biogenic processes (cf. Fig. 2.21). Closure of the volcanic cycle via oceanic subduction and magma transformation is very slow, thus we talk in geological time-scales. It is clear that volcanic activity during the early history of the earth (degassing period) was orders of magnitude higher than it is today (cf. Chapter 2.2.1). Before subduction, volcanic gases enrich the ocean. Assuming 10 Tg yr^{-1} as mean volcanic sulfur over the last 4 Gyr, it would correspond to a mass about 3000 times larger than the sulfur dissolved in the ocean (cf. Tables 2.9 and 2.12). Taking for HCl a volcanic emission of 1–10 Tg yr^{-1} (Table 2.41) over the same period (4 Gyr) results in only one-third to one-thirtieth of the mass of chlorine dissolved in the

Table 2.40 Estimates of global SO_2 emissions by volcanic activities, in Tg S yr^{-1}.

emission	author
0.75	Kellog et al. (1972)
17	Bartels (1972)
2	Friend (1973)
5	Stoiber and Jepsen (1973)
3.75	Cadle (1975)
23.5	Naughton et al. (1975)
3	Granat et al. (1976)
2.5	Cullis and Hirschler (1980)
20–30	Cadle (1980)
15	Várhelyi and Gravenhorst (1981)
7.5	Berresheim and Jaeschke (1983)
2	Möller (1983,1984a)
9.3	Stoiber et al. (1987)
4.5	Warneck (1988)
25	Lambert et al. (1988)
9.6	Spiro et al. (1992)
20	Graf et al. (1997)
7	Chin and Jacob (1997)
6.5	Andres and Kasgnoc (1998)
3.8–5.2	Halmer et al. (2002)
10 (± 5)	Möller (2003)
0.75–25	Textor et al. (2004)
12.6	Diehl et al. (2007)

Table 2.41 Emission of volcanic gases (in Tg yr^{-1}); – no value given.

substance	Cadle (1975)	Textor et al. (2004)[b]	Halmer et al. (2002)[b]
H_2O[a]	–	?	–
CO_2	–	75	–
SO_2	3.75	1.5–50	7.5–10.5
H_2S	–	1–2.8	–
HCl	0.75	0.4–11	1.5–37.1
HF	0.038	0.06–6	0.7–8.6
HBr	–	0.0078–0.1	0.0026–0.0432
CO	–	–	–
COS	–	0.006–0.1	0–0.32
CS_2	–	0.007–0.096	0–0.044

[a] based on the concentration ratio H_2O/Cl (Table 2.37) to be > 100, the water emission should be larger then 100–1000 Tg yr^{-1}
[b] means, based on values from Symonds et al. (1988, 1994), Cadle (1980), Chin and Davis (1993)
[c] compiled on a global data set on about 360 volcanoes over the past 100 yr of volcanic degassing during both explosive and quiescent volcanic events

ocean. This simple example should only show that it makes no sense to assume constant emission fluxes and ratios between different compounds over geological epochs.

Volcanoes regulate the climate through CO_2 emissions. Carbon dioxide emissions from volcanoes are given as 75 Tg yr^{-1} by Textor et al. (2004) and as 200−500 Tg yr^{-1} by Bickle (1994). The Mt Etna CO_2 plume emission and diffuse emission combined to amount to 25 Mt yr^{-1} (Gerlach 1991).

Volcanoes are the only net source (remember: biospheric actions provide closed cycles due to emission = uptake globally) of reduced substances (mainly sulfur species) which influence the oxidation capacity of the atmosphere. The most widespread effects will be derived from dust and volcanic gases, sulfur gases being particularly important. This gas is converted into sulfuric acid aerosols in the stratosphere and layers of aerosol can cover the global atmosphere within a few weeks to months. These remain for several years and affect atmospheric circulation, causing surface temperature to fall in many regions (Mather 2008). The effects include temporary reductions in light levels and severe and unseasonable weather (including cool summers and colder-than-normal winters).

Every now and again the earth experiences a tremendous, explosive volcanic eruption, considerably greater than the largest witnessed in historic times. Those yielding more than 450 km^3 of magma have been called super-eruptions (Self 2006). The record of such eruptions is incomplete; the most recent known example occurred 26 000 years ago. According to the Toba catastrophe theory, a supervolcanic event at Lake Toba, on Sumatra, 70 000 to 75 000 years ago reduced the world's human population to 10 000 (Ambrose 1998). The eruption was about 3000 times greater than the 1980 eruption of Mount St Helens in the USA. It is hypothesized that the Toba explosion may have reduced the average global temperature by 3−5 °C for several years. According to Robock et al. (2009), the Toba incident did not initiate an ice age, but rather exacerbated an ice age that was already underway.

It is more likely that the earth will experience a super-eruption than an impact from a large meteorite greater than 1 km in diameter. Depending on where the volcano is located, the effects will be felt globally or at least by a whole hemisphere. Large areas will be devastated by pyroclastic flow deposits, and the more widely dispersed ash falls will be laid down over continent-sized areas. The 1883 Krakatau eruption was the largest explosion ever observed. The 1815 Tambora eruption produced the "Year Without a Summer" in 1816. The 1963 Agung eruption produced the largest stratospheric dust veil in more then 50 years in the northern hemisphere, and inspired many modern scientific studies (Robock 2003). The subsequent 1982 El Chichón and 1991 Mt Pinatubo eruptions have been extensively studied from the beginning of the eruptions; they also led to very large stratospheric aerosol clouds and climatic effects. Labitzke and McCormick (1992) found a significant increase in stratospheric temperatures at 30 and 50 mbar levels in the Northern Hemisphere to be 2.5 °C higher than the 26−year mean, with some daily zonal mean increases of almost 3 °C. They believe this warming is due to the absorption of radiation by the new aerosols produced from the June 1991 eruptions of the volcano Pinatubo. The largest emissions from past which are estimated are from the Roza member of the Columbia River bassalt flow (in the north-west of the USA) about 15 Myr ago (Table 2.39). Over 10 years annual fluxes (in Tg element

yr^{-1}) have been assessed to be ~ 800 (SO_2), ~ 200 (HF), and ~ 100 (HCl) (Thordarson and Self 1996). Thus, the atmospheric perturbations associated with the Roza eruption may have been of the magnitude predicted for a severe "nuclear" or "volcanic" winter, but lasting up to a decade or more.

The composition of volcanic gases depends on a number of factors, the most important being the composition of the magma and the depth of gas-melt segregation prior to eruption; this latter parameter has proved difficult to constrain in the past, yet is arguably the most critical for controlling eruptive style. Some fundamental processes for how gases evolve from magma depth in the earth were discussed in Chapter 2.2.1.1. The principal components of volcanic gases are water vapor, nitrogen and carbon dioxide (Robock 2003, Robock and Oppenheimer 2003). Among the trace gases are sulfur species, halogens (HCl, HF, HBr) and in very small concentrations H_2, CH_4 and NH_3 (Table 2.38). However there are volcanoes where some trace gases have never been detected (Ryan et al. 2006). Sulfur (S) is a common trace element in many volcanically-emitted compounds. Sulfur dioxide (SO_2) and hydrogen sulfide (H_2S) are the primary S-containing gases emitted. Other S-containing species (COS, CS_2, S_2) usually occur in smaller quantities. The distribution between these various species differs not only between volcanoes, but also during a single volcanic episode at one volcano. The chemical composition of volcanic gases varies so widely (Tables 2.9 and 2.38) that it makes no sense to establish global mean volcanic gas compositions. Moreover several types of volcanoes differ considerably. Recently HNO_3 emission in the plumes of the Lascar and Villarrica volcanoes (Chile) has been reported (Mather et al. 2004a); molar emission ratios HNO_3/SO_2 of 0.01–0.07 have been found. At Masaya volcano (Nicaragua), NO and NO_2 are intimately associated with volcanic aerosol, such that NO_x levels reach as much as an order of magnitude above local background (Mather et al. 2004b). Volcanic sulfate aerosol is also emitted directly from volcanic vents (e. g., Allen et al. 2002, Mather et al. 2003). Andres and Kasgnos (1998) report on a mass-ratio of 0.006 for sulfate/SO_2.

The only published estimate attempting to quantify NH_3 emissions from a volcano appears to be that of Uematsu et al. (2004) for Mijahama volcano, in the south of Japan. They measured plume concentrations of NH_3 up to 5 ppb (~ 3 g m^{-3}) approximately 100 km downwind of the source, and reported an emission ratio of 1 ammonia : 1 ammonium : 1 sulfate : 10 sulfur dioxide. Based on the estimate that NH_x emissions were 15 % of SO_2 emissions, they inferred an NH_x release of 340 kt NH_3—N per year for the period since 2000. This is by far the largest NH_3 point source emission ever reported, being similar to the total annual NH_3 emission of the UK. Improved quantification of global volcanic NH_3 and NH_4^+ emissions must therefore be a priority (Sutton et al. 2008).

Volcanoes emit with the dust many metals: Cd, Hg, Ni, Pb, Zn, Ca, Sb, As and Cr (Nriagu and Pacyna 1988); for Cd and Hg metals, volcanoes contribute 40–50 % to global emission and for the latter 20–40 %. Based on global emissions estimates (Nriagu 1989), the emission of heavy metals amounts to 2–50 kt yr^{-1}. The total volcanic dust emission has been estimated between 25 and 300 Tg yr^{-1} (Peterson and Junge 1971, Jonas et al. 1995, Pueschel 1995).

2.6.4.4 Lightning

Liebig (1827) with his theory of nitrogen fixing by lightning probably stimulated interest in the importance of electric discharges for atmospheric trace gas formation. Later in the nineteenth century, the formation of ozone, hydrogen peroxide and nitric acid in air was attributed generally to electric discharges, "... variations in the electrical condition of the atmosphere ..." (Fox 1873). The general discourse was of "ozone-producing conditions" (quiet electric discharges of air). As mentioned in Chapter 1.3.2, at the end of the eighteenth century it was in vogue to study this subject, and *van Marum* noted in 1785 "the odor of electrical matter". *Cavendish* first used electric sparks for air (chemistry) studies in 1784. He and *Priestley*, noted in 1788 that the HNO_3 formation in moist air under the influence of electric discharges. NO formation has been suggested by Noxon (1976) and Hill (1979). Lightning (Rakov and Uman 2007, Betz et al. 2009) provides − depending on the flash energy − for molecule dissociation, similar to photochemistry (see Chapter 4.2.3). Therefore ambient air molecules are dissociated and subsequent new molecules are produced according the air chemical conditions. Depending on the flash energy and the molecule dissociation energy, there is no other limit to decompose (and subsequently to synthesize) any molecule in the atmosphere; as an example, CO production has been reported (Chameides 1979), and decomposition of CFCs by lightning (Cho and Rycroft 1997). But surely the most dominant dissociation is the oxygen splitting, starting a chain of subsequent reactions:

$$O_2 \rightleftarrows O + O \tag{2.95}$$

The subsequent atomic oxygen reacts with the main constituents N_2 and O_2 to produce NO and O_3. Ozone however easily decays back to $O + O_2$ by electrical discharges and radiation (2.98). However, direct ozone formation by lightning (Griffing 1977, Egorova et al. 1999) is small compared with the downward transport of stratospheric ozone. Indirectly lightning contributes to ozone formation in the upper tropospheric background due to the NO produced from lightning (Hauglustaine et al. 2001, see Chapter 5.4.1). The specific reaction rate k (Baulch et al. 1980) is given in terms of cm^3 molecule^{-1} s^{-1} (bimolecular) and cm^6 molecule^{-2} s^{-1} (trimolecular) respectively; and the activation energy (if available) in kJ mol^{-1} (see Chapter 4.2.1)

$$O + N_2 \rightarrow NO + N \qquad k_{2.96} = 1.8 \cdot 10^{12} \exp(-319/\boldsymbol{R}T) \tag{2.96}$$

$$N + O_2 \rightarrow NO + O \qquad k_{2.97} = 6.4 \cdot 10^{9} \exp(-26/\boldsymbol{R}T) \tag{2.97}$$

$$O + O_2 \rightleftarrows O_3 \tag{2.98}$$

A reaction chain follows where radical reactions (2.99 and 2.100) are rapid whereas thermal molecular reactions (2.101−2.103) are slow; the radical reaction rates of (2.104) and (2.105) are slow due to less reaction probability:

$$NO + N \rightarrow N_2 + O \tag{2.99}$$

$$NO + O \rightarrow N + O_2 \tag{2.100}$$

$$NO \rightleftarrows N + O \tag{2.101}$$

Table 2.42 Recent estimates of global NO production by lightning, in Tg N yr^{-1}.

emission	author
12.2 (5−25)	Price et al. (1997)
5 (2−20)	Pickering et al. (1998)
2 (1−8)	Lawrence et al. (1997)
0.2−22	Huntrieser et al. (1998)
5	Jourdain and Hauglustaine (2001)
7	Zhang et al. (2003)
2.8 (0.8−14)	Beirle et al. (2004)
3.5−7	Tie et al. (2004)
1.1−6.4	Boersma et al. (2005)
5.7	Mari et al. (2006)
6 ± 2	Martin et al. (2007)
5 ± 3	Schumann and Huntrieser (2007)

$$NO + NO \rightarrow N_2O + O \tag{2.102}$$

$$NO + NO + O_2 \rightarrow 2\,NO \qquad k_{2.103} \sim 2 \cdot 10^{-19} \tag{2.103}$$

$$N_2 + O + M \rightarrow N_2O + M \tag{2.104}$$

$$N_2O + O \rightarrow NO + NO \tag{2.105}$$

This mechanism is known as "Thermal NO Mechanism" or "*Zeldovich* Mechanism". The further oxidation of NO is described in Chapter 5.4.1. The range of NO formation is cited as between 0.3 and 230 Tg N yr^{-1}; $(1-30) \cdot 10^{16}$ molecules of NO per joule of input energy are produced (see citations in Lawrence et al. 1995, Rakov and Uman 2007). The key in estimating NO production by lightning is the flash rate distribution and the NO production per flash (Pickering et al. 2009).

Among the various NO_x sources, the contribution from lightning probably represents the largest uncertainty, but it seems that recent estimates (Table 2.42) agree with a global value of 5 ± 3 Tg N yr^{-1}. Interestingly many of the recent NO estimates are based on modelling and fitting emission assumptions by comparison with observations. As an example, Labrador et al. (2005) state that 0 and 20 Tg N yr^{-1} production rates of NO_x from lightning are too low and too high, respectively.

2.6.4.5 Biomass burning

Biomass burning is the oldest known man-made source and was responsible for significant local and regional air pollution even before the present era of fossil-fuel combustion (Andreae 1991, 2007). Several source categories are included with biomass burning (percentages in parenthesis according to Andreae 2004); savanna (35%), forests (21%; tropical 14% and extra-tropical 7%) agricultural residues (13%), charcoal making and burning (2%), and biofuel burning (29%), totaling 9200 Tg dry mass yr^{-1} (based on the late 1990s). The term "biofuel" used in combination with biomass burning considers only primary biomass, such as wood, but also peat, savanna grass and animal excrements (to avoid misunderstanding it should be renamed in biomass-fuel). Biofuel in a technical sense is defined as solid, liquid, or gaseous fuel obtained from living organisms or from metabolic by-products (organic or food waste products). Global biofuel production consists of ethanol (90%) and diesel (10%); the production tripled from $1.7 \cdot 10^{10}$ L in 2000 to about $5.6 \cdot 10^{10}$ L in 2007 (roughly 50 Tg C yr^{-1}), but still accounts for less than 3% of the global transportation fuel supply (Worldwatch Institute 2006). This amount is negligible compared with wood burning.

In contrast to open (wild) fires, domestic fires (mainly burning wood from farmland trees, indigenous forests, woodlands and timber off-cuts from plantations; also termed firewood) for cooking and (less importantly) heating are an important source of air pollutants. Little is known about the contribution of domestic fires to global biomass-burning emissions; the only estimate is given by Andreae (2004), which assessed this source to be one-third of the total burning. Three-quarters of the world's population use wood as their main source of energy. Of the global woodfuel production (1996: $1.9 \cdot 10^{15}$ g), 50% is used in Asia, 27% in Africa, 10% in Southern America and 13% in the rest of the world (FAO 2001, 2003). The FAO estimates that woodfuel consumption rose by nearly 80% between 1961 and 1998, slightly trailing world population growth of 92% over the period. There are indications that as much as two-thirds of woodfuel worldwide probably comes from non-forest sources (woodlands, roadside verges, residues, wood recovered, and waste packaging); however this amount is finally also turned in the cycle of harvested forest wood.

Biomass burning is known to be a significant source of many air pollutants, as estimated from a large number of measurements (reviewed by Andreae and Merlot 2001, Simoneit 2002, and Koppman et al. 2005). Biomass burning caused by fires is an important source of soot, particulate matter and gases. Generally there is a distinction between different types of biomass: wild fires (savanna, forest), agricultural burning, wood and charcoal burning (for cooking and heating), and waste (agricultural residues) burning. Almost all wild biomass burning (90%) is caused by humans (whether careless or intentional), of which 80% happens in the inner-tropical zone and 45% is due to savanna fires (Cachier 1998). Biomass burning shows large interannual variations: between 1997 and 2004 the variation was estimated from 2.0 Pg C yr^{-1} (minimum) and 3.2 Pg C yr^{-1} (maximum in 1998) with an average of 2460 ± 366 Tg C yr^{-1} (Werf et al. 2006), see also Table 2.43. The factors influencing the emissions from biomass burning are the combustion completeness (CC), fuel loads, moisture and burned area, integrated over the time and space scale of interest. Burned area is usually considered to be the most uncertain parameter in estimates of these emissions (it has become available on a global scale only recently using satellite data). Total carbon emission tracks burning in forested areas (including deforestation fires in the tropics, see Chapter 2.8.3.1), whereas burned area is largely controlled by savanna fires that responded to different environmental and human factors. Care should be taken when using and comparing values from the literature because dimensions are given in mass, in DM (dry matter), in volume and in carbon; conversion factors are often not given (density, C content).

Table 2.43 Global estimates of burned biomass and released carbon into the atmosphere.

source	burned biomass (Tg dry mass yr^{-1})		total carbon release (Tg C yr^{-1})		
	Dignon (1995)	Malingreau and Zhuang (1998)	Dignon (1995)	Malingreau and Zhuang (1998)	Mouillot et al. (2006)
tropical forest	1230	1230–2430	550–1090	570	1260
savanne	3470	1190–3690	540–1660	1660	1650
boreal forest	520	280–1620	130–230	130	205
temperate forest	–	–	–	–	185
wood fuel	1880	620–1880	280–850	640	–
charcoal	–	21	30	30	–
agricultural waste	1360	280–2020	300	910	–
tundra	4	–	2	–	–
total[b]	7660	3625–11665	1800–4740	3940	3300[a]

[a] 1700–4100 Tg C yr^{-1} after Denman et al. (2007), corresponds to 3–8 % of total terrestrial NPP
[b] 9200 Tg DM after Andreae (2004)

The biomass combustion process is generally divided into four basic combustion phases; ignition, flaming, smoldering, and glowing. During the initial heating period, large quantities of VOCs are released. Once the fuel is sufficiently dry, combustion proceeds from the ignition phase to the flaming phase (325–350 °C). During the flaming process, hydrocarbons are volatized from the thermally decomposing biomass and are rapidly oxidized in a flame. Products of complete combustion are CO_2 and H_2O. Incomplete combustion leads to emission of CO and a large variety of organic compounds. Biomass (wood) consists of two main constituents; cellulose and hemicellulose (50–65 %) and lignin (16–35 %); minor constituents are extractives (0.2–15 %) and trace minerals. During a low-temperature phase (< 100 °C) the functional groups of the main constituents decompose and methanol, and C_1–C_2 aldehydes and acids are the dominant emissions. Above 220 °C the polymer structure of the wood is decomposed; in this stage methane, aldehydes, methanol, furanes and aromatic compounds are emitted. The emitted gases will burn in the flame[16] more or less completely, be cracked and will synthesize new products, especially aromatic compounds. The smoldering phase is characterized by flameless combustion (solid phase oxidation) of the charcoal. Charcoal consists of approximately 90 % carbon, 5 % oxygen and 3 % hydrogen. As the fuel oxidizes, it burns with a characteristic glow, until the temperature is reduced so much that combustion cannot be continued, or all combustible material is consumed. The compound predominantly emitted is CO. Since the emission rates of all the products of incomplete combustion tend to be correlated, CO is often taken as a

---

[16] Production of gases from wood heating (called dry distillation) was known since ancient times to produce charcoal and tar. Methanol was first produced as a by-product in the manufacture of charcoal through the destructive distillation of wood, with yields of 12–24 L per ton of wood. With the beginning of the nineteenth century, wood distillation became a manufacturing process.

Table 2.44 Emission factors for biomass burning (after Andreae and Merlot 2001, Randerson et al. 2007), in g kg^{-1} (note that uncertainties are large and not given in this Table).

substance	savanna	tropical forest	extratropical forest
CO_2	1664	1580	1568
CO	63	102	106
CH_4	2.2	6.8	4.8
CH_3OH	1.3	2.0	2.0
CH_3COOH	1.3	2.1	3.8
CH_3CHO	0.5	0.65	0.48
HCHO	0.26	1.4	2.2
HCOOH	0.7	1.1	2.9
NMVOC	3.4	8.1	6.7
H_2	0.99	3.8	1.81
NO_x	2.35	1.85	3.00
NH_3	1.3	1.3	1.3
N_2O	0.21	0.20	0.26
SO_2	0.35	0.57	1.0
K	0.34	0.29	0.05
$PM_{2.5}$	4.9	9.1	13.0
TPM[a]	8.5	8.5	17.6
TC[b]	0.37	6.0	8.3
OC	3.2	5.2	9.1
BC	0.46	0.63	0.56

[a] total particulate matter
[b] total carbon in PM (OC + BC + inorganic)

surrogate for hydrocarbons and particulate carbon (Koppmann et al. 2005, Statheropoulos and Goldammer 2007).

Many laboratory and field studies have been carried out to estimate specific emission factors (e. g. Levine et al. 1993, Andreae and Merlot 2001, Koppmann et al. 2005). Andreae and Merlot (2001) list factors and global emissions for 92 species; they also separate from savanna and forest burning other sources: biofuel burning, charcoal making, charcoal burning, and agricultural residue burning (see Table 2.44). Despite the increase in research in biomass burning and climate interactions during the 1990s the current estimates of biomass consumption by fire remain the same, according to Innes (2000); it seems that this statement is still valid today, another decade later (Table 2.46).

Besides the species listed in Table 2.46, many other compounds have been detected in biomass plumes, for example, CH_2Cl_2 (F-l2) (Radke et al. 1991), CH_3Br (Mano and Andrea 1994), H_2O_2 (Lee et al. 1997), CHCN (Malingreau and Zhang 1998), HNO_2 (Keene et al. 2006, Stavrakou 2009b), COS (Crutzen et al. 1979, Nguyen et al. 1995), PAHs (Ballentine et al. 1996), and more than 80 different NMVOCs (Smith et al. 1992, Koppmann 2007). The biomass-burning emissions of CH_3CN may well dominate the global source of this compound, which thus might well be a unique tracer for biomass burning (Holzinger et al. 1999). It is remarkable

Table 2.45 Comparison between emission factors for biomass burning, in g kg^{-1}.

substance	Ward et al. (1992)	Ferek et al. (1998)	Andreae and Merlot (2001)	Yokelson et al. (2007)
CO_2	1614 ± 40	1599	1580 ± 90	1615 ± 40
CO	101 ± 20	105	102 ± 20	101 ± 20

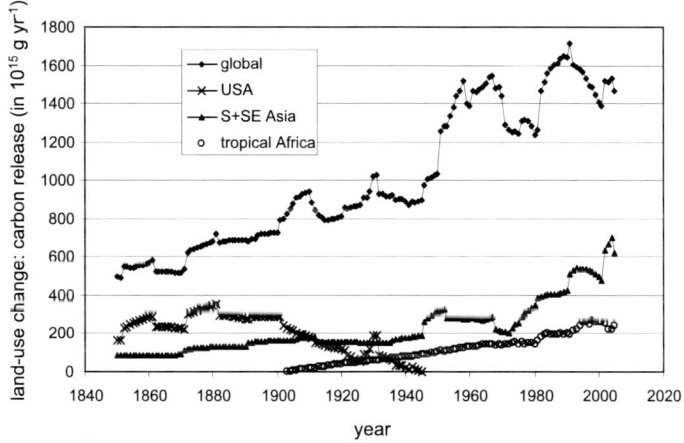

Fig. 2.43 Carbon release due to land-use change (1850−2005), after Houghton (2005b).

that emission factors (EF) for CO_2 and CO lie within a close range (Table 2.45). Forest fires contribute approximately 25 % of the total biomass burning, savanna burning 40 %, and woodfuel (firewood) 25 % (Table 2.47).

It is important to note that, over a long time scale, biomass burning is neutral in the atmospheric CO_2 budget, but in the last 200 years a steadily increasing amount of biomass burning has been occurring (Dignon 1995). It is likely that only savanna biomass burning lies within a timely closed carbon cycle because of rapid bush and grass growth in the next growing season. Therefore, forest burning (so-called carbon flux from land use change) contributes to the increase of atmospheric CO_2. The estimated global total net flux of carbon from changes in land use increased from approx. 500 Tg C in 1850 to a maximum of approx. 1700 Tg C in 1991, then declined to 1400 Tg C (1.4 Pg C) in 2000; see Fig. 2.43. It seems that the annual flux slightly decreases. Houghton et al. (2012), however, notes that the uncertainty is large ($\pm$ 0.3 Pg C) and that several modelling studies (giving an "average" of about 1.1 ± 0.2 Pg C for the period 1980−2009) do not reflect that. The global net flux during the period 1850−2000 was 146 Pg C, about 55 % of which was from the tropics. There have been major changes in the regional distribution of emissions from fires in the last hundred years, as a consequence of (a) increased burning in tropical savannas and (b) a switch of emissions from temperate and boreal forests towards the tropics. Therefore, biomass burning contributes significantly to the "greenhouse" effect, estimated to be up to 25 % (Andreas 1991).

Table 2.46 Estimates of global biomass burning emissions (in Tg yr^{-1}); note that some values are given based on the elements and compound, respectively. A: Lobert et al. (1991), B: Dignon (1995), C: Malingreau and Zhuang (1998), D: Andreae and Merlot (2001), Andreae (2004) E: Ito and Penner (2004), F: Randerson et al. (2007)[a], G: Schultz et al. (2008).

substance	A (1991)	B (1995)	C (1998)	D[i] (2001/04)	E (2005)	F (2007)	G (2008)
CO_2 (as C)	1860–4012	3200	3500	2237	2290	2418	2078
CO (as C)[h]	120–370	350	350	177	496	184	330
CH_4 (as C)	8–30	38[c]	–	12.9	32.2	14.1	15.4
HCHO[g] (as C)	–	–	–	2.6	11.5	–	3.9
CH_3OH[f] (as C)	–	–	–	2.8	9.2	–	–
HCOOH (as C)	–	–	–	1.3	–	–	–
CH_3CHO (as C)	–	–	–	1.7	–	–	–
CH_3COOH (as C)	–	–	–	3.6	21.7	–	–
acetone (as C)	–	–	–	1.3	–	–	–
CH_3CN (as C)	–	–	–	0.4	–	–	–
C_3H_6 (as C)	1–5	–	–	–	–	–	–
C_2H_6 (as C)	1–4.5	–	–	–	–	–	–
C_3H_8 (as C)	0.3–1.3	–	–	–	–	–	–
C_2H_2 (as C)	–	–	–	–	–	–	–
benzene	1.2–6.1	–	–	–	–	–	–
toluene	0.6–3.1	–	–	–	–	–	–
NMVOC[b]	19–96	24	38	24.8	38	27.7	–
CH_2Cl_2 (F12)	–	–	–	–	–	–	–
N_2O (as N)	0.2–0.7	0.8	0.8	0.71	–	0.73	–
NO_x (as N)	4–12	9	8.5	5.0	–	–	4.6
NH_3 (as N)	1.1–5.4	5.3	5.3	4.9	–	–	–
HCN (as N)	0.6–3.2	–	–	0.2	–	–	–
HNO_3 (as N)	0.3–0.5	–	–	–	–	–	–
SO_2 (as S)	–	2.2	2.8	1.2	–	–	2.2
COS (as S)	–	0.09	0.09	0.07	–	–	–
CH_3Cl (as Cl)	–	–	0.51	0.26	–	–	–
H_2	–	–	19	9.4	–	10.6	–
OC[c]	–	60–80	69	23.4	–	23.7	17.6
BC[d]	80–600	7–10	19	2.7	–	2.84	2.2
total PM[e]	–	100–200	104	48.8	–	50.7	–
PM2.5	–	–	–	36.4	38.3	37.8	–

[a] mean between 1997–2005 (SD 0.15–0.26)
[b] further estimates (in Tg C yr^{-1}): 28 including an uncertainty factor of 4 (Crutzen et al 1979), 42 (Lobert et al. 1991), 25–48 including about 30% OVOC (Koppmann et al. 1997, 2005), ~95 (Werf et al. 2006, Andreae 2007)
[c] further estimate (in Tg yr^{-1}): 54 (45–80), Guenther et al. (2006)
[d] further estimate (in Tg yr^{-1}): 5.7 (5–6), Guenther et al. (2006)
[e] further estimates (in Tg yr^{-1}): 29–72 (Pueschel 1995), 80 (Jonas et al. 1995)
[f] further estimate (in Tg yr^{-1}): 4 (Holzinger et al. 1999)
[g] further estimate (in Tg yr^{-1}): 34 (Holzinger et al. 2004); estimates for acetaldehyde and acetone (in Tg yr^{-1}) to be 10 and 6, respectively (Holzinger et al. 1999)
[h] further estimate (in Tg CO yr^{-1}): 663–807 (Holzinger et al. 2004)
[i] sum of savanna and forest burning (Andreae and Merlot also provide date for charcoal making, charcoal burning and agricultural residual burning)

Table 2.47 Biomass burning emissions in Tg C yr^{-1} (related to the late 1990s), values adapted and rounded from Andreae (2004).

substance	forest and wood		savanna burning
	total[a]	forest fire	
CO_2	2000	850	1400
CO	260	85	85
CH_4	25	9	5
NMVOC	60	22	20
particulate C	40	14	12
total C	2386	980	1522

[a] includes charcoal making and burning (amounts ~ 2% of total burned carbon)

2.6.4.6 Atmospheric chemistry: Secondary sources

We have already defined secondary sources (substances and hence emissions in terms of matter flux) as being chemical conversion processes in the atmosphere, but there must also be a primary source at the earth's surface. Therefore we exclude the uncountable number of intermediates and products from chemical processes in the atmosphere not having a primary source (Chapter 5); the best known atmospheric substance without a known primary source is ozone. In some estimates of biogenic and biomass burning emissions based on plume measurements, it is often impossible to separate primary from secondary sources, i. e., the values represent the sum. For example, Andreae and Merlot (2001) give for glyoxal a value from biomass burning of 5.9 Tg yr^{-1} and Stavrakou et al. (2009b) give a similar value of 7−9 Tg yr^{-1} for primary and secondary pyrogenic source, distinguished into 4−6 Tg yr^{-1} for primary and 3 Tg yr^{-1} for secondary formation (Table 2.48).

On the other hand, the number of chemicals theoretically is endless. It is simple to assume that from all the chemical compounds ever produced by humans and by nature, a certain amount of the substances escape to the atmosphere. Every chemical compound that has been described in the literature carries a unique numerical identifier, its CAS number, given by the Chemical Abstracts Service (CAS), a division of the American Chemical Society. CAS also maintains and sells a database of these chemicals, known as the CAS Registry. About 23 million compounds have received a CAS number so far, with about 4000 new ones being added each day. However only a small number of chemicals is commercially used; the EPA (the US Environmental Protection Agency) registers all environmentally relevant chemicals in an inventory, the Toxic Substances Control Act Chemical Substance Inventory (TSCA) (about 87 000), but only about 10% is in actual commercial use in significant amounts.

For the climate system, all chemicals are of importance which have a significant effect on changing physico-chemical properties; the effect results from the specific property (e. g. climate forcing) and the mass (emission flux). It remains a continuous research work to identify new substances having intolerable impacts on the

Table 2.48 Global secondary production of gaseous compounds; in Tg yr^{-1}.

substance	process	emission	author
NO	oxidation of NH$_3$	< 1	Möller (1996a)
NO	oxidation of N$_2$O	< 1	Möller (1996a)
CO	from NMVOC	350–2000 (?)	Möller (2003)
	from CH$_4$	362	Lelieveld et al. (1998)
CO$_2$	oxidation of CO and VOC	2000–4000 (?)	Möller (2003)
SO$_2$	oxidation of reduced sulfur	< 10	Möller (1996a)
HCHO	from CH$_4$	960	Stavrakou et al. (2009b)
	from isoprene	480	Stavrakou et al. (2009b)
	from anthropogenic NMVOC	112	Stavrakou et al. (2009b)
	from other biogenic NMVOC	48	Stavrakou et al. (2009b)
	total	1600	Stavrakou et al. (2009b)
CHOCHO	from biomass plumes	3	Stavrakou et al. (2009b)
	from anthropogenic NMVOC	12–13	Stavrakou et al. (2009b)
	from UVOC	54	Stavrakou et al. (2009b)
	from other biogenic NMVOC	32	Stavrakou et al. (2009b)
	total	40–100	Stavrakou et al. (2009b)
CH$_3$OCH$_3$	from anthropogenic alkanes	20 ± 10	Jacob et al. (2002)[a]
	from isobutene, isopentene	1 (1–2)	Singh et al. (2000)
	from monoterpenes	7 ± 4	Jacob et al. (2002)
	from methylbutenol	1.8 ± 1.8	Jacob et al. (2002)
	from NMVOC, biomass burning	25–31	Holzinger et al. (2004)
	total	29 (±50 %)	Jacob et al. (2002)
	total	38 (±100 %)	Singh et al. (2000)
CH$_3$OH	from NMVOC, biomass burning	29–35	Holzinger et al. (2004)

climate system. According to our present knowledge, the following reaction chains are significant in producing secondary substances from precursors:

$$VOC \rightarrow OVOC \text{ (oxygenated volatile organic compounds)} \rightarrow CO \rightarrow CO_2$$
$$CH_4 \rightarrow HCHO \rightarrow CO \rightarrow CO_2$$
$$VOC \rightarrow SOA \text{ (secondary organic aerosol)}$$
$$S_{red} \rightarrow SO_2 \rightarrow \text{sulfuric acid (sulfate aerosol)}$$
$$NH_3 \rightarrow NO$$
$$N_2O \rightarrow NO$$
$$SO_2, NH_3, NO \rightarrow \text{atmospheric aerosol}$$

As seen, the final oxidation state of volatile organic compounds (VOCs) is CO_2, but probably only a small fraction of emitted VOCs are totally decomposed by oxidative processes in the atmosphere. However, once organic fragments (C_1 and C_2 molecules) are produced these will be dominantly photolyzed and oxidized by OH radicals. Carbon monoxide is by far the most important secondary product from which again CO_2 is derived by further oxidation consuming OH radicals. C_1 and C_2 compounds are the most abundant decay products from VOC oxidation: formaldehyde (HCHO), methanol (CH_3OH), formic acid (HCOOH), acetone

Table 2.49 Global secondary aerosol production (in $Tg\ yr^{-1}$ as S or N, respectively).

substance	source	Penner et al. (2001)	Wolf and Hidy (1997)	Pueschel et al. (1995)[a]
sulfate from SO_2	anthropogenic	34 (19−60)		47
	biogenic	16 (8−33)	40	30
	volcanic	3 (0−13)		4
nitrate from NO	anthropogenic	3.2 (2.2−4.3)	9	8
	natural	0.9 (0.4−1.7)		5
ammonium from NH_3	anthropogenic	15 (8−26)	−	−
	natural	9 (5−20)	−	−
SOA^b	anthropogenic	0.6 (0.3−1.8)	4	10
	biogenic	16 (8−40)	−	55

[a] adapted from IPCC 2004 (see Jonas et al. 1995)
[b] secondary organic aerosol

(CH_3OCH_3), glyoxal ($OHC-CHO$), and oxalic acid ($HOOC-COOH$). Many oxygenated VOCs (aldehydes, ketones and acids) are water-soluble and are going into wet deposition. A significant percentage of atmospheric aerosol is produced from gaseous primary emitted nitrogen (NO and NH_3) and sulfur coumpounds (SO_2 and to a less share reduced sulfur), Table 2.49.

Because of their high solubility in water, multifunctional secondary compounds derived from the gas-phase oxidation of VOCs are suspected to be key contributors to the WSOC. Measurements show that 20−60% of the carbon mass present in fine atmospheric particulate matter consists of water soluble organic compounds (WSOC). However, only 5−20% of this WSOC has been identified, mainly as dicarboxylic acids. Speciation of WSOC was found to be mainly as tri- or higher multifunctional hydroxy-carbonyl species and hydroxy-hydroperoxide-carbonyl species, in urban and rural environments, respectively. However, it was also found that taking into account only the absorption of secondary VOC does not bring the mass concentration of carboxylic acids into agreement with the measurements. An attempt was made to explain this discrepancy by introducing chemistry occurring within deliquescent aerosols, promoting the role of water-soluble organic aerosol particles (Aumont et al. 2000). A large fraction (mainly from biogenic VOCs such as isoprene and monoterpenes) transforms into organic aerosol which has an important role in the climate system. Once in the aerosol phase, those compounds react further by photochemical reactions that alter particulate composition (e. g. Molina et al. 2004). Finally they will be removed by deposition processes from the atmosphere. The formation of inorganic salty aerosol by gas-to-particle conversion (see Chapter 4.3.4) from the precursors (and important anthropogenic emissions) ammonia (NH_3), sulfur dioxide (SO_2), and nitrogen oxides (NO_x) provides globally distributed background aerosol, probably producing the most important CCN and transporting essential bioelements around the globe and returning them by wet deposition to the biosphere.

Aldehydes belonging to secondary-produced OVOC, especially formaldehyde (HCHO), are most important as very reactive species (Altshuller 1993, Friedfeld et al. 2002). OVOCs are important secondary products in aged biomass burning plumes; the secondary production of methanol and acetone is probably more important than their primary sources. An example of an important secondary compound of which no primary sources are known (Volkamer et al. 2005) is glyoxal (CHO—CHO), the smallest dicarbonyl, which has recently been observed from space (Sinreich et al. 2007). The global anthropogenic glyoxal emission has been estimated by Olivier et al. (2003) to be 3.2 Tg yr^{-1}, almost negligible compared with the secondary emission. The chemical production of glyoxal is calculated to equal about 56 Tg yr^{-1} with 70 % being produced from oxidation of biogenic hydrocarbons, 17 % from acetylene, 11 % from aromatic chemistry and 2 % from ethene and propene (Myriokefalitakis et al. 2008). Current modeling studies underestimate substantially the observed glyoxal satellite columns, pointing to the existence of an additional source of glyoxal of biogenic origin on land (Stavrakou et al. 2009a); the largest columns of glyoxal are found in the tropical and sub-tropical regions, associated with high biological activity and the plumes from vegetation fires.

2.6.5 Anthropogenic sources

It is beyond the focus of this book to present a complex overview of man-made sources and emissions; the reader is referred to the literature (e. g. Klingenberg 1996, Ebel et al. 1997, Power and Baldasano 1998, Borrell and Borrell 1999, Lenz and Cozzarini 1999, Hutchinson 2003, Jackson 2003, Friedrich and Reis 2004, Olivier et al. 1998, 1999a, 1999b, 2001, 2002, 2003, 2005, Grauier et al. 2011); Table 2.50 shows generally the relation between anthropogenic sources and pollutants. An excellent guide for source characteristics and preparation of emissions inventories is given in EMEP (2006). The third edition of the emission inventory guidebook prepared by the UNECE/EMEP Task Force on Emissions Inventories and Projections provides a comprehensive guide to state-of-the-art atmospheric emissions inventory methodology. Its intention is to support reporting under the UNECE Convention on Long-Range Transboundary Air Pollution and the EU directive on national emission ceilings. Fig. 2.44 shows schematically the chain from natural resources to emissions via production and consumption of goods. It is remarkable that the number of substances being emitted into air in significant volumes is very limited; however, behind NMVOC and PM are a myriad of substances.

Table 2.50 Contribution of anthropogenic sources to main air pollutants; x – significant, (x) – minor importance, – negligible or zero.

source		CO_2	CO	CH_4	BC	NMVOC	SO_2	NO	NH_3	N_2O	dust
fossil	coal	x	–	(x)	–	–	x	x	–	(x)	x
fuels	oil/liquids	x	x	–	x	x	(x)	x	–	–	–
	gas	x	–	x	(x)	–	–	x	–	–	–
agri-	cropland	x[a]	–	–	–	(x)	–	(x)	x	x	x
culture	rice-fields	–	–	x	–	–	–	–	–	–	–
	animals	–	–	x	–	(x)	–	–	x	–	–

[a] burning of residues (see biomass burning)

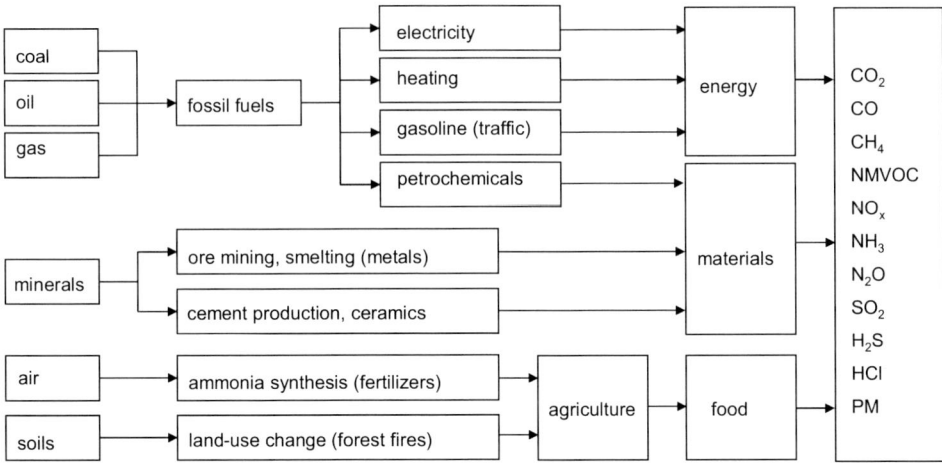

Fig. 2.44 Scheme of the air pollutants origin from natural resources through production and consumption.

It is also possible to create a myriad of subcategories in technology and product use. In an understanding of the (changing) chemistry of the climate system we can focus on only two categories of sources; fossil fuel use for energy supply and agriculture for food supply. A third category, materials, is important to create the technical fundamentals for the energy and food sector. With the exception of cement production, an important CO_2 source, all other production technologies play a minor role as emission sources compared with the combustion of fossil fuels and agricultural activities. In Chapter 2.8.1 the historical dimension will be described briefly in bridging the past with future developments.

2.6.5.1 Fossil fuel use: The energy problem

Fossil fuels, coal, crude oil (petroleum) and natural gas are a non-renewable source of energy and materials (see also Chapter 2.3.6); other fossil fuels are being investigated, such as bituminous sands and oil shale. The difficulty is that they need expensive processing before we can use them. It was estimated by EIA (2009) that in 2006 primary sources of energy consisted of petroleum 36.8 %, coal 26.6 %, and natural gas 22.9 %, amounting to an 86 % share for fossil fuels in primary energy production in the world; the rest is made up of nuclear and renewable energy sources (for historic energy statistics see Rethinaraj and Singer 2010).

The ASA-ASTM system[17] established four classes or ranks of coal; anthracite, bituminous, sub-bituminous, and lignite, based on fixed-carbon content and heat-

---

[17] ASTM International (ASTM), originally known as the American Society for Testing and Materials, is an international standards organization that develops and publishes voluntary consensus technical standards for a wide range of materials, products, systems, and services. The American Supply Association (ASA) is the US national organization serving wholesale-distributors and manufacturers.

Table 2.51 Chemical composition of coal; after Teichmüller and Teichmüller (1968).

coal type[a]	carbon (% of ash free)	volatile com- pounds (%)	water content (%)	caloric value (MJ kg^{-1})
lignite (brown coal)	60–71	49–53	35–75	17–23
sub-bituminous	71–77	42–49	10–35	23–29
bituminous	77–91	8–42	0–10	29–36
anthracite	>91	<8	0	36

[a] according to the ASTM classification

ing value measured in British thermal units per pound (Btu lb^{-1}). Anthracite, a hard black coal that burns with little flame and smoke, has the highest fixed-carbon content, 86–98% and a heating value of $(9.2–10.7) \cdot 10^8$ J kg^{-1} (Speight 2006, see Tables 2.51 and 2.55). This classification was already used 100 years ago (Robertson 1919); and Seyler (1900) used the C/H ratio for the coal classification, from lignite ($\sim$ 10) to anthracite (> 26). Based on raw coal, the chemical composition varies widely (in %):

Carbon	15–90
water	0–50
ash	0–30
hydrogen	1–5
sulfur	0–5
nitrogen	0–1

Coal combustion – except in open fires, where conditions similar to biomass burning are valid – is carried out at high temperature to achieve a complete oxidation into CO_2 and H_2O as main products. The flue gas (at about 55 °C and water-saturated) of coal-fired power stations consists roughly[18] of N_2 (79%), CO_2 (12–15%), O_2 (6–8%), SO_2 (20–100 ppm), NO (20–50 ppm), and PM_{10} (10–20 mg m^{-3}) as well as traces of HCl, NH_3, N_2O, H_2SO_4 and other. In spite of today's high-efficiency filters, fine PM from the ash will be emitted with a wide range of compounds (heavy metals, among them Hg, As, Cd; Ca, K, Mg, almost as oxides) are emitted; mercury is partly volatile and partitioning between the gas and solid phase. Since the end of the 1980s technologies for flue-gas desulfurization have been introduced stepwise, based on stripping SO_2 by limestone solution (with a removal efficiency of 90–95%). Nitrogen oxides (NO_x) are removed in the DeNOx process by selective catalytic reduction; commonly known as SCR, which uses ammonia (NH_3) to reduce NO_x by injecting it into flue gases while passing over a catalyst. NO_x cleaning is achieved at levels between 50% and 70% (Theloke et al. 2007). As a consequence, small amounts of N_2O and NH_3 have been found in the cleaned gas. Modern coal-fired power plants now just contribute to N_2O emissions (see Chapter 2.7.1.2).

Coal has been known and used since Antiquity, but the "era of fossil fuels" began with the Industrial Revolution in the mid nineteenth century (Fig. 2.45). The

---

[18] Based on our own studies in different German power plants, fired with Australian hard coal.

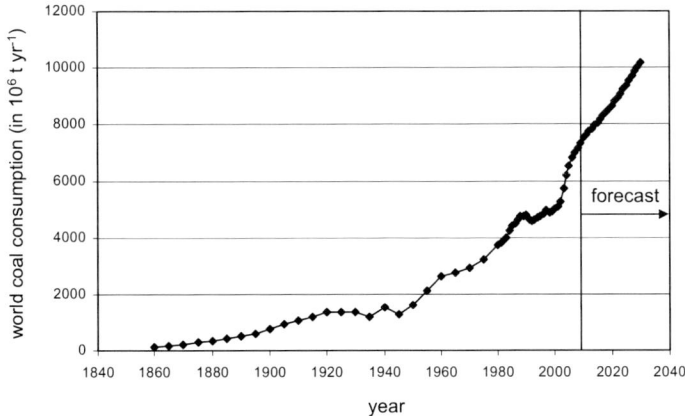

Fig. 2.45 World coal consumption in Mt yr^{-1}. Data from 1860 until 1975 after Möller (2003); data since 1980 after EIA (2009).

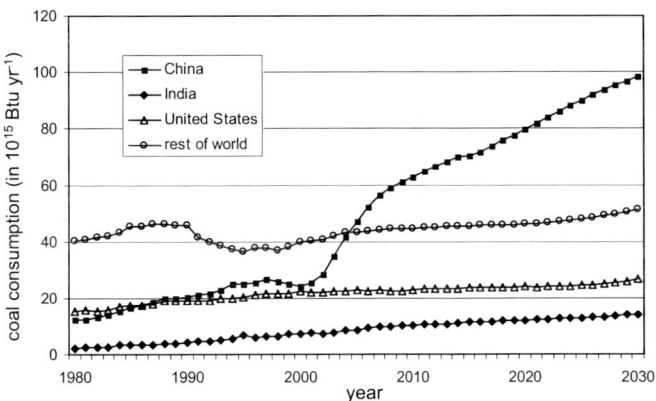

Fig. 2.46 Coal consumption in selected regions if the world in 10^{15} Btu yr^{-1}; data from EIA (2009). 1 Btu (British thermal unit) = 1.055 kJ.

period 1860–1914 is characterized by an exponential growth in coal use in the industrial countries; between 1920 and 1945 we clearly see the depression phase, and then after 1950 (until 1990) a second, nearly exponential growth phase is observed, interrupted by the "oil crisis" in the early 1970s. The rapid growth since 2000 is exclusively explained by the economic growth of China (Fig. 2.46). A doubling of the world's coal consumption by 2030 is expected compared with the consumption level of the 1990s (EIA 2009). The consequences – if the CCS technology, the flue-gas removal of CO_2, is not introduced early – are apparent (see also Chapters 2.8.4.2). The expected duration for continued coal use is more than 100 years (known resources will allow further energy supply by coal for several hundred years) – hence coal remains the most important energy carrier in the future until transfer to the "solar era" (Chapter 2.8.4.4). Möller (2003) wrote that the "peak

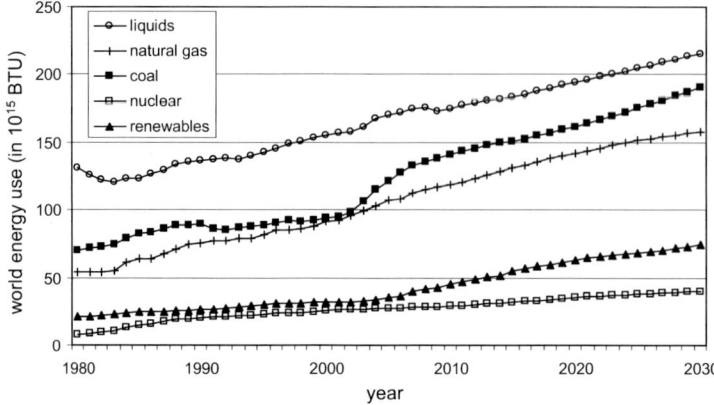

Fig. 2.47 World production of fossil fuels, in Mt oil equivalent (toe); data from EIA (2009). The IEA/OECD defines one toe to be equal to 41.868 GJ or 11630 kWh.

oil" will be expected between 2010 and 2020. Peak oil is the point in time when the maximum rate of global petroleum extraction is reached, after which the rate of production enters terminal decline. Nowadays it is controversially discussed whether there is/was a peak, or will be or not (Hirsch et al. 2005). The forecast in Fig. 2.47 shows a further slight increase in oil consumption; the known world fossil energy reserves increased continuously from 1996 ($8.9 \cdot 10^9$ toe) to 2008 ($11.1 \cdot 10^9$ toe), according to BP (2009). Of the world coal reserves, the USA holds 29%, Russia 19%, China 14%, Australia 9%, India 7%, and Ukraine, Kazakhstan and South Africa about 4% each (BP 2009). About 60% of coal is used for electricity production (Fig. 2.48); some 20% of coal is used for steel and cement production and 10% for heat for industrial processes. In 2006, world production was 5.5 Gt of hard coal and 0.95 Gt of brown coal. The efficiency of a conventional thermal power station, considered as saleable energy (in MWe) produced at the plant as a percentage of the heating value of the fuel consumed, is typically 33% to 48% efficient. Most modern coal-fired power plants have an electric efficiency of 48%, and there is work to achieve over 50%.

Petroleum (from Latin petra and Greek πέτρα – rock, and Latin oleum and Greek έλαιον – oil), mineral oil, or crude oil, is a thick, dark brown or greenish flammable liquid. Table 2.52 shows the elemental composition of crude oil; in contrast to coal, the variation of carbon is much less. Four principal groups of compounds are distinguished; hydrocarbons, non-hydrocarbons, organometallic compounds, and inorganic salts. Sulfur (mainly as H_2S) is removed before chemical treatment of the components (oil together with gas becomes an important sulfur source). The component chemicals of petroleum are separated by distillation (Table 2.53). The chains in the C_{5-7} range are all light, easily vaporized, clear naphthas. They are used as solvents, dry-cleaning fluids, and other quick-drying products. The chains from C_6 to C_{12} are blended together and used for gasoline (petrol). Kerosene is made up of chains in the C_{10} to C_{15} range, followed by diesel fuel/ heating oil (C_{10} to C_{20}) and heavier fuel oils such as those used in ship engines. Lubricating oils and semi-solid greases range from C_{16} up to C_{20}. Chains above

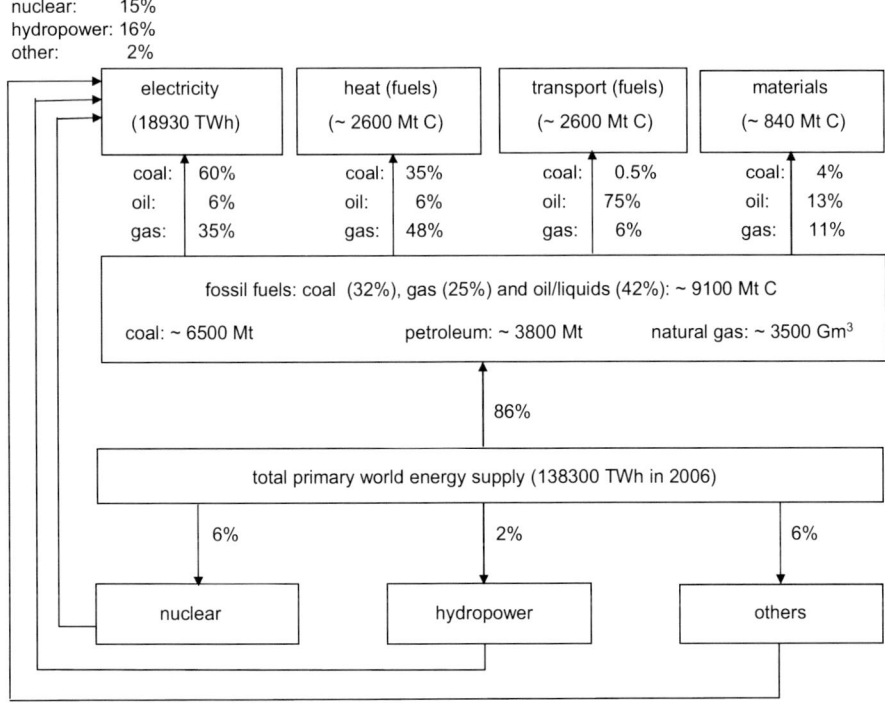

Fig. 2.48 World use of energy carriers (in %); data adapted from EIA (2009), WCI (2005). Note that $1\,W = 0.315 \cdot 10^8$ J yr^{-1}; 138300 TWh in 2006 corresponds to 15.8 TW (cf. Fig. 2.17). Percentage are based on out-fluxes; only sources of electricity are in-flux percentages. "Other" includes biofuels, geothermal heat, wind etc.

Table 2.52 Chemical composition of crude oil; data adapted from Leverson (1967), Simanzhenkov and Idem (2003), and Jones and Pujadó (2003); type of compounds after Manning and Thompson (1995), Hatch and Matar (1977).

compounds	content (in %)	type of compounds
carbon	82–87	alkanes, aromatics, napthenes
hydrogen	10–15	
sulfur[a]	0.1–4.5	organic aliphatic and aromatic sulfides, di- and polysulfides, thiophene and homologs
nitrogen	0.1–1.5	pyridines, quinolines, acridines, pyroles, indoles, carbazoles, phorphyn
oxygen	0.1–1.7	organic acids, phenols, cresols, esters, ketones, furans
metals	0.0–0.1	Na, Ca, Mg (soaps)
		V, Ni, Co, Fe (phorphyrins)

[a] also elementary and evolved to H_2S around 200°C (Eccleston et al. 1952) and as aliphatic sulfur compounds

Table 2.53 Crude oil fractions according to distillation temperature; after Simanzhenkov and Idem (2003).

fraction	T (°C)	named	use
light	40–70	petrol ether	used as solvent
(20–60%)	60–100	light petrol	automobile fuel
	100–140	heavy petrol or light kerosene	household solvent and fuel
heavy	140–180	heavy naphtas	jet engine fuel
(10–20%)	180–240	kerosene fraction	
	240–350	diesel fraction[a]	diesel fuel, heating
	> 350	atmospheric residues	= heavy fraction
vacuum	300–400	vacuum oil	engine oil
(~ 50%)	400–450	medium oil	
	450–490	heavy oil	
	> 500	vacuum residues	tar, asphalt, residual fuel

[a] also named gas oil (250–350 °C) and lubrication oil (> 300 °C)

Table 2.54 Chemical composition of natural gas (in vol-%); data adapted from Matar and Hatch (2001), Guibet and Faure-Birchem (1999), Leverson, A. I. (1967).

compound		mean	range
CH_4	(methane)	~ 95	62–97
C_2H_6	(ethane)	~ 2.5	1–15
C_3H_8	(propane)	~ 0.2	0–7
C_4H_{10}	(butane)	~ 0.2	0–3
C_5H_{12}	(pentane)	~ 0.03	< 0.2
C_6H_{14}	(hexane)	~ 0.01	< 0.1
N_2		~ 1.3	1–25
CO_2		~ 0.02	1–9
H_2S		~ 0.2	< 3
He		~ 0.1	< 2
H_2		~ 0.01	< 0.02

C_{20} form solids, starting with paraffin wax, then tar and asphaltic bitumen. The boiling ranges of petroleum atmospheric pressure distillation fractions are up to 350 °C; for higher boiling temperature (up to 500 °C) vacuum distillation is used. Petroleum is roughly used as fuels for transport (75%), heating (10%), petrochemicals (10%), asphalt (3%) and electricity (2%) (McKetta 1992, Jones and Pujadó 2003). With the exception of vacuum residues it is possible to convert the heavier fraction into lighter (through catalytic cracking) to increase the yield of gasoline (petrol). From 1996 to 2000, the share of fuel oil decreased from 26% to 11% whereas the percentage of the light and middle fraction increased from 58% to 68% (BP 2009); it is expected that in future it will be used exclusively for engine fuels.

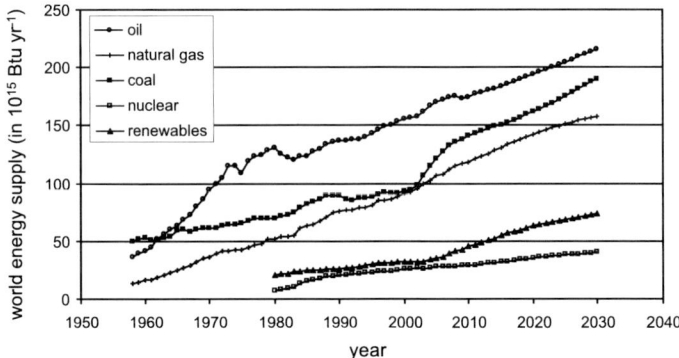

Fig. 2.49 World energy supply 10^{15} Btu yr^{-1}; data adapted from EIA (2009) and BP (2009). Note that „renewable" is almost hydropower until the year 2000.

Natural gas consists primarily of methane (Table 2.54). It is found in association with fossil fuels, in coal beds, and together with oil. Gas present in contact with and/or dissolved in crude oil is coproduced with it. Non associated gas is found in reservoirs containing no oil (in dry wells but also as clathrates under frozen conditions or under high pressure at the deep sea bed). Before natural gas can be used as a fuel, it must undergo extensive processing to remove almost all materials other than methane. The by-products of that processing include ethane, propane, butanes, pentanes and higher molecular weight hydrocarbons, elemental sulfur, and sometimes helium and nitrogen. After further removal of CO_2 and water, the gas for heating is almost CH_4. This methane is converted temporarily to liquid form (atmospheric pressure at a temperature of $-162\,°C$, $-260\,°F$ or 111 K) for ease of storage or transport (called LNG). The gross heat of combustion of one normal cubic meter of commercial quality natural gas is around $39 \cdot 10^6$ J ($\approx$ 10.8 kWh). Heating and electricity generation have been the main traditional uses of natural gas. Natural gas remains a key energy source for industrial sector uses (heat for processing) and electricity generation. In the electric power sector, natural gas is an attractive choice for new generating plants because of its relative fuel efficiency and low carbon dioxide intensity. Electricity generation is predicted to account for 35 % of the world's total natural gas consumption in 2030, up from 32 % in 2006; industrial use is expected to be 40 % in 2030, the rest is residential use (heating and cooking), BP (2009).

Still more than 85 % of the total primary energy supply is provided by fossil fuels (Fig. 2.48). It seems (Fig. 2.49) that in the next 20 years all growth in energy carriers will be linear (in % yr^{-1}): oil (1.25), coal (3.25), gas (2.25), renewable (6.75), and nuclear (1.75); see also discussion in Chapter 2.8.4.2 for limits. From the world electricity production of 18.9 GWh in 2006 (only 16.4 GWh was the world net electricity consumption, i. e., 13 % of electricity produced was lost before use), 67 % is due to fossil fuels (Fig. 2.48), but it corresponds to a total fossil-fuel energy supply of about 41.1 GWh; in other words, the global conversion efficiency (not to be confused with the efficiency of power plants) is only 30 %. It is obvious that increasing efficiency leads to reduced fuel consumption and specific emissions. All fossil fuels consumed in 2006 provided about 9.1 Gt of carbon (Fig. 2.48); a mere 9 % were converted into materials, such as chemicals, plastics, elastomers etc., which delays its return as CO_2 into the atmosphere after its final combustion. Therefore, more than 90 % of fossil-fuel carbon is released as CO_2 (8.2 Gt in 2006) into the air. Around 30 % of fossil-fuel carbon is used as engine fuels (transport) and the same amount (30 %) for heat and industrial processing (10 %, or about 1 Gt C for steel and cement production). The different uses of fossil-fuel carbon also show that in future carbon cannot be fully replaced by renewable energy sources; about 1.5−2 Gt C is needed for non-energetic (materials) and process-specific use.

The reserve-to-production (R/P) ratio denotes a number of years of continued use of fossil fuels when the reserves remaining at the end of any year are divided by the production in that year. From the year 2012 the world R/P amounts to 110 for coal, 60 for gas and 50−60 for oil (BP 2013), little change compared with the year 2008 (in the first edition of this book). Therefore, the limit of fossil fuels to serve energy carriers is one side in approaching the transfer from the fossil fuel era to the solar era; but it seems that an amplifying force is climate change due to CO_2 emissions. In 2008, the world electricity (19.2 TWh) was produced from 41 % coal, 20 % gas, 19 % renewable energy (90 % of which is hydropower), 15 % nuclear energy, and 5 % oil. In 2030 (expected 31.8 TWh electricity), there is no significant change in the share of coal and gas, but a reduction in oil (to 3 %) and nuclear energy (to 12 %) and a slight increase in renewable energy (to 21 %) (BP 2009). From the data we learn that the ultimate future source, direct solar energy use, is totally negligible and will not rise before 2030.

The calorific value or heat of combustion or heating value of a sample of fuel is defined as the amount of heat evolved when a unit weight (or volume in the case of a sample of gaseous fuels) of the fuel is completely burnt and the products of combustion cooled to a standard temperature of 298 °K. Net calorific value assumes the water leaves with the combustion products without fully being condensed. Fuels should be compared based on the net calorific value (Table 2.55). The calorific value of coal varies considerably depending on the ash, moisture content and the type of coal, while calorific values of fuel oils are much more consistent. Table 2.55 shows that oil and gas have about 30 % less specific CO_2

Table 2.55 Carbon content and net caloric values of fossil fuels (global averages based on 2006); based on raw material. Calculated from data in IEA (2009), BP (2009) and date from Table 2.51.

raw fuel[a]	carbon content (%)	caloric value (MJ kg^{-1})	caloric value (MJ kg^{-1} carbon)	specific CO_2 emission (kg carbon kJ^{-1})
lignite	30 (25−35)	10.0	~ 33	30
bituminous	70 (45−85)	25.0	~ 36	28
anthracite	90 (86−98)	34.0	~ 38	26
crude oil	85	42.7	~ 50	20
natural gas[b]	75	38.8	~ 52	19

[a] includes water and ash (large variations in coals in contrast to oil and gas)
[b] 1 m^3 natural gas $\approx$ 1 kg natural gas

emission than coal, based either on carbon content or energy output; the only reason for this difference is the water content of the fuel and the energy loss due to water evaporation. Water condensation and heat recovery seems a way to increase the net efficiency and hence reduce emissions.

2.6.5.2 Agriculture: The food problem

More than 99.7% of human food comes from the terrestrial environment while less than 0.3% comes from the oceans and other aquatic ecosystems (FAO 2001). Worldwide, of the total land area on earth, 11% is used for growing crops and 27% is pasture land (Table 2.56). Around 21% of the land area is unsuitable for crops, pasture, and/or forests because the soil is too infertile or shallow to support plant growth, or the climate and region are too cold, dry, steep, stony, or wet. As discussed concerning biogenic sources in Chapter 2.6.2, we have to consider the vegetation (crop), soils (farmland and pasture) and animals (grazing and livestock farming). Arable land may be regarded as a "high-performance" ecosystem, but only because of targeted crops and fertilizer application. Therefore, agricultural activities contribute to emission of carbon and nitrogen compounds, where CH_4 and NH_3 are the only two important compounds, and agriculture itself represents the dominant source of these compounds. Nitrogen release (beside NH_3, the greenhouse gas N_2O is important but also NO from soils is not negligible) is the result of fertilizing to increase yield but at the expense of high nitrogen loss into soil, groundwater and

Table 2.56 Global land-use areas, in 10^{12} m²; see also Table 2.18.

Land type	area	source	land type	Loveland et al. (2000)	IPCC (2001)
closed forest	38.8	FAO (2006)			
fragmented forest	4.8	FAO (2006)			
other woodland	12	FAO (2006)	forest[d]	40.1	41.7
croplands[a]	14.7	Goldwijk (2001)	croplands[a]	14.0	16.0
pastureland	34.5	Goldwijk (2001)	grassland[c]	54.8	44.5
wetlands	5.5	Mattews (1993)	wetlands	1.3	3.5
deserts	13.1	Nishigami et al. (2000)	non-vegetated	18.5	45.5
urban areas	0.4	WRI	urban	0.3	–
total[b]	130.7	FAO (2006)	sum	129.0	–
Antarctica	13.9	Loveland et al. (2000)	snow and ice	16.6	–
Greenland	2.2				
sum continental	146.8		total	145.7	151.2

[a] cultivated land
[b] total is not equal to sum
[c] natural cropland, savannas, scrublands, wooded scrublands
[d] all types

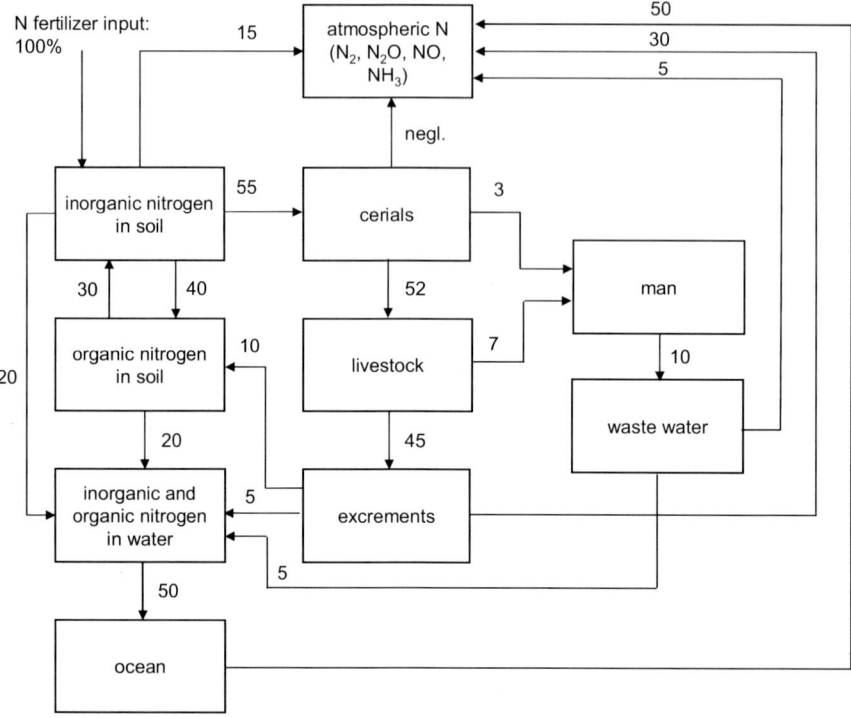

Fig. 2.50 Fate of applied nitrogen fertilizers to farmland; data from Möller und Schiefer-decker (1982).

air (Fig. 2.50). Finally, only 10% of applied N fertilizers are transferred into food. The general conclusion, shown in Fig. 2.50, that only half of the nitrogen applied to croplands was incorporated into plant biomass and that less than 10% was consumed by humans, is also shown by Bodinsky et al. (2012). FAO/IFA (2001), estimates that globally from nitrogen fertilizers 3.3% is lost from cropped fields as N_2O (3.5 Mt N yr^{-1}) and NO (2.0 Mt yr^{-1}) and 14% is lost as NH_3 (11.2 Mt yr^{-1}); an additional 7.8 Mt NH_3-N is emitted from animal manure application (corresponds to 23% of manure N); all data related to 1995. Raery et al. (2012) estimate the global N_2O emission from agriculture to be 5.6 Tg N yr-a (year 2000) having an increase up to 7.3 Tg N yr^{-1} by 2030. An increase of N_2O emission from agriculture is also modelled by Bodinsky et al. (2012), from 3 Tg N_2O-N in 1995 up to 7–9 Tg by 2045. Of global anthropogenic emissions, in 2005 agriculture accounted for about 60% of N_2O and 50% of CH_4 (Smith et al. 2007), but 90% of NH_3. Agriculture accounted for an estimated CO_2 emission of 1.4 to 1.7 Gt C yr^{-1}, but the net flux is estimated to be approximately balanced, with CO_2 emissions around 0.01 Gt C yr^{-1} only (Smith et al. 2007); emissions from energy and fuel use are covered in the fossil fuel sector. Another important agricultural source is biomass (residues) burning. Unfortunately, there are large residues in agriculture (primarily crop stalks and roots), estimated for 1997/99 to be 4.7 Gt and increasing to 7.4 Gt in 2030 (Bruinsma 2003), hence two to three times more than the harvested crop. Using a mean carbon content of agricultural residues of 29%, another 1 Gt C is taken out of the food chain, and mainly burned (see Chapter 2.6.4.5); forest residues are in the some order of magnitude. Andreae and Merlot (2004) estimated burning of agricultural residue to be 1.19 Gt dm yr^{-1} or about 0.55 Gt C yr^{-1}, respectively. Table 2.50 indicates the compounds emitted by agricultural activities. It should be mentioned that dust evolution from croplands by wind erosion but also by ploughing contributes to PM_{10} loading of the air.

According to FAO estimates, the world population will be increasingly well-fed by 2030, with 12.7 MJ available per person, compared with 9.8 MJ per person per day in the mid 1960s and 11.6 MJ today. It follows that the world population in 2006 ($6.5 \cdot 10^9$ humans) consume $0.27 \cdot 10^{20}$ J yr^{-1} (or 0.9 TW), corresponding to 5.5% of the world energy supply (Fig. 2.48). Taking the values from Fig. 2.17, this amount of biomass energy corresponds to about 0.7 Gt C and is equal to about 25% of yearly NEP. The annual production of cereals (2007) amounts to 2082 Mt (1033 Mt grains, 626 Mt wheat, and 423 Mt rice) (OECD-FAO 2009). With the mean carbon content of cereals at 40% (Metzger and Benford 2001), it follows there is 830 Mt carbon, very close to the above rough estimate of 700 Mt. On average each human consumes 130 kg C yr^{-1} which is about ten times the human's carbon content (19%, and 70 kg weight taken into account). Globally, some 660 Mt of cereals are used as livestock feed each year. This represents just over one-third of total world cereal use.

Agricultural N_2O emissions are projected to increase by 50−60% up to 2030 due to increased nitrogen fertilizer use and increased animal manure production (Smith et al. 2007). Combined CH_4 emissions from enteric fermentation and manure management will increase by 21% between 2005 and 2020.

2.6.5.3 Land-use change and deforestation: The population problem

Land use is the human modification of natural environment or wilderness into built environment such as fields, pastures, and settlements. Land-use change represents the anthropogenic replacement of one land-use type by another, e.g., forest to cultivated land (or the reverse), as well as subtle changes of management practices within a given land-use type, e.g., intensification of agricultural practices; both of which are effecting 40% of the terrestrial surface (Fischlin et al. 2007, see Table 2.56). Agriculture has been the greatest force of land transformation on earth (Lambin and Geist 2006). A global increase of cropland area from $265 \cdot 10^{10}$ m^2 in 1700 to $1471 \cdot 10^{10}$ m^2 in 1990, while the area of pasture has increased more than six times from $524 \cdot 10^{10}$ m^2 to $3451 \cdot 10^{10}$ m^2 have been estimated (Goldewijk 2001). Another estimate by Forster et al. (2009) is similar; for the year 1750 the global area of cropland and pasture has been assessed to be $(7.9−9.2) \cdot 10^{12}$ m^2 (corresponding to 6−7% of the global land) and for the year 1990 to be $(45.7−51.3) \cdot 10^{12}$ m^2 (35−39% of global land: $131 \cdot 10^{12}$ m^2 excluding Antarctica, cf. Table 2.2). This land-use change was partly due to deforestation; the forest area decreased in this time period "only" by $11 \cdot 10^6$ km^2 (9% of global land), Forster et al. (2007). The forest change has been estimated by IPCC (2007) to be -8.9% (1990–2000) and slowing down to -7.3% (2000–2005). Interestingly the ratio 6.7 of the world population increase from 1750 to 1990 (cf. Fig. 2.58) is only somewhat higher than the increase of the agricultural area of 5.8 (5.0–6.5) from the data cited above, suggesting that the crop yield per area only slightly increased globally.

The major effect of land use since 1750 has been deforestation of temperate regions (only within the last few decades has tropical forest become more important). Of particular concern is deforestation, where logging or burning are followed by the conversion of the land to agriculture or other land uses. Even if some forests are left standing, the resulting fragmented landscape typically fails to support many species that previously existed there. It is estimated that roughly 50 % of the original forest (existing ca. 8000 years ago) has been lost (Billington et al. 1996). In addition to the intrusion of humans and their activities throughout the earth's land area, degradation of soil has emerged as a critical agricultural problem (Pimentel and Kounang 1998). Erosion on agricultural land is estimated to be 75 times greater than that occurring in natural forest (Pimentel and Pimentel 2007). About 50 % of the global land is devoted to agriculture. Of this, about one-third ($15 \cdot 10^{12}$ m²) is planted with crops and two-thirds is pastureland. As a result of erosion, during the last 40 years about 30 % of world's arable land has become unproductive (roughly $200 \cdot 10^{10}$ m²). Throughout the world, current erosion rates are greater than ever before. Further, people in developing countries have been forced to burn crop residues for cooking and heating and this exposure of the soil to wind and rainfall energy intensifies soil erosion as much as 10 times. Most of the additional cropland needed yearly to replace lost land now comes from the world's forest areas (roughly $4000 \cdot 10^{10}$ m² in 2000). Forests[19] cover about 30 % of land surface and hold almost half of the world's terrestrial carbon; if only vegetation is considered (ignoring soils), forests hold about 75 % of the living carbon (Houghton 2005b, see Table 2.57). Per unit area, forests hold 20 to 50 times more carbon in their vegetation than the ecosystem that generally replaces them. The above-ground living biomass has been estimated by Keith et al. (2009) to be about 200 t C ha^{-1} for tropical and temperate forest and 60 t C ha^{-1} in boreal forests. The global estimate by Kindermann et al. (2008) results in 60 t C ha^{-1} globally (see Table 2.58). According to different scenarios (Nabuurs et al. 2007), forest area in industrialized regions will increase between 2000 and 2050 by about 60 to 230 million ha. At the same time, the forest area in the developing regions will decrease by about 200 to 490 million ha.

Fig. 2.43 shows that deforestation is a significant CO_2 source to the atmosphere. The deforested area was calculated to be $8.9 \cdot 10^{10}$ m^{-2} yr^{-1} in the period 1990–2000 and $7.3 \cdot 10^{10}$ m^{-2} yr^{-1} between 2000 and 2005 (Denman et al. 2007). According to different estimates, the tropical deforestation released $(1-2) \cdot 10^{15}$ g C yr^{-1}; the area of deforestation was $(8-15) \cdot 10^{10}$ m² yr^{-1} in the 1980s and

---

[19] The definition of a forest according to FAO (2001) is: an ecosystem that is dominated by trees (defined as perennial woody plants taller than 5 m at maturity), where the tree crown (or equivalent stocking level) exceeds 10 % and the area is larger than 0.5 ha. The term includes all types of forests, but excludes stands of trees established primarily for agricultural production and trees planted on agroforestry systems.

Table 2.57 Global forest, after FAO (2003); C&N − Central and North, S − South (values rounded).

region	landarea, in 10^{12} m^2	forest, in stock				production[a], in 10^{15} g yr^{-1}	
		area, in 10^{12} m^2	volume, in 10^9 m^3	mass, in 10^{15} g	density, in m^3 ha^{-1}	wood-fuel	total wood
Africa	29.8	6.5	46.5	70.9	72	0.81	0.93
Asia	30.8	5.5	34.5	45.1	63	1.02	1.56
Europa	22.6	10.4	116.4	61.1	112	0.06	0.48
C&N America	21.4	5.5	67.3	52.4	123	0.12	0.95
Oceania	8.6	2.0	10.8	12.6	55	0.01	0.09
S America	17.5	8.9	110.8	180.2	125	0.30	0.65
world	130.7	38.8	386.4	422.3	100	2.32[b]	4.66

[a] data by FAO (2003) given in m^3 yr^{-1}; recalculated using the regional mass-volume ratio
[b] given in 1.78 · 10^9 m^3 yr^{-1} by FAO (2003); by using world mean mass/volume ratio it follows only 1.94 · 10^{15} g yr^{-1}. This slightly inconsistency may be explained by different averaging, but reflects the uncertainty of global values, which are "masked" when using "virtual correct" statistical data given in digits down to kt yr^{-1} by FAO (2003). As seen for an example, the values of the world forest area are different in Table 2.57 (38.8), Table 2.58 (39.8), and Table 2.18 (42.3)

Table 2.58 Global forest biomass, after Kindermann et al. (2008).

forest area (in 10^{12} m^2)	39.8	
forest volume (in 10^{11} m^3)	4.7	
mass (in 10^{15} g)	biomass	carbon
above-ground	470	234[b]
below-ground	130	62
dead	80	41
litter	−	23
soil	−	398
total	680[a]	758

[a] total is not equal to the sum of above-ground, below-ground and dead biomass
[b] The changes have been estimated to be (in 10^{15} g C) 299 (1990), 288 (2000), and 282 (2005), after Denman et al. (2007). These values (in the range 230−290 Pg C) are considerable lower than other estimates: 359−536 (different estimates in IPCC 2001), 335−365 (IPCC 2007), 422 (Nieder and Benbi 2008). Interestingly, all authors state a percentage of global forest to global land vegetation of about 80% (77−82%); Chapin et al. (2000) states that 50% are in tropical forests and the remaining 30% in other forests. Schlesinger (1997) state 560 Pg as global plant carbon stock; hence derived about 450 Pg in forests.

$(10-17) \cdot 10^{10}$ m^2 yr^{-1} in the 1990s (Houghton 2005b, see Table 2.59). The lower estimates are based on satellite observations and the higher estimates (according to Houghton (2005a) the more reliable ones) are based on the data base from the FAO (the Food and Agriculture Organization of the United Nations). It is important to note that carbon release is due to two principal processes: first by the vegetation replacement (mainly by fires − biomass burning) and secondly by release from the

Table 2.59 Carbon emission from tropical land-use change (deforestation) in 10^{15} g C yr^{-1}.

carbon release	author
1.5 ($\pm$1.2)	Gurney et al. (2002)
2.4	Fearnside (2000)
0.9 (0.5−1.4)	DeFries et al. (2002)
2.2 ($\pm$0.8)	Houghton (2003)
1.1 $\pm$ 0.3	Archard et al. (2004)

Table 2.60 Approximated global terrestrial carbon fluxes, based on forest inventories, in 10^{15} g C yr^{-1} (or Gt yr^{-1}); − release from ecosystems (harvesting and emission into atmosphere), + uptake by ecosystem (NEP − CO_2 sequestration), + emission into the atmosphere (source) and − uptake from the atmosphere (sink).

process	fluxes, total	fluxes, forests	fluxes, atmosphere
land-use change, total	−2.0[a]		
soil emission[b]			+1.0
forest fires		−1.0[c]	+1.0[c]
wood harvesting[d]		−2.4	+?[e]
woodfuel use			+1.4
vegetation growth, total	+3.3[f] (1.7−4.1)		−3.3[f] (1.7−4.1)
forest growth		+1.6[g]	
budget (net flux)	+1.3 (0.9)[g]	−1.8	$\pm$ 0

[a] from Table 2.59
[b] difference between total land-use change and forest fires
[c] from Table 2.47
[d] includes woodfuel use
[e] wood products (paper, panels etc.) return back
[f] Mouillot et al. (2006)
[g] Denman et al. (2007)

soil's carbon pool after land-use change. Table 2.59 shows a range of 1−2 Gt yr^{-1} carbon release due to tropical land-use. Taking into account a value for tropical forest burning to be about 0.8 Gt C yr^{-1} (Table 2.43), it follows that about the same amount of carbon is emitted from soil into the atmosphere; a considerable amount additional to other anthropogenic sources (cf. Fig. 2.10). As noted in Table 2.12 and Table 2.58, the ratio between above-ground carbon in vegetation and organic carbon in soils is globally 0.4−0.6, where the upper estimate represents tropical forests; or in other words, about twice the carbon in living biomass is stored in the soil. The historic loss of soil carbon (to be emitted as CO_2 to the atmosphere) from cultivated cropland soils of the world ranges from 41 to 55 Pg C (Amundson 2001). The global organic carbon amount in soil ranges from 1500 to 2000 Pg C (to 1 m depth) and is in various forms, from recent plant litter to charcoal and very old humified compounds. Inorganic carbon (carbonate) amounts to 800−1000 Pg C.

Table 2.60 shows a very rough assessment of the budget of forest carbon fluxes. The estimated net flux (~ 1 Gt C yr^{-1}) suggests that the global forest is a sink for atmospheric CO_2. However, this value, based on forest inventories, is only a fraction of the sinks inferred from atmospheric data and models, estimated to be 2.1 (± 0.8) Pg C yr^{-1}, referred to the northern mid-latitudes and suggesting that more than half of the northern terrestrial carbon sink is in non-forest ecosystems (Houghton 2005a). The uncertainties are large, however.

Historically, global land-use emissions were predominantly caused by the expansion of cropland and pasture, while wood harvesting (for timber and fuel wood) only played a minor role. These findings are robust even when changing some of the driving forces like the extent of historical land-use changes (Minnen et al. 2009). Due to land-use change, large amounts of carbon have been emitted into the atmosphere (see Figs. 2.43 and 2.56). The high uncertainty in land-based CO_2 flux estimates is thought to be mainly due to uncertainty in not only quantifying historical changes among forests, croplands, and grassland, but also due to different processes including calculation methods. Inclusion of a nitrogen cycle in models is fairly recent and strongly affects carbon fluxes (Jain et al. 2013). Land use change between 1750 and 2011 (including deforestation, afforestation and reforestation) contributed a 180±80 Pg CO_2-C release into the atmosphere; vegetation biomass and soils not affected by land use change stored 150±90 Pg C (IPCC 2013). Hence, the net effect was an emission of 30 Pg C into the atmosphere druing the period of 1700−2011 (35 Pg C for the period 1750−200 after Minnen et al. 2013).

2.7 Emission of atmospheric substances

An *emission inventory* is a database that lists, by source, the amount of air pollutants discharged into the atmosphere of a community during a given time period. As already emphasized several times, emissions as the flux (mass per time) of substances in the atmosphere are the key parameter for atmospheric chemistry − they determine the initial air concentration (the key parameter for the reaction rate) and the composition (the key parameter for the reaction mechanisms). Beginning with estimates of global values of a high degree of uncertainty 30−40 years ago, now emission (or better, source) processes are described with the time function for diurnal, weekly and annual cycles with the aim to achieve a better temporal resolution for global inventories (Table 2.61). By using geographical statistical data, the sources and emissions are spatially resolved, from the 1° · 1° normally used in global models down to 1 · 1 km^2 in local models. It is another question how reliable the temporal and spatial resolved estimates are. The *Global Emissions Inventory Activity* (GEIA) aims to provide global gridded emissions inventories to science and policy communities for all trace gases. GEIA was created in 1990 within the frame of IGAC (*International Global Atmospheric Chemistry Program*) to encourage the development of global emissions inventories of gases and aerosols emitted into the atmosphere from natural and anthropogenic sources. The long-term goal is to provide inventories of all trace species relevant to global atmospheric chemistry (Graedel et al. 1993). Emission inventories are available for the following species:

Table 2.61 List of selected global and regional emission inventory activities. RIVM – Rijksinstituut voor de Volksgezondheid (Bilthoven, Netherland), MPI – Max-Planck-Institute (Mainz, Germany), LSCE – Laboratoire des Sciences du Climat et l'Environnement (Paris, France), IER – Institut für Energiewirtschaft und Rationelle Energieanwendung (Stuttgardt, Germany).

inventory	spatial cover and resolution	temporal cover and resolution	remarks
GEIA (IGAC)	global, $1° \cdot 1°$ or per country	1985,1999	natural and anthropogenic (EDGAR) emissions
EDGAR 3.2 (RIVM)	global, $1° \cdot 1°$ or per country	1990–1995	Only anthropogenic emissions
AEROCOMN (MPI/LSCE)	global, $1° \cdot 1°$	2000, 1750	natural, anthropogenic and "effective" secondary aerosol
POET	global, $1° \cdot 1°$	1990–2000	POET emission WP extended EDGAR and GEIA
RETRO	global, $0.5° \cdot 0.5°$	Global emissions for the ERA-40 period	RETRO is about modelling intra annual trends in tropospheric chemistry
EMEP	Europe, $50 \cdot 50$ km^2	1990–2002, 1 hour	anthropogenic emissions, based on official reporting of countries and experts assessment
GENEMIS (IER)	Europe, Germany, $10 \cdot 10$ km^2	1 hour	production on demands for selected episodes
TNO	Europe, $25 \cdot 25$ km^2	1 hour	
ABBI	Asia, $1° \cdot 1°$	2000–2001	Asian biomass burning inventory
REAS	Asia	1983–2003, 2010, 2020	regional emission inventory for Asia[a]

[a] For China, three emission scenarios have been developed: REF (reference case), PSC (Policy Success Case), and PFC (Policy Failure Case)

– acidification: NO_X from soils and lightning; NH_3 from natural soils, oceans and wild animals,
– aerosol formation: SO_2 from volcanoes; DMS from oceans,
– climate change: CH_4 from wetlands, termites, oceans/hydrates; N_2O from natural soils and oceans,
– tropospheric ozone: CO from vegetation and oceans; CO soil sink; NMVOC from vegetation,
– major reactive chlorine compounds: CH_3Cl, $CHCl_3$, CH_2Cl_2, C_2HCl_3, and C_2Cl_4 from oceans; CH_3Cl and $CHCl_3$ from land-based sources; HCl and $ClNO_2$ from sea salt dechlorination.

The EDGAR (*Emission Database for Global Atmospheric Research*) project is a comprehensive task carried out jointly by the Netherlands National Institute for Public Health and the Environment (RIVM) and the Netherlands Organization for Applied Scientific Research (TNO) (Bouwman et al. 1997, Olivier et al. 1999a, 1999b, 2005, Olivier and Berdowski 2001, Olivier 2002, Janssens-Maenhout et al.

2012). This set of inventories combines information on all different emission sources, and it has been used over the past few years as a reference database for many applications. The work is linked into and part of the Global Emissions Inventory Activity (GEIA) of IGBP/IGAC. The last version of EDGAR v4.2 (November 2011) provides calculated global annual man-made emissions for direct greenhouse gases (CO_2, CH_4, N_2O, and 22 halogenated compounds), ozone precursor gases (CO, NO_x, CH_4, NMVOC), acidifying gases (NH_3, SO_2, NO_x) and PM_{10} for the period 1970−2008. This new database structure allows the calculation of emissions by country, sector and includes specific technologies for combustion/processing and emission abatement measures. Furthermore, to facilitate both the use of EDGAR data in air pollution and climate modelling on different scales, the country emissions are allocated to a $0.1° \cdot 0.1°$ grid using newly developed $0.1°$ grid maps for a large variety of emission sources. EDGAR datasets have also been used in IPCC Assessments, both on source strengths and on spatial distribution of emissions in the development of emission scenarios (Nakicenovic et al. 2000). These SRES scenarios, as they are often called, were used in the IPCC Third Assessment Report (TAR), published in 2001, and in the IPCC Fourth Assessment Report (AR4), published in 2007 (IPCC 2001, 2007).

Beside the large international activities in emissions inventories (Table 2.61) there are several substance- or source-related emissions estimates; for example, for mercury in Europe (Pacyna et al. 2001) and the world (Pacyna and Pacyna 2002, Pacyna et al. 2003, 2006), global carbonaceous aerosol inventory 1860−1997 (Junker and Liousse 2006), and trace metals (Nriagu 1979, 1989). Moreover, many countries (almost all developed) provide national emission inventories.

There are two general methodologies used to estimate regional to global emissions; bottom-up and top-down. Bottom-up methodologies apply the following general equation to estimate emissions:

$$E_i = A_i \, (EF)_i \, P_{1i} \, P_{2i} \, \text{.......} \tag{2.106}$$

where E_i are emissions (for example, kg sulfur hr^{-1}), A_i is the activity rate for a source (or group of sources i, for example, kg of coal burned in a power plant), $(EF)_i$ is the *emission factor* (amount of emissions per unit activity, for example, kg sulfur emitted per kg coal burned), and P_{1i}, P_{2i} ... are parameters that apply to the specified source types and species in the inventories (for example, sulfur content of the fuel, efficiency of the control technology). Top-down methodologies, also known as *inverse modeling*, derive emission estimates by inverting measurements in combination with additional information, such as the results of atmospheric transport and transformation models.

Table 2.62 shows the significance of emitted substances to key properties of the climate system; all substances have natural as well as anthropogenic sources. Abatement strategies have been applied successful but to very different degrees for all substances with the only exception being CO_2 and with only limited reduction for NH_3, CH_4, N_2O. Some substances have been completely controlled within the last few decades − or at least to an extent that they no longer play role (or more than a negligible role) in the climate system or as local pollutants, such as trace metals, smell-intensive substances and CFCs. Behind NMVOC are so many organic compounds with different properties that an assessment of the climate system relevance

Table 2.62 Relevance of emitted substances to the climate system. + and – positive and negative significant, respectively, (+) less significant. CCN cloud condensation nuclei, i. e. influencing cloud and precipitation formation.

species	global warming	global cooling	tropospheric ozone formation	stratospheric ozone depletion	secondary aerosol formation	acidifying	CCN
SO_2	–	+[b]	–	–	+	+	+[b]
DMS[f]	–	+[b]	–	–	+	(+)[e]	+[b]
NO_x	–	(+)[b]	+	(+)[d]	+	+	+[b]
N_2O	+	–	–	+	–	–	–
NH_3	–	+[b]	–	–	+	+	+[b]
CH_4	+	–	+	–	–	–	–
PM	–	(+)	–	–	–	(+)[h]	(+)
seasalt	–	(+)	(+)[j]	–	–	–	+
NMVOC	(+)[a]	(+)[b]	+	–	+	–	(+)[b]
CO_2	+	–	–	–	–	(+)[c]	–
CO	(+)[a]	–	–	–	–	(+)[a]	–
BC	+	–	–	–	–	–	–
HCFC's[g]	(+)	–	–	+	–	–	–
HFC's	(+)	–	–	+	–	–	–

[a] via CO_2 formation
[b] via (secondary) aerosol formation
[c] weak acidity in global background
[d] from high-altitude aircrafts
[e] via oxidation to SO_2 and further to H_2SO_4
[f] only natural emitted
[g] synonym with CFC's
[h] as NH_3, partly acid-neutralizing if alkaline such as carbonates
[j] in sense of ozone depletion via heterogeneous processes

needs a compound-specific approach; however, as already seen (Tables 2.30 and 2.33) natural sources become more and more dominant with a further abatement of anthropogenic sources.

Table 2.63 summarizes global natural emissions of gases. As seen the uncertainties are large. Although the citation is from the year 2000 (which means that the data results from research around and before 1995) there is no reason not to use this data in comparison with anthropogenic sources for climate change. In recent years, the values of natural NMVOC emissions have been corrected continuously to larger figures (Friedrich 2009). It is worth noting again that man-made climate change (deforestation, warming etc.) has a feedback on natural emissions too. For example, the global NMVOC emission has been estimated to be 717 Tg C yr^{-1} in 1986 and 778 Tg C yr^{-1} in 1995, i. e., the increase amounts 8.5%.

As seen from Table 2.50, only three anthropogenic activities contribute substantially to emissions on a global scale: stationary combustion of fossil fuels (CO_2 and SO_2), mobile combustion of fossil fuels (CO_2, CO, and NMVOC), and agriculture (CH_4, NH_3, N_2O).

Table 2.63 Estimation of global natural emission of gases (in Tg element yr^{-1}). Source: Watts (2000); reduced sulfur modified after Möller (2003).

source	NH_3	NO	N_2O	CO_2	DMS	CS_2	SO_2	COS	H_2S
ocean	10–15	–	2	–[c]	15–25	0.2	–	0.25	1.8
aquatic ecosystems	–	–	–	–	≤ 0.5	0.02	–	0.03	0.7
soils and plants	3–5	4–8	8–10	–[c]	2–3	0.07	–	0.02	0.8
wild life	1–3	–	–	–	–	–	–	–	–
biomass burning[a]	5–7	9	1	3500	–	–	2–3	0.07	?
volcanisms	–	–	–	–	–	< 0.2	5–10	0.2	1–3
lightning	–	4	–	–	–	–	–	–	–
secondary sources	–	1–2	–	–	–	–	5–20	0.6	–
total	25 (± 5)	20 (± 2)	10 (± 3)	3500[c]	25 (± 5)	0.3 (± 0.1)	12 (± 6)	0.4 (± 0.1)	4.5 (± 1)

source	CO	CH_4	C_2–C_4 alkanes	C_2–C_4 alkenes	terpenes	other NMVOC
ocean	20–100	10[d]	< 1 (?)	1	–	?
aquatic ecosystems	–	125	–	–	–	–
soils and plants	100	–	?	3	300–1000	3–5
wild life	–	4	–	–	–	2
biomass burning[a]	300–350	40	6	15	–	10–20
volcanisms	–	–	10[b]	–	–	–
lightning	–	–	–	–	–	–
secondary sources	700–2400	–	–	–	–	–
total	450 (± 100)	150 (± 30)	15 (± 5)	20 (± 10)	600 (± 300)	20 (± 10)

[a] including man-made percentage (see text)
[b] earth emanation
[c] NEP

2.7.1 Nitrogen compounds

2.7.1.1 Ammonia (NH_3)

Our knowledge of biogenic ammonia emission is still surprisingly small. Although there has been interest in ammonia research since 1800, and the recognition of its importance for plant growth and the changes in atmospheric ammonia have been well documented (see Chapter 1.3.3), almost all studies deal with agricultural issues (see Chapter 2.6.5.2). Modern estimates of the global terrestrial NH_3 emission are in the range 3–12 Tg N yr^{-1} (Table 2.64). In most studies, such values are derived from budget estimations, supported by up-scaled experimentally determined values of grassland emission. However, a number of sources are not generally included in such values (Sutton et al. 2008).

Table 2.64 Global emission of ammonia (in Tg N yr^{-1}).

author	natural emission		biomass burning	man-made emission
	terrestrial	oceanic		
Söderlund and Svenson (1976)	2–6	–[a]	–	24–47
Jaffe (1992)	–	–	–	–
Schlesinger und Hartley (1992)	10	13	5	–
Dentener and Crutzen (1994)	7.6[b]	–	–	–
Möller (1996)	8	15	–	30
Bouwman et al. (1997)	7.2	8.2	4.1	34.1
Schlesinger (1997)	–	13	–	52
Friedrich and Obermeier (2000)	3	10	7.2	–
Watts (2000)	4–8	10–15	5–7	20–40
Brasseur et al. (2003)	12	–	–	40
best estimate	8 ± 2	10–15	5 ± 1	40 ± 5

[a] no specification
[b] separated into 5.1 (soils) and 2.5 (wildlife)

Table 2.65 NH$_3$ emission factors for different animals and humans (in g N species^{-1} yr^{-1}). A: Buismann et al. (1986), B: Möller and Schieferdecker (1989), C: Klaassen (1992), D: Battye et al. (1992), E: Strogies and Kallweit (1996), F: Ryaboshapko (2001), G: Misselbrook et al. (2001).

source	A	B	C	D	E	F	G
cattle	13.64	22.1[a]	12.5/35.5[b]	22.9	23.04	14.3/28.5[b]	5.6/22[b]
pigs	3.87	5.2	5.1	4.0/9.2[c]	5.36	6.4	4.0/4.3[c]
poultry	0.23	0.22	0.18/0.32[c]	0.18	0.25	0.32	0.2/0.4[c]
horses	–	15.0	12.5	–	12.20	–	–
sheeps	0.46	3.0	2.1	3.4	1.70	1.34	0.6
humans	–	1.3	0.3	0.25	–	1.3 (± 50%)	–

[a] milk cow 35.0, fattening cow 15.4
[b] other cows / milk cows
[c] different husbandry (fattening pig / sow, cock / laying hen)

The bacterial decomposition of animal excreta is the largest source of NH$_3$ (Table 2.65). For example, colonies of wild animals, such as seabirds and seals, although they produce a small fraction of the global emission, have recently been recognized as hotspots of NH$_3$ emission in remote areas with otherwise low NH$_3$ emissions. Seabird colonies are found to represent the largest point sources of ammonia globally (up to ~0.006 Tg yr^{-1} per colony). Measured colony emissions ranged from 1 to 90 kg h^{-1}, and equated to 16 and 36% volatilization of excreted nitrogen for colonies dominated by ground/burrow nesting and bare-rock nesting birds, respectively. These ammonia "hot spots" explain significant perturbations of the nitrogen cycle in these regions and add ~20% to oceanic ammonia emissions south of latitude 45° S (Blackall et al. 2007). Recent global estimate amounts 0.22 Tg N yr^{-1} (Riddick et al. 2012).

It is likely that natural ecosystems (forest, grassland) emit no or only small amounts of ammonia because normally there is a deficit of fixed nitrogen in landscapes. Reported emissions factors over forests span three orders of magnitude and are likely be influenced by re-emission of wet deposited ammonium. Older publications considerably overestimated emission by using simple models considering soil ammonium concentrations obtained from relative decomposition and nitrification rates, where *Henry's* law gives the equilibrium concentration of ammonia gas in the soil, and a simplified diffusion equation yields the flux to the atmosphere, for example, Dawson (1977) calculated it to be about 47 Tg N yr^{-1}.

Therefore, (dry) deposition exceeds emission, for example, in forests the soil may emit NH$_3$ but it is up taken by the canopy, and hence remains within the ecosystem. This has been shown in the 1980s, when the European interest in ammonia motivated the first measurements of ammonia dry deposition to semi-natural ecosystems (e. g., Horváth 1983, Duyzer et al. 1987, Erisman and Wyers 1993, Sutton et al. 1993). While the earlier compensation point studies had identified plant stomata as the key exchange site, these European measurements noted large rates of dry deposition that could only be explained by the fact that most of the NH$_3$ was deposited directly onto leaf surfaces.

There is reason to assume that wildlife, as shown in Chapter 2.6.2.2 for domesticated animals, is the dominant natural source of NH$_3$ (Table 2.32). A completely different pathway of emissions is the decay of urea or uric acid to ammonia in animal manure (from mammals or birds). This pathway may also lead to N$_2$O formation. Emissions are much more pronounced for domestic animals, however, where manure is actually collected and kept liquid for longer periods of time, or other sites where animals live in a very dense populations (point emissions from bird breeding colonies on small islands, e. g., in the North Sea).

Although the ocean was proposed as a source of ammonia by Boussingault (1856), it was principally assumed until 1980 that the ocean emits no NH$_3$ (Lenhard and Georgii 1980). When comparing the photosynthesis flux between continents and the oceans, and taking into account river run-off as a mineral source for the sea, it is logical that ammonium is abundant. Although fluxes are small, seas may represent a significant contribution to the nitrogen budget of near-coast land areas (Barret 1998). It has been found that DMS and NH$_3$ fluxes had a similar strength with an average NH$_3$ to DMS molar flux ratio of 1−1.5 (Dentener and Crutzen 1994 and literature therein). This conception assumes only physico-chemical processes in sea-air exchange and similarly uni-directional flux (DMS is not readily soluble in water and is fast oxidized in air, and ammonia is fast converted into particulate ammonium, which is no longer in equilibrium with dissolved ammonia). Based on a lower DMS emission flux of only 16 Tg S yr^{-1} (see Chapter 2.6.3), Dentener and Crutzen (1994) calculated the NH$_3$ emission flux to 7 Tg N yr^{-1}. Möller (1996) calculates the oceanic NH$_3$ emission to be 15 Tg N yr^{-1}, adopting a higher DMS value (cf. Table 2.64). This figure is very close to the estimate of 16.8 Tg N yr^{-1} by Duce and Liss (1991). By using modern models of sea-air exchange an ammonia emission in the range of 10−15 Tg N yr^{-1} is likely (Table 2.64). This concept has recently been revised (Johnson and Bell 2008) by introducing the concept of "co-emission" of the gases, where DMS emission controls the rate of emission of ammonia from the ocean by acidifying the atmosphere (due to sulfurous particle formation).

To summarize the knowledge into a "best estimate", the global NH_3 emission is calculated to be 65 ± 10 Tg N yr^{-1}, of which anthropogenic activities (including biomass burning) contribute 70%.

2.7.1.2 Dinitrogen monoxide (N_2O)

N_2O budgets are associated with considerable uncertainties (Bouwman et al. 1995, 2002a, b, Nevison et al. 1996). As seen from Figs. 2.26 and 2.27, denitrification is the direct way of producing molecular nitrogen (N_2), dinitrogen monoxide (N_2O) and nitrogen monoxide (NO). This way is partly parallel to the formation of ammonia/ammonium (ammonification), and therefore it is assumed that all these compounds appear together, but in different quantities. Soil structure and pH, oxygen content, humidity and temperature, but also radiation, are important parameters in determining emissions (Meixner and Neftel 2000). N_2O emissions from soils under natural vegetation are significantly influenced by vegetation type, soil organic C content, soil pH, bulk density and drainage, while vegetation type and soil C content are major factors for NO emissions (Stehfest and Bouwman 2006). A soil emission similar to those of ammonia is proposed. The emission of N_2O is perhaps twice that, which is supported by the more direct chemical formation pathway during denitrification (Fig. 2.27). In analogy to ammonia, both gases are assimilated by microorganisms living in soils and plants. As for ammonia, emission of NO and N_2O is considered to be a loss for the organisms, in contrast to emission of N_2 by denitrification, closing the atmospheric cycle (> 100 Tg N yr^{-1} after Schlesinger 1997). Sometimes emission of NO_2 and HNO_2 is reported. Again similar to ammonia, soil-emitted NO circulates within local circuits, namely in tropical rainforest (Delmas and Servant 1987). Additionally, NO is fast converted into NO_2 and further in nitric acid (HNO_3) which will be effectively deposited in or close to the original NO source. An uptake of NO_2 seems favorable because of fast metabolization via electron transfer into nitrite (see Fig. 2.27).

Whereas emissions from agriculture are relatively well understood, no studies have calculated global emissions from soils under natural vegetation due to lack of data for many vegetation types (Stehfest and Bouwman 2006). Table 2.66 shows the high uncertainties in natural (but also in anthropogenic) global emission estimates. Newer studies can revise some values but they do not lead to any greater certainty; for example, the estimated N_2O emission from global tropical rainforest is 1.3 (0.9−2.4 uncertainty) Tg N yr^{-1} compared with older estimates in the range 1.2−3.6 Tg N yr^{-1} (Werner 2007). Globally, tropical soils (primarily wet forest soils, but also savannas) are estimated to produce 6.3 Tg of N_2O annually and oceans are thought to add around 4.7 Tg of N_2O annually to the atmosphere (IPPC 2001). Latest estimates (Saikawa et al. 2013) for global average soil N_2O emission (1878−2000) amounts to 7.5−8.9 Tg N_2O-N yr^{-1}. The estimate of global N_2O emission from natural ecosystem soils is much lower at 3.4 (2.0−4.6) Tg N_2O-N yr^{-1} (Zhang et al. 2012). Most previous studies of nitrous oxide had focused on the

Table 2.66 Uncertainties in global emission estimates; value in Tg N yr^{-1}.

compound	natural emission[a]	natural emission[b]	anthropogenic emission[b]
NH_3	25 ± 5	15 ± 12	40 ± 20
N_2O	10 ± 3	12 ± 5	4 ± 3.5
NO	20 ± 2	20 ± 15	30 ± 15

[a] Watts (2000), see Table 2.63
[b] Olivier at al. (1999)

open ocean, which led to an underestimation of the oceanic source (Bange 2006). Not only are the fluxes of N_2O from coastal waters higher, but the processes responsible for its production also appear to be different from those in the open sea in that denitrification can sometimes produce large amounts of N_2O in shallow suboxic waters (Naqvi et al. 2008). A large amount of N_2O is produced by bacteria in the oxygen-poor parts of the ocean using nitrites (Trimmer et al. 2006). Newer studies suggest a much lower oceanic N_2O emission in the range $1-2$ Tg N yr^{-1} (Rhee et al. 2009) as already proposed by Khalil and Rasmussen (1992). Duce et al. (2008) were the first to show that the open ocean is also an anthropogenic source of N_2O due to atmospheric deposition of nitrogen compounds, amounting 1.6 Tg N yr^{-1}.

Agricultural activities and animal production systems are the largest anthropogenic sources of N_2O emissions. Recent calculations using IPCC 1996 revised guidelines indicate that N_2O emission from agriculture is 6.2 Mt N as N_2O per year (IPCC 2007). About one-third is related to direct emissions from the soil, another third is related to N_2O emission from animal waste management, and the final third originates from indirect N_2O emissions through ammonia (NH_3), nitrogen oxides (NO_x), and nitrate losses.

To summarize the knowledge into a "best estimate", the global N_2O emission is calculated to be 12 ± 5 Tg N yr^{-1}, of which anthropogenic activities (including biomass burning) contribute 40%.

2.7.1.3 Nitrogen monoxide (NO)

As discussed in Chapter 2.6.4.4 for NO production during lightning, similar conversion processes occur in all combustion and high-temperature processes (see Chapter 5.4.1). To a minor percentage the fuel nitrogen content contributes to NO formation, but this pathway is relatively more important for biofuels. In soils, denitrification of nitrate is the source of NO formation (Chapter 2.4.3.1). On cultivated soils, as for NH_3, the primary NO source is N fertilizer application (cf. Fig. 2.51); natural soils have much less specific emissions. The soil NO emission (now often termed SNO) is roughly half of the N_2O emission (Galloway et al. 1995, Stehfest and Bouwman 2006): 1.4 Tg N yr^{-1} for fertilized cropland and 0.4 Tg N yr^{-1} for grassland; data from natural soils are too scarce to create global values. Older estimates of the soil NO sources range between 9.7 Tg N yr^{-1} (Potter et al. 1996a) and 21 Tg N yr^{-1} (Davidson and Kingerlee 1997), whereas estimates that include

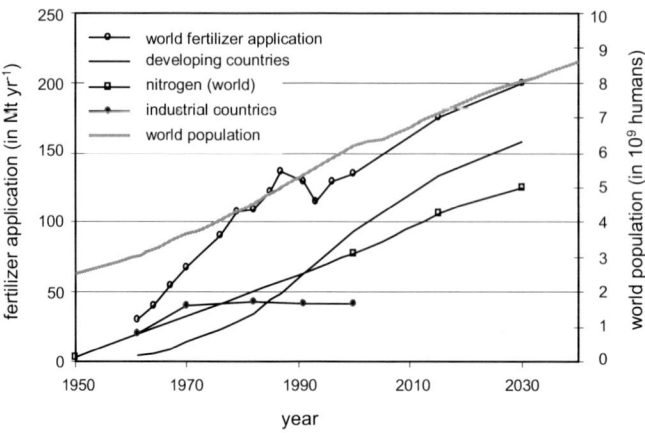

Fig. 2.51 Trend in fertilizer application (from Möller 2003).

the role of canopy deposition, are about 50 % lower (Yienger and Levy 1995, Gan-zeveld et al. 2002). The inventory by Yienger and Levy (1995) presents a NO soil emission flux of 11 Tg N yr^{-1} (below canopy, and consequently their atmospheric source estimate is 5.5 Tg N yr^{-1}) which is slightly larger compared with the estimate of 9.7 Tg N yr^{-1} by Potter et al. (1996a). Recently, Hudman et al. (2012) estimate above-soil NO$_x$ emission in the same order to be 10.7 Tg N yr^{-1}, including 1.8 Tg N from fertilizer input (0.68% of applied N) and 0.5 Tg N from atmospheric N deposition. Thus it remains uncertain – mainly because of local nitrogen cycles between soil and canopy – what the global (and regional) natural soil NO emissions are. NO from lightning has been estimated to be relatively consistent at 5 ± 3 Tg N yr^{-1} (Chapter 2.6.4.4). Without having information on natural soils, but assuming 50 % of N$_2$O (corresponding to ~ 3 Tg N yr^{-1}), the global natural NO emission is ~ 10 Tg N yr^{-1}.

There is no less uncertainty in estimating the anthropogenic NO emission (Table 2.66). While the fossil-fuel source is estimated at 20–25 Tg N yr^{-1} (e. g., Delmas et al. 1997 and citations therein), total NO emissions are about 44 (23–81) Tg N yr^{-1} (Lee et al. 1997). Martin et al. (2003) derive a top-down NO$_x$ emission inventory from the GOME data by using the local GEOS-CHEM rela-tionship between NO$_2$ columns and NO$_x$ emissions. The resulting NO$_x$ emissions for industrial regions are seasonal, despite large seasonal variation in NO$_2$ columns, providing confidence in the method. Top-down errors in monthly NO$_x$ emissions are comparable with bottom-up errors over source regions. Annual global a posteri-ori errors are half of a priori errors. Their global a posteriori estimate for annual land surface NO$_x$ emissions (37.7 Tg N yr^{-1}) agrees closely with the GEIA-based a priori (36.4) and with the EDGAR 3.0 bottom-up inventory (36.6), but there are significant regional differences. This estimate includes natural and anthropogenic sources; thus, considering about 10 Tg N yr^{-1} from natural sources, the remaining ~ 25 Tg N yr^{-1} is attributed to anthropogenic sources, which is close to the esti-mates of Graedel et al. (1995), Lawrence et al. (1995), and Benkowitz et al. (1996, 2004). International ship traffic significantly contributes to anthropogenic NO emission: $6-7$ Tg N yr^{-1} (Corbett and Fischbeck 1997, Corbett and Köhler 2003, Eyring et al. 2005) where the emission only was 1.6 Tg N yr^{-1} in 1951 and is forecast to rise to 11.8 Tg N yr^{-1} in 2050.

To summarize the knowledge into a "best estimate", the global NO emission is calculated to be 35 ± 5 Tg N yr^{-1}, of which anthropogenic activities contribute 70%.

2.7.2 Sulfur compounds

2.7.2.1 Sulfur dioxide (SO$_2$)

The development of a reliable regional emission inventory of sulfur as a function of time is an important first step in assessing the potential impact on the climate system. Whereas in the past the contribution of SO$_2$ to acidity ("acid rain") was the main reason for making SO$_2$ inventories, in the late 1980s the question of sulfate aerosol on climate (cooling) came into focus. SO$_2$ was the key pollutant for centuries (see Chapter 2.8.1) and the first interest in global cycling arose in the 1970s (Kellog et al. 1972, Friend 1973, Granat et al. 1976, Bettelheim and Littler 1979, Ryaboshapko 1983, Möller 1983, Langner and Rodhe 1991, Pham et al. 1995, Möller 1995a, Örn et al. 1996, Lefohn et al. 1999, Brimblecombe 2003, Smith et al. 2004). Table 2.67 shows anthropogenic SO$_2$ estimates from different authors over the last 50 years. Fig. 2.52 shows that the SO$_2$ emission (before the introduction of air pollution control) was closely connected with economic growth, especially reflecting exponential increase before World War I, and from World War II until the end of the 1970s. Large discrepancies among the approximations probably reflect the many uncertainties associated with estimating sulfur emissions. In addition to the global estimates summarized in Table 2.67, there are also continental and national estimates of sulfur emissions (e. g. Gschwandtner et al. 1986, Fujita et al. 1991, Kato and Akimoto 1992, Mylona 1996). The maximum historical SO$_2$ emission with around 70 Tg SO$_2$-S yr^{-1} was reached around 1970. While global sulfur

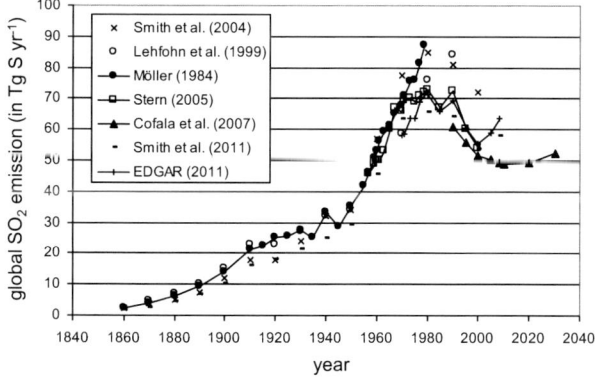

Fig. 2.52 Global SO$_2$ emission trend.

Table 2.67 Global anthropogenic SO_2 emission estimates (in Tg S yr^{-1}).

emission	base year	author
34.4	1937	Magill et al. (1956)
38.7	1943	Magill et al. (1956)
39	1959	Ericksson (1959, 1960)
73	1965	Robinson and Robbins (1970)
39	1940	Katz and Gale (1971)
50	1970	Koide and Goldberg (1971)
50	1965	Kellogg et al. (1972)
65	1972	Friend (1973)
80	1973	Almquist (1974)
65	?	Granat et al. (1976)
77	1974	Möller (1977)
93.6	1974	Cullis and Hirschler (1980)
98	?	Ryaboshapko (1983)
79.2	1979	Várhelyi (1985)
87.2	1979	Möller (1984)
103.6	1980	Cullis and Hirschler (1980)
103	1980	Warneck (1988)
88.4	1980	Spiro et al. (1992)
62.8[a]	1980	Dignon and Hameed (1989)
65.4	1980	Smith et al. (2011)
71.5	1980	EDGAR v4.2 (2011)
72	1990	Smith (2001)
72.3	1990	Stern (2005a, b)
61	1990	Cofala et al. (2007)
63.9	1990	Smith et al. (2011)
77.5	1995	Benkovitz et al. (2004)[b]
60	1995	Stern (2005a, b)
54	2000	Stern (2005a, b)
51.5	2000	Cofala et al. (2007)
53.5	2000	Smith et al. (2011)
57.8	2005	Smith et al. (2011)
59	2005	EDGAR v4.2 (2011)
63.5	2008	EDGAR v4.2 (2011)

[a] only from fossil fuels [b] EDGAR 3.2

dioxide emissions have generally declined since the mid 1970s, the upturn in emissions from 2000 to 2005 (Smith et al. 2011) raised the possibility of a short period that increasing aerosol forcing might be offsetting recent warming from increasing greenhouse gas concentrations. Declining emissions from both China (after 2005), still the world's largest source of sulfur dioxide, the planned decrease in shipping emissions, and continued decreases in industrialized countries are likely to lead to a further net decrease in global sulfur dioxide emissions in the future (Klimont et al. 2013).

The key to reliable global emission estimations are the *EF* as function from time (changing technology and introduction of flue gas desulfurization) as well as distinction between countries. Eq. (2.107) shows a simple relationship between the emission E_i (of the process i), the emission factor EF and the parameters determining the sulfur release: sulfur content S (in %), ash bonding degree α (or sulfur non-release) and flue gas desulfurization (FGD) degree β; M mass of fuel or processed sulfur-containing material.

$$E_i = EF_i \cdot M_i = \frac{S_i}{100} (1 - \alpha)(1 - \beta) M_i \qquad (2.107)$$

More than 90% of SO_2 emission is related to the combustion of fossils fuels (coal and oil with an approximate share of two-thirds and one-third, respectively), primarily metal smelting (Cu, Zn, Mn, and Ni) and sulfuric acid production (only in the past). Since 1980, flue gas desulfurization has been introduced stepwise with different degrees in industrialized countries. FDG works with an efficiency of about 95%; hence the country-based SO_2 emission can be reduced by about one order of magnitude as has occurred in Germany after instalment of FGD at all large SO_2 sources (Fig. 2.53). Germany probably remains an extreme example of changing SO_2 emissions in a very short period; in 1992 the ratio between Eastern and Western Germany in SO_2 emission was still 5.1 and dropped 10 years later to 0.25. Such country-based emission changes (expressible in specific emission densities too) also result without any doubt in changes of atmospheric chemistry.

Global anthropogenic sulfur emissions increased until around 1980. (Fig. 2.52). The different estimates in the period 1990–2000 show large variations between 65 and 85 Tg S yr^{-1}; Möller's (1984b) forecast did not include FGD – hence this estimate shows the further emission without abatement which became important in Western Europe from the beginning of the 1980s and in Eastern Europe in the middle of the 1990s (Japan had already introduced FGD fully in the late 1970s). Combustion of fossil fuels has increased continuously until the present (Fig. 2.47). Newer estimates (Stern 2005, Cofala et al. 2007, Smith et al. 2011) show relative stability throughout the decade of the 1980s and a 25% decline from 1990 to 2000 to a level not seen since the early 1960s. The decline is evident in North America, but most significant in Europe. Recently, Cofala et al. (2007) revealed that changes

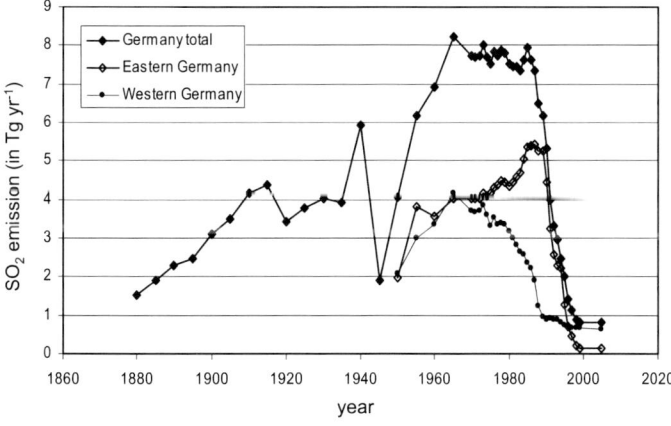

Fig. 2.53 German SO_2 emission trend; data from Möller (2003), after 2000 from UBA.

in the pattern of global sulfur emission have been more dramatic than previously believed (Fig. 2.52). As Möller (2003) pointed out, the scenario of future SO_2 development depends primarily on the introduction of FGD technologies in China. Through the drastic increase in China's energy consumption (see also Chapter 2.7.3.1) a further rise of all emissions from coal combustion must be expected (Carmichael et al. 2002, Dasgupta et al. 2002). Streets and Waldorf (2000) predicted that sulfur dioxide emission would increase from 12.6 Tg S in 1995 to 15.2 Tg S in 2020, provided emission controls are implemented on major power plants; if this does not happen, emissions could increase to as much as 30.4 Tg S by 2020. Between 2000 and 2005 alone coal consumption in China increased by a factor of about three, corresponding to a global coal combustion increase of about 45% (Fig. 2.46). China started after 1995 with the implementation of different abatement strategies, and after the year 2000 the percentage of FGD in operation drastically increased (in GW installed FGD): 2005 (53), 2006 (166), 2007 (374), and (2008 (379). China's official annual SO_2 emission in 1990 amounted to about 8 Tg S, and increased to 10 Tg S by 2000. This is similar to the last estimates by Smith et al. (2011), who state that China's maximum SO_2 emission was in 2005 (16.5 Tg SO_2-S) and decreased to 14 Tg in 2010. Controversially, SO_2 emission from India is rising, from 1.8 Tg SO_2-S in 1990 to 5.0 Tg SO_2-S in 2010 (Smith et al. 2011). Adopting the coal consumption increase and no FGD, in 2008 an emission of about 45 Tg S would be expected. Officially, in 2008 an SO_2 emission of about 12 Tg S has been reported and an FDG instalment of 66% (China Desulfurization Industry Report 2009); this roughly corresponds to the above estimate without FGD. After increasing until about 2006, Klimont et al. (2013) estimate a declining trend continuing until 2011. However, there is strong spatial variability, with North America and Europe continuing to reduce emissions, and Asia and international shipping playing an increasing role. China remains a key contributor, but the introduction of stricter emission limits followed by an ambitious program of installing flue gas desulfurization in power plants resulted in a significant decline in emissions within the energy sector and stabilization of total Chinese SO_2 emissions.

2.7.2.2 Reduced sulfur compounds (H_2S, DMS, COS)

Biogenic sources emit so-called reduced sulfur compounds, being in the oxidation state of S^{2-} such as hydrogen sulfide (H_2S), dimethyl sulfide (DMS), and dimethyl disulfide (DMDS) carbon disulfide (CS_2) and carbonyl sulfide (COS). It is likely that other organic sulfur compounds (in analogy to organic nitrogen), especially sulfides, are emitted too; for example, from the plant *Allium ursinum* (a type of garlic): dipropenyldisulfide, methylpropenyldisulfide, diallylsulfide, and cis-propenyldisulfide (Puxbaum and König 1997). An uncountable number of publications exist on biogenic sulfur emission. The range of estimates is from 13 to 280 Tg S yr^{-1} (Möller 2003 and citations therein), whereas higher values before 1975, which were almost all due to H_2S, have been proposed.

Plants emit DMS and COS (similarly to other organic compounds) predominantly by respiration (Kesselmeier and Hubert 2002). Table 2.68 summarizes global natural sulfur emissions; see Chapter 2.4.3.2 for details on DMS emission. One of

Table 2.68 Natural sulfur emission; after Möller (2003).

source	emission (Tg S a^{-1})	emitted compound
volcanism	10 ($\pm$ 5)	SO_2, H_2S, COS (?)
soils and plants	1–4	H_2S, DMS, COS, CH_3SH
wetlands	2	H_2S, DMS, COS, CS_2
biomass burning	2–3	SO_2, COS
ocean	36 ($\pm$ 20)	DMS, H_2S, COS, CS_2

compounds	emission (Tg S a^{-1})
DMS	35 ($\pm$ 20)
SO_2	12 ($\pm$ 6)
H_2S	3 ($\pm$ 2)
CS_2	1 ($\pm$ 1)
COS	1 ($\pm$ 0,5)

total	50 ($\pm$ 25)

the major estimated sources of COS[20] is the atmospheric oxidation of CS_2 and it is also formed by oxidation of DMS. It seems that there has been no progress in estimating global natural sulfur emissions since the last published inventory used in modeling by Chin and Jacob (1996). The best-guess global annual-integrated COS net flux estimate does not differ from zero within the range of estimated uncertainty, consistent with the observed absence of long-term trends in atmospheric COS loading. Interestingly, the hemispheric time-dependent monthly fluxes are very closely in phase for the northern and southern hemispheres. The monthly variation of the northern hemisphere flux seems to be driven primarily by high COS vegetation uptake in summer, while the monthly variation of the southern hemisphere flux appears to be driven mostly by high oceanic fluxes of COS, CS_2, and DMS in summer (Kettle et al. 2002).

The anthropogenic emissions of reduced sulfur have been regarded over time to be negligible compared with SO_2. Because of its long residence time, only COS from anthropogenic sources may be of interest; however, the only global estimate is given by Turco et al. (1980) who proposed a flux between 1 and 5 Tg S yr^{-1} due to coal combustion, which is between 2% and 12% of SO_2-S emitted from coal combustion. From the combustion chemistry of fuel sulfur it is simple to assume COS formation in all combustion processes; biomass burning has been identified (Crutzen et al. 1979). From an analysis of the correlations between measured emission rates and environmental parameters, the sources of COS and CS_2 are estimated here to be 1.23 (0.83–1.71) Tg COS yr^{-1} and 0.57 (0.34–0.82) Tg CS_2 yr^{-1}, respectively. The results of Chin (1992) and Chin and Davis (1993, 1995) indicate that nearly 30% of the atmospheric COS source is derived from the oxidation of CS_2, while emissions from the ocean and other natural terrestrial sources contribute 28% and 25%, respectively. Harnisch et al. (1992) identified Al production to be the most important COS source (65%): 0.08±0.06 Tg S yr^{-1}.

---

[20] Some authors use the formula OCS, which also represents the molecule structure (O=C=S).

Blake et al. (2008) have shown that COS is very highly correlated with CO_2 in continental plumes. That shows a close interrelation with plant assimilation, as a source as well as a sink for COS. A correlation of COS with fossil-fuel sources would expect a trend similar to SO_2 and CO_2. Surprisingly, no COS concentration trends over the long term could be detected by analyzing spectral data dating back to 1951 (Rinsland et al. 2008). Also findings from the Antarctic air samples measured at Siple Dome with air ages between 1616 and 1694 yielded a mean mixing ratio of 373 ± 37 ppb (Aydin et al. 2002). Deutscher et al. (2006), from air samples taken in Australia in the 1990s, found a small but significant downward trend of 0.3% yr^{-1} and attributed it to decreasing industrial emissions, almost in the NH. Contrarily, adopting no trend over the last few decades before 1990 as found by several authors (see in Rinsland et al. 2008) is not consistent with the increase of all industrial emissions. In contrast to SO_2 it is not to expected that FGD will significantly abate COS, hence − when there is a correlation with coal combustion − emissions should continuously increase (cf. Fig. 2.46). There are two possibilities to explain the data: either there is no, or no significant, industrial COS emission, or the anthropogenic COS emission is buffered by vegetation uptake. Recent work has highlighted the potential of atmospheric COS measurements to constrain gross primary production (Montzka et al. 2007). Berry et al. (2013) state that biomass burning contributes much more to COS release and canopy and soils uptake COS than previously assumed.

Barkley et al. (2008) found globally tropospheric and stratospheric means to be 433 ppt and 330 ppt, slightly lower than previous measurements of free tropospheric mixing ratio of 480−520 ppt (Notholt et al. 2003) and stratospheric mixing ratios of 380 ppt (Chin and Davies 1995); they estimated the stratospheric lifetime to be 64 ± 21 yr, resulting (remember: $F = M/\tau$) in stratospheric turn-over between 63 and 124 Gg yr^{-1}.

Recently, Lejeune et al. (2012) presented remarkable results on atmospheric COS trends through Jungfraujoch FTIR measurements. The 7.1 km column density trend from 1995−2002 is negative (0.7% yr^{-1}) whereas the trend from 2002−2008 is positive (1.1% yr^{-1}). The mixing ratio growth by 5 ppt yr^{-1} from 450 to 485 ppt (2002−2008) and further remains further constant. The COS concentration is considerably less at that altitude than in Mace Head (around 500 ppt) where a positive trend from 2002−2008 was also encountered ($+1.19$ ppt yr^{-1}). They argue that anthropogenic sources may be responsible (e. g. strong increase of coal consumption in China). The importance of man-made sources is supported by the COS interhemipheric ratio of 1.12 ± 0.07 (Deutscher et al. 2008) and the finding by Aydin et al. (2008) that COS measurements coming from firn air and ice core suggest that COS levels of the late twentieth century are significantly larger than the preindustrial levels.

Concerning the debate on COS contribution to the stratospheric sulfate layer (SSA), Lejeune at al. (2012) argue that the recent trend evolution of SSA layer suggests that stratospheric COS plays a minor role in maintaining the SSA background level. After the eruption of Mount Pinatubo (1991), SSA increased and finally declined to the background level reached between 1998 and 2002 but since 2002 a systematic increase of the SSA level is observed. This can be explained by the increase of coal burning in China (Hofmann et al. 2009) and a series of moderate but intense volcanic eruptions in tropical latitudes and the transfer of sulfur particles by the *Brewer-Dobson* circulation to higher latitudes (Vernier et al. 2010).

2.7.3 Carbon compounds

2.7.3.1 Carbon dioxide (CO_2)

About 95% of all anthropogenic industrial CO_2 emission is caused by fossil fuel use; 4% is from cement production (limestone burning and CO_2 release from past carbonate sediments). China is the world's largest hydraulic cement producer. In 2006 China produced over 1.2 billion metric tons of hydraulic cement, or roughly 47% of the world's production. Emissions from cement production account for 9.8% of China's total industrial CO_2 emissions in 2006.

It lies in the nature of the combustion processes of fossil and biofuels (gaseous, liquid and solid) that all carbon will be oxidized into CO_2 to obtain a maximum heat of reaction. Naturally, depending on the completeness of the oxidation process (see discussion in Chapter 2.6.4.5), unburned carbon (BC), volatile organics, and CO are emitted too. Besides wood, coal was the dominant carbon carrier for centuries; the rise of coal-released CO_2 after the Industrial Revolution around 1850 is expressive (Fig. 2.54). Only after 1900 liquid fuels (dominant for transport) become significant and after 1950 natural gas was substituted for coal in stationary combustion processes (also due to reduced SO_2 emissions). In 2005 the percentage of the global CO_2 emission (in parenthesis the 2012 figure) was about 39% for liquids, 38% for solids and 18% for gas, not very much changed since 1980 (Boden et al. 2009, 2013). According to reported energy statistics, coal production and use in China has increased since the early 1960s by a factor of ten. As a result, Chinese fossil-fuel CO_2 emissions have grown a remarkable 79.2% since 2000 alone. At 1.66 billion metric tons of carbon in 2006, China has surpassed the United States as the world's largest emitter of CO_2 due to fossil-fuel use and cement production (Fig. 2.55)

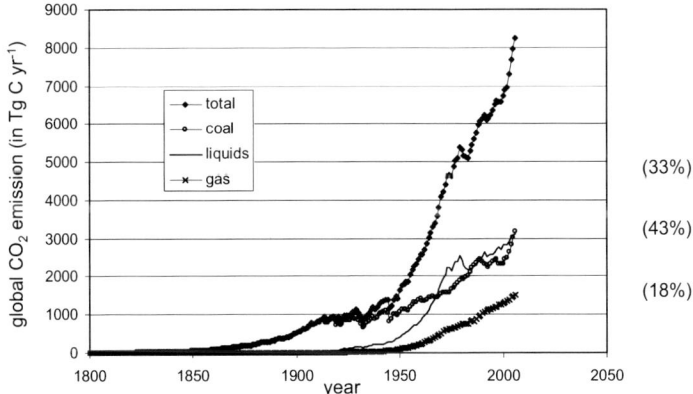

Fig. 2.54 Global CO_2 emission Trend; data from Boden et al. (2009).

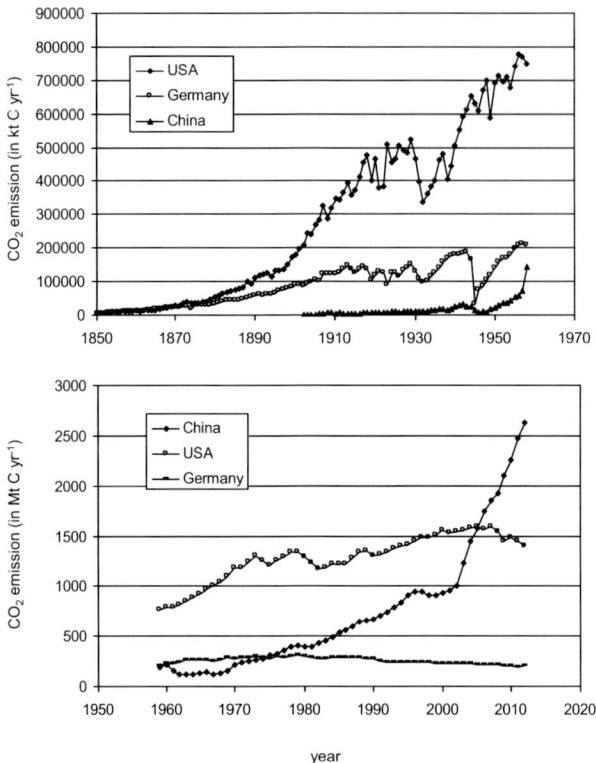

Fig. 2.55 Historical record of CO_2 Emission in Germany, the USA, and China (1850–1958 and 1959–2012, respectively); data from Boden et al. (2013).

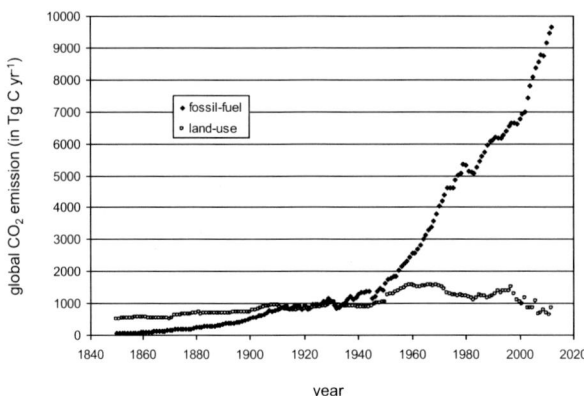

Fig. 2.56 Global CO_2 emission trend; fossil fuel use data from Boden et al. (2013) and land-use change date from Houghton (2012).

Despite the large uncertainties in CO_2 from land use change (see IPCC 2013 for recent results), this source contribution has decreased continuously in the last 15 years and contributes to only about 10% to total anthropogenic CO_2 emission today. Before 1910, land-use change was the dominant anthropogenic CO_2 source and remained significant until 1960 (Fig. 2.56).

In 2013, the global CO_2 emission increase is smaller at 2.1% than the average of 3.1% since 2000 but larger than in 2012, reaching a level that is 61% above emission in 1990 (CDIAC 2013); it will reach 9.8 Gt C from fossil fuel use. The annual average growth rates of the historical emission estimates are as follows (Peters et al. 2013); in % yr^{-1}: 1.75 (1985−2012), 1.96 (1990−2012), 2.91 (2000−2012), and 2.10 (2005−2012).

Fig. 2.56 shows the global CO_2 emission trend 1959−2012 (Boden et al. 2013). Additionally to fossil fuel CO_2 release, cement production increasingly contributes to the global CO_2 emission (in Tg C yr^{-1}): 43 (1960), 157 (1990), 320 (2005), and 509 (2012). Land-use change (see also Chapters 2.6.5.3 and 2.8.3.1) is another important anthropogenic source but with decreasing source strength: average 2005−2012 amounts 0.77 Tg C yr^{-1} comparing to the 1975−1995 average of 1.24 Tg C yr^{-1} (Houghton et al. 2012, Houghton and Hackler 2013). Altogether, the present anthropogenic CO_2 emission reaches about 11 Tg C yr^{-1}. For 2012, remarkable trends were seen in the top three emitting countries/regions, which accounted for 55% of total global CO_2 emissions (Olivier et al. 2013). Of these three, China (29% share) increased its CO_2 emissions by 3%, which is low compared with annual increases of about 10% over the last decade. In the United States (16% share) and the European Union (11% share), CO_2 emissions decreased by 4% and 1.6%, respectively. In addition, emissions in India and Japan increased by 7% and 6%. The small increase in emissions in 2012 of 1.1% may be the first sign of a slowdown in the increase in global CO_2 emissions, and ultimately of declining global emission.

However, the latest carbon dioxide emissions continue to track the high end of emission scenarios, making it even less likely global warming will stay below 2° C. A shift to a 2° C pathway requires immediate significant and sustained global mitigation, with a probable reliance on net negative emissions in the longer term (Peters et al. 2013).

2.7.3.2 Carbon monoxide (CO)

As mentioned in the previous section, CO is the product of incomplete combustion or oxidation. Both the burning of biofuels and biomass and the combustion of fuels in mobile engines contribute dominantly ($\sim 80\%$) to primary CO emission (Tables 2.30, 2.46−2.48, and 2.63). Most of the CO accounted in global balances, however, is indirect from the atmospheric oxidation of hydrocarbons (Tables 2.48 and 2.69). Direct biogenic CO formation is only very small. Wilks (1959) first reported the direct CO emission from green plants when they were exposed to sunlight. Two main biogenic sources have been considered: vegetation, with an estimate of ~ 100 Tg C yr^{-1} globally (IPCC 2001, Lelieveld and Dentener 2000, Marufu et al. 2000, Schade and Crutzen 1999, Warneck 1999, Lelieveld and Dorland 1995, Seiler and Conrad 1987), and ocean contribution (~ 50 Tg C yr^{-1}) (Prather et al. 1995, Khalil and Rasmussen 1995, Bates et al. 1995). The uncertainty is

Table 2.69 Global CO_2, CH_4, N_2O, and CO emissions in the base year (1990) by source (IPCC 2007).

substance	total emission	emission by source	
CO_2 (in Gt C yr^{-1})	6.0–8.2 (anthropogenic)	5–6	fossil fuel combustion
		0.16	cement production
		0.6–2.6	land use change
CO (in Gt C yr^{-1})	0.3–0.5 (anthropogenic)	0.13–0.21	technical
		0.13–0.30	biomass burning
	0.03–0.14 (biogenic)	0.02–0.04	biogenic
		0.01–0.1	oceanic
	0.3–0.7 (secondary)	0.2–0.4	CH_4 oxidation
		0.1–0.3	NMVOC oxidation
CH_4 (in Gt C yr^{-1})	0.1–0.4 (anthropogenic)	0.03–0.07	enteric fermentation
		0.01–0.07	rice paddies
		0.01–0.02	animal waste
		0.01–0.05	biomass burning
		0.01–0.05	landfills
		0.01–0.05	domestic sewage
		0.05–0.07	fossil fuel use
	0.07–0.14 (natural)		
N_2O (in Mt N yr^{-1})	3.7–7.7 (anthropogenic)	1.8–5.3	cultivated soils[a]
		0.2–0.5	cattle and feed lots
		0.2–1.0	biomass burning
		0.7–1.8	industrial

[a] 4.1 according to Stehfest and Bouwman (2006)

probably in the order of 100% (Kanakidou and Crutzen 1999). As for the case of other reduced gases, soils appear mostly as sink, but can also contribute to CO emissions depending on the conditions; globally, however, this source seems be negligible (~ 10 Tg yr^{-1} according to Potter et al. 1996b).

When we consider that ~ 200 · 10^{15} g C of atmospheric CO_2 is globally assimilated annually (Fig. 2.10) and that the same amount of carbon is roughly turned back into air by respiration, almost in the form of CO_2, the estimate of 0.03–0.1 · 10^{15} g CO-C plant release is very small and amounts to less than 0.1% of CO_2 not used by metabolism (see reaction 2.53 and Fig. 2.10), or lost during respiration before oxidizing into CO_2. As seen later (Table 2.69), the metabolic release of non-methane volatile organic compounds (NMVOCs) is much larger (0.4–4.4 Gg C) than that of CO but (note the large uncertainty in the order of one magnitude) is also relatively small compared with respiratory CO_2 (<2% of the carbon turnover).

2.7.3.3 Methane (CH_4)

Atmospheric methane is produced by both natural sources (e. g. wetlands) and human activities. Present-day global natural CH_4 emissions are estimated to be

145−260 Tg yr^{-1} (Spahni et al. 2013) which is 25−50% of total CH_4 emissions (503−610 Tg yr^{-1} according to Denman et al. 2007 or IPCC 2007, resp.); Table 2.69. Newer estimates (Höglund-Isaksson 2012) assume an increase of anthropogenic global CH_4 from 323 Tg (2005) to 414 Tg in 2030. The 2005 figure is very close to the EDGAR (2011) estimate (324 Tg), whereas global CH_4 emissions show approximately a linear increase from 231 Tg CH_4 in 1970 to 324 Tg CH_4 in 2005 and 346 TG in 2008. The anthropogenic sources are for the year 2005 (Höglund-Isaksson 2012): fossil fuel extraction and use 131 Tg (40%), agriculture 126 Tg (39%), solid waste and waste water 57 Tg (17%), and combustion 11 Tg (4%). Methane emission rates from rice paddies (an anthropogenic source) have been estimated to be 20 to 120 Tg CH_4 yr^{-1} (Yan et al. 2009) with an average of 60 Tg CH_4 yr^{-1} (Denman et al. 2007). Meng et al. (2012) estimates rice paddy emissions in the year 2000 to be 42 Tg CH_4 yr^{-1}.

Although most sources and sinks of methane have been identified, their relative contributions to atmospheric methane levels are highly uncertain. As such, the factors responsible for the observed stabilization of atmospheric methane levels in the early 2000s, and the renewed rise after 2006, remain unclear. Kirschke et al. (2013) construct decadal budgets for methane sources and sinks between 1980 and 2010 and conclude − despite uncertainties in emission trends − that the observed stabilization of methane levels between 1999 and 2006 can potentially be explained by decreasing-to-stable fossil fuel emissions, combined with stable-to-increasing microbial emissions, and that a rise in natural wetland emissions and fossil fuel emissions most likely account for the renewed increase in global methane levels. All estimates are considerably smaller than those provided by IPCC (1995).

Methane is produced in anaerobic conditions by methanogenic bacteria. As seen in Fig. 2.42, CH_4 is one of the (theoretically) final products in the hydrogenation of CO in the process of photosynthesis. It is an unwanted pathway because the "sense" of plant metabolism consists in the synthesis of C_n molecules (see discussion in Chapter 2.6.2.1). Metabolic processes, especially in the intestines of animals but also processes in their excretions are responsible for gas formation. One important pathway leading primarily to methane formation is the anaerobic degradation of plant cellulose by symbiotic microflora (methanogenic and acetogenic bacteria) in the intestines. Major kinds of animals known to emit methane are mammals (primarily ruminants and rodents) and termites. Termites contribute 20 ± 2 Tg C yr^{-1} (Sanderson 1996). Wildlife generates significant specific emissions compared with those of ammonia (Table 2.32). The largest contribution (but highest uncertainty) in natural methane emission comes from wetlands (Matthews 1993, Meng et al. 2012, Bridgham et al. 2013). The annual methane emission from wetlands is 100−200 Tg (Matthews and Fung 1987), well within the range of older estimates (11−300 Tg). Using a baseline set of parameter values, model-estimated average global wetland emissions by Meng et al. (2012) for the period 1993−2004 were 256 Tg CH_4 yr^{-1} (including the soil sink). Tropical wetlands contributed 201 Tg CH_4 yr^{-1}, or 78% of the global wetland flux. Tropical/subtropical peat-poor swamps from 20°N to 30°S account for ∼30% of the global wetland area and ∼25% of the total methane emission. About 60% of the total emission comes from peat-rich bogs concentrated around 50−60°N, suggesting that the highly seasonal emissions from these ecosystems is a major contributor to the large annual oscillations observed in atmospheric methane concentrations at these latitudes. The seasonal thawing of permafrost could be a "new" CH_4 source (Walter et al. 2007) and might explain variations in CH_4 concentration.

One surprising result was the (hypothetical) finding that higher plants produce and emit CH_4. Keppler et al. (2006) estimated a methane source strength of $62-236$ Tg yr^{-1} for living plants and $1-7$ Tg yr^{-1} for plant litter, which could be to blame for $10-45\%$ of the world's methane emissions (Nisbet 2006). However, Nisbet et al. (2009) found that plants were only ever a passive transmitter of the methane present in other places – for example, methane in water, soaked into the soil and could be taken up by a plant and released – but the methane was not produced by the plant. In fact, it is soil-based bacteria that manufacture methane. However, under high UV stress conditions, there can be a spontaneous breakdown of plant material, which releases methane. In addition, plants take up and transpire water containing dissolved methane, leading to the observation that methane is released.

Total estimated anthropogenic methane emissions rose from 79.3 Tg in 1860 to 371.0 Tg in 1994 (Stern and Kaufmann 1998). Between 1860 and 1994, the relative importance of the various component sources changed, with fossil fuels increasing and agriculture – although still dominant – declining in dominance. Within the agricultural sector, livestock replaced rice as the leading component.

2.7.3.4 Non-methane volatile organic compounds (NMVOC)

Volatile organic compounds (VOCs) are among the trace gases that have a crucial role as ozone precursors (see Chapter 5.3.2.2) and thus in controlling the distribution of chemical species in the atmosphere. Global emissions of VOCs are dominated by natural (e. g. biogenic) sources. Taking into account the huge number of organic substances in the metabolism of plants and microorganisms, it is not surprising that all types of organic compounds have basically been found in emissions from vegetation and soils, including alkanes, alkenes, organic acids, alcohols, esters, ethers, aldehydes and ketones (Fig. 2.42). In addition to plant emissions, biomass burning (Table 2.47) and secondary atmospheric chemical reactions (Table 2.48) produce a variety of NMVOCs (Table 2.70).

Scientific investigations into biogenic VOC emissions were initiated in the Soviet Union in 1928, and a considerable VOC flux measurement database had been produced by the 1950s (Isidorov et al. 1985, Isidirov 1990). Biogenic VOC emissions research in North America began in the 1950s (Went 1960) and reached a period of peak activity in the mid-1970s to early 1980s because of speculation that these compounds had a role in urban ozone formation (Altshuller 1983). These efforts included the systematic enclosure survey of North American vegetation species and above-canopy flux measurements of emissions from entire ecosystems (Rasmussen 1972, Zimmerman 1979, Winer et al. 1982, Arnts et al. 1982).

Scientific literature is full of studies estimating the emission factors for different types of vegetation and soils independent of controlling parameters such as temperature and humidity. To establish regional, global and (the most sophisticated approach) grid-resolved emission inventories, numerical assessment is used. The mod-

Table 2.70 Global emissions of NMVOC (after Williams and Koppmann 2007).

source	compound	emission (in Tg C yr^{-1})
fossil fuel use	alkanes	28 (15−60)
	alkenes	12 (5−25)
	aromatics	20 (10−30)
biomass burning	alkanes	15 (7−30)
	alkenes	20 (10−30)
	aromatics	5 (2−19)
terrestrial plants	Isoprene	460 (200−1800)
	Σ monoterpenes	140 (50−400)
	Σ other NMVOCs	580 (150−2400)
oceans	alkanes	1 (0−2)
	alkenes	6 (3−12)
total anthropogenic		100 (49−194)
total biogenic		1187 (403−4614)
total		1287 (452−4808)

els calculate emission fluxes based on (experimentally estimated) ecosystem-specific biomass and emission factors and algorithms describing the light and temperature dependence of NMVOC emissions. Advanced models include a general circulation model that can estimate factors as a function of geophysical variables and use satellite data. Numerical models often claim to provide accurate values but the calculated flux for monoterpenes (194.66 Tg yr^{-1}; RETRO 2008) includes two problems. First, the value is mass-related to the compound but the chemical mixture or averaged molar mass is not given. Second, an estimate in the format 200 Tg yr^{-1} seems more appropriate and suggests the "roughness" of this value. In other words, such numerical figures include errors of an unknown dimension, but at least 50%. One of the first modelling approaches to estimate annual global NMVOC fluxes was presented by Guenther et al. (1995) and Guenther (1999). Large uncertainties exist for each of these estimates, especially for compounds other than isoprene and monoterpenes. Tropical woodlands (rainforest, seasonal, drought-deciduous and savannah) contribute about half of all global natural NMVOC emissions. Croplands, shrublands and other woodlands contribute 10−20% each. Isoprene emissions calculated for temperate regions are as much as five times higher than older estimates. In 1995, total global anthropogenic emissions of NMVOCs were estimated to be 160 Tg, which corresponds to an increase of about 5 Tg since 1990 (van Aardenne et al. 2001).

Isoprene and monoterpenes

Most of the hydrocarbon flux from the biosphere to the atmosphere is just one compound, isoprene. Isoprene (HC=C(CH$_3$)HC=CH$_2$) is the building block of monoterpenes that has been found in all structures known from organic molecules.

The first ideas about emissions of hydrocarbons from plants were discussed by Went (1955). Despite the more obvious emissions of pleasant smells such as pine scent and lemon scent (resulting from monoterpenes), isoprene emission is the predominant biogenic source of hydrocarbon in the atmosphere, roughly equal to the global emission of methane from all sources (Guenther et al. 2006, Kesselmeier and Staudt 1999). In forests, more than 70 organic compounds derive from different families (Isidorov et al. 1985). The uncertainties in global estimations, however, are very large (Table 2.70).

This surprising finding of such a large influx of isoprene from plants into the atmosphere raises a number of questions, including what happens to the isoprene in the atmosphere and why plants emit isoprene, or rather what advantage does isoprene emission provide to the plant that makes it. The energy cost of isoprene emission is significant (Sharkey et al. 2008). Thermotolerance has most often been discussed as the advantage plants gain by synthesising isoprene. A second role for isoprene is its tolerance of ozone and other reactive oxygen species (ROS). Isoprene can prevent visible damage caused by ozone exposure. By contrast, isoprene emission can increase ozone production when NO_x is present, and simultaneously help plants tolerate ozone; however, ecosystem composition could change as isoprene-emitting species lead to high levels of ozone that they can better tolerate (Lerdau 2007). In addition, the mechanism by which isoprene protects against heat flecks and ROS is unknown.

In plant species emitting isoprene under illumination, this process is closely related to photosynthesis. Subsequent studies have shown that leaves are also capable of releasing isoprene in darkness, although at a rate two orders of magnitude lower than that in illuminated leaves. It is presently known that the isoprene is not emitted by all plant species from various taxonomic groups, whereas the dark release of isoprene occurs in cells of all living organisms (Sanadze 2004). The general view on isoprene emission is that it results from regulated conversions of carbon and free energy in a series of photosynthetic reactions under stressful conditions caused by CO_2 deficit inside illuminated autotrophic cells. This stress generates an energy overflow far in excess of the energy-consuming capacity. The necessity of discharging this energy excess is dictated by the fact that the living cell is a dissipative structure.

In addition to the important role biogenic terpenes play in gas-phase chemistry, their impact also extends to heterogeneous air chemistry. Although Went (1960) linked the formation of the "blue haze" over coniferous forests to the biogenic emission of 20 monoterpenes over 40 years ago, it was not until recently that terpenes received their due attention with respect to their role in secondary organic aerosol (SOA) formation. O'Dowd et al. (2002) reported that nucleation events over a boreal forest were driven by the condensation of terpene oxidation products. Formaldehyde (HCHO) is a high-yield product of isoprene oxidation. The short photochemical lifetime of HCHO allows the observation of this trace gas to help constrain isoprene emissions (Shim et al. 2005).

The annual global emission for isoprene and monoterpenes ranges from 250 to 450 and 128 to 450 Tg per year, respectively (Steinbrecher 1997, Möller 2003, and citations therein). By using HCHO column observations from the Global Ozone Monitoring Experiment (GOME) Shim et al. (2005) estimate that an annual global isoprene emissions figure of 566 Tg C yr^{-1} is $\sim 50\%$ larger than the *a priori* estimate, which was within the range given above. Guenther et al. (2006) estimate ~ 520 Tg C yr^{-1} as the global isoprene emissions figure. Meanwhile, in a recent estimation, Müller et al. (2007) stated that annual global isoprene emissions ranged between 374 Tg (in 1996) and 449 Tg (in 1998 and 2005), for an average of ca. 410 Tg yr^{-1} over the whole period, i.e. about 30% less than the standard MEGAN estimate (Guenther et al. 2006). This difference is largely because of the impact of the soil moisture stress factor, which is found to decrease global emissions by more than 20%. In qualitative agreement with past studies, high annual emissions are found to be generally associated with El Niño events. Current isoprene emission estimates are highly uncertain because of a lack of direct observations.

Williams and Koppman (2007) state a large range of uncertainty (Table 2.70). By contrast, Arneth et al. (2008) analyzed 15 modeling studies published between 1995 and 2008 and they showed similar "mean" global emissions data, but much smaller ranges of estimates (in Tg C yr^{-1}): 32–127 for monoterpenes and 412–601 for isoprene. Hence, the ratio of maximum to minimum emissions estimate for monoterpenes is between 4 (Arneth et al. 2008) and 8 (Williams and Koppmann 2007), but for isoprene it is only 1.5 (Arneth et al. 2008) compared with 9 (Williams and Koppmann 2007). Arneth et al. (2008) argue that most of the isoprene models are based on similar emissions algorithms, using fixed values for emission capacity. Arneth et al. (2013) show that most bottom-up model simulations to date converge on global isoprene emission strength of around 500±100 Tg C yr^{-1}, despite not only large differences in the values of the assigned emission capacities and in the modelled processes, but also in how vegetation is represented and which climatology is used in the experiments. By contrast, the model-to-model variation is significantly larger for monoterpenes. Arneth et al. suggest there is no evidence to improve the reliability of the emission mechanisms. There is no apparent reason why the spread in monoterpenes emission rates should be so much larger than isoprene emission rates because both are based on similar model set experiments.

In modeling the carbon cycle and the climate, Le Quéré (2006) identified three main phases in model development: the illusion, the chaos and the relief. Arneth et al. (2008), adopting this view, state that modeling BVOC emissions is in the illusion phase. Because of the lack of observations different models tend not to depart greatly from previously published estimates, doing so the wrong impression of being in the "relief" phase already. In fact, the biome-specific emission capacities are unknown rather than uncertain. The high importance, especially of monoterpenes, lies in the formation of SOAs. Against a backdrop of current discussions on the further reduction of PM levels (does the policy set up limits below the natural emissions? see Chapter 4.3.5) and the possible role of SOAs in the hydrologic cycle (formation of CCN and ice nuclei) and radiation budget (light scattering and hence, cooling the atmosphere) it is a challenge to improve global terrestrial emission data.

Other hydrocarbons

Oxygenated volatile organic compounds (OVOCs) have been found to be ubiquitous and abundant components of the global troposphere (Singh et al. 2000, 2001, Guenther 2013). Among the myriad of such chemicals present, acetone (propanone CH_3COCH_3) and methanol (CH_3OH) are the most dominant (Jacob et al. 2002,

Tie et al. 2003, Lathière et al. 2005). That is not surprising based on the CO stepwise reduction scheme shown in Fig. 2.42. The uncertainty in such emission values is best expressed by the statement given by Tie et al. (2003) that methanol has a relatively large surface emission with an estimated global methanol emission of 70 to 350 Tg yr^{-1}. Plant growth represents 60−80% of methanol sources and hence, results in a strong seasonal variation (Galbally and Kristine 2002, Jacob et al. 2005). It is again remarkable that today's estimates are double those previously given for acetone (10−16 Tg yr^{-1}) by Singh et al. (1994). HCHO, the most abundant carbonyl in the atmosphere, is directly emitted by fossil fuel combustion and biomass burning, but is mainly formed from the oxidation of methane and NMVOCs. By far the largest NMVOC emission source (excluding terpenes) on the global scale is of biogenic origin (ca. 85%), followed by anthropogenic technical (9%) and biomass burning emissions (6%; Table 2.70). As discussed before (Chapter 2.6.4.5), biomass burning must be included within anthropogenic sources. Newer biomass burning estimates range in the upper range of 95 Tg Y yr^{-1} (Werf et al. 2006, Andreae 2007).

Concerning natural organic emissions, secondary products (for example, OVOCs) often play *the* important role in air chemistry and climate, but are difficult to separate from primary compounds using global biosphere-atmosphere models. OVOCs, however, are important secondary products from the oxidation of primary emitted biogenic-organic compounds, mainly terpenes. Care must be taken on modeling results with the aim of estimating fluxes from global budget calculations. Direct measurements of the air/sea flux of acetone were made over the North Pacific Ocean by Marandino et al. (2005) and demonstrated that the net flux of acetone is into, rather than out of, the oceans. The authors argued that extrapolated global ocean uptake of 48 Tg yr^{-1} requires a major revision of the atmospheric acetone budget (Marandino et al. 2005). Using a global 3−D chemical transport model, Fischer et al. (2012) find that the Northern Hemisphere oceans are a net sink for acetone while the tropical oceans are a net source; on a global scale the ocean is in near-equilibrium with the atmosphere. Global total acetone sink has been estimated to be 101 Tg yr^{-1} (Marandino et al. 2005) and because (assuming a global steady state) sink = source is valid, but source = primary and secondary sources, it is a challenge to separate primary and secondary sources. Jacob et al. (2002) estimated the global total acetone source to be 95 Tg yr^{-1}, in agreement with the estimate by Marandino et al. (2005) when taking into account source = sink. Jacob et al. (2005) estimated among the terrestrial vegetation source (33 Tg yr^{-1}) an additional 8 Tg yr^{-1} (1.1 anthropogenic, 4.5 biomass burning and 2 plant decay), i. e. a total of 41 Tg yr^{-1}, which was consistent with other estimates. Consequently, 50 Tg yr^{-1} can be attributed to secondary sources, and thereby the result by Marando et al. (2005) not only suggests that the ocean does not emit acetone but that the budget needs no revision.

2.8 The human problem: A changing earth system

In recent decades, humans have become a very important force in the earth system, demonstrating that emissions and land use change are the cause of many of our environmental issues. These emissions are responsible for the major global reorganizations of biogeochemical cycles. With humans as part of nature and the evolution of a man-made changed earth's system, we also have to accept that we are unable to remove the present system into a preindustrial or even prehuman state because this means disestablishing humans. The key question is which parameters of the climate system allow the existence of humans under which specific conditions. The chemical composition of air is now contributed by both natural and man-made sources. As discussed in previous sections, large uncertainties in the estimations of global emissions (and subsequent regional gridded emissions patterns) remain. Nevertheless, major regional and global environmental issues, such as acid rain, stratospheric ozone depletion, pollution by POPs, and tropospheric ozone pollution, resulting in adverse effects on human health, plant growth and ecosystem diversity, have been identified and controlled to different extents by various measures. Some key issues remain unsolved, such as the further increase of greenhouse gases (GHG), most importantly that of CO_2. With the growth of "megacities", local pollution will have a renaissance, and this will inevitably contribute to regional and subsequently global pollution by large plumes, such as "brown clouds". Thus, it is important to find answers to the following questions:

a) What is the ratio between natural and man-made emissions?
b) What are the concentration variations on different time scales?
c) What are the true trends of species by man-made origin?
d) What are the concentration thresholds for the effects we cannot tolerate?

The chemical composition of air has been changing since the settlement of humans. In addition to the scale problem (from local to global), we have to regard the time scale. Natural climate variations (e. g. due to ice ages) had a minimum time scale of 10 000 years. The man-made changes in our atmosphere over the last 2000 years were relatively small before the 1850s. In the past 150 years (but almost all after 1950), however, the chemical composition has changed drastically. For many atmospheric compounds anthropogenic emissions have grown to the same or even larger order of magnitude than natural ones. Because of the huge population density, the need (or consumption)[21] of materials and energy has drastically forced the earth's system.

The time scale of the adaptation and restoration of natural systems is much larger than the time scale of man-made stresses (or changes) to the climate system. We should not forget that "nature" cannot assess its own condition. In other words, the biosphere will accept all chemical and physical conditions, even worse (catastrophic) ones. Only humans possess the facility to evaluate the situation, accepting it or not, and coming to the conclusion of making it sustainable. Under the aspect

---

[21] This is an interesting question: do we need all this consumption? What consumption do we need to realize a *cultural* life? Of course we move from natural (earth sciences) to a social and political dimension (life sciences) in answering these questions. But there is a huge potential to economize and save resources in answering these questions and implementing it. Karl Marx wrote: "The philosophers have only managed to interpret the world in various ways. The point is to change it" (this is still fixed in the main hall of the central building of the Humboldt University in Berlin, Germany). However, the key point is *how* and in *which* direction we have to change the world to receive *sustainability*.

of chemical evolution, this section briefly outlines the role of humans in forcing the climate system. Let us define a *sustainable society* as one that balances the environment, other life forms and human interactions over an indefinite time period.

2.8.1 Human historic perspective: From the past into the future

The value in identifying current trends and viewing them in a historical light is that the results can be used to inform ongoing policy and investment decisions.

As discussed in Chapter 2.6.5, food and energy are the key limiting factors for human development (Table 2.71). Humans have relied on various sources of power for centuries with solar power providing most of the essential energy. Solar energy is vital to all natural ecosystems. The energy sources have ranged from human, animal, wind, tidal and water energy to wood, coal, gas, oil and nuclear sources for fuel and power. The planet could not support the six to seven billion people that exist today without the commercialization of first coal, and then oil and gas. If these energy sources were necessary for the historically rare and unprecedented population growth that has occurred over the past 300 years, then this growth might be correlated (and modeled), in some way, after the pattern of consumption of these energy sources (Figures 2.45−2.47 and 2.58).

With the increase of the world's population, however, the per capita consumption of energy has increased (Table 2.71). Humans convert energy from less desirable to more desirable forms, i.e. from grass to meat, wood to heat and fossil fuels to electricity. Throughout history, humans have developed ways to expand their ability to harvest energy. The primitive man found in east Africa 1 000 000 years ago, who had yet to discover fire, had access only to the food he ate, and his daily energy consumption has been estimated to have been 2000 kcal (Table 2.71). This corre-

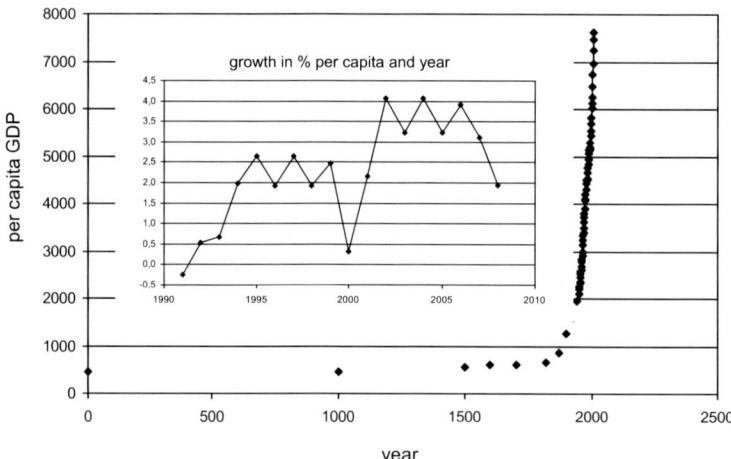

Fig. 2.57 Historical growth of GDP per capita (in million 1990 International Geary-Khamis dollars) and in % per capita and year (in 1990 International Geary-Khamis dollars) after Maddison, A. (2009), http://www.ggdc.net/maddison/oriindex.htm.

Table 2.71 Daily consumption of energy per capita, in 10^3 kcal (note: 1 kcal $\approx$ 4.19 kJ). Adopted from Cook (1976).

use	primitive man	hunting man	primitive agricultural man	advanced agricultural man	industrial man	techno-logical man
food	2	3	4	6	7	10
home and commerce	0	2	4	12	32	66
industry and agriculture	0	0	4	7	24	91
transport	0	0	0	6	14	63
total	2	5	12	31	77	230

sponds to the minimum daily dietary energy requirement (1800 kcal), given by the Food and Agriculture Organization of the United Nations, the average daily energy is about; the real consumption is between 1600 and 3800 kcal worldwide. Recommended are about 40–50 kcal/kg human body. The energy consumption of the hunting man found in Europe about 100 000 years ago was about 2.5 times that of the primitive man because he had better methods of acquiring food and also burned wood for both heating and cooking. The modern man consumes ten times more energy for food because of transportation, cooling, heating, losses etc. Roughly 10 000 years ago, the increasing population pressure on wild food resources led to a shift from food gathering (hunter/gatherers) to food production (agriculturists) in several parts of the world; however, the population density was still small (0.25 man km^{-2} rising to 250 km^{-2}; Guderian 2000). This led to demand-induced technologies and sources of energy supply, such as "water power for flow irrigation, animal draft power, iron tools and fire for land clearing and for improvement of hunting and pastoralism" (Boserup 1981, p. 46). Energy consumption increased again by almost 2.5 times as man evolved into the primitive agricultural humans of about 5000 years ago, people who harnessed draft animals to aid crop growth. The advanced agricultural man of AD 1400 in north-western Europe again doubled the amount of energy consumption as he began inventing devices to tap wind and water power, utilising small amounts of coal for heating and harnessing animals to provide transportation. Not only did the total and specific energy consumption increase but so did (and much stronger) the energy density (or power output) per unit (Table 2.72). For stationary energy units power output increased by six orders of magnitude from the first waterwheel 3000 years ago to the modern power plant; for mobile engines within the last 200 years the power output has increased between two and four orders of magnitude (referring to aircraft turbines and rocket propulsion, respectively).

Population pressures in many parts of Europe in the seventeen's and eighteen's centuries led to serious shortages of wood, which in turn led to many of the technological innovations that fuelled the Industrial Revolution. Coal's replacement of wood as the most important source of energy in Western Europe was "a classic example of demand-induced innovation ... promoted by population pressures on

Table 2.72 Chronological advances in power output. Adopted from Cook (1976).

primer mover	date	power output (in MW)
man pushing a lever	3000 BC	0.00004
ox pulling a load	3000 BC	0.0004
water turbine	1000 BC	0.0003
vertical waterwheel	350 BC	0.002
turret windmill	1600 AD	0.01
Savery's steam pump	1697 AD	0.0007
Newcommen's steam pump	1712 AD	0.004
Watt's steam pump	1800 AD	0.03
steam engine (marine)	1837 AD	0.6
steam engine (marine)	1843 AD	1.1
water turbine	1854 AD	0.6
steam engine (marine)	1900 AD	6
steam engine (land)	1900 AD	15
steam turbine	1921 AD	30
aircraft Ilyushin IL-18[a]	1947 AD	7.2
aircraft A 380[a]	2008 AD	80 (maximum)
aircraft Eurofighter[a]	2006 AD	120 (maximum)
steam turbine	1943 AD	200
coal-fired steam power plant	1973 AD	1 000
nuclear power plant	1974 AD	1 200
coal-fired steam power plant[a]	1990 AD	4 000
S-IC of Saturn V rocket[a]	1992 AD	20 000

[a] internet information

forested land in western and central Europe" (Boserup 1981, p. 109). The world's coal output per year was less than 10 million metric tons. Then, things began to change, thanks mainly to steam-powered pump engines, which allowed coal miners to drain the water that tended to accumulate in mine shafts and tunnels. In 1860, the world produced about 130 million tons. The dawn of the age of industrialization, ushered in by the invention of the steam engine, caused a threefold increase in energy consumption by 1875. Among other things, the steam engine allowed man to unlock the earth's vast concentrated storage deposits of solar energy − coal, gas and oil − so he was no longer limited to natural energy flows. In 1900, production rose to an astonishing one billion tons, and coal provided 90 % of the world's energy consumption.

Whereas historical increases in energy consumption had been gradual, once industrialization occurred the rate of consumption increased dramatically over a period of just a few generations. The technological man of 1970 in the Unites States consumed approximately 230 000 kcal of energy per day (~ 115 times that of primitive man) with about 26 % of that electrical energy. Of that electrical energy only about 10 % resulted in useful work, whereas the remaining 16 % was wasted by inefficiencies in electrical generation and transmission (Cook 1971, 1976).

Since about 1700, abundant fossil fuel energy supplies have made it possible to augment agricultural production to feed an increasing number of humans, as well as improve the general quality of human life in many ways. In essence, ample energy

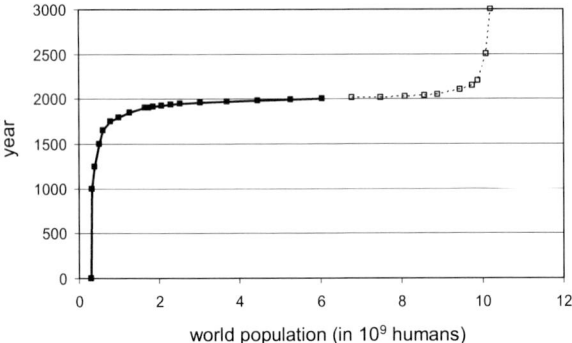

Fig. 2.58 Growth and trend of world population (data from UN Population Division).

supplies, especially fossil energy, have supported the rapid population growth and increased agricultural production.

In 1750, the world's population was approximately 720 million people. Over the previous 1000 years, this population had been growing very slowly at an average annual rate of about 0.13%. At this rate population doubles every 500 years, and it would have taken over 1500 more years (sometime around year 3250) to reach our current population of six billion people.

But sometime during the eighteens century, circumstances changed and the population began growing rapidly (Fig. 2.58). The population explosion corresponds with continued increase of the gross domestic product (GDP) per capita (Fig. 2.57). The relative growth (in percentage per capita and year), however, shows large variations from year to year and between (not shown here) countries. Moreover, the growth remains small and rose slowly over a long period (from the year 1000–1820 it lay between 0.03 and 0.07% $yr^{-1}head^{-1}$). In 1870, it was already 0.6% $yr^{-1}head^{-1}$ and only with the exception of post war time (1945–1950), was always above 1% (as seen in Fig. 2.57 up to 4%), but for some countries negative and for China with a maximum of about 15% $yr^{-1}head^{-1}$ in 2003 (Maddison 2009). The most common explanation for this change is that a *mortality revolution* reduced the rate at which people died, and that this mortality revolution was brought about by the Industrial Revolution. The Industrial Revolution changed everything. It was an economic revolution that spawned revolutions in science, technology, transportation, communication and agriculture. As a consequence, humanity began to experience improvements in health, nutrition, food variety, medicine and quality of life. More people survived infancy and childhood and they carried on to live longer lives. Because people were dying less quickly, populations grew more quickly.

In fact, the rate of energy use from all sources has been growing even faster than world population growth. Thus, from 1970 to 1995 energy use increased at a rate of 2.5% per year (doubling every 30 years) compared with a worldwide population growth of 1.7% per year (doubling every 40 years). During the next 20 years, energy use is projected to increase at a rate of 4.5% per year (doubling every 16 years) compared with a population growth rate of 1.3% per year (doubling every 54 years). Although about 50% of all the solar energy captured by worldwide photosynthesis is used by humans, this amount is still inadequate to meet all human needs for food and other purposes (Pimentel and Pimentel 2007). Hence, only the direct conversion of solar energy (heat and radiation) into electricity and chemically stored energy can overcome the future gap. As discussed above, two factors drive resource and environmental problems; global population growth and per-capita growth. Without any doubt, sustainable economy means zero-growth; in other words, growth must de-accelerate and come to saturation levels, expressing steady-state conditions.

The slowdown of population growth after 1995 might lead to an inverse population development within the next 250 years shifting in a steady state (Fig. 2.58). Large and sustained population growth is a contemporary phenomenon; until recent times it was rare to non-existent. Preindustrial populations grew when times were good (favourable climatic, agricultural, political and economic conditions) and shrank when times were bad (droughts, famines, wars, plagues, bad weather). Population growth was at all times restricted by the amount of land and food available. Land was needed to grow food for humans, fodder for animals and trees for building and fuel. As populations grew and occupied prime land, people were forced onto less productive land and the competing interests of food, fodder and fuel grew stronger. This pressure on land led to a number of different consequences: rising prices, under-nourishment, hunger, migration, territorial expansion through aggression and war and internal revolt. Populations became more susceptible to famine, disease, plague and death.

Population pressures led to the commercialization of a new energy source (water and wind power, animal draft power, coal, oil and natural gas), which in turn led to technical advances. Unlike today, many parts of the world were sparsely populated. There were frontier regions where land was sparsely populated and into which populations could migrate when domestic situations became too crowded or life too miserable. By contrast, it is well known that a large number of people in rural areas in developing countries do not have access to commercial energy because of a lack of purchasing power or other reasons. These people depend for survival on non-commercial energy sources, principally firewood, dung and agricultural wastes, which they gather at a negligible monetary cost. In many developing countries, non-commercial energy accounts for a significant proportion of total energy consumption (7500 kcal d^{-1} and capita) is considered a representative value (Goldemberg 2000).

As a consequence of increasing emissions from energy and food supplies (Figures 2.43, 2.51, 2.52 and 2.54), air pollution control measures have been stepwise introduced, from administrative local orders (ban of coal use) and technical applications (increasing chimneys and use of filters) to exhaust gas treatment (catalytic, thermal and washing) including technological improvements (increased efficiency, reduced energy demand) as well as alternative technologies. The first regulations concerning pollution from coal combustion were known as far back as the thirteen's century, but the first nationwide regulations were only introduced in England in the mid-nineteen's century (see also Chapter 1.3.4). Britain not only underwent the first Industrial Revolution, it also "invented" pollution (Thorsheim 2006). Soon this coupled phenomenon had been recognized in Germany and other western European countries, and by the beginning of the twenties century in the United States.

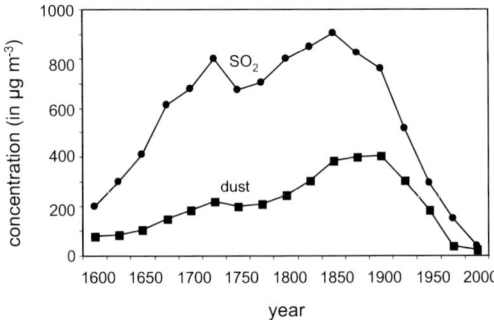

Fig. 2.59 Historical development of air pollution in London, after Lomborg (2001).

Much of the coal consumption, and the smoke that accompanied it, was concentrated in urban areas – large towns and cities. But in the second half of the nineteen's century the population started to accept that the smoke plague was not longer the price of industrialization and began to regard it as a problem. Fig. 2.59 shows the continuous increase of pollution in London over the centuries and its decrease at the turn of the twenties century because of regulatory measures. In most parts of the region until the 1960s, coal was the primary source of energy for electricity generation, industry and domestic heating. Air purification devices were practically non-existent. This led to high levels of air pollution, particularly in cities, with soot, dust, sulfur dioxide and nitrogen oxides. Winter smog's, particularly the notorious episodes in London during the 1950s, had serious effects on health as well as on building materials and historic monuments.

In Western Europe, after World War II, industry was restructured and oil, gas and nuclear power were increasingly used for energy production. This, together with the introduction of low sulfur fuels, natural gases, electricity and district heating schemes for domestic heating, contributed to the virtual disappearance of winter smog. Road transport, by contrast, has grown inexorably and is now the main source of urban air pollution. Between 80 % and 90 % of "classical" air pollution had been reduced by 1980 without the introduction of technological measures – almost all because of fuel replacement (Fenger 2009).

Over the past 20 years, the levels and patterns of air pollution in Europe have changed because of the adoption of important agreements aimed at reducing emissions and the dramatic changes occurring in central and Eastern Europe and central Asia. Emissions of acidifying substances in the region as a whole have decreased substantially. Between 1985 and 1994, SO_2 emissions in western, central and eastern Europe fell by 50 % as a result of the convention on *Long-range Transboundary Air Pollution* protocols (GEO 2000). The main reasons for these reductions were the installation of low sulfur coal and flue-gas desulfurization equipment at large point sources in Western Europe and the renewal of power plants and economic restructuring in Eastern Europe. The most drastic nationwide emission reduction happened in the former East Germany after the political changes in 1990 (Fig. 2.60). In two steps, first cutting down large parts of the industrial and agricultural sector (1990/1991) and second introduction of flue-gas desulfurization

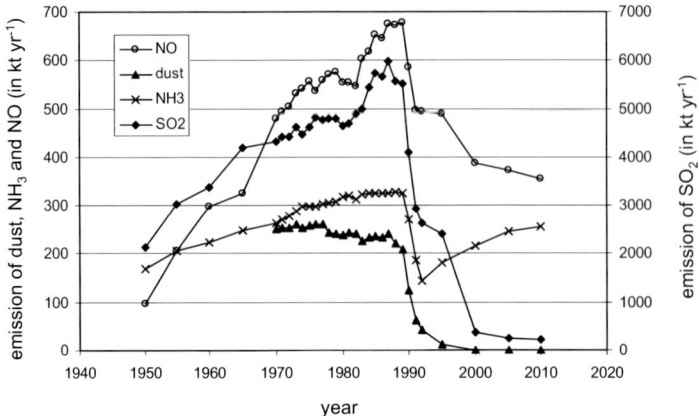

Fig. 2.60 East German emission trend; data from Möller 2003, after 2000 from UBA.

and other gas treatments (1995–1997), the emissions of all primary pollutants dropped by between 50% and more than 90% compared with 1988. The ratios between pollutants have also changed significantly, leading to another "chemical climate sub-system".

2.8.2 Changing the chemical composition of the atmosphere: Variations and trends

The chemical composition of the atmosphere is determined by the emission of compounds from marine and terrestrial biospheres, anthropogenic sources and their chemistry and deposition processes. Biogenic emissions depend on physiological processes and the climate, and the atmospheric chemistry is governed by the atmospheric composition and climate involving feedbacks. Anthropogenic emissions depend on technical processes and the willingness of humans to control the climate. Predicting the future climate needs an understanding of biogeochemistry. Nowadays, after more than 50 years of intensive research in atmospheric chemistry and air pollution control, GHG remain out of control. Climate research shows strong evidence that global warming will change the "physical" climate and thereby biogenic emissions (but also geogenic when we consider, for example, CH_4 from thawing permafrost soils in Siberia), which subsequent lead to the changing of the "chemical" climate.

With the aim of establishing a chemical climatology (see also Chapter 3.4) it is important to analyse and understand the reasons for concentration variations on different times scales to the determine trends of *climate change*, and separate natural- from man-made-caused influences to establish natural reference values of concentrations.

As pointed out in Chapter 1.3.4 and 2.8.1, urban pollution in towns and cities must be regarded at the end of the nineteen's century tremendously, likely by about

Table 2.73 Historical air pollutant concentrations (in µg m^{-3}); data from Cohen (1895), Rubner (1907) and Junge (1963).

City	CO	SO$_2$	NO$_2$	O$_3$[a]
Manchester (1882), winter		2000−3500		
Berlin (1906)	2000	1000−1500		
London (~ 1900), foggy		6000−11 000		
London (~ 1900)	2000	1300−2000		
London, suburb (1930)			1−6	
London (1930)			16	
London (1940−1950)	6000	600	20	20
Los Angeles (1950−1955)	120−1000	60−600	200−600	100−300
Frankfurt/M. (~ 1955)		87/443[b]	50/89[b]	
New York (1957)		280−1500		
USA, rural (~ 1955)		23		
Florida and Hawaii (~ 1955)		1−3		
Sweden, background (1954/55)		12 (2−19)		
Central Europe, background (1950s)	200−300		2−3	

[a] daytime values
[b] summer/winter

two orders of magnitude higher in concentration than at present concerns the "classical" pollutants (soot, dust, SO$_2$; see Fig. 2.59; Möller 2009b). Strong-smelling substances (likely organic sulfides) and ammonia (NH$_3$) must have been dominant in cities before the establishment of sewerage systems in the second half of the nineteen's century. *The Great Stink* or *The Big Stink* was a time in the summer of 1858 during which the smell of untreated sewage was very strong in central London. Nowadays, in a more political discussion of urban pollution concerning PM$_{10}$ and NO$_x$, these historical aspects are almost forgotten. Table 2.73 shows values from around 1950 that also support the data shown in Fig. 2.59. It should be recognized that measurement methods (air sampling and analytical procedure) changed over time, and it is often impossible to characterize the accuracy of early published data.

Distinguishing between local, regional and global pollution is mainly by using the atmospheric residence time (Chapter 4.5) of the compounds. However, in the 1970s SO$_2$ pollution rose to at least a regional problem because of the widely distributed sources of coal combustion around the world. The oxidation product, sulfuric acid in the form of sub micron particulate matter with a residence time 5−10 times greater than that of SO$_2$, had already become a hemispheric problem. Ozone, a secondary product with a variable residence time in the atmosphere of 2−60 days, can be regarded as a problem on hemispheric scale. With respect to the climate, GHG are of most interest. Therefore, in the following subsections the historic trends of man-made gases such as CO$_2$, CH$_4$, N$_2$O and chlorofluorocarbons (CFCs) will also be presented. Table 2.74 summarizes the increase since 1850, around the times of the beginning of the Industrial Revolution, which also can be seen as the preindustrial background situation.

Table 2.74 Concentrations of "greenhouse" gases (at Mace Head) and their changes; data from IPCC (2007).

substance	mixing ratio		GWP (100 year time horizon)	radiative forcing (W m^{-2})	residence time (years)
	pre-1750	2005			
in ppm					
CO_2	280	385	1	1.66	~ 100
in ppb					
CH_4	700	1857	25	0.48	12
N_2O	270	321	298	0.16	114
O_3	25	34	–	0.35	0.01–0.1
in ppt					
CFC-11 (CCl_3F)	0	246	4 750	0.063	45
CFC-12 (CCl_2F_2)	0	541	10 900	0.17	100
CFC-113 ($CCl_2F_2ClF_2$)	0	77	6 130	0.024	85
HCFC-22 ($CHClF_2$)	0	197	1 820	0.033	12
HCFC-141b (CH_3CCl_2F)	0	21	720	0.0025	9.3
HCFC-142b (CH_3CClF_2)	0	20	2 310	0.0031	17.9
Halon 1211 ($CBrClF_2$)	0	4.4	1 890	0.001	16
Halon 1301 ($CBrClF_3$)	0	3.2	7 140	0.001	65
HFC-134a (CH_2FCF_3)	0	49	1 430	0.0055	14
CCl_4	0	90	1 400	0.012	26
CH_3CCl_3	0	12.7	146	0.0011	5
SF_6	0	6.4	22 800	0.0029	3200

2.8.2.1 Fundamentals: Why concentration fluctuates?

The amount of a substance in atmospheric air can be expressed by different quantities, but the most useful is the concentration and mixing ratio (see Chapter 4.1.2). There is a basic problem in establishing the quantified chemical air composition. In Table 1.1, the sum of the first four listed constituents (N_2, O_2, Ar and CO_2) amounts to 100.00006%. The next eight listed compounds (Ne, He, CH_4, Kr, H_2, N_2O, CO and Xe) contribute 0.02732%, i. e. the sum of the first four compounds must be less than 100%, namely 99.973%. Additionally, trace constituents further contribute about 0.00042%. We must conclude that the substances denoted in Table 1.1 as "constant" are not really constant (we do not consider the variation of water vapor, which also contributes to a changing percentage in the range from negligible to about 3% to the total gas mixture; therefore, it is important to relate all concentration values to dry air). With the increase of minor (CO_2) and trace (CH_4, N_2O, CO and others) gases the relative percentage of all other gases must decrease. Moreover, because of the oxidation of reduced compounds used in raw materials (mainly carbon and nitrogen) an equivalent amount of atmospheric oxygen is chemically fixed and the abundance of O_2 in air also decreases absolutely (Keeling and Shertz 1992). For example, the actual atmospheric CO_2 level is now (year 2013 average) 298 ppm, 18 ppm more than listed in Table 1.1, but the CO_2 increase is at the cost of oxygen only, i. e. the O_2 mixing ratio decreases by the same amount. With increasing CH_4 and N_2O (all other trace gases are too small to measurably influence the level of the main constituents), however, the percentage of main substances must be corrected. *August Krogh* was the first to use a new gas analysis apparatus in Copenhagen to show that oxygen varies inversely with carbon dioxide (however, the data do not provide a stoichiometrical balance because of still missing precision, see discussion below). He wrote that "there is usually an oxygen deficit corresponding to the increase in carbon dioxide" (Krogh 1919, p. 12), but the air is rapidly renewed by atmospheric ventilation and that urban CO_2 emission contributed less than 0.001 % to the ventilation.

The precise measurement of oxygen is only recently possible with a precision of ± 1 ppm (Manning et al. 1999). Using paramagnetic oxygen analyzers precision was previously no better than ± 60 ppm (Keeling 1988). With regard to the CO_2 mixing ratio in the range of 320−380 ppm and comparing the CO_2 increase with an equivalent O_2 decrease it was previously impossible to constrain a budget. Moreover, chemical methods of O_2 determination in air must be assessed with a precision of less than 500 ppm (however, this contributes only to an error of the absolute O_2 concentration of 0.2 %)[22]. Table 2.75 shows a series of (at that time) precise measurements. Without discussing the variation of CO_2 (Chapter 2.8.2.3), according to current understanding, the sum of $O_2 + CO_2$ should be constant or reflect the uncertainty, respectively. As seen, the deviation in total concentration ($O_2 + CO_2$) amounts to 560 ppm (0.056 %), which is only the variation in the O_2 mixing ratio. The variation of CO_2 only amounts to 50 ppm (corresponding to 12 % of the absolute CO_2 value); thus, the relative precision of O_2 measurements was much less than that of CO_2.

What does the mixing ratio (or the corresponding concentration depending on pressure and temperature) express at a given site and a given time? It relates to positive (sources F_+) and negative (sinks F_-) volume-based (an air box or column with different spatial values x, y, and z; Chapter 4.1.1) fluxes, as well as the residence (removal) time of the substance regarded. There is no concentration changes or fluctuation when $F_+ = F_-$, i. e. $dc/dt = 0$ (this status we call steady state). This does not mean that dc/dx (or correspondingly in the vertical axis dc/dz) is zero. The time scale of concentration changes ranges from less than seconds (because of turbulent mixing) to years (because of slow chemical transformation or other removal). But transport and chemical reactions are always ongoing simultaneously but with different characteristic times. Even from (even time and spatially resolved) concentration measurement it is impossible to derive the individual processes that determine the concentration changes because of the large number of processes that determine the fluxes. According to the convention based on a time scale of several decades, it should be adequate to use annual averages to understand climate change. These, however, are normally based on monthly or daily means. Daily means, by

---

[22] The state-of-the-art of gas analysis at the end of the nineteen's century is best described by Hempel (1913), a pioneer of gas analysis; see also Haldane (1918).

Table 2.75 Simultaneous measurements of oxygen and carbon dioxide (given in % of dry air) in Dresden, November 8–18, 1884 (made by *Felix Oettel*, assistant of *Walther Hempel*); data from Benedict (1912). and carbon dioxide (1912).

substance	O_2	CO_2	sum
	20.840	0.0360	20.8760
	20.850	0.0370	20.8870
	20.870	0.0390	20.9090
	20.890	0.0410	20.9310
	20.880	0.0500	20.9300
	20.860	0.0550	20.9150
	20.900	0.0389	20.9389
	20.910	0.0391	20.9491
	20.740	0.0400	20.7800
	20.750	0.0440	20.7940
	20.770	0.0440	20.8140
	20.750	0.0480	20.7980
	20.810	0.0490	20.8590
	20.780	0.0540	20.8340
	20.820	0.0380	20.8580
	20.790	0.0410	20.8310
	20.800	0.0416	20.8416
	20.830	0.0430	20.8730
	20.860	0.0400	20.9000
	20.840	0.0370	20.8770
	20.920	0.0400	20.9600
	20.920	0.0440	20.9640
standard variation	0.056	0.0050	0.0560
mean	20.835	0.0430	20.8780

contrast, are constructed according to the sampler's and/or analyzer's technical time resolution (varies from less than a second up to a day). For an understanding of concentration values (and finally to analyze time series) it is important to know all influences on the concentration with regard to different time scales.

The following list should give an idea of which factors can influence the concentration value (note that we talk mainly on surface-near concentrations regarding the biosphere−atmosphere interaction):

Transport: The turbulent and advective transport of air parcels (horizontally and vertically) is often the most important factor (mixing and dilution). Transport processes show another wide range of causes, but are mainly driven by density gradient (hence, T gradients). Thus, it is clear that characteristic daily and seasonal patterns can be observed (for example, the vertical down-mixing of air in the morning after a nocturnal inversion).

Emission: This flux from the earth's surface into the lower air layer is the most important input of trace substances, changing the mass and thereby the concentration. This flux shows strong variations with time, e. g. daily and seasonal concerning different biogenic processes (photosynthesis, respiration), as well as varying with other parameters (temperature, light, water), rush hour traffic (but there is also a weekly traffic cycle) and domestic fuel use (seasonal cycle).

Deposition: Regarding the deposition opposite to the emission as the ultimate sink of atmospheric substances, similar factors (but almost inversely correlated) in biological uptake have to be taken into account. Wet removal depends on precipitation (it is occasionally flux only) and dry deposition depends on the surface and atmospheric conditions, again showing timely variations.

Chemical conversion: Finally, chemical reactions can produce (secondary) substances (e. g. CO, O_3, H_2O_2) but also destroy them and all primary emitted compounds, but on very different time scales (from minutes to years).

Therefore, all processes occurring simultaneously but on different time and spatial scales are in "competition" to establish the budget of a substance and hence, its mass in the atmosphere and its three-dimensional spatial distribution. Substances with large residence time (in other words, the fluxes of deposition and chemical conversion are very small) are mostly globally distributed (e. g. GHG). Substances with an equal source and sink term (e. g. all biogenic-emitted substances – but disturbed by the anthropogenic emission) over the regarded time period (e. g. year) will not accumulate in the atmosphere and show a constant average value in time but not naturally in space because of the possible heterogeneous distribution of source and sink terms. Discontinuous or even catastrophic emission can lead to concentration spots (e. g. volcanic eruptions). Substances with a small residence time (less than a few days, e. g. SO_2, NO_x) are only of local and regional importance, but with globally distributed sources have become a global problem with increasing emissions.

2.8.2.2 SO_2, NO_2 and dust: Classic for local to regional up-scaling

Air pollution in the industrialized world has undergone drastic changes in the past 50 years (Table 2.73). Until World War II the most important urban compound was sulfur dioxide combined with soot from the use of fossil fuels in heat and power production (Möller 1977). The data listed in Table 2.73 support the approximated record of air pollution in London (Fig. 2.59). Fig. 2.61 is a rare example of historical seasonal SO_2 cycles. Manufacture (attributing them to the summer figure) led to 2−4 mg m^{-2} and household fires in winter to an additional 4−8 mg m^{-3} in Manchester by the 1890s. By the turn of the century almost everything that was known in the 1950s (Halliday 1961) about the causes of smoke and their elimination had already been said, but hardly anything had been done to reduce the smokiness of cities. The seasonal cycle from 1905 in Manchester (Fig. 2.61) shows on average half as much SO_2 as in 1891/1892. Interestingly, both cycles show minima in February and September, which are likely to be meteorologically originated. Compared with Manchester, in the former German city of Königsberg, very low SO_2 concentration (summer 30−40 µg m^{-3} and winter 300−400 µg m^{-3}) were measured, probably because there was no industrial sector and mixing with pristine air from the

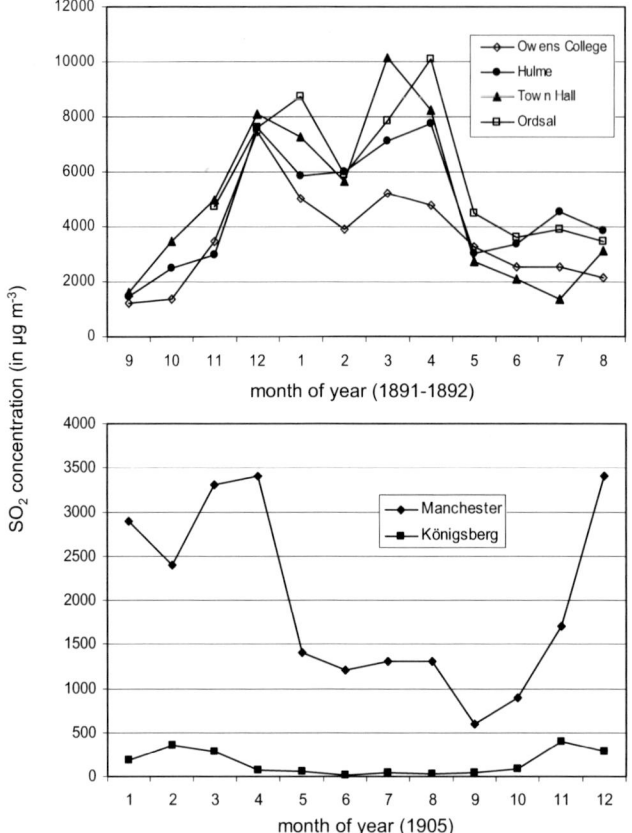

Fig. 2.61 Historical seasonal cycles of SO$_2$ concentration in Manchester (different sites 1891−1992 and 1905) and Königsberg (1905); data from Rubner et al. (1907), pp. 390−391.

sea. Although SO$_2$ concentration in cities were in the range of 1−10 mg m^{-3} around 1900, they dropped to the upper range of μg m^{-3} (some hundreds) by the 1950s and nowadays are in the lowest range of μg m^{-3} (5−10), close to the remote background values of the order of 0.5 to 4 μg m^{-3} (Georgii 1970). Table 2.73 also shows the change in the "type" of pollution (from London to Los Angeles smog). Los Angeles still shows large SO$_2$ concentration as well as very high concentrations of NO$_2$ and O$_3$ caused by car traffic. Without doubt the large concentrations of CO (and CO$_2$), SO$_2$ (and soot) before 1950 were caused by domestic coal combustion.

When that problem was partly solved by cleaner fuels, higher stacks and flue-gas cleaning in urban areas, the growing traffic gave rise to nitrogen oxides and VOCs and in some areas photochemical air pollution, which can be abated by catalytic converters. Lately, interest has centered on small particles and more exotic organic compounds that can be detected with new sophisticated analytical techniques.

In the early 1970s, annual average sulfur dioxide concentrations in urban areas of the United States tended to range from 10 to 80 μg m^{-3} (EPA 1972). In 1970 and 1971, the annual average level among all urban monitoring sites was about

Table 2.76 Air pollutants annual concentrations (in μg m^{-3}) from about 50 cities in the word (between 2000 and 2005); data after WHO (2006).

compounds	average	range
SO_2	~ 20	7–100
NO_2	~ 30	18–83
PM_{10}	~ 75	25–225

25 μg m^{-3}. At nonurban NASN monitoring sites, annual mean sulfur dioxide concentrations tended to be around 5 to 15 μg m^{-3}, slightly above estimated natural background levels (EPA 1972). Between 2000–2005, the annual NO_2 average in the United States in 125 cities was 36 μg m^{-3} (WHO 2006), similar to the value found in Europe (variation 14–44 μg m^{-3}). Higher SO_2 concentrations were reported in some of the megacities in developing countries, but in Europe and the United States the annual urban SO_2 concentration varied from 10 to 30 μg m^{-3} (Table 2.76). In eastern Germany, besides electricity production that was predominantly based on lignite-fired power plants (but equipped with high stacks), domestic coal fuels contributed to large local pollution (annual average 80–100 μg m^{-3} SO_2 but in winter periods in Leipzig up to 1500 μg m^{-3}). The concentration decrease of about one order of magnitude of the emission decrease after 1990 is shown in Figures 2.62 and 2.63. Fig. 2.62 also shows a scattering of the SO_2 concentration between different cities (as a result of the local emission values), whereas after 2000 SO_2 has been homogeneously distributed at around 5 μg m^{-3} and decreased further to 3 μg m^{-3} in 2009, remaining later constant. The Wahnsdorf record, one of the longest continuous measurements in the world, also shows the permanent increase of O_3 concentration

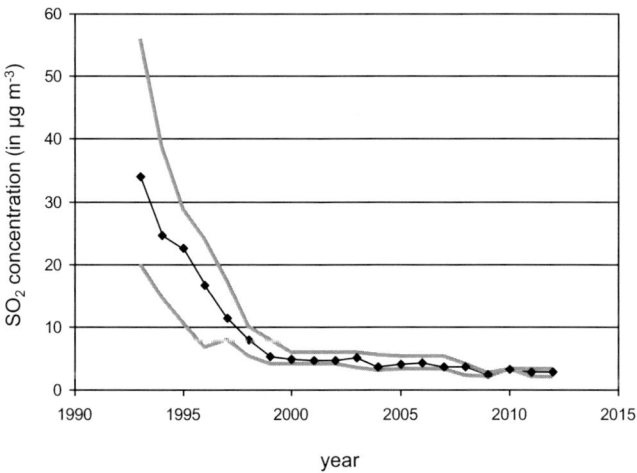

Fig. 2.62 Historical record of SO_2 concentration (annual means in μg m^{-3}) from five different cities in Brandenburg (Germany); grey lines: min and max, resp., representing the spatial variation. Data from LUGV (Landesamt für Umwelt, Gesundheit und Verbraucherschutz, Land Brandenburg, 2012).

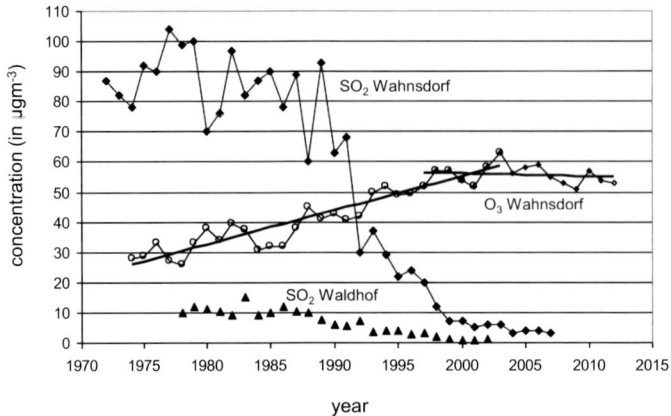

Fig. 2.63 Historical records of SO$_2$ and O$_3$ concentration (annual means in µg m^{-3}) at Wahnsdorf (former Meteorological Observatory near Dresden, Germany); data from LfULG (Sächsisches Landesamt für Umwelt, Landwirtschaft und Geologie); SO$_2$ monitoring was discontinued in 2008. UBA Background station Waldhof (Umwelbundesamt).

(see Chapter 2.8.2.7 for more discussion). This curve shows that a long measurement is needed to avoid misinterpretation because of shorter time fluctuations.

Dust concentrations in the air were in past several times larger than nowadays. Large global differences still were observed (Table 2.76) but only on a local scale, in other words influenced by local (primary) dust sources. Fig. 2.64 shows an example on long-term PM$_{10}$ monitoring at two sites in eastern Germany. The decrease by a factor of more than two between 1989 and 1999 is because of reduced flue-ash emissions and the introduction of desulfurization in coal-fired power stations in this period. Today's values of between 15 and 20 µg m^{-3} are typical for European background sites. Because two-thirds of PM$_{10}$ (the percentage of PM$_{2.5}$) is formed from secondary sources (about half ammonium, nitrate and sulfate and half SOAs), air mass characteristics and subsequent air chemistry conditions are responsible for the inter-annual variations being on average 15–20%. As a consequence, to validate PM trends due to anthropogenic abatement measures, long-term monitoring (minimum 10 years) is essential for any source-receptor relationship analysis of the PM constituents (the Melpitz station near Leipzig in Germany is the only site in Europe with long-term monitoring of chloride, sodium, sulfate, nitrate, ammonium, calcium, magnesium and potassium). Fig. 2.64 shows the decreasing trend of PM$_{10}$ in Melpitz between 1993 and 2000, representing the regional background in East German lowland, still influenced by East European Countries (Poland, Czech Republic, and the Ukraine). In the last 12 years, the PM$_{10}$ concentrations remain constant at 22.4 µg m^{-3} with inter-annual variation at ±2.9 µg m^{-3} (Spindler et al. 2013). The PM$_{2.5}$/PM$_{10}$ ratio increased slightly from 0.71 (1995) to 0.84 (2012), being higher in summer than in winter.

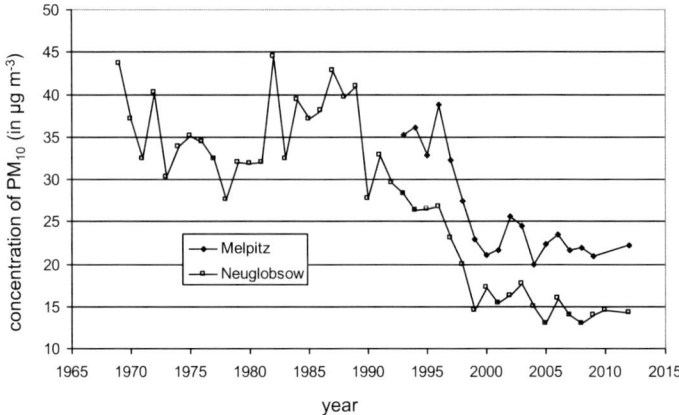

Fig. 2.64 Long-term PM_{10} at Melpitz and Neuglobsow (Germany), Melpitz data from *Gerald Spindler*, Institute for Tropospheric Research, Leipzig (see also Spindler et al. 2004, 2009, 2013), Neuglobsow data from Umweltbundesamt (Federal Environmental Agency), before 1990 as TSP from GDR Environmental Authorities. Melpitz is located close to Leipzig and characterizes agricultural background whereas Neuglobsow represents the rural background in Germany.

2.8.2.3 CO_2: The fossil fuel era challenge

The preindustrial CO_2 level derived from historical measurements

Being aware of the pre-industrial CO_2 level is important for two reasons: first we need this figure to estimate how much CO_2 humans have added to the atmosphere since the beginning of the Industrial Revolution, and second, this value is the reference for all human caused impacts dependant on the actual CO_2 level, e. g. climate forcing and oceanic acidification. This "reference" value has been estimated by several authors to be in the 243–290 ppm range (Holmén 1992 and citations therein, p. 242). Today's most accepted "pre-industrial" value is 280±10 ppm where mainly data from bubbles in ice cores has been employed (Neftel et al. 1985, Friedli et al. 1986, Etheridge et al. 1998) and from historical ocean CO_2 data a smaller value of 268±13 ppm was derived (Chen and Poisson 1984). A value around 290 ppm (Callendar 1938, 1940) and 292 ppm (From and Keeling 1986) was derived from selected historical CO_2 measurements in the 19th century respectively.

Joseph Black was the first who found it in the air of Edinburgh, most probably between the years 1752 and 1754. He writes (Black 1803, p 74): *"... From this it was evident that the sort of air with which the lime is disposed to unite in a particular species which is in a small quantity only with the air of the atmosphere. To this particular species I gave the name of fixed air"*. In 1787, *Horace Bénédict De Saussure* (*Saussure* the elder) employed lime water as a test for CO_2 in the air when he was at the summit of Mont Blanc (Saussure 1796). First attempts to determine CO_2 in atmospheric air were done at the end of the 18th century by *Alexander von Humboldt*, and *John Dalton*, only showing, however, that its level was less than 0.1%. There was a group of famous scientists (*Louis-Jacques Thénard, Theodore*

De Saussure, Jean Baptiste Boussingault, Hugo von Gilm) who measured CO_2 levels around 400 ppm (and more), between 1815 and 1850, at different sites in Central Europe. All later researchers agreed that such values were much too high. *Albert Edmund Letts* (1852–1918), professor of chemistry in Queen's College, Belfast, and *Robert F. Blake* (1867–1944) together published two remarkable papers (1900, 1901), giving a new accurate CO_2 determining method and the most extensive critical overview on CO_2 observations in the 19th century, citing 198 authors. They wrote (1900, p. 172): "*From the time of De Saussure whose researches extended over the period 1809–1830, and who made some hundreds of observations ... until about the year 1870, the normal amount of atmospheric carbonic anhydride appears to have been taken as 4 in 10,000 vols. ... but it is interesting to notice, as Blochmann has pointed out, that his figures almost steadily decreased as his work progressed.*"

It is remarkable that many authors agreed with 0.03 % (according to 300 ppm) as "natural reference value" (Schoesing 1880, Blochmann 1886, Renk 1886, Letts and Blake 1900, Brown and Escombe 1905a, Friedheim 1907, Lode 1911, Benedict 1916, Krogh 1919, Loewy 1924, Quinn and Jones 1936, Mellor 1940, D'Ans and Lax 1943, Remy 1965). *Humboldt* wrote: "*The relative quantities of the substances composing the strata of air accessible to us have, since the beginning of the nineteenth century, become the object of investigations, in which Gay-Lussac and myself have taken an active part...*" (Humboldt 1850, p. 311). He accurately cites the concentration of oxygen (20.8 %) and nitrogen (79.2)[23], and named as a minor species "carbonic acid gas". Citing *Boussingault* and *Lewy*, *Humboldt* mentioned in a footnote on p. 311 that "*the proportion of carbonic acid in the atmosphere ... varied only between 0.00028 and 0.00031 in volume*" (280–310 ppm).

What is the "error" of such analytical figures? It is known from chemical practice that a precision (being the standard deviation from several measurements of a sample with the same concentration) of ±3 % is excellent. However, a high precision does not automatically mean that the value is "true"; the trueness is the degree of match with the "true value". A high precise measurement can be far away from the true value, as found in inter-comparisons for rain water analysis; today's wet chemical analysis of ions have a precision of 0.02 mg Ca L^{-1} and an inter-network bias of 15 % for calcium[24]. However, when the analysis has a high accuracy, the mean from many measurements can fit the "true" value. *Letts* and *Blake* write:

"These variations, if actually small, are *relatively* large, and correspond with fluctuations of at least 10 per cent of the total quantity, and according to some observers to a much higher proportion (1901, p. 436) ... and that very few of these have been tested as regards their degree of absolute accuracy; and also that the observers have collected their samples at different heights above the ground ... Müntz and Aubin are an exception to this rule (1901, p. 139).

---

[23] Only at the end of the 19th century was it known that about 1 % of this figure is due to novel gases.
[24] Assuming this being also similar for barium; however, later we discuss that not the (excellent) precision of chemical analysis (likely already provided in the 19th century) is important but the errors during CO_2 sampling/absorbing (incompleteness) and alkaline titration (additional CO_2 absorption) are essential. Such precision (0.02 mg Ba L^{-1}) would result only in a 1 % relative error when sampling CO_2 from 6 L air.

CO_2 has been analyzed wet-chemically by being absorbed from air, exclusively in alkaline solutions (almost baryt water, i. e. BaOH); after precipitating of $BaCO_3$ after 1−2 hours, an extracted part of the clear solution is titrated (mostly by oxalic acid). Many modifications have been made (apparatus such as bottles or tubes with an aspirator, instead of titration gravimetric estimation, other titrating acids such as HCl solution, different indicators for the neutralization point etc.). Blochmann (1886), Walker (1900) and Letts and Blake (1900, 1901) found that *Pettenkofer's* ordinary process ("Flaschenmethode") and the "aspiration method" systematically produces inflated values due to the baryt water additionally absorbing CO_2 (e. g. when handling solutions in laboratory air). Letts and Blake (1901) constructed careful simultaneous sampling and analysis with the *Pettenkofer's* and its own new process, showing that the *Pettenkofer* ordinary process results in 25−35% to high values and so explaining the difference between "4" and "3" parts of 10,000.

It is clear to any chemist that many sources of errors can appear in the procedure, and a lot of experience or practice by the operating person is necessary to obtain a high accuracy. Fresenius (1875) states that a precision of ±0.002% absolute CO_2 level (according to ±20 ppm) is excellent for air analysis and Haldane (1918) gives ±10 ppm as precision (3−7 % relatively). Hempel (1913) first cites results from inter-comparisons; the accuracy is no better than ±100 ppm for three different methods (50−100 ppm for ambient air analysis and 100−200 ppm for indoor CO_2 analysis) and ±(30−50) ppm by comparing *Pettersson's* and *Pettenkofer's* method. Today, it is self-evident that measurements are corrected to standard conditions and related to dry air. In many older papers (before the 1870s) this is not clearly expressed and we have to consider another source of "variation".

When discussing historical values, we have to take into account that CO_2 in laboratories air can go up to 0.2% (Hempel 1913), and therefore any uncareful analytical operation might lead to higher analytical figures. The next fact which must be taken into account when comparing different historical CO_2 values is the town influence, which was estimated between 30 and 200 ppm higher than in rural surroundings (Renk 1886, Lode 1911), with today's values of city dome CO_2 (see below). Today, higher town CO_2 is nearly almost a result of traffic, whereas in the past domestic coal combustion (heating, cooking, manufactures) and locomotives were the CO_2 sources. Möller (2009b) found that in cities around 1900 (shown for Dresden), the soot pollution was likely to be a factor of 50 and more higher than today. Of course, CO_2 is not proportional to soot emission, but is supported by this finding to be an important local pollutant during that time.

Moreover, seasonal CO_2 variation at remote continental sites amounts to 10−20 ppm and at Mauna Loa only about 5 ppm (see below) whereas inter-annual CO_2 variation is small with 2 ppm. Only very few of the historical measurement series were able to reflect a seasonal cycle; accuracy with 10−20 ppm and a too small number of measurements mask this natural variation. However, the seasons influence on CO_2 was already stated by *De Saussure*, who found it higher in summer than in winter, supported by Fittbogen and Hässelbarth (1879), who found a minimum in December and Petermann and Graftiou (1891) with a very small variation of only a few ppm. Marie-Davy (1880) found a CO_2 maximum in December. Whereas seasons and towns influences on the CO_2 level may be excluded or to be assessed of minor importance in evaluation of historic background levels, there remain two significant factors responsible for large variations among the observers; the sampling height above ground and the time of sampling. The most serious effect of CO_2 variation is the diurnal cycle, which amounts to 30–80 ppm (as maximum-minimum, and often more; there are many measurements in literature, not cited here) and appears as sinusoidal curve. Hence, the difference between continental daytime-means and nocturnal means could range between 10–30 ppm. Today, CO_2 is measured continuously ("automatically" giving a mean daily value) resulting in representative monthly means and subsequent annual mean figures (when the time gaps of measurements are not too large). In the past, often only one value (most often daytime) per day or only few values per month were produced. According to the time of day, such figure cannot reflect a daily mean; only a large number of measurements made at different times can reflect a mean daytime value. The majority of CO_2 measurements have been carried out at business times; hence, a "corrected" daily mean must be around or higher than 10 ppm. The diurnal amplitude strongly depends on the local situation. Principally, this was known at the end of the nineteenth century after a few more intense studies; but extensive sampling over longer periods was impossible due to the length of time needed for sampling and single analysis (several hours) and hence the need for large man-power. Letts and Blake (1900) wrote "… *and also that the observers have collected their air samples at different heights above the ground*". More CO_2 is found in the ground air from lower than from higher levels; the gradient has a definite connection with the season and rainfall has a marked influence, Letts and Blake (1900, p. 216) summarize, citing Fodor "… *the fluctuations in the amount of atmospheric carbonic anhydride are mainly due to the absorbing action of soil on the one hand, and the evolution of ground air on the other*". Fodor (1879) found soil CO_2 emission to be 0.175 L m^{-2}d^{-1}.

In summary of the discussed facts affecting the CO_2 measurement, it is no longer surprising that historically published values after 1870, between 270 and 360, can be seen as not principally erroneous but not representing a "true" mean CO_2 figure. As found from the few inter-comparisons, the scattering of only single or a few CO_2 measurements can due to different reasons be ±(50–100) ppm: methodical errors by different authors and/or methods, timely influences (almost diurnal cycle), and different local influences (but these are likely of minor influence). Consequently, it makes no sense to list and to interpret published historical CO_2 data based on very few measurements. Therefore, only results from "long-term" observations and/or more than 50 single measurements will be regarded here.

Additionally, some of the longer times series show large timely (but not periodically as we know it today) variations of the single measurements which can be interpreted as the scattering of the analytical figure by errors and likely predominant by the different time of the single measurements and the subsequent average. This is best illustrated by the well-known Montsouris Observatory monitoring (1876–1910), Fig. 2.65. First, the inter-annual (5–15 ppm) and monthly variations (>10 ppm) cannot be explained by natural physical reasons (Waterman 1983) and secondly, the marked rise in 1890 (by 27 ppm) suggests a change in the sampling and/or analytical procedure. Here are the characteristics of two periods (period average based on mean annual values, standard variation, min-max, number of mean values):

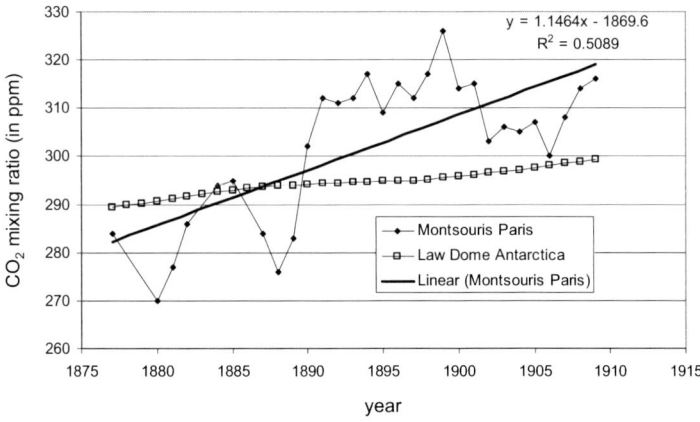

Fig. 2.65 Historical records of CO_2 measurements at Montsouris Observatory near Paris (data from Stanhill 1982), compared with CO_2 data from Law Dome in Antarctica (see Fig. 2.66 for data source).

Table 2.77 Monthly means at Mountsouris station (unknown single measurements) after Marie-Davy (1880), in ppm. Mean from all monthly data: 312±39 and from annual averages: 319±25.

	01	02	03	04	05	06	07	08	09	10	11	12
1876	–	–	–	269	249	256	261	–	–	313	307	280
1877	280	282	276	270	278	280	277	267	280	269	308	344
1878	333	335	322	331	359	351	342	350	347	353	354	355
1879	356	357	357	358	356	356	346	333	330	304	255	244

1876−1889: 283±8 (270−294) 9
1890−1910: 311±6 (300−325) 20

The obvious problems with the Montsouris analysis (mentioned already by Reiset 1880) is illustrated by the data published by Marie-Davy (1880), despite his explanation of different air mass influences (without doubt, the Montsouris station in a park north of Paris which today is within Paris, is much less suitable then the Dieppe station for background estimates). Table 2.77 shows the monthly means from April 1876 until December 1879; the standard variation of total monthly means ranges between 35−56 ppm and that for annual means ranges between 12−41 (monthly minimum-maximum 244−359 ppm). This data illustrates the above discussed uncertainties.

Lévy and Miquel (1891) gave a mean from the Mountsouris series (1881−1890), being 287±8 ppm (277−302) based on annual means whereas the mean from all monthly means amounts to 293±2.5 ppm (289−297), showing only very small variations. Despite the large general uncertainties of Montsouris data, the last figure is close to the following cited mean values (and the "selected pre-industrial value" by From and Keeling 1986).

In Gembloux, a rural site in Belgium, Petermann and Graftiau (1893) found for the period 1889–1891, a mean of 294±18 ppm from 525 measurements (daytime). Similar values have been found by Schulze (1871) in Rostock (1868–1871, $n = 36$) to be 291±12 (272–323) ppm; however, in an earlier measurement series (1863–1864, $n = 41$) *Schulze* estimated 360±21 (320–405) ppm, but he doubts himself. As examples of other unbelievably high values owing to systematic errors, the following figures are given by Gilm (1857) from Innsbruck (November 1856 – March 1857, $n = 18$) to be 415±23 (381–458) ppm and by Farsky (1877) from Tabor, Bohemia (1874–1875, $n = 27$) to be 342±8 (328–362) ppm. Farsky (1877), a professor of chemistry, denotes himself as a newcomer ("*Als Neuling dieser Art von Untersuchungen* ..."), giving to us, as an example, that handling and long-term experience is needed for "correct" CO_2 atmospheric air analysis as Blochmann (1886) emphasizes. On the other hand, there are a few historic measurements with means significantly less than 290 ppm. Remarkable measurements have been done in Dorpat (today Tartu, Estonia) belong the Medical Faculty published in two doctor thesis (Heimann 1888, Frey 1889). Frey (1889) found a mean of 262 ppm ($n = 556$) with 189 (min) and 336 (max) whereas Heimann (188) found 269 ($n = 601$) with 182 (min) and 375 (max); variation among monthly means are 250–283 ppm. When reading both works, the reader gets a strong feeling of reliability and that the measurements are accurately done but having a systematic error which (of course) cannot identify. The interesting finding by Heimann (1888) is the day-night difference to be on average 28 ppm: day-time average 258 ppm ($n = 379$) and night-time average 286 ppm ($n = 222$); obviously about 30 ppm (10 % absolute error) too low.

I fully agree with the statement by From and Keeling (1986) that *Jules Reiset's* (1818–1896) analysis shows the highest level in analytical science for that time and which seems to be most reliable. Reiset (1879) carried out a first series at a rural station 8 km from Dieppe (September 1872 until August 1873) with analysis during day and night-time ($n = 92$) giving a mean of 294 ppm; additional parallel measurements above different crops and under free field conditions show differences up to 10 ppm. Less above vegetation and near a herd of sheep (300 head), the CO_2 increases to 318 ppm. In a later measurement campaign from June until November 1879, using improved sampling and described in detail, Reiset (1880) estimates a mean of 298 ppm ($n = 91$), including daytime (mean 289 ppm) and nighttime measurement (mean 308 ppm), but unfortunately not mentioning the number of analysis. During fog, he found a mean of 317 ppm (up to a maximum of 342 ppm in intense fog). All data has been related by *Reiset* to standard conditions (dry air, 0 °C and 760 torr). Besides *Reiset*, Müntz and Aubin (1882) made very careful measurements at different sites; unfortunately only for a relatively short time. In Vincennes, suburb of Paris, they found ($n = 37$) 284±13 (270–317) ppm in April and May 1881, in Paris (December 1881 – February 1882, $n = 20$) 319±26 (301–422) ppm and in April/June 1882 ($n = 8$) 293±6 (288–306) ppm, at a rural site in 1881 (agricultural institute, $n = 8$) 300±15 (273–329) ppm, and at Pic du Midi (2887 m a.s.l) in August 1881 as mean 286±9 (269–301) ppm ($n = 14$). The number of single measurements is too low to receive robust figures; the standard variations are, therefore, larger than expected from natural variations. Petermann and Graftiou (1891) compare his result with other measurements (in 10^{-2}%):

Rostock (Schulze), 1871	2.92
Dieppe (Reiset), 1872−1873	2.96
Plaine de Vincennes (Müntz et Aubin), 1881−1881	2.84
Montsouris (Marié-Davy et Lévy), ca. 1880	2.93
Gembloux (Petermann et Graftiau), 1889−1891	2.94

More findings around "3" can be added:

Halle, Germany (Marchard), 1845	3.10
Atlantic Ocean (Thorpe), 1866	3.00
Leeds, UK (Armstrong), 1879	3.13
Belfast, Ireland (Letts and Blake), 1897	3.00
Kew, UK (Brown and Escombe),1898−1901	2.94
Boston, USA (Benedict), 1909−1912	3.07
Stockholm (Selander), 1885−1886	3.03
Copenhagen (Krogh), 1917	3.00
Sweden, costal site (Lundegårdh), 1920−1923	3.04

Richard Felix Marchand (1813−1850), professor for chemistry in Halle, was the first to estimate a mean from 150 observations (around 1845) in Halle, significantly less than the "4" figure to be 310 ppm. His finding is only cited by very few following scientists. Petermann and Graftiou (1891, p. 30 and 29) wrote: " … *acide carbonique de l'atmosphère est en general d'une grade constance … se rapproche trés sensiblement de 3 litres par 10,000 litres d'air à 0° et 760 millimétres de pression.*" From the cited data, we can assess the town contribution to CO_2 (remarkable only in winter) to be as mean 40±20 ppm and the day−night difference of 20−30 ppm. Assuming that 285−295 ppm is the range for a "best" daytime average, it follows a daily mean in the range of 295−310 ppm with the most probable mean between 300 and 310 ppm for continental background. Dumas (1882) states as "*la grande moyennes*" 294−310 ppm CO_2 as "normal value" (background in today's sense). According to NOAA, Mauna Loa CO_2 represents the global CO_2 average level within 1 ppm range. Baumann (1893) states in an excellent overview that the (mean) CO_2 content of the atmosphere varies only slightly. However, according to Baumann (1893), the absolute minimum and maximum from all these measurements was 260 ppm and 354 ppm, respectively. This is much more than it can be explained by natural timely variations from such relatively remote sites

Brown and Escombe (1905b) estimated in Kew (UK) from 1898−1901 ($n = 94$) a mean of 294 ppm CO_2 (243−360, were only 9 values where larger than 320 ppm); they give an error of only ±1% (±3 ppm) which is unlikely low. The measurement series carried out in Boston by *Francis Benedict* (1780−1957) between April 1909 and January 1912 is probably the longest and most precise of that time. Benedict (1912) tried carefully to avoid town influences, and by separating the values shown in Table 2.78 it is possible to derive a seasonal amplitude of about 9 ppm and a town contribution (adding to the rural background) of about 45 ppm; the most

Table 2.78 Statistics of the CO_2 measurement by Benedict (1912) carried out in Boston at the Nutrition Institute from April 1909 to January 1912 (assessed imprecision larger than 10 ppm).

	all data	summer[a]	winter[b]	all < 330 ppm	all > 320 ppm
mean (ppm)	307	302	311	298	343
standard deviation	24	19	26	15	21
min	210	260	260	210	330
max	460	460	430	320	460
n	670	313	356	534	123

[a] April–September
[b] October–March

likely background mean is 307 ppm. Krogh (1919) states from *Benedict's* analysis, the background averages for CO_2 of 0.030 % (0.001 % accuracy) were identical with his own measurements (adding 10–70 ppm by town influence from Copenhagen). To give an example how different values become when averaging, Callendar (1958) cites *Benedict's* analysis as mean of 317.5 ppm ($n = 645$) – much higher than shown in Table 2.78 – for all cases but correct *Krogh's* analysis to be 300 ($n = 40$). Callendar (1938) states a CO_2 mean for the USA to be 310 ppm in the 1930s.

Callendar (1938, 1949, 1958) and From and Keeling (1986) state that the "pre-industrial" CO_2 level derived from historical measurements is 290 or 292 ppm, respectively. However, the limited data sets (despite the huge total number of measurements) do not allow any trend construction. We can assume that most data was accumulated during daytime (this value would be lower compared to a daily mean) and during the warm season (this value would be lower compared to an annual mean). Hence, a certain value (10 ppm or so) could be added to represent annual mean values. On the other hand, local CO_2 influence from heating can mostly be excluded in the warm season, and the measurement sites have been selected to avoid industrial and town CO_2. The "best guess" means from measurements between 1870 and 1910 are between 290–300 ppm. Without going into speculations how large the corrective is, this Section ends with the conclusion that historical analyses gives evidence for a "pre-industrial" CO_2 (period 1870–1910) mean level to be around 300 ppm. This is slightly higher than values derived from ice cores (see below) but may be within the uncertainties of chemical air analysis.

The preindustrial CO_2 level derived from ice core data

Ice cores are unique with their entrapped air inclusions enabling the direct recording of past changes in atmospheric trace gas composition. The CO_2 record presented in Fig. 2.66 is derived from DE08 ice cores obtained at Law Dome, East Antarctica from 1849 to 1978. The Law Dome site satisfies many of the desirable characteristics of an ideal ice core site for atmospheric CO_2 reconstructions including the negligible melting of the ice sheet surface, low concentrations of impurities, regular stratigraphic layering undisturbed at the surface by wind or at depth by ice flow, and a high snow accumulation rate.

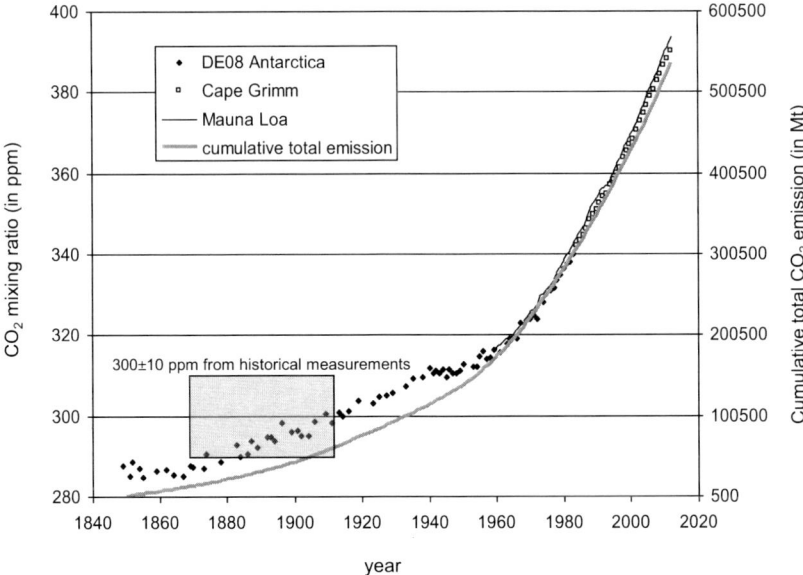

Fig. 2.66 Historical CO_2 records derived from Law Dome ice core DE08 (2000-year), Antarctica, 1849–1978 (data from Etheridge et al. 1996 and MacFarling Meure et al. 2006) and atmospheric annual CO_2 levels (in ppm) derived from in situ air samples collected at Cape Grimm, 1980–2006 (data from *L. P. Steele, P. B. Krummel* and *R. L. Langenfelds*, Commonwealth Scientific and Industrial Research Organization (CSIRO), Atmospheric Research, Aspendale, Victoria, Australia, 3195), Cape Grimm 2007–2012 (http://csiro.au/green-house-gases/GreenhouseGas/data/CapeGrim_CO2_data_download.txt) and Mauna Loa 1959–2012 (date were obtained from the Scripps website (scrippsco2.ucsd.edu); *Pieter Tans*, NOAA/ESRL (www.esrl.noaa.gov/gmd/ccgg/trends/) and *Ralph Keeling*, Scripps Institution of Oceanography (scrippsco2.ucsd.edu/); data from March 1958 through April 1974 was obtained by *C. David Keeling* of the Scripps Institution of Oceanography (SIO). See also www.esrl.noaa.gov/gmd/ccgg/trends/ for additional details. Cumulative global total anthropogenic CO_2 emission (fossil-fuel use, cement production, and land-use change); data from CDIAC (Carbon Dioxide Information Analysis Center), http://cdiac.ornl.gov/.

Ice cores recovered from the Antarctic ice sheet reveal that the concentration of atmospheric CO_2 at the Last Glacial Maximum (LGM) at 21 000 years ago was about one third lower than during the subsequent interglacial (Holocene) period started 11.7 ka ago (Delmas et al. 1980, Neftel et al. 1982, Monnin et al. 2001). Longer (to 800 ka) records exhibit similar features, with CO_2 values of ~180–200 ppm during glacial intervals (Petit et al., 1999). Prior to 420 ka, interglacial CO_2 values were 240–260 ppm rather than 270–290 ppm after that date (Lüthi et al. 2008). The variations in atmospheric CO_2 over the past 11 000 years preceding industrialization are more than five times smaller than the CO_2 increase observed during the Industrial Era. During three interglacial periods prior to the Holocene, CO_2 did not increase, and this led to a hypothesis that pre-industrial anthropogenic CO_2 emissions could be associated with early land use change and forest clearing (Ruddiman 2003, 2007). The conclusion of the above studies was that cumulative Holocene carbon emissions as a result of pre-industrial anthropogenic land use and

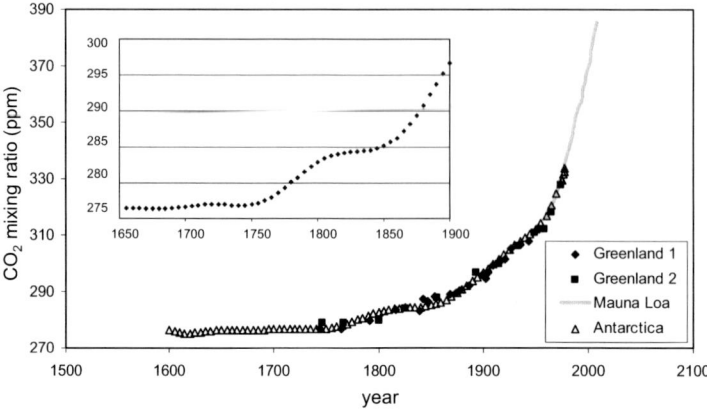

Fig. 2.67 Historical CO_2 records from different ice cores and Mauna Loa in situ monitoring (in ppm): Law Dome, Antarctica (Fig. 2.66), Mauna Loa annual means (cf. Fig. 2.68) and two different ice cores from Greenland (core 1 see Neftel et al. 1985), core 2 see Friedli et al. 1986); data from A. Neftel, H. Friedli, E. Moor, H. Lötscher, H. Oeschger, U. Siegenthaler, and B. Stauffer (1994) Historical CO_2 record from the Siple Station ice core. In Trends: A Compendium of Data on Global Change. Carbon Dioxide Information Analysis Center, Oak Ridge National Laboratory, U.S. Department of Energy, Oak Ridge, Tenn., U.S.A.

land cover change were not large enough (~50–150 Pg C during the Holocene before 1850) to have had an influence larger than an increase of ~10 ppm on late Holocene observed CO_2 concentration increases (IPCC 2013).

Air bubbles were extracted using the "cheese grater" technique. Ice core samples weighing 500–1500 g were prepared by selecting crack-free ice and trimming away the outer 5–20 mm. Each sample was sealed in a polyethylene bag and cooled to −80 °C before being placed in the extraction flask where it was evacuated and then ground into fine chips. The released air was dried cryogenically at −100 °C and collected cryogenically in electropolished stainless steel "traps", cooled to about −255 °C. The ice cores were dated by counting the annual layers in oxygen isotope ratio ($\delta^{18}O$ in H_2O), ice electroconductivity measurements and hydrogen peroxide (H_2O_2) concentrations. For these three parameters, each core displayed clear, well-preserved seasonal cycles allowing a dating accuracy of ±2 years at AD 1805 and ±10 years at AD 1350. Further details on the site, drilling and cores as well as on the extraction technique, are provided in Etheridge et al. (1996, 1988, 1992) and Morgan et al. (1997).

These atmospheric CO_2 reconstructions offer a record of atmospheric CO_2 mixing ratios from AD 1006 to 1978 (Fig. 2.67), having an excellent overlapping with the Mauna Loa CO_2 record between 1959 and 1978 (Fig. 2.68). Etheridge et al. (1996) reported the uncertainty of the ice core CO_2 mixing ratios as 1.2 ppm. Preindustrial CO_2 mixing ratios were in the range of 275–284 ppm, with lower levels between AD 1550 and 1800, probably because of the colder global climate (Etheridge et al. 1996). Law Dome ice core CO_2 records show major growth in atmospheric CO_2 levels over the industrial period, except during 1935–1945 A.D. when levels stabilized or decreased slightly (Fig. 2.66).

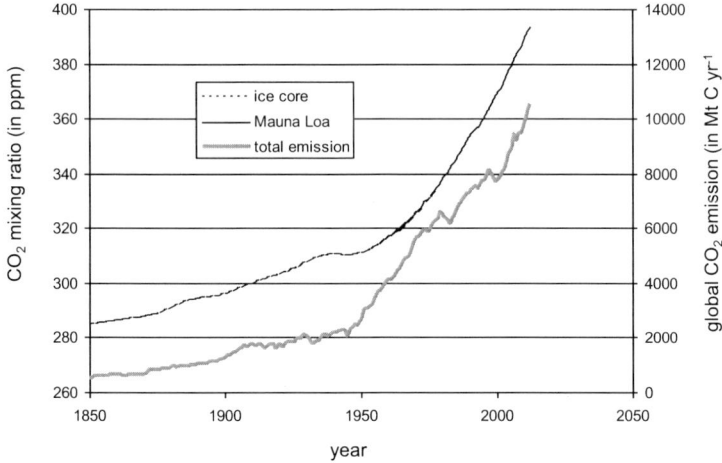

Fig. 2.68 Historical CO_2 records (yearly averages) derived from Law Dome ice core 20 year-smoothed (1850−1978) and Mauna Loa (1959−2012) and global total CO_2 emission (1850−2012); data source see Fig. 2.66.

As stated in the last Section, air chemical analysis gives evidence for atmospheric CO_2 levels being around 300 ppm before 1920. This is in excellent agreement with the value derived from the Law Dome ice core (Antarctic) of (1909−1917 average) 300.7±0.9 ppm (Fig. 2.66). It is remarkable that between 1939 and 1946, no CO_2 increase has been observed (310.2±0,1 ppm).

Derived from the ice core, the AD 1010−1850 average amounts to 280.9±2.6 ppm the while 1850−1900 average was 290.6±3.8 ppm. Hence, there is evidence to set the "pre-industrial" CO_2 level to ~285 ppm.

The twenties century increase

The Mauna Loa record (Fig. 2.69), also known as the *Keeling* curve, is almost certainly the best-known icon illustrating the impact of humanity on the planet as a whole (Keeling et al. 1976). These measurements have been independently confirmed at many other sites around the world. When *Charles David Keeling* started his CO_2 measurements in the 1950s, he almost always derived the same value of 310 ppm. Previous measurements of CO_2 in the atmosphere did not show such constancy, but these measurements had been made by wet chemical methods, which were considerably less accurate than the dry manometric method he deployed. He concluded that a) the earth's system might behave with surprising regularity and b) there was a need to make highly accurate measurements to reveal that regularity. By the early 1970s this curve was gaining serious attention and played a key role in launching a research program into the effect of rising CO_2 on the climate. Since then, the rise has been relentless and shows a remarkably constant relationship with fossil fuel burning (Fig. 2.66 and 2.68). It can be well accounted for based on the simple premise that 57 % of fossil fuel emissions remain airborne. Measurements of the changes in atmospheric molecular oxygen using a new interferometric technique

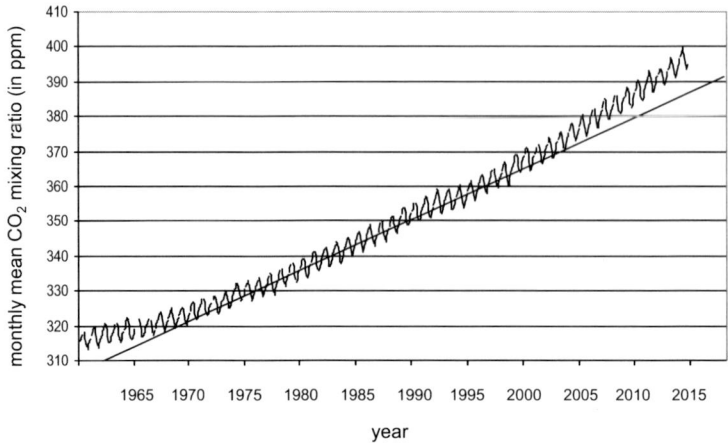

Fig. 2.69 The *Keeling*-curve: January 1959 – November 2013; atmospheric monthly mean CO_2 values (ppm) derived from in situ air samples collected at Mauna Loa, Hawaii, USA. Data source: *Pieter Tans*, NOAA/ESRL (www.esrl.noaa.gov/gmd/ccgg/trends/) and *Ralph Keeling*, Scripps Institution of Oceanography (scrippsco2.ucsd.edu/). Data through April 1974 was obtained by *C. David Keeling* of the Scripps Institution of Oceanography (SIO) and from the Scripps website (scrippsco2.ucsd.edu). The line represents a linear regression fit between 1970 and 2000.

(Keeling and Shertz 1992) show that the O_2 content of air varies (inversely with CO_2) seasonally in both the northern and southern hemispheres because of plant respiration and photosynthesis. The seasonal variations provide a new basis for estimating global rates of biological organic carbon production in the ocean, and the inter-annual decrease constrains estimates of the rate of anthropogenic CO_2 uptake by the oceans. In addition, during the process of combustion O_2 is removed from the atmosphere. Recent work by Manning and Keeling (2006) indicates that atmospheric O_2 is decreasing at a faster rate than CO_2 is increasing, which demonstrates the importance of the oceanic carbon sink.

For discussing the robustness and precision of CO_2 measurements it is worth briefly describing the measurement procedure and kinds of fluctuations. Since 1958, CO_2 concentrations at the Mauna Loa Observatory have been obtained using a non-dispersive, dual detector, infrared gas analyzer. Air samples are obtained from air intakes at the top of four 7 m towers and one 27 m tower. Four samples are collected every hour from air intakes on the taller tower and from one of the 7 m towers. Air is sampled from one tower intake for 10 minutes, followed by a second tower intake for 10 minutes, and then from a reference gas for 10 minutes. Air flow through the intakes registers a voltage on the infrared gas analyzer, which then records the concentrations on a strip chart recorder. The gas analyzer is calibrated by standardized CO_2-in-nitrogen reference gases twice daily. Flask samples are taken twice a month for comparison to the data recorded using the infrared gas analyzer. The imprecision in measuring reference gases approaches 0.1 ppm and is rarely greater than 0.2 ppm. However, agreement differences of less than 0.5 ppm between flask and analyzers, or between different analyzers on a short-term basis,

Table 2.78 Mean CO_2 mixing ratios (1981−1992 average) for different remote measurement stations; data from Carbon Dioxide Information Analysis Center (CDIAC), http://cdiac.ornl.gov/trends/co2/contents.htm.

station	geographic coordinates	altitude (m above MSL)	CO_2 mixing ratio (ppm)
Alert, Canada	82° 27' N, 62° 31' W	210	344.7 ± 13.1
Barrows, Alaska	71° 19' N, 156° 36' W	11	349.9 ± 5.9[a]
Mauna Loa, Hawaii	19° 32' N, 155° 35' W	3397	348.4 ± 5.7
Guam, Mariana Islands, Pacific	13° 43' N, 144° 78' E	2	348.8 ± 5.7
Seychelles, Mahe Island, Indian Ocean	4° 67' S, 55° 17' E	7	348.1 ± 5.9
Samoa, Indian Ocean	14° 15' S, 170° 34' W	30	347.1 ± 5.6
Amsterdam Island, Indian Ocean	37° 47' S, 77° 31' E	70	345.7 ± 5.7
Cape Grimm, Tasmania	40° 41' S, 144° 41' E	94	348.3 ± 4.1
Palmer, Antarctica	64° 92' S, 64° W	10	346.7 ± 5.5
South Pole	89° 59' S, 24° 48' W	2810	347.1 ± 5.4
mean			347.4 ± 1.7[b]

[a] standard variation based on the annual values
[b] standard variation based on the stations mean values

are difficult to obtain. Monthly averages from May 1964 to January 1969 might have been in error by as much as 1.0 ppm, but since 1970 systematic error probably does not exceed 0.2 ppm. The precision of monthly averages is approximately 0.5 ppm. In summary, monthly and annual averages of the Mauna Loa data are statistically robust and serve as a precise, long-term record of atmospheric CO_2 concentrations. Daily, monthly and annual averages are computed for the Mauna Loa data after the deletion of contaminated samples and readjustment of the data.

The steady rise in atmospheric CO_2 concentration shown by this record has been widely interpreted as a global trend. It is remarkable that all CO_2 measurements taken at different remote sites in the world and the ice core data from Antarctica and Greenland show significant overlap in averages, reflecting the global increase of the CO_2 burden (Fig. 2.66). Despite differences in seasonal cycles among the stations, which reflect the timely concentration variation (carbon budget concerns sources and sinks; Chapter 2.8.2.1), as well as very small differences in the absolute mixing ratios (Table 2.78), the trends are similar. All variations are smaller than the precisions, i. e. are not significant.

The annual mean rate of growth of CO_2 in a given year is the difference in concentration between the end of December and the start of January of that year. This represents the sum of all CO_2 added to, and removed from, the atmosphere during the year by human activities and natural processes (Fig. 2.70; Conway et al. 1994); the estimated uncertainty in the global annual mean growth rate is 0.07 ppm yr^{-1}. The relative yearly global CO_2 emission increase (based on the 1959 value set equal to one) looks like a smoothed CO_2 concentration growth rate (Fig. 2.70), again emphasizing the general close relationship between anthropogenic carbon dioxide increases in the atmosphere and rising emissions. However, on a shorter

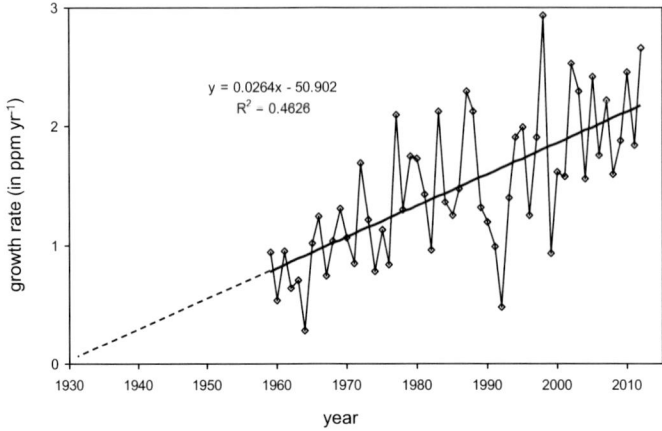

Fig. 2.70 Annual mean rate of growth of atmospheric CO_2 at Mauna Loa and global total anthropogenic CO_2 emission (1959–2012); data source see Fig. 2.66.

time scale (less than 2–3 years) the atmospheric CO_2 response is not linear to the emissions rate, expressing the complexity of the carbon cycle.

The atmospheric CO_2 growth rate averages continuously increase between 1960 and 2008 at 0.026 ppm yr^{-1}. As Fig. 2.70 clearly shows, the growth rate itself increases, reflecting an exponential growth. The *Keeling* curve (Fig. 2.69) shows three periods with subsequent increasing growth rates but relative constant rates within the period (in ppm CO_2 yr^{-1}):

	from linear fit (Fig. 2.69)	from growth rate (Fig. 2.70)
01/1959–12/1969	0.76 ($r^2 = 0.62$)	0.85 ± 0.31
01/1070–12/1999	1.47 ($r^2 = 0.97$)	1.46 ± 0.53
01/2000–11/2013	2.03 ($r^2 = 0.93$)	2.03 ± 0.41

The overall averaged growth rate amounts to 1.47 ± 0.61 ppm yr^{-1}. The 1974–1985 period of continuous atmospheric CO_2 measurements from the NOAA GMCC (Geophysical Monitoring for Climate Change) program at the Mauna Loa Observatory in Hawaii shows the average growth rate of CO_2 was 1.42 ± 0.02 ppm yr^{-1}, and the fraction of CO_2 remaining in the atmosphere from fossil fuel combustion was 59% (Thoning et al. 1989). Hence, with constant increases expected the atmospheric CO_2 mixing ratio is predicted to grow to 500 ppm in 2050 and to 700 ppm in 2100. There is no doubt that CO_2 has been increasing since the Industrial Revolution and has reached a concentration unprecedented for over more than 400 000 years (Chapter 3.4).

At this point, let us come back to the question of the "pre-industrial" atmospheric CO_2 level. The *Keeling* curve is non-linear but shows a remarkable highly correlated fitted polynomial function (Fig. 2.71). The curve before 1959 is hypothetically constructed according to the mathematical function to "extent" the *Keeling* curve back. Purely mathematically, it follows for the beginning ("zero-year" of CO_2

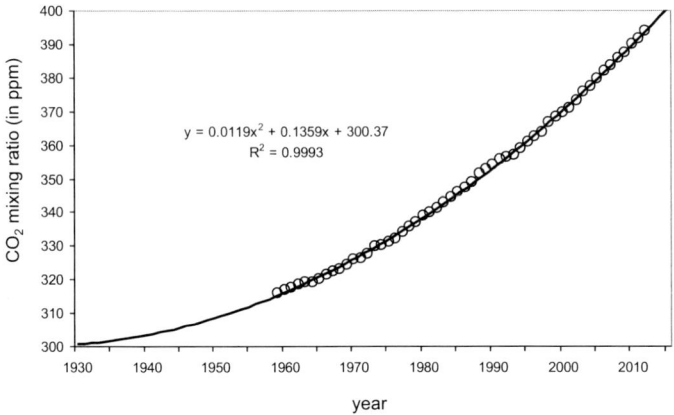

Fig. 2.71 The fitted *Keeling* curve, based on yearly means.

increase) in the year 1930 with a reference level of 300 ppm. We receive the same result from the linear fit of the growth rate trend (Fig. 2.70); despite the large inter-annual variations of the growth rate (due to different air mass characteristics), it must be assumed that the linear fit is robust (apart from the low r^2). Note that this does not mean "zero-emission" in 1930. If this extrapolation would be true, then all CO_2 emitted earlier must be taken up by oceans and/or terrestrial ecosystems, i. e. the air-borne fraction was zero. However, the ice-cores tell us that the CO_2 level in air further decreases over time, but obviously (Fig. 2.66) more smoothly. Excluding the correctness of the ice-core data, we must conclude that the atmospheric CO_2 level growth in different periods after different models. Between 1850 (and assuming that this time corresponds to the "pre-industrial" references level of 286.3±1.3 ppm) and 1940 (310.0±1.0 ppm), the ice-core record also represents a significant ($r^2 = 0.95$) polynomial fit, but with a slower growth rate compared to the *Keeling* curve (cf. Fig. 2.66). The period 1937−1949 (mean 310.2±0.1 ppm) shows no CO_2 increase from the ice-core data. Comparing the CO_2 concentration with the global CO_2 emission (Fig. 2.71, see also Fig. 2.66) it is apparent that in the period 1910−1950 that there is no coincidence in contrast to the periods before 1910 and after 1950.

Latitudinal variation

There are large seasonal cycles, and a major hemispheric gradient: CO_2 is only well mixed on a multi-year timescale. The seasonal cycle is large in northern latitudes (10−20 ppm) and small in southern (2−3 ppm).

The natural or unperturbed component is equivalent to that part of the atmospheric CO_2 distribution, which is controlled by non-anthropogenic CO_2 fluxes from the ocean and terrestrial biosphere. Taylor and Orr (2000) found the following key features of the natural latitudinal distribution: (a) CO_2 concentrations in the Northern Hemisphere that were lower than those in the Southern Hemisphere, (b) CO_2 concentration differences that are higher in the tropics (associated with outgassing of the oceans) than those currently measured, and (c) CO_2 concentrations over the Southern Ocean that are relatively uniform.

Anthropogenic CO_2 sources are almost located between 30°N and 60° N and negligible in the Southern Hemisphere. From baseline CO_2 measurement stations it is found that the North-South gradient increases: 5 ppm in 1980 (338 ppm / 333 ppm) and 7 ppm in 2009 (385 ppm/378 ppm) for Alert (82° N) and South Pole, respectively.

Timely variations

The annual fluctuation in carbon dioxide is caused by seasonal variations in carbon dioxide uptake by land plants. Since many more forests are concentrated in the northern hemisphere, more carbon dioxide is removed from the atmosphere during northern hemisphere summers than southern hemisphere summers. This annual cycle is shown in Fig. 2.72 by taking the average concentration for each month across a year; interestingly, there is no significant change in the seasonal amplitude over the years (the amplitude amounts to 1.008, corresponding to 6−7 ppm only). Taguchi et al. (2002) show that on average the seasonal variation of atmospheric CO_2 at Mauna Loa is influenced mostly by the Siberian CO_2 flux, followed by temperate Asia and North America. The inter-annual variability (less than 2 ppm) of the seasonal cycle is caused mainly by the inter-annual variation in the transport of the Siberian signal to Mauna Loa.

Cyclical changes over shorter time scales are harder to spot in the records because they are usually much weaker than seasonal oscillations and can be masked by random variations in the data. Daily cycles have not been observed but possible ambient error sources at Mauna Loa include volcanic, vegetative and man-made effects (e. g. vehicular traffic and industry). Daily peaks in measured concentrations occur because of complex wind currents. Downslope winds often transport CO_2 from distant volcanic vents, causing elevations in measured CO_2 concentrations. Upslope winds during afternoon hours are often low in CO_2 because of photosynthetic depletion occurring in sugarcane fields and forests. Recently, a weekly cycle has been found at Mauna Loa by Cerveny and Coakley (2002). They found that the measurements rise to a peak on Mondays and then decline steadily to a minimum on Saturdays. Crucially, they find no such cycle in carbon dioxide records from the Amundsen-Scott South Pole Station in Antarctica, which is far from any sources of pollution. The Antarctic measurements show the same yearly trend and seasonal cycle, but there is no significant difference between average daily values. Such short-term variations have evened out by the time CO_2 pollution reaches Antarctica.

In Germany, several CO_2 monitoring series are available on "urban background" stations, showing a wider range of mean CO_2 mixing ratios and different seasonal cycles (Fig. 2.72). It is remarkable that the growth rate for the 1981−1992 period is exactly the same (1.5−1.7) ppm yr^{-1} as at Mauna Loa. The German average CO_2 level amounts to 352 ± 9 (Mauna Loa average 348.4 ± 5.7), about 4 ppm larger than in the remote background. From the data it can by summarized that a) seasonal cycles are smoothed with increasing altitude above sea level, b) the concentra-

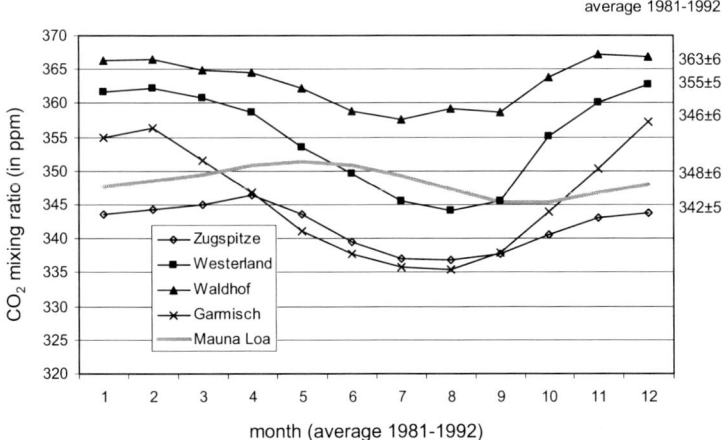

Fig. 2.72 Seasonal variation of CO_2 at Mauna Loa (data source see Fig. 2.66) and German stations (data from Carbon Dioxide Information Analysis Center (CDIAC) and German Weather Service DWD, Observatory Hohenpeissenberg, Dr. Wolfgang Fricke).

tion differences between the sites amounts to up to 20 ppm and c) the inter-annual variations amount to about 2 ppm (similar to that found at remote background stations).

The city dome CO_2

As already mentioned, in the vicinity of large CO_2 sources (big cities, industrial areas) elevated concentrations are measured at a level, strongly depending on meteorological parameters, which influence dispersion. This difference between rural and town sites had already been recognized 100 years ago (Rubner et al. 1907) within a range of 33 to 133 ppm. In recent years, an increasing number of studies on CO_2 observations in urban areas have been published. Concentrations of CO_2 measured in cities reflect a complex response of both the local anthropogenic and biogenic surface-atmosphere exchange of CO_2 and the concentrations accumulated, diluted or advected over time controlled by meteorological factors. This makes the interpretation of urban CO_2 observations challenging[25]. The situation is usually different in urban areas where patterns are heavily influenced by anthropogenic emissions, which often cause strong short-term variations, but less visible seasonal patterns over the year.

Measurements of CO_2 concentration carried out in Rome (Gratani and Varone 2005) showed a mean yearly value increase from 1995 (367 ± 29 ppm) to 2004 (477 ± 30 ppm). The daily trend had a peak in the early morning when traffic was highest and the atmosphere was more stable. The annual trend showed a peak in winter, 18% greater than the summer one, which correlated to traffic density. The

---

[25] By contrast, because it is now clear that a) the city dome CO_2 concentrations result in no adverse health effect and b) do not contribute significantly to a local "greenhouse" effect, more studies do not make any sense either from an academic or environmental point of view.

Table 2.79 Carbon dioxide offset in cities comparing to the rural environment (in ppm).

location and year	additional CO_2	author
Phoenix, USA (2000)	100–200	Idso et al. (2002)
Rome	80–100	Gratani and Varone (2005)
Essen, Germany (2003)	20–40	Henninger and Kuttler (2004)
Tokyo	15–60	Moriwaki et al. (2006)
Paris	20–550	Widory and Javoy (2003)

weekly trend had lowest values (414 ± 19 ppm) during the weekends when traffic density was 72% lower. Miyaoka et al. (2007) reported on measurements made in Sapporo in 2005 that showed significant diurnal variation (360–400 ppm in summer and 390–400 ppm in winter for lowest daytime values) as well seasonal differences, i. e. southern air mass (urban plume) 379 ± 7 (summer) and 416 ± 12 (winter) and for northern air masses (remote) 374 ± 4 (summer) and 396 ± 7 (winter). In Essen (Germany), a similar pattern has been found (Henninger and Kuttler 2004), showing lower daytime values (402 ± 20 ppm in winter and 369 ± 16 ppm in summer) and higher values at night (427 ± 23 ppm in winter and 417 ± 20 ppm in summer); average values amount to 415 ± 21 ppm in winter and 393 ± 18 ppm in summer. Similar results have been found by Idso et al. (2002) in Phoenix (United States) that emphasize that the character of the city's urban CO_2 dome is almost exclusively a product of vehicular emissions and the region's distinctive meteorology. According to results by Balling et al. (2002), the CO_2 concentration of the air over Phoenix drops off rapidly with altitude, returning to a normal non-urban background value of approximately 378 ppm at an air pressure of 800 hPa. Consequently, Phoenix's urban CO_2 dome did not have much of an impact on its near-surface air temperature, creating a calculated warming of just 0.12 °C at the time of maximum CO_2-induced warming potential. Table 2.79 shows the available data on the city dome CO_2; it is surprising that the supplementary CO_2 to the background value is roughly in the same range as the historical measurement of 100 years ago despite the different CO_2 sources (today traffic and in past coal combustion). From all these studies we can conclude that anthropogenic CO_2 emissions are the primary source of the urban CO_2 dome. The dome is generally stronger in city centers, in winter, on weekdays, at night, under conditions of heavy traffic, close to the ground, with little to no wind and in the presence of strong temperature inversions.

2.8.2.4 CH_4 and N_2O: Permanent agricultural associates

Besides carbon dioxide and the problem of fossil fuel combustion, two more GHGs are of anthropogenic (among natural) origin: methane (CH_4) and nitrous oxide (N_2O). Whereas anthropogenic N_2O is absolutely dominated by nitrogen fertilizer applications (percentage of total global emission is about 40%), CH_4 has several sources (Chapter 2.7.3.3) with agricultural activities contributing about 40% of total anthropogenic CH_4 emissions (Table 2.69). Anthropogenic CH_4 represents about 70% of the total CH_4 emission amounts compared with an anthropogenic CO_2 percentage of "only" 4%.

Global anthropogenic methane emissions are estimated at 323 Mt methane in 2005, with an expected increase to 414 Mt methane in 2030 when assuming no further implementation of control measures than those currently adopted or prescribed by implemented legislation (Höglund-Isaksson 2012). The full technical mitigation potential for CH_4 in 2030 is estimated at 195 Mt CH_4, i.e. 47 percent below baseline or 33 percent below the 2005 emission level.

Methane (CH₄)

The atmospheric CH_4 reconstructions presented here offer a record of atmospheric CH_4 concentrations since AD 1008 (Fig. 2.73). Because of the high rate of snow accumulation, the air enclosed in the three ice cores from Law Dome, Antarctica, has unparalleled age resolution and extends into recent decades. Etheridge et al. (1998) reported that the uncertainty of the ice core CH_4 concentrations is about 10 ppb. The recent EPICA Dome C (Antarctica) record, which shows the methane-tracked climate over the past 650 000 years, highlights that the current concentration of methane in the atmosphere is far outside the range experienced with lower methane concentrations in glacials than interglacials, and lower concentrations in cooler interglacials than in warmer ones (Wolff and Spahni 2007). There is controversy about whether changes in the preindustrial Holocene are natural or anthropogenic in origin. Changes in wetland emissions are generally cited as the main cause of the large glacial-nterglacial change in methane. However, changing sinks must also be considered, and the impact of possible newly described sources evaluated.

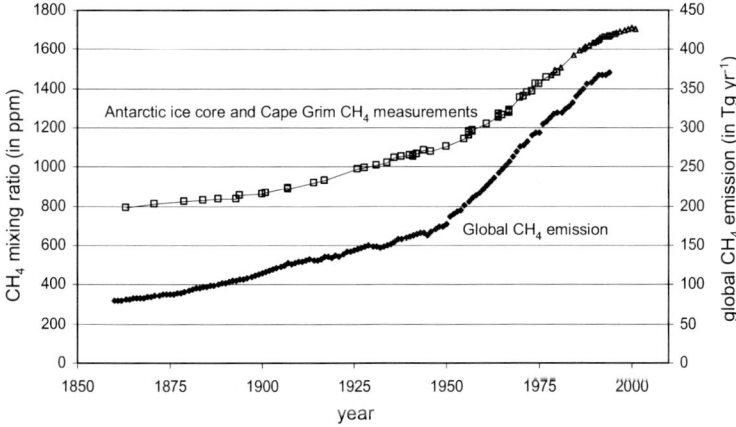

Fig. 2.73 Historical records (1863–1980) of CH_4 mixing ratio derived from ice cores (for data source, see Fig. 2.66), CH_4 annual data (1978–2001) from Cape Grim (see Fig. 2.74), and global anthropogenic CH_4 emission 1860–1994 (data from: D. I. Stern and R. K. Kaufmann (1998) Annual Estimates of Global Anthropogenic Methane Emissions: 1860–1994. Trends Online: A Compendium of Data on Global Change. Carbon Dioxide Information Analysis Center, Oak Ridge National Laboratory, U.S. Department of Energy, Oak Ridge, Tenn., U.S.A. doi:10.3334/CDIAC/tge.001).

Recent isotopic data seem to finally rule out any major impact of clathrate releases on methane at these time scales (Wolff and Spahni 2007). Any explanation must take into account that, at the rapid *Dansgaard-Oeschger* warmings[26] of the last glacial period, methane rose by around half its glacial-interglacial range in only a few decades.

From the ice core record (Fig. 2.73) a preindustrial (reference) level of 680 ± 11 ppb can be derived, which increased from around 1750 until the late 1990s up to around 1730 ppb (Fig. 2.74). It is remarkable that in contrast to CO_2 (where an increase has been observed since 1850) the CH_4 increase had already begun around 1750, according to world population growth (Fig. 2.58). Fig. 2.73 shows the remarkable general relationship between emission and concentration increase; however, there is no simple relationship. CH_4 removal (hence, residence time) is mainly determined by the reaction with the OH radical (see Eq. 5.42). Possible soil-emitted CH_4 is largely removed by soil uptake; therefore, the seasonal cycles seen in Fig. 2.74 are well correlated with the solar radiation seasonal variation. The 1985−2001 average from Cape Grim amounts to 1660 ± 30 ppb with a seasonal summer minimum (1643 ppb) and winter maximum (1672 ppb). The long-term CH_4 trend shows different periods with various growth rates. The largest continuous growth has been observed between 1952 and 1977 (about 16 ppb yr^{-1}), which continuously slowed down until 1999 and remained unchanged almost a decade. Since 2007, a renewed global growth has been observed (Rigby et al. 2008) with about +7 ppb yr^{-1} and this continued until the end of 2013 (Fig. 2.74); see also discussion in Chapter 2.7.3.2.

Column-averaged infrared solar measurement between 1983 and 2003 at Kit Peak Observatory (2090 m a.s.l., Colorado, United States) resulted in an averaged tropospheric volume mixing ratio of 1814 ± 48 ppb with trend equal to 8.3 ± 2.2 ppb yr^{-1} in 1983 decreasing to 1.9 ± 3.7 ppb yr^{-1} in 2003 (Rinsland et al. 2006). A similar trend has been observed in the lower stratosphere where at the 120−10 hPa (~ 16−30 km altitude) at 25° N−35° N from 2004−2008 a value of ~ 1370 ppb was estimated using spectroscopic measurement (Rinsland et al. 2009). In 1985, this value was estimated to be 1151 ppb.

The seasonal amplitude between 1977 and 1995 was 30 ppb, very close to the peak-to-peak seasonal cycle found at Cape Grim. It is obvious that the NH concentration of CH_4 is about 100 ppb larger than the SH value because the main sources are in the NH and the transport through the ITCZ is slow. The mean difference in CH_4 between the most northern station Alert and the South Pole amounts to ~ 120 ppb (Rigby et al. 2008).

The reasons for the decrease in the atmospheric CH_4 growth rate (Dlugokencky et al. 2003) and the implications for future changes in its atmospheric burden are not understood. Recent estimates by WMO (2009), including 2006, show large variations in the growth rate (compared with the discussion about CO_2 above). But despite the large growth rates in 1998 and 2002/2003 (during El Niño events), these estimates also showed a relatively stable global mean mixing ratio between 1999

---

[26] *Dansgaard-Oeschger* events are rapid climate fluctuations that occurred 25 times during the past glacial period. The processes behind the timing and amplitude of these events (as recorded in ice cores) are still unclear.

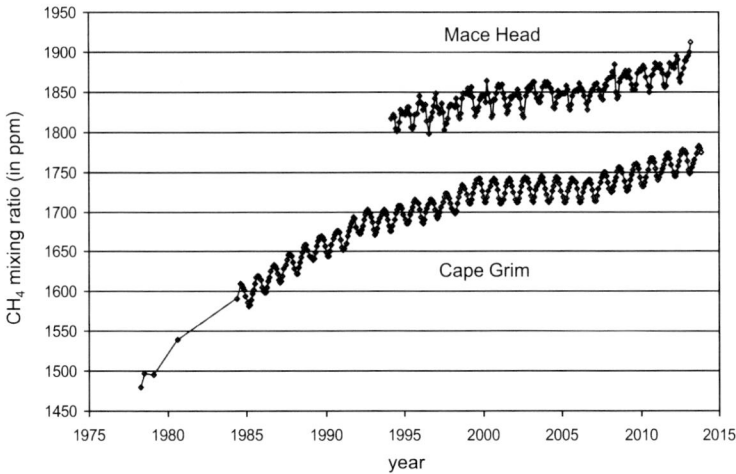

Fig. 2.74 Historical records of CH_4 mixing ratios (in ppb) as monthly summary averages of the individual measurements at Cape Grim (April 1974 – November 2013) and Mace Head (March 1994 – November 2013) Data for Cape Grim: from CSIRO Marine & Atmospheric Research and Cape Grim Baseline Air Pollution Station/Australian Bureau of Meteorology, see also Rigby et al. (2008). Date source for Mace Head: ALE/GAGE/AGAGE global network program: http://cdiac.ornl.gov/ndps/alegage.html; see also Prinn et al. (2000).

and 2006 (globally averaged CH_4 in 2006 was 1782 ppb, a decrease of 1 ppb since 2005 and a decrease of 2 ppb since 2003). However, the growth rates during 2007 increased significantly throughout the entire monitoring network (Rigby et al. 2008). Between 1999 and 2007 the average growth rate was 2.1 ± 5.2 ppb yr^{-1} compared with ~ 15 ppb yr^{-1} before 1985 and increased between 2007 and 2013 to ~ 5 ppb yr^{-1}.

Etheridge et al. (1998) suggest that high growth rates follow high global temperature anomalies. Rigby et al. (2008) discuss the relationship with the OH radical. Modeling results by Bousquet et al. (2006) indicate that wetland emissions dominated the inter-annual variability of methane sources, whereas fire emissions played a smaller role, except during the 1997−1998 El Niño event. Over longer time scales, the results show that the decrease in atmospheric methane growth during the 1990s was caused by a decline in anthropogenic emissions (Lelieveld et al. 1998). Since 1999, however, anthropogenic emissions of methane have risen again. The effect of this increase on the growth rate of atmospheric methane has been masked by a coincident decrease in wetland emissions, but atmospheric methane levels might increase in the near future if wetland emissions return to their mean 1990s levels (Bousquet et al. 2006).

Nitrous oxide (N_2O)

The results from weekly global measurements of nitrous oxide from 1981 to 1996 show that there is more N_2O in the northern hemisphere by about 0.7 ± 0.04 ppb, and that the Arctic to Antarctic difference is about 1.2 ± 0.1 ppb (Khalil et al. 2002; Fig. 2.75). Concentrations at locations influenced by continental air are higher than

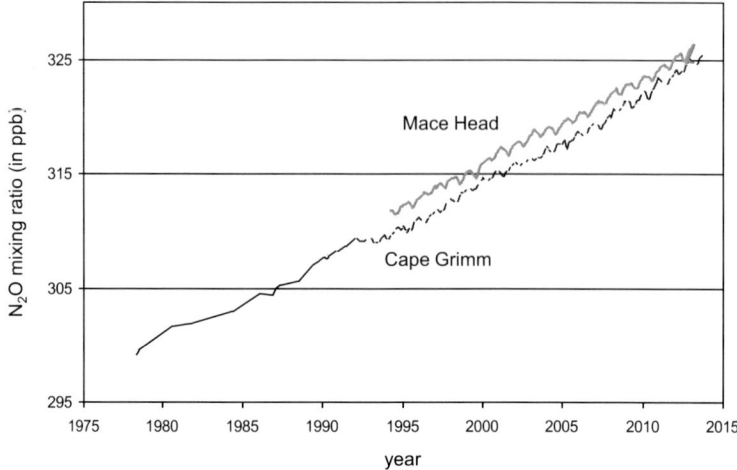

Fig. 2.75 Historical records of N_2O mixing ratios (in ppb) as monthly averages of the individual measurements at Mace Head (grey line: August 1994 to September 2008) and Cape Grim (black line: April 1978 − November 2013). Data source for Cape Grimm: CSIRO Marine & Atmospheric Research and Cape Grim Baseline Air Pollution Station/Australian Bureau of Meteorology. Data source for Mace Head: ALE/GAGE/AGAGE global network program: http://cdiac.ornl.gov/ndps/alegage.html; see also Prinn et al. (2000).

at marine sites, showing the existence of large land-based emissions. For the second period (Fig. 2.75), N_2O increased linearly at an average rate of about 0.73 ppb yr^{-1} ($r^2 = 0.99$), although there were periods when the rates were substantially different.

Using ice core data (Sowers et al. 2003), a record of N_2O that goes back about 106 000 years, suggests minimal changes in the ratio of marine to terrestrial N_2O production. It shows preindustrial levels of about 287 ± 1 ppb and that concentrations have now risen by about 27 ppb or 9.4 % over the past century. The ice core data show that N_2O only started increasing during the twenties century (Pearman et al. 1986), which supports the conclusion that N_2O is the result of synthetic nitrogen fertilizer application. During the past glacial termination, both marine and oceanic N_2O emissions increased by 40 ± 8 %.

2.8.2.5 Halogenated organic compounds: Sit out problem

Almost all long-living halogenated organic compounds are ozone-depleting substances (ODSs) in the stratosphere. The Montreal Protocol (2000), based on meetings held in London (1990), Copenhagen (1992), Vienna (1995), Montreal (1997) and Beijing (1999), sets regulations on different classes of ozone depletion potentials (ODPs), subdivided into several groups concerning the stepwise reduction and final banning of these compounds: 34 hydrobromofluorocarbons (HBFCs) until 1996, 15 CFCs and 3 halons until 2010, 40 hydrochlorofluorocarbons (HCFCs) until 2030 and three individual compounds namely carbon tetrachloride (CCl_4), methyl chloroform (CH_3CCl_3) and bromochloromethane (CH_2BrCl). As of Sep-

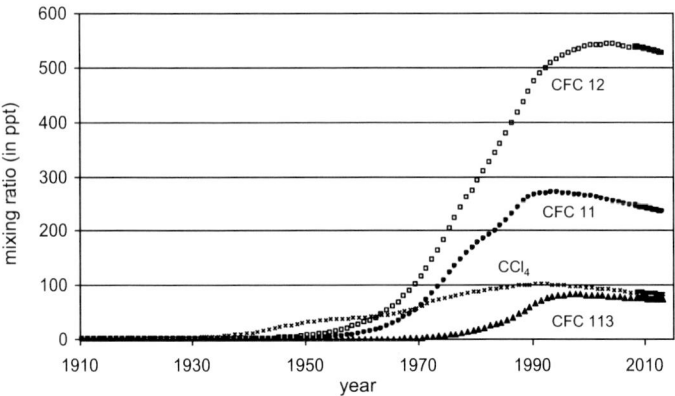

Fig. 2.76 Historical records of CFC 11, CFC 12, CFC 13, and CCl₄ mixing ratios (in ppt). Data from 1908 − 1999 is from Walker et al. (2000), reconstructed histories of the annual mean atmospheric mole fractions, compiled by *John Bullister* (NOAA/PMEL); the values from 1999 − 2013 are monthly global mean of baseline data from AGAGE GC-MD data: http://agage.eas.gatech.edu/data_archive/global_mean/.

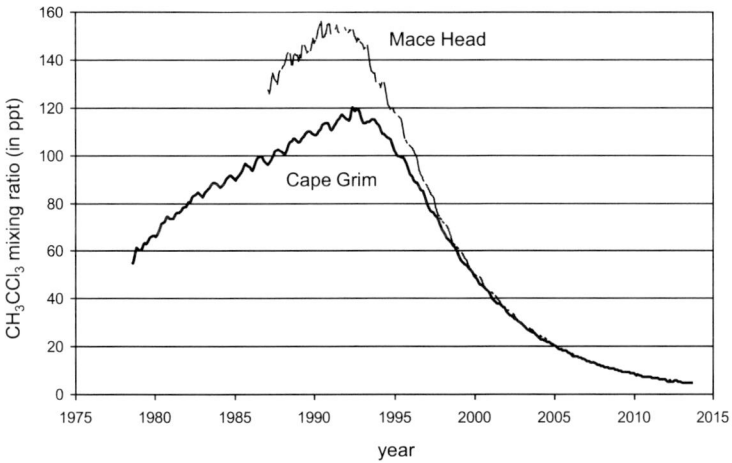

Fig. 2.77 Historical records of methyl chloroform (CH₃CCl₃) mixing ratios (in ppt) at Cape Grim (July 1978 – September 2013) and Mace Head (January 1987 – September 2013). Data source for Cape Grim data from CSIRO Marine & Atmospheric Research, Cape Grim Baseline Air Pollution Station/Australian Bureau of Meteorology and the Advanced Global Atmospheric Gases Experiment (AGAGE), see also Prinn et al. (2005). Data source for Mace Head: ALE/GAGE/AGAGE global network program: http://cdiac.ornl.gov/ndps/alegage.html; see also Reimann et al. (2004).

tember 16, 2009, all countries in the United Nations have ratified the original Montreal Protocol. The reference gas CFC-11 is defined as having an ODP of 1.0.

CFCs, along with other chlorine- and bromine-containing compounds (Table 2.74), have been implicated in the accelerated depletion of the ozone in the

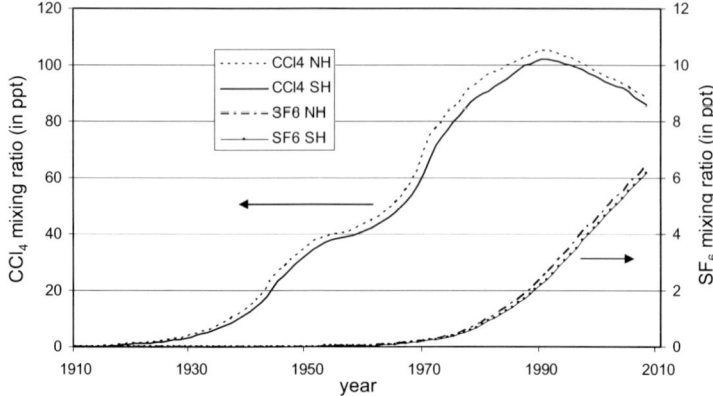

Fig. 2.78 Historical records of CCl_4 and SF_6 mixing ratios (in ppt); CCl_4 data from Fig. 2.76 and SF_6 data are based on production/release data from Maiss and Brenninkmeijer (1998)

earth's stratosphere in the 1970s. Although CFCs are GHG, they are regulated by the Montreal Protocol, which was motivated by their contribution to ozone deple-tion rather than global warming. CFCs were developed in the early 1930s and are used in a variety of industrial, commercial and household applications. These substances are non-toxic, flammable and non-reactive with other chemical com-pounds. These desirable safety characteristics, along with their stable thermody-namic properties, make them ideal for many applications, such as coolants for commercial and home refrigeration units, aerosol propellants, electronic cleaning solvents and blowing agents.

Replacement compounds for CFCs are HCFCs and HBFCs. Although HBFCs were not originally regulated under the Clean Air Act, subsequent regulation added them to the list of class I substances and they were phased out in 1996 because of their high ODP. HCFCs (ODP < 0.1) still contain chlorine atoms but the presence of hydrogen makes them reactive with chemical species in the troposphere. This greatly reduces the prospects of the chlorine reaching the stratosphere, as chlorine will be removed by chemical processes in the lower atmosphere. Hence, HCFCs have the longest allowed use (to be phased out in 2030).

Nevertheless, a significant proportion of HCFCs do break down in the strato-sphere and they have contributed to more chlorine build-up there than originally predicted. Later alternatives, such as hydrofluorocarbons (HFCs) that contain no chlorine, have even shorter lifetimes in the lower atmosphere. One of these com-pounds, HFC-134a, is now used in place of CFC-12 in automobile air conditioners. HFCs were only introduced in 1990 but they are among the GHG listed in the Kyoto Protocol (besides CO_2, CH_4, N_2O, PFCs, and SF_6).

According to the IPCC, sulfur hexafluoride (SF_6) is the most potent greenhouse gas (not listed in the Montreal Protocol), with a GWP of 22 800. However, because of its high density relative to air, SF_6 flows to the bottom of the atmosphere and this limits its ability to heat the atmosphere. SF_6 is very stable (its lifetime is 3200 years) and its mixing ratio in the atmosphere is very low, but has steadily increased (Fig. 2.78). Of the 8000 tons of SF_6 produced per year, 6000 tons is used as a gaseous dielectric medium in the electrical industry. The measured concentration of SF_5CF_3 increased from zero in 1965–1966 to about 0.12 ppt in 1999, with a current growth rate of about 0.008 ppt yr^{-1} (about 6% yr^{-1}). Given the similarity of the growth curves of SF_5CF_3 and SF_6 (which increased from 0.18 ppt in 1970 to 4.0 ppt in 1999), Sturges et al. (2000) speculate that the former might originate as a breakdown product of the latter in high-voltage equipment. Although the current radiative forcing of SF_5CF_3 might be minor, the high growth rate and long atmospheric residence time suggest that the greenhouse significance of this gas could increase markedly in the future. Conversely, SF_5CF_3 seems not to have any natural sources, so control might be feasible once the sources are identified.

2.8.2.6 CO: The biomass burning problem

CO is primarily emitted through incomplete combustion processes (biomass burning, domestic combustion and traffic) and is likely to be biogenically produced from soils as well as being an important secondary product from NMVOC oxidation in the atmosphere. It has a lifetime of about 2–3 months, strongly depending on the OH concentration and thereby shows inversely correlated seasonal concentration profiles with OH (Fig. 2.79). It also plays a role in background ozone formation (Chapter 5.3.6.3). A well-defined annual cycle of CO is evident (Fig. 2.80), largely because of an increase in its destruction by the OH radical during the summer months. The Mace Head 1993–2008 averaged CO concentration amounts to 123 ± 26 ppb, whereas that from Cape Grim is only 51 ± 8 ppb. In the marine boundary layer, mixing ratios were greatest in the northern winter (200–220 ppb) and lowest in the southern summer (35–45 ppb). The inter-hemispheric gradient showed strong seasonality with a maximum difference between the high latitudes of the northern and southern hemispheres (160–180 ppb) in February and March and a minimum in July and August (10–20 ppb; Novelli et al. 1998). Higher CO

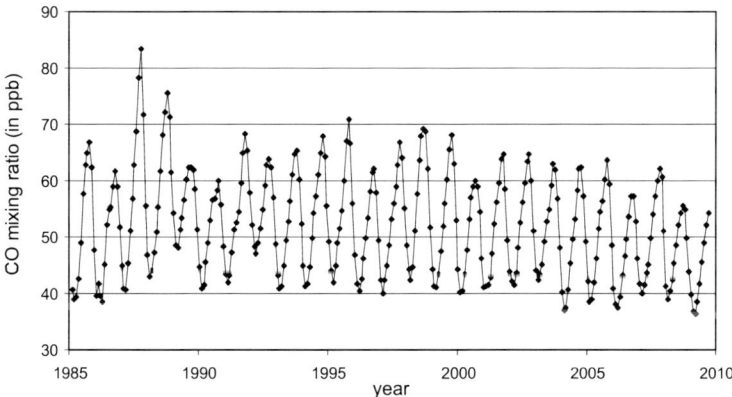

Fig. 2.79 Atmospheric carbon monoxide record (monthly means, calculated as the mean of daily values obtained from flask air samples; February 1985 – December 2013) from Cape Grim, Australia; data from CSIRO Marine & Atmospheric Research and Cape Grim Baseline Air Pollution Station/Australian Bureau of Meteorology; see also Steele et al. (2007).

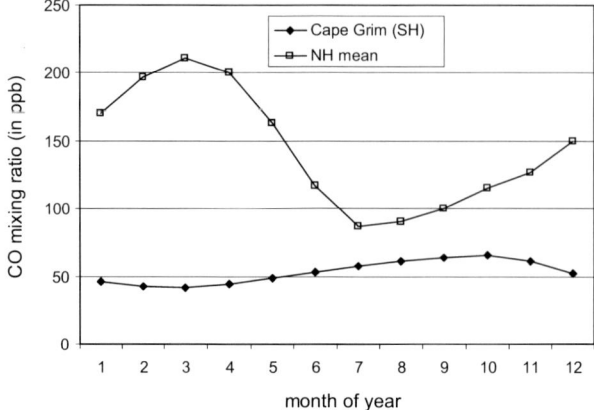

Fig. 2.80 Seasonal variation of carbon monoxide (monthly means); average of the 1985–2001 record from Cape Grim (see Fig. 2.79) – lower curve; Northern Hemispheric mean 1991–1994 after Yurganov (2000) – upper curve.

was found in regions near human development than in more remote areas (Table 2.73). Although tropospheric concentrations of CO exhibit periods of increases and decreases, the globally averaged CO mixing ratio between 1990 and 1995 decreased by approximately 2 ppb yr^{-1} (Novelli et al. 1998), with a further decrease in values to the end of 2009 (Fig. 2.79).

Published total vertical column amounts of CO deduced from infrared solar spectra recorded during 1950/1951 in the Swiss Alps have been reanalyzed based on improved spectroscopic parameters for CO and a curve of growth. From a comparison of the 1950/1951 results and modern measurements, an average increase of $\sim 2\%$ yr^{-1} in the free tropospheric concentration of CO above Europe is estimated for 1950–1977 (Curtis et al. 1985).

Annual average CO concentrations at Cape Grim have remained 50–55 ppb between 1985 (beginning of the recordings) and 2002; except for 1987/1988 when mixing ratios were 58.2 ppb and in 1989 and 1998 when mixing ratios of around 56 ppb occurred. Elevated mixing ratios were observed on a global scale during 1998 (Langenfelds et al. 2002). It seems that recently at Cape Grim a trend of decreasing mixing ratios has started (about 1% yr^{-1}; Fig. 2.79). Sampled air at Cape Grim is well mixed, having recently travelled over the Southern Ocean, and is far removed from variations in source strength characteristic of land areas.

CO mixing ratios derived from the time series of solar absorption spectra on Kitt Peak (United States) at a 2.09 km altitude have shown no long-term CO trend since the late 1970s (mean volume mixing ratio of 102 ± 3 ppb below 10 km altitude), suggesting that the global average long-term decline reported from 1990–1995 has not continued in the free troposphere (Rinsland et al. 2007).

2.8.2.7 O_3: Locally believed to be solved but regional unsolved

Ozone (O_3) and hydrogen peroxide (H_2O_2) are secondary products, and their likely formation in combustion processes and during lightning (Chapter 2.6.4.4) might be of only small importance (see Chapter 5.3.1 concerning the formation in the atmosphere). It has been concluded that within the past 100 years ground-based ozone concentration has risen by about a factor of two. Models also show a doubling of the tropospheric ozone content because of human activities (Möller 2002a, 2004 and citations therein). Very likely, the mean ozone concentration will further increase. By contrast, episodes with exceeded ozone levels ("summer smog") have become rare since the end of the 1990s in central Europe. This controversial figure is explained by a reduction of NO_x (the "catalyst" in ozone formation) as well as reactive ozone precursors such as NMVOCs because of catalytic converters in cars. By contrast, a further increase of methane emissions and probably CO is responsible for the regional and even global background increase of ozone. The principal formation mechanisms are well understood despite some open questions concerning the contribution of specified organic substances, especially from biogenic sources. Less known, however, are chemical sinks, especially heterogeneous processes (Acker at al. 1995, Möller et al. 1999).

Trend (troposphere and stratosphere)

The long-term trend given by Marenco et al. (1994) is well known (Fig. 2.81). Its exponential increase is significant, showing a yearly increase by 1.54%. However, there has been some criticism because data from different sites and various altitudes were included in this diagram. Some additional values have now been introduced by Arosa and Zugspitze, suggesting that there is no continuous exponential slope over the entire period. There is also some belief that the old data from Pic du Midi are too low. Values around 10 ppb and less were typical only for urban areas at sea level (known from Paris and Moscow around 1900). Generally, the strong vertical

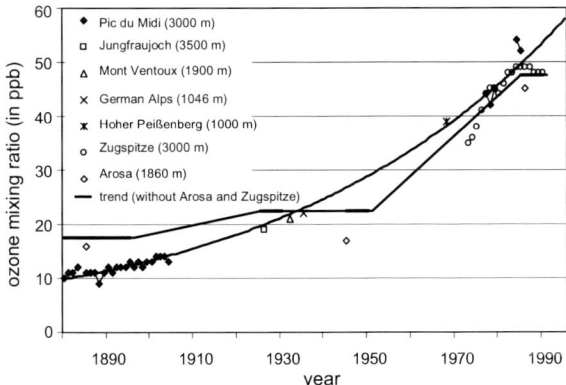

Fig. 2.81 Composite of different historic European Mountain ozone measurements ("*Marenco*" curve changed, Marenco et al. 1994); additional data (not included in the trend line) from Arosa (Staehelin et al. 1994) and Zugspitze (UBA 2000). Grey line: trend $[O_3] = 9.67\exp(0,0154 \cdot x)$ with x number of years (beginning at 1885), $r^2 = 0.99$. Black staircase-shaped line represents corrected larger ozone before 1900 and likely ozone evolution (Möller 2003).

Table 2.80 Ozone concentration in different periods and seasons in the surrounding of Arose (Switzerland), after Staehelin et al. (1993).

period	winter (in ppb)	summer (in ppb)	increase (in ppb)	increase factor	ratio summer/ winter
1889–1891	~15	15–20	–	–	1.2
1950–1951 and 1954–1958	10–12	20–25	5	1.3	~2
1989–1991	35–40	45–50	25	2.2	1.3

Table 2.81 Global tropospheric sources and sinks of ozone (in Tg a^{-1}).

process	sources			sinks		
	Crutzen et al. (1999)		Wang et al. (1998)	Crutzen et al. (1999)		Wang et al. (1998)
	present	pre-industrial	present	present	pre-industrial	present
photochemistry	3940	1780	4100	3120	1630	3700
stratosphere	480	480	400	–	–	–
dry deposition	–	–	–	1300	670	800
total	4420	2260	4500	4420	2300	4500

dependence of ozone concentration has to be taken into account (see next subsection). Staehelin et al. (1994) have shown from Swiss data that ozone might have risen by a factor of 2.5 in the past century (Table 2.80), as also supported by global models (Table 2.81).

Other widely unknown ozone data from the Soviet Union measured in Moscow and in the Caucasus Mountains (Konstantinova-Schlesinger 1937a, 1937b, 1938), support ozone values in the lower mountain region between 20 and 30 ppb for the 1930s (see Fig. 2.84). Thus, we assume a stepwise increase shown in Fig. 2.81 by the fat black line with periods of increase (1905–1930 and 1950–1990) and periods of stagnation in between. The first period of increase, however, was likely to have been very small. Thus, after 1950 – similar to other "classical" air pollutants (SO$_2$, NO$_x$) – ozone significantly rose. The following values, typical for an altitudes 2000 ± 1000 m a.s.l., are most likely:

    1885–1900: 15–20 ppb
    1930–1935: 20–25 ppb
    1950–1955: 20–25 ppb
    1990–1995: 45–50 ppb

Before World War II ozone formation was believed to only be occurring in the stratosphere. In 1924, *Gordon Dobson* designed a spectrometer, which became the standard instrument for measuring both total column ozone and the profiles[27] of ozone. The total column ozone (mass per square) is expressed in the *Dobson* unit (1 DU is defined as 0.01 mm thickness at stp)[28]. With *Dobson* spectrophotometers *Joe Farman* and his team on the British Antarctic survey discovered the ozone hole in 1984 (Farman et al. 1985). It is worth noting that the view of *Farman* on discovering the ozone hole was based on long-term monitoring (BBC broadcast on July 6, 1999):

> The British Antarctic survey set up stations in Antarctica. And so we'd been monitoring very many things in Antarctica for a long while. And suddenly in 1985 it dawned on us that we were sitting on top of one of the biggest environmental discoveries of the decade, I suppose, or perhaps even of the century. We saw this little dip appearing, and then it just accelerated so rapidly that, within three or four years, we were talking about a 30 per cent in the thickness of the ozone above us, which was an enormous amount. We can be slightly proud of the fact. This was the first time that anyone had shown that ozone levels had changed since the measurements began, way back in 1926 or thereabouts, when Dobson made his original pioneering measurements. The long-term monitoring of the environment is a very difficult subject. There are so many things you can monitor. And basically it's quite expensive to do it. And, when nothing much was happening in the environmental field, all the politicians and funding agencies completely lost interest in it.

In a recent commentary in *Nature*, Nisbet (2007) called environmental monitoring "science's Cinderella, unloved and poorly paid" and noted that sustained, long-term, ground-based measurements are underappreciated and underfunded because they are seen neither as basic measurements to test scientific hypothesis nor as challenging high-tech opportunities for profit by commercial interests[29]. However, high-quality long-term monitoring is the platform for understanding climate change, providing insights into processes and data for model validation. Experimental environmental science is based on tripod field campaigns, laboratory studies and long-term monitoring. Only from all three approaches can we identify processes and quantify them as well as implement and validate models and (in sense of a feedback) use them to interpret experimental data.

Some modern versions of the *Dobson* spectrophotometer exist and continue to provide data, e. g. the *Brewer* spectrophotometer in the WMO-GAW ozone observ-

---

[27] The vertical distribution of ozone is derived by the Umkehr method. Direct solar intensity is measured at two different wavelengths, one being more absorbed by ozone than the other. At sunrise and sunset, the intensities decrease at different rates. The ratio shows an inversion. This is called the Umkehreffekt and gives information about the vertical distribution of ozone in the atmosphere. An Umkehr measurement takes about three hours, and provides data up to an altitude of 48 km, with the most accurate information for altitudes above 30 km.

[28] For example, 300 DU of ozone brought down to the surface of the earth at 0 °C would occupy a layer only 3 mm thick. One DU is $2.69 \cdot 10^{16}$ ozone molecules cm^{-2}, or $2.69 \cdot 10^{20}$ m^{-2}. This is 0.4462 millimoles of ozone m^{-2}.

[29] The author (DM) had same experience with the Mt. Brocken Cloud Chemistry Climate Monitoring (BROCCMON) station with German Funding Agencies: The station was only funded from 1992 until 1995 but remains in operation (now 18 years with more than 25000 one-hour analyzed cloud-water samples). We discovered in 1994 the cloud ozone hole (Acker et al. 1995). We closed the station in 2011.

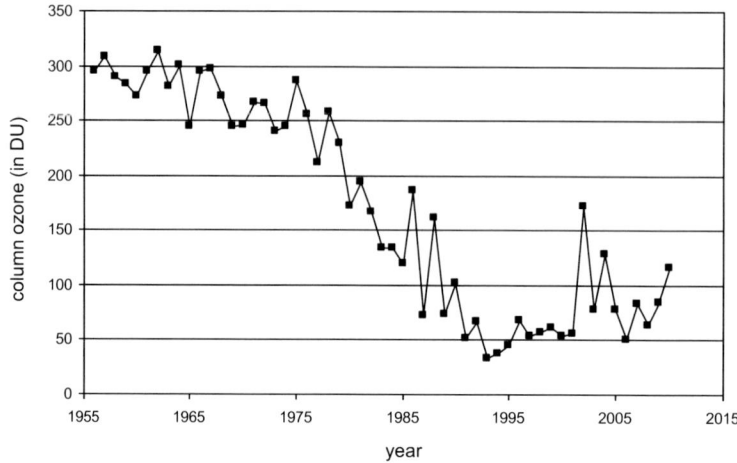

Fig. 2.82 Total column ozone amounts (October mean) as measured at Halley, Antarctica, by the British Antarctic Survey from 1956 − 2010. Data after Kummel et al. (2012).

ing system comprises more than 100 stations worldwide that measure total column O_3 and O_3 profiles in the troposphere and stratosphere. Fig. 2.82 shows the total ozone record from Antarctica, showing two main features: a) the drastic drop in ozone after 1975, b) the minimum O_3 level in 1993, which has been attributed to residual volcanic effects (Pinatubo, 1991) and c) the significant ozone growth (recovery) of 3.2 ± 1.0 DU yr^{-1} (Kummel et al. 2012) as the result of the Montreal Protocol.

Long-term tropospheric ozone measurements show that the ozone over Europe has increased by more than a factor of two between World War II and the early 1990s, which is consistent with the large increase in anthropogenic ozone precursor emissions in the industrialized world. However, the further increase in background ozone over Europe and North America since the early 1990s cannot be solely explained by regional ozone precursor changes because anthropogenic ozone precursor emissions decreased in the industrialized countries as a consequence of air pollution legislation. Measurements also indicate large increases in ozone in the planetary boundary layer over the tropical Atlantic since the late 1970s, which have been attributed to large increases in fossil fuel-related emissions. Measurements at southern mid-latitudes, which are limited in number, show a moderate increase in tropospheric ozone since the mid-1990s (Staehelin et al. 2007).

In 1952, the first long-term measurements of the European ozone started in Wahnsdorf near Dresden (in a former observatory of the weather service; Fig. 2.63). Measurement was based on iodometry with a strong interference of SO_2. Thus, between 1952 and 1971 a "constant" O_3 concentration of 25 ± 5 µg m^{-3} was registered because of rising SO_2 concentrations. Only in 1967, when SO_2 concentrations remained constant (Fig. 2.63), and in 1972, after the introduction of a prefilter, was the data considered representative. The following regression is valid for the mean summer season values (in µg m^{-3}):

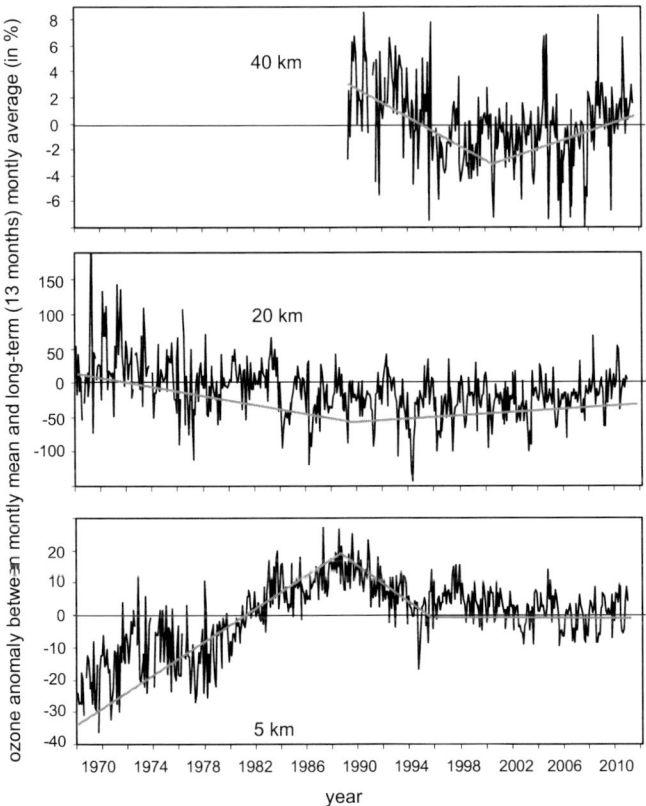

Fig. 2.83 Historical record (November 1967–October 2009) of the relative change of ozone in three different atmospheric layers measured at the Meteorological Observatory Hohenpeissenberg; until 30 km altitude balloon sonding and above 30 km lidar measurement. The changes are calculated as the difference from the monthly average and the long-term monthly mean (13 month, traversing means). Data from Hans Claude (Deutscher Wetterdienst, DWD), see also Claude et al. (2008).

$$[O_3] = 38.9 + 1.4 \cdot x \, (r^2 = 0.80; 1974-2001), x = 0 \text{ for } 1974 \qquad (2.108)$$

Fig. 2.83 shows the different trends of ozone at different altitudes. The whole troposphere has been characterized by an ozone increase since measurement started in 1967, where the average decadal growth rate amounts to about +10 % (Claude et al. 2004, 2008). However, the growth rates decreased between 1988 and 1993 and showed no further change until the end of 2009. By contrast, in the lower stratosphere (20 km level) ozone decreased until 1988 and remained constant or slightly increased in subsequent years. Similarly, at the 40 km altitude ozone decreased until end of the 1990s but then slightly increased, probably showing the slow recovery of the ozone layer.

It is worth noting that the ozone trends at various mountain sites vary largely. At the beginning of the 1990s some stations in Europe showed no further O$_3$ increases. The number of days with high ozone concentrations (exceeding the thresholds of 180 µg m^{-3} and 240 µg m^{-3}) significantly decreased in Germany (and all across Europe). This coincides with the reduction of some precursor emissions (NMVOCs and NO). However, it seems that there are differences between seasons and site locations (background, rural, urban). In assessing ozone levels between different sites and different periods it is essential to regard the "typical" variation of ozone with altitude, season and daytime.

At Mt. Brocken, we found an increase in winter values from 15 ppb in 1992/1993 to 29 ppb in 1997/1998 and no change or only little variation after the end of the 1990s (Fig. 2.85). This clearly demonstrated that such change was connected with the changing air pollution in eastern Germany (Möller et al. 1996, 1999). The long-term ozone monitoring at Mt. Brocken shows that the ozone concentration has remained approximately constant since 2000 and has decreased slightly since 2005. The Brocken site represents the background processes where CO and CH$_4$ contribute to O$_3$ formation; CO more in densely populated NH regions and CH$_4$ more in the SH. Thus, it seems a clear hypothesis that significantly decreasing CO (Fig. 2.79) since 2000 and stagnant CH$_4$ (Fig. 2.74), at least until 2007, could explain this small decrease on background ozone.

The picture of long-term tropospheric ozone changes is a varied one in terms of both the sign and magnitude of trends and the possible causes for the changes (Oltmans et al. 2006). At the mid-latitudes of the SH, three time series of ~ 20 years all agree, showing increases that are strongest in the austral spring (August–October). Profile measurements show this increase extending through the mid-troposphere but not into the highest levels of the troposphere. In the NH in the Arctic, a period of declining ozone in the troposphere through the 1980s and into the mid-1990s has reversed and the overall change is small. Decadal variations in the troposphere in this region are related in part to changes in the lowermost stratosphere. At mid-latitudes in the NH, continental Europe and Japan showed significant increases in the 1970s and 1980s. Over North America, rises in the 1970s were less than those seen in Europe and Japan, suggesting significant regional differences. In all three of these mid-latitudes, regional tropospheric ozone amounts seem to have levelled off or in some cases declined in recent decades. Over the North Atlantic, three widely separated sites have shown significant increases since the late 1990s, and that might have peaked in recent years. In the NH tropics, both the surface record and the ozonesondes in Hawaii have shown a significant increase in the autumn months in the past decade compared with earlier periods, which drives the overall increase in the 30-year record. This seems to be related to a shift in the transport pattern during this season with more frequent flow from higher latitudes in the latest decade.

From 1979 until the late 1990s, all available data show a clear decline of ozone by 10–15% near 40 km (Steinbrecht et al. 2009). This decline has not continued in the past 10 years. At some sites, ozone at 40 km seems to have increased since 2000, consistent with the beginning decline of stratospheric chlorine. The phase out of CFCs after the Montreal Protocol in 1987 has been successful, and is now showing positive effects on the ozone in the upper stratosphere. The upper stratospheric ozone shows three main features (Steinrecht et al. 2006): (a) a decline of 10–15% since 1980 because of chemical destruction by chlorine; (b) two to three year's fluctuations by 5–10 % because of the quasi-biennial oscillation (QBO); and (c) an 11-year oscillation of about 5 % because of the 11-year solar cycle. Negative ozone anomalies (the magnitude is ~ 20 % and much larger than any previously reported solar cycle effect) are strongly correlated with the flux of energetic electrons in the radiation belt (Sinnhuber et al. 2006).

The 1979 to 1997 ozone trends are larger at the southern mid-latitude station Lauder (45° S), reaching −8 % decade^{-1} compared with only −6 % decade^{-1} at Table Mountain (35° S) and Hohenpeissenberg (47° N). At all stations, ozone residuals after the subtraction of QBO and solar cycle effects have levelled off in recent years, or are even increasing. Assuming a turning point in January 1997, the change of trend is largest at southern mid-latitudes (+11 % decade^{-1}), compared with +7 % decade^{-1} at northern mid-latitudes. This point is the beginning of a recovery of upper stratospheric ozone. However, chlorine levels are still very high and ozone will remain vulnerable (Steinbrecht et al. 2009). At this point, the most northern mid-latitude station Hohenpeissenberg (Fig. 2.83) differs from the other stations and shows much less clear evidence for the beginning of a recovery, with a change of trend in 1997 by only +3 % decade^{-1}.

Ozone change with altitude

Generally, an increase of ozone is found with increasing height above sea level (note that O_3 is not a primary emission produced photochemically within the boundary layer and transported downwards from the stratosphere). In the first few 100 meters (within the mixing height) the strong diurnal variation (see next subsection) can influence the mean value (daily average) by lowering. Thus, Kley et al. (1994) suggest to regard the around noon averaged ozone concentration in summer as a "reference" compared with data from other stations. These authors found a linear regression with altitude (between 0.9 and 2.4 km):

$$[O_3] = 44 + 3.8 \cdot z \ (r^2 \text{unknown}) \tag{2.109}$$

where z is altitude in km and concentration is in ppb. An increase in concentration with altitude has been known for many years. From data from Staehelin et al. (1994) it follows that:

$$[O_3] = 41.7 + 5 \cdot z \ (r^2 = 0.99) \tag{2.110}$$

and that between 0.8 and 3.45 km there is a somewhat stronger increase with height. Historical data show a much stronger gradient $d[O_3]/dz \approx 11$ ppb km^{-1}. This is because of the missing anthropogenic ozone production in the lower troposphere, i. e. the earth's surface acts as a sink (via dry deposition) and the upper troposphere as a source (O_3 downward transport from the stratosphere). Former Soviet data (Fig. 2.84) from the mid-1930s result in the equation (0.1–14 km):

$$[O_3] = 9.5 + 10.9 \cdot z \ (r^2 = 0.99) \tag{2.111}$$

Remarkably, this is the same gradient as from Swiss data from the 1920s (Fig. 2.84):

$$[O_3] = 3.5 + 11 \cdot z \ (r^2 = 0.90) \tag{2.112}$$

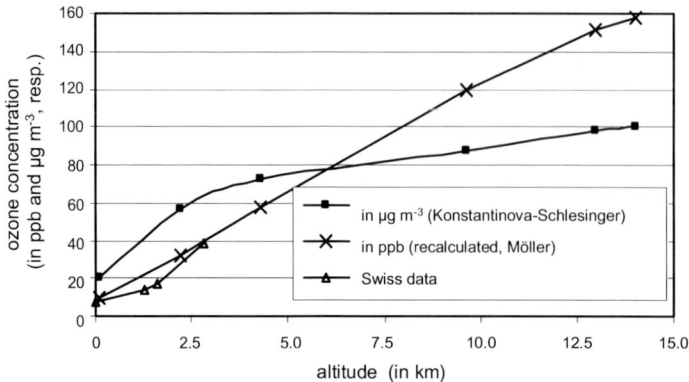

Fig. 2.84 Historical data on the dependence of ozone from altitude. Data from Konstanti-nova-Schlesinger (1937a, 1937b, 1938) from the Soviet Union in the 1930s years (Moscow 100 m, Elbrus 2200 m and 4300 m, aircraft measurements 9620 m, 13 000 m and 1400 m); Swiss date: (Genf 200 m, Zermatt 1650 m, Rochers de Naye 2045 m and Gornergrat 3200 m) from Gmelin (1943) and Staehelin et al. (1994).

It does not seem reasonable to derive an ozone figure for $z = 0$ from the equations, which is too large for the present period (Eqs. 2.107 and 2.110) and too small for the period before World War II (Eqs. 2.111 and 2.112), i.e. such relationships are only valid within a given range of altitudes. The equations clearly show how the man-made ozone formation in the lower troposphere reduces the vertical gradient but increases the concentration.

Timely variations (cycles)

A "typical" diurnal ozone variation concurs with the intensity of solar radiation, whereas the maximum is shifted later to the afternoon. The diurnal variation, how-ever, represents the budget (sources − sinks). Beginning at night (no photochemical production), O_3 is removed by dry deposition and chemical reaction with NO_x. In case of nocturnal inversion layer, no ozone transport from the residual layer (above mixing height) occurs and, consequently, O_3 concentration depletes (even to zero in urban areas). After sunrise, the inversion layer is broken down and, by vertical mixing, O_3 is transported down. Additionally, photochemical production increases. The daytime maximum O_3 concentration represents the well-mixed boundary layer atmosphere. In summer, the daily maximum net photochemical O_3 production is around 15 ppb; the ozone formation rate is proportional to the photolysis rate, assuming no variation in precursors (NMVOCs). Thus, O_3 photochemical produc-tion shows a maximum at noon. After having the concentration maximum (which represents sources = sinks) in late afternoon, the inversion layer builds up again (no more vertical O_3 mixing in) and finally dry deposition and surface-based chemi-cal removal reactions reduce the ozone. Thus, vertical transport is dominant in determining the diurnal variation, and, consequently, wind speed and temperature (again linked with radiation) are key meteorological parameters showing a correlation with O_3. Mean O_3 is again well correlated with mean temperature (Treffeisen et al. 2002).

Similar to diurnal variation, seasonal variation is driven by photochemical ozone production with a maximum in summer. The winter minimum represents a reduced photochemical activity (Fig. 2.85) but might also show the chemical ozone depletion into cloud droplets. As seen in Fig. 2.85, daily variation might even be larger because of changes in the air masses with different characteristics in ozone production and depletion. Differences from year to year depend on the large-scale weather situation as well as varying stratosphere–troposphere exchange processes (STE), the main source for tropospheric ozone. This stratospheric ozone often reaches a maximum in spring and has been made responsible for the maximum in annual ozone variation found in northern latitudes (Bazhanov 1994, Oltsman and Levy 1994). Wang et al. (1998) have shown by modeling that this maximum, also found in remote areas out of the tropics, is based on overlapping stratospheric O_3 (which is transported in late winter into the troposphere) and O_3 photochemically produced in the free troposphere (with a maximum in spring). This might also be explained by the seasonal variation of the tropopause with its minimum in January. With increasing tropopause after January, vertical mixing arises where ozone will be transported downwards (surface-near maximum originated) and, finally, the same ozone amount is distributed over a higher troposphere, resulting in lower surface-near concentrations (seasonal minimum).

The seasonal amplitude (ratio between summer and winter concentration) represents a measure for annual photochemical ozone production (neglecting the first possible ozone sinks). Before the 1950s, seasonal ozone variation was small. Considering the O_3 concentration as an expression for the source-sink budget, even the summer-winter ratio must be a sensible parameter, but this also depends on altitude, showing a characteristic increase with height.

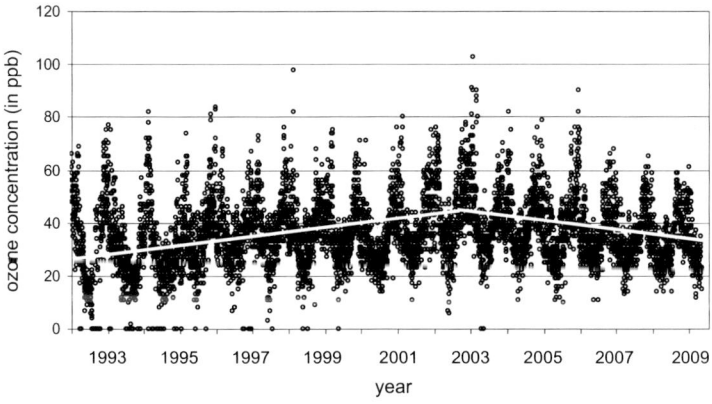

Fig. 2.85 Historical record of ozone at Mt. Brocken, 1142 m a.s.l. (daily means June 1, 1992–December 31, 2009); data from: Saxony-Anhalt State Environmental Protection Agency, Air Quality Monitoring and Information System. Trend: 1992–2002: $[O_3] = 28.8 + 1.0 \cdot yr^{-1}$ (in ppb), 2003–2009: $[O_3] = 38.8 - 1.7 \cdot yr^{-1}$ (in ppb).

Budget − sources versus sinks

At a given site ozone concentration can have two different sources: the local *in situ* photochemical formation and the *onsite* transportation (vertical and horizontal) of ozone. Local sinks of ozone are dry deposition (irreversible flux to ground) and chemical reactions in the gas phase, on aerosol particles and in droplets. Air masses have different potential to form and destruct the ozone. Depending on air mass type, the past and fate of ozone (in relation to the site) can be characterized by production and/or destruction (positive or negative budget). With changing air masses, the ozone level can also drastically change. The atmospheric ozone residence time, depending only on sink processes, is extremely variable. It amounts to a few days near the ground and several months in the upper troposphere. Thus, ozone can be transported over long distances in the free troposphere. The advective transport to a receptor site is therefore an important source of local ozone.

The mean (annual) background O_3 concentration for central Europe (reference year 1995 ± 5) was 32 ± 3 ppb, where the following "sources" can contribute (Möller 2004):

− 10 ± 2 ppb stratospheric ozone with small seasonal variation (spring peak);
− 6 ± 2 ppb natural biogenic ozone from natural VOCs with seasonal variation (0−12 ppb); and
− 16 ± 2 ppb man-made ozone from CH_4 and CO.

This base ozone of 32 ppb shows a seasonal variation between 26 ppb (winter) and 38 ppb (summer), where about 50 % is anthropogenic. Additional to the base ozone is 5 ppb of man-made hot ozone from NMVOCs with seasonal variation (0−15 ppb) and strong short-term variation (0−70 ppb).

In total, the "typical" European annual mean figure amounts to 37 ppb (winter 25 ppb and summer 47 ppb). It has been clearly shown that the "acute" air pollution problem is given from NMVOC precursors (hot ozone), which contributed only 5 ppb (25 % of man-made contribution and 14 % of total ozone) to the long-term average in the 1990s. The "chronic" air pollution problem is given by methane oxidation (and partly by CO), contributing to a likely continuously increasing base ozone (50 %). In Germany, and stepwise in all other European countries, the hot ozone problem has been solved. The problem of background ozone, connected with CH_4, cannot be solved. The reduction of "hot" ozone led to constant ozone levels or only minor ozone increases in the 1990s, but in the future ozone will again rise according to the CH_4 emissions increase.

2.8.2.8 H₂O₂: Mysterious

One of the most interesting and important species in air is hydrogen peroxide (H_2O_2). Kok et al. (1978) suggested that H_2O_2 is an index of the HO_2 radical concentration and better represents oxidation capacity than ozone. However, the relatively complicated and labor-intensive measurement technique limits H_2O_2 measurements to very few sites and even then only occasionally (probably because of fixing to ozone as the oxidant leader species).

Despite the decreasing emissions of many pollutants (SO_2, dust, NO_x, VOCs), key impact factors such as oxidation capacity, acidity, toxicity and climate forcing have not been reduced to the same extent but have remained constant or even increased. It is likely that our incomplete understanding of such relationships is because of insufficient budget estimation (i. e. taking into account *all* sources and sinks), which again strongly depends on time and space.

H_2O_2 is produced in the atmospheric gas phase only through a single pathway, namely the HO_2 radical recombination. Its main role is oxidizing the SO_2 dissolved in hydrometeors to sulfate. Thus, aqueous-phase chemistry has been considered a main sink (apart from dry deposition and scavenging) but rarely a source of H_2O_2 despite early findings of its heterogeneous and aqueous-phase production.

Over the past 20 years, many H_2O_2 measurements have been carried out but almost always (with a very few exceptions) covering short measurement periods. The main findings of these studies are summarized as follows (Möller 2002b, 2009a, Lee et al. 2000, Gunz and Hoffmann 1990, and citations therein):

- the H_2O_2 concentration is generally lower in the boundary layer (< 2 ppb) than in the free troposphere (2−8 ppb), whereas "typical" surface-near concentrations in summer are between 0.2 and 0.8 ppb;
- the H_2O_2 concentration increases from the North Pole to the equator and rises drastically around the equator and above Africa;
- the H_2O_2 concentration drops in the gas-phase during rain and within clouds (*interstitial air*);
- increased H_2O_2 concentration has often been found close to clouds;
- in remote areas different diurnal variations have been found with maxima between late afternoon and midnight;
- in winter the H_2O_2 concentration is almost always significantly lower (factor ≤ 15);
- the H_2O_2 concentrations correlate with UV radiation;
- in "NO poor" areas an inverse correlation exists between H_2O_2 and O_3 (in diurnal variation, not in seasonal variation);
- the H_2O_2 concentration in urban areas is lower than in rural areas; and
- in the aqueous phase an inverse correlation between H_2O_2 and S(IV) has been found.

Hydrogen peroxide has been found with increasing concentrations in Greenland ice cores over the past 200 years (Fig. 2.86). Most of the increase has occurred over the past 20 years (Sigg and Neftel 1991, Anklin and Bates 1997) because of human activities in general (Sigg and Neftel 1991), the increase of emission rates of NO_x, CH_4 and CO (Sigg and Neftel 1991, Anklin and Bates 1997, Thompson 1992) and the possible increased radiation because of decreasing stratospheric ozone (Thompson 1992). Möller (1999a, 2002b) proposed an explanation based on an inverse correlation between atmospheric SO_2 and H_2O_2 during long-range transportation and chemical interaction.

Looking into details of the smoothed H_2O_2 ice core record (Fig. 2.86) it seems obvious to distinguish between several time periods connected with different chemical climate, namely:

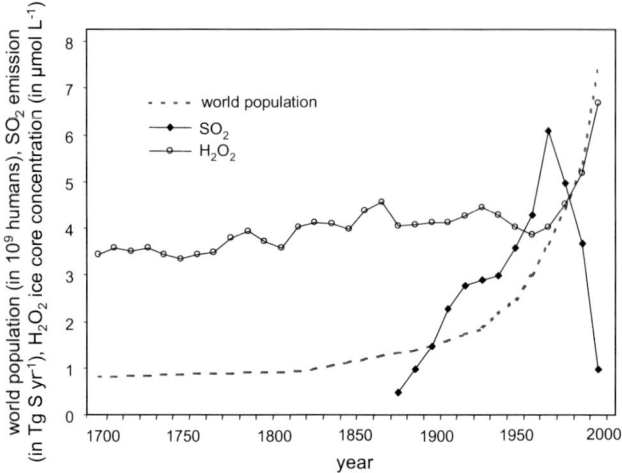

Fig. 2.86 Historical records of 10−years averages of H_2O_2 concentration (in μMol L^{-1}),in Greenland ice cores (before 1988 after Sigg und Neftel, 1991 and 1988 to 1995 after Anklin and Bales, 1997), world population (cf. Fig. 2.58) and total SO_2 emission from Northern America (after Gschwandtner et al. 1986) and Western Europe (after Mylona 1996) in 10^7 t S yr^{-1} (the year 2000 value has been adopted from Stern 2005). Typical errors: $\pm 10\%$ ice core concentration, $\pm 20\%$ and 50% SO_2 emission before 1995 and from 1995, respectively.

- 1300−1750: with no trend in H_2O_2 (mean of 3.6 ± 0.3 μM) before significant population growth;
- 1750−1850: with a small increase ($\sim 0.2\%$ yr^{-1}), proportional to the human population (i. e. wood burning as the main human energy source) and before significant SO_2 emissions (hypothesized H_2O_2 formation by photocatalytic inter-facial processes, see Chapter 5.3.4);
- 1850−1975: with stagnant or somewhat decreasing H_2O_2 (mean of 4.0 ± 0.6 μM) due to the "masking" of the formation term by the consumption of H_2O_2 be-cause of highly cloud-dissolved SO_2 from fossil fuel combustion; and
- 1975−1995: with the rapid increase by about a factor of two ($\sim 5\%$ yr^{-1}) because of the widespread use of flue-gas desulfurization and consequently decreasing SO_2 emissions (note that the Greenland area is mainly influenced by air masses coming from North America and western Europe).

Hence, it is likely that during the decades before significant SO_2 emissions (before 1850), atmospheric H_2O_2 concentrations were larger than at present because of missing consumption ("masking") by atmospheric sulfur dioxide. Now, 150 years later, the atmospheric aqueous phase after desulfurization might again have the natural potential for photocatalytic oxidant formation and we see a "demasking" in the H_2O_2 budget (Möller 1999a, 2002b, 2009a).

As a consequence, it could be possible that within the past few decades the atmospheric H_2O_2 level has already increased (and might further rise in future). This hypothesis, however, is impossible to prove because of missing long-term mon-itoring. To my knowledge, there have been only two medium-term measurement series. Dollard and Davies (1993) found a large increase in H_2O_2 between 1988 and

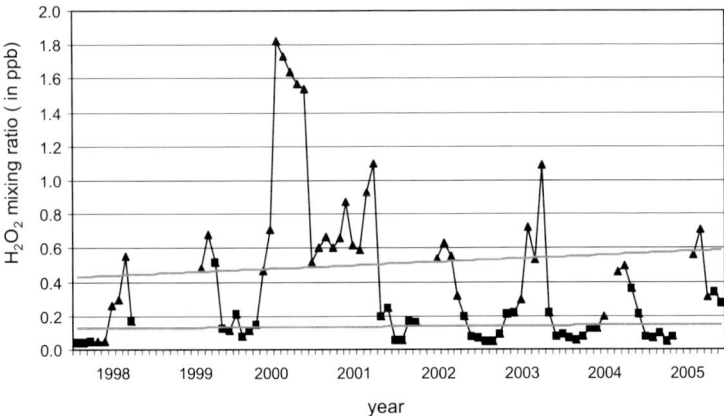

Fig. 2.87 Variation of monthly means of H_2O_2 (in ppb) at Mt. Hohenpeissenberg. Upper grey line: linear trend of summer values (May–September). Lower grey line: linear trend of winter values (October–April); data from Dr. Stefan Gilge (Deutscher Wetterdienst, DWD).

1991 from 0.15 to 0.3 ppb with +0.028 ppb yr^{-1} ($r^2 = 0.83$) in Harwell (England), i. e. a doubling in only four years. However, they measured H_2O_2 at the same site again in 1993/1994 and found higher values (even though the concentration mean in the second period was the same as in the first period) and provided no explanation. A much longer and careful H_2O_2 monitoring study was carried out over many years between January 1998 and October 2005 at the Mt. Hohenpeissenberg (988 m a.s.l.), GAW station and German Meteorological Observatory, about 60 km south of Munich. Fig. 2.87 shows all data based on monthly means (the data availability lies between 30 and 80%). It is clear that there are high excess (compared with the trend line) values in summer 2000, winter 2000/2001 and summer 2001. The extremely hot summer of 2003 led only to an increase of H_2O_2 by a factor of two above the trend line. The winter values at Hohenpeissenberg show no trend (mean of 0.14 ppb), whereas the summer values (mean of 0.51 ppb) show a trend of about +0.02 ppb yr^{-1}, remarkably similar to the trend found in Harwell (see above). Hence, the increase (based on monthly means) corresponds to 4% yr^{-1}, similar to the discussed increase in Greenland ice cores. The trend, however, is not significant and the few measurements support increasing atmospheric summer H_2O_2 concentrations.

2.8.2.9 OH: The key oxidant

The most important chemical cleaning agent in the climate system is the hydroxyl radical (Ehhalt 1999) destroying about $3.7 \cdot 10^{15}$ g of trace gases each year (Prinn et al. 2005). OH is very short-living and is thereby produced by very few reactions (Chapter 5.3.2) and destroyed by a myriad of reactions within short-term photostationary states. Hence, selected long-living trace substances, which are removed only

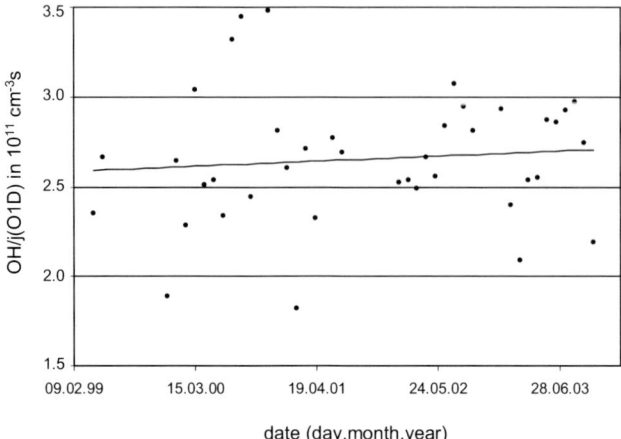

date (day.month.year)

Fig. 2.88 OH trend between 1999 and 2003 at Hohenpeissenberg; data from Franz Rohrer (Institute of Atmospheric Chemistry Jülich) and Harald Berresheim (Deutscher Wetterdienst DWD), see also Rohrer and Berresheim (2006).

via OH attack such as 1,1,1-trichloroethane (CH_3CCl_3, methyl chloroform), provide an accurate method for determining the global and hemispheric behavior of OH.

Different spectroscopic techniques have been developed since the end of the 1980s to realise local OH measurements. At the Hohenpeissenberg observatory, the first *in situ* long-term OH monitoring started in 1999 (Berresheim et al. 2000, Rohrer and Berresheim 2006) and showed no observable trend until 2003 (Fig. 2.88) apart from strong seasonal and diurnal cycles. Rohrer and Berresheim (2006) found that OH concentration can be described by a surprisingly linear dependency on solar UV radiation throughout the measurement period, despite OH being influenced by thousands of reactants. Essentially, this means that the whole OH data set can be described by a single variable, $j(O^1D)$. This also is confirmed by two-year OH measurements at Mace Head (Berresheim et al. 2013). From the Mace Head measurements a summer-winter ratio was found to be 10.8 for noontime OH values (summer maximum $2.26 \cdot 10^6$ cm^{-3}); nighttime OH is close to the detection limit ($\sim 9 \cdot 10^4$ cm^{-3}). The variance of OH is dominated by the diurnal cycle (76 %; Fig. 2.89) and the seasonal cycle (23 %), and several field campaigns support the strong linear OH-$j(O^1D)$ correlation (Rohrer and Berresheim 2006, and citations therein):

$$[OH] = a \left(\frac{j(O^1D)}{10^{-5}s^{-1}} \right)^b + c \quad (r^2 = 0.85 \ldots 0.95) \tag{2.113}$$

The exponent b reflects the combined effects of all photolytic processes producing OH either directly or indirectly via HO_2 recycling ($b = 1$, with the exception of the MINOS campaign where $b = 0.68$). The dependence of OH on reactants is condensed into a single pre-exponential factor a (varying from 1.4 to 7.2 for different sites). Finally, the coefficient c (varying from 0.01 to 0.20 for different sites) includes

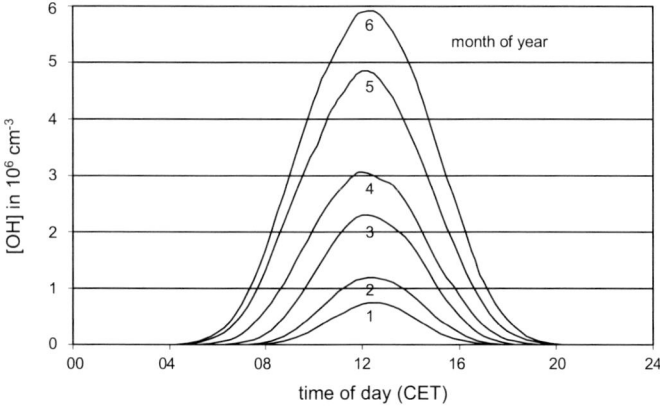

Fig. 2.89 Mean OH diurnal variation between 1999 and 2003 at Hohenpeissenberg (data source see Fig. 2.88).

all processes that are light-independent, e. g. night time OH formation. The findings of Rohrer and Berreshcim (2006) allows us to define an "OH index" characterized by a simple set of coefficients for months or even years to describe observable impacts and trends in the oxidation efficiency of the troposphere in different chemical regimes. The five-year monitoring at Hohenpeissenberg has resulted in an average OH of $1.97 \cdot 10^6$ molecules cm^{-3} with a maximum of $11.6 \cdot 10^6$ molecules cm^{-3}.

Analysis of the methyl chloroform observations (Fig. 2.77) shows that global average OH levels (since 1978; average until 2004 about $1.1 \cdot 10^6$ molecules cm^{-3}) had a small maximum around 1989 and a larger minimum around 1998, with OH concentrations in 2003 being comparable to those in 1979. This post-1998 recovery of OH contrasts with the situation four years ago when reported OH was decreasing. The 1997–1999 OH minimum coincides with, and is likely to be caused by, major global wildfires and an intense El Nino event at this time (Prinn et al. 2005). Prinn et al. (2001) also discussed a strong OH decline in the 1990s; hence, care is taken when estimating trends. Certainly, variations are linked with those of CO, the major sink of OH and CH_4, the second major sink (Chapters 2.8.2.4 and 2.8.2.6). The OH concentration in 2003/2004 is indistinguishable from that in 1979–1981 (Prinn et al. 2005). It should be noted here without going into chemical details (Chapter 5) that the same class of pollutants (CH_4, CO, NO_x, NMVOCs), promoting ozone as the key OH source via its photolysis, is responsible for OH destruction and the experimental findings suggest this balance of OH concentration. In the case of no pollutants, OH itself destructed with O_3 (so-called catalytic ozone destruction mechanism).

2.8.2.10 H_2: Light but problematic

First Gautier (1900, 1901) detected hydrogen to be permanent present in air. As the second most abundant reactive gas in the troposphere after CH_4, hydrogen is present at about 500 ppb. In contrast to the distribution of most anthropogenically influenced gases hydrogen is 3% more abundant in the SH than in the NH (Khalil and Rasmussen 1989, Novelli et al. 1999). Comparing the 1993−2008 averaged H_2 concentrations from Mace Head (487.3 ± 84.8 ppb) and Cape Grim (517.3 ± 67.6 ppb), SH hydrogen is 6% more abundant than NH. Khalil and Rasmussen (1990b) have shown that between 1985 and 1989 the concentration of H_2 increased at an average rate of 3.2 ± 0.5 ppb yr^{-1} (a relative increase of 0.6 ± 0.1% yr^{-1}). Obviously, this trend (with a slower growth rate of about 2 ppb yr^{-1}) was continuous in the 1990s at Cape Grim (Fig. 2.90); the 1992/1993 mean amounts to 517 ppb and reaches 530 ppb in 2000/2001. These increases originate from anthropogenic sources.

Surface emissions include technological sources (industry, transportation and other fossil fuel combustion processes), biomass burning, nitrogen fixation in soils and oceanic activity and totals 39 Tg yr^{-1}. The photochemical production (31 Tg yr^{-1}) from HCHO photolysis accounts for about 45% of the total source of H_2. Soil uptake (55 Tg yr^{-1}) represents a major loss process for H_2 and contributes 80% of the total destruction. H_2 oxidation by OH in the troposphere contributes the remainder. The global burden of H_2 in the atmosphere is 136 Tg. Its overall lifetime in the atmosphere is 1.9 years (Hauglustaine and Ehhalt 2002). H_2 is rather well mixed in the free troposphere. However, its distribution shows a significant seasonal variation (Fig. 2.90) in the lower troposphere where soil uptake dominates. This loss process shows a strong temporal variability and is maximal during summer.

Higher levels of hydrogen will add more water vapor to the stratosphere, where it can affect stratospheric ozone. Ultimately, hydrogen comes from water via thermal dissociation deep in the earth (Chapters 2.1.2) and via water splitting by photosynthesis. It is lost into space (Chapter 2.2.1.3) and buried in the form of fossil fuels

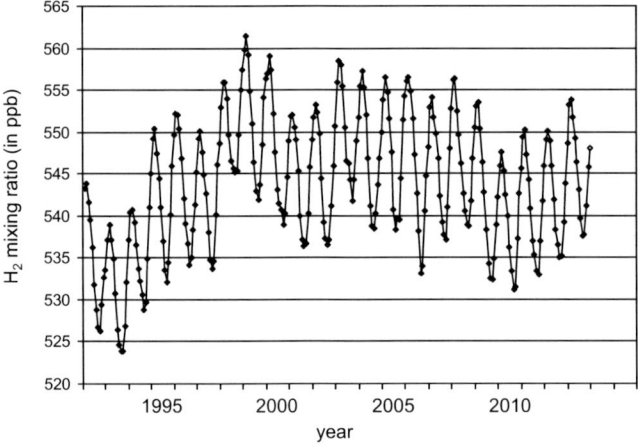

Fig. 2.90 Monthly mean hydrogen concentration (gaslab flask H_2 data): January 1992 – December 2013 (in ppb). Data from: *Paul Steele, Paul Krummel, Ray Langenfelds* and *Marcel van der Schoot*, CSIRO Marine and Atmospheric Research, Aspendale, Australia.

(Chapter 2.2.2.5). It returns by combining with the OH radical in air back to water. Ideas on a future so-called hydrogen technology could lead to a permanent hydrogen leakage; a small loss of the order of 1 % from the hydrogen-based energy industry would increase hydrogen sources twofold (Wuebbles et al. 1997).

2.8.2.11 Volatile acid and OH precursor: HNO_2

Nitrous acid (HNO_2) is an important source of the hydroxyl radical (OH), the most important daytime oxidizing species contributing to the formation of ozone as well as of other reactive oxygen species in the troposphere. Understanding the sources and sinks of HNO_2 is of crucial interest for accurately modelling the chemical composition of the troposphere and predicting future trace gas concentrations. Nitric acid (HNO_3) is the final product of NO_x oxidation (see Chapter 5.4.4.1) and the most important gaseous acid in air, in equilibrium with NH_3 and particulate ammonium nitrate and responsible for chlorine degassing from sea salt (next Chapter 2.8.2.12).

Nitrous acid and a number of other atmospheric components and variables were continuously measured during complex field experiments at seven different suburban and rural sites in Europe from which a short summary is here reported (Acker et al. 2000, 2001, 2004, 2005a, 2005b, 2006, 2007, Acker and Möller 2007). For measurement of soluble gases (HNO_2, HNO_3, HCl, NH_3) and separation from its particulate phase (nitrite, nitrate, chloride, ammonium) we adopted and modified a wet effluent diffusion denuder system (WEDD) in parallel-plate design, coupled with online ion chromatographic analysis (Acker et al. 2005).[30]

HNO_2 is mainly formed by heterogeneous processes on surfaces (see Chapter 5.4.4.2) and often accumulated in the night time boundary layer. We also found HNO_2 formation in clouds; Fig. 2.91 shows that HNO_2/NO_2 rises significantly during passing clouds but highly fluctuates, likely due to the LWC variation (grey area in upper Fig. 2.91) which represents the surface for heterogeneous processes.

Our results confirm that the photolysis of HNO_2 is an important source of the OH radical, not only in the early morning hours but also throughout the entire day and often comparable with the contribution of ozone and formaldehyde photolysis. We were the first who reported that HNO_2 photolysis contributes more strongly to daytime primary OH production than O_3 photolysis, both under urban and rural condition (Acker et al. 2005); Fig. 2.92 shows that the diurnal variation of HNO_2 and OH is similar but OH is a bit phase-delayed.

At all research sites, unexpected high HNO_2 mixing ratios were observed during daytime (up to several hundred ppt, correspond to pmol mol^{-1}). Assuming a quickly established photo-equilibrium between the known significant gas phase reactions, only a few ppt HNO_2 should be present around noon. The ratio of known sources to sinks indicates a missing daytime HNO_2 source of 160−2600 ppt h^{-1} to make up the balance. Moreover surprisingly, the HNO_2 mixing ratio at the three

---

[30] We can not confirm the criticisms of aqueous-phase scrubbing techniques having interferences in HNO_2, esp. at low NO_x conditions (see literature cited in Elshorbany et al. 2012) with our WEDD technique (to our knowledge no other group applied such sophisticated wet sampling technique).

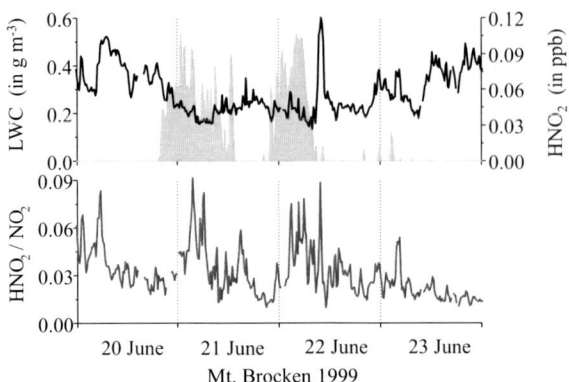

Fig. 2.91 Time series of the gas phase HNO_2 mixing ratio, the liquid water content (LWC) of clouds (upper graph, grey area) and the HNO_2 to NO_2 ratio (lower graph) at Mt. Brocken in 1999.

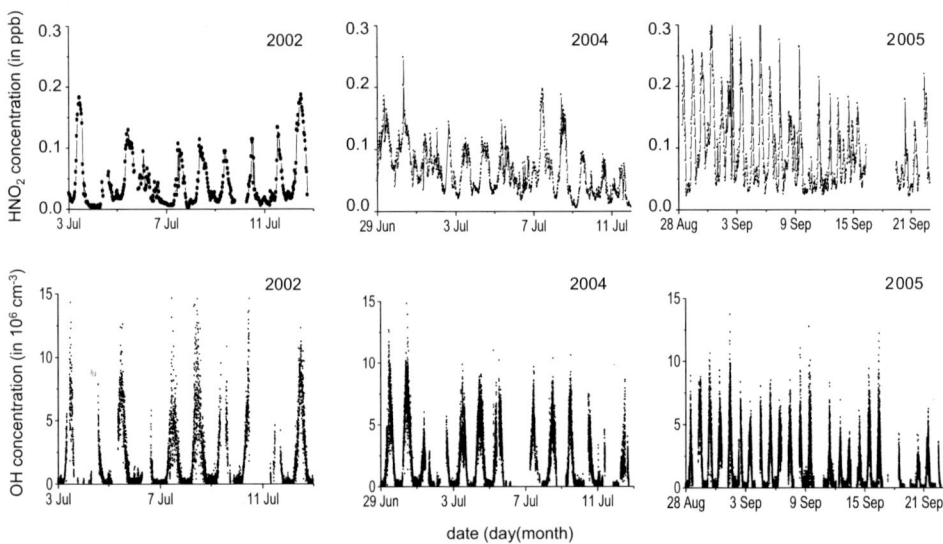

Fig. 2.92 Time series of HNO_2 mixing ratios (upper diagrams) and OH (lower diagrams) concentrations measured simultaneously during three experiments (2002, 2004 and 2005) at the Meteorological Observatory Hohenpeissenberg.

mountain sites often showed a broad maximum or several distinct peaks at midday and lower mixing ratios during the night. Based on these values and on production mechanisms proposed in literature, we hypothesize that the daytime mixing ratio levels may only be explained by a fast photoenhanced electron transfer onto adsorbed NO_2.

Gaseous HNO_2 in ambient air was measured for the first time by Perner and Platt (1979) using Differential Optical Absorption Spectroscopy (DOAS). Existing techniques to detect HNO_2 were reviewed by Genfa et al. (2003), Clemitshaw (2004) and Liao et al. (2005). Generally, the HNO_2 mixing ratios are found to be higher in (polluted) urban than in rural areas and often found to scale with mixing ratios of NO_2; on average the HNO_2 to NO_2 ratios reach a few percent. Nitrous acid mixing ratios mostly showed a diurnal variation dominated by night-time accumulation and photolytic decay during the day.

Several studies have been carried out to estimate the source strength of nitrous acid from direct emissions such as combustion engines. On average 0.3 to 0.8 % of the total NO_x emitted from combustion processes was observed to be direct nitrous acid emission, but these observations may be overlapped by heterogeneous HNO_2 formation (reaction 5.205, see also Fig. 5.23) on wetted soot, asphalt and walls (see literature cited in Acker and Möller 2007). In the atmosphere significantly higher HNO_2 to NO_2 ratios are mostly observed; even in Milan and Rome. Therefore, the contribution of direct emission to the measured atmospheric HNO_2 levels was found to be only very locally relevant (e. g., in urban canyons, traffic channels).

We obtained more than 200 diurnal profiles of nitrous acid (measured by identical equipment, measurement and personal conditions) under different pollution and meteorological conditions; for each site average diurnal cycles were derived, and partly found to be similar to those reported by other authors with heterogeneous formation and accumulation of HNO_2 over night and a rapid decrease after sunrise due to efficient photolysis (5.190) and vertical mixing processes. Beside expected diurnal cycles at the flat country station, surprising results were obtained for the mountain sites for the first time (Mt. Brocken/Harz, Goldlauter/Thuringia forest and Mt. Hohenpeissenberg/Bavaria).

Summarizing all Mt. Brocken HNO_2 data, no marked diurnal variation was found. On average, about 0.060 ppb occurred over the whole day (Fig. 2.93), mainly due to high horizontal advection at the isolated summit which is often above the planetary boundary layer during nights. During and after cloudy episodes, often higher HNO_2/NO_2 ratios (0.03−0.06) were found, see Fig. 2.91. Besides gaseous HNO_2, the amount of nitrite dissolved in the cloud water was also determined (100−5700 nmol L^{-1}). As expected, the distribution between (interstitial) gas and liquid phase was found to be a function of the pH. In most cases, cloud water pH was below 5 and bulk samples were supersaturated with respect to the atmosphere from which the samples were drawn, suggesting an interfacial HNO_2 formation.

Goldlauter is windward of the Mt. Schmücke (937 m a.s.l.), the nearest town is Suhl (~5 km). In autumn, the mixing ratios of nitrous acid were expectedly low under rural conditions but showed unexpected diurnal cycles. Only in a few cases (6 days), an increase of HNO_2 was observed in the evening hours. For a night time formation through reaction (5.205), an average rate constant of 0.006 ± 0.006 h^{-1} was calculated from the linear increase of the HNO_2 to NO_2 ratio, which varied between 0.015 and 0.04 on average. Few days did not show any clear structure in diurnal HNO_2 variation. But surprisingly, a daytime HNO_2 maximum was often found between 0900 and 1500 CET with values between 0.1−0.43 ppb. The average diurnal variation given in Fig. 2.93 reflects this behaviour.

Measurements of nitrous acid were made by our group in the summers of 2002, 2003 and 2005 at the Meteorological Observatory Hohenpeissenberg. The highest HNO_2 mixing ratios were consistently measured during daytime in all campaigns,

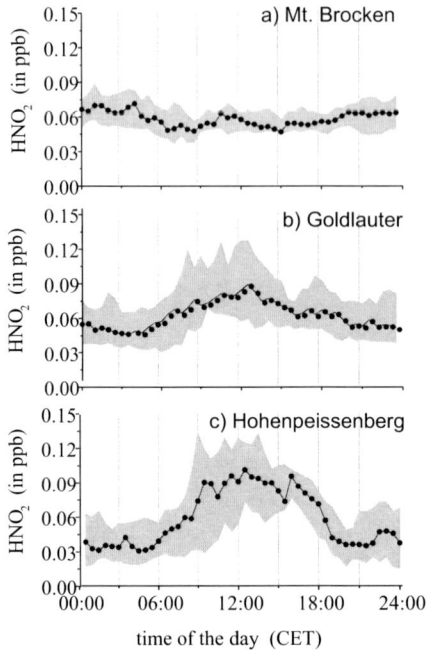

Fig. 2.93 Average diurnal variation of HNO$_2$ at different mountainous rural sites (averaged over the measurement campaign, based on Central European Time (CET)), given as median and 25 and 75 quartiles): Mt. Brocken (19 June – 04 July 1999, Goldlauter (03 October – 07 November 2001) and Hohenpeissenberg (29 June – 14 July 2004).

showing a broad maximum occurring between 0800 and 1600 CET. The average diurnal cycle of nitrous acid measured in June/July 2004 at Hohenpeissenberg is shown in Fig. 2.93. The noontime levels reached on average 0.110 ppb whereas night time values were significantly lower (~0.040 ppb). In the 2005 campaign at Hohenpeissenberg, we simultaneously measured at the summit and foothill (Fig. 2.94). The foothill site, a patch of grass, was often covered by dew until noon, an ideal condition for interfacial HNO$_2$ formation. Daily mean HNO$_2$ levels are higher at foothill than at summit but only the summit shows a strong diurnal variation with a maximum around noon.

In spring 2000 as well as in summer 2006, field studies including HNO$_2$ measurements were performed near Leipzig, at the research station Melpitz. The site is surrounded by agriculturally used land, mainly grassland for hay harvest, and mixed forests. The nitrous acid mixing ratios varied between 0.02 and 0.35 ppb. The HNO$_2$ mixing ratios increased more or less every night reaching up to 1.4 ppb, followed by effective photodecomposition after sunrise until minimum values (0.03−0.2 ppb) were reached at about 1600 CET. A complex field experiment, (INTERREG) was performed with numerous field sites around Strasbourg (France) in late spring 2003. We found a very similar behaviour at Rossfeld (Fig. 2.95) as in the 2006 Melpitz experiment (not shown here). Night time HNO$_2$ mixing ratios of up to 1.3 ppb were observed – mainly caused by heterogeneous formation processes

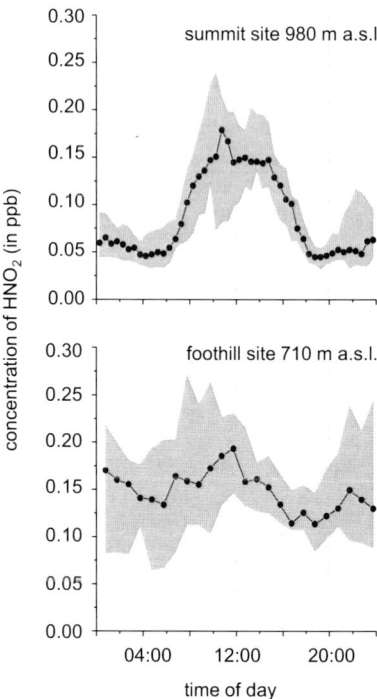

Fig. 2.94 Averaged diurnal variation of HNO_2 at the summit and the 270 m lower foothill site at Hohenpeissenerg in 2004.

and accumulation of HNO_2 in the night time boundary layer. From the increasing HNO_2 to NO_2 ratio, a conversion rate constant of 0.022 ± 0.017 h^{-1} on average was determined. Average HNO_2 to NO_2 ratios of $0.016-0.06$ were found for the single periods during the monotonic increase of nitrous acid in the evening hours. The experiment ESCOMPTE took place in the summer of 2001 in the urban environment of Marseille (France), Fig. 2.95. Highest HNO_2 mixing ratios (up to 1.2 ppb) were observed during episodes of strong pollution accumulation when sea breeze transported industrial, traffic and urban pollution land-inwards. In spring 2001, our group was involved in the European research program NITROCAT to study the influence of nitrous acid on the oxidation capacity of the atmosphere in Rome (Fig. 2.95). The atmosphere of Rome can be generally described as polluted: up to 90 ppb NO, 55 ppb NO_2, 75 mg m^{-3} TSP, but on average only 2 ppb SO_2, were detected. Heterogeneous formation on ground, urban, and aerosol surfaces seems to be a significant source of nitrous acid during the night; up to 2 ppb were found and a mean NO_2 to HNO_2 formation rate constant of 0.01 ± 0.003 h^{-1} was derived. On average, the HNO_2 to NO_2 ratio increased during the nights from 0.02 to 0.05. Table 2.82 summarizes the mean HNO_2/NO_2 ratios found in pour campaigns; obviously it does not depend on NO_2 concentration during daytime.

Other campaigns at different sites of the world also show similar HNO_2/NO_2 ratios (and variations) as listed in Table 2.82 (see citations in Elshorbany et al.

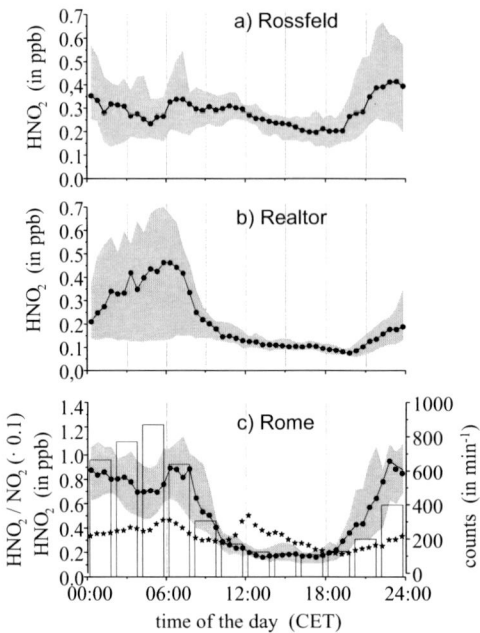

Fig. 2.95 Average diurnal variation of HNO_2 at different flat country urban and suburban sites (averaged over the measurement campaign, based on Central European Time (CET)), given as median and 25 and 75 quartiles): Rossfeld, near Strasbourg (22 May − 16 June 2003); Realtor, near Marseille (13 June − 13 July 2001); Rome, Villa Ada (19 May − 05 June 2001). Additional, for the Rome site the variation in natural radioactivity (columns) and in the HNO_2 to NO_2 ratio (stars) are shown.

Table 2.82 Mean HNO_2 characteristics at different European sites (altitude a.s.l in m).

site	ratio	HNO_2 (in ppt)[a]	
	HNO_2/NO_2	nocturnal	noon
Mt. Brocken, 1142 m	0.04±0.02	60	60
Melpitz, 0 m	0.09±0.05	150	50
Realtor (Marseille), 0 m	0.06±0.05	500	100
Rome, 0 m	0.03±0.01	1000	200
Goldlauter (Mt. Schmücke), 605 m	0.02±0.01	50	100
Rossfeld (Strasbourg), 0 m	0.08±0.08	500	100
Hohenpeissenberg, 980 m	0.03±0.03	20	100
direct emission	0.008±0.001	−	−

[a] Schneefernerhaus (2650 m): 5/15−25 (Kleffmann et al. (2002)
  Whitfeace Mt. (1483 m): 10/30−40 (Zhou et al. 2002)

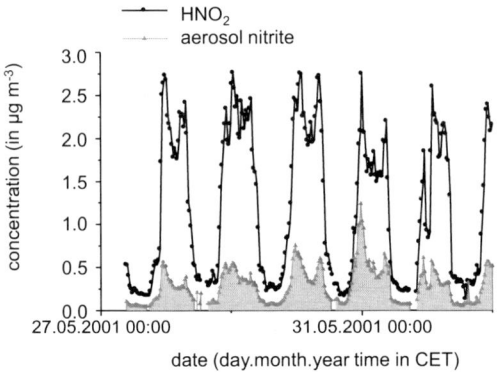

Fig. 2.96 Times series of HNO_2 and particulate nitrite (NO_2^-) in Rome (May 2001).

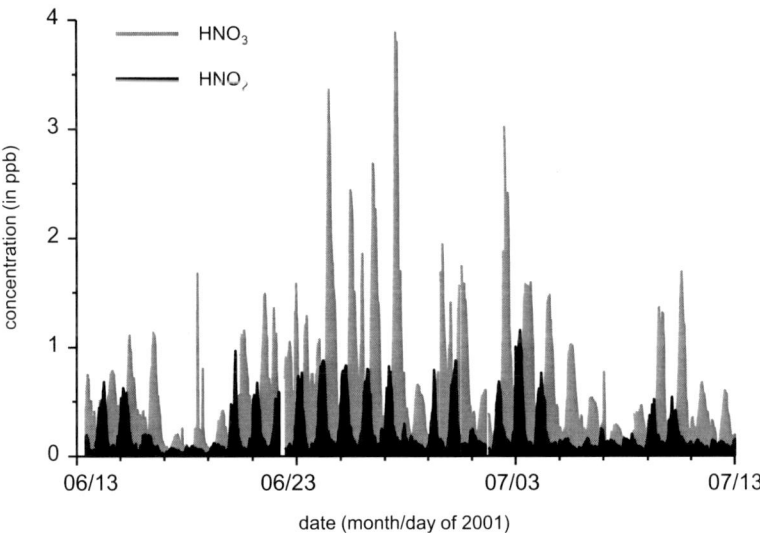

Fig. 2.97 Times series of HNO_2 and HNO_3 in Realtor near Strasbourg (June – July 2001).

2012); these authors also suggest by modelling that realistic HNO_2 levels should be accounted for by global models and that simulated HO_x, O_3 and secondary oxidation products should be revised accordingly. However, the assumption of a globally robust value of 0.02 for HNO_2 / NO_2 ratio seems to be wrong. Despite still hypothetical mechanisms of HNO_2 formation (see for details Chapter 5.4.4.2), we may assume that the formation is photocatalytic interfacial and that HNO_2 decomposition is photolytic, i.e. both depend on actual solar radiation but the formation also on surface characteristics (e.g. surface to volume ratio, catalytic activity) it follows under steady-state conditions

$$[HNO_2]/[NO_2] = k_{het}/j = const \cdot \alpha(surface).$$

The relative high variance of the HNO_2/NO_2 ratio at each site (Table 2.82) could show that variable parameters such as surface wetness and specific radiation parameters may influence formation and photolysis of HNO_2; Elshorbany et al. (2012) derived for all investigated measurement campaigns only 0.02 ± 0.002.

The relationship between gaseous HNO_2 and particulate NO_2^- shows Fig. 2.96 for two European sites. Remarkable that the diurnal slope for both compounds is synonymic, not surprising when assuming that the particulate phase is a source of HNO_2 (compare it below with the discussion on HNO_3-nitrate diurnal variation shown in Fig. 2.99). Interesting is the relationship between diurnal variation of HNO_2 and HNO_3 (Fig. 2.97). Regularly HNO_2 maxima are reached before approaching the HNO_3 maxima, suggesting that HNO_2 formation occurs earlier before noon when maximum photochemical HNO_3 formation happens – also from NO_2 produced from HNO_2 photolysis.

2.8.2.12 Sea salt degassing: HCl and the role of HNO_3

Hydrochloric acid (HCl) in the gas phase, chloride and sodium in the particle phase were measured for the first time with a high time-resolution and simultaneously with a number of other atmospheric components (in gas and particulate phase) as well as meteorological parameters during a campaign at the Melpitz research station (Germany, 51°32 N, 12°54 E; 87 m a.s.l.) to study the Cl partitioning (Möller and Acker 2007) in the summer of 2006 (12−30 June). On most of the 19 measurement days, the HCl concentration showed a broad maximum around noon/afternoon (on average 0.1 µg m^{-3}) and much lower concentrations during night (0.01 µg m^{-3}) with high correlation to HNO_3. The data supports that (a) HNO_3 is responsible for Cl depletion, (b) there is an increase in the Na/Cl ratio due to faster HCl removal during continental air mass transport, and (c) on average 50% of total Cl is as gas-phase HCl. The time series for HCl and HNO_3 are shown in Fig. 2.98, exhibiting a pronounced diurnal variation with usually low nocturnal

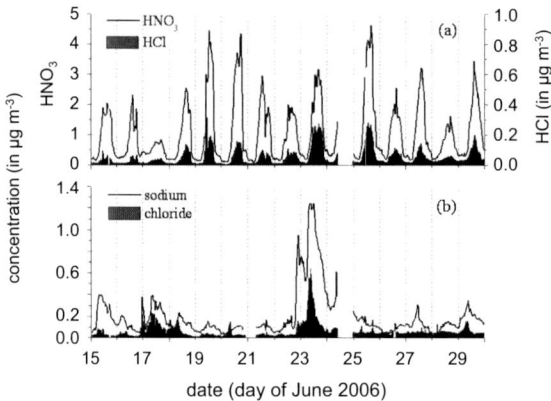

Fig. 2.98 Times series of HNO_3, HCl and particulate sodium (Na^+) and chloride (Cl^-) at Melpitz (June 2006).

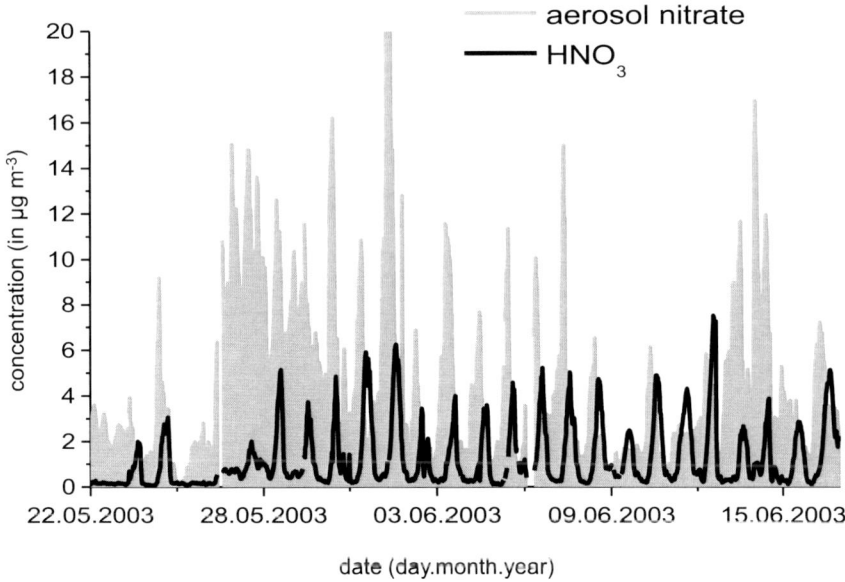

Fig. 2.99 Times series of HNO₃ and particular nitrate (NO₃⁻) in Realtor near Strasbourg (June – July 2001).

values and high values at daytime, reaching a maximum around noon. Remarkable, that even the fine structure in temporal variation is identical for HCl and HNO_3, resulting in high correlation:

$$[HCl] = 0.001 + 0.05[HNO_3]; \; r^2=0.79; \; n = 800.$$

Other gases (O_3, SO_2 and NH_3) also show a pronounced diurnal cycle with usually low nocturnal values and high values at daytime reaching a maximum around noon. Such a cycle is typical for many compounds having direct or indirect photo-chemical sources (we may exclude SO_2 and NH_3) and are produced in the free troposphere (what does not mean that the boundary layer (BL) is excluded being a chemical source). Hence, reasons for the diurnal cycle are (a) due to intensive mixing from the free troposphere into the BL during daytime, (b) deposition processes at night due to formation of a stable inversion layer, (c) higher temperature and lower humidity during daytime which enhances the evaporation of gases (HCl, HNO_3, NH_3) from PM phase and (d) daytime photochemical induced formation processes. Local sources of NH_3 and SO_2 can be excluded during the sampling period. As shown in other campaigns (Möller 2004), the above described vertical mixing processes contribute significantly to the daily amplitude. As for corresponding gas species, diurnal but less pronounced patterns were found for particulate NO_3^-, Cl^- and Na^+, showing mostly higher concentration (with exceptions) at daytime. High air temperatures and low relative humidity enhance evaporation of gases (HCl, HNO_3) from particulate matter phase. This is shown as an example for HNO_3 and particulate nitrate in Fig. 2.99; regularly gaseous HNO_3 approaches its

Table 2.83 Air composition at Mace Head (June 25−29, 2006; continuous denuder measurements with 30 min sampling/analysing); in µg m^{-3}.

	maritime air (n = 240)	continental air (n = 144)
sulfate	1.28	1.07
nss sulfate	0.40	0.84
nitrate	0.13	0.77
calcium	0.13	0.06
magnesium	0.42	0.10
potassium	0.07	0.01
sodium	3.62	0.94
chloride	5.52	0.90
chloride loss (in %)	16	54
HNO$_3$	0.02	0.11
HCl	0.12	0.08

daily maximum after the aerosol nitrate peak. Obviously, this expresses that nitrate formation occurs via photochemical HNO$_3$ formation, where simultaneous HNO$_3$ sticking on particulate matter occurs (as above discussed with HCl release) and later in the day (with temperature increase, see Chapter 4.3.4) with HNO$_3$ thermal degassing from nitrate PM.

The very good correlation

$$[Mg] = 0.13[Na], r^2 = 0.86,$$

with early identical bulk seawater ratio (0.12), suggests the only origin of sea-salt for both cations. On the other site, $[K] = 0.055[Na]$, $r^2 = 0.89$ also shows an excellent correlation but suggesting excess potassium, likely soil derived. On average, 83 % Cl depletion, calculated by $Cl_{depl} = 1 - R_{sea}/R_{sample}$, was observed in PM with only a small variation (Fig. 2.98) mainly given by diurnal cycle. The results demonstrate the important role of continental Cl degassing by an acid replacement process, very likely driven only by gaseous HNO$_3$. To a large extent (50−75%), sea salt in Cl already is depleted when entering the continent. During daytime (and consequently in the free troposphere), 2/3 of total chlorine is as HCl within the gas phase. The data supports that HCl deposition led to an increase of Na/Cl ratio. In summer 2008, we carried out a campaign at Mace Head to study the maritime HCl climatology (Table 2.83). HCl degassing is small (16%) in maritime air masses compared to continental air (54%); the data (HCl/HNO$_3$ ratio) suggest that replacement of chloride by sulfate via SO$_2$ oxidation (preferable on small particles) in maritime sea salt occurs whereas in continental particulate matter HNO$_3$ led to HCl degassing.

2.8.3 The carbon problem: Out of balance

From the analysis of the history of anthropogenic trace compounds in the atmosphere, its impacts on the climate system − so far we understand the processes of climate change − and recognizing the introduced abatement, few atmospheric environmental problems remain that are connected with the compounds N_2O, CH_4 and CO_2. The carbon dioxide problem is by far the most serious. There is no or only very little hope that CO_2 emissions will decrease in the next two or three decades. In contrast to N_2O and CH_4, the atmospheric residence time of "anthropogenic" CO_2 is orders of magnitudes larger (Chapter 2.8.3.3); in other words, even when there is a zero CO_2 emission world, the subsequent "greenhouse" effect will still last several hundreds of years. However, there are ways to solve the problem (Chapter 2.8.4). That part of N_2O and CH_4 linked with fossil fuels will "automatically" be solved together with the CO_2 problem. Hence, because the remaining sources of N_2O and CH_4 are linked with agriculture and food production, they are likely to become the dominant residual problem in 100 years, but with a far lower impact factor then today. By then there will be another residual air pollution problem, the formation of NO_x in all high-temperature processes. At present, the catalytic treatment of NO_x containing exhaust gases will abate these emissions only between 50 % and 70 %.

In the following three subsections, the "carbon problem" is characterized in terms of the budget, residence time and global equilibrium.

2.8.3.1 The carbon budget

Since the beginning of the Industrial Revolution, humans have emitted about $(365\pm30) \cdot 10^{15}$ g CO_2-C from the combustion of fossil fuels and cement production, and about $(180\pm80) \cdot 10^{15}$ g CO_2-C from land-use change, mainly deforestation (period 1750−2011, IPCC 2013). The atmospheric increase amounts $(240\pm10) \cdot 10^{15}$ g CO_2-C and the oceans take up $(155\pm30) \cdot 10^{15}$ g CO_2-C and the residual terrestrial uptake amounts to $(150\pm90) \cdot 10^{15}$ g CO_2-C. The world carbon stocks around 1990 are (Scurlock and Hall 1991) in 10^{15} g C:

    560 in plant biomass (80 % in trees),
    725 in ocean 75 m surface layer,
  1515 soil carbon content,
38000 deep ocean carbon, and
  5900 fossil fuel resources (92 % as coal).

In 1750, the atmospheric CO_2 level was 278 ± 5 ppm, which increased to 390.5 ± 0.1 ppm in 2011 (Ballentyne et al. 2012). Since 1950, these sources have amounted to about $400 \cdot 10^{15}$ g CO_2-C, i. e. 70 % of the total carbon release. Measurements and constructions of carbon balances, however, reveal that less than half of these emissions remain in the atmosphere (Prentice et al. 2001). The anthropogenic CO_2 that did not accumulate in the atmosphere must have been taken up by the ocean, by the land biosphere or by a combination of both.

It is easy to calculate the amount of CO_2 accumulated in the atmosphere. Between mass and atmospheric mixing ratio x the following relationship is valid. The changing air concentrations of trace compounds do not influence the value of the total mass of the atmosphere in any detectable way.

$$\text{mass of } CO_2 \;=\; x \text{ (in ppm) } 10^{-6}\frac{\text{molar mass of } CO_2}{\text{molar mass of air}} \text{ mass of the atmosphere} \qquad (2.114)$$

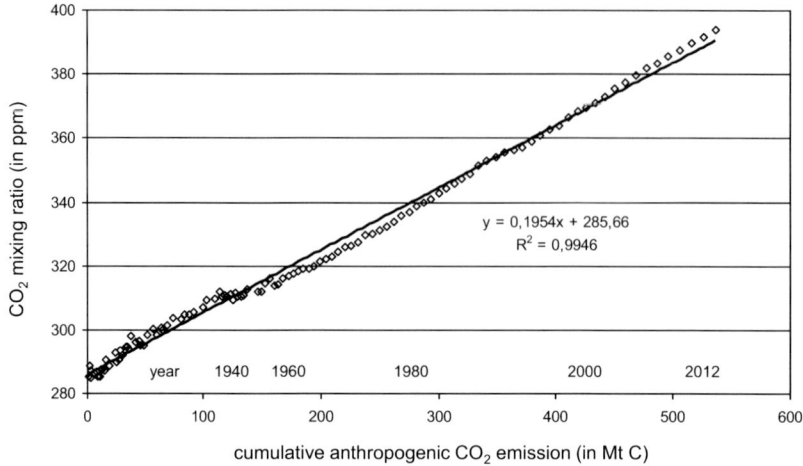

Fig. 2.100 Relationship between cumulative total anthropogenic CO_2 (fossil fuel, cement, and land-use) and atmospheric CO_2 mixing ratio (data source see Fig. 2.66).

The mass of human cumulated CO_2 to the atmosphere at a given year i is

$$m = (1 - \alpha) \sum_{i=1}^{t} Q_i \qquad (2.115)$$

Where Q emission for year $i=1$ until $i=t$ and α fraction $(0...1)$ of up taken anthropogenic CO_2. It follows for the atmospheric CO_2 mixing ratio (taking into account the values for the parameters in Eq. 2.114); Q in Pg CO_2-C (Gt):

$$x(t) = x(0) + \frac{1 - \alpha}{2.15} \sum_{i=1}^{t} Q_i \qquad (2.116)$$

This is equivalent to the regression shown in Fig. 2.100. The point of intersection for zero emission amounts to $x(0) = 286$ ppm in sense of a "pre-industrial level" (see Chapter 2.8.2.3). The slope $(1 - \alpha) \approx 0.20$ represents the atmospheric CO_2 increment (in ppm) per Pg (10^{15} g) of total emitted CO_2. From this it follows that from Eq. (2.116) $\alpha = 0.57$, i.e. 57% of emitted anthropogenic CO_2 is removed by the biosphere; $(1 - \alpha)$ represents the airborne fraction (0.43). As seen from Fig. 2.100, the slope (which corresponds to the airborne fraction) varies over the whole period 1850–2010, giving $(1 - \alpha)$ between 40% and 60%. On average, 50% of the cumulative emissions have been absorbed by the environment assuming, of course, that the carbon source is entirely human (of which there is no doubt). This means that the emitted carbon dioxide is directly "partitioned among reservoirs".

The year 2012 CO_2 mixing ratio (394 ppm) corresponds to $847 \cdot 10^{15}$ g CO_2-C: taking into account the total mass of the atmosphere ($5.2 \cdot 10^{21}$ g; Table 2.2) it follows that[31]

---

[31] Expressed in other terms, consequently the atmospheric increase amounts to 2.150 Gt CO_2-C per ppm CO_2; Prater et al. (2012) use a "conversion factor" to be 2.120 Pg per ppm.

atmospheric CO_2-C mass $= 0.0394[\text{vol}-\%] \cdot 10^{-2} \cdot \dfrac{12}{29} \cdot 5.2 \cdot 10^{21} = 847 \cdot 10^{15}$g,

where 12 is the molar mass of carbon and 29 the molar mass of air. Hence, the "reference" level of 286 ppb (related to about 1850) is equivalent to $615 \cdot 10^{15}$ g CO_2-C; the total added CO_2 mass since 1850 is, therefore, $232 \cdot 10^{15}$ g CO_2-C, less than half (43%) of the total emitted carbon ($536 \cdot 10^{15}$ g). In other words, the airborne fraction amounts as mean over the period 1850–2011 to about 0.57. Without ocean and land uptake, the atmospheric CO_2 concentration would have increased to 540 ppm.

Bolin et al. (1982) estimated a "pre-industrial" atmospheric CO_2 content of 614 Pg (290 ppm) assuming a constant airborne fraction of 0.54. The airborne fraction will increase if emissions are too fast for the uptake of CO_2 by the carbon sinks (Raupach 2013). It is thus controlled by changes in emissions rates, and by changes in carbon sinks driven by rising CO_2, changes in climate and all other biogeochemical changes. However, the percentage of CO_2 injected into the atmosphere from human activities that remains in the atmosphere has remained pretty much constant for the last 50 years (Ballantyne et al. 2012). Knorr (2009) extended his analysis back 150 years, and concluded that the airborne fraction of carbon dioxide had remained constant over that longer period as well. Thus, identifying the mechanisms and locations responsible for increasing global carbon uptake remains a critical challenge in constraining the modern global carbon budget and predicting future carbon-climate interactions.

From Eq. (2.115), we can simply calculate the future CO_2 concentration assuming different CO_2 based on the slope[32] given by Fig. 2.100 and assuming (which is not self-evident) that the airborne fraction (~ 50%) also remains constant in future. In Fig. 2.101, two different emission scenarios are presented:

a) A further but slowed down increase of CO_2 emission (3% until 2020, 2% until 2030 and a further 1% growth); land use change and biomass burning 3 Gt CO_2-C yr^{-1} constant until 2040 and further 2 Gt CO_2-C yr^{-1}, no carbon capture; and
b) The continuous increase of CO_2 emission, but with a faster slowing down (3% until 2010, then 2% until 2015 and 1% until 2020), from 2020 carbon capture (1% yr^{-1} increase until 2015 and then 5% yr^{-1} until 2050).

Scenario b) seems to be very optimistic – it results in a constant CO_2 mixing ratio of 465 ppm after 2050. It is more likely that carbon capture and sequestration/storage (CCS) technology (Chapter 2.8.4) will only become important after 2030 and will capture a maximum of 50% of the fossil fuel-released CO_2. It is also unlikely that the yearly consumption of fossil fuels will be further reduced before 2050. This is due to the increasing alternative energy source percentage of the

---

[32] It is remarkable that the linear fit ($r^2 = 0.999$) only begins in 1850. The period before (Fig. 2.66) is characterised by several distinguished positive and negative CO_2 trends. Between 1600 and 1800, there is a CO_2 minimum plateau (~ 270 ppm). This period is called the little ice age (LIA). It is generally agreed that there were three minima, beginning around 1650, 1770 and 1850, each separated by intervals of slight warming. Beginning around 1850, the climate began warming and the LIA ended. We may assume that natural CO_2 exchange was dominant and that the anthropogenic signal masked before 1850.

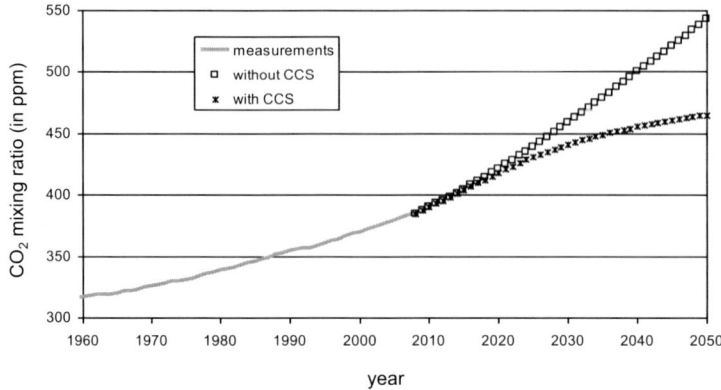

Fig. 2.101 Two scenarios of future atmospheric CO_2 development (see text for assumptions).

total energy consumption. Hence, in 2050, a value of around 500 ppm CO_2 seems more likely.

The most important reservoir is the backmixed surface layer of the ocean. The anthropogenic emissions are added to the atmosphere, continuously increasing the equilibrium carbon content of the surface layer as a result of the increasing partial pressure of the carbon dioxide in the gas phase (see next section, Chapter 2.8.3.2). The further transport of anthropogenic carbon from the surface to the ocean bulk (deep sea) is believed to be extremely slow (thousands of years) and hence the limiting step. A first quantification of the oceanic sink for anthropogenic CO_2 is based on a huge amount of measured data from two international ocean research programs; the cumulative oceanic anthropogenic CO_2 sink for 1994 was estimated to be $(118\pm19) \cdot 10^{15}$ g CO_2-C (Sabine et al. 2004). The cumulative uptake for the 1750 to 2011 period is $\sim (155\pm30) \cdot 10^{15}$ g CO_2-C from data-based studies (IPCC 2013), about 30 % of the total cumulative anthropogenic CO_2 emission.

The total oceanic dissolved carbonate carbon (Table 2.9) corresponds to 0.028 g L^{-1} as carbon in seawater, taking into account the volume of the world's oceans (Table 2.3). The experimentally estimated seawater standard carbonate carbon is 0.0244 g L^{-1} seawater (Dickson et al. 2007). In the first 200 m of the ocean, the total deposited anthropogenic CO_2 (Table 2.84 and assuming that 30 % is within this layer) only contributes to 3 % of dissolved inorganic carbon (DIC). Hence, it is very difficult to measure trends in the DIC because of man-made changes (see Fig. 2.105).

The kinetic processes of exchange and transport for the system atmosphere − the mixed surface layer of the ocean − are in the area of a few years whereas the exchange time with deep sea is between 500 and 2000 years (Wagener 1979). Variations in surface concentration are related to the length of time that the waters have been exposed to the atmosphere and the buffer capacity for seawater, as expressed by the *Revelle* factor. This factor represents the ratio of the instantaneous fractional change in the partial pressure of CO_2 ($\Delta p/p_0$) exerted by seawater to the fractional change in total CO_2 dissolved in the ocean waters ($\Delta\Sigma CO_2/(CO_2)_0$), where the subscript 0 denotes the reference status and ΣCO_2 denotes all forms of DIC (Chap-

Table 2.84 Global carbon budgets.

process	In 10^{15} g C yr^{-1} (Denman et al. 2007)		In 10^{15} g C (Sabine et al. 2004)		total
	1990s	2000–2005	1800–1994	1980–1999	
industrial CO_2 emission[a]	6.4 ± 0.4	7.2 ± 0.3	244 ± 20	117 ± 5	329[d]
land-use change	1.6[b]	1.6[b]	100–180	24 ± 12	156[e]
sum	8.0	8.9	344–424	141	485
atmospheric increase	3.2 ± 0.1	4.1 ± 0.1	165 ± 4	65 ± 1	225[f]
difference (biospheric uptake)	4.8	4.8	179–259	76	260
ocean uptake	2.2 ± 0.4	2.2 ± 0.5	118 ± 19	37 ± 8	165[g]
terrestrial uptake	2.6[c]	2.6[c, e]	61–141	39 ± 18	95

[a] from fossil fuel use and cement production
[b] range 0.5–2.7
[c] range 0.9–4.3
[d] 1751–2005 (Boden et al. 2009)
[e] 1850–2005 (Houghton 2008)
[f] calculated from the difference 384 to 280 ppm CO_2
[g] to be assumed 50% of cumulative industrial CO_2 emission
[e] forest sequestration for the period 1993–2003 has been estimated to be $0.3 \cdot 10^{15}$ g C yr^{-1} (IPCC 2007)

ter 2.8.3.2). Changes between ocean sediment and deep water play no role on a time scale of a few hundred years of anthropogenic perturbation of the surface layer CO_2. But over thousands of years nearly all anthropogenic CO_2 will be captured by the ocean.

The impact of global climate change on future carbon stocks is particularly complex. These changes might result in both positive and negative feedbacks on carbon stocks. For example, increases in atmospheric CO_2 are known to stimulate plant yields, either directly or via enhanced water use efficiency, and thereby enhance the amount of carbon added to soil. Higher CO_2 concentrations can also suppress the decomposition of stored carbon because C/N ratios in residues might increase and because more carbon might be allocated below ground. Predicting the long-term influence of elevated CO_2 concentrations on the carbon stocks of forest ecosystems remains a research challenge (Prentice et al. 2001). The severity of damaging human-induced climate change depends not only on the magnitude of the change but also on the potential for irreversibility. Solomon et al. (2009) show that climate change that takes place because of increases in carbon dioxide concentration is largely irreversible for 1000 years after emissions stop. The question of how large the residence time of anthropogenic CO_2 in the atmosphere is will be discussed in Chapter 2.8.3.3. There are strong arguments that the anthropogenic-caused CO_2 increase is largely irreversible; hence, stopping emissions will not solve (though might smooth) climate change problems. As a consequence, CO_2 capture from the atmosphere remains the challenge for climate sustainability (Chapter 2.8.4). The oceans have certainly been identified as the final sink of anthropogenic CO_2 but after thousands of years; moreover, the seawater uptake capacity will decrease and oceanic acidification will result in serious ecological consequences (see next section, Chapter 2.8.3.2).

2.8.3.2 The CO_2-carbonate system

In water, the following chemical carbon-IV species exist in equilibrium: carbon dioxide (CO_2), carbonic acid (H_2CO_3), bicarbonate (HCO_3^-) and carbonate (CO_3^{2-}). Additionally, the phase equilibriums with gaseous CO_2 and a possible solid body such as $CaCO_3$ and $MgCO_3$ have to be considered. Free carbonic acid is not isolated but the structure $O{=}C(OH)_2$ in aqueous solution has been confirmed. Often the expression $CO_2 \cdot H_2O$ is also used for carbonic acid. The sum of the dissolved carbonate species is denoted as total DIC and is equivalent with other terms used in literature:

$$DIC \equiv \Delta\Sigma\, CO_2 \equiv TCO_2 \equiv C_T = [CO_2] + [H_2CO_3] + [HCO_3^-] + [CO_3^{2-}]$$

The carbon dioxide (physically) dissolved in water − we denote it as $CO_2(aq)$ − is in equilibrium with gaseous atmospheric carbon dioxide $CO_2(g)$. There is no way to separate non-ionic dissolved $CO_2(aq)$ and H_2CO_3; therefore, it is often lumped into $CO_2^*(aq)$. Analytically, DIC can be measured by acidifying the water sample, extracting the CO_2 gas produced and measuring. The marine carbonate system represents the largest carbon pool in the atmosphere, biosphere and ocean, meaning it is of primary importance for the partition of atmospheric excess carbon dioxide produced by human activity.

The ocean is saturated with $CaCO_3$, which represents the largest carbon reservoir in sediments in forms of calcite and aragonite (Zeebe and Gattuso 2006). The higher carbonate (in terms of DIC) solubility is because of dissolved CO_2, which converts carbonate (CO_3^{2-}) into higher soluble bicarbonate (HCO_3^-). It follows that the capacity of the ocean for CO_2 uptake is still very large − the system is far away from saturation in DIC (or total carbonate, respectively) but rather in equilibrium. With increasing atmospheric CO_2 the seawater CO_2/carbonate concentration increases, and vice versa, i. e. in the case of decreasing atmospheric CO_2 concentrations the ocean will degas CO_2, thereby leading to a new equilibrium. That means, even in the case of carbon capture from the atmosphere (Chapter 2.8.4.2), the reduction of atmospheric CO_2 is not equivalent to (or less than) the captured CO_2. The history of anthropogenic CO_2 is the story of the coupled ocean-atmosphere reservoir.

Seawater is slightly alkaline (pH $\approx$ 8.2) because of the equilibrium between $CaCO_3$ bottom and dissolved carbonate; subscripts s denote solid phase, aq dissolved phase and g gaseous phase (note that ions are generally dissolved chemical species):

$$CaCO_3(s) \xrightleftharpoons{H_2O} Ca^{2+} + CO_3^{2-} \quad \text{solubility product } L \text{ (or } K_s) \quad (2.117)$$

$$CO_3^{2-} + H_2O \rightleftharpoons HCO_3^- + OH^- \quad \text{hydrolysis constant 1} \quad (2.118)$$

Carbonate acts as a base (Chapter 4.2.5.2). The solubility of $CaCO_3$ at 20 °C in water is only about 0.007 g L^{-1} as carbon. $CaCO_3$ water solubility (in mg L^{-1})

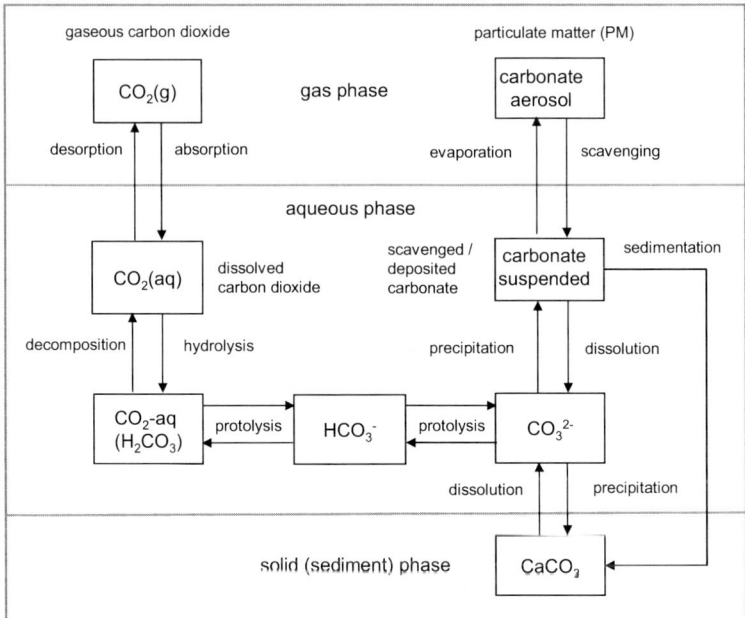

Fig. 2.102 Scheme of the multiphase CO_2-carbonate system.

decreases linearly with increasing temperatures $- [CaCO_3] = 80.3 - T (r^2 = 0.997)$, T in °C $-$ and increases slightly with increasing CO_2 partial pressure $- [CaCO_3] = 56.5 + 0.0219 \cdot [CO_2 (g)]$ $(r^2 = 0.986; 20\,°C)$, valid in the range $20-1000$ ppm CO_2 (data from D'Ans and Lax 1943).

Processes (2.117) and (2.118) describe the solid-liquid equilibrium at the oceanic bottom (sediment-seawater interface) with suspended matter in seawater (calcareous organisms). Moreover, similarly the processes of heterogeneous nucleation and droplet formation from CCN as well as in general the solid-water interfacial process are described (Fig. 2.102). Hence, seawater is an excellent solvent for acidic gases such as SO_2 (used for flue-gas desulfurization at some coastal site power stations) and atmospheric CO_2. The ratio between atmospheric and oceanic (water dissolved) CO_2 is described by the *Henry* equilibrium (2.119; see also Chapter 4.3.2):

$$CO_2 (g) \rightleftharpoons CO_2 (aq) \qquad \text{true } \textit{Henry} \text{ constant } H_{CO_2} \qquad (2.119)$$

This "physical" equilibrium depends only on the temperature. As mentioned above, it is not possible to measure $CO_2 (aq)$ but only the sum $CO_2 (aq) + H_2CO_3$ hence, all listed equilibrium constants are related to an *apparent Henry* constant:

$$H_{ap} = \frac{[CO_2 (g)]}{[CO_2 (aq)] + [H_2CO_3]} \approx \frac{[CO_2 (g)]}{[CO_2 (aq)]} = H \qquad (2.120)$$

As seen later, taking into account the hydrolysis constant of H_2CO_3, the percentage of carbonic acid is very small ($\sim 0.3\,\%$), so that $H_{ap} \approx H$ is valid within the measurement errors. Fig. 2.103 shows the temperature dependency of the *Bunsen*'s absorption coefficient α (data from D'Ans and Lax 1943). This coefficient is directly related to the *Henry* constant via $\alpha = H \cdot RT$ (R gas constant), where H has the dimension mol L^{-1} atm^{-1}. An empirical fit results in ($r^2 = 0.999$), but is slightly different for two ranges of temperature:

$$\alpha = 0.0008 \cdot T^2 - 0.0585 \cdot T + 1.6992 \qquad (0-25\,°C)$$

$$\alpha = 0.0002 \cdot T^2 - 0.0256 \cdot T + 1.283 \qquad (25-60\,°C)$$

Carrol and Mather (1992) have shown that between 0 °C and 100 °C the dependency of H from T is nearly linear. From the empirical data the following can be derived:

$$H = \frac{1}{11.39 + 0.81 \cdot T}, \; T \text{ in } °C$$

A "true" equilibrium is given through subsequent chemical reactions, leading to the higher solubility of CO_2 in water. Dissolved carbon dioxide forms bicarbonate via different steps:

$$CO_2(aq) + H_2O \rightleftharpoons HCO_3^- + H^+ \qquad \text{apparent first dissociation constant } K_{ap} \qquad (2.121)$$

$$CO_2(aq) + H_2O \underset{k_{-2.121}}{\overset{k_{2.121}}{\rightleftharpoons}} H_2CO_3 \qquad \text{hydration constant } K_h \qquad (2.122)$$

$$H_2CO_3 \rightleftharpoons H^+ + HCO_3^- \qquad \text{true first dissociation constant } K_1 \qquad (2.123)$$

$$HCO_3^- \rightleftharpoons H^+ + CO_3^{2-} \qquad \text{second dissociation constant } K_2 \qquad (2.124)$$

CO_2 hydration (2.122) is relatively slow and in comparison with total dissolved CO_2 the concentration of H_2CO_3 is very low (negligible). Reaction (2.122) occurs for pH < 8; for pH > 10 reaction (2.125) is dominant, in the pH region 8–10 both reactions are parallel and it is complicated to study the kinetic. Hence, reliable kinetic data are valid only for pH < 8 and pH > 10, respectively.

$$CO_2(aq) + OH^- \underset{k_{-2.122}}{\overset{k_{2.122}}{\rightleftharpoons}} HCO_3^- \qquad \text{hydrolysis constant 2} \qquad (2.125)$$

However, reaction (2.125) plays basically no role in natural waters with the exception of the initial state of cloud/fog droplet formation from alkaline CCN (e. g. flue ash and soil dust particles). The reaction rate constants (minus prefix means the inverse reaction) have been "best" estimated to be $k_{2.122} = 0.03$ s^{-1} (25 °C) and $k_{-2.122} = 20$ s^{-1} (25 °C) as pseudo-first order rates and $k_{2.125} = 8400$ L mol^{-1} s^{-1} and $k_{-2.125} = 2 \cdot 10^{-4}$ s^{-1}. All K and k data given here are taken from Gmelin (1973); these data are also adapted by Sigg and Stumm 1996, Stumm and Morgan 1996).

The equilibrium constant of the apparent first dissociation (2.121) is given by:

$$K_{ap} = \frac{[HCO_3^-][H^+]}{[CO_2(aq)]} = K_1 K_h \qquad (2.126)$$

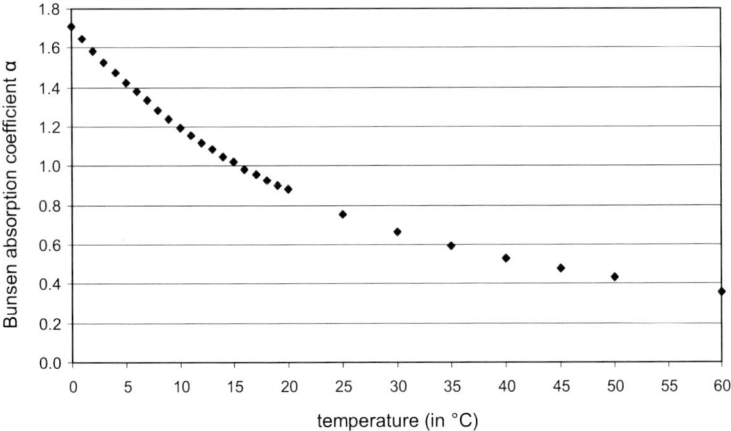

Fig. 2.103 Solubility of CO_2 in water (at 1 atm CO_2) in terms of the dimensionless *Bunsen* absorption coefficient α in dependence from temperature (in volume dissolved CO_2 / volume of solution); data from Gmelin (1943).

Table 2.85 Equilibrium constants in the aqueous CO_2-carbonate system (data from Gmelin 1973 if not other noted).

T (in °C)	0	5	10	15	20	25
H (in 10^{-2} atm^{-1} mol L^{-1})	7.70	–	5.36	–	3.93	3.45
K_{ap} (in 10^{-7} mol L^{-1})[a]	2.64	3.04	3.44	3.81	4.16	4.45
K_1 (in 10^{-4} mol L^{-1})	–	1.56	–	1.76	1.75	1.72
$K_h \cdot 10^3$ ([b])	–	1.96	–	2.16	2.52	2.59
K_2 (in 10^{-11} mol L^{-1})	2.36	2.77	3.24	3.71	4.20	4.29
$K_{ap} = K_s \cdot K_h$ (in 10^{-7} mol L^{-1})	–	3.06	–	3.80	4.41	4.45

[a] Lide (2005)
[b] $K_h = [H_2CO_3]/[CO_2(aq)]$

K_{ap} can be relatively easily estimated from equilibrium concentration measurements and the hydration constant K_h is calculated according to Eq. (2.125); Table 2.85.

The adjustment of equilibriums (2.123) and (2.124) is fast; the direct estimation of the dissociation constants K_1 and K_2 is not possible from concentration measurements (only indirectly through potentiometric and/or conductometric measurements). The true first dissociation constant K_1 (p$K = 3.8$) is three orders of magnitude larger than the apparent dissociation constant K_{ap}. Hence, carbonic acid is 10 times stronger than acetic acid but acetic acid can degas CO_2 from carbonic solutions because H_2CO_3 is decomposed to about 99 % into CO_2 (which escapes from the water) and H_2O as it follows from K_h. This makes the aqueous carbonic system unique: carbonic acid exists as well (but in very low concentrations) as H_2CO_3 (in kinetic-inhibited equilibriums) and largely as $CO_2 \cdot H_2O$ where CO_2 degassing is also inhibited. The second dissociation constant characterizes bicarbonate as a very weak acid ($pK_2 = 10.4$). The aqueous-phase concentrations of different C(IV) species can be calculated from the equilibrium expressions:

$$[CO_2\,(aq)] = H \cdot [CO_2\,(g)] \tag{2.127}$$

$$[H_2CO_3] = H \cdot K_h[CO_2\,(g)] \tag{2.128}$$

$$[HCO_3^-] = H \cdot K_1K_h[CO_2\,(g)]\,[H^+]^{-1} \tag{2.129}$$

$$[CO_3^{2-}] = H \cdot K_1K_2K_h[CO_2\,(g)]\,[H^+]^{-2} \tag{2.130}$$

At a typical surface seawater pH of 8.2, the speciation between $[CO_2]$, $[HCO_3^-]$ and $[CO_3^{2-}]$ is 0.5%, 89% and 10.5%, showing that most of the dissolved CO_2 is in the form of HCO_3^- and not CO_2. For the description of the overall gas-aqueous equilibrium the so-called effective *Henry* constant is used (Chapter 4.3.2):

$$H_{eff} = \frac{[CO_2\,(g)]}{DIC} = \frac{[CO_2\,(g)]}{[CO_2\,(aq)] + [H_2CO_3] + [HCO_3^-] + [CO_3^{2-}]} \tag{2.131}$$

Using the expression (2.127) to (2.130), it follows for the total DIC:

$$DIC = H[CO_2\,(g)]\,(1 + K_h(1 + K_1[H^+]^{-1} + K_1K_2[H^+]^{-2})) \tag{2.132a}$$

or

$$DIC \approx H[CO_2\,(g)]\,(1 + K_{ap}[H^+]^{-1} + K_{ap}K_2[H^+]^{-2}) \tag{2.132b}$$

The true *Henry* constant strongly depends on temperature. From data listed in *Gmelin* (including more up to 50 °C, not listed in Table 2.85) it can be derived ($r^2 = 0.99$):

$$H = 0.061 \cdot \exp(-0.023 \cdot T) \quad T \text{ in } °C \tag{2.133}$$

Expressions for K_1 and K_2 independent of temperature T (in K) and salinity S (in ‰) are given by Lueker et al. (2000); $k^0 = 1$ mol kg^{-1} seawater:

$$\log\,(K_1^*/k^0) = \frac{-3633.86}{T} + 61.2172 - 9.6777 \cdot \ln T$$
$$+ 0.011555 \cdot S - 0.0001152 \cdot S^2$$

$$\log\,(K_2/k^0) = \frac{-471.78}{T} + 25.929 + 3.16967 \cdot \ln T$$
$$+ 0.01781 \cdot S - 0.0001122 \cdot S^2$$

In Lueker et al. (2000) and Dickson et al. (2007), the equilibrium constant K_1^* is expressed as:

$$K_1^* = \frac{[H^+]\,[HCO_3^-]}{[CO_2\,(aq)] + [H_2CO_3]}$$

which corresponds to:

$$\frac{1}{K_1^*} = \frac{[CO_2\,(aq)]}{[H^+][HCO_3^-]} + \frac{[H_2CO_3]}{[H^+][HCO_3^-]} = \frac{1}{K_{ap}} + \frac{1}{K_1} \approx \frac{1}{K_{ap}}$$

Hence, care is taken in adopting constants from different studies to avoid confusing similar symbols that are based on different equilibriums. Moreover, using the same name for different definitions presents another source of confusion.

It follows that pK_1^* ($\approx pK_{ap}$) $= 5.8472$ and $pK_2 = 8.966$ ($pK = -\log(K/k^0)$ whereby $S = 35‰$ and $25\,°C$. From the values listed in Table 2.83 it would follow that pK_{ap} ($\approx pK_1^*$) $= 6.3$ and $pK_2 = 11.63$, considerably different from the values for seawater. This certainly explains the different calculated concentrations based on given atmospheric CO_2 partial pressures (see below). At $25\,°C$ and water having pH $= 8.1$, the equilibrium concentration of total DIC can be calculated from Eq. (2.129) as (in mol L^{-1} whereas $CO_2(g)$ in atm):

$$DIC_{seawater} = 7.0 \cdot [CO_2\,(g)] \tag{2.132c}$$

using K values given for seawater, respectively. It follows with 380 ppm atmospheric CO_2 for DIC in water 11 mg C L^{-1} and in seawater 26.6 mg C L^{-1}. The measured DIC in surface seawater (Dickson et al. 2007) amounts to 27.6 mg L^{-1} as carbon or 0.0023 mol L^{-1} and is in good agreement with the above estimate. Assuming that the relationship is valid, it can be tuned by taking the measured values (28 mg L^{-1} carbon in surface seawater and 384 ppm atmospheric CO_2) to get:

$$DIC_{seawater} = 6.1 \cdot [CO_2\,(g)] \tag{2.132d}$$

or generally

$$DIC_{seawater} = a\,(T,\,pH) \cdot [CO_2\,(g)] \tag{2.132e}$$

It is important to note that the "factor" a depends on the seawater pH and (to a lesser extent) T; in (2.132d) $a = 6.1$ is valid only for pH $= 8.1$ but for pH $= 7.9$ and pH $= 8.2$ the factor ranges twofold, namely 3.8 and 7.7, respectively. This means that with decreasing seawater pH the uptake capacity for atmospheric CO_2 decreases significantly. Increasing sea surface temperature (SST) also leads to decreasing DIC; from Eq. (2.132b), taking into account the temperature dependency of H and K_{ap}, it follows in the range $0-30\,°C$ a linear fit ($r^2 = 0.99$) with a slope of -0.13 mg L^{-1} carbon. However, the relationship between atmospheric CO_2 is more complicated because of the buffer capacity of seawater (besides carbonate in a more exact treatment all buffering chemical species – for example borate – have to be considered). The buffer capacity of carbonized water (here seawater) is given to complete the acid-based reaction:

$$CO_2(aq) + CO_3^{2-} + H_2O \rightleftharpoons 2\,HCO_3^- \qquad \text{primarily buffering} \tag{2.134}$$

Anthropogenic CO_2 dissolves in seawater, produces hydrogen ions and neutralizes carbonate ions. Hence, H^+ concentration (and pH) will not change in small ranges depending on the CO_2 partial pressure increase and the available carbonate in seawater. But when seawater pH declines as a result of rising CO_2 concentrations, the concentration of CO_3^{2-} will also fall (see reaction 2.134) reducing the calcium carbonate saturation state. Marine carbonates also react with dissolved CO_2 through the reaction:

$$CO_2(aq) + CaCO_3 + H_2O \rightleftharpoons 2\,HCO_3^- + Ca^{2+} \qquad \text{secondary buffering} \qquad (2.135)$$

This reaction depends on the calcium carbonate saturation state Ω; the latter is expressed by:

$$\Omega = \frac{[Ca^{2+}][CO_3^{2-}]}{L_{CaCO3}} \qquad (2.136)$$

where L_{CaCO3} is the solubility product. Unless sufficient carbonate is present, $CaCO_3$ will dissolve back into the surrounding seawater. Because the ocean is in contact with carbonate sediments, both on shelves and in the deep sea, the ocean as a first approximation is roughly saturated with respect to calcite. If we assume that the concentration of Ca^{2+} has remained nearly constant, this is equivalent to a roughly constant carbonate-ion concentration. In regions where $\Omega > 1.0$, the formation of shells and skeletons is favored. Below a value of 1.0, the water is corrosive and the dissolution of pure aragonite and unprotected aragonite shells will begin to occur. Equation (2.136) is a bit curious because from reaction (2.118) it follows that:

$$K_s = \frac{[Ca^{2+}][CO_3^{2-}]}{[CaCO_3]} \quad \text{and} \quad K_s[CaCO_3] = [Ca^{2+}][CO_3^{2-}] = L_{CaCO_3}, \qquad (2.137)$$

where the $CaCO_3$ activity of the solid body is set by convention to be one. Therefore, $K_s \equiv L$ and hence, $\Omega = 1$. Supersaturation and undersaturation remain, therefore, in small limits. However, water bodies with insufficient solid $CaCO_3$ are generally undersaturated.

Oceanic uptake of anthropogenic CO_2 leads to a shift between carbonate and bicarbonate: in other words, to a larger solubility of mineral $CaCO_3$. It has been argued that, as a consequence of higher CO_2 partial pressures, the surface layers of the ocean will become undersaturated with calcium carbonate, according to equation (2.135), with possible catastrophic biological consequences for a variety of marine organisms (for example, coral reefs, shells and skeletons of other marine-calcifying species). However, from general aspects, it seems unlikely that $CaCO_3$ supersaturation is a precondition for carbonate biomineralisation for the following three reasons: a) life processes are far out of equilibrium; b) biomineralisation depends on the active transport of ions through biomembranes driven by metabolic energy and thereby do not depend on the free energy of any carbonate reaction; and c) calcium carbonate structures of living organisms can be covered by organic tissue, representing a barrier against fast exchange processes (Wagener 1979). The ecological consequences of a possible change in the calcium carbonate supersaturation, with respect to calcareous organisms, can only be examined experimentally.

Furthermore, CO_2 uptake leads to so-called oceanic acidification. With a very good approximation (i. e. neglecting the carbonate concentration compared with that of bicarbonate), equation (2.132) can be reduced to:

$$[CO_2(g)] = \frac{[H^+][DIC]}{HK_{ap}} \qquad (2.138)$$

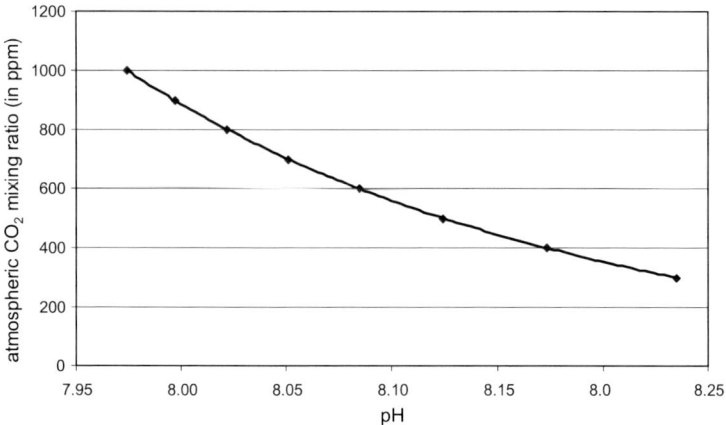

Fig. 2.104 Relationship between atmospheric CO_2 mixing ratio and seawater pH assuming a pH of 8.25 at a CO_2 mixing ratio of 280 ppm and constant seawater carbonate concentration.

and further transformed into:

$$pH = \log DIC - \log HK_{ap} - \log [CO_2 (g)] \qquad (2.139)$$

Unfortunately, DIC is a function of pH and, therefore, no simple relationship between atmospheric CO_2 concentration and seawater pH can be found. Pearson and Palmer (1999) held DIC constant to estimate the relationship between changing atmospheric CO_2 and pH, assuming a preindustrial pH of 8.25 at the preindustrial CO_2 value of 280 ppm. However, there is no serious argument to hold DIC constant over time; by contrast Sabine et al. (2004) have shown that the oceanic DIC reflects the accumulated atmospheric CO_2. Caldeira and Berner (2000) held carbonate constant and used the formula (simply derived from Eq. 2.130):

$$[CO_2 (g)] = \frac{[H^+]^2 [CO_3^{2-}]}{HK_{ap} K_2} \qquad (2.140)$$

from which:

$$\begin{aligned} pH &= 0.5 \log [CO_3^{2-}] - 0.5 \log HK_{ap}K_2 - 0.5 \log [CO_2(g)] \\ &\equiv \text{const} - 0.5 \log [CO_2 (g)] \end{aligned} \qquad (2.141)$$

follows and const $= 6.474$, assuming a preindustrial pH of 8.25 at the preindustrial CO_2 value of 280 ppm; Fig. 2.104 shows this relationship.

It follows that by 2100, using the scenario without CCS technology (Fig. 2.92) for about 1000 ppm CO_2, the seawater pH would decrease to 7.96. From the reactions (2.134) and (2.135), it follows that increasing CO_2 would lead to increasing bicarbonate, whereas carbonate is transformed into bicarbonate and the carbonate concentration is adjusted through bicarbonate dissociation (2.124) and $CaCO_3$ dissolution (2.117) but the relation is strong non-linear. From reaction (2.124) it follows that increasing $[H^+]$ leads to decreasing the carbonate ion concentration. Hence, it is more likely that oceanic acidification is faster than predicted from equation (2.141) and Fig. 2.94, which can be concluded from the measurements of

Table 2.86 Mean data sets from the deep-water Station ALOHA (A Long-Term Oligo-trophic Habitat Assessment; 22° 45' N, 158° 00' W) located 100 km north of Oahu, Hawaii. Data from: Dore, J. E. (2009) Hawaii Ocean Time-series surface CO_2 system data product, 1988−2008, http://hahana,soest,hawaii,edu/hot/products/products,html, SOEST, University of Hawaii, Honolulu, HI.

parameter	1988−2008	1992−1998	2003−2008
pH (measured)	−[h]	8.10 ± 0.01	8.09 ± 0.01
T, in ° C	24.9 ± 1.1	24.8 ± 1.3	24.9 ± 1.1
n	203	52	57
TA^a in µeq kg^{-1}	2306 ± 13	2300 ± 14	2308 ± 11
DIC^b in µmol kg^{-1}	1975 ± 15	1966 ± 12	1980 ± 11
$[CO_2(aq)]^c$, in µmol kg^{-1}	9.8 ± 0.4	9.6 ± 0.2	10.0 ± 0.3
$[CO_3^{2-}]^d$, µmol kg^{-1}	234 ± 6	235 ± 5	231 ± 5
$[HCO_3^-]^e$, µmol kg^{-1}	1731	1721	1737
pCO_2, in ppm^f	345 ± 16	337 ± 13	356 ± 14
$[CO_2(g)]$, in ppm^g	366 ± 10	360 ± 3	380 ± 3

[a] total alkalinity, measured by open cell titration; $TA = [HCO_3^-] + 2[CO_3^{2-}]$ + others
[b] measured by coulometry
[c] free seawater CO_2 concentration
[d] seawater carbonate ion concentration
[e] calculated as difference between DIC and $CO_2(aq) + CO_3^{2-}$
[f] mean seawater partial pressure, calculated from DIC and TA
[g] mean atmospheric CO_2 mixing ratio (from Mauna Loa)
[h] measured pH values only from the selected periods available density of seawater (S = 35‰, 25° C): 1.023343 kg L^{-1} (Dickson et al. 2007)

surface water pH (Marsh 2009, Caldeira and Rau 2000). According to the Ocean Station Aloha, current pH (2008) is about 8.08 but from Eq. (2.141) it follows that pH = 8.18. Finally it makes no sense to adopt relationships such as (2.141) or (2.139) to calculate historical and future seawater pH. There is no scientific evidence to set the preindustrial seawater pH to 8.25. When using the reference 384 ppm CO_2 and pH = 8.08 as measured values for 2008, it follows from Eq. (2.141) pH = 8.15 as "preindustrial", i. e. related to 280 ppm CO_2. Exact measurements of pH in natural waters are fully uncertain to the second decimal place.

Dore et al. (2003, 2009) found that over two decades of observation (Station Aloha, see Table 2.86) the surface ocean grew more acidic at exactly the rate expected from the chemical equilibration with the atmosphere. However, that rate of change varied considerably in terms of seasonal and inter-annual time scales, and even reversed for a period of nearly five years (Fig. 2.105). The concentrating/diluting effect of salinity changes on DIC can be removed from the measured data through the normalization to a constant S ($= 35‰$ where nDIC $= 35$ DIC$/S)^{33}$. Whereas S has no seasonal pattern, DIC (and nDIC) have distinct annual cycles, largely driven by the input of DIC from below (via winter mixing) and biological drawdown of CO_2 (via photosynthesis).

---

[33] Fig. 2.105 is an interesting example of how to make misinterpretations when not considering collateral factors.

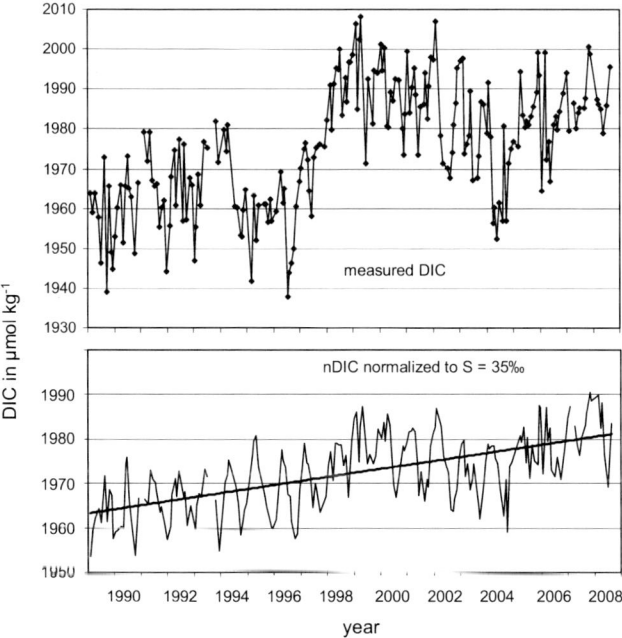

Fig. 2.105 Trend of dissolved inorganic carbon (DIC) at Station ALOHA (see Table 2.86 for details); data from: Dore, J,E, 2009, Hawaii Ocean Time-series surface CO_2 system data product, 1988−2008, SOEST, University of Hawaii, Honolulu, HI, http://hahana.soest. hawaii.edu/hot/products/products.html. (see also Dore et al. 2003, 2009).

Year-to-year changes seem to be driven by climate-induced changes in ocean mixing and attendant biological responses to mixing events. Evidence has been found for the upwelling of corrosive "acidified" water onto the continental shelf (Feely et al. 2008). It is not possible to fit the measured data in Table 2.84 with the equilibrium equations derived here, suggesting that a) either different equilibrium constants are valid in seawater compared with pure water and/or b) that the seawater DIC concentration is not fully described by equilibrium conditions. The latter is possible because of the fact of the mixing and varying biological activity in respect to assimilation and respiration.

Over the period 1990−2008, it is derived from Table 2.86 an increase of DIC by about 15 μmol C L^{-1} or (taking into account the atmospheric CO_2 increase from 354 to 386 ppm) about 6 μg C per ppm CO_2. In other terms, the oceanic DIC increased yearly by about 10 mg C m^{-3} seawater or about 4 Tg C within the 1-m-surface layer of the oceans, corresponding to only 0.3 Gt within the 75-m-layer. The yearly ocean uptake, however, is assumed to be about 2 Gt C yr^{-1} (cf. Tab. 2.82). Hence, the subtropical station ALOHA can not reflect the global seawater chemistry. On the other hand, using Eq. (2.132d), it would result in a DIC increase for the mentioned period, corresponding to 3.5 Gt within the 75-m-layer. This result is much closer to the estimates, but as discussed with Eq. (2.132e), changing T and pH have to be taken into account.

Moreover, this calculated inorganic carbon concentration does not incorporate the fact that carbon is continuously supplied into the atmosphere and oceans by degassing from metamorphism and magmatism and by the weathering of carbonate minerals and organic carbon, and is continuously consumed by the production of carbonate and organic carbon sediments (Fig. 2.9). Hence, the total DIC load of the ocean can be expected to vary over time. Finally, the uptake of atmospheric CO_2 by the sea surface is a dynamic process that depends on oceanic and wind circulation interlinked. Observations suggest that the Southern Ocean sink of CO_2 has weakened between 1981 and 2004 by $0.08 \cdot 10^{-15}$ g decade^{-1} relative to the trend expected from the large increase in atmospheric CO_2 and this trend is attributed to the observed increase in wind (Le Quére et al. 2007).

2.8.3.3 Atmospheric CO_2 residence time[34]

As mentioned already, the CO_2 cycle has one major problem in the atmosphere − there is no direct chemical sink. In nature, CO_2 can only be assimilated by plants (biological sink) through conversion into hydrocarbons (Chapter 2.2.2.3) and stored in calcareous organisms, partly buried in sediments but almost completely turned back into CO_2 by respiration; hence, CO_2 partitionates between the biosphere and atmosphere. The only definitively carbon sink is the transport of DIC to Deep Ocean − when the ocean-atmosphere system is not in equilibrium, i. e. in case of increasing atmospheric CO_2 levels (due to anthropogenic and volcanic activities). As mentioned in Chapters 2.2.2.5 and 2.6.4.3, the CO_2 source term by volcanic exhalations is very uncertain but is likely to be a value much less than $0.1 \cdot 10^{15}$ g yr^{-1} carbon. Hence, with respect to time periods being of interest for humankind (from decades to hundreds of years) this natural biogeochemical recycling can be regarded to be closed or, in other words, the net flux is zero. Consequently, all concentrations (pools) in the biosphere and atmosphere remain constant (we should not consider short-term variations because of seasonal and interannual fluctuations).

The only driving forces behind removing CO_2 from the atmosphere are dry deposition (including plant uptake) and wet deposition (CO_2 scavenging). As discussed, the marine and terrestrial earth's surface can be assumed to be carbonate saturated. This equilibrium is only disturbed by the yearly increase of the CO_2 level due to anthropogenic emissions. Hence, the physical (not the biological) surface resistance is zero as is the physical dry deposition flux (Chapter 4.4.1 for details). Therefore, the only abiogenic removal pathway from the atmosphere is CO_2 scavenging by clouds and finally precipitation. We can easily calculate the DIC in precipitation (assuming equilibrium) using Eq. (2.132b). Thus, we get 0.21 mg L^{-1} and 0.28 mg L^{-1} carbon for 280 and 383 ppm CO_2, respectively (pH = 5.6 and at 10 °C). By using the global precipitation (Table 2.22), it results in a very small total of wet removal fluxes (in 10^{15} g yr^{-1} carbon):

---

[34] In Chapter 4.5, the theoretical conception of atmospheric residence time will be developed.

- 280 ppm CO_2 (preindustrial): 0.08 and 0.02 for marine and terrestrial precipitation, respectively and
- 400 ppm CO_2 (today): 0.10 and 0.03 for marine and terrestrial precipitation, respectively.

The river run-off (Tables 2.28 and 2.29) is about $0.46 \cdot 10^{15}$ g yr^{-1} carbon and is much larger than the total wet deposited carbonate ($0.13 \cdot 10^{15}$ g yr^{-1} carbon). The global volcanic CO_2 emission is uncertain and there is given a value (Table 2.41) of $0.02 \cdot 10^{15}$ g yr^{-1} carbon. We state that the global carbon wet removal is significantly larger (by about a factor of 6–7) than the annual volcanic CO_2 release. Hence, it is likely that biogenic CO_2 is precipitated but this amount is extremely small compared with the assimilation flux (about $0.1 \cdot 10^{15}$ g yr^{-1} carbon). Moreover, the river run-off is much larger than the total continental wet removal flux (by a factor of ~ 15). It is likely that it comprises biospheric carbonate from soils but we cannot exclude anthropogenic CO_2. We can summarize that the maximal physical removal fluxes are $0.13 \cdot 10^{15}$ g yr^{-1} carbon wet deposition and $0.46 \cdot 10^{15}$ g yr^{-1} carbon river run-off.

The residence time τ in the sense of the turnover from one reservoir (atmosphere) to another (biosphere) is generally described by (cf. Eq. 4.339 and 4.334):

$$\tau = \frac{m}{F_{sink}} \tag{2.142}$$

where m mass of the compound in the reservoir. The CO_2 mass in the preindustrial atmosphere amounts to about $600 \cdot 10^{15}$ g yr^{-1} carbon with a global assimilation rate (Fig. 2.10) of about $200 \cdot 10^{15}$ g yr^{-1} carbon, meaning the turnover time (a "pseudo-residence time") of natural CO_2 amounts to about three years. Some climate skeptics (Segalstad 2009) use this time-quantity (enlarged because of rising atmospheric CO_2 mass to about four years) to explain that after the cessation of anthropogenic CO_2 emissions the recovery of the atmospheric CO_2 concentration will soon be expected (within less than 10 years). However, this is misinterpreting the conception of budgets and fluxes (Prather 2007). As discussed above, it follows that the natural removal is balanced with the new yearly input by respiration. In Chapter 4.5 we will define the residence time mathematically and see that Eq. (2.142) is valid only for removal processes which can be described by a first-order rate equation (cf. Eq. 2.143):

$$F_{sink} = \frac{dm}{dt} = k \cdot m$$

Because almost all chemical reactions can be described as pseudo-first order (see Chapter 4.2.1) and dry deposition (so far it is driven by physico-chemical sorption) as well wet deposition can also mathematically described as first-order rate, removal of most atmospheric trace constituents can be described by Eq. (2.142). However CO_2 assimilation must be considered as a zero-order process, i.e. the removal rate is constant and (largely) independent from the atmospheric CO_2 concentration. This becomes reliable when considering the global biosphere as heterogeneous uptake process, only depending from the plant amount. This consideration should not be fully valid but explains that application of Eq. (2.142) is invalid. The above estimated "pseudo-residence time" of CO_2 is the time when all atmospheric CO_2

(assuming no CO_2 source via respiration) is completely consumed and the "reaction" (photosynthesis) abrupt stops. However because respiration brings about at the same rate CO_2 back to the atmosphere, the "pseudo-residence time" of CO_2 becomes infinite. Only burial of organic matter (lignin-derived organic matter, relatively non-biodegradable) in sediments, representing a small excess of photosynthesis over respiration, is important over million of years for control of CO_2 and O_2 (Berner 2005, Beerling and Berner 2005).

The CO_2 measurement clearly shows that the accumulative CO_2 increase is due to anthropogenic emission without fully balancing it by a sink. We have discussed in Chapter 2.8.3.1 that only about 50 % of annual anthropogenic CO_2 is taken up by the biosphere and oceans. The remaining 50 % builds up the carbon stock in the atmosphere ("airborne fraction") and can only be removed physically.

Taking the present anthropogenic CO_2 mass in the atmosphere (Table 2.82) of about $225 \cdot 10^{15}$ g carbon, and assuming that the river run-off represents the maximum physical removal rate, it follows from Eq. (2.142) that there is a residence time of about 500 years. Taking into account only the atmospheric wet removal flux ($0.13 \cdot 10^{15}$ g yr^{-1} carbon), the residence time is 1700 years. Moreover, the residence will increase with increasing airborne CO_2. Having a mixing ratio of 500 ppm (corresponding to about $300 \cdot 10^{15}$ g carbon) by 2050, the residence time increases to 650 and 2300 years, respectively. It is, therefore, likely that the removal capacity of our climate system for the recovery of anthropogenic atmospheric CO_2 is in the order of 1000 years. Then, the removal flux amounts to $0.2 \cdot 10^{15}$ g yr^{-1} carbon, only 2 % of the present man-made emission flux.

2.8.4 Climate change mitigation: Global sustainable chemistry

We can state that humans became a global force in the chemical evolution with respect to climate change by interrupting naturally evolved biogeochemical cycles. But humans also do have all the facilities to turn the "chemical revolution" into a sustainable chemical evolution. That does not mean "back to nature". At the beginning of Chapter 2.8, we defined a sustainable society as being able to balance the environment, other life forms and human interactions over an indefinite time period. "The Great Acceleration is reaching criticality. Whatever unfolds, the next few decades will surely be a tipping point in the evolution of the Anthropocene" write Steffen et al. (2007).

A global sustainable chemistry first needs a paradigm change, namely the awareness that growth drives each system towards a catastrophe. Sustainable chemistry, also known as green chemistry, is a chemical philosophy encouraging the design of products and processes that reduce or eliminate the use and generation of hazardous substances. Anastas and Warner (1998) developed 12 principles of green chemistry:

1. *Prevention*: It is better to prevent waste than to treat or clean up waste after it has been created.
2. *Atom Economy*: Synthetic methods should be designed to maximise the incorporation of all materials used in the process into the final product.

3. *Less Hazardous Chemical Syntheses*: Wherever practicable, synthetic methods should be designed to use and generate substances that possess little or no toxicity to human health and the environment.
4. *Designing Safer Chemicals*: Chemical products should be designed to affect their desired function while minimizing their toxicity.
5. *Safer Solvents and Auxiliaries*: The use of auxiliary substances (e. g. solvents, separation agents, etc.) should be made unnecessary wherever possible and innocuous when used.
6. *Design for Energy Efficiency*: Energy requirements of chemical processes should be recognized for their environmental and economic impacts and should be minimised. If possible, synthetic methods should be conducted at ambient temperature and pressure.
7. *Use of Renewable Feedstocks*: A raw material or feedstock should be renewable rather than depleting whenever technically and economically practicable.
8. *Reduce Derivatives*: Unnecessary derivatization (use of blocking groups, protection/ deprotection, temporary modification of physical/chemical processes) should be minimized or avoided if possible, because such steps require additional reagents and can generate waste.
9. *Catalysis*: Catalytic reagents (as selective as possible) are superior to stoichiometric reagents.
10. *Design for Degradation*: Chemical products should be designed so that at the end of their function they break down into innocuous degradation products and do not persist in the environment.
11. *Real-time Analysis for Pollution Prevention*: Analytical methodologies need to be further developed to allow for real-time, in-process monitoring and control prior to the formation of hazardous substances.
12. *Inherently Safer Chemistry for Accident Prevention*: Substances and the form of a substance used in a chemical process should be chosen to minimize the potential for chemical accidents, including releases, explosions and fires.

With respect to the climate system, principles 1, 2, 7 and 11 are most important. The basic principle of global sustainable chemistry, however, is to transfer matter for energetic and material use only within global cycles without changing reservoir concentrations above a critical level, which is "a quantitative estimate of an exposure to one or more pollutants below which significant harmful effects on specified sensitive elements of the environment do not occur according to present knowledge" (Nilsson and Grennfelt 1988).

To keep the 2 degrees C climate warming goal, the cumulative CO_2 emission must be limited to about 1000 Gt carbon (nowadays we are approaching 600 Gt). Hence, the atmospheric CO_2 increase must be stopped within the next few decades and the "energy turnaround" ("Energiewende" − renewable energy revolution) must be completed by the end of this century. There are several ways as a consequence of the nearly irreversible accumulation of anthropogenic CO_2 in the air and the cognition that carbon from fossil fuels is limited but the findings that carbon compounds are optimal carriers for energy conversion and materials:

− To reduce fossil fuel combustion (replacement by solar energy);
− to capture CO_2 from exhaust gases and storage (CCS technology);

- to capture CO_2 from ambient air (DAC) air (and/or other reservoirs such as seawater[35]); and
- to recycle CO_2 into utilizable carbon compounds (or sequestration achieving a negative flux).

From the long-term perspective, the transmission to renewable energies will solve the problem of the depletion of fossil fuels but is surely too late to minimize climate change. Moreover, although future oil consumption might stagnate or even decrease, coal consumption is expected to rise. Carbon capture and storage (CCS) technology is considered the only practical solution for early CO_2 control. CO_2 storage (and sequestration) is the ultimate technology for "neutralizing" the CO_2 budget through the combustion of fossil fuels but also the only way to make the biosphere-atmosphere CO_2 budget "negative", i. e. reducing atmospheric CO_2 loading. CCS technology is now widely accepted as a solution for meeting the climate protection targets (CO_2 emission reduction) worldwide and a precondition for further use of coal-fired power plants beyond 2020 in Europe. However, the reduction in energy efficiency by about 25 % because of carbon capture (based on CO_2 washing by monoethanolamine (MEA) and subsequent thermal CO_2 stripping) needs an equivalent increase in coal combustion to maintain the energy budget. Moreover, CO_2 storage is accompanied by several problems that have not yet been fully solved. The main objection, however, lies in the one-way function: the one time use of fossil fuels by transforming them into carbon dioxide, which is at best deposited for a million or more years in geological stocks.

The "safety" of CO_2 storage is realized by a small diffusive loss rate compared with the capture rate (catastrophic events of CO_2 eruptions from geological stocks postulated not to occur). Hence, a CO_2 residence time of only a few hundred years in geological stocks is sufficient to effectively compensate climate impacts from CO_2. Moreover, CO_2 storage is a resource for future carbon use. But even if CCS technology is deployed at all large industrial facilities, more than half of global CO_2 emissions would remain.

Instead of storing CO_2, it is suggested that it should be recycled by transformation into different reduced carbon species (CH_4, CO, C and many hydrocarbons), now called *solar fuels* (see Chapter 2.8.4.4).

2.8.4.1 Growth and steady state economy

In nature, many processes follow this simple law (2.143), which expresses that the change of a quantity N (for example population, mass, energy) is proportional to

---

[35] The desorption of dissolved CO_2 from seawater (being 90 % HCO_3^- with a mean total DIC concentration of ~ 28 mg C L^{-1} − by a factor of 1000 larger than atmospheric carbon) could be another approach to closing the carbon cycle (we currently call it *seawater capture*). Hence, in the case of a technology with 60 % desorption efficiency, "only" 150 km^3 seawater must be globally processed daily to attain the production mentioned above (1 Gt yr^{-1}). In our laboratory, an ultrasonic-based CO_2 desorption technique has been developed as an alternative to the thermal stripping in the CCS process, and it is not difficult to believe that this technology could be applied to seawater decarbonization.

the quantity itself. In other words, exponential growth occurs when some quantity regularly increases by a fixed percentage. The proportionality coefficient λ characterizes the process (biology, chemistry, physics, economy, etc.) as follows:

$$\frac{dN}{dt} = \lambda \cdot N = F \quad \text{and} \quad N(t) = N_0 \exp\ (\lambda t) \tag{2.143}$$

We see that dN/dt denotes a flux F; according to the sign, it could result in a growth (positive sign) or decline (negative sign). It is clearly seen that a negative flux will end with $N(t) = 0$ with $t \to \infty$ when there is no permanent source (positive flux) of N to maintain a pool of this quantity.

Growth (positive sign), however, is first a mathematical problem but bacterial growth is the best example. Crichton (1969) wrote: "The mathematics of uncontrolled growth is frightening. A single cell of the bacterium E. coli would, under ideal circumstances, divide every twenty minutes. That is not particularly disturbing until you think about it, but the fact is that bacteria multiply geometrically: one becomes two, two become four, four become eight, and so on. In this way it can be shown that in a single day, one cell of E. coli could produce a super-colony equal in size and weight to the entire planet earth." Fortunately, mathematics describes natural processes but does not control them.

The growth of a population is only possible if nutrients or food are available to allow the increase in N. As discussed, this concerns the super-exponential growth of the human population from the eighteen's century onwards. Any (exponential) growth of non-human population will soon be limited through nutrition, water and the subsequent limits of the habitat. As a consequence of the increasing population and the limits in its sustainability, a decline (negative growth) starts, in other words the death of individuals. Basically, this is also valid for humans without solving the supply problems concerning energy, food and materials. Scientific and technical innovation, however, has (so far) overcome the limits. But there is a fundamental difference to non-human populations; in nature (without humans) producers and consumers are balanced – the mass budget is zero[36]. Moreover, any growth is limited through the flux of renewable (solar) energy to provide process energy.

Humans exceeded such natural limitations by exploiting raw materials from the lithosphere (fossil fuels and minerals) and biosphere (wood), disregarding (or shifting it into future) that the extraction fluxes are larger than the recovery fluxes. For abiogenic matter (natural deposits recovery), the latter one is zero over human time scales. Hence, the limits are well defined and replacement strategies have been developed or designed. The problem of why they are not yet or only hesitantly introduced is simply the economy, or more correctly the price. For so long traditional technologies (especially mining, transportation and the combustion of fossil fuels) have produced more profit, with solar energy (which first needs a new infrastructure) secondary. Hence, the economic paradigm has to be changed: the costs of energy must include the costs of sustaining the climate and/or the cost of climate change on long-term basis. This done, only solar energy (direct or indirect) is much cheaper.

---

[36] That is not fully true. Permanent climate change occurs but over long time scales allowing adaption through evolution.

Another much more serious problem than resource limitation is the waste accumulation in the climate system. The challenge is clear: to adopt the natural model of cycling matter. Each produced good (from nature-extracted matter) must be recycled to the chemical status of the primary compound or (compromising) transferred to a pool without climate impact.

Another message to traditional economists and politicians is that exponential growth is far away from any idea of a sustainable society. Growth leads to crisis: financial, pollution and supply. A sustainable society is characterized by equilibrium:

$$\left(\frac{dN}{dt}\right) = \left(\frac{dN}{dt}\right)_{product} - \left(\frac{dN}{dt}\right)_{waste} = 0 \tag{2.144}$$

This does not exclude growth but growth (production $\approx$ consumption assumed[37]) must be compensated by removal. Self-limitation of this system is naturally given if waste cannot be turned back into products.

The observation that economic growth has limits led to the development of "steady state economics" (sometimes also called ecological economics or full-world economics). Ecological economists also observe that an economy is structured like an ecosystem. Permanent growth – as stated by politicians – will not solve life problems such as employment; this is a question of reorganizing society. After productivity (expressed as constant annual turnover) satisfies social consumption needs, stationary conditions are then achievable, i.e. λ becomes zero in Eq. (2.143). Before 1800, global growth was less than 0.1% yr^{-1} and increases to more than 3% yr^{-1} in the second half of the 20th century (Fig. 2.57). Naturally, the human population will (and must) tend to a constant number. This limitation process is likely to go on over the next 200 years. Another limitation must be set through per capita consumption to provide social and cultural standards. The growth, however, is going on this century. Without revolutionary technological changes, the climate will become out of control.

The steady state economy is an entirely physical concept. Any non-physical components of an economy (e.g., knowledge) can grow indefinitely. But the physical components (e.g. supplies of natural resources, human populations, and stocks of human-built capital) are constrained and endogenously given. An economy could reach a steady state after a period of growth or after a period of downsizing or degrowth.

However, the technical man also uses materials (non-organic such as minerals and elements) with life-cycles being extremely small comparing to its geochemical recycling. Even in a steady state economy, simple reproduction consumes materials not being within the time-scale of natural reproduction. Hence, also sustainable economy lead on long-term scale (thousands of years) to an irreversible degradation. The technical society therefore, always remains a factor in global chemical weathering (cf. Chapter 2.5.5). The challenge of sustainable chemistry is looking for replacement of inorganic through organic (hence reproducible) materials and substances; the carbon-based economy is described in Chapter 2.8.4.3.

---

[37] In reality, production > consumption because of losses. It is a challenge to achieve production $\approx$ consumption both from resource management and climate control.

2.8.4.2 Direct air capture

The idea of *air capture* (CO_2 extraction from air) as climate control strategy is now accepted and seriously considered in global ecological (for example Cao and Caldeira 2010) and economic models (e. g. Edenhofer et al. 2006). Keith (2009) writes: "Air capture is an industrial process for capturing CO_2 from ambient air; it is one of an emerging set of technologies for CO_2 removal that includes geological storage of biotic carbon and the acceleration of geochemical weathering. Although air capture will cost more than capture from power plants when both are operated under the same economic conditions, air capture allows one to apply industrial economies of scale to small and mobile emission sources and enables a partial decoupling of carbon capture from the energy infrastructure, advantages that may compensate for the intrinsic difficulty of capturing carbon from the air."

A complete air capture system requires both a contactor and a system for regenerating the absorbing solution. However, with the exception of CCS, which is presently transferred to larger technical equipment being tested in pilot plants, DAC (direct air capture) and CCU (carbon capture and utilization) still only exist within the laboratory or only on conceptual levels, characterized by different approaches[38].

The idea of CO_2 capture from ambient air using alkaline solution is not new (Tepe and Dodge 1943, Spector and Dodge 1949, Greenwood and Pearce 1953) and was used as a pre-treatment before cryogenic air separation. In general, air capture includes all processes of CO_2 fixing and sequestration. In the past, it focused on biomass (Marchetti 1977, Keith 2000, Metzger and Benford 2001), but it remains an option today and for the future too. Bio-energy with carbon storage (BECS) is the term referring to a number of biofuel technologies, which are followed by carbon sequestration and yielding "negative emission energy" (Read and Lermit 2005). However, the key factor in CO_2 removal from the atmosphere is the specific carbon flux per time and square. Plant assimilation needs time and a large area, whereas bringing biomass (almost always wood) to biofuel power plants also needs energy. However, it is important to study all practical measures for avoiding abrupt climate change (ACC) and ensuring the safety of risky geo-engineering (Lenton and Vaughan 2009).

The large-scale scrubbing of CO_2 from ambient air was first suggested by Lackner et al. (1999, Zeman and Lackner 2004). They wrote: "It is not economically possible to perform significant amount of work in air, which means one cannot heat or cool it, compress it or expand it. It would be possible to move the air mechanically but only at speeds that are easily achieved by natural flows as well. Thus, one is virtually forced into considering physical or chemical adsorption from natural airflow passing over some recyclable sorbent." The basic principles of CO_2 capture from ambient air with respect to a climate strategy were described in Elliott et al. (2001), Dubey et al. (2002), Keith et al. (2005) and Keith (2009). Almost all of these authors suggested techniques based on sodium hydroxide, whereas sodium carbonate is converted back into NaOH by "causticization," one of the oldest proc-

---

[38] Criticism of the American DAC Report came also from the Climeworks Company which is doing solar-thermal CO_2 capture and conversion in cooperation with the Professorship of Renewable Energy Carriers, Institute of Energy Technology at ETH Zurich (Switzerland).

esses in the chemical industry. Different absorbers have been proposed such as large convective towers (Lackner et al. 1999), packed scrubbing towers (Zeman 2007) and a fine spray of the absorbing solution in open towers (Stolaroff et al. 2008). CaO-CaCO$_3$ cycles have also been proposed using solar reactors (Nikulshina et al. 2009).

Holmes and Keith (2012) adapt technology used in large-scale cooling towers and waste treatment facilities, which are designed to efficiently bring very large quantities of ambient air into contact with fluids. The design they present assumes that absorber fluid is an aqueous solution that absorbs CO_2 from ambient air (typically of a $1-2$ M NaOH solution) with flux across the surface of the liquid film of order 1mg m^{-2} s^{-1}, and that, under typical operating conditions, each kilogram of solution absorbs about 20 g of CO_2 before it is returned for regeneration.

Generally, it is a huge challenge to believe that direct CO_2 extraction from air can be achieved in quantities approaching an order of several Gt C yr^{-1}. Remember that about 50 % (about 4 Gt C yr^{-1}) of technically emitted CO_2 comes from small and mobile units, a percentage likely to increase further in the future. Additionally, about $1-2$ Gt C yr^{-1} comes from land use change and wood fuel use, which are categories that should diminish in the future. Some 4 Gt C yr^{-1} is absorbed by the biosphere (ocean and forest) but with an anticipated decreasing capacity. This "uptake capacity" is not constant but at a certain percentage (likely non-linearly) of the total CO_2 release into the air. Hence (assuming full CO_2 capture from stationery large sources), there is a requirement of at least 2 Gt C yr^{-1} air capture. A compensation of atmospheric CO_2 buildup through engineered chemical sinkage was proposed by Elliott et al. (2001). They calculated the CO_2 removal from air by asking for the area needed if this was a perfect, flat sink with a dry deposition velocity (Chapter 4.4.1) of 1 cm s^{-1}. It is a hundred thousand square kilometer value, which constitutes an upper limit for absorbing the annual anthropogenic CO_2 input. Roughness elements and vertical fences could increase the transfer velocity (by reducing the atmospheric residence) and increase the specific absorbing area per horizontal air column surface. A total square reduced by a factor of 10 might be able to be reached.

Technically, CO_2 is extractable from air by cryogenic techniques. However, based on 400 ppm CO_2, an air volume of about 10 km^3 must be processed daily to get 0.1 Mt C d^{-1} (this rate corresponds to about 30 "capture units" globally to achieve a yearly capture of 1 Gt C). Today's high-performance cryogenic air separation plants have an air capacity of about 0.02 km^3 d^{-1}. In other words, more than 10 000 such plants would have to be in use to provide CO_2 capture of 1 Gt C yr^{-1} from air. The non-use of other gases from air separation would also not conform to a sustainable approach. However, new air separation techniques will eventually make it possible to generate only carbon dioxide and water from air and to increase the daily capacity by a factor of 10. In that case, only 1000 such plants will be needed for the extraction of 1 Gt C yr^{-1} from air, less than the global number of coal-fired power stations.

Carbon dioxide capture can be applied both in closed technical plant systems as well as in an open-field technology (geo-engineering). The processed air volume is large (10^7 km^3 because of about 40 t CO_2-C km^{-3}) but corresponds to the air volume passing through about 100 cooling towers of large power plants.

Assuming CO_2 solvents having a surface resistance being zero, the atmospheric (dry deposition) flux is determined only by the quasi-laminar and atmospheric resistance (see Chapter 4.4.1), and a value between 0.4 and 1.2 kg CO_2-C $m^2 d^{-1}$ can be estimated. This corresponds to an uptake rate of about 2000 t C $ha^{-1} yr^{-1}$; at least 50 times more than most manipulated algal aquacultures will yield[39]. To "capture" 1–2 Gt CO_2-C yearly, a square (assuming 50 % scrubbing efficiency) of 10^4 km^2, smaller than the State Brandenburg, is ("only") needed. However, in contrast to CCS in this approach it is not the aim, to extract CO_2 in a short time from a given volume of gas (air) but to reach a saturation of the CO_2 solvent for further desorption and solvent cycling.

Design and synthesis of new CO_2 absorbing materials is the key for application of DAC. The argument that large costs will limit the technology is overcoming when including the costs for climate change into energy price. It is self-evident that only solar energy is used for DAC. In the next Chapter 2.8.4.3, we will see that DAC is a basic technology for the carbon economy similar to biogenic assimilation.

From today's perspective, it seems that as a result of the extremely low concentration of CO_2 in the air, the technical and economic solution of direct atmospheric CO_2 reuse is not very likely (DAC 2011). However, any technical solution in our concept is based on the paradigm change to establish a zero-carbon budget (not zero emissions!), and to no longer measure the effect on energy efficiency (solar energy is in "excess") but on budget, with respect to climate sustainability. The "price" of CO_2 emitted from fossil fuels (and hence fossil fuel costs) must include climate change affects; this would encourage energy transition and DAC technologies also.

2.8.4.3 The carbon economy: CO_2 cycling

Mining and combustion of fossils fuels now results in geological reservoir redistribution of carbon close to (or even passing?) the "tipping point". It is assumed that in the near future the acceleration of CO_2 release still further increases due to economic growth. The large CO_2 residence times in air and seawater avoid reaching a steady-state (global cycle in-time) and a recovery (climate restoration) also after full stop of fossil fuel use.

Therefore, much more forced by climate change and its uncertain but very likely catastrophic impacts after reaching the "tipping points" than by fossil resource limits, we need the transfer into the "solar era" as soon as possible. Nuclear power may concern as "bridging technology" but risks may not any longer be accepted

---

[39] Algal productivity rates between 5 and 10 g C m^{-2} d^{-1} have usually been cited (Drapcho and Brune 2000), but have been reported to be up to 15 C m^{-2} d^{-1} in highly modern farming systems (Shelef et al. 1978). Again, to achieve 0.1 Mt C d^{-1}, a farming area of about 7000 km^2 is needed or 210 000 km^2 globally, which corresponds to an area roughly 50 % of the size of Germany. Surely there is a research need to optimize (and maximize) CO_2 capture by industrial biofarming in sunbelt countries. For example, nutrients for biofarming could be taken from municipal wastewater of nearby "solar cities" and/or recycled from the biomass conversion process into CO_2 (note that fixed CO_2 is the aim rather than biofuel).

by society. Secondary "renewable" energy, for a long time already in use (and we should not forget, it was the only significant source of energy before the first Industrial Revolution), such as water and wind, will probably never contribute on a global scale to fit the energy demand (this does not exclude national and regional solutions, nowadays proposed for Germany). Hence, only direct use of solar energy as it is proposed, for example, by the *desertec* conception, can realistically solve the global energy problem and fully replace fossil-fuels. Without any doubt, electricity is the unique form of energy in future and its direct application (also for mobility and heating) will increase − and can replace to a large percentage traditional fuels based on fossil resources. The *desertec* technology is not unlikely to apply within the next few decades to remarkably replace fossil fuels − if political (and thus financial) willingness is given. However, there are some questions which have to be answered and transferred into technical solutions to establish the solar era.

− Electricity will be produced not constantly over time and not correlated with the demand for energy, hence it must be stored, most likely best by transfer into "chemical energy", to manage energy supply.
− Due to safety reasons, excess energy must be stored (for example in water reservoirs, but this way is limited), again best way seems to transfer electricity into "chemical energy".
− There are technical applications (for example air traffic, long-distance street traffic, ship traffic, metallurgy) where electricity can not be taken directly from nets or storage units and will be neither ecological nor economic.
− Humans always need synthetic organic materials (polymers, drugs, chemicals etc.). These can be produced from remaining fossil resources, but also from biomass − and from CO_2.

Möller (2012) put forward an option to create a globally closed anthropogenic carbon cycle by using only solar energy to (a) stop further increase of CO_2 emission and to get a global zero-carbon budget, (b) to solve the problem of electricity storage based on CO_2 utilization, (c) to provide carbon-based materials only form CO_2 utilization, and (d) to use further the infrastructure developed for the fossil fuel era, called SONNE[40] conception ("Sonne" is the German word for sun). The SONNE conception will interlink solar electricity conceptions such as *desertec* with CO_2 utilization to overcome the above mentioned open problems after the fossil fuel era. In other words, SONNE build a man-made carbon (CO_2) cycle in analogy to the natural assimilation-respiration carbon cycle (Fig. 2.106). CO_2 is recycled within hybrid power plants (see Fig. 2.107) and captured from ambient air. It is replaced from waste (emission) to resource; process energy is taken from solar energy.

The specific approaches put together in this "CO_2 economy" are already known and/or proposed, but to my knowledge, the creation of a man-made carbon cycle in such an integrative approach and with such rigorousness in linking energy with material economy adopting the principle of natural cycling but not copying natural

---

[40] SOlar-based maN-made carboN cyclE.

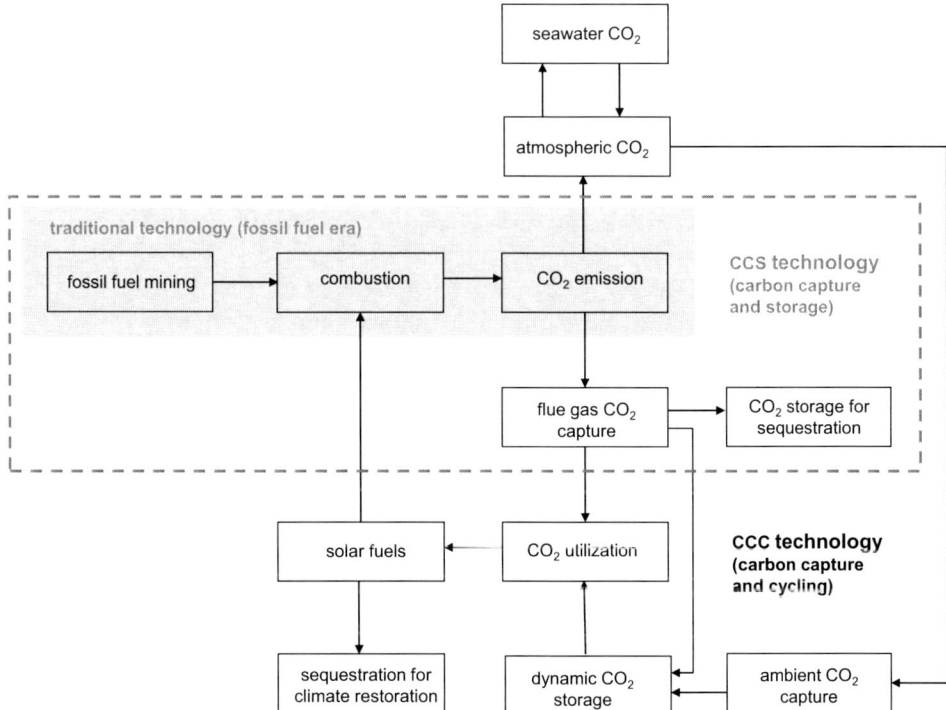

Fig. 2.106 Scheme of energy transition from fossil to solar era including a CO_2 economy (SONNE conception). Three overlapping systems: fossil fuel burning without (grey box) and with carbon capture (dotted box) as well as solar fuel production/use and global carbon cycling. The driving force is exclusively solar radiation; hence the CO_2 economy is interdepended with solar electricity conceptions such as desertec. Elements of this concept can be introduced concurrent with further use of fossil fuels aimed as its stepwise replacement.

processes[41], as suggested here, is world-wide unique and new (even much more complex than the "methanol economy").

CO_2 is unique[42],

- as final oxidation product of all organic matter and materials,
- because of its globally cycling and homogeneous distribution in the atmosphere (but keeping a level before "tipping points"),
- as resource for organic materials concerns carriers of energy and functional materials,

---

[41] For illustration, some scientists' dream from the artificial leaves to transform CO_2 into (solar) fuels. Our approach consists in "secondary" use of solar energy in terms of electricity and heat in large industrial operational units, which principally already are known.

[42] In a certain sense hydrogen (H_2) can also play this role when we adopt the natural water splitting process, which was already proposed as "hydrogen technology" in the early 1980s. But there are several problems, a) safety in storage and transport, b) leakage and atmospheric implications and c) missing material supply. Water electrolysis will play an important role in SONNE for oxy-fuel combustion (O_2 supply) and CO_2 reduction (H_2 supply).

- carbon is the only element forming complex molecules and substances and being within a global dynamic[43] cycle and gaseous compounds on lowest (CH_4) and highest oxidation state (CO_2),
- the only environmental problem of CO_2 is its rise in atmosphere (and seawater) with climatic implications; hence controlling its level on acceptable values will overcome the environmental problem.

It is evident that through the realization of these principles a CO_2 "zero-budget world" rather than a "CO_2 free world" can be reached because there is a closed anthropogenic carbon cycle (we call it CO_2 economy). All CO_2 still emitted – and also cannot be captured in future from mobile and small equipment's – will be captured and cycled for reuse; I call it "**C**arbon **C**apture and **C**ycling" (CCC) technology. With this in mind, CCS technology (carbon capture and storage/sequestration) makes (more) sense – despite of the controversially discussed CO_2 storage problems – and provides considerable incentive because CO_2 storage is now only temporary (we call it dynamically) until it is recycled from waste to feedstock. The proposed CCC technology allows a stepwise replacement of coal and other fossil fuels by solar fuels but keeping the carbon based infrastructure such as pipelines, tankers, storage facilities, engines and allows the continuous use of other available technical applications developed within the last hundred plus years, but within a CO_2 neutral closed loop.

A closure of the carbon cycle, however, is only possible when CO_2 is extracted from natural reservoirs such as the atmosphere and seawater (DAC, see previous Chapter 2.8.4.2) because a complete "industrial" CO_2 capture will be impossible when taking into account many small and mobile sources.

Principally, the SONNE conception is not aimed for the very near future but for the solar era with "unlimited" access to useable solar energy, likely after 2050. However, because CCS will be an essential technology within the "internal cycle" of hybrid-type power plants (Fig. 2.107), CCC could be introduced to some extent parallel with further use of fossil fuels and step-wise replacing them until fully establishing the SONNE cycle. The principal scheme of CO_2 use in solar electricity storage (valid also for other "renewable" energy such as wind) could soon be realized. So-called oxy-fuel combustion would provide high purity CO_2 as exhaust gas which can be recycled without energy-intensive capture (Fig. 2.107). We can set ten mission statements or principles:

1. Further use of fossil fuel combustion in large stationary units but only with CO_2 capture (CCS technology) until the full transfer into the solar fuel world: capture CO_2 from combustion units as much as possible;
2. Replacement of fossil fuel use in small stationary and mobile units as far as possible (electricity-based and hybrid techniques): reduce carbon carriers as fuels as much as possible;
3. Sequestration of carbon (not CO_2) on medium and long-term scale for buffering the further CO_2 emission increase within the next decades and climate sanitation in far future;

---

[43] In (biogeochemical) cycles move all elements and its compounds – but often on geological time scale (beside carbon only sulfur and nitrogen are in similar dynamic cycles).

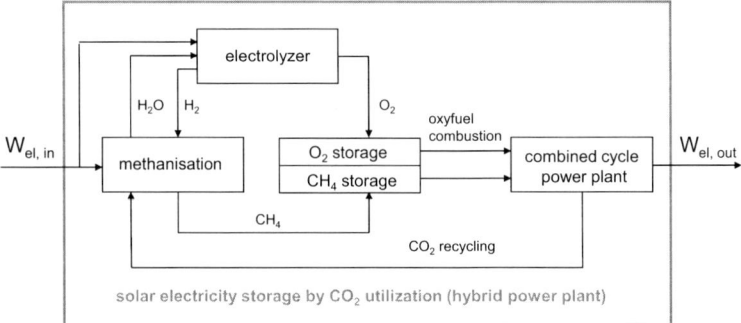

Fig. 2.107 Schema of a "hybrid power plant": chemical storage of "renewable" energy (preferably solar electricity) by CO_2 utilization (probably by methanisation) and internal CO_2 recycling (probably by oxyfuel combustion); $W_{el, in}$ – direct solar electricity, $W_{el, out}$ – indirect energy solar energy (electricity/heat) "on demand". The energy efficiency is negative (for example in the case of CH_4 production from CO_2, only 30% of electricity input can be re-used).

4. Developing technologies for CO_2 extraction from natural reservoirs (ambient air, seawater) to achieve a global man-made carbon cycle while allowing CO_2 emissions into the atmosphere from mobile and small sources: atmospheric CO_2 is considered as the only carbon reservoir for chemical CO_2 utilization (CCU);

5. Developing technologies for CO_2 reduction but applications only with renewable energy, namely solar radiation (solar fuel production);

6. Introduction of large solar thermal power plant units for electricity generation;

7. Developing technologies for electricity conversion into chemical energy carriers (solar fuels used in hybrid power plants);

8. Build-up a solar fuel infrastructure (widely on the basis of the existing fossil fuel infrastructure);

9. Developing technologies for electricity conversion into large central heat storage units (based on molten minerals); and

10. Economic paradigm change: solar energy is "in excess" (compared with the global human demand) and is naturally dissipated in the atmosphere; hence, large energy consuming conversion processes and direct air capture can be carried out for resource generation and climate sustainability: a new economy-thinking based on sustainability (or closed carbon cycle) is needed. In other terms, not energy but material efficiency becomes the key factor.

It is evident that through the realization of these principles, a CO_2 "zero budget world" rather than a CO_2 free world can be reached because there is a closed anthropogenic carbon cycle. Human's evolutionary responsibility should also consider the retransfer of emitted CO_2 into geological stocks, for example, as elemental carbon for safe sequestration and step-wise but long-lasting climate recovery.

All CO_2 still emitted will be captured and cycled for reuse. Naturally, the energy needed for CO_2 reduction comes from renewable sources. The proposed CCC technology allows a stepwise replacement of coal and other fossil fuels by solar fuels.

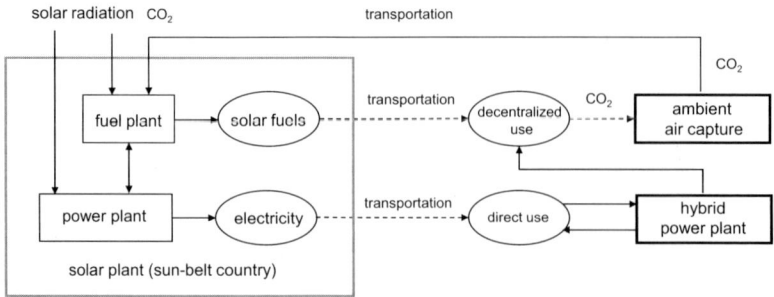

Fig. 2.108 Solar plant complex (solar-to-electricity and CO_2-to-fuel conversion in sun-belt countries), interlinked with transportation of fuels, chemicals and materials, and electricity to Northern Hemispheric countries (to be used there) and back transport of CO_2 captured from ambient air to solar plants (getting a geo-economics equilibrium and political interdependence as win-win-situation). Note that the whole system dissipates solar energy as it happens in the climate system by natural processes. Therefore, the energy efficiency plays a minor role (but should be maximized in singular technical process); the key for a sustainable economy (socio-ecological system) is the closed cycling of matter (here carbon, but this is true for all elements to be used).

Finally, there is a closed carbon cycle similar to the natural photosynthesis, a respiration cycle (Fig. 2.106). Carbon-based solar fuels solve the problem of energy storage and allow the continuous use of available technical applications to provide products for materials, and are within a CO_2 neutral closed loop.

Olah (2005) proposed the "methanol economy" but in the SONNE concept, CH_3OH is only one possible product among C_1 chemicals; the *Fischer-Tropsch* synthesis (from $CO + H_2$) basically offers a wide range of organics including liquid fuels. Our "CO_2 economy" includes the "CH_3OH economy". Recently it has been shown that the energetic efficiency of the overall energy conversion-storage system (see. Fig. 2.107) including CH_3OH as storage medium, is only 17.6 % in contrast to 29.7 % for CH_4 (Rikho-Struckmann et al. 2010). However, taking into account ambient CO_2 capture, the overall energetic efficiency will drastically lower. As does nature (the photosynthesis efficiency concerning solar light is only 2−3%), we realize a closed carbon loop only with a large energy input (in other words low energetic efficiency), but based on incoming solar radiation, 1000 times higher than present global human energy demand. Still unanswered is the question, however, of what are the limits of solar use without getting additional climate implications.

It is remarkable that establishing the SONNE conception (CO_2 economy), we first see that sun-belt countries, many of them being privileged with natural oil and gas reservoirs, will provide "solar sites" for electricity generation and likely CO_2 processing (Fig. 2.108). On the other hand, future use of fossil fuels will mainly be in the non-sun countries in the Northern Hemisphere, which should become responsible for ambient CO_2 capture (there are good reasons to more importantly establish DAC units in the north because CO_2 absorption processes needs low temperature and DIC in seawater is significantly higher in cold areas). For example, Northern Europe will capture ambient CO_2 and transport them to Northern Africa as "fuel feedstock" for solar processing (Fig. 2.108). Thus, a social win-win situation with many positive political and educational effects may be created.

At this point, it is important to state that SONNE is based on ideas already known and further investigated (for example CCS, CCR[44], CCU, DAC) at many scientific institutions worldwide. As mentioned, a key idea of CCC technology is the capture of CO_2 from the atmosphere (and its dynamic storage) to close the man-made global carbon cycle analogously to the biosphere. The CO_2-carbon economy is the adaption of the biospheres' assimilation-respiration cycle by humans, the only long-term sustainable way of surviving.

From today's perspective, it seems that due to the extremely low concentration of CO_2 in air, the technical and economic solution of direct atmospheric CO_2 reuse is not very likely (DAC 2011). However, any technical solution in our conception is based on the paradigm change to establish a zero carbon budget (not zero emission!) and measuring the effect no longer on the energy efficiency (solar energy is in "excess") but on the matter budget with respect to climate sustainability. The "price" of CO_2 emitted from fossil fuels (and hence fossil fuel costs) must include climate change affects; this would force the energy transition and also DAC technologies.

2.8.4.4 Solar fuels: Carbon as a material and energy carrier

The idea of using CO_2 as a chemical raw material is not new (Aresta and Forti 1987, Edwards 1995, Aresta and Aresta 2003, Park et al. 2004, Olah 2005, Aresta 2010). However, when using CO_2 from fossil-fuel gases, it is only climate-sustainable if the products are "sequestrated", for example by long-term use in carbon materials such as polyurethanes. CO_2 captured from fossil-fired power plants and "utilized" for storage of excess electricity (for example from wind power) may help to improve the energy efficiency (because the excess electricity cannot be used on demand) but not solve the climate problem. Zeman and Keith (2008) also suggested synthesizing carbon neutral hydrocarbons (CNHCs) from air-captured CO_2. This term (in fact CO_2 neutral) is inconsistent, and in its place we will use "solar fuels" to express that the supply of process energy for the chemical reduction of captured CO_2 must be based on *solar* energy processing instead of fossil (geothermal heat is another option).

Within the last years, considerable progress has been achieved in the catalytic hydrogenation of CO_2 (methanisation). The possible synthesis of C_1-chemicals (CO, C, CH_4, CH_3OH, and HCHO) from CO_2 and directly further to C_3 in analogy to the assimilation process (see also Fig. 2.42) leads to a variety of important basic chemicals being available for either direct combustion or material use (industrial synthesis in organic chemistry); we now call them *solar fuels*. However, a global CO_2 economy must not only provide chemicals in the order of hundred million tons but $1-2$ orders of magnitude more gaseous and liquid fuels. By using high-temperature chemical processes (which have been known for many years but due to the high energy consumption have hardly been mentioned before) based on solar thermal energy, it is also possible to remake "coal chemistry" (gasification and

---

[44] R stands for recycling.

liquefaction) via CO_2 reduction. Namely, carbon monoxide (CO) and elemental carbon may be produced and inversely transformed. For example, elemental carbon could be stored better than carbon dioxide (sequestration) but could also be reused directly in an early stage of the CCC technology. It is known that in high-temperature processes of conversion of carbon compounds to elemental carbon, the yield of polymeric carbon structure (fullerenes) are large and unforeseen changes in creating new carbon materials are made possible.

However, within recent years, considerable progress has been achieved in the catalytic hydrogenation of CO_2 (methanisation). The possible synthesis of methanol and formic acid from CO_2 leads to important basic chemicals for industrial synthesis in organic chemistry. For example, elemental carbon could be stored better than carbon dioxide but could also be reused directly in an early stage of CCC technology. It is known that in high-temperature processes of converting carbon compounds to elemental carbon the yield of polymeric carbon structures (fullerenes) is large, and unforeseen changes in creating new carbon materials can be possible. It is out of the focus of this book to review the chemical processes of CO_2 utilization – this section will only present the basic ideas of chemical conversion in the sense of the chemical evolution sustained by humans – and the reader is referred to (as examples) Aresta and Aresta (2003), Park et al. 2004, Grimes et al. (2007), Rajeshwar et al. (2008) and Varghese et al. (2009).

It is self-evident that all energy for processing is solar-based. By using high-temperature chemical processes (which have been known for many years but because of high energy consumption have hardly been mentioned before) based on solar thermal energy, it is also possible to remake "coal chemistry" (gasification and liquefaction) via CO_2 reduction. Namely, carbon monoxide and elemental carbon can be produced and inverse transformed. Thus, a variety of basic chemicals (CO, C, CH_4, CH_3OH, HCHO, etc.) would be available for either direct combustion or material use as solar fuels.

There are several modular technical systems for realizing the CCC conception (Fig. 2.108) where the solar (desert) site in sun-belt countries (module A) plays a key function by providing solar-based electricity (and heat for on-site chemical processing). Principally, it is possible to combine the solar site with other modules (B – air capture and CO_2 supply and C – CO_2 processing), but there are arguments to localize air capture close to CO_2 storage (module E) to avoid transportation, to sites where climatic conditions support large CO_2 absorption (see above) and/or to sites of CO_2 processing. Solar power plants are decentralized close to urban and industrial areas (module D). Solar fuel (in contrast to CO_2) can be easily transported and stored using traditional infrastructure for liquids and gases. Processing sites (module C) must have access to solar (or non-fossil) energy. Then, by using high voltage direct current transmission lines (*desertec* conception), they can be sited several thousands of kilometers from the solar thermal power plants. An interesting site would be Iceland, which might be able to provide power from geothermal heat and cold carbon-rich seawater for CO_2 extraction as well as air capture because of the large temperature differences in rich and lean CO_2 loading. Captured CO_2 could be reduced on site to solar fuels which are transported by tank ships to Central Europe.

There is no need for the long-range transportation of hydrogen (which is one feedstock for CO_2 reduction) because electricity (according to the *desertec* conception) can be transported advantageously in comparison to H_2. Hydrogen can also be produced at any site by water electrolysis (also using renewable sources other than direct solar electricity). There are clear advantages to CO_2 reduction close to the solar fuel consumers including the avoidance of the expensive transportation of CO_2 back to the sun-belt countries and the possibility of mixing solar fuels directly with oxygen from the water electrolysis to create "oxyfuels". Oxyfuels can be burned in stationary power plants, with the result that the flue gas is almost all pure CO_2. However, there is an interesting advantage to the back transportation of CO_2 from industrial to sun-belt countries. The solar site depends on the delivery of feedstock CO_2. Thus, a geopolitical equilibrium can be reached in the sense of a win/win situation.

As mentioned, the world's infrastructure is based on fossil fuels and products derived from coal and petrochemistry. The following groups of substances are delivered from fossils fuels and can also be produced from CO_2 using solar-based processes:

- Gaseous hydrocarbons (C_nH_{2n+2} with $n < 6$); methane as a key substance; all substances with $n > 2$ are easily liquefiable;
- liquid hydrocarbons (C_nH_{2n+2} with $n > 5$) such as alkanes but also oxygenated liquid compounds from C_1 upwards (for example, methanol);
- gaseous CO (the typical town gas in the past); and
- elemental carbon (C in different modifications).

Energy use is simply combustion in different burners and engines with, ideally, transformation back into CO_2. Reaction enthalpies are different (in the sequence $CH_4 > C_nH_{2n+2} > C > CO$), and thereby provide possible usable energy (for comparison in terms of kJ mol^{-1} C).

$$CH_4 + 2\,O_2 \rightarrow CO_2 + 2\,H_2O \tag{2.145}$$

$$C_nH_{2n+2} + (2\,n + 2)\,O_2 \rightarrow n\,CO_2 + (2\,n + 2)\,H_2O \tag{2.146}$$

$$CO + \frac{1}{2}\,O_2 \rightarrow CO_2 \tag{2.147}$$

$$C + O_2 \rightarrow CO_2 \tag{2.148}$$

Despite the lower energy yield, it is evident that hydrogen-free carbon carriers such as CO and C have the large advantage of consuming a lot less oxygen and only producing CO_2, when considering oxyfuel technology in the future. That could be an important feature in establishing the carbon cycle and turning CO_2 back into the feedstock for solar fuels. In the production of group one and group two compounds in the list above, some progress has already been made towards CO_2 hydrogenation (reactions 2.149 and 2.150). It has already been reported that catalytic CO_2 methanization has been carried out at 160 °C and below 100 °C at pressures of 80 kPa (Abe et al. 2009). Reaction (2.149) has already been discovered by *Paul Sabatier* in the 19th century.

$$CO_2 + 3\,H_2 \rightarrow CH_4 + H_2O + 90 \text{ kJ mol}^{-1} \tag{2.149}$$

$$CO_2 + 3\,H_2 \rightarrow CH_3OH + H_2O \text{ and } CO + 2\,H_2 \rightarrow CH_3OH \tag{2.150}$$

What must be considered in reactions (2.149) and (2.150) is the source of H_2. It is clear that "traditional" reactions (water-gas shift and CH_4 reforming) cannot be used and that H_2 must be generated via water electrolysis using renewable energy sources such as solar radiation. Thus, the argument that it is preferable to use H_2 directly as a fuel is inconsistent with our aim of carbon cycling. Moreover, carbon carriers provide a range of products that are better suited to the available infrastructure than H_2. Overall, the photosynthesis-like reactions (2.151) and (2.152) are carried out:

$$CO_2 + 2\,H_2O + \text{solar energy} \rightarrow CH_3OH + 1.5\,O_2 \qquad (2.151)$$

$$n\,CO_2 + n\,H_2O + \text{solar energy} \rightarrow (CH_2O)_n + n\,O_2 \qquad (2.152)$$

The formation of methanol from CO_2 hydrogenation is known as the CAMERE process (Melián-Cabrera et al. 1998, Joo et al. 1999). In the past, a thermochemical heat pipe application has been proposed (Edwards 1995), which is based on Reaction (2.153a), where CO_2/CH_4 reforming (using solar energy) gives CO/H_2 gas, which can be converted back (2.153b) in an exothermic reactor with an equivalent energy output. However, this technology needs an entirely new infrastructure.

$$CH_4 + CO_2 + \text{energy} \rightleftarrows 2\,CO + 2\,H_2 \qquad (2.153a,b)$$

The CO/H_2 gas (also called water-gas) − depending on the $CO:H_2$ ratio − can also be converted (using the *Fischer-Tropsch* synthesis) to alkanes, alkenes and alcohols. Preferably, substances should be generated for application in known technical systems such as liquefied petroleum gas or low pressure gas (LPG) and gasoline as well as natural gas. The production of solar substitutes for diesels and oils ($C > 8$), that is petrol products from the fractional distillation of crude oil between $200\,°C$ and $350\,°C$, is also possible, but offers no advantages in the solar fuel cycle and its stepwise replacement by gases and gasoline should be foreseen.

An interesting way of generating solar fuels can be seen in two "classical" inorganic carbon carriers: CO and carbon itself. The gasification of carbon (2.154a) produces CO (generator gas) in the so-called *Boudouard* equilibrium:

$$C + CO_2 + 171 \text{ kJ mol}^{-1} \rightleftarrows 2\,CO \qquad (2.154a,b)$$

The inverse reaction (2.154b) opens the way to finally convert CO_2 to elemental carbon via reactions (2.149) and (2.153a). Alternatively, CO can be produced by the high-temperature electrolysis of CO_2:

$$CO_2 + \text{energy} \rightarrow CO + 0.5\,O_2, \qquad (2.155)$$

where the overall reaction represents the inverse reaction (2.148). High-temperature electrolysis using solid oxide electrolytic cells offers absolute new synthesis pathways. In contrast to reaction (2.155), the electrolysis of CO_2/H_2O leads to CO and CH_4:

$$CO_2 + H_2O + \text{energy} \rightarrow CO + H_2 + 1.5\,O_2 \qquad (2.156)$$

$$CO_2 + 2\,H_2O + \text{energy} \rightarrow CH_4 + 2\,O_2 \qquad (2.157)$$

For long-term space missions, such reactions have been considered to provide a closed cycle production of oxygen and consumption of respiratory CO_2. A final pyrolysis reaction (2.158) recycles hydrogen, but more interesting for earth applications is the formation of elemental carbon according to the gross conversion process shown in equation (2.159).

$$CH_4 + energy \rightarrow C + 2\,H_2 \qquad (2.158)$$

Reactions (2.156) and (2.158) provide the following overall reaction:

$$CO_2 + H_2O + energy \rightarrow C + 2\,H_2 + 2\,O_2 \qquad (2.159)$$

This set of reactions based on solid high-temperature electrolytic CO_2 reduction shows that more advantages are likely at desert solar sites than the conversion processes in detail decisions will be ever made. The general budget is described by:

$$CO_2 + H_2O + solar\ energy \rightarrow solar\ fuels\ (C, CO, CH_4, H_2, O_2, etc.), \qquad (2.160)$$

independent of the detailed conversion processes. In the inverse reaction (2.160), solar fuels reconvert into CO_2 and H_2O via the combustion or electricity in fuel cells and set the energy free, primarily as heat, for energetic use.

3 Climate, climate change and climate system

Our understanding of complex systems is naturally limited for three main reasons. First, the number of parameters, processes and compartments is always larger than our facilities to observe them simultaneously and continuously (otherwise it would not be a *complex* system). Secondly our research has historically developed within distinct disciplines. *Leibniz* and *Humboldt* were two of the last great polymaths[1], and nobody nowadays is able to recognize (and to understand) different scientific fields at a consistently high level. Thirdly the necessity of subdividing sciences (as a tool for understanding nature) led to an increasing misunderstanding between scholars due to the use of different (scientific) language. An anonymous reviewer[2] of *Charles Lyell*'s *Principles of Geology* (published in 1835 and subtitled, "An attempt to explain the former changes of the earth's surface by reference to causes now in operation") wrote to the point:

> Just as the dairy-maid believes the moon to be a great cheese, so the astronomer fancies our globe a condensed nebula; the chemist, an oxidized ball of aluminum and potassium; the mineralogist, a prodigious crystal – "one entire chrysolite"; and the zoologist, an enormous animal – a thing of life and heat, with volcanoes for nostrils, lava for blood, and earthquakes for pulsations.

It is remarkable that the English term *science* includes the sciences in general *and* the *natural* sciences (it is different in the German language, for example, where the analogous term only means science in general). There are three *natural* sciences; namely physics, chemistry and biology. As discussed in Section 1.2, there is no need in principle to use prefixes such as "environmental", "geo-" or "bio-" because the aim of these sciences is a priori the study of material and life or, in other words, *nature*. In so far as "life" is quite different from just a simple application of physics and chemistry, as we have discussed in Chapter 2.2.2.1, biology seems to be a science naturally based on physical and chemical laws but additionally having "biological" laws. Therefore, to draw a sharp border between physics and chemistry makes absolutely no sense in the attempt to understand the climate system. However, without understanding biological laws – and probably we can enlarge that by including social laws when we consider that the biosphere was long ago transformed into the noosphere – we will neither understand climate change nor find solutions for *climate control*.

---

[1] The term "polymath" (an artificial word) was surely unknown at the time of *Leibniz*, and the German word *Universalgelehrter* express this facility much better.

[2] "Principles of Geology: being an inquiry how far the former changes of the earth's surface are preferable to causes now in operation". *London Quarterly Reviews*, Vol. LIII, April 1835, pp. 213–235 (at p. 217).

The expression *climate* is currently often used in connection with *climate change* (a term widely popularized in recent years). Synonymously, the expression *climate alteration* (a more scientific term) is used. In the *United Nations Framework Agreement on Climate Change*, which was signed by 34 countries and was approved in New York on 9 May 1992, *climate change* is defined as "changes in climate, that are directly or indirectly attributable to anthropogenic activities, which change the composition of the atmosphere and which add up to the natural climatic changes observed over comparable periods of time". This definition is circumscribed, as natural processes are included in the sense of climate change but not in the sense of climate alteration. However, it can only be observed that climate change is due to *all* factors. Possible anthropogenic factors having an effect on climate (and therefore on its alteration) are manifold and stretch from an alteration in the use of land (altered surface properties) to emissions, whereas direct energetic influences (e.g. warmth islands) are at most (first) of local importance. Thus, a climatic change is the difference between two states of climate. A *state of climate* is described through the static condition of the climatic system. A *climatic fluctuation* can therefore be described as a periodic climatic change, independent of the respective scale of time. Climate is a function of space and time – and is continuously changing (Schneider-Carius 1961).

The climate system is too complex to be clearly mathematically described. In all likelihood, this will be valid in the future as well. Our knowledge about many single processes in this system is still imperfect. Nevertheless, immense progress has been made in the past twenty years and the description of the *principal* processes is considered as assured (Hansen et al. 2007).

Both climate researchers and historians of climate science have conceived of climate as a stable and well defined category, but such a conception is flawed. In the course of the nineteenth and twentieth centuries, the very concept of climate changed considerably. Scientists came up with different definitions and concepts of climate, which implied different understandings, interests, and research approaches. Understanding of climate shifted from a timeless, spatial concept at the end of the nineteenth century to a spaceless, temporal concept at the end of the twentieth. Climatologists in the nineteenth and early twentieth centuries considered climate to be a set of atmospheric characteristics associated with specific places or regions. In this context, while the weather was subject to change, climate remained largely stable. Of particular interest was the impact of climate on human beings and the environment. In modern climate research, at the close of the twentieth century, the concept of climate lost its temporal stability. Instead, climate change has become a core feature of the understanding of climate and a focus of research interest. Climate has also lost its immediate association with specific geographical places and become global. Interest is now focused on the impact of human beings on climate (Heymann 2009).

In summarizing Chapter 2, the following major phases in the evolution of the earth can be identified:

1. The accretion of the earth and its primitive atmosphere from the primordial solar nebula.
2. The differentiation of the interior of the planet and the associated outgassing of volatile materials.

3. The chemical era of abiotic photochemical transformation of the primordial atmosphere and oceanic water to form the organic molecules from which life could spring.
4. The microbial era during which the first simple life-forms evolved, proliferated, and forever modified the atmosphere and environment.
5. The geological era in which the physical reconfiguration of oceans and continents caused major deviations in the evolution of life and the atmosphere.
6. The recent age, when humans appeared with the intelligence to exploit fully all of the capability to alter significantly the global atmosphere and environment.

Each of those epochs started with rapid climate change and (with the exception of phase 1, which ended soon after the formation of the earth) continued in more gradual change due to progressive evolution concerning an adapting climate system. To some extent, the disturbances of biogeochemical cycles caused by anthropogenic activities since the Industrial Revolution already exceed the natural matter flow, but many are still within the natural variability of the air composition over geological epochs. The time periods of natural variability and changes, however, are far beyond the dimensions of the last 150 years. Our atmospheric environmental problem – climatic change – is therefore a matter of adaptation in time of various subsystems with the global system in the first place. The humans causing alterations may disappear, but another climate and climate system will remain.

3.1 Climate and climatology: A historical perspective

The term *climate* has been used over time in different senses. The ancient Greek word κλίμα (*clima*) means area or region (clime) and is first found in the New Testament (Benseler and Schenkl 1900); the Greek philosophers did not use the word (Gilbert 1907). It is derived from κλίνω (*klinein*) which means "to incline" (Latin: *inclino*) and was probably first used in French in the late fourteenth century (*climat*) in the sense of "zone". The ancient geographers and travellers used the term *climata* (plural of *clima*) to divide parallel zones of the earth from equator to pole (Brown 1949, Sanderson 1999). The *Encyclopædia Britannica* (1771) denotes "climate" as belonging to a geographical category: "A space upon the surface of the terrestrial globe, contained between two parallels, and so far distant from each other, that the longest day in one differs half an hour from the longest day in the other parallel".

The concept of a spherical world is attributed to the Greek philosopher *Pythagoras* in the sixth century B.C. (see also Chapter 1.3.1). Based on this mathematical-astronomical definition[3], the ancient geographers, namely *Ptolemy*, introduced 24 *climata*, being zones between two parallel circles, for which the length of the longest

---

[3] In the same sense: slope at a given site against rotations of the axis of earth (depending on geographical latitude).

day increases from the equator to the polar circle by half an hour, stepwise. It is easy to see that these zones correspond to different "climates"[4].

The first indication of shifting from that "calculated solar or geographic climate" to the modern concept of climate, describing the physical characteristics of a location, and referring beside heat (temperature) to many other aspects (wind, precipitation, evaporation, exhalations)[5] is found in Gehler (1789). Remarkably he notes "that the many local phenomena made it difficult to attribute the observations to a general theory" (Gehler, 1789, p. 765). A first short example of the modern usage of the term climate is given by *Wilhelm August Lampadius* (Lampadius 1806, p. 45) who wrote: "climate is behind the type of weather of a location, which should be studied in more detail by climatology" (... *die Art der Witterung eines Ortes, welche in der Climalehre näher soll bestimmt werden*). In Gehler (1830) "the most important conditions of the climate" (now called climatic elements) are listed (seven); temperature, humidity, earth surface type, relative position of the location to its surrounding, winds, altitude above sea level, and active volcanoes.

Our modern understanding of "climate" in the sense of "typical weather of specific locations", however, has long been used to describe the physical situation using meteorological elements such as temperature (heat−cold), precipitation (flood−drought), etc. without using the term *climate* (Khrgian 1970). Early historians and geographers, blending natural and human scientific exploration and description, lent a scholarly basis to determinist views. They described vegetation, animal, and even human populations as adapted to climatic constraints (Riebsame 1985). For example, *Hippocrates of Kos* (*circa* 460−377 B.C.) wrote the first book we know of on the relationship between air (climate), water and soil properties and its impacts on the psychological and physiological constitutions of the inhabitants.[6] The Greek and Latin titles of Hippocrates' work are: *Kōou Peri aerōn, hydatōn, topōn. Peri physōn Coi* "De aëre, aquis, et locis libellus eiusdem de flatibus"[7] or *Book of airs, waters and localities* (often simple entitled: *On the environment*)[8].

Hippocrates cast climate as the determinant of health and disease, and claimed that the microclimates of cities affect the civility and personality of their inhabitants. The ideas of Greek philosophers remained unchanged and unimproved (not even by the Romans) by new observations until end of the Middle Ages. *Johann Gottfried von Herder* (1744−1803) noted in the seventh book of his series "Ideen zur Philosophie" (Ideas on Philosophy), *Geschichte der Menschheit* (*History of Mankind*) in footnote 123:

---

[4] The astronomer *Hipparchus* in 140 B.C. stated that countries lying beyond the "climate" with a longest day of 17 h were uninhabitable on account of the cold and, thus, were of no interest.
[5] "... die im Luftkreise vorgehenden Verbindungen, Zersetzungen und Niederschläge, die Wirkung der Ausdünstung der Erdoberfläche, die Mittheilung der Temperatur anderer Orte durch Winde" (Gehler 1789, p. 765). This clearly expresses that atmospheric chemical processes, surface-air exchanges and circulations are parts ("elements") of the climate.
[6] *Poseidonius* (135−51 B.C.) was the first geographer to relate climate to man, saying that the people in the torrid zone, as a consequence of the heat and lack of rain, were born with woolly hair and protruding lips, their extremities being, as it were, gnarled (Brown 1949).
[7] Hippocratis: Graece & Latine Iano Cornario Zviccaviense Interprete. Basel: Hieronymus Froben und Johannes Herwagen, August 1529.
[8] "Airs, Waters, Places" in *Hippocrates*. Volume I. Translated by W. H. S. Jones. London: Heinemann, 1962.

S. Hippokrates, »De aëre, locis et aquis«, vorzüglich den zweiten Teil der Abhandlung. Für mich der Hauptschriftsteller über das Klima. (... the second part of treatment [is] excellent. For me the main writer on climate).

Alexander von Humboldt still based his work on climatology as "geographic meteorology" but with the aim being to "understand the entity in the diversity and to study the commonness and inner context of telluric phenomena" as the "ultimate scope of a description of the physical earth" (Humboldt 1845, Vol. I, pp. 55–56). His often cited definition of climate shows three remarkable particularities: the relation of *changes* instead of a mean status (as defined later in the nineteenth century), the inclusion of the chemical status of the atmosphere (by using the terms cleanness and pollution) and the restriction on parameters affecting human organisms but also the whole biosphere:

> The term *climate*, taken in its most general sense, indicates all the changes in the atmosphere which sensibly affect our organs, as temperature, humidity, variations in the barometric pressure, the calm state of the air or the action of opposite winds, the amount of electric tension, the purity of the atmosphere or its admixture with more or less noxious gaseous exhalations, and, finally the degree of ordinary transparency and clearness of the sky, which is not only important with respect to the increased radiation from the earth, the organic development of plants, and the ripening of fruits, but also with reference to its influence on the feeling and mental condition of men. (Humboldt 1850–1852, Vol. I, pp. 317–318)

> Der Ausdruck Klima bezeichnet in seinem allgemeinsten Sinne alle Veränderungen in der Atmosphäre, die unsere Organe merklich afficieren: die Temperatur, die Feuchtigkeit, die Veränderungen des barometrischen Druckes, den ruhigen Luftzustand oder die Wirkungen ungleichnamiger Winde, die Größe der elektrischen Spannung, die Reinheit der Atmosphäre oder die Vermengung mit mehr oder minder schädlichen gasförmigen Exhalationen, endlich den Grad habitueller Durchsichtigkeit und Heiterkeit des Himmels: welcher nicht bloß wichtig ist für die vermehrte Wärmestrahlung des Bodens, die organische Entwicklung der Gewächse und die Reifung der Früchte, sondern auch für die Gefühle und ganze Seelenstimmung des Menschen (Humboldt 1845, Vol. 1, p. 340).

Although recognizing that humankind's mental powers provide some independence from environmental factors, *von Humboldt* argued that culture is essentially a product of adaptation to the physical world, a key element of which is climate. *Humboldt's* definition was the basis for *Carl Dorno* establishing the modern (in contrast to the ancient idea of *Hippocrates*) *bioclimatology* in 1906, further developed by *Adolf Loewy* (1924). *Loewy* defined climate (in a physiological sense) as "the sum of all atmospheric and terrestrial states, typical for a given location and influencing human feeling" (behind "terrestrial state" he understood a mental category). Rubner (1907) defined climate as "all through the special location caused influences on human health". *Ellsworth Huntington* (1915) extended this view, claiming that climate is all-pervasive in molding social structure, settlement patterns, and human behavior (Huntington 1915).

It is remarkable that all these later definitions no longer include *Humboldt's* statement of *variations* of the "states", but set the climate based on mean or averaged values. It seems that *Humboldt's* definition, with his focus on biosphere-atmosphere interaction (we will see below that nowadays we accept this definition again in the sense of earth system research) was too broad and — with the rapid mathematical development of *meteorology* as a physical discipline in the second half of the nineteenth century — it was soon related only to atmospheric properties. But it is worth noting that Loewy (1924) includes not only the physical climate factors (which he considered to be the most important) but also *chemical* climate factors (in terms of constituents in air).

The father of the *physical* definition of climate is *Julius Hann* who defined the climate as "the entirety of meteorological phenomena which describe the mean status of the atmosphere at any given point of the earth's surface"[9] (Hann 1883, p. 1). He further stated that "climate is the entirety of the weathers[10] of a longer or shorter period as occurring on average at a given time of the year". The aim of climatology is to establish the mean states of the atmosphere over different parts of the earth's surface, including describing its variations (anomalies) within longer periods for the same location (Hann 1883). *Hann* also introduced the term *climatic element* emphasizing that measurements must characterize these through numerical values. He also freed the term "climate" from the close relation to humans and plants and related it to the time before life appeared on the earth (without using the term palaeoclimate). The climatic elements *Hann* listed in this order: temperature (air and soil, radiation), atmospheric humidity (including water vapour and precipitation), cloudiness, winds, air pressure (being less important in contrast to weather), evaporation, air composition. Measurements of minor constituents in air were very rare and uncertain at that time, but *Hann* noted explicitly dust, organics, ozone, hydrogen peroxide, in reference to the hygienic state of the air.

Köppen (1906) adopted *Hann's* definition ("mean weather at a given location") but stated in contrast to *Hann* that it is meaningless to define a climate without focus on human beings; hence only factors (elements) influencing "organic life" should be considered. *Köppen* explicitly presents a "second definition" of climate "... as the entirety of atmospheric conditions, which make a location on earth more or less habitable for men, animals and plants" (Köppen 1906, p. 8). *Hann* (1908) emphasized the importance of climatology not only to describe the "mean states of the atmosphere" but also the variations from this mean. In the first decade of the twentieth century, the hygienic aspect becomes very important (see Chapter 1.3.4). Before his definition of *bioclimatology*, Rubner (1907) defined climate in a similar way to *Köppen's* second definition of climate as: "... all influences on health given by the location". Köppen (1923, p. 3) wrote that "with the progress in knowledge new subjects will be included in the number of climatic elements when

---

[9] " ... Gesamtheit meteorologischer Erscheinungen, welche den mittleren Zustand der Atmosphäre an irgendeiner Stelle der Erdoberfläche charakterisieren."

[10] There is no direct English translation of German *Witterung*. The translation as "weather" (German *Wetter*) is not fully correct because *Witterung* denotes short-term averaged weather (a weather period).

their geographical characteristics are unveiled"[11]. Köppen (1923, 1931) also noted the importance of variation of the climatic elements, but stated the constancy of climate and did not discuss any climate change.

Around the middle of the nineteenth century increasing scientific observation of the problems of polluted air began (Smith 1845). An approach to *acid rain* occurred in a remarkably insightful way as early as 1852 by *Robert Angus Smith* who writes (Smith 1852, p. 213) "I do not mean to say that all rain is acid – it is often found with so much ammonia in it as to overcome the acidity; but in general, I think, the acid prevails in the town". The term *acid rain* he shaped indirectly. Then he remarked (Smith 1852, p. 216) relative to that city's air that "We may therefore find easily three kinds of air, – *that with carbonate of ammonia in the fields at a distance,* – *that with sulphate of ammonia in the suburbs,* – *and that with sulphuric acid, or acid sulphate, in the town.*" (*Smith*'s italics). In his book *"Air and Rain: The Beginning of a Chemical Climatology"*, published in London in 1872, *Smith* coined the term "chemical climatology".

Between 1857 and 1864, Prestel (1865) systematically investigated the ozone concentration in Emden. *Prestel*'s "ozonometric wind rose" we now call *air pollution climatology*[12]. Prestel (1872) wrote that "... the determination of the periodic and unperiodic occurrence of ozone be a relevant moment for the *climatology*".

During the last 60 years the idea of climate has broadened in so far that, in the definition of climate, apart from the mean value, higher statistical moments are included. According to the new definition, climate describes the "statistical behavior of the atmosphere, which is characteristic for a relatively large temporal order of magnitude" (Hantel et al. 1987). The climatic variables, also called climatic elements, are given as statistical figures, like, for example, yearly or monthly means or probabilities and frequency of events. The World Meteorological Organization (WMO) has fixed the period of time that forms the basis for the calculation of mean values of the weather to 30 years. Hence, the times from 1931 to 1960 and 1961 to 1990 are frequently used for climatic comparisons. However, for the calculation of mean values, other periods of time are found in literature. According to the WMO (1992), climate is defined as follows:

> Synthesis of weather conditions in a given area, characterized by long-term statistics (mean values, variances, probabilities of extreme values, etc.) of the meteorological elements in that area (WMO 1992, p. 112).

Likewise, according to the WMO, weather is defined as: "The state of the atmosphere mainly with respect to its effects upon life and human activities. As distinguished from climate, weather consists of the short-term (minutes to about 15 days) variations of the atmosphere state." That is, the terms weather and climate are

---

[11] "Mit dem Fortschritt des Wissens werden neue Gegenstände in die Zahl der klimatischen Elemente aufgenommen, wenn deren geographische Züge entschleiert werden."
[12] In German, *Immissionsklimatologie.* The term *immission* – not known in English – is deduced in analogy to *emission* (entry of matter into the atmosphere), however, it does not describe the discharge (which is referred to as deposition), but the concentration of a matter at the effective location, a somewhat spongy definition.

ascribed to the atmosphere and its *state*. The WMO also gives a very nice definition of atmosphere (and implicitly for *meteorology*):

> The envelope of air surrounding the earth and bound to it more or less permanently by virtue of the earth's gravitational attraction; the system whose chemical properties, dynamic motions, and physical processes constitute the subject matter of meteorology.

The reader's attention will instantly be attracted by the point, that here, in the description of the system, "chemical properties" are mentioned as well. Additionally, a (logical) definition of the discipline of meteorology is given by the inclusion of atmospheric chemistry. Also, a differentiation between the concepts of air and atmosphere can be made out in the sense that air is seen as a substantial (that is, chemical) composite, which behaves within the atmosphere according to geo-physical laws. Now, the WMO becomes inconsistent with its description of "meteorological elements" (that is, the system parameters of the state of the atmosphere), as no (atmospheric) chemical properties are numerated:

> Any one of the properties or conditions of the atmosphere which together specify the weather at a given place for any particular time (for example, air temperature, pressure, wind, humidity, thunderstorm and fog).

Hitherto we have understood "climate" as

> the sum of meteorological factors (elements) or
> the summary of weather or
> the mean (averaged) weather,

describing the mean status of the atmosphere at a given site of the earth's surface, represented by the statistical total properties (mean values, frequencies, durations etc.) of a long enough time period. As climate changes with both space and time, for a specification of climatic elements an inclusion of location and average time period, for which the statistical characteristics are given, is needed.

3.2 Climate and the climate system

Within climatology, as the rather descriptive science of climate, the meteorological definition of climate has proved itself. For an understanding of the dynamics of climate, that is, the processes that determine the average state and the variability of the atmosphere over longer periods, the meteorological definition is inadequate, as over longer periods changes in the atmosphere are considerably affected by interdependencies of the atmosphere, the ocean, vegetation and ice masses (Claußen 2006). For this reason, in climate dynamics, climate is defined by the state and the statistical behaviour of the *climatic system*, as can be read in modern textbooks on meteorology and climate physics (e. g., Peixoto and Oort 1992, Kraus 2004, Lutgens et al. 2009). Claußen (2006) distinguishes between a meteorological and a system-analytical definition. Note that it is essential to include the *statistics* into the idea of climate, i. e., climate means not simply the "mean weather". It follows that

- climate is a function of space and time;
- climate cannot be described as a single unit.

With our increasing understanding of global environmental processes, it has been concluded that the mean atmospheric status (over several years) not only depends on processes occurring in the atmosphere itself but also the oceanic circulation, the glacier movement, the spread of vegetation, etc. Consequently, Gates (1975) defined climate with three categories, namely the *climate system*, *climate states* and *climate change*. The climate system consists of the atmosphere, hydrosphere, cryosphere, lithosphere and biosphere. A climate state is determined by the full description of the statistics of the inner climate system. A climate change is the difference between two climate states of same kind.

In the history of mankind and the exploration of air and the atmosphere, the concept of climate has been subject to change, but also various descriptions have existed at the same time. It is beyond the scope of this book to address them here. Here, one can conclude that different definitions of climate are also in use. A priori, this is a contradiction, as there is only one climate system on earth. Obviously, this results from a pragmatic approach to the cognition and description of the climatic system by

a) diverse disciplinary points of view
b) different objectives (e. g. description of subsystems) and/or
c) differentiated knowledge of the system relationships.

The climate system (Fig. 3.1) consists of various subsystems: the atmosphere, the hydrosphere (which includes oceans, rivers, lakes, rain, groundwater), the cryosphere (ice sheets, sea ice, snow, permafrost), the marine and terrestrial biosphere, soil, and − when considering the development of climate over many millennia − the earth's crust and the upper mantle. Basically, this classification is carried out by means of the matter involved (gaseous, liquid and solid) and the timescales that can be observed for typical changes in the subsystems. The subsystems are linked to each other through flows of energy, momentum and matter. To the flows of matter, the transport of chemical substances and the processes of their transformation need to be added, as far as these substances − e. g., greenhouse gases or nutrients of the biosphere − are directly or indirectly related to the energy budget. The definition of the climate system is not derived from superior principles, but is a pragmatic restriction of the subject to be examined by classification in subsystems and interpretation of the respective system environment. The separation of the climate system from its environment is carried out in that way, as no significant flow of matter between the system and its environment occurs on timescales relevant for examination.

For Kraus (2004), the anthroposphere, the world of human action, belongs to the climatic system as well. On one hand, this seems to be reasonable, as human beings as a part of nature have altered the "natural system" to a large extent (see discussion in Chapter 2.4.1). On the other hand, it is not pragmatic, as human action, especially culture and psychology, elude thermodynamic description. In the literature, the total of climate system and anthroposphere is defined as the earth system (Schellnhuber and Wenzel 1998, Schellnhuber 1999, Claußen 1998, 2001). Hantel (2001) presents a very pregnant definition:

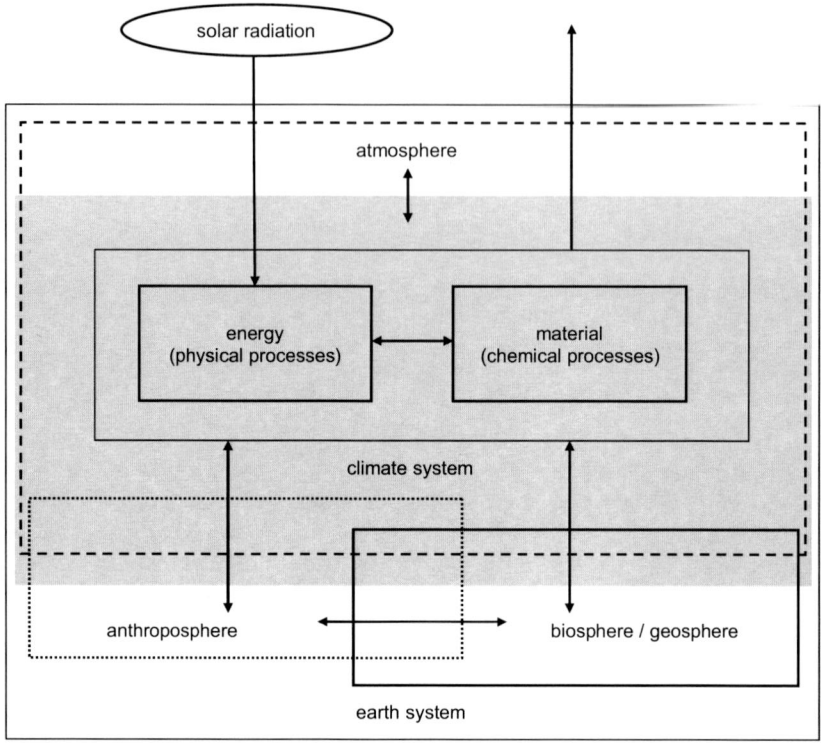

Fig. 3.1   The climate system (cf. also Fig. 1.1).

The climate is not a subject-matter but a property. Its carrier is the climate system. The climate is the entirety of the properties of the climate system.[13]

The climate system can be described (and widely quantified through measurements) by

1. the natural energy system,
2. the hydrologic cycle,
3. the carbon cycle and
4. the other biogeochemical cycles.

In a broader sense, the climate system can be seen as an interlayer within the earth system, buffering a habitable zone from uninhabitable physical and chemical conditions in altitude (upper atmosphere) and depth (deep lithosphere) (see Fig. 3.2). In a more narrow sense, the human-habitable zone is limited to the gas-solid interface (earth-surface/atmosphere) with a very small extension of a few tens of metres. The anthroposphere (or noosphere), however, is permanent spatially ex-

---

[13] Das Klima ist kein Gegenstand, sondern eine Eigenschaft. Ihr Träger ist das Klimasystem. Das Klima ist die Gesamtheit der Eigenschaften des Klimasystems.

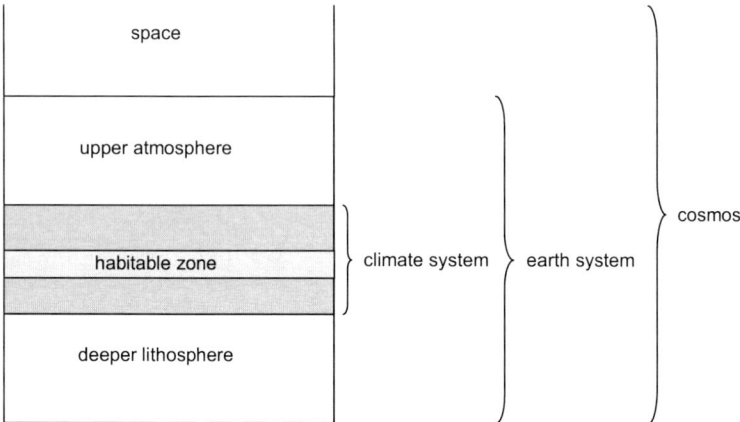

Fig. 3.2 Layer structure of the climate system

pending – also out of the climate system – due to the fact that humans are creating inhabited and uninhabited closed habitable systems in an uninhabitable surrounding.

3.3 Climate change and variability

Using the definition "the climate is the entirety of the properties of the climate system" (Hantel 2001), it is logical to define *climate change* as any change of the status of the climate system. According to Bärring (1993), the distinction between the terms *climate change*, *climate variation*, *climate fluctuation* and *climate variability* is still unclear, or was in 1993. He argued that this, perhaps, is not a problem in the scientific community but that difficulties may arise when climatological information is interpreted by non-specialists, for example, in a political or socio-economic context. Already one hundred years ago (Hann 1908, Alt 1916) the term climate change(s)[14] had been separated into changes, ongoing continuously of the same tenor and others, varying over more or less long periods around a mean value, called periodic variability.

Climate variations within a defined period (for example, 30 years) are not climate change but belong to the climate value, expressed as the periodic mean *and* its variance including statistics of extreme values. Human societies are highly adapted to the ordinary *annual* rhythms of climate, which we call the seasons. Moreover, almost all climates display *interannual* differences, which have hitherto appeared to be unpredictable (Hare 1985). *Climate change* is said to occur when the differences between successive averaging periods exceed what variance can account for, i. e., when a distinct signal exists that is visible above the variance. There are three principal causes of climate change and variability's:

---

[14] In German: *Klimaänderungen*.

- changing solar radiation, namely the solar constant (due to solar activities and/ or astronomical factors),
- changing chemical composition of the atmosphere (due to emission changes) and
- changing characteristics of the earth's surface (due to land-use change and/or geological as well as biological changes).

This includes also catastrophic events such as volcanic eruptions and the impact of celestial bodies. It is beyond the focus of this book to describe the physics of such alterations. However, with an understanding of the chemical evolution of the climate system (Chapter 2), it is evident that different chronological processes are superposed and that the different causes are interlinked in the sense of climatic feedbacks which make it very difficult to quantify climate changes and variations. Temperature and precipitation, as the most important climatic factors, are interrelated, but they are also interrelated with atmospheric composition and surface characteristics which again are interrelated.

Let us briefly look into the past climate, the *palaeoclimate*. With plant land settlement, the rate of photosynthesis quickly increased and induced in the *Carboniferous* era a decrease of atmospheric CO_2 due to the storage of carbon in biomass. On the other hand, CO_2 regeneration via oxidation of organic carbon was unable to balance the atmospheric CO_2 pool. As a consequence, due to lower absorption of terrestrial radiation by atmospheric CO_2, the earth's surface cooled. Lower temperatures reduced the photosynthesis and lowered the O_2 content. This explains the ice age in the *Permian* (280 Myr before present). By this controlling mechanism, oxygen and carbon dioxide did oscillate, being inversely correlated, but very smoothed and within a period of about 100 Myr (Budyko et al. 1987). The changing $CO_2 : O_2$ ratio was probably not a consequence but the cause of the *Pleistocene* ice ages. Photosynthetic assimilation works as a "regulator". Ice ages lasted over tens of thousands of years. With the beginning of an ice age, CO_2 concentration fell to 200 ppm and at the end (the beginning of a warm period) it rose to 270 ppm. The last *Würm* glaciation era ended 11 000 years ago with a reduced mean temperature of 3 °C. The global climate − or *earth climate* − is also influenced by external factors (variations of solar radiation and orbit), which are probably responsible for the 10 000−50 000-year oscillation of ice ages within the last 2 Myr. Internal factors concern changes in surface albedo or solar scattering due to volcanic eruptions, variation of oceanic circulation and changes of atmospheric chemical composition. Solar models show that at the beginning of the earth's evolution (4.5 Gyr ago), solar radiation must have been reduced by 25−30 % compared with the present time. The subsequent lower earth surface temperature (8 %) would freeze all oceans for 2 Gyr. However, total freezing never happens and it is assumed that temperature compensation occurred due to an increasing greenhouse effect (Wayne 1996). Despite alternating ice ages and warmer eras, our climate has to be considered as quite stable over the last 3.5 Gyr. Ice ages affected the earth's surface in areas above 45° N and below 45° S − about 30 % of the earth's surface was permanently partially frozen. All biogeochemical findings suggest that the ocean has never been completely frozen or boiling. Very likely the mean surface temperature ranged between 5 °C and 50 °C. If there were not any infrared absorbing gases in our atmosphere and if the surface had an albedo − just as Mars − the mean near surface temperature would measure about 23 °C lower. These components, called greenhouse gases, are mainly H_2O, CO_2 and CH_4. The radiation budget (and therefore the heat budget) is modified by atmospheric aerosol and clouds. The presence of H_2O and CO_2 in the secondary atmosphere originated more than 4 Gyr ago and was a requirement for the oceans not to freeze. Natural climate variations in periods ranging from decades to hundreds of years have been found; the causes are not yet clear. Variations of the earth's rotation − caused by core dynamics or shifts in mass at the surface − may be mechanisms. Sunspot cycles (11 and 22 years) as well as cycles of orbits of the outer planets (Uranus, Neptune and Pluto) (208 years) are other causes.

The conclusion that the climate is permanently but slowly changing is true. Now, however, we have the likely situation of man-made *abrupt climate change*. An abrupt climate change occurs when the climate system is forced into transition to a new state at a rate that is determined by the climate system itself, and which is more rapid than the rate of change of the external forcing. Moreover, with this background we understand *climate change* better, because it means that one or more of the earth-system determining climate elements is changing by a qualitative jump[15]. Despite the linguistic simplification, we should not forget the complexity of the system: even the most advanced mathematical models running on the biggest (and networked) computers are still unable to produce a true picture of the climate system sufficient to draw *reliable* conclusions. Experimental validation of the model outputs, however, is only possible *after* climate change − that is the dilemma! Parts of the model (or the system) can be validated with relevant experimental approaches, for example, cooling of the atmosphere by particulate matter after large volcanic eruptions.

Volcanism has long been implicated as a possible cause of weather and climate variations; *Benjamin Franklin* and *William Jackson Humphrey* were pioneers in their association of volcanic eruptions with climate change. Franklin (1784) put forward a theory linking volcanic dust to climatic change:

> During several of the summer months of the year 1783, when the effect of the sun's rays to heat the earth in these northern regions should have been greater, there existed a constant fog over all Europe, and great part of North America. This fog was of a permanent nature; it was dry, and the rays of the sun seemed

---

[15] In nature there is no jump, as nature itself composes of jumps. The idea of interactive changes from quality to quantity and vice versa is a main principle of dialectics. The fact that our subjective thinking and the objective world underlies the same laws and therefore cannot be contradictory in their results, but correspond, rules our complete theoretical thinking. It is an unconscious and unconditional premise. The materialism of the eighteenth century studied this premise according to its vitally metaphysical character on its content only. It limited itself to the proof that the content of all thinking and knowledge descends from sensual experience, and reconstructed the sentence: *Nihil est in intellectu, quod non fuerit in sensu* (Nothing is in the mind, which has not before been in the senses). Dialectics as the science of the common laws of all movement was first scientifically postulated by *Friedrich Engels* (s. *Karl Marx / Friedrich Engels − Werke. Band 20. Dialektik der Natur*, Berlin: (Karl) Dietz Verlag, 1962).

to have little effect towards dissipating it, as they easily do a moist fog, arising from water... The cause of this universal fog is not yet ascertained. ... or whether it was the vast quantity of smoke, long continuing; to issue during the summer from Hecla in Iceland, and that other volcano which arose out of the Sea near that island, which smoke might be spread by various winds, over the northern part of the world, is yet uncertain... It seems however worth the enquiry, whether other hard winters, recorded in history, were preceded by similar permanent and widely extended summer fogs.

Remarkably, in 1783 it was not Hekla (Iceland's most active volcano) that erupted, but another volcano, Laki (Stothers 1996). Humphrey (1913) was inspired by the Kalmai (Alaska) eruption in 1912, which led to a decrease in temperature, to study records following all volcanic eruptions dating back to the year 1759, and found that all the eruptions were connected with a cooling of the earth's surface air temperature. Furthermore, in the discussion concerning the changes resulting from sunspot activities, *Humphrey* shows that its influence is small compared with the transparency of the atmosphere and explained the small solar influence on surface temperature by increased UV absorption by oxygen and ozone formation in the upper atmosphere.

However, only in recent decades has it became clear (Wexler 1952, Lamb 1970, Hammer 1981, Sigurdsson 1982, Rampino and Self 1982, Handler 1989, Robock 1991, Lacis et al. 1992) that sulfate aerosol particles from SO_2 emitted into high altitudes, which have an effective radius of about 0.5 μm, equivalent to the wavelength of visible light, interact more strongly with the shortwave solar radiation than the longwave terrestrial radiation (~ 10 μm). Some absorption occurs, but mostly scattering of solar radiation and thus cooling the planet (Robock 2003).

Hann (1897) first addresses climate changes with a separate section in his handbook in the sense of periodic variations ("the distribution of climatic elements is not [an] absolute constant") but also in the sense of continuous changes over the earth's evolution. A first systematic description of climate changes is found in Hann (1908). Brückner (1890) first identified climatic cycles by analyzing meteorological records (the so-called *Bruckner cycle* of 35 years).

At the end of the nineteenth century, almost "astronomical theories" (Adhémar 1842, Croll 1864, 1875) were cited to explain climate variation, for example, glacial periods. Marchi (1895, p. 207) arrived at the conclusion that all these hypotheses must be rejected. Hann (1897, 1908) also found more arguments against the astrononomical theory than for it. Today this theory (*Milankovic cycles*) is largely supported by findings from oceanic sediments. *James Croll* introduced the idea of changes in the earth's orbital elements as probably a periodical and extraterrestrial mechanism for initiating multiple glacial epochs. *Joseph Adhémar* had considered only the climatic effects of the present amount of eccentricity, not the effect of its changes (Fleming 2006). The astronomical theory re-emerged from eclipse and was formulated into the mathematical theory of insolation by *Milutin Milankovic* between 1920 and 1941 (Milankovic 1920, 1930, 1941). *Milankovic cycles* are cycles in the earth's orbit that influence the amount of solar radiation striking different parts of the earth at different times of year. Besides the astronomical theory, the carbon dioxide theory developed in the nineteenth century; today both theories are among the mainstream.

The carbon dioxide theory emerged as a consequence of the experimental work of *John Tyndall*, who wrote in 1861 that slight changes in the amount of any of the radiatively active constituents of the atmosphere – water vapor, carbon dioxide, ozone or hydrocarbons – may have produced "all the mutations of climate which the researches of geologists reveal ... they constitute true causes, the extent alone of the operation remaining doubtful of climate which the researches of geologists reveal ..." (Tyndall 1861). *Luigi de Marchi* first explains glacial periods through changes of atmospheric properties such as transmission (Marchi 1895). At the same time, *Svante Arrhenius* first established the carbon dioxide theory. He wrote (Arrhenius 1896):

> The air retains heat (light or dark) in two different ways. On the one hand, the heat suffers a selective diffusion on its passage through the air; on the other hand, some of the atmospheric gases absorb considerable quantities of heat. These two actions are very different (p. 238) ...
> The selective absorption of the atmosphere is, according to the researches of *Tyndall, Lecher* and *Pernter, Röntgen, Heine, Langley, Ångström, Paschen,* and others, of a wholly different kind. It is not exerted by the chief mass of the air, but in a high degree by aqueous vapour and carbonic acid, which are present in the air in small quantities. Further, this absorption is not continuous over the whole spectrum, but nearly insensible in the light part of it, and chiefly limited to the long-waved part, where it manifests itself in very well-defined absorption-bands, which fall off rapidly on both sides (p. 239) ...
> ... that there exists as yet no satisfactory hypothesis that could explain how the climatic conditions for an Ice Age could be realized in so short a time as that which has elapsed from the days of the glacial epoch... a preliminary estimate of the probable effect of a variation of the atmospheric carbonic acid on the belief that one might in this way probably find an explanation for temperature variations of 5−10 °C. (p. 267) ...
> One may now ask how much must the carbonic acid vary according to our figures, in order that the temperature should attain the same values as in the Tertiary and Ice Ages respectively? A simple calculation shows that the temperature in the arctic regions would rise about 8 ° to 9 °C, if the carbonic acid increased to 2.5 or 3 times its present value. In order to get the temperature of the ice age between the 40th and 50th parallels, the carbonic acid in the air should sink to 0.62−0.55 of its present value (lowering of temperature 4−5 °C.). (p. 268)

This idea could only answer the riddle of the ice ages, however, if such large changes in atmospheric composition really were possible. For that question *Arrhenius* turned to a colleague, *Arvid Högbom*. It happened that *Högbom* had compiled estimates for how carbon dioxide cycles through natural geochemical processes, including emission from volcanoes, uptake by the oceans, and so forth. Arrhenius (1896) cited Högbom (1894) in the following:

> The world's present production of coal reaches in round numbers 500 millions of tons per annum. (p. 270)...
>   ... that the most important of all the processes by means of which carbonic acid has been removed from the atmosphere in all times, namely the chemical

weathering of siliceous minerals, is of the same order of magnitude as a process of contrary effect, which is caused by the industrial development of our time, and which must be conceived of as being of a temporary nature. (p. 271) ...

Carbonic acid is supplied to the atmosphere by the following processes:

1. volcanic exhalations and geological phenomena connected therewith,
2. combustion of carbonaceous meteorites in the higher regions of the atmosphere,
3. combustion and decay of organic bodies,
4. decomposition of carbonates, and
5. liberation of carbonic acid mechanically enclosed in minerals on their fracture or decomposition.

The carbonic acid of the air is consumed chiefly by the following processes:

6. formation of carbonates from silicates on weathering, and
7. the consumption of carbonic acid by vegetative processes.

The ocean, too, plays an important role as a regulator of the quantity of carbonic acid in the air by means of the absorptive power of its water, which gives off carbonic acid as its temperature rises and absorbs it as it cools. The processes named under (4) and (5) are of little significance, so that they may be omitted. So too the processes (3) and (7), for the circulation of matter in the organic world goes on so rapidly that their variations cannot have any sensible influence. From this we must except periods in which great quantities of organisms were stored up in sedimentary formations and thus subtracted from the circulation, or in which such stored-up products were, as now, introduced anew into the circulation. The source of carbonic acid named in (2) is wholly incalculable. Thus the processes (1), (2), and (6) chiefly remain as balancing each other. As the enormous quantities of carbonic acid (representing a pressure of many atmospheres) that are now fixed in the limestone of the earth's crust cannot be conceived to have existed in the air but as an insignificant fraction of the whole at any one time since organic life appeared on the globe, and since therefore the consumption through weathering and formation of carbonates must have been compensated by means of continuous supply, we must regard volcanic exhalations as the chief source of carbonic acid for the atmosphere. But this source has not flowed regularly and uniformly. Just as single volcanoes have their periods of variation with alternating relative rest and intense activity, in the same manner the globe as a whole seems in certain geological epochs to have exhibited a more violent and general volcanic activity, whilst other epochs have been marked by a comparative quiescence of the volcanic forces. It seems therefore probable that the quantity of carbonic acid in the air has undergone nearly simultaneous variations, or at least that this factor has had an important influence. (p. 272)

Twenty years later, Alt (1916) noted that since Dufour (1870) "... the question of climate change in historic time is still completely open and the statement of the majority of meteorologists that the climate does not change is neither proved nor rejected". The carbon dioxide theory of climate change was in deep eclipse in 1938 when *Guy Stewart Callendar* revived it and placed it on a firm scientific basis.

Callendar (1938, 1949) documented a significant upward trend in temperatures for the first four decades of the twentieth century and noted the systematic retreat of glaciers. He compiled estimates of rising concentrations of atmospheric CO_2 since pre-industrial times and linked the rise of CO_2 to the combustion of fossil fuel. Finally, he synthesized information newly available concerning the infrared absorption bands of trace atmospheric constituents and linked increased sky radiation from increased CO_2 concentrations to the rising temperature trend. Today this is called the *Callendar* Effect. The next great contributor to the carbon dioxide theory was *Gilbert Norman Plass* who established connections between the physics of infrared absorption by gases, the geochemistry of the carbon cycle, feedback loops in the climate system and computer modeling (Plass 1953, 1956a, b). Using recent measurements of the influence of the 15-μm CO_2 absorption band, he calculated a 3.6 °C surface temperature increase for doubling of atmospheric carbon dioxide and a 3.8 °C degree decrease if the concentration were halved. Contrary to the assumptions of many scientists at the time, the effect of water vapor absorption did not mask the carbon dioxide effect by any means. He used these results to argue for the applicability of the carbon dioxide theory of climate change for geological epochs and in recent decades (Fleming 2010). It is worth mentioning in this line *Charles David Keeling* who commenced continuous measurement of atmospheric CO_2 at Mauna Loa (Hawaii) in 1958, showing the rise of atmospheric CO_2 over the decades.

However, it has only become possible in the last few decades to look back into the past through different sediment and ice-core investigations, to record data indicating climate change. In January 1998, the collaborative ice-drilling project between Russia, the United States and France at the Russian Vostok station in East Antarctica yielded the deepest ice core ever recovered, reaching a depth of 3623 m (Petit et al. 1997, 1999). Preliminary data indicates that the Vostok ice-core record extends through four climate cycles, with ice slightly more than 400 000 years old (Petit et al. 1997, 1999). Because air bubbles do not close at the surface of the ice sheet but only near the firn-ice transition (that is, at ~ 90 m below the surface at Vostok), the air extracted from the ice is younger than the surrounding ice (Barnola et al. 1991). Using semi-empirical models of densification applied to past Vostok climate conditions, Barnola et al. (1991) reported that the age difference between air and ice may be ~ 6000 years during the coldest periods instead of ~ 4000 years, as previously assumed. Ice samples were cut with a bandsaw in a cold room (at about −15 °C) as close as possible to the centre of the core in order to avoid surface contamination (Barnola et al. 1983). Gas extraction and measurements were performed with the Grenoble analytical setup, which involved crushing the ice sample (~ 40 g) under vacuum in a stainless steel container without melting it, expanding the gas released during the crushing in a pre-evacuated sampling loop, and analyzing the CO_2 concentrations by gas chromatography (Barnola et al. 1983) (see Fig. 3.3). The analytical system, except for the stainless steel container in which the ice was crushed, was calibrated for each ice sample measurement with a standard mixture of CO_2 in nitrogen and oxygen. For further details on the experimental procedures and the dating of the successive ice layers at Vostok, see Barnola et al. (1987, 1991), Lorius et al. (1985), and Petit et al. (1999). The changes shown in Fig. 3.3 cover three glacial terminations (Fischer et al. 1999) and are most notably

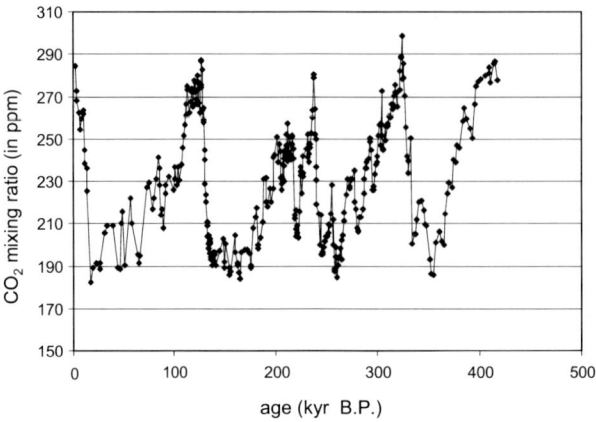

Fig. 3.3 Historical CO_2 record from the Vostok Ice Core, Antarctica; data from Barnola et al. (2003).

an inherent phenomenon of change between glacial and interglacial periods. Besides this major 100 000-year cycle, the CO_2 record seems to exhibit a cyclic change with a period of some 21 000 years (Etheridge et al. 1996).

There is a close correlation between Antarctic temperature (Fig. 3.4) and atmospheric concentrations of CO_2 (Barnola et al. 1987, Blunier et al. 2005). The extension of the Vostok CO_2 record shows that the main trends of CO_2 are similar for each glacial cycle. Major transitions from the lowest to the highest values are associated with glacial-interglacial transitions. During these transitions, the atmospheric concentration of CO_2 rises from 180 to 280–300 ppm (Petit et al. 1999). The extension of the Vostok CO_2 record shows the present-day levels of CO_2 are unprecedented during the past 420 000 years. Pre-industrial Holocene levels (~ 280 ppm) are found during all interglacials, with the highest values (~ 300 ppm) found approximately 323 000 years BP. When the Vostok ice-core data was compared with other ice-core data (Delmas et al. 1980, Neftel et al. 1982) for the past 30 000–40 000 years, good agreement was found between the records. All show low CO_2 values (~ 200 ppm) during the Last Glacial Maximum and increased atmospheric CO_2 concentrations associated with the glacial-Holocene transition. According to Barnola et al. (1991) and Petit et al. (1999), these measurements indicate that, at the beginning of the deglaciations, the CO_2 increase either was in phase or lagged by less than ~ 1000 years with respect to the Antarctic temperature, whereas it clearly lagged behind the temperature at the onset of the glaciations. The main significance of the new data lies in the high correlation between GHG concentrations and temperature variations over 420 000 years and through four glacial cycles. However, because of the difficulty in precisely dating the air and water (ice) samples, it is still unknown whether GHG concentration increases precede and cause temperature increases, or vice versa, or whether they increase synchronously. It is also unknown how much of the historical temperature changes have been due to GHGs, and how much has been due to orbital forcing, i.e., increases in solar radiation, or perhaps long-term shifts in ocean circulation.

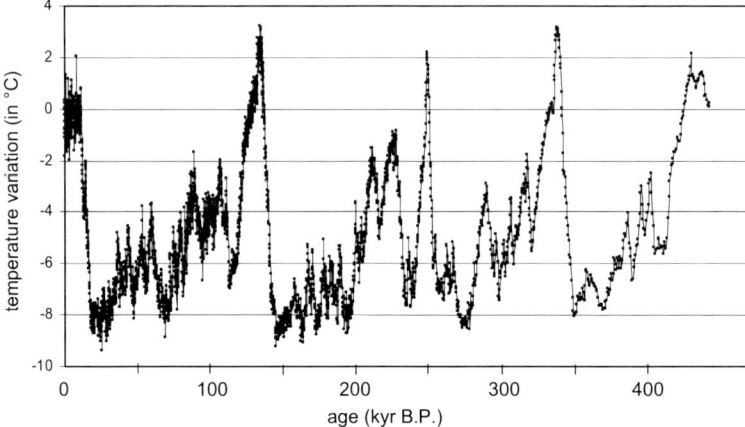

Fig. 3.4 Historical isotopic temperature record from the Vostok Ice Core. Source: Petit et al. (2000). Temperature variation as a difference from the modern surface temperature value of −55.5 °C.

Whether the ultimate cause of temperature increase is excess CO_2, or a different orbit, or some other factor, probably does not matter much. It could have been one or the other, or different combinations of factors at different times in the past. The effect is still the same. Nevertheless, the scientific consensus is that GHGs account for at least half of temperature increases, and that they strongly amplify the effects of small increases in solar radiation due to orbital forcing. Given all the new ice-core data, what changes can we anticipate for our climate? If CO_2 has increased over the past 150 years as much as it normally increases over thousands of years leading up to an interglacial phase (about 80 ppm), then we could expect as much as a corresponding 10−12 °C increase in temperature. But if half the historical temperature increases have been due to orbital forcing and other factors, then we should expect an increase of "only" about 5−6 °C.

There have been two interruptions in the rise of global average temperature since 1956 (Fig. 3.5), and of course, the earth's climate is influenced by more than just CO_2. Other trace gases and black carbon warm the climate, and aerosols cool it. On a larger scale, the astronomical theory of orbital influences was revived and climate variation attributed to such factors as ENSO, the Pacific Decadal Oscillation, and solar activities (or the lack thereof) are now being widely discussed. Still, more than 50 years later, scientists agree that the uncontrolled experiment pointed out by *Plass* in 1956 has been verified, and a warmer future caused by the radiative effects of CO_2 is in store (Fleming 2010).

Whereas the *Milanković cycles* affect directly the temperature through redistribution of insolation, carbon dioxide changes begin timely delayed. Due to the uncertainties of differences in the ages of ice and included air bubbles, this delay (long assumed by climatologists) could not be proved before the exploration of the Vostok ice core. It means that the beginning (and ending) of glacial periods was caused without CO_2 but only the *Milanković cycles*. CO_2 amplifies warming and cooling according to the increase and decrease of its atmospheric concentration. The feed-

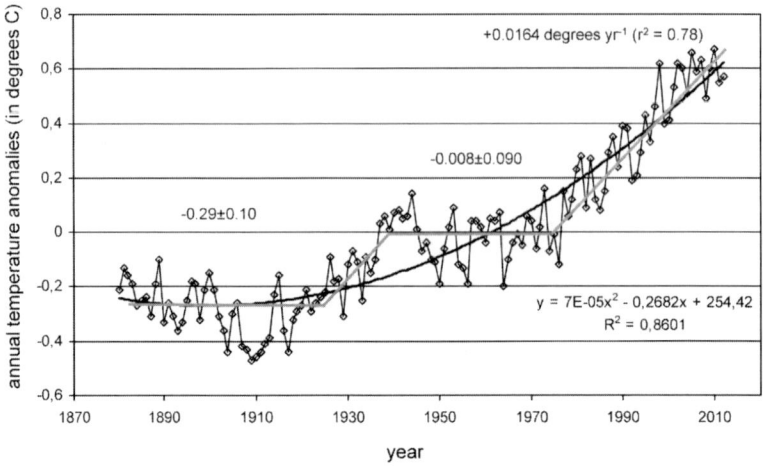

Fig. 3.5 Line plot of global mean land-ocean temperature index, 1880 to present, with the base period 1951–1980. Source: NASA Goddard Institute for Space Studies, New York, NY (see also Hansen et al. 2006). The curve represents a polynomial fit, the grey lines a stepwise linear approach.

back is as follows: increasing temperature exhausts CO_2 from oceans and increasing atmospheric CO_2 amplifies warming. With decreasing temperature, oceans absorb more atmospheric CO_2 and less atmospheric CO_2 reduces air temperature. What starts in this cycle depends simply on the trigger: temperature or carbon dioxide.

The prime indicator of global warming is, by definition, global mean temperature (Fig. 3.5). Time series of global temperature show a well-known rise since the early 20th century, and most notably since the late 1970s. From 1900 to 1940, there is an increase of 0.3 °C, between 1940 and 1975 it remains static, and it increases again by a further 0.6 °C until 2009. This widespread temperature increase is corroborated by a range of warming-related impacts: shrinking mountain glaciers, accelerating ice loss from ice sheets in Greenland and Antarctica, shrinking Arctic sea ice extent, sea level rise, and a number of well-documented biospheric changes like earlier bud burst and blossoming times in spring (IPCC 2007). Whereas greenhouse gas increase has been shown to cause the centennial trend of global temperature rise since the industrial revolution (see Figs. 3.5 and 3.6), global temperature has remained flat for the past 15 years. Several hypothetical explanations have been published (see Benestad 2012 and citations therein). However, as seen from Fig. 3.5, periods of more or less warming (or cooling, respectively) changed; but on average, every decade was warmer than the previous one. The decade 2000–2009 was warmer than the previous decade, with a surface temperature about 0.54 °C above the long-term twentieth-century average.

The warming goes on (Foster and Rahmstorf 2011) but the ocean (likely due to changing exchange and current processes) stores more heat, cooling the air surface temperature (Kosaka and Xie 2013). The globally-averaged temperature for 2013

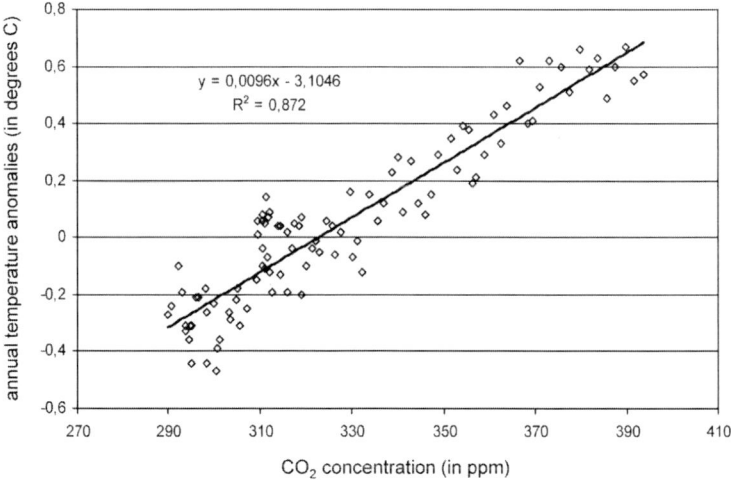

Fig. 3.6 Scatter plot between annual temperature anomalies (data source see Fig. 3.5) and atmospheric CO_2 concentration, derived from Antarctic ice core and Mauna Loa records (data source see Fig. 2.66).

tied as the fourth warmest year since record keeping began in 1880[16]. It also marked the 37th consecutive year with a global temperature above the 20th century average. The last below-average annual temperature was 1976. Including 2013, all 13 years of the 21st century (2001−2013) rank among the 15 warmest in the 134-year period of record. The three warmest years on record are 2010, 2005, and 1998.

According to the report of the IPCC (2007, 2013) the following statements concerning climate change can be made:

- The globally averaged combined land and ocean temperature data as calculated by a linear trend, show a warming of 0.85 (0.65 to 1.06) °C, over the period 1880−2012.
- The 2000s was the warmest decade and 2013 was the warmest year since 1861.
- The rise in temperature over 100 years was the greatest in the last 1000 years.
- In some areas of the southern hemisphere, especially in Antarctica and over the oceans, no warming has been registered.
- It is virtually certain that the upper ocean (above 700 m) warmed from 1971 to 2010, and likely that it has warmed from the 1870s to 1971.
- There is very high confidence that the Arctic sea ice extent (annual, multi-year and perennial) decreased over the period 1979−2012; between 3.5 and 4.1% per decade (range of 0.45 to 0.51 million km² per decade).
- It is very likely that the annual Antarctic sea ice extent increased at a rate of between 1.2 and 1.8% per decade (0.13 to 0.20 million km² per decade) between 1979 and 2012.

---

[16] State of the Climate. Global Analysis. December 2013. National Oceanic and Atmospheric Administration (NOAA), USA. National Climatic Data Center.

- Antarctic ice sheet has been losing ice during the last two decades The average rate of ice loss from Antarctica likely increased from 30 (37 to 97) Gt yr^{-1} (sea level equivalent, 0.08 (−0.10 to 0.27) mm yr^{-1}) over the period 1992−2001, to 147 (72 to 221) Gt yr^{-1}over the period 2002−2011 0.40 (0.20 to 0.61) mm yr^{-1}).
- There is very high confidence that glaciers world-wide are persistently shrinking as revealed by the time series of measured changes in glacier length, area, volume, and mass. During the last decade (2001−2010), the largest contributions to global glacier ice loss were from glaciers in Alaska, the Canadian Arctic, the periphery of the Greenland ice sheet, the Southern Andes, and the Asian mountains. Together, these areas account for more than 80 % of the total ice loss.
- There is very high confidence that maximum global mean sea level (GMSL) during the last inter-glacial period (129 to 116 kyr) was, for several thousand years, at least 5 m higher than present. GMSL has risen by 0.19 m (0.17 m to 0.21 m), estimated from a linear trend over the period 1901−2010.
- Satellite and in situ observations show significant reductions in the NH snow cover extent over the past 90 years, with most of the reduction occurring in the 1980s. Snow cover decreased the most in June when the average extent decreased very likely by 53 % (40 to 66 %) over the period 1967 to 2012.
- Observations from surface stations, radiosondes, global positioning systems, and satellite measurements indicate increases in tropospheric water vapour at large spatial scales. It is very likely that tropospheric specific humidity has increased since the 1970s. The magnitude of the observed global change in tropospheric water vapour of about 3.5 % in the past 40 years is consistent with the observed temperature change of about 0.5 °C during the same period, and the relative humidity has stayed approximately constant.
- At present, there is medium confidence that there has been a significant human influence on global scale changes in precipitation patterns, including increases in Northern Hemisphere mid-to-high latitudes. Precipitation in the twentieth century rose in most areas of the middle and higher latitudes of the northern hemisphere by 0.5−1 % per decade, while it sank over tropical areas (10° N−10° S) by 0.2−0.3 %. The frequency of heavy rains in middle and higher latitudes of the northern hemisphere rose by 2−4 % in the last 20−30 years. Cloud coverage in middle and higher latitudes of the northern hemisphere probably rose by 2 %
- Between 1750 and 2011, CO_2 emissions from fossil fuel combustion and cement production are estimated from energy and fuel use statistics to have released 375 (345 to 405) Pg C. In 2002−2011, average fossil fuel and cement manufacturing emissions were 8.3 (7.6 to 9.0) Pg C yr^{-1}, with an average growth rate of 3.2 % yr^{-1}. This rate of increase of fossil fuel emissions is higher than during the 1990s (1.0 % yr^{-1}). In 2011, fossil fuel emission derived from land cover data and modelling, is estimated to have released 180 (100 to 260) Pg C. Of the 555 (470 to 640) Pg C released to the atmosphere from fossil fuel and land use emissions from 1750 to 2011, 240 (230 to 250) Pg C accumulated in the atmosphere, as estimated with very high accuracy from the observed increase of atmospheric CO_2 concentration from 278 (273 to 283) ppm in 1750 to 390.5 (390.4 to 390.6) ppm in 2011.
- The concentration of CH_4 has increased by a factor of 2.5 since pre-industrial times, from 722 (697 to 747) ppb in 1750 to 1803 (1799 to 1807) ppb in 2011.

- Since pre-industrial times, the concentration of N_2O in the atmosphere has increased by a factor of 1.2. The atmospheric concentration of N_2O (46 ppb) rose by 17% from 1750.

Solomon et al. (2009) stated that following the cessation of emissions, the removal of atmospheric carbon dioxide will decrease radiative forcing, but that is largely compensated by the slower loss of heat to the ocean, so that atmospheric temperatures do not drop significantly for at least 1000 years. Among illustrative irreversible impacts that should be expected if atmospheric carbon dioxide concentrations increase from current levels (~ 400 ppm) to a peak of $\geq$ 500 ppm over the coming century are irreversible dry season rainfall reductions in several regions comparable to those of the "dust bowl" era and inexorable sea level rise. The thermal expansion of the warming ocean provides a conservative lower limit to irreversible global average sea level rise of at least 0.4−1.0 m if twenty-first century CO_2 concentrations exceed 600 ppm and 0.6−0.9 m for peak CO_2 concentrations exceeding 1000 ppm. Additional contributions from glaciers and ice sheet contributions to future sea level rises are uncertain but might equal or exceed several metres over the next millennium or longer.

3.4 Climate and chemistry

At this point, we have seen that the term *climate* inevitably includes chemistry and the climate system "works" because of chemical processes. Chemistry (in terms of chemical compounds and reactions) and transport are inextricably linked with each other. Physics (in terms of energy conversions) is the driving force of chemical processes. However, the climate system does not work without biology (in terms of biochemical processes). Predicting the changing chemical state is an essential task of climate research and without long-term chemical monitoring (concentrations of chemical compounds in air, within particulate matter, rain and cloud water) we have no insight in climatological processes and no opportunity to test adequate mathematical models. Monitoring of compounds in air was presented in Chapter 2.8.2 and some key results from our precipitation and cloud chemistry monitoring is now presented in Chapters 3.4.2 and 3.4.3.

3.4.1 Chemical weather and climate

Climatologists agree that 'climate' refers to the long-term *state of the atmosphere* (characterized by its mean as well as by variations thereof). Atmosphere, in turn, is characterized by air (in the sense of a mixture of substances) and the processes occurring therein. It follows that, in the attempt to describe the atmosphere, a separation between chemistry and physics would be fatuous. The chemical composition of air contains, among its main components (nitrogen and oxygen) and its minor components (noble gases and water), a virtually "unlimited" number of trace gases and solid components (aerosol particles) as well as (temporarily) dissolved substances (in hydrometeors). In analogy to the *meteorological* definition of weather, Lawrence et al. (2005) define the *chemical weather* as:

> Local, regional, and global distribution of important trace gases and aerosols and their variabilities on time scales of minutes to hours to days, particularly in light of their various impacts, such as on human health, ecosystems, the meteorological weather, and climate.

Certainly, the alert observer will instantly notice that (in hydrometeors) dissolved trace matters are missing, and then the question arises about what are *important* and what are *unimportant* trace matters. In the term "weather" (which describes the short term time scale of the state of the atmosphere) with respect of effects on life − actually unavoidably − a priori atmospheric chemical species is included. Hence, without further discussion, atmospheric chemistry can be incorporated into the definition of weather (although in all likelihood no meteorologist would have thought so).

Thus, the definition of Lawrence et al. (2005) can be adjusted to:

> The chemical state of the atmosphere mainly with respect …

When we leave out the short-term time scale, that is, we use the latter sentence concerning the definition of climate, we arrive at a formal definition of chemical climate as:

> The synthesis of chemical weather conditions in a given area, characterized by long-term statistics (mean values, variances, probabilities of extreme values, etc.) of the chemical substances in that area.

Instead of "meteorological elements", we have "chemical substances". When − as implicitly mentioned earlier − meteorology feels responsible as a discipline for the description of the chemical state of the atmosphere, the list of "meteorological elements" can simply be expanded to accommodate all relevant chemical variables as well. So as not to raise misunderstandings, with the extension of the term climate chemical aspects are dealt with, not with "chemistry", just as climatology is not a subdiscipline of physics. The variation over time (intraday, trend, etc.) of the concentration of an atmospheric trace matter is not "chemistry". Chemistry deals with the transformation of the trace substance and the corresponding atmospheric conditions, that is, it investigates − in analogy to the physics of the atmosphere − the process and the change of states. Here, the chemical substance (like, for example, temperature as physical "element") is understood as a "state variable" in the sense of a geographical (time-space dependence) and a meteorological (atmospheric-phenomenological) property. I would like to define climate in a general sense as follows:

> Climate describes the mean status of the atmosphere at a given site of the earth's surface, represented by the statistical total properties (mean values, frequencies, durations etc.) of a long enough time period.

It is understood (and therefore no differentiation between a *meteorological* and a *system-related* definition of climate according to *Claußen* shall be undertaken), that the atmosphere is only one part of the climatic system and therefore, climate can only correctly be described in a physical-chemical manner taking into consideration material and energetic interdependencies with the other subsystems. Advantageously, a *climatic state* (i. e., the scientific, broadly mathematical description of the climate) can be defined:

A climate state is given by the whole description of the statistical status of the internal climate system.

Why – apart from an academic point of view – is the inclusion of the "chemical dimension" into the description of climate so important? Why is the observation of a "chemical weather" phenomenon of no or only little importance? The current state of the atmosphere (weather) in relation to its physical component has an enormous relevance for humans and society and any further rationale is superfluous.

Using the definitions of chemical weather above, the meaning is limited to selected "weather elements", like the chemical composition of even selected species, e. g., ozone concentration concerns the exceeding of threshold values (ozone forecast and daily information). Some institutions prepare forecasts of "chemical weather". There is not much sense in doing this, as trace gas concentrations presently (when there is not an accident) are almost always below acute impact levels and there is neither any way nor any need to respond to chemical weather. Such information only makes sense when (a) a possible hazard is associated with it and (b) the possibility of prevention and riposte, respectively, exists. An effective *acute* countermeasure is not known (e. g. bans or restraints on driving in particular cities have been shown to be ineffective – which with deeper knowledge of atmospheric and chemical processes should have been known before). However, prior warning (e. g. of winter and summer smog)[17] would have been of importance to enable sensitive individuals to stay indoors. The acute potential of chemical variables in the atmosphere to affect (see further below), compared with physical variables, is negligibly low, and therefore only circumstantial to society.

The record of the current chemical state of the atmosphere (that is, the chemical weather) is a condition precedent to the long-term record, so as to be in a position to compile a *chemical climatology*. The alteration of the chemical composition of the atmosphere has become a continuous process since the Industrial Revolution[18]. Only with the concept of sustainability (which brought on the necessity of "green" chemistry and solar power as technological alternatives) can the resource and energy industry be integrated into (anthropogenically strongly modified) biogeochemical cycles and – as we hope – lead to a stability of the atmospheric-chemical composition. The alteration of the chemical composition (greenhouse gases and aerosols, to mention just two of the components) is generally accredited as the cause of ongoing climate change. With our new, comprehensive definition, the alteration of the chemical composition itself is climate change.

The awareness that certain chemical substances (which are in interdependency with radiation and thus change the energetic balance of the atmosphere) alter *physi-*

---

[17] For the sake of completeness, the necessity of warnings related to chemical disasters should be mentioned as well.

[18] More exactly, since the start of large-scale forest clearing.

Table 3.1 Climatic elements (examples – not complete).

atmospheric-physical	atmospheric-chemical
radiation and heat (temperature)	atmospheric aerosol (number, mass, size etc.)
humidity (absolute and relative)	concentration (of gaseous compounds)
cloudiness	oxidation capacity
precipitation	acidity potential
wind (direction and speed)	deposition

cal components of climate by internal feedback is a "simple" scientific interrelation. Conversely, changed physical variables (cloud coverage and radiation amongst others), by providing changed reaction conditions, change the chemical turnover (spatiotemporal concentration distribution), and thus the chemical setting. Climate is then determined by internal feedback within the atmosphere as well as by external feedback with processes in other subsystems (which, from the point of view of the climatic system can be seen as internal relations as well). From a viewpoint of the earth system, extraterrestrial as well as magmatic influences can be seen as external feedback too.

The record of chemical components in climate (i. e., the development of a *chemical climatology*) is of great importance for societies. Long-term forecast of the chemical climate is extremely important due to the close interaction between the physical and chemical system (climate change) and due to the fact that any feedback response to establish abatement strategies needs many years and even decades to take effect. Climate modelers have lately realized that only a better description of chemical processes will lead to greater soundness regarding important physical statements (e. g. concerning the water cycle). On the other hand, the description of the change of the chemical composition of the atmosphere (as a reminder, we understand this as a component of climate) is a self-contained value, for example, concerning the toxic potential of the atmosphere (further rising mean ozone concentrations cause stress for plants), the oxidative potential (decomposition rates of harmful trace matters) and the acid-base balance (geochemical erosion, availability of nutrients, mobility of heavy metals). These slowly ongoing changes require an extensive global and careful monitoring, analysis and evaluation, as well as (only possible on a long-term basis) technological changes and adaptations with the aim of securing a sustainable development. The parameters shown in Table 3.1 will be included in the list of climatic elements, which is not exhaustive.

3.4.2 Precipitation chemistry climatology

We have seen (Chapter 1.3.4) that sampling and chemical analyzing of rain water is among the first activities in atmosphere chemistry: the first half-quantitative rain water analysis was carried out in Berlin by *Andreas Sigismund Marggraf* in 1753. However, there are known older but alchemistic precipitation studies (Möller 2008); earliest "scientific" studies begun in 1840; seven rain, snow, and hail investigations until 1850 are known of. From that time, the number of rain sampling and chemical

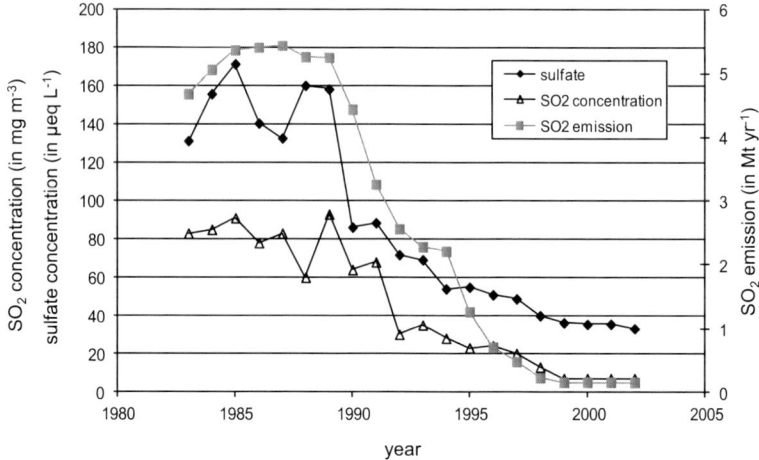

Fig. 3.7 Historical trend of East German SO_2 emission (see also Fig. 2.53), SO_2 concentration measured in Wahnsdorf (see also Fig. 2.63) and sulfate concentration in precipitation in Seehausen, based on annual means.

analysis boomed. In the second half of the 19th century, five monitoring series longer than two years (and partly over more than 10 years) are known of. The first long-term monitoring was started in 1853 in Rothamsted, an agricultural site in the UK. After World War II, precipitation chemistry monitoring begun in Scandinavia, first to study chemical cycling (nitrogen, chlorine and sulfur) and later air pollution. The target was a) investigation of trends (pH and compounds) and b) regional differences (see Chapter 2.5.3 concerning parameters of precipitation chemical climatology). In the 1980s, another booming period started, mainly because of the EMEP program (long-range air pollutant transport).

As described in Chapter 4.4.2, wet deposition is simple to measure but interpretation of precipitation chemical composition is not that simple due to the complex processes from emission to deposition, including in-cloud and below-cloud scavenging (cf. Figs. 4.8 and 4.9). In other words, at the point of precipitation and its sampling, long-range transportation of precursors in air, atmospheric aerosol, and clouds with a complex of physico-chemical history determines the composition in the precipitating clouds (in-cloud scavenging) and local processes determines the below-cloud scavenging.

With this in mind, it is remarkable that from long-term monitoring in Europe, approximately linear relationships with high correlation have been found between emission and concentration of SO_2 and the sulfate concentration in precipitation (Figs. 3.7 and 3.8). The linear relationship, however, is only valid in the period of changing SO_2 emission (1990−2000); before 1990 and after 2000, the emission was about constant (Table 3.2). The variation of sulfate (upper plateau of the grey line in Fig. 3.8) only shows the inter-annual meteorological variations while emission remains constant. The same is seen for low emission regime after year 2000.

Whereas the SO_2 emission decreased within 10 years by a factor of about 30 and the SO_2 concentration by a factor of 13, the rainwater sulfate only decreased by a factor of 4 (Table 3.2). Nowadays, we can almost neglect the anthropogenic SO_2

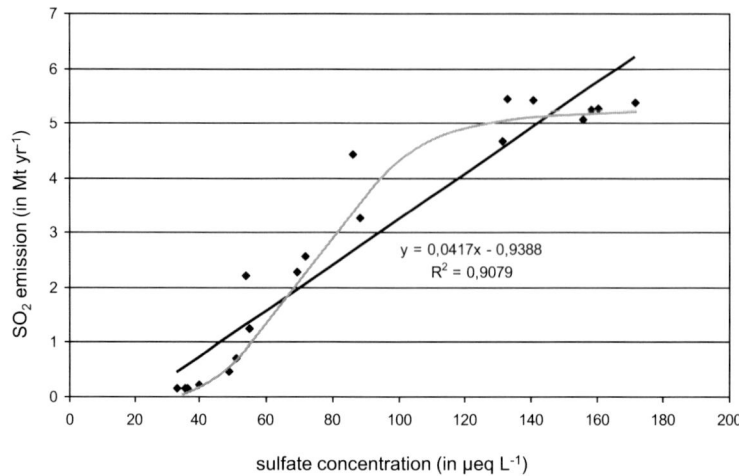

Fig. 3.8 Scatter plot between East German SO_2 emission (in Mt S yr^{-1}) and sulfate concentration in precipitation in Seehausen (in μeq L^{-1}), based on annual means 1982–2002. Linear regression and approximated relationship (grey line).

Table 3.2 Period means for SO_2 emission, SO_2 concentration and rain water sulfate in Eastern Germany.

period	SO_2 emission (in Mt yr^{-1})	SO_2 concentration (in mg m^{-3})	sulfate concentration[a]	
			(in μeq L^{-1})	(in mg L^{-1})
< 1990	5.2 ± 0.3	81 ± 13	150 ± 14	14.4 ± 1.3
> 2000	0.15	6 ± 1	35 ± 1	3.4 ± 0.1

[a] precipitation weighted means

emission in Germany and no further decrease in rain water sulfate is expected – the actual 3–4 mg L^{-1} represent a regional background; it is worth noting here that in cloud water (see next Chapter) the "background" amounts now to 6–7 mg L^{-1} and recalculated on the air volume 2–3 μg m^{-3} (the ground-based sulfate background from the Melpitz station is ~ 3 μg m^{-3} – from which we can deduct that "background" sulfate is homogeneously distributed within the PBL). The measured SO_2 concentrations are no longer influenced by local sources and also represent the regional pollution situation (it is fact that SO_2 concentration – as well as other pollutants – increases with eastern air masses). This rigorous pollution change (we now term it change of chemical climate), however, is unique for Eastern Germany and much more smooth for Europe.

Having a view on the trends of rain water constituents in Europe over the last 100–150 years (Table 3.3), we state that sulfate increased from the beginning of the 19th century until 1980 by a factor of 3 and decreased relative to the 1980 period by factor of 6 until now, hence now about on a 50% level of the 19th century value (remember that the 19th century already was the coal combustion period with sulfur

Table 3.3 Changing chemical composition (European average) by factors, relative to 1 (1900); before 1980 citations see in Möller (2003).

compound	~ 1900	1900–1960	1960–1980	1980–2000
nitrate	1	4–6	6–12	~ constant
sulfate	1	~ 2	~ 3	≤ 0.5
ammonium	1	~ constant	~ constant	~ 0.8
pH[a]	1	1	1	~ 1.2

[a] pH remains probably constant (4.0–4.5) between 1860 and 1980 due to neutralization of acidic by alkaline compounds and increased (in other terms, the acidity decreased) between 1990 and 200 to about 5.0–5.5[b] since 2000 no more change observed

Table 3.4 Historical comparison of precipitation chemical composition.

site	pH	SO_4^{2-}	NO_3^-	NH_4^+	Cl^-	sum[a]	Δ[b]	author
		in mg L^{-1}				in µeq L^{-1}		
London 1870	4.8	57.7	8.9	10.6	10.5	1029	1014	Smith (1872)
Beijing 1981	6.5	16.6	5.0	4.0	2.1	254	254	Zhao et al. (1988)
Seehausen 1985	4.3	13.4	4.3	2.7	2.9	274	229	Möller (2003)

[a] $[SO_4^{2-}] + [NO_3^-] + [Cl^-] - [NH_4^+]$

[b] $[SO_4^{2-}] + [NO_3^-] + [Cl^-] - [NH_4^+] - [H^+]$ = missing neutralizing cations such as Ca^{2+}

pollution). In contrast, nitrate increased by about one order of magnitude from 1900 – and had not well abated until now. This reflects the 20th century as the mobile fuel-based engine period. Remarkable that ammonium did not change over the 1900–1980 period (for more information and discussion see Sutton et al. 2008). The pH (in terms of acidity) also remains nearly constant and only increased slightly after 1990 due to the almost complete SO_2 abatement. To assess anthropogenic modifications of biogeochemical cycles it is important to know the "natural" pH or background acidity (Horváth and Möller 1987).

As seen from Table 3.4, rain water chemical composition reflects the extreme city air pollution in the 19th century. The London values also show the high importance of NH_3 emissions in that time from missing waste water management. As a result, the pH was not different from late 20th century values. Interesting are the Beijing rain water values which were not significantly different from German background values but had a very high pH due to alkaline soil dust from the Gobi desert. Later, however, the pH value became less than 5 due to the extreme increase of Chinese SO_2 emission. We see that the balance between acidic and alkaline constituents determines the rain water acidity (see also Chapter 4.2.5).

The meteorological station Seehausen (52° 06′ N, 12° 24′ E), about 100 km north-westerly of Magdeburg, is representative for the Northern German lowland, situated close to the former border between East and West Germany. In 1982, rainwater chemistry monitoring (Table 3.5) was begun to study the different acidity from

Table 3.5 Annual weighted-mean concentrations (in µeq L^{-1}) of chemical composition of precipitation at Seehausen; RR − annual precipitation amount (in mm or $Lm^{-2}yr^{-1}$, resp.), n − number of samples (based on 4-hour wet-only sampling).

year	n	RR	pH	H^+	Na^+	K^+	NH_4^+	Ca^{2+}	Mg^{2+}	Cl^-	NO_2^-	NO_3^-	SO_4^{2-}
1982[a]	17	32	4.45	35.7	47.4	5.2	92.4	132.7	16.3	102.2	6.3	25.3	207.2
1983	100	212	4.55	28.4	36.4	8.3	75.7	58.0	13.5	71.8	3.8	29.8	131.5
1984	152	320	4.42	38.0	29.7	6.5	83.2	54.6	12.9	55.5	4.0	34.1	146.8
1985	165	289	4.56	27.4	36.0	8.3	114.8	57.6	15.0	77.6	3.5	47.4	171.9
1986	127	408	4.21	62.1	22.9	4.7	89.5	52.2	10.1	57.1	3.1	47.2	139.8
1987	201	434	4.21	61.4	19.3	3.5	84.2	43.7	8.6	39.5	1.5	43.5	132.8
1988	189	317	4.32	48.1	44.9	6.5	73.7	71.1	12.5	64.0	1.1	48.6	160.5
1989	142	265	4.40	40.2	29.2	5.1	85.1	79.6	12.2	57.9	0.6	55.5	158.5
1990	196	412	4.43	37.0	27.5	4.3	69.3	43.4	9.5	53.9	−	48.6	86.2
1991	118	222	4.39	40.6	32.3	6.1	69.0	43.1	7.2	42.5	−	50.2	89.0
1992	230	475	4.30	49.5	41.1	5.8	59.9	28.0	7.5	45.3	−	44.5	71.8
1993	225	386	4.35	44.4	17.1	4.9	63.8	22.4	5.4	34.3	−	46.2	69.2
1994	229	600	4.40	39.4	28.4	4.0	43.7	19.6	7.4	41.3	−	38.1	54.0
1995	189	384	4.62	24.1	38.0	2.8	49.4	35.0	10.8	41.5	−	43.1	54.9
1996	170	392	4.68	20.7	21.3	1.2	60.6	16.8	6.6	27.3	−	47.6	50.9
1997	153	158	4.91	12.3	28.6	1.3	73.1	18.6	7.7	29.6	−	43.8	51.3
1998	266	542	5.03	9.4	17.0	1.5	46.0	20.9	3.8	26.9	−	36.7	40.3
1999	198	429	4.84	14.6	12.2	0.9	44.7	15.5	4.1	19.6	−	40.1	34.4
2000	148	357	4.80	15.7	8.6	1.3	62.1	11.4	2.6	14.3	−	44.6	35.5
2001	241	221	4.82	15.1			50.0					42.0	35.6
2002	212	261	4.94	11.4			47.6					29.8	33.2

[a] Begin of monitoring in October 1982

western and eastern air masses[19]. In that time, depending on the trajectory, high concentrations of sulfate in southern Sweden were accompanied by greater or lesser carbon as black or white episode, respectively (Rahn et al. 1982). The former Eastern Germany (GDR) was known to be the second largest SO_2 emitter (after the UK) in Europe but based on lignite combustion instead of hard coal, used dominantly in Western Europe. It was expected that long-range transport from the GDR would be associated with high sulfate but also with high calcium, and therefore less acidic than in air masses from the UK to Scandinavia. Therefore, wet-only sampling on 4-hours basis was established to attribute chemical composition to air masses characterized by backward trajectories (Marquardt and Ihle 1988, Möller et al. 1996, Marquardt et al. 1996, 2001), Table 3.5.

The data set imposingly shows the changes in chemical precipitation composition due to emission changes in the period before and after 1990 as well as significant differences between "west" and "east" (Tables 3.6 and 3.8). Within a period of

[19] Established by the former Institute for Energetics (Leipzig), continued from 1992 by the Institute for Troposheric Research (Leipzig) and since 1996 by the Chair for Atmospheric Chemistry and Air Pollution Control (Cottbus) until ending the monitoring in 2002. The chemical analysis was carried out from 1982−1995 by Dr. Erika Brüggemann (Leipzig) and from 1996 to 2002 by Dr. Renate Auel (Berlin) with highest quality standards. No change of any sampling procedure and data classification has been made over the whole period.

Table 3.6 Weighted-mean concentrations (in µeq L^{-1}) of precipitation constituents at See-hausen in the period before 1990 and after 1990 (note: the year 1990 has been not considered as year of "political change"); change in % between periods (+ means increase).

trajectory east; n = 169 (a), 46 (b), 97 (c)

period	H$^+$	SO$_4^{2-}$	Ca^{2+}	Mg^{2+}	Cl$^-$	NH$_4^+$	Na$^+$	NO$_3^-$	K$^+$
1983−89 (a)	45.5	273.5	158.5	21.8	82.1	151.0	33.9	69.2	11.0
1991−94 (b)	87.2	118.6	39.3	6.4	31.0	76.5	14.2	78.7	5.3
1995−00 (c)	29.5	72.3	30.7	5.2	21.9	80.9	11.2	64.0	1.5
change (b-a)	+91	57	75	71	62	49	58	+14	52
change (c-a)	35	74	81	76	73	46	68	8	86

trajectory west; n = 142 (a), 146 (b), 135 (c)

period	H$^+$	SO$_4^{2-}$	Ca^{2+}	Mg^{2+}	Cl$^-$	NH$_4^+$	Na$^+$	NO$_3^-$	K$^+$
1983−89 (a)	44.9	132.9	48.8	9.0	47.9	85.9	22.1	47.6	4.2
1991−94 (b)	35.5	47.6	22.6	4.4	31.2	49.3	19.6	39.4	3.7
1995−00 (c)	10.2	34.7	13.9	4.3	21.5	46.8	14.6	35.1	0.9
change (b-a)	21	64	54	51	35	43	11	17	12
change (c-a)	77	74	72	51	45	46	34	26	79

all samples; n = 1185 (a), 561 (b), 768 (c)

period	H$^+$	SO$_4^{2-}$	Ca^{2+}	Mg^{2+}	Cl$^-$	NH$_4^+$	Na$^+$	NO$_3^-$	K$^+$
1983−89 (a)	46.3	147.9	58.0	11.7	58.5	86.8	29.9	44.2	5.8
1991−94 (b)	34.8	51.8	23.1	7.2	36.8	45.2	26.4	38.5	3.7
1995−00 (c)	14.4	41.2	16.8	4.6	23.2	41.9	16.2	41.2	1.0
change (b-a)	25	35	60	38	37	48	12	13	36
change (c-a)	69	72	67	61	60	52	46	7	83

almost five years (1990−1995), changes happened in an order of magnitude never before and will be observed at any site in the world: the "political change" (in German: die "Wende") in the former GDR was associated with an "economic collapse", and hence linked with emission decrease and subsequent changing "chemical climate".

Obviously, potentially high acidity derived from sulfate was neutralized by calcium from alkaline fly ash emission in "eastern air masses". Hence, in the period before 1990, no difference in acidity (expressed as pH) can be seen in different trajectories. In the period after 1995, the pH rose from about 4.3 (before 1990) to 4.8 (east trajectories) and 5.0 (west trajectories). The changes are mainly due to decrease of sulfate and calcium but also other "neutralizing" compounds such as nss-magnesium and nss-sodium decreases. It is remarkable that ammonium decreased between the two periods by a factor of two whereas nitrate remained constant. The most significant change is seen in the sulfate/nitrate ratio from 3.4 to 1.0. Concerning total inorganic nitrogen (nitrate + ammonium), the S/N ratio was between 1.0 and 1.2 before 1990 (east to west) and decreased to 0.4 to 0.5 after 1995. In other words, the chemical climate changed from sulfur to a nitrogen regime.

Table 3.7 Estimations and change of *excess* chloride in precipitation (in $\mu mol\ L^{-1}$) at Seehausen, based on different sea salt reference values for $[Na]_{sea}/[Cl]_{sea}$ (Eq. 2.83).

period	$[Na^+]$	$[Cl^-]$	$R = [Na]/[Cl]$	Cl_{excess} calculated on	
				$R_{sea} = 0.86$	$R_{sea} = 1.16$
1983–1989	33 ± 9	63 ± 18	0.52 ± 0.11	25	35
1990–1994	31 ± 7	39 ± 6	0.79 ± 0.15	3	12
1995–2002	21 ± 11	26 ± 9	0,75 ± 0.16	2	8

Table 3.8 Change of equimolar concentration ratios in precipitation at Seehausen before and after 1990; $[NH_4^+]/[cations]] = [NH_4^+]/([Ca^{2+}] + [Mg^{2+}] + [Na^+] + [K^+])$, $[N]/[S] = ([NO_3^-] + [NH_4^+])/[SO_4^{2-}]$

concentration ratio	before 1990[a]		after 1990[b]	
	trajectory east	all data	trajectory east	all data
$[N]/[S]$	0.80 ± 0.06	0.88 ± 0.07	1.70 ± 0.57	1.90 ± 0.53
$[NH_4^+]/[cations]$	0.67 ± 0.10	0.82 ± 0.19	1.61 ± 0.51	1.22 ± 0.60

[a] $[N]/[S]$ 1983–1989 and 1983–1991
[b] $[N]/[S]$ 1991–2000 and $[NH_4^+]/[cations]$ 1992–2000

There are other interesting changes in chemical composition of precipitation. Substances, normally associated with natural source, such as potassium (soil dust), magnesium (sea-salt and soil dust), and sodium and chloride (both sea-salt) have been found in large excess in precipitation before 1990. Obviously, Mg and Na (beside Ca) were emitted with flue-ashes from lignite-fired power plants. We can not exclude it for Cl also (the so-called west-Elbe coal was salty) but much of HCl was emitted by the potash industry. Excess chloride accounted to about 50% to total chloride before 1990 and decreased significantly until 2002 but obviously still exists (Table 3.7). Regarding Table 3.7, it can be deduced that sea salt NaCl is much lower in the period after 1995 than in the period 1990–1994 (and before 1990). The molar ratio [Na]/[Mg] is much lower (2.6 before 1990 and 3.8 after 1990, Table 3.6) than expected (molar 9.8) assuming that all Mg is sea salt. It must be concluded that Mg found in precipitation comes also from other primary sources of dust (like Ca).

Despite general air pollution control, in the period from 1991–1995, an acidity increase was observed (Tables 3.5 and 3.6), followed by a sharp decrease after 1995. The simple explanation is that air pollution abatement first PM emission decrease – containing alkaline compounds – from power plants had been fulfilled in Eastern Germany, and flue-gas desulfurization was only completed until after 1995. We again see that not the absolute concentration (or emission), but the relative percentage of compounds having acidifying and/or neutralizing properties is essential in determination of the budget.

Since about the year 2000, changes in emission structure, and therefore "chemical climate" became negligible – emission abatement finished and "energy transition" (third industrial revolution) is not yet seen. In Germany and Central Europe, we

Table 3.9 Chemical composition of precipitation in different regions of the world (in μeq L^{-1}, rounded).

site	H$^+$	Na$^+$	K$^+$	NH$_4^+$	Ca^{2+}	Mg^{2+}	Cl$^-$	NO$_3^-$	SO$_4^{2-}$
China (1981−1984)[a]	0.2	44	21	160	460	−	39	34	162
India (1988−1996)[b]	0.1	18	8	40	56	−	32	18	36
Cape Grimm (1977−1985)[c]	1	1167	30	2	78	247	1372	3	152
Central Australia (1980−1984)[d]	17	4	1	3	2	1	8	4	4
Lancaster (NW England)[e]	34	57	4	55	15	15	79	28	89
Central Bohemia[f]	42	7	3	62	30	−	−	51	116
Hungary[g]	32	23	9	61	85	16	26	41	119
Seehausen (1983−1989)	44	33	6	85	59	12	63	45	150
North Sweden (1983−1987)[h]	23	4	2	13	3	2	5	11	33
South Sweden (1983−1987)[h]	46	24	2	36	10	7	30	36	67
Seehausen, Germany (1996−2002)[k]	14	20	2	52	16	5	25	39	40
Melpitz, Germany (2000)[m]	13	8	1	44	13	3	8	31	35
Neuglobsow (2000)[n]	13	14	2	39	18	4	14	31	33
Staudinger, Germany (2007−2008)[o]	10	14	3	44	18	5	14	36	31
Peitz, Germany (2011−2012)[p]	6	30	17	78	30	9	28	58	47
Radewiese, Germany (2011−2012)[r]	4	31	7	52	483	12	20	41	469

[a] suburb of Beijing (Zhao et al. 1988)
[b] suburb in semiarid area (Kumar et al. 2002)
[c] Tasmania, south Pacific air (Ayers and Ivey 1988)
[d] Katherine, annual rainfall 75−136 mm (Likens et al 1987)
[e] Harrison und Pio (1983)
[f] Hradec, mean of 483 samples (Moldan et al. 1988)
[g] mean from 6 stations (Horváth and Mészáros 1983)
[h] mean from 10 stations (Granat 1988)
[k] mean from 20 stations (Granat 1988)
[m] rural station near Leipzig, weekly wet-only samples (UBA)
[n] background station 60 km north of Berlin, weekly wet-only samples (UBA)
[o] Hanau (within power station Staudinger) 25 km east of Frankfurt, $n = 61$ (Möller et al., unpublished)
[p] 15 km north of Cottbus, close to power station Jänschwalde, $n = 123$ (Möller et al., unpublished)
[r] 15 km northeast from Cottbus, direct under the plume of Jänschwalde power station, $n = 128$

see a relative homogeneous distribution of atmospheric constituents (in air, precipitation, and particulate matter); concentrations variation in time and space are small and reflect meteorological variation ("physical climate") and not longer variations in emissions pattern. Precipitation composition in Germany is not very different from Scandinavia 20 years ago (Table 3.9, last 6 lines). Local "pollution" (soil dust, agriculture, power plants) however, may contribute to higher concentration; remarkable is the below-cloud scavenging of sulfate and calcium at Radewiese (Table 3.9), located 5 km downwind of the largest lignite-fired power station in Germany (Jänschwalde).

Still higher concentrations found in precipitation (and also particulate matter) with easterly trajectories (Table 3.5) are not a result of the often cited still higher emissions in southeastern countries (Poland, Czech Republic, and Ukraine) but a result of long-range transportation. Clouds and precipitation are far less with eastern air masses than with western air masses; in other terms, the time interval between "wet events" is large. Hence, concentrations increase due to missing wet deposition; especially particulate matter (secondary aerosol), exceeding EU limits of 50 μgm^{-3}. In case of precipitation, associated with easterly trajectories, higher wet deposition is found. In western air masses, many precipitation (and cloud) events occur during transport from the Atlantic or North Sea which remove (and transform) gases and particles until they can be deposited in Eastern Germany (and further east). A strong correlation between duration of the "dry period" and concentrations of rain water constituents has been found (Marquardt et al. 1988).

3.4.3 Cloud chemistry climatology

In contrast to precipitation chemistry, it was much more difficult to collect water from fog or clouds; due to missing appropriate sampling techniques (from analytical point of view), low occurrence of fog/cloud, and laborious sampling on mountains. First chemical analyses of fog water were carried out by *Boussingault*[20] in Alsace (at Mt. Liebfrauenberg), Paris and Rhine valley (1857−58). He found nitrate to be between 2 mg L^{-1} (at mountain site) and 138 mg L^{-1} (Paris) and ammonium in Paris to be 10 mg L^{-1}. Unfortunately, *Boussingault* was only interested in nitrogen from an agricultural point of view, and therefore did not analyse any other substance (also in rain, snow and dew). There are no known other cloud studies in the 19th century (except for town fog); only after World War 2 did cloud studies begin; first in the USA (Houghton 1956), later in Germany (Mrose 1966) and Russia (Petrenchuk and Drozdova 1966) as well as Japan (Okita 1968). Systematic studies[21] around the world began in the 1980s: for example in Italy (*Sandro Fuzzi*), in the USA (*Daniel Jacob, Jed Waldman, William Munger, Michael Hoffmann, Volker Mohnen, Vin Saxena, Jeffrey Collet* and others), in Austria (*Hans Puxbaum*), in Japan (*Manabu Igawa* and others), and in Germany (*Hans-Walther Georgii, Wolfgang Jaeschke, Otto Klemm*); intensive cloud and fog studies also were started in Taiwan and China.

The first continuous cloud chemistry program was established in the middle 1980's in the eastern U.S., which was primarily focused on studying the reasons for new-type forest decline. The whole *Mountain Cloud Chemistry Project*, however, resulted in several important findings which were the basis for the first establishment of chemical cloud climatology (e. g., Mohnen and Kadlecek 1989, Mohnen and Vong 1993, DeFelice and Saxena 1991, Schemenauer et al. 1995, 1998). Mostly, stratus type clouds were observed during this program. But it was also shown that

---

[20] *Jean Baptiste Joseph Dieudonne Boussingault* (1802−1887) French agricultural chemist, was the author of *Traite d'economie rurale* (1844), which was remodelled as *Agronomie, chimie agricole, et physiologie* (5 vols., 1860−1874; 2nd ed., 1884), translated into many languages. In German "Die Landwirthschaft in ihren Beziehungen zur Chemie, Physik und Meteorologie", Halle (1851) 442 pp. and supplement „Beiträge zur Agricultur-Chemie und Physiologie" (1856) 312 pp., with abundant data on ammonia and nitrate in air.

[21] Some of them (and others not cited) were focussed on fogs and/or mixed mountain clouds.

a greater data base and more detailed information (e. g., determination of exact cloud base height for each event) are required to deduce a climatology that is truly representative of a region. Fog chemistry in the Po Valley (Italy) area has been studied over a fifteen year period and high levels of pollutants have been reported within fog droplets (Fuzzi et al. 1996). Continuous monitoring of cloud and rain samples (time resolution weekly) at three sites in the UK over one year has allowed consideration of the impact of the enhancement of wet deposition of pollutants by orographic effects (Fowler et al. 1995), specifically the scavenging of cap cloud droplets by rain falling from above (the seeder-feeder effects).

In Europe, an intensive phase in organising cloud field experiments was given within the EUROTRAC-GCE project between 1989 and 1993. 13 research groups from 7 European countries participated at 3 large field campaigns. In principle, it was the aim of GCE (Ground based Cloud Experiment) to characterise the interaction between droplets (clouds and fog) with the gaseous and solid particulate phase. Using different measurement sites, the scavenging and aqueous-phase transformation was studied in different polluted air masses and cloud types, as well as fog (Po valley campaign in 1989 in an agricultural area; Kleiner Feldberg, near Frankfurt/M campaign in 1990 in polluted urban area; Great Dun Fell experiment 1993 in mostly marine influenced orographic clouds). During the Kleiner Feldberg experiment, mainly the formation of cloud drops, the time-dependent evolution of chemical species in the multiphase system and studying of different sampling systems were done (Wobrock et al., 1994). Both experiments (Po valley and Kleiner Feldberg) were conducted in autumn and can be chemically characterised by oxidant limitation. In contrast, the Great Dun Fell experiment took place in spring, with low pollution but high oxidant concentrations (Chourlarton et al. 1997, Colvile et al. 1997, Swietlicki et al. 1997, Wells et al. 1997, Cape et al. 1997, Lüttke et al. 1997, Sedlak et al. 1997, Bower et al. 1997, Wiedensohler et al. 1997, Pahl et al. 1997, Laj et al. 1997).

Vertical distributions of aerosol particles and CCN in clean air around the British Isles were obtained during several flights through research fields of small cumulus clouds over the sea; results are given in Raga and Jonas (1995). Also, as a result of research flights within the CLEOPATRA project, the distribution of trace substances inside and outside of clouds were determined (Preis et al. 1994). Cloud events with high pollutant loading were observed during the EASE field campaigns in the so-called black triangle region at Mt. Szrenica in the Sudeten Mountains (Poland) in 1995 and 1996 (ApSimon 1997, Acker et al. 1999b, Kmiec et al. 1998). The idea that came from the Great Dun Fell experiment to study cap-cloud chemistry has been continued within the joint project FEBUKO, where several campaigns at Mt. Schmücke (Thuringian Forest, Germany) have been carried out (e. g. Acker et al. 2003, Herrmann et al. 2005). In the area of focus was the microphysics and chemistry of different types of aerosols, the role of aerosol chemical composition for cloud formation as well as the chemical transformation, where behaviour of organic substances were of special interest (e. g., organic acids, peroxides, organic carbon, soot).

The first continuous long-time mountain site cloud monitoring in Europe started in 1992 at Mt. Brocken (1142 m a.s.l., Harz Mountains, 51.80°N, 10.67°E), located at the former border between East and West Germany. Due to the high occurrence of clouds (30–50% of the time from April to October, in winter time up to 80%

Table 3.10 Annual and total LWC weighted mean concentrations (in µeq L^{-1}) of cloud water constituents at Mt. Brocken, n – number of samples (total 22 841), LWC – liquid water content (annual mean).

year	n	LWC	Cl^-	NO_3^-	SO_4^{2-}	NH_4^+	Na^+	K^+	Ca^{2+}	Mg^{2+}	H^+	pH
1992	36	266	75	321	257	336	80	–	115	27	36	4.44
1993	1054	277	51	193	189	290	43	–	33	17	70	4.15
1994	1079	309	90	280	231	319	87	10	100	28	129	3.89
1995	1343	391	76	232	215	300	80	2	40	21	105	3.98
1996	2026	377	75	244	224	327	74	1	37	17	92	4.04
1997	1269	346	109	234	198	300	119	5	45	25	57	4.24
1998	1507	353	80	220	180	243	72	2	31	16	87	4.06
1999	809	332	122	265	168	282	125	10	61	31	51	4.29
2000	1533	387	79	272	182	328	81	5	40	19	65	4.19
2001	1598	376	94	258	189	295	105	5	33	22	67	4.17
2002	1370	359	74	158	139	224	75	4	23	17	41	4.39
2003	815	290	129	265	209	362	137	7	37	31	40	4.40
2004	1615	295	87	185	146	261	88	5	24	24	41	4.39
2005	1143	268	70	184	154	282	72	5	26	19	34	4.47
2006	1304	307	81	238	168	338	85	6	28	22	44	4.36
2007	1843	345	84	183	139	250	80	5	23	21	48	4.31
2008	1236	314	44	175	122	225	46	3	25	14	54	4.27
2009	1261	340	71	180	121	232	77	7	21	19	37	4.43
mean	–	338	82	221	172	285	84	4	35	21	64	4.19

of the time), the Mt. Brocken summit is an ideal platform for sampling of cloud water. Simulations of surface wind fields in the Harz Mountains for different synoptic situations (Adrian and Fiedler, 1991) showed that the surface wind at the Mt. Brocken summit itself is relatively uninfluenced by the orography and represents the predominant low tropospheric wind (air masses streaming over the mountain top). The summit plateau of the Mt. Brocken is normally situated in the upper atmospheric boundary layer (ABL) or even in the free troposphere (Beyrich et al. 1996) due to the steep slopes at all of its edges. The wind field at the station is dominated by westerly/south-westerly winds. Gas and liquid phase data from Mt. Brocken are, therefore, representative of the regional air mass pollution situation (Acker et al. 1996). The Mt. Brocken station was designed and implemented in 1991, and the first cloud water results for a whole frost free period (April/May−October/November) were available in 1992. We adopted experiences from the *Mountain Cloud Chemistry Project* in the eastern United States (Mohnen and Vong 1993); for almost 18 years (until 2009). We analysed more than 20 000 water samples based on 1 hour collection (for details see Möller et al. 1993, 1996 and Acker et al. 1996, 1998, 2002), Table 3.10. A full set of meteorological data and gases (SO_2, NO, NO_2, O_3) were continuously measured and occasionally many other compounds during campaigns, such as organics (Lüttke et al. 1997, 1999), heavy metals (Plessow et al. 2001), HNO_2 (Acker et al. 2001), and hydrogen peroxide (unpublished). There are several "typical" characteristics for cloud chemistry climatology:

- Concentrations of all dissolved compounds are several times higher in cloud than in rain water;
- concentration variation between cloud events is huge (much more than for rain events),
- concentration variation within a cloud event (passing cloud) is also huge and ir-regular;
- LWC increases linearly with height above cloud base (up to about 80% of cloud thickness);
- total ionic content (TIC) of cloud water depends nonlinearly from LWC; gener-ally, concentration tends to increase with small LWC (less than 100 mg m^{-3}) but there are large variations;
- near the cloud base, concentration of all compounds increase exponentially (due to small LWC because of evaporation);
- in small cloud droplets, significantly higher concentrations of all compounds are found, especially organics (a result from our campaigns);
- in precipitating clouds, the concentration of all constituents decreases (dilution effect due to rain water evolution from higher cloud levels);
- the interstitial air of cloud contains less gases then during cloud-free atmos-phere: highly soluble gases such as H_2O_2, HNO_2 and HNO_3 decrease to the detection limit but also O_3 can be significantly reduced due to cloud chemical processes.

As an example, the statistics for sodium (Na) in terms of arithmetic mean, standard deviation, minimum and maximum value and LWC weighted mean (in µeq L^{-1}) for the whole period 1993–2009 are presented; note that the sample number for weighted mean is smaller because of missing LWC data:

arithmetic mean: 124.7 ± 310.2 (0.0 − 7065.2) $n = 23\,743$
weighted mean: 86.0 $n = 21\,803$

It is remarkable that (as a "mathematical" result of the large variation of LWC and concentration) the difference between the arithmetic and the weighted mean is considerable. In literature, often only arithmetic means are used (because of missing information on LWC and/or RR) but only the weighted mean represents the "true" chemical averaged in the sense of collecting samples continuing over the whole period, in a "bottle", and analyzing them. Another important point is that the volume-based cloud water concentration (cf. Eq. 4.55; the product of LWC and aqueous-phase concentration) is more robust and shows less variation than the liquid-phase concentration (cf. Fig. 3.13)[22]. It represents the residual or − assuming that the regarded substance is totally dissolved (scavenging efficiency = 1) − the aerosol (PM) concentration.

Without discussing the relationships here in detail (see cited literature), it is evi-dent that cloud microphysics and dynamic (including cloud transport and cycles,

---

[22] Note that each single value (based on the water sample) must be calculated from LWC and aqueous-phase concentration with subsequent averaging according to the regarded period. This mean is different from the product of averaged LWC and averaged concentrations, based on differ-ent averaging periods. It is a mathematical phenomenon: LWC only can be arithmetically averaged but the annual mean concentration is based on single LWC weighted values.

Table 3.11 Comparison of cloud (Brocken) and rain water (Seehausen) chemical composition; weighted means (in µeq L^{-1}). LWC – liquid water content (in mg m^{-3}) and RR – yearly rain fall amount (in L m^{-2}) as mean over the period; n – number of samples.

period	n	LWC/ RR	Cl$^-$	NO$_3^-$	SO$_4^{2-}$	NH$_4^+$	Na$^+$	K$^+$	Ca^{2+}	Mg^{2+}	H$^+$
cloud water											
1993–1995	3476	331	73	236	213	303	72	6	57	22	103
1996–2002	8086	362	92	235	176	279	96	4	38	22	61
rain water											
1993–1995	643	529	39	42	60	53	26	4	24	7	38
1996–2002	1388	475	25	40	40	52	20	2	16	5	14

i. e. evaporation, precipitation, nucleation) dominantly determine the chemical composition (likely through the droplet size distribution which we only measured during campaigns). Of secondary importance are the air mass characteristics (maritime, continental, pollution level, etc.). For any cloud chemistry monitoring, it is crucial to include a minimum of cloud physics, such as LWC, height above cloud base, duration of cloud and cloud-free events. It would be desirable to also monitor the droplet size distribution (nowadays, a robust and inexpensive technique is available).

It must, therefore, be concluded that any climatological statements can only be drawn after long-term monitoring having a large timely representativeness (we usually collected from April to October every year, gathering about 80–90% of all cloud events at the summit; note that the overall cloudiness is larger due to higher clouds). As already said, cloud chemical composition is much more regionally (meso-scale) representative than wet deposition (precipitation chemistry).

Even with the large number of available cloud water samples and the lengthy period (over 17 years), it is not simple to classify "cloud chemistry" according to trends and variation of air pollution, trajectories and cloud physical parameters because all "effects" overlay and varies often for different reasons. We see that the emission changes after 1990 almost happened in 1990 and were completed around 1995/1996 (Fig. 3.7). To compare cloud with rain water chemistry, two periods in Table 3.11, namely 1993–1995 and 1996–2002, have been selected. As can be seen, changes in cloud water are less compared with rain water; it is remarkable that there are no changes for Mg (and K) and the decrease for ammonium is insignificant.

It is interesting that sea salt (as NaCl) significantly increased by about 40% but no change in the Na/Cl ratio (Fig. 3.9) and it is also notable that this ratio is much larger than in rain water, suggesting that in cloud, there is no excess chloride found, indeed missing chloride, as discussed in Chapter 2.4.3 for "clean" maritime air masses passing the continent (see also Chapter 2.8.3.12). The NaCl increase might be interpreted as an increase in the percentage of maritime air masses passing the Brocken (Central Germany).

Interannual variation of LWC is relatively small (insignificant), see Table 3.12 and Fig. 3.10. The years 1993 and 2003 (both characterized by hot summers) show

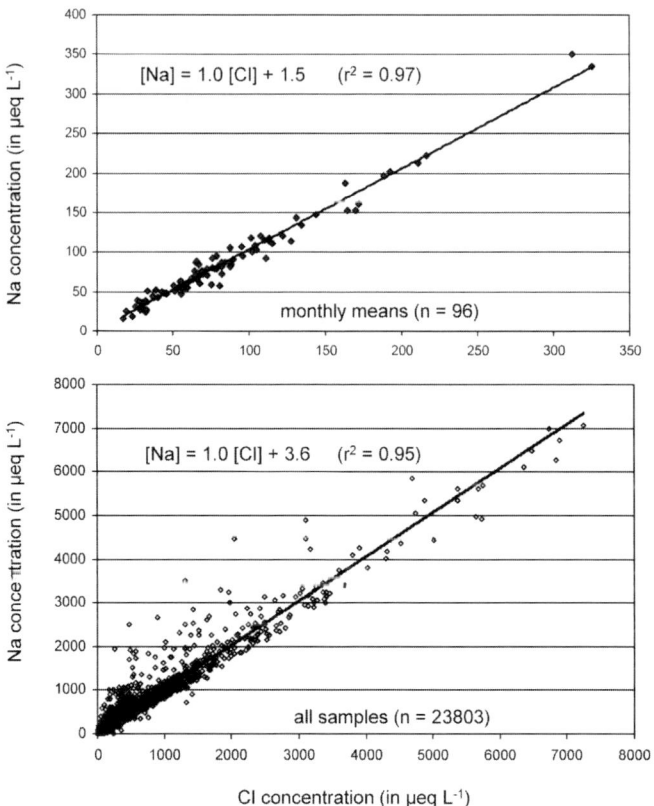

Fig. 3.9 Correlation between sodium and chloride in Brocken cloud water (in μeq L^{-1}) based on monthly weighted means ($n = 96$) and all data (1993–2009, $n = 23803$ one-hour values).

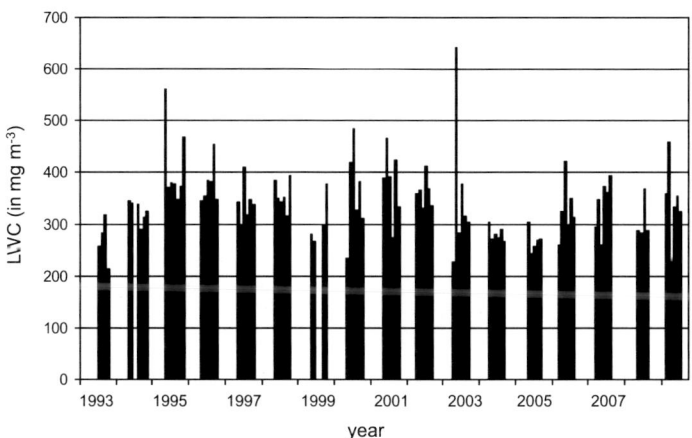

Fig. 3.10 Variation of monthly LWC means at Brocken from 1993–2009 (April/May–October/December yearly variable); $n = 96$.

Table 3.12 Statistics of LWC in Mt. Brocken cloud monitoring program (in mg m^{-3}), based on sampling periods (about 80% of all cloud events).

period	arithmetic mean	weighted mean
1993–1995	326 ± 59	331
1995–2002	358 ± 20	362
1993–2002	348 ± 37	353
1992–2009	330 ± 41	338
2003–2009	308 ± 27	312

Table 3.13 Trend in cloud sulfate at Mt. Brocken.

year	air-volume related concentration (in µg m^{-3})		aqueous-phase concentration (in mg L^{-1})	
	from all samples	from annually weighted means	from all samples[a]	
			weighted mean	arithmetic mean
1993	2.6 ± 2.5	2.6	189	262 ± 365
1994	3.4 ± 3.9	3.5	231	264 ± 326
1995	4.1 ± 4.0	4.1	215	287 ± 409
1996	4.1 ± 3.1	4.1	224	291 ± 380
1997	3.7 ± 2.9	3.4	198	285 ± 296
1998	3.6 ± 3.1	3.1	180	276 ± 294
1999	3.6 ± 4.0	2.7	168	273 ± 334
2000	3.5 ± 2.9	3.4	182	268 ± 327
2001	3.4 ± 2.2	3.5	189	265 ± 256
2002	3.4 ± 2.5	2.4	139	258 ± 249
2003	3.4 ± 2.5	3.0	209	257 ± 228
2004	3.2 ± 1.7	2.1	146	250 ± 167
2005	3.1 ± 1.8	2.0	154	245 ± 199
2006	3.0 ± 2.8	2.5	168	242 ± 220
2007	3.0 ± 2.2	2.4	139	238 ± 249
2008	2.9 ± 1.7	1.9	122	233 ± 192
2009	2.9 ± 1.8	2.0	121	229 ± 151

[a] This data set comprises only samples where LWC is available to compare the weighted mean exactly with the arithmetic mean; in Fig. 3.13. arithmetic mean is calculated from all data available including samples with missing LWC (1994: 300 ± 396 and 2004: 179 ± 166).

significantly lower LWC compared to average; characterizing the air masses by the sea salt loading, 1993 was "continental" (lowest NaCl) and 2003 was "maritime" (highest NaCl), Table 3.10. If this explanation is true, however, the low concentrations of "pollutants" (sulfate, calcium, and nitrate) in 1993 are hard to explain in continental (i. e. easterly) air masses.

A trend in NaCl is not seen, however, it seems that after the year 2003, LWC is significantly lower than in all periods prior (Table 3.12). To conclude on "climate change" is not applicable because LWC is an integrated parameter, whereas droplet

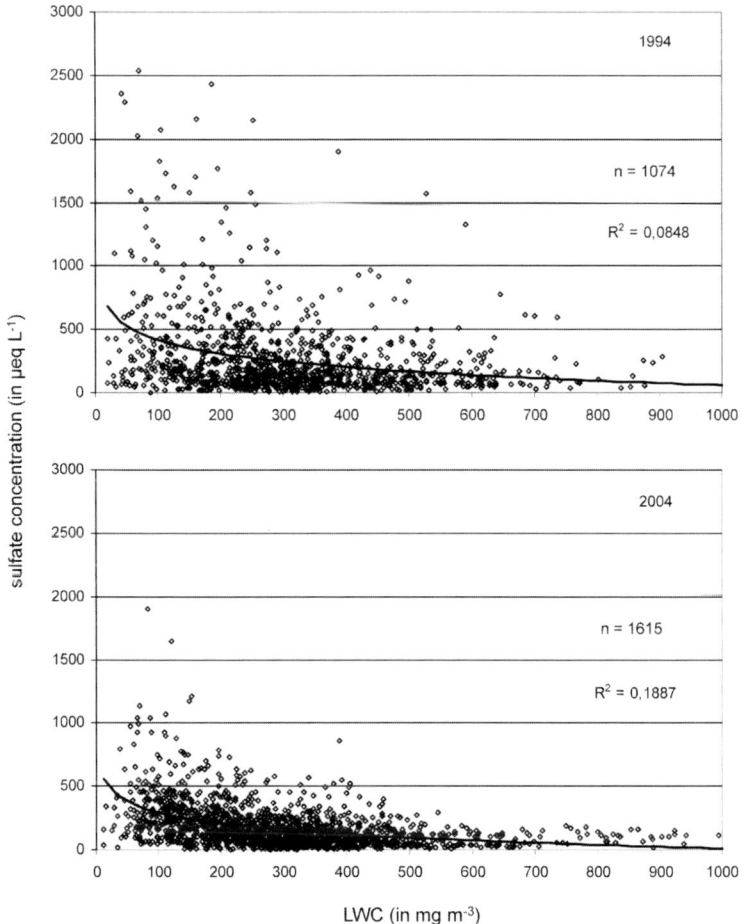

Fig. 3.11 Scatter plot of sulfate and LWC for two selected years.

size and droplet number are directly climate relevant. It is discussed (e. g. Junkermann et al. 2011) that due to "warming" and air pollution abatement the number of potential CCN increases, especially below 0.1 μm (nanoparticles), but the droplet sizes decreases. As a possible consequence, more cloudiness and less precipitation could result. However, any conclusion on changing LWC is impossible. Assuming (and there is no reason not to assume it) that the number of activated CCN (and newer results show that natural sources are likely dominant) remain unchanged but the absolute humidity for nucleation decreased, the number of cloud droplets will not change but its size decreases – this would mean that LWC decreases.

The most significant change in cloud chemistry at Mt. Brocken is the decrease of calcium (by 33 %) and H^+ (by 41 %), Table 3.11, Fig. 3.13. Sulfate only decreased by 17 % (in rain water by 72 %), Table 3.11 and Fig. 3.14 shows the sulfate climate statistics. The trend in sulfate decrease continues despite no change in SO_2 emission being observed since the year 2000 (see Fig. 3.7). The trend consists due to different

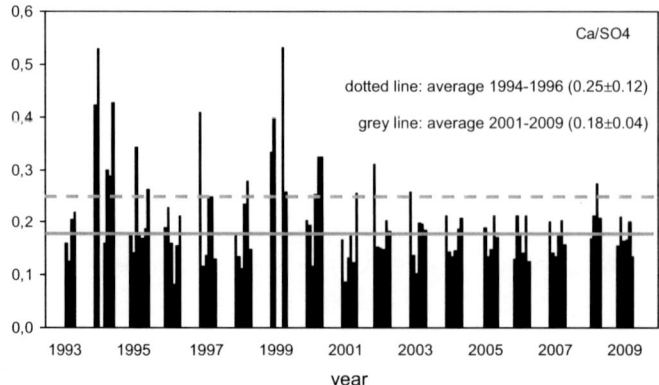

Fig. 3.12 Variation of monthly means of calcium to sulfate ratio at Brocken from 1993−2009 (April/May−October/December yearly variable); $n = 96$.

components; the yearly mean and the standard variation as well as the number of exceedance (extreme values). From Figs. 3.11 and 3.14, it is shown that the number of samples with very large sulfate significantly decreases. Fig. 3.14 only shows cloud water sulfate values larger than 300 µeq L^{-1} (the average over the whole period amounts 200−300 µeq L^{-1}, depending on the type of averaging. Concentrations larger than 800 (up to 2800) µeq L^{-1} after the year 2000 likely originated from cloud microphysical and dynamic processes. Only trajectory analysis (which is presently under work) can (likely) show an air pollution influence. Without any doubt, short-term concentration increase by a factor of ten and more is caused by meteorological factors (transport phenomena). There are also fluctuations between the years, and only any long-term trend is climate change. Further work must be done to separate the meteorological (and air chemical) factors from the air pollution level on the number and the magnitude of extreme events (Fig. 3.14).

Based on period averages, nitrate shows the same degree in decrease (in contrast to rain water at Seehausen) as sulfate: from 250 µeq L^{-1} (1994−2001) to 190 µeq L^{-1} (2001−2009), i. e. to 75 %. This can be explained through the decreasing NO_x emissions by power plants which emit into higher atmospheric altitudes whereas the NO_x emission (no significant changing over the last two decades) from traffic is ground-released (and to a large percentage fast deposited after chemical conversion).

Calcium in cloud (and rain) water is originated from dust (here flue ash from lignite fired power plants), emitted in the period before 1995 from high stacks in Eastern Germany, and therefore directly "injected" into long-range transport. This also is the explanation why nitrate in easterly air masses, crossing the border to the west, was larger than in westerly air masses. The "calcium abatement " was very efficient in the early 1990s and directly resulted in equivalent reduction of air pollution (concentration in PM). As seen in Fig. 3.12, despite larger Ca abatement than for SO_2 in the period from 1990−1995, the mean ratio calcium/sulfate is significantly larger in that period than after 2000. This is not a coincidence with the (mean) result that acidity (in terms of H^+) was higher in this period. However, taking into account the large standard variation (monthly variation, seen in

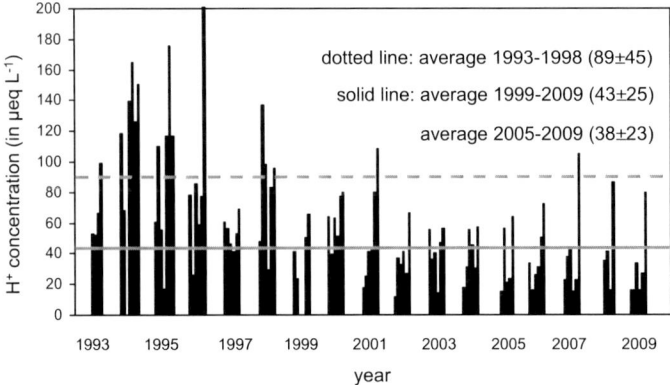

Fig. 3.13 Variation of monthly weighted H^+ means at Brocken from 1993–2009 (April/May–October/December yearly variable); $n = 96$.

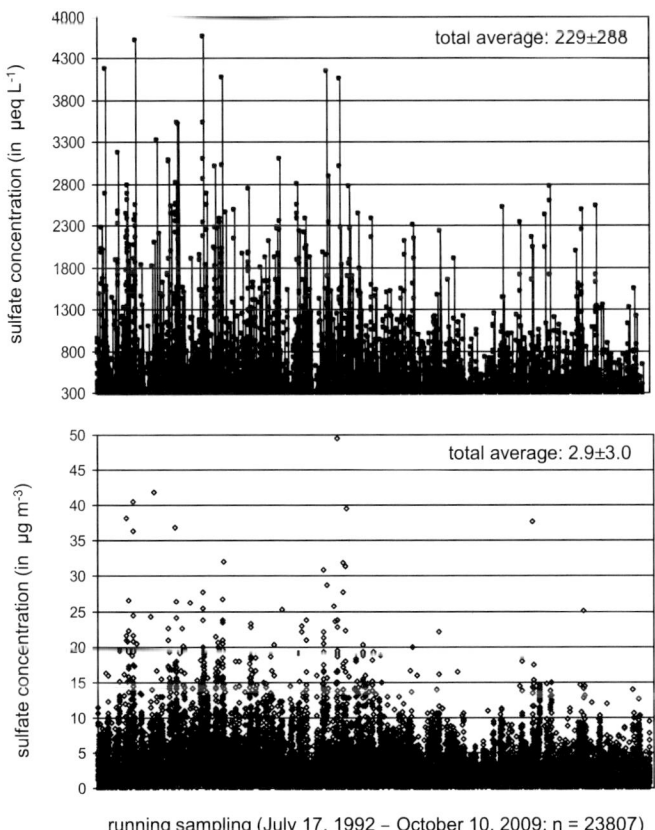

Fig. 3.14 Trend of sulfate in Brocken cloud water (in μeq L^{-1}) and the "residual" (in μg m^{-3}) based on all 1-hour samples (1992–2009), $n = 23807$.

Fig. 3.13) and the pH distribution (Fig. 3.15) one can draw the conclusion that in the period before 1996, different air masses occurred having different chemical climate. It shows that averaging might result in masking different effects.

Sulfate in cloud water is originated a) from CCN, and therefore of large-scale characteristic and b) from SO_2 scavenging (local to meso-scale characteristic). As seen above, SO_2 abatement was also very efficient (but from 1990−1995 less than for dust) − it is logical that consequent sulfate change in cloud is smoother and smaller than in rain where below-cloud scavenging additionally increases sulfate loading. Concerning acidity change, we first look at the budget (in µeq L^{-1}):

1993−1995: $[Cl^-] + [NO_3^-] + [SO_4^{2-}] - ([NH_4^+] + [Ca^{2+}] + [Mg^{2+}] + [K^+])$
            $= 522 - 460 = 62$
1993−1995: $[H^+] = 103$
1996−2009: $[Cl^-] + [NO_3^-] + [SO_4^{2-}] - ([NH_4^+] + [Ca^{2+}] + [Mg^{2+}] + [K^+])$
            $= 503 - 439 = 64$
1996−2009: $[H^+] = 62$

The difference corresponds to $[H^+]$ − when there are no other constituents. We see that this is true for the second period whereas in the first period there are missing anions for 41 µeq L^{-1}. It is not unlikely that this was bicarbonate (HCO_3^-), associated with alkaline flue ash particles.

As discussed for precipitation chemistry at Seehausen, the years 1994−1995 also show an increased acidity in Brocken cloud water (Fig. 3.13), very likely due to stronger abatement of alkaline dust than of acidic gases (where only SO_2 is important). This effect, however, as obviously only measurable in easterly trajectories.

Fig. 3.15 shows frequency distributions of pH for selected years. Climate parameters normally are unimodally distributed (*Gauss* distribution) but often showing a skewness. Most years here show a clear bimodal pH distribution; only a few years (1994, 2002, 2004, 2005, 2006 and 2009) show a skewed distribution and the year 2007 approximately a symmetric distribution. The pH minimum was reached in 1994 (pH 3.9) and then increased until 2002 (pH 4.4), from then on being constant. However, more significant than change of mean pH is the change of pH classes (Fig. 3.15). The number of very acid cloud events drastically decreased: from $11.5 \pm 1.8\%$ (230 h yr^{-1}, resp.) in 1994−1996 to $2.6 \pm 1.5\%$ (35 h yr^{-1}, resp.) in 1999−2009. In cases of bimodal pH distributions, the first maximum is around pH 4−5 (only in the years 1994 and 1995 being slightly less at around pH 4) and the second pH maximum moves from 5.5−6.0 in the period from 1993−1997 to around 6−7 after 1998 but showing a second maximum only in 2008 whereas in the years 2002−2007 and 2009 no maximum is seen. The rain water pH shifted in the last few years (see Table 3.9) to around 6 (from about 4.5 before 1995). We conclude that clouds are on average more acidic than rain (due to cloud processing and acid formation) and that the second (less frequent) maximum represents the "clean" background; when clouds precipitate, the pH evolution (pH increase) is due to dilution by growing droplets in the cloud−rain transformation and subsequent below-cloud scavenging of (less acidic) dust particles.

A very strong correlation between ions only exists for Na and Cl (Fig. 3.9) but weaker correlations are found for the following relationships (note that in cloud events and short-term periods, the correlation coefficient is much higher; based on

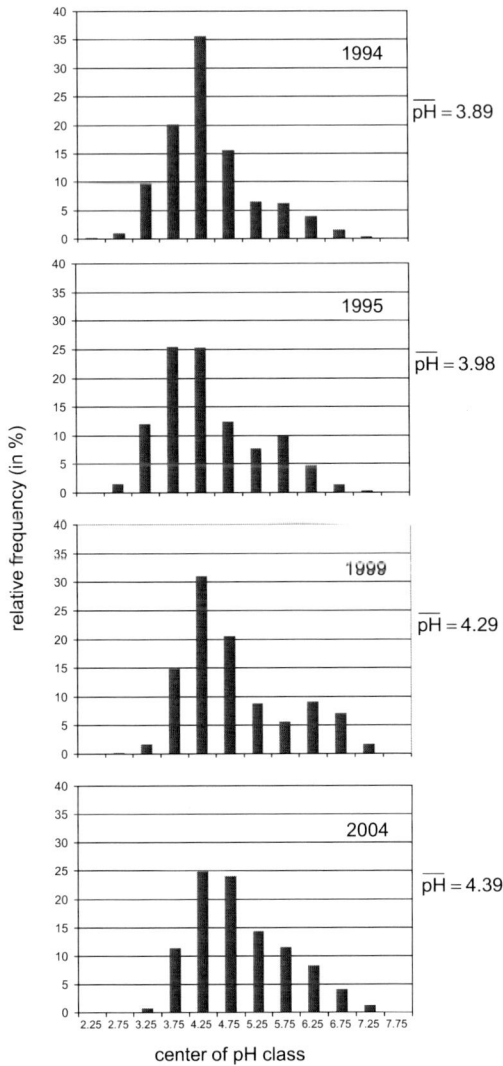

Fig. 3.15 Frequency distribution of pH in cloud water for selected years.

microequivalents) We can conclude that magnesium is in large part of sea salt origin and (what is not surprising) ammonium, sulfate and nitrate are interlinked via particle formation as described in Chapter 4.3.4 and providing an important percentage of CCN.

$[Na^+]$ = 4.9 $[Mg^{2+}]$ − 18.5 r^2 = 0.86
$[SO_4^{2-}]$ = 0.5 $[NO_3^-]$ + 82.5 r^2 = 0.67
$[NH_4^+]$ = 1.7 $[SO_4^{2-}]$ + 2.8 r^2 = 0.81
$[NH_4^+]$ = 0.9 $[NO_3^-]$ + 90.8 r^2 = 0.80

Taking into account the molar cloud water composition, we are able to "construct" the salty composition to be (in parenthesis "volatile acid", see also Chapter 2.8.3.12):

$NaCl + MgCl_2 + NaNO_3 + (NH_4)_2SO_4 + CaSO_4 + NH_4NO_3 (+ HNO_3)$.

From the mass budget, the volatile HNO_3 amounts to around 30 μeq L^{-1}, or in other terms, only 3 eq-% of cloud water composition. The total inorganic salty mass, recalculated on the air volume, did not changed over the period from 1993–2009 and amounts to 12 μg m^{-3}. This corresponds to the water-soluble inorganic percentage of PM_{10} measured at the Fronauer Tower (320 m a.s.l.) in Berlin in 2001–2002 (Beekmann et al. 2007).

4 Fundamentals of physico-chemistry in the climate system

The transport and transformation of *chemical species*[1] is ongoing permanently in the atmosphere. The air constituents (gases, solid and liquid particles) change permanently through chemical reactions, transfers among the physical states and transfers to (deposition) and from the earth's surface (emissions). Direct solar radiation and scattered and re-emitted radiation interacts with the air constituents, resulting in changes of the energetic state and photochemical conversions. This chapter mainly focuses on the chemical fundamentals of the climate system. As often already mentioned in this book, we cannot separate chemical and physical processes. Hence, it is inevitable to outline briefly the physical fundamentals in this regard. However, there are many excellent books on atmospheric physics and meteorology on the market and the reader is referred to them: atmospheric physics and thermodynamics (Peixoto and Oort 1992, Andrews 2000, Zdunkowski and Bott 2004, Hewitt and Jackson 2003, Tsonis 2007), atmospheric dynamics (Gill 1982, Holton 2004, Vallis 2006, Marshall and Plumb 2007), meteorology (Garratt 1992, Kraus 2004, Wallace and Hobbs 2006, Ackerman and Knox 2006, Ahrens 2007, Lutgens et al. 2009) and radiation (Kyle 1991, Zdunkowski et al. 2007, Wendisch and Yang 2012).

Despite important achievements in understanding of the key aspects of aerosols, cloud and precipitation chemistry and physics, clouds and aerosols remain the largest source of uncertainty in the two most important climate change metrics: radiative forcing and climate sensitivity (IPPC 2007, 2013). This uncertainty is the "manifestation of the lack of our understanding of how aerosol-cloud-precipitation processes act in the climate system" (Heintzenberg and Charlson 2009). An understanding of this complex matter needs an interdisciplinary and integrated approach as suggested by AQCPC (2009)[2]. Contemporary science has had to adopt a new way of thinking because of the emergence of a rapid accumulation of knowledge at different levels. It seems, however, that progress in understanding nature is slow and infinite. Wise sayings illustrate this predicament[3]: "Nature is simple, but scientists are complicated", said *Francisco Torrent-Guasp*, Spanish cardiologist (1931−2005) and *Jean-Jacques Rousseau* said, "Nature never deceives us; it is we who deceive ourselves" *Albert Einstein* said that "everything should be made as simple as possible, but not simpler". Stewart and Cohen (1994) also stated:

---

[1] The term *species* is normally only used in biology as a basic unit of biological classification. In recent years, the term *chemical species* has come into use as a common name for atoms, molecules, molecular fragments, ions, etc.

[2] Aerosols, Clouds, Precipitation and Climate. Science plan & implementation strategy (Melbourne 2009. Joint initiative of IGBP and WCRP. This group has proposed a regime-based approach to study specific cloud regimes.

[3] Citations from Buckberg (2005).

The role of science is to seek simplicity in a complex world. This is a comfortable picture, which encourages a view of the relation between laws and their consequences − between cause and effect, i. e. simple rules imply simple behaviour, therefore complicated behaviour must arise from complicated rules.

4.1 Physical basics

The fundamental law of nature is the law of the conservation of matter. Matter occurs in two states: flowing energy (Chapter 2.3) and cycling material (Chapter 2.4). The climate system, however, is an open system and is in permanent exchange with its surrounding earth system and thereby space (Fig. 3.2). Therefore, we generally expect changes and variations of internal energy and mass over time. From a chemical point of view, the interest lies in the quantification of the amount − in terms of the number of particles and thereby mass − of different chemical species in a volume regarded. This quantity, called *density* (an equivalent term to *concentration*) is investigated as a function of time and space. Hence, mass, time and distance are the fundamental quantities of the climate system, and all its quantities can be expressed in terms of meters, kilograms and seconds (Appendix A.2). Motion is based on *Newton's* equations:

$$\left(\frac{d\vec{v}}{dt}\right)_m = \frac{1}{m} \cdot f \quad \text{and} \quad \left(\frac{dm}{dt}\right)_{\vec{v}} = \frac{1}{\vec{v}} \cdot f$$

$$\text{and} \quad \frac{d}{dt}(m\vec{v}) = \sum_i f_i = ma \tag{4.1}$$

where m is mass, t is time, $\vec{v}$ is velocity (as vector), a is acceleration and f is force. The last equation of motion ($m\vec{v}$ is the momentum) characterizes the *fluid* atmosphere (force = mass · acceleration = power / velocity and work = power · distance). *Force* is the change of energy by distance (1 J m^{-1} = 1 N = 1 kg m s^{-2}). *Energy* is the capacity to do work (or produce heat); there are different kinds of energy: energy ≡ heat ≡ radiation ≡ work (1 J = 1 N m = 1 kg m^2 s^{-2}). *Heat* is the transfer of thermal energy from one object to another. *Power* is the rate of energy change (1 J s^{-1} = 1 W = 1 kg m^{-2} s^3). *Climate forcing* is expressed by energy flux (1 W m^{-2} = 1 J s^{-1}m^{-2}).

The *rate R* is a ratio between two measurements, normally per unit of time: $R = \Delta\varepsilon/\Delta t$. However, a rate of change can be specified per unit of length (Δl) or mass (Δm) or another quantity, for example $\Delta\varepsilon/\Delta x$ where Δx is the *displacement* (the shortest directed distance between any two points). In chemistry and physics, the word "rate" is often replaced by (or synonymously used with) *speed* (see below about the difference to *velocity*), and can be the distance covered per unit of time (i. e. *acceleration,* the rate of change in speed) or the change in speed per unit of time (i. e. *reaction rate,* the speed at which chemical reactions occur). In physics, *velocity* $\vec{v}$ is the rate of change of position. This is a vector physical quantity, and both *speed* and *direction* are required to define it. The scalar absolute value (magnitude) of velocity is *speed*. The *average* velocity $\bar{v}$ of an object moving through a displacement (Δx) during a time interval (Δt) is described by $\bar{v} = \Delta x/\Delta t$.

A *flux* F is defined as the amount that flows through a unit area per unit of time: $F = \Delta\varepsilon/q\Delta t$ (q area). Flux in this definition is a vector. However, in general "flux" in earth system research relates to the movement of a substance between compartments. This looser usage (Chapter 2) is equivalent to a rate (change of mass per time); sometimes, the term *specific* rate is used in the more exact sense of time- *and* area-related flux. Generally, the terms "rate" and "flux" as well as "velocity" and "speed" are often not separated in the literature in such an exact physical sense, but used synonymously.

The *gradient* of a scalar field (for example, air temperature, wind speed, density) is a vector field that points in the direction of the greatest rate of increase of the scalar field, and whose magnitude is the greatest rate of change. The gradient (or gradient vector field) of a scalar function $\varepsilon(x_1, x_2, \dots x_n)$ is denoted as $\vec{\nabla}\varepsilon$ where $\vec{\nabla}$ (the nabla symbol) denotes the vector differential operator (*del*). The notation grad(ε) is also used for gradient. The gradient of f is defined to be the vector field whose components are the partial derivatives of ε. That is:

$$\vec{\nabla}\varepsilon = \left(\frac{\partial\varepsilon}{\partial x_1}, \dots, \frac{\partial\varepsilon}{\partial x_n} \right).$$

4.1.1 Properties of gases: The ideal gas

Without doubt, one of the most important models in physical chemistry is that of an *ideal gas* assuming the gas as the congregation of particles existing in omnidirectional stochastic motion.

4.1.1.1 Fluid characteristics

The velocity of moving air, more exactly those of an air parcel, is the wind velocity. It is not the velocity of molecules and particles contained in the air. Wind is another expression for flow. The wind velocity $\vec{v}$ is a vector given by a norm (*wind speed*) and a direction in space (*wind direction*). An air parcel is a volume large enough to allow a statistic macroscopic consideration of the gas; this can even be given for a small volume of about 1 mm in diameter. In the Cartesian (rectangular) coordinate system (in the atmosphere where the direction of the x-axis is east, the y-axis north and the z-axis upward), the wind velocity is separated into three directions, named u, v and w:

$$\vec{v} = u \cdot \vec{i} + v \cdot \vec{j} + w \cdot \vec{k} \tag{4.2}$$

where $\vec{i}$, $\vec{j}$ and $\vec{k}$ are the unit vectors in directions of x, y and z. The wind speed (absolute value of velocity) follows from:

$$|\vec{v}| = \sqrt{u^2 + v^2 + w^2} \quad \text{or} \quad |\vec{v}| = \sqrt{v_x^2 + v_y^2 + v_z^2} \tag{4.3}$$

Because of the time-depending wind velocity, the wind vector is described by a complicated four-dimensional field $\vec{v}(x, y, z, t)$ with the components $u(x, y, z, t)$,

$v(x, y, z, t)$ and $w(x, y, z, t)$. The vertical component is much smaller (by a factor 10^3 to 10^4) than the horizontal components, and the typical wind speed at a 1 km altitude is 10 m s^{-1}. The velocity describes the motion of an air parcel, i. e. the distance $\vec{l}$ that is passed in the time dt:

$$\vec{v} = \frac{d\vec{l}}{dt} = \frac{dx}{dt}\vec{i} + \frac{dy}{dt}\vec{j} + \frac{dz}{dt}\vec{k} \qquad (4.4)$$

This means that $dx/dt = u$, $dy/dt = v$ and $dz/dt = w$ are the directional components of wind velocity and that any property ε of this air parcel can be described as:

$$\frac{d\varepsilon}{dt} = \frac{\partial\varepsilon}{\partial t} + u\frac{\partial\varepsilon}{\partial x} + v\frac{\partial\varepsilon}{\partial y} + w\frac{\partial\varepsilon}{\partial z} = \frac{\partial\varepsilon}{\partial t} + \vec{v} \cdot \vec{\nabla}\varepsilon \qquad (4.5)$$

This is the total derivative of ε over time, where the partial derivatives have been included. The quantity changed (for example, velocity, density or momentum) is expressed through a mean value (convective transport) and a jitter (*turbulent transport*):

$$\varepsilon(t) = \bar{\varepsilon}(t_R, \Delta t) + \varepsilon'(t) \qquad (4.6)$$

Turbulence or turbulent flow is a fluid regime characterized by chaotic, stochastic property changes as a self-organizing process; the information entropy serves as a criterion for the degree of its self-organization. The degree of turbulence (Tu) in a fluid characterizes the ratio between the velocity jitter and averaged speed $\bar{u}$; it is a parameter of the laminar to turbulent foldover:

$$Tu = \frac{\sqrt{\frac{1}{3}(u^2 + v^2 + w^2)}}{\bar{u}} \qquad (4.7)$$

The change along a particle path, i. e. the trajectory of the air parcel, is called *Lagrangian*. By contrast, the *Eulerian* approach considers changes at a given site or fixed volume element, i. e. local changes of $\partial\varepsilon/\partial t$. The atmospheric fluid is always turbulent with the exception of quasi-laminar interfaces. Turbulence is produced dynamically and thermally; it grows with increasing wind, surface roughness and a vertical negative temperature gradient. In contrast to the molecular diffusion, the turbulent diffusion coefficient is not a physical constant. The transport of mass and convective heat is phenomena of turbulence. Flows are triggered through local temperature differences and thereby density differences.

4.1.1.2 The gas laws

The *gas laws* developed by *Robert Boyle, Jacques Charles* and *Joseph Gay-Lussac* are based upon empirical observations and describe the behavior of a gas in macroscopic terms, that is, in terms of properties that a person can directly observe and experience. The kinetic theory of gases describes the behavior of molecules in a gas based on the mechanical movements of single molecules. A gas is defined as a collection of small particles (atom- and molecule-sized) with the mass m_i (subscript i denotes a defined substance):

- occupying no volume (that is, they are points),
- where collisions between molecules are perfectly elastic (that is, no energy is gained or lost during the collision),
- having no attractive or repulsive forces between the molecules,
- traveling in straight-line motion independent each other and not favoring any direction, and
- obeying *Newton*'s Laws.

While colliding they exchange energy and momentum. Collisions lead macroscopically to the *viscosity* of a gas and *diffusion* of molecules. Considering only the translation of particles (the aim of the kinetic theory of gases), there is no need for quantum mechanical description. However, when describing rotation and vibration in molecules (the other parts of the internal energy), quantum mechanics is essential. The size and speed of molecules with a mass m_m can be different. Therefore, the gas is considered macroscopic by averaging individual quantities. The molecule number density c_N is defined as the ratio between the number of gas molecules N and the gas volume V. Hence, $c_N = N/V$ (we will later see that c_N in pure gases denotes the *Loschmidt* constant n_0). Please note that in the scientific literature the number density is normally termed with n but here that would be confusing with the mole number n.

Let us now consider the motion of a molecule in x direction onto a virtual wall. Because of the three spatial directions and each positive and negative direction, the particle density in each direction amounts to $c_N/6$. For this, molecules $v = v_x$ is valid because the speed in all other directions (y, z) is zero (according to the agreement they only move in x direction). Within the time dt a molecule passes the distance $v\,dt$. At the wall with the area q, $c_N q v\,dt/6$ molecules collide totally. Therefore, each molecule transfers the momentum $2 m_m v$ because of the action-reaction principle (to and from the wall). All collisions, given by the collision number z (collision frequency or rate is z/dt), transfer in a given time period dt a momentum that generates the force f according to Eq. (4.1):

$$f = \frac{z \cdot 2 m_m v}{dt} = \frac{c_N}{6} q v \cdot 2 m_m v = \frac{1}{3} c_N q \cdot m_m v^2 \tag{4.8}$$

For the *pressure* p (force/area) results:

$$p = \frac{f}{q} = \frac{1}{3} c_N \cdot m_m v^2 \tag{4.9}$$

With the definition of the gas density ϱ as a product of particle density c_N and molecule mass m_m, the fundamental equation of the kinetic theory of gases follows:

$$p = \frac{1}{3} \varrho v^2 \tag{4.10}$$

Expressing the gas density ϱ as the ratio between total mass m ($m = N\,m_m$) and gas volume V the *Boyle-Mariotte* law then follows. pV represents the quantity of an energy, or more specifically the pressure-volume work of this gaseous system:

$$pV = \frac{1}{3} m \cdot v^2 = \text{constant} \tag{4.11}$$

Empirically, it has been found that (T is the absolute temperature):

$$\frac{pV}{T} = \text{constant} \tag{4.12}$$

This is the combined gas law that combines *Charles's* law ($V = \text{constant} \cdot T$ where c_N, $p = \text{constant}$), *Boyle's* law ($pV = \text{constant}$ where c_N, $T = \text{constant}$) and *Gay-Lussac's* law ($p/T = \text{constant}$ where m, $V = \text{constant}$). *Avogadro's* law expresses that the ratio of a given gas volume to the amount of gas molecules within that volume is constant (where T, $p = \text{constant}$):

$$\frac{V}{c_N} = \text{constant} \tag{4.13}$$

Combining Eqs. (4.12) and (4.13) leads to the *ideal gas law*, where $\boldsymbol{R}$ is the universal gas constant. The amount of gas molecules we will express by the mole number $n = m/M$:

$$pV = n\boldsymbol{R}T \tag{4.14}$$

Now, we can derive an expression for the work W done by the gas volume (that is equivalent to the kinetic energy). Using the definitions $f = dW/dx$, $a = dv/dt$ and $v = dx/dt$ it follows from Eq. (4.1):

$$E_{kin} = \Delta W = \int_0^x f dx = \int_0^x m_m a\, dx = \int_0^v m_m \frac{dx}{dt}\, dv = \int_0^v m_m v\, dv = \frac{1}{2} m_m v^2 \tag{4.15}$$

Now, combining Eqs. (4.10) and (4.14) and taking into account that $\varrho = m/V$ and $n = m/M$, we derive an important relationship for the mean velocity of molecules:

$$v_{mean} = \sqrt{\frac{3p}{\varrho}} = \sqrt{\frac{3\boldsymbol{R}T}{M}} = \sqrt{\frac{3\mathbf{k}t}{m_m}} \tag{4.16}$$

More exactly the velocity derived in Eq. (4.16) is the *root mean square velocity* v_{rms} (see equation 4.35b). Remember that m_m is the molecule mass, $\mathbf{k}$ is the *Boltzmann* constant, which is in the following relationship to the gas constant (hence, the equations with $\mathbf{k}$ are related to averaged single molecule properties and those with $\boldsymbol{R}$ to the gas being the collection of molecules):

$$\frac{\mathbf{k}}{\boldsymbol{R}} = \frac{m_m}{M} = \frac{1}{N_A} \tag{4.17}$$

N_A *Avogadro* constant[4] (in the past called *number*), that is the number of "elementary entities" (usually atoms or molecules) in one mole, is (from the definition of the mole) the number of atoms in exactly 12 grams of C^{12}, which is about $6.023 \cdot 10^{23}$. From Eqs. (4.16) and (4.15) another expression for the kinetic energy of molecules follows (remember that $M = m_m N_A$):

$$E_{kin} = \frac{1}{2} m_m v^2 = \frac{1}{2} \frac{M}{N_A} v^2 = \frac{3}{2} \frac{\boldsymbol{R}T}{N_A} = \frac{3}{2} \mathbf{k}T \tag{4.18}$$

---

[4] *Johann Josef Loschmidt* first calculated the value of *Avogadro's* number; often referred to as the *Loschmidt* number in German-speaking countries (*Loschmidt* constant now has another meaning).

The average kinetic energy of a molecule is $3kT/2$ and the molar kinetic energy amounts to $3RT/2$. It is important to distinguish between molecular (denoted by the subscript m here) and molar quantities. Furthermore, it is valid (V_m is the molar volume):

$$\frac{p}{RT} = \frac{n}{V} = \frac{1}{V_m} \tag{4.19}$$

Another (original) expression for the gas law then follows:

$$pV_m = RT = \frac{1}{N_A}kt \tag{4.20}$$

The *Loschmidt* constant n_0 is the number of particles (atoms or molecules) of an ideal gas in a given volume V (the number density), and is usually quoted at standard temperature and pressure:

$$n_0 = \frac{N}{V} = \frac{p}{kT} = \frac{p}{RT}N_A = \frac{N_A}{V_m} \tag{4.21}$$

From Eqs. (4.14), (4.17), (4.20) and (4.21) the following gas equations are derived:

$$pV = \frac{m}{m_m}kT \quad \text{and} \quad p = n_0 kT \tag{4.22}$$

The model of an ideal gas can be applied with sufficient accuracy[5] to air with a mean pressure of 1 bar (variation about 0.65−1.35 bar) and a mean temperature near the earth's surface of 285 K (variation about 213−317 K). For a description of air as a gas mixture, *Dalton*'s law (also called *Dalton*'s law of partial pressures) is important. This law states that the total pressure p exerted by a gaseous mixture is equal to the sum of the partial pressures p_i of each individual component in a gas mixture: $p = \Sigma p_i$. It follows that $V = \Sigma V_i$ and $n = \Sigma n_i$. The gas density ϱ (note the difference to the number density η and n_0, respectively) can also be expressed by several terms:

$$\varrho = \frac{m}{V} = \eta m_m = \frac{N}{V}m_m = \frac{pM}{RT} \tag{4.23}$$

In the gas mixture, we can now introduce all equations related to a specific compound (molecule) i or to the sum of all gases; hence, it also valid that $m = \Sigma m_i$ and $\varrho = \Sigma \varrho_i$. A useful quantity for describing mixtures (not only gaseous) is the mole fraction $x_i = n_i/n$, which generally characterizes a *mixing ratio* (Chapter 4.1.2). From the different forms of the gas law it follows (the subscript i denotes *partial* quantities) that:

$$x_i = \frac{n_i}{n} = \frac{V_i}{V} = \frac{p_i}{p} = \frac{m_i}{m} = \frac{\varrho_i}{\varrho} \tag{4.24}$$

A mean molar mass $\overline{M}$ of air (or generally a gas mixture) can be defined based on mole-weighted fractions of individual molar masses:

---

[5] Because of the complexity of air, which is a multicomponent and multiphase system, we are only able to roughly model this system. It makes no sense to describe single processes with a complexity several orders of magnitudes more than the process with the lowest accuracy.

$$\overline{M} = \frac{\Sigma n_i M_i}{\Sigma n_i} = \Sigma x_i M_i \qquad (4.25)$$

Eq. (4.23) states that the molar mass of a gas is directly proportional to its densities in the case of same pressure and temperature. Based on standard values (normally $0\,°C$ and 1 bar), the density for different pressures and temperature can be calculated according to:

$$\varrho_{T,p} = \varrho_0 \frac{p}{p_0} \frac{T_0}{T} \qquad (4.26)$$

Furthermore, it is practicable to introduce a *relative* density ϱ_r, which is the ratio of a gas density to the density of a gas selected as standard (ϱ_s) because the ratio of densities does not depend on pressure and temperature. Oxygen and dry air are often selected as standard, which have the following densities (at 1 bar and 273.15 K):

$$\varrho_{O_2} = 0.001429 \text{ g cm}^{-3}$$
$$\varrho_{air} = 0.0012928 \text{ g cm}^{-3}.$$

The relative density is also equivalent to the ratio of the relevant molar masses:

$$\varrho_r = \frac{\varrho}{\varrho_s} = \frac{M}{M_s} \qquad (4.27)$$

Air has a mean relative mole mass $M_{r,\,air} = M_{r,O_2} \cdot \varrho_{air}/\varrho_{O_2} = 28.949$.

4.1.1.3 Mean-free path and number of collisions between molecules

To understand heat conduction, diffusion, viscosity and chemical kinetics the mechanistic view of molecule motion is of fundamental importance. The fundamental quantity is the *mean-free path*, i. e. the distance of a molecule between two collisions with any other molecule. The number of collisions between a molecule and a wall was shown in Chapter 4.1.1.2 to be $z = c_N q v \, dt / 6$. Similarly, we can calculate the number of collisions between molecules from a geometric view. We denote that all molecules have the mean speed $\overline{v}$ and their mean *relative speed* with respect to the colliding molecule is $\overline{g}$. When two molecules collide, the distance between their centers is d; in the case of identical molecules, d corresponds to the effective diameter of the molecule. Hence, this molecule will collide in the time dt with any molecule centre that lies in a cylinder of a diameter $2d$ with the area πd^2 and length $\overline{g} \, dt$ (it follows that the volume is $\pi d^2 \overline{g} \, dt$). The area πd^2 where d is the molecule (particle) diameter is also called *collisional cross section* σ. This is a measure of the area (centered on the centre of the mass of one of the particles) through which the particles cannot pass each other without colliding. Hence, the number of collisions is $z = c_N \pi d^2 \overline{g} \, dt$. A more correct derivation, taking into account the motion of all other molecules with a *Maxwell* distribution (see below), leads to the same expression for z but with a factor of $\sqrt{2}$. We have to consider the relative speed, which is the vector difference between the velocities of two objects A and B (here for A relative to B):

$$(\bar{g})^2 = \overline{(\vec{v_A} - \vec{v_B})^2} = \overline{\vec{v_A}^2} - 2\overline{\vec{v_A}\vec{v_B}} + \overline{\vec{v_B}^2} = \overline{\vec{v_A}^2} + \overline{\vec{v_B}^2} \approx 2\overline{\vec{v}^2} \qquad (4.28)$$

Since $\overline{\vec{v_A}\vec{v_B}}$ must average zero, the relative directions being random, the average square of the relative velocity is twice the average square of the velocity of A + B and, therefore, the average root mean square velocity is increased by a factor $\sqrt{2}$ (remember that $\sqrt{\overline{v^2}} = \bar{v}$), and the collision rate is increased by this factor ($\bar{g} = \sqrt{2}\,\bar{v}$). Consequently, the mean-free path decreases by a factor of $\sqrt{2}$ when we take into account that all the molecules are moving:

$$z = \sqrt{2}\,c_N \pi d^2 \bar{v}\,dt \qquad (4.29)$$

To determine the distance travelled between collisions, the *mean-free path l*, we must divide the mean molecule velocity by the collision frequency or, in other words, the mean travelling distance $\bar{v}\,dt$ by the number of collisions. Furthermore, we must find an expression for the collision frequency ($v = z/dt$) and the *mean-free time* ($\tau = 1/v$):

$$l - \frac{\bar{v}\,dt}{z} = \frac{1}{\sqrt{2}c_N \pi d^2} \quad \text{and} \quad \tau = \frac{1}{\sqrt{2}c_N \pi d^2 \bar{v}} \quad \text{and} \quad \bar{v} = \frac{l}{\tau} \qquad (4.30)$$

That fraction of gas molecules is of interest when velocity is within v and $v + dv$. This fraction $f(v)dv$ is time-independent from the exchange with other molecules. The *Boltzmann* distribution states that at higher energies ($= \mathbf{k}T$) the probability to meet molecules is exponentially less:

$$f(v)dv = \text{const} \cdot \exp\left(-\frac{m_m v^2}{2\mathbf{k}T}\right)dv = \text{const} \cdot \exp\left(-\frac{Mv^2}{2RT}\right)dv \qquad (4.31)$$

The *Maxwell* distribution describes the distribution of the speeds of gas molecules at a given temperature:

$$f(v)dv = \sqrt{\frac{2}{\pi}}\left(\frac{m_m}{\mathbf{k}T}\right)^{\frac{3}{2}} v^2 \cdot \exp\left(-\frac{m_m v^2}{2\mathbf{k}T}\right)dv \qquad (4.32)$$

The fraction $f(v)\,dv$ denotes the probability dW and is also represented by dz/z ($= d\ln z$):

$$f(v)dv = dW = \frac{dz}{z} \qquad (4.33)$$

To find the *most probable speed v_p* of molecules, which is the speed most likely to be possessed by any molecule (of the same mass m_m) in the system (in other words, it is the speed at maximum of probability), we calculate $df(v)/dv$ from Eq. (4.32), set it to zero and solve for v, which yields:

$$v_p = \sqrt{\frac{2\mathbf{k}t}{m_m}} = \sqrt{\frac{2RT}{M}} \qquad (4.34)$$

By contrast, the *mean speed* follows as the mathematical average of the speed distribution, equivalent to the weighted arithmetic mean of all velocities (N number of molecules):

Table 4.1 Useful quantities in molecular kinetics (V – gas volume, d – molecule diameter, q – square).

quantity	meaning	description
m_m	molecule mass	$= M/N_A$
m	mass	$= N\,m_m$
M	molar mass	$= N_A m_m$
n	mole number	$= m/M$
N	number of molecules	
N_A	*Avogadro* constant [a]	$= M/m_m$
n_0	*Loschmidt* constant	$= N/V$
ϱ_N	density, number related [b]	$= N/V$
ϱ_m	density, mass related [c]	$= m/V = n_0 m_0$
c	concentration	$= m/V = n_0 m_0$
V_m	molar volume	$= V/n$
p	pressure	$= F/q = n_0 kT = nRT = \varrho_m v^2/3$
$\mathbf{k}$	*Boltzmann* constant	$= R(m_m/M) = R/N_A$
R	gas constant	$= \mathbf{k}(M/m_m) = \mathbf{k}\,N_A$
$\bar{v}$	mean molecule velocity [d]	$= \sqrt{v_{rms}^2}$
v_{rms}	root-mean-square speed	$= \sqrt{3\mathbf{k}T/m_m}$
a	acceleration	$= dv/dt$
F	force	$= p \cdot q = m_m(dv/dt) = v(dm/dt) = ma$
F_d	diffusion flux	$= dn/dt$
z	collision number	$= \sqrt{2}\,n_0 \pi d^2 v\,dt$
l	mean free path	$= v\,dt/z = 1/\left(\sqrt{2}\,n_0 \pi d^2\right)$
η	(dynamic) viscosity	$= (1/\pi d^2)\sqrt{m_m \mathbf{k}T/6}$
D	diffusion coefficient	$= (\bar{v} \cdot l)/3$

[a] identical with a number density
[b] identical with a number concentration
[c] identical with a mass concentration
[d] also transport velocity

$$\bar{v} = \frac{1}{N}\int_0^N v\,dN = \int_0^\infty vf(v)dv = \sqrt{\frac{2}{\pi}}\left(\frac{m_m}{\mathbf{k}T}\right)^{\frac{3}{2}}\int_0^\infty \exp\left(-\frac{m_m v^2}{2\,\mathbf{k}T}\right)v^3 dv$$

$$= \sqrt{\frac{8\,\mathbf{k}T}{\pi m_m}} = \sqrt{\frac{8\,RT}{\pi M}} \tag{4.35a}$$

The *root mean square speed* v_{rms} is the square root of the average squared speed:

$$v_{rms} = \left(\int_0^\infty v^2 f(v)dv\right)^{\frac{1}{2}} = \sqrt{\frac{3\,\mathbf{k}T}{m_m}} = \sqrt{\frac{3\,RT}{M}} \tag{4.35b}$$

All three velocities are interlinked in following ratio:

$$v_p : \bar{v} : v_{rms} = \sqrt{2} : \sqrt{\frac{8}{\pi}} : \sqrt{3} \tag{4.36}$$

Now, we can set the expression (4.34) for the mean molecular velocity in Eq. (4.29) and ascertain the mean molecular number of collisions z and the mean-free path l, replacing n by p/kT (r molecule radius) in Eq. (4.30):

$$z = 2c_N\pi d^2 dt\sqrt{\frac{kT}{m_m}} \quad \text{or collision frequency} \quad \nu = 2c_N\pi d^2\sqrt{\frac{kT}{m_m}} \tag{4.37}$$

$$l = \frac{kT}{\sqrt{2}\,\pi d^2 p} = \frac{kT}{4\sqrt{2}\,\pi r^2 p} \tag{4.38}$$

Note that these equations are valid only for like molecules. In the case of air, many different (unlike) molecules have to be considered with different radiuses or diameters, molar masses and molecular speeds. For the example of gas mixture A and B, the following expressions must be applied for mean molecule diameter and mass:

$$d_{AB} = (d_A + d_B)/2 \tag{4.39}$$

$$m_{AB} = m_A m_B/(m_A + m_B) \tag{4.40}$$

In air, the mean-free path has an order of 10^{-7} m and does not depend on temperature but is inversely proportional to pressure. The expressions given for collision number and mean-free path are useful for understanding chemical reactions (see collision theory in Chapter 4.1.1.2) but have only limited worth for applications because the molecular diameter (or radius) is not directly measurable. However, the molecular diameter is typically determined from viscosity measurements.

4.1.1.4 Viscosity

Air is a *viscose* medium; hence, we observe *friction* or *drag*. Friction converts kinetic energy into heat. The internal friction between two moving thin air layers in x direction results in a gradient of the speed in y direction (perpendicular to the flow); f_f is the *frictional force*, q the area between the moving layers:

$$f_f = q\eta\frac{dv}{dz} = q \cdot \tau \tag{4.41}$$

where η is the *dynamic* or *absolute* viscosity (dimension kg/m · s) and the *shear stress* τ in Newtonians fluids is defined by $\tau = q\,(dv/dz) = f_f/q$. A *kinematic* viscosity v is defined as the ratio of the dynamic viscosity to the density of the fluid: $v = \eta/\varrho$. The derivative (dv/dz) is the *shear velocity*, also called friction velocity. When compared with *Newton's* Eq. (4.1), the meaning of τ is clearly seen to be a flux density of momentum. The momentum flows between moving layers in the direction of decreasing shear velocity (from higher layers down to the earth's surface). More exactly, the frictional force must be regarded in all directions, i.e. being a vector.

The *viscosity* of a gas is in direct relation to the mean-free path (Gombosi 1994); mass density $\varrho_m = c_N m_m = N m_0/V = m/V$ (note the difference to the number density):

$$\eta = \frac{1}{3}\varrho_m \bar{v} l = \frac{1}{3}c_N m_m \bar{v} l = \frac{1}{\pi d^2}\sqrt{\frac{m_m \mathbf{k}T}{6}} \tag{4.42}$$

However, this equation is an approximation; with respect to intra-molecular interactions, a more exact relation $\eta = 0.499\varrho_m \bar{v} l$ is obtained. Viscosity is independent of pressure but increases with increasing temperature because of rising molecule speed. The temperature dependence of viscosity is described by an empirical expression; the C *Sutherland* constant and the constant B follow from Eq. (4.42):

$$\eta = \frac{B\sqrt{T}}{1 + C/T} \tag{4.43}$$

4.1.1.5 Diffusion

In summarizing, we can state that no preferential direction exists in molecular motion. Hence, there is no transport of any quantity (Brownian motion). Any flux (transport of material, heat or momentum) is caused either by turbulent diffusion (air parcel advection) or by laminar diffusion. Diffusion (when using this term in chemistry only laminar transport is meant) is the flux (dn/dt) due to concentrations gradients (dc/dz). Such concentration gradients occur near interfaces very close to the earth's surface (see dry deposition in Chapter 4.4.1) and between air gases and solid aerosol or aqueous particles (cloud and raindrops) through phase transfer processes. Physically diffusion means that the mean-free path of molecules (or particles[6]) increases in the direction of decreasing number concentration; therefore, diffusion is directed Brownian motion. The diffusion flux $F_d = dn/qdt$ (mole per time and area) is proportional to the concentration gradient dc/dx; D is the *diffusion coefficient*, V volume ($q \cdot dz$) and n mole number, known as *Fick's* first law:

$$F_d = \frac{1}{q}\frac{dn}{dt} = -D\frac{1}{V}\frac{dn}{dz} = -D\frac{dc}{dz} \tag{4.44}$$

Combining the basic *Newton's* equation (4.1) with (4.41) in terms of the general quantity ε (Eq. 4.5), which can be a mass m_m, velocity v, heat etc. and where β means a proportionality coefficient, namely a *characteristic transfer coefficient*, it follows that:

$$\frac{d\varepsilon}{dt} = -\beta \cdot \varepsilon = -\frac{1}{\varepsilon}F = -q\frac{\eta}{m_m}\frac{d\varepsilon}{dz} \tag{4.45}$$

This is a *general flux equation* for material, energy and momentum. The negative sign indicates a decrease in the quantity ε in the direction of the gradient $d\varepsilon/dz$ (z is for vertical direction and x, y for any horizontal one). In terms of velocity from Eq. (4.44), an expression for the *fluidity* φ follows:

---

[6] The diffusion of large particles (PM) compared with molecules is described in a different way. The term *particle* here involves molecules, atoms and molecule clusters.

$$\frac{d\vec{v}}{dx} = -\frac{1}{\eta \cdot q} \, m_m \frac{d\vec{v}}{dt} = -\varphi \cdot \tau \tag{4.46}$$

where $\varphi = 1/\eta$ (τ shear stress). Hence, viscosity means the transport of momentum against a velocity gradient, and diffusion is the transport of material against a concentration gradient. Comparing Eqs. (4.44) and (4.45) generates a simple expression for the gas diffusion coefficient D_g (which is different from the liquid-phase diffusion coefficient). Table 4.1 summarizes the most important quantities for describing the molecular kinetics.

$$D_g = \frac{1}{3}\bar{v}l = \frac{1}{\pi d^2}\frac{1}{\varrho_m}\sqrt{\frac{m_m \mathbf{k}T}{6}} = \frac{1}{\pi d^2}\frac{1}{n_0}\sqrt{\frac{N_A}{M}}\sqrt{\frac{\mathbf{k}T}{6}} \tag{4.47}$$

4.1.2 Units for chemical abundance: Concentrations and mixing ratios

Air is a multiphase and multicomponent system or, in other words, a mixture of gases and particles. The latter we distinguish into liquid droplets and solids. Solid particles are also often mixtures and always contain water to different extents, i. e. they are humid. Within the troposphere, droplets are exclusively aqueous solutions (hydrometeors) but in the stratosphere, droplets can be formed from sulfuric and nitric acid.

Despite the proposal by scientific organizations (IUPAC and IUPAP) to use only SI units (Appendix A.2), practical purposes mean different units for describing the abundance of air traces are in use. In chemistry, concentration is the measure of how much of a given substance is mixed with another substance. This can apply to any sort of chemical mixture, but most frequently the concept is limited to homogeneous solutions, where it refers to the amount of solute in a substance. Air always provides a heterogeneous mixture. We can define a unit of air volume (in the sense of a space), let's say 1 m³, which is filled with an amount (how much there is or how many there are of something that you can quantify) of a substance (gas and condensed matter). Amount (mass, volume, number) denotes an extensive quantity, i. e. the value of an extensive quantity increases or decreases when the reference volume changes. The amount of a substance in a given gas volume is strongly determined by the gas laws. By contrast, an intensive quantity (pressure, temperature) does not change with volume. We can quantify the amount of a substance in different terms:

- *mass m* (related to a prototype made from iridium and stored in Paris), measured in kg
- *volume V* (how much three-dimensional space is occupied by a substance), measured in m³
- *number N* (the sum of individuals such as molecules, particles, droplets), dimensionless
- *mole n* (1 mol corresponds to the molecular weight or to a fixed number of molecules expressed by the *Loschmidt* constant n_0), dimensionless

Table 4.2 Recalculation between mass concentration c_i (in µg m^{-3}) and mixing ratio x_i (in ppb) under standard conditions (25 °C and 1 bar).

compound		$x(v)_i$ in ppb	$c(m)_i$ in µg m^{-3}
ozone	O_3	1	1.96
hydrogen peroxide	H_2O_2	1	1.39
sulfur dioxide	SO_2	1	2.63
nitrogen monoxide	NO	1	1.22
nitrogen dioxide	NO_2	1	1.89
nitrous acid	HNO_2	1	1.92
nitric acid	HNO_3	1	2.56
ammonia	NH_3	1	0.69
hydrogen chloride	HCl	1	1.63
benzene	C_6H_6	1	3.23

Mass, volume, mole and pressure (note that pressure is an intensive quantity) are linked in the general gas equation (which, however, is only valid for diluted mixtures, which is true when we consider traces in air), as seen in Eq. (4.14). We also can apply the gas law based on the partial quantities $m = \Sigma m_i$, $V = \Sigma V_i$, $p = \Sigma p_i$ and $n = \Sigma n_i$ in two forms:

$$p_i V = n_i RT \quad \text{as well as} \quad pV_i = n_i RT \tag{4.48}$$

Hence, for any mixing ratio x_i by taking into account the gas equation for a given substance i, it follows that: $x_i = \dfrac{n_i}{n} = \dfrac{V_i}{V} = \dfrac{m_i}{m} = \dfrac{p_i}{p}$ (Eq. 4.24). These ratios are called (dimensionless) *mixing ratios* (or *fractions*) by the amount of moles, volume, mass and pressure, respectively. The large advantage in its use compared with concentrations[7] (moles or mass per volume) lies in its independence from p and T. Depending on the magnitude of the mixing ratio the most convenient units can be:

− percent: %, where 1 % = 10^{-2}
− parts per million: ppm, where 1 ppm = 10^{-6}
− parts per trillion: ppt, where 1 ppt = 10^{-9}

Normally, it must be decided on which quantity the mixing ratio is based (volume or mass) and then written as ppmm or ppmV (also written as to ppm(m) or ppm(V)), respectively (Table 4.2).

A standard volume of air (e. g. 1 m^3) is well defined but hard to measure when taking into account the volume of condensed matter (hydrometeors and particulate matter; PM). Normally, we can neglect the volume (fraction) of the condensed matter occupying a volume of air because of the small values in the order of 10^{-6} (Table 1.1), which is orders of magnitude smaller than the best gas volume measurement facilities. The volume of condensed matter, however, can be well measured using optical methods. It must also be considered when soluble gases are distributed

---

[7] However, whenever measuring an atmospheric substance, p and T must be co-measured to allow standard recalculations for exact averaging and intercomparisons.

among the gas and the droplet phases in air. Highly soluble gases (such as HCl and HNO_3) will be transferred quantitatively in a cloud from the gas into the liquid phase (scavenging efficiency equal to one). Partially soluble gases (such as SO_2 and CO_2) are distributed among the phase according to the *Henry* law. Hence, the partial pressure is according to:

$$p_i = p_i(g) + p_i(aq) \qquad (4.49)$$

where $p(g) = n(g)RT/V(g)$ and $p(aq) = n(aq)RT/V(aq) = p(g)H_{eff}RT \cdot V(aq)/V(g)$ because of $n(aq) = p(g)V(aq)H_{eff}$. The reservoir distribution δ (ratio of molar mass in the aqueous phase to the molar mass in gas phase) is given by:

$$\delta = \frac{n_{aq}}{n_g} = \frac{V_{aq}}{V_g} H_{eff} RT \approx LWC \cdot H_{eff} \cdot RT \qquad (4.50)$$

where H_{eff} is the effective *Henry* constant (Chapter 4.3.2). The volume ratio between the phases is given by $V = V(aq)/V(g) \approx V(aq)/\text{air volume} = LWC$ (liquid water content), whereas air volume $= V(aq) + V(g) \approx V(g)$. The total partial pressure is now described by:

$$p_i = p_i(g) \cdot (1 + LWC \cdot H_{eff} \cdot RT) = p(g) \cdot (1 + \delta) \qquad (4.51)$$

In all cases, when $\delta \ll 1$ (i.e. $p(g) < p$) the partial pressure of a substance must be corrected by the factor $1/(1 + \delta)$ in equations. The concentration can be defined based on mass (e.g. g L^{-1} or g m^{-3}), number (e.g. cm^{-3}) and mole (molarity, e.g. mol L^{-1} or molar or M):

$$c(m)_i = m_i/V, \quad c(N)_i = N_i/V, \quad c(n)_i = n_i/V \qquad (4.52)$$

Using the gas law (Eq. 4.14), and taking into account Eqs. (4.24) and (4.48), a recalculation between mass concentration and the mixing ratio is based on

$$x_i = c(m)_i RT/pM_i \qquad (4.53)$$

where M_i is the molar mass of the substance i. Gaseous trace species were often measured in mass concentration (e.g. µg m^{-3}). It is obligatory to also measure pressure and temperature because otherwise there is no way for a recalculation between the mass concentration and mixing ratio. For standard conditions (1 atm $= 1.01325 \cdot 10^5$ Pa and 25 °C $= 298.15$ K) the mass concentration is listed as 1 ppb in Table 4.2.

Only molecules in very small concentrations (especially reactive radicals such as OH and HO_2) are quantified in number concentrations, e.g. the number of radicals in a volume (cm^{-3}). This concentration measure is also used for the particle number of condensed matter (droplets and solid particles)

According to heterogeneous reactions and optical properties, another quantity − the surface to volume (of air) ratio − is useful for describing the condensed phase. For droplets, it is simple to define the volume of an individual droplet based on its diameter assuming a spherical form. The mass, number and surface quantities of particles show a very different (but characteristic) behavior when related to the size distribution.

Concerning the chemical composition of the condensed matter, there are two ways to describe the abundance. First, chemically analyzing the matter (either single particle or collected mass of particles) provides the concentrations (or mixing ratios) related to the volume (or mass) of the condensed phase, e. g. moles (or mass) of a substance per liter of cloud water and mass of a substance per total mass of PM. When sampling the condensed phase, normally for hydrometeors, all droplets are collected in a bulk solution (however, multistage cloud water impactors are also available). However, for solid PM, different particle size fractions are in use. TSP means the total suspended matter (i. e. it is sampled over all particles sizes). A subscript denotes the cut-off during sampling (for example, PM_{10}, $PM_{2.5}$ and PM_1), i. e. the value denotes the aerodynamic diameter in μm of (not sharp) separation between particles smaller than the given value. Second, the specific concentration of a substance within the condensed matter can be related to the volume of air. To do so, the volume (or mass) concentration of the total condensed matter must be known. This is given by the LWC for hydrometeors and the total mass concentration of PM. The resulting value of the mass dissolved matter or PM in a volume of air is the atmospheric abundance, and this air quality can be compared between different sites. The concentration of a substance in cloud water, however, also depends on the cloud's physical properties. This is also valid for fog and precipitation. Hence, with hydrometeor sampling the simultaneous registration of the LWC and rainfall amount is obligatory:

$$c_i\,(air) = m_i\,(\text{PM}) \cdot \text{PM} \tag{4.54}$$

where $\sum m_i\,(\text{PM}) = m\,(\text{PM})$ and PM denotes the total mass of PM in an air volume. In analogy to species dissolved in cloud water, the aqueous phase concentration must be multiplied by the LWC:

$$c_i\,(air) = m_i\,(aq) \cdot \text{LWC} \tag{4.55}$$

When using averaged concentration values it is strongly recommended to calculate them from so-called weighted values, namely LWC for cloud and fog water and precipitation amount for rain and snow. Only such weighted values represent the correct total abundance of trace substances in air. The averaged precipitation composition, e. g. as annual mean results from:

$$\overline{c_i(aq)} = \frac{\sum_i R_i\, c_i(aq)}{R} \tag{4.56}$$

where R_i denotes the precipitation amount for sample i, and R denotes the annual (or any other timescale) precipitation amount (rainfall rate). In analogy for cloud water chemical composition:

$$\overline{c_i(aq)} = \frac{\sum_i^n c_i(aq)\text{LWC}_i}{n\,\overline{\text{LWC}}} = \frac{\sum_i^n c_i(aq)\text{LWC}_i}{\sum_i \text{LWC}_i} \tag{4.57}$$

where $\overline{\text{LWC}}$ is the arithmetic mean of samples LWC_i and n is the number of samples.

Another concentration measure is important for ions in hydrometeors and soluble substances in PM: the normality (N) and the equivalent (eq). Because of the condition of electroneutrality, the sum of the equivalent concentration of all cations must be equal to the sum of the equivalent concentration of all anions in droplets and solid particles.

Normal is one gram equivalent of a solute per liter of solution. The definition of a gram equivalent varies depending on the type of chemical reaction being discussed − it can refer to acids, bases, redox species and ions that will precipitate. More formally, a gram equivalent of a substance taking part in a given reaction is the number of grams of the substance associated with the transfer of N_A electrons or protons or with the neutralization of N_A negative or positive charges, where N_A is *Avogadro*'s number. The expression of concentration in equivalents per liter (or more commonly, microequivalents per liter, µeq L^{-1}) is based on the same principle as normality. A normal solution is one equivalent per liter of solution (eq L^{-1}), where $c_i(n)$ denotes the molar concentration (e. g. in mol L^{-1}) and e symbolizes the ionic charge.

$$c_i(eq) = c_i(n) \cdot e \tag{4.58}$$

4.1.3 Thermodynamic: The equations of state

Thermodynamics was originally the study of the energy conversion between heat and mechanical work, but now tends to include macroscopic variables such as temperature, volume and pressure. In physics and chemistry, and thereby the atmosphere and more generally the climate system, thermodynamics includes all processes of equilibrium between water phases, including hydrometeors, trace gases and aerosol particles (APs). These processes occurring in energetic changes are the key factor for understanding atmospheric states and thereby climatic changes. Changes in heat and kinetic energy can be measured in the work carried out.

The *internal energy* of a system or body (for example, a unit of air volume) with well-defined boundaries, denoted by U, is the total kinetic energy due to the motion of particles (translational, rotational and vibrational) and the potential energy associated with the vibrational and electric energy of atoms within molecules or any matter state. This includes the energy in all chemical bonds and that of free electrons (for example, hydrated electrons in water and photons in air).

This quantity (and some still following) is defined as a *variable of state* because it only depends on the state of the system and, inversely, describes the state of the system. The change $\Delta U = U_2 - U_1$ as a result of a state change means, according to the law of energy conservation, that energy is either taken up from the environment ($\Delta U > 0$) or released into the surroundings ($\Delta U < 0$). The first case is called *endothermic* (for example, the evaporation of water) and the second case *exothermic* (for example, oxidation). *Heat Q* takes a special place among different kinds of energy (which are summarized behind the term *work W*). Hence, the change of internal energy is defined by:

$$\Delta U = W + Q \tag{4.59}$$

In the gas phase, work is carried out primarily[8] as *pressure-volume work* (in the condensed phase such as droplets and solid particles, surface work, electrical work and expansion work also occur). The change of volume occurs at a constant pressure (*isobaric* change of state) and so it is valid that

$$-W = p\Delta V \tag{4.60}$$

If no other work is carried out, the differential change of internal energy is described by

$$dU = dQ - p\Delta V \tag{4.61}$$

The internal energy of an ideal gas depends only on temperature (the *Gay-Lussac* law). During an isothermal expansion, when air performs positive work through overbearing external pressure, it must uptake an equivalent amount of heat to meet a constant temperature ($-W = Q > 0$). Is there no heat exchange with the surroundings ($Q = 0$); this is defined as an *adiabatic* change of state. Consequently, the gas (air) cools and the internal energy decreases by the amount equivalent to the work performed ($-W = -\Delta U$). In air parcels, pressure and volume change with each shift in height. With an ascent, the volume increases (expansion) and pressure decreases. As long as there is no heat exchange with the surrounding air, the internal energy remains constant and the altitude change is adiabatic. Therefore, adiabatic air mass changes are an important condition for the condensation of water vapor into the cloud condensation nuclei. While adiabatic, rising air cools by 0.98 °C per 100 m; this is called the *dry* adiabatic *lapse rate* (DALR), or dT/dz. As soon as the air parcel is saturated by water vapor, it partly condenses and is then heated by the released heat. Then, the *wet adiabatic temperature gradient* (lapse rate) is observed, which lies between 0.4 °C at large temperatures and 1 °C for low temperatures. During adiabatic changes, the *potential temperature* remains constant, i.e. an air parcel with 10 °C in 1000 m altitude contains about the same heat as a surface-near air parcel at 20 °C. The temperature gradient determines the atmospheric layering. It is called a stable atmospheric boundary layer (SBL) if the air temperature decreases less with altitude than in the case of adiabatic layering. The lifting air becomes cold faster than its environment and sinks down again so that only small vertical displacements occur. By contrast, if the air temperature decreases faster than the adiabatic lapse rate (the rising air is warmer than the environment) another buoyant force evolves — it becomes a *labile* layering.

The heat needed to heat one mole of a gas at constant volume is called molar heat capacity C_V; it is defined as $Q = C_V\Delta T$. The temperature dependency of C_V is:

$$C_V = \left(\frac{\Delta Q}{\Delta T}\right)_V = \left(\frac{\Delta U}{\Delta T}\right)_V \tag{4.62}$$

Now, we can formulate the *first law of thermodynamics*: In a closed system, the internal energy remains constant. For the calculation of processes with a constant pressure, the internal energy is replaced by the quantity H, denoted *enthalpy*:

$$H = U + pV = U + n\boldsymbol{R}T \tag{4.63a}$$

---

[8] Further transformed into accelerational and frictional work.

In differential form:

$$dH = dU + pdV + Vdp \tag{4.63b}$$

The last equation can be integrated for isobaric changes ($dp = 0$):

$$\Delta H = \Delta U + p\Delta V = Q \tag{4.64}$$

The heat capacity at a constant pressure is defined:

$$C_p = \left(\frac{\Delta H}{\Delta T}\right)_p \tag{4.65}$$

From Eqs. (4.61) and (4.65) we can derive the relationships between the change of enthalpy and the (infinitesimal) change of temperature at a constant pressure in the case of constant heat capacity within a certain range of temperature:

$$dH = C_p dT \quad \text{and} \quad \Delta H = C_p \Delta T \tag{4.66}$$

The quantities ΔU und ΔH might also be expressed as heat because all kinds of energy, which the system exchanges with its surroundings, can be completely transferred into heat, in agreement with the law of energy conservation. The general driving force can be quantified as *entropy*. With such quantification, it can be studied whether a process runs voluntarily. Voluntary processes in the atmosphere or generally in the climate system are of crucial interest because it is nearly[9] impossible to trigger the intended changes of pressure and temperature. Only *voluntary* chemical processes can be observed in nature. The finding that all processes can be grouped into voluntary and non-voluntary processes leads to the second law of thermodynamics. It is impossible to carry out a process with an uptake of heat from a reservoir and its complete transfer into work (there is no *perpetuum mobile*). In a slightly different phrasing, heat is low-grade energy, i. e. whereas heat always degrades, heat of a lower grade will always remain (for example in the form of infrared radiation). This "loss" can be called dissipated work. The "value" of heat is determined by the temperature of the system; the more elevated the temperature, the larger the part of heat that is transferable into work (useful energy). This property of heat is characterized by the entropy:

$$dS = \frac{dQ_{rev}}{T} \tag{4.67}$$

In the case of isothermal processes we can rewrite it as follows:

$$\Lambda S = \frac{Q_{rev}}{T} \tag{4.68}$$

Q_{rev} is the reversible heat taken up by the system at a given temperature. In a closed system,[10] the total energy remains constant and thereby the direction of a process

---

[9] It is not impossible, for example in weather modification (rain making, hail prevention and fog dissipation), but almost through "catalytic" triggering.

[10] In nature a closed system is a fiction or a model approximation. The atmosphere is open to space and the earth's surface. The earth system is open regarding energy flux and only apparently closed regarding mass, when not considering cosmic epochs.

is associated with the redistribution of energy. Experience has shown that voluntary processes always result in a larger disorder of the system ($\Delta S > 0$). This is a condition for an irreversible process, which cannot return either the system or surroundings to their original conditions. Consequently, a reversible process in an adiabatic isolated system, as characterized by $\Delta S = 0$. This is an ideal abstraction because all processes in nature are spontaneous and thereby voluntary and irreversible. This does not exclude cycling processes, for example biogeochemical material cycles where all single processes are irreversible but directed in a cycle, not returning the system to their original conditions but keeping a stationary state. This also does not exclude small intermediate steps that are reversible.

Considering spontaneous processes from the view of probability f (with a range of values 0 … 1), irreversible processes operate as transfers from a less probable in a more probable state. *Boltzmann* derived the equation:

$$S = \mathbf{k}\ln f + \text{constant} \tag{4.69}$$

With the assumption by *Max Planck* that the constant is zero, Eq. (4.69) can also be written as:

$$\Delta S = \mathbf{k} \cdot \ln \frac{f_2}{f_1} \tag{4.70}$$

where f_1 and f_2 denote the probabilities of the initial and final state, respectively. To characterize a process as "voluntary", two more thermodynamic variables are introduced: *free energy* (also called *Helmholtz* energy) F and *free enthalpy* (also called *Gibbs* energy) G:

$$F = U - TS \tag{4.71}$$

$$G = H - TS = F + pV \tag{4.72}$$

Changes in the state at a constant temperature can be written as:

$$dF = dU - TdS \quad \text{and} \quad dG = dH - TdS.$$

With the condition of "voluntariness" of the process, that is $dS \geq 0$, another important thermodynamic condition follows:

$$dF_{T,V} \leq 0 \quad \text{and} \quad \Delta G_{T,p} \leq 0, \text{ respectively.}$$

In a spontaneous operating process, the change of free energy is negative, whereas in equilibrium $dW_{T,V} = 0$ is valid. The change of free energy corresponds to the maximum possible work that can be carried out ($dW = -p\Delta V$). A more general criterion for the "voluntariness" of processes, however, is the attempt to garner a maximum from the sum of entropy changes of the system (dS) and the surroundings ($-dU/T$) or, in other words, to gain a small total entropy. The criterion $\Delta G_{T,p} \leq 0$ is in chemistry in this sense interpreted as a chemical reaction at a constant temperature and constant pressure if it is connected with a decrease in free enthalpy. Hence, it makes sense to introduce *free standard enthalpies* $\Delta_R G^\ominus$ to calculate reactions and equilibriums:

$$\Delta_R G^\ominus = \Delta_R H^\ominus - T\Delta_R S^\ominus \tag{4.73}$$

It is logical to treat the change of free enthalpy as a function of p and T:

$$dG = dH - TdS - SdT. \tag{4.74}$$

Because $H = U + pV$ it is $dH = dU + pdV + Vdp$ and using the fundamental equation $dU = TdS - pdV$ it follows finally that:

$$dG = Vdp - SdT \tag{4.75}$$

and

$$\left(\frac{\partial G}{\partial T}\right)_p = -S \quad \text{and} \quad \left(\frac{\partial G}{\partial p}\right)_T = V \tag{4.76}$$

Because S takes positive values, G must decline if T is increasing in a system at a constant pressure and constant composition. In gases, G responds more sensibly to pressure variation than in condensed phases (because gases have a large molar volume). From Eq. (4.76), the temperature dependency of free enthalpy can be derived. Owing to $S = (H - G)/T$, after a few steps we get the well-known *Gibbs-Helmholtz equation*:

$$\left(\frac{\partial}{\partial T}\left(\frac{G}{T}\right)\right)_p = -\frac{H}{T^2} \tag{4.77}$$

Relating this equation to the initial and final state of a chemical reaction or physical change of state, it follows because $\Delta G = G_2 - G_1$

$$\left(\frac{\partial}{\partial T}\left(\frac{\Delta G}{T}\right)\right)_p - -\frac{\Delta H}{T^2} \tag{4.78}$$

According to *Gibbs,* the partial free enthalpy $(\partial G/\partial n)$ of any chemical species is called its chemical potential μ. The chemical potential denotes how the free enthalpy of a system changes with changing chemical composition. Therefore, the equilibrium of all kinds can be clearly described, especially in mixed or multiphases. For pure substances (because of $dn_i = 0$) it is valid that $\mu = G$. The molar-free enthalpy for solids and liquids depends little on pressure, but for gases this dependency is large. For ideal gases, according to the definition $G = H - TS$, the total derivative dG follows from Eq. (4.74) with consideration of the gas equation for molar quantities ($V = RT/p$; we disclaim here the exact marking as V_m):

$$dG = Vdp = RTd \ln p \tag{4.79}$$

where $dp/p = d \ln p$ ($- d \ln (p/p_0)$ more exactly; p_0 is set to 1 bar but might earn any reference value. After integration and $\mu = G$ we get:

$$\mu = \mu^\ominus + RT \ln p \tag{4.80}$$

where $\mu^\ominus$ denotes the *chemical standard potential*. The difference $\mu - \mu^0$ is equal to the molar work when transferring the ideal gas reversible and isothermal from standard pressure on p.

In an open system — its chemical composition does not need to be constant — the change of G must be described with variation of p and T as well with n, the composition. Besides the terms $(\partial G/\partial p)_{T,n} = V$ und $(\partial G/\partial T)_{p,n} = -S$ it further follows (Eq. 4.75) as a general definition of the chemical potential:

$$\left(\frac{\partial G}{\partial n_i}\right)_{p,T,n_j} - \mu_i \tag{4.81}$$

With n_j (besides p and T) the constancy of the chemical composition is expressed. From the thermodynamic fundamental equation the following now follows for H and U:

$$\mu_i = \left(\frac{\partial U}{\partial n_i}\right)_{S,V,n_j} \quad \text{and} \quad \mu_i = \left(\frac{\partial H}{\partial n_i}\right)_{S,p,n_j} \tag{4.82}$$

In an ideal mixing, i. e. the components do not interact through intramolecular forces, the chemical potential of each component is equal to that of the pure component if its pressure is identical to the partial pressure in the composition:

$$\mu_i = \mu_i^{\ominus} + RT \ln p_i \tag{4.83}$$

Because of $p_i = p \cdot x_i$ it follows that:

$$\mu_i = \mu_i^{\ominus} + RT \ln p_i + \ln x_i = \mu_i^{\ominus} + RT \ln x_i \tag{4.84}$$

Similar to gases for diluted solutions (otherwise activities a_j must be used) we write $\mu_i^{\ominus}$. This denotes a new standard potential of the dissolved compound i at a concentration 1 mol L^{-1}:

$$\mu_i = \mu_i^{\ominus} + RT \ln c_i \tag{4.85}$$

4.1.4 Equilibrium

We stated above that a reversible process between two states characterizes equilibrium A $\rightleftarrows$ B, whereas no limiting conditions are expressed for the states A and B. Thus, it can be a chemical (reversible) reaction or any phase transfer (gas-liquid, solid-liquid, solid-gas). There are, however, only a few types of chemical reactions representing equilibrium: the acid-base reaction, adduct formation, complexation and addition-dissociation. Phase equilibrium is often observed in nature: dissolution-precipitation, evaporation-condensation and absorption-desorption. All non-chemical processes in multiphase systems can be subdivided into partial steps considering each chemical species separately transferring to the interface. In equilibrium, each substance has the same chemical potential in all phases. When the state variable changes (pressure, temperature, mole fraction) but the equilibrium remains, the chemical potentials change. However, it is valid that the changes are the same in all phases: $d\mu_i = d\mu_i'$. Any chemical reaction or phase transfer including the chemical species X_i and Y_j is described by (ν stoichiometric constant):

$$\nu_1 X_1 + \nu_2 X_2 + \nu_3 X_3 + ... \nu_m X_m \underset{k_-}{\overset{k_+}{\rightleftarrows}} \nu_{m+1} Y_1 + \nu_{m+2} Y_2$$

$$+ \nu_{m+3} Y_3 + ... \tag{4.86}$$

The law of mass action establishes the relationship between states A and B via the *equilibrium constant K*; the brackets denote the concentration (or for non-ideal systems the activity and fugacity, respectively). By convention the products form the numerator:

$$K = \frac{[Y_1]^{v_{m+1}} [Y_2]^{v_{m+2}} [Y_3]^{v_{m+3}} \cdots}{[X_1]^{v_1} [X_2]^{v_2} [X_3]^{v_3} \cdots} \tag{4.87}$$

In Eq. (4.86), k_+ und k_- represent the rate constants of the partial processes and the forward (k_+) and back (k_-) reaction or transfer. Equilibrium also means that the fluxes of the forward and backward processes are equal: $F_+ = F_-$ because of $d\mu_i = d\mu_i'$ and therefore $dW_{T,V} = 0$ and $\Delta G_{T,p} = 0$. Hence, it is valid that:

$$F_+ = \left(\frac{dn}{dt}\right)_+ = k_+ [X_1]^{v_1} [X_2]^{v_2} \cdots = F_- = \left(\frac{dn}{dt}\right)_-$$
$$= k_-[Y_1]^{v_{m+1}} [Y_2]^{v_{m+2}} \cdots . \tag{4.88}$$

It follows that:

$$K - \frac{k_+}{k_-} \tag{4.89}$$

In 1886, *Jacob van't Hoff* derived thermodynamically the law of mass action based on the work from the initial substances X_i and final substances (products) Y_j as well as the following relationship between free energy and enthalpy and the equilibrium constant either for constant pressure or constant volume but always at constant temperature:

$$(\Delta_R F^{\ominus})_{V,T} = (\Delta_R G^{\ominus})_{p,T} = -RT \ln K \tag{4.90}$$

From the general definition of the chemical equilibrium ($d\mu_i = d\mu_i'$), it follows when arriving at the equilibrium (depending on the rate constants, time is needed to achieve the equilibrium)[11] the conditions (lower cases denote mass-related and not molar quantities as for capital letters):

$$(dg)_{p,T} = 0 \text{ (isobar)} \quad \text{and} \quad (df)_{V,T} = 0 \text{ (isochoric−isothermal)} \tag{4.91}$$

Another general condition for equilibrium follows:

$$(g)_{p,T} = (f)_{V,T} = \Sigma \mu_i dn_i = 0 \tag{4.92}$$

Because of $\Sigma dn_i = v_i$ it also follows that $\Delta_R G = \Sigma v_i \mu_i = 0$. Now, the thermodynamic derivation of K becomes coherent:

$$0 = \Delta_R G = \Sigma v_i \mu_i = \Sigma (v_i \mu_i + RT \ln c_i^{v_i}) = \Delta_R G + RT \ln K \tag{4.93}$$

We now transfer from the infinitesimal change d to the difference Δ and it follows from $\Delta_R G^{\ominus} = \Delta_R H^{\ominus} - T\Delta_R S^{\ominus} = -RT \ln K$ through transformation with respect

---

[11] This is sometimes forgotten by modellers treating fast equilibrium according to the integrating time. However, when the time to achieve the equilibrium is larger than the numeric time step, nonsense is calculated.

to ln K and the derivative with respect to T. Using the *Gibbs-Helmholtz* equation (4.78), we derive the universal *van't Hoff*'s reaction isobar:

$$\frac{d \ln K}{dT} = \frac{\Delta_R H^{\ominus}}{RT^2} \tag{4.94}$$

Transforming this equation, a practicable expression for the temperature-dependent equilibrium constant then follows:

$$K(T) = K_{298} \exp\left[\frac{\Delta_R H^{\ominus}}{R}\left(\frac{1}{298} - \frac{1}{T}\right)\right] \tag{4.95}$$

4.1.5 Steady state

As noted before, one of the general conditions for equilibrium and steady state[12] is that the forward and backward fluxes of a process are equal ($F_+ = F_-$). The specific condition to fulfill a chemical equilibrium is $\mu_+ = \mu_-$ (Eq. 4.92). Although a chemical equilibrium occurs when two or more reversible processes occur at the same rate, and such a system can be said to be in steady state, a system that is in steady state might not necessarily be in a state of equilibrium, because some of the processes involved are not reversible. A system in a steady state has numerous properties that do not change over time. The concept of steady state has relevance in many fields, in particular thermodynamics. Hence, steady state is a more general situation than dynamic equilibrium. If a system is in steady state, then the recently observed behavior of the system will continue into the future. In stochastic systems, the probabilities that various different states will be repeated will remain constant. We will generalize Eq. (2.144) as follows (see also 4.88):

$$\left(\frac{dn}{dt}\right) = \left(\frac{dn}{dt}\right)_+ - \left(\frac{dn}{dt}\right)_- = 0 \quad \text{or} \quad F = F_+ - F_- = 0 \tag{4.96}$$

In many systems, steady state is not achieved until some time has elapsed after the system is started or initiated. This initial situation is often identified as a transient state, start-up or warm-up period.

The term steady state is also used to describe a situation where some, but not all, of the state variables of a system are constant. For such a steady state to develop, the system does not have to be a flow system. Therefore, such a steady state can develop in a closed system where a series of chemical reactions take place. Literature on chemical kinetics usually refers to this case, calling it *steady-state approximation*. Steady-state approximation, occasionally called stationary-state approximation, involves setting the rate of change of a reaction intermediate in a

---

[12] Often also called *stationary* state. But in physics, especially in quantum mechanics, a stationary state is an eigenstate of a Hamiltonian or, in other words, a state of definite energy. It is called *stationary* because the corresponding probability density has no time dependence. For all kinds of "stationary and/or steady state" in German (and likely in other languages too) there is only one term "*stationärer Zustand*".

reaction mechanism equal to zero. Steady-state approximation does not assume the reaction intermediate concentration is constant (and therefore its time derivative is zero). Instead, it assumes that the variation in the concentration of the intermediate is almost zero. The concentration of the intermediate is very low, so even a big relative variation in its concentration is small, if considered quantitatively.

These approximations are frequently used because of the substantial mathematical simplifications this concept offers. Whether or not this concept can be used depends on the error the underlying assumptions introduce. Therefore, even though a steady state, from a theoretical point of view, requires constant drivers (e. g. constant inflow rate and constant concentrations in the inflow), the error introduced by assuming steady state for a system with non-constant drivers might be negligible if the steady state is approached fast enough (relatively speaking).

This is often the case for very reactive chemical species, especially radicals such as hydroxyl (OH). Steady state OH concentration is than expressed by the condition $d[OH]/dt = 0$ from which follows:

$$\sum_i k_i[A_i][B_i] = [OH]\sum_j k_j[C_j] \tag{4.97}$$

where the terms on the left side represent all reactions producing OH and those on the right side all reactions consuming OH. However, in truth this is valid only for a short time, depending on the error being taken into account (a few minutes, but this time is large compared with the atmospheric OH lifetime in the order of seconds). The approximation of OH steady state is also useful for adopting a mean "constant" OH concentration (independent of diurnal cycles, Chapter 2.8.2.9) to reduce secondary-order reactions to pseudo-first-order reactions (see next Chapter).

The principle of stationary (or instationary) is also applied to the atmospheric budget of trace species, regarding F_+ the source term (emission Q) and F_- the total removal term R (deposition and chemical conversion). With the definition of the residence time (see Chapter 4.5 for more details) it follows from Eq. (4.96) that:

$$\ln\frac{n_0}{n(\Delta_t)} = \frac{\Delta t}{\tau} \tag{4.98}$$

where n_0 is the (initial) amount of a substance in air at $t = 0$. To fit the steady-state condition $Q = R$ over a given unit of time (for example a year) there is no condition concerning the quantity of the residence time. Normally, it is believed that a short residence time (let's say $\tau \ll 1$ yr) corresponds to a large removal capacity; if $\tau = 1$ day (for example formic acid), after seven days 99.9% of the initial amount n_0 is removed from the atmosphere independent of the absolute amount. Consequently, the yearly removal capacity is some 50 times higher. As another consequence, the mean atmospheric concentration of this substance remains very low (but depends on the influx and the emission). Regarding the example of a large residence time $\tau = 10$ years (for example methane), after one year only about 10% of n_0 is removed from air as it follows from Eq. (4.98). Yet we have to consider that n_0 represents the total amount of the substance at an arbitrary time (here set to $t = 0$); this consists of the actual emission flux and past cumulative emission versus the removal rate. Let's say an experiment with an emission process starts (year 1) with continuous emission over time (100 units per year); after a year 90 % remains in air due to the slow removal capacity ($\tau = 10$ yr). In year 2, the fresh emission (100 units) will be added to the remaining emission from year 1, but the absolute removal (10 % relatively) is larger: $(100 + 90) \cdot 0.1 = 19$ units. Hence, the atmospheric amount (and concentration) increases from year to year but the absolute removal amount also rises until reaching the equilibrium, when the yearly removal becomes equal to the emission (100 units) – that is, for the example of CH_4, in 10 years. In all following years, the stationary $Q = R$ remains and a stationary concentration is achieved. The higher the residence time (and emission) the larger the atmospheric amount of the substance. Because of $dn/dt = F_- = R = n/\tau$ (Eq. 4.96) and the condition $Q = R$ it follows that:

$$n_{atm} = \tau \cdot Q \tag{4.99}$$

However, this experiment simply implies that a steady state is achieved some time after the system is started and that Q is constant, which is the normal case for natural processes on a climatological timescale. With yearly rising emissions, as is typical for anthropogenic processes such as fossil fuel combustion, the system remains out of stationary and the time to achieve the steady state is endless while the emissions continue to increase. Hence, the atmospheric concentration increases (as seen for the greenhouse gases). This increase is simply calculated from:

$$\text{concentration increase (in \%)} = 100 \cdot \frac{Q_i - R_i}{n_i + Q_i} \tag{4.100}$$

where the subscript i denotes the year regarded, n total amount in air and Q and R the yearly amount of emission and removal, respectively. For the example of CO_2, it follows using the actual values from Fig. 2.10 and Table 2.84 that:

$$100 \, \frac{8.9 - 4.8}{800 + 8.9} = 0.6 \, \%.$$

This is in full agreement with the measurements from Fig. 2.70, which shows that the increase for the period 2000–2005 is about 2.2 ppm yr^{-1} and that this results in a 0.6 % increase based on a total of 370 ppm. Despite its complexity and the fact that it cannot fully understanding single atmospheric processes this approximation shows that nature follows simple generic principles.

Let us return to CO_2 residence time. Eq. (4.99) cannot be applied because the system is still far from the steady state ($F_+ > F_-$). Hence, the condition $Q = R$ is invalid. It is stated that from a total of 8.9 Gt emitted carbon, about 50 % is removed by absorption through the sea and biosphere. Hence, the anthropogenic CO_2 uptake is biogenic – as discussed in Chapter 2.8.3.3 – and cannot be described by a first-order law (4.98). The key question remains as to what controls the biogenic uptake of anthropogenic CO_2. Fig. 2.100 suggests that it is (so far) a linear process with 50 % yearly uptake of emitted CO_2. Again, this discussion emphasizes that budget and time-dependent properties of trace species cannot be understood as purely physically and/or chemically. We conclude that 50 % of emitted anthropogenic CO_2 remains in air in the absence of any considerable removal process and thereby has an indefinite residence time.

4.2 Chemical reactions

A *chemical reaction* is a process that results in the interconversion of chemical species; Table 4.3 summarizes the base reaction types. Chemical reactions might be elementary reactions or stepwise reactions. A stepwise reaction consists of at least one reaction intermediate and involves at least two consecutive elementary reactions. Parallel reactions are several simultaneous reactions that form different respective products from a single set of reactants (Svehla 1993, Muller 1994).

Thermodynamics is important for describing chemical reactions. As seen before, it explains whether a reaction is "voluntary" and in which direction and in what state equilibrium is. Hence, thermodynamics also describes the mechanisms but without providing information on the explicit pathway. Many reactions can be parallel and thereby competitive. In quantifying the overall chemical processes (production and/or decay of substances), because of the substantial mathematical simplifications, it is important to delete reactions of minor importance (i. e. those with reaction rates about two orders of magnitude less than the fastest reaction). The task of chemical kinetics is to describe the speed at which chemical reactions occur. The reactions rate is the change in the number of chemical particles per unit of time through a chemical reaction. This is the term:

$$R_N = \frac{dN}{dt} \text{ [number/time]}.$$

Because $n = N/N_A$ (N_A is the *Avogadro* constant) the rate (or speed) also can be expressed in mole per time:

$$R_n = \frac{dn}{dt} \text{ [mole/time]}.$$

Table 4.3 Types of chemical reactions; A, B and C any atomic or molecular entity, H hydrogen, e^- electron.

term	reaction scheme
addition	$A + B \rightarrow AB$
insertion	$A + BC \rightarrow AB + C$
abstraction	$AB + C \rightarrow A + BC$
dissociation[a]	$AB \rightarrow A + B$
radioactive decay[b]	$A \rightarrow B + C$
proton transfer	$A + H^+ \rightarrow AH^+$ or $A + BH \rightarrow AH^+ + B^-$
electron transfer	$A + e^- \rightarrow A^-$ or $A + B^- \rightarrow A^- + B$

[a] thermal or irradiative
[b] A, B and C are atoms only

Finally, because $c = N \cdot M / V$ (M is the molar mass) is the change of concentration per time:

$$R_c = \frac{dc}{dt} \text{ [mass / volume} \cdot \text{time]}.$$

In chemistry, the last concentration-related term is normally used, and Eq. (4.101) shows the recalculation:

$$R = \frac{dN}{dt} = N_A \frac{dn}{dt} = \frac{V\, dc}{M\, dt} \tag{4.101}$$

In nature, especially the climate system, the investigation of chemical reactions can be limited. All early attempts to estimate reaction rates were unsuccessful because too many variables existed. The concentration measurements of different trace species – together with meteorological parameters that describe transport, mixing and phase state conditions (in so-called complex field experiments) – are extremely helpful in several directions. These directions include: a) establishing empirical relationships between substances and other atmospheric parameters to gain insights into mechanisms, b) recognizing still unidentified or not yet considered substances under the specific measurement conditions and c) providing data sets for model evaluations.

Kinetics, specifically studying rate laws and measurement of rate constants, can only be done under laboratory conditions, whereas reaction conditions should be simulated in special reactors ("smog chambers") closely resembling the atmospheric one. Once established, the k-value of an elementary reaction is universally applicable, or in other words, "pure chemistry" is independent of meteorological and geographical specifics but the conditions for reactions (pressure, temperature, radiation, humidity) and the concentration field depends from location. This is the difference to "air chemistry". For more detailed information on chemical kinetics, see Zumdahl (2009), Atkins (2008) and Houston (2006).

4.2.1 Kinetics: The reaction rate constant

Even complex chemical reaction mechanisms can be separated into several definite *elementary reactions*, i. e. the direct electronic interaction process between molecules and/or atoms when colliding. To understand the total process $A \rightarrow B$ – for example the oxidation of sulfur dioxide to sulfate – it is often adequate to model and budget calculations in the climate system to describe the *overall reaction*, sometimes called the *gross reaction*, independent of whether the process $A \rightarrow B$ is going via a reaction chain $A \rightarrow C \rightarrow D \rightarrow E \rightarrow ... \rightarrow Z \rightarrow B$. The complexity of mechanisms (and thereby the rate law) is significantly increased when parallel reactions occur: $A \rightarrow X$ beside $A \rightarrow C$, $E \rightarrow Y$ beside $E \rightarrow F$. Many air chemical processes are complex. If only one reactant (sometime called an educt) is involved in the reaction, we call it a unimolecular reaction, that is the reaction rate is proportional to the concentration of only one substance (first-order reaction). Examples are all radioactive decays, rare thermal decays (almost autocatalytic) such as PAN decomposition and all photolysis reactions, which are very important in air. The most frequent are bimolecular reactions (A + B); for example, simple molecular reactions ($NO + O_3 \rightarrow NO_2 + O_2$) are frequent radical reactions ($NO + OH \rightarrow HNO_2$), but these are second-order reactions. However, trimolecular reactions are common in the atmosphere ($O + O_2 + N_2 \rightarrow O_3 + N_2$) because the primary collision complex $A \cdots B$ is energetically instable and must transfer excess energy onto a third collision partner (third-order reaction). Collision partners are the main constituents of air (N_2 and O_2) because of a collision's probability, but also H_2O (for example $HO_2 + HO_2 + H_2O \rightarrow H_2O_2 + O_2 + H_2O$), probably because of its specific role in the transfer complex (charge transfer, hydrogen bonding or sterical reasons). The number of reactants from which the concentration of the reaction rate depends determines the reaction order. The third-order reactions might be considered elementary (despite the fact that the collision partner certainly does not collide at the same time with A and B). All reactions with higher or non-integer numbers principally reflect the kinetics of complex mechanisms. The bimolecular reaction A + B can be written in three different schemes):

$$A + B \rightarrow C + D \quad \text{or} \quad A \xrightarrow[-D]{B} C \quad \text{or} \quad B \xrightarrow[-C]{A} D \qquad (4.102)$$

The following rate law is valid (remember that the brackets denote the concentration of the substance A etc. and k reactions rate constant:

$$-\frac{d[A]}{dt} = -\frac{d[B]}{dt} = \frac{d[C]}{dt} = \frac{d[D]}{dt} = k[A][B] = R \qquad (4.103)$$

An important quantity to describe the state of the chemical reaction (4.102) is the *reaction quotient Q*. In contrast to Eq. (4.87), the concentrations are not expressed as an equilibrium (when the reaction reaches an equilibrium the condition $Q = K$ is fulfilled):

$$Q = \frac{[C][D]}{[A][B]} \qquad (4.104)$$

To solve the differential equation (4.103) analytically, we substitute time-dependent concentrations by ($[A]_0 - x$) and ($[B]_0 - x$) where subscript 0 denotes the initial concentration for $t = 0$ and x expresses the concentration of any of the products (assuming that neither C nor D are included in other reactions).

$$\int_0^x \frac{dx}{([A]_0 - x)([B]_0 - x)} = k \int_0^t dt \qquad (4.105)$$

By partial fraction decomposition we get:

$$k = \frac{1}{t([A]_0 - [B]_0)} \ln \frac{[A]_0([B]_0 - x)}{[B]_0([A]_0 - x)} \qquad (4.106)$$

or, for the simple case that A = B (2 A $\rightarrow$ products), we get:

$$k = \frac{1}{t}\left(\frac{1}{[A]_0 - x} - \frac{1}{[A]_0}\right) = \frac{1}{t}\frac{1}{[A]_0([A]_0 - x)} \qquad (4.107)$$

Often in nature there are situations where the concentration of the second reactant B can be considered constant in the given period or its relatively timely change $\Delta[B]/[B]$ can be neglected. This we assume most when the concentration of B is larger than that of A and thereby the numerical error stays small; this is the case in many reactions when O_2 is the partner. Another case is the steady state of B ($d[B]/dt = 0$). Then, the concentration of B can be included in the rate constant and the reaction type is reduced to a pseudo-first-order reaction with mathematical simplifications in further treatment:

$$-\frac{d[A]}{dt} = k[A][B] = k'[A] \tag{4.108}$$

Similar trimolecular reactions (remember that the third collision partner is often in high excess) can be simplified to pseudo-second-order and eventually even to pseudo-first-order. The solution of a first-order rate law is simple:

$$\int_{[A]_0}^{[A]} \frac{d[A]}{[A]} = \int_{[A]_0}^{[A]} d\ln[A] = k\int_0^t dt \tag{4.109}$$

It follows that the exponential time dependency of concentration of A is:

$$[A] = [A]_0 \exp(-kt) \tag{4.110}$$

From Eq. (4.110) useful characteristic times such as the half-life $\tau_{1/2}$ and residence or lifetime τ can be derived. The half-life is defined as the time when the initial concentration is decreased to the half: $[A] = 0.5[A]_0$:

$$\tau_{1/2} = \frac{1}{k}\ln 2 \tag{4.111}$$

More important for the climate system is the residence time, which is simply the reciprocal of k, but only in the case of a first-order process (Chapter 4.5): $\tau = 1/k$.

The residence time τ corresponds to the condition $[A]/[A_0] = e$ (this follows from Eq. 4.110), where e is a mathematical constant, sometimes called the *Euler* number (e $\approx$ 2.7183). In other words, the initial concentration of A is decreased to about 37%.

The rate-determining step (RDS), sometimes also called the limiting step, is a chemistry term for the slowest step in a chemical reaction. In a multistep reaction, the steps nearly always follow each other, so that the product(s) of one-step is/are the starting material(s) for the next. Therefore, the rate of the slowest step governs the rate of the whole process. In a chemical process, any step that occurs after the RDS will not affect the rate (see the discussion on net ozone formation in Chapter 5.3.6.3) and, therefore, does not appear in the rate law.

It has been empirically found that between the logarithm of the reaction rate k and the reciprocal temperature a linear relation exists, which is similar to *van't Hoff*'s reaction isobar (Eq. 4.94) by *Arrhenius*:

$$\frac{d\ln k}{dT} = \frac{E_A}{\boldsymbol{R}T^2} \tag{4.112}$$

and termed now the *Arrhenius* equation, E_A activation energy:

$$k = k_m \exp\left(-\frac{E_A}{RT}\right) \qquad (4.113)$$

The term $\exp(-E_A/RT)$ simply expresses the fraction of chemical species with a per mole higher energy than E_A. The *Boltzmann* distribution follows from Eq. (4.31), which shows the probability of molecules having (or exceeding) a certain speed or energy, when taking into account that $3RT/M$ is equivalent (Eq. 4.15) to the mean molar kinetic energy of the molecules. Thus, E_A terms the (molar) kinetic energy required for the transfer from A + B to the transition state $AB^{\ddagger}$. The factor k_m expresses the maximum reaction rate constant ($E_A \rightarrow 0$ and/or $T \rightarrow 0$), also called the *Arrhenius* constant. The interpretation of k_m is possibly both from the collision theory and the theory of the transition state; in reality k_m depends on T (often $\beta = 1/2$):

$$k_m = BT^{\beta} \qquad (4.114)$$

Exceptions of the validity of Eq. (4.113) are trimolecular reactions where k is decreasing with increasing T. This become clear when we consider that the transition complex $AB^{\ddagger}$ can decompose thermal back to A + B instead of transforming to C + D:

$$A + B \underset{k_b}{\overset{k_a}{\rightleftharpoons}} [AB^{\ddagger}] \xrightarrow{+M(k_c)} C + D + M \qquad (4.115)$$

The reaction scheme (4.115) is principally valid for all bimolecular reactions:

$$A + B \underset{k_b}{\overset{k_a}{\rightleftharpoons}} [AB^{\ddagger}] \xrightarrow{k_c} C + D \quad \text{and} \quad A + B \xrightarrow{k} C + D \qquad (4.116)$$

According to the kinetic theory of gases, the reaction rate (here expressed as the number N of molecule changes per time) is equal to the number of collisions z

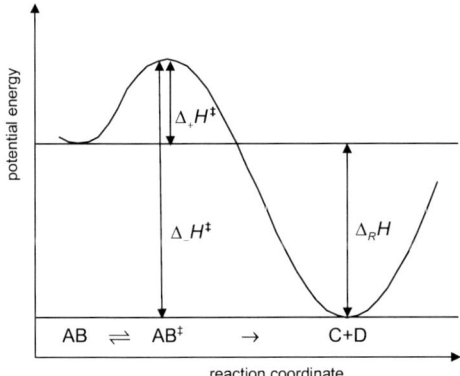

Fig. 4.1 Energetic scheme of a bimolecular reaction; instead of enthalpies (ΔH) also free enthalpies can be used (ΔG) in this scheme.

between A and B (Eq. 4.37) where the factor 2 means that at each collision two molecules disappear:

$$R_N = \frac{dN}{dt} = k_m N_A N_B = \frac{2z}{dt} \tag{4.117}$$

Now, expressing z in terms of two different molecules (Eq. 4.40) with a radius r_A and r_B as well as a molar mass M_A and M_B it follows that:

$$z = 4(r_A + r_B)^2 n_A n_B \, dt \sqrt{\pi RT \left(\frac{1}{M_A} + \frac{1}{M_B}\right)} \tag{4.118}$$

and

$$k_m = 8(r_A + r_B)^2 \sqrt{\pi RT \left(\frac{1}{M_A} + \frac{1}{M_B}\right)} \tag{4.119}$$

Thus, in $k_m = f(T^{1/2})$, k_m clearly represents the maximum reaction rate constant. Not all collisions result into a successful reaction, i.e. turnover according to Eq. (4.102) because the transition state AB is formally in equilibrium with the reactants with a pseudo equilibrium constant $K^\ddagger = [AB^\ddagger]/[A][B]$. This circumstance is described by the introduction of a probability factor γ (also called a steric factor, but not to be confused with the accommodation coefficient despite some similarities) in the *Arrhenius* equation:

$$k = \gamma \cdot k_m \exp\left(-\frac{E_A}{RT}\right) = k_0 \exp\left(-\frac{E_A}{RT}\right) \tag{4.120}$$

The rate law concerns the transition state is:

$$\frac{d[AB^\ddagger]}{dt} = k_a[A][B] - k_b[AB^\ddagger] - k_c[AB^\ddagger] \tag{4.121}$$

Assuming a steady state for the short-living intermediate $d[AB^\ddagger]/dt = 0$ (*Bodenstein* principle) and further assuming that $AB^\ddagger$ once formed will preferably react to C + D because of the much larger free reaction enthalpy $\Delta_- H^\ddagger$ compared with $\Delta_+ H^\ddagger$ (Fig. 4.1) and therefore $k_c \gg k_b$ it follows that:

$$\frac{[AB^\ddagger]}{[A][B]} = \frac{k_a}{k_b + k_c} \approx \frac{k_a}{k_c} \tag{4.122}$$

This reliable approach, however, is in contradiction to the *Eyring* approach, which assumes that $AB^\ddagger$ is within an equilibrium with the initial reactants. This is derived by using the general relationships (4.87) and (4.90) and treating the transition state in pseudo equilibrium (Eq. 4.115) as follows:

$$\frac{k_a}{k_b} = K^\ddagger = \frac{[AB^\ddagger]}{[A][B]} = \exp\left(-\frac{\Delta_+ G^\ddagger}{RT}\right)$$

$$= \exp\left(-\frac{\Delta_+ H^\ddagger}{RT}\right) \exp\left(\frac{\Delta_+ S^\ddagger}{R}\right) \tag{4.123}$$

where the state variables are related to the transition state $AB^{\ddagger}$. Note the difference between (4.122) and (4.123). It would follow from both equations that $k_b = k_c$, which means that each 50% of $AB^{\ddagger}$ will turn back to the reactants and turn forward to the products – an unlikely condition. The rate of product formation ("successful" reaction) is equal to the rate of $AB^{\ddagger}$ right-hand transformation in Eq. (4.116) but also to the disappearance (turnover) of A (or B); τ represents the residence time of the transition state:

$$R = \frac{d[\text{products}]}{dt} = k[A][B] = k_c[AB^{\ddagger}] = k_c K^{\ddagger}[A][B]$$
$$= [AB^{\ddagger}]\tau^{-1} . \tag{4.124}$$

From Eq. (4.124) it follows that:

$$k = k_c \frac{k_a}{k_b} = k_c K^{\ddagger} = k_c \exp\left(-\frac{\Delta_+ H^{\ddagger}}{RT}\right) \exp\left(\frac{\Delta_+ S^{\ddagger}}{R}\right) \tag{4.125}$$

According to kinetics theory, the residence time $\tau = 1/k_c$ represents the ratio of the mean translation velocity of $AB^{\ddagger}$ and the displacement $v\,dt$ the transition state passes; with $\tau - h/kT$, where h is *Planck's* constant. It follows:

$$k = \frac{kT}{h} \exp\left(-\frac{\Delta_+ H^{\ddagger}}{RT}\right) \exp\left(\frac{\Delta_+ S^{\ddagger}}{R}\right) \tag{4.126}$$

Fig. 4.1 shows schematically the chemical reaction of type (4.116). It follows that for the reaction enthalpy of the total process (A + B = C + D), $\Delta_R H = \Delta_- H^{\ddagger} - \Delta_+ H^{\ddagger}$ is the difference of the transition state enthalpies of the backward and forward reactions (Fig. 4.1). $\Delta_+ H^{\ddagger}$ is the difference between the enthalpy of the transition state and the sum of the enthalpies of the reactants in the ground state. This is called activation enthalpy (Fig. 4.1).The transition state $AB^{\ddagger}$ is formed at a maximum energy. $\Delta_R G$ is the free reaction enthalpy of the overall process or, in other words, the formation of (C + D). When comparing the right-hand term of Eq. (4.120) with the middle term of Eq. (4.123) and applying Eq. (4.90) to the total process, it follows that:

$$k = k_c \exp\left(-\frac{\Delta_+ G^{\ddagger}}{RT}\right) = k_0 \exp\left(-\frac{E_A}{RT}\right) = \exp\left(-\frac{\Delta_R G}{RT}\right) \tag{4.127}$$

and

$$E_A = \Delta_+ G^{\ddagger} - RT \ln\frac{k_0}{k_c} \tag{4.128}$$

Hence, the activation energy is not (fully) identical to the activation enthalpy $\Delta_+ H^{\ddagger}$ as is often reported in the literature. The *Eyring* plot $\ln(k/T)$ versus $1/T$ gives a straight line with the slope $-\Delta_+ H^{\ddagger}/R$ from which the enthalpy of activation can be derived and with intercept $\ln(k/h) + \Delta S^{\ddagger}/R$ from which the entropy of activation is derived.

The reaction scheme (4.116) is valid but with the most likely belief that $k_c \gg k_b$ there is no flux from the transition state $AB^{\ddagger}$ back and thereby no equilibrium. The *Arrhenius* factor simply explains that the formation of $AB^{\ddagger}$ is *inhibited*, i. e. only a fraction of collisions A + B turn to $AB^{\ddagger}$ despite the reacting molecules providing the needed activation energy E_A. Let us compare the following reactions between the OH radical and a simple organic compound (k in cm^3 molecule^{-1} s^{-1}):

$$OH + CH_4 \rightarrow H_2O + CH_3 \qquad k_{CH4} = 6.9 \cdot 10^{-15} \qquad (4.129)$$

$$OH + HCHO \rightarrow H_2O + HCO \qquad k_{HCHO} = 1 \cdot 10^{-11} \qquad (4.130)$$

Both reaction rate constants are considerably different (ratio more than 10^3). Although the residence time of methane is about 10 years, that of formaldehyde is only a few days. Both transition states represent a hydrogen bonding and H abstraction − $H_3CH \cdots OH$ and $HO \cdots HCHOH$ − and it is not likely that in the case of methane the transition state turns with a probability of more than a factor of 10^3 back to the initial substances compared with formaldehyde. The reason of rate differences lies in the activation energy which is much higher for the methane than for the formaldehyde reaction.

The *Arrhenius* plot $\ln k$ versus $1/T$ give a straight line with a slope $-E_A/R$ from which the activation energy can be derived and with intercept $\ln k_0$. Normally, in chemical standard reference books or tables rate constants are given for 25 °C (about 298 K) and the following equation can be used for recalculations:

$$\ln \frac{k(T)}{k_{298}} = \frac{E_A}{R}\left(\frac{1}{298} - \frac{1}{T}\right) \quad \text{or} \quad k(T) = k_{298} \exp\left(\frac{E_A}{R}\left(\frac{1}{298} - \frac{1}{T}\right)\right) \qquad (4.131)$$

The activation energy of chemical reactions lays in the range 20–150 kJ mol^{-1}. The time of processing $A + B \rightarrow C + D$ is in the order of only 10^{-12} s. This corresponds to $k = k_m$ and thereby $\gamma = 1$ (sometimes in the literature the steric or probability coefficient is denoted by A); this means that each collision will turn the state A + B into C + D. Controversially, when $\gamma = 0$, no collision results into a chemical conversion. The reality is somewhere in between.

In the case of (pseudo-) first-order reactions, the dimension of the reaction rate constant k is a reciprocal time (for example 1/s). To obtain a better understanding of the rate of disappearance of a pollutant, it is often useful to take a rate r in % h^{-1}; the recalculation is then made with the following expression:

$$r = 100[1 - \exp(-k \cdot 3600)] \quad \text{and} \quad k = \frac{-\ln\left(1 - \frac{r}{100}\right)}{3600} \qquad (4.132)$$

For illustration, a reaction with $k = 10^{-5}$ s^{-1} has a specific conversion of $r = 3.5$ % · h^{-1} and a residence time of about 1 day (exactly 27.7 h). Therefore, after one day of air mass travelling, which is equivalent to (typical wind speed 5 m s^{-1}) a distance of about 500 km, the initial concentration has decreased to about 37%.

It is important not to confuse the reaction rate constant (sometimes referred to as the specific reaction rate) with the absolute reaction rate $R (= dc/dt = kc)$, which is often called the simple turnover flux (mass per volume and time). In air chemistry, this reaction rate is often given in ppb · h^{-1}. Recalculation between concentration c and mixing ratio x is then given by Eq. (4.53).

Finally, there exist zero-order reactions where the reaction rate does not depend on the concentration of the reactant and remains constant until the substance has disappeared. Zero-order reactions are always heterogeneous.

4.2.2 Radicals

We have seen in previous chapters that T (and to a lesser extent p) is the key parameter in determining the inner energy of a system and thereby the ability to proceed chemical conversion. Thermal reactions are therefore limited in the atmosphere (in industrial chemical reactors, temperature increase is the most important trigger to initiate and accelerate conversions). In nature, where the modification of reaction conditions is limited and not under the control of the chemical system itself, reactive *radicals* play a crucial role (often referred to as free radicals, but there is no difference in use of the terms). Radicals are atoms, molecules or ions with unpaired electrons on an open shell configuration. The unpaired electrons cause them to be highly chemically reactive. Although radicals are generally short-lived because of their reactivity, long-lived radicals exist. The prime example of a stable radical is molecular dioxygen O_2 in the triplet state. Oxygen is also the most common molecule in a diradical state. Multiple radical centers can exist in a molecule. Other common atmospheric substances of low-reactive radicals are nitrogen monoxide NO and nitrogen dioxide NO_2.

Three radicals are involved in reactions (4.129) and (4.130): hydroxyl OH as the primary reactant as well as the very reactive radicals methyl CH_3 and formyl HCO[13]. According to IUPAC recommendations, the formulas should be given as $^{\bullet}OH$, $^{\bullet}CH_3$ and $HC^{\bullet}O$ where the dot symbolizes the unpaired electron and should be placed to indicate the atom of highest spin density, if possible. In this book, we cancel out the dots for clarity. Table 4.4 lists the most important atmospheric radicals.

Depending on the core atom that possesses the unpaired electron, the radicals can be described as carbon-, oxygen-, nitrogen- or metal-centered radicals. If the unpaired electron occupies an orbital with considerable s or more or less pure p character, the respective radicals are termed σ- or π-radicals.

Radicals exist is the gas and liquid (aqueous) phase. They play an important role in many chemical transformations. In natural waters, radical ions, a radical that carries an electric charge, exist. Those positively charged are called radical cations and those negatively charged radical anions, but the most important is the superoxide anion O_2^-. Free radicals are produced under high temperature (combustion,

Table 4.4 Important reactive radicals.

source element	radicals
oxygen	O (O^1D, O^3P), OH, HO_2, O_2^-, O_3^-, RCH_2O, RCH_2O_2
hydrogen	H
nitrogen	NO_3
sulfur	HSO_3 (SO_3^-), HSO_5 (SO_5^-), SO_4^-
carbon	CH_3, HCO, RCO
halogens	Cl, ClO, Br, BrO

---

[13] Note that the symbol for formyl is also different used in literature: HOC and CHO (it is the aldehyde group $H-C=O$).

lightning) and through the photodissociation of source molecules. There are presently two fields where a more detailed knowledge of the thermodynamic properties of radicals would be extremely useful.

The first is biomedicine. The discovery of superoxide dismutase and nitrogen monoxide as messengers has led to an explosive growth in articles in which one-electron oxidations and reductions have been explored. Organic radicals play an important role in the treatment of cancers. The other is atmospheric chemistry where the modeling of reactions requires accurate reduction potentials (Stanbury 1989, Wardman 1989).

Radicals play a key role in chain reactions, in which one or more reactive reaction intermediates (frequently radicals) are continuously regenerated, usually through a repetitive cycle of elementary steps (the "propagation step"). The propagating reaction is an elementary step in a chain reaction in which one chain carrier is converted into another. The chain carriers are almost radicals. Termination occurs when the radical carrier reacts otherwise. An example of one of the possible ozone destructions is shown below (R-Cl − chloro-organic compound):

$$R\text{-}Cl \xrightarrow{\text{radiation}} R + Cl \qquad \text{initiating}$$

$$O_3 + Cl \rightarrow O_2 + ClO$$
$$O_3 + ClO \rightarrow 2\,O_2 + Cl \qquad \text{propagating (chain)}$$

$$Cl + H \rightarrow HCl$$
$$ClO + NO_2 \rightarrow ClNO_3 \qquad \text{termination}$$

Without termination, the gross propagating step results in $2\,O_3 \rightarrow 3\,O_2$ and can very often be cycled depending on parallel reactions. In the example above, the products of termination (HCl, ClNO$_3$) can act as source molecules and provide Cl radicals through photodissociation.

4.2.3 Photochemistry: The photolysis rate constant

Photochemistry is the branch of chemistry concerned with chemical reactions caused by the absorption of light (far UV to IR). There are many excited states of a molecule, which results in chemical conversions without photodissociation. Photochemical paths offer the advantage over thermal methods of forming thermodynamically disfavored products, overcoming large activation barriers in a short time and allowing reactivity otherwise inaccessible by the thermal method. In atmosphere, however, photodissociation (often also termed photolysis), which is the cleavage of one or more covalent bonds in a molecular entity resulting from an absorption of light, is of importance, especially for producing radicals.

Fig. 4.2 shows schematically a photochemical process. It is beyond the scope of this book to go into detail about photoexcitation. According to the quantum structure of the electronic molecular system, such photons are absorbed corresponding to existing bands of rotational and vibrational states (Fig. 4.3). The excited molecule AB* can turn back to the ground state through light emission (fluorescence and/or phosphorescence) but also via collisions with air molecules, called quench-

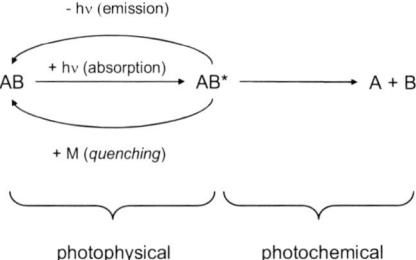

Fig. 4.2 Scheme of the photophysical and photochemical process.

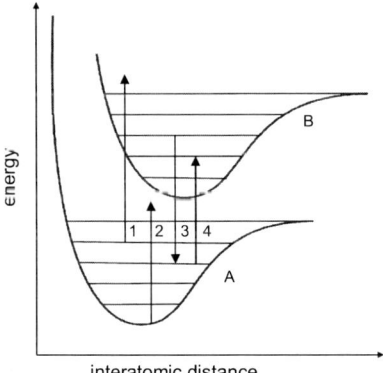

Fig. 4.3 Scheme of electronic excitation of a diatomic molecule, also called potential energy curve (inharmonic oscillator). A ground state, B first excited state, 1 direct dissociation from exited ground state, 2 direct dissociation from ground state, 3 radiation transfer from exited state, 4 excitation from ground state.

ing. Only when the absorbed light energy (corresponding to a photolysis threshold wavelength) is large enough to overcome the intermolecular distance can the excited state turn into breakdown products A + B. The photolysis is represented by:

$$AB \xrightarrow{h\nu} A + B \qquad (4.133)$$

The term $h\nu$ symbolizes the energy of a photon (h is *Planck*'s quantum and ν is frequency; remember that $c = \nu \cdot \lambda$, where λ is the wavelength and c the speed of light). A first-order law describes the rate of process (4.133):

$$\frac{d[AB]}{dt} = -j_{AB}[AB] \qquad (4.134)$$

where j is the *photolysis rate coefficient*, often also called the *photolysis rate constant* (sometimes termed *photolysis frequency* – which physically is the most correct term), but in contrast to the reactions rate constant k (which only depends on T), j depends on many parameters and is not constant but changes permanently over time. The dimension of j is reciprocal time, called photolytic residence time

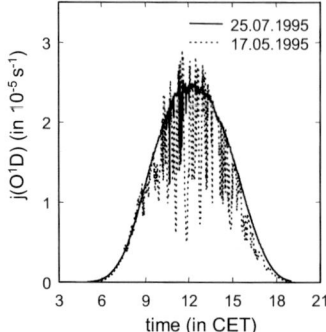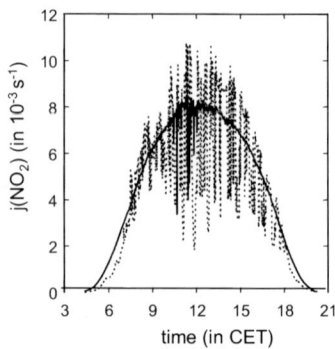

Fig. 4.4 Diurnal variation of j-O(^{1}D) and j-NO$_2$ at two different days in summer 1995, measured at the airport Munic (Germany). July 25 represents a cloud-free day with few cumulus at noon whereas May 17 was cloudy between 4/8 and 7/8; after Reuder (1999).

$\tau = j^{-1}$. For a daily mean estimate of the photochemical conversion rate according to Eq. (4.134), it is useful to determine an average rate coefficient as:

$$\bar{j} = \frac{1}{t_{sunset} - t_{sunrise}} \int_{t_{sunrise}}^{t_{sunset}} j(\Delta t)\, dt \qquad (4.135)$$

where Δt is the time interval of the integration step. Fig. 4.4 shows two examples of j in its diurnal variation. The photolysis rate coefficients j are calculated by integrating the product of the *spectral actinic flux* $S(\lambda)$, *spectral absorption cross section* $\sigma(\lambda)$ and the photodissociation *quantum yield* $\Phi(\lambda)$ over all relevant wavelengths (Madronich 1987):

$$j(\lambda, T) = \int_{\lambda_{min}}^{\lambda_{max}} S_\lambda(\lambda)\, \sigma(\lambda, T)\, \Phi(\lambda, T)\, d\lambda \qquad (4.136)$$

Because of the quantum characteristic of light absorption in the consideration of photochemistry we use here *spectral* quantities, i. e. per unit of wavelength interval (in nm normally given despite the SI recommendation of m). In Chapter 2.3.1.2, quantities of radiation transfer through the atmosphere were described. The *actinic flux* S_λ in terms of photons per time, called in traditional terms light intensity, photon flux, irradiance or radiant flux and simply radiation, which has caused some confusion (Table 2.14), is the quantity of light available to molecules at a particular point in the atmosphere and which, on absorption, drives photochemical processes in the atmosphere. Actinic flux describes the number of photons (or radiation) incident on a spherical surface, such as the molecule of the atmospheric species, and is the suitable radiation quantity for photolysis rate coefficient determination. From Eq. (2.66) we can then integrate the (spectral actinic) flux with respect to the spherical surface, which is characterized by the angles θ and φ (remember that the solid angle $d\Omega$ is defined as $\sin(\theta)d\theta d\varphi = d\Omega$, Chapter 2.3.1.2):

$$S_\lambda(\lambda) = \int_0^{2\pi} \int_0^{\pi/2} hc \frac{L(\lambda, \theta, \varphi)}{\lambda} \sin(\theta) d\theta d\varphi \tag{4.137a}$$

$$(\text{Dimension: photons} \cdot s^{-1} \cdot cm^{-2} \cdot nm^{-1})$$

Note: Instead of "photons" the equivalent term "quanta" is sometimes used. Because each photon of a given wavelength is characterized by a specific energy, $S_\lambda(\lambda)$ is also equivalent to a value given in $J \cdot s^{-1} \cdot cm^{-2} \cdot nm^{-1}$. A photon is a particle of zero charge, zero rest mass, with spin quantum number 1 and energy $h\nu$; it is the carrier of electromagnetic force:

$$S_\lambda(\lambda) = \int_0^{2\pi} \int_0^{\pi/2} L(\lambda, \theta, \varphi) \sin(\theta) d\theta d\varphi \tag{4.137b}$$

$$(\text{Dimension: } J \cdot s^{-1} \cdot cm^{-2} \cdot nm^{-1})$$

Sometimes the actinic flux is only expressed in terms of the solid angle:

$$S_\lambda(\lambda) = \int_{4\pi} L(z, \Omega, \lambda) d\Omega \quad (\text{Dimension: } J \cdot s^{-1} \cdot cm^{-2} \cdot nm^{-1}) \tag{4.137c}$$

Note that the actinic flux is always related to a given altitude (depth z, Fig. 2.11) in the atmosphere (earth's surface at $z = 0$). It is important to distinguish the actinic flux $S_\lambda(\lambda)$ from the *spectral irradiance* $I(\lambda)$, which refers to energy arrival on a flat surface with a fixed spatial orientation (Table 2.14) and which is the most common measured radiation quantity:

$$I(\lambda) = \int_0^{2\pi} \int_0^{\pi/2} L(\lambda, \theta, \varphi) \cos(\theta) \sin(\theta) d\theta d\varphi \tag{4.138}$$

$L(\lambda, \theta, \varphi)$ is the spectral radiance, and θ and φ are the zenith and azimuth angles, respectively. The irradiance $I(\lambda)$ describes the radiance on a horizontal surface integrated over the whole upper hemisphere, weighted with the cosine of the incidence angle. From the definition of irradiance (Eq. 2.64) it follows that:

$$I = \frac{dP}{da} = \int_0^{2\pi} L \cos \theta d\Omega \tag{4.139}$$

The spectral irradiance (and analog of other *spectral* quantities) is given by:

$$I(\lambda) = \int_{\lambda_1}^{\lambda_2} E(\lambda) d\lambda \tag{4.140}$$

The actinic flux and the irradiance are split into a diffuse and a direct part. Actinic flux measurements are not trivial. They require spectroradiometers with especially configured optics to enable measurements of radiation equally weighted from all directions (Ruggaber et al. 1993, Kazadzis et al. 2004, Trebs et al. 2009). The actinic flux does not refer to any specific orientation because molecules are oriented randomly in the atmosphere. This distinction is of practical relevance: the actinic flux (and thereby a *j*-value) nears a brightly reflecting surface (e. g. over snow or above a thick cloud) can be a factor of three times higher than that near a non-reflecting surface. Hence, the presence of clouds can change drastically the actinic flux throughout the atmosphere.

The *absorption cross section* $\sigma(\lambda)$ is a measure of the area (dimension: cm^2 molecule^{-1}) of the molecule (given by the electron density function), through which the photon cannot pass without absorbing, when its energy is equivalent to a molecular quantum term (otherwise it will be reflected) in contrast to the definition of a collisional cross section. Furthermore, the dimensionless *quantum yield* $\Phi(\lambda)$ is the ratio between the number of excited (or dissociated) molecules and the number of absorbed photons (dimensionless with values 0 ... 1). This quantity depends on wavelength and approaches one at the so-called *threshold wavelength*. The quantum yield is numerically dimensionless but formally denotes in molecules per photon.

The expression (4.136) for the *j*-value can be transformed (through integration) into a more applicable form by using the mean values of σ and Φ for given wavelength intervals $\Delta\lambda$ (as tabulated in standard books and sources (for example, Finlayson-Pitts and Pitts 2000):

$$ j = \sum_{\lambda = 290 \text{ nm}}^{\lambda_i} \overline{S_\lambda}(\lambda)\, \overline{\sigma}(\lambda)\, \overline{\Phi}(\lambda) \tag{4.141}$$

The values for $\overline{\sigma}(\lambda)$ and $\overline{\Phi}(\lambda)$ have been determined from laboratory investigations (Finlayson-Pitts and Pitts 2000). For the actinic flux with respect to the solar zenith angle θ, height z and wavelength λ either measurements and/or calculations are needed. For fundamentals see Chandrasekhar (1960), Lenoble (1985), Liou (1992) and Thomas and Stamnes (1999), and for some examples of application see Ruggaber et al. (1993), Los et al. (1997) and Kay et al. (2001). All radiative transfer modeling is ultimately based on the fundamental equation of radiative transfer, or the transfer of energy as photons. Eq. (4.142) describes the net photon flux:

$$ dQ = L(\lambda, \theta, \varphi)\cos(\theta) \cdot da \cdot d\Omega \cdot dt \cdot d\lambda \tag{4.142}$$

Analytic solutions to the radiative transfer equation (RTE) exist for simple cases; however, for more realistic media with complex multiple scattering effects, numerical methods are required. The equation of radiative transfer simply states that as a beam of radiation travels, it loses energy to absorption, gains energy by emission and redistributes energy by scattering. The differential form of the equation for radiative transfer is:

$$ \frac{1}{c}\frac{\partial L}{\partial t} = -\Omega\nabla L - (\kappa_a + \kappa_s)L + \varepsilon + \frac{1}{4\pi c}\kappa_s\int_\Omega L d\Omega \tag{4.143}$$

where κ_a and κ_s are extinction coefficients for absorption and scattering, respectively (Chapter 2.3.2.1), c is the speed of light and ε is the emission coefficient. Analytical solutions need strong simplifications; numerical solutions differ in the forms of the coefficients. Another useful expression for *j* is given by Hough (1988) in parameterized form:

$$ j = a_i \exp(b_i \cdot \sec\theta) \tag{4.144}$$

Table 4.5 Some important photolytic reactions in air.

no.	reactant	products	photolysis rate (in s^{-1}); Hough (1988)	τ (in h)[a]
(1)	O_3	$O(^1D)$	$2 \cdot 10^{-4} \exp(1.4 \sec \theta)$	0.3
(2)	NO_2	$O(^3P) + NO$	$1.45 \cdot 10^{-2} \exp(0.4 \sec \theta)$	0.01
(3)	HNO_2	$OH + NO$	$0.205 \cdot j_2 \, (3 \cdot 10^{-3} \exp(0.4 \sec \theta)$	0.06
(4)	HNO_3	$OH + NO_2$	$3 \cdot 10^{-6} \exp(1.25 \sec \theta)$	38
(5)	NO_3	different	$3.29 \cdot j_2$	0.003
(6)	$HCHO$	$H + HCO$	$6.65 \cdot 10^{-5} \exp(0.6 \sec \theta)$	2.1
(7)	$HCHO$	$CO + H_2$	$1.35 \cdot 10^{-5} \exp(0.94 \sec \theta)$	6.9
(8)	CH_3CHO	HO_2	$= j_6$	2.1
(9)	$ClONO_2$	$ClO + NO_2$	$2.9 \cdot 10^{-5}$	9.4
(10)	Cl_2	$2\,Cl$	$= j_1$	0.3

[a] calculated for $\theta = 30°$ (about maximum value in Central Europe)

where a_i and b_i are substance-specific constants (Table 4.5). It is clear that Eq. (4.144) reflects the *Lambert-Beer* law and is analog to the *Arrhenius* equation. Fig. 4.4 shows two examples of photolysis frequency for two days with different cloudiness. As mentioned, clouds scatter solar radiation and can reduce as well as enhance photodissociation.

4.2.4 Oxidation and reduction (the redox processes)

Oxidation and reduction are key processes in the chemical evolution and thereby the climate system (Chapter 2.2.2.3). Oxidation is always coupled with reduction (therefore, *redox* as shorthand for the reduction-oxidation reaction):

$$A_{red} + B_{ox} \rightarrow A_{ox} + B_{red} \tag{4.145}$$

The reaction (4.145) describes all chemical reactions in which atoms have their oxidation number (oxidation state) changed. Oxidation state is defined as the charge an atom might be imagined to have when electrons are counted according to the following agreed set of rules (Table 4.6):

1. the oxidation state of a free element (uncombined element) is zero;
2. for a simple (monatomic) ion, the oxidation state is equal to the net charge on the ion;
3. hydrogen has an oxidation state of +1 and oxygen has an oxidation state of −2 when they are present in most compounds[14]; and
4. the algebraic sum of the oxidation states of all atoms in a neutral molecule must be zero, whereas in ions the algebraic sum of the oxidation states of the constituent atoms must be equal to the charge on the ion.

---

[14] Exceptions to this are that hydrogen has an oxidation state of −1 in hydrides of active metals, e. g. LiH, and oxygen has an oxidation state of −1 in peroxides.

Table 4.6 Oxidation states of elements in stable compounds in the climate system (in radicals further states occur).

−4	−3	−2	−1	0	+1	+2	+3	+4	+5	+6	+7
			Hᵃ	H	H						
C						C		C			
	N				N	N	N	N	N		
	P				P		P	P	P		
		O	Oᵇ	O							
		S		S				S		S	
			Cl	Cl	Cl		Cl	Cl	Cl		
			Br	Br	Br			Br	Br		
			I	I	I			I	I		I
				Fe		Fe	Fe				
				Mn	Mn	Mn					
				Cu	Cu	Cu					

ᵃ in metal hydrides which are unstable under normal conditions
ᵇ only in peroxide

Hence, oxidation and reduction can be defined by these three criteria, where all oxidations meet criteria 1 and 2 and many meet criterion 3 but this is not always easy to demonstrate.

1. The complete, net removal of one or more electrons from a molecular entity (also called "de-electronation");
2. An increase in the oxidation number of any atom within any substrate; and
3. The gain of oxygen and/or loss of hydrogen of an organic substrate.

The redox process (4.145) can formally divide into oxidation reduction:

$$A_{red} \rightarrow A_{ox} + e^- \quad \text{and} \quad A_{ox} + e^- \rightarrow A_{red} \qquad \text{(4.146a) and (4.146b)}$$

$$B_{ox} + e^- \rightarrow B_{red} \quad \text{and} \quad B_{red} \rightarrow B_{ox} + e^- \qquad \text{(4.147a) and (4.147b)}$$

The reducing agent (also called a reductant or reducer) is thereby an electron donator and the oxidizing agent (also called an oxidant or oxidizer) is an electron acceptor. With the extension of criterion 3, oxygen is an oxidizing agent and hydrogen a reducing agent according to

$$O_2 + e^- \rightarrow O_2^- \qquad (4.148)$$

$$H \rightarrow H^+ + e^- \qquad (4.149)$$

Because of the general principle of electroneutrality and mass conservation, oxidation and reduction must be balanced within the whole earth system. The gross "reaction" of sequences (4.148) and (4.149) is $H_2O \rightleftarrows H_2O$. Water is the source substance of oxidizing agents (oxygen and related substances) and reducing agents (hydrogen and related substances), whereas in the biosphere hydrogen is enriched and oxygen depleted and vice versa in the atmosphere. This simple difference (gradients in "reservoirs'" chemical potential) creates biogeochemical driving forces. As biology is relatively ubiquitous in the hydro- and lithosphere, the most ubiquitous elements that readily undergo redox transformations – iron in the ferrous/ferric (Fe^{2+}/Fe^{3+}) system and oxygen – are commonly encountered. Many microbial species when starved of oxygen (anoxic conditions) turn to metals to shuttle their electron transport chains along and store energy via organophosphates for later use in living processes, such as reproduction or building protective coatings (Gentleman 2010).

In atmospheric chemistry, the term *oxidant* is frequently used. This is a qualitative term that includes any and all trace gases with a greater oxidation potential than oxygen (for example, ozone, peroxoacetyl nitrate, hydrogen peroxide, organic peroxides, NO_3, etc.). Hence, oxygen is excluded (therefore, the term *reactive oxygen species* (ROS) has been introduced). This "definition" of an oxidant is incorrect and should not be used in this narrow sense. As reaction (4.145) suggests, the determination of whether a chemical species is a reducing or oxidizing species is relative and depends on the specific redox potential of the pairs of half-reactions (4.146 and 4.147). If a half-reaction is written as a reduction (4.146b and 4.147a), the driving force is the *reduction potential*. If a half-reaction is written as an oxidation (4.146a and 4.147b), the driving force is the *oxidation potential* related to the reduction potential by a sign change. So the redox potential is the reduction-oxidation potential of a compound measured under standard conditions against a standard reference half-cell. In biological systems, the standard redox potential is defined at pH $= 7.0$ versus the hydrogen electrode and partial pressure of $p_{H_2} = 1$ bar.

However, the concept of current flow is only applicable to aqueous systems[15]. In the gaseous phase of air, electron exchange occurs within the transition state of two molecular entities, basically in a wider sense of charge transfer complexes. Any substance in a specified phase has an electrochemical potential $(\mu_i)_{el}$ consisting of the chemical potential μ_i (partial molar free enthalpy) and a specified electric potential:

$$(\mu_i)_{el} = \mu_i + z\,F\,E \tag{4.150}$$

where z is the number of elementary charges (e) exchanged in the oxidation-reduction process and F is the *Faraday* constant (molar electric charge $F = e \cdot N_A$). The *electromotive force* E (the equivalent of the term redox potential) is the energy supplied by a source divided by the electric charge transported through the source: $-z\,FE = \Delta_R G$. By contrast, the *electric potential* V is the work required to bring a charge from infinity to that point in the electric field divided by the charge (for a galvanic cell it is equal to the electric potential difference for zero current through the cell). The molar charge $z\,F \cdot n$ (n mole number) is defined as the *electric charge* Q (base unit ampere) and the current I is defined as the rate of charge flow:

$$I = \frac{dQ}{dt} = z\,F\,\frac{dn}{dt} . \tag{4.151}$$

---

[15] In the earth climate system, only aqueous systems occur, but under laboratory conditions any other solution can be regarded (for example, organic solvents), thereby providing carriers for electrons.

The current has the rate dimension of moles per unit of time. This equality is important for understanding the equivalence between mass conversion (dn/dt) and current flow (dQ/dt). By using Eqs. (4.88) and (4.90), we can describe the free enthalpy of the redox process (4.145) by:

$$\Delta_R G = -\boldsymbol{R}T \ln K + \boldsymbol{R}T \ln \frac{[A_{ox}][B_{red}]}{[A_{red}][B_{ox}]} = -zFE \qquad (4.152)$$

whereas $-\boldsymbol{R}T \ln K = \Delta_R G^{\ominus}$ is the standard free enthalpy (in equilibrium). The important *Nernst* equation then follows:

$$E = E^{\ominus} + \frac{\boldsymbol{R}T}{zF} \ln \frac{[A_{ox}][B_{red}]}{[A_{red}][B_{ox}]} \quad \text{or} \quad E = E^{\ominus} + \frac{0.05916}{z} \log \frac{[A_{ox}][B_{red}]}{[A_{red}][B_{ox}]} \qquad (4.153)$$

Standard redox potentials are listed in Table 4.7. From Eq. (4.153), when only considering the half-reaction oxidation or reduction, an adequate reduction potential E_{red} and oxidation potential E_{ox} can be derived:

$$E_{ox} = E_{ox}^{\ominus} + \frac{\boldsymbol{R}T}{zF} \ln \frac{[A_{ox}]}{[A_{red}]} \qquad (4.154)$$

$$E_{red} = E_{red}^{\ominus} + \frac{\boldsymbol{R}T}{zF} \ln \frac{[B_{red}]}{[B_{ox}]} \qquad (4.155)$$

It follows for the coupled process that:

$$E = E_{red} + E_{ox} . \qquad (4.156)$$

Eq. (4.154) describes the (oxidation) reaction (4.146a); therefore, and the reverse (reduction) process (4.146b) is expressed by Eq. (4.155) when substance B is replaced by A. Because $E_{ox}(A) = E_{red}(A)$ it also follows then that $E_{red}^{\ominus} = E_{ox}^{\ominus}$. This explains why "only" reduction standard potentials are listed in the literature, almost termed the electrode potentials of half-reactions (Table 4.7).

For solutions in protic solvents (water is the most important one) the universal reference electrode for which, under standard conditions, the standard electrode potential is zero by definition all temperatures is the hydrogen electrode (H^+/H_2). In Table 4.7 are the absolute electrode potentials for hydrogen, which can be interpreted in the following way. A redox couple more negative than 0.414 V should liberate hydrogen from water and a couple more negative than 0.828 V should liberate H_2 from 1 mol L^{-1} OH^- solution.

In atmospheric chemistry, the term atmospheric oxidation capacity is frequently used and mostly attributed to hydroxyl radicals (OH), primarily formed through the photodissociation of ozone (for example, Lelieveld et al. 2008) but also defined as being "essentially the global burden of these oxidants" (Thompson 1992), which includes O_3, OH and H_2O_2. We state that atmospheric oxidation capacity is not well defined in any thermodynamic and/or kinetic phrasing. Becker et al. (1999) and Möller (2003) proposed that the sum of pseudo-first-order oxidation rates (4.159) could be used to describe the atmospheric oxidation capacity. Let us use the general gas phase reaction scheme (*i* different substances to be oxidized and *j* different oxidizing agents) as follows:

Table 4.7 Standard reduction potentials (in V) of selected half-reactions in the aqueous H_xO_y system (Milazzo and Carol 1978, Bard et al. 1985, Stanbury 1989, Wardman 1989, Holze 2007), at 25 °C. e^- electron transferred from electrode, e_{aq}^- hydrated electron, (g) gaseous, (aq) dissolved.

reactants		products	E	
e^-	$\rightarrow$	e_{aq}^-	-2.89	
$H^+ + e^-$	$\rightarrow$	$H(aq)$	-2.32	
$\frac{1}{2}H_2 + e^-$	$\rightarrow$	H^-	-2.25	
$2\,H_2O + 2\,e^-$	$\rightarrow$	$H_2 + 2\,OH^-$	-0.828	pH = 14
$2\,H^+ + 2\,e^-$	$\rightarrow$	H_2	-0.414	pH = 7
$O_2 + 4\,H^+ + 4\,e^-$	$\rightarrow$	$2\,H_2O$	$+1.229$	pH = 0
$O_2 + 4\,H^+ + 4\,e^-$	$\rightarrow$	$2\,H_2O$	$+0.816$	pH = 7
$O_2(g) + 2\,H^+ + 2\,e^-$	$\rightarrow$	H_2O	$+2.430$	acid
$O_2 + 2\,H^+ + 2\,e^-$	$\rightarrow$	H_2O_2	$+0.695$	acid
$O_2 + 2\,H_2O + 4\,e^-$	$\rightarrow$	$4\,OH^-$	$+0.401$	pH = 14
$O_2(g) + H_2O + 2\,e^-$	$\rightarrow$	$2\,OH^-$	$+1.602$	pH = 14
$O_2(g) + e^-$	$\rightarrow$	O_2^-	-0.35	pH = 7
$O_2(g) + H^+ + e^-$	$\rightarrow$	HO_2	-0.07	acid
$O_2(aq) + H^+ + e^-$	$\rightarrow$	HO_2	$+0.10$	acid
$O_2(aq) + e^-$	$\rightarrow$	O_2^-	-0.18	pH = 7
$O_2 + e^-$	$\rightarrow$	O_2^-	-0.563	pH = 14
$O_2 + H_2O + 2\,e^-$	$\rightarrow$	$OH^- + HO_2^-$	-0.0649	pH = 14
$O + e^-$	$\rightarrow$	O^-	$+1.61$	pH = 7
$O^- + H_2O + e^-$	$\rightarrow$	$2\,OH^-$	$+1.59$	pH = 14
$O_2^- + H^+ + e^-$	$\rightarrow$	HO_2	-0.105	pH = 14
$O_2^- + 2\,H^+ + 2\,e^-$	$\rightarrow$	H_2O_2	$+0.695$	pH = 14
$O_2^- + 4\,H^+ + 4\,e^-$	$\rightarrow$	$2\,H_2O$	$+1.229$	pH = 14
$O_3 + 2\,H^+ + 2\,e^-$	$\rightarrow$	$O_2 + H_2O$	$+2.075$	pH = 0
$O_3 + H^+ + e^-$	$\rightarrow$	$O_2 + OH$	$+1.77$	acid
$O_3 + H_2O + 2\,e^-$	$\rightarrow$	$O_2 + 2\,OH^-$	$+1.246$	pH = 14
$O_3 + H_2O + 2\,e^-$	$\rightarrow$	$O_2 + OH + OH^-$	-0.943	neutral
$OH + H^+ + e^-$	$\rightarrow$	H_2O	$+2.85$	pH = 0
$OH + e^-$	$\rightarrow$	OH^-	$+1.985$	pH = 0
$OH + H^+ + e^-$	$\rightarrow$	H_2O	$+2.30$	pH = 0
$HO_2 + H^+ + e^-$	$\rightarrow$	H_2O_2	$+1.495$	pH = 0
$HO_2^- + H_2O + 2\,e^-$	$\rightarrow$	$3\,OH^-$	$+0.867$	pH = 14
$H_2O_2 + H^+ + e^-$	$\rightarrow$	$OH + H_2O$	-1.14	pH = 0
$H_2O_2 + 2\,H^+ + 2\,e^-$	$\rightarrow$	$2\,H_2O$	$+1.763$	pH = 0

$$(A_{red})_i + OX_j \rightarrow (A_{ox})_i + \text{products} \tag{4.157}$$

The total reaction rate is given by:

$$(R_{ox})_{total} = \sum_i \sum_j k_{i,j} [(A_{red})_i][OX_j] \tag{4.158}$$

and the oxidizing capacity is given by ($OX = O_3$, OH, etc.):

$$(C_{ox})_{total} = \sum_i \sum_j k_{i,j} [OX_j] \tag{4.159}$$

and when we reduce it only to reactions between A and OH:

$$(C_{OH})_{total} = \sum_i k_i[OH] \tag{4.160}$$

From reaction (4.157) we can also define the atmospheric oxidation potential (in the thermodynamic sense) to be:

$$(P_{OX})_{total} = \sum_i \Delta_R G_i \tag{4.161}$$

4.2.5 Acid–base reactions: Acidity and alkalinity

4.2.5.1 Environmental relevance of acidity

The term *acid rain* denotes one of the most serious environmental problems, i. e. the *acidification* of our environment. *Acidity* is a chemical quantity that is essential for biological life. It is a result of the budget between *acids* and *bases* existing in the regarded reservoir, which finally is an equilibrium state because of the interaction of all biogeochemical cycles including the water cycle. Consequently, the anthropogenic disruption of biogeochemical cycles leads to changing acidity (Stumm et al. 1983). *Acidification* is used to describe a process by which a given environment is made more acidic (Odén 1976).

The term *acidity* is often used to characterize the ability of a compound to release hydrogen ions (H^+) to water molecules as a measure, expressed by pH value, but sometimes denoted *acid capacity* or *acid strength*, not to be confused with the *acid constant* K_a, although K_a is a measure for acid strength. By contrast, the *alkalinity* (can also be called *basicity* or *basic capacity*) is used to characterize the ability of a compound to be a proton acceptor. This definition, limited only to free H^+, is not appropriate for the atmospheric multiphase system because of the gas-liquid equilibrium of acids and bases in the pH range of interest. Therefore, Waldman et al. (1992) defined *atmospheric acidity* to be the "acidity" in the aqueous, gaseous and aerosol phases, representing the sum of individual compounds which are measured. They wrote, "The net acidity, however, is measured in solution, following eluation or extraction of the sample. Measurements of sample acidity are performed with a pH electrode".

Not all these definitions help clarify what we have to understand about *atmospheric* acidity. The term *acidifying capacity* (Möller 1999b) is only of qualitative value and meets the same basic problems as for defining the oxidizing capacity of the atmosphere. The problem lies in the different points of view between the analytical chemist and impact researcher. *Svante Odén* (1976) wrote, "The problem of air pollution or the impact of a specific pollutant can only be understood when reactions and interactions between all reservoirs are taken into account". The impact (acidification) is caused by acid deposition, which is a result of atmospheric acidity or, in other words, the *acidifying capacity* of the atmosphere. The changing acidifying capacity of the atmosphere is only part of a larger problem − changes of the chemical climate caused by a variety of emissions into the atmosphere. The importance of atmospheric acidity, and especially *acid fog*, had already been identified at the end of the last century to be a cause for forest damage (Wislicenus et al. 1916) as well as by *Robert Angus Smith* while analyzing rainwater in the middle of the 19th century (Chapter 3.4).

However, it is worth briefly describing the formation and function of natural acidity. There are only three strong acids (HCl, HNO_3 and H_2SO_4) that are not directly emitted in nature (except a small amount of HCl from volcanoes probably). Gaseous HCl produced from sea salt particles is equivalent to the consumption of other strong acids (HNO_3 and H_2SO_4) because of heterogeneous reactions and HCl degassing. Hence, from NO and reduced sulfur emissions globally strong acids will be produced via oxidation (Chapters 5.4.4 and 5.5.2). Assuming a 50 % conversion of naturally emitted NO and reduced sulfur into acids (which surely represents a maximum), about 1.5 Teq yr^{-1} acidity is provided. Direct emissions are known for organic acids (HCOOH and CH_3COOH corresponding to about 0.15 Teq yr^{-1}; Kesselmeier et al. 1997) and indirectly via HCHO oxidation; assuming that only 10 % of HCHO is converted into HCOOH via the cloud water phase (Chapter 5.7.2), an additional 3 Teq yr^{-1} is produced. The quantification of other organic acids (oxalic, propionic, butyric and many others) is impossible due to missing source information. The large concentrations (1−2 ppb) found in coniferous forests (Kesselmeier et al. 1997) for formic and acetic acid, and its relationship to plant physiological parameters (photosynthesis, transpiration and stomatal conductance) suggest an ecological controlling function. However, most acidity in ecosystem budgets is produced in soils from humic substances. By contrast, atmospheric acidity is moveable and thereby influences other ecosystems through weathering. From a budget point of view, carbon dioxide provides a global precipitation (assuming pH = 5.6 and taking into account the global rainfall from Table 2.22) of about 2 Teq yr^{-1} in the form of HCO_3^-. For global weathering, carbonic acid seems to be dominant but other acids are not negligible. Moreover, only organic acids, sulfuric acid and nitric acid can produce acidic rain down to pH ~ 4 locally. By contrast, formic acid / formate and acetic acid / acetate are efficient buffers not only in the laboratory.

4.2.5.2 Acid–base theories

Arrhenius (1887) and Ostwald (1894) provided the oldest scientific definitions. Acids (A) are species that produce during their aquatic dissolution hydrogen ions (H^+), whereas bases (B) produce hydroxyl ions (OH^-):

$$A + H_2O \rightarrow H^+ \quad + product_1 \tag{4.162}$$

$$B + H_2O \rightarrow OH^- + product_2 \tag{4.163}$$

This definition excludes ions to be acids. Brønsted (1934) modified the *Arrhenius* definition in the way that acids are chemical species that separate H^+, whereas bases uptake H^+:

$$A \rightleftarrows B + H^+ \tag{4.164}$$

Table 4.8 Atmospheric acids (A) and bases (B).

	strong acids	week acids	strong bases	weak bases
gases	H_2SO_4 HCl HBr HNO_3 HNO_2	$RCOOH$ $ROOH$ HO_2 H_2O_2	NH_3	none
particulate matter and solute	HSO_4^-	HCO_3^- NH_4^+ HSO_3^-	CO_3^{2-} HCO_3^- OH^- $[O^{2-}]^a$ $RCOO^-$	SO_4^{2-} HSO_4^- SO_3^{2-} HSO_3^- NO_3^- NO_2^-

[a] metal oxides

with $K_a = [B][H^+]/[A]$ the acidity constant. According to this definition, H^+ is not an acid, whereas OH^- is a base. The *Brønsted* theory includes the *Arrhenius-Ostwald* theory and is the most useful for the atmospheric application of diluted aqueous systems with the gas–liquid interaction. According to the *Brønsted* definition, Table 4.8 lists the most abundant acids and bases. Note that the listed species in PM (aerosols) do not occur in ionic form or free acids (e. g. H_2SO_4) but only as salts (e.g. NH_4NO_3 or NH_4HSO_4); O^{2-} denotes oxides (this "ion" does not exist in aqueous solution) and $RCOOH$ organic acids. Hydroperoxides are weak acids because of following equilibriums forming radical ions: $H_2O_2 \leftrightarrow H^+ + HO_2^-$, $HO_2 \leftrightarrow H^+ + O_2^-$ and $ROOH \leftrightarrow H^+ + ROO^-$.

The advantage of the *Brønsted* theory is that the formation of acids and bases occurs via protolysis reactions including the corresponding acids and bases. The *dissociation degree* α describes the position of the equilibrium (4.164) and therefore the *acid strength*:

$$\alpha = \frac{[B]}{[A] + [B]} \tag{4.165}$$

For $\alpha = 0.5$ it follows $[A] = [B]$. For binary solutions (acid + water), the dissociation degree was originally defined to be:

$$\alpha' = \frac{[H^+]}{[A]} = \frac{[B]}{[A]} \tag{4.166}$$

So, it follows from Eqs. (4.165) and (4.166) that:

$$\alpha = \frac{\alpha'}{1 - \alpha'} \tag{4.167}$$

The solvent H_2O itself is amphoteric, that is, reacts both as an acid and as a base. We can specify Eq. 4.164 as follows:

$$H_2O \rightleftharpoons OH^- + H^+ \qquad (4.168)$$

$$H_3O^+ \rightleftharpoons H_2O + H^+ \qquad (4.169)$$

The *acid constant,* often called the *acidity*[16] constant K_a, is defined as $([B][H^+])/[A]$. Including the corresponding base reaction:

$$H_2O + B \leftrightarrow A + OH^- \qquad (4.170)$$

defined as $K_b = [A][OH^-]/[B]$. Using the equilibrium expressions of Eqs. (4.164) and (4.168), the relationships:

$$K_a K_b = K_w \quad \text{or} \quad pK_a + pK_b = pK_w \qquad (4.171)$$

are valid, where the water *ion product* $K_w = [H^+][OH^-] \approx 10^{-14}$ or, written in logarithmic form $pK_w = -\lg K_w \approx 14$. In contrast to the hydrogen ion (H^+), the oxonium ion H_3O^+ ($H^+ \cdot H_2O$ – also called protonized water) is an acid (hydronium is an old but still frequently used term). With $pK_a(H_3O^+) = -1.74$ the ion H_3O^+ is the strongest acid that exists in aqueous solution. That means, acids with $K_a > K_a(H_3O^+) \approx 55$ are totally protolyzed ($\alpha \to 1$) and do not exist as acids in aqueous solutions. K_a values of these "very strong" acids are only inexactly detectable because $K_a \to \infty$ (Table 4.9). With $pK_a(H_2O) = ([H^+][OH^-])/[H_2O] \approx 15.74$ is OH^- the strongest base existing in aqueous solutions; note that $[H_2O] \approx 55$ mol L^{-1} has been used in the equations above. Adopting the general reaction equation for corresponding acids and bases:

$$A_1 + B_2 \leftrightarrow B_1 + A_2 \qquad (4.172)$$

it follows for water that:

$$H_2O + H_2O \leftrightarrow OH^- + H_3O^+ \qquad (4.173)$$

Eq. (4.168) does not represent a chemical reaction since the hydrogen ion (or proton) H^+ does not exist *free* in aqueous solutions. Instead, it "reacts" with H_2O according to Eq. (4.169) to the hydronium ion in a first step and then hydration occurs, e. g. to $H_9O_3^+$ (Wicke et al. 1954). However, according to recommendations given by IUPAC, only the symbol H^+ should be used. The high mobility of the proton in aqueous solutions (contribution to the conductivity) is provided by *tunnel transfer* belong hydrogen bridges within the H_2O clusters. The dissociation of neutral water is very low ($\alpha' = 1.8 \cdot 10^{-9}$) and that is why the water *activity* $\lg[H_2O] = 1.745$ is constant and is included $[H_2O]$ in all equilibrium constants. According to the pK_a value rank, acids could be subdivided (this is not an objective ranking) into strong and weak acids (Table 4.9):

very strong	pK_a	$\leq pK_a(H_3O^+) = -1.74$
strong	$-1.74 < pK_a$	≤ 4.5
weak	$4.5 \leq pK_a$	≤ 9.0
very weak	pK_a	≥ 9.0

---

[16] It is recommended to use only the term *acid* constant to avoid mistakes with the term *acidity,* which is different from the equilibrium constant K_a based on the mass effect law (Chapter 4.1.4).

Table 4.9 pK_a values (in Mol-L units) of different atmospheric acids (298 K); for literature see Möller and Mauersberger (1995) and citation therein; Graedel and Weschler (1981), Jacob et al. (1983).

acid		H^+ + base	pK_a
HCl	$\rightleftarrows$	$H^+ + Cl^-$	−6.23
HO_2NO_2	$\rightleftarrows$	$H^+ + {}^-O_2NO_2$	−5
HNO_3	$\rightleftarrows$	$H^+ + NO_3^-$	−1.34
$H_2O_2^+$	$\rightleftarrows$	$H^+ + HO_2$	−1
$HOCH_2SO_3H$	$\rightleftarrows$	$H^+ + HOCH_2SO_3^-$	<0
HSO_4^-	$\rightleftarrows$	$H^+ + SO_4^{2-}$	1.92
$SO_2\,(+H_2O)$	$\rightleftarrows$	$H^+ + HSO_3^-$	1.76
$Fe(H_2O)_6^{3+}$	$\rightleftarrows$	$H^+ + Fe(H_2O)_5OH^{2+}$	2.2
HNO_2	$\rightleftarrows$	$H^+ + NO_2^-$	3.3
$Fe(H_2O)_5OH^{2+}$	$\rightleftarrows$	$H^+ + Fe(H_2O)_4(OH)^+$	3.5
$CO_2\,(+H_2O)$	$\rightleftarrows$	$H^+ + HCO_3^-$	3.55
HCOOH	$\rightleftarrows$	$H^+ + HCOO^-$	3.74
CH_3COOH	$\rightleftarrows$	$H^+ + CH_3COO^-$	4.75
HO_2	$\rightleftarrows$	$H^+ + O_2^-$	4.7
HSO_3^-	$\rightleftarrows$	$H^+ + SO_3^{2-}$	7.2
H_2S	$\rightleftarrows$	$H^+ + HS^-$	7.2
NH_4^+	$\rightleftarrows$	$H^+ + NH_3$	9.23
HCO_3^-	$\rightleftarrows$	$H^+ + CO_3^{2-}$	10.3
H_2O_2	$\rightleftarrows$	$H^+ + HO_2^-$	11.7
$HOCH_2SO_3^-$	$\rightleftarrows$	$H^+ + {}^-OCH_2SO_3^-$	11.7
HS^-	$\rightleftarrows$	$H^+ + S^{2-}$	12.9

The definition of acids can extend according to the solvent theory as follows (this definition is symmetric concerning the acid-base relationship in contrast to the *Brønsted* theory): acids/bases increase/decrease H^+ or increase/decrease OH^-.

$$A \leftrightarrow H^+ + B \qquad \text{(donator acid)} \qquad (4.174)$$

$$A + OH^- \leftrightarrow B \qquad \text{(acceptor acid)} \qquad (4.175)$$

As we have seen, $[H^+]$ is an important chemical quantity in the diluted aqueous phase; Sørensen (1909) defined the pH to be:

$$pH = -\lg[H^+] \qquad (4.176)$$

Today, the pH as an *acidity measure* is defined as pH ($\equiv$ paH) $= -\lg a_{H^+}$. Single activities, however, are not measurable. Therefore, using mean activities $a_\pm$, a conventional pH scale is defined based on fixed buffer solutions with known pH values. Another acidity measure is the *Hammett* acidity function H_0 (Hammett 1940):

$$H_0 = pK_a + \lg[B] - \lg[A] \qquad (4.177)$$

where $H_0 = -\lg[H^+]$. In the following, no difference is made between various pH definitions based on the assumption that in most cases in atmospheric chemistry $a_{H^+} = a_\pm = [H^+]$ and $\gamma = 1$ (activity coefficient). A generalized acid−base equation can be derived (Eqs. 4.162, 4.169 and 4.170):

$$A + H_2O \leftrightarrow H_3O^+ + B \tag{4.178}$$

Therefore, it follows that:

$$pK_a = pH - \lg([B]/[A]) \quad \text{or} \quad pH = pK_a - \lg([A]/[B]) \tag{4.179}$$

which is also called *Henderson-Hasselbalch* equation and describes the derivation of pH as a measure of acidity (using pK_a) in biological and chemical systems. The equation is also useful for estimating the pH of a buffer solution and finding the equilibrium pH in acid-base reactions. The ratio [A]/[B] or [acid]/[conjugate base] is termed in the buffer ratio. Weak acids such as almost all organic acids can play the role of buffering strong acids in nature. Solutions containing acid and its salt (for example, acetic acid and acetate) convert H^+ from strong acids (nitric and sulfuric acid) and from CH_3COO^- into undissociated CH_3COOH without changing the pH (in certain limits).

The (mean) hydrogen ion activity in terms of pH can be estimated directly using the hydrogen electrode (against a reference electrode) and measuring the electric potential. Based on the *Nernst* equation (see Eq. 4.153), $E_H = RT \ln a_\pm/F$ it follows that (F is the *Faraday* constant):

$$pH = \frac{F(E - E_B)}{2.303\, RT} \tag{4.180}$$

where E_B is the reference electrode potential ($E = E_H - E_B$); for details of pH measurements see Schwabe (1976) and Galster (1991). Eq. (4.180) does not reflect the diffusion potentials that make pH measurements more complicated (Schwabe 1959). The accuracy of pH estimations in atmospheric water samples in the pH range 4−5 is no better than ± 0.1 pH units despite the *quality assurance* of the pH measurement against standards to be ± 0.02 in this range of pH.

4.2.5.3 Atmospheric acidity

In atmospheric waters (cloud, fog and raindroplets), the following 10 main ions must be taken into account: SO_4^{2-}, NO_3^-, Cl^-, HCO_3^-, NH_4^+, Ca^{2+}, Mg^{2+}, Na^+, K^+ and H^+ (Granat 1972, Möller and Zierath 1986, Möller and Horváth 1989). Of minor importance are HSO_4^-, HSO_3^-, SO_3^{2-}, NO_2^-, CO_3^{2-}, F^- and OH^-. Because of the *electroneutrality condition*, the following:

$$\sum_i [\text{cation}_i] = \sum_i [\text{anion}_j] \tag{4.181}$$

must be valid[17]. In this and the following equations, only the equivalent concentrations will be used, otherwise the stoichiometric coefficients must be considered, for example, 1 eq $SO_4^{2-} \equiv 0.5$ mol SO_4^{2-}. The relationship (4.181) must not be confused

---

[17] This equation can also be used as a quality control measure for the analytical procedure in atmospheric water: if the deviation of the quotient $\Sigma[\text{anions}]/\Sigma[\text{cations}]$ is $\geq 20\%$ from unit (1), the analysis must be repeated or the samples should be rejected.

with the definition of an acid $[H^+] = [A] - [B]$, where A is acids and B is bases. It follows from (4.181):

$$[H^+] = \sum_i [anion_i] - \sum_j [(cation\ without\ H^+)_j] = [A] - [B] \qquad (4.182)$$

Werner Stumm (see, for example, Stumm and Morgan 1981, 1995, Stumm 1987, Sigg and Stumm 1996, Zobrist 1987) introduced the acidity as a *base neutralizing capacity* (BNC), corresponding to the equivalent of all acids within the solution, titrated to a given reference point:

$$BNC \equiv [Acy] = [A] + [H^+] - [OH^-] \qquad (4.183)$$

and a corresponding alkalinity as an *acid neutralizing capacity* (ANC):

$$ANC = [Alk] = [B] + [OH^-] - [H^+] \qquad (4.184)$$

Eqs. (4.183) and (4.184) are, therefore, not based on the electroneutrality equation (4.181) because they also include weak acids and bases which are not protolyzed in solution at the given pH (that means in non-ionic form), but contribute to *acid* or *base titration*. The first attempt to determine airborne acidity was done by careful titration inserting microliter quantities of a NaOH solution (Brosset 1976) using *Gran's* titration method (Gran 1950). The reference points, however, are not objective criteria[18]. Zobrist (1987) extended this definition to the general equations:

$$[Acy]_{total} = [H^+] + [A] - [B] \approx [Acy]_{H^+}$$
$$= [H^+] + [A]_{strong} - [B]_{strong} \qquad (4.185)$$

where Acy_{total} denotes the *total acidity* (including weak acids) and Acy_{H^+} the *free acidity*.

These definitions might be useful for bulk water bodies such as rivers and lakes. Atmospheric water, however, as mentioned before, consists of single droplets that are mixed up while sampling. Moreover, individual samples collected in the field will furthermore mix, e. g. when getting a time-averaged sample. All that means that individual droplets or samples with different acidity/alkalinity have been mixed, resulting in acid-base reactions and a shifting liquid-gas equilibrium with a new reference point with an averaged "final" acidity. Therefore, it is important to define acidity and alkalinity as conservative parameters, i. e. they are independent of pressure, temperature and ionic strength as well as CO_2 gas exchange. The acidity is then defined as (Liljestrand 1985, Sigg and Stumm 1996):

$$[Acy] = [H^+] - [HCO_3^-] - [CO_3^{2-}] - [OH^-] = -[Alk] \qquad (4.186)$$

and, using Eq. (4.178):

$$[Acy] = [A^*] - [B] \qquad (4.187)$$

---

[18] The German DIN 38406 (1979) defined the acid capacity (in German *Säurekapazität*) to be equivalent to the hydrochloric acid consumption of pH = 4.3 and the sum of all carbonaceous bonded cations: $[Acy]_{H^+} = [HCO_3^-] + 2[CO_3^{2-}] + [OH^-] - [H^+]$.

where A* = anions without HCO_3^-, CO_3^{2-} and OH^-. It follows, using the equilibrium expressions for the ions (see Chapter 2.8.3.2 concerning carbonate protolysis chemistry, especially Eqs. 2.127–2.130):

$$[Acy] = [H^+] - (K_{ap}H_{CO_2}[CO_2(g)] + K_w)[H^+]^{-1}$$
$$- K_{ap}K_2H_{CO_2}[CO_2(g)][H^+]^{-2} \tag{4.188}$$

Adopting standard conditions (340 ppm CO_2, 298 K) it follows that:

$$[Acy] \approx [H^+] - 4.42 \cdot 10^{-12}[H^+]^{-1} - 2.2 \cdot 10^{-22}[H^+]^{-2} \tag{4.189}$$

Eq. (4.188) is only valid without changing partial pressure (p_0), i.e. neglecting reservoir distribution between the gas and aqueous phases. The whole mass of the gas (expressed as p_0), however, is distributed between the gas phase (expressed as equilibrium partial pressure $p_{eq} = p_0 - p' = n(g)RT/V(g)$) and the aqueous phase, expressed as $n(aq) = V(aq)H_{eff}RT$, where $p' = p_{eq}H_{eff}RT$ and:

$$H(CO_2)_{eff} = \frac{[CO_2(aq)] + [H_2CO_3] + [HCO_3^-] + [CO_3^{2-}]}{[CO_2(g)]} \tag{4.190}$$

The reservoir distribution ratio ς (mass in aqueous phase/mass in gas phase) is given by:

$$\varsigma = \frac{n(aq)}{n(g)} = H_{eff}V_rRT \tag{4.191}$$

where V_r denotes the volume ratio between the phases (volume aqueous phase/volume gas phase): $V(aq)/V(g) \approx V(aq)/air\text{-}volume = LWC$. In cases if not $\varsigma \ll 1$, in Eq. (4.188) the gas phase concentration must be expressed by the equilibrium gas phase concentration:

$$[CO_2(g)] = (1 + \varsigma)[CO_2(g)]_{eq} \tag{4.192}$$

Fig. 4.5 shows that under "normal" atmospheric conditions with a water pH between 4 and 5.7 almost all HNO_2 and NH_3 are dissolved into droplets, whereas SO_2 and CO_2 remain in the gas phase. Within the pH range 6–8, SO_2 is effectively scavenged but CO_2 is measurably transferred from the gas to the aqueous phase only above a pH of about 8. In Fig. 4.6, the "acidity" parameters $[H^+]$, $[OH^-]$, [Acy] and [Alk] are calculated based on Eqs. (4.186) and (4.188), and are then expressed as a logarithm with a dependency on pH. Fig. 4.6 represents some characteristic points:

pH	[Acy]	[Alk]
< 5.68	$\approx [H^+]$	< 0
= 5.68	$\to 0$	$\to 0$
> 5.68 (but < 10.3)	< 0	$\approx [HCO_3^-]$

From this table we cannot conclude that pH = 5.68 (Acy $\to$ 0) represents the reference point for "neutral" atmospheric water, i.e. water with a pH < 5.68 can be characterized as "acid". This reference point is valid only for the binary system CO_2/H_2O. To quantify changes in the amounts of acids or bases that cause acidifi-

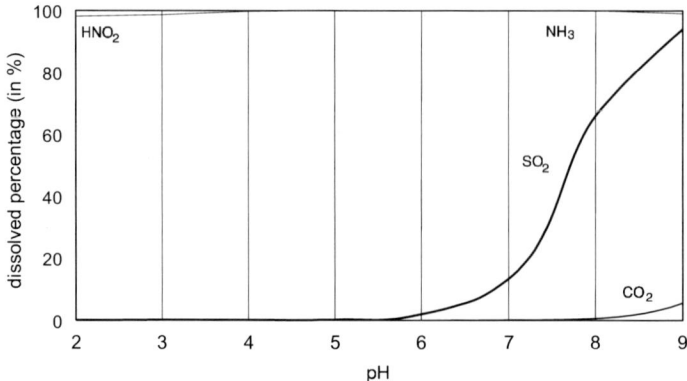

Fig. 4.5 Reservoir distribution of gases between gas and liquid phase.

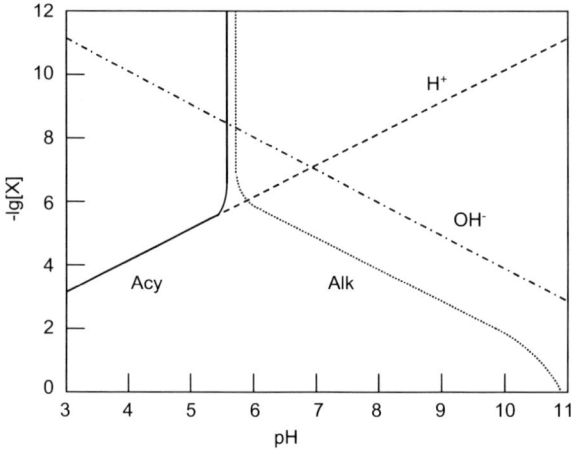

Fig. 4.6 pH dependent H_2O protolysis, acidity and alkalinity in a gas-aqueous atmospheric system.

cation, we also have to compare one state with another reference state. In natural systems, the most appropriate reference state is that which relates to the system itself. Other reference states have no meaning (physically or chemically), but might be useful or even necessary from a computational point of view (Odén 1976). Other atmospheric ions originating from the natural emission of gaseous species and PM lead to a natural acidity $[Acy]_{nat}$ or alkalinity, which varies in time and space (e. g. discussions in Möller and Zierath 1986, Möller and Horváth 1989). Therefore, the definition of *acid* rain or clouds gains by:

$$[Acy]_{measured} = [Acy]_{man-made} + [Acy]_{natural} \geq [Acy]_{natural}, \qquad (4.193)$$

where subscript "measured" denotes the experimental determined acidity in samples. For pH > 10.3, CO_3^{2-} must be taken into the budget. However, this alkaline range is unlikely in clouds and rain; for pH = 10.7 it follows that [Alk] = 1 mol

L^{-1}, i.e. in solution must be 1 mol L^{-1} cations (e.g. Ca^{2+}). Such high-concentrated solutions occur only in transition states (condensation/evaporation). Neglecting the CO_3^{2-} contribution to alkalinity, Eq. (4.188) transforms into (β constant):

$$[H^+] = 0.5[Acy] + \sqrt{0.25[Acy]^2 + \beta} \qquad (4.194)$$

There is another complication. The samples of collected cloud and/or rainwater are most probably not in equilibrium with atmospheric CO_2. In this case, Eq. (4.186) is still valid for the acidity calculation; however, $[HCO_3^-]$ cannot substitute using *Henry*'s law. The only way to solve this problem is an analytical estimation of $[HCO_3^-]$, e.g. using titration or electroanalytical methods. Note that traditional titration methods result in the figure ($[H_2CO_3/CO_2$-aq$] + [HCO_3^-]$), where $[HCO_3^-]$ must be calculated using the equilibrium equation. Assuming a total carbonaceous concentration of 100 µeq L^{-1} at pH = 4.5, it follows that $[HCO_3^-]$ = 90 and $[H_2CO_3/CO_2$-aq$]$ = 10 µeq L^{-1}.

CO_2 plays a special role in the formation of atmospheric acidity because of its high and constant concentration. Fig. 2.93 shows that an important pathway in *alkalinity* (carbonate) formation goes via the CCN (nucleation and droplet formation) as well as aerosol scavenging. The last process, however, is of minor importance for clouds but contributes significantly to sub-cloud scavenging into falling raindrops. There is a variety of PM acting as a ANC including flue ash, soil dust and industrial dust. Carbonate particles could lead to an initial high alkaline aqueous phase that is an efficient absorber for gaseous acids and acidic precursors, especially SO_2. The solution becomes oversaturated with CO_2(aq) and, consequently, CO_2 desorption occurs from the droplet. The neutralization stoichiometry follows the overall reaction:

$$CO_3^{2-}(solid) + 2\,H^+ \rightarrow CO_2\,(g) + H_2O \qquad (4.195)$$

For rainwater in equilibrium with atmospheric CO_2, the dissolved $[HCO_3^-]$ amounts to only between 0.2 and 0.5 µmol L^{-1} in the pH range 4.0 to 4.5. However, in rainwater at different stations in eastern Germany a background concentration of 98 µmol L^{-1} and for polluted areas 130−260 µmol L^{-1} have been found in the past (Zierath 1981). These several orders of magnitude higher bicarbonate concentrations suggest the high importance of sub-cloud scavenging on carbonaceous PM (in the case of its abundance such as in polluted areas) and − this is a very important conclusion − the *non-equilibrium state* of collected rainwater with atmospheric CO_2. By contrast, the possible high contribution of bicarbonate to the ionic balance in rainwater samples implies "errors" in the quality check between measured conductivity (where HCO_3^- is included) and the analyzed ions (where HCO_3^- is excluded). Moreover, all acidity conceptions based on equilibrium become questionable. Other important neutralizing species are oxides, which are frequently found in flue ash and form hydroxyl ions at a first step:

$$CaO\,(solid) + H_2O \rightarrow Ca^{2+} + 2\,OH^- \qquad (4.196)$$

The only gaseous neutralizing species is ammonia:

$$NH_3\,(g) + H^+ \rightarrow NH_4^+ \qquad (4.197)$$

Only under limited atmospheric conditions the inverse process, that is ammonia degassing, could be possible, e. g. while precipitating remote cloud over areas, polluted by "alkaline" aerosols (Möller and Mauersberger 1990a) via:

$$CO_3^{2-} + NH_4^1 \rightarrow HCO_3^- + NH_3 \tag{4.198}$$

As we have seen, acidity is a chemical quantity defined only in exclusively aqueous solutions in the climate system (similar to the redox potential). However, it makes sense to introduce a *total acidity* for the whole volume of air including the droplet, gaseous and particulate phases. Fachini and Fuzzi (1993) defined atmospheric acidity in the sense of a multiphase concept as the BNC per unit of volume of the atmospheric system. They stated that "in analogy with solution chemistry the addition of acidic or basic components to an atmospheric system can be viewed as a titration of the global system". This basic idea is acceptable because it suggests a potential neutralization of a reference reservoir (for example, soil or lake) and the *acidification* (or basification) of this reference reservoir. This idea is based on the acidity concept first presented by Stumm and Morgan (1981), see Eq. (4.182). Fachini and Fuzzi (1993) applied this concept to the *Po Valley Fog Experiment* and neglected the aerosol reservoir, probably based on the assumption that during fog episodes all soluble PM is scavenged through fog droplets. This assumption is credible; our own experiments in clouds at Mt. Brocken (Harz Mt., Germany) suggest that the interstitial PM of acidifying species, listed at the beginning of this Chapter 4.2.5.3, is negligible. They describe then atmospheric acidity by:

$$[Acy]_{atm} = [A]_{fog} - [B]_{fog} + [\text{organ. acids (g)}] + [HNO_3 \text{ (g)}]$$
$$+ [SO_2 \text{ (g)}] - [NH_3 \text{ (g)}] \tag{4.199}$$

where A and B are the sum of anions and cations dissolved in the fog water, respectively. In Eq. (4.199), it is hard to understand why gaseous HNO_3 and NH_3 are considered, since $\varsigma \gg 1$, i. e. these species can be neglected in the gas phase in the pH range 2–8 (Fig. 4.5 and Eq. 4.191). By contrast, SO_2 dissolves only to a small degree. Why does gaseous SO_2 contribute to atmospheric acidity at all (remember, SO_2 is not an acid but the anhydride of sulfurous acid)? SO_2 can contribute to soil acidity after dry deposition (Chapter 4.4.1). An *overall titration* of SO_2 per unit of air volume makes no sense: this is an artificial idea based on laboratory practice with no reason for atmospheric application. The time- and space-dependent fate of atmospheric SO_2 (dry deposition, dissolution, oxidation, wet deposition), quantitatively parameterized, can lead only to acidity contributors such as sulfite and sulfate. Only the termination, SO_2 being an *acidifying species*, is correct. Möller (1999b) introduced the atmospheric *acidifying capacity* as the sum of the *potential* acidity in gas $[Acy]_g$ and aerosols $[Acy]_a$ as well as the acidity in aqueous phase $[Acy]_{aq}$ as follows:

$$[Acy]_{atm} = [Acy]_g + [Acy]_a + [Acy]_{aq} \tag{4.200}$$

Whereas $[Acy]_{aq}$ is defined according to Eq. (4.182), the *potential aerosol acidity* can be defined similar to Eq. (4.187):

$$[Acy]_a = [A^*] - [B] \approx [SO_4^{2-}] + [NO_3^-] + [Cl^-]$$
$$- [NH_4^+] - [Ca^{2+}] - [Mg^{2+}] - [K^+] \tag{4.201}$$

The ions listed in Eq. (4.201) represent the main constituents in cloud and rainwater and thereby in the CCN. This list can be extended by minor species such as carbonate, bicarbonate, sulfite, nitrite, iron and others. Based on the electroneutrality condition, the soluble PM (ions listed in Eq. 4.201) must be balanced with H^+, HCO_3^-, CO_3^{2-} and OH^-.

The definition of *potential gaseous acidity* is more complicated. Gas molecules form acidity only after dissolution in cloud, fog and raindrops and subsequent protolysis reactions. The degree of dissolution − or in other words − phase partitioning depends on the initial droplet acidity (i. e. aerosol acidity) and from the gas phase composition of acidifying gases themselves. Therefore, it does not make sense to only add gaseous acids according to Table 4.8. With a good approximation we can assume that strong gaseous acids and bases are completely transferred to the aqueous phase and dissociated therein ($\alpha \to 1$, $\varsigma \to 1$, see Eqs. 4.165 and 4.191), whereas weak acids (e. g. carbonic acids) contribute less to acidity. However, in the case of missing strong acids they contribute significantly to acidity, e. g. Galloway et al. 1982). Another problem is *anhydrides*, gases that form acids after reaction with water (SO_2, SO_3 and N_2O_5)[19] only. Other gases that are not direct anhydrides (e. g. NO_2) can also produce acids in reaction with water. Taking into account only SO_2 besides strong gaseous acids and NH_3, the following equation represents the potential gaseous acidity, where $\varepsilon = 1/(1 + \varsigma)$:

$$[Acy]_g \approx [HNO_3] + [HNO_2] + [HCl] + \varepsilon[SO_2] - [NH_3] \qquad (4.202)$$

Generally, the term *acid deposition* could be better used to describe the acidity phase transfer (e. g. from the atmosphere to the biosphere).

4.2.5.4 pH averaging

As we have seen from Fig. 4.6, we can adopt $[Acy] = [H^+] = 10^{-pH}$ for samples with pH < 5.7. Note that only the pH is (Eq. 4.180) the directly measurable value. The determination of [Acy] has been tried by titration against different reference points (Zobrist 1987); however, has the problem of separation between *free* and *total* acidity. The problem of differentiation between $[H^+]$ and [Acy] arises if samples occur with pH > 5.7 (Fig. 4.6). After mixing two samples, one with pH < 5.7 and another with pH > 5.7, the *physically meaningful* averaged $[H^+]$ does not follow from $[H^+]_i$ based on precipitation-weighted means because H^+ is not a conservative value. Let's first consider the procedure for the *physically meaningful* averaging of concentrations from conservative species (such as Ca^{2+}, SO_4^{2-}). The averaged precipitation composition of annual mean results from

$$\overline{c_{rain}} = \frac{\sum\limits_{i} R_i c_i}{R} \qquad (4.203)$$

---

[19] CO_2 is also an anhydride, however, because of its constant concentration the acidity contribution only depends on solution pH.

where R_i denotes the precipitation amount for samples i, and R denotes the annual precipitation amount (rainfall rate), i. e. $R = \Sigma R_i$. Similarly, we derive for cloud water an expression for LWC_i, which is the mean LWC during collecting an individual cloud water sample:

$$\overline{c_{cloud}} = \frac{\sum_i c_i LWC_i}{\sum_i LWC_i} \tag{4.204}$$

These equations can be applied also to [Acy] taking into account standard conditions (Eq. 4.189) where const $= 4.42 \cdot 10^{-12}$ mol^2 L^{-2}:

$$[Acy]_i \approx 10^{-pH_i} - const \cdot 10^{pH_i} \tag{4.205}$$

Finally, we get an averaged pH for rainwater (for cloud water R_i has to be replaced by LWC_i and R by ΣLWC_i):

$$\overline{pH} = -\lg\left(-0.5\frac{\sum R_i [Acy]_i}{R} + \sqrt{0.25\left(\frac{\sum R_i [Acy]_i}{R}\right)^2 + const}\right) \tag{4.206}$$

It is important to view this relatively complicated formula as an estimation of averaged pH and/or [H$^+$]. This average represents the resulting acidity of the "physically mixed" samples taking into account the neutralization of acids and bases. The following sequence summarizes the steps:

- pH measurement of sample i, transformation into [H$^+$]$_i$ according to Eq. (4.176),
- calculation of [Acy]$_i$ according to Eq. (4.205);
- calculation of $\overline{[Acy]}$ according to Eqs. (4.203) or (4.204);
- recalculation of $\overline{[H^+]}$ based on Eq. (4.194); and
- transformation into $\overline{pH}$ (based on Eq. 4.176).

4.3 Multiphase processes

The climate system is simplified as the water-soil-air system, representing the physical states of liquid, solid and gaseous. We have already discussed the air (or atmosphere) as a multiphase system and despite the volume-based dominance of gases, the solid matter (APs) and liquid matter (cloud droplets) play a crucial role in weather formation and thereby climate phenomena. Similarly, the hydrosphere (oceans) is considered a multiphase system where solid carbonates play a key role in the biogeochemical status. It is remarkable that in each reservoir the minor phases are obviously the driving forces (water in air, solids in water, air/gases and water in soils). Therefore, studying the interfacial processes (solid-gas, solid-liquid and gas-liquid interfaces concerning the mass transfer and chemical reaction) is of high importance for understanding the climate system. This short chapter can only present a brief overview; for more details the reader is referred to special literature:

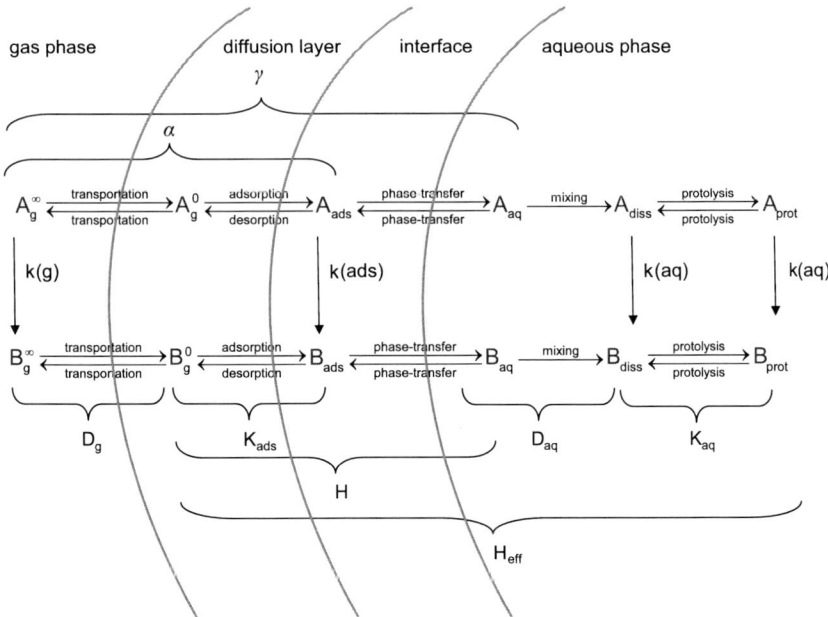

Fig. 4.7 Scheme of the processes while phase transferring. Indices: ∞ far from interface, 0 beginning of diffusion layer, ads outer surface, aq inner surface and aqueous phase, diss – dissolved, prot – protolyzed form. k reactions rate coefficients in gas phase (g), at surface (*ads*) and in aqueous phase (*aq*). D_g diffusion coefficient gas phase, K_{ads} sorption equilibrium coefficient, D_{aq} diffusion coefficient aqueous phase, K_{aq} equilibrium coefficient of protolysis, H *Henry* coeffient, H_{eff} effective *Henry* coefficient. α accommodation coefficient, γ uptake coefficient.

aerosols, clouds and chemistry (Jaeschke 1986, Fuzzi and Wagenbach 1997, Pruppacher and Klett 1997, Guderian 2000, Hobbs 2000, McFiggans et al. 2005a, Barnes and Kharytonov 2008) and multiphase transport and chemistry (Helmig and Schulz 1997, Faghri and Zhang 2006). Whereas Fig. 4.7 shows the processes on a micro-scale, Figs. 4.8 and 4.9 represent the whole system on a macro-scale, to which we turn first to discuss the limits of our understanding of the multiphase system.

In air, permanent solid particles (atmospheric aerosols) occur either from primary sources out of the atmosphere or from *gas-to-particle conversion* within the atmosphere. All trace matter shows a high variability in concentration[20] because of chemical and microphysical processes. Liquid particles (cloud, fog and rain droplets), however, are not permanent in air and form and exist only under specific physicochemical conditions (the presence of condensation nuclei and water vapor saturation). The transition from molecules to droplets comprises many steps:

1. homogeneous nucleation of molecules: formation of nuclei and growth ("new particle formation");
2. activation of potential condensation nuclei through water vapor adsorption;

---

[20] Excluding trace species with a very long residence time such as some halogenated hydrocarbons.

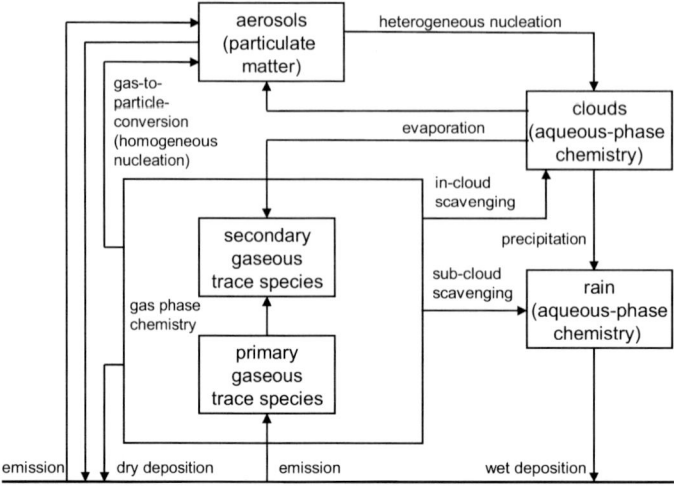

Fig. 4.8 Scheme of the atmospheric multiphase system.

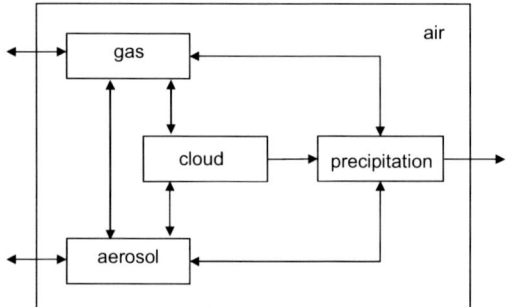

Fig. 4.9 The evolution scheme: From gas to aerosol, cloud and precipitation; ⟶ non-reversible pathway, ⟷ pathways in both direction, but not mandatory equilibrium.

3. heterogeneous nucleation: formation of droplets through water vapor condensation onto condensation nuclei;
4. adsorption of molecules onto particles and surface chemical reactions (heterogeneous chemistry);
5. absorption of molecules by droplets and droplet phase chemistry (aqueous phase chemistry;
6. cloud droplet growth and formation of precipitation elements (microphysical and dynamic processes);
7. desorption of molecules from particles and droplets;
8. evaporation of droplets: formation of solid residual particles and back transfer of dissolved gases; and
9. precipitation and scavenging of gases and particles in the sub-cloud layer.

The characterization of the size-depending chemical composition of particles and droplets is an important issue of atmospheric chemistry. *In-cloud scavenging* (often

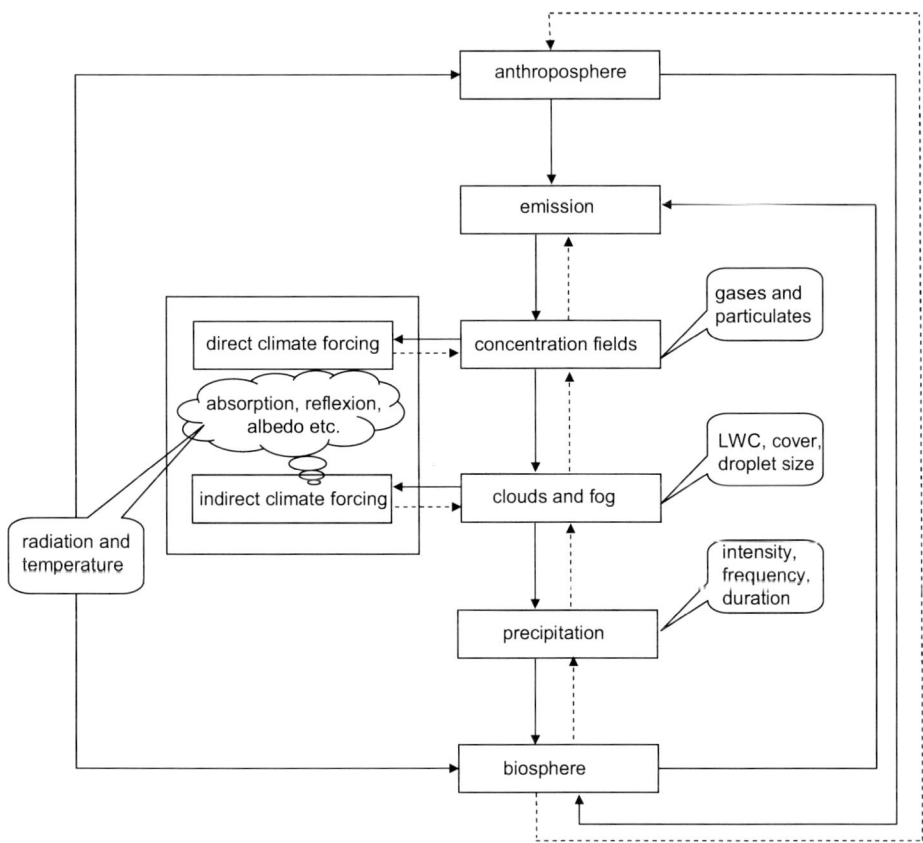

Fig. 4.10 The climate impact system. ⟶ direct impacts, ---➤ dotted line: feedbacks and indirect impacts.

called rain-out in the past) comprises the processes 3, 4 and 5, which all accumulate material in cloud droplets, either turned back to air (processes 7 and 8) or removed via precipitation (process 9). Together with this *sub-cloud scavenging* (often called washout in the past), the chemical composition of wet deposition (precipitation) is determined. *Precipitation chemistry* in a more narrow sense deals with the chemical composition of rain and snow in a climatological sense (see Chapter 3.4.2), i.e. studying the space- and time-dependency (location, season) and meteorological influences (air masses, rainfall characteristics). *Cloud chemistry,* however, cannot be reduced on aqueous phase chemistry (droplet chemistry) because it describes the chemistry in clouds, i.e. the chemical interactions in the multiphase system containing APs, cloud droplets and interstitial gases (Fig. 4.10), see Chapter 3.4.3.

4.3.1 Aerosols, clouds and precipitation: The climate multiphase system

Interactions among aerosols, clouds and precipitation are critical in shaping the climate system. Aerosols, cloud and precipitation are intrinsically linked (Figs. 4.8

and 4.9). APs are the condensation and freezing nuclei on which cloud particles form (heterogeneous nucleation process). APs are either directly emitted or produced from gaseous precursors via homogeneous nucleation (gas-to-particle conversion). Clouds occasionally produce precipitation, which removes trace species via in-cloud and sub-cloud scavenging back to the earth's surface (wet deposition). In most cases, however, clouds evaporate and turn chemical species back to air in the form of aged APs and secondary gaseous matter. This cloud processing (nucleation-evaporation cycle) is important for the transport, redistribution and conversion of pollutants. Cloud processing ultimately leads to particle growth and generally to an increase of water-solubility of inorganic particles. Aqueous phase droplet chemistry, however, can also produce organic polymers as an important part of secondary organic aerosols (SOA). Cloud processing has long been identified as the main source of climate active sulfate aerosols.

We have described the crucial role played by cloud and precipitation (Chapter 2.5.2) in the earth's energy budget and water cycle (Fig. 4.10). Despite many important achievements, clouds and aerosols remain the largest source of uncertainties in the two most important climate change quantities: radiative forcing and climate sensitivity (IPCC 2007, 2013). Whereas the role of greenhouse gases on the absorption of terrestrial radiation is well understood and the interaction between APs and radiation has been well described (direct climate forcing), the dominant uncertainty among radiative forcing of climate change is the impact of aerosols on clouds (indirect climate forcing). Fig. 4.10 illustrates the direct cause-impact relationships and feedbacks. Changing radiation, temperature and precipitation are essential parameters for plant growth and, therefore, the biosphere's emission capacity and biogeochemical cycling. The climate system is anthropogenically influenced through atmospheric emissions and land use changes. The concentration fields of gases and PM are a result of emission patterns, atmospheric chemistry and transport processes. All these processes depend on climate. Qualitatively, the feedbacks are well known, and for small-scale relationships also quantitatively described such as temperature influence on biogenic emissions, parameters determining wind-blown mineral dust and condensation nuclei activation. But on a large-scale, when atmospheric (and oceanic) circulation and the whole multiphase (and multicomponent) chemistry must be considered as an interlinked system, cloud formation and precipitation as essential climate variables are only poorly quantified. Moreover, the feedbacks of aerosols, clouds, precipitation and climate as an interlinked system, and its sensitivity to perturbations, are not well understood (Andreae et al. 2009, Heintzenberg and Charlson 2009).

There are simple and complex feedbacks or relationships, for example:

– increasing humidity → increasing OH concentration
– increasing T → increasing stability → decreasing wind speed → decreasing PBL → increasing concentrations
– decreasing SO_2 (and sulfate, resp.) → decreasing negative forcing → increasing T
– decreasing dust pollution (PM_{10}) → increasing nanoparticulates → decreasing precipitation

A main reason for the difficulties in understanding the aerosol-cloud-precipitation system is the large range of spatial and temporal scales on which the involved processes act (Fig. 4.11). Important microphysical processes such as the formation

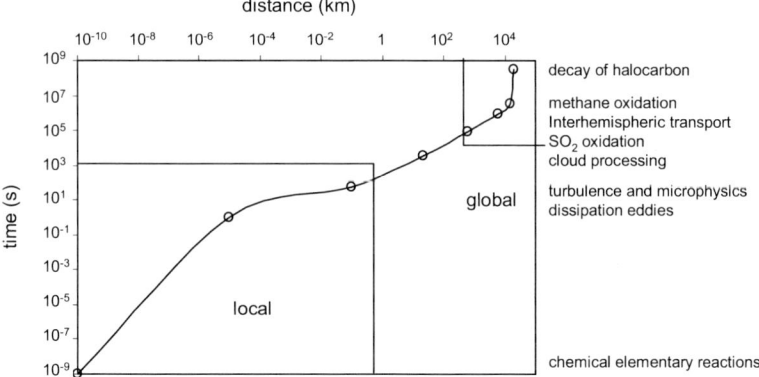

Fig. 4.11 Scheme of different processes having a characteristic time and hence spatial characteristics (from local to global).

of particles through homogeneous nucleation are in the range of nanometers and nanoseconds and of cloud droplet formation in the ranges of micrometers and seconds. Cloud droplets exist from minutes to hours and the cloud system from hours to days. Precipitation is within the range of millimeters (drops interaction) and a few hundred of meters (sub-cloud scavenging) and minutes (falling raindrops). The organization of cloud systems, however, is a large-scale dynamic taking place on spatial scales of the order of hundreds to thousands of kilometers and temporal scales between days and weeks. At the end of the upper scale are long-lived pollutants in the stratosphere and global climate feedbacks between tens and hundreds of years.

Disciplinary research on many individual aspects of the climate system has advanced substantially. But what is needed to study the system as a whole with attention to all the pertinent scales? Another permanent question is to what extent we are able to understand the whole system. It is more likely to believe that a full understanding remains the goal in science and philosophy. Our present limited understanding of the climate and aerosol−cloud−precipitation system, however, is sufficient to establish an urgent challenge for climate control measures as suggested in Chapter 2.8.4.3.

4.3.2 Gas-liquid equilibrium (*Henry* equilibrium)

Each gas in contact with a liquid develops an equilibrium with the dissolved component as found first by *William Henry* in 1803 (Henry 1803):

$$A(g) \rightleftarrows A(a\,q). \tag{4.207}$$

The equilibrium constant is named *Henry*'s law constant H (also reciprocal ratios are in use):

$$H_c = \frac{[A(aq)]}{[A(g)]} \quad \text{or} \quad H_p = \frac{[A(aq)]}{p_A} \tag{4.208}$$

The temperature dependency is described by the *van't Hoff* equation (4.94) where the reaction enthalpy must be replaced by the enthalpy of solution $\Delta_{sol}H$. The derivation of (4.209) follows from the equality of the chemical potentials in the gas and aqueous phases in equilibrium, $\mu^\ominus(g) + RT \ln p_A = \mu^\ominus(aq) + RT \ln [A]$ from which $[A] = p_A \cdot$ constant follows. In close relation to the *Henry's* law constant are the *Bunsen* absorption coefficient α (the volume of gas absorbed by one volume of water at a pressure of 1 atmosphere) and *Ostwald's* solubility β (the quantity of solvent needed to dissolve a quantity of gas at a given temperature and pressure) (dimensions of $[A]$ in mol L^{-1} and of p in Pa):

$$\frac{[A(aq)]}{[A(g)]} = H_c = \alpha RT = \beta = \frac{H_p p_A}{[A(g)]} = H_p RT \tag{4.209}$$

Hence, the *Bunsen* coefficient is identical to the partial pressure-related *Henry* coefficient H_p. Another quantity used is the gas solubility S (the mass of a gas dissolved in 100 g of pure water under standard conditions, that is the partial pressure of the gas and the water saturation pressure is equal to 1 atm or 101.325 Pa). Approximately (without considering the density of the solution) it follows (M molar mass of the gas, H_p in mol L^{-1} Pa^{-1}) that:

$$S = M \cdot H_p 10.1325 \text{ (exactly in g of solute per 100 g of water)} \tag{4.210}$$

It is important to note that few simplifications have been considered in application of *Henry's* law (sometime also called *Henry-Dalton's* law): the validity of the ideal gas equation, ideal diluted solution and that the partial molar volume of the dissolve gas is negligible compared with that in the gas phase. However, the range of its validity in the climate system is appreciable. It is clear that during heterogeneous nucleation (CCN to cloud droplet formation) *Henry's* law is not valid. Absolute values of solubility cannot be found from thermodynamic considerations. Nevertheless, general rules are valid for all gases:

- decreasing solubility with increasing temperature;
- decreasing solubility with increasing salinity of waters (same ratio for all gases); and
- increasing volume of aqueous solution with gas dissolution.

The equilibrium (4.207) only describes the physical dissolved gas species (Fig. 4.7) without considering the subsequent protolysis equilibrium as discussed for the example of CO_2 absorption in Chapter 2.8.3.2. In Eq. (2.119), we introduced an *apparent Henry* coefficient, where the dissolved matter comprises the anhydride (for example CO_2 or SO_2) and the acid (H_2CO_3, H_2SO_3). The acid can dissociate according to Eq. (4.174) and thereby increases the total solubility of gas A, as described by the *effective Henry* coefficient H_{eff}:

$$A + H_2O \rightleftharpoons H_2AO \underset{-H^+}{\rightleftharpoons} HAO^- \underset{-H^+}{\rightleftharpoons} AO^{2-} \tag{4.211}$$

$$H_{eff} = \frac{[A(aq)] + [H_2AO] + [HAO^-] + [AO^{2-}]}{[A(g)]} \tag{4.212}$$

Excluded from the total solubility or effective *Henry*'s law are subsequent reactions, for example, the oxidation of dissolved SO_2 into sulfuric acid (sulfate). Such oxidation increases the flux or phase transfer into water and thereby the phase partitioning, but cannot be described by equilibrium conditions (Chapter 4.1.4).

For very soluble gases such as HCl and HNO_3, the "physical" dissolved molecule does not exist (or is in immeasurably small concentrations) because of full dissociation. Therefore, the *Henry* coefficient for such gases represents the equilibrium between gas and the first protolysis states:

$$H = H_{eff} = \frac{[HAO^-]}{[A(g)]} \tag{4.213}$$

More limitations must be considered in the application of *Henry*'s law under atmospheric conditions. In clouds, because of the different chemical composition of the CCN and thereby initial droplets after its formation through heterogeneous nucleation, different equilibriums occur on a micro-scale. It is important to note that in clouds *Henry*'s law is valid only for each droplet with its gaseous surrounding, i. e. taking into account the gas concentration close to the droplet's surface and the aqueous phase concentration (neglecting here further limitations through mass transport). This likely is the main reason (besides others such as surface active components influencing the gas-liquid equilibrium) why in bulk experimental approaches (integral collecting the droplets and analyzing the cloud water) deviations from *Henry*'s law have always been found. In modeling, the spectral resolution of particles and droplets (concerning size and chemical composition) is the only way to come closer to the reality. Few field experiments with multistage cloud water samplers have shown significant differences in size-resolved chemistry.

4.3.3 Properties of droplets

In Chapter 2.5.3, the phenomena of atmospheric water are described from a meteorological and hydrological point of view. Here, the physicochemical properties of droplets that are important for an understanding of cloud physics and chemistry will briefly be described. Atmospheric droplets are always solutions of gases and salts and partly suspended particles. Key properties such as the size and salinity of droplets are dominantly determined by the CCN (Chapter 4.3.5). Although droplets can normally be considered diluted solutions, during nucleation and evaporation processes highly concentrated solutions occur and instead of concentrations in thermodynamic equations (if they are still valid) activities have to be used.

With the presence of cloud droplets the inner energy U of the system (Eq. 4.61) is enlarged by several new types of energy. The work of mixing $\mu\,dn$, surface work $\gamma\,ds$ (s surface) and electrical work $\varphi\,dQ$ (Q charge) are the most important forms:

$$dU = T\,dS - p\,dV + \mu\,dn + \gamma\,ds + \varphi\,dQ + \dots \tag{4.214}$$

Considering only the phase transfer and change of temperature, pressure and amount (expressed in mol n), Eq. (4.72) transforms into:

$$G = G(T, p, n_i) = H(T, p, n_i) - TS = U + pV - TS \tag{4.215}$$

and

$$dG(T, p, n) = \frac{\partial G(T, p, n)}{\partial T} dT + \frac{\partial G(T, p, n)}{\partial p} dp$$

$$+ \sum_i \frac{\partial G(T, p, n_i)}{\partial n} dn_i \tag{4.216}$$

It follows at a constant pressure and temperature:

$$dG_{p,T} = \Sigma \mu_i dn_i \tag{4.217}$$

Again, the system is in equilibrium (Eqs. 4.91 to 4.93) when $dG = 0$. Without further discussion, it is clear that such a condition is hardly achievable in clouds.

4.3.3.1 Vapor pressure change: The *Kelvin* equation

In Chapter 2.5.3.1, we considered water vapor as a gaseous constituent of air. Here, we discuss the vapor-droplet equilibrium in clouds. We can consider each liquid as a condensed gas. At each temperature a part of the liquid-water molecule transfers back to the surrounding air, consuming energy (enthalpy of evaporation). The droplet is in equilibrium with air, when the flux of condensation is equal to the flux of evaporation. The equivalent vapor pressure p (in a closed volume or close to the droplet surface) is the vapor pressure equilibrium. In a closed system, it corresponds to the saturation vapor pressure. The vapor pressure equilibrium depends neither on the amount of liquid nor vapor but only on temperature and droplet size.

Condensation and evaporation occurs at any vapor pressure. When the vapor pressure becomes smaller than the equilibrium value, the droplets are thermodynamically instable and evaporate. In equilibrium in both phases exists the same chemical potential:

$$- S_{aq} dT + V_{aq} dp = - S_g dT + V_d dT \tag{4.218}$$

From this we derive the molar evaporation enthalpy at temperature T ($\Delta_V H / T = \Delta_V S$):

$$\frac{dp}{dT} = \frac{\Delta_V S}{\Delta T} = \frac{\Delta_V H}{T \Delta_V V} \tag{4.219}$$

which is called *Clapeyron*'s equation. Since the molar volume of air is much larger than that of droplets, we can approximate $\Delta_V V \approx V_m \boldsymbol{R} T / p$ and get the *Clausius−Clapeyron* equation, describing the change of vapor pressure with temperature:

$$\frac{d \ln p}{dT} = \frac{\Delta_V H}{\boldsymbol{R} T^2} \quad \text{or} \quad p_2 = p_1 \exp \left\{ \frac{\Delta_V H}{\boldsymbol{R}} \left(\frac{1}{T_1} - \frac{1}{T_2} \right) \right\} \tag{4.220}$$

In laboratory praxis, from the *Clausius-Clapeyron* plot of $\ln p$ against $1/T$, the enthalpy can be derived.

In air, we have to consider droplets and not a bulk solution. From experience, we know that dispersed small droplets combine to larger drops. That is because the enthalpy also depends on the surface: the aqueous amount in the form of droplet possesses a higher chemical potential than the same amount of liquid after coalescence (bulk solution). Another consequence consists in the higher partial pressure droplets have ($\overline{p} + \Delta p = p$) compared with a bulk volume with a flat surface ($\overline{p}$). Assuming spherical particles, the change in vapor pressure Δp can be simply derived. The change in free enthalpy $dG_{T,n} = dW_V + dW_s$ is expressed (T, n = constant) by the changing pressure-volume work $dW_V = dp\,dV = \Delta p \cdot \frac{4}{3}\pi r^3$ and the surface energy change dW_s (r particle radius, γ surface tension, s surface) $dW_s = \gamma ds = \gamma 4\pi r^2$.

Equilibrium gains when $\partial \dfrac{dW_V}{\partial r} = \partial \dfrac{dW_s}{\partial r} = 0$. It follows that $\Delta p\, 4\pi r^2 = \gamma 8\pi r$ and finally $\Delta p = \dfrac{2\gamma}{r}$.

Now, we calculate the change in the chemical potential with changing droplet size, where $d\mu = RT \ln p$ (Eq. 4.80) and molar volume, defined by $V_m = RT/p$ (Eq. 4.19), through the dispersion of a bulk liquid on droplets:

$$\int_{p^\infty}^{p} d\mu = \int_{p^\infty}^{p} RT \ln p = V_m \int_{p^\infty}^{p} pd \ln p = V_m \int_{p^\infty}^{p} dp \tag{4.221}$$

After integration and using the general definition of the chemical potential (Eq. 4.83) we obtain:

$$\Delta\mu = RT \ln \frac{p}{p^\infty} = V_m \Delta p \tag{4.222}$$

Now replacing the expression for Δp we get the *Kelvin* equation; The equation is named in honor of *William Thomson*, commonly known as Lord Kelvin:

$$\ln \frac{\overline{p + \Delta p}}{p^\infty} = \ln \frac{p}{p^\infty} = \frac{2\gamma V_m}{rRT} = \ln \mathfrak{S} \tag{4.223}$$

where $\mathfrak{S}$ is the saturation ratio. This equation[21] is valid only for pure water, but in air, we always meet diluted aqueous solutions. Later, in combination with *Raoult's* law we consider the influence of dissolved matter on lowering the vapor pressure. Eq. (4.223) says that the formation of droplets is possible only for immense supersaturation; a droplet with $r = 10$ nm is stable only if supersaturation is 120% ($p/p^\infty = 1.12$). The small water droplets are thermodynamically instable because of their large vapor pressure. This agrees with the observation that droplets in air are formed only through condensation onto nuclei. By contrast, when droplets exist in air, the *Kelvin* equation says that larger droplets grow via vapor condensation at the expense of smaller droplets, which evaporate.

---

[21] This equation is also called *Gibbs-Thompson* equation and the effect (surface curvature, vapor pressure and chemical potential) is also called the *Gibbs-Kelvin* effect or *Kelvin* effect.

4.3.3.2 Surface tension and surface active substances

An important property of the surfaces of droplets is the surface tension that expresses the cohesion of water molecules (Chapter 2.5.1.1). On molecules existing close to the droplet surface, forces are directed to the inner of the droplet. Therefore, each liquid has the tendency to form spherical particles (if they are not counteracting forces such as gravitation and other outer forces). The reason is simple: a sphere of a given volume has the smallest surface of all bodies. Thus, a growing droplet needs to overcome the molecular cohesion. There are two equivalent definitions of surface tension:

$$\gamma = \frac{f}{l} \quad \text{and} \quad \gamma = \frac{dW_s}{ds} \tag{4.224}$$

The force f concentrates a surface band of width l and a dimension of surface tension is N m^{-1} or kg s^{-2}). The other definition is the ratio of the surface energy to the surface (dimension: energy/surface but reduced on metric units to kg s^{-2}). Water has a surface tension (298 K) of $72.85 \cdot 10^{-3}$ N m^{-1}.

Some organic substances dissolved in the droplets or transported from the gaseous surrounding to the surface can accumulate at the surface (Fig. 4.12) when having hydrophilic and hydrophobic properties in one molecule (for example, aliphatic alcohols, aldehydes and acids). They form a liquid film and reduce the surface tension according to *Gibbs* equation:

$$d\gamma = -\boldsymbol{R}T\,\Gamma_s d \ln c \tag{4.225}$$

where c is the concentration of the surface active substance and Γ_s surface excess is $n_s(s)/s$, i.e. the amount of matter per square unit. The importance of such films becomes clear when considering that all processes, linked with the free enthalpy G of the droplet (evaporation, adsorption, desorption, surface reactions), result generally in a change of G through a change of T, p, S or n (do not mix entropy S with surface s):

$$dG = -S\,dT + V\,dp + \gamma\,ds + \Sigma\,\mu_i\,dn_i \tag{4.226}$$

In cloud and rainwater, polycarboxylic acids, having a molecular structure analogous to that of humic-like substances (HULIS) are the most effective surface active species within the droplets. In addition, monocarboxylic acids and polyaromatic hydrocarbons can play an important role in the atmospheric aquatic system because

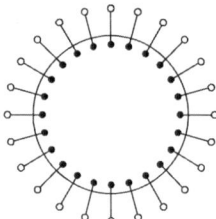

Fig. 4.12 Scheme of surface active substances in a droplet.

of their surface active potential (Fachini et al. 2000, Ćosović et al. 2007). Seidl and Hänel (1983) found in rainwater that soluble surface active materials had concentrations of the order of 2 μmol L^{-1} and that this was too small to have a significant influence on cloud's physical processes. However, organic aerosol constituents can influence the surface tension of nucleating cloud droplets and thereby modify the critical supersaturation necessary to activate APs. Kiss et al. (2005) found HULIS that accounted for 60% of the water-soluble organic carbon present in rural aerosols. The isolated organic matter present in a concentration of about 1 g L^{-1} decreased the surface tension of the aqueous solutions by 25–42% compared with pure water.

4.3.3.3 Vapor pressure lowering: *Raoult*'s law

Solutions have two fundamental property changes compared with pure water: lowering the vapor pressure and freezing point depression. In cloud microphysics, these changes are crucial for droplet growth and precipitation formation.

In about 1886, *Raoult* discovered that substances have lower vapor pressures in solution than in pure form and that the freezing point of an aqueous solution decreases in proportion to the amount of a non-electrolytic substance dissolved. The ratio of the partial vapor pressure of substance i in solution to the vapor pressure of the pure substance (subscript 0 denotes the pure substance) is equal to the mole fraction x of i:

$$p_i = x_i p_i^0 \tag{4.227}$$

This law is strictly valid only under the assumption that the chemical interaction between the two liquids is equal to the bonding within the liquids: the conditions of an ideal solution. In the atmosphere, water is the solvent and dissolved matter is predominantly non-volatile. The vapor pressure of water p_w in solution is smaller than that of pure water p_w^0, whereas the vapor pressure of the solution p_S is $p_S = p_w + p_i$. Because of $x_w + x_i = 1$ we rewrite *Raoult*'s law as follows:

$$p_S = p_w^0 + (p_i^0 - p_w^0) x_i \tag{4.228}$$

Assuming that $p_i^0 \to 0$ (the dissolved substance is non-volatile) we find for the *relative* lowering of vapor pressure of the solution ($\Delta p / p$), i. e. equal to the mole fraction of i:

$$\Delta p = p_w^0 - p_S = x_i p_w^0 \tag{4.229}$$

We can now transform the *Kelvin* equation (4.223), valid for pure water, by consideration of *Raoult*'s equation for a droplet with dissolved matter. The total change in vapor pressure consists of the vapor change above the curved surface and the vapor change of the solution:

$$\Delta p = \Delta p_{\text{Kelvin}} + \Delta p_{\text{Raoult}} = \frac{2\gamma}{r - x_i p^\infty} \tag{4.230}$$

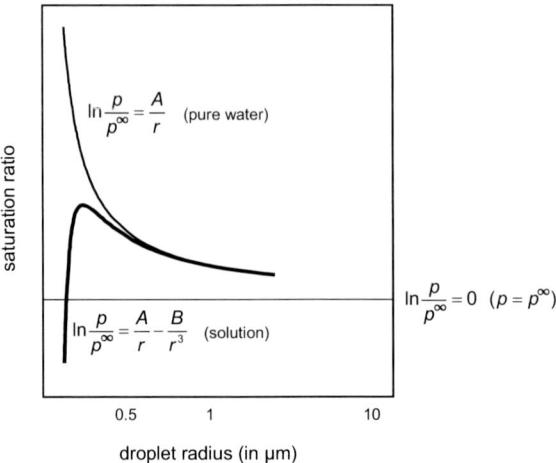

Fig. 4.13 The *Köhler* curve.

We set the vapor pressure of pure water p_w^0 equal to the equilibrium vapor pressure p^∞. From Eq. (4.223), after some reforming,[22] it follows that:

$$\ln\frac{p}{p^\infty} = \ln \mathfrak{S}_{solution} = \frac{A}{r} - \frac{B}{r^3} \qquad (4.231)$$

where $A = 2\gamma V_m / RT$ and $B = 3n_i / 4\pi$ (n_i mole number of dissolved substance).

Curves resulting from Eq. (4.231) are named *Köhler* curves (Fig. 4.13). The first term on the right side of the *Köhler* equation is called the *Kelvin* term and the second is called the *Raoult* term. From Eq. (4.231) follows the droplet radius for $\mathfrak{S} = p/p^\infty = 1$:

$$r_{saturation} = \sqrt{\frac{B}{A}} = \sqrt{\frac{3n_i}{8\pi\gamma}}\, p^\infty \qquad (4.232)$$

The maximum in the *Köhler* curve is given for $d\ln\mathfrak{S}/dr = 0$:

$$r_{critical} = \sqrt{\frac{3B}{A}} \qquad 4.233)$$

Finally, it follows for the supersaturation ratio that:

$$\ln\mathfrak{S} = \sqrt{\frac{4A^3}{27B}} \qquad (4.234)$$

---

[22] We consider that in a binary solution $x_i = n_i/(n_i + n_w)$ and the droplet volume results from the partial molar volume; furthermore, for diluted solutions the volume of dissolved matter i is negligible; thereby $4\pi r^3/3 = n_i V_i + n_w V_w \approx n_w V_w$. Because of $n_i \ll n_w$ it follows that $x_i \approx n_i/n_w \approx n_i 4\pi r^3/3V_m$. Finally, $V_m p^\infty = RT$.

The exponential function can be approximated by a progression ($\exp x \approx 1 + x$) and we see that the solution effect (*Raoult* term) is dominant when $\mathfrak{S} = p/p^{\infty} \leq 1 + (4A^3/27\,B)^{1/2}$, whereas $\mathfrak{S}$ strongly decreases by further reducing the radius. By contrast, for $\mathfrak{S} > 1 + (4A^3/27\,B)^{1/2}$ the *Köhler* curve approaches the curve for pure water (Fig. 4.13). Each point on the *Köhler* curve represents the equilibrium vapor ratio of water, i. e. the relative humidity (RH) at which a droplet with a given radius and amount of dissolved matter exists. For $r < r_{critical}$ the equilibrium is stable; for constant $\mathfrak{S}$ the droplet would evaporate water (because of the momentum of the larger pressure) if there is a slight growth through the adsorption of water molecules and return to equilibrium. For $r > r_{critical}$ the supersaturation $\mathfrak{S}$ becomes smaller for larger droplets; therefore, water molecules would permanently absorb and enlarge the droplet. Conversely, smaller droplets are quickly undersaturated when losing water and tend to evaporate. This is one principle of cloud droplet growth that is at the expense of smaller droplets.

4.3.3.4 Freezing point depression

The freezing point depression follows from the lowering of vapor pressure. From the *Clausius-Clapeyron* equation and *Raoult*'s law, it follows ($\Delta_{sm}H$ enthalpy of smelting) that:

$$\Delta T = \left(\frac{RT^2}{\Delta_{sm}H}\right)x_i \qquad (4.235)$$

Taking the molality m (defined as the ratio of the amount of dissolved matter and the mass of water, expressed in mol kg^{-1} and in contrast to the molarity not depending on T; in dissolved solution molarity is proportional to molality) we obtain:

$$\Delta T = K_f m_i \qquad (4.236)$$

where K_f is the cryoscopic constant, which is empirical and can be determined experimentally; for water $K_f = 1.86$ K kg mol^{-1}.

In clouds, the important effect is the existence of supercooled droplets, i. e. despite a certain freezing point depression droplets remain at lower temperatures liquid because of the kinetic inhibition of crystallization. The homogeneous process is that spontaneous freezing occurs only for ≤ 232 K or $-41°C$ (T_H) and saturation near that of liquid water (Koop et al. 2003). However, the temperature when water becomes ultimate a solid (T_S) may be deeper; computer studies based on classical nucleation theory using experimental data suggest that the crystallization rate of water reaches a maximum around 225 K ($-48°C$), below which ice nuclei form faster than liquid water can equilibrate (Moore and Molerino 2011). The heterogeneous process requires the presence of ice nuclei (IN) with a hexagonal crystal structure similar to that of water ice, which allows freezing at temperatures as high as $-5°C$ (Vali 2008). Heterogeneous ice nucleation in clouds with supercooled water results in the subsequent efficient growth of the ice crystals because of the *Bergeron-Findeisen* process. It is assumed that this is the main initiation process of precipitation in the midlatitudes (Pruppacher and Klett 1997).

It is remarkable that biological particles such as bacteria or pollen might be active as both the CCN and heterogeneous IN. These biological particles are much larger than the inorganically or organically derived CCN (but probably smaller in number density) and are important in the development of precipitation through giant CCN and IN. However, there is yet no direct evidence that bacterial IN participate in cloud processes to any significant degree (Möhler et al. 2007).

Although recognized for hundreds of years, artificial weather modification was not applied before the turn of the 20th century. It started by simply introducing artificial CCN to the cloud layers by guns, later rockets and aircrafts. By end of the 1940s the introduction of "crystallization nuclei" (e. g. AgI, dry ice and many others) affected the ice phase processes in supercooled clouds and larger hygroscopic particles (artificial precipitation embryos about 30 μm in diameter) to stimulate collision-coalescence processes in warm clouds. About 10 countries applied different techniques with variable success.

4.3.4 Gas-to-particle formation: Homogeneous nucleation

According to the *Köhler* equation, the formation of droplets only from water vapor is impossible in air. However, some products of gas phase reactions such as SO_3/ H_2SO_4 (sulfuric acid), CH_3HSO_3 (methanesulfonic acid) and many oxygenated organic compounds have small vapor pressures but partly a high affinity to H_2O. Such molecules can accommodate each other (single component nucleation) and among different species (multicomponent nucleation) and form a cluster of molecules as a metastable phase. After reaching a critical radius, they become stable. It is believed that in the atmosphere, *gas-to-particle conversion* occurs not by condensation of single species but rather by involving at least two, probably more, different species. Students well know from laboratory praxis that condensed fine matter forms when opened near bottles of aqueous ammonia solution and sulfuric acid, nitric acid and/or hydrochloric acid. Clearly single and combined reactions occur with different numbers of species, depending on its gaseous concentration and the stability of the embryo formed (g − gaseous, p − particulate):

$$NH_3(g) + HCl(g) \rightleftharpoons NH_4Cl(p) \tag{4.237}$$

$$NH_3(g) + HNO_3(g) \rightleftharpoons NH_4NO_3(p) \tag{4.238}$$

$$NH_3(g) + H_2SO_4(g) \rightleftharpoons NH_4HSO_4(p) \tag{4.239}$$

$$2\,NH_3(g) + SO_3(g) + H_2O \rightleftharpoons (NH_4)_2SO_4(p) \tag{4.240}$$

$$SO_3(g) + H_2O(g) \rightleftharpoons H_2SO_4(p) \tag{4.241}$$

$$NH_3 + HNO_3 + SO_3 + H_2O \longrightarrow (NH_4)_2NO_3HSO_4 \tag{4.242}$$

Interestingly, it has been found that the formation of SOA is accelerated in the presence of SO_2 and sulfuric acid (Jang et al. 2002, Edney et al. 2005, Surrat et al. 2007). The first step in atmospheric SO_2 oxidation is OH addition (Chapter 5.5.2.1) and this radical can react with alkoxy radicals (RO) to form sulfonic acid and further with organic peroxo radicals to form dialkyl sulfates:

$$SO_2 \xrightarrow{OH} HSO_3 \xrightarrow{RCH_3O} ROHSO_3 \xrightarrow[-HO_2]{RCH_3O_2} ROS(O_2)OR \qquad (4.243)$$

These results strongly suggest the importance of particle phase reactions that lower the volatility of organic species, via accrecation (oligomerization) processes (Carlton et al. 2009). Organosulfates, including nitrated derivatives (e. g. nitroxy organosulfates) have been detected in ambient aerosols (Surrat et al. 2007, 2008, Gómez-González et al. 2008). Jacobson et al. (2000) and Carlton et al. (2009) have also presented reviews on organic atmospheric aerosols.

The main concern of the classical homogeneous nucleation theory has been a thermodynamic description of the initial stage of nucleation from embryo to nucleus with a little larger size over the critical one (Seinfeld 1986, Pruppacher and Klett 1997, Seinfeld and Pandis 1998, Kulmala et al. 2000). The change of the free enthalpy of the cluster is at first positive because the decrease of entropy is initially larger (regular structure formation) than the decrease in enthalpy:

$$\Delta G = G_{cluster} - G_{vapor} \qquad (4.244)$$

Applying Eq. (4.226) as a special case for T, n = constant on Eq. (4.244), it follows (see also Eq. 4.75 concerning volume-pressure change):

$$dG_{T,n} = dVdp + \gamma ds \qquad (4.245)$$

It is $dV = V_{cluster} - V_{vapor}$ and because $V_{cluster} \ll V_{vapor}$ under all conditions, $dV = -V_{vapor}$. The vapor phase is assumed to be ideal, thereby $V_{vapor} = nRT/p$. The mole number n corresponds to the number of molecules transferred from vapor into the cluster so $n = V_{cluster}/V_m$ and with $V_{cluster} = 4\pi r^3/3$; just it follows:

$$dG_{T,n} = -\frac{4\pi r^3}{3V_m} RT \frac{dp}{p} + \gamma ds \qquad (4.246)$$

Now transferring from infinitesimal to difference expressions, we integrate dp/p $(= d\ln p)$ and replace p/p^{∞} by the saturation term (see *Kelvin* equation):

$$\int_{p^{\infty}}^{p} \frac{dp}{p} = \ln \frac{p}{p^{\infty}} = \ln \mathfrak{S}$$

Finally, using the standard expression for the surface (of the cluster) and replacing ds, we get (V_m is the molar volume of the cluster phase):

$$\Delta G_{T,n} = -\frac{4\pi r^3 RT}{3V_m} \ln \mathfrak{S} + 4\pi r^2 \gamma \qquad (4.247)$$

As a result, the nucleation curve (free energy change versus nucleus size) passes through a well-known single maximum point corresponding to the critical size of the nucleus (Fig. 4.14). The process is at first reversible (as expressed in Eqs. 4.237−4.241). The radius of the critical cluster can be derived from Eq. (4.246) under the condition $\partial(\Delta G/\partial r) = 0$ (compared with the radius derived directly from the *Kelvin* equation) to be:

$$r_{critical} = \frac{2\gamma V_m}{RT \ln \mathfrak{S}} = \frac{2\gamma V}{kT \ln \mathfrak{S}} \qquad (4.248)$$

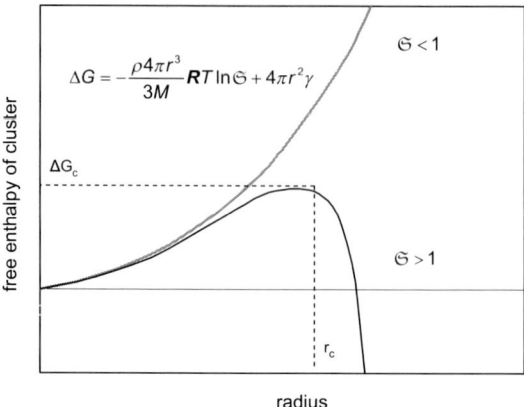

Fig. 4.14 Change of free enthalpy while forming clusters from a vapor (homogeneous nucleation).

Table 4.10 Critical size of H_2O clusters in dependence from supersaturation.

supersaturation	radius (in nm)	N_{H_2O}
1.02	54	$2.2 \cdot 10^{10}$
1.1	11	$2.0 \cdot 10^5$
2	1.5	500
10	0.5	18

Note that $V_m/V = R/k$ and $V_m = M/\varrho$ where M and ϱ are the molar mass and density of the nucleating substance, respectively (Eq. 4.19). Table 4.10 shows the critical radius of the pure water cluster is independent of supersaturation. We can clearly see that under normal atmospheric conditions the formation of water droplets is impossible.

In atmospheric modeling the nucleation rate and number of clusters produced with a critical radius per volume is simple to approach from kinetics theory (Möller 2003) but is still highly controversial after almost a century of research. The weakness of the classical homogeneous nucleation theory is the importance of energy transfer (as opposed to the mass transfer) along the chain of the growing molecular cluster (Bakhtar et al. 2005, Wasai et al. 2007). Altman et al. (2008) suggested that, to bring theory in consistency with experiments, certain fundamental propositions of the theory of nucleation should be revised. The inclusion of an additional contribution to the *Gibbs* energy of a cluster caused by the size dependence of the specific heat capacity of the cluster decreases the critical cluster size compared with the value calculated by the *Kelvin* equation.

In the following, we will consider the example of ammonium chloride formation in air. More exactly, reaction (4.237) most split into gaseous (monomolecular) and particulate (multimolecular) ammonium chloride:

$$NH_3(g) + HCl(g) \rightleftharpoons NH_4Cl(g) \rightleftharpoons NH_4Cl(p) \qquad (4.249)$$

Pure NH_4Cl aerosols can grow up to $1\,\mu m$ (Gmelin 1936). The particles are less hygroscopic than $NaCl$ but very soluble in water. The equilibrium constant is given by:

$$K_p = \frac{p_{NH_3} p_{HCl}}{p_0^2} \approx p_{NH_3} p_{HCl} \quad \text{and} \quad K_c = \frac{c_{NH_3} c_{HCl}}{c_{NH_4Cl}} \tag{4.250}$$

where p_0 is the atmospheric pressure (assumed to be 1 bar). *Gibbs* equation (4.251) and the *Kelvin* equation (4.252) describe the temperature dependency of the equilibrium as well as the cluster critical radius independent of the partial pressures of NH_3 and HCl:

$$\boldsymbol{R}T \ln K_p = \mu_{NH_4Cl(p)}(T) - \mu_{NH_3}(T) - \mu_{HCl}(T) \tag{4.251}$$

$$-\ln \frac{p_{NH_3} p_{HCl}}{p_0^2} = \frac{2\,\gamma_{NH_4Cl}\, M_{NH_4Cl}}{\boldsymbol{R}T\varrho_{NH_4Cl}\,r} = \ln \mathfrak{S} \tag{4.252}$$

Hence, $p_{NH_3} p_{HCl}$ must exceed K_p to obtain a critical cluster size. Seinfeld and Pandis (1998) used an expression:

$$\ln K_p = 34.266 - \frac{21.196}{T},$$

which shows that K_p is relatively independent of T ($K_{298} = 7.08 \cdot 10^{14}$). Although $K_p = p_{NH_3} p_{HCl}$ and in equilibrium $p_{NH_3} = p_{HCl}$ we now estimate the condition for NH_4Cl formation to be $1/\Sigma K_p = 0.37 \cdot 10^{-7}$ atm ($= 37$ ppb), an impossible condition under free tropospheric conditions. Therefore, particulate NH_4Cl belongs to the minor constituents in air. However, to follow this example given by Seinfeld und Pandis (1998), we calculate the critical radius by using Eq. (4.252) and the data given $M = 52.49$ g mol^{-1}, $\gamma = 150 \cdot 10^{-5}$ N cm^{-1} and $\varrho = 1.527$ g cm^{-3} (Countess und Heicklen 1973). It follows that $r_{critical} \approx 1$ nm for this numerical example (37 ppb gas concentration). Baek and Aneja (2005) estimated the reaction rate constant of $NH_3 + HCl \rightarrow NH_4Cl$ to $k = 3.44 \cdot 10^{-4}$ m^3 μmol^{-1} s^{-1} and (recalculated for 298 K) $k = 1.4 \cdot 10^4$ atm^{-1} s^{-1}. The most likely particle formation will go first through sulfate production (4.241) and subsequent NH_3 absorption onto the sulfate clusters/APs. Harrison und Kitto (1992) found pseudo-first-order reaction rate constants for absorbing gaseous ammonia of between $4 \cdot 10^{-6}$ s^{-1} and $4.1 \cdot 10^{-4}$ s^{-1}, corresponding to residence times between less than one hour and a few hours.

This process of homogeneous nucleation, nowadays also called *new particle formation*, *Charles Darwin* describes in his book "The Voyage of the Beagle" in 1839 as a phenomenon we now call *blue haze*:

> During this day I was particularly struck with a remark of Humboldt's, who often alludes to the thin vapour which, without changing the transparency of the air, renders its tints more harmonious, and softens its effects. This is an appearance which I have never observed in the temperate zones. The atmosphere, seen through a short space of half or three-quarters of a mile, was perfectly lucid, but at a greater distance all colours were blended into a most beautiful haze, of a pale French grey, mingled with a little blue. The condition of the atmosphere

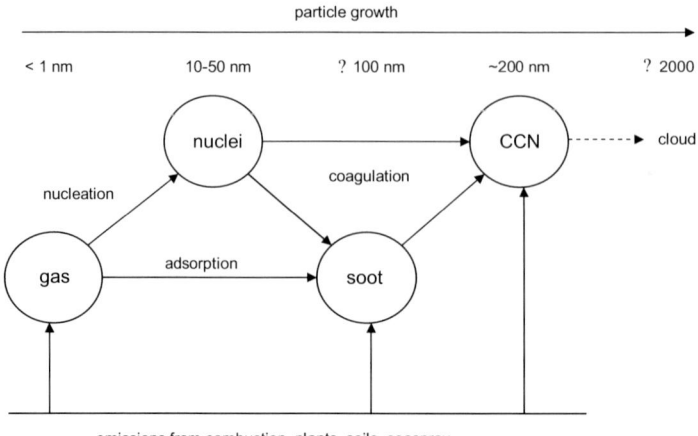

Fig. 4.15 Scheme of particle growth and formation of CCN.

> between the morning and about noon, when the effect was most evident, had undergone little change, excepting in its dryness. In the interval, the difference between the dew point and temperature had increased from 7.5 degrees to 17 degrees.
> (Darwin 1839, p. 33)

The formation of SOA, leading to the formation of nanoparticles of blue haze over forested areas, is highly complex and not fully understood. Zhang et al. (2009) showed that the interaction between biogenic organic acids and sulfuric acid enhances nucleation and the initial growth of those nanoparticles. The climate impact by forests is still controversially discussed in the chain new particle formation and cloud formation (Spracklen et al. 2008). Modeling with the so-called partitioning method, which gives a more realistic estimate of SOA formation, produced 15.3 Tg yr^{-1} with the biogenic fraction dominant (13.6 Tg yr^{-1}) (Lack 2003).

The fate of nuclei is particle coagulation, a process in which small particles (assumed to be spherical) collide with each other and coalesce completely to form larger spherical particles. Small particles are indeed spheroidal and the assumption of spherical particles seems to be reasonable (Mitchell and Frenklach 2003, Balthasar et al. 2005). After a certain size, however, the particles cannot coalesce completely and start to form long chains, which eventually grow into three-dimensional fractal-like structures. Fig. 4.15 shows the new particle formation and growth pathways. A great role is played by the surface coating of primary hydrophilic soot particles, likely via SO_2 absorption and sulfuric acid formation but also via NMVOC absorption, with its oxygenation possibly providing a matrix for the stabilization of very fine SOA particles.

4.3.5 Atmospheric aerosols and the properties of aerosol particles

In physics, an *aerosol* is defined as the dispersion of solids or liquids (the dispersed phase) in a gas (the dispersant). More generally, dispersion is a heterogeneous mixture of at least two substances that are not soluble within each other (in contrast to molecular dispersion). However, the classical scientific terms are *colloid* and *colloidal system*. Colloids exist between all gas, solid and liquid combinations with the exception of gas-gas (all gases are mutually miscible); L − liquid, S − solid, G − gaseous:

S + S	L + S	G + S
S + L	L + L	G + L
S + G	L + G	

As well as the terms *sol* and *gel,* the terms colloid and *colloidal condition of matter* were introduced by *Thomas Graham* (Graham 1861). Over 100 years ago it had already become clear that the sciences to study colloids − chemistry and physics − must deal with the colloidal state and not only with the colloid, the dispersed phase itself (Ostwald 1909). The colloidal system was characterized as multiphase or heterogenic. For colloids such as (G + S), cigarette smoke, atmospheric dust and for (G + L), atmospheric fog was given as examples (Ostwald 1909). As mentioned in Chapter 1.3.4, the term *aerosol* was introduced by the German meteorologist *Schmauß* in 1920 (Schmauß 1920, Schmauß and Wigand 1929). Remember that the role of dust particles in atmospheric optics and water condensation was known since the discoveries of *Tyndall* and *Aitken* (1870−1880). Schmauß and Wigand (1929) considered the three categories of dispersed particles behind atmospheric aerosols: ions, dust and hydrometeors.

Large ions (electric charged molecular clusters) were first observed in air by Langevin (1905) and later Pollock (1909) as being produced photochemically from atmospheric trace gases as first proposed by Lenard (1900) and directly from combustion processes (Wigand 1913). Later studies (Pollock 1915a, 1915b, Wigand 1919, 1930) showed that different classes of larger ions exist in the air under ordinary conditions. Nowadays we call these (nanometer-sized) ultrafine atmospheric particles, produced via homogeneous nucleation. The formation rate of 3 nm particles is often in the range $0.01-10$ cm^{-3} s^{-1} in the boundary layer. However, in urban areas formation rates are often higher (up to 100 cm^{-3} s^{-1}), and rates as high as 10^4-10^5 cm^{-3} s^{-1} have been observed in coastal areas and industrial plumes. Typical particle growth rates are in the range $1-20$ nm h^{-1} in the mid-latitudes depending on the temperature and availability of condensable vapors (Kulmala et al. 2004).

The confusion in using different terms for classifying atmospheric disperse systems was discussed by Grimm (1931). The WMO classification (manual of codes) of horizontal obscuration into categories of fog, ice fog, steam fog, mist, haze, smoke, volcanic ash, dust, sand and snow shows that there is little scientific understanding of the atmospheric colloidal system behind it. The atmospheric aerosol always comprises the whole system, i. e. suspended particles in air; therefore, in the air itself. The particles can be named APs; thereby only the APs can be collected but not the aerosol (particles are separated from air). Nowadays, the term bioaerosol or biological aerosol has been introduced, which is scientifically unsound[23].

---

[23] The phrase "Bioaerosol is defined as airborne particles, large molecules that are living …" (Gelencsér 2004, p. 70) contains two more errors: Aerosol is more than a particle and even the largest existing molecule is not living (Chapter 2.2.2.1).

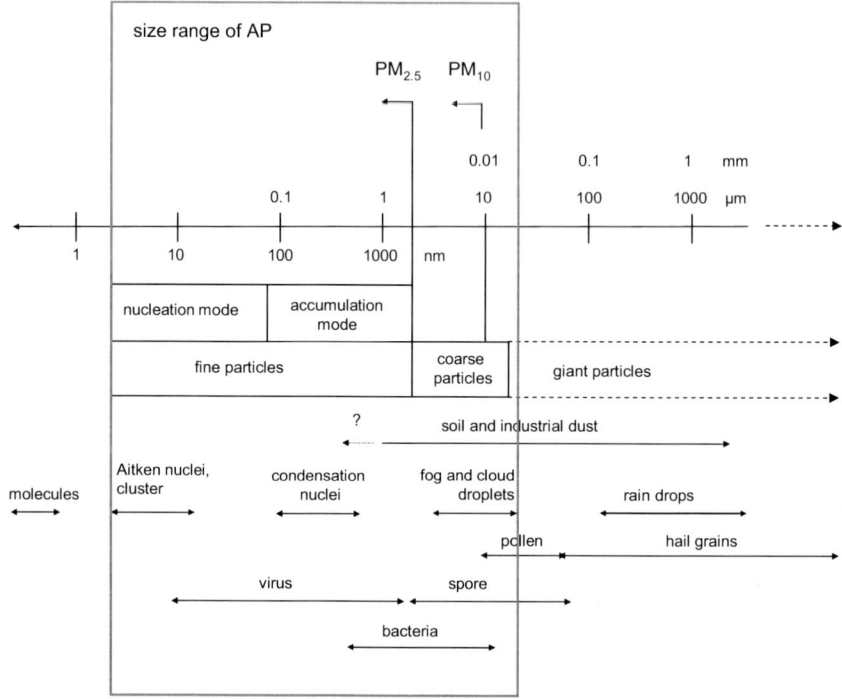

Fig. 4.16 Size ranges and types of particles in air.

Clearly, the role of biogenically derived APs (to call them *biological particles* is correct) is reconsidered (after the insights given by *Pasteur* 150 years ago) in a new light, i.e. not as a transmitter of diseases but playing a likely role as CCN and IN (Möhler et al. 2007).

Despite the term *aerosol* implying solids in air, the scientific community remains divided into those, calling hydrometeors (cloud and fog droplets), that belong to atmospheric aerosols and those don't. There is no question that atmospheric APs contain water, and that the transfer from potential CCN (CN) via activated CCN to haze particles (some scientists separate them between dry and wet haze) is more or less fluent. However, when forming mist[24], the particle suddenly growths (heterogeneous nucleation) by several orders of magnitude in volume and becomes almost a water droplet, containing some dissolved matter from the CCN. The physical and chemical properties of hydrometeors, whether solid (icy particles) or liquid (water droplets), are different from non-aqueous particles in the atmosphere. Therefore, behind atmospheric aerosols we understand the entity of non-hydrometeor particles in the air from a few nanometers to a few micrometers (Fig. 4.16). The limits are not fixed; the smallest particles might comprise a few molecules (embryo) or a

---

[24] If we can see less than 1 km through the cloud of water droplets, it is known as fog. If we can see between 1 and 2 km, we call it mist. A fog is a cloud that has come into contact with the earth's surface. Mist is very thin (concerning number density) fog.

Table 4.11 Origin and types of atmospheric aerosol particles (further classification possible concerning biogenic, geogenic and anthropogenic origin).

source characteristics		particle characteristics	
direct	wind blow	inorganic	soil dust and sea salt
		organic	plant debris, degradation products
		biological	bacteria, viruses, pollen
	combustion	inorganic	ash
		organic	smoke, soot (BC)
	industrial	inorganic	dust
	volcanic	inorganic	ash
	extraterrestrial	inorganic	meteoric dust
indirect	gas emissions	inorganic	salts (sulfate, nitrate, ammonium, chloride)
		organic	SOA

macromolecule and the largest particles might be biological species such as pollen. The transfer to giant particles (which might have radii of hundreds of μm) is also running[25]. Such large particles[26] no longer fit the scientific understanding of an aerosol (or colloidal system) because of its very small number concentration.

In contrast to hydrometeors, APs are always in the air, but not all types of APs can be observed at any time because of different sources and formation conditions (Table 4.11). The entity of APs is called PM or simply *dust*. Suspended matter in air was separated by *Carl Wilhelm von Nägeli* in 1879 into the three classes: a) coarse (visible through human eyes), b) sun-dust[27] (visible through light scattering) and c) non-visible, only through water vapor condensation detectable (cited after Rubner et al. 1907).

There is another fundamental difference between hydrometeors and APs: the size-depending chemistry. Size distribution of hydrometeors (cloud droplets) cover about 1 to 40 μm, where a maximum is observed at 5–10 μm. They are well approximated by spherical form. Typical differences in cloud droplet chemical composition have been found but each hydrometeor consists of > 99.999 % H_2O. The size range of APs is three orders of magnitude larger, having several maxima (modes) and showing extreme differences in chemical composition and particle form from spheres to fibers and crystalloids. APs are ubiquitous in the earth's atmosphere and they can grow by several processes and age, and thereby change size and chemical composition.

Although we do not consider cloud drops as APs, they play a crucial role in AP aging: after heterogeneous nucleation, the CCN dissolves[28] in the aqueous phase

---

[25] It has been suggested to set a threshold for particles with significant sedimentation velocity according to *Stoke*'s law.

[26] It is remarkable that hail particles can be produced in the upper atmosphere with diameters of many centimetres. It has been reported that icy chunks (beside its extraterrestrial origin) are likely to be produced in upper atmosphere too.

[27] Called *Sonnenstäubchen* in German.

[28] This is valid only for the water-soluble CCN; water-insoluble particles only coating the CCN-type will form a suspension in drops and only the surface layer is dissolved (as observed for soot and SOA particles).

and further gas uptake as well aqueous phase reactions provide mass for a residual after droplet evaporation (Fig. 4.9). We will see later that typical AP substances (such as sulfate, organic acids and macromolecular substances) are produced through cloud processing. Let us state that each AP is unique and singular despite several classes of particles being established. As a logical consequence, scientific understanding needs single particle sampling and analysis, a crestfallen approach. By contrast, each PM sample contains more different molecules than we normally consider for the external phase, the gas mixture (let us say thousands). For understanding the timely and spatial behavior of gases (Chapter 2.8.2), we need thousands of measurements of a single component. There is no further need to debate that PM climatology is only at the very beginning.

The role of APs within the climate system was outlined at the beginning of Chapter 4.3 and is summarized again here (Ramanathan et al. 2001, Carslaw et al. 2010):

1. optical function: changing visibility (by reducing it)
2. radiation function: changing atmospheric heat budget (cooling tendency)
3. water cycle function: cloud and thereby precipitation formation (increasing and decreasing); and
4. chemical function: providing the surface for heterogeneous chemistry.

Without a detailed explanation it is self-evident that a quantified description of atmospheric aerosols is extremely important for understanding the climate system. However, it is so complex that any final story will not be seen in the near future. Much is known about the chemical composition and sources of APs but modeling their climate impact functions still provides most uncertainties. This seems to be a dilemma but there are two simple answers. First, it is a normal process in understanding complex systems that it takes time. Second, our current understanding of aerosols, clouds and climate is good enough to pass the available scientific results and conclusions as management tools into the hands of engineers and politicians. It is simple: end the era of fossil fuel combustion (see proposed solution in Chapter 2.8.4.3).

It is out of the scope of this book to describe the AP mechanics, i. e. microphysics and dynamics (Friedlander 1977, Hinds 1882, Kouimtzis and Samara 1995, Harrison and van Grieken 1998, Mészáros 1999, Spurny 1999, 2000, Baron and Willeke 2001). Here, we only summarize the important topic of atmospheric aerosol size distribution (Jaenicke 1999). Fig. 4.16 shows that the size range covers several orders of magnitude. Therefore, the common logarithm of the radius[29] is useful to describe the different distribution functions: $dN(r)/d\lg r = f(\lg r)$ or $dN(r)/dr = f(\lg r)/2.302 \cdot r$. $N(r)$ cumulative number size distribution (or the integral of radii) having dimension cm^{-3}, r radius of particle:

---

[29] The logarithm of a physical quantity instead of the quantity itself is mathematically incorrect because the logarithm must be always a figure. The origin of the logarithm of a physical quantity is based on the integral $\int \frac{dx}{x} = \int d\ln x = \ln x + c$; finally, the definition is valid $\ln x \equiv \ln (x/x_0)$ where the reference value $x_0 = 1$ (for example pressure, mol fraction).

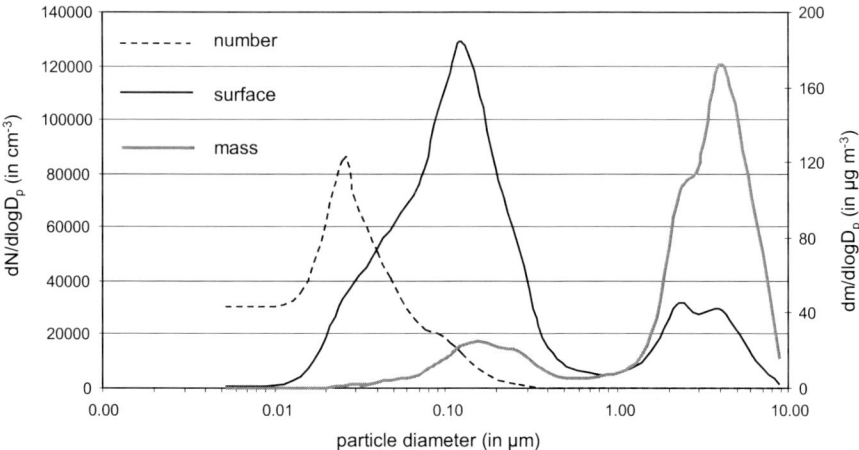

Fig. 4.17 Particle size distributions from an inner kerbside site in Stockholm (surface in arbitrary units) Data from Mohnen (1995).

$$N(r) = \int_{o}^{\infty} n(r)\, dr \tag{4.253}$$

$n(r)$ differential number size distribution, which denotes a number density (not number concentration) having dimension $cm^{-3}\, cm^{-1}$:

$$n(r) = \frac{dN(r)}{dr} \tag{4.254}$$

Their sizes range from a few nanometers to a few micrometers often with pronounced concentration modes around a few tens of nm (*Aitken* mode), in the range $\sim 100-500$ nm (accumulation mode) and at a few μm (coarse mode); see Fig. 4.17. Coarse mode particles typically originate from the dispersion of solid and liquid matter such as soil, dust and sea spray (Chapter 2.6.4). Submicron particles are usually mixtures of primary particles (almost combustion products from biomass smoke and diesel soot) and secondary produced particles via the gas-to-particle conversion (most important is sulfuric acid from SO_2 oxidation and organics from oxidation of VOCs).

APs must also satisfy the fundamental thermodynamic principles (which are important in chemical analysis and studying particle origin):

- $\prod_{i} x_i \leq K_S$ (solubility equility)
- $\sum_{i} x_i^+ = \sum_{j} x_j^-$ (electroneutrality)
- $\sum_{i} x_i + x_{H_2O} = 1$ (closure)

APs generally comprise the following three types of matter or fractions (Table 4.12):

- highly water soluble inorganic salts;
- insoluble mineral dust; and
- carbonaceous material.

Table 4.12 Principal chemical composition of particulate matter of different origin.

origin	≥ 95%	1–5%	< 1%
soil dust	O, Si, Al	Fe, Ca, K	all other elements
sea salt	Na, Cl	SO_4^{2-}, Mg	all other elements
industrial dust	Ca, O, C	Fe, elements	all other elements
secondary inorganic	SO_4^{2-}, NH_4^+, NO_3^-	Cl	–
secondary organic	C, O	H	S, N

The water-soluble fraction (mostly a mixture of sulfate, nitrate and chloride compounds) has been studied extensively, and several models of the geographical distribution and climatic effects of this group of compounds in the aerosol have appeared. In this group are sea salt and secondary inorganic aerosols (fully water
soluble) but also a small fraction of soil dust (sulfate, chloride, potassium, calcium)
and SOA (organic acids). Because in cloud and rainwater we also find many dissolved metal ions (mainly Fe, Cu and Mn but in extremely small concentrations no
metal can be excluded), their origin is from crustal matter and sea salt. Insoluble
inorganic matter is predominantly soil dust but also occasionally volcanic ash and
industrial dust. Oxygen and silicon (silicates) are the dominant elements. Whereas
soil dust is only inconvenient, volcanic particles emitted into the upper troposphere
can influence the climate for 1–2 years. Industrial dust, predominantly flue ash
from coal combustion, initially provides a certain acid neutralization potential into
the air (within its plume); however, from the emission of heavy metals only mercury
plays a global role because of its volatility and different chemical species. Transition
metal ions (Fe, Mn and Cu) play a huge role in aqueous phase redox processes
(Chapter 5.3.5).

Whereas secondary produced APs are < 1 μm, soil dust is > 1 μm. Sea salt
particles can range to the size of the CCN (~ 0.2 μm) but are dominantly in the
lower μm range. Industrial dust can range from nm particles (soot from combustion) to coarse particles. Biological particles also range from sub μm (bacteria) to
10–30 μm (pollen; Matthias-Maser 1998). The three fractions (insoluble minerals,
carbonaceous and soluble salts) contribute each roughly one-third of total PM,
however, this can vary according to the source characteristics and – for secondary
matter – depending on the reaction conditions. Hence, the plural term *aerosols* is
attributed to special source and chemical regimes characterizing the aerosol type:
marine aerosol, rainforest or tropical aerosol, arctic aerosol, desert dust aerosol
and so on.

Carbonaceous material includes organic compounds ranging from very soluble
to insoluble, plus elemental carbon and biological species. Organic compounds
cover a very wide range of molecular forms, solubilities, reactivities and physical
properties, which makes complete characterization extremely difficult, if not impossible. The reader is referred to review the following papers: Jacobson et al. (2000),
Decesari et al. (2000), Gelencsér (2004), McFiggans et al. (2005b), Kanakidou et
al. (2005), Sun and Ariya (2006), Facchini and O'Dowd (2007) Timonen et al.
(2008) and Hallquist (2009).

Table 4.13 Composition of particulate matter PM_{10} in Berlin and environment (daily samples 1 year: 2001/2002); in $\mu g\ m^{-3}$; data from Möller (2009b).

species	kerbside	urban	rural
total PM_{10}	34.5	24.4	20.1
residual[a]	15.3	9.4	6.8
OC	4.3	3.4	2.8
BC	4.3	2.2	1.3
sulfate	4.1	3.6	3.6
nitrate	3.4	3.0	3.0
ammonium	2.0	1.8	1.8
chloride	0.50	0.19	0.21
sodium	0.34	0.29	0.34
calcium	0.42	0.19	0.16
potassium	0.20	0.14	0.09
magnesium	0.05	0.04	0.05
iron	0.7	0.2	0.1

[a] insoluble minerals as difference PM_{10} and measured species

A large fraction of PM is soot, the historic *symbol* of air pollution (Chapter 1.3.4). There has been a long dispute in the literature on the definition of soot, which is also called elemental carbon (EC), black carbon (BC) and graphitic carbon (see Cachier 1998, Gelencsér 2004). Surely soot is the best generic term that refers to impure carbon particles resulting from the incomplete combustion of a hydrocarbon (EM – elemental matter is also found in literature and might "integrate" EC and BC). The formation of soot depends strongly on the fuel composition. It spans carbon from graphitic (EC) through BC to organic carbon fragments (OC). Each of the available methods (optical, thermal, thermo-optical) refers to a different figure; it remains a simple question of definition. Hence, the comparison of different soot methods is basically senseless. The atmospheric implications of soot:

– provide the largest surface-to-volume ratio for heterogeneous processes;
– form most complex structural nanoparticles;
– carry (toxic) organic substances; and
– warm the atmosphere through light absorption.

New research has found that the sooty *brown clouds*, caused primarily by the burning of coal and other organic materials in India, China and other parts of south Asia, might be responsible for some of the atmospheric warming that had been attributed to greenhouse gases (Ramanathan and Crutzen 2003, Ramanathan et al. 2007).

In Europe, carbonaceous matter ranges from $0.17\ \mu g\ m^{-3}$ (Birkenes, Norway) to $1.83\ \mu g\ m^{-3}$ (Ispra, Italy) for EM and for OC from $1.20\ \mu g\ m^{-3}$ (Mace Head, Ireland) to $7.79\ \mu g\ m^{-3}$ (Ispra, Italy). The percentage of TC to PM_{10} in rural backgrounds amounts to 30 %: 27 % OM and 3.4 % EM, respectively (Yttri et al. 2007). Within this range are values measured in Berlin and surrounding areas: $1.3–2.2\ \mu g\ m^{-3}$ EM and $2.8–3.4\ \mu g\ m^{-3}$ OC, whereas TC contributes 20 % to PM_{10} (Table 4.13). Roughly one-third of the contribution share is found (Table 4.14).

Table 4.14 Classified composition (in %) of particulate matter PM_{10} in Berlin and environment.

species	kerbside	urban	rural
insoluble minerals[a]	44	38	34
OC + BC	25	23	20
sulfate + nitrate + ammonium	27	34	42
soluble minerals	4	5	4
total	100	100	100

[a] residual (see Table 4.13)

4.3.6 Formation of cloud droplets: Heterogeneous nucleation

As seen from the *Köhler* equation (4.223), the condensation of water vapor into preexisting droplets is only possible when the saturation ratio is larger than one. However, the mass transfer of water molecules is possible for $\mathfrak{S} < 1$ when the solid (aqueous particles are not existing) particle surface provides a large enough affinity to H_2O. This property is called *hygroscopicity*. Each crystalline water-soluble surface dissolves or deliquesces at a certain *RH* (Table 4.15).

Surfactants apparently decrease the deliquescence point (Chen and Lee 2001) by up to 8% *RH*. Fig. 4.18 shows the behavior of ammonium sulfate with increasing humidity; the dry particle with a diameter of 100 nm does not grow over a wide range of humidity changes (range a) but suddenly, shortly before reaching the deliquescence point, (range c) in changes by about 240% of the particle volume by water absorption. With decreasing humidity, however, the particle will not lose all the water again but only according to the *Köhler* theory (hysteresis curve), i. e. it exists in a metastable state (range c). Particles that attain the deliquescence point are called *activated* CCN Beyond this deliquescence point, the particle takes up continuous water with increasing *RH* according to the following equilibrium equation:

$$RH = \frac{p_{H_2O}}{p_{H_2O}^{\infty}} = a_{H_2O} \tag{4.255}$$

Table 4.15 Deliquescence point (*RH* in %) for pure salts.

substance	deliquescence point
$MgCl_2 \cdot 2\,H_2O$	33
NH_4HSO_4	40
$K_2CO_3 \cdot 2\,H_2O$	43
NH_4NO_3	62
NaCl	76
$(NH_4)_2SO_4$	80
$NaNO_3$	80
$KHSO_4$	86

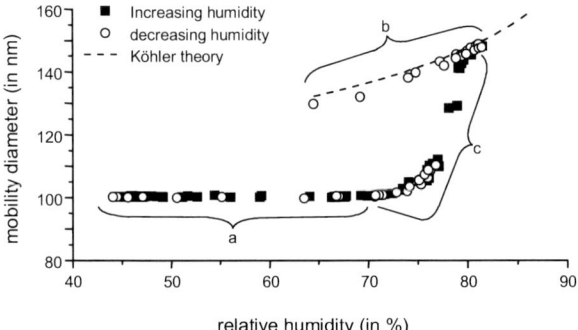

Fig. 4.18 Experimental estimated particle growth with increasing and decreasing humidity (humidogram) for synthetic $(NH_4)_2SO_4$ with $r = 50$ nm at $20\,°C$, after Weingartner et al. (2002).

where a is the activity of water in the salt solution, p is the water partial pressure and p^∞ is the saturation pressure (T depending). Optically, these water-coated particles (we call them haze particles) show light scattering similar to cloud droplets (the key difference is that the number of haze particles is larger but the size is much smaller than for cloud droplets); therefore, the sky becomes milky and visibility reduces.

When $p_{H_2O} > p_{H_2O}^\infty$, water condenses onto the activated CCN and the particle volume roughly increases by at least two orders of magnitude to form mist, fog and/or cloud droplets. This process is called *heterogeneous nucleation*. The supersaturation is always a result of the adiabatic cooling of the air parcel. Because of the release of the condensation heat and the entrainment of dryer air from surrounding droplet formations (or cloud extensions) is limited.

From a mass transfer point of view, the heterogeneous nucleation is also a *nucleation scavenging* of PM, the first process of in-cloud scavenging, following by a gas uptake. Junge (1963) described it by the simple equation:

$$c_{aq} = \varepsilon \frac{c_p}{LWC} \tag{4.256}$$

where c_{aq} is the concentration of dissolved particles in the droplet phase and ε is the fraction $(0 \ldots 1)$ of the scavenged (washout) PM in air (c_p gas phase PM concentration) and LWC. Under normal boundary layer conditions, $\varepsilon = 0.9 \ldots 1.0$ is estimated (Mészáros 1981). Despite only a small percentage of the total number of particles serving as CCN (and thereby being scavenged), these few particles contribute $> 90\%$ of PM mass, whereas smaller non-CCN particles contribute $< 10\%$ of PM mass but $>99\%$ of number concentration (Fig. 4.17).

4.3.7 Scavenging: Accommodation, adsorption and reaction (mass transfer)

4.3.7.1 Mass transfer: General remarks

Let us now consider the mass transfer from the gas phase into the droplet or onto an AP according to the scheme presented in Fig. 4.7. The process includes the following steps (here seen from the site of particle surrounding, but the transfer can by reversible):

- turbulent transport of molecules close to the particle (to the diffusion layer interface);
- molecular diffusion to the particle surface (interface);
- adsorption onto the surface (namely solid particles but for droplet surfactants as an example) or absorption (uptake) into the particle (mainly droplets but not to exclude for solids);
- chemical surface reactions (if possible);
- desorption into the gas phase;
- diffusion and mixing within the particle; and
- simultaneous chemical reactions (if possible) within the condensed phase (e. g. aqueous phase or in-droplet reactions).

From the gas kinetic theory, the number of collisions of a gas molecule with a particle can be derived. In Chapters 4.1.1.2 and 4.1.1.3, we derived different expressions for the number of collisions (z) between molecules and a wall ($c_N qv\, dt/6$) and among molecules ($\sqrt{2}\, c_N v\, dt$), where c_N is the molecule density (N/V), the area q has different definitions and v is the mean molecular speed. In the above expressions, the subscript 0 refers to the *Loschmidt* constant, i. e. to the mean gas phase density. In the following, the subscript 0 refers to the surface (interfacial layer), the subscript g close to the surface layer (about one mean-free path) and the subscript ∞ to the gas phase concentration far from the surface. According to Fig. 4.20, the number of striking molecules is $z = (c_N)_g\, qv_x\, dt$, with $q = \pi r^2$ (r – particle radius). We now have to consider that only a fraction of molecules have the speed v_x expressed by $v_x f(v_x)\, dv_x$ (Eq. 4.31). From the gas kinetic theory (Gombosi 1994), it can be derived that $v_x = \bar{v}/4$. Finally, the gross (maximum) flux F_{coll} onto the particle surface is (= number of collisions in terms of number per unit of time and unit area) is given by Eq. (4.257); $(c_N)_\infty$ – gas concentration far from the particle, i. e. $x_\infty \gg l$ (Fig. 4.20):

$$F_{coll} = \frac{1}{4}\, (c_N)_g \bar{v} = \frac{z}{\pi r^2 dt} = \frac{1}{\pi r^2}\, R_{coll} \tag{4.257}$$

where R is the rate and n_g is the gas concentration close to the particle. It follows that:

$$F_{net} = \frac{1}{4}\, \gamma\, (c_N)_g \bar{v} = \frac{1}{4}\, \gamma_{eff}\, (c_N)_\infty \bar{v} \tag{4.258}$$

Distinguishing between γ and γ_{eff} (introduced by Pöschl et al. (2006) but not used by other authors) is meaningful because it reflects the transport from the bulk gas phase close to the particle surface ($\infty \rightarrow g$) which might have limitations, expressed by the actual surface collision flux F_{coll} and the average gas kinetic flux $\overline{F_{coll}}$. From equations (4.257) and (4.258) it follows that:

$$\gamma = \frac{F_{net}}{F_{coll}} \tag{4.259}$$

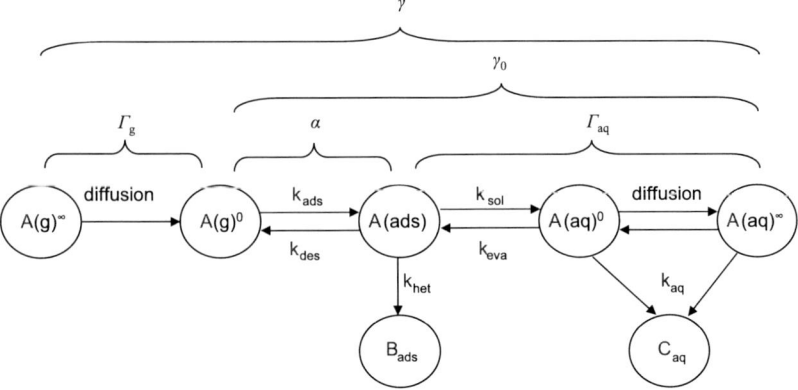

Fig. 4.19 Resistance model of mass accommodation.

$$\frac{\gamma_{eff}}{\gamma} = \frac{(c_N)_g}{(c_N)_\infty} = \frac{F_{coll}}{F_{coll}} \quad \text{and} \quad \gamma_{eff} = \frac{F_{net}}{F_{coll}} \tag{4.260}$$

If there is no concentration gradient in the particle surrounding ($n_g \approx n_\infty$) it follows that $\gamma = \gamma_{eff}$. The uptake coefficient γ refers to the net (sometime also called effective) uptake of gas molecules by the droplets. The ratio γ_{eff}/γ represents a gas diffusion correction factor. The mean molecular speed $\bar{v}$ is given by Eq. (4.35), the "weighted arithmetic mean of all velocities". Care must be taken in the terminology of different "averaged" velocities (remember the discussion on different uses of velocity/speed in Chapter 4.1). This maximum flux (Eq. 4.258) occurs only in the absence of the gas phase diffusion and solubility limitations and in the absence of chemical reactions. When chemical reactions occur in the interfacial region (Fig. 4.7), reactive loss at the surface competes with mass accommodation and subsequent reaction in the liquid phase (see later).

The gas uptake by liquid droplets is widely described by the so-called resistance model, where substance A must overwhelm layers with a specific resistance (Fig. 4.19). The idea of this model is to describe the single processes (transport, adsorption, dissolution and reaction) as decoupled processes. The dimensionless parameter γ ($0 \le \gamma \le 1$) is called the net uptake coefficient, which describes the total process including all physical and/or chemical partial processes, here given by the gas phase diffusion ("resistance" $1/\Gamma_g$ or "conductance" Γ_g) and net uptake (resistance $1/\gamma$):

$$\frac{1}{\gamma_{eff}} = \frac{1}{\Gamma_g} + \frac{1}{\gamma} . \tag{4.261}$$

Again, when the gas phase diffusion has a low resistance, it becomes $\gamma = \gamma_{eff}$ or, in other words, if the rate of diffusion of gas to the surface is large, then $1/\Gamma_g$ can be neglected. The net uptake coefficient γ_{eff} represents the major observable parameter in laboratory studies; it is also sometimes called the measurable and apparent uptake coefficient. Table 4.16 lists some γ values for common gas species.

Table 4.16 Measured net uptake coefficient γ on water surface at 273 K; data from Davidovits et al. (2006).

gas-phase species	γ
SO_2	0.34
H_2O_2	0.24
H_2O	0.22
HCl	0.18 ... 0.23
NH_3	0.12 ... 0.20
HNO_3	0.15
HBr	0.079 ... 0.26
DMS	0.13
HI	0.091
CH_3OH	0.056
C_2H_5OH	0.049
CH_3OOH	0.012

The most common formula used to calculate the gas transport coefficient is (Pöschl et al. 2005):

$$\Gamma_g \approx \frac{4 D_g}{\bar{v} r} = \frac{4}{3} \frac{l}{r} = \frac{4}{3} Kn \tag{4.262}$$

The ratio between the mean-free path l and droplet radius r is defined as the *Knudsen* number. Another empirical formulation is known (Davidovits et al. 2006):

$$\Gamma_g \approx \frac{0.75 + 0.238 \, Kn}{Kn \, (1 + Kn)} \tag{4.263}$$

With regard to the mass transfer to particles, we have to distinguish between three so-called transport regimes:

- *kinetic regime*: the radius of the particle is small compared with the mean-free path of the molecule (this is valid for particles in the nucleation mode); $r \ll l$ or $Kn \gg 1$
- *continuum regime*: the radius of the particle is large compared with the mean-free path of the molecule (this is valid for particles in the accommodation mode and thereby for all hydrometeors); $r \gg l$ or $Kn \ll 1$
- *transition regime*: the radius of the particle is in the order of the mean-free path of the molecule (this is valid for the CCN); $r \approx l$ or $Kn \approx 1$

In the transfer regime, which is important for CCN formation and activation, several approaches are known to solve the *Boltzmann* equation (*Fuchs* theory, *Fuchs* and *Sutugin* approach, *Dahneke* approach, *Sitarski* and *Nowakowski* approach; see Seinfeld and Pandis (1998), Pruppacher and Klett (1997)). In the following, we will consider only the case uptake by cloud droplets, i. e. the continuum regime. For more details, see Pöschl et al. (2005), Davidovits et al. (2006) and Morita and Garrett (2008). It should be noted that termination and symbols are different to those used in the scientific literature (Pöschl et al. 2005). The mean-free path is

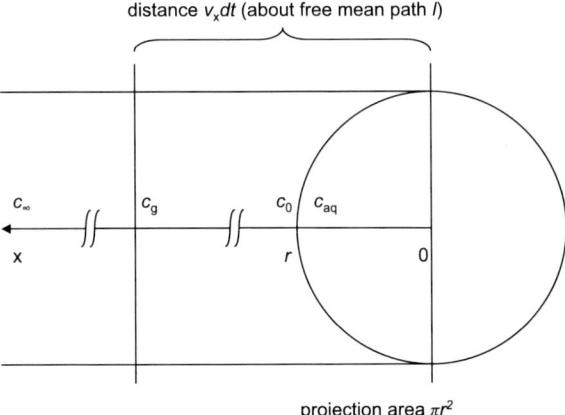

distance $v_x dt$ (about free mean path l)

projection area πr^2

Fig. 4.20 Schema of mass transfer into droplets.

about 60 nm for most atmospheric gases under standard conditions (Pöschl et al. 2005).

Fig. 4.20 shows schematically the accommodation along the x-axis. Mass accommodation occurs when a gas molecule strikes the particle surface; for solid particles adsorption and surface chemistry almost occurs and for hydrometeors absorption (dissolution) and aqueous phase chemistry must be considered (Fig. 4.19). The uptake of molecules into droplets lowers the surface concentration $((c_N)_o < (c_N)_g < (c_N)_\infty)$ and results in a subsequent gas phase diffusion to the surface because of the concentration gradient. The mass accommodation coefficient[30] α is the probability that a molecule that strikes the particle surface is adsorbed or enters the liquid. In a further application of the resistance model, we split the net uptake into two processes in series: adsorption and dissolution (with possibly aqueous phase chemical reaction; see later). Pöschl et al. (2005) proposed the term "surface accommodation coefficient" for α.

In Eq. (4.264), surface chemistry is excluded because it would be a parallel process to adsorption (α must be replaced in the case of heterogeneous chemistry by another adequate parameter because this would increase the incoming flux):

$$\frac{1}{\gamma} = \frac{1}{\alpha} + \frac{1}{\Gamma_{aq}} \qquad (4.264)$$

The mass accommodation process can also be defined as the fraction of the collision flux to the surface which is adsorbed; no adsorption means collision where the gas molecule is "reflected" back to the gas phase. Therefore, it is valid that:

---

[30] In the scientific literature similarly defined parameters have been called: condensation coefficient, sticking coefficient, sticking probability, trapping probability, adsorptive mass accommodation coefficient, accommodation coefficient and thermal accommodation coefficient (Pöschl et al. 2005). The term "sticking" is often used for the case of "chemisorption", i. e. the binding energy between the gas molecule and the (solid) surface is large and "trapping" for "physisorption" when the binding energy (more exactly adsorption energy) is small (< 50 kJ mol^{-1}); in the case of liquid droplets, "sticking" means absorption/dissolution.

$$F_{ads} = \alpha\, F_{coll} = \frac{1}{4}\, \alpha\, (c_N)_g\, \bar{v} \equiv k_{ads}\, (c_N)_g \tag{4.265}$$

where k_{ads} is the rate constant of adsorption (see 4.3.7.2). Similarly, we can also define the desorption flux $F_{des} = k_{des}\, n_0$. The mass accommodation coefficient is also defined through:

$$\alpha = \frac{\text{number of molecules, absorbed by the particle}}{\text{number of molecules which collide with the particle surface}}$$

From Eq. (4.265) it follows that:

$$k_{ads} = \frac{1}{4}\, \alpha\, \bar{v} \tag{4.266}$$

After the molecule enters the liquid, it diffuses away from the surface into the inner droplet; this net flux into the liquid is given by $F_{sol} = k_{sol}\, (c_N)_0$. We just completed the transfer chain of a molecule A:

$$A(\infty) \to A(g) \to A(0) \to A(aq)$$

The net flux (gas molecule uptake by droplets) is given by the difference of adsorption and desorption:

$$F_{net} = F_{ads} - F_{des} \quad \text{or} \quad \frac{1}{4}\gamma\, (c_N)_g\, \bar{v} = \frac{1}{4}\alpha\, (c_N)_g\, \bar{v} - k_{des}\, (c_N)_0 \tag{4.267}$$

Setting the net incoming flux from the gas phase equal to the net flux into the liquids F_{sol} gives:

$$F_{net} = F_{sol} \quad \text{or} \quad \frac{1}{4}\, \gamma\, (c_N)_g\, \bar{v} = k_{sol}\, (c_N)_0 \tag{4.268}$$

Combining both equations (4.267) and (4.268) leads to:

$$\alpha = \gamma\left(1 + \frac{k_{des}}{k_{sol}}\right) \quad \text{or} \quad \frac{1}{\gamma} = \frac{1}{\alpha} + \frac{k_{des}}{\alpha k_{sol}} = \frac{1}{\alpha} + \frac{1}{\Gamma_{sol}} \tag{4.269}$$

If $k_{sol} \gg k_{des}$ it follows that $\gamma = \alpha$; Γ_{sol} denotes the process of the solvation and/or dissolution of adsorbed gas molecule. Comparing Eq. (4.269) with (4.265) we see that $\Gamma_{aq} = \Gamma_{sol}$ in the absence of chemical reactions (Eq. 4.287). In general, the aqueous phase conductance Γ_{aq} can include dissolution (limited by the slow diffusion into the droplet and by saturation), aqueous phase chemical reaction and heterogeneous surface reactions. The coefficient Γ_{sol} is also a pure number but its value (in contrast to α and γ) is not restricted to numbers less than one. The solubility limited uptake coefficient Γ_{sol} varies with time t. The gas is exposed to the droplet and is given by the relationship (Fogg 2003) D_{aq} with the aqueous phase diffusion coefficient ot the dissolved substance:

$$\frac{1}{\Gamma_{sol}} = \frac{1}{4}\frac{\bar{v}}{H \cdot RT}\sqrt{\frac{t}{D_{aq}}} \tag{4.270}$$

If the solubility is low or the exposure time has been so long that the droplet is saturated, then Γ_{sol} can be neglected. According to the resistance model, more parallel processes and processes in series can be added or split (for example, surface or heterogeneous chemical reactions and bulk phase chemical reactions). For solid particles without phase dissolution, the last term in Eq. (4.269) is omitted.

4.3.7.2 Adsorption

A measure for the adsorption of a substance on the surface is the coverage degree θ:

$$\theta = \frac{\text{number of occupied adsorption sites}}{\text{number of available adsorption sites}}$$

In the case of multilayer formation, it is useful to express the coverage degree by $\theta = V/V_m$ where V_m denotes the volume of the adsorbed substance in a monolayer. Adsorption and desorption can be described as well as the kinetic and equilibrium process. Equilibrium is described by the general condition $f(n, p, T) = 0$. The process is described either at a constant pressure (adsorption isobar) or constant temperature (adsorption isotherm). The most simple adsorption isotherm after *Freundlich* and *Langmuir* is based on the equilibrium

$$A\,(g) + P_{surface} \underset{k_{des}}{\overset{k_{ads}}{\rightleftarrows}} A\,(ads)\text{-}P_{surface} \tag{4.271}$$

or simplified $A\,(g) \rightleftarrows A\,(ads)$. It is assumed that the coverage degree changes over time are proportional to the partial pressure of A and the number N of free adsorption sites; similarly we set Eq. (4.273) for the desorption kinetic:

$$\left(\frac{d\theta}{dt}\right)_{ads} = k_{ads}\,p_A\,N\,(1 - \theta) \tag{4.272}$$

$$\left(\frac{d\theta}{dt}\right)_{des} = k_{des}N\theta \tag{4.273}$$

In equilibrium ($K = k_{ads}/k_{des}$) is $d\theta/dt = 0$ and we get:

$$\theta = \frac{Kp_A}{1 + Kp_A} \tag{4.274}$$

When $Kp_A \gg 1$ goes $\theta \to 0$. Normally, however, it is the condition $Kp_A \ll 1$ (low gas concentration) that leads to $\theta = Kp_A$ and $\theta \ll 1$. As a consequence Eq. (4.272) simplifies to $R_{ads} = k'_{ads}\,p_A$; the adsorption site number N we include in the adsorption coefficient. Note that the expression for the adsorption flux $F_{ads} = k_{ads}\,n_g$ (Eq. 4.266) looks similar but the dimensions of k_{ads} are different and it must be taken into account that $p = n\,\mathbf{k}T$ (Eq. 4.22; note that n is not the mole number but the gas phase molecule density or molecule number concentration, respectively); flux is the rate per unit of area ($F = R/q$).

Because $\theta = m/m_{max}$ (m – molality of the adsorbed substance) and setting $1/K = \beta$ (in sense of an adsorption coefficient, see Eq. 4.209) we finally obtain:

$$\frac{p_A}{m} = \frac{\beta}{m_{max}} + \frac{p_A}{m_{max}} \tag{4.275}$$

Using the derived expression for adsorption and desorption flux and setting $F_{ads} = F_{des}$ for the equilibrium, we get (as we already have derived from Eq. (4.266) above):

$$k_{ads} = \frac{1}{4}\alpha\bar{v} = \alpha\sqrt{\frac{RT}{2\pi M}} \qquad (4.276)$$

Another adsorption isotherm according to *Brunauer, Emmet* and *Teller* (the so-called BET isotherm) is useful for the formation of multilayer adsorption, i. e. there is no principal saturation or coverage degree limitation:

$$\frac{p_A}{(p_A^\infty - p_A)V} = \frac{1}{EV_m} + \frac{(E-1)p_A}{EV_m p_A^\infty} \qquad (4.277)$$

where p^∞ denotes the saturation pressure of the pure liquid phase of the adsorbed gas A, V is the total volume of the adsorbed substance A (V_m that of the mono-layer) and E is a constant. The adsorption of soluble gases on the water surface cannot be described by any type of adsorption isotherm because the net flux is given by the dissolution flux:

$$F_{net} = F_{ads} - F_{des} + F_{sol} \approx F_{sol} \qquad (4.278)$$

The gas-liquid equilibrium is given by *Henry*'s law (Chapter 4.3.2). In the case of surface and/or bulk chemical reactions the flux equation (mass budget) is expressed by:

$$F_{net} = F_{ads} - F_{des} + F_{het} + F_{chem} \approx F_{het} + F_{chem} \qquad (4.279)$$

The next two chapters deal briefly with the mass transfer through surface chemistry and bulk chemistry.

4.3.7.3 Surface chemistry: Kinetics of heterogeneous chemical reactions

The substance A can undergo at the droplet (or particle) surface after adsorption chemical conversion: A(ads) $\rightarrow$ B(ads) (Fig. 4.19). This chemical flux is given by $F_{het} = k_{het}(c_N)_0$. Using the adsorption equilibrium condition $K = (c_N)_0/(c_N)_g$ it follows that:

$$F_{het} = k_{het} K (c_N)_g = k'_{het} (c_N)_g \qquad (4.280)$$

In the case that each surface striking of a gas molecule goes into surface chemical conversion – this is the steady state with $F_{het} = F_{ads}$ – we get:

$$k_{het} = k_{ads} = \frac{1}{4}\gamma(c_N)_g\bar{v} \qquad (4.281)$$

This also represents the maximum heterogeneous chemical conversion rate, i. e. it is limited by the adsorption flux; $(c_N)_\infty$ is the gas phase molecule density, M is the molar mass:

Table 4.17 Expressions for gas-particle mass transfer; n_∞, n_g, n_0 and n_{aq} molecule density of the same substance far from the particle, close to the particle, at particle surface and within the particle (droplet); p − gas partial pressure far from the droplet, c_{aq} − aqueous-phase concentration. k − mass transfer coefficient (recalculable into specific rate constant); g − gas-phase, aq − aqueous-phase, het − interfacial layer (chemistry), in − interfacial layer (transport), $coll$ − collision, ads − adsorption (surface striking), sol − dissolution, $diff$ − diffusion in gas-phase, des − desorption.

flux	parameter
$F_{coll} = \dfrac{1}{4} n_g \bar{v}$	$\bar{v} = \sqrt{\dfrac{8RT}{\pi M}}$
$\overline{F_{coll}} = \dfrac{1}{4} n_\infty \bar{v}$	
$F_{ads} = \alpha F_{coll} = \dfrac{1}{4} \alpha n_g \bar{v} = k_{ads} n_g$	$\alpha = \dfrac{F_{ads}}{F_{coll}} \qquad k_{ads} = \dfrac{1}{4} \alpha \bar{v}$
$F_{des} = k_{des} n_0 = k_{des} K_{des} n_g$	
$F_{net} = \dfrac{1}{4} \gamma n_g \bar{v} = \dfrac{1}{4} \gamma_{eff} n_\infty \bar{v} \equiv \dfrac{\gamma}{\alpha} k_{ads} n_g$	$\gamma = \dfrac{F_{net}}{F_{coll}} \qquad \gamma_{eff} = \dfrac{F_{net}}{F_{coll}} \qquad \dfrac{\gamma_{eff}}{\gamma} = \dfrac{n_g}{n_\infty}$
$F_{sol} = k_{sol} n_0$	
$F_{het} = k_{het} n_0$	$k_{het}^\infty = \alpha \dfrac{\gamma_{eff}}{\gamma} \dfrac{1}{m_m^{3/2}} \sqrt{\dfrac{kT}{\pi}}$
$F_{chem} = k_{aq} n_{aq}$	
$F_{diff}^{aq} = k_{diff}(c_\infty - c_g) = \dfrac{k_{diff}}{RT}\left(p - \dfrac{c_{aq}}{H_{eff}}\right)$	$k_{diff} = \dfrac{3}{r} D_g$
$F_{chem}^{aq} = -k_{aq} c_{aq}$	$\dfrac{p}{c_{aq}} = \dfrac{1}{H_{eff}} - \dfrac{k_{aq}}{k_g} RT$
$F^{aq} = \dfrac{k_g}{RT}\left(p - \dfrac{c_{aq}}{H_{eff}}\right)$	$\dfrac{1}{k_g} = \dfrac{r^2}{3 D_g} + \dfrac{4r}{3 \alpha \bar{v}}$
$R_{air}^g = \dfrac{dp}{dt} = \sum_i k_i p - LWC \cdot k_g\left(p - \dfrac{c_{aq}}{H_{eff}}\right)$	
$R_{air}^{aq} = \dfrac{dc_{aq}}{dt} = \sum_i (k_{aq})_i c_{aq} - \dfrac{k_g}{RT}\left(p - \dfrac{c_{aq}}{H_{eff}}\right)$	

$$F_{ads} = F_{het}^{max} = \alpha \dfrac{\gamma_{eff}}{\gamma}(c_N)_\infty \sqrt{\dfrac{RT}{2\pi M}} = k_{het}''(c_N)_\infty \qquad (4.282)$$

Note that this flux is normalized per unit area, and to convert it to the "more chemical" heterogeneous chemical rate R_{het} (dimension concentration per time) we first have to include the total particle surface per volume of air S ($\sum_i s_i$, where s is the single particle surface) and then convert the molecule density $(c_N)_\infty$ into the gas phase concentration c_g of species A using $c_N = \dfrac{N}{V} = \dfrac{nN_A}{V} = \dfrac{mN_A}{MV} = c\dfrac{N_A}{M}$.

$$R_{het} = -\left(\frac{dc_\infty}{dt}\right)_{het} = S\frac{N_A}{M}k''_{het}c_\infty = k^\infty_{het}S \cdot c_\infty \tag{4.283}$$

For the gas bulk phase, the heterogeneous-specific rate constant follows (m_m – mass of molecule):

$$k^\infty_{het} = \alpha\frac{\gamma_{eff}}{\gamma}\frac{1}{m_m^{3/2}}\sqrt{\frac{kT}{\pi}} \tag{4.284}$$

The important conclusion is that atmospheric heterogeneous chemical transformation depends on the available particle surface. Table 4.17 summarizes the different expressions for fluxes and mass transfer parameters.

Let us consider the change of surface density n_0 over time and assuming a steady state:

$$\frac{d(c_N)_0}{dt} = k_{ads}(c_N)_g - k_{des}(c_N)_0 - k_{het}(c_N)_0 - k_{sol}(c_N)_0 = 0. \tag{4.285}$$

It follows, assuming that $k_{ads} \gg k_{des}$, that:

$$(c_N)_0 = (c_N)_g\frac{1}{\dfrac{1}{K_{ads}} + \dfrac{k_{het} + k_{sol}}{k_{ads}}} = (c_N)_g\frac{k_{ads}}{k_{het} + k_{sol}}. \tag{4.286}$$

4.3.7.4 Mass transfer into droplets with chemical reaction

After gas phase diffusion onto the droplet surface, in addition to dissolution limitation, the enhancement of the mass-transferred substance by chemical removal is also possible. The resistance model equation, combining (4.262) and (4.265), is then enlarged to (Γ_{rxn} denotes chemical processes):

$$\frac{1}{\gamma_{eff}} = \frac{1}{\Gamma_g} + \frac{1}{\alpha} + \frac{1}{\Gamma_{sol} + \Gamma_{rxn}} \tag{4.287}$$

The reactive uptake coefficient Γ_{rxn} for the non-reversible reaction is given by the relationship (Fogg 2003):

$$\frac{1}{\Gamma_{rxn}} = \frac{1}{4}\frac{\bar{v}}{H \cdot RT}\frac{1}{\sqrt{D_{aq}k_{aq}}} \tag{4.288}$$

where k_{aq} is the aqueous phase reaction rate constant. In the case of simultaneous surface chemistry (heterogeneous chemistry) and bulk-chemistry, the resistance model leads to:

$$\frac{1}{\gamma_{eff}} = \frac{1}{\Gamma_g} + \frac{1}{\alpha} + \frac{1}{\frac{1}{\frac{1}{\Gamma_{sol}} + \frac{1}{\Gamma_{chem}}} + \frac{1}{\Gamma_{het}}} \tag{4.289}$$

where Γ_{chem} and Γ_{het} refer to in-droplet and droplet surface chemistry, respectively. Sometimes only one of the four stages shown in Eq. (4.288) is the significant RDS, and so it is then possible to assume that:

$\gamma_{eff} \approx \Gamma_g$ (fast absorption and/or chemical reaction: gas diffusion limited)
$\gamma_{eff} \approx \alpha$ (fast gas phase diffusion and reaction: sticking limited)
$\gamma_{eff} \approx \Gamma_{sol}$ (low solution or saturated droplet, no reaction: dissolution limited)
$\gamma_{eff} \approx \Gamma_{rxn}$ (fast gas diffusion and effective accommodation but slow reaction and saturation)

From the measurements of the effective uptake coefficient, the aqueous phase reaction rate constant can be calculated, as long as the gas phase and liquid phase diffusion coefficients and the *Henry* constant are known. As mentioned, a gas transfer into droplets is characterized by the continuum regime. The unsteady-state diffusion flux (it means that F_{diff} depends on t as well as on x) of species A along the x-axis to the stationary droplet (Fig. 4.20) was described by Seinfeld and Pandis (1998), where $c(x, t)$ is the concentration, depending on time and location:

$$\frac{\partial c}{\partial t} = -\frac{1}{x^2} \frac{\partial}{\partial x}(x^2 F_{diff}) \tag{4.290}$$

Remember that $F_{diff} = -D_g (dc/dx)$ is given by *Fick*'s first law. This molar flux is given in moles per area and time at any radial position. Now, after derivation and assuming that D_g is constant, it follows that:

$$\frac{\partial c}{\partial t} = D_g \left(\frac{\partial^2 c}{\partial x^2} + \frac{2}{x} \frac{\partial c}{\partial x} \right) \tag{4.291}$$

After some time steady state is reached ($\partial c/\partial t = 0$). Seinfeld and Pandis (1998) have shown that under atmospheric conditions for almost all chemical species of interest this relaxation time is about 10^{-3} s or smaller. The integration of:

$$\frac{d^2 c}{dx^2} + \frac{2}{x} \frac{dc}{dx} - 0$$

with the boundary conditions (r is the droplet radius) (note that we use c_g as the gas phase concentration close to the surface (where $l \ll r$) in contrast to c_0, which denotes the surface concentration of the adsorbed species):

$$c(x, t) = c_\infty \ (x > r)$$
$$c(\infty, t) = c_\infty$$
$$c(r, t) = c_g$$

to the simple solution:

$$c(x) = c_\infty - \frac{r}{x}(c_\infty - c_g) \quad \text{or} \quad \frac{c(x) - c_\infty}{c_\infty - c_g} = \frac{r}{x} \tag{4.292}$$

The total flow (or rate) R_S onto the droplet surface ($4\pi r^2$) is given by:

$$R_{diff}^S = 4\pi r^2 (F_{diff})_{x=r} = -4\pi r^2 D_g \frac{dc}{dx} = 4\pi r D_g (c_\infty - c_g) \tag{4.293}$$

because the derivation of Eq. (4.292) leads to $(dc/dx) = -(r/x^2)(c_\infty - c_g)$ and $x = r$ (droplet surface). The total flux onto the droplet surface is described by:

$$F_{diff}^S = \frac{D_g}{r}(c_\infty - c_g) \tag{4.294}$$

The flux to the droplet per volume is obtained by dividing the molar flow R_S by the droplet volume ($4\pi r^3/3$); note that we now consider a volume-based flux instead of surface-based flux, which is equivalent to the change of concentration per time:

$$\left(\frac{dc}{dt}\right)_{aq} = F_{diff}^{aq} = \frac{3D_g}{r^2}(c_\infty - c_g) = k_{diff}(c_\infty - c_g) \tag{4.295}$$

Now, substituting the molar concentration c ($p = c\mathbf{R}T$) by gas phase partial pressure and the surface-near concentration by the aqueous phase concentration according to *Henry*'s law, we obtain (p is the gas phase partial pressure far from the droplet):

$$F_{diff}^{aq} = \frac{3D_g}{r^2 \mathbf{R}T}\left(p - \frac{c_{aq}}{H_{eff}}\right) \tag{4.296}$$

It is seen that the diffusion flux into the droplet depends strongly from the droplet radius: smaller droplets under otherwise constant conditions will have a larger gas uptake than larger drops. Eq. (4.296) represents the flux into a single droplet, with the dimension mass per droplet volume and time, e. g. mol L^{-1} s^{-1}. To gain the mean flux into a droplet in a cloud we must integrate the single droplet flux over all droplets ($N(r)$ – droplet size distribution):

$$\overline{F_{diff}^{aq}} = \frac{4}{3}\pi \int_0^\infty N(r)\, F_{diff}^{aq}(r)\, r^3 dr \tag{4.297}$$

In the case of a monodisperse cloud, $\overline{F_{diff}^{aq}} = F_{diff}^{aq}$ is valid. Of interest is the total flux into all droplets in a cloud, i. e. in the liquid water volume. From the volume of air, we obtain for the air-volume-related flow, expressed as change of gas phase partial pressure per time (e. g. ppb s^{-1}), the following:

$$(\overline{F_{diff}^{aq}})_{air} = \mathbf{R}T \cdot \text{LWC} \cdot \overline{F_{diff}^{aq}} \tag{4.298}$$

To consider the flux into the droplet, a so-called two-layer model is used, which consists of the diffusion layer ($x \to r$) and an interfacial layer at the droplet surface (between c_o and c_{aq}) (Fig. 4.20). The interfacial flux corresponds to the adsorption flux (Eq. 4.265) but now based on the droplet volume, thereby (S and V surface and volume of the droplet, respectively):

$$F_{in}^{aq} = \frac{S}{V} F_{ads} = \alpha \frac{3}{4} \frac{\bar{v}}{r} (c_N)_g = \alpha \frac{3}{4r} \frac{1}{RT} \sqrt{\frac{8RT}{\pi M}} \, p = \frac{3\alpha}{4r\sqrt{2\pi MRT}} \, p \qquad (4.299)$$

The interfacial mass transfer coefficient is $k_{in} = 3\alpha\bar{v}/4r$. Schwartz and Freiberg (1981), Schwartz (1986) and Seinfeld (1986) also introduced a mass transfer coefficient k_g and set the overall mass transfer equation to be (cf. Eq. 4.296):

$$F^{aq} = k_g (c_\infty - c_g) = \frac{k_g}{RT} \left(p - \frac{c_{aq}}{H_{eff}} \right) \qquad (4.300)$$

where:

$$\frac{1}{k_g} = \frac{1}{k_{diff}} + \frac{1}{k_{in}} \qquad (4.301)$$

It follows for the overall mass transfer coefficient, now split in two partial steps, that the diffusion to the surface and the transfer through the surface is:

$$\frac{1}{k_g} = \frac{r^2}{3D_g} + \frac{4r}{3\alpha\bar{v}} \qquad (4.302)$$

We will later see when discussing the dry deposition (Chapter 4.4.1) that a similar conception is applied to explain the "partial conductance"; the first term on the right side in Eq. (4.302) denotes the resistance of diffusion and the second the interfacial transfer. The steps that follow after interfacial transfer are the (fast) salvation and/or protolysis reactions until reaching the equilibrium, the diffusion within the droplet (until reaching steady-state concentrations or, in other words, a well-mixed droplet) and finally the aqueous phase chemical reactions. Let us first consider a pseudo-first-order reaction:

$$F_{chem}^{aq} = -k_{aq} c_{aq} \qquad (4.303)$$

Because there are no other sources of the substance transferred into the droplet than the mass transfer from the gas phase into the droplet (i. e. chemical production in aqueous phase is excluded), we get under steady-state conditions the following:

$$F^{aq} = \frac{k_g}{RT} \left(p - \frac{c_{aq}}{H_{eff}} \right) = F_{diff}^{aq} = \frac{k_{diff}}{RT} \left(p - \frac{c_{aq}}{H_{eff}} \right)$$
$$= F_{chem}^{aq} = -k_{aq} c_{aq} \qquad (4.304)$$

It follows that $k_g = k_{diff}$ and a "modified" *Henry* law including aqueous phase reactions:

$$\frac{p}{c_{aq}} = \frac{1}{H_{eff}} - \frac{k_{aq}}{k_g} RT \qquad (4.305)$$

The mass budget of a soluble and chemical active species in a cloud is now given by the following (considering Eqs. 4.295, 4.303 and 4.304) and assuming that LWC is constant and k_i is the reaction rate coefficient in the gas phase:

$$\frac{dp}{dt} = \sum_i k_i p - \text{LWC} \cdot k_g \left(p - \frac{c_{aq}}{H_{eff}} \right) \tag{4.306}$$

$$\frac{dc_{aq}}{dt} = \sum_i (k_{aq})_i c_{aq} - \frac{k_g}{RT} \left(p - \frac{c_{aq}}{H_{eff}} \right) \tag{4.307}$$

4.4 Atmospheric removal: Deposition processes

Deposition is the mass transfer from the atmosphere to the earth's surface; it is the opposite of emission (the escape of chemical species from the earth's surface into the air). It is a flux given in mass per unit of area and unit of time. According to the various forms and reservoirs of atmospheric chemical species (molecules in the gas, particulate and liquid phases), different physical and chemical processes are distinguished:

- *Sedimentation* of matter because of the earth's gravitational force (this is valid only for particles of a certain size);
- sorption of molecules and small particles at the earth's surface with subsequent vertical transport process, called *dry deposition*;
- sorption of molecules and impaction of particles by falling hydrometeors, called *wet deposition*; and
- *impaction* of particles from an airflow at surfaces.

Gravitational settlement is only of interest for large particles in the upper range of the coarse mode (Chapter 4.3.5). In addition, hydrometeors (raindrops, hail and snow particles) settle because of gravitation but we do not consider water to be deposited (it is physically sedimented, but rather we call the process precipitation) but only the scavenged chemical trace species. Therefore, we consider three reservoirs of trace substances: gaseous (molecules), mixed in APs and dissolved in hydrometeors. Despite the fact that sedimentation and impaction can be important removal processes under specific conditions, they play no significant role in the regional and global budget of fluxes in the climate system. On local sites, however, such removal processes can be important, for example fog droplet impaction by animals (e. g. *Stenocara* beetle) and montane cloud forests, and the sedimentation of dust after volcanic eruptions or soil dust after storm events (for details, see Mark (1999), Pahl (1996), Herckes et al. (2002)). Particle removal is often parameterized as not sharply separating between dry deposition and sedimentation but including deposition processes such as turbulent transfer, Brownian diffusion, impaction, interception, gravitational settling and particle rebound (Zhang et al. 2001). Unfortunately, in the scientific literature the term "dry" for all "non-wet" deposition processes is widely distributed. Fig. 4.21 shows the removal process of gases and particles from the atmosphere (impaction is not shown but it is a simple collision impaction of solid particles and cloud droplets). Without further comments, it is clear

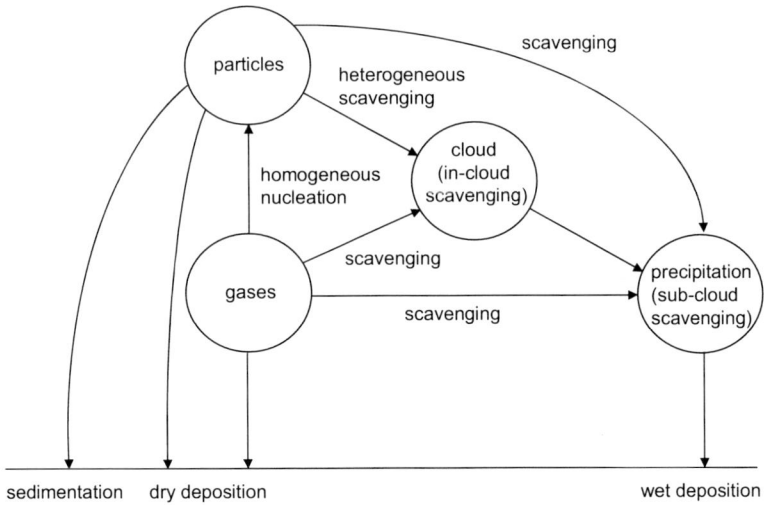

Fig. 4.21 Scheme of deposition processes.

that dry deposition also occurs during precipitation; for soluble gases, dry deposition velocity increases because of the reduced soil resistance (see below). Total deposition onto the earth's surface (normally in sense of an artificial collector surface) is called *bulk deposition*:

$$\text{bulk deposition} = \text{dry deposition} + \text{wet deposition} + \text{sedimentation}$$

From a measurement point of view, there are difficulties in approaching the physically correct removal processes. Dry deposition strongly depends on the surface characteristics; correct estimation is only possible by flux measurements (eddy correlation and gradient methods). Any so-called deposition sampler can be installed with a wet-only/dry-only cover-plate to avoid bulk sampling, but it is never possible to avoid the collection of sedimentation dust in dry-only samplers or dry deposition or sedimentation by wet-only samplers. However, with enough accuracy all other deposition processes contribute negligibly to the deposition flux under wet-only and dry-only conditions, respectively.

4.4.1 Dry deposition

Dry deposition is generic and very similar process to the mass transfer by scavenging described in Chapter 4.3.7 where the transport axis is vertically downwards and the uptaking object is the fixed earth's surface. The situation, however, is complicated by the possible upward flux of the substance A due to plant and soil emission. Only the net flux is measurable (Foken et al. 1995). The case that downward and upward fluxes (so-called bidirectional trace gas exchanges) are equal ($F_{dry} = F_Q$) is called the *compensation point*. It only depends on the concentration gradient ($c_h - c_0$). Additionally, the net flux can be influenced by a fast gas phase reaction within the diffusion layer. Let us now consider the surface as sink, independent of the specific process, which can be:

- surface adsorption;
- interfacial transfer (absorption including chemical reaction); or
- stomatal uptake.

Chamberlain (1953) first introduced the concept of dry deposition and the dry deposition flux is considered of first-order concerning atmospheric concentration (Slinn 1977):

$$F_{dry} = -v_d c \tag{4.308}$$

where v_d is the dry deposition velocity (dimension: distance per time). Because c is a function of the height z, v_d must also related be to a reference height at which c is specified. It is also valid that:

$$F_{dry} = \frac{1}{q}\frac{dn}{dt} = \frac{V dc}{q dt} = \frac{q \cdot h}{q}\frac{dc}{dt} = h\frac{dc}{dt} = -v_d c \tag{4.309}$$

where h is the reference height from which the dry deposition flux starts. The rate of dry deposition follows as:

$$R_{dry} = \left(\frac{dc}{dt}\right)_{dry} = -\frac{v_d}{h}c \tag{4.310}$$

The reference height h is normally considered the mixing height (there are other meteorological definitions):

$$h = \frac{1}{\bar{c}}\int_0^{\infty} c(z)\,dz \tag{4.311}$$

The advantage of introducing the deposition velocity is to avoid a microphysical treatment of vertical diffusion and surface interfacial processes in a single mass transfer coefficient, which is measurable. The limiting application of Eq. (4.309) is because of the dependence of v_d from various parameters and states of the atmosphere and the earth's surface. Besides Eq. (4.309), the general diffusion equation is valid:

$$F_{dry} = K_z \left(\frac{dc}{dz}\right)_{z=h} = v_d c \tag{4.312}$$

where K_z is the turbulent vertical diffusion coefficient. From this equation, it can be seen that v_d is experimentally quantifiable through $v_d = K_z\,(d \ln c / dz)$. The dry deposition process consists of three steps:

- aerodynamic (turbulent) transport through the atmospheric surface layer to the molecular (diffusion) boundary layer close to the surface;
- molecular diffusion transport onto the surface (quasi-laminar sub-layer); and
- uptake by the surface as the ultimate process of removal.

Each step contributes to v_d or the dry deposition resistance $r = 1/v_d$. According to the three layers, the total resistance r is given from the partial resistances (Fig. 4.22), the aerodynamic r_a, the quasi-laminar r_b and the surface resistance r_c:

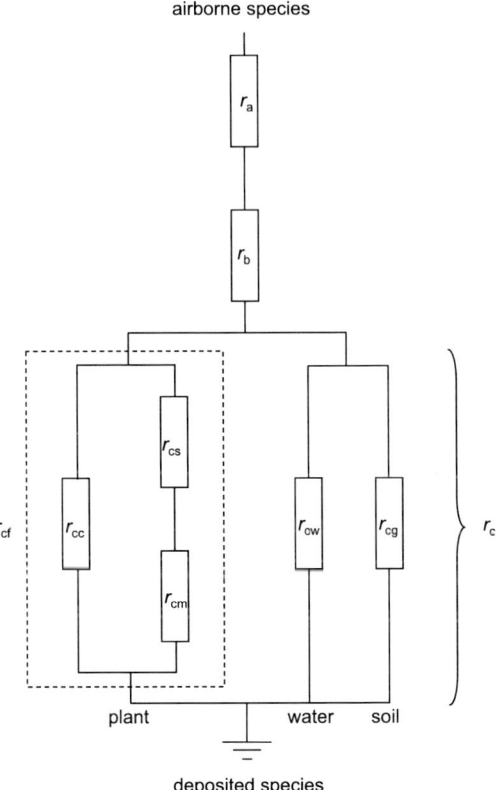

airborne species

plant water soil

deposited species

Fig. 4.22 Resistance model of dry deposition. r_a – aerodynamic resistance, r_b – quasi-laminar resistance, r_c – soil resistance, r_{cw} – soil resistance (water), r_{cg} – soil resistance (other ground), r_{cf} – foliar resistance (weighted by leaf area index), r_{cs} – stomatal resistance, r_{cc} – cuticular resistance, r_{cm} – mesophylic resistance.

$$\frac{1}{v_d} = r = r_a + r_b + r_c, \tag{4.313}$$

This equation follows from the flux through different layers under steady-state conditions with the boundary condition $c_0 = 0$, rearranging after c_1, c_2 and c_3:

$$F = \frac{c_3 - c_2}{r_a} = \frac{c_2 - c_1}{r_b} = \frac{c_1 - c_0}{r_c} = \frac{c_3 - c_0}{r}. \tag{4.314}$$

The limitation of the resistance model lies in its application only for sufficient homogeneous surfaces such as forests, lakes and grasslands. Therefore, in dispersion models dry deposition can be described by using partial areas or weighted partial areas within the grid. The aerodynamic resistance through the upper layer can be calculated using Eq. (4.312) and an approach for the turbulent diffusion coefficient (eddy diffusivity) $K_z = \kappa u_* z$ (valid only for neutral conditions):

$$F_a = \frac{c_3 - c_2}{r_a} = K_z \frac{\partial c}{\partial z} = (c_3 - c_2) \left(\int_{z_2}^{z_3} \frac{dz}{\kappa u_* z} \right)^{-1} \tag{4.315}$$

where κ is the *von Kármán* constant (~ 0.4), u_* is the friction velocity and F_a is the partial aerodynamic flux. The integral in Eq. (4.315) is evaluated from the bottom of the constant flux layer (at z_0, the roughness length of the surface) to the top (z, the reference height implicit in the definition of v_d); therefore, it follows that:

$$r_a(z_3, z_2) = \frac{1}{\kappa u_*} \ln \frac{z}{z_0} \tag{4.316}$$

The roughness length varies from 0.003 m for a flat uncovered surface to 1–6 m for forests and 1.5–10 m for urban areas (Oke 1987). Using the logarithmic wind law

$$u = \frac{u_*}{\kappa} \ln \frac{z}{z_0}$$

it follows from Eq. (4.316) that

$$r_a(z_3, z_2) = \frac{u(z_0)}{u_*^2} \tag{4.317}$$

The aerodynamic resistance grows proportionally with wind speed at the reference height and decreases with increasing roughness of the surface. The quasi-laminar resistance r_b is based on the idea that close to the surface exists a molecular diffusion sub-layer where the transport only depends on the diffusivity of the molecule and the surface characteristics of the surface but not on atmospheric parameters (such as wind speed). The flux under steady-state conditions is parameterized by (B dimensionless transfer coefficient):

$$F_b = Bu_* (c_2 - c_1) \tag{4.318}$$

A useful expression for B in terms of the *Schmidt* number ($Sc = v/D_g$); v is the kinematic viscosity (Chapter 4.1.1.4) is (Wesely 1989):

$$r_b = \frac{5 Sc^{2/3}}{u_*} \tag{4.319}$$

The surface or canopy resistance r_c is almost the dominant resistance in dry deposition and depends on vegetation characteristics, building materials, soil, water and snow surface with different humidity, pH and reactivity concerning the deposited substance A (Hosker and Lindberg 1989; Table 4.18). It is the most difficult of the three resistances to describe (Weseley 1989). Baldochi (1991) defined the plant canopy as a complex consisting of a plant community, plant litter located on the soil surface and incorporated in the upper layers of the soil and the rhizosphere consisting of roots, soil organic matter, microbes and soil fauna. The canopy control of trace gas exchange operates on the interconnected vegetation-litter-soil complex. According to Fig. 4.22, three parallel surface resistances exist: plant (or foliar), water and ground resistance (the resistances in Eq. 4.313 are in series):

Table 4.18 Averaged surface resistances (in s m^{-1}) for different gases during daytime, with stomatal resistance also given; after Erisman and Pul (1997).

		wet, moderate temperature		dry, summer	
		coniferous forest	grassland	coniferous forest	grassland
r_c	SO_2	1[a]	1[a]	300	50
	NH_3	1[a]	1[a]	500[b]	500[b]
	NO_2	320	100	240	80
	O_3	320	100	240	80
	HNO_3	1[a]	0[b]	1[a]	1[a]
	r_{cb}	200	60	150	50

[a] zero but set to 1 to avoid high v_d at low r_a and r_b
[b] under these conditions NH_3 will be emitted by vegetation

$$\frac{1}{r_c} = \frac{1}{r_{cf}} + \frac{1}{r_{cw}} + \frac{1}{r_{cg}} \tag{4.320}$$

The foliar resistance consists again of two parallel resistances: the cuticular resistance r_{cc} and the stomatal resistance r_{cs}, which is in series with the mesophyll resistance r_{cm}.

$$\frac{1}{r_{cf}} = \frac{1}{r_{cs} + r_{cm}} + \frac{1}{r_{cc}} \tag{4.321}$$

Investigations show that mass transfer through the cuticle can be neglected in comparison to stomatal exchange (Kerstiens et al. 2006). The stomata are the leaf openings used in photosynthesis and respiration, respectively. The air spaces in the leaf are saturated with water vapor, which exits the leaf through the stomata (this is known as transpiration). Therefore, plants cannot gain carbon dioxide without simultaneously losing water vapor. Stomatal resistance can, therefore, be calculated from the transpiration rate and humidity gradient. Afterwards the mesophyll (the layer between the upper and lower epidermis) resistance is the final step up gas uptake. In the mesophyll, the palisade cells are exposed directly to the air spaces inside the leaf, which is primary site of photosynthesis in a plant's leaves. The gas transfer through the stomata is by molecular diffusion. An inverse dependence on molecular diffusivity is generally accepted (Grünhage et al. 2000). The surface resistance reduces to a particular resistance in the following cases (Erisman and Pul 1997):

- water surface: $r_c = r_{cw}$
- bare soil: $r_c = r_{cg}$
- snow cover: $r_c = r_{snow}$

The detailed processes in trace gas exchange based on the photosynthesis budget equation are as follows:

$$F_{net.photo} = F_{photo.sun} + F_{photo.shadow} - F_{resp.sun} - F_{resp.shadow}$$

Grünhage and Haenel (2008) provided an explicit parameterization, while Sehmel (1980), Voldner et al. (1985), Hanson and Lindberg (1991), Padro (1996), Wesely

Table 4.19 Averaged dry deposition velocities for SO_2 (in cm s^{-1}); after Möller (2003).

	farmland		grassland		dec. forest		con. forest		urban		water	
	su	wi	su	wi	su	wi	su	wi	su	wi	su	wi
wet	1.0	1.0	1.0	1.0	3.0	1.5	2.0	2.0	1.0	1.0	0.5	0.5
dry	0.7	0.5	0.6	0.4	1.5	0.5	0.7	0.5	0.1	0.1	0.5	0.5
snow	–	0.	–	0.1	–	0.2	–	0.2	–	0.1	–	0.1

Table 4.20 Averaged dry deposition velocities (in cm s^{-1}) above land; after Möller (2003).

substance	SO_2	NO	NO_2	HNO_3	O_3	H_2O_2	CO	NH_3
v_d	0.8	<0.02	0.02	3	0.6	2	<0.02	1

and Hicks (2000) and Petroff et al. (2008) all give reviews on dry deposition. Pioneering work was done by Chamberlain (1953), Hicks and Liss (1976), Slinn (1977), Wesely and Hicks (1977), Garland (1978), Weseley (1989) and many others. Table 4.19 shows the averaged dry deposition velocities for sulfur dioxide for several surface conditions and Table 4.20 the typical v_d for several gases on land.

A special case of large importance is dry deposition onto water surfaces, especially oceans (see for details Danckwerts 1970, Hicks and Liss 1976, Liss and Slinn 1983, Schwartz 1986). The dry deposition flux is described basically by Eq. (4.300):

$$(F_{dry})_{aq} = k_g\left(c - \frac{c_{aq}}{H}\right) \tag{4.322}$$

where c and c_{aq} are the concentrations in the gas and liquid phase, close to the surface, respectively and H is the dimensionless *Henry* constant. It is evident from this equation that for there to be a deposition flux, c_{aq} must be less than $c\,H_{eff}$. For $c_{aq} \to 0$, this equation leads to Eq. (4.309): $F = k_g c \equiv v_d c$. Similarly to the gas transfer described in Chapter 4.3.7.1, we can consider three fluxes: gas phase diffusion close to the water surface (g), interfacial transfer (i) and dissolution including chemistry (L) within the water body. The index i denotes the concentrations immediately adjacent to the interface and the coefficient β represents mass transfer enhancement due to chemical reactions in water; when there is no reaction $\beta = 1$:

$$(F_{dry})_d = k_d(c - c^i) \tag{4.323}$$

$$(F_{dry})_i = \frac{1}{4}\alpha\bar{v}(c^i - c_{aq}^i) \tag{4.324}$$

$$(F_{dry})_L = \beta k_L(c_{aq}^i - c_{aq}) \tag{4.325}$$

Under steady-state conditions, the flux is constant and equal in both media and across the interface. By equating the several expressions (4.322–4.325) for the flux, one obtains the overall mass transfer coefficient k_g as the inverse sum of the mass transfer coefficients in two media and the interface:

$$\frac{1}{k_g} = \frac{1}{k_d} + \frac{4}{\alpha \bar{\upsilon}} + \frac{1}{\beta k_L H} \qquad (4.326)$$

According to Schwartz (1986), large magnitudes of mass transfer coefficients appear sufficient when the interfacial resistance is negligible in the natural environment. In the absence of any aqueous phase, the chemical reaction of the dissolved gas $\beta = 1$ and it follows for non-reactive gases that:

$$\frac{1}{k_g} = \frac{1}{k_d} + \frac{1}{k_L H} \qquad (4.327)$$

It is seen that there is a critical value for the *Henry* coefficient $H_{crit} = k_d / k_L$ for which the gas and liquid phase resistances are equal. For $H \ll H_{crit}$ liquid phase, the mass transport is controlling and dry deposition flux is linearly dependent from H. For $H \gg H_{crit}$, the gas phase transport is controlling and the dry deposition flux is independent of H. For reactive gases, the dry deposition flux is enhanced and it is βH that is compared with H_{crit}.

4.4.2 Wet deposition

As seen from Fig. 4.21, wet deposition is a complex process beginning with heterogeneous scavenging during cloud droplet formation, cloud processing (in-cloud scavenging) and precipitation (sub-cloud scavenging; sometimes called below-cloud scavenging). Hence, several microphysical processes and chemical reactions must be considered. It is remarkable that wet deposition is extremely difficult to model in contrast to dry deposition but is easier to measure by rain gauges (water sampling and analyzing the dissolved substances), whereas dry deposition measurements (vertical flux quantifications) are laborious. The number of precipitation chemistry measurements in uncountable; a huge number of networks has existed since the 1950s. In contrast to dry deposition, wet deposition is occasional. In remote areas, the average fluxes concerning dry and wet deposition are similar but because precipitation occurs for only 5–10 % of the year, the event-based wet flux is considerably larger than the dry flux within a comparable time. Table 4.21 shows characteristic data of cloud and rain events. Despite the complexity of the specific steps in gaining wet deposited substances, we can parameterize the wet deposition flux F_{wet} similar to the approach concerning dry deposition:

$$F_{wet} = -\left(\frac{dc}{dt}\right)_{wet} = k_{wet}\, c \qquad (4.328)$$

By contrast, it is valid from the simple balancing of deposited rainwater that:

$$F_{wet} = R \cdot c_{aq} \qquad (4.329)$$

where c_{aq} is the rainwater concentration of the substance (dimension: mass per liter) and R is the rainfall amount (dimension: liter per unit of time and unit of area). From both equations, λ is called the scavenging coefficient (dimension: time^{-1}):

Table 4.21 Averaged values characterizing wet deposition or precipitation, respectively, typical for Germany; after Möller (2003).

parameter	deep clouds	rain
duration of the event (in h)	6.6	2
duration of dry period (in h)	14	30
volume occupying the mixing layer (in %)	34	100
rain amount (in mm yr^{-1})	–	800
LWC (in mg m^{-3})	0.3	0.05
life time of droplets (in min)	60	10
droplet radius (in μm)	5	500
droplet number	$2 \cdot 10^5$	0.03

$$k_{wet} = \frac{R \cdot c_{aq}}{c} \equiv \lambda \tag{4.330}$$

The older literature is full of "experimental estimations" of λ, also called the washout coefficient, by the measurement of the gas and rainwater concentration of soluble gases. However, that approach is wrong because a) λ is a function of height (it should measured as the vertical gas phase concentration profile and not the surface concentration) and b) a dominant part of the dissolved matter arises from in-cloud scavenging. For sub-cloud scavenging, assuming the washout process to be a first-order process $(dc/dt) = \lambda c$, we can describe the sub-cloud process for gases as well as particles; indexes g and p denote the gas and particle, respectively:

$$F_{sub}(g) = \int_0^\infty \lambda_{sub}^g(z,t) c(x,y,z,t) dz \tag{4.331}$$

$$F_{sub}(p) = \int_0^\infty \lambda_{sub}^p(r,z,t) c_N(r,x,y,z) dz \tag{4.332}$$

The wet deposition velocity was defined by Seinfeld (1986) as:

$$v_{sub} = \frac{F_{sub}}{c(x,y,0,t)} \tag{4.333}$$

For the homogenous distribution of gases within the sub-cloud layer of height h we derive:

$$v_{sub} = \int_0^h \lambda(z,t) dz = h\overline{\lambda_{sub}} = \frac{F_{sub}}{c(x,y,0,t)} \tag{4.334}$$

An in-cloud scavenging coefficient follows by using Eq. (4.298) and (4.300); t_c lifetime of the cloud. Table 4.22 shows the relationship between λ values for SO_2 and H_2O_2 as an example for enhancement due to chemical reactions.

Table 4.22 Mean scavenging coefficients λ (in 10^{-4} s^{-1}) of H_2O_2 in dependence on the SO_2 gas phase concentration c (in ppb), calculated from a cloud model with coupled gas-aqueous chemistry; after Möller (1995b).

species	$c(SO_2)$ and $\lambda(SO_2, H_2O_2)$			
	0.1 ppb	1.0 ppb	2.0 ppb	5.0 ppb
H_2O_2	4.1	13.5	69.0	269.0
SO_2	27.9	8.8	3.6	0.8

$$\lambda_{in-cloud} = \boldsymbol{R}T \cdot \text{LWC} \int_o^{t_c} \frac{F^{aq}(t)}{c(t)} dt \qquad (4.335)$$

Now considering the raindrop distribution, we get an averaged rain event scavenging coefficient:

$$\overline{\lambda_{sub}} = \boldsymbol{R}T \cdot \text{LWC} \int_o^{\infty} \pi r^2 \lambda_r \cdot N(r) dr. \qquad (4.336)$$

4.5 Characteristics times: Residence time, lifetime and turnover time

The *residence time* (often interchangeable with the term *lifetime* but later we will distinguish between the two terms) is the average amount of time that a particle spends in a particular system. This definition is adapted to fit with groundwater, the atmosphere, glaciers, lakes, streams and oceans. It was first introduced by Barth (1952) as the ratio of the total amount m of an element in the ocean to the rate of input (flux) F to the ocean: $\tau = m/F$. Junge (1963, 1974), Bolin and Rodhe (1973) and Bolin et al. (1974) first applied this conception to atmospheric trace gases.

We now consider the various flux equations in the general form (the removal flux can be given by different specific processes such as deposition, decomposition or transportation):

$$\begin{aligned} F_{rem} = k_r c &= -\left(\frac{dc}{dt}\right)_r = -\frac{1}{V}\left(\frac{dm}{dt}\right)_r = -\frac{M}{V}\left(\frac{dn}{dt}\right)_r \equiv \sum_i F_i \\ &= \sum k_i c \end{aligned} \qquad (4.337)$$

where k_r is the removal coefficient (dimension: time^{-1}) and c is the concentration (remember we always regard the substance A and don't use the subscription c_A to avoid confusing with other subscripts). The flux, therefore, has the dimension mass per unit of time and unit of volume, corresponding to the rate definition in chemistry. The reciprocal removal coefficient denotes a *characteristic time* τ_c:

$$\tau_c = \frac{1}{k_r} \tag{4.338}$$

Formally, it leads to:

$$\tau_c = \frac{c}{F_{rem}} \tag{4.339}$$

It is important to note that $c(t)$ is not constant over time and thereby the removal flux depends on time (it decreases over time if c is decreasing) or, in other words, τ depends on concentration or reservoir mass. Moreover, the removal coefficient is not obligatorily constant, i. e. $k_r(t)$ leads to $\tau(t)$. Only in the case of constant k_r (we call it the first-order removal process), it follows from Eq. (4.337) that:

$$c = c_0 \exp\left(-\frac{t}{\tau_c}\right) \tag{4.340}$$

from which we define a *turnover time* τ_t as:

$$\tau_t = -\ln\left(\frac{c}{c_0}\right)\tau_c \tag{4.341}$$

From Eq. (4.340), it follows that $c \to 0$ for $t \to \infty$. At $t = \tau_c$, $c/c_0 = \ln(-1) \approx 0.37$ or, in other words, the initial amount of A in the volume (or reservoir) decreases to about 37 % (1/e). We have seen that the condition $c = 0$ is undefined because of the exponential removal law. However, we can set the condition $c \ll c_0$ and it follows for $c = 0.01 \cdot c_0$ that:

$$\tau_t = 4.6\,\tau_c.$$

The turnover time of a substance denotes that time when the substance in the volume regarded becomes "practically" zero. Only in the case of the zero-order removal process, i. e. the removal rate is constant over time and depends not on the concentration and τ_t marks exactly the time when $c = 0$. The basic equation (4.337) changes into (note that the removal coefficient k_r^0 becomes identical to the removal flux and gets another dimension):

$$F_{rem} = k_r^0 = -\left(\frac{dc}{dt}\right) = -\frac{1}{V}\left(\frac{dm}{dt}\right) = -\frac{M}{V}\left(\frac{dn}{dt}\right) \tag{4.342}$$

The solution of Eq. (4.342) is:

$$c_0 - c = F_{rem}\,t \tag{4.343}$$

It follows (F_{rem}^V is the reservoir- or volume-based flux in mass per unit of time):

$$\tau_t = \frac{c_0}{F_{rem}} = \frac{1}{V}\frac{m_0}{F_{rem}} = \frac{m_0}{F_{rem}^V} \tag{4.344}$$

The often used equation in the form $\tau = m/F$ (see Eq. 2.142)[31] referring the life time and here given in terms of Eq. (4.344) with *constant* m and F represents the turnover time for zero-order removal (constant removal rate); it is not a residence time! However, when we consider time-depending mass and flux, we will later see that it corresponds to the residence time:

---

[31] We use m instead of M for mass to avoid confusing with the molar mass.

$$\tau = \frac{m(t)}{F(t)} \equiv \frac{m(t)}{dm} dt = \frac{dt}{d \ln m}.$$

The mass of A in a reservoir is given by the three-dimensional concentration distribution within the reservoir:

$$m = \int_0^{x_x} \int_0^{y_y} \int_0^{z_z} c(x, y, z) \, dx \, dy \, dz \qquad (4.345)$$

The residence time τ is generically defined as the average time a substance spends within a specified region of space, such as a reservoir. We now define τ as the arithmetic mean of the individual lifetimes of all molecules of substance A in the reservoir. It must be noted that there is much confusion in the scientific literature over the different characteristic times. The *lifetime t* of a molecule (it is similar to *life expectancy at birth* of a biological species) is the time the molecule spends in the reservoir from entering to leaving it:

$$\tau - \frac{1}{N_0} \int_{N_0}^{0} t \, dN \qquad (4.346)$$

The specific expression for residence time depends on the kinetic law applicable to $N(t)$. In the simple case of (pseudo-)first-order law $dN = -k_r N_0 \exp(-k_r t) \, dt$ and replacing the number by the concentration ($N \sim c$), it follows that:

$$\tau = k_r \int_0^{\infty} t \cdot \exp(-k_r t) \, dt \qquad (4.347)$$

The solution[32] is:

$$\tau = \frac{1}{k_r} \qquad (4.348)$$

Compared with Eq. (4.338), we conclude that the characteristic time τ_c corresponds to the residence time τ for removal processes of the first-order rate. Using the *residence time distribution* approach, we define:

$$\tau = \int_0^{\infty} t \, df_A(t) \qquad (4.349)$$

where f_A denotes the normalized distribution function of the age of the molecules of A $\int_0^{\infty} f_A(t) \, dt = 0$, which is in the following relationship to the concentration:

---

[32] Because $\int x \cdot \exp(ax) \, dx = \dfrac{\exp(ax)}{a^2} (ax - 1)$ it follows $\tau = k_r \dfrac{\exp(-k_r t)}{k_r^2} (-k_r t - 1) \Big|_0^{\infty} = \dfrac{k_r}{k_r^2}$

$$f_A = \frac{c_0 - c(t)}{c_0} = 1 - \frac{c(t)}{c_0} \tag{4.350}$$

With $df_A(t) = -\frac{1}{c_0}\left(\frac{dc}{dt}\right)dt$, a general definition of residence time, not depending on the specific kinetic law of concentration decrease (depletion of A), follows from Eq. (4.349):

$$\tau = -\int_0^\infty t\left(\frac{dc}{dt}\right)\frac{1}{c_0}\,dt \tag{4.351}$$

From Eq. (4.351) follows Eq. (4.347) when using a first-order rate law $(dc/dt) = -k_r c_0 \exp(-k_r t)$ and, therefore, $\tau = 1/k_r$. With the logical assumption that $c(t) \to 0$ for $t \to \infty$, Eq. (4.351) simplifies to (the second term becomes zero):

$$\tau = -\int_0^\infty t\left(\frac{dc}{dt}\right)\frac{1}{c_0}\,dt = \int_0^\infty \left(\frac{dc}{dt}\right)\frac{1}{c_0}\,dt - \left(\frac{dc}{dt}\right)\frac{1}{c_0}t\bigg|_0^\infty = \int_0^\infty \left(\frac{dc}{dt}\right)\frac{1}{c_0}\,dt \tag{4.352}$$

Now, we assume the general time law $\dfrac{dc(t)}{dt} = -k_r c(t)^\alpha$; for $\alpha < 1$ follows a hyperbolic characteristics or, in other words, $c \to 0$ for $t \to \infty$ (infinite time) and for $\alpha > 1$ it follows parabolic characteristics, i.e. $c \to 0$ is obtained in a finite time. For these cases, the following solution from Eq. (4.352) is valid:

$$\alpha > 1 : c(t) = \frac{\theta}{(tg + t)^{\frac{\alpha}{\alpha - 1}}}$$

$$\alpha > 1 : c(t) = \frac{\theta}{(tg - t)^{\frac{\alpha}{1 - \alpha}}}\,,$$

where $\theta = \left(|\alpha - 1|k_r\right)^{\frac{-1}{\alpha - 1}}$ and $tg = \left(c_0^{\alpha-1}|\alpha-1|k_r\right)^{-1}$. In both cases, the solution of the integral results as $\alpha < 2$ to:

$$\tau = \int_0^\infty \frac{c(t)}{c_0}\,dt = \frac{c_o^{1-\alpha}}{(2 - \alpha)k_r} \tag{4.353}$$

This equation represents the general expression for the residence time of $1 \leq \alpha \leq 2$. The case $\alpha = 1$ represents the first-order rate law as already described and Eq. (2.142) becomes identical to Eq. (4.338). Now, we see that the residence time is independent of time only if k_r is constant. With $F_{rem} = -(dc/dt)$ and $k_r = F_{rem}/c$ it follows that:

$$\tau = \frac{c(t)}{F_{rem}(t)} \equiv \frac{dt}{d\ln c} \tag{4.354}$$

An example for the application of Eq. (4.354) is shown in Fig. 4.23, where the long-term measurement of CH_3CCl_3 data (Fig. 2.77) results in a linear function $\lg c$

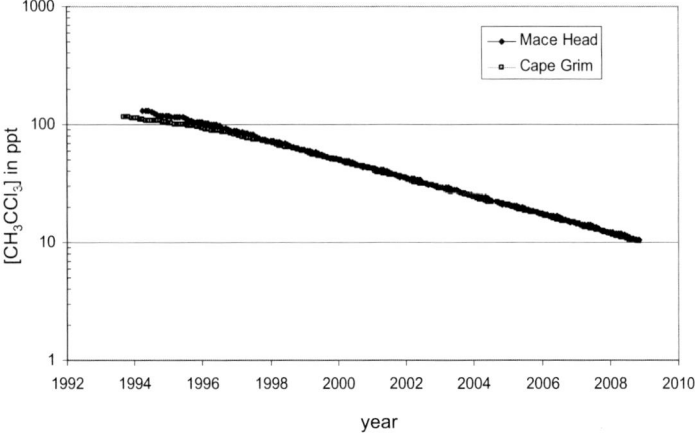

Fig. 4.23 Historical record of methyl chloroform (CH_3CCl_3) in logarithmic relationship; data from Fig. 2.77.

$= f(t)$. It follows from the diagram that $d \ln c / dt = 0.315$ and, therefore, $\tau = 3.2$ years.

According to the resistance model, we have to consider parallel and serial removal processes. Without consideration of source terms (see later), we have the budget equation (first-order law) of:

$$F_{rem} = \left(\frac{dc}{dt}\right)_r = \sum_{i=1}^{n} \left(\frac{dc}{dt}\right)_i = \sum_{i=1}^{n} k_i c \tag{4.355}$$

from which a relationship between the overall residence time τ and process-specific residence times τ_i follows, for example concerning dry deposition τ_d, wet deposition τ_{wet} and chemical reactions τ_{chem}:

$$\frac{1}{\tau} = \sum_{i=1}^{n} \frac{1}{\tau_i} \tag{4.356}$$

In the case of subsequent (in serial) processes, the residence times sum up as:

$$\tau = \sum_{j=1}^{m} \tau_j \tag{4.357}$$

when the fluxes branch out and pass several reservoirs (number o) parallel with the fluxes F_k.

Where $\Sigma F_k = F_{rem}$, the residence follows as the weighted mean:

$$\tau = \frac{1}{F_{rem}} \sum_{k=1}^{o} F_k \, \tau_k \tag{4.358}$$

In all cases, it is valid that the overall residence time = the total mass (or mean concentration) / total removal flux (Eq. 4.354) at a given time. The removal flux normally splits into dry and wet removal as well as chemical transformation. These flux equations have been shown in previous chapters and we now use the reciprocal characteristic mass transfer coefficients in the sense of specific residence times concerning dry deposition, wet deposition and chemical decomposition:

$$F_{rem} = k_r c = F_{dry} + F_{wet} + F_{chem} = \left(\frac{k_d}{h} + k_{wet} + \sum_i k_i \right) c \qquad (4.359)$$

We are often interested in determining the relative importance of different sinks contributing to the overall removal of a species. For example, the fraction f removed by the process i is given by:

$$f_i = \frac{F_i}{F_{rem}} \qquad (4.360)$$

So-called wet events such as wet deposition and aqueous phase chemistry in clouds and precipitation are episodic, i. e. the specific residence time during the dry period is not defined (flux is zero). Regarding averaging over longer periods containing several alternating dry periods and wet events, Rodhe (1978) developed an expression for the averaged wet removal residence time:

$$\overline{\tau_{wet}} = \tau_a \frac{\tau_a}{\tau_a + \tau_b} + \frac{\tau_b + \tau_a}{\tau_a} \tau_{wet} = A + B \tau_{wet} \qquad (4.361)$$

The subscripts a and b characterize the mean time of the dry periods (non-rainy and/or cloudy-free) and wet periods (clouds, rain), respectively. τ_{wet} represents the residence time during the wet event, either according to Eq. (4.329) for wet deposition or Eq. (4.303) for aqueous phase chemistry. The constant A corresponds to a fractional time of the dry period time τ_a, which becomes smaller when the ratio between the dry and wet periods decreases. The coefficient B is normally larger than one (if $\tau_a > \tau_b$). Most likely $\tau_a \gg \tau_b$ but $\tau \ll \tau_b$ follows $\tau \rightarrow \tau_a$. Eq. (4.361) represents the flow through two reservoirs according to Eq. (4.357), first a space without wet events (dry period) and then a space with a wet event. For a given atmospheric volume, we might also define a transport residence time as:

$$\tau_{advection} = \frac{2 \bar{u}}{x} \qquad (4.362)$$

where $\bar{u}$ is the mean wind velocity and x is the length of the atmospheric reservoir. It follows for the overall atmospheric residence time of a substance that:

$$\frac{1}{\tau} = \frac{1}{\tau_{chem}^g} + \frac{1}{\tau_{dry}} + \frac{1}{\tau_{wet} + \tau_{chem}^{aq}} + \frac{1}{\tau_{advection}} \qquad (4.363)$$

Similar to Eq. (4.360), we can calculate the fractions of specific removal processes by:

$$f_i = \frac{\tau}{\tau_i} \qquad (4.364)$$

because $\Sigma \left(\tau / \tau_i \right) = 1$. Different, i. e. process-related residence times for sulfur dioxide are listed in Table 4.23.

Table 4.23 Mean residence time of gaseous and dissolved SO_2; after Möller (1995a, b, 1996).

chemical regime	residence time
gas phase concerns OH oxidation:	
averaged over one year	~ 20 d
sunny days (day time)	≤ 4 d
liquid phase:	
within the droplet (day time)	~ 2 min
averaged over the mixing layer:	
cloud	~ 3 h
rain	~ 6 h
averaged over one year:	
cloud	~ 0.8 d
rain	~ 2.9 d
overall average over one year	1−2 d

5 Substances and chemical reactions in the climate system

As mentioned in Chapter 1.2, this book will treat the chemistry of the *climate system* according to the elements and its compounds depending on the conditions of reactions and whether the substance exists in the gaseous or condensed (aqueous and solid) phase (Fig. 5.1).

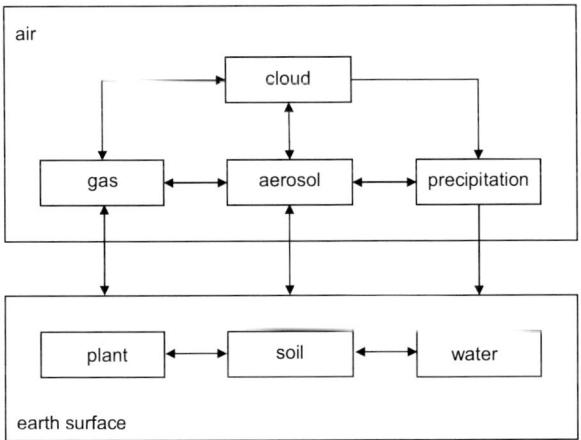

Fig. 5.1 The chemical reservoirs and mass-transfers in the climate system.

5.1 Introduction

5.1.1 The principles of chemistry in the climate system

The chemistry of the climate system is more than simply air or atmospheric chemistry, but *it is* to a large extent atmospheric chemistry. The atmosphere is the only reservoir connecting all other reservoirs such as the biosphere (plants), hydrosphere (waters) and pedosphere (soils). As chemistry is the science of matter, we consider substances moving through the climate system. We have already mentioned that elementary chemical reactions are absolutely described by a mechanism and a kinetic law and, therefore, are applicable to all regimes in nature if the conditions (temperature, radiation, pressure, concentration) for this reaction correspond to those of the natural reservoir. For example, the photolytic decay of molecules is only possible in the presence of adequate radiation. Hence, in darkness there is no

Table 5.1 Chemical regimes in the climate system, characterized by temperature T, pressure p, H_2O vapor pressure, radiation (wavelength λ) and trace concentrations.

T	p	H_2O	λ (in nm)	traces	regime
low ($< 200\,°C$)	very low	very low	< 300	very low	stratosphere and upper atmosphere
normal ($\sim 285 \pm 30\,°C$)	normal	normal	> 300	normal	troposphere, waters and plants
high ($\leq 1000\,°C$)	normal	large	$-$	high	biomass burning
very high ($> 1300\,°C$)	normal-high	normal	> 300	normal	lightning

photolysis and within the troposphere there is no O_2 photolysis because of the lack of radiation lower than a wavelength of 300 nm.

Table 5.1 classifies how chemical regimes meet in the climate system. We see that almost "normal" conditions occur and extreme low and high temperatures border the climate system. The chemistry described in the following chapters concerns almost these normal conditions of the climate system. We focus on the troposphere and the interfaces. For example, aqueous phase chemistry in cloud droplets does not differ principally from surface water chemistry (aquatic chemistry) and much soil chemistry does not differ from aerosol chemistry (colloidal chemistry). Plant chemistry, however, is different and only by using the generic terms (Chapter 2.2.2.3) of inorganic interfacial chemistry can we link it. The chemistry of the atmosphere is widely described (Seinfeld and Pandis 1998, Warneck 1999, Finlayson-Pitts and Pitts 2000, Wayne 2000, Brasseur et al. 2003) and the branches in atmospheric chemistry are well defined (Fig. 5.2).

The approach of this book is to use the classical chemical textbook concept based on the elements and its compounds. However, it is not the aim of this chapter to include as many substances and reactions as possible. Two criteria always help make the chemistry as simple as possible. First, we only consider the main reactions of a species A with competitive chemical pathways; all reactions that contribute less than 1 % to species A conversion are neglected[1]. Let us regard as an example the two reactions ($A = O(^3P)$):

$$O(^3P) + O_2 \rightarrow O_3; \quad \text{and} \tag{a}$$

$$O(^3P) + SO_2 \rightarrow SO_3. \tag{b}$$

There are hundreds of reactions of $O(^3P)$ described[2] but under normal conditions in the climate system (Table 5.1), the pathway (a) is the only relevant reaction because the overall rate is many orders of magnitude larger than all other competitive reactions because of the size of the oxygen concentration compared with all other

---

[1] This condition is reliable because any quantification of fluxes, budgets and pool concentrations in the climate system has a large uncertainty; an error of only 10 % is an excellent result.
[2] See, for example, the IUPAC subcommittee for gas kinetic data evaluation (http://www.iupac-kinetic.ch.cam.ac.uk/; Atkinson et al. 2004, 2005, 2007, 2008).

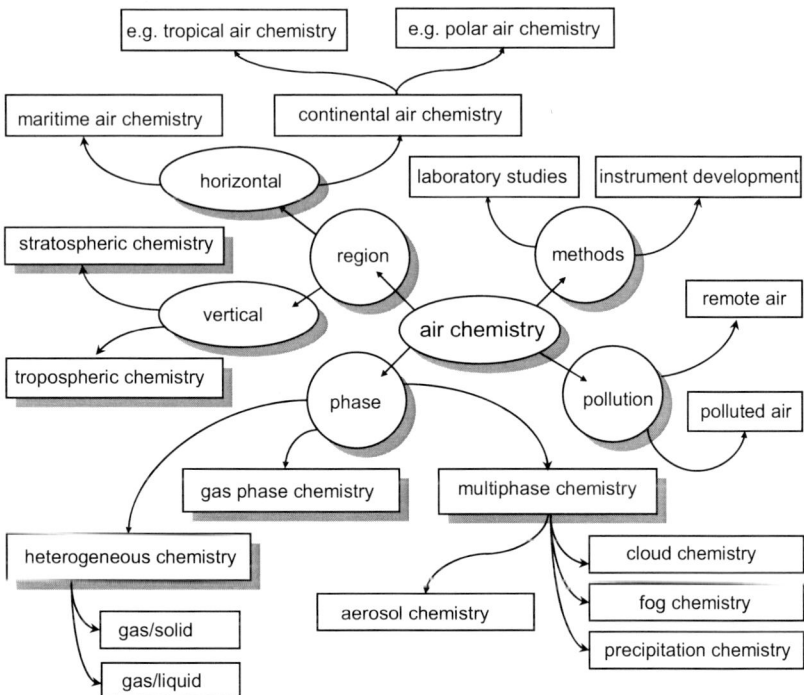

Fig. 5.2 Sub-disciplines of atmospheric chemistry.

trace gas concentrations. Second, we only consider a limited number of chemical species (next chapter) that play a significant role in the climate system. We do not regard the huge number of organic compounds of different origins and different fates in the environment (this is a special task for environmental chemistry). It is possible to understand the climate system chemistry by "lumping" together characteristic groups with specific "roles". A key question, however, is their specific and/or significant roles within the climate system. We focus on chemical species contributing to:

- acidity/alkalinity (atmospheric acidifying potential);
- oxidizing/reducing agents (atmospheric oxidation capacity);
- particulate matter formation (global cooling);
- greenhouse effect (global warming); and
- ozone depletion (UV radiation effects).

Climate system chemistry is driven by biogeochemical cycling where the emissions from plants and microorganisms (in soils and waters) in terms of rate and substances specify the chemical regime. Behind the chemistry of the climate system stands the chemistry in the lower atmosphere with interfaces *to* soils, waters and plants. According to the scheme shown in Table 5.2, we classify this as multiphase chemistry. Discussing the fate, transport and transformation of chemicals *in* soils, waters and plants is the task of biogeochemistry; for instance, marine biogeochemistry

Table 5.2 Scheme of chemical interactions.

multiphase	homogeneous	gas-phase aqueous-phase solid phase
	heterogeneous	gas-solid gas-aqueous solid-aqueous

(Turner and Hunter 2002, Hansell and Carlson 2002, Fasham 2003, Libes 2009), trace elements (Braids and Cai 2002, Prasad et al. 2005, Huang and Gobran 2005), wetlands (Bianchi 2006, Reddy and DeLaune 2008), soils (Bohn et al. 2001, Sposito 2008, Konya and Nagy 2009), soils and water (Essington 2003, Franzle 2009, Tan 2010), water chemistry (Gianguzza et al. 2002, Jensen 2003, Yves 2006, Kuwajima et al. 2009, Jensen 2015), plant chemistry (Haas and Hill 1929, Heldt 2004) and space chemistry (Lewis 2004, Rehder 2010).

5.1.2 Substances in the climate system

Life determines the global biogeochemical cycles of the elements of biochemistry, especially C, N, P and S (Schlesinger 2003), but the key process of plant life is the water splitting process into H and O (Chapter 2.4.4) and thereby creating oxic (oxidizing) and anoxic (reducing) environments. Fig. 5.3 shows schematically the principal three groups of compounds and their biogeochemical cycling. Hydrides (methane, ammonia, hydrogen sulfide and water) represent the lowest oxidation state of the elements C, N, S and O because they bond in the biomass and are then released into the abiogenic surroundings (soil, water and air). In the air, these substances (and many more such as hydrocarbons, amines and sulfides) are oxidized, finally to oxides of C, N and S. Many oxides represent anhydrides and form oxoacids and subsequent salts (such as carbonate, sulfate, nitrate) when reacting with water. The important role of acidity in weathering has already been discussed in Chapter 2.5.5. This last group provides the stock chemicals for the biosphere that are reduced to hydrides and close the cycle. There are other important elements. First, hydrogen, which in its molecular form (H_2) is relatively non-reactive but exists atomically as short-lived transient and stable hydronium ion in solutions. Its role (Fig. 2.31) alters between water, hydrides and acids. Second, phosphorous is also a key bioelement. P has few and instable volatile compounds and plays only a minor role in atmospheric chemistry. Moreover, there are halogens (F, Cl, Br, I) that we could also include in Fig. 5.3 because they form hydrides (acids such as HCl), oxides (e. g. ClO) and oxoacids (e. g. HOCl). As discussed in Chapter 2.4.3.3, the role of organic Cl compounds in soils has only recently been investigated. Finally, many trace elements are biogeochemically important, either because of their roles as redox elements or as bioelements with very specific properties. However, some trace elements (for example Cd and Hg) have no biological functions but are extremely poisonous.

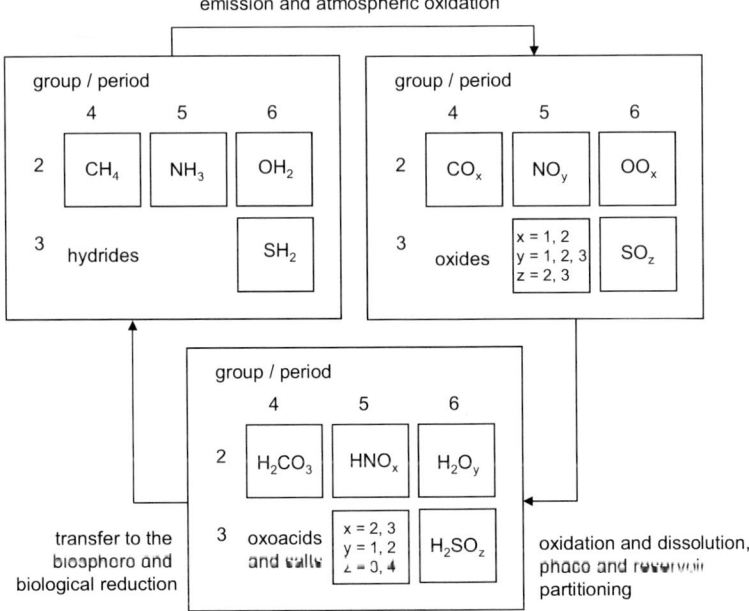

Fig. 5.3 Characteristic groups for C, N, O and S and its role in biogeochemical cycling.

Table 5.3 Principle main compounds in the climate system. PM – particulate matter, SOA – secondary organic aerosol, OA – (primary) organic solid matter (biogenic), EC – elemental carbon, red. S – H_2S, COS, CS_2, DMS and others.

gas	liquid	dissolved[a]	solid
H_2O	H_2O (water)	$H^+ + OH^-$	H_2O (ice, snow)
CO_2	–	$H^+ + HCO_3^-$	PM
HCl	–	$H^+ + Cl^-$	(chloride,
NH_3	–	$NH_4^+ + OH^-$	ammonium,
NO_y	–	$H^+ + NO_2^- / NO_3^-$	carbonate,
SO_2	–	} $H^+ + HSO_3^- / SO_4^{2-}$	nitrate,
red. S	–		sulfate) PM
NMVOC	–	NMVOC[b]	SOA
–	–	OA[b]	OA
–	–	–	EC + BC (soot)
–	–	$Na^+ + Cl^-$[c]	seasalt (NaCl)[c]
		$K^+ + Ca^{2+} + Mg^{2+}$[d]	dust[e]
CH_4	–	–	–

[a] partly after oxidation
[b] if soluble, organic acids protolysed
[c] about 90% NaCl but including other compounds (sulfate, magnesium, calcium, potassium and many other)
[d] and many others, including anions (carbonate, sulfate)
[e] insoluble main components: Si, Al

Table 5.4 Main-group elements in gaseous compounds in the climate system (and Hg from transition elements). Note: elementary in air only exist novel gases, nitrogen (N_2), oxygen (O_2), hydrogen (H_2), and in traces halogen atoms, atomic oxygen (O).

	4	5	6	7	8
1				H	He
2	C	N	O	F	Ne
3	Si	P	S	Cl	Ar
4		As	Se	Br	Kr
5				I	Xe
6					Ra

Table 5.5 Principal main educts and products (simplified).

element / group	educts and intermediates	final products
oxygen	O_3, OH, HO_2, H_2O_2	O_2, H_2O
inorganic carbon	CO, CO_2	CO_2, H_2CO_3, HCO_3^-
sulfur	H_2S, DMS, COS, CS_2, SO_2	H_2SO_4, SO_4^{2-}
nitrogen	NO, NO_2, NO_3, N_2O_5, HNO_2	HNO_3, NO_3^-
clorine[a]	Cl_2, Cl, HCl	HCl, Cl^-
organic carbon	RCH_3, RCHO	RCOOH, $RCOO^-$

[a] representative for other halogens (F, Br, I)

Table 5.3 shows the distribution of main compounds in the climate system among different state of matter. It is remarkable that only one substance (H_2O) exists in all three phases. Not all gases exist in dissolved form in the aqueous phase but the main inorganic ones (essential for life such as sulfurous, nitrous, carbonaceous, chloride, ammoniacial) exist gaseous, dissolved, and particulate. All dissolved species also exist in particulate matter. All insoluble species exist either only as gases or particulates. Insoluble but oxidable substances exist dissolved and hence in PM too. However, some organic substances can be transformed into solids (homogeneous nucleation process) but need not be soluble.

Excluding novel gases, the number of gaseous elements is very limited: N, O, H, Cl (Table 5.4). All these elements form many gaseous compounds. It is remarkable that the number of further elements (which are not gaseous) forming gaseous compounds is also limited: C, N, S, Br and I (Table 5.4). Table 5.4 lists more elements (Si, P, As, Se) that also form gaseous compounds but play no role in the climate system (P may be an exception, see Chapter 5.6). In addition, many compounds and a few other elements (such as Br, I and Hg) can be detected as gaseous in the atmosphere despite having boiling points larger than the temperature at the earth's surface and having volatile properties. The number of primarily emitted important substances, intermediates and final products is relatively limited (excluding organic compounds; Tables 5.5 and 4.4). All kinetic data given in the following chapters are from the IUPAC subcommittee for gas kinetic data evaluation if not otherwise cited (also Atkinson et al. 2004, 2005, 2007, 2008).

5.2 Hydrogen

The troposphere has an estimated 155 Tg of hydrogen gas (H_2), with approximately a two-year lifetime (Chapter 2.8.2.10). Many sources of hydrogen gas and a few major sinks account for this relatively short lifetime. The main pathway in the production of hydrogen atoms in the air is the methane (CH_4) conversion by the OH radical and subsequent photolysis of formaldehyde (HCHO); see reactions (5.42) to (5.48). This process accounts for about 26 Tg H yr^{-1} (Novelli et al. 1999).

$$CH_3 \xrightarrow{\text{OH}} HCHO \xrightarrow{hv} H$$

Once H is formed in the troposphere it combines with O_2 to produce the important HO_2 radical (Chapter 5.3.2): $k_{5.1} = 5.4 \cdot 10^{-32} \, (T/300)^{-1.8} [N_2] \, cm^3 \, molecule^{-1} \, s^{-1}$ over the temperature range 200−600 K:

$$H + O_2 + M \rightarrow HO_2 + M \tag{5.1}$$

Hugh Taylor first proposed the formation of HO_2 (in the photochemical $H_2 + O_2$ reaction) in 1925[3] and it was detected by mass spectroscopy in 1953 (Foner and Hudson 1953). The relatively fast reaction with the HO_2 radical, producing OH, can be of interest only in the stratosphere; $k_{5.2} = 7.2 \cdot 10^{-11} \, cm^3 \, molecule^{-1} \, s^{-1}$ (298 K):

$$H + HO_2 \rightarrow 2\,OH \tag{5.2a}$$

This reaction splits in two further product pathways of minor importance:

$$H + HO_2 \rightarrow H_2 + O_2 \tag{5.2b}$$

$$H + HO_2 \rightarrow H_2O + O \tag{5.2c}$$

The main sink (besides soil uptake) of molecular hydrogen is the relatively slow reaction with OH; $k_{5.3} = 7.7 \cdot 10^{-12} \exp(2100/T) \, cm^3 \, molecule^{-1} \, s^{-1}$ ($6.7 \cdot 10^{-15}$ at 298 K):

$$H_2 + OH \rightarrow H + H_2O \tag{5.3}$$

Hence, if there is more H_2 in the stratosphere it will react with hydroxyl radicals so there will be more H_2O in the stratosphere. Modeling studies show that this increase in H_2O cools the lower stratosphere. There are several excited species of molecular hydrogen (Bozek et al. 2006) but there is no direct photodissociation of molecular hydrogen in the interstellar medium because atomic hydrogen depletes the spectrum above 13.6 eV, which corresponds to a wavelength smaller than 90 nm (Balashev et al. 2009). At temperatures of 2000 K, only 0.081% of H_2 dissociates and this fraction increases to 7.85% at 3000 K (95.5% at 5000 K). In Chapter 4.2.5.2, we met the species H^+ (proton) in the aqueous phase; however, this does not freely exist and is only present in combination with H_2O as the hydronium ion H_3O^+. In the gas phase, H^+ (but not under atmospheric conditions because the ionization energy is very high: 1311 kJ mol^{-1}) can be produced from atomic hydrogen. Atomic hydro-

---

[3] Trans. Faraday Soc. 21, 560

gen has a large affinity to electrons (H⁻) but this species exists only in hydrides. Most elements produce hydrides from very stable – the best known is water OH_2 – to very unstable. It is likely that at an early state of chemical evolution (as assumed for nitrides; Chapter 2.1.3) many unstable metal hydrides (such as NaH) are formed. In the accretion phase of the earth, they then decompose to hydrogen (H_2) in contact with water (reaction 5.68). Hydrogen undergoes several reactions in the aqueous phase (Chapter 5.3.4).

5.3 Oxygen

Oxygen is the most abundant element in the climate system. The atmosphere consists of 21% oxygen, the oceans 85% and the earth's crust 46%. The mass of the atmosphere is negligible and the oceans and crust contribute about 10% and 90%, respectively (in mass %). With respect to the climate system that comprises air, waters, plants and soils, oxygen is found to be 99% in the form of water (Table 2.12). With respect to the earth's system, oxygen is found to be 99% in the form of silicates (Table 2.13). In the air, oxygen (O_2) is found to be about 98% (the remaining form of oxygen is water vapor and that held in CO_2 is negligible). From a climate system point of view, oxygen is the carrier of carbon (in the form of CO_2 and carbonate) and hydrogen (as H_2O), the main building blocks of life. Atmospheric oxygen in its different forms (O, O_2 and O_3) provides oxidizing agents for the decomposition of organic compounds (symbolized in Fig. 5.4 as $(CH_2O)_n$) where it turns back into H_2O via several reactive (and biologically and atmospherically important) H_yO_x species (Chapter 5.3.2). Oxygen is only known in electronegative bondings, with the exception of peroxides (−1) almost in the oxidation state −2. Only fluorine is more electronegative than oxygen but elemental F does not exist in nature. Therefore, it forms oxides with all other elements[4]. In the past few decades, the role of molecular oxygen to form metal complexes (superoxo and peroxo), which play the role of oxygen carriers in biochemistry, have been studied. Molecular oxygen in its $^3\Sigma_g^-$ ground state has triplet multiplicity but not singlet multiplicity unlike most natural compounds. The unpaired electrons in two different molecule orbitals account for the paramagnetism of molecular oxygen. The high O_2 dissociation energy (493.4 kJ mol⁻¹) does not allow, under tropospheric conditions, photolytic decay into oxygen atoms. Triplet multiplicity is the reason why most reactions of oxygen with organic substances, although exergonic, do not proceed at normal temperature but upon heating or in the presence of catalysts. Thus, reactions of organic compounds with dioxygen are kinetically inhibited. The other main and very stable oxygen compound is water (H_2O) because the O—H bonding cannot be broken by photodissociation within the climate system.

Therefore, oxygen comprises four chemical groups from O_1 to O_4 (Fig. 5.5). Transfers between these groups provide very few special reactions. The variety of

---

[4] Xenon (Xe) also forms oxides (and fluorides); from He, Ne and Ar stable compounds are unknown.

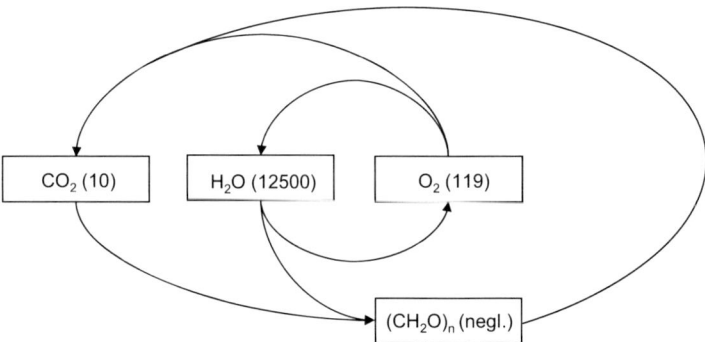

Fig. 5.4 Oxygen in chemical reservoirs and chemical cycling; numbers represents total mass of oxygen in 10^{19} g (CO_2 is almost as carbonate dissolves in ocean and O_2 molecular in air), $(CH_2O)_n$ represents organic compounds including biomass.

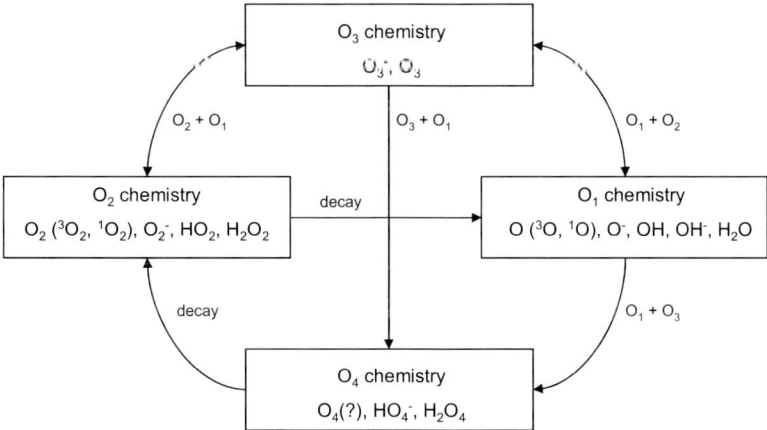

Fig. 5.5 Chemical species and schematic relationships between $O_1 - O_4$ chemistry.

species within groups O_1 and O_2 is large, whereas group O_3 plays the role of connecter between groups O_1 and O_2. Group O_4 only plays a (hypothetical) transient state from O_3 to O_2. Hydrogen radicals (H) and ions (H^+) are closely connected with oxygen chemistry; in Chapter 4.2.4, we stated that H and O are the "symbols" for reduction and oxidation, respectively.

In the troposphere, reactive oxygen species (ROS) are primarily only produced in the gas phase via ozone photolysis and from atomic hydrogen, which is produced from the photolysis of some aldehydes and which turns immediately into HO_2. In the past two decades, our understanding of the photocatalytic formation of ROS in the aqueous phase and at interfaces has quickly grown. The effect of relatively stable molecular oxygen (and low ROS concentrations) enables our life in an oxygen-containing atmosphere. However, it becomes clear that changing climate system chemistry to more ROS would result in so-called oxidative stress.

The chemistry of oxygen species is very complex; all oxygen species are interlinked with all other chemical species found in the climate system. To gain an understanding of oxygen chemistry and its role in oxidizing other chemical species, we will first discuss pure O_x chemistry in Chapter 5.3.1 and then O_xH_y chemistry in an atmosphere free of other trace gases (Chapter 5.3.2.1) and C, N and S species concerning the cycling of oxygen species in Chapter 5.3.2.2. Aqueous oxygen chemistry is even more complex than gas phase chemistry and will be presented separately in Chapter 5.3.4.

5.3.1 Atomic, molecular oxygen and ozone: O, O_2 and O_3 (O_x)

As mentioned, O_2 cannot be photochemically destructed in the troposphere because of its missing short wavelength (below and Chapter 5.3.7). The main fate of O_2 in the troposphere is the addition of several radicals X to produce peroxo radicals (reaction 5.1 and below), an important step in the oxidation of trace gases:

$$X + O_2 \rightarrow XO_2 \qquad (X = O, Cl, Br, R, \text{ and other}) \tag{5.4}$$

The role of singlet oxygen ($^1\Delta_g$ and $^1\Sigma_g^+$), where two electrons in the outer shell ($2p^4$) are in the degenerated π-symmetric orbitals with the antipodal spin (total zero spin), is well known in organic chemistry. Singlet oxygen is a metastable species and because of its excitation energy of 94 kJ mol^{-1} it is chemically extraordinary reactive. In the gas phase, photoexcitation forming singlet molecules from triplet are quantum chemically forbidden. There are several ways to form singlet oxygen. In the gas phase, the photolysis of ozone provides different excited O_2 molecules (reaction 5.7, however, only occurs in the upper stratosphere):

$$O_3 + h\nu \rightarrow O(^3P) + O_2(^1\Sigma_g) \qquad \lambda_{\text{threshold}} = 611 \text{ nm} \tag{5.5}$$

$$O_3 + h\nu \rightarrow O(^1D) + O_2(^1\Delta_g) \qquad \lambda_{\text{threshold}} = 310 \text{ nm} \tag{5.6}$$

$$O_3 + h\nu \rightarrow O(^1D) + O_2(^1\Sigma_g^+) \qquad \lambda_{\text{threshold}} = 267 \text{ nm} \tag{5.7}$$

As for many excited molecule states, quenching (collisional deactivation) is the most important fate; $k_{5.8} = 1.6 \cdot 10^{-18}$ cm^3 molecule^{-1} s^{-1} (M = O_2) at 298 K:

$$^1O_2 + M \rightarrow O_2 + M \tag{5.8}$$

Singlet oxygen is directly produced from triplet oxygen by collision with so-called photosensitizers A^* (electronically excited molecules), which transfer energy onto O_2 in the ground state; the third collisional partner M stabilizes the transfer complex:

$$O_2(^3\Sigma_g^-) + A^* (+ M) \rightarrow O_2(^1\Sigma_g^+) + A (+ M) \tag{5.9}$$

This reaction can occur in the air (for example with excited SO_2) but is of high importance in condensed phases, especially cells. Once singlet oxygen forms, it undergoes interconversion reactions:

$$O_2(^1\Sigma_g^+) + A^* (+ M) \rightarrow O_2(^1\Delta_g^+) + A (+ M) \tag{5.10}$$

$$O_2(^1\Delta_g^+) + O_2(^1\Delta_g^+) \rightarrow O_2(^1\Sigma_g^+) + O_2(^3\Sigma_g^-) \tag{5.11}$$

Finally, singlet oxygen can be produced chemically in the aqueous phase:

$$H_2O_2 + OCl^- \rightarrow {}^1O_2 + H_2O + Cl^- \tag{5.12}$$

Hypochlorite plays a minor role in the atmosphere (Chapter 5.8.2) but hydrogen peroxide is ubiquitous in all natural systems and hypochlorite is enzymatically produced in cells. $^1O_2$ is probably not important in initiating chemical change in the lower atmosphere, at least in the gas phase; excited molecules dissolved in water droplets can promote chemical change under special circumstances. In the stratosphere and mesosphere, each of the bound excited states gives rise to characteristic emission features of the airflow, both by day and night (Schiff 1970, Wayne 1994). In the past, it was assumed that singlet oxygen played a role in urban smog chemistry (Pitts 1970).

In the troposphere, only the photodissociation of ozone and nitrogen dioxide produces atomic oxygen. The photodissociation of O_3 produces several oxygen species including the triplet O (reaction 5.5) and the single O (reaction 5.6); the formation of $O(^3P)$ and O_2 in the ground state is already possible at very large wavelengths:

$$O_3 + hv \rightarrow O(^3P) + O_2(^3\Sigma_g) \qquad \lambda_{threshold} = 1180 \text{ nm} \tag{5.13}$$

The different absorption bands are called *Chappuis* bands (440–850 nm), *Huggins* bands (300–360 nm) and *Hartley* bands (200–310 nm); the strongest absorption occurs at 250 nm. These different absorption bands through to O_3 play a key role in the atmosphere a) to prevent radiation in the lower atmosphere with wavelengths lower than about 300 nm, which destroys life and b) to provide $O(^1D)$ from O_3 photolysis (reaction 5.6), which subsequently forms other ROS (Chapter 5.3.2). Once $O(^3P)$ is formed, it ultimately combines with oxygen to produce O_3; $k_{5.14} = 5.6 \cdot 10^{-34}[N_2]$ cm^3 molecule^{-1} s^{-1} at 298 K:

$$O(^3P) + O_2 + M \rightarrow O_3 + M \tag{5.14}$$

In all such reactions with molecular oxygen, we can assume a constant O_2 concentration of $5.6 \cdot 10^{18}$ molecules cm^{-3} under standard conditions. Furthermore, taking into account the constant N_2 concentration ($2.08 \cdot 10^{19}$ molecules cm^{-3}), a pseudo-first-order rate constant $k_{5.14} = 6.4 \cdot 10^4$ s^{-1} follows and a lifetime of only $1.5 \cdot 10^{-5}$ s for $O(^3P)$. About 90% of $O(^1D)$ turns by quenching into the triplet oxygen; $k_{5.15} = 4.4 \cdot 10^{-11}$ cm^3 molecule^{-1} s^{-1} (M = N_2) at 298 K:

$$O(^1D) + M \rightarrow O(^3P) + M \tag{5.15}$$

The only other (and this is one of very few initial reactions producing ROS) reaction of $O(^1D)$ is with H_2O − producing OH radicals (reaction 5.21). There are known as faster reactions of singlet O (with N_2O and H_2) but the water vapor concentration is orders of magnitude larger in the lower atmosphere. Thus, only in the upper atmosphere, where extremely low H_2O is found, do such reactions become important (Chapter 5.3.7).

Finally, we consider the photodissociation of molecular oxygen:

$$O_2 + hv \rightarrow O(^3P) + O(^3P) \qquad \lambda_{threshold} = 242 \text{ nm} \tag{5.16}$$

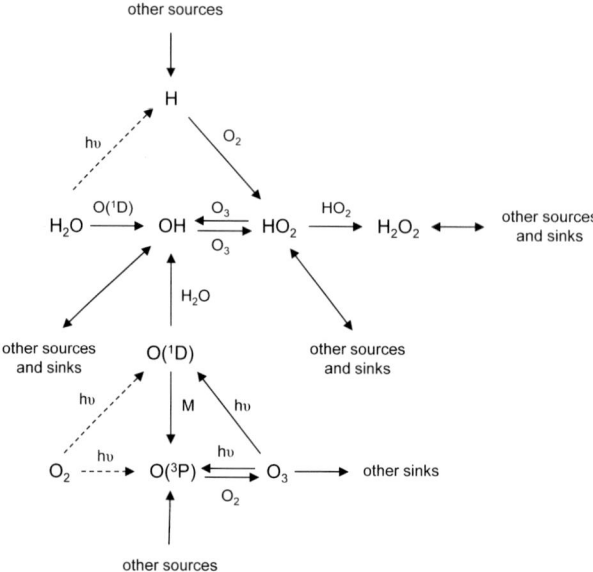

Fig. 5.6 Scheme of O_xH_y gas-phase chemistry; dotted lines denote photolysis only in the stratosphere, other sources and sinks occur in presence of other species than H_yO_x.

$$O_2 + hv \rightarrow O(^3P) + O(^1D) \qquad \lambda_{\text{threshold}} = 175 \text{ nm} \qquad (5.17)$$

$$O_2 + hv \rightarrow O(^1D) + O(^1D) \qquad \lambda_{\text{threshold}} = 137 \text{ nm} \qquad (5.18)$$

Reaction (5.16) refers to the *Herzberg* continuum (230−280 nm), which consists of three bands. It is clear that O_3 formation (reaction 5.14) as well as O_3 photolysis occurs. Hence, at 30 km (above the peak in the ozone layer), we observe substantial reductions in the amount of radiation received between 225 and 275 nm. The O_2 photodissociation in the *Schumann-Runge* continuum (125−175 nm) directly produces $O(^1D)$ and opens many radical reactions (but almost all molecules photodissociates at such hard radiation); this is in altitude of 100−200 km. The solar H *Lyman*-α line (121.6 nm) is, through O_2 photodissociation, an important source of $O(^1D)$ production throughout the mesosphere and lower thermosphere.

The *Lyman*-α line also dissociates H_2O at altitudes between 80 and 85 km, where the supply of water vapor is maintained by methane oxidation even for very dry conditions at the tropospheric−stratospheric exchange region. An increase in HO_x follows and thereby a sharp drop in the ozone concentration nears the mesopause (Chapter 5.3.7 concerns stratospheric O_3 cycles). The ozone concentration rapidly recovers above 85 km because of the rapid increase in O produced by the photodissociation of O_2 by the absorption of ultraviolet solar radiation in the *Schumann-Runge* bands and continuum. Above 90 km, there is a decrease in ozone because of photolysis and because the production of ozone through the three-body recombination of O_2 and O becomes slower with decreasing pressure.

The O_2 photodissociation and subsequent O_3 formation in the lower stratosphere is the most important source of O_3 in the troposphere via the stratosphere−troposphere exchange. The tropospheric net O_3 formation (Chapter 5.3.7) is another important source, especially because of anthropogenic enhancement. Fig. 5.6 shows all principal oxygen reactions; note that O_2 photolysis can be excluded in the climate system in the more narrow sense close to the earth's surface. The exit pathways (sinks) to products (Fig. 5.6) are only possible if there are additional elements and compounds (Chapter 5.3.2.2). In a pure oxygen gas system, we only consider four reactions (5.5, 5.6, 5.13 and 5.14), forming a steady state:

$$O(^3P) \xleftarrow{\text{M}} O(^1D) \xleftarrow{h\nu} O_3 \underset{O_2}{\overset{h\nu}{\rightleftharpoons}} O(^3P).$$

It follows for the steady-state concentration that:

$$[O_3] = \frac{k_{5.14}}{j_{5.5} + j_{5.6}} [O_2][O(^3P)]. \tag{5.19}$$

Besides the O_3 photolysis (5.5 and 5.13), there is only one other important $O(^3P)$ source to provide the only net source for O_3 formation in the troposphere.

$$NO_2 + h\nu \rightarrow O(^3P) + NO \qquad \lambda_{\text{threshold}} = 398 \text{ nm} \tag{5.20}$$

The NO_3 photodissociation also produces $O(^3P)$ but NO_3 concentrations are very small compared with NO_2. Without discussing here in detail, O_3 has several chemical sinks, where reactions with NO and NO_2 are the most important (Chapter 5.4.4), but alkenes also react with O_3 (Chapter 5.7.4) and heterogeneous loss (Chapter 5.3.6) is important.

5.3.2 Reactive oxygen species I: OH, HO_2 and H_2O_2 (O_xH_y species)

5.3.2.1 Atmosphere, free of trace species

We now extend the chemical system from pure oxygen by adding first H_2O as the second abundant oxygen species in the air and learn the formation of reactive O_xH_y species (Table 5.6). Besides reaction (5.15), $O(^1D)$ reacts in another channel (to about 10%) with H_2O to hydroxyl radicals (OH); $k_{5.21} = 2.2 \cdot 10^{-10}$ cm³ molecule⁻¹ s⁻¹ at 298 K:

$$O(^1D) + H_2O \rightarrow 2OH \tag{5.21}$$

The OH radical is the most important species in the atmosphere. This is often referred to as the "detergent" of the troposphere because it reacts with many pollutants, often acting as the first step in their removal. Since OH (this is the neutral form of the hydroxyl ion OH⁻, i.e. OH is an extremely weak acid) is from H_2O, it returns to water by the abstraction of H from nearly all hydrocarbons RH; this pathway is the definitive sink for OH. However, the organic radical R combines with O_2 (reaction 5.4) and in many subsequent reactions produces organic oxygen radicals and recycles HO_2 (Chapter 5.3.2.2):

$$OH \xrightarrow[-R]{RH} H_2O$$

A second important pathway is the addition of OH on different species (SO_2, NO, NO_2) producing oxo acids:

Table 5.6 All important O_xH_y components (O_x special case when y = 0).

oxidation state	species	formation/destruction	name
−3	H_2O^-	$H_2O + e^-$	hydrated electron
−2	H_2O	$OH + H / OH^- + H^+$	water
	$H^+(H_3O^+)$	−	hydrogen ion (hydrogenium)
	OH^-	$OH + e^-$	hydroxyl ion
	$[O_2^-]$	$O_2 + e^-$	oxide
−1	H_2O^+	$H_2O - e^- / OH + H^+$	water anion
	O^-	$O^- + H^+$	oxo anion
	OH	$O + H / H_2O - H$	hydroxyl radical
	HO_2^-	$HO_2 + e^-$	hydrogen peroxide anion[a]
	H_2O_2	$OH + OH$	hydrogen peroxide
−1/±0	O_2^-	$O_2 + e^-$	hyperoxide anion[b]
	HO_2	$O_2 + H / O + OH$	hydroperoxo radical[c]
	O_3^-	$O_3 + e^-$	ozonid anion
	HO_3	$OH + O_2$	hydrogen trioxide[d]
	HO_4^-	$O_3 + OH^-$	hydrogen superoxide anion
	H_2O_4	$HO_2 + HO_2 / 2 OH + O_2$	hydrogen superoxide
±0	H	−	hydrogen
	O	−	oxygen, atomar
	O_2	$O + O$	oxygen, molecular

[a] also named: perhydroxyl radical
[b] also named: superoxide anion
[c] also named: perhydroxyl radical
[d] also named: hydrogen ozonide
Note: the —O—O— is the peroxo group (also called peroxide group); in past termed peroxy (in organic compounds the term peroxy is preferred). Peroxyl, however, is a radical R—O—O, derived from peroxide.

$$OH \xrightarrow{X} XOH$$

A third pathway is OH transformation into the hydroperoxo radical HO_2:

$$OH \xrightarrow[-O_2]{O_3} HO_2$$

HO_2 can turn back into OH (with O_3 but later we will that NO is most important), resulting in the scheme:

$$O_3 \xrightarrow{h\nu} OH \rightleftarrows HO_2 \longrightarrow H_2O_2$$

and thereby in an overall ozone destruction in the clean atmosphere. Later we see that the presence of other trace gases will enhance O_3 destruction in NO_x free air.

The following two reactions provide recycling between OH and HO_2 (they are in a fixed ratio of about $1:10$ as OH/HO_2) and an ozone destruction as the most important O_3 sink in the remote atmosphere. In the presence of other trace gases, other dominant pathways exist, and in NO-containing air, a net formation of O_3

rather than destruction occurs; $k_{5.22} = 7.3 \cdot 10^{-14}$ cm^3 molecule^{-1} s^{-1} and $k_{5.23} = 2.0 \cdot 10^{-15}$ cm^3 molecule^{-1} s^{-1} at 298 K:

$$OH + O_3 \rightarrow HO_2 + O_2 \tag{5.22}$$

$$HO_2 + O_3 \rightarrow OH + 2O_2 \tag{5.23}$$

The net result of this reactions sequence is O_3 destruction according to $2O_3 \rightarrow 3O_2$, whereas the radicals cycle $OH \rightleftarrows HO_2$. HO_2 produces hydrogen peroxide H_2O_2: $k_{5.24} = 1.6 \cdot 10^{-11}$ cm^3 molecule^{-1} s^{-1} and $k_{5.25} = 5.2 \cdot 10^{-32}$ [N$_2$] cm^3 molecule^{-1} s^{-1} (M = N$_2$) at 298 K:

$$HO_2 + HO_2 \rightarrow H_2O_2 \tag{5.24}$$

$$HO_2 + HO_2 + M \rightarrow H_2O_2 + M \tag{5.25}$$

The dimerisation of HO_2 proceeds around 298 K by two channels: one bimolecular and the other termolecular. A cyclic hydrogen bond intermediate $\cdots$H—O—O$\cdots$H—O—O$\cdots$ is very likely, which explains the "abnormal" negative dependence from T and the pressure dependence (Atkinson et al. 2004). Enhancement in the presence of H_2O (M = H_2O) was observed; $k_{5.26} = k_{5.24}[1 + 1.4 \cdot 10^{-21}[H_2O] \exp(2200/T)]$ cm^3 molecule^{-1} s^{-1}, for 1 Vol-% H_2O. It follows that $k_{5.26} = 1.6\,k_{5.24}$ at 298 K (exceeded H_2O_2 concentration where observed above cloud layers).

$$HO_2 + HO_2 + H_2O \rightarrow H_2O_2 + H_2O. \tag{5.26}$$

Again assuming steady states, we can formulate different expressions for the radical concentrations and production rates (taking into account all reactions until now listed):

$$[HO_2] \approx \sqrt{\frac{k_{5.21}}{k_{5.24}}[H_2O][O(^1D)]} \approx \sqrt{\frac{k_{5.21}}{k_{5.24}}\frac{j_{5.6}}{k_{5.15}\frac{[M]}{[H_2O]} + k_{5.21}}[O_3]} \tag{5.27}$$

$$[OH] \approx \frac{k_{5.23}}{k_{5.22}}[HO_2] + \frac{j_{5.6}}{k_{5.22}}\left(\frac{2}{1 + \frac{k_{5.15}[M]}{k_{5.21}[H_2O]}}\right) \tag{5.28}$$

$$\frac{d[OH]}{dt} = \left(\frac{2}{1 + \frac{k_{5.15}[M]}{k_{5.21}[H_2O]}}j_{5.6} + k_{5.23}[HO_2] - k_{5.22}[OH]\right)[O_3]$$

$$\approx \frac{2}{1 + \frac{k_{5.15}[M]}{k_{5.21}[H_2O]}}j_{5.6}[O_3] \tag{5.29}$$

Note that $[M]/[H_2O] \geq 100$. Hence, OH production is proportional to O_3 singlet oxygen photolysis (see Section 2.8.2.9). The second term in Eq. (5.28) contributes a maximum value of about 10^8 molecules cm^{-3} to [OH], whereas the first term suggests that [OH] is about 0.03 [HO$_2$]. These values differ considerably from measured values, where $[OH] \approx 0.1 \, [HO_2] \approx 5 \cdot 10^6$ molecules cm^{-3} at the maximum. That is not remarkable because the OH chemistry is much more complex, including other sources and many more sinks (below and Fig. 5.6).

H_2O_2 is rather stable (Finlayson-Pitts and Pitts 1986) in the gas phase, i. e. it does not undergo fast photochemical and gas phase reactions. The only important sinks in the boundary layer are dry deposition and scavenging by clouds (with subsequent aqueous phase chemistry) and precipitation (wet deposition). In the free troposphere (FT), H_2O_2 photolysis is regarded as an important radical feedback (Crutzen 1999). The photodissociation of H_2O_2 into two OH radicals was first shown by Urey et al. (1929).

To complete the O_xH_y chemistry, further reactions should be noted, which are unimportant near the earth's surface but become of interest in the FT and upper atmosphere. The photolysis of H_2O_2 is very slow and heterogeneous sinks (scavenging by clouds, precipitation and dry deposition) can be neglected.

$$H_2O_2 + h\nu \rightarrow 2\,OH \qquad\qquad \lambda_{threshold} = 360 \text{ nm} \qquad\qquad (5.30a)$$

$$H_2O_2 + h\nu \rightarrow H_2O + O(^1D) \qquad \lambda_{threshold} = 359 \text{ nm} \qquad\qquad (5.30b)$$

$$H_2O_2 + h\nu \rightarrow H + HO_2 \qquad\qquad \lambda_{threshold} = 324 \text{ nm} \qquad\qquad (5.30c)$$

It has long been assumed that reaction channel (5.30a) is the only significant primary photochemical channel at $\lambda > 200$ nm. The quantum yield is very low for $\lambda > 300$ nm, resulting in a slow photodissociation with j in the order of 10^{-6}–10^{-5} s^{-1} (a residence time of around one week). More important in the upper atmosphere is the reaction with OH; $k_{5.31} = 1.7 \cdot 10^{-12}$ cm^3 molecule^{-1} s^{-1} at 298 K. Reactions (5.31) and (5.24) together represent a net radical sink: $OH + HO_2 \rightarrow H_2O + O_2$.

$$H_2O_2 + OH \rightarrow H_2O + HO_2 \qquad\qquad\qquad\qquad\qquad\qquad (5.31)$$

The reactions (5.32) and (5.33) only play a role in the stratosphere when there is no significant competition with $OH + A$; $k_{5.32} = 1.48 \cdot 10^{-12}$ cm^3 molecule^{-1} s^{-1} and $k_{5.33} = 1.1 \cdot 10^{-10}$ cm^3 molecule^{-1} s^{-1} at 298 K.

$$OH + OH \rightarrow H_2O + O \qquad\qquad\qquad\qquad\qquad\qquad (5.32)$$

$$OH + HO_2 \rightarrow H_2O + O_2 \qquad\qquad\qquad\qquad\qquad\qquad (5.33)$$

5.3.2.2 Atmosphere with trace species

Let us now consider the most important reactions of the O_xH_y species (Table 5.6) in the air containing other trace gases[5]. As mentioned, the most important species is the hydroxyl radical OH, which reacts with almost all gases but at very different

---

[5] A complete description of radical reactions with the citation of original sources can be found in Leigthon (1961). Many radicals have been proposed while studying atomic reactions in combustion and explosive processes since about 1925 but fast kinetic observations and intermediate detection became possible only in the 1940s with spectroscopic techniques. The stimulus of radical reactions was the recognizing of photochemical air pollution in Los Angeles towards the end of the 1940s. Scientific literature in the early 1950s is full of studies on this topic.

rates. The decomposition of trace gases is not the aim of the discussion here, but rather we will consider the role of other molecules in odd oxygen recycling. Under odd oxygen we understand the sum of all ROS: O_3, OH, HO, H_2O_2, NO_2 (later NO_3 and N_2O_5 will be added); in the aqueous phase, more species must be added (Table 5.6 and Chapter 5.3.5).

The simplest example of radical recycling is given by the atmospheric oxidation of carbon monoxide (there are no other important CO reactions in the air); $k_{5.34}$ − $2.2 \cdot 10^{-13}$ cm^3 molecule^{-1} s^{-1} at 298 K:

$$OH + CO \rightarrow H + CO_2 \quad \text{or} \quad OH + CO + O_2 \rightarrow HO_2 + CO_2 \quad (5.34)$$

H rapidly combines with O_2 to HO_2 (reaction 5.1). Reaction (5.34) is the only known OH splitting reaction. The mechanism goes via an additive intermediate:

$$OH + CO \rightleftarrows HOCO \rightarrow H + OCO$$

Because reaction (5.34) is faster than OH + O_3 (5.22), ozone decay is enhanced (reaction 5.23 becomes relatively more important) in the presence of carbon monoxide, an almost abundant gas (Chapter 2.8.2.6). This statement is only virtually in contradiction with the experience that biomass burning leads to O_3 formation because CO (and organics as we will see later) in the presence of NO_x results in net ozone formation. In competition with (5.23), HO_2 turns back to OH, whereas NO is oxidized: $k_{5.35} = 8.8 \cdot 10^{-12}$ cm^3 molecule^{-1} s^{-1} at 298 K:

$$HO_2 + NO \rightarrow OH + NO_2 \quad (5.35)$$

NO is also quickly oxidized by ozone to NO_2: $k_{5.36} = 1.8 \cdot 10^{-14}$ cm^3 molecule^{-1} s^{-1} at 298 K:

$$O_3 + NO \rightarrow O_2 + NO_2 \quad (5.36)$$

These two reactions together with (5.20) show the role of NO_x in oxygen chemistry, OH−HO_2 recycling and the net formation of O_3 (Fig. 5.7). There are two cycles, OH ↔ HO_2 and NO ↔ NO_2, which are interlinked, consuming so-called ozone precursors (CO, CH_4, RH) and producing O_3. However, reactions (5.36, 5.22 and 5.23) are in competition with O_3 replacing the other trace species (dotted lines in Fig. 5.7), which results in a net decay of ozone. Both cycles are interrupted when $R_{5.22} \gg R_{5.35}$, what is given for a few tenths of ppb NO:

$$[NO] \ll \frac{k_{5.22}}{k_{5.35}} [O_3] \approx 10^{-2} [O_3]$$

This concentration is the threshold between the ozone-depleting chemical regimes (in the very remote air) and ozone-producing regimes (in NO_x air). Other important reactions of the HO_2 radical (with the exception of H_2O_2 formation via reaction 5.24) are unknown. It also follows that H_2O_2 formation is favored (and vice versa) in low NO_x and low O_3 environments, which are relatively restrictive conditions.

As mentioned at the beginning of Chapter 5.3.2.1, OH can be lost by several reactions. The first type is the abstraction of H and forming reactive intermediates without recycling radicals (but possible further radical consumption in subsequent reactions to the final oxidation product):

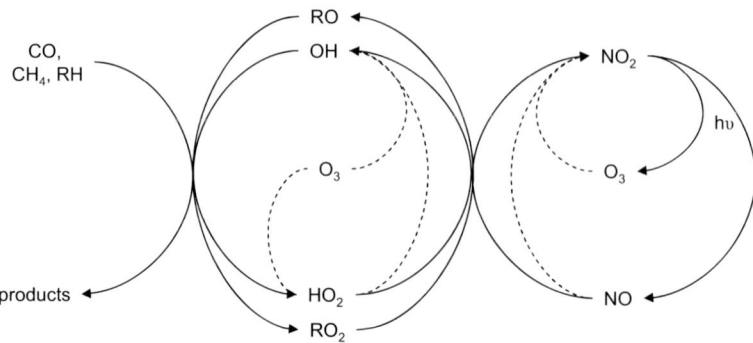

Fig. 5.7 Scheme of gas-phase net O_3 production: HO_x, RO_x cycle and NO_x cycle; dotted lines denote competitive reaction $HO_2 + O_3$ and $NO + O_3$ (for large NO_x).

$$OH + HA \rightarrow H_2O + A \quad (HA = NH_3, HNO_2, H_2S) \tag{5.37}$$

$$A \rightarrow products \ (NO, NO, SO_2) \tag{5.38}$$

Another type of reaction is the addition of OH with the possible photolytic decomposition of the product back to OH but in competition with other pathways of HO-C:

$$OH + B \rightarrow HO{-}B \xrightarrow{h\nu} OH + B$$
$$(B = NO, NO_2; HOB = HNO_2, HNO_3) \tag{5.39}$$

The addition of OH is possible while producing molecules without radical recycling (not to confuse with formyl HCO):

$$OH + C \rightarrow HO{-}C \rightarrow products \quad (C = CS_2) \tag{5.40}$$

Most important are the reactions between OH and organic compounds where the first step also leads to H abstraction according to (5.37), but in subsequent steps to the formation of HO_2 and other radicals, thereby continuing the radical chain:

$$OH + RH \xrightarrow[-H_2O]{} R \xrightarrow{O_2} HO_2 + products \tag{5.41}$$

Let us now consider in more detail reaction (5.41), beginning with the most simple but also most abundant hydrocarbon methane CH_4. Reaction (5.42) is very slow ($k_{5.42} = 6.4 \cdot 10^{-15}$ cm^3 molecule^{-1} s^{-1} at 298 K), giving a residence time of about 10 years:

$$OH + CH_4 \rightarrow H_2O + CH_3 \tag{5.42}$$

A low specific rate (large residence time), however, does not mean that this pathway is unimportant. On the contrary, because of the large CH_4 concentration in the air, nearly homogeneously distributed in the whole troposphere (Chapter 2.8.2.4), CH_4 controls to a large extend the background OH concentration and tropospheric net O_3 formation. The methyl radical CH_3 rapidly reacts with O_2, producing the methylperoxo radical; $k_{5.43} = 1.2 \cdot 10^{-12}$ cm^3 molecule^{-1} s^{-1} at 298 K):

$$CH_3 + O_2 \rightarrow CH_3O_2 \tag{5.43}$$

Similar to HO_2 (reaction 5.35), CH_3O_2 oxidizes with NO (hence, we see that the first HO_x radical cycle in Fig. 5.7 overlaps with another RO_x cycle, turning two molecules NO into NO_2 per one molecule of RH): $k_{5.44} = 7.7 \cdot 10^{-12}$ cm^3 molecule^{-1} s^{-1} at 298 K):

$$CH_3O_2 + NO \rightarrow CH_3O + NO_2 \tag{5.44}$$

There are some competing reactions with (5.43), which we will discuss in Chapter 5.3.3. But when the methoxy radical CH_3O is produced, reaction (5.45) rapidly proceeds and gives formaldehyde, HCHO and HO_2, closing the HO_y cycle; $k_{5.45} = 1.9 \cdot 10^{-15}$ cm^3 molecule^{-1} s^{-1} at 298 K):

$$CH_3O + O_2 \rightarrow HCHO + HO_2 \tag{5.45}$$

Formaldehyde is the first intermediate in the CH_4 oxidation chain with a lifetime longer than a few seconds (Table 2.48 shows the huge amount of secondary HCHO). Formaldehyde is removed by photolysis (5.46) and also reacts with OH (5.47); $k_{5.47} = 8.5 \cdot 10^{-12}$ cm^3 molecule^{-1} s^{-1} at 298 K):

$$HCHO + h\nu \rightarrow H + HCO \tag{5.46}$$

$$HCHO + OH \rightarrow H_2O + HCO \tag{5.47}$$

The formyl radical HCO rapidly reacts with O_2 (HCO is a carbon radical and not oxygen):

$$HCO + O_2 \rightarrow HO_2 + CO \tag{5.48}$$

We see that CH_4 oxidation results in a net gain of radicals when HCHO is photolyzed and turns equivalent OH into HO_2 when HCHO oxidizes by OH. The gross budgets are:

$$CH_4 + OH + 4\,O_2 \xrightarrow{h\nu} CO + 3\,HO_2 + 2\,H_2O \tag{5.49}$$

$$CH_4 + 2\,OH + 3\,O_2 \longrightarrow CO + 2\,HO_2 + 3\,H_2O \tag{5.50}$$

Moreover, CO reacts faster with OH than CH_4 to provide the first cycle in Fig. 5.7; the overall "reactions" are:

$$CH_4 + 2\,OH + 4\,O_2 \xrightarrow{h\nu} CO_2 + 4\,HO_2 + 2\,H_2O \tag{5.51}$$

$$CH_4 + 3\,OH + 3\,O_2 \longrightarrow CO_2 + 3\,HO_2 + 3\,H_2O \tag{5.52}$$

All these reactions turn directly or in multiple steps OH $\rightarrow$ HO_2. When we now extend the budget by including the second NO_x cycle shown in Fig. 5.7, we can establish the budget equation for net O_3 formation. The first oxidation step of CH_4 to HCHO results in:

$$CH_4 + 4\,O_2 + 2\,h\nu \rightarrow HCHO + H_2O + 2\,O_3 \tag{5.53}$$

Concerning formaldehyde, the budget amounts to:

$$HCHO + 2\,O_2 + h\nu \rightarrow CO + H_2O + O_3 \tag{5.54}$$

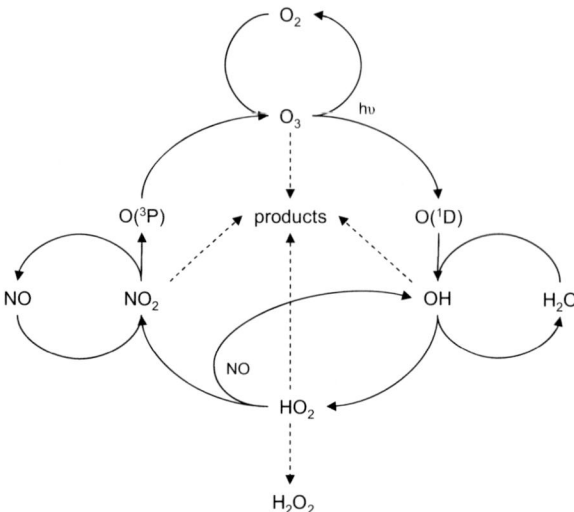

Fig. 5.8 Cycle of odd-oxygen; dotted lines denote competitive reactions not shown.

Similarly, the CO oxidation leads to the formation of one molecule O_3:

$$CO + 2\,O_2 + h\nu \rightarrow CO_2 + O_3 \tag{5.55}$$

Therefore, adding all reactions in the conversion $CH_4 \rightarrow CO$ (remember that after the initial step $OH + CH_4$ the following reactions are all much faster) we gain:

$$CH_4 + 6\,O_2 + 4\,h\nu \rightarrow CO_2 + 2\,H_2O + 4\,O_3 \tag{5.56}$$

Equation (5.56) represents the maximum yield of ozone from CH_4 oxidation. But because of radical chains and recycling, no radical loss is given and instead CH_4 drops the radical concentration to lower values compared with an atmosphere without trace gases (Chapter 5.3.2.1). Definitive OH sinks are given by reactions of the type (5.37)–(5.39); see sulfur and nitrogen chemistry later. The "perfect" ozone formation cycle (Fig. 5.7) is based on the odd oxygen cycling shown in Fig. 5.8. Not shown in Fig. 5.8 are the O_3 precursors (CO, CH_4 and RH), which maintain the OH ↔ HO_2 cycle without consuming O_3. Because of the very slow photolysis, H_2O_2 does not maintain the odd oxygen cycle; its formation already interrupts this cycle. Moreover, many competitive reactions (denoted by the dotted arrows in Fig. 5.8) reduce net ozone formation.

5.3.3 Reactive oxygen species II: RO, RO_2 and ROOH

In the CH_4 oxidation chain are the organic homologs to HO_x radicals (Table 5.7). Any hydrocarbon can react analogously to CH_4; we use the formula RCH_3 (also in terms of RH):

$$RCH_3\ (RH) + OH \rightarrow RCH_3\ (R) + H_2O \tag{5.57}$$

Table 5.7 Organic radicals.

name	symbol	formula
alkyl	R	$-C^\bullet(H_2)$
alkoxo	RO	$-C(H_2)O^\bullet$
peroxoalkyl	RO_2	$-C(H_2)O-O^\bullet$
formyl	HCO	$H-C^\bullet=O$
acyl	RCO	$-C^\bullet=O$
peroxoformyl	$HC(O)O_2$	$H-\overset{\displaystyle O}{\overset{\|}{C}}-O-O^\bullet$
peroxoacyl	$R(O)OO$	$-\overset{\displaystyle O}{\overset{\|}{C}}-O-O^\bullet$
formate	$HC(O)O$	$H-\overset{\displaystyle O}{\overset{\|}{C}}-O^\bullet$
acylate	$RC(O)O$	$-\overset{\displaystyle O}{\overset{\|}{C}}-O^\bullet$

The fate of the alkyl radical R is combining with O_2 to form the alkyl peroxo radical RO_2 ($R-O-O^\bullet$):

$$RCH_3 \text{ (R)} + O_2 \rightarrow RCH_3O_2 \text{ (RO}_2) \tag{5.58}$$

Similar to HO_2, RO_2 oxidizes NO to NO_2 to form the alkoxy radical $R-O^\bullet$ (sometimes also called alkoxyl and acyl radical; acyl denotes the group $R-C-O-$):

$$RCH_3O_2 \text{ (RO}_2) + NO \rightarrow RCH_3O \text{ (RO)} + NO_2 \tag{5.59}$$

There are a number of reactions competing with reaction (5.59), mostly with other peroxo radicals. Combining with HO_2 leads to organic peroxides; the simplest is methylhydroperoxide CH_3OOH:

$$RCH_3O_2 \text{ (RO}_2) + HO_2 \rightarrow RCH_3OOH \text{ (ROOH)} + O_2 \tag{5.60}$$

Two organic peroxo radicals combine to give an alcohol and aldehyde:

$$RCH_3O_2 + RCH_3O_2 \rightarrow RCH_3OH + RCHO + O_2 \tag{5.61}$$

However, it should be noted that $[NO] \gg [HO_2] > [RO_2]$ when discussing the percentages of different pathways. Organic peroxides (the most important are CH_3OOH and C_2H_5OOH) are permanently found in the air besides H_2O_2 but in smaller concentrations.

The alkoxy radical gained in (5.59) reacts with a comparable rate as in reaction (5.58) in the order of $10^{-11}-10^{-12}$ cm^3 molecule^{-1} s^{-1} giving an aldehyde and recycling HO_2:

$$RCH_3O \text{ (RO)} + O_2 \rightarrow RCHO + HO_2 \tag{5.62}$$

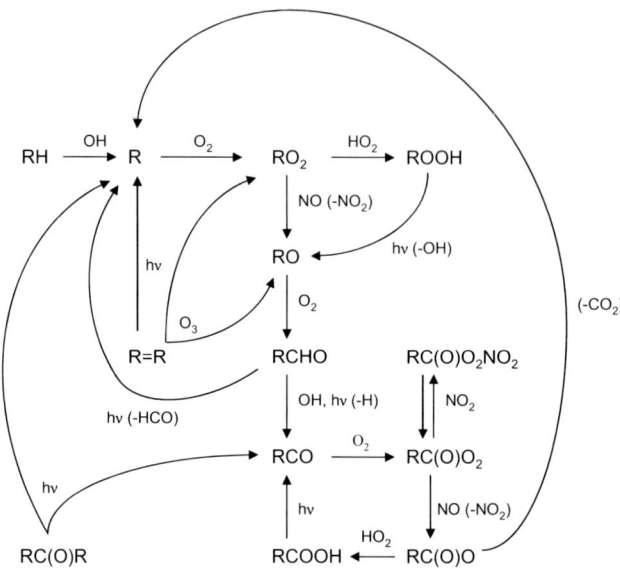

Fig. 5.9 Scheme of organic radical chemistry and fate of characteristic organic groups. RH hydrocarbon, R organic radical, RO_2 organic peroxo radical, ROOH organic peroxide, RO organic oxo radical, RCHO aldehyde, RCOOH carbonic acid, $RC(OC)R$ keton, R=R olefin. Reaction between RO or RO_2 and NO or NO_2 are not included in this scheme.

Aldehydes are very reactive substances. They react with OH by the abstraction of H (almost from the chain) to produce finally bicarbonyls, e. g. dialdehydes and ketoaldehydes (the carbonyl group denoted by $>C=O$). Nevertheless, more important is the photolysis of aldehydes and ketones that initiates radical chains. We have seen that the simplest aldehyde HCHO is photolyzed (reaction 5.46) and gains H (which turns into HO_2) and the formyl radical HCO (a carbon radical), which reacts with O_2 to HO_2 and CO. Therefore, HCHO is an efficient radical source. In analogy, higher aldehydes are photolyzed and produce H atoms and acyl radicals RCO, for example acetyl CH_3CO, which do not decompose like the formyl radical HCO and add O_2, thereby giving acyl peroxo radicals $R—C(O)O_2$:

$$RCH_3O + h\nu \rightarrow RCO + H \tag{5.63}$$

$$RCO + O_2 \rightarrow RC(O)OO \tag{5.64}$$

$$RC(O)OO + NO \rightarrow RC(O)O + NO_2 \tag{5.65}$$

This acylperoxo radical reacts in analogy to the RO_2 radical, thereby oxidizing NO and giving acyl oxoradicals ($RC(O)O$), reacting with HO_2 and forming acyl peroxide or adding NO_x (Chapter 5.4.5). Many of these species are radical reservoirs (especially peroxides but also nitro and nitroso compounds), which can be transported from polluted areas away and release radicals after photolysis, starting new radical chains, oxidizing trace species and decomposing O_3. Acyl radicals are also given by the photolysis of ketones:

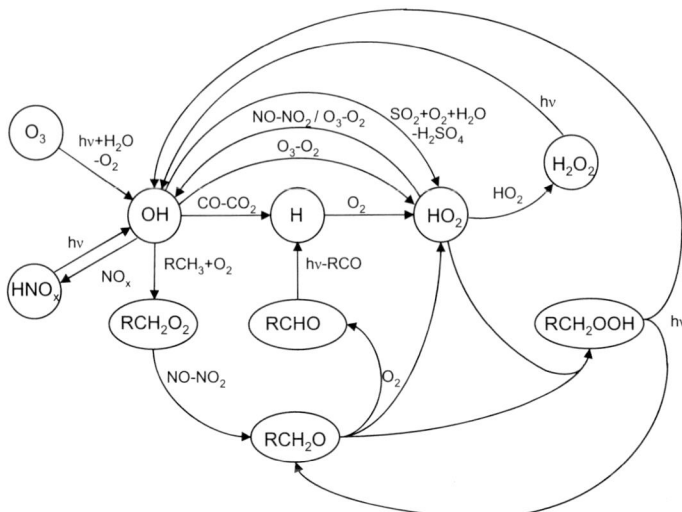

Fig. 5.10 Scheme of O_xH_y gas-phase chemistry including all relevant trace gases (cf. with Fig. 5.6).

$$RC(O)R + h\nu \rightarrow RCO + R \qquad (5.66)$$

The fate of this radical (which is chemically similar to the formyl radical HCO) is not decomposition but O_2 addition onto the carbon radical, giving the peroxo acyl radical RC(O)OO, which is a precursor to peroxoacylnitrates (below). This can be similar to other peroxo radicals oxidizing NO and transforming it into alkyl oxo radical R(O)O, which converts the carbonic acid by reacting with HO_2 (Fig. 5.9). For HO_2 recycling, the photolysis of olefins (R=R) and aldehydes (RCHO) is very important. Aldehydes produce a variety of primary radicals (HCO, RCO and H) that retransform in multisteps back to HO_2 (and to OH; Fig. 5.10). As seen, low reactive hydrocarbons (RH), after oxidation by OH, produce many more reactive intermediates that amplify net ozone formation (Fig. 5.7). All organic oxo and peroxo radicals can also add NO and NO_2 and form organic nitroso (—N=O) and nitro (—NO_2) compounds (Chapter 5.4.5). These compounds present a radical reservoir and can be photolyzed back to the original compounds.

5.3.4 Water and the hydrated electron: H_2O and H_2O^- (e^-_{aq})

We stated that 99 % of oxygen is fixed in H_2O in the climate system but 99 % is fixed in SiO_2 in the earth's system. Oxygen is the most abundant element in the climate system (Table 5.8) but also on earth (excluding the core, where iron is more abundant but only because of its molar mass) and in space (excluding hydrogen and novel gases). Although C and N are the first four elements (again excluding H_2 and novel gases) in space, they are much less abundant on earth because of its high volatility (Chapter 2.1). The enrichment of oxygen on earth is because of its

Table 5.8 Ranking list of elements, based on mass-% (cf. Table 2.13 and Fig. 2.2).

reservoir	elements			sub sum %	elements					total sum %
space	O	C	Fe	84.7	N	Si	Mg	S		99.9
earth crust	O	Si	Al	82.5	Fe	Ca	Na	Mg	K	99.6
earth core	Fe	Ni	S	99.1						99.1
earth	Fe	O	Si	77.3	Mg	S	Ni	Ca	Al	98.8

[a] excluding H_2 (comprises 98 % of all elements) and novel gases

ability to form stable oxides with all other elements (only the core consists of elemental iron and nickel).

As stated in Chapter 2.1.2, water (H_2O) and hydrocarbons (C_xH_y) are the most abundant molecules in space and both were delivered to earth from space (i. e. are not produced on earth). In Chapter 2.5.1, we discussed the unique physical and chemical properties of water. Table 2.23 shows data that explain the chemical stability. However, H_2O is chemically reactive, and Fig. 2.31 and reaction (5.4) show the central role of H_2O in (biogeo-)chemical cycling between oxygen (O_2), carbon dioxide (CO_2) and hydrocarbons (C_xH_y). As discussed in Chapter 2.4.4, the water-splitting process proceeds in plants under the influence of light and enzymes. Nevertheless, we have seen that H_2O is chemically decomposed (reaction 5.21, see below for more examples) and produced in many reactions (5.2c, 5.3, 5.31, 5.32 and 5.33). We have also already mentioned that H_2O is photolyzed in the upper atmosphere (outside the climate system):

$$H_2O + h\nu \rightarrow OH + H \quad \lambda_{threshold} = 242 \text{ nm } (\Delta_R H = 498 \text{ kJ mol}^{-1}) \quad (5.67)$$

However, there is neither the loss nor formation of water on earth since the budget is zero (neglecting possible H_2 loss in the upper atmosphere to space through diffusion). The abovementioned gas phase radical reactions giving H_2O are unimportant in the climate system because of competing reactions between the radicals (H, OH, HO_2) and other trace gases. However, in the upper atmosphere these reactions occur but are balanced with the back formation of H_2O. Only in a chemical regime with excess hydrogen (such as in space and likely only at an early time of chemical evolution and – of course – in a manmade chemical reactor), H_2O is net produced, gaining a thermodynamic equilibrium between H_2, O_2 and H_2O. Pioneering work in studying the reactions between H_2 and O_2 was made by Hinshelwood and Williamson (1934). It is worth summarizing here the most important reactions. The gross reaction $2 H_2 + O_2 = 2 H_2O$ is highly exothermic but does not proceed at a measurable rate at ambient temperatures because of the high activation energy needed to break the O—O (493 kJ mol^{-1}) and H—H (436 kJ mol^{-1}) bonding:

starting phase	$O_2 + \text{energy} \rightarrow O + O$
chain propagation	$O + H_2 \rightarrow OH + H$
	$H + O_2 \rightarrow HO_2$
	$H + HO_2 + M \rightarrow H_2O + O + M$
	$H + HO_2 + M \rightarrow 2 OH + M$

$$OH + OH + M \rightarrow H_2O_2 + M$$
$$OH + H_2 \rightarrow H_2O + H$$
$$HO_2 + HO_2 \rightarrow H_2O_2$$
$$HO_2 + H_2 \rightarrow H_2O_2 + H$$
$$H_2O_2 + OH \rightarrow H_2O + HO_2$$
$$H_2O_2 + h\nu \rightarrow 2\,OH$$
$$H_2O_2 + h\nu \rightarrow H_2O + O$$
$$H_2O_2 + h\nu \rightarrow H + HO_2$$

termination
$$H + HO_2 + M \rightarrow H_2 + O_2 + M$$
$$H + H + M \rightarrow H_2 + M$$
$$H + OH \rightarrow H_2O$$
$$OH + HO_2 \rightarrow H_2O + O_2$$
$$O + O \rightarrow O_2$$

There are several chemical reactions irreversibly consuming H_2O. Each student knows the experiment when one places elemental sodium (or potassium) into liquid water:

$$2\,Na + H_2O \rightarrow H_2 + Na_2O \quad \text{or} \quad 2\,Na + 2\,H_2O \rightarrow H_2 + 2\,Na^+ + 2\,OH^-$$

In detail, however this reaction goes heterogeneous when Na comes into contact with H^+. Liquid water dissociates to a small percentage (0.00000556% – one liter of water contains 55.6 mol H_2O and produces 10^{-7} mol L^{-1} H^+):

$$H_2O \rightleftarrows OH^- + H^+ \tag{5.68}$$

The hydroxyl ion OH^- (also termed hydroxide ion) is the strongest known base in water and a one-electron donor according to $OH^- + X \rightarrow OH + X^-$. For simplification, we disregard that $H^+ + H_2O \rightarrow H_3O^+$ (it is convention to use the symbol H^+). The elemental step of the above reaction is $Na\,(s) + H^+ \rightarrow Na^+ + H\,(g)$, an electron transfer from Na into H^+; below, we will see that first an electron transfer into H_2O occurs. This reaction is the reason that alkali and earth alkali metals do not exist elementarily on earth. It is likely that at the beginning of the earth's chemical evolution metal hydrides such as NaH existed, produced by solar chemistry (such as metal nitrides). By contrast, the hydride ion H^- does not exist freely. Let us assume its existence is short-lived and transient:

$$NaH\,(s) \xrightarrow{\ H_2O\ } Na^+ + H^- \tag{5.68a}$$

$$H^- + H^+ \rightarrow H_2\,(g) \tag{5.68b}$$

Therefore, the budget equation is $NaH\,(s) + H_2O \rightarrow H_2\,(g) + Na^+ + OH^-$. Sodium and calcium are abundant elements and at the beginning of chemical evolution, their most abundant forms were oxides: Na_2O and CaO. Oxygen in oxides is similar to the hydride ion that is non-existent in aqueous solution, but let us again assume it existed as a short-lived transient anion O^{2-}. Hence, the overall reactions were:

$$Na_2O + H_2O \rightarrow 2\,Na^+ + 2\,OH^- \quad \text{and} \quad CaO + H_2O \rightarrow Ca^{2+} + 2\,OH^-,$$

which would proceed via:

$$Na_2O\,(s) \xrightarrow{\;H_2O\;} 2\,Na^+ + O^{2-} \quad \text{and} \quad CaO\,(s) \xrightarrow{\;H_2O\;} Ca^{2+} + O^{2-};$$
$$\text{and} \quad O^{2-} + H^+ \rightarrow OH^- \tag{5.69}$$

In Chapter 4.2.5, we said that OH^- is the strongest existing base. According to the above reaction, it would be the corresponding weakest acid and, as a consequence, the hypothetical ion O^{2-} would be the strongest base. All these reactions can produce alkalinity from water according to the budget reaction:

$$H_2O \xrightarrow{\;e^-\;} H_2O^- \longrightarrow H + OH^- \tag{5.70}$$

In the early state of the climate system, huge amounts of carbon dioxide (CO_2) were found in the atmosphere as well as in the earth because of degassing processes. This CO_2 would react with alkaline aqueous solution (reactions 2.122 and 2.131) in the sequence:

$$CO_2\,(g) + OH^- \rightarrow HCO_3^- \tag{5.71}$$

$$HCO_3^- + OH^- \rightarrow CO_3^- + H_2O \tag{5.72}$$

$$CO_2\,(g) + CO_3^- + H_2O \rightleftarrows 2\,HCO_3^- \tag{5.73}$$

This inorganically produced carbonate could have provided the first feedstocks for photosynthesis in water. Moreover, bicarbonate buffers acid deposition (HCl and SO_2) exhalations deposited in early oceans:

$$HCO_3^- + SO_2\,(g) \rightarrow HSO_3^- + CO_2$$

$$HCO_3^- + HCl\,(g) \rightarrow H_2CO_3 + Cl^- \rightleftarrows H_2O + CO_2 + Cl^-$$

In all the reactions listed above, we see that the two fundamental species of water, the ions H^+ and OH^-, act as reactive species. Their concentrations in neutral water (10^{-7} mol L^{-1}) correspond to an order of magnitude in the ppm range, thereby by a factor of 10^3 larger than the comparable gas phase concentrations of trace gases. This is one reason why aqueous phase chemistry in the air (clouds, fog and precipitation) is so important for the quantification of air/volume-based total chemical conversion.

Fig. 5.11 comprises most of the important oxygen species and reactions in water. The small box with the continuous line comprises the chemical regime reduced in oxygen-free water. Because in natural waters oxygen is always dissolved, ROS such as peroxides are produced and destroyed in waters under the influence of light (larger dotted box in Fig. 5.11). But besides hydrogen (H^+) and hydroxide ion (OH^-) another fundamental species of aqueous solutions exists. This is the hydrated electron e_{aq}^-, also written as H_2O^-, and first postulated by radiation chemists (*Gabriel Stein*)[6] in 1952 and characterized in 1962 by recording its absorption spectrum (Stein 1968, Hart and Anbar 1970, Hughes and Lobb 1976). The solvated electron[7] and especially the hydrated electron have since been found to be an ex-

---

[6] Stein, G. (1952) *Disc. Faraday Soc.* 12, 227
[7] Many papers refer to the first observation of a solvated electron in 1864 by Weyl (*Pogg. Ann.* 123, 350) who studied the dissolution of some alkali and earth alkali metals in liquid ammonia and who observed the characteristic deep blue color of the solution (owing to the light absorption of solvated electrons) and formation of hydrogen, called "hydrogen ammonia". However, Kraus (1908) first stated the existence of "an electron, surrounded by an envelope of solvent molecules".

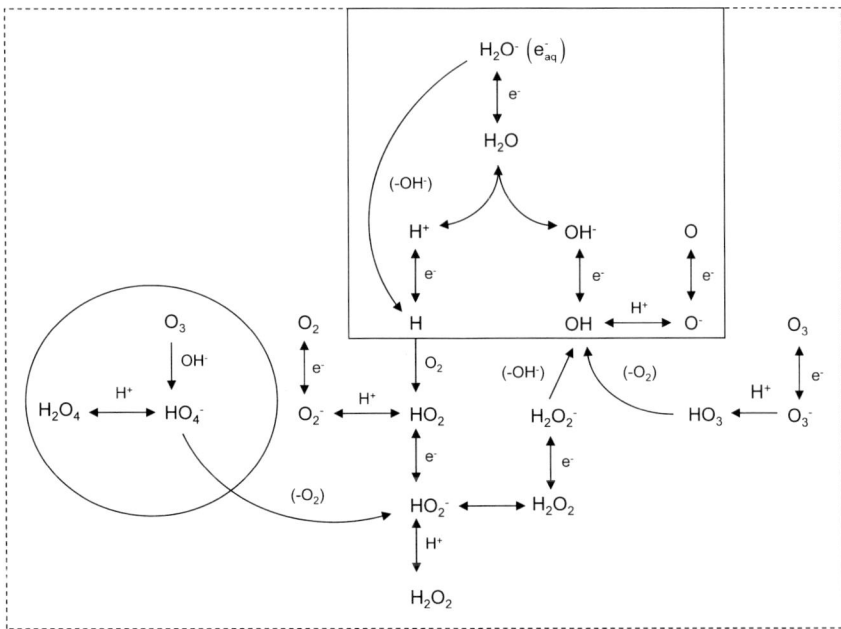

Fig. 5.11 Aqueous oxygen chemistry; HO_3, HO_4^- and H_2O_4 hypothetical intermediates. The box comprises the oxygen-free water chemistry under radiolysis/radiation influence and the dotted box includes oxygen containing water. e− electron donor; note that some pathways strongly depend on the pH, i.e. whether the reactions occur in alkaline or acid solution. The circle box represents a strong alkaline regime (pH > 10).

tremely important reactive species. When the solvent medium is water, the hydrated electron becomes essential to myriad physical, chemical and biological processes. In a simple picture of an electron in a cavity, the description of the hydrated electron state structure is analogous to that of a hydrogen atom, with a ground state of s-type and an excited state of p-type character. They absorb between 600 nm and 800 nm and appear blue. However, the hydrated electron is far more complex because of the ultrafast dynamics of structural change, solvation and recombination (Paik et al. 2004). It is the simplest electron donor and all of its reactions are in essence electron transfer reactions (Logan 1967). What is known is that their presence enhances the reactivity of water molecules with other molecules in a number of important chemical, physical and biological processes. Hydrated electrons form when an excess of electrons are injected into liquid water.

Studying water radiolysis is well established like the discovery of radioactive radiation (radium emanation). Without knowing the transient processes, the generic reaction equation:

Liquid ammonia dissociates similar to H_2O: $2\,NH_3 \rightleftarrows NH_2^- + NH_4^+$. The reaction of potassium with liquid ammonia is thereby very similar to that of sodium with water (see above in text): $K + NH_3\,(l) \rightarrow KNH_3 \rightarrow KNH_2 + H$. Amides ($-NH_2$) are well-known compounds; today, we know that K reacts with NH_4^+ via electron transfer, forming the ammonium radical NH_4, which decomposes into $NH_3 + H$.

$$2\,H_2O \xrightarrow{\text{radiation}} H_2O_2 + H_2$$

was established by Kernbaum (1909a) while investigating the influence of β radiation (from $RaCl_2$) on pure water. *Mirosław Kernbaum*, who worked at the research school of *Marie Curie* in the Paris faculty, also studied the influence of UV radiation (Kernbaum 1909b), electric discharges (Kernbaum 1910) and metals (Kernbaum 1911) on water decomposition and was one of the pioneers of H_2O_2 formation from water. The formation of H_2O_2 due to electric discharges in air containing water vapor was first proposed by Boehe (1873)[8] and confirmed by Fischer and Ringe (1908). Duane and Scheuer (1913) first studied H_2O_2 formation in water during irradiation with α radiation (from Rn ampulla). Risse (1929) first studied the influence of X-rays on water. In all these studies, H_2O_2 formation was confirmed but hydrogen often did not escape and was instead dissolved in water with sometimes oxygen even being evolved (Fricke 1934). However, X-rays do not produce H_2O_2 in oxygen-free water (Risse 1929, Bonét-Maury and Lefort 1948). All these early observations are consistent with the modern view of H_2O_2 radiolysis through radioactive radiation. Haïssinsky and Magat (1951) first studied analysis of products such as H, OH, O^- and suggested reaction 5.74. Reaction 5.75 was then confirmed years later:

$$H_2O + \gamma \rightarrow H_2O^+ + e^- \tag{5.74}$$

$$e^- + n\,H_2O \rightarrow (H_2O)_n^- \tag{5.75}$$

$$H_2O^+ \rightarrow H^+ + OH \tag{5.76}$$

Hydrogen and hydrogen peroxide formation goes subsequently via $H^+ + e^- \rightarrow H$ ($2\,H \rightarrow H_2$) and $OH + OH \rightarrow H_2O_2$. Furthermore, H_2O_2 again dissociates by radiation. Obviously, X-rays do not produce H_2O^+ and without O_2 there is no peroxide (O_2^-) formation; for historic overviews on radiation chemistry and inorganic free radicals in solutions see Uri (1952), Hart (1959), Allen (1961) and Zimbrick (2002). Reaction (5.74) needs 8.8 eV ionization energy and cannot occur under natural conditions in the climate system (Pozdnyakov et al. 2000). Alkaline solutions produce OH under UV irradiation and electrons (which in a subsequent step fixes in finite-sized water clusters; reaction 5.75):

$$OH^- + h\nu \rightarrow OH + e^-, \qquad \lambda_{\text{threshold}} = 195 \text{ nm}. \tag{5.77}$$

However, within the climate system there is no liquid water where radiation smaller than 195 nm exists. Liquid water photolysis under the formation of H_2O_2 was first proven by Tian (1911) for wavelengths < 190 nm. Other early studies on the influence of sunlight on aqueous solutions will be cited in Chapter 5.3.5. In the following, we will term the hydrated electron with the symbol e_{aq}^- or H_2O^- unless it exists only in H_2O clusters where $n = 6-50$ (reaction 5.75). In the climate system, there are only two possibilities to produce e_{aq}^- either chemically or under solar light conditions via so-called photosensitizers. The only known direct chemical production (Hughes and Lobb 1976) is:

---

[8] The first evidence of H_2O_2 in rain during a thunderstorm was provided by *Georg Meissner* in 1862 (Chapter 5.3.6).

$$H + OH^- \xrightleftharpoons{aq} H_2O + e_{aq}^- \qquad (5.78)$$

Therefore, in anoxic medium and (for example biogenic) *in situ* hydrogen production, this is an important pathway to initiate reduction chains by electron transfer processes. The hydrogen atom and hydrated electron are interconvertible. Reaction (5.78) represents a conjugated acid-base pair, in other terms H is a (very weak) acid in an equilibrium where $K = 2.3 \cdot 10^{-10}$ mol L^{-1} ($pK_a = 9.6$); $k_{5.78} = 2 \cdot 10^7$ L mol^{-1} s^{-1} and $k_{-5.78} = 16$ L mol^{-1} s^{-1}. The back reaction (5.78) is so slow that it can be neglected compared with other reactions of the hydrated electron. Reaction (5.79) shows that the electron is the ultimate *Lewis* base:

$$H \rightleftarrows H^+ + e_{aq}^- \qquad (5.79)$$

The most studied and likely formation of e_{aq}^- under normal climatic conditions goes via photosensitization by reversible electron transfer. Photoinduced electron transfer has been demonstrated in many molecules where the donor and acceptor are linked together intramolecularly (Kavarnos and Turro 1987). The efficiency of intramolecular electron transfer is strongly influenced by the separation distance between the donor and acceptor and the structure of the molecular link. The donor-acceptor groups can form collision complexes or exchange an electron at a long distance. In "electron hopping" mechanisms, an electron proceeds from the donor to acceptor via a series of consecutive "hops" to various acceptor groups. This mechanism plays a crucial role in the primary stages of photosynthesis where consecutive electron transfer takes place (Chapter 2.2.2). The generic reaction scheme is given here, where W represents a chromophoric substance and A an electron acceptor:

$$W + h\nu \longrightarrow W^* \xrightarrow{A} (W^+ - A^-) \rightarrow products \qquad (5.80)$$

There are several pathways in product formation:

$W^+ + A^-$	(radical separation)
$W + A$	(ground state reversible reaction)
WA	(coupling reaction)
$W^+ + A + e^-$	(sensitization)

It has been shown that in all natural waters chromophoric substances exist (e. g. chlorophyll) that can produce hydrated electrons after illumination and, consequently, reduce dissolved O_2 (Faust and Allen 1992, Anastasio et al. 1997):

$$W \xrightarrow{h\nu} W^+ + e^- \xrightarrow{aq} e_{aq}^- \qquad (5.81a)$$

$$W \xrightarrow{h\nu} W^* \xrightarrow{O_2} (W^+ - O_2^-) \longrightarrow W^+ + O_2^- \qquad (5.81b)$$

Later we will see that many inorganic semiconductors (for example TiO_2, ZnO, Fe_2O_3) produce hole-electron pairs (for example Hoffmann 1990) when absorbing photons with energy equal to or greater than the band gap energy E_g of the semiconductor:

$$TiO_2 \xrightarrow{h\nu} e_{cb}^- + h_{vb}^+ \qquad (5.82)$$

In the absence of suitable electron and hole scavengers adsorbed to the surface of a semiconductor particle, recombination occurs within 1 ns (5.83a):

$$W^+ (h_{vb}^+) + e^- (e_{cb}^-) \longrightarrow W. \tag{5.83a}$$

However, when appropriate scavengers are present, the valence band holes h_{vb}^+ function as powerful oxidants, whereas the conduction band electrons e_{cb}^- function as moderately powerful reductants. In aqueous solution, e_{cb}^- can transfer to H_2O, gaining the hydrated electron H_2O^- or e_{aq}^-, which moves along the water structure until scavenging by electron receptors. Much less is known on the nature and fate of W^+ and h_{vb}^+, respectively. We speculate that they can recombine again with free electrons or react with main anions in the solution, producing radicals (and thereby amplifying the oxidation potential; Ollis et al. 1989, Turchi and Ollis 1990, Litter 1999):

$$W^+ (h_{vb}^+) + OH^- \longrightarrow W + OH \tag{5.83b}$$

$$W^+ (h_{vb}^+) + Cl^- \longrightarrow W + Cl \tag{5.83c}$$

Nonetheless, it is clear that the rate of the oxidative half reaction involving h_{vb}^+ is closely related to the effective removal of the partner species, i.e. e_{cb}^-, by suitable electron scavengers, which must be present in the solution and available at the semiconductor interface. By contrast, in processes aiming at the oxidative destruction of a target organic substrate, any hole scavengers present in the same environment are potential competitors for the consumption of h_{vb}^+. As far as electron scavenging is concerned, the most common electron scavenger, in aerated aqueous solution, is O_2 and its role has already been emphasized (Litter 1999).

Many organic substances have been found to act as photosensitizers such as phenones, porhyrines, naphtalenes, pyrenes, benzophenones and cyano compounds as well as inorganic species such as transition metal complexes and many others (Kavarnos and Turro 1987). It is well known that the formation of singlet dioxygen in biochemistry (which can afterwards add on olefins) generates:

$$W^* + {}^3O_2 \longrightarrow W + {}^1O_2 \tag{5.84}$$

As a transient step, (5.84) can first occur with the following subsequent reaction:

$$W^* + {}^3O_2 \longrightarrow W^+ + O_2^- \longrightarrow W + {}^1O_2 \tag{5.85}$$

For natural waters and hydrometeors, reaction (5.81a) is the key formation process of hydrated electrons. However, little is known about the rate of formation of superoxides in natural aqueous solution. It is clear that different photosensitizers (and mixtures of them) provide a large variety of photocatalytic oxygen activation. The formation of hydrated electrons can explain many processes such as autoxidation and corrosion. This looks at first confusing because e_{aq}^- works as a reducing species but the key oxidizing species in solution is the OH radical (a strong electron acceptor similar to the atmospheric gas phase), which is produced in a chain of electron transfer processes:

$$O_2 \xrightarrow{e^-} O_2^- \xrightarrow{H^+} HO_2 \xrightarrow{e^-} HO_2^- \xrightarrow{H^+} H_2O_2 \xrightarrow{e^-} OH + OH^-$$

Table 5.9 Aqueous chemistry of the hydrated electron (data from Hart and Anbar 1970, Hughes and Lobb 1976, Buxton 1982), X − halogen.

reaction			k (in L mol^{-1} s^{-1})
$e_{aq}^- + O_2$	$\rightarrow$	O_2^-	$1.9 \cdot 10^{10}$
$e_{aq}^- + H_3O^+$	$\rightarrow$	$H + H_2O$	$2.3 \cdot 10^{10}$
$e_{aq}^- + e_{aq}^-$	$\rightarrow$	$H_2 + 2\,OH^-$	$0.54 \cdot 10^{10}$
$e_{aq}^- + OH$	$\rightarrow$	OH^-	$3.0 \cdot 10^{10}$
$e_{aq}^- + H$	$\rightarrow$	$H_2 + OH^-$	$2.5 \cdot 10^{10}$
$e_{aq}^- + CO_2$	$\rightarrow$	CO_2^-	$7.7 \cdot 10^{9}$
$e_{aq}^- + NO_3^-$	$\rightarrow$	$NO_3^{2-} \xrightarrow{H_2O} (NO_2)_{aq} + 2\,OH^-$	$1.0 \cdot 10^{10}$
$e_{aq}^- + N_2O$	$\rightarrow$	$N_2 + O^-$	$0.87 \cdot 10^{10}$
$e_{aq}^- + RX$	$\rightarrow$	$RX^- \rightarrow R + X^-$	
$e_{aq}^- + Mn^{2+}$	$\rightarrow$	Mn^+	$3.8 \cdot 10^{7}$
$e_{aq}^- + Fe^{3+}$	$\rightarrow$	Fe^{2+}	$3.5 \cdot 10^{8}$
$e_{aq}^- + Cu^{2+}$	$\rightarrow$	Cu^+	$3 \cdot 10^{10}$
$e_{aq}^- + HCOOH^a$	$\rightarrow$	products	$1.4 \cdot 10^{8}$
$e_{aq}^- + CH_3COOH$	$\rightarrow$	products	$1.8 \cdot 10^{8}$

[a] alcohols are unreactive

The e^- symbolizes either charge-transfer complexes or electron transfer by e_{aq}^-. These processes will be discussed in Chapter 5.3.5, but it is clearly seen in Fig. 5.11 that oxygen water chemistry is enhanced in the presence of O_2 and O_3 via parallel inputs of reactants. The hydrated electron reacts rapidly with many species with a reduction potential more positive than −2.9 V. Tunneling between solvent traps can also explain the mobility of e_{aq}^- since it is much higher than expected for a singly charged ion with a radius of 0.3 nm (Buxton 1982). Once e_{aq}^- is produced, then the timescale for the formation of subsequent reactive species (H, HO_2, OH, H_2O_2) is in the order of 10^{-7} s. Table 5.9 lists some reactions of e_{aq}^-; many reactions are so fast that they are at the diffusion controlled limit ($\sim 10^{10}$ L mol^{-1} s^{-1}). The lifetime of e_{aq}^- is in the order of 1 ms, but e_{aq}^- does not react with OH^- and H_2. The back reaction of (5.79) leads to atomic hydrogen: $k_{5.86} = 2.26 \cdot 10^{10}$ L mol^{-1} s^{-1} (Tojima et al. 1999):

$$e_{aq}^- + H_3O^+ \rightarrow H + H_2O \tag{5.86}$$

This reaction only proceeds in acid solution, whereas in neutral and alkaline solutions e_{aq}^- reacts with O_2 (reaction 5.96). Atomic hydrogen is the major reducing radical in some reactions[9]. It effectively reacts as an oxidant forming hydride intermediates such as:

$$H + I^- \rightarrow [H \cdots I^-] \xrightarrow{H^+} H_2 + I \tag{5.87}$$

$$H + Fe^{2+} \rightarrow [Fe^{3+} \cdots H^-] \xrightarrow{H^+} H_2 + Fe^{3+} \tag{5.88}$$

---

[9] Fast radical−radical reactions (in parenthesis k in 10^{10} L mol^{-1} s^{-1}) such as $H + H \rightarrow H_2$ (1.3), $OH + OH \rightarrow H_2O_2$ (0.53) and $OH + H \rightarrow H_2O$ (3.2) play no role in natural waters because of other available reactants in concentrations orders of magnitude higher.

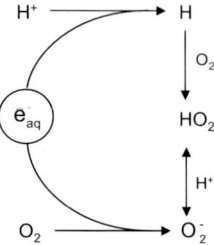

Fig. 5.12 Simplified scheme of the hydrated electron fate in natural water (possible scavenging by ozone not shown, see Fig. 5.14).

In its reactions with organic compounds the hydrogen atom generally abstracts H from saturated molecules and adds to the centers of unsaturated; the fate of the organic radicals is well known (Chapter 5.3.3):

$$H + CH_3OH \rightarrow H_2 + CH_2OH \tag{5.89}$$

$$H + H_2C{=}CH_2 \rightarrow CH_3CH_2 \tag{5.90}$$

In the presence of oxygen, similar to the gas phase reaction (5.1), follows the fast recombination with O_2: $k_{5.91} \sim 2 \cdot 10^{10}$ L mol^{-1} s^{-1} (Bielski et al. 1985):

$$H + O_2 \rightarrow HO_2 \tag{5.91}$$

Reaction (5.91) occurs in an acid solution, whereas $e_{aq}^- + O_2 \rightarrow O_2^-$ (reaction 5.96) proceeds in neutral and alkaline solutions. Interestingly, the above discussed reaction of elemental sodium with water is now considered a source of hydrated electrons (Hughes and Lobb 1976):

$$Na + H_2O \rightarrow Na^+ + e_{aq}^- \tag{5.92}$$

Again, considering our principle of main pathways according to the overall rate, the fate of e_{aq}^- in natural waters and hydrometeors is the formation of superoxide/hydroperoxide radicals (Fig. 5.12), but whether the electron transfer goes directly onto O_2 or via H^+ plays no role because all reactions proceed very fast. The gross water splitting process is given by:

$$H_2O \xrightarrow{\ e_{aq}^-\ } H + OH^- \tag{5.93}$$

where the hydrated electron is produced by reaction (5.81a). We see that this primary process of photosynthesis occurs in the plant cell as well as in abiotic environments. Only the oxygen content determines the father fate of hydrogen, either excess hydrogen for a reducing environment or the formation of O_xH_y in an oxic environment. In natural waters, electron scavengers reduce the lifetime of e_{aq}^- (see 5.123–5.125 for fate of O^-):

$$e_{aq}^- + NO_3^- \rightarrow [NO_3^{2-}] \rightarrow NO_2 + 2\,OH^- \tag{5.94}$$

$$e_{aq}^- + N_2O \xrightarrow{\ e_{aq}^-\ } N_2 + O^- \tag{5.95}$$

5.3.5 Aqueous phase oxygen chemistry

Figs. 5.11 and 5.14 show that in water, under light influence and contact with air all the oxygen species listed in Table 5.6 can be produced. It can also clearly be seen that the presence of dioxygen (O_2) and ozone (O_3) opens competing pathways for gaining peroxo species. In hydrometeors, however, we have to take into account that all, including short-lived, oxygen species (OH, HO_2) will be scavenged. Ta-

Table 5.10 Basic O_xH_y reactions in aqueous phase.

catalase	$2\,H_2O_2$	$\rightarrow$	$2\,H_2O + O_2$
Fenton reaction	$H_2O_2\,(+\,Fe^{II})$	$\rightarrow$	$OH + OH^-\,(+Fe^{III})$
dismutase	$HO_2 + O_2^-\,(+\,H^+)$	$\rightarrow$	$H_2O_2 + O_2$
enzymatic reduction	$O_2\,(+\,e^-)$	$\rightarrow$	O_2^-
desactivation	$O_2^-\,(+\,Fe^{III})$	$\rightarrow$	$O_2\,(+\,Fe^{II})$
ozone decay	$O_3 + O_2^-\,(+\,H^+)$	$\rightarrow$	$OH + 2\,O_2$

Table 5.11 *Henry*'s law coefficients for selected gases.

species	H (in mol L^{-1} atm^{-1})	$-d\ln H/d\,(1/T)$ in K	source
H_2O_2	$1.0\cdot10^5$	6300	Hanson et al. (1992)
HO_2	$4.0\cdot10^3$	5900	Hanson et al. (1992)
OH	$3.0\cdot10^1$	4500	Hanson et al. (1992)
O_2	$1.3\cdot10^{-3}$	1500	Lide and Frederikse (1995)
O_3	$0.94\cdot10^{-3}$	2400	Seinfeld (1986)
H_2	$7.8\cdot10^{-4}$	500	Lide and Frederikse (1995)
HNO_3	$2.1\cdot10^5$	8700	Leliefeld and Crutzen (1991)
N_2O_5	∞		Sander and Crutzen (1996)
HNO_2	$5.0\cdot10^1$	4900	Becker et al. (1996)
NO_3	1.8		Thomas et al. (1998)
N_2O_4	1.4		Schwartz and White (1981)
N_2O_3	0.6		Schwartz and White (1981)
N_2O	$2.5\cdot10^{-2}$	2600	Lide and Frederikse (1995)
NO_2	$1.2\cdot10^{-2}$	2500	Lide and Frederikse (1995)
NO	$1.9\cdot10^{-3}$	1400	Lide and Frederikse (1995)
N_2	$6.1\cdot10^{-4}$	1300	Kavanaugh and Trussel (1980)
NH_3	$6.1\cdot10^1$	4200	Clegg and Brimblecombe (1989)
SO_2	1.2	2900	Lide and Frederikse (1995)
HOCl	$6.6\cdot10^2$	5900	Huthwelker et al. (1995)
Cl_2O	$1.7\cdot10^1$	1700	Lide and Frederikse (1995)
ClO_2	1.0	3300	Lide and Frederikse (1995)
Cl	0.2		Lide and Frederikse (1995)
Cl_2	$9.5\cdot10^{-2}$	2100	Mozurkevich (1986)
$ClNO_3$	∞		Sander and Crutzen (1996)
NH_2Cl	$9.4\cdot10^1$	4800	Holzwarth et al. (1984)
H_2S	$8.7\cdot10^{-2}$	2100	De Bryun et al. (1995)
Hg	$9.3\cdot10^{-2}$		Brimblecombe (1996)

ble 5.11 shows that H_2O_2 and HO_2 are very soluble and OH also shows significant solubility. Despite the low solubility of O_2, the large concentration saturated in natural waters and in the large aqueous phase concentration compared with trace species favors the recombination of many radicals with O_2 as in air. The solubility of O_3 is low but we have to consider that the uptake is not limited by dissolution but through aqueous phase reactions (Chapter 4.3.7.4). In natural surface waters, however, the scavenging of oxo radicals from air is very limited because of the diffusion controlled uptake and the short lifetime in air. The absorption of O_2 and O_3 with subsequent light-induced radical formation is dominant. It is likely that interfacial and/or surface water formation of hydrogen peroxide is more evident than possible dry deposition and also gas phase production (Möller 2009a). For plants, however, the uptake of H_2O_2 might be an important impact of oxidative stress as ascribed by Möller (1988, 1989). Table 5.10 shows the main oxygen chemistry in biological systems (enzymatic processes) but also occurring in abiotic environmants (catalytic processes).

Let us now consider the oxygen chemistry stepwise from O_2 to H_2O_2 via HO_2 (Chapter 5.3.5.1) and then discuss O_3 chemistry (Chapter 5.3.5.2); where OH will play a central role as the "detergent" of water or, in other words, in the oxidation of "contaminants" such as organics, sulfur and nitrogen species. The combination of aqueous chemistry with air chemistry is described in Chapter 5.3.6.

5.3.5.1 From dioxygen to peroxide (O_2 chemistry)

The most important initial reaction is (5.81), which we can also write in the generic form, where e^- means any kind of electron donor. The reduction potential of the donor must be smaller than -0.33 V in neutral solution or -0.125 V in acid solution (Fig. 5.13):

$$O_2 + e^- \longrightarrow O_2^-$$ (5.96)

The limiting step (in terms of reduction potentials; Fig. 5.13) is the electron transfer to O_2 from the photosensitizer. This means that an electron source adequate for the reduction of O_2 will produce all the other reduced forms of dioxygen (O_2^-, HO_2, HOOH, HO_2^-, OH) via reduction, hydrolysis (or proteolysis) and disproportionation steps (Fig. 5.14). Thus, the most direct means to activate O_2 is the addition of an electron (or hydrogen atom), which results in significant fluxes of several ROS. The idea that an electron transfer occurs in oxygen was proposed by Evans and Uri (1949) and was accepted by Leighton (1961) as a mechanism likely to be important for particulate matter:

$$O_2 \text{ (ads)} \xrightarrow{e^-} O_2^- \text{ (ads)}$$

The superoxide anion O_2^- is the proteolytic dissociation product of the weak acid HO_2, the perhydroxyl radical ($pK_a = 4.9$). O_2^- is dominant in cloud and rainwater because of pH values normally around 4.5:

$$HO_2 \rightleftarrows H^+ + O_2^-$$ (5.97)

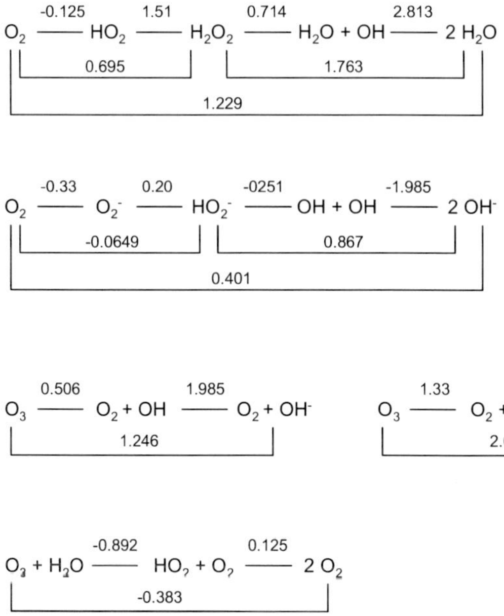

$$O_2 \xrightarrow{-0.125} HO_2 \xrightarrow{1.51} H_2O_2 \xrightarrow{0.714} H_2O + OH \xrightarrow{2.813} 2\,H_2O$$

0.695   1.763

1.229

$$O_2 \xrightarrow{-0.33} O_2^- \xrightarrow{0.20} HO_2^- \xrightarrow{-0251} OH + OH \xrightarrow{-1.985} 2\,OH^-$$

-0.0649   0.867

0.401

$$O_3 \xrightarrow{0.506} O_2 + OH \xrightarrow{1.985} O_2 + OH^-$$

1.246

$$O_3 \xrightarrow{1.33} O_2 + OH \xrightarrow{2.813} O_2 + H_2O$$

2.075

$$O_3 + H_2O \xrightarrow{-0.892} HO_2 + O_2 \xrightarrow{0.125} 2\,O_2$$

-0.383

Fig. 5.13 Standard reduction potentials among H_xO_y species in aqueous solution (see also Table 4.7).

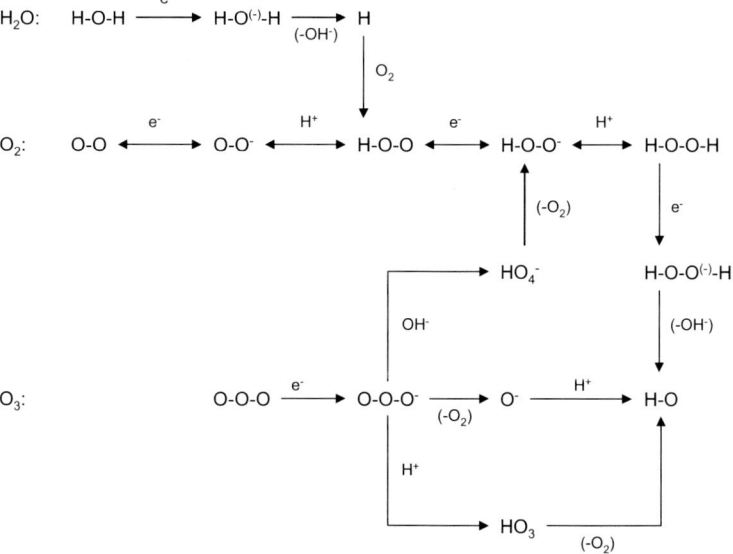

Fig. 5.14 Water, oxygen and ozone redox chemistry, depending on pH.

The uptake of HO_2 from air (Schwartz 1984) is an important source of O_2^- besides photocatalytic formation (5.82). It should be considered that almost all gas phase reactions described in Chapter 5.3.2.1 also proceed in aqueous solutions, for example the H_2O_2 photolysis to OH radicals (5.30a) as first described by Urey et al. (1929); $\lambda_{threshold} < 379$ nm. Another spontaneous radical formation in alkaline solution was suggested (Schroeter 1963, Walling 1957):

$$O_2 + OH^- \rightleftarrows O_2^- + OH \qquad (5.98)$$

Finally, the following interesting pathway of O_2^- formation from photolysis of iron oxalate complexes was described by Zuo and Hoigné (1992):

$$Fe^{III}(C_2O_4) \xrightarrow[-Fe^{II}]{h\nu\,(\lambda < 350\,nm)} [C_2O_4] \xrightarrow[-2\,CO_2]{O_2} O_2^- \qquad (5.99)$$

If the solution contains formate (the anion of formic acid, one of the most abundant organic acids in the air; Chapter 5.7.2) and is saturated with O_2, which is likely for cloud drops, and there is a source of OH radicals (which is also likely for cloud drops under light influence), then the following process produces peroxide:

$$HC(O)O^- + OH \rightarrow C(O)O^- + H_2O \qquad (5.100)$$

$$C(O)O^- + O_2 \rightarrow CO_2 + O_2^- \qquad (5.101)$$

Although superoxide ion is a powerful nucleophile in aprotic solvents, in water it has less reactivity, presumably because of its strong hydration. Hence, within water, superoxide anions are rapidly converted to dioxygen and peroxide:

$$O_2^- \xrightarrow[(-OH^-)]{H_2O} HO_2 \xrightarrow[(-O_2)]{O_2^-} HO_2^- \xrightarrow[(-OH^-)]{H_2O} H_2O_2 \qquad (5.102)$$

The reaction with water can be viewed as a polar-group-transfer reaction:

$$OO^- + HOH \rightarrow [OO\cdots H\cdot\cdot\ ^-OH] \rightarrow OOH + {}^-OH$$

This multistep process can be considered an overall equilibrium: $k_{5.103} \approx 2.5 \cdot 10^8$ atm (Sawyer 1991):

$$2\,O_2^- + 2\,H_2O \rightleftarrows O_2(g) + H_2O_2 + 2\,OH^- \qquad (5.103)$$

Such a proton-driven disproportionation process means that O_2^- can deprotonate acids much weaker than water. The first step is that O_2^- reacts with the proton source to form HO_2, which disproportionates with another O_2^-; $k_{5.104} = 1.0 \cdot 10^8$ L mol^{-1} s^{-1}:

$$O_2^- + HO_2 \rightarrow O_2 + HO_2^- \qquad (5.104)$$

In competition with (5.104) is the deactivation (radical chain termination) of O_2^- by reaction (5.107); HO_2^- is also produced from unprotonized HO_2 via electron transfer:

$$HO_2 + Fe^{2+} \rightarrow HO_2^- + Fe^{3+} \qquad (5.105)$$

HO_2^- is the anion of the extremely weak acid H_2O_2 (p$K_a \approx 11.4$):

$$HO_2^- + H^+ \rightleftarrows H_2O_2 \tag{5.106}$$

O_2^- undergoes competing reactions with ozone, where the ozonide anion (O_3^-) is produced:

$$O_2^- + O_3 \rightarrow O_2 + O_3^- \tag{5.107}$$

Furthermore, it reduces transition metal ions (ubiquitous in natural waters):

$$O_2^- + Fe^{3+} (Cu^{2+}, Mn^{2+}) \rightarrow O_2 + Fe^{2+} (Cu^+, Mn^+) \tag{5.108}$$

The HO_2 radical undergoes a rapid homolytic disproportionation, but much slower than the process with O_2^- in reaction (5.101); $k_{5.109} = 8.6 \cdot 10^5$ L mol^{-1} s^{-1}:

$$HO_2 + HO_2 \rightarrow H_2O_2 + O_2 \tag{5.109}$$

The homogeneous disproportionation of HO_2 seems to involve a "head-to-tail" dimer intermediate that undergoes H-atom transfer (Sawyer 1991):

$$2\,HOO \rightarrow [HOO \cdots HOO] \rightarrow HOOH + \bar{O}_2$$

Before we go onto OH and $\bar{O}_3$ chemistry, we now turn to hydrogen peroxide H_2O_2 or HOOH, which is seen in (5.102) as the final product of oxygen reduction but which is also considered an intermediate because of its reactivity in aqueous solution (in contrast to the gas phase). In Chapter 1.3.3 we cited *Prout's* notable idea on the existence of H_2O_2 in air (just after its discovery by *Thénard* in 1818) and the "bleaching qualities of dew". Dew, surface droplet water in intensive contact with the surrounding air and the underlying surface, often provides plants with chromophoric substances and is an ideal medium for photocatalytic oxygen chemistry according to the process:

$$2\,O_2 \xrightarrow[(-W^+)]{W+h\nu} 2\,O_2^- \xrightarrow{2\,H^+} H_2O_2$$

First, Traube (1887) showed that water electrolysis in acid solution (never in alkaline solution) always first produces H_2O_2 when oxygen is bubbling around the cathode and then hydrogen. He explained it by a reaction between *in statu nascendi* hydrogen (H) and molecular oxygen (reaction 5.91) and concluded the formula H—O=O—H (2 H + O_2) in contrast to H—O—O—H for H_2O_2. That could be interpreted as the first (not yet clear) idea on the reaction H + O_2 → HO_2. Based on our current knowledge, we can formulate the following reaction chain (in contrast to 5.102 this chain proceeds in acid solution):

$$H^I \xrightarrow{e^-} H \xrightarrow{O_2} H\bar{O}_2 \xrightarrow{e^-} HO_2^- \xrightarrow{H^+} H_2O_2$$

The proof that all the oxygen in H_2O_2 comes from molecular oxygen (O_2) was shown by Calvert et al. (1954) using O^{18} as a tracer. Weiss (1944) published the first general radical theory, but the earliest clear idea on the participation of free radicals (H and OH) in oxidation mechanisms in aqueous solution was given by Risse (1929).

The formation of H_2O_2 and ROS via radiolysis and strong UV radiation (Chapter 5.3.4) has been known for about 100 years. In 1967, *Akira Fujishima* discovered

Table 5.12 Electrochemical series of metals (standard electrode potential in V).

metal	potential	reduced species
Mg	−2.372	Mg^{2+}
Al	−1.662	Al^{3+}
Ti	−1.63	Ti^{2+}
Mn	−1.185	Mn^{2+}
Cr	−0.913	Cr^{2+}
Zn	−0.762	Zn^{2+}
Fe	−0.447	Fe^{2+}
Cd	−0.403	Cd^{2+}
Co	−0.280	Co^{2+}
Ni	−0.257	Ni^{2+}
Pb	−0.126	Pb^{2+}

the photocatalytic water decomposition (water photolysis by sunlight conditions) in the process in the TiO_2 surface, later called the *Honda-Fujishima* effect (Fujishima 1972). However, the phenomena that finely divided ZnO and TiO_2 are capable of acting as photosensitizers for a number of chemical reactions were known long before. The formation of H_2O_2 was found in water under UV radiation (< 400 nm) containing suspended ZnO (Baur and Perret 1924, Baur and Neuweiler 1927). Water also decomposes by zinc dust under the formation of H_2O_2 but only in contact with air or free oxygen (Kernbaum 1910). It has been suggested that all metals work in a similar process:

$$Zn + 2\,H_2O + O_2 \rightarrow Zn\,(OH)_2 + H_2O_2$$

Merckens (1906) stated that H_2O_2 is produced according to the reduction potential of metals (Mg, Al, Zn, Cd, Ni, Co, Pb: decreasing from left to right; Table 5.12). Therefore, Baur (1918), combining the ideas of electrolysis with those of photolysis, suggested the theory of photoinduced release of electrons:

$$M \xrightarrow{\;h\nu\;} M^+ + e^- \tag{5.110}$$

It was observed that paints, containing binders and TiO_2, show no durability under the influence of weather and light (Wagner 1939). It was also known that bleaching light-resistant color paints occurred because of adding TiO_2 (Keidel 1929, Wagner 1939), and similar effects were seen when TiO_2 was used as a delustering agent for rayon (Keiner 1934). Keidel (1929) had already proposed a photochemical action where TiO_2 acts as a sensitizer. Goodeve (1937) suggested that the photosensitization through TiO_2 results in an exciton transport, consisting of a negatively charged electron and a positively charged hole, bound. Schossberger (1942) found that rutile shows no photosensitization but only anatase with disorders in the crystal lattice. It seems that older literature was unknown to many researchers (who were unable to read German) to refer that photocatalysis and the formation of ROS was discovered decades before. It is likely that Schumb et al. (1955), the only monograph on

H_2O_2 in English[10], cited the other two earlier monographs written in German (Machu 1937, Kausch 1938) but did not cite one of the abovementioned papers on H_2O_2 formation from "surface-water-sunlight" pathways. It is remarkable that Leighton (1961) cited the early papers on ZnO and TiO_2 photocatalysis. Kormann et al. (1988) reported on H_2O_2 formation in aqueous suspensions containing oxides (TiO_2, ZnO, desert soil) under atmospheric conditions (without citing these older papers).

Dunstan, Jowett and *Goulding* studied the rusting of iron and published (Dunstan et al. 1905) the "hydrogen peroxide theory". It was shown, however, that iron exposed to water and oxygen, with the exclusion of carbon dioxide, underwent rusting. They concluded that Whitney (1903) was correct in assuming that pure oxygen and liquid water alone are essential to the corrosion of iron, the presence of an acid being unnecessary. The detection of hydrogen peroxide during the corrosion of many metals gave rise to the idea that it acted as an intermediary in corrosion processes. First, Schönbein (1861) showed that H_2O_2 is formed during the slow oxidation of metals in atmospheric air. Dunstan et al. (1905) wrote the reaction according to

$$Fe + H_2O \rightarrow FeO + H_2 \quad \text{and} \quad 2\,FeO + H_2O_2 \rightarrow Fe_2O_2(OH)_2 \text{ (rust)}.$$

In modern terms we write this sequence as:

$$Fe \xrightarrow[(-Fe^+)]{H_2O} e_{aq}^- \xrightarrow{O_2} O_2^-$$

or (Me metal)

$$Me \xrightarrow{O_2} [Me^+\!\!-\!O\!-\!O^-] \xrightarrow[(-OH^-)]{H_2O} [Me^+\!\!-\!O\!-\!O\!-\!H]$$

$$\rightarrow [Me^{2+}\cdots\,^-O\!-\!O\!-\!H] \xrightarrow[(-OH^-)]{H_2O} Me^{2+} + H_2O_2$$

Autoxidation, defined as the slow oxidation process where dioxygen adds on some bodies, can be written as:

$$B\,(s) + O_2 \rightarrow B\!-\!O\!-\!O\,(s) \xrightarrow{O_2} B\!-\!O\,(s) + O_3$$

and

$$B\!-\!O\!-\!O\,(s) + A \rightarrow B\!-\!O\,(s) + AO$$

H_2O_2 formation in autoxidation processes was proposed by Traube (1882) despite nothing being known on formation mechanisms. Years before, Schönbein (1866) had detected the formation of H_2O_2 from essential oils and turpentine oil under sunlight. Wieland and Franke (1929) studied the kinetics of H_2O_2 formation in autoxidation processes. Because of the significance of biological processes, H_2O_2 formation was also found (in the late 1930s) under visible and UV light influence

---

[10] These books are relatively rare. There are wartime photo-lithoprint reproductions by Edwards Bros, Ann Arbor, Michigan, under the authority of the Alien Property Custodian (Kausch 1943 and Machu 1944) because H_2O_2 production was important for torpedo engines. Machu's monograph was edited later by Springer-Verlag, Berlin, 1951. All older literature is cited on oxygen species in Gmelin (1966, 1969).

in different aqueous suspensions containing plant parts (Gmelin 1966). Water containing organic substances produces H_2O_2 under (visible) light influence (Blum and Spealman 1933). Between 1876 and 1890, several authors believed they could prove the existence of H_2O_2 in plants[11]. However, because of insufficient analytical methods, there were doubts about such claims (Machu 1937). After the turn of the 19th century, Bach and Chodat (1902) and Gallagher (1923) found evidence for H_2O_2 in living plants and Tanaka (1925) proved H_2O_2 was a primary product in respiration.

However, the role of superoxide anions in autoxidation processes only became clear after the 1950s, interestingly because of studying bleaching processes. The bleaching and germicidal properties of H_2O_2 had been found very early and opened the door to many industrial, wastewater treatment and medical applications. However, unbalancing the oxidation potential could result in oxidative stress (Sies 1986). Thus, Möller (1989) proposed that increasing atmospheric H_2O_2 was responsible for the declining forests in Europe, which was first recognized in the second half of the 1970s. Even the redox behavior of H_2O_2 produces during the photosynthesis in the formation of oxygen from water proved to be an important intermediate (Halliwell and Gutteridge 1984). In cells, it is both a source of oxidative stress and a second messenger in signal transduction (Georgiou and Masip 2003).

The "aquatic surface chemistry" (Stumm 1987) mechanism, now often termed as interfacial and photocatalytic, was not understood before the beginning of the 1960s (Gmelin 1966). Since then the reduction and oxidation of H_2O_2 has been well described in the sense of electron transfers depending on the reduction potential and pH of the aqueous solution (electron donators and/or acceptors), i. e. combining equations (5.81) and (5.103):

$$W \xrightarrow[(-W+)]{W\,(+O_2)} O_2^- \underset{}{\overset{H^+}{\rightleftharpoons}} HO_2 \xrightarrow[(-O_2)]{O_2^-} HO_2^- \xrightarrow{H^+} H_2O_2. \tag{5.111}$$

As seen in Fig. 5.13, such high positive potentials places H_2O_2 in the group of the most powerful oxidizing agents known. The H_2O_2/H_2O couple has such a high potential that in many instances the reduced species of the oxidizing agent is oxidized back to its original state. The result of this behavior is the decomposition of peroxide according to $2\,H_2O_2 \rightarrow O_2 + 2\,H_2O$.

Oxidative damage (stress) *in vivo* is often ascribed to the *Fenton* reaction. In 1876, *Henry Fenton* described a colored product obtained by mixing tartaric acid with hydrogen peroxide and a low concentration of a ferrous salt (Fenton 1876) and found that iron acts catalytically (Fenton 1894). Haber and Weiss (1932, 1934) proposed[12] the involvement of free hydroxyl radicals in the iron(II)/hydrogen peroxide system (called *Fenton* chemistry):

$$H_2O_2 + Fe^{2+} \rightarrow OH + OH^- + Fe^{3+} \tag{5.112}$$

which was then followed by (5.113); $k_{5.113} = 4.5 \cdot 10^7$ L mol^{-1} s^{-1} (Hughes and Lobb 1976):

$$H_2O_2 + OH \rightarrow H_2O + O_2^- + H^+ \quad \text{or} \quad OH + H_2O_2 \rightarrow H_2O + HO_2 \tag{5.113}$$

---

[11] Schönbein (1864) was the first person to detect H_2O_2 in the human body.
[12] They did not mention *Fenton* (Koppenol 2001).

Reaction (5.112) also proceeds with hydrated electrons; $k_{5.114} = 1.36 \cdot 10^{10}$ L mol^{-1} s^{-1} (Hughes and Lobb 1976):

$$H_2O_2 + e_{aq}^- \rightarrow OH + OH^- \tag{5.114}$$

Solutions, containing H_2O_2 and Fe^{2+} (ferrous ion) are called *Fenton* reagents and are used for the oxidation of organic compounds as well as oxidizing contaminants of wastewaters. *Fenton*-like chemistry goes on with other TMI such as Cu and Mn:

$$H_2O_2 + Cu^+ (Mn^+) \rightarrow OH + OH^- + Cu^{2+} (Mn^{2+}) \tag{5.115}$$

The high importance of this pathway lies in OH generation. When H_2O_2 acts as a reducing species, the elemental step can be regarded as electron transfer into H_2O_2, whereas $H_2O_2^-$ decays in a non-reversible manner in acid medium (it is assumed to be in equilibrium $H_2O_2^- \rightleftarrows OH + OH^-$) with OH generation:

$$H_2O_2 \xrightarrow{\ e^-\ } H_2O_2^- \xrightarrow{\ H^+\ } OH + H_2O \tag{5.116}$$

In polluted air, the main fate of aqueous H_2O_2 is the fast oxidation of dissolved SO_2 (Chapter 5.5.2.2), which limits the lifetime of both species. In biological systems, besides inorganic *Fenton* chemistry, three oxygenic base processes occur (Haber and Willstätter 1931):

$$O_2 + e_{aq}^- \rightarrow O_2^- \qquad \text{enzymatic reduction}$$
$$2\,O_2^- + 2\,H^+ \rightarrow O_2 + H_2O_2 \qquad \text{dismutase}$$
$$2\,H_2O_2 \rightarrow 2\,H_2O + O_2 \qquad \text{catalase}$$

Hydrogen peroxide (which is between the oxygen state -2 and 0) can also act as a reducing agent and thereby turn oxidized metals (reaction 5.117) back to lower oxidation states; however, at a slow rate, $k_{5.117}$ (Fe^{3+}) $= 6.0 \cdot 10^2$ L mol^{-1} s^{-1} (Graedel et al. 1986) and $k_{5.117}$ (Mn^{2+}) $= 7.3 \cdot 10^4$ L mol^{-1} s^{-1} (Davies et al. 1968):

$$H_2O_2 + Fe^{3+} (Cu^{2+}, Mn^{2+}) \rightarrow HO_2 + H^+ + Fe^{2+} (Cu^+, Mn^+) \tag{5.117}$$

John Baxendale and coworkers (Barb et al. 1949) suggested that superoxide reduces (5.108) the iron(III) formed on reaction (5.105) to explain the catalytic of the metal.

5.3.5.2 From ozone to hydroxyl (O_3 and O_1 chemistry)

Ozone is unstable in water. The major product formed from ozone decomposition in aqueous solution is the OH radical. Hence, the ozonation of drinking water has been widely used after the recognition of the germicidal properties of ozone, first shown by *Eugen Gorup-Besanez* (1859) and later applied to treat organically polluted water by *Fritz Emich* in 1885[13]. Only after the manufacture of ozone generators was the first water treatment plant established 1892 in Martinikenfelde[14] near Berlin by the firm *Siemens* and *Halske*. Many studies have since been carried out at this plant, supported by the German government, to investigate the efficiency of ozonization as a basis for developing larger water treatment plants (Fonrobert

---

[13] Monatshefte der Chemie 6, 89 (cited after Fonrobert 1916).
[14] This former village is now an urban region at the northwestern edge of downtown Berlin.

1916). However, before the 1980s nothing was known about detailed chemical mechanisms (below).

In the gas phase, evidence of OH radicals was spectroscopically shown (despite the characteristic bands seen in the 1880s) in 1924 (Watson 1924). In the aqueous phase, its existence was first proposed by Haber and Willstädter (1931) but definitively detected only in 1965 (Thomas 1965). From the kinetic measurements of H_2O_2 decomposition, Haber and Willstädter (1931) concluded that:

$$H_2O_2 + OH \rightarrow H_2O + H^+ + O_2^- \tag{5.113}$$

$$O_2^- + H^+ + H_2O_2 \rightarrow O_2 + OH + H_2O$$

or $$\tag{5.118}$$

$$O_2^- + H_2O_2 \rightarrow O_2 + OH + OH^-$$

Haber and Weiss (1932, 1934) added two more reactions, reaction (5.112) as the first, (5.113) and (5.118) as numbers 2 and 3 and as number 4 the following (5.119):

$$Fe^{2+} + OH + H^+ \rightarrow Fe^{3+} + H_2O \quad \text{or} \quad Fe^{2+} + OH \rightarrow Fe^{3+} + OH^- \tag{5.119}$$

These four reactions were terms the *Haber-Weiss* cycle despite two reactions being proposed by Haber and Willstädter (1931). Indeed, the only paper by Haber and Weiss (1934) was written in English and the other papers, published in German by Haber et al., were almost ignored (Koppenol 2001). Moreover, the third reaction in this cycle (5.118) was considered inefficient (George 1947, Barb et al. 1949), and Weiss and Humphrey (1949) proposed to replace (5.118) by (5.108) to maintain the cycling between ferrous and ferric ions:

$$O_2^- + Fe^{3+} \rightarrow O_2 + Fe^{2+} \tag{5.108}$$

OH radicals are less scavenged than HO_2 (which is an important fact when considering the gas phase ozone formation cycle; Chapter 5.3.6). Note that the OH yield in reaction (5.116) is stoichiometric to H_2O_2 and thereby provides OH concentrations that are orders of magnitudes larger than by the phase transfer of gaseous OH (remember that $[HO_2]/[OH] \sim 10$ and $[H_2O_2]$ in gas phase is orders of magnitude higher). The OH radical has a standard reduction potential of +2.8 V in acidic solution and is therefore a strong oxidant. A main sink in aqueous solution is similar to the gas phase reaction (5.41) and the abstraction of the H atom from dissolved organic compounds (DOC):

$$OH + RH \xrightarrow[(-H_2O)]{} R \xrightarrow{O_2} HO_2 + \text{products} \tag{5.120}$$

However, recycling between OH and HO_2 has not always given (5.100) and (5.101). OH also reacts with many inorganic ions (such as ferric ions in reaction 5.119) as an electron acceptor:

$$OH + A^{n+} \rightarrow OH^- + A^{(n+1)+} \tag{5.121}$$

Another pathway is the addition with anions to radical ions (not to be confused with hypochloric acid; HOCl), which decompose into hydroxyl ions and new radicals:

$$OH + Cl^- (Br^-) \rightarrow HOCl^- (HOBr^-) \rightarrow OH^- + Cl (Br) \tag{5.122}$$

Reaction (5.119) with metal ions is slower ($\leq 10^8$ L mol^{-1} s^{-1}) because of the replacement of the H_2O ligand:

$$OH + Me^{n+}(H_2O)_x \xrightarrow[(-H_2O)]{} Me^{n+}(H_2O)_{x-1}OH \xrightarrow{H_2O} Me^{(n+1)+}(H_2O)_x + OH^-$$

In strong alkaline solution, the OH radical dissociates (it is a weak acid with $pK_a = 11.9$): $k_{5.123} = 1.2 \cdot 10^{10}$ L mol^{-1} s^{-1} and $k_{-5.123} = 9.3 \cdot 10^7$ L mol^{-1} s^{-1}:

$$OH + OH^- \rightleftarrows O^- + H_2O \quad \text{or} \quad OH \rightleftarrows O^- + H^+ \tag{5.123}$$

The standard electrode potential of OH in alkaline solution is smaller than in acidic solution with 1.4 V. The O$^-$ radical reacts quickly with organic compounds under H abstraction:

$$O^- + RH \rightarrow OH^- + R \tag{5.124}$$

However, O$^-$ and OH are interconvertible:

$$O^- + H^+ \rightarrow OH \tag{5.125}$$

As in the gas phase, in solution the H atom is also abstracted from organics:

$$OH + RH \rightarrow H_2O + R \tag{5.126}$$

The dominant pathway in alkaline solution containing oxygen ($k_{5.127} = 2.6 \cdot 10^9$ L mol^{-1} s^{-1}) is:

$$O^- + O_2 \rightarrow O_3^- \tag{5.127}$$

and the reaction with water is $k_{5.128} = 10^8$ s^{-1} (Elliot and McCracken 1989). The likely intermediate $H_2O_2^-$ in reaction (5.128) has already been met in reaction (5.116):

$$O^- + H_2O \rightarrow OH + OH^- \tag{5.128}$$

O$^-$ is produced from the reaction between e_{aq}^- and N$_2$O in aqueous solution (which is the most important scavenger for hydrated electrons): $k_{5.129} = 2.6 \cdot 10^9$ L mol^{-1} s^{-1}:

$$e_{aq}^- + N_2O \rightarrow O^- + N_2 \tag{5.129}$$

The fate of O$^-$ (if it is ever produced in natural waters corresponding to OH) is transformation into the ozonide anion under oxide conditions (5.127), which is an important intermediate in alkaline solution with a lifetime of about 10^{-3} s. Ozonide anion (not to be mixed up with the olefin-ozone adduct, Chapter 5.7.4) is easily produced through electron transfer onto dissolved ozone; the electron affinity of O$_3$ is several times (2.1 eV) that of O$_2$ (0.44 eV; Pichat et al. 2000, Addamo et al. 2005).

Before we summarize current knowledge about ozone decay in natural water, let us turn for a moment to early observations. It has long been known that ozone decays in alkaline solution. Schönbein (1844) found that ozone is removed from air after bubbling through alkaline solutions. This was proven quantitatively by Soret (1864) but Cossa (1867) found that O$_3$ will not be destroyed in pure KOH solution free of any organic substance. However, a mechanism was not known (only the formation of O$_2$) before Weiss (1935) first proposed the following three reactions, based on the observation that ozone decay is effective only in alkaline solution:

$$O_3 + OH^- \rightarrow O_2^- + HO_2 \tag{5.130}$$

$$O_3 + HO_2 \rightarrow 2O_2 + OH \tag{5.131}$$

$$O_3 + OH \rightarrow O_2 + HO_2 \tag{5.132}$$

With the exception of reaction (5.132), we have discussed as an important gas phase process (5.23) − and nowadays not regarded as an important aqueous phase reaction − the equations (5.130) and (5.131) do not represent elementary steps. Hoigné and Bader (1976) re-evaluated the importance of OH in the water ozonation process. Bahnemann and Hart (1982) and Staehelin and Hoigné (1982) estimated the reaction rate of (5.132) to be between $1 \cdot 10^8$ and $3 \cdot 10^9$ L mol^{-1} s^{-1}; $k_{5.133} = 3.6 \cdot 10^{10}$ L mol^{-1} s^{-1} (Bahnemann and Hart 1982):

$$O_3 + e_{aq}^- \rightarrow O_3^- \tag{5.133}$$

Staehelin and Hoigné (1982), Sehested et al. (1984) and Staehelin et al. (1984) added several reactions to the *Weiss* mechanism:

$$O_3 + OH^- \rightarrow O_2 + HO_2^- \qquad k_{5.134} = 70 \text{ L mol}^{-1} \text{ s}^{-1} \tag{5.134}$$

$$O_3 + HO_2^- \rightarrow O_2 + OH + O_2^- \qquad k_{5.135} = 2.8 \cdot 10^6 \text{ L mol}^{-1} \text{ s}^{-1} \tag{5.135}$$

$$O_3 + O_2^- \rightarrow O_3^- + O_2 \qquad k_{5.136} = 1.6 \cdot 10^9 \text{ L mol}^{-1} \text{ s}^{-1} \tag{5.136}$$

It is clear that only reaction (5.136) is fast enough for further consideration; the spontaneous alkaline ozone decay (non-radical and non-photochemical) according to (5.130) and (5.134) is too slow to obtain any atmospheric importance. The formation of H_2O_2 in ozone decay has been controversial discussed since *Schönbein*, where H_2O_4 ($H_2O \cdot O_3$), the hypothetically ozone acid, was first proposed by Gräfenberg (1902, 1903): H—O—O—O—O—H (also called hydrogen superoxide according to Ardon 1965) being a so-called spontaneous decay reaction:

$$O_3 \xrightarrow{\text{OH}^-} HO_4^- (\xleftarrow{\text{H}^+} H_2O_4) \xrightarrow[(-O_2)]{} HO_2^- \xrightarrow{\text{H}^+} H_2O_2 \tag{5.137}$$

In summarizing all the proposed reactions of O_3 decay in solution, the intermediate O_3^- formation either via (5.133) or (5.136) where the fate of O_3^- is quickly decayed according to (Sehested et al. 1984): $k_{5.138} = 5 \cdot 10^{10}$ L mol^{-1} s^{-1} and $k_{-5.138} = 3.3 \cdot 10^2$ s^{-1}: $k_{5.139} = 1.4 \cdot 10^5$ s^{-1}:

$$O_3^- + H^+ \rightleftarrows HO_3 \tag{5.138}$$

$$HO_3 \rightarrow OH + O_2 \tag{5.139}$$

The overall process is given by:

$$O_3 + e_{aq}^- \longrightarrow O_3^- \xrightarrow[(-O_2)]{\text{H}^+} [HO_3] \longrightarrow OH \tag{5.140}$$

The intermediates HO_3 (also called hydrogen trioxide) was detected by Cascade et al. (1999). H_2O_4 (H_2O_3, produced from OH + HO_2, has also been suggested) have never been identified (see for example Marshall and Rutledge 1959, Skorckho-

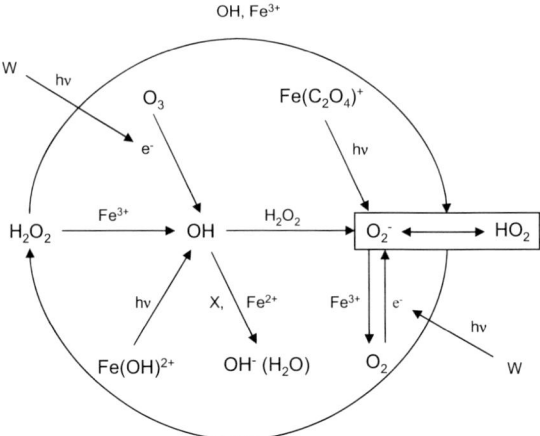

Fig. 5.15 Simplified main aqueous oxygen chemistry; W photosensitizer, X oxidizible species RH, N(III), S(IV), Cl⁻.

dov et al. 1962, Kobozev et al. 1959); nevertheless, the formation of H_2O_2 from O_3 in alkaline medium has been proved to affect the direct formation of OH radicals. Thus, in the presence of O_3 and electron donors it can produce OH radicals with much higher yields than in reaction $O_2 + e^-$ (5.96) as shown by Pichat et al. (2000) and Zhang et al. (2003) and finally H_2O_2 in subsequent steps, likely via degradation of DOC[15]:

$$OH \xrightarrow{\text{RH}} HO_2 \xrightarrow{e^-} HO_2^- \xrightarrow{H^+} H_2O_2 \tag{5.141}$$

Thus, *Cossa*'s remarkable early observations support the essential role of water-dissolved organics (DOC or NOM) in reaction (5.120). It seems very likely that reaction (5.133) apart from (5.111) takes place on all wetted surfaces. For example, dew provides a medium where all needed reactants are available. Hence, organic matter plays a crucial role as an electron donor as well as a converter of OH into HO_2. Recently, humic-like substances (HULIS, also called macromolecular compounds) have been found in atmospheric aerosols (Havers et al. 1998) and cloud water (Feng and Möller 2004), which were likely to have been produced in atmospheric chemical processes from primary aromatic compounds and have chromophoric properties. Fig. 5.15 summarizes the most important aqueous phase oxygen reactions. Similar to the gas phase, OH plays a central role as a key oxidant for many dissolved inorganic and organic species. Peroxides (H_2O_2 and HO_2/O_2^-) play the role of OH generation and cycling. Organic photosensitizers in the presence of sunlight produce hydrated electrons or directly O_2^- – the key role of natural water, mainly at the interface to air, is the formation of ROS, thereby providing oxidation processes, including corrosion and autoxidation.

---

[15] It is self-evident that this reaction provides efficient water treatment and cleaning.

5.3.6 Multiphase oxygen chemistry

In the previous chapters, we considered either homogeneous gas phase or aqueous phase chemistry. However, the condensed aqueous phase exists only in interaction with the gas phase. As just mentioned, at the interface reactants meet from the aqueous and gaseous sites and build up high concentrations to provide favorable conditions for chemical reactions; in recent years this has brought the term "interfacial chemistry" into fashion.

Concerning the earth's surface water, it is relatively simple with respect to the plane interface and fluxes in both directions as emission and deposition. In a cloud, however, up to a few hundred droplets are within one cubic centimeter. Thus, in a cloud, the specific surface-to-volume ratio is in the order of $0.3 \ m^2 \ m^{-3}$ but the aqueous volume-to-gas volume ratio (LWC) amounts to only $10^{-6} \ m^3 \ m^{-3}$. The volume between the droplets is termed the interstitial air. In Chapter 4.3.7, we discussed that despite the low aqueous solution volume, the chemical turnover can be significant because of much faster reactions in solution and the fact that highly soluble gases are almost completely transferred into the droplet phase. Depending on the present chemicals, each phase provides different budget terms (source sinks). The chemistry within the aqueous phase depends on the concentrations of the reactants before droplet formation and its solubility. Because of different gas solubilities, the ratios among gas phase species changes and thereby so does the significance between different competing pathways. One of the most important feedbacks of clouds onto the net ozone formation is the decoupling between $OH-HO_2$ and $NO-NO_2$ (Fig. 5.7). HO_2 is far more scavenged than the other species, which results in OH depletion (not recycling) and reduced NO to NO_2 transformation because of O_3, which could result in a net ozone depletion in clouds. From the aqueous phase of O_3 chemistry, we have learned that in-droplet O_3 formation is unlikely and, once air is in contact with water, O_3 is scavenged and decomposed.

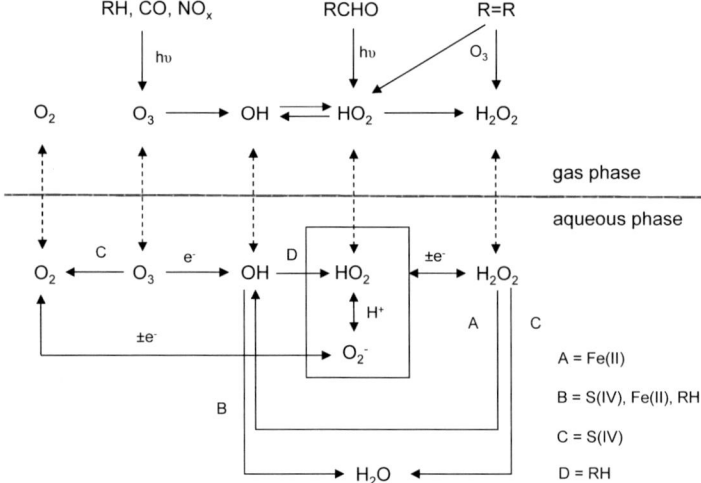

Fig. 5.16 Chemical relationships between O_x and O_xH_y species.

HO_2 and OH radicals are also scavenged and consumed as well as produced in many aqueous phase reactions. Whether the aqueous phase is a net sink or source of ROS depends on the presence of consuming species (organics, sulfur, nitrogen and halogen compounds). The general relationships among stabile oxygen species (O_2, O_3, H_2O_2 and H_2O) are not different between the gas and aqueous phase (Fig. 5.16). The short-life radicals (HO_2/O_2^-, OH) are the key oxidants, generated from and turned back into the "stable" oxygen species.

5.3.6.1 Historical remarks

Junge and Ryan (1958) first pointed out the importance of the atmospheric aqueous phase for SO_2 oxidation. Although acid rain (Smith 1852), the bleaching properties of dew (Prout 1834) and hydrogen peroxide in rain (Meissner 1863) were known as phenomena many years ago, atmospheric aqueous phase chemistry has long been ignored compared with gas phase chemistry. Only the large air pollution relevance of sulfur dioxide in winter smog episodes stimulated relevant research in the 1950s (Chapter 5.5.2.2). However, recognizing the complex oxygen chemistry in solution (which has been known for more than 60 years) in its atmospheric relevance did not happen before around 1990.

It is remarkable that the existence of H_2O_2 in air (as a gas as well as dissolved in hydrometeors) was definitely established before 1880 but the existence of O_3 was still being discussed around then. Definite proof of the existence of O_3 in the atmosphere was not provided until the first spectrometric measurements at the end of the 19th century (Möller 2004). *Georg Meissner* provided the first evidence of H_2O_2 in rain during a thunderstorm[16] in 1862 (Meissner 1863). Schönbein (1869) confirmed this observation and *Heinrich Struve* detected it in snow soon after (Struve 1869). *Struve* even proposed in 1870 that H_2O_2 is produced during all burning processes in air (Struve 1871). The German chemist *Emil Schöne*, however, was the first scientist to study atmospheric H_2O_2 in detail in rain, snow and air near Moscow (Petrowsko-Rasumowskaja, an agricultural research station) in the 1870s (Schöne 1874, 1878, 1893, 1894). *Schöne* noted the following remarkable observations:

- in showers the H_2O_2 concentration was higher than in drizzle;
- in rain from southern air masses H_2O_2 was higher than in rain from polar air masses;
- the H_2O_2 concentration decreases strongly from July until November; and in snow less H_2O_2 is found than in rain during the same season

Schöne also used an artificial dew and frost sampler made from glassware and found H_2O_2, which had to be co-deposited from the gas phase. In this way, he detected a diurnal variation and found H_2O_2 to be largest at noon. Under comparable conditions, he also found that the amount of deposited H_2O_2 is largest when the temperature is highest, there is more sunshine and less relative humidity. During

---

[16] Zuo and Deng (1999) are believed to be the first authors to establish that lightning can induce H_2O_2 production (discovered during a Florida thunderstorm).

rain, he found no H_2O_2 in his "sampler" and concluded correctly that "rain washes out all H_2O_2 which is vaporous in air". He further concluded "... *dass bei der Entstehung des atmosphärischen Wasserstoffhyperoxyds das Sonnenlicht eine hervorragende Rolle spielt*" (that sunlight plays a singular role in the formation of atmospheric hydrogen peroxide; Schöne 1874, p. 1708). Kern (1878) later confirmed these results. *Schöne* was the first to state that H_2O_2 is a permanent natural constituent of our atmosphere. Thiele (1907), Tian (1910), Chlopin (1911) and Kernbaum (1911) proved that the existence of H_2O_2 in air is because of the influence of solar radiation (onto water). These authors' experiments show that the ordinary moist air of a room, after subjection for a few minutes to the action of UV light, shows the presence of ozone, hydrogen peroxide and nitrogen trioxide. Hence, the formation of H_2O_2 during electric discharges in the air and water vapor was established at the beginning of the 20th century, which explains its excess occurrence in rain from thunderstorms.

At that time, nothing was known about the atmospheric formation reactions of O_3 and H_2O_2 and which chemical mechanisms existed between both species in the gas and aqueous phases. The formation of ozone, hydrogen peroxide and nitric acid in air had been attributed generally to "*variations in the electrical condition of the atmosphere*" (Fox 1873). The period 1850–1880 of uncounted ozone measurements using the *Schönbein* paper has often been criticized, and it is likely that apart from ozone, nitrous acid[17] and hydrogen peroxide were also responsible for the blue coloring of the paper[18]. The objection to *Schönbein*'s ozonometer (potassium iodide on starch paper) and to *Houzeau*'s ozonometer (potassium iodide on red litmus paper) lies in the fact that their materials were hygroscopic and their indications varied widely with air moisture (for instance, the intensive "ozone reaction" near places with intensive water evaporation (e. g. waterfalls) were wrongly interpreted as an ozone source). Later, *Schöne* provided the results of an extended series of experiments on the use of thallium paper for estimating the oxidizing material in the atmosphere, whether hydrogen peroxide alone or mixed with ozone or even other constituents hitherto unknown. His findings indicated an oxidizing species, but with our present knowledge it is possible to conclude that the "ozone reaction" was because of hydrogen peroxide in the air. Whereas no definite reagents were available for ozone detection, H_2O_2 was clearly detectable in solution (e. g. rainwater) in the presence of ozone and nitrous acid (Schöne 1893).

Gottfried Wilhelm Osann found in 1853 that each snowflake falling on a paper coated with potassium iodide is followed by a "Schönbein reaction". Which substance other than H_2O_2 could it be? Thus, many so-called ozone observations can be attributed to H_2O_2. Boehm (1858) observed in Prague that the majority of thunderstorms are accompanied by a simultaneous increase in the depth of the color in tests[19]. Remarkably, it was also observed that thunderstorms without or with little

---

[17] Before 1850, this acid was wrongly attributed to be NO_2 or N_2O_4; note that sulfurous acid was SO_2 and no difference with the acid anhydride was made (e. g. $SO_2 + H_2O$).

[18] Owing to its high sensitivity to humidity it is likely that heterogeneous surface reactions (e. g. $O_3 + H_2O$) forming oxidants also produced the color.

[19] Fox (1873, pp. 70–72) cited many other observers concerning increased "ozone" in connection with rain, fog and snow.

rain did not show an "ozone reaction" (Schiefferdecker 1855). From today's knowledge, these findings indicate aqueous phase production pathways; the conclusion that thunderstorms do not produce O_3 but only NO_y has not yet been drawn. *Augustin Reslhuber* stated that "*with thunderstorms, the amount of ozone* [we should turn into hydrogen peroxide] *is dependent upon the amount and kind of the aqueous precipitations which accompany them*" (Reslhuber 1856). Finally, it is cited by *Auguste Houzeau* (after Fox 1872, p. 72) that "… that ozone is more frequently present in the air during days of rain than during days of fine weather". Moreover, *Robert Scoutetten* found that rain and fog determine different affects, according to their conditions of production of "ozone reactions": "*If rain follows a storm, and returns after a temporary reappearance of blue sky, the test-paper exhibits deep tints. If, on the contrary, the rain is fine and continuous, and the temperature is slightly elevated, there is little Ozone*" (Scoutetten 1856). However, we can deduce from the humidity dependence of ozonometers that the "ozone reaction" is a result of the interfacial chemistry of O_3 and H_2O_2 onto wetted surfaces.

5.3.6.2 Hydrogen peroxide

Heikes et al. (1982) first suggested the idea of atmospheric aqueous phase H_2O_2 formation in cloud droplets. The in-droplet chemical formation of H_2O_2 occurs much faster than in the gas phase. However, one must consider the liquid volume fraction (LWC) ratio as well as the occurrence of clouds to assess the share of in-droplet production in the total atmospheric budget. Liquid phase H_2O_2 is also formed from gas phase scavenging of H_2O_2 and HO_2 (Schwartz 1984) as well as via chemical reactions within the liquid phase, all based on the general reactions (5.96) and (5.102) where electron donors are given by different species, e. g. O_2^- itself (McElroy 1986) and transition metal ions (Gunz and Hoffmann 1990). Related to the volume of air, however, the aqueous phase production concerning this radical mechanism is small (about 10^{-7} ppb s^{-1}; Möller and Mauersberger 1990b, 1992) compared with the gas phase production rate of about 10^{-5} ppb s^{-1} (Martin et al. 1997). All these findings emphasize the central role of H_2O_2 in radical chains during biochemical, combustion and air chemical processes. The fundamental role in H_2O_2 environmental chemistry lies in the simple redox behavior among water and oxygen depending on the available electron donors or acceptors:

$$H_2O \xleftarrow{\text{reduced}} H_2O_2 \xrightarrow{\text{oxidized}} O_2 \tag{5.142}$$

Reaction (5.142) is valid for the aqueous and gaseous phases in all environmental media. The oxo and peroxo radicals provide the intermediates. In Chapter 2.8.2.8, we summarized the characteristic behavior of atmospheric H_2O_2 resulting from its multiphase chemistry.

Under atmospheric conditions, the self-reaction of the HO_2 radical according to Eq. (5.24) is the most important gas phase formation of H_2O_2 (reviews: Gunz and Hoffmann 1990, Lee et al. 2000, Vione et al. 2003). H_2O_2 is rather stable (Finlayson-Pitts and Pitts 1986) in the gas phase, i. e. it does not undergo fast photochemical and gas phase reactions. The only important sinks in the boundary layer are dry deposition and scavenging by clouds (with subsequent aqueous phase chemistry) and precipitation (wet deposition).

Thus, the transport of H_2O_2 over long distances is possible during essentially cloud-free conditions. As mentioned, H_2O_2 will be completely and quickly consumed under SO_2-rich conditions. Hence, in the presence of clouds and rain only under SO_2-poor conditions (above) is the net formation of H_2O_2 likely to occur in the aqueous phase. By contrast, the HO_2 radical (and in the aqueous phase, O_2^- too) can undergo many competitive reactions that limit H_2O_2 formation. Under NO-rich conditions, HO_2 will be retransformed into OH ($NO_2 + HO_2 \rightarrow NO_2 + OH$) with decreasing H_2O_2 yield. However, under NO-poor conditions ($\ll 1$ ppb), reaction $HO_2 + O_3 \rightarrow OH + 2O_2$ occurs and results in a diurnal inverse correlation (Ayers et al. 1996) between the short-term concentrations of O_3 and H_2O_2. However, under NO-poor conditions, hydrocarbon oxidation by OH leads to the formation of organic peroxides by HO_2 consumption. Thus, only under "medium NO" does gas phase H_2O_2 production reach its optimum. Gilge et al. (2000) have shown that under NO limited ozone formation conditions, i.e. in each time step more HO_2 is produced than transformed by NO into NO_2 and OH, H_2O_2 concentration increases. It is important to note that with increasing tropospheric ozone the concentration of hydrogen peroxide can − but not must − increase.

Another pathway in the formation of HO_2 and probably direct H_2O_2 is via the ozonolysis of alkenes, which has been investigated for many years. However, several open questions remain about the initial and product relationship. It is well known that the excited *Criegee* radical can produce OH and HO_2 radicals (Neeb et al. 1998). The stabilized *Criegee* radical can react with water vapor to form α-hydroxyalkyl peroxides (RR′C(OH)OOH); Gäb et al. 1985, Becker et al. 1993). Recently, it was found[20] that substituted *Criegee* radicals could directly produce H_2O_2, i.e. without the HO_2 intermediate. The yield strongly depends on the alkene species and the water vapor pressure (Sauer et al. 1999). This way is independent of radiation and can be important under all situations when competition with the OH attack on alkenes is small, e.g. under cloudy conditions, during night-time and in winter. It can also be important in urban areas where high alkene emission occurs. Weinstein-Lloyd et al. (1998) believed that the increase of peroxides during photochemical periods in the USA is because of the ozonolysis of biogenic and man-made alkenes.

The most important H_2O_2 sink is usually the atmospheric aqueous phase, via wet deposition and first through the oxidation of dissolved SO_2 (Chapter 5.5.2.2). Modeling studies have shown that at low atmospheric SO_2 concentrations (around 1 ppb) the droplet phase acts as a net source of H_2O_2 (Möller and Mauersberger 1992), even without taking into account the photocatalytic formation according to (5.102). As discussed in Chapter 2.8.2.8 regarding the H_2O_2 increase in Greenland's ice cores, atmospheric SO_2 definitely controls the source-sink relationship of the aqueous phase. Möller (2009) has shown for the first time by longer simultaneous H_2O_2 monitoring in rain and air that in summer the net formation of H_2O_2 occurs via the aqueous phase. Fig. 5.17 shows the coupled air-aqueous oxygen chemistry with a focus on H_2O_2. Whereas H_2O_2 in the gas phase is more like an oxidant

---

[20] The formation of H_2O_2 when boiling organic ozonides was described by Harries (1916).

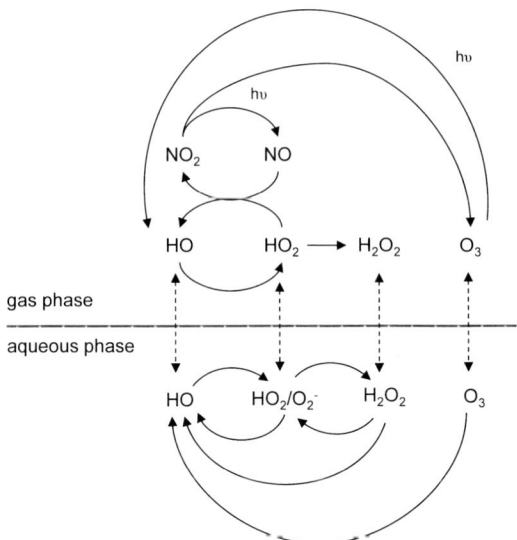

Fig. 5.17 Scheme of H_2O_2 multiphase chemistry. Note that OH, HO_2/O_2^- and O_3 can undergo competitive reactions with many substances (organic, sulfur and nitrogen compounds), not shown here.

reservoir (the slow photolysis only plays a role in the upper troposphere), it is included in intensive cycling between OH, HO_2 and H_2O_2 in the aqueous phase. Thus, depending on the pH the aqueous phase provides a variety of ROS, partly scavenged from the gas phase and also produced in solution where oxygen photocatalysis in the presence of photosensitizers occurs in a unique way for oxidation processes. As for the gas phase, sunlight is essential but in the absence of light transition, metal ions carry out the electron transfer processes. It is likely that the morning wetted earth's surface (dew formation) provides an interface for photochemical reactive oxygen production.

Summarizing the atmospheric H_2O_2 sources, we can separate between the following processes:

a) direct emission (e. g. biomass burning);
b) photochemical-initiated formation via the HO_2 radical as a product of the photolysis of O_3 and aldehydes; HO_2 can either recombine to H_2O_2 in the gas phase or react after scavenging in the liquid phase;
c) thermal direct formation in the gas phase via the ozonolysis of alkenes;
d) decay of O_3 in the (alkaline) aqueous solution; and
e) photosensitized formation ("photocatalysis") via electron transfer onto O_2 in the aqueous phase.

In addition to direct emission in combustion processes, most secondary formation pathways in the gas and aqueous phases depends on ozone. By contrast, there is only one way that is light independent (ozonolysis) and another way that is independent of oxidant precursors (aqueous phase electron transfer onto oxygen; Eqs. 5.96 and 5.102).

5.3.6.3 Ozone

The net rate of ozone formation is given by three terms: the chemical production $P(O_3)$ in the gas phase, the sink term $S(O_3)$ by chemical conversion and deposition and the net transport term $T(O_3)$, given by influx and outflow:

$$\left(\frac{d[O_3]}{dt}\right) = P(O_3) - S(O_3) + T(O_3) \tag{5.143}$$

The chemical production term is given by the $OH-HO_2$ conversion reaction shown in Fig. 5.7 of the whole ozone formation cycle as the rate-determining step:

$$P(O_3) = \left(k_{5.34}[CO] + k_{5.41}[CH_4] + \sum_i k_i[NMVOC]\right)[OH] \tag{5.144}$$

Considering mean concentrations, it follows that O_3 formation rates of 0.2−2 ppb d^{-1} concerns CO and about 0.5 ppb d^{-1} concerns CH_4, respectively. This is close to the global mean net ozone formation rates found by modeling 1−5 ppb d^{-1} (Wang et al. 1998), suggesting that long-life carbon monoxide and methane are responsible for the global background ozone concentration. However, local and regional O_3 net production rates of about 15 ppb d^{-1} are observed under cloud-free conditions in summer (Möller 2004, Beck and Grennfelt 1994). The *in situ* formation rate can quickly reach up to 100 ppb h^{-1} (Volz-Thomas et al. 1997). It is often neglected that the transport term in Eq. (5.143) can be dominant in the morning after the breakdown of the inversion layer and vertical O_3 mixing, and in the case of changing air masses vertical fluctuations can "produce" O_3 changes up to 10 ppb in minutes. During night, dry deposition might reduce near-surface ozone to zero. As discussed in Chapter 5.3.2.2, ozone production abruptly stops when NO becomes very small and the "catalytic" O_3 destruction cycle (reactions 5.22 and 5.23) works.

The ozone loss in clouds was almost disregarded before the 1990s. Ozone is less soluble in water compared with H_2O_2. The discussed aqueous phase production of O_3 goes through nitrate ion photolysis (5.197). The aqueous phase is an effective sink for O_3 through reaction (5.107) and S(IV) oxidation (5.301−5.303). As seen in Fig. 5.16 and 5.17, the gas phase ozone formation cycle is disturbed mainly because of HO_2 scavenging. Only at the beginning of the 1990s, modelers (Lelieveld and Crutzen 1991, Möller and Mauersberger 1992, Liu et al. 1997) showed that clouds can effectively reduce gas phase ozone. Acker et al. (1995) showed by cloud chemistry monitoring that − depending on the transport characteristics − the interstitial air of clouds can have "ozone holes". The main pathway of ozone decomposition in solution goes via OH formation, which turns either to H_2O_2 (we discussed net H_2O_2 formation in the previous chapter) or oxidizes organic substances when they are present. This is similar to the discussed "catalytic ozone destruction" in the remote atmosphere. Nevertheless, in the polluted atmosphere, dissolved sulfur dioxide can react directly with O_3 (5.5.2.2).

Another indirect but very efficient pathway of irreversible O_3 consumption via the aqueous phase goes with NO_y species, which are highly soluble and will transfer into dissolved nitrate (next chapter). Fig. 5.18 shows the destruction of O_3 where

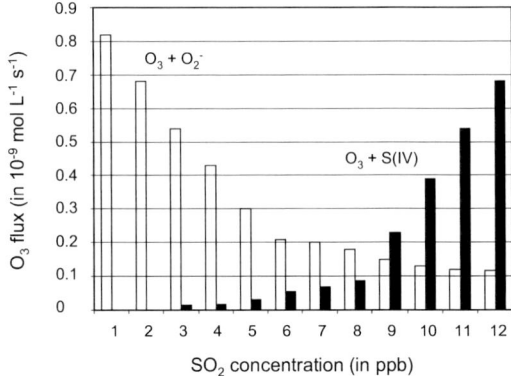

Fig. 5.18 Modeled ozone flux (in 10^{-9} mol L^{-1} s^{-1}) into the aqueous phase of a monodisperse stratiform cloud (droplet radius 5 µm, LWC $0.3 \cdot 10^{-6}$), 30 ppb O_3 for different gaseous SO_2 concentrations, using a complex cloud chemistry model (Möller and Mauersberger 1995).

Table 5.13 Statistical parameters for mean ozone concentrations in summer (mid April until mid October) and winter (mid October until mid April) and for cloud-free and cloudy condition ("station-in-cloud") at Mt. Brocken 1992−1997 (Harz Mt., Germany) 1142 m a.s.l., 51.80° N and 10.67° E, based on monthly means (in ppb).

	winter	summer	year
all events (ppb)	26.3 ± 4	44.0 ± 3	34.2 ± 3
cloud-free (ppb)	31.1 ± 5	47.1 ± 2	37.4 ± 4
station-in-cloud (ppb)	21.1 ± 4	33.5 ± 3	26.8 ± 4
LWC (in mg m^{-3})	272 ± 22	272 ± 27	263 ± 63
station-in-cloud (%)	59 ± 16	28 ± 10	45 ± 2
difference cloud-free − cloudy (ppb)	10.0	13.6	10.6
ratio cloudy / cloud-free	0.68	0.71	0.72

the cycling through OH dominates in alkaline solution (here adjusted because of the low availability of SO_2). In an SO_2-polluted atmosphere, the solution becomes acidic because of sulfurous acid formation and oxidation to sulfuric acid, but the indirect O_3 decay via OH plays no role. With decreasing atmospheric SO_2, the solution becomes less acidic and the rate of the O_3—S(IV) reaction decreases. Fig. 5.18 shows modeling results illustrating this relationship. It can clearly be seen that in the remote air, by reaction (5.107) droplets consume more O_3 than the polluted atmosphere through S(IV) oxidation.

Other modeling groups (Johnson and Isakson 1993, Matthijsen et al. 1997, Liang and Jacobs 1997) have confirmed that clouds can decompose O_3 up to 100%, depending on pH, LWC and the extension of the cloud system (in other words, the air volume the cloud occupies). Mauldin et al. (1997) carried out direct measurements in clouds and found that the OH concentration above the cloud is 3−5 times higher $(8 \ldots 5) \cdot 10^6$ cm^{-3} than under cloud-free conditions within the same altitude. Simultaneously, they found O_3 concentration in clouds is about 25% smaller than above the clouds (30 ppb). Based on long-term monitoring at Mt. Brocken (Harz, Germany), we generally found smaller O_3 concentrations under cloudy conditions than under comparable cloud-free conditions. However, significantly smaller interstitial O_3 concentrations only under special conditions favoring "freezing" cloud chemical conditions such as no mixing, no precipitation and long-scale cloud transport have been found (Acker et al. 1995). As seen in Table 5.13, under cloudy conditions the ozone concentration is on average 30% less than under cloud-free conditions.

5.3.7 Stratospheric oxygen chemistry

Chemistry in the stratosphere (which goes up to about 50 km in altitude) differs from that in the troposphere for three reasons: (i) radiation below 300 nm is available for photodissociations that do not occur in the troposphere (Fig. 5.19), (ii) not all trace gases (those with a short lifetime) are either available or found in very small concentrations (especially water – the stratosphere is very dry) and (iii) there is no precipitation and no liquid cloud (but special *polar stratospheric clouds;* PSCs). The following reactions are of crucial importance for the climate system because they result in the continuous absorption of radiation below 300 nm (Fig. 5.20):

$$O_3 \xrightarrow{h\nu \, (\lambda \leq 300 \text{ nm})} O(^1D) + O_2 \qquad\qquad (5.6 \text{ and } 5.7)$$

$$O_2 \xrightarrow{h\nu \, (\lambda \leq 242 \text{ nm})} O(^3P) + O(^3P) \qquad\qquad (5.16)$$

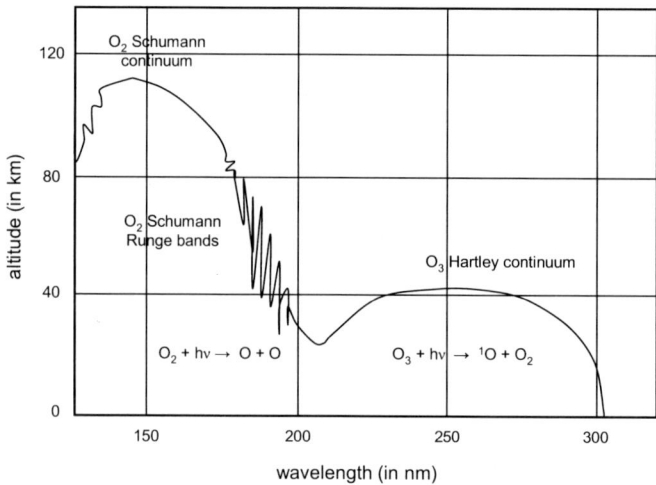

Fig. 5.19 Penetration of solar UV radiation into the atmosphere as a function of wavelength and absorption by oxygen and ozone. The curve indicates the altitude at which incoming radiation is attenuated to about 1/10 of its initial intensity.

$$O_3 \xrightarrow{\ h\nu\ (\lambda \leq 175\ \text{nm})\ } O(^1D) + O_2 \tag{5.17}$$

Evidently, O_3 formation and reaction (5.145) follows, which have no importance in the troposphere but do form *the* natural O_3 removal pathway in the stratosphere.

$$O(^3P) + O_2 \rightarrow O_3 \tag{5.14}$$

$$O + O_3 \rightarrow O_2 + O_2 \tag{5.145}$$

There are other important differences between the stratosphere and troposphere. Whereas the troposphere is heated from the bottom (i. e. the earth's surface), the stratosphere heats from the top (i. e. by incoming solar radiation). This positive temperature gradient results in an extremely stable layering. Therefore, mixing and transport are much weaker than in the troposphere. At the bottom of the stratosphere, close to the tropopause, the lowest temperatures are found (200–220 K and at the poles down to ~ 180 K), whereas temperature rises up to 270 K can be found at 50 km altitude. At each point of the stratosphere the temperature is determined by adiabatic radiation processes: O_3 absorbs UV radiation and heats the air and CO_2 absorbs IR radiation and cools the air. The chemical composition (mixing ratios) concerning the main constituents (N_2, O_2, CO_2 and CH_4) remains constant but the pressure strongly decreases from about 200 hPa at 12 km to 1 hPa at 50 km altitude. The most important trace species is O_3; in the layer between 15 and 30 km altitude about 90% of atmospheric ozone is available. This is called the *ozone layer*.

The photolysis of oxygen was described many years ago by Chapman (1930). The attention to the stratospheric ozone was drawn by Bates and Nicolet (1950), who presented the idea of catalytic O_3 decay. However, only through the implications of manmade influences on the stratospheric ozone cycle by Crutzen (1971), Johnston (1971), Molina and Rowland (1974) as well as Stolarski and Cicerone (1974) was our attention to the stratospheric ozone layer drawn.

Two of the above listed reactions comprise the O_3 catalytic decomposition:

$$\begin{aligned} O + O_3 &\rightarrow O_2 + O_2 \\ O_3 + h\nu &\rightarrow O + O_2 \\ \hline 2\,O_3 &\rightarrow 3\,O_2 \end{aligned}$$

By contrast, this decay is balanced with O_3 formation according to:

$$\begin{aligned} O + O_2 &\rightarrow O_3 \\ O_2 + h\nu &\rightarrow O + O \\ \hline 3\,O_2 &\rightarrow 2\,O_3 \end{aligned}$$

Reaction (5.7) does not lead to a net destruction of ozone. Instead, O is almost exclusively converted back to O_3 by reaction (5.14). However, because O_2 dissociates to free oxygen atoms above about 30 km, below 30 km reaction (5.145) results in a net loss of odd oxygen (if the odd oxygen concentration is defined as the sum of the O_3 and O concentrations). The budget between the photodissociation of O_3 and its formation via (5.14) is zero (steady state). Because the rate of reaction (5.14) decreases with altitude, whereas that for reaction (5.7) increases, most of the odd

Table 5.14 Reactions rate constants for X + O_3 ($k_{5.146}$) and O + XO ($k_{5.147}$) at 22 km altitude and 222 K (in 10^{-12} cm^3 molecule^{-1} s^{-1}); after DeMore et al. (1997).

	O	H	OH	NO	F	Cl	Br	I
$k_{5.146}$	8	140	1.6	2	22	29	17	23
$k_{5.147}$	–	22	30	6.5	27	30	19	120
XO	–	OH	HO$_2$	NO$_2$	FO	ClO	BrO	IO

oxygen below 60 km is in the form of O_3, whereas above 60 km it is in the form of O. Odd oxygen is produced by reaction (5.16). It can be seen that reactions (5.14) and (5.7) do not affect the odd oxygen concentrations but merely define the ratio of O to O_3.

A significant fraction of the removal is caused by the presence of chemical radicals X, such as nitric oxide (NO), chlorine (Cl), bromine (Br), hydrogen (H) or hydroxyl (OH), which serve to catalyze reaction (5.145), termed cycle 1:

$$X + O_3 \rightarrow XO + O_2 \tag{5.146}$$
$$O + XO \rightarrow X + O_2 \tag{5.147}$$
$$\overline{O + O_3 \rightarrow O_2 + O_2}$$

When $k_{5.146} > k_{5.145}$ and substance X is not quickly lost by any termination reaction, the ozone decomposition flux according to cycle 1 becomes larger than it follows only from reaction (5.145). Reaction (5.147) is generally faster than (5.146) because of the stronger radical characteristics (with the exception of hydrogen atoms; Table 5.14). The species H and OH are natural (in contrast to halogens and nitrogen species) constituents of the stratosphere (but with lower concentrations than in the troposphere). Only the hydroxyl radical OH (no other XO species) can directly react with O_3 (reaction 5.22), providing another effective O_3 decay cycle 2:

$$OH + O_3 \rightarrow HO_2 + O_2$$
$$HO_2 + O_3 \rightarrow OH + 2O_2$$
$$\overline{2O_3 \rightarrow 3O_2}$$

The reactive species, called ozone-depleting substances, come from different source gases such as H_2, H_2O, N_2O, CH_4 and halogenated organic compounds. Noteworthy that influencing the stratosphere with water (from aircrafts) and hydrogen (H_2 technology in discussion) results in ozone depletion. An important source of H_2O is methane, and a H_2O concentration maximum found between 50 and 70 km altitude results from CH_4 oxidation:

$$CH_4 \xrightarrow[(-H_2O)]{OH} CH_3 \xrightarrow{O_2} CH_3O_2 \xrightarrow[(-O_2)]{HO_2} CH_3OOH \xrightarrow[(-OH)]{h\nu}$$

$$CH_3O \xrightarrow[(-HO_2)]{O_2} HCHO \xrightarrow[(-HO_2)]{h\nu(+O_2)} HCO \xrightarrow[(-HO_2)]{O_2} CO \xrightarrow[(-HO_2)]{OH(+O_2)} CO_2$$

An additional source of radicals (OH and HO_2) generates the photolysis of O_2 (5.16 and 5.17), O_3 (5.7) and H_2O (5.67) as well as CH_4, giving H (and subsequently HO_2):

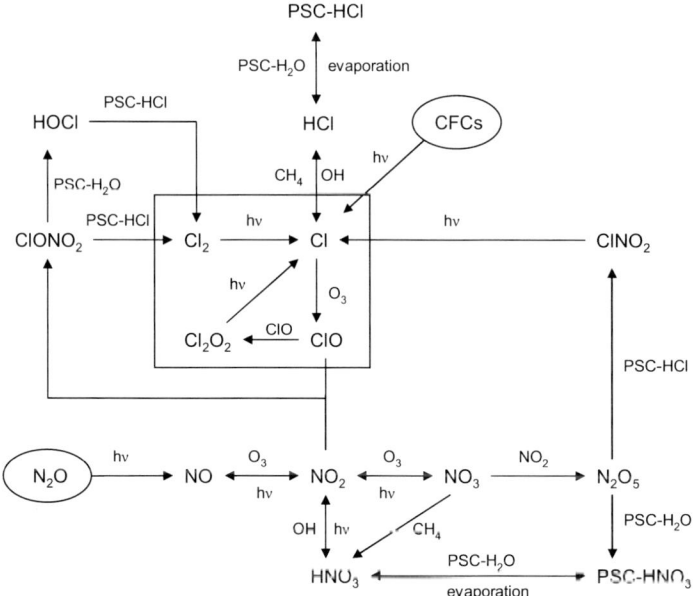

Fig. 5.20 Scheme of stratospheric multiphase chemistry.

$$CH_4 + h\nu \rightarrow CH_3 + H \ (\lambda > 230 \text{ nm}) \tag{5.148}$$

Further considering the H_2O production reactions $OH + OH$ (5.32) and $OH + HO_2$ (5.33) the gross reaction accounts for:

$$CH_4 + 2 O_2 \rightarrow CO_2 + 2 H_2O \tag{5.149}$$

About 90 % of N_2O entering the stratosphere is photolyzed to $N_2 + O$ (5.156); the remaining 10 % reacts with $O(^1D)$ to NO (5.167), which can then enter the ozone decay cycle.

Halogens are most important for ozone depletion. Below we will see that NO_y and halogens produce condensed reservoir species such as $ClONO_2$ and $BrONO_2$, which play a role in the "ozone hole" (Fig. 2.82). Precursors, such as halogenated organic compounds, are photolyzed but also react with the oxygen atoms X = H, Br, F and Cl.

$$R—CX_2Cl + h\nu \rightarrow R—CHX + Cl \tag{5.150}$$

$$R—CX_2Cl + O(^1D) \rightarrow R—CHX_2O + Cl \tag{5.151a}$$

$$R—CX_2Cl + O(^1D) \rightarrow R—CHX_2 + ClO \tag{5.151b}$$

The formation of ClO is dominant (about 60 %) because reactions with OH are too slow. Cycle 2 can run several thousand times before competing products (HOCl, HOBr, HOI and nitrates) are produced. The ODP cycles cannot explain the dramatic O_3 depressions observed in Antarctica every spring (September to October) in the layer between 12 and 24 km because gas phase chemistry (which stops in the arctic winter like at night) is a continuous process. The simple explanation consists of an accumulation of radicals in a condensable matter, called PSC, which can form in different types at $T < -80°C$. Type I consists of pure water, type II nitric acid trihydrate ($HNO_3 \cdot 3H_2O$) and type III a mixture ($HNO_3/H_2SO_4/H_2O$). These "clouds" provide a surface for heterogeneous chemistry and for the absorption and storage of so-called reservoir gases HCl, HBr, HI, $ClONO_2$ and others. Easily photodissociable species are Cl_2, $ClNO_2$ and HOCl (similar to the other halogens). For example, the following reactions occur:

$$HCl(s) + ClONO_2 \rightarrow Cl_2 + HNO_3(s) \tag{5.152}$$

$$HCl(s) + N_2O_5 \rightarrow ClNO_2 + HNO_3(s) \tag{5.153}$$

$$ClONO_2 + H_2O(s) \rightarrow HOCl + HNO_3(s) \tag{5.154}$$

$$HOCl + HCl(s) \rightarrow Cl_2 + H_2O(s) \tag{5.155}$$

In the Antarctic spring, PSCs evaporate and set free "active" halogens (Cl_2 + Cl + ClO + Cl_2O_2; Fig. 5.20). Most NO_x is stored as HNO_3 in solid PSCs and contributes only little to O_3 depletion. After ending the ozone hole period (at the end of November), PSCs almost evaporate and remain only as liquid sulfuric acid particles. Meanwhile, the "normal" gas phase cycles of ozone depletion control the steady-state concentration.

5.4 Nitrogen

A total of 78.1 Vol-% of air contains dinitrogen (N_2) and this corresponds to about 99% of all nitrogen on earth. The other main forms of nitrogen are nitrate deposits and proteins in the biomass (Fig. 2.26), both representing the most oxidized N(+V) and most reduced N(−III) form of nitrogen and cycling through diverse oxides of nitrogen (Fig. 2.27). The dissociation energy of the N≡N triple bond is very large (945.33 kJ mol^{-1}); hence, molecular nitrogen is the most stable nitrogen compound. The formation of ammonia $N_2 + 3H_2 \rightleftarrows 2NH_3$ is exothermic (92.28 kJ mol^{-1}) but the initial N_2 dissociation needs large activation energy. The photodissociation of N_2 occurs at wavelengths smaller than 100 nm in the upper mesosphere (Richards et al. 1981, Wu et al. 1983). The reaction $N_2 + O_2 \rightleftarrows 2NO$ is endothermic (180.6 kJ mol^{-1}) but in the gross budget $N_2 + 2.5O_2 \rightarrow 2HNO_3$ exothermic (30.3 kJ mol^{-1}); again, the activation energy to split N_2 under normal conditions in the climate system is not available (otherwise all atmospheric O_2 would be converted into nitric acid and dissolved in the ocean).

Therefore, nitrogen fills the atmosphere as chemical non-reactive "buffer substances" (too high an oxygen concentration would result in burning), but is the only source for vital nitrogen compounds via biological and chemical fixation (Chapter 2.4.3.1). Natural chemical fixation is because of lightning where NO is produced, similar to industrial high-temperature processes (Chapter 5.4.1). This source is small compared with biological fixation from N_2 (Table 2.19) and is negligible for the biological nitrogen cycle but important for the atmospheric NO budget. The industrial ammonia synthesis $N_2 + 3H_2 \rightarrow 2NH_3$ (at about 200 bar and 500°C) is a copy of the enzyme-based reduction of molecular nitrogen by nitrogen-fixing

Table 5.15 Inorganic nitrogen species; note: All H atoms can be partly and fully replaced by organic groups (R) forming analog organic nitrogen compounds (cf. Table 5.16).

	hydrides and oxides	name	oxo acids	name
−3	NH_3	ammonia		
−2	$N_2H_4{}^g$ (H_2N-NH_2)	hydrazine		
−1	$N_2H_2{}^g$ ($HN=NH$)	diazene[a]	$HONH_2{}^g$	hydroxyl amine
	$NH^{e, g}$	nitrene[h]		
0	N_2 ($N\equiv N$)	nitrogen		
+1	N_2O	dinitrogen monoxide[c]	$HNO^{e, g}$	hydrogen oxonitrate[f]
			$HON=NOH^g$	hyponitrous acid[k]
+2	NO^g	nitrogen monoxide[d]	$HONO^g$	nitrous acid
+3	N_2O_3	dinitrogen trioxide	$HOONO^e$	peroxonitrous acid
+4	NO_2	nitrogen dioxide		
	N_2O_4	dinitrogen tetroxide		
+5	NO_3	nitrogen trioxide[b]	$HONO_2$	nitric acid
	N_2O_5	dinitrogen pentoxide		
			$HOONO_2{}^e$	peroxonitric acid

[a] as azo group ($-N=N-$) in organics, also called diimine and diimide
[b] nitrate radical
[c] nitrous oxide
[d] nitric oxide
[e] instable (intermediates)
[f] nitroxyl radical $H-N=O$; other name: nitrosyl hydride (IUPAC name: azanone)
[g] likely intermediates in plants
[h] imidogene
[k] IUPAC name: diazenediol

bacteria. Nitrogen occurs in all oxidation states from −3 to +5 (Table 5.15), which makes the compounds important for global redox processes; however, many substances only exist as intermediates. In reaction (5.20), we saw the only source of $O\,(^3P)$ for atmospheric ozone formation in the troposphere. In air, NH_3 (NH_4^+), N_2O, NO, NO_2, NO_3, HNO_2 (NO_2^-) and HNO_3 (NO_3^-) are only of interest among inorganic nitrogen. Most organic nitrogen compounds are derived from some of the above inorganic species, such as amines ($-NH_2$), nitroso (or nitrosyl) compounds ($-N=O$) and nitro compounds ($-NO_2$).

5.4.1 Thermolysis of nitrogen: Formation of NO

At high temperatures ($T > 1000\,°C$), molecular nitrogen from air converts into NO. This can happen during lightning (biomass combustion do not provide such high T) and in industrial combustion processes. Because of the large dissociation energy in N_2, the initial step is dioxygen thermal dissociation with a subsequent reaction of oxygen atoms with N_2:

$$N_2 + O \rightarrow NO + N \tag{5.156}$$

$$N + O_2 \rightarrow NO + O \tag{5.157}$$

In steady state ($d[N]/dt = 0$), it follows that:

$$\frac{d[NO]}{dt} = k_{5.157}[O][N_2] \tag{5.158}$$

For a small percentage, the formation of N_2O is possible, which can also effect oxidization:

$$N_2 + O + M \rightarrow N_2O + M \tag{5.159}$$

$$N_2O + O \rightarrow NO + NO \tag{5.160}$$

As termination reaction occurs:

$$N + NO \rightarrow N_2 + O \tag{5.161}$$

After all, the primary product is NO, which than can go into further oxidation processes, and thereby the early detection of nitric acid in thunderstorms is the final product in NO oxidation.

5.4.2 Ammonia (NH_3)

All nitrogen-fixing organisms are prokaryotes (bacteria). Some of them live independent of other organisms − the so-called free-living nitrogen-fixing bacteria. Others live in intimate symbiotic associations with plants or other organisms (e. g. protozoa). Biological nitrogen fixation can be represented by the following equation, in which two moles of ammonia are produced from one mole of nitrogen gas, at the expense of 12 moles of ATP (adenosine triphosphate, empirical formula: $C_{10}H_{16}N_5O_{13}P_3$; ADP adenosine diphosphate) and a supply of electrons and protons, using an enzyme complex termed nitrogenase. This reaction is performed exclusively by prokaryotes (the bacteria and related organisms):

$$N_2 + 6H^+ + 6e^- + 12\,ATP \rightarrow 2\,NH_3 + 12\,ADP + 12\,phosphate \tag{5.162}$$

There is stepwise hydrogenation via diimine N_2H_2, an unstable intermediate, and diazene N_2H_4 (hydrazine) to ammonia:

$$N_2 \xrightarrow{2H} N_2H_2 \xrightarrow{2H} N_2H_4 \xrightarrow{2H} 2\,NH_3$$

Hydrazine is found in natural waters as a pollutant from industrial manufacturing in small concentrations. Warburg (1914) first suggested the photodissociation of NH_3:

$$NH_3 + h\nu \rightarrow NH_2 + H \tag{5.163}$$

The dissociation energy of ammonia using photolysis at 205 nm was determined to be (4.34 ± 0.07) eV (Amorim et al. 1996); thereby it is out of interest in the atmosphere (but was discussed in the early atmosphere). The only gas phase reaction is that with OH forming the amidogen (amide) radical NH_2; $k_{5.164} = 1.6 \cdot 10^{-13}$ cm^3 molecules^{-1} s^{-1}:

$$NH_3 + OH \rightarrow NH_2 + H_2O \qquad (5.164)$$

Ennis et al. (2009) first detected the water amidogen radical complex (H_2O-NH_2) as a reactive intermediate in atmospheric ammonia oxidation. The reaction of NH_2 with O_2 is slow ($k < 6 \cdot 10^{-21}$ cm^3 molecule^{-1} s^{-1} at 298 K), making it unimportant in the atmosphere. The products are not specified (NH_2O_2, NO + H_2O, OH + HNO). The most likely fate (Finlayson-Pitts and Pitts 1999 provide a lifetime of about 2–3 s) is reaction with NO_x: $k_{5.165}$ ($= k_a + k_b + k_c$) $= 1.6 \cdot 10^{-11}$ cm^3 molecule^{-1} s^{-1} at 298 K with $k_a/k_{5.165} = 0.9$ and $(k_b + k_c)/k_{5.165} = 0.1$:

$$NH_2 + NO \rightarrow N_2 + H_2O \qquad (5.165a)$$

$$NH_2 + NO \rightarrow N_2H + OH \qquad (5.165b)$$

$$NH_2 + NO \rightarrow N_2 + H + OH \qquad (5.165c)$$

In the reaction with NO_2, channels (a) and (c) are the most probable; no evidence has been found for the occurrence of channel (b) or the other exothermic channels leading to N_2 + 2OH and/or 2HNO: $k_{5.166}$ ($= k_a + k_b + k_c$) $= 2.0 \cdot 10^{-11}$ cm^3 molecule^{-1} s^{-1} at 298 K where $k_a/k_{5.166} = 0.25$ and $k_c/k_{5.166} = 0.75$ over the temperature range 298–500 K:

$$NH_2 + NO_2 \rightarrow N_2O + H_2O \qquad (5.166a)$$

$$NH_2 + NO_2 \rightarrow N_2 + H_2O_2 \qquad (5.166b)$$

$$NH_2 + NO_2 \rightarrow H_2NO + NO \qquad (5.166c)$$

In summary, ammonia oxidation is negligible (5%) compared with the main fate of NH_3, deposition (40%) and particle formation (55%); the numbers in parenthesis provide the percentage of emitted NH_3 (Möller 2003). Thus, ammonia remains in its oxidation state −3, is mainly seen as ammonium (NH_4^+) in air and returns to soils and waters as ammonium, where it moves between the amino group ($-NH_2$) in the biomass and nitrate through nitrification and ammonification.

5.4.3 Dinitrogen monoxide (N$_2$O)

This biologically important gas is rather stable in the troposphere and only undergoes either uptake by soils and vegetation or transport to the stratosphere where it photodissociates up to 90%, however, the photolysis is effective only for $\lambda <$ 240 nm:

$$N_2O + h\nu \ (\lambda < 240 \text{ nm}) \rightarrow N_2 + O \qquad (5.167)$$

Another pathway is via ($k_{5.168} = 1.2 \cdot 10^{-10}$ cm^3 molecules^{-1} s^{-1})

$$N_2O + O(^1D) \rightarrow 2NO \qquad (5.168a)$$

$$N_2O + O(^1D) \rightarrow N_2 + O_2 \qquad (5.168b)$$

5.4.4 Nitrogen monoxide (NO), nitrogen dioxide (NO_2) and oxo acids

The chemistry of different nitrogen oxides with oxidation states larger than +1 is so closely interlinked − also with the oxo acids − that any treatment in separated chapters would diminish the importance of the climate system view. As for oxygen, nitrogen shows a complex aqueous phase chemistry, which is extensively studied because of its importance in biology.

5.4.4.1 Gas phase chemistry

NO_2 and NO play crucial roles in the ozone formation cycle. The former provides the source of atomic oxygen (5.20) and the latter cycles the HO_2 radical back to OH (5.35) for the continuous "burning" of the ozone precursors CO, CH_4 and NMVOC (Fig. 5.7).

$$NO_2 + h\nu \rightarrow NO + O \tag{5.20}$$

$$NO + HO_2 \rightarrow NO_2 + OH \tag{5.35}$$

$$NO + O_3 \rightarrow NO_2 + O_2 \tag{5.36}$$

In Fig. 5.7, the competing reaction to (5.35) is included:

$$NO + RO_2 \rightarrow NO_2 + RO \tag{5.169}$$

It is useful to distinguish between groups (the termination of NO_x is also practical). Most *in situ* analyzers based on chemiluminescence measure the sum of $NO + NO_2$ and only by using a two-channel technique is it possible to detect NO and NO_x, where the difference is interpreted to be NO_2.

$$NO_x = NO + NO_2$$

$$NO_y = NO_x + NO_3 + N_2O_5 + HNO_2 + HNO_3 + \text{organic N}$$
$$+ \text{ particulate N};$$

and

$$NO_z = NO_y - NO_x$$

Therefore, NO_y represents the sum of all nitrogen with the exception of ammonia (and amines), N_2O and N_2. The aging of air masses in terms of oxidation state is often described by ratios such as NO_y / NO_x. It can clearly be seen that in daytime the ratio NO_2 / NO depends on radiation and O_3 concentration. Close to NO sources (e. g. traffic), O_3 can be totally depleted through (5.36). This is also called "ozone titration". NO_2 carries the oxygen and releases it via photodissociation. This was observed by the fact that O_3 concentration in suburban sites is often larger than in urban sites, thereby defining O_x as:

$$O_x = O_3 + NO_2$$

O_x represents the sum of odd oxygen and should be roughly constant for suburban and urban sites, as found by Kley et al. (1994). NO and NO_2 form equilibrium with dimers:

$$NO + NO_2 \rightleftarrows N_2O_3 \tag{5.170}$$

$$NO_2 + NO_2 \rightleftarrows N_2O_4 \tag{5.171}$$

Both substances (they are more soluble than NO and NO_2) have been discussed as precursors to the formation of acids in solution (Möller and Mauersberger 1990a). However, compared with other pathways and because of their very low gas phase concentrations they have been assessed as negligible; N_2O_3 might still play a role as a interfacial intermediate (below). N_2O_3 is the anhydride of nitrous acid (N_2O_3 + H_2O = $2\,HNO_2$). The gas phase equilibrium constant $K_{N_2O_3}$ = $[NO][NO_2]/$ $[N_2O_3]$ = 1.91 atm at 298 K suggests that N_2O_3 is negligible in air. By contrast, $K_{N_2O_4}$ strongly depends on temperature (0.0177 at 273 K and 0.863 at 323 K; Chao et al. 1974) but remains for the atmosphere (out of plumes) without consideration. The biological role of nitrogen oxides will be briefly discussed in Chapter 5.4.4.2.

Under atmospheric conditions, the formation of NO_2 is relevant only through reactions (5.35) and (5.36). Reactions involving direct dioxygen might have importance only under extremely large NO_x concentrations (like in exhaust systems and biological cells) and are negligible: $k_{5.172}$ = $2.0 \cdot 10^{-38}$ cm^6 molecule^{-2} s^{-1} at 298 K:

$$NO + NO + O_2 \rightarrow 2\,NO_2 \tag{5.172}$$

Equation (5.172) does not represent an elementary reaction; it is a multistep mechanism involving NO_3 or the dimer $(NO)_2$. This NO_3 (O=NOO) is an isomer to the nitrate radical O=N=O(O) and the first step in NO oxidation. It is clear that this very instable peroxo radical will almost decompose by quenching to NO + O_2, which results in a slow reaction probability (we will meet this reaction later in biological systems):

$$NO + O_2 \rightarrow O{-}O{-}N{=}O \tag{5.173}$$

Since NO_2 is the precondition for near surface ozone formation and thereby beginning radical (oxygen) chemistry, any possible direct NO_2 emission is of large importance for changing the atmospheric ROS budget because there is no ROS consumption in the NO—NO_2 conversion. The next higher nitrogen oxide, only a short-life but extremely important atmospheric intermediate, is nitrogen trioxide NO_3 (directly derived from nitrate ions via electron loss): $k_{5.174}$ = $3.5 \cdot 10^{-17}$ cm^3 molecule^{-1} s^{-1} at 298 K:

$$NO + O_3 \rightarrow NO_3 + O_2 \tag{5.174}$$

The lifetime of NO_3 is very short and is much shorter in daytime because of effective photodissociation (quantum yield 1.0 for $\lambda \leq 587$ nm); at wavelengths less than 587 nm, NO_3 radical dissociation is dominantly to NO_2 + O(^{3}P):

$$NO_3 + h\nu \rightarrow NO + O_2 \qquad \lambda_{\text{threshold}} \geq 714 \text{ nm} \tag{5.175a}$$

$$NO_3 + h\nu \rightarrow NO_2 + O(^3P) \qquad \lambda_{\text{threshold}} = 587 \text{ nm} \tag{5.175b}$$

Orlando et al. (1993) suggested photodissociation rates at the earth's surface, for an overhead sun, and the wavelength range 400–700 nm, of $j(NO_2 + O)$ = 0.19 s^{-1} and $j(NO + O_2)$ = 0.016 s^{-1}. Moreover, collision with NO_x removes NO_3: $k_{5.176}$ = $2.6 \cdot 10^{-11}$ cm^3 molecule^{-1} s^{-1} at 298 K and $k_{5.177}$ = $3.6 \cdot 10^{-30}\,(T/300)^{-4.1}\,[N_2]$ cm^3 molecule^{-1} s^{-1} over the temperature range 200–300 K:

$$NO_3 + NO \rightarrow 2\,NO_2 \tag{5.176}$$

$$NO_3 + NO_2 \rightarrow N_2O_5 \tag{5.177}$$

Hence, at night NO_3 is accumulated (daytime concentrations are negligible) and plays a role similar to OH in H abstraction, whereas stable HNO_3 is produced as the final product of NO oxidation:

$$NO_3 + RH \rightarrow HNO_3 + R \tag{5.178}$$

The specific reaction rates with different hydrocarbons are generally lower (about 3−4 orders of magnitude) compared with OH + RH, but the large night-time NO_3 concentration can balance it and provide absolute rates comparable with OH. Of high importance is the fast reaction of NO_3 with isoprene and α-pinene ($k >$ $3 \cdot 10^{-12}$ cm^3 molecule^{-1} s^{-1}; see Wayne et al. 1991 for more details on atmospheric NO_3 chemistry). Reaction (5.178) is only one way of producing nitric acid. N_2O_5 is the anhydride of HNO_3 but it reacts absolutely negligibly in the gas phase with H_2O: $k_{5.179} < 1 \cdot 10^{-22}$ cm^3 molecule^{-1} s^{-1}:

$$N_2O_5 + H_2O \rightarrow 2\,HNO_3 \tag{5.179}$$

In a certain sense, N_2O_5 is in (dynamic) equilibrium with NO_2 and NO_3 according to the fast reaction (no kinetic is known). It is remarkable that until now no measurements of N_2O_5 but extensive measurements of NO_3 have existed, from which indirect conclusions on N_2O_5 have been drawn:

$$N_2O_5 + M \rightarrow NO_2 + NO_3 + M \tag{5.180}$$

NO_3 and N_2O_5 will be quantitatively scavenged by clouds and precipitation to form HNO_3 and NO_3^-, respectively (next chapter). Some reactions of N_2O_5 onto sea salt have been studied, producing gaseous nitrile halogenides from NaCl (and similar with NaBr and NaI; Finlayson-Pitts and Pitts 1999):

$$N_2O_5 + NaCl\,(s) \rightarrow ClNO_2\,(g) + NaNO_3\,(s) \tag{5.181}$$

NO_2 can react with sea salt to produce gaseous nitrosyl chloride:

$$2\,NO_2\,(g) + NaCl\,(s) \rightarrow ClNO\,(g) + NaNO_3\,(s) \tag{5.182}$$

In both reactions, the significance consists of the possible subsequent photolytic release of Cl (or related halogen) radicals, which can go, for example, in O_3 destruction cycles:

$$ClNO + h\nu \rightarrow NO + Cl \tag{5.183}$$

$$ClNO_2 + h\nu \rightarrow NO_2 + Cl \tag{5.184}$$

In summary, we present the NO_x—NO_y chemistry in the following line:

$$NO \underset{h\nu}{\overset{O_3}{\rightleftarrows}} NO_2 \underset{NO,\,h\nu}{\overset{O_3}{\rightleftarrows}} NO_3 \underset{M}{\overset{NO_2}{\rightleftarrows}} N_2O_5$$

We now turn to the formation of oxo acids. As noted, nitric acid is the final product that is formed from NO_3 and N_2O_5 via phase transfer as well as from NO_2 in a fast reaction with OH; $k_{5.185} = 3.3 \cdot 10^{-30} \, (T/300)^{-3.0} \, [N_2] \, cm^3 \, molecule^{-1} \, s^{-1}$ over the temperature range 200–300 K (about $6.0 \cdot 10^{-11} \, cm^3 \, molecule^{-1} \, s^{-1}$ at 298 K):

$$NO_2 + OH + M \rightarrow HNO_3 + M \tag{5.185}$$

This reaction is the ultimate OH sink in the air; the photodissociation of HNO_3 is negligible and the fate of nitric acid is scavenging, dry deposition and particle formation according to (4.238) and (4.242), which also finally deposited. The reaction with OH is too slow to be important in the lower troposphere; $k_{5.186} = 1.5 \cdot 10^{-13} \, cm^3 \, molecule^{-1} \, s^{-1}$ at 298 K and 1 bar of air:

$$HNO_3 + OH \rightarrow NO_3 + H_2O \tag{5.186}$$

NO_2 also reacts with HO_2 to produce peroxonitric acid $HOONO_2$; $k_{5.187} = 1.8 \cdot 10^{-31} \, (T/300)^{-3.2} \, [N_2] \, cm^3 \, molecule^{-1} \, s^{-1}$ over the temperature range 220–360 K; $k_{5.188} = 1.3 \cdot 10^{-20} \, [N_2] \, cm^3 \, molecule^{-1} \, s^{-1}$ at 298 K. It follows formally that the equilibrium constant is $2 \cdot 10^{-11}$ at 298 K:

$$NO_2 + HO_2 + M \rightarrow HNO_4 + M \tag{5.187}$$

$$HNO_4 + M \rightarrow NO_2 + HO_2 + M \tag{5.188}$$

This acid is very instable and decomposes immediately; only in the upper troposphere at low temperatures is a certain accumulation and formation of $NO_2 + HO_2$ possible. Furthermore, photolysis produces all breakdown species: OH, HO_2, NO and/or NO_2.

Let us now turn to the formation of nitrous acid. In the atmosphere, the formation by OH radicals is quickly followed by photodissociation (similar to HNO_4); $k_{5.189} = 7.4 \cdot 10^{-31} \, (T/300)^{-2.4} \, [N_2] \, cm^3 \, molecule^{-1} \, s^{-1}$ over the temperature range 200–400 K and quantum yield 1.0 throughout the wavelength range 190–400 nm; $k_{5.191} = 6.0 \cdot 10^{-12} \, cm^3 \, molecule^{-1} \, s^{-1}$ at 298 K:

$$NO + OH + M \rightarrow HNO_2 + M \tag{5.189}$$

$$HNO_2 + h\nu \rightarrow NO + OH \qquad \lambda_{threshold} = 578 \text{ nm} \tag{5.190a}$$

$$HNO_2 + h\nu \rightarrow H + NO_2 \qquad \lambda_{threshold} = 361 \text{ nm} \tag{5.190b}$$

$$HNO_2 + OH \rightarrow NO_2 + H_2O \tag{5.191}$$

Gaseous HNO_2 in ambient air was first measured by Perner and Platt (1979) using differential optical absorption spectroscopy. Recent measurements indicate that HNO_2 also plays a much larger role in the reactive nitrogen budget of rural sites than previously expected (Acker and Möller 2007 and citations therein). The formation of nitrous acid (HONO) via heterogeneous and interfacial pathways (next chapter) provides a source (especially in the morning after sunrise) to produce OH radicals parallel to the O_3 (5.5) and HCHO (5.46) photolysis. From measurements it has been derived (Acker et al. 2005) that HNO_2 accounts for about 30–42% of the radical production in the air close to the ground, similar to contributions from photolysis of HCHO and O_3.

Fig. 5.21 shows the significant difference between nocturnal and daytime NO_y chemistry. Note that HNO_3 formation goes through very different pathways. The daytime chemistry can be characterized as a $NO \rightleftarrows NO_2 \rightarrow HNO_3$ interrelationship and the night-time chemistry as a $HNO_2 \leftarrow NO_2 \rightleftarrows NO_3 \rightarrow HNO_3$ chain.

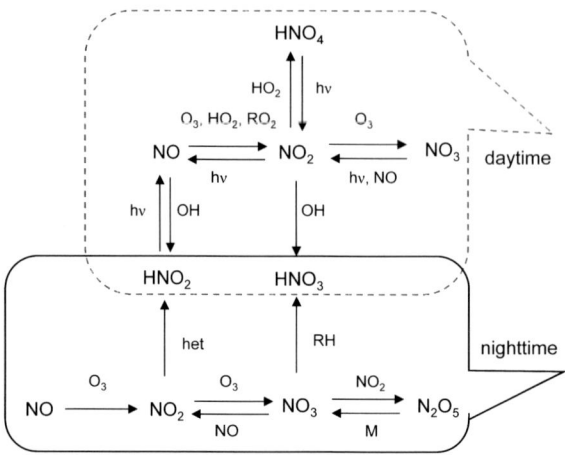

Fig. 5.21 Gas-phase NO_y chemistry at daytime and night-time.

5.4.4.2 Aqueous phase and interfacial chemistry

NO_3, N_2O_5 and HNO_3 will be quantitatively scavenged by natural waters. Nitric acid is a strong acid, and thereby fully dissociated (Table 4.9), whereas nitrous acid is roughly 50% dissociated in hydrometeors.

$$HNO_3 \rightarrow NO_3^- + H^+ \tag{5.192}$$

$$HNO_2 \rightleftarrows NO_2^- + H^+ \qquad pK_a = 3.3 \tag{5.193}$$

When N_2O_5 sticks to the water's surface, it is completely and quickly converted into nitrate ions ($N_2O_5 + H_2O \rightarrow 2\,NO_3^- + 2\,H^+$). The nitrate radical NO_3 can react with all electron donors according to (5.194) and is, therefore, a strong oxidant:

$$NO_3 + e^- \rightarrow NO_3^- \tag{5.194}$$

It reacts with dissolved hydrocarbons (5.178) but is likely to be dominated by chloride and sulfite. The fate of Cl radicals is described in Chapter 5.8.2 and that of sulfite radicals in Chapter 5.5.2.2; $k_{5.195} = 9.3 \cdot 10^6$ L mol^{-1} s^{-1} and $k_{5.196} = 1.7 \cdot 10^9$ L mol^{-1} s^{-1}:

$$NO_3 + Cl^- \rightarrow NO_3^- + Cl \tag{5.195}$$

$$NO_3 + HSO_3^- \rightarrow NO_3^- + HSO_3 \tag{5.196}$$

Nitrate ions can be photolyzed. However, in the bulk water phase the reaction is very slow ($j \approx 10^{-7}$ s^{-1}). The photodecomposition of NO_3^- into OH and NO_x species within and upon ice has been discussed over the past decade and can be crucial to the chemistry of snowpacks and the composition of the overhead atmospheric boundary layer (Dubowski et al. 2002, Chu and Anastasio 2003 and citations therein); $\lambda > 300$ nm and pH < 6:

$$NO_3^- + h\nu \rightarrow NO_2^- + O(^3P) \tag{5.197a}$$

$$NO_3^- + H^+ + h\nu \rightarrow NO_2 + OH \qquad (5.197b)$$

Warneck and Wurzinger (1988) found that channel (a) and (b) contribute 10% and 90%, respectively, but that nitrate photolysis is generally too slow (a lifetime of about seven days) to be significant in the atmosphere. In the presence of O_2, following (5.197a) O_3 formation follows. However, nitrite will be photolyzed, yielding OH; thereby independent of the channels, OH is produced according to the following budget equation:

$$NO_2^- + H^+ + h\nu \rightarrow NO + OH \qquad (5.198)$$

In the absence of radical scavengers, nitrate can react with oxygen; $k_{5.199} = 2 \cdot 10^8$ L mol^{-1} s^{-1}:

$$NO_3^- + O(^3P) \rightarrow NO_2^- + O_2 \qquad (5.199)$$

Nitrite is quickly converted back to NO_2 by OH radicals; $k_{5.200} = 2 \cdot 10^{10}$ L mol^{-1} s^{-1}:

$$NO_2^- + OH \rightarrow NO_2 + OH^- \qquad (5.200)$$

Later, we will see that NO_2 will again be converted to nitrite (5.206) via electron transfer processes independent of the reaction chain (5.201) to (5.203), thereby establishing dynamic equilibria. In the presence of formate HCOO$^-$, which is an efficient OH radical scavenger, it not only prevents nitrite losses via reaction (5.190), but actually converts NO_2, the major product of nitrate photolysis, into additional nitrite via the following reaction sequence; $k_{5.201} = 4.3 \cdot 10^9$ L mol^{-1} s^{-1} and $k_{5.202} = 2.4 \cdot 10^9$ L mol^{-1} s^{-1}:

$$HCOO^- + OH \rightarrow CO_2^- + H_2O \qquad (5.201)$$

$$CO_2^- + O_2 \rightarrow CO_2 + O_2^- \qquad (5.202)$$

$$NO_2 + O_2^- \rightarrow NO_2^- + O_2 \qquad (5.203)$$

For exactly 200 years, it has been known that nitrous gases produce nitrous acid in contact with water (see Chapter 1.3.3). Raschig (1904) showed that when passing N_2O_3 into liquid water, only HNO_2 is formed:

$$NO + NO_2 + H_2O \rightarrow 2\,HNO_2 \qquad (5.204)$$

When passing NO_2 through water, nitrous as well as nitric acid is produced; thereby N_2O_4 is interpreted as a "mixed" anhydride:

$$NO_2 + NO_2 + H_2O \rightarrow HNO_2 + HNO_3 \qquad (5.205)$$

Reaction (5.205) has been proposed to proceed at wetted surfaces, first proposed by Akimoto et al. (1987) as an artefact OH source via HONO photolysis in smog chamber studies. Kessler (1984) first studied the nocturnal increase of HNO_2 due to surface formation. Acker et al. (2001) suggested HONO formation in clouds because of the large surface provided for heterogeneous processes. The heterogeneous hydrolysis of NO_2, which is believed to occur with the same mechanism during the day as at night, has been investigated by numerous field and laboratory studies on many different surfaces (citations in Acker and Möller 2007). Soils, buildings,

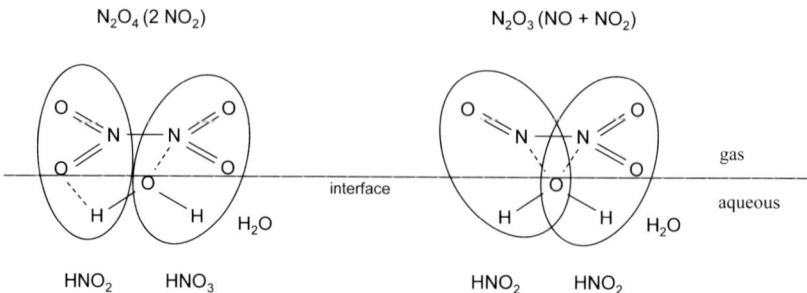

Fig. 5.22 Scheme of N_2O_4 and N_2O_3 interfacial reaction to HNO_2 and HNO_3 (adapted and changed from Möller 2003).

roads and vegetation provide similar solid support and should hold water in sufficient amounts to promote heterogeneous reactions during the day. Accordingly, different heterogeneous formation pathways have been postulated. For example, HNO_2 can be formed from NO_2 suspended on a reducing soot surface as indicated by Ammann et al. (1998). However, only a very small fraction of the surface of fresh soot particles acts, reducing the reaction found with a high reaction probability and a very fast termination (e. g. see discussion in Trick 2004). A proposed heterogeneous reaction of NO and NO_2 on surfaces and the reaction of NO with adsorbed nitric acid (HNO_3) were also found to be unimportant as a potential HNO_2 source. Until now, the chemical formation mechanism has not been well understood and no kinetic is available. According to (5.205), nitrous acid is released to the gas phase and can accumulate during the night, whereas nitric acid remains adsorbed on the surface. Although formation on ground surfaces is expected to be the main source, it seems likely that under circumstances − when atmospheric aerosol or fog droplets provide sufficient surface area − they could also supply an additional surface for the heterogeneous formation of HNO_2. A process with dinitrogen tetroxide (N_2O_4) after the adsorption of NO_2 and steric rearrangement as a key intermediate has been proposed by Finlayson-Pitts et al. (2003) and Möller (2003); see Fig. 5.22.

Whereas the interpretation of observed night-time HNO_2 formation rates is mainly based on (5.205), this "classical" heterogeneous HNO_2 formation via NO_2 disproportionation is too slow to account for the observed atmospheric daytime HNO_2 mixing ratios. From midday measurements above a mixed deciduous forest canopy near Jülich (Germany), Kleffmann et al. (2005) calculated a missing daytime source of ~500 ppt h^{-1}. For the potential candidate (5.205), the authors estimated a mean HNO_2 source strength of 8 ppt h^{-1}, 64 times less efficient than required. It is surprising, that the idea to transfer reaction mechanism (5.96) as a simple electron transfer onto NO_2 in atmospheric chemistry was not proposed before 2005[21]:

---

[21] In air pollution treatment, it was long before it was discovered that NO_x is photocatalytically converted into the TiO_2 catalyst surface and then into HNO_2 and HNO_3 (Hoffmann et al. 1995, Fujishima et al. 1999).

$$NO_2 + e^- \rightarrow NO_2^- \overset{H^+}{\rightleftharpoons} HNO_2 \qquad (5.206)$$

George et al. (2005) found that the photoinduced conversion of NO_2 into HNO_2 on various surfaces containing photosensitive organic substrates might exceed the rate of reaction (5.205) by more than one order of magnitude, depending on the substrate. A high conversion yield for the NO_2 reduction by photochemically activated electron donors (5.206) being present in films of humic acid was observed by Stemmler et al. (2006). Meanwhile, Acker and Möller (2007) discussed that this photochemically driven conversion is common to many surfaces "rich" in partly oxidized aromatic structures and, therefore, is a major candidate for a daytime HNO_2 source.

Vertical measurements using a small aircraft (and other measurements using towers and long path absorption techniques) all showed significant vertical HONO gradients within the lower portion of the boundary layer, indicating the ground surface is a significant HONO source (Zhang et al. 2009 and citations therein). However, for the first time HONO concentration in the FT was measured, with figures ranging from 4 to 17 pptv in the FT and from 8 to 74 pptv in the boundary layer, thereby suggesting that an *in situ* HONO production rate of 57 ppt h^{-1} would be required in the FT. Because the concentration varied little with the time of day, this suggests that *in situ* HONO production was photochemical in nature and that the ambient HONO concentration was in a near photo-steady state. However, the net budget of reactions (5.189) and (5.190) only contributes less than 10 % to the concentration measured. The remaining 90 % of the needed HONO *in situ* production could be derived from photoenhanced processes on aerosol particles involving NO_2. He et al. (2006) and Zhou et al. (2003) proposed the photolysis of HNO_3/nitrate but this process seems to be too limited (discussion above concerns HNO_3 photolysis on ice). Other proposed sources, such as photolysis of nitrophenols (Bejan et al. 2006), cannot explain the HONO concentration because of the very small ambient concentrations of nitrophenol (Zhang et al. 2009, Acker and Möller 2007). Kleffmann (2007) confirmed that a hitherto unknown photochemical HONO source explains the large daytime HONO concentrations and proposes the photolysis of nitroaromatics. Nevertheless, with respect to the evidence of strong photochemical *in situ* HONO production in the FT, there would be significant implications with respect to the cycling of reactive nitrogen in the troposphere and the OH budget.

Other more "exotic" formation pathways of HNO_2 in the condensed phase include possible inorganic reactions similar to biogenic denitrification and nitrification processes, e. g. from NO (which is the anhydride) via the formation of "hydronitrous acid" (H_2NO_2):

$$NO \xrightarrow{H_2O} H_2NO_2 \xrightarrow{H_2NO_2} HNO_2 + HNO \qquad (5.207)$$

The nitroxyl radical HNO can form, via dimerisation, hyponitrous acid ($H_2N_2O_2$ and this can oxidize to HNO_2, likely via the not freely existing hyponitric acid (known as *Angeli*'s salt):

$$N_2O_3^{2-} \xrightarrow{H^+} HNO + NO_2^- \qquad (5.208)$$

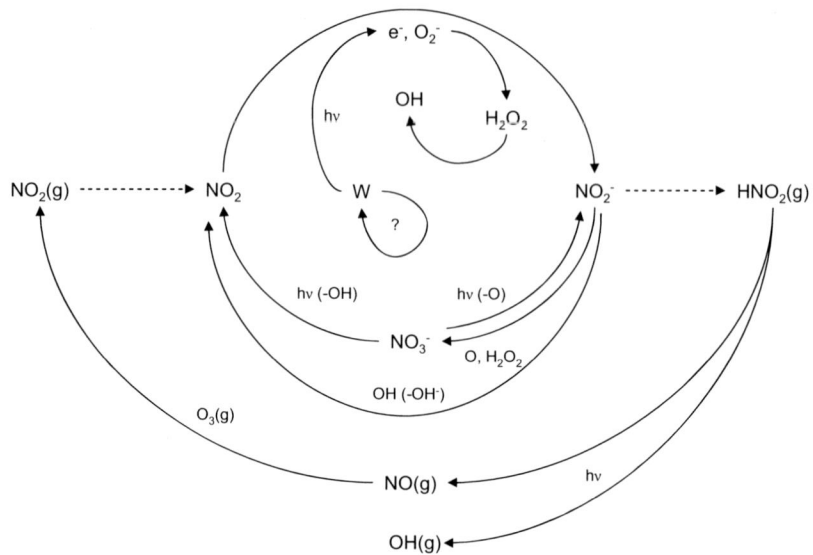

Fig. 5.23 Photocatalytic NO_2 conversion at condensed phases (natural waters, hydrometeors and aerosol particles); (g) gas phase.

Hence, it cannot be excluded that NO is directly associated with HNO_2 formation via the aqueous phase despite its very low solubility, which can be increased because of N_2O_3 formation ($NO + NO_2$); see below for a speculative transfer of NO biological chemistry to atmospheric interfacial chemistry. The atmospheric importance would be that directly emitted NO could be produced via the condensed phase HNO_2 and finally OH radicals.

Gustafsson et al. (2006) showed experimentally under laboratory conditions the reduction of NO_2 to HONO when a TiO_2 aerosol was present. As mentioned before, H_2O_2 is produced in desert sand, which contains TiO_2 (Kormann et al. 1988). Beaumont et al. (2009) detected H_2O_2 when reducing NO_2 into TiO_2 particles. Combining oxygen and nitrogen chemistry via heterogeneous (or interfacial) photochemistry would explain all experimental results. The aerosol particles provide organic matter with chromophoric properties and/or inorganic matter with photochromic properties. Many oxides (and partly sulfides) of transition state metals (Wo, No, Ir, Ti, Mn, V, Ni, Co, Fe, Zn and others) have been characterized to be able to support photosensitizing (Granquist 2002). All these observations lead to the conclusion (Acker et al. 2005, 2008, Acker and Möller 2007, Möller 2009a) that NO_2 undergoes photosensitized conversion by single electron transfer according to (5.196) and similar to those of O_2 (5.96) and O_3 (5.129) at all wetted surface containing photosensitizers. Despite photosensitizers seeming to be ubiquitous, its specific abundance in natural waters and aerosol particles varies, thereby the potential to provide electrons can vary and this explains the different ratios of $HONO/NO_x$ (Acker and Möller 2007) found in different regions. Moreover, the importance of relative humidity as a precondition to forming water films on the surface can clearly be seen. Fig. 5.23 shows the cycle between NO_x and HONO via the condensed phase.

In detail, the following reactions proceed in the NO_2 reduction process (Fig. 5.23); g denotes the gas phase and the adsorbed phase (particulate and/or aqueous) is not indexed; $k_{5.210} = 4.5 \cdot 10^9$ L mol^{-1} s^{-1} (Logager and Sehested 1993) and $k_{5.211} = 1.2 \cdot 10^7$ L mol^{-1} s^{-1} (Clifton et al. 1988):

$$NO_2(g) \rightleftarrows NO_2$$

$$NO_2 + e^- \rightarrow NO_2^- \xrightarrow{H^+} HONO \rightleftarrows HONO(g) \xrightarrow{h\nu} NO + OH \qquad (5.209)$$

$$NO_2 + O_2^- \rightarrow NO_2^- + O_2 \qquad (5.210)$$

$$NO_2 + HSO_3^- \rightarrow NO_2^- + HSO_3 \qquad (5.211)$$

From this sequence the budget equation $NO_2 + H^+ \xrightarrow{e^- + h\nu} NO + OH$ follows, which shows a smaller stoichiometry to OH than the budget from the pure gas phase cycle of reactions (5.20), (5.14), (5.6) and (5.21): $NO_2 + H_2O \xrightarrow{h\nu} NO + 2\,OH$.

The overall budget also depends on the interfacial conditions. Acid solutions favor HONO formation and desorption, and thereby reduce nitrite oxidation (note that HONO formation is a reducing step in an aerobic environment). Therefore, alkaline conditions and high oxidation potential lead to nitrate formation; $k_{5.212} = 4.6 \cdot 10^3$ L mol^{-1} s^{-1} and $k_{5.213} = 5.0 \cdot 10^5$ L mol^{-1} s^{-1} (Damschen and Martin 1993)[22]:

$$NO_2^- + H_2O_2 \rightarrow NO_3^- + H_2O \qquad (5.212)$$

$$NO_2^- + O_3 \rightarrow NO_3^- + O_2 \qquad (5.213)$$

Both reactions are relatively slow compared with the oxidation of dissolved SO_2, which has to be assumed to be in the presence of NO_x (Chapter 5.5.2.2), H_2O_2 will exclusively react with HSO_3^- and O_3 with SO_3^{2-}. Recently, Mudgal et al. (2007) studied the very slow autoxidation ($NO_2^- + O_2$) of nitrous acid according to:

$$2\,HNO_2 + O_2 \rightarrow 2\,HNO_3 \qquad (5.214)$$

It is proposed a HNO_2 dimerization with intermediate peroxo formation and hence complicated steric arrangement (cf. Fig. 5.22), likely explaining the very high acceleration of the reaction rate under freezing conditions (Takenaka et al. 1996, 1998):

$$2\,HNO_2 \rightleftarrows (HNO_2)_2 \xrightarrow{O_2} HO_2N{-}O{-}O{-}NO_2H \rightarrow 2\,HNO_3$$

Experimental studies also show that nitrite in rainwater and cloud water samples exists for hours and even days, whereas S(IV) is oxidized in minutes. Therefore, it is very likely that – especially at interfaces – nitrite is accumulated and transferred back to the gas phase as HONO. Nitrite can be quickly oxidized back to NO_2 by OH; $k_{5.215} = 4.5 \cdot 10^9$ L mol^{-1} s^{-1} (Logager et al. 1993); thereby NO_2 reduction (5.211) and oxidation (5.214) are in "equilibrium":

$$NO_2^- + OH \rightarrow NO_2 + OH^- \qquad (5.215)$$

---

[22] A good compilation of aqueous phase chemical reactions and equilibria can be found in Williams et al. (2002) and Herrmann et al. (2005).

The very fast reaction (5.211) is dominant in the presence of low SO_2 concentrations and has not been considered until now to budget for the NO_2 transfer to HONO. In Fig. 5.23, the nitrate chemistry is included; however, it is likely that we can neglect this and conclude the following main pathway:

$$NO\,(g) \xrightarrow[(-O_2)]{O_3} NO_2(g) \xrightarrow{ads} NO_2(ads) \xrightleftharpoons[OH\,(-OH^-)]{e^-,\,O_2^-} NO_2^- \xrightarrow{H^+}$$

$$HONO\,(g) \xrightarrow{h\nu} NO + OH$$

The budget equation follows, which represents a water splitting process similar to the gas phase OH production from O_3 (reactions 5.6 and 5.21) $O_3 + H_2O \xrightarrow{h\nu} O_2 + 2\,OH$:

$$O_3 + e^- + H_2O \rightarrow O_2 + OH + OH^- \tag{5.216}$$

It is fascinating that this overall conversion is also given by the aqueous phase ozone decay, which here is given in a slightly different reaction pathway (5.141a). The overall budget represents Eq. (5.216) independent of the intermediates HO_3, O^- and/or $H_2O_2^-$, which can vary in dependence from pH:

$$O_3 \xrightarrow{e^-} O_3^- \xrightarrow[(-O_2)]{} O^- \xrightarrow{H_2O} H_2O_2^- \xrightarrow[(-OH^-)]{} OH$$

Hydrogen peroxide can be generated parallel to the photoenhanced processes from adsorbed oxygen:

$$O_2 \xrightleftharpoons{e^-} O_2^- \xrightleftharpoons{H^+} HO_2 \xrightarrow{e^-} HO_2^- \xrightarrow{H^+} H_2O_2 \left(\xrightarrow{e^-} H_2O_2^- \xrightarrow[(-OH^-)]{} OH \right)$$

As shown in Chapter 5.3.5, from ozone via OH, H_2O_2 is also produced through the *Fenton* reaction. Moreover, it is seen that photocatalytic dioxygen reduction is less effective compared with O_3 because many reactions scavenge superoxide (return it to O_2) and the OH formation needs a three-electron transfer ($O_2 + 3\,e^- + 2\,H_2O \rightarrow 3\,OH^- + OH$) compared with the one-electron transfer step into O_3. Therefore, the presence of O_3 will enhance all interfacial oxidation processes.

Fig. 5.24 shows the key photochemical and electron transfer reactions in the multiphase oxygen-nitrogen system. The link from nitrogen to oxygen goes only in the gas phase via photolysis of NO_2 (to O_3) and HNO_2 (to OH) – there is no ROS formation in aqueous phase from dissolved NO_y (neglecting the already discussed nitrate photolysis) but NO_y plays a crucial role in both phases in cycling of ROS.

Let us now turn to an interesting pathway in combining ammonium and nitrite chemistry. Takenaka et al. (2009) studied the chemistry of aqueous solution of salts while drying with special respect to dew. We already mentioned that dew water is very common and likely hitherto an underestimated interfacial chemical pathway in linking biosphere and atmosphere. Beside photosensitized oxidation processes (see the schemata above) under drying (in other terms evaporation) conditions, high concentration of solutes occur. Without any doubt, ammonium (NH_4^+) and nitrite (NO_2^-) are important species in dew. Likely in the early 19th century, it has been known that ammonium nitrite is a "natural" substance found in air (Chap-

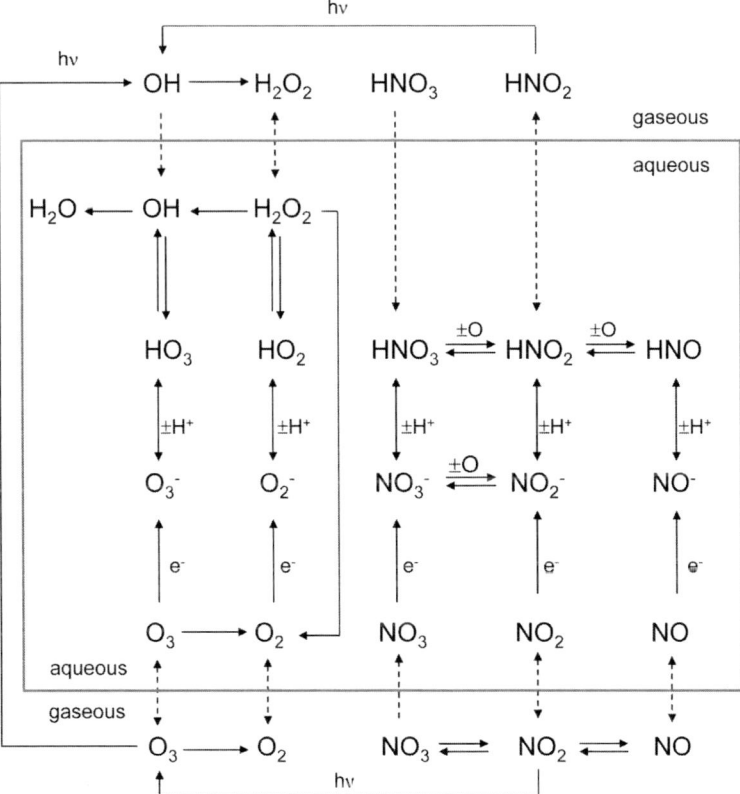

Fig. 5.24 Scheme of principal reactions among oxygen and nitrogen multiphase chemistry. Dotted lines: phase equilibrium and transfer, resp.; double arrow: protolytic equilibrium; e⁻ electron transfers, O symbol for odd oxygen in terms of different oxidants, note: behind the solid arrows are mostly complex multistep pathways.

ter 1.3.3)[23]. In 1812, *Jöns Jakob* Berzelius found that aqueous solutions of NH_4NO_2 decompose, and in 1875, *Marcellin Berthelot* stated that concentrated solutions decompose quickly, under formation of N_2 where acid accelerates this. Takenaka et al. (2009) found that drying dew droplets, containing NH_4^+ and NO_2^-, not only evaporate HNO_2 (which turns into the OH production as discussed above) but also N_2, NO and NO_2. The gross reactions already have been described in the 1930s (Gmelin 1936, see also for citations):

$$NH_4^+ + NO_2^- \rightarrow N_2 + H_2O \tag{5.217}$$

$$2\,HNO_2 + NO_2^{\,-} \rightarrow NO_3^{\,-} + 2\,NO + H_2O \tag{5.218}$$

$$2\,HNO_2 \rightarrow NO + NO_2 + H_2O \tag{5.219}$$

---

[23] Alchemists did collect dew (see Mutus Liber) in large amounts and distilled it to find the "materia prima"; it is likely that they also mentioned the "explosive" character of the residual salt − ammonium nitrite (Möller 2008).

Audrieth (1930) proposed as intermediates HN=NOH (hydroxyl diimide or hydroxy[1,1 or 1,2]diazene) or the tautomer $H_2N—NO$ (nitrosamide). This compound (N_2H_2O) has been described *ab-inito* in form of 9 isomers (Casewit and Goddard 1982), but none has been isolated and characterized; several are believed to be intermediates in processes involving the reduction of nitrogen oxides to molecular nitrogen and water, namely the species mentioned by Audrieht (1930) and HN=N(O)H (diimide-N-oxide). The reaction

$$NH_2 + NO \rightarrow N_2 + H_2O \tag{5.165a}$$

is considered as a key step in the thermal DeNOx process and nitrosation of amines (see also next Chapter 5.4.5.1). Here, we propose the following hypothetical mechanisms:

$$NH_4^+ + NO_2^- \rightleftarrows NH_3 + HNO_2 \rightarrow [H_3N—NO(OH)]$$
$$\xrightarrow[-H_2O]{} H_2N—NO \rightarrow N_2 + H_2O \tag{5.217a}$$

$$NH_4^+ + NO_2^- \rightleftarrows NH_3 + HNO_2 \rightarrow [H_3N—NO(OH)]$$
$$\xrightarrow[-H_2O]{} HN—NOH \rightarrow N_2 + H_2O \tag{5.217b}$$

$$2\,HNO_2 \xrightarrow[-H_2O]{} N_2O_3 \text{ (note: O=N—NO(O))} \rightleftarrows NO + NO_2 \tag{5.219}$$

$$HNO_2 \xrightarrow[-H_2O]{H^+} NO^+ \xrightarrow{NO_2^-} ONNO_2\ (N_2O_3) \rightarrow N_2O + O_2 \tag{5.220}$$

Note that nitrous oxide N_2O is tautomer: $N^-{=}N^+{=}O$ and $N{\equiv}N^+—O^-$. Reaction (5.219) is the "back" reaction (5.204); cf. also Fig. 5.23 for steric arrangement. Takenaka et al. (2009) found that droplet size (smaller accelerate decomposition), pH, the nitrite/ammonium ratio and the dying speed determine the reactions channels and finally, the yield of N_2, N_2O, NO and NO_2 versus evaporation of simply NH_3 and HNO_2, or residual formation of NH_4NO_3.

Nothing is found in the air chemistry literature on the aqueous phase chemistry of nitric oxide (NO). In textbooks of inorganic chemistry (e. g. Wiberg et al. 2001), it is noted that NO does not react with water. However, in older literature (Gmelin 1936) we see that NO slowly reacts with water with the formation of HNO_2, N_2 and N_2O. It was speculated that NO combines with OH^- and via $H_2N_2O_2$ (hyponitrous acid, an isomere of nitramide) formation, which decays to N_2O, or that NO directly combines with H_2O to generate $H_2N_2O_3$ (oxo hyponitrous acid), which decays to HNO and HNO_2. The dimerisation of HNO to $H_2N_2O_2$ was also proposed but detailed mechanisms were unknown (see below). The intermediate existence of HNO in HNO_2 reduction as well as NH_3 or NH_2OH (hydroxylamine) oxidation is well established (Heckner 1977).

With the findings that NO (one of the smallest and simplest molecules) is an important signaling molecule in biological chemistry and its inactivation is unique with respect to other signaling molecules in that it depends solely on its non-enzymatic chemical reactivity with other molecules (Miranda et al. 2000), an explosion in NO solution chemistry began in the 1990s. Until recently, most of the biological effects of nitric oxide have been attributed to its uncharged state (NO), yet NO can also exist in a reduced state as nitroxyl (HNO and its protolytic form, the nitroxyl anion NO^-) and in an oxidized form as nitrosonium ion (NO^+; Hughes 1999, Fukuto et al. 2005, Doctorovich et al. 2011):

$$NO^+\ [N{\equiv}O]^+ \leftrightarrow NO\ [N{=}O] \leftrightarrow NO^-\ [N{=}O^-] \leftrightarrow HNO\ [H{-}N{=}O]$$

Thus, unlike NO, HNO (not N—OH) can target cardiac sarcoplasmic ryanodine receptors to increase myocardial contractility, can interact directly with thiols, and is resistant to both scavenging by superoxide (O_2^-) and tolerance development (Irvine et al. 2008).

Nitrosonium (sometimes also termed nitrosyl cation) is very short lived in aqueous solution. Nitrosonium ions react with secondary amines to generate nitrosamines, many of which are cancer-inducing agents at very low doses (5.224). NO^+ is formed in very small concentrations from nitrous acid in strong acid solution, when HNO_2 reacts as a base:

$$HNO_2 + H^+ \rightleftarrows H_2NO_2^+ \xrightleftharpoons{H^+} NO^+ + H_3O^+ \ (pK_a = -7) \qquad (5.221)$$

or

$$HONO \rightleftarrows OH + NO^+ \qquad (5.222)$$

If such a reaction is possible under atmospheric conditions (at highly concentrated particle interfaces), the subsequent step would be the formation of nitrosyl chloride ClNO, an important species in gas phase chemistry. The likely importance of this pathway lies in the photolysis of ClNO gaining Cl radicals:

$$NO^+ + Cl^- \rightarrow ClNO \qquad (5.223)$$

$$NO^+ + R_2{-}NH \rightarrow R_2{-}N{-}NO + H^+ \qquad (5.224)$$

We cannot exclude that NO^+ is produced from electron-holes:

$$NO + W^+\ (h_{vb}^+) \rightarrow NO^+ \qquad (5.225)$$

See reactions (5.220) for another fate of NO^+ in high concentrated nitrite solutions. Nitroxyl anion NO^- is the base of nitroxyl HNO, which is a very weak acid; $pK_a = 11.4$, when both species are in their ground states, ^1HNO and $^3NO^-$ (Shafirovich and Lymar 2002):

$$HNO \rightleftarrows NO^- + H^+ \qquad (5.226)$$

HNO quickly dimerizes to hyponitrous acid $H_2N_2O_2$, including electronic rearrangement and H shift. $H_2N_2O_2$ (HO—N=N—OH) slowly decompose in aqueous solution into nitrous oxide and water (cf. Fig. 2.27 and discussion in Chapter 2.4.3.1 with respect to nitrification):

$$H_2N_2O_2 \rightarrow N_2O + H_2O \qquad (5.227)$$

Hence, N_2O can be regarded as anhydride of hyponitrous acid. HNO also adds NO and generates the anionic radical of the NO dimer (Lymar et al. 2005); $k_{5.228} = 5.8 \cdot 10^6$ L mol^{-1} s^{-1}, which further reacts with NO ($k_{5.229} = 5.4 \cdot 10^9$ L mol^{-1} s^{-1}) to the long-lived $N_3O_3^-$, which finally decomposes into N_2O and nitrite ($k_{5.230} = 3 \cdot 10^2$ s^{-1}):

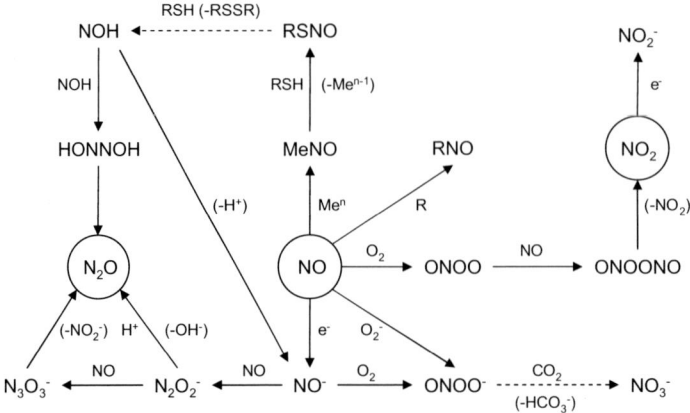

Fig. 5.25 Biological NO_x chemistry; adapted and simplified from Fukuto et al. (2000). Dotted arrow: multistep process, Me transition metal ion, RSH organic sulfide (−SH thiol group), RSSH disulfide, ONOO is an isomeric form of the nitrate radical NO_3, $ONOO^-$ (ONOOH) peroxonitrous acid, HON=NOH ($H_2N_2O_2$) hyponitrous acid, $N_2O_2^-$ ($O=N-N-O^-$) radical anion of NO dimer, $N_3O_3^-$ ($O=N-N-O-NO^-$) trimer radical anion (not to confuse with *Angeli* salt $N_3O_4^{2-}$: $[O_2N-N-NO_2]^{2-}$).

$$HNO + NO \rightarrow O=N-N=O^- (N_2O_2^-) + H^+ \quad (5.228)$$

$$N_2O_2^- + NO \rightarrow O=N-N(O)-N=O^- (N_3O_3^-) + H^+ \quad (5.223)$$

$$N_3O_3^- \rightarrow N_2O + NO_2^- \quad (5.230)$$

As for other electron receptors (such as NO_2 and NO_3), NO is easily reduced and we assume that the electron source is the hydrated electron and/or the conduction band electron (Lymar et al. 2005):

$$NO + e^- \rightarrow NO^- \quad (5.231)$$

NO^- forms peroxonitrite (not to be confused with nitrate NO_3^-); $k_{5.232} = 2.7 \cdot 10^9$ L mol^{-1} s^{-1} (Shafirovich and Lymar 2002), which converts with CO_2 via nitroperoxo carbonate ($ONOOCOO^-$) to nitrate (Fukuto et al. 2000); $k_{5.233} = 3 \cdot 10^4$ L mol^{-1} s^{-1}:

$$NO^- + O_2 \rightarrow O=N-OO^- \quad (5.232)$$

$$ON-OO^- + CO_2 \rightarrow O=N-O-O-COO^- \xrightarrow{H_2O}$$
$$HCO_3^- + NO_3^- + H^+ \quad (5.233)$$

Biochemical routes for the formation of nitroxyl ions are shown in Fig. 5.25 but without considering inorganic or non-enzymatic solution chemistry. In cells, reactions of NO with haems[24], thiols (R—SH) and metals to form metal nitrosyl com-

---

[24] Compounds of iron complexed in a porphyrin (tetrapyrrole) ring differ in side chain composition. Haems are the prosthetic groups of cytochromes and are found in most oxygen carrier proteins.

plexes (Me—NO) as carriers of NO^+ at physiological pH and nitrosation (formation of S-nitrosothiols) is of large biological significance. NO can add onto iron complexes and go into internal electron transfer, thereby forming electropositive nitrosyl ligands:

$$Me^n + NO \rightarrow [Me - NO] \leftrightarrow [Me^{-1} - NO^+]. \qquad (5.234)$$

The nitro(ly)sation of thiol compounds goes directly via N_2O_3 and indirectly through metallic nitrosyl ligands:

$$RSH + N_2O_3 \rightarrow RS\text{—}NO + HNO_2 \qquad (5.235)$$

$$RSH + Me^n\text{—}NO \rightarrow RS\text{—}NO + Me^{n-1} + H^+ \qquad (5.236)$$

The nitrosothiol is quickly decomposed by cuprous, thereby transforming nitroxyl back to nitric oxide (Nelli et al. 2000):

$$RS^-NO + Cu^+ \rightarrow RS\text{—} + NO + Cu^{2+} \qquad (5.237)$$

An important role of NO lies in the interchange between oxyhemoglobin $Hb(Fe\text{—}O_2)$ and deoxyhemoglobin $Hb(Fe^{III})$, which is the primary mechanism by which the movement and concentration of NO are controlled *in vivo*. Because of NO lipid solubility, it can enrich high concentrations. The following reactions illustrate the conversion among NO, NO_2 and nitrite and nitrate (Fukuto et al. 2000):

$$Hb(Fe^{II}\text{—}O_2) + NO \rightarrow Hb(Fe^{III}) + NO_2^-; \qquad (5.238a)$$

$$Hb(Fe^{II}\text{—}O_2) + NO_2 \xrightarrow{H^+} Hb(Fe^{III}\text{—}OOH) + NO_2^-; \qquad (5.238b)$$

$$Hb(Fe^{III}\text{—}OOH) + NO_2 \rightarrow Hb(Fe^{IV}=O) + HNO_2; \qquad (5.238c)$$

$$Hb(Fe^{IV}=O) + NO_2^- \xrightarrow[(-H_2O)]{2\,H^+} Hb(Fe^{III}) + NO_2; \qquad (5.238d)$$

$$Hb(Fe^{IV}=O) + NO \rightarrow Hb(Fe^{III}) + NO_2^-; \qquad (5.238e)$$

$$Hb(Fe^{III}) + H_2O_2 \rightarrow Hb(Fe^{IV}=O) + H_2O; \text{ and} \qquad (5.238f)$$

$$Hb(Fe^{II}\text{—}O_2) + NO \rightarrow Hb(Fe^{III}) + NO_3^-. \qquad (5.238g)$$

Fig. 5.26 summarizes the main routes in the multiphase NO_x—NO_x chemistry. In air, odd oxygen tends to accumulate in NO_x and NO_y but is in a photo-steady state between oxygen and nitrogen species. This condensed phase acts as a sink of oxygen gas phase species, mainly in the form of NO_z (HNO_3, NO_3, N_2O_5). It is obvious that NO_x can transfer either non-photochemical (via $NO + NO_2$ reaction, not shown in Fig. 5.24) or photosensitized to nitrite, which might return to the gas phase as HNO_2 and photolyzed to OH and NO. N(III) oxidation in solution to nitrate is relatively slow and not important as an oxidant consumer. Not shown in Fig. 5.33 is the decomposition of NH_4NO_2 (comproportionation of NH_3 and HNO_2) into gases (N_2, NO, NO_2, N_2O) while evaporating droplets (dew, fog, cloud).

The nitrate photolysis is too slow in bulk solution but can have significance under special climatic condition (snow and ice surface) and at aerosol particle surfaces.

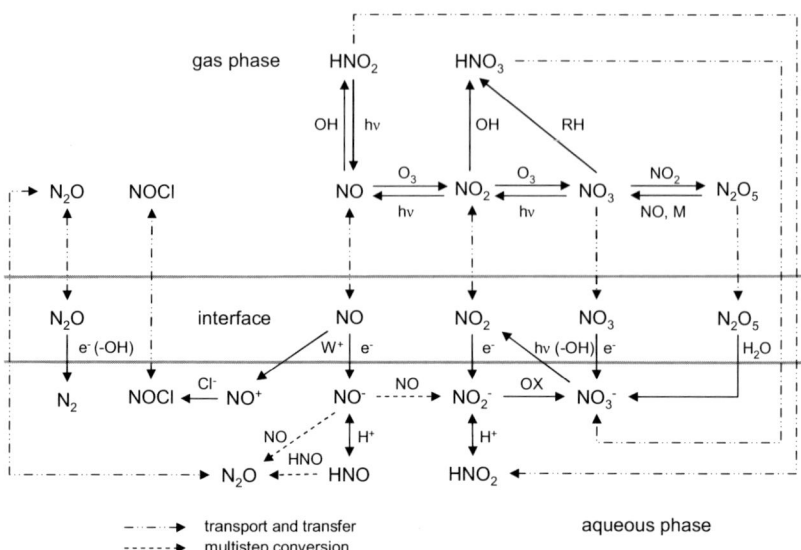

Fig. 5.26 Scheme of multiphase NO_x—NO_y chemistry (main pathways); OX oxidants (OH, H_2O_2); not included is the N_2O_3 chemistry, which might have importance during night for nitrite formation in solution and at interfaces.

Interestingly, a new aspect has potential significance for nitrosyl anions NO^+ that mainly produce nitrosyl chloride (ClNO) but could also react with organics (nitrosation).

The role of aqueous phase oxidant formation, namely OH radicals through photosensitized electron transfer processes onto different nitrogen oxides and oxygen species, is shown in Fig. 5.27 (to avoid overloading the scheme, the OH-consuming reactions and intermediates are not shown; Figs. 5.14, 5.15 and 5.17). Photosensitizers (W) have a key role to provide surface electrons and/or hydrated electrons after illumination for the reduction of the molecules transferred from the gas phase onto the wetted surface (natural waters, hydrometeors and aerosol particles). N_2O provides a direct way for OH production (not yet included in atmospheric models). Similarly, ozone reduction goes directly to OH. The O_2 reduction is less efficient and goes through peroxides. As shown in Chapter 5.3.5, all ROS (O_2^-, HO_2, H_2O_2, OH) can be found in solution but dependent of pH and redox potentials. A new pathway − adapted from biological chemistry − could be the NO-nitrite cycle with the relevance of permanent OH production. Much experimental evidence for HONO formation from NO_2 condensed phase photoconversion is available but this pathway is not a net source of OH because in the first step of NO to NO_2 conversion, ozone (or HO_2) is consumed (note the budget equation in the gas phase: $2\,O_3 + H_2O \rightarrow OH + HO_2 + 2\,O_2$):

$$NO + O_3\,(or\,HO_2) \xrightarrow[(-O_2\,or\,OH)]{} NO_2 \xrightarrow{e^- + H^+} HONO \xrightarrow{h\nu} NO + OH$$

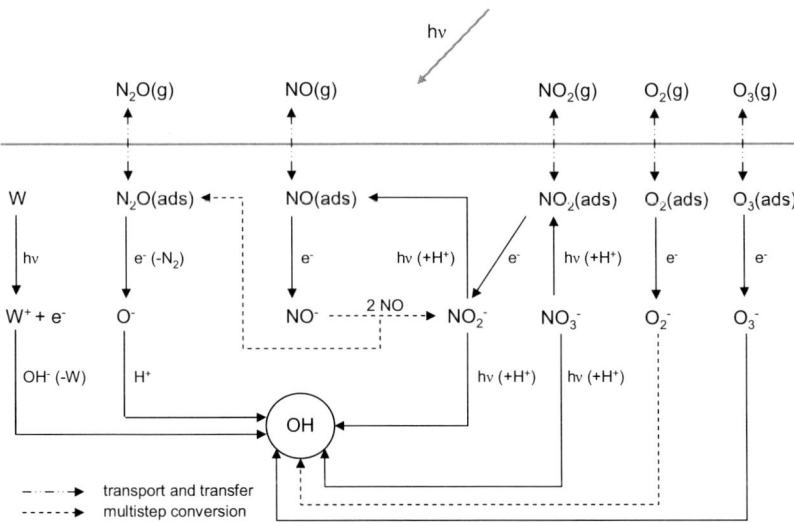

Fig. 5.27 Scheme of solution and interfacial photosensitized electron transfer processes of N_2O, NO_x and O_x giving OH radicals; not included are oxidant consuming processes W any photosensitizer such as organic chromophoric substances and inorganic semiconductors.

This widely discussed process only shifts (because of much faster photolysis of HONO compared with that of O_3) the photo-steady states and species reservoir distribution.

5.4.5 Organic nitrogen compounds

Nitrogen is an essential element in life and thereby in biogeochemical cycling (Figs. 2.23, 2.25 and 2.26 and Chapter 2.4.3.1). In the biomass, nitrogen in organic molecules is almost in a reduced state, derived from ammonia NH_3. The main building block is the amine group NH_2, occurring in its simplest organic compounds, amines (Chapter 5.4.5.1 and Table 5.16). In the biomass, amino acids $=C(NH_2)C(O)OH$ are the building blocks of peptides (which are polymers of amino acids with the $=C-N(H)-C=$ central group) and next are proteins, which are polymers of more than 50 amino acids and have enormous diversity in structure and functions. Additionally, nitrogen is found heterocyclically (mainly in nucleic acids) in biomolecules Oxidized nitrogen (such as organic nitrites and nitrates) should be excluded and expected only as a result of reactions between organics and NO_x / NO_y (Chapter 5.4.5.2). Because of the large N-containing biomolecules, almost non-volatile and after degradation highly water soluble, direct emissions are less probable. Biomass burning has been identified as the source of organic nitrogen but this is probably already in a decomposed form such as HCN or CH_3CN (Table 2.46). The simplest amine is methylamine CH_3NH_2, whose global emissions from animal husbandry was estimated by Schade and Crutzen (1995) in 1988 to be 0.15 ± 0.06 TgN.

Table 5.16 Organic nitrogen species: Functional groups of environmental importance.

structure	name	formula symbol
—NII$_2$	amino group (primary amine)	RNH$_2$
=NH	secondary amine	R$_2$NH
≡N	tertiary amine	R$_3$N
—N=N—	azo group	RNNR
—C≡N	cyano group (nitrile)	RCN
—C(=O)NH$_2$	amide	RC(O)NH$_2$
—N=O and $^+$N=O	nitroso group (nitrosyl)	RNO
—O—N=O	nitrite	RONO
—N(=O)O or —NO$_2$	nitro group (nitryl)	RNO$_2$
—O=NO$_2$	nitrate	RONO$_2$
=N—N=O	nitrosamin	R$_2$NNO

Almost three-quarters of these emissions consisted of trimethylamine-N. Other sources were marine coastal waters and biomass burning.

Nothing is known on the natural emissions of amines. Several anthropogenic sources (traffic, fertilizer production, paper mills, rayon manufacturing, the food industry) have been identified. Possible amine emissions along with the scrubbed flue gas from CO_2 capture facilities (using amine washing solutions) were recently studied. Other simple degradation products of amino acids and peptides are amides, which consist of the building block R—C(=O)NH$_2$. The simplest molecule is formamide HC(=O)NH$_2$. Amides are derivates of carbonyl acids where the hydroxyl group (—OH) is replaced by the amino group (—NH$_2$). Nitriles with the cyano group —CN are potential biomass combustion products from degraded biomolecules.

5.4.5.1 Amines and nitriles

Amines have been detected in ambient air, namely close to industrial areas, livestock homes, and animal waste and wastewater treatments. For example, monomethylamine, trimethylamine, isopropylamine, ethylamine, n-butylamine, amylamine, dimethylamine and diethylamine have all been found (Schade and Crutzen 1995 and citations therein). In southern Sweden, the major amines in both air and rain were trimethylamine and methylamine, but dimethylamine, diethylamine and triethylamine were also detected and individually quantified. The total amine concentration was $0.16-2.8$ nmol m^{-3} in air and $30-540$ nM in rain (Grönberg et al. 1990). Amides (dimethyl formamide) were also detected in air (Howard 1990), whereas other amides have been found to be emitted in industrial processes (acetamide, acrylamide). Under the aspect of CCS (carbon capture and storage, see Chapter 2.8.4), intensive studies on amine emission and degradations have been carried out (e. g. Nielsen et al. 2010, Ge et al. 2011) recently. Amines until C_3 are gaseous and highly volatile. They are *Lewis* bases (somewhat more basic than NH$_3$) and water-soluble (without longer carbon chains), thereby forming salts; the cation NR$_3$H$^+$ is called aminium:

$$NR_3(g) + HNO_3(g) \rightleftarrows NR_3HNO_3(s) \qquad (5.239)$$

Therefore, the acidity of individual particles can greatly affect gas/particle partitioning, and the concentrations of amines, as strong bases, should be included in estimations of aerosol pH (Pratt et al. 2009).

The general fate of amines is similar to that of ammonia, i. e. reactions with OH. They do not photolyze. Because the C—H bond energy is smaller than the N—H bond, predominantly OH abstracts H from C—H (see Chapter 5.7.2) forming bifunctional compounds (e. g. amides $OC(H)NH_2$, see below for further fate). In the less probable case of N—H abstraction, the following radicals are formed:

$$CH_3NH_2 + OH \rightarrow CH_3NH + H_2O \qquad (5.240a)$$

$$(CH_3)_2NH + OH \rightarrow (CH_3)_2N + H_2O \qquad (5.240b)$$

The further fate is not explicitly known, but molecule rearrangements and/or the addition of O_2, NO_x and NO_3 is possible.

$$CH_3NH + O_2 \rightarrow CH_3N(H)O_2 \xrightarrow[(-NO_2)]{NO} CH_3N(H)O \xrightarrow[(-HO_2)]{O_2}$$

$CH_3N{=}O$ (Nitrosyl)

From secondary amines, azo compounds or organic hydrazines (derived from hydrazine N_2H_4) can be given:

$$2\,(CH_3)_2N \rightarrow (CH_3)_2NN(CH_3)_2 \qquad (5.241a)$$

$$(CH_3)_2N + NO \rightarrow (CH_3)_2N{-}NO \text{ (nitrosamines: } R_2NNO) \qquad (5.241b)$$

$$(CH_3)_2N + NO_2 \rightarrow (CH_3)_2N{-}NO_2 \text{ (nitramines: } R_2NNO_2) \qquad (5.241c)$$

They also fast photolyze back to R_2N and NO_x in gas phase (hence they are enriched in air during night time). Nitrosamines are of high interest because of their cancerogenous properties. They have been found to be everywhere, but almost always in small concentrations.

As good electron donors, they are easily oxidable in solution, gaining derivates of hydroxylamine (NH_2OH):

$$(CH_3)_2N + H_2O_2 \rightarrow CH_3N{-}OH + H_2O \qquad (5.242)$$

In water, tertiary amines form quaternary ammonium compounds, e. g. with methyl halogenides. In aqueous solution, secondary (aliphatic and aromatic) amines react with nitric acid HONO (namely the nitrosonium ion NO^+) as already mentioned (reaction 5.224) to generate nitrosamines:

$$(CH_3)_2NH + HONO \rightarrow CH_3N{-}NO + H_2O \qquad (5.243)$$

In water, amines undergo decomposition more than in the gas phase and are accumulated in the condensed phase. Even simpler are formed aromatic nitroamines (Ar—aryl, e. g. C_6H_5):

$$Ar{-}NHR + HNO_2 \xrightarrow[(-H_2O)]{} Ar{-}NR{-}NO \qquad (5.244)$$

They reform by internal rearrangement (*Fischer-Hepp* rearrangement) of the nitroso group into nitroamine:

$$Ar—NR—NO \rightarrow ON—Ar—NHR.$$

The rate of all nitrosation (nucleophilic substitution) increases with basicity of released group X:

$$X—NO \xrightarrow{\text{Y}} [X \cdots NO \cdots Y]^- \rightarrow Y—NO + X^-$$

For amines ($HY = HNR_2$), this reaction is adequate to (5.243) – there is an equilibrium $HNR_2 \rightleftarrows H^+ + R_2N^-$; however, not nitrous acid (ON—OH) is the direct nitrosation agent but the protonized form $ON—OH_2^+$ with measurable rates for pH < 1 only. Other excellent nitrosation agents are nitrosamine itself (OH—NR$_2$) and dinitrogen trioxide N_2O_3 (ON—NO$_2$):

$$NO—NO_2 (N_2O_3) + RNH_2 \rightarrow R—NH—NO + HNO_2 \tag{5.245}$$

Dimethyl formamide is expected to exist almost entirely in the vapor phase in ambient air and react with photochemically produced hydroxyl radicals in the atmosphere (Howard 1990):

$$(CH_3)_2C(O)NH_2 + OH \rightarrow (CH_3)_2C(O)NH + H_2O \tag{5.246}$$

Further dimerization led to imides or imido compounds (—CONH$_2$), which react quickly with OH (Barnes et al. 2010). Formamide HC(O)NH$_2$ is the simplest amide in air. The following listed compounds have been identified as main products of amine oxidation.

isocyanic acid	O=C=NH
methyl isocyanate	O=C=NCH$_3$
N-formylformamide	HN(C(O)H)$_2$
N-formyl-N-formamid	N—CH$_3$(C(O)H)$_2$

In the aqueous phase, tertiary amines react with H_2O_2 to amine oxides (Singh et al. 2006), which act as surfactants and can be of importance in cloud droplets:

$$R_3N + H_2O_2 \rightarrow R_3N^{(+)}—O^{(-)} + H_2O \tag{5.247}$$

Organic cyanides (nitriles) react with OH according to the general C–H abstraction, probably leading ultimately to the formation of CN radicals, which turn to HCN by reaction with hydrocarbons (Goy et al. 1965):

$$CH_3CN + OH \rightarrow CH_2CN + H_2O \tag{5.248}$$

$$CN + RH \rightarrow HCN + R \tag{5.249}$$

HCN is a highly soluble but weak acid ($pK_a = 9.2$), forming many salts (cyanides):

$$HCN \rightleftarrows H^+ + CN^- \tag{5.254}$$

Hydrogen cyanide or hydrocyanic acid (HCN) is found in air, waters, and soils (Ghosh et al. 2006). It is emitted from several thousand plant species, including many economically important food plants, which synthesize cyanogenic glycosides and cyanolipids. Upon tissue disruption, these natural products are hydrolyzed, liberating the respiratory poison hydrogen cyanide. This phenomenon of cyanogenesis accounts for numerous cases of acute and chronic cyanide poisoning of animals including man (Poulton 1990). The —C≡N group (as well as thiocyanate or, in older terminology, rhodanide —S—C≡N) is the simplest organic nitrogen bond, and is already produced in interstellar chemistry (Chapter 2.1.1) and a building block of biomolecules. In air, HCN is insignificantly destructed by OH ($k = 3 \cdot 10^{-14}$ cm^{-3} molecule^{-1} s^{-1}) into the CN radical. HCN is cycled back to vegetation and soils through dry and wet deposition and biologically taken up.

5.4.5.2 Organic NO$_x$ compounds

The direct natural emission of oxidized organic nitrogen compounds is unknown. Therefore, organic nitrites and nitrates are produced in the decomposition process of hydrocarbons (Chapter 5.7) where different oxo and carbon (alkyl) radicals can add NO (leading to nitrites) and NO$_2$ (leading to nitrates; Fig. 5.9):

$$CH_3O + NO + M \rightarrow CH_3ONO + M \qquad (5.250)$$
$$k_{5.250} = 3.3 \cdot 10^{11} \text{ cm}^3 \text{ molecule}^{-1} \text{ s}^{-1}$$

$$CH_3O + NO_2 + M \rightarrow CH_3ONO_2 + M \qquad (5.251)$$
$$k_{5.251} = 2.1 \cdot 10^{11} \text{ cm}^3 \text{ molecule}^{-1} \text{ s}^{-1}$$

Organic peroxo radicals also add NO, where the intermediate CH$_3$OONO rearranges to CH$_3$ONO(O):

$$C_2H_5OO + NO + M \rightarrow CH_3ONO_2 + M \qquad (5.252)$$
$$k_{5.252} \leq 1.3 \cdot 10^{13} \text{ cm}^3 \text{ molecule}^{-1} \text{ s}^{-1}$$

The peroxonitrate (not to be confused with peroxoacyl nitrates) decomposes by collision:

$$C_2H_5OO + NO_2 + M \rightarrow CH_3OONO_2 + M \qquad (5.253)$$

$$CH_3OONO_2 + M \rightarrow CH_3O_2 + NO_2 + M \qquad k_{5.254} = 4.5 \text{ s}^{-1} \qquad (5.254)$$

NO reacts with the methyl radical to form formaldehyde and nitroxyl (remember: in air, CH$_3$ + O$_2$ is absolute predominant):

$$CH_3 + NO \rightarrow HCHO + HNO \qquad (5.255)$$
$$k_{5.255} = 2 \cdot 10^{13} \text{ cm}^3 \text{ molecule}^{-1} \text{ s}^{-1}$$

The methoxy radical can in contrast to (5.251) react to generate nitrous acid; however, both reactions are very slow and therefore not interesting in air:

$$CH_3O + NO_2 \rightarrow HCHO + HNO_2 \qquad (5.256)$$
$$k_{5.256} = 2.1 \cdot 10^{13} \text{ cm}^3 \text{ molecule}^{-1} \text{ s}^{-1}$$

Alkyl nitrite is photolyzed to give the initial molecules in reaction (5.251); the photolytic lifetime is only 10−15 min (Seinfeld and Pandis 1998). Organic nitrates also photodissociate, where (5.258a) is the dominant pathway (reaction 5.258b including internal rearrangement):

$$CH_3ONO + h\nu \rightarrow CH_3O + NO \qquad (5.257)$$

$$RONO_2 + h\nu \rightarrow RO + NO_2 \qquad (5.258a)$$

$$RONO_2 + h\nu \rightarrow RCHO + HONO \qquad\qquad (5.258b)$$

$$RONO_2 + h\nu \rightarrow RONO + O \qquad\qquad (5.258c)$$

Peroxoacylnitrates have been intensively studied as NO_x reservoirs because they form in photochemical processes but are relatively stable and decompose thermally, thereby providing the long-range transport of nitrogen and regaining organic radicals. They are yielded from the peroxoacyl radical $RC(O)OO$, which is produced after the photolysis of aldehyde (reactions 5.63 and 5.64):

$$RC(O)OO + NO_2 \rightleftarrows RC(O)OONO_2 \qquad\qquad (5.259)$$

Peroxoacyl nitrates also can decompose as follows:

$$RC(O)OONO_2 \rightarrow RC(O)O + NO_3 \qquad\qquad (5.260a)$$

$$RC(O)OONO_2 \rightarrow RC(O) + O_2 + NO_2 \qquad\qquad (5.260b)$$

In the condensed phase (droplets and particles), many nitration reactions proceed (Heal et al. 2007). The nitrating properties of N_2O_3 ($NO + NO_2$) and N_2O_5 are seen from the structure, transferring $-NO_2$, $-N{=}O$ and $-O-N{=}O$:

$$N_2O_3 \quad O{=}N-O-N{=}O$$

$$N_2O_5 \quad (O)O{=}N-O-N{=}O\,(O)$$

In clouds (Mt. Brocken monitoring station), we found nitro phenols (Lüttke et al. 1997).

5.5 Sulfur

In the climate system, sulfur ranks concern total mass on the fourth site (Table 5.17). After oxygen and nitrogen, sulfur (and later phosphorous) is an important constituent of biomolecules with specific properties in the form of functional groups. As for the other life-essential elements occurring dominantly in its most stable chemical compound (Table 5.17), sulfur occurs as sulfate dissolved in oceans. The ocean is also the deposit for oxygen and hydrogen (both in the form of H_2O) and carbon (as carbonate). All other reservoirs and chemical forms can be neglected with regard to mass but not chemical relevance. Sulfur is the oldest known element (cited in the bible but known long before) because it was found residing close to volcanoes and the people quickly recognized its specific properties such as burning with a penetrative odor. In *Homer's Odyssey*, Ulysses said … "Bring my sulphur, which cleanses all pollution, and fetch fire also that I might burn it, and purify the cloisters" (translated by S. Butcher, Orange Street Press, 1998, p. 278). However, despite sulfur being the main chemical (among mercury) of the alchemists, it was regarded until 1809 (by *Humphrey Davy*) as a composite body even though *Lavoisier* in 1777 had already recognized it as an element (simple body). *Gay-Lussac* and *Thénard* confuted still in 1809 this mistake (Kopp 1931, pp. 310−311) and from that time sulfur has been seen as an element.

Table 5.17 Reservoirs or pools of elements in the climate system (cf. Table 2.12).

element	mass (in 10^{19} g element)	atmosphere	ocean	lithosphere
oxygen	12 620	O_2 (1%)	H_2O (99%)	
hydrogen	1 500		H_2O (100%)	
nitrogen	442	N_2 (100%)		
sulfur	128		sulfate (100%)	
carbon	4[a]		carbonate (100%)	
phosphorous	15[b]			phosphate (100%)

[a] without consideration of the lithosphere
[b] based on O/P ratio for the lithosphere after Clarke (1920)

As for nitrogen, biogenic redox processes are essential to convert sulfur from its largest oxidation state +VI (sulfate) to its lowest −II (sulfide). Sulfate is the most stable compound in the atmosphere; once it has been produced there is no abiotic reduction possible in the climate system. Organisms need sulfur in the form of thiols RSH (simple thiols are called mercaptanes) and sulfides R_2S. Thiols are (similar to alcohol ROH) weak acids (but much stronger). Thiols are easy oxidized into sulfides (this reaction is the main function in biological chemistry; Fig. 5.25), thereby thiols provide hydrogen for the reduction of other molecules (the dissociated form RS^- acts as an electron donor):

The following scheme shows the biological cysteine sulfur turnover (Gupta and Carrol 2013); see Table 5.18 for names of compounds:

$$RSO_2H \leftarrow RSOH \leftarrow RSH \rightarrow RSSR \rightarrow RSS(=O)R$$
$$\downarrow \qquad\qquad\qquad\qquad\qquad\qquad\qquad\qquad\qquad \downarrow$$
$$RSO_3H \leftarrow RS(=O)_2S(=O)_2R \leftarrow RS(=O)S(=O)_2R \leftarrow RSS(=O)_2R$$

There are many organic sulfur compounds in nature but all probably exist in a higher oxidation state than S(−II) such as sulfides (including disulfides) and thiols because they are oxidized after their release from organisms (see Table 5.18 for different functional sulfur groups). As mentioned in Chapter 2.4.3.2, carbonyl sulfide $O=C=S$ (commonly written as COS) is the most abundant sulfur compound naturally present in the atmosphere. Hydrogen sulfide H_2S was believed before 1970 to be the major emitted species but it is almost all oxidized before escaping from anoxic environments into the atmosphere. In the natural climate system, dimethyl sulfide $(CH_3)_2S$ (DMS) is the most important sulfur species because of its large emissions and contribution to climate-relevant atmospheric particulate sulfates. With industrialization, sulfur dioxide SO_2 became the key pollutant and the most important sulfur species in air. Nowadays, after the introduction of flue gas desulfurization, SO_2 has been replaced in air chemical significance by NO_x. However, concerning the atmospheric background submicron aerosol, sulfates remain dominant and important in climate control. It is a subtle irony of too much SO_2 abatement that cooling sulfate aerosol disappears, demasking the greenhouse effect.

Table 5.18 Organic sulfur compounds. Sulfenes and sulfenic acid are extremely instable and convert into disulfides and sulfinic acids. The open bond — is connected with organic rest R.

structure	formula	name of compound	formal oxidation state
—S—H	RSH	thiol (thio group SH)[a]	−2
—S—	R_2S	sulfide or sulfane	−2
—S—S—	R_2S_2	disulfide	−1/−1
—S(=O)—	RSO	sulfine or sulfoxide (sulfinyl group SO)	0
—S(=O)$_2$	RSO_2	sulfone (sulfonyl group SO_2)	+3
=S(=O)$_2$	RSO_2	sulfene[b]	+6
—S(=O)—S—	R_2S_2O	thiosulfinic acid	0/−1
—S—S(=O)$_2$—	$R_2S_2O_2$	thiosulfanate	−1/+3
—S(=O)—S(=O)$_2$—	$R_2S_2O_3$	disulfide trioxide	+1/+3
—S(=O)$_2$—S(=O)$_2$—	$R_2S_2O_4$	disulfone	+3/+3
—S—OH	RSOH	sulfenic acid[c]	0
—S(=O)—OH	RSO_2H	sulfinic acid	+2
—S(=O$_2$)—OH	RSO_3H	sulfonic acid	+4
—O—S(=O$_2$)—OH	RSO_4H	sulfane oxide	+4

[a] also called mercaptane (old)
[b] note that S is in double bond with a C atom in difference to sulfone
[c] tautomer: —S(=O)—H

The atmospheric chemistry of sulfur is simpler than that of nitrogen in two aspects. First, the number of stable species in air is smaller[25] and second, the variety of interactions in the climate system is less − almost all effects come from sulfate, such as acidity and radiation scattering. In the oxidation line to sulfate, oxidants are consumed:

$$H_2S \xrightarrow[(-2\,H_2O)]{3\,O} SO_2 \xrightarrow{O} SO_3 (\xrightarrow{H_2O} H_2SO_4)$$

Because SO_2 is produced (and emitted) from fuel sulfides in combustion processes, four O atoms are needed to produce one sulfate. This is one atom of oxygen more than for nitrate (HNO_3) formation from air nitrogen (N_2). It has always been out of consideration that SO_2 plays a role in the ozone formation cycle in transferring OH into HO_2 (reactions 5.278 and 5.279); however, the rate of OH $\rightarrow$ HO_2 conversion through SO_2 is small compared with the other O_3 precursors of interest. There is also another irony of fate: when SO_2 concentration in urban areas has been so significant in the past (Table 2.73) that it contributed to O_3 formation, nothing was known about the mechanism of tropospheric ozone formation.

5.5.1 Sulfides (H_2S, CS_2, COS and RSH): Reduced sulfur

In chemical textbooks, carbon disulfide and carbonyl sulfide are treated among carbon compounds. In the climate system, the contribution of CS_2 and COS to the

---

[25] The number of possible intermediates (often hypothetical) is huge (Tables 5.18 and 5.19), especially in aquatic environments.

carbon budget is negligible and both species play no role in carbon chemistry. Moreover, COS only decomposes in the stratosphere because of its chemical stability. Most tropospheric COS is removed by dry deposition (plant assimilation and uptake by microorganisms in soils). Crutzen (1976) suggested COS to be a contributor to the stratospheric sulfate layer (*Junge* layer). However, nowadays it is assumed that it plays a minor role in the stratospheric sulfur budget. Stratospheric chemistry is simple. Either it photodissociates into CO and S, which subsequently form CO_2 and SO_2, or it reacts slowly with OH to CO + SO or CO_2 + HS. All sulfur species finally oxidize to H_2SO_4, which is important in the formation of PSCs (see Chapter 5.3.7). SO and HS are short-lived intermediates, which also occur in the oxidation of CS_2 and H_2S: $k_{5.262} = 7.6 \cdot 10^{-17}$ cm^3 molecule^{-1} s^{-1}, $k_{5.263} < 4 \cdot 10^{-19}$ cm^3 molecule^{-1} s^{-1} and $k_{5.264} = 3.7 \cdot 10^{-12}$ cm^3 molecule^{-1} s^{-1}.

$$SO + O_2 \rightarrow SO_2 + O \tag{5.262}$$

$$HS + O_2 \rightarrow products\ (HSO_2, HO_2 + S, H + SO_2) \tag{5.263}$$

$$HS + O_3 \rightarrow HSO + O_2 \tag{5.264}$$

The radicals HSO and HSO_2 quickly decompose (with O_3 and O_2) into several other radicals (HS, SO, HSO_2 and OH; $k = 6 \cdot 10^{-14}$ cm^3 molecule^{-1} s^{-1}); finally, SO_2 is produced. Radical sulfur fast oxidizes; $k_{5.265} = 2.1 \cdot 10^{-12}$ cm^3 molecule^{-1} s^{-1}).

$$S + O_2 \rightarrow SO + O \tag{5.265}$$

The only important (and very fast) reaction of CS_2 goes via OH radicals; $k_{5.266} = 2.5 \cdot 10^{-11}$ cm^3 molecule^{-1} s^{-1} and $k_{5.267} = 2.8 \cdot 10^{-14}$ cm^3 molecule^{-1} s^{-1}.

$$CS_2 + OH + M \rightarrow HOCS_2 + M \tag{5.266}$$

$$HOCS_2 + O_2 \rightarrow products\ (CO, CS, SO, COS) \tag{5.267}$$

The CS radical reacts in two channels; $k_{5.268} = 2.9 \cdot 10^{-19}$ cm^3 molecule^{-1} s^{-1}.

$$CS + O_2 \rightarrow CO + SO \tag{5.268a}$$

$$CS + O_2 \rightarrow COS + O \tag{5.268b}$$

It is self-evident that O reacts with O_2 to give O_3 and CO forms CO_2 via OH attack. Finally, the oxidation of H_2S is also relatively fast (2−3-day lifetime):

$$H_2S + OH \rightarrow HS + H_2O \tag{5.269}$$

Analog OH reacts with methanethiol (methyl mercaptan) CH_3SH ($k = 3.3 \cdot 10^{-11}$ cm^3 molecule^{-1} s^{-1}). We can summarize that H_2S is 100 % converted into SO_2 and CS_2 with each 50 % transformed into SO_2 and COS.

The most important reduced sulfur species is DMS. Many studies have been carried out to clear the oxidation mechanisms shown in Fig. 5.28. The fate of radical intermediates such as CH_3O, SO and CH_3 have been described above, but many other radicals can only be grouped together within the term "product" on the right-hand side of the equation. Many molecules found in the reaction scheme can be identified from Table 5.18. Fig. 5.28 only shows the OH attack; in addition, NO_3 (nitrate radical) has been described as a very effective oxidizer. OH initiates two

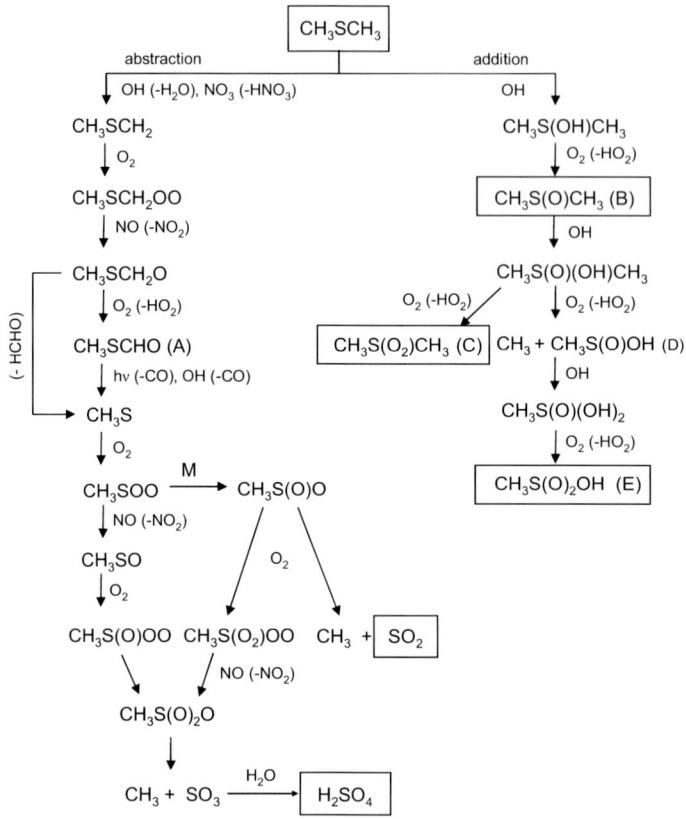

Fig. 5.28 Scheme of DMS oxidation; simplified on the main routes; after Berresheim et al. (1995), Berresheim (1998) and Finlayson-Pitts and Pitts (2000).

channels: the H abstraction and the addition: $k_{5.270} = 4.8 \cdot 10^{-12}$ cm^3 molecule^{-1} s^{-1} and $k_{5.271} = 2.2 \cdot 10^{-11}$ cm^3 molecule^{-1} s^{-1}.

$$\text{DMS} + \text{OH} \rightarrow \text{CH}_2\text{SCH}_3 + \text{H}_2\text{O} \tag{5.270}$$

$$\text{DMS} + \text{OH} \rightarrow \text{CH}_3\text{S(OH)CH}_3 \tag{5.271}$$

Two more channels that break down the DMS molecule are described into CH$_3$ + CH$_3$OH and CH$_3$ + CH$_3$SOH. Because of the nocturnal NO$_3$ attack on DMS (H abstraction channel), its lifetime is short (only about one day). It is remarkable that via the abstraction channel only sulfate is seen at the end, whereas in the OH addition channel many stable organic sulfur species, such as methanesulfonic acid or methanesulfonate (MSA) and dimethylsulfoxide (DMSO) as well as dimethylsulfone (DMSO$_2$), can be identified.

The further fate of these compounds is transfer into the aqueous phase (clouds and precipitation). It is known that organic sulfide (here DMS) can also be simply oxidized to sulfoxide (here DMSO) and further to sulfone in solution:

$$\text{CH}_3\text{SCH}_3 \,(\text{DMS}) + \text{H}_2\text{O}_2 \rightarrow \text{CH}_3\text{S(O)CH}_3 \,(\text{DMSO}) + \text{H}_2\text{O} \tag{5.272}$$

$$CH_3S(O)CH_3 \xrightarrow{\quad O \quad} CH_3(O)S(O)CH_3 \ (DMSO_2) \tag{5.273}$$

Although DMSO is the major product of aqueous phase DMS oxidation (and is ubiquitously found in all aquatic environments), minor amounts of methanethiol CH_3SH, dimethyldisulfide CH_3SSCH_3 (DMDS), MSA and $DMSO_2$ have been observed (Lee and de Mora 1999 and citations therein). DMSO was first detected in the marine environment by Andreae (1980).

DMSO$_2$ will oxidize in droplets like MSA, which is found (together with DMSO) in clouds and precipitation water with a distinct seasonal variation (Munger et al. 1986, Ayers et al. 1986, Brimblecombe and Shooter 1987, Adewyui 1989, Sciare et al. 1998). The rate of DMS oxidation has been observed to increase in the presence of humic substances as well as model substances that could act as photosensitizers. This again is a remarkable observation supporting that in aquatic environments oxidizing agents (where OH plays the key role) are photoenhanced (reaction 5.96). It has been found (see citation in Lee and de Mora 1999) that besides the formation of DMSO$_2$, DMSO also reacts in the following chain (CH_3SO_2H − methanesulfuric acid):

$$DMSO \xrightarrow{OX} CH_3SO_2H \xrightarrow{OX} MSA$$

A number of studies on the oxidation of H_2S with O_2 in natural waters have been conducted in the laboratory and the field (Chen and Morris 1972, Hoffman and Lim 1979, Millero 1986, Millero et al. 1987, Zhang and Millero 1993, Brezonik 1994) but only one study is known to have an atmospheric relevance (Hoffmann 1977). With respect to the sediment chemistry of natural waters and the pollution treatment of wastewater, H_2S (aut)oxidation was of interest long before. It has been found that the process is in the first order with respect to H_2S. Trace metals (especially Fe and Mn) increase the rate of oxidation, and sulfate (SO_4^{2-}), thiosulfate ($S_2O_3^{2-}$) and elemental sulfur have been detected as products. From the water bottom to the surface, the H_2S oxidation rate increases, which is attributed to a larger amount of dissolved oxygen concentration as well as additional oxidants such as hydrogen peroxide (Hoffmann 1977, Millero 2006) and ozone closer to the interface. The oxidation of intermediate sulfite (HSO_3^-) is discussed in Chapter 5.5.2.2. Atmospheric H_2S concentrations are significant only near their sources. Therefore, considering the much larger significance of DMS and SO_2 in the sulfur cycle, aqueous phase H_2S oxidation has never been considered in cloud and precipitation chemistry models. H_2S is only slightly soluble (Table 5.11) and partly dissociates: $pK_a = 7$ (the second dissociation with $pK_a \approx 13$ can be neglected).

$$H_2S \rightleftarrows HS^- + H^+ \tag{5.274}$$

In the studies cited above, only gross reaction equations are given. However, with modern knowledge we can speculate about aqueous oxygen chemistry and photo-catalytic-enhanced redox processes (Chapter 5.3.5) that reactive oxidants such as OH, O_3 and H_2O_2 will elementarily react with HS^- (in the autoxidation process all these species are slowly produced from O_2), similar to the sulfite oxidation (Chapter 5.5.2.2).

Table 5.19 Oxoacids of sulfur and its isomeric forms (hypothetically); after Hollemann-Wiberg (2007). Note: many are instable intermediates or exist only as anions in solution. The —S—S— group is called sulfane, from which polysulfane is derived.

structure	formula	name (acid)	name (salt)
HO—S(H)=O	H_2SO_2	sulfinic acid	sulfonate[a]
(HO)$_2$S=O	H_2SO_3	sulfurous acid	sulfite
(HO)$_2$S(=O)$_2$	H_2SO_4	sulfuric acid	sulfate
HO—S(=O)$_2$—OOH	H_2SO_5	peroxomonosulfuric acid[b]	peroxomono-sulfate
HS—S(H)=O	H_2S_2O	thiosulfoxyl acid	thiosulfinate
HS—S(=O)—OH	$H_2S_2O_2$	thiosulfurous acid[c]	thiosulfite
(HO)$_2$S=S(=O)	$H_2S_2O_3$	thiosulfuric acid	thiosulfate
HO—S(=O)—S(=O)—OH	$H_2S_2O_4$	hyposulfurous acid[d]	hyposulfite[d]
HO—S(=O)$_2$—O—S(=O)—OH	$H_2S_2O_5$	disulfurous acid	disulfite
HO—S(=O)$_2$—S(=O$_2$)—OH	$H_2S_2O_6$	dithionic acid[f]	dithionate
HO—S(=O)$_2$—O—S(=O$_2$)—OH	$H_2S_2O_7$	disulfuric acid[h]	disulfate
HO—S(=O$_2$)—OO—S(=O$_2$)—OH	$H_2S_2O_8$	peroxydisulfuric acid[k]	peroxydisulfate

[a] only as organic compounds:
[b] also called *Caro*'s acid
[c] disulfur(I)acid, anhydride: S_2O (disulfur monoxide); tautomers: HO—S—S—OH (dihydroxy disulfane) and S—S(OH)$_2$ (thiothionyl hydroxide)
[d] also called dithionous acid (salt: dithionite or hydrosulfite)
[e] also called pyrosulfurous acid
[f] generic name: thionic or polythionic acids $H_2S_nO_6$ (n = 2 … 6), also named disulfanic acid
[g] or hypodithionic acid[h]generic name: polysulfuric acids[k] also called *Marshall*'s acid

$$HS^- + OH\ (O_3, H_2O_2) \rightarrow HS + OH^-\ (O_3^-, H_2O_2^-) \tag{5.275}$$

The HS radical (in solution it dissociates to a very small extent into $H^+ + S^-$; thereby instable S^- could donate its electron and provide sulfur) can react in analogy to the gas phase reaction (5.264) with O_3 gaining HSO but also with OH forming S:

$$HS\ (S^-) + OH \rightarrow S + H_2O\ (OH^-) \tag{5.276}$$

$$HS + O_2^- \rightarrow S + HO_2^- \tag{5.277}$$

The elemental sulfur becomes colloidally stable in S_8 molecules but can also add to sulfite and form thiosulfate, which also loses the sulfur via disproportionation into H_2S and sulfate:

$$S + SO_3^{2-} \rightarrow S_2O_3^{2-} \xrightarrow{H_2O} H_2S + SO_4^{2-}$$

In hydrometeors and interfacial waters, however, sulfur reacts with oxygen (reactions 5.265 and 5.262) via SO_2 to form hydrogen sulfite HSO_3^-. Thiosulfate is an important intermediate in biological sulfur chemistry from both sulfate reduction and sulfide oxidation. Many hypothetical so-called lower sulfuric acids (Table 5.19) might appear as intermediates or in the form of radicals (such as SOH, HSO, HSO$_2$, HSS and HS as seen from the structure formulas) in the oxidation chain from sulfide to sulfate:

$$HS^- \rightarrow S \rightarrow HSO_3^- \rightarrow SO_4^{2-}$$

The role of such intermediates and organosulfur compounds as antioxidants (due to radical scavenging) has been known for many years (Kulich and Shelton 1978, 1991).

5.5.2 Oxides and oxoacids: SO_2, H_2SO_3, SO_3, H_2SO_4

Out of biological chemistry, sulfur dioxide (and sulfurous acid) and sulfur trioxide (and sulfuric acid) are the dominant species; remember that the only natural source of SO_2 is given by volcanic activities[26]. Apart from minor global (but locally it can be a significant source) contributions from metallurgic processes, anthropogenic SO_2 is exclusively from the combustion of fossil fuels, mainly coal. In the air, we have to consider gas phase SO_2 oxidation and the following gas-to-particle conversion to sulfuric acid and sulfate particulate matter (Chapter 4.3.4). However, the main route of S(IV) oxidation goes via the aqueous phase (Chapter 5.5.3), where cloud droplet evaporation provides sub-μ aerosol particles containing sulfate as illustrated in the scheme below.

$$SO_2 \left(\underset{}{\overset{H_2O}{\rightleftharpoons}} H_2SO_3 \right) \overset{OX}{\longrightarrow} SO_3 \left(\underset{}{\overset{H_2O}{\rightleftharpoons}} H_2SO_4 \right) \rightarrow \text{particulate sulfate}$$

$$\text{aqueous-phase S(IV)} \overset{OX}{\longrightarrow} \text{aqueous-phase S(VI)}$$

5.5.2.1 Gas phase SO_2 oxidation

From all gas phase reactions studied since the 1950s (see Möller 1980 and citations therein), OH remains the sole component for SO_2 oxidation: $k_{5.278} = 1.3 \cdot 10^{-12}$ cm^3 molecule^{-1} s^{-1}: $k_{5.279} = 4.3 \cdot 10^{-13}$ cm^3 molecule^{-1} s^{-1} and $k_{5.280} = 5.7 \cdot 10^4$ s^{-1} (50 % RH). Therefore, it is seen that (5.278) determines the overall rate of conversion.

$$SO_2 + OH + M \rightarrow HOSO_2 + M \tag{5.278}$$

$$HOSO_2 + O_2 \rightarrow HO_2 + SO_3 \tag{5.279}$$

$$SO_3 + H_2O \rightarrow \text{products (sulfate clusters)} \tag{5.280}$$

Assuming a mean OH radical concentration of 10^6 molecule cm^{-3}, the SO_2 lifetime is about 10 days. Therefore, dry deposition and uptake by clouds and precipitation are important removal pathways (Chapter 5.5.3).

---

[26] Recent field measurements have provided evidence that acid sulfate soils, when drained, oxidize sulfides and might emit SO_2, globally contributing to about 3 Tg S yr^{-1} (Macdonald et al. 2004).

5.5.2.2 Aqueous sulfur chemistry

As seen from Fig. 4.5, the reservoir distribution of SO_2 strongly depends on the aqucous phase pH. The sulfurous acid H_2SO_3 is not known as a pure substance; several isomeric forms have been spectroscopically detected. Therefore, the symbol $SO_2 \cdot aq$ is often used; the equilibrium (5.281) lies full on the left-hand side ($K_{5.281} \ll 10^{-9}$). Sulfurous acid, however, is largely dissociated with $pK_{5.282} = 2.2$ and $pK_{5.283} = 7.0$, $k_{5.283} = 3.1 \cdot 10^3$ s^{-1}.

$$SO_2 + H_2O \rightleftarrows SO_2 \cdot aq \rightleftarrows H_2SO_3 \tag{5.281}$$

$$H_2SO_3 \rightleftarrows HSO_3^- + H^+ \tag{5.282}$$

$$HSO_3^- \rightleftarrows SO_3^{2-} + H^+ \tag{5.283}$$

Because of the complication with H_2SO_3, dissolved SO_2 can be directly associated with bisulfite: $k_{5.284} = 6.3 \cdot 10^4$ s^{-1} (Graedel and Weschler 1981).

$$SO_2 + H_2O \rightleftarrows HSO_3^- + H^+ \tag{5.284}$$

SO_2 directly reacts with hydroxyl ions: $k_{5.285} = 1.1 \cdot 10^{10}$ L mol^{-1} s^{-1} (Boniface et al. 2000):

$$SO_2 + OH^- \rightarrow HSO_3^- \tag{5.285}$$

It is useful to name with S(IV) all dissolved sulfur species in this oxidation state: $SO_2 + H_2SO_3 + HSO_3^- + SO_3^{2-}$, whereby $[H_2SO_3] \rightarrow 0$. Sulfite forms adducts with dissolved aldehydes, of which the most important is α-hydroxymethanesulfonate (HMS), which is stable against oxidation: $pK_{5.286} = 3.4$ (Warneck 1989).

$$HCHO + HSO_3^- \rightarrow HCH(OH)SO_3^- \text{ (HMS)} \tag{5.286a}$$

$$HCHO + SO_3^- \rightarrow HCH(O)^-SO_3^- \xrightleftharpoons{H^+} HMS \tag{5.286b}$$

However, competitive formaldehyde gives a hydrate, which is unable to add onto sulfite.

$$HCHO + H_2O \rightleftarrows CH_2(OH)_2 \tag{5.287}$$

HMS is the anion of the strong hydroxymethanesulfone acid (HMSA), full dissociated (Munger et al. 1984). The second dissociation step to $HCH(O)^-SO_3^-$ corresponds to a weak acid with $pK_a \approx 10$ (Sörensen and Anderson 1970). HMS slowly decomposes into the initial substances, thereby sometimes describing (5.286) and (5.288) as equilibrium with $K_{5.286/5.288} = 6.6 \cdot 10^9$, $k_{5.288} = 7.7 \cdot 10^{-3}$ s^{-1} (Möller and Mauersberger 1995).

$$HCH(OH)SO_3^- \text{ (HMS)} \rightarrow HCHO + HSO_3^- \tag{5.288}$$

Often the formaldehyde hydration is also included in the equilibrium (Deister et al. 1986):

$$K = [CH_2(OH)SO_3^-]/[CH_2(OH)_2][HSO_3] = 3.6 \cdot 10^6 \text{ L mol}^{-1} \tag{5.289}$$

In alkaline solution, HMS decomposes: $k_{5.289} = 3.7 \cdot 10^{-3}$ L mol^{-1} s^{-1}.

Table 5.20 Reaction rate constants for $HSO_3^- + H_2O_2$.

k in 10^7 L^2 mol^{-2} s^{-1}	author
35.0	Hoffmann and Edwards (1975)
2.6	Penkett et al. (1979)
4.7	Martin and Damschen (1981)
2.4	McArdle and Hoffmann (1983)
8.0	Kunen et al. (1983)
9.5	Lee at al. (1986)
9.1 ± 0.5	Maaß et al. (1999)

$$HCH(OH)SO_3^- + OH^- \rightarrow CH_2(OH)_2 + SO_3^- \tag{5.290}$$

The only oxidation of HMS goes via OH attack (Buxton et al. 1996):

$$HCH(OH)SO_3^- + OH \rightarrow CH_2(OH)_2 + SO_3^- \tag{5.291}$$

From all three HMS sinks (oxidation, alkaline decomposition and decay), only (5.288) is considered important (Möller and Mauersberger 1995). Sulfite also forms adducts with other aldehydes such as benzaldehyde, methylglyoxal, acetaldehyde and hydroxyacetaldehyde (Olson and Hoffmann 1988a, 1988b, 1989; Olson et al. 1988).

Because of the importance of SO_2 and sulfate in the atmosphere, the oxidation pathways in solution have been studied extensively and typically been subdivided as follows:

- by peroxides (H_2O_2 and ROOH)
- by ozone O_3
- by oxygen (autoxidation)
- by oxygen with participation of TMI
- by radicals (OH, NO_3, Cl and others)
- by other oxidants (e. g. HNO_4, HOCl)

The decades of SO_2 research have given almost only gross reaction rates such as those first studied by Mader (1958) and later recognized as being atmospherically important by Hoffmann and Edwards (1975) and Penkett et al. (1979); $k_{5.292} = (5.3 \pm 2.7) \cdot 10^7$ L^2 mol^{-2} s^{-1} recommended by Möller and Mauersberger (1995), see Table 5.20.

$$-(d[S(IV)]/dt = (d[S(VI)]/dt = R_{H_2O_2} = k_{5.292}[H^+][HSO_3^-][H_2O_2] \tag{5.292}$$

Thénard recognized that one O is only weakly bonded in H_2O_2 and that the molecule easily decomposes[27]. Remarkably, soon after its discovery it was known that H_2O_2 oxidizes sulfurous acid into sulfuric acid without the formation of free oxygen (Gmelin 1827), a mechanism recognized to be important in air chemistry almost 150 years later (Möller 1980) as the most important pathway in the oxidation of

---

[27] This (wrong) statement is based on the often generally accepted reaction scheme A + H_2O_2 → AO + H_2O. However, H_2O_2 cannot transfer O (like O_3); the only way is decomposing according to $H_2O_2 + e^- + H^+$ → OH + H_2O; in a subsequent reaction OH is the oxidizer.

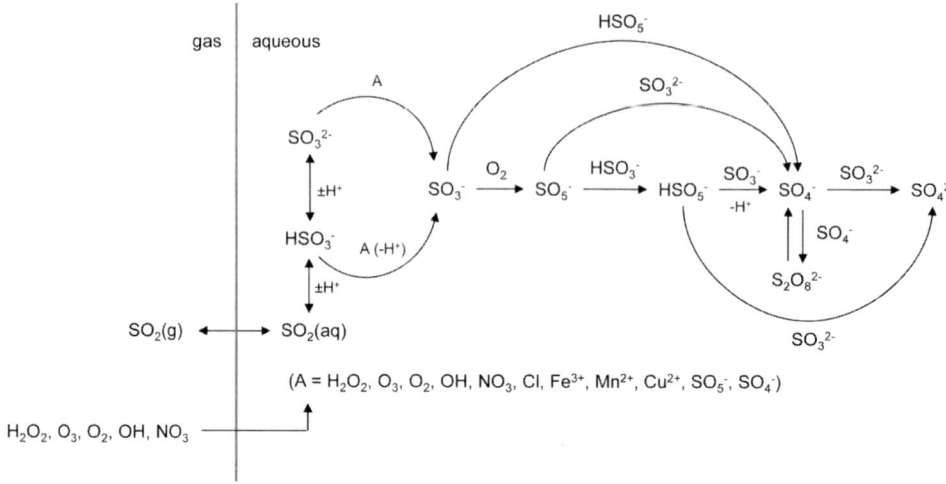

Fig. 5.29 Scheme of general S(IV) oxidation in alkaline and acidic solution. Electron acceptors: H_2O_2, O_3, O_2, OH, NO_3, Cl, Fe^{3+}, Mn^{2+}, Cu^{2+}, electron donors: Fe^{2+}, Mn^+, Cu^+, HSO_3^-, SO_3^{2-}, OH^-.

dissolved SO_2 in hydrometeors. It is remarkable that the detailed mechanism of the S(IV)—H_2O_2 reaction is not known; it was proposed as the formation of peroxomonosulfite (HO—S(O)—OO^-), an unknown substance (Möller 2003). The subsequent (slow) proton-catalyzed rearrangement into sulfate could explain the pH-dependence (rate increase with H^+). It is hard to understand in which way HOOH (H_2O_2) might transfer a single O atom or the whole peroxo group (−O—O−).

$$HO-S\overset{O^-}{\underset{O}{<}} + H_2O_2 \longrightarrow \left[HO-S\overset{O^{\cdots}O^{\diagdown H}}{\underset{O^{\cdots}O}{<}} \right] \longrightarrow HO-S\overset{O^-}{\underset{O-O}{<}} (HSO_4^-) + H_2O$$

Drexler et al. (1991) and Lagrange et al. (1993) proposed the formation of peroxomonosulfite in an equilibrium (what is hard to assume because of the many structural intermolecular rearrangements) with an acid-catalyzed slow conversion of peroxomonosulfite into sulfate. Möller (2009a) proposed that a single electron transfer occurs in the sense of another *Fenton*-like reaction. The fate of $H_2O_2^-$ is well known (5.116). Note that the OH yield is stoichiometric to H_2O_2 and thereby provides OH concentrations orders of magnitudes larger than by the phase transfer of gaseous OH (the "radical" oxidation by OH has been described as slow; Möller 1980).

$$HSO_3^- + H_2O_2 \rightarrow HSO_3 + H_2O_2^- \tag{5.293}$$

Fig. 5.29 shows a generalized scheme of S(IV) oxidation, depending on pH. Ermakov et al. (1997) considered that the radical chain mechanism is most likely in

Table 5.21 Rate constants of reactions of sulfite with radicals.

reaction			k in L mol^{-1} s^{-1}	
$OH + HSO_3^-$	$\rightarrow$	$H_2O + SO_3^-$	$2.7 \cdot 10^9$	Buxton et al. (1996)
$OH + HSO_3^-$	$\rightarrow$	$OH^- + SO_3^-$	$4.6 \cdot 10^9$	Buxton et al. (1996)
$OH + SO_3^{2-}$	$\rightarrow$	$OH^- + SO_3^-$	$5.5 \cdot 10^9$	Neta and Huie (1985)
$SO_4^- + HSO_3^-$	$\rightarrow$	$SO_4^{2-} + SO_3^- + H^+$	$6.8 \cdot 10^8$	Buxton et al. (1996)
$SO_4^- + SO_3^{2-}$	$\rightarrow$	$SO_4^{2-} + SO_3^-$	$3.1 \cdot 10^8$	Buxton et al. (1996)
$SO_5^- + HSO_3^-$	$\rightarrow$	$HSO_5^- + SO_3^-$	$8.6 \cdot 10^3$	Buxton et al. (1996)
$SO_5^- + SO_3^{2-} + H^+$	$\rightarrow$	$HSO_5^- + SO_3^-$	$2.15 \cdot 10^5$	Buxton et al. (1996)
$SO_5^- + HSO_3^-$	$\rightarrow$	$SO_4^- + H^+ + SO_4^{2-}$	$3.6 \cdot 10^2$	Buxton et al. (1996)
$SO_5^- + SO_3^{2-}$	$\rightarrow$	$SO_4^- + SO_4^{2-}$	$5.5 \cdot 10^5$	Buxton et al. (1996)
$NO_3 + HSO_3^-$	$\rightarrow$	$NO_3^- + SO_3^- + H^+$	$5.6 \cdot 10^3$	Exner et al. (1992)
$NO_3 + SO_3^{2-}$	$\rightarrow$	$NO_3^- + SO_3^-$	$3.0 \cdot 10^8$	Exner et al. (1992)
$Cl_2^- + HSO_3^-$	$\rightarrow$	$2\,Cl^- + SO_3^- + H^+$	$1.7 \cdot 10^8$	Jacobi (1996)
$Cl_2^- + SO_3^{2-}$	$\rightarrow$	$2\,Cl^- + SO_3^-$	$6.2 \cdot 10^7$	Herrmann et al. (1996)

S(IV) oxidation. The existence of the sulfite radical and its role in biological damage (whereby the subsequently produced peroxosulfate radical SO_5 is a much stronger oxidant) has long been known (Neta and Huie 1985). Many molecules, radicals and metal ions react with sulfite and bisulfite in a one-electron oxidation (Table 5.21); A − electron acceptor (see also Fig. 5.29):

$$HSO_3^- + A \rightarrow SO_3^- + H^+ + A^- \qquad (5.294a)$$

$$SO_3^{2-} + A \rightarrow SO_3^- + A^- \qquad (5.294b)$$

The sulfite radical SO_3^- reacts quickly with oxygen to form the peroxosulfate radical; $k_{5.295} = 2.5 \cdot 10^9$ L mol^{-1} s^{-1} (Buxton et al. 1996):

$$SO_3^- + O_2 \rightarrow SO_5^- \qquad (5.295)$$

The peroxosulfate radical SO_5^- almost reacts with sulfite (HSO_3^- and SO_3^{2-}) in different pathway to peroxomonosulfate (HSO_5^-) and the final product sulfate (SO_4^{2-}), generating sulfur radicals (SO_3^- and SO_4^-), see Fig. 5.29); $k_{5.296a} = 5.5 \cdot 10^5$ L mol^{-1} s^{-1}, $k_{5.296b} = 3.6 \cdot 10^2$ L mol^{-1} s^{-1}, $k_{5.296c} = 8.6 \cdot 10^3$ L mol^{-1} s^{-1} (Buxton et al. 1996):

$$SO_5^- + SO_3^{2-} \rightarrow SO_4^- + SO_4^{2-} \qquad (5.296a)$$

$$SO_5^- + HSO_3^- \rightarrow HSO_5^- + SO_3^- \qquad (5.296b)$$

$$SO_5^- + HSO_3^- \rightarrow HSO_5^- + SO_3^- \qquad (5.296c)$$

At this state, the pathway splits from sulfate radical SO_4^- and peroxomonosulfate (HSO_5^-); the latter decomposes with sulfite (HSO_3^- and SO_3^{2-}) to sulfate; $k_{5.297} = 7.1 \cdot 10^6$ L mol^{-1} s^{-1} (Betterton and Hofmann 1988):

$$HSO_5^- + HSO_3^- \text{ (or } SO_3^{2-}) \rightarrow 2\,SO_4^{2-} + 2 \text{ (or 1) } H^+ \qquad (5.297)$$

The sulfate radical reacts similarly with sulfite but regenerates sulfite radicals; $k_{5.298} = 3.2 \cdot 10^8$ L mol^{-1} s^{-1} (Betterton and Hofmann 1988):

$$SO_4^- + HSO_3^- \text{ (or } SO_3^{2-}) \rightarrow SO_4^{2-} + SO_3^- + H^+ \tag{5.298}$$

Hence, a radical chain has been established where in the initial step sulfit ions will also be transformed into sulfit radical by other sulfur radicals (SO_4^- and SO_3^-). Of less importance are radical-radical reactions; we cited first dimerizations to relative stable dithionate ($S_2O_6^{2-}$) and peroxydisulfate ($S_2O_8^{2-}$); $k_{5.299} = 1.8 \cdot 10^8$ L mol^{-1} s^{-1}, $k_{5.300a} = 1.8 \cdot 10^8$ L mol^{-1} s^{-1}, $k_{5.300b} = 7.2 \cdot 10^8$ L mol^{-1} s^{-1} (Herrmann et al. 1995):

$$SO_4^- + SO_4^- \rightarrow S_2O_8^{2-} \tag{5.299}$$

$$SO_5^- + SO_5^- \rightarrow S_2O_8^{2-} + O_2 \tag{5.300a}$$

$$SO_5^- + SO_5^- \rightarrow 2\,SO_4^- + O_2 \tag{5.300b}$$

Peroxydisulfate (it is produced as a strong oxidant) decays back in aqueous solution by homolysis in sulfate radicals. There are several competing reactions in this radical sulfur oxidations mechanism (not noted here) because the chain

$$SO_5^- \xrightarrow{e^- + H^+} HSO_5^- \xrightarrow[-H_2O]{e^- + H^+} SO_4^- \xrightarrow{e^-} SO_4^{2-}$$

also proceeds by all available electron donors (such as reduced TMI, O_2^-, OH^- and Cl^-); see reactions (5.309), (5.310), (5.311), (5.314), (5.315), (5.317), (5.319). Furthermore, there are described transfers of sulfur radicals by ROS (H_2O_2, HO_2, and OH; see reactions 5.313, 5.316, 5.318), whereby OH acts as H abstractor (formation of H_2O) and the ionic species acts as electron donor according to

$$H_2O_2 \rightleftarrows HO_2^- \xrightarrow{-e^-} HO_2 \rightleftarrows O_2^- \xrightarrow{-e^-} O_2.$$

Remember (see Chapter 5.3.5) that other ROS, oxidizing sulfite in the initial step, undergo conversions and provide the important superoxide O_2^- ($pK_{HO2/O2-} = 4.5$):

$$H_2O_2 \xrightarrow{e^-} H_2O_2^- \rightarrow OH + OH^-$$

$$O_3 \xrightarrow{e^-} O_3^- \xrightarrow{H^+} HO_3 \rightarrow OH + O_2$$

$$O_2 \xrightarrow{e^-} O_2^- \xrightarrow{H^+} HO_2 \xrightarrow[(-O_2)]{O_2^-} HO_2^- \xrightarrow{H^+} H_2O_2$$

Note the large differences in the reaction rate constants are independent of whether sulfite or bisulfite is the reagent. This makes the overall process strongly dependent on pH. This pH dependence, however, is much more subtle because the sulfur radicals (sulfite, sulfate and peroxomonosulfate) undergo acid-base equilibrium, which likely is on the left-hand side because the hydrogenated species are strong acids:

$$SO_3^- \xrightleftharpoons{H^+} HSO_3, SO_4^- \xrightleftharpoons{H^+} HSO_4, \text{ and } SO_5^- \xrightleftharpoons{H^+} HSO_5$$

Parallel to all possible starting reactions of the type (5.294), the direct nucleophilic attack of oxygen species (O_3, OH and HO_2) might be possible:

$$O_3 + SO_3^{2-} \rightarrow [OOO-SO_3]^{2-} \rightarrow O_2 + SO_4^{2-} \tag{5.301}$$

Table 5.22 Reaction rate constants for S(IV) + O_3 for the nucleophilic mechanism.

$10^4 \, k_0$	$10^5 \, k_1$	$10^9 \, k^2$	author
0	3.1	2.2	Erickson et al. (1977)
2.2	3.2	1.0	Hoigné et al. (1985)
2.4	3.7	1.5	Hoffmann (1986)

Table 5.23 Reaction rate constants for S(IV) + O_3 for the radical mechanism.

k in $10^{-4} \, \text{L}^{1/2} \, \text{mol}^{-1/2} \, \text{s}^{-1}$	author
1.45	Penkett et al. (1979)
1.4 ... 3.5	Maahs (1983)
1.9	Martin (1984)
1.23	Nahir and Dawson (1987)
2.99	Lagrange et al. (1992)
1.27	Botha at al. (1994)

$$O_3 + HSO_3^- \rightarrow [OOO-SO_3H]^{2-} \rightarrow O_2 + HSO_4^- \tag{5.302}$$

$$O_3 + SO_2 \rightarrow [OOO-SO_2] \xrightarrow[(-O_2)]{} SO_3 \xrightarrow[(-H^+)]{H_2O} HSO_4^- \tag{5.303}$$

$$OH + SO_3^{2-} \rightarrow [HO-SO_3]^{2-} \xrightarrow[(-HO_2)]{O_2} SO_4^{2-} \tag{5.304}$$

$$HO_2 + SO_3^{2-} \rightarrow [HOO-SO_3]^{2-} \xrightarrow[(-OH)]{} SO_4^{2-} \tag{5.305}$$

The rate law (5.292) has been confirmed by many experimental studies. The reaction of ozone with S(IV) is assumed as a nucleophilic attack with all S(IV) species (with the exception of the study by Erickson et al. 1977; Table 5.22).

$$R_{O_3} = (k_a[SO_2] + k_b[HSO_3^-] + k_c[SO_3^{2-}])[O_3] \tag{5.306}$$

Using the expressions for the protolysis equilibrium and simplification for pH > 3, we get:

$$R_{O_3} = (k_b + k_c K_b[H^+]^{-1})[S(IV)][O_3] \tag{5.307}$$

where K_b is the equilibrium constant of the second dissociation ($HSO_3^- \rightleftarrows SO_3^{2-}$). This expression is identical to the rate law proposed by Maahs (1983) even though he proposed the radical mechanism instead of the nucleophilic. A general rate law can be derived from the studies suggesting the radical mechanism (Table 5.23):

$$R_{O_3} = k[H^+]^{-1/2}[S(IV)][O_3] \tag{5.308}$$

Now comparing the pseudo-second-order rate constants for the nucleophilic (5.307) and radical (5.308) mechanism $R_{O_3}(nuc) = {}^1k[S(IV)][O_3]$ and $R_{O_3}(rad) = {}^2k[S(IV)][O_3]$:

$${}^1k = k_b + k_c K_b[H^+]^{-1}$$

$${}^2k = k[H^+]^{-1/2},$$

Table 5.24 Comparison of pseudo-second order rate constants of S(IV) + O_3 reaction for the nucleophilic (^1k) and radical (^2k) mechanisms in dependence from pH.

k in L mol^{-1} s^{-1}	pH = 3	pH = 4	pH = 5
^1k	$4.3 \cdot 10^5$	$1.3 \cdot 10^6$	$1.0 \cdot 10^7$
^2k	$6.0 \cdot 10^5$	$1.9 \cdot 10^6$	$0.6 \cdot 10^6$

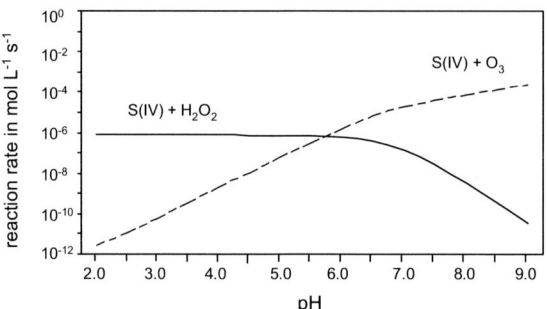

Fig. 5.30 Dependence of S(IV) by H_2O_2 and O_3 from pH.

we come to the noteworthy conclusion that the difference is numerically insignificant (Table 5.24) and that all experimental studies strongly agree with the kinetics, although without clarifying the mechanism. Fig. 5.30 shows the strong influence of pH on both the H_2O_2 (dominant in acidic solution) and O_3 (dominant in alkaline solution) pathways.

Accepting the radical mechanism theory, the reaction rate of the sulfite radical formation (5.294) determines the overall rate. Amplifying the S(IV) oxidation is given by the subsequent formation of radicals (OH, O_2^-, SO_4^-, SO_5^-) that further react with sulfite or bisulfite (Fig. 5.29). It is impossible to regard the S(IV) oxidation under natural conditions (i.e. outside laboratory conditions) in the sense of a definite mechanism because all reactive species (providing A to E in the following scheme) are available, but in different concentrations and superposed, making sulfate formation very complex and depending on the redox state and the pH of the solution as well as the radiation and photosensitizers.

$$SO_3^{2-}(HSO_3^-) \xrightarrow{\text{A}} SO_3^- \xrightarrow{\text{O}_2} SO_5^- \begin{array}{c} \xrightarrow{\text{B}} SO_4^- \xrightarrow{\text{C}} \\ \searrow_{\text{D}} \quad \nearrow_{\text{E}} \end{array} SO_4^{2-} \\ HSO_5^-$$

All conversions through species A to E can proceed through sulfur radicals (sulfite, peroxomonosulfate and sulfate) as well as via many other species, whereas some of them are permanently abundant and others are produced via photochemical processes. Only coupled gas-aqueous phase models comprising the whole chemistry can

Table 5.25 Ions and radicals of oxoacids of sulfur (instead of hydrogen- also bi- is used to term prononized forms).

ion	radical	name	acid	name
HSO_3^-	HSO_3	hydrogensulfite	H_2SO_3	sulfurous acid
SO_3^{2-}	SO_3^-	sulfite		
HSO_4^-	HSO_4	hydrogensulfate	H_2SO_4	sulfuric acid
SO_4^{2-}	SO_4^-	sulfate		
HSO_5^-	HSO_5	hydrogenperoxomonosulfate	H_2SO_5	peroxomonosulfuric acid
SO_5^{2-}	SO_5^-	peroxomonosulfate		

describe the very complex S(IV) oxidation in hydrometeors. The following list of reactions only compiles the most important steps, for further reactions see for example CAPRAM; citations for reactions rates, given in L mol^{-1} s^{-1}, see Möller and Mauersberger (1995) or Möller (2003).

$$SO_4^- + Fe^{2+}\ (Mn^{2+}, Cu^+) \rightarrow SO_4^{2-} + Fe^{3+}\ (Mn^{3+}, Cu^{2+})$$
$$k_{5.309}(Fe) = 4.1 \cdot 10^9 \qquad (5.309)$$

$$SO_4^- + Cl^- \rightarrow SO_4^{2-} + Cl \qquad k_{5.310} = 2.5 \cdot 10^8 \qquad (5.310)$$

$$SO_4^- + O_2^- \rightarrow SO_4^{2-} + O_2 \qquad k_{5.311} = 4 \cdot 10^9 \qquad (5.311)$$

$$SO_4^- + HO_2 \rightarrow SO_4^{2-} + H^+ + O_2 \qquad k_{5.312} = 3.5 \cdot 10^9 \qquad (5.312)$$

$$SO_4^- + H_2O_2 \rightarrow SO_4^{2-} + H^+ + HO_2 \qquad k_{5.313} = 1.2 \cdot 10^7 \qquad (5.313)$$

$$SO_4^- + Cl^- \rightarrow SO_4^{2-} + Cl \qquad k_{5.314} = 2.5 \cdot 10^8 \qquad (5.314)$$

$$SO_4^- + OH^- \rightarrow SO_4^{2-} + OH \qquad k_{5.315} = 1.4 \cdot 10^7 \qquad (5.315)$$

$$SO_4^- + OH \rightarrow HSO_5^- \qquad k_{5.316} = 1 \cdot 10^{10} \qquad (5.316)$$

$$SO_5^- + Fe^{2+}\ (Mn^{2+}, Cu^+) + H^+ \rightarrow HSO_5^- + Fe^{3+}\ (Mn^{3+}, Cu^{2+})$$
$$k_{5.317}(Fe) = 4.3 \cdot 10^7 \qquad (5.317)$$

$$SO_5^- + HO_2 \rightarrow HSO_5^- + O_2 \qquad k_{5.318} = 1.7 \cdot 10^9 \qquad (5.318)$$

$$SO_5^- + O_2^- + H_2O \rightarrow HSO_5^- + OH^- \qquad k_{5.319} = 2.3 \cdot 10^8 \qquad (5.319)$$

Dithionate ($S_2O_6^{2-}$) und peroxodisulfate ($S_2O_8^{2-}$) have never been detected in cloud and rainwater. Dithionate slowly disproportionates (into $SO_4^{2-} + SO_2$) and peroxodisulfate hydrolyzes in acid solution (into $HSO_5^- + SO_4^{2-}$). Table 5.25 lists all higher oxidized sulfur species of interest in the climate system.

5.5.3 Multiphase sulfur chemistry

Because of the relatively slow gas phase of SO_2 oxidation, aqueous phase oxidation in clouds contributes to 80–90 % of sulfate formation in the northern mid-latitudes.

The in-droplet S(IV) oxidation rate is related to the volume of air by using the general equation (4.278) in terms of:

$$\left(\frac{d\,[\text{sulfate}]}{dt}\right)_{\text{air}} = k\,[\text{SO}_2]_{\text{gas}} = \boldsymbol{R}T \cdot \text{LWC} \cdot k_{aq}\,[\text{S(IV)}] \tag{5.320}$$

In a cloud under normal daytime conditions of LWC and concentrations of H_2O_2 and O_3, the SO_2 lifetime is ≤ 1 h and sulfate production can reach 8 ppb h^{-1}. This is by a factor of 100 higher than the maximum gas phase production (Table 5.26). On a yearly basis, however, we need the statistical information on the occupancy of the PBL by clouds and occurrence of clouds to calculate the mean aqueous phase S(IV) oxidation. Any error in k_{aq} is insignificant compared with the uncertainty of cloud statistics. The atmospheric SO_2 residence time strongly depends on the event-related cloud and precipitation statistics. As shown in Fig. 4.5, scavenging of no other gas than sulfur dioxide depends so strongly on changing pH in the range 6 to 8; hydrometeors normally have pHs between 4 and 6 and natural waters between 7 and 8. The key questions with respect to climate-relevant sulfate formation are:

- What is the percentage of gas-to-particle sulfate formation in the very low submicron range?
- What percentage do cloud processes contribute to sulfate formation in the upper submicron range?
- How much non-oxidized sulfur is deposited (dry as SO_2 and wet as S(IV))?
- How much sulfur is on average dissolved in clouds and thereby not climate relevant?

The last question is important because not only does the percentage of atmospheric SO_2 oxidation (often global climate models assume a simple conversion factor in percentage of SO_2 emission) determine the "climate active" sulfate but so does the particulate sulfate in air. Fig. 5.31 shows a scheme of the multiphase atmospheric sulfur chemistry. The figures are derived from many field studies and modeling attempts, and are representative of Europe. It is noteworthy that the number of studies with *in situ* measurements of S(IV) in hydrometeors (rain, cloud and fog) is very limited (because of the instability of S(IV) and a complicated measurement technique; Möller 2003). Normally in rainwater only sulfate is detected because

Table 5.26 SO_2 oxidation (percentage of pathway[a]) and sulfate formation rates (in % h^{-1}) for central European conditions; data from Möller (1995a), Möller and Mauersberger (1992).

pathway	summer day	summer night	winter day	winter day
liquid phase	75.8	8.1	11.3	4.4
gas phase	1.4	0.0013	0.22	0.0002
O_3	0.3	2.2	65.5	24.1
H_2O_2	99.6	87.3	26.7	34.7

[a] the difference to 100% is given by other not listed pathways such as radicals and TMI catalytic

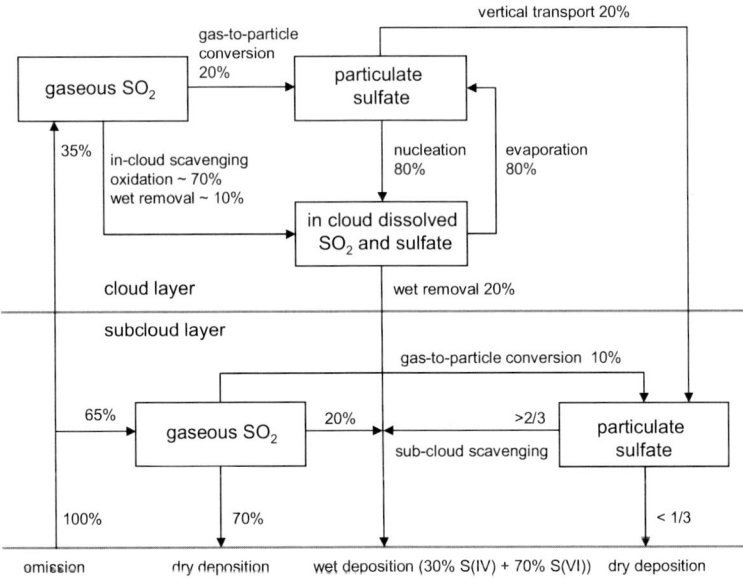

Fig. 5.31 Scheme of SO_2 multiphase chemistry (after Möller 1995a, 2003); % related to each box as budget.

much time is taken from sampling to analysis. *In situ* field measurements, however, support the theoretical assumption that in fresh rainwater S(IV) contributes about 30% to total sulfur. In other words, the percentage of wet deposited SO_2 (in-cloud and sub-cloud) is significant but might vary depending on the gas phase SO_2 concentration and droplet pH. These findings support the conclusion that on average less than 50% of SO_2 is converted into sulfate in air.

5.6 Phosphorus

In line with oxygen, nitrogen and sulfur, phosphorus is the next important life-essential element but in contrast to the other listed elements, it does not naturally occur in nature. Moreover, P *only* (according to textbook knowledge) occurs in derivates of phosphorus acid in nature. Phosphorus (white) was first produced as an element (but not recognized as an element) by the German alchemist *Hennig Brand* (c. 1630–1692) in Hamburg in 1669 from the heating of distilled urine remaining with sand. *Bernhardt Siegfried Albinus* (1697–1770) isolated phosphorus from the charcoal of mustard plants and cress, confirmed by *Marggraf* in 1743 and likely *Scheele*, in 1769 found it from bone (Kopp 1931). *Lavoisier* recognized P as an element. It is believed that all original phosphates came from the weathering of rocks. Phosphate ion enters into its organic combination largely unaltered. From phosphorus acid H_3PO_4 many esters are formed simply by the exchange of H through organic residual.

It was thought that phosphorus can cycle in the atmosphere only as phosphate bound to aerosol particles such as pollen, soil dust and sea spray. Similar to nitrogen, chemical forms of P have been found in an oxidation state -3 to $+5$. It is generally accepted that in contrast to O, S, N and C, the phosphorus cycle does not contain redox processes (i.e. remains on the state of phosphate) or volatile compounds (i.e. the atmosphere is excluded from the P cycle). In continental rainwater, Lewis et al. (1985) found phosphate in such concentrations that it could not be explained by the scavenging of particulate phosphorus and concluded that it was a terrestrial source of a volatile P compound. Recently, Glindemann et al. (2005a and citations therein) detected monophosphane PH_3 (formerly phosphine, also termed hydrogen phosphide) in air in the range pg m^{-3} to ng m^{-3}. Close to identified emission sources (paddy fields, water reservoirs and animal slurry) reported concentrations are significant higher. Glindemann et al. (2003) found for the first time PH_3 in remote air samples (low ng m^{-3} range) in the high troposphere of the north Atlantic. In the lower troposphere, PH_3 is observed at night in the 1 ng m^{-3} range, with peaks of 100 ng m^{-3} in populated areas. During the day, the concentration is much lower (in the pg m^{-3} range).

Monophosphane[28] has a low water solubility, $H = 8.1 \cdot 10^{-3}$ L mol^{-1} atm^{-1} (Williams et al. 1977), which is about three times less than the *Henry* coefficient for NO_2. PH_3 cannot be photolyzed in the troposphere but reacts quickly with OH; $k_{5.321} = 1.4 \cdot 10^{-11}$ cm^3 molecule^{-1} s^{-1} (much faster than OH + NH_3).

$$PH_3 + OH \rightarrow PH_2 + H_2O \qquad (5.321)$$

The literature is absent of studies on atmospheric PH_3 oxidation in detail. In PH_3 + O_2 explosions, the radicals PH_2 and PO have been detected (Norrish and Oldershaw 1961), and PO has also been found in interstellar clouds. According to Glindemann et al. (2003, 2005a), the final product of PH_3 oxidation in air is phosphate ion but nothing is known about the reaction steps. We can only further speculate that the oxidation proceeds in solution and/or interfacial, i.e. in the cloud and aerosol layer and this might explain why PH_3 is found in the upper troposphere. As we have largely discussed, it is not the solubility that controls the "washout" but the interfacial chemistry for low-soluble species.

PH_3 is a very weak base (p$K_b \approx 27$ compared with 4.5 for NH_3) but the phosphonium ion is known in solid salts, for example with chloride (PH_3 + HCl $\rightleftarrows$ PH_4Cl). Because the PH_3 mixing ratio is $10-100$ times lower than that of NH_3, it is likely that during inorganic gas-to-particle conversion in the late morning ($SO_2 \rightarrow H_2SO_4$) ammonium and phosphonium are taken up by the evolving particulate phase. The equilibrium is generally on the left-hand side (PH_3) but because of acid excess and dynamic processes it might be shifted to salt formation. Moreover, in the condensed phase a strong oxidation regime can stepwise oxidize PH_3/PH_4^+ by OH to phosphate. This process can better explain the observed daytime decrease of PH_3 concentrations and the relatively high PH_3 concentrations in the upper and remote atmosphere.

---

[28] P forms many compounds with hydrogen according the general formula P_nH_{n+m} (n in whole numbers, m = 2, 0, -2, -4, ...) but few compounds have been isolated, of which the most important are PH_3 and P_2H_4, which are both volatile and self-igniting at high concentrations.

However, to gain the PO_4^{3-} ion a very complex P chemistry has to be assumed including all oxidants of interest (O_2^-, OH, O_3, H_2O_2). We can speculate that alternate OH attacks abstract H from P and subsequent O_3 attacks add O onto P according to the known chain of oxo acids:

$$\begin{bmatrix} H \\ | \\ H-P-H \\ | \\ H \end{bmatrix}^+ \rightarrow \begin{bmatrix} H \\ | \\ O-P-H \\ | \\ H \end{bmatrix}^0 \rightarrow \begin{bmatrix} O \\ | \\ O-P-H \\ | \\ H \end{bmatrix}^- \rightarrow \begin{bmatrix} O \\ | \\ O-P-O \\ | \\ H \end{bmatrix}^{2-} \rightarrow \begin{bmatrix} O \\ | \\ O-P-O \\ | \\ O \end{bmatrix}^{3-}$$

Fine particulate matter might be transported and release PH_3, whereas particulate acidity decreases and thereby explains why PH_3 seems to be ubiquitous in air. Moreover, PH_2 probably returns to PH_3 via H abstraction from hydrocarbons in the gas phase:

$$PH_2 + RH \rightarrow PH_3 + R \tag{5.322}$$

In the biomass, phosphorus is as important as ATP and ADP for providing energy transfers and in DNA and RNA where phosphate interlinks the nucleosides. In bones and teeth, it is found as calcium phosphates (approximately 85% of phosphorus in the body). In contrast to nitrogen, plants do not reduce phosphate. However, PH_3 (and likely P_2H_4) exist in air and are emitted from the decomposing biomass under anaerobic conditions. Even exotic formation via lightning, shown by simulated lightning in the presence of organic matter, which provides a reducing medium, was suggested by Glindemann et al. (2004) but this is hard to accept under atmospheric conditions. Natural rock and mineral samples release trace amounts of phosphine during dissolution in mineral acid. Strong circumstantial evidence has been gathered on the reduction of phosphate in the rock via mechanochemical or "tribochemical" weathering at quartz and calcite/marble inclusions (Glindemann et al. 2005b). However, this can also be because of traces of phosphides in rocks, which produce PH_3 in acidic hydrolysis. Although phosphorus could be expected to occur naturally as a phosphide, the only phosphide in the earth's crust is found in iron meteorites as the mineral schreibersite (Fe, Ni)$_3$P, in which cobalt and copper might also be found (WHO 1988).

Many papers have suggested that PH_3 forms by microbial processes in soils and sediments but no bacteria responsible or any chemical mechanism is known (Zhu et al. 2007). This question rose 40 years ago but Burford and Bremner (1972) could not confirm earlier evidence for the evolution of phosphine through the microbial reduction of phosphate in waterlogged soils. They also showed that phosphine is sorbed by soil constituents and might not escape to the atmosphere if produced in soils. Recently, PH_3 has been detected at surprisingly high concentrations in the marine atmosphere (Zhu et al. 2007). The measurement technique to detect such low PH_3 concentrations became available around 1993; therefore, the absence of evidence of PH_3 before is no longer evidence of PH_3 absence. It seems that only yet unknown microbial processes where phosphate is used as an oxygen source under anaerobic conditions (similar to the sulfate reduction) might explain PH_3 formation. This is still an open field of research and needs more specific microbial studies. Organophosphanes PR_3, where R = alkyl or aryl (for example triphenyl-phosphine oxide $OPPh_3$ and trimethylphosphine oxide $OP(CH_3)_3$) are known to form metal complexes and have been found in fungi (Holland et al. 1993). Organophosphanes are easily oxidized to the corresponding phosphane oxide OPR_3, which are considered the most stable organophosphorus compounds. The identification that organophosphines are responsible for the "typical smell" when touching metals (Glindemann et al. 2006) might give evidence for such compounds in the human body. For century's ignis fatuus, a phosphorescent light seen over marshy ground and around graveyards at night (ghostly lights) is now known to be caused by the spontaneous combustion of gases emitted by decomposing organic matter where phosphanes would act as a "chemical match".

5.7 Carbon

Thousands of organic compounds used as chemicals by man are described in relation to properties and environmental fate in air, soil and water (Howard 1990, 1993, Mackay et al. 2006, Verschueren et al. 2009). However, the detailed chemistry is almost unknown and only studied for a few hundred of substances. We have already presented chapters on carbon, its origin (2.1.2.2), cycling (2.2.2.5), emissions into air (2.7.3) and the "carbon problem" (2.8.3). We have also discussed the unique chemical properties of carbon and its compounds to create life (2.2.2.1, 2.2.2 and 2.4.4). It is the harmonic balanced affinity of carbon to electropositive and electronegative elements as well, which provides the largest quantity of different chemical compounds, and the huge reservoir of CO_2 (including water-dissolved bicarbonate) supports the maintenance of life in the form of omnipresent plants and animals and carbon cycling. By contrast, silicon as an element with similar properties, provides non-volatile SiO_2 with the tendency to form polymers and cannot provide global turnover rates compared with carbon but is the fundament of inorganic "life", the rocky world.

In Chapter 5.3.3, we introduced some extremely important reactive oxygen-carbon species in air. The separation between inorganic and organic carbon chemistry (and compounds) is not strongly fixed. In nature, the synthesis of organic compounds only occurs in living cells of plants and animals, where only plants are able, through photosynthesis, to link the organic world with the inorganic, i.e. use CO_2 as feedstock. Nature provides organic matter in a large variety for food, materials and energy carriers. So far, the extraction of such compounds has been limited to the carbon cycle (i.e. limited to renewable sources), and problems have only arisen because of local limits of carbon supply and local waste loadings. Only because of the exhaustion of fossil fuels do humans again meet the same general problems, but now on a global scale. An idea for a solution the carbon resource limitation and the pollution problem is proposed in Chapter 2.8.4.3 by creating the manmade closed carbon cycle.

The general fate of organic carbon in the climate system is mineralization, i.e. oxidation back to CO_2 and/or carbonate and water. In Chapter 5.3.2.2, we presented the elementary steps in CH_4 oxidation, finally to CO_2:

$$CH_4 \xrightarrow[(-H_2O)]{OH} CH_3 \xrightarrow{O_2} CH_3O_2 \xrightarrow[(-NO_2)]{NO} CH_3O \xrightarrow[(-HO_2)]{O_2} HCHO \xrightarrow[(-H_2O \text{ or } -H)]{OH \text{ or } h\nu}$$

$$HCO \xrightarrow[(-H_2O)]{OH} CO \xrightarrow[(-H)]{OH} CO_2$$

Table 1.1 shows that CO_2, CH_4 and CO are the main carbon compounds in air, roughly in a ratio of $1000:10:1$. It is noteworthy that these ratios also express roughly the ratios of the chemical lifetime of these species in the atmosphere ($\tau_{CH_4} \approx 10$ years). In this reaction chain, there are two competitive pathways. At low NO concentrations, CH_3O_2 will react (with HO_2) to form methyl peroxide CH_3OOH, which mainly transfers into the aqueous phase but can also be photolyzed (into CH_3O + OH). Formaldehyde HCHO will also be scavenged; thereby the hydrometeors (and finally natural waters) are an important sink for reactive carbon intermediates. Organic acids probably only form in aqueous solutions. For the CO_2 budget these play no role on how much the percentage of CH_4 and NMVOC conversion to CO_2 is. The natural atmospheric CO_2 concentration is determined by the difference between global photosynthesis and respiration whereby the ocean plays the role of a controller (see Chapter 2.8.3.2). It was discussed in Chapter 5.3.6 that both CO and CH_4 (because of its OH $\rightarrow$ HO_2 conversion, see scheme above) determine the regional and global tropospheric net O_3 production, whereas NMVOC contributes to local and mesoscale additional O_3 formation. Within these photochemical conversions, organic oxidized nitrogen compounds are formed (see Chapter 5.4.5), which (at high concentrations) have been attributed to human health problems.

Without life there would be no carbon cycle on earth. But this is also true for all other elements: as well as carbonate, nitrate, sulfate and phosphate all also cannot be reduced by simple chemical processes in the climate system. Deep in the earth, however, we cannot exclude − even hypothesizing the existence of elemental carbon − reducing chemical regimes, turning elements on geological timescales.

In the following subchapters, we will summarize the basic principles of carbon chemistry; the reader is recommended to refer textbooks on organic chemistry, biogeochemistry and biochemistry.

5.7.1 Elemental carbon and soot

Elemental carbon exists naturally in nature as graphite (hexagonal C structure) and diamond (tetrahedral C structure). In graphite, very small amounts of fullerenes, where C_{60} molecules are most known, have been detected. Elemental carbon is chemically extremely stable. Only at high temperatures does C react with other elements and burns with O_2 (well known for centuries as a coal dust explosion). Another phenomenon is the self-ignition of coal, but locally the neccessary increased temperature must rise and several processes have been suggested (for example Stepanov and Andrushchenko 1990). This process of self-oxidation until self-ignition needs time and is only possible in condensed coal stocks − when burning in deposits can occur over centuries. However, this ignition was always initiated by humans who interrupted the chemical regime of the deposit through contact with

Table 5.27 Soot types.

soot origin	characteristics
wood combustion soot	large OC fraction, usually only 20% BC; lignine-derived substances with OC
biomass burning soot	similar to wood soot but OC fraction larger up to 90%
coal combustion soot[a]	different from biomass burning soot; large BC and EC fraction, mainly in the coarse mode
diesel soot	OC may approaches 50%, the remainder is BC and EC; smallest size fraction 3–20 nm consists of oil nanodroplets, accumulation mode (50–250 nm) contains EC and OC and the coarse mode is EC due to coagulation
aviation soot	includes undefined OC fraction up to 300 nm

[a] This soot was important in past from household coal heating and steam locomotives; coal-fired power plants with low efficiency such as largely still in use in India and China may also produce large soot emissions.

atmospheric oxygen. In the climate system, soot is a phenomenon that is as old as the culture of fire. Remember that biomass burning is dominantly caused by men, and thereby "natural" burning caused by lightning strikes has always been negligible. Therefore, soot is largely an artefact to nature (Table 5.27).

In Table 2.34, the total BC emission was given between 11 and 17 Tg C yr^{-1} where about 50% is contributed by biomass and fossil fuel burning. Other more recent data suggest a lower emission of 5.8–8.0 Tg C yr^{-1} (Haywood and Boucher 2000, Bond et al. 2004). Bond et al. (2004) estimated 8.0 Tg for black carbon and 33.9 Tg for organic carbon where the contributions of fossil fuel, biofuel and open burning are estimated as 38%, 20% and 42%, respectively, for BC and 7%, 19% and 74%, respectively. The uncertainty ranges of 4.3–22 Tg yr^{-1} for BC and 17–77 Tg yr^{-1} for OC.

The answers on the question "what is soot" (Chapter 4.3.5) are as different as different people will ask this question. A general definition was given by Popovicheva et al. (2006): "soot is a carbon-containing aerosol resulting from incomplete combustion of hydrocarbon fuel of varying stoichiometry, defined by the ratio of fuel to oxygen". Soot not only addresses the properties of BC and EC fractions commonly associated with *soot*, but also includes the organic fraction (OC). The chosen combustion conditions control the soot properties to a large extent. Soot aerosol consists of harmful substances, such as adsorbed PAHs as well as their hydroxylated and nitro-substituted congeners, which have significant carcinogenic and mutagenic potential.

A lot is known on the direct and indirect climate impact of atmospheric soot, the absorption of gases, possible heterogeneous processes and water-soot interactions, but nothing is known on the fate of soot, especially elemental carbon. Studies on the chemistry of NO$_y$, O$_3$, SO$_2$ and many other species on soot have been carried out over recent decades, showing that soot might provide a reactive surface in air. However, such surface chemistry has been assessed to be insignificant in the budget of chemical species compared with gas phase and liquid phase processes; mainly because of the limited PM surface-to-air volume ratio. Many studies suggest that direct ozone loss on soot aerosol is unlikely under ambient conditions in the troposphere (Disselkamp et al. 1999, Nienow and Roberts 2006).

Without doubt, the large OC fraction in "soot" will undergo "aging" by oxidation. Decesari et al. (2002) showed that the WSOC produced from the oxidation of soot particles increased rapidly with ozone exposure and consisted primarily of aromatic polyacids found widely in atmospheric aerosols and which are frequently referred to as macromolecular humic-like substances (HULIS).

However, we only can speculate that reactive oxygen species such as O_3 and OH can react with carbon similar to CO:

$$C + OH \rightarrow CO + H \tag{5.323}$$

$$C + O_3 \rightarrow CO + O_2 \tag{5.324}$$

This process is extremely slow and can result in the chemical lifetime of hundreds or more years under climate system conditions. It is known that surfaces covered with photocatalytic active TiO_2 obviously remain "clean" with respect to soot pollution, whereas reference surfaces become black. As discussed in Chapter 5.3.5, under such photocatalytic conditions large OH concentrations might locally produced.

The fate of the about 8 Tg BC yr^{-1} widely dispersed on the globe is deposition into oceans and soils. It is known that coal can survive in soils for hundreds of years and can improve soil structure and water budget. The survival of atmospheric coal combustion soot from the Middle Ages can still be seen at old churches and it is cultural question to regard it as patina with respect or unwanted pollution.

5.7.2 C_1 chemistry: CO, CO_2, CH_4, CH_3OH, HCHO, HCOOH

Most of the gas phase C_1 chemistry is elsewhere presented (Chapters 5.3.2.2 and 5.3.3 concerns organic and 2.8.3.2 concerns CO_2 and carbonate dissolution). Two species from the heading above, methanol CH_3OH and formic acid HCOOH we met as emissions from biomass burning (Table 2.44). Because C—O and O—H bonds are much stronger than the C—H bond, OH attack goes preferable onto C—H (at higher carbon chains preferably at the α C—H); $k_{5.325} = 7.7 \cdot 10^{-13}$ cm^3 $molecule^{-1}$ s^{-1} and $k_{5.326} = 1.3 \cdot 10^{-13}$ cm^3 $molecule^{-1}$ s^{-1}, i.e. about 85% of methanol goes via (5.325):

$$CH_3OH + OH \rightarrow CH_2OH + H_2O \tag{5.325}$$

$$CH_3OH + OH \rightarrow CH_3O + H_2O \tag{5.326}$$

The fate of the methoxy radical CH_3O is known (5.45) and also CH_2OH gives the same products: $k_{5.327} = 9.7 \cdot 10^{-12}$ cm^3 $molecule^{-1}$ s^{-1}:

$$CH_2OH + O_2 \rightarrow HCHO + HO_2 \tag{5.327}$$

Hence, the methanol oxidation yields formaldehyde according to the budget:

$$CH_3OH \xrightarrow[(-H_2O - HO_2)]{OH + O_2} HCHO$$

Formaldehyde (IUPAC name methanal) quickly converts at daytime to CO (5.46–5.47) but also transfers into the aqueous phase where it hydrates (5.287) or reacts with S(IV) (reaction 5.286). In aqueous solution, methanol quickly oxidizes similar to the gas phase mechanisms to formaldehyde, which (together with scavenged HCHO) further oxidizes to formic acid (methane acid), likely via an OH adduct (the following reaction is speculative):

$$HCHO + OH \rightarrow H_2C(O)OH \xrightarrow[(-HO_2)]{O_2} HC(O)OH \qquad (5.328)$$

Numerous measurements show that in the gas phase [HCHO] > [HCOOH] and in hydrometeors [HCHO] < [HCOO)]. Formic acid was first isolated in 1671 by the English researcher *John Ray* (1627–1795) from red ants (its name comes from the Latin word for ant, *formica*). Considering methane acid as a transient between inorganic and organic carbon, the carbonyl group —C(O)OH provides the huge class of organic acids RCOOH and the formate HCOO⁻ gives the class of esters HCOOR and RCOOR.

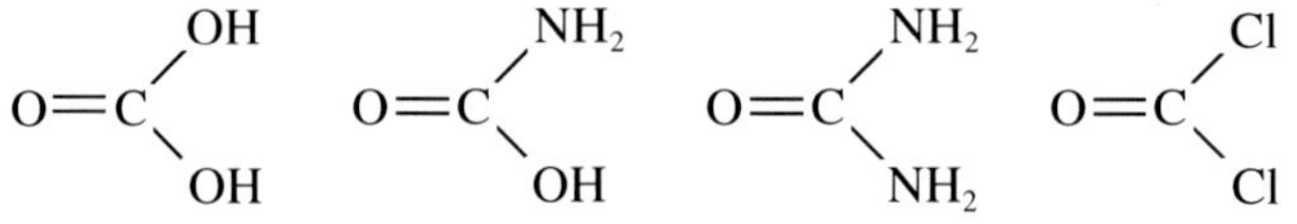

$$O=C\overset{OH}{\underset{OH}{<}} \qquad O=C\overset{NH_2}{\underset{OH}{<}} \qquad O=C\overset{NH_2}{\underset{NH_2}{<}} \qquad O=C\overset{Cl}{\underset{Cl}{<}}$$

carbonic acid   carbamic acid   urea   phosgene

From carbonic acid (CO_2 + H_2O) important derivates are derived: carbamic acid (or carbamates[29], which also provides a class of organic carbamines substituting H for organic R), urea and halogenated substitutes (such as phosgene). Urea is the "symbol" linking inorganic with organic chemistry; thereby here are two IUPAC names: diaminomethanal (as organic compound) and carbonyl diamide (as inorganic compound).

In aqueous solution, especially in cellular environments, the carbonate radical anion (CO_3^-) is produced from the reaction between the ubiquitous carbon dioxide and peroxonitrite ($ONOO^-$), which is an instable intermediate (Fig. 5.25) in biological NO reduction and first forms as an CO_2 adduct nitrosoperoxocarboxylate, which then decomposes:

$$CO_2 + ONOO^- \rightarrow ONOOCO_2^- \rightarrow CO_3^- + NO_2 \qquad (5.329)$$

Carbonate radicals react with many organic compounds (Chen and Hoffmann 1973, Umschlag and Herrmann 1999) in the general H abstraction reaction (competing with OH); $k_{5.330} = 10^4 \ldots 10^7$ L mol^{-1} s^{-1}:

$$CO_3^- + RH \rightarrow HCO_3^- + R \qquad (5.330)$$

It is a strong acid (i. e. the radical HCO_3 fully dissociates) with p$K_a = -4.1$ and a strong oxidizing agent with $E°(CO_3^-/CO_3^{2-}) = 1.23 \pm 0.15$ V (Czapski et al. 1999,

---

[29] They are formed while CO_2 capture from (flue) gases using aqueous amine solutions.

Armstrong et al. 2006, Medinas et al. 2007), which likely exists as the dimer $H(CO_3)_2^-$ (Wu et al. 2002). The importance in natural waters is likely limited because it is produced in radical reactions such as listed following ($k_{5.331a} = 3.9 \cdot 10^8$ L mol^{-1} s^{-1}, $k_{5.331b} = 1.7 \cdot 10^7$ L mol^{-1} s^{-1}, $k_{5.332} = 4.1 \cdot 10^7$ L mol^{-1} s^{-1}, $k_{5.333} = 2.6 \cdot 10^6$ L mol^{-1} s^{-1}; CAPRAM model) and reacts back to carbonate.

$$CO_3^{2-} + OH \rightarrow CO_3^- + OH^- \tag{5.331a}$$

$$HCO_3^- + OH \rightarrow CO_3^- + H_2O \tag{5.331b}$$

$$HCO_3^- + NO_3 \rightarrow CO_3^- + H^+ + NO_3^- \tag{5.332}$$

$$CO_3^{2-}(HCO_3^-) + Cl_2^- \rightarrow CO_3^- + 2\,Cl^- \, (+\, H^+) \tag{5.333}$$

$$CO_3^{2-} + SO_4^- \rightarrow CO_3^- + SO_4^{2-} \tag{5.334}$$

It reacts with TMI ($k_{5.335} = 2 \cdot 10^7$ L mol^{-1} s^{-1}) and with peroxides ($k_{5.336} = 6.5 \cdot 10^8$ L mol^{-1} s^{-1}), which represents radical termination in one-electron transfers. Reactions (5.336a) and (5.337) also represent H abstraction reactions: $k_{5.337} = 4.3 \cdot 10^5$ L mol^{-1} s^{-1} (Draganic et al. 1991).

$$CO_3^- + Fe^{2+}\,(Mn^{2+},\,Cu^+) \rightarrow CO_3^{2-} + Fe^{3+}\,(Mn^{3+},\,Cu^{2+}) \tag{5.335}$$

$$CO_3^- + HO_2 \rightarrow HCO_3^- + O_2 \tag{5.336a}$$

$$CO_3^- + O_2^- \rightarrow CO_3^{2-} + O_2 \tag{5.336b}$$

$$CO_3^- + H_2O_2 \rightarrow HCO_3^- + HO_2 \tag{5.337}$$

The following fast reaction obviously transfers O$^-$ (adequate reactions concerns $NO \rightarrow NO_2^-$ and $O_2 \rightarrow O_3$ are not described in literature): $k_{5.338} = 1 \cdot 10^9$ L mol^{-1} s^{-1} (Lilie et al. 1978).

$$CO_3^- + NO_2 \rightarrow CO_2 + NO_3^- \tag{5.338}$$

A reaction with ozone is slow and implies the intermediate O_4^- ($\xleftarrow{\;H^+\;}$ HO_4): $k_{5.339} = 1 \cdot 10^5$ L mol^{-1} s^{-1} (Sehested et al. 1983).

$$CO_3^- + O_3 \rightarrow CO_2 + O_2 + O_2^- \tag{5.339}$$

Another interesting species is given by the carbon dioxide anion radical CO_2^-, produced from aquated electrons (Hart and Anbor 1970), which is an efficient reducing agent in two ways, electron transfer and radical addition; $k_{5.340} = 4 \cdot 10^9$ L mol^{-1} s^{-1} (CAPRAM):

$$CO_2 + e_{aq}^- \rightarrow CO_2^- \tag{5.340}$$

$$CO_2^- + H_2O_2 \rightarrow CO_2 + H_2O_2^- \tag{5.341}$$

$$CO_2^- + O_2 \rightarrow CO_2 + O_2^- \tag{5.342}$$

The CO_2^- radical is also given from the oxidation of formate ions (Todres 2008).

$$HCOO^- + OH \rightarrow CO_2^- + H_2O \tag{5.343}$$

It adds onto organic radicals and double bonds (Morkovnik and Okhlobystin 1979).

$$CO_2^- + RCH(OH) \rightarrow RCH(OH)COO^- \qquad (5.344)$$

$$CO_2^- + R\!-\!CH\!=\!CH\!-\!R \rightarrow RCH(COO^-)\!-\!CH\!-\!R \qquad (5.345)$$

It disproportionates and dimerizes to oxalate (Morkovnik and Okhlobystin 1979).

$$CO_2^- + CO_2^- \rightarrow CO_3^{2-} + CO \qquad (5.346)$$

$$CO_2^- + CO_2^- \rightarrow {}^-O(O)C\!-\!C(O)O^- \qquad (5.347)$$

Basically, these processes represent a way of sustainable chemistry in the future for CO_2 air capture and CO_2 reduction to fuels (Fujita 1999, Wu 2009). Two-electron steps onto adsorbed CO_2 are favored compared with (5.340), whose reduction potential amounts -1.0 V:

$$CO_2 \xrightarrow{2\,H^+ + 2e^-} HCOOH \qquad\qquad E = -0.61 \text{ V}$$

$$CO_2 \xrightarrow{2\,H^+ + 2e^-} CO + H_2O \qquad\qquad E = -0.53 \text{ V}$$

$$CO_2 \xrightarrow{6\,H^+ + 6e^-} CH_3OH + H_2O \qquad E = -0.38 \text{ V}$$

In aqueous solution, some metal-ligand complexes form CO_2 adducts, which internally undergo a two-electron step:

$$M^I\!-\!L + CO_2 \rightarrow M^I\!-\!L(CO_2) \rightleftarrows$$
$$M^{III}\!-\!L(CO_2^{2-}) \xrightarrow{H^+} M^{III}\!-\!L + CO, HCOOH, H_2O \qquad (5.348)$$

Thus, one of the best routes to remedy the CO_2 problem is to convert it to valuable hydrocarbons using solar energy in "photocatalytic farms".

Finally, another radical is given from CO, forming the carbon monoxide anion CO^- (Hart and Anbor 1970), which very quickly reacts with water to give the formyl radical, which undergoes very rapid hydration to $CH(OH)_2$ and further then dimerizes to glyoxal $(HCO)_2$.

$$CO + e_{aq}^- \rightarrow CO^- \qquad (5.349)$$

$$CO^- + H_2O \rightarrow HCO + OH^- \qquad (5.350)$$

$$2\,HC(OH)_2 \rightarrow (HCO)_2 + 2\,H_2O \qquad (5.351)$$

The formyl radical in solution can be considered the conjugated acid of CO^-. In the next chapter, we will see that C_2 species are produced in the aqueous phase by carbonylation (reactions with HCO) and carboxylation (reactions with CO_2^-). Raef and Swallow (1966) showed that directly from (5.349) and (5.350) the hydrated form is given:

$$CO + e_{aq}^- + H_2O \rightarrow HC(OH)O^- \underset{}{\overset{H^+}{\rightleftharpoons}} HC(OH)_2 \qquad (5.352)$$

The hydrated formyl radical $HC(OH)_2$ is a strong reducing species. It has been shown as an intermediate in the reaction of hydrated formaldehyde with OH and is oxidized by H_2O_2 to form formic acid (or the formate anion, respectively).

$$OH + H_2C(OH)_2 \rightarrow HC(OH)_2 + H_2O \qquad (5.353)$$

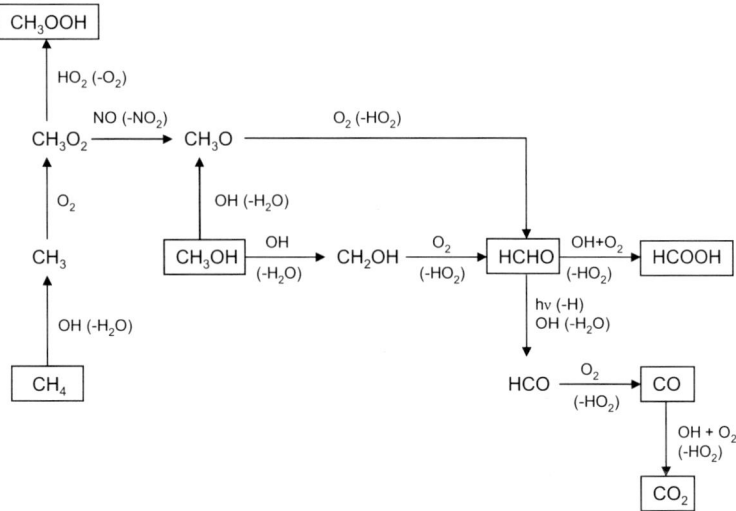

Fig. 5.32 Scheme of C_1 gas-phase chemistry.

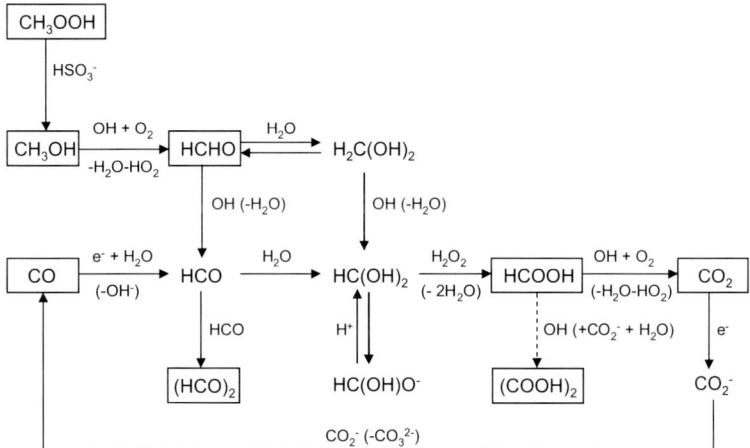

Fig. 5.33 Scheme of C_1 aqueous-phase chemistry. $(HCO)_2$ – glyoxal, $(COOH)_2$ – oxalic acid; dotted line – multistep process.

$$H_2O_2 + HC(OH)_2 \rightarrow HCOOH + 2\,H_2O \tag{5.354}$$

A possible propagation mechanism involves:

$$HC(OH)_2 + OH^- \rightarrow HCOOH + e_{aq}^- + H_2O \tag{5.355}$$

Figs. 5.32 and 5.33 summarize the C_1 chemistry in the gas and aqueous phase. Whereas CH_4 is very slowly oxidized in air (but provides methyl peroxide as a ubiquitous compound), methanol and particularly formaldehyde quickly oxidizes (the latter also photodissociates) to inorganic CO and finally CO_2. As shown in

Chapter 5.3.3, HCHO provides a net source of radicals. The formation of formic acid is likely to be negligible in the gas phase. But methanol, formaldehyde and formic acid (all produced and/or emitted in huge quantities from biogenic sources) will be scavenged and provide more efficient oxidation, finally with an accumulation of formic acid, but partly until its mineralization to CO_2 and, most interestingly, a pathway in the formation of highly reactive bicarbonyls such as glyoxal and oxalic acid.

5.7.3 C_2 chemistry: C_2H_2, C_2H_4, C_2H_6, C_2H_5OH, CH_3CHO, CH_3COOH, $(COOH)_2$

From a single carbon, according to the scheme in Fig. 2.42, C_2 (and higher) hydrocarbons are built up using the building blocks CH_3 and CO. Alkanes (also known as paraffins), with the general formula C_nH_{2n+2}, represent CH_2 chain blocks with methyl ends. They generally produce first an aldehyde by the OH oxidation of the terminal carbon atom and then a bicarbonyl at the other end of the chain (here C_2).

$$H_3C-CH_3 \xrightarrow[\ (-H_2O\ -\ NO_2\ -\ HO_2)\ ]{OH\ +\ NO\ +\ O_2} H_3C-C(O)H \xrightarrow[\ (-H_2O\ -\ NO_2\ -\ HO_2)\ ]{OH\ +\ NO\ +\ O_2} (C(O)H)_2$$

$$(5.355)$$

Acetaldehyde CH_3CHO (similar to formaldehyde) photolyzes and oxidizes through the OH attack. Other photolysis pathways (to CH_4 + CO or CH_3CO + H) are important above 300 nm in the troposphere.

$$CH_3CHO + h\nu \rightarrow CH_3 + HCO \tag{5.357}$$

$$CH_3CHO + OH \rightarrow CH_3CO + H_2O \tag{5.358}$$

CH_3CO (it is not RO but a carbon radical $^\bullet CO$) adds O_2 (see 5.64) and produces PAN (5.259) but also in a competitive reaction as RO_2 radical it oxidizes NO and afterwards decomposes (from the methylperoxo radical CH_3O_2 finally formaldehyde is given).

$$CH_3C(O)O_2 \xrightarrow[\ (-NO_2)\ ]{NO} CH_3C(O)O \tag{5.359}$$

$$CH_3C(O)O \xrightarrow[\ (-CO_2)\ ]{O_2} CH_3O_2 \xrightarrow[\ (-NO_2)\ ]{NO} CH_3O \xrightarrow[\ (-HO_2)\ ]{O_2} HCHO \tag{5.360}$$

Alcohols give (acetic) acid; $k_{5.361} = 2.8 \cdot 10^{-12}$ cm^3 molecule^{-1} s^{-1}. Acetic acid is likely stable in the gas phase and transferred into hydrometeors.

$$C_2H_5OH + OH \xrightarrow[\ (-H_2O)\ ]{} CH_3CHOH \xrightarrow[\ (-HO_2)\ ]{O_2} CH_3C(O)OH\ (90\%) \tag{5.361a}$$

$$C_2H_5OH + OH \xrightarrow[\ (-H_2O)\ ]{} CH_2CH_2OH\ (5\%) \tag{5.361b}$$

$$C_2H_5OH + OH \xrightarrow[\ (-H_2O)\ ]{} CH_3CH_2O \xrightarrow[\ (-HO_2)\ ]{O_2} CH_3CHO\ (5\%) \tag{5.361c}$$

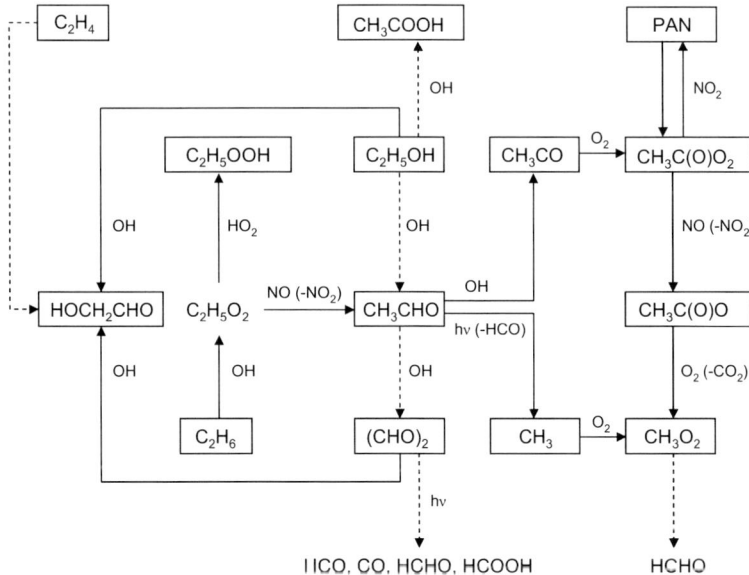

Fig. 5.34 Scheme of C_2 gas-phase chemistry. C_2H_5OOH – acethyl peroxide, $HOCH_2CHO$ – glycolaldehyde, C_2H_6 – ethane, C_2H_4 – ethene; dotted line – multistep process. OH reactions involve subsequent step with O_2 ($-HO_2$).

The radical adduct CH_2CH_2OH reacts the same as in the ethene + OH reaction (see below) and forms glycolaldehyde (IUPAC name 2-hydroxyethanal):

$$CH_2CH_2OH \xrightarrow[(-H_2O-NO_2-HO_2)]{OH+NO+O_2} HOCH_2CHO \tag{5.362}$$

The lifetime of glycolaldehyde in the atmosphere is about one day for reaction with OH, and > 2.5 days for photolysis, although both wet and dry deposition are other important removal pathways (Bacher et al. 2001). The primary products of OH attack and photolysis are mainly HCO (for further fate, see 5.48) and CH_2OH (5.327). We later will see that glycolaldehyde is the main product of the OH + C_2H_2 (ethine) reaction. However, the hydroxyalkoxy intermediate might also decompose (the oxidation of glycolaldehyde gives HCHO and CO as final products).

$$HOCH_2CH_2O \rightarrow HCHO + CH_2OH \tag{5.363}$$

Finally, the CH_2OH radicals react with O_2 to give HCHO and HO_2 (5.327). Thus, C_2 is broken down into C_1 species. Fig. 5.34 shows schematically the C_2 gas phase chemistry. It is obvious that there is no ethanol formation and acetic acid decomposition, whereas acetaldehyde provides many pathways back to C_1 chemistry. Glycolaldehyde is a highly water-soluble product from several C_2 species (ethene, acetaldehyde and ethanol); other bicarbonyls, however, are likely to be produced preferably in solution. The aqueous phase produces other C_2 species but also decomposes them (Fig. 5.35). By contrast, in aqueous solution from C_1, C_2 species can be given as shown by the formation of glyoxal from the formyl radicals (5.351); the latter is

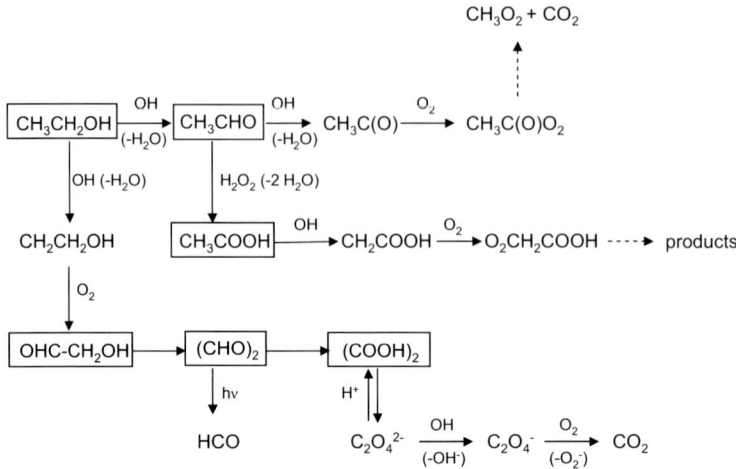

Fig. 5.35 Scheme of C_2 aqueous-phase chemistry.

an often found species in solution. From methanol and formaldehyde dicarbonyls are produced via carbonylation in the aqueous phase:

$$CH_3OH \xrightarrow[(-H_2O)]{OH} CH_2OH \xrightarrow{HCO} OHC{-}CH_2OH \quad \text{(glycolaldehyde)} \tag{5.364}$$

$$HCHO \xrightarrow[(-H_2O)]{OH} HCO \xrightarrow{HCO} OHC{-}CHO \quad \text{(glyoxal)} \tag{5.365}$$

Furthermore, from methanol and formic acid the corresponding acids gained through carboxylation (in reaction with the CO_2^- radical) are:

$$CH_3OH \xrightarrow[(-H_2O)]{OH} CH_2OH \xrightarrow{CO_2^-} HOCH_2{-}CO_2^- \quad \text{(glycolic acid)} \tag{5.366}$$

$$HCOOH \xrightarrow[(-H_2O)]{OH} COOH \xrightarrow{CO_2^-} COOH{-}CO_2^- \quad \text{(oxalic acid)} \tag{5.367}$$

Ethanol C_2H_5OH oxidizes in a multistep process in solution to acetaldehyde.

$$CH_3CH_2OH \xrightarrow[(-H_2O\ -\ HO_2)]{OH\ +\ O_2} CH_3CHO \tag{5.368}$$

Such starting H abstraction can also go with other H acceptors such as SO_4^-, NO_3, Cl_2^-, Br_2^- and CO_3^- (see, for example, the CAPRAM model). Further oxidation of the aldehyde is similar in elementary steps to acetic acid.

$$CH_3CHO \xrightarrow[(-H_2O\ -\ HO_2)]{OH\ +\ O_2} CH_3COOH \tag{5.369}$$

Acetic acid further oxidizes in a first step to give the CH_2COOH radical; in the absence of O_2, it can dimerize to succinic acid $COOH(CH_2)_2COOH$ (Garrison et al. 1959).

$$CH_3COOH \xrightarrow[(-H_2O)]{OH} CH_2COOH \xrightarrow{O_2} O_2CH_2COOH \qquad (5.370)$$

This peroxo radical can then give glycolic acid, glyoxylic acid, oxalic acid, HCHO and CO_2 (Garrison et al. 1959). The acetylperoxo radical (ACO_3) − the precursor of PAN − is given from the oxidation of acetaldehyde. It forms with HO_2 peroxoacetic acid, also detected in air in small concentrations.

$$CH_3CHO \xrightarrow[(-H_2O)]{OH} CH_3C(O) \xrightarrow{O_2} CH_3C(O)O_2 \xrightarrow[(-O_2)]{HO_2} CH_3C(O)OOH \qquad (5.371)$$

In aqueous solution, peroxo radicals can also react with bisulfite, gaining the sulfite radical:

$$RO_2 \xrightarrow[(-SO_3^-)]{HSO_3^-} ROOH \qquad (5.372)$$

As for other peroxo radicals, it can transfer O onto other dissolved species ($NO \rightarrow NO_2$) and then decompose them (finally, the methylperoxo radical forms HCHO).

$$CH_3C(O)O_2 \xrightarrow[(NO_2)]{NO} CH_3C(O)O \xrightarrow{O_2} CH_3O_2 + CO_2 \qquad (5.373)$$

Glyoxal (CHOCHO or $(CHO)_2$) is one of the simplest multifunctional compounds found in the atmosphere and is produced by a wide variety of biogenic- and anthropogenic-emitted organic compounds (Ortiz et al. 2006). One current model estimates global glyoxal production to be 45 Tg yr^{-1}, with roughly 50% due to isoprene photooxidation, whereas another estimates 56 Tg yr^{-1} with 70% produced from biogenic precursors (Galloway et al. 2009 and citations therein). CHOCHO is destroyed in the troposphere primarily by reaction with OH radicals (23%) and photolysis (63%), but it is also removed from the atmosphere through wet (8%) and dry deposition (6%; Myriokefalitakis et al. 2008). Glyoxal sulfate has also been detected in filter samples:

$$CHO-C(OH)-OSO_3^- \qquad \text{(glyoxal sulfate)}$$

$$COOH-C(H)-SO_3^- \qquad \text{(glycolic acid sulfate)}$$

In diluted aqueous solution, glyoxal exists as a dihydrate $CH(OH)_2CH(OH)_2$, which is fast and reversibly formed (Schweitzer et al. 1998). The aqueous phase photooxidation of a glyoxal is a potentially important global and regional source of oxalic acid and secondary organic aerosol (SOA; Fu et al. 2009) and could account for a missing SOA source in budgeting.

The gas phase photolysis of glyoxal produces two HCO radicals as the most important pathway under atmospheric conditions. Tadić et al. (2006) estimated $j_{obs} = 1.0 \cdot 10^{-4}$ s^{-1}. Although glyoxal has a very low effective quantum yield ($\Phi_{eff} = 0.035 \pm 0.007$), photolysis remains an important removal path in the atmosphere.

Oxalic acid is the most abundant dicarboxylic acid found in the troposphere, yet there is still no scientific consensus concerning its origins or formation process. Concentrations of oxalic acid gas at remote and rural sites range from about 0.2 ppb to 1.2 ppb with a very strong annual cycle with high concentrations during

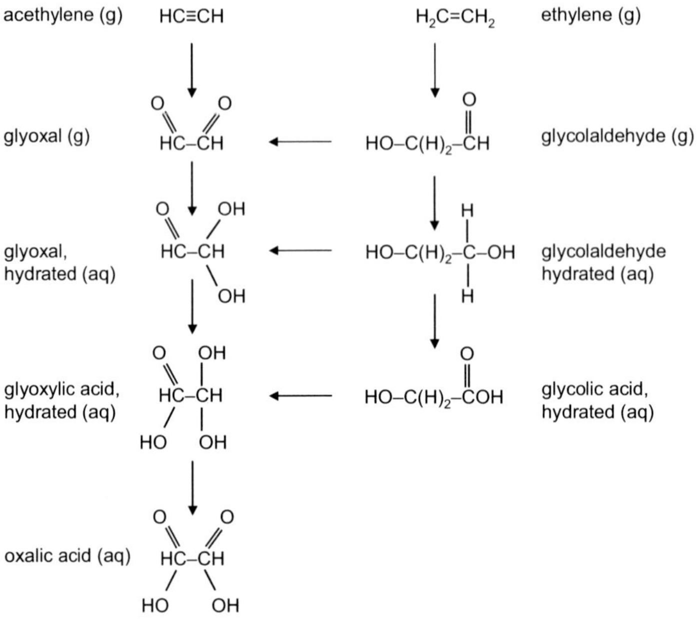

Fig. 5.36 The proposed reactions pathway for the formation of oxalic acid in cloud water; after Warneck (2003)

the summer periods. Oxalate was observed in the clouds at air-equivalent concentrations of 0.21 ± 0.04 g m^{-3} to below-cloud concentrations of 0.14 g m^{-3}, suggesting an in-cloud production as well (Crahan et al. 2004). Oxalic acid is the dominant dicarboxylic acid (DCA) and it constitutes up to 50% of total atmospheric DCAs, especially in non-urban and marine atmospheres. The large occurrence in the condensed phase led Warneck (2003) to suggest that oxalate does not solely originate in the gas phase and condense into particles (Fig. 5.36). Hence, hydroxyl radicals might be responsible for the aqueous phase formation of oxalic acid from alkenes. Among different dicarboxylic acids (oxalic, adipic, succinic, phthalic and fumaric) only the dihydrate of oxalic acid, enriched in particles in the upper troposphere, acts as a heterogeneous ice nucleus (Zobrist et al. 2006). Ubiquitous organic aerosol layers above clouds with enhanced organic acid levels have been observed and field data suggest that aqueous phase reactions to produce organic acids, mainly oxalic acid, followed by droplet evaporation are the source (Sorooshian et al. 2007).

Organic acids (Table 5.28) are ubiquitous components of the troposphere in urban and remote regions of the world. Organic acids contribute significantly to rainwater acidity in urban areas (Pena et al. 2002) and account for as much as 80−90% of the acidity in remote areas (Andreae et al. 1988). Keene and Galloway (1986) were the first to demonstrate that formic and acetic acid concentrations are highly correlated in rainwater. Organic acid concentrations in rainwater vary both spatially and temporally. Concentration variations in the gas and aqueous phases have been attributed to seasonal variations in biogenic emissions.

Besides alkanes, C$_2$ carbon comprises two other classes: alkenes and alkynes. The double bond C=C is stronger than the simple C—C but paradoxically is more

Table 5.28 C$_1$ and C$_2$ carbon acids, after Holleman-Wiberg (2007).

formula	structure	name of acid	name of salt
monocarbon acids			
H$_4$CO	H$_3$COH	methanol[a]	methanolate
H$_2$CO$_2$	HC(O)OH	formic acid	formate
H$_2$CO$_3$	C(O)(OH)$_2$	carbonic acid	carbonate
H$_2$CO$_4$	C(O)(OH)OOH[b]	peroxocarbonic acid	peroxocarbonate
dicarbon acids			
H$_4$C$_2$O$_2$	CH$_3$C(O)OH	acetic acid[e]	acetate
H$_4$C$_2$O$_3$	HOCH$_2$C(O)OH	glycolic acid[f]	glycolate
H$_2$C$_2$O$_2$	HOC≡COH[b, d]	dihydroxyacetylene	dihydroxyacetylate
H$_2$C$_2$O$_3$	HOC—C(O)OH	glyoxylic acid[g]	glyoxalate
H$_2$C$_2$O$_4$	HO(O)C—C(O)OH	oxalic acid	oxalate
H$_2$C$_2$O$_5$	HO(O)C—O—C(O)OH[b]	dicarbonic acid	dicarbonate
H$_2$C$_2$O$_6$	HO(O)C—O—O—C(O)OH[b]	peroxodicarbonic acid	peroxodicarbonate

[a] very weak acid forming
[b] only as salts
[d] isomer with the non-acidic glyoxal O=CH—CH=O (ethandial)
[e] IUPAC name: ethanoic acid; other names: methanecarboxylic acid, acetyl hydroxide
[f] IUPAC name: 2-hydroxyethanoic acid, other name: hydroxoacetic acid
[g] IUPAC name: oxoethanoic acid; other names: oxoacetic acid, formylformic acid (ubiquitous in nature in berries)

reactive. The most simple is ethylene (ethene) CH$_2$=CH$_2$. Alkenes are produced in so-called elimination reactions, mainly by the dehydration ($-$H$_2$O) of alcohols and dehydrohalogenation ($-$HX) of alkyl halides:

$$H_3C\text{—}CH_2OH \rightarrow H_2C\text{=}CH_2 + H_2O \tag{5.374}$$

Alkenes add OH and provide a radical adduct (which reacts further as shown in reaction 5.359). More reaction principles of alkenes are described in Chapter 5.7.4.

$$H_2C\text{=}CH_2 + OH \rightarrow CH_2CH_2OH \tag{5.375}$$

In contrast to alkenes (also known as olefins), which are important compounds in nature, alkynes (also termed alkines) $-$C≡C$-$ are relatively rare in nature but highly bioactive in some plants. The triple bond is very strong with a bond strength of 839 kJ mol^{-1}. Ethylene C$_2$H$_2$ (commonly known as acetylene) is generally considered to be produced only by human activities with an average tropospheric lifetime of the order of two months, allowing this compound to reach remote areas as well as the upper troposphere. Thus, the presence of C$_2$H$_2$ in open oceanic atmosphere is commonly explained by its long-range transport from continental sources together with CO originating from combustion. Both species are strongly correlated in atmospheric observations, offering constraints on atmospheric dilution and chemical aging. In effect, its mixing ratio is typically in the range 500$-$3000 ppt in inhabited countries compared with 50$-$100 ppt in remote oceanic areas. Xiao et al. (2007) estimated by modeling the global source of 6.6 Tg yr^{-1}. However, Kanakidou et al. (1988) do not exclude an oceanic source.

The destruction of acetylene in the atmosphere occurs only by reaction with OH radicals (Siese and Zetzsch 1995). Typical products are HO_2, glyoxal (CHO—CHO), but SOA formation (Volkamer et al. 2009) is also found under laboratory conditions. The polymerization of C_2H_2 after radiolysis has long been known.

5.7.4 Alkenes, ketones and aromatic compounds

In the previous chapter, we introduced the OH addition to alkenes. Alkenes are the only class of organic compounds that react in the gas phase with ozone (the often referred DOC decomposition by O_3 is almost a result of secondary ROA formation from O_3 in solution). This occurs by the addition of ozone, a reaction called *ozonolysis* and has been known for more than 100 years.

$$\begin{array}{c} R_1 \\ \diagdown \\ \diagup C = C \diagdown \\ R_2 \qquad R_4 \end{array} \xrightarrow{O_3} \begin{array}{c} R_1 \quad \overset{O}{\underset{|}{\overset{|}{O}}} \quad \overset{O}{\underset{|}{\overset{|}{O}}} \quad R_3 \\ \diagdown C - C \diagup \\ R_2 \qquad R_4 \end{array} \longrightarrow R_1R_2C{=}O + R_3R_4{}^\bullet COO^\bullet \qquad (5.376)$$

$$\longrightarrow R_1R_2{}^\bullet COO^\bullet + R_3R_4C{=}O$$

The reaction rate increases with increasing carbon numbers, ranging between 10^{-18} and 10^{-14} cm^3 molecule^{-1} s^{-1}. Considering the O_3 concentration is larger by a factor of $> 10^5$ than that of OH, the absolute rate of ozonolysis, even for lower alkenes (C_1–C_4), is about 10% of the OH addition. Hence, at night the ozonolysis is an important pathway. For higher alkenes such as isoprene and terpene, the atmospheric lifetime is only in the range of minutes. Because of the steric consideration, the probability of each pathway a) and b) amounts to 50%. The ozonide intermediate decomposes with the formation of ketone and biradicals, called *Criegee* radicals, which then stabilize and decompose. For the example of propene, the following products are given (Finlayson-Pitts and Pitts 2000, Neeb et al. 1998).

$$(C(H)HOO)^* \xrightarrow{M} HCO + OH \quad (37-50\%) \qquad (5.377a)$$

$$(C(H)HOO)^* \xrightarrow{M} CO + OH \quad (12-23\%) \qquad (5.377b)$$

$$(C(H)HOO)^* \xrightarrow{M} CO_2 + H_2 \quad (23-38\%) \qquad (5.377c)$$

$$(C(H)HOO)^* \xrightarrow{M} CO_2 + 2H \quad (0-23\%) \qquad (5.377d)$$

$$(C(H)HOO)^* \xrightarrow{M} HCOOH \quad (0-4\%) \qquad (5.377e)$$

The produced OH and HO_2 (as a subsequent product from H) react additionally with the alkenes and provide a huge spectrum of products. Most important is SOA formation (known as a blue haze from biogenic emissions). The stabilized *Criegee* radical reacts with major species (H_2O, SO_2, NO, NO_2, CO, RCHO and ketones). In the reaction with water vapor direct H_2O_2 can also be formed (Sauer et al. 1999).

$$RC(O)OH + H_2O \qquad (5.378a)$$

$$RH\dot{}COO\dot{} \xrightarrow{H_2O} RHC(OH)OOH$$

$$RHC{=}O \ (RCHO) + H_2O_2 \qquad (5.378b)$$

This is an important source of secondary organic acids (reaction 5.378a is an intra-molecular rearrangement via H_2O-collision intermediate). The hydroxyperoxide has been identified as an intermediate. Competing are the following conversions to aldehyde.

$$RH\dot{}COO\dot{} + (NO, NO_2, CO) \rightarrow RCHO + (NO_2, NO_3, CO_2) \qquad (5.379)$$

The reaction with SO_2 goes via an adduct, which decays to sulfuric acid.

$$RH\dot{}COO\dot{} + SO_2 \rightarrow R(H)C \overset{O-O}{\underset{O}{\diamond}} S{=}O \xrightarrow{H_2O} RCHO + H_2SO_4 \qquad (5.380)$$

Nitrate radicals also react with alkenes by addition, which results in a variety of different compounds such as hydroxynitrates, nitrohydroperoxides and hydrocarbonyls (Wayne et al. 1991).

Another class of atmospherically important compounds is ketones, which are almost of biogenic origin. The most simple is acetone, which is ubiquitous in air. With the exception of acetone, higher ketones react preferably with OH at any carbon atom by H abstraction. With subsequent O_2 addition the fate of RO_2 is known, either forming aldehyde (HCO) at terminal carbon or ketone (C=O) in the middle of the chain. The photolysis of acetone (Emrich and Warneck 2000) provides the known radicals, which finally convert into HCHO and form the acetylperoxo radical (see 5.46 and 5.47).

$$CH_3C(O)CH_3 + h\nu \rightarrow CH_3C(O) + CH_3 \qquad (5.381)$$

Finally, we can show the principal reactions of aromatic compounds. The most simple aromatic (aryl) compound is benzene C_6H_6, which is similar to the simplest alkane, methane CH_4, namely it is very stable against OH attack. Substituted aryl compounds react faster with OH, for example methylbenzenes such as toluene, p-xylene and 1,3,5−trimethylbenzene. The degradation chemistry of aromatic VOC remains an area of particular uncertainty. What is clear is the attack onto the methyl group because it turns into the aldehyde (toluene → benzaldehyde). However, about 90 % of the OH attack goes to the ring, generating an adduct called the hydroxycyclohexadienyl radical (Becker et al. 1999).

This can further react in different channels; first it can give benzeneoxide C_6H_5O, which is in equilibrium with oxepine − from which through photolysis phenols are gained, for example α-cresol from toluene.

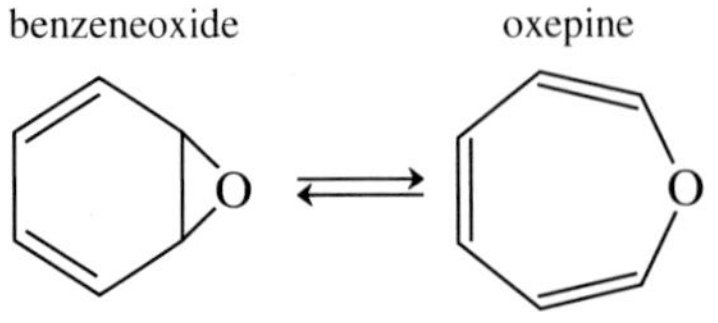

benzeneoxide   oxepine

Because benzeneoxide has a conjugated but not localized radical carbon atom, O_2 addition is another pathway; toluene gives the following peroxide, which is linked within the ring:

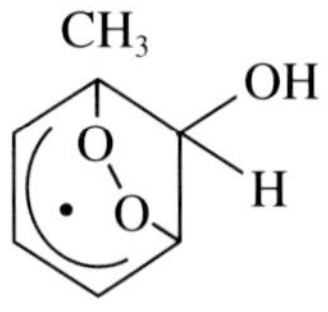

This intermediate peroxide decays while opening the ring into butendial and methylglyoxal. Methylglyoxal and glyoxal have been identified to be the main components (up to 3% only).

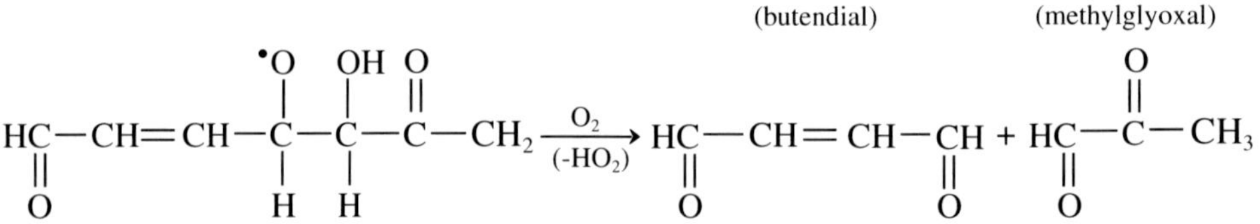

$$HC-CH=CH-\overset{\bullet O}{\underset{O}{C}}-\overset{OH}{\underset{H}{C}}-\overset{O}{\underset{H}{C}}-CH_2 \xrightarrow[(-HO_2)]{O_2} HC-CH=CH-CH + HC-C-CH_3$$

(butendial)   (methylglyoxal)

(5.382)

The variety of products is large; often bi- and polycarbonyls are found, which have been suggested to produce SOA and which also transfer to the aqueous phase. However, it has also been found that during new particle formation in forested areas OM mass fraction is significantly increasing but that the CCN efficiency is reduced by the low hygroscopicity of the condensing material (Dusek et al. 2010).

5.7.5 Is the atmospheric fate of complex organic compounds predictable?

In recent years, the availability of kinetic and mechanistic data relevant to the oxidation of VOCs has increased significantly and various aspects of the tropospheric chemistry of organic compounds have been reviewed extensively. We have a good generic VOC mechanism for the OH-initiated oxidation of organics that becomes progressively less accurate with increasing carbon number, functionalization (with oxygen) and especially at low NO_x. What we do know is summarized in the master chemical mechanism[30], which is a crucial link between laboratory kinetics and the reduced mechanisms used in large-scale models.

The simple enumeration of the possible branching pathways in the oxidation of large (C_5 and larger) organic compounds is bound to fail. One reason is simply the number of possible reaction pathways open to a large molecule, including radical attack at numerous sites as well as photolysis and isomerization. However, it is often argued that we need these mechanisms to predict ambient radical budgets and ratios, to predict ozone generation and removal and to predict organic aerosol production as well as organic aerosol aging in both vapor and condensed phases. Do we really need it? When 90 % of the radical budget and ozone formation is well described by the available mechanisms and models from the past 10−15 years, what will bring any improvement when considering the *main* uncertainty, the quantity of the emission of organic species (which is within a factor of two or more)? Nevertheless, science is unlimited and new generations of atmospheric chemists will (probably) compile more and more insights into atmospheric organic chemistry. The question of whether we need better mechanisms is therefore wrong. But the answer to whether atmospheric organic chemistry is predictable is definitively no.

5.8 Halogens (Cl, Br, F and I)

There is a reservoir of chlorine (and other halogens) in the ocean and sea salt in the form of NaCl from which HCl (and other halogens) can be released (Fig. 2.29). However, the ocean has been identified as source of many organic halogen compounds (Table 2.21) and newer insights provide evidence for halogenated compounds in soils (Chapter 2.4.3.3). Most spectacular attention on halogens was given by the implications on the stratospheric ozone layer (Chapter 5.3.7).

Table 5.29 lists the halogen compounds that have been detected in nature but not all exist in all phases (natural waters, hydrometeors and the air-gas phase). In the gas phase, acids such as HCl, HBr, HI, HClO, HBrO, HIO_3, $HClO_4$, $HBrO_4$ and HIO_4 exist. In the aqueous phase, perhalogenic acid, where O is exchanged by the peroxo group O_2, also exist as an intermediate in redox processes. Several interhalogens are known and might exist in air (such as iodine chloride). HF exists in air only in the condensed phase (particulate matter) despite it being primarily emitted as a gas. SF_6 is exclusively from anthropogenic sources (it is used for air dispersion tracer experiments). This is extremely stable with the longest known lifetime of 3200 years.

The ocean acts as both a source and a sink for methyl halides, where algae are responsible for the production of halogenated organic compounds with surface photochemical processes most likely breaking them down to simple species (similar to the DMS production − thereby correlations have often been found). The natural

---

[30] The Master Chemical Mechanism (MCM − University of Leeds) is a near-explicit chemical mechanism describing the degradation of CH_4 and 142 VOCs in the troposphere. The organic component of the version MCMv3 contains in the region of 12 600 reactions and 4 500 chemical species.

Table 5.29 Important halogen compounds in the climate system.

formula[a]	names[b]	generic name
F^-, Cl^-, Br^-, I^-	chloride	halogenide
HCl, HBr, HI	hydrogen chloride[c]	hydrogen halogenide[g]
Cl, Br, I	chlorine atom	halogen radical or atom
Cl_2, Br_2, I_2	chlorine	halogen molecule
Cl_2^-, Br_2^-, I_2^-	dichlorine anion radical	dihalogen anion radical
ClO, HOBr, IO	chlorine monooxide	halogen monoxide
HOCl, HOBr, HOI	hypochlorous acid	hypohalogenic acid
OCl^-, OBr^-, OI^-	hypochlorite	hypohalogenite
ClO_2, BrO_2, IO_2	chlorine dioxide	halogen dioxide
$HClO_2$, $HBrO_2$	chlorous acid	halogenic acid[g]
ClO_2^-, BrO_2^-	chlorite	halogenite
$HClO_3$, $HBrO_3$, HIO_4	chloric acid	halogen acid[g]
ClO_3^-, BrO_3^-	chlorate	halogenate
$ClONO_2$, $BrONO_2$	chlorine nitrate	halogen nitrate
ClNO, BrNO	nitrosyl chloride	nitrosyl halogenide
$ClNO_2$, $BrNO_2$	nitryl chloride	nitryl halogenide
CCl_4, CF_4	carbon tetrachloride	carbon tetrahalogenide
$CHCl_3$, $CHBr_3$	trichloromethane[d] (TCM)	trihalogen methane
CH_2Cl_2, CH_2Br_2	dichloromethane[e] (DCM)	dihalogen methane
CH_3Cl, CH_3Br, CH_3I	chloromethane[f]	methyl halide
SF_6	sulfur hexafluoride	—

[a] are listed only substances found to be significant in nature
[b] given as example for Cl compounds
[c] common name: hydrochloric acid (gas)
[d] common names: chloroform, bromoform
[e] common name: methylen chloride
[f] common name: methyl chloride
[g] in English (in contrast to German where generic specific meanings exist and are here proposed in English terms) halogen or halogenic acid is the general name for all acids (HCl, HOCl, $HClO_2$, $HClO_3$)

emission of volatile organic chlorine (CH_3Cl is dominant) lies in the Tg range per year (Table 2.21) and those of CH_3Br only in the kt range (Yvon-Lewis et al. 2004) but the net CH_3I oceanic emission to the atmosphere is 0.2 Tg (Bell et al. 2002), whereas rice paddies, wetlands and biomass burning only contribute small amounts (Lee-Taylor et al. 2005). Unexpectedly, during Saharan dust events, methyl iodide mixing ratios have been observed to be high relative to other times, suggesting that the dust-stimulated emission of methyl iodide has occurred (Williams et al. 2007)[31].

---

[31] Of course, experiments (Williams et al. 2007) with adding collected dust to seawater as well as adding H_2O_2 rapidly produced CH_3I; another example in line with photoenhanced radical aqueous chemistry.

5.8.1 Gas phase chemistry

The photolysis of organic halogenated compounds in the troposphere (i. e. at wavelengths > 300 nm) is insignificant. The only possible decomposition pathway goes via OH attack, but at relatively slow rates leading to lifetimes of about two years for CH_3Cl. Nevertheless, according to Fabian et al. (1996), less than 10% of the amount of methyl chloride emitted reaches the stratosphere. The OH pathway $(k_{CH_3Cl} = 3.6 \cdot 10^{-14}$ cm^{-3} molecule^{-1} s^{-1} and $k_{CH_3F} = 2.1 \cdot 10^{-14}$ cm^{-3} molecule^{-1} s$^{-1})$ is as follows:

$$CH_3Cl \xrightarrow[(-H_2O)]{OH} CH_2Cl \xrightarrow{O_2} O_2CH_2Cl \xrightarrow[(-NO_2)]{NO} HCHO + Cl \qquad (5.383)$$

Methyl halides with $\geq$ 3 halogens react too slowly ($k < 10^{-16}$ cm^{-3} molecule^{-1} s^{-1}) to give measurable decomposition in the troposphere. As seen from this pathway, the halogen atoms (F, Cl, Br and I) first appear in air. The Cl atom (and other halogens with decreasing rates from F over Cl to Br; no reactions with I are described) reacts similar to OH (e. g. in the oxidation of volatile organic compounds; Cai and Griffin 2006).

$$CH_4 + F \rightarrow CH_3 + HF \qquad k_{5.384} = 6.3 \cdot 10^{-11} \text{ cm}^{-3} \text{ molecule}^{-1} \text{ s}^{-1} \quad (5.384)$$

$$CH_4 + Cl \rightarrow CH_3 + HCl \qquad k_{5.385} = 1.0 \cdot 10^{-13} \text{ cm}^{-3} \text{ molecule}^{-1} \text{ s}^{-1} \quad (5.385)$$

$$HCHO + Cl \rightarrow HCO + HCl \qquad k_{5.386} = 7.2 \cdot 10^{-11} \text{ cm}^{-3} \text{ molecule}^{-1} \text{ s}^{-1} \quad (5.386)$$

$$HCHO + Br \rightarrow HCO + HBr \qquad k_{5.387} = 1.1 \cdot 10^{-12} \text{ cm}^{-3} \text{ molecule}^{-1} \text{ s}^{-1} \quad (5.387)$$

Whereas the role of halogens in the depletion of stratospheric ozone has been studied for several decades, the role of halogens in tropospheric O_3 reactions in marine and polar environments has only been studied in the past 15 years. The competitive fate of halogen atoms (F, Cl, Br and I) is their reaction with ozone; $k_{5.388} = 1.2 \cdot 10^{-11}$ cm^{-3} molecule^{-1} s^{-1}.

$$Cl + O_3 \rightarrow ClO + O_2 \qquad (5.388)$$

ClO can photodissociate (Table 5.30) and react with O_3 (as BrO, IO); $k_{5.389} < 1.5 \cdot 10^{-17}$ cm^{-3} molecule^{-1} s^{-1}, $k_{5.390} < 2 \cdot 10^{-17}$ cm^{-3} molecule^{-1} s^{-1}, $k_{IO} < 10^{-15}$ cm^{-3} molecule^{-1} s^{-1}.

$$ClO + O_3 \rightarrow ClO_2 + O_2 \qquad (5.389)$$

$$BrO + O_3 \rightarrow Br + 2O_2 \qquad (5.390a)$$

$$BrO + O_3 \rightarrow BrO_2 + O_2 \qquad (5.390b)$$

The atoms can also react with HO_2 to give HX (X = F, Cl, Br, I); $k_{5.391} = 3.5 \cdot 10^{-11}$ cm^{-3} molecule^{-1} s^{-1}.

$$Cl + HO_2 \rightarrow HCl + O_2 \qquad (5.391a)$$

$$Cl + HO_2 \rightarrow ClO + OH \qquad (5.391b)$$

Table 5.30 Photodissociation of inorganic halogen compounds in the troposphere.

HOCl	$\rightarrow$	OH + Cl
ClO		Cl + O
OClO	$\rightarrow$	ClO + O
Cl_2O	$\rightarrow$	Cl + ClO or Cl_2 + O
Cl_2O_2	$\rightarrow$	2 ClO
ClONO	$\rightarrow$	Cl + NO
$ClNO_2$	$\rightarrow$	Cl + NO_2
$ClONO_2$	$\rightarrow$	ClO + NO_2
Cl_2	$\rightarrow$	2 Cl

also for Br and I

HBr	$\rightarrow$	H + Br
Br_2	$\rightarrow$	2 Br
HI	$\rightarrow$	H + I

Monoxides (ClO, BrO and IO) react with NO_x:

$$ClO + NO \rightarrow Cl + NO_2 \qquad k_{5.392} = 1.7 \cdot 10^{-11} \text{ cm}^{-3} \text{ molecule}^{-1} \text{ s}^{-1} \qquad (5.392)$$

$$ClO + NO_2 \xrightarrow{M} ClONO_2 \qquad k_{5.393} = 7 \cdot 10^{-11} \text{ cm}^{-3} \text{ molecule}^{-1} \text{ s}^{-1} \qquad (5.393)$$

The reaction with HO_2 gives hypohalogenic acids (HOCl, HOBr and HOI; structure: H—O—X); $k_{5.394} = 6.9 \cdot 10^{-12} \text{ cm}^{-3} \text{ molecule}^{-1} \text{ s}^{-1}$.

$$ClO + HO_2 \rightarrow HOCl + OH \qquad (5.394a)$$

$$ClO + HO_2 \rightarrow HCl + O_3 \qquad (5.394b)$$

An interesting formation pathway of peroxohypochlorous acid in the gas phase has been suggested and proposed that ClOOH might be relatively stable (Lee and Rendell 1993). The reaction is too slow in the troposphere but could be of interest in the stratosphere (in solution we will meet ClOOH below).

$$ClO + OH \rightarrow ClOOH \rightarrow Cl + OOH \qquad (5.395)$$

The chlorine chemistry is too slow for ozone depletion in the troposphere but bromine and, in particular, iodine provide – combining with heterogeneous chemistry (see below) – lead to ozone depletion in marine environments by the following catalytic cycles (Barrie and Platt 1995, Platt and Mortgaat 1999); $k_{5.398} < 5 \cdot 10^{-12}$ cm^{-3} molecule^{-1} s^{-1}.

$$BrO + BrO \rightarrow 2 Br + O_2 \qquad k_{5.396} = 2.7 \cdot 10^{-12} \text{ cm}^{-3} \text{ molecule}^{-1} \text{ s}^{-1} \qquad (5.396)$$

$$Br + O_3 \rightarrow BrO + O_2 \qquad k_{5.397} = 1.2 \cdot 10^{-12} \text{ cm}^{-3} \text{ molecule}^{-1} \text{ s}^{-1} \qquad (5.397)$$

$$IO + IO \rightarrow I_2 + O_2 \qquad (5.398a)$$

$$IO + IO \rightarrow 2 I + O_2 \qquad (5.398b)$$

$$IO + IO \rightarrow I + IO_2 \qquad k_{5.398c} = 3.8 \cdot 10^{-11} \text{ cm}^{-3} \text{ molecule}^{-1} \text{ s}^{-1} \qquad (5.398c)$$

$$IO + IO \rightarrow I_2O_2 \qquad k_{5.398d} = 9.9 \cdot 10^{-11} \text{ cm}^{-3} \text{ molecule}^{-1} \text{ s}^{-1} \qquad (5.398d)$$

$$I_2 + OH \rightarrow HOI + I \qquad k_{5.399} = 2.1 \cdot 10^{-10} \text{ cm}^{-3} \text{ molecule}^{-1} \text{ s}^{-1} \qquad (5.399)$$

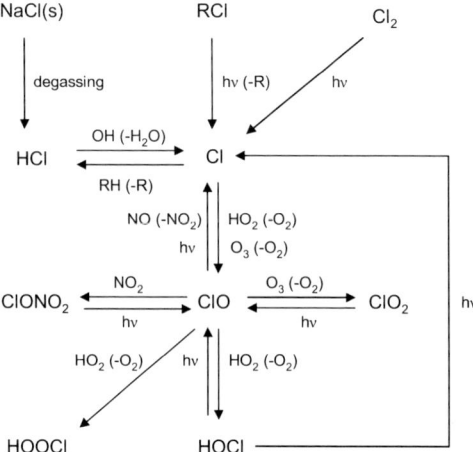

Fig. 5.37 Simplified scheme of gas-phase chlorine chemistry.

$$IO + ClO \rightarrow ICl + O_2 \qquad k_{5.400} = 2.4 \cdot 10^{-12} \ cm^{-3} \ molecule^{-1} \ s^{-1} \qquad (5.400)$$

The hydrogen halogenides (HCl, HBr, HI) react with OH but relatively slowly (only HI reacts quickly): $k_{5.401} = 7.8 \cdot 10^{-13} \ cm^{-3} \ molecule^{-1} \ s^{-1}$, $k_{HI} = 7.0 \cdot 10^{-11} \ cm^{-3} \ molecule^{-1} \ s^{-1}$. The photolysis of HCl is too slow (even in the stratosphere) to provide atomic Cl. Thus, the only direct Cl source from HCl is because of its reaction with OH. Fig. 5.37 shows a simplified tropospheric chlorine chemistry (those of bromine and iodine are similar).

$$HCl + OH \rightarrow H_2O + Cl \qquad (5.401)$$

5.8.2 Aqueous and interfacial chemistry

A significant source of hydrogen and nitrosyl halogenides is the surface of sea salt. Sea salt aerosol (SSA) has multiple impacts (besides other particulate matter categories) on atmospheric properties: responding to climate by optical properties (Mahowald et al. 2006), providing cloud condensation nuclei (McGovern et al. 1994; Clarke et al. 2006), being a heterogeneous surface for multiphase chemical reactions, e. g. SO_2 oxidation (Luria and Sievering 1991) and being a source for reactive chlorine (Finlayson-Pitts et al. 2003).

$$NaCl(s) + H_2SO_4 \rightarrow NaHSO_4 + HCl(g) \qquad (5.402)$$

$$NaCl(s) + N_2O_5 \rightarrow NaNO_3 + ClNO_2 \qquad (5.403)$$

Here, we present the following speculative scheme of multiphase chemical production of atomic Cl from dissolved HCl (its oxidation by H_2O_2 is described in Gmelin 1927):

$$\left[HCl \xrightarrow{H_2O_2} HOCl \underset{H_2O}{\overset{H^+ + Cl^-}{\rightleftarrows}} Cl_2) \right]_{particulate} \xrightarrow{degassing} Cl_2 \xrightarrow{h\nu} Cl\,(+\,Cl)$$

$$\xrightarrow{RH} HCl\,(+\,R) \qquad (5.404)$$

Assuming photocatalytic H_2O_2 formation in liquid films, the above scheme would represent another radical source via the aqueous phase without the consumption of gas phase-produced oxidants and a recycling between HCl and atomic Cl. This scheme would also support the findings of increasing chlorination together with (we know: TiO_2 containing) desert dust deposition onto oceans. O_2^- as a precursor of H_2O_2 has recently been measured directly in the ocean (Heller and Crott 2010).

Only in the past few years has evidence been found that SSA also occurs in the very fine fraction below 250 nm (down to 10 nm) depending on the sea state (Nilsson et al. 2001, Geever et al. 2005, Clarke et al. 2006). In Chapter 2.4.3.3, we presented knowledge of HCl degassing from sea salt. Brimblecombe and Clegg (1988) described the equilibrium:

$$HX(g) \rightleftarrows HX(aq) \rightleftarrows H(aq)^+ + X(aq)^- \quad (X = HSO_4^-, SO_4^{2-}, NO_3^-) \qquad (5.405)$$

$$H(aq)^+ + Cl(aq)^- \rightleftarrows HCl(g) \qquad (5.406)$$

and argued that only for strong acids with $H_{eff} > 10^{-3}$ mol L^{-1} atm^{-1} the reaction:

$$NaCl(p) + H(p)^+ \rightarrow Na(p)^+ + HCl(g) \qquad (5.407)$$

takes place. Only in the past decade has HCl degassing been observed in continental PM, mainly believed by gaseous HNO_3 sticking onto the particles (Pakkanen 1996, Keene and Savoie 1998, Zhuang et al. 1999, Yao et al. 2003, Xiu et al. 2004, Wai and Tanner 2004). This acid displacement has been studied by Guimbaud et al. (2002) in the laboratory to be diffusion limited in the gas phase. In other words, after uptake all HNO_3 chloride is readily displaced as HCl into the gas phase. Because of recrystallization, nitrate can replace all chloride even in the deeper layers of sea salt crystals (Finlayson-Pitts et al. 2003).

In aqueous solution, hypochlorous acid HOCl (and HOBr and HOI) is produced by Cl_2 hydrolysis: $k_{5.408} = 0.4$ cm^3 molecule^{-1} s^{-1}.

$$Cl_2 + H_2O \rightarrow HOCl + H^+ + Cl^- \qquad (5.408)$$

The acid and its salts are unstable; HOCl is a very weak acid ($pK_a = 7.5$; $pK_{HOBr} = 7.7$ and $pK_{HOI} = 10.6$), thereby is almost protolyzed in hydrometeors. Because dichloroxide is regarded as the anhydride, the following equilibrium is valid ($K = 3.55 \cdot 10^{-3}$ L mol^{-1}; Holleman-Wiberg 2007):

$$2\,HOCl \rightleftarrows Cl_2O + H_2O \qquad (5.409)$$

HOCl is a strong oxidizing agent ($E_{HOCl/Cl^-} = 1.49$ V, pH $= 0$) in acid but not in alkaline solution. Hence, oxidation goes only from HOCl or its protonized form HOClH$^+$ with the direct transfer of Cl (chlorination), similar to the reaction of HOBr and HOI. It disproportionates into chloride and chlorate; singlet oxygen is then partly formed, which can oxidize organic compounds (the biocide application of hypochlorite is based on this):

$$2\,HOCl \rightarrow 2\,Cl^- + 2\,H^+ + O_2 \tag{5.410}$$

$$3\,HOCl \rightarrow 2\,Cl^- + 2\,H^+ + ClO_3^- \tag{5.411}$$

The formation reaction (5.408) goes reverse in acid solution:

$$HOCl + HCl \rightarrow Cl_2 + H_2O \tag{5.412}$$

The last three reactions are explained through first the homolytic dissociation (which also occurs in surface water and clouds through photodissociation; Table 5.30):

$$HOCl \rightarrow OH + Cl \tag{5.413}$$

Chlorine radicals are also produced in solution from chloride by other radicals such as nitrate (NO_3), sulfate (SO_4^-) and hydroxyl (OH) with $k \approx 10^7 \dots 10^8$ cm^3 molecule^{-1} s^{-1}.

$$Cl^- + X \rightarrow Cl + X^- \tag{5.414}$$

The process including OH is also described as equilibrium (CAPRAM): $K_{5.412} = 0.7$ and $K_{5.413} = 1.6 \cdot 10^7$.

$$Cl^- + OH \rightleftarrows ClOH^- \tag{5.415}$$

$$ClOH^- + H^+ \rightleftarrows Cl + H_2O \tag{5.416}$$

This forms an adduct chlorine radical in equilibrium: $K = 1.9 \cdot 10^5$ (CAPRAM). Therefore, all reactions go from Cl_2^-.

$$Cl + Cl^- \rightleftarrows Cl_2^- \tag{5.417}$$

The presence of peroxides leads to radical termination: $k_{5.418} = 1.3 \cdot 10^{10}$ cm^3 molecule^{-1} s^{-1} (CAPRAM).

$$Cl_2^- + HO_2 \rightarrow 2\,Cl^- + H^+ + O_2 \tag{5.418}$$

This decays in water and gives the hydroxyl radical: $k_{5.419} = 6$ cm^3 molecule^{-1} s^{-1} (CAPRAM).

$$Cl_2^- + H_2O \rightarrow 2\,Cl^- + H^+ + OH \tag{5.419}$$

Hypochlorous acid directly reacts with organic nucleophilic compounds (X = O, N, S) or via the intermediate formation of the chlorine cation Cl^+ (HOCl $\rightarrow$ OH$^-$ + Cl$^+$; Gallard and Gunten 2002, Hanna et al. 1991). Such reactions might also explain the formation of halogenated organic compounds in soils from OM such as humic material where the volatile fraction (for example chloroform) emits into air. All olefinic hydrocarbons and aromatic compounds add Cl and OH from HOCl (see for example reaction 5.422).

$$HOCl + RX^- \rightarrow OH^- + RXCl \tag{5.420}$$

$$Cl^+ + RXH \rightarrow RXCl + H^+ \tag{5.421}$$

$$HOCl + CH_3CH{=}CHCH_3 \rightarrow CH_3CH(Cl){-}CH(OH)CH_3 \tag{5.422}$$

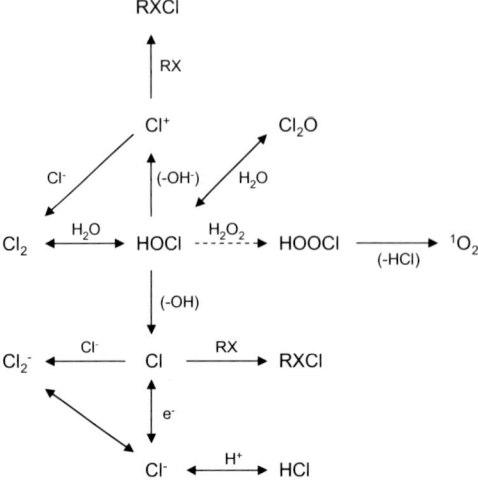

Fig. 5.38 Simplified scheme of aqueous-phase chlorine chemistry; dotted line – multistep process.

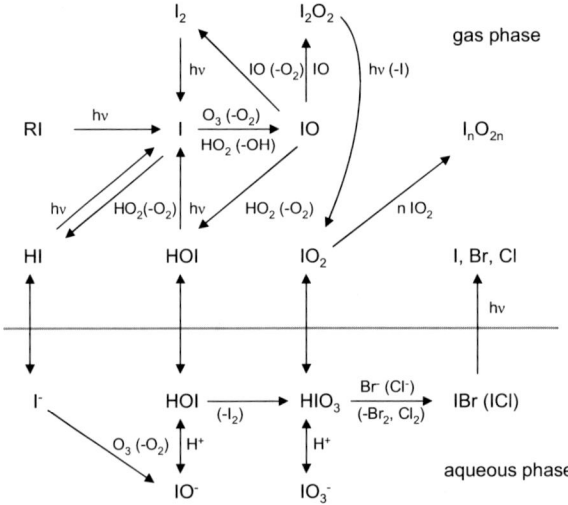

Fig. 5.39 Simplified scheme of multiphase iodine chemistry (no NO_x chemistry included).

Recently, a reaction between hypochlorite and hydrogen peroxide was proposed in alkaline solution (Castagna et al. 2008) as an electron transfer where the OCl radical can couple together with OH to form peroxohypochlorous acid HOOCl, which decomposes with a yield of singlet oxygen.

$$ClO^- + H_2O_2 \rightarrow ClO + OH + OH^- \tag{5.423}$$

$$ClO + OH \rightarrow ClOOH \tag{5.424}$$

$$ClOOH + OH^- \rightarrow {}^1O_2 + H_2O + Cl^- \tag{5.425}$$

However, this pathway of chemical singlet oxygen formation from MO_2 molecules (such as in the reaction between hypochlorite and hydrogen peroxide) where O_2 eliminates with the conservation of its total spin has long been known (Holleman-Wiberg 2007; oxygen formation in this reaction has been known since 1847, see Gmelin 1927). Fig. 5.38 summarizes the aqueous phase chlorine chemistry (that of bromine is similar) and Fig. 5.39 shows the multiphase chemistry of iodine. In contrast to the other halogens, iodine forms polymeric oxides that can provide effective CCN in the marine environment (O'Dowd et al. 2005, O'Dowd and de Leeuw 2007).

5.9 Other elements

So far we have discussed chemistry of almost all volatile compounds from oxygen, hydrogen, carbon, nitrogen, sulfur, phosphorus and halogens. We have also mentioned silicon, earth, earth-alkaline elements and others in composition of the ocean, soils and rocks. In aqueous phase oxygen and sulfur chemistry, we have emphasized the importance of redox and states of transition metal ions such as iron, copper and manganese. We have mentioned the role of aqueous semiconductor metal oxides (Ti, Zn, Fe, Cu, Sn and others), which have been recognized in photosensitized electron transfers onto important gas molecules such as O_2, O_3, NO, NO_2, CO and CO_2. Elements with an impact on biogeochemical redox processes are (in line with increasing atom number) Se, Fe, Mn, Co, Cu, Cr, Hg, Tc, As, Sb, U and Pu (Borch et al. 2010); some of them are only anthropogenic released (Hg, Tc, U, Pu). The basic elements in soil dust and rainwater are Na, K, Ca and Mg, but in minor and trace concentrations all stable elements can be detected such as (in decreasing concentrations) Fe, Zn, Mn, Cu, Ti, Al, Pb, Ni, Sb, V, Cr, Sb, Hg and Th (called crystal elements and thereby found in dust emissions from coal-fired power plants, soil dust and sea salt). Vital elements occurring in some but not all organisms are metals such as Li, Be, Na, Mg, K, Ca, V, Cr, Mn, Fe, Co, Ni, Cu, Mo, Sb and W and non-metals such as O, H, C, N, S, P, halogens and B, Si, As, Se and Te. Those likely without vital functions and thereby poisoning (unlike each substance they can become toxic with concentration) are Be, Cd and Hg. The interested reader is referred to Braids and Cai (2002), Turner and Hunter (2002), Fasham (2003), Prasad et al. (2005), Huang and Gobran (2005) and Likes (2009). Most literature is found on iron, the fourth most abundant element on earth. Many metals form complexes, highly dependent on pH, soluble and insoluble minerals, and thereby vary in their availability for life (essential or not) and redox processes.

6 Final remark

Recently, some of the younger generation of atmospheric chemists (Abbat et al. 2013) emphasized the importance of the three-legged stool to be balanced in atmospheric chemistry: laboratory, ambient observations, and modelling studies to address the most pressing issues of our time. They further state that each leg of the stool is only as stable as the fundamental chemistry that underpins it, is further studied. Unfortunately, that "fundamental chemistry" (studying properties and reactivity's of molecules) is weakening within the last 15 years. On the other hand, the three-legged stool (naturally not "tipping") is rose to misbalancing due to more and more modelling studies relative to experimental work (*theoria cum praxi* drift apart). The reason is clear, because "fundamental research" needs much more funding than modelling studies. I agree with Abbat et al. (2012) that atmospheric chemistry (now to avoid the adjective "fundamental") is increasingly focussed on much more complex chemical systems. That is the answer on my question (academically) asked at this point in the first edition four years ago: But why write 600 pages on climate chemistry?

At this point, the reader might come to the conclusion that this book is in the long line with other books on atmospheric chemistry (Table 6.1). A total of 90% of this book deals with air chemistry but the atmosphere represents 90% of the climate system. However, it is certainly not my aim to add another book belonging to *Finlayson-Pitts* and *Pitts* and *Seinfeld* and *Pandis*. *Barbara Finlayson-Pitts* is a great experimentalist and *John Seinfeld* a great theoretic chemist. Nonetheless, both combine *theoria cum praxi*. More than two columns, flanking the entrance into the science of air chemistry, namely experiment (practice) and model (theory), we do not need. If this book contributes something to atmospheric chemistry, which legitimates its publication in line with those books listed in Table 6.1, then it is the presentation of deeper insights into atmospheric aqueous and interfacial chemistry. Moreover – and I did try it to emphasize even more in the 2nd edition – the reader should understand that chemistry takes place in our climate system (or simply said our environment) simultaneous in all phases, respectively state of matter (aqueous, gaseous and solid) and reservoirs, respectively sphere (atmosphere, hydrosphere, lithosphere, biospheres): this is *climate system chemistry*.

Tom Graedel and *Paul Crutzen*, if they could rewrite those books (Graedel and Crutzen 1993, 1995) would possibly rename *global chemistry* and *earth chemistry*. In this line down (global system → earth system → climate system) is *climate chemistry*. This is parallel to climate physics and both are the fundamentals of the very observational physical and chemical climatology. At the beginning of this book, we emphasized that it makes no sense to draw strong borderlines between the disciplines and this is valid too for the systems because they overlap and the most important processes can happen at their interfaces. Therefore, chemistry of the climate system comprises atmospheric with water, soil and biological chemistry. At the end, however, it is chemistry solely.

However, research on our climate system and climate change will not end while climate change now proceeds with increasing rate. At present, on the other side, we are deep enough in understanding our climate system to conclude that further "business-as-usual" brings us to different "tipping points" with social feedbacks out of economic rationality. So, why decisions makers are unable to draw the right decisions? The answer may be simple because of the "Renish basic law" (here in High German: *Es kommt wie es kommt − es ist noch immer gut gegangen* (whatever will be, will be − so far, everything has always gone well in the end). But this erroneous belief neglects the evolution of our climate system ...

Table 6.1 Books (monographs and textbooks) on air composition and chemistry (not listed are books, written mainly on meteorological and physical aspects of air); note that all books published before 1960 are either on selected gases only or air composition but little or not on chemical processes in air. First edition given.

year	author(s)	biography
1774	Joseph Priestley	Experiments and observations on different kinds of Air. W. Bowyer and J. Nichols, London (Vol. 1)[a]
1777	Carl Wilhelm Scheele	Chemische Abhandlung von der Luft und dem Feuer. Nebst einem Vorbericht von Torbern Bergman. Upsala & Leipzig Magn. Swederus, 155 pp.
1872	Robert Angus Smith	Air and rain – the beginning of chemical climatology. Longmans, London, 600 pp.
1873	Cornelius B. Fox	Ozone and Antozone. Churchill, London, 329 pp.
1891	A. Petermann and Jean Graftiau	Recherches sur la composition de l'atmosphère. Acide carbonique, combinaisons azotées contenues dans l'air atmosphèriquè. In: Mèmoires Courinnès et Autres Mèmoires publies part L'Academie Royale des Sciences, des Lettres et des Beaux-Arts de Belgique. Bd. XLVII, Brüssel. 77 pp.
1896	H. Henriet	Les gaz de L'atmosphère. Encylopedie Scientifique des Aide-Memoire, Guthier-Villars et Masson, Malicorne (France), 192 pp.
1896	William Ramsay	The gases of the atmosphere. The history of their discovery. London, Macmillian and Co., 240 pp.
1900	Hans Blücher	Die Luft. Ihre Zusammensetzung und Untersuchung, ihr Einfluss und ihre Wirkung sowie ihre technische Ausnutzung. Leipzig, Verlag Otto Wigand, 322 pp.
1912	Francis G. Benedict	The composition of the atmosphere with special references to its oxygen content. Carnegie Institution, Washington, 115 pp.
1913	Anthon John Berry	The atmosphere. Cambridge Univ. Press, 146 pp.
1915	William Ramsay	The gases of the atmosphere. The history of their discovery. Fourth edition. London, Macmillian and Co., 306 pp.
1961	Philipp A. Leighton	Photochemistry of air pollution. Academic Press, New York, 300 pp.

Table 6.1 (continued)

year	author(s)	biography
1963	Christian E. Junge	Air chemistry and radioactivity. Academic Press, New York, 382 pp.
1972	Samuel S. Butcher and Robert J. Charlson	An introduction to air chemistry. Academic Press, New York, 241 pp.
1976	Julian Heicklen	Atmospheric chemistry. Academic Press, New York, 406 pp.
1979	John D. Butler	Air pollution chemistry. Academic Press, New York, 408 pp.
1981	Ernö Mészáros	Atmospheric chemistry. Fundamental Aspects. Akadémiai Kiadó, Budapest, 201 pp.
1986	John H. Seinfeld	Atmospheric chemistry and physics of air pollution. John Wiley & Sons, New York, 738 pp.
1986	Barbara J. Finlayson-Pitts and James N. Pitts	Atmospheric chemistry: Fundamentals and experimental techniques. John Wiley & Sons, New York, 1098 pp.
1988	Peter Warneck	Chemistry of the natural atmosphere. Academic Press, San Diego, 753 pp.
1990	Valerii A. Isidorov	Organic chemistry of the earth's atmosphere. Springer-Verlag, Berlin, 215 pp.
1995	Peter V. Hobbs	Basic physical chemistry for the atmospheric sciences. Cambridge University Press, 206 pp.
1996	Peter Brimble-combe	Air composition and chemistry. Cambridge University Press, 253 pp.
1996	Richard P. Wayne	Chemistry of the atmospheres. Clarendon Press, Oxford, 447 pp.
1998	John H. Seinfeld and Spyros N. Pandis	Atmospheric chemistry and physics – from air pollution to climate change. John Wiley & Sons, New York, 1326 pp.
1999	Daniel J. Jacob	Introduction to atmospheric chemistry. Princeton University Press, 264 pp.
2000	Barbara J. Finlayson-Pitts and James N. Pitts	Chemistry of the upper and lower atmosphere. Academic Press, San Diego, 969 pp.
2000	Peter V. Hobbs	Introduction to atmospheric chemistry. A companion text to basic physical chemistry for the atmospheric sciences. Cambridge University Press, 206 pp.
2003	Detlev Möller	Luft: Chemie, Physik, Biologie, Reinhaltung, Recht. DeGruyter, Berlin, 750 pp.
2010	Detlev Möller	Chemistry of the climate system. DeGruyter, Berlin, 722 pp.

[a] It is a six-volume work (1774–1786) in different editions. First printing: Observations on different kinds of air. In: *Philosophical Transactions*, Vol. 62, 1772, pp. 147–264

Appendix

A.1 List of acronyms and abbreviations found in literature[1]

ACPC	aerosols, clouds, precipitation, climate program
AOD	aerosol optical depth
AOT	aerosol optical thickness
AOT	accumulated dose over threshold
ANC	acid neutralizing capacity
AP	aerosol particle
BBOA	biomass burning organic aerosol
BC	black carbon (= soot)
B.C.	(also BC) before Christ, from the Latin *Ante Christum*, an epoch based on the traditionally reckoned year of the conception or birth of Jesus
BECS	bio-energy with carbon storage
BL	boundary layer
BNC	base neutralizing capacity
BOVOC	biogenic oxygenated volatile organic carbon
BP	before present (also B.P., e. g. kyr BP – thousand years before present)
Btu	British thermal unit (sometimes also BTU)
BVOC	biogenic volatile organic carbon
CAPRAM	chemical aqueous phase radical mechanism
CCC	carbon capture and cycling
CCM	chemistry-climate model
CCN	cloud condensation nuclei
CCS	carbon capture and storage (or: sequestration)
CFC	chlorofluorocarbons (contain carbon and some combination of fluorine and chlorine atoms)
CLIVAR	climate variability and predictability program
CN	condensation nuclei
CORINAIR	core inventory air (emission estimation program)
CSP	concentrating solar thermal power (plant)
DALR	dry adiabatic lapse rate
DIC	dissolved inorganic carbon
DM	dry mass or dry matter (sometemes also dm)
DMDS	dimethyldisulfide
DMS	dimethylsulfide
DMSO	dimethylsulfoxide

---

[1] Often the plural is formed by adding „s", for example: PSCs – polar stratospheric clouds.

DNA	deoxyribonucleic acid
DOC	dissolved organic carbon
DU	*Dobson* unit (total ozone column-concentration)
EC	elemental carbon (= soot), no difference to BC
EDGAR	emission database for global atmospheric research
EF	emission factor
EF	enrichment factor
EM	elemental matter (carbon related)
EMEP	co-operative programme for monitoring and evaluation of the long range transmission of air pollutants in Europe
ENSO	el Niño southern oscillation
EOS	earth observing system
EPA	U.S. environmental protection agency
EU	European Union
FAO	food and agriculture organization of the United Nations
FGC	flue gas cleaning
FGD	flue-gas desulfurization (plant)
FT	free troposphere
GAW	global atmospheric watch (WMO monitoring program)
GCM	general circulation model
GEIA	global emission inventory activity (within IGAC)
GEP	gross ecosystem production
GEWEX	global energy and water cycle experiment (project of WCRP)
GHG	greenhouse gases
GVM	global vegetation monitoring
GPCP	global precipitation climatology project (of GEWEX)
GPP	gross primary production
GWP	global warming potential
HALOE	halogen occultation experiment
HCFCs	hydrochlorofluorocarbons (contain hydrogen, chlorine, fluorine and carbon atoms)
HVDC	high voltage direct current
HFC	hydrofluorocarbon (contain hydrogen, fluorine, and carbon (no chlorine)
HMS	hydroxymethanesulfonate
HMSA	hydroxymethanesulfonic acid
HOA	hydrocarbon-like organic aerosol
HULIS	humic-like substances
IEA	international energy agency
IGAC	international global atmospheric chemistry (project within IGBP)
IGBP	international geosphere-biosphere program
IN	ice nuclei
IPCC	intergovernmental panel on climate change (supported by WMO and UNEP)
IR	infrared
ITCZ	inter-tropical convergence zone
IWC	cloud ice water content

LHB	late heavy bombardment
LIA	little ice age
LNG	liquefied natural gas (CH_4)
LPG	liquefied petroleum/propane gas, also low pressure gas (C_3/C_4 hydrocarbons)
LRTAP	convention on long-range transboundary air pollution
LS	lower stratosphere
LWC	liquid water content
MCE	modified combustion efficiency
MCM	master chemical mechanism
MCS	mesoscale cloud system
MEGAN	model of emissions of gases and aerosols from nature
MODIS	moderate resolution imaging spectroradiometer
MOPITT	measurement of pollution in the troposphere
MSA	methansulfonic acid
Mt	megaton (10^{12} g)
NASA	national aeronautics and space administration (USA)
NCAR	national center for atmospheric research (USA)
NDACC	network for the detection of atmospheric composition change
NCEP	national centers for environmental prediction
NEP	net ecosystem production
NH	northern Hemisphere
NMHC	nonmethane hydrocarbons
NMVOC	nonmethane volatile organic carbon
NOAA	national oceanic and atmospheric administration (USA)
NOM	natural organic matter
NPP	net primary production
NSS	non-seasalt
OA	organic aerosol (total, see also HOA, BBOA)
OC	organic carbon
ODP	ozone depletion potential
ODS	ozone depleting substances
OECD	organisation for economic co-operation and development
OM	organic matter
OOA	oxygenated organic aerosol
OVOC	oxygenated volatile organic carbon
PAH	polycyclic aromatic hydrocarbons
PAN	peroxyacetyl nitrate
PAR	photosynthetic active radiation
PBL	planetary boundary layer
Pg	picogram (10^{15} g)
PFC	perfluorocarbon
PM	particulate matter
PM_1	particulate matter with aerodynamic particle diameter ≤ 1 μm
PM_{10}	particulate matter with aerodynamic particle diameter ≤ 10 μm
$PM_{2.5}$	particulate matter with aerodynamic particle diameter ≤ 2.5 μm
POA	primary organic aerosol

POCP	photochemical ozone creation potential
POM	particulate organic matter (part of PM)
POP	persistent organic pollutant
PSC	polar stratospheric cloud
QBO	quasi-biennal oscillation
R	organic carbon radical
RDS	rate-determining step
REE	rare earth element
RH	relative humidity
RH	organic compound
RCHO	aldehyde
RCOOH	organic acid
RNA	ribonucleic acid
RO	organic oxy radical (alkoxy or alkoxyl radical)
RO_2	organic hydroperoxy radical
ROOH	organic peroxide
ROS	reactive oxygen species
RTE	radiative transfer equation
SAGE	stratospheric aerosol and gas experiment
SBUV	solar backscatter ultra-violet
SCIAMACHY	scanning imaging absorption spectrometer for atmospheric cartography
SD	standard deviation
SBL	stable atmospheric boundary layer
SH	southern Hemisphere
SIC	soil inorganic carbon
SOA	secondary organic aerosol
SOC	soil organic carbon
SOM	soil organic matter
SPARC	stratospheric processes and their role in climate (an international research project)
SPM	suspended particulate matter (= dust, atmospheric aerosol)
SSA	sea-salt aerosol
SST	sea surface temperature
STE	stratosphere-troposphere exchange
stp	standard temperature and pressure
SVOC	secondary volatile organic carbon
TC	total carbon (sum of organic and inorganic)
TCL	tropospheric chlorine loading
Tg	terragram (= Mt)
TIC	total ionic content
TMI	transition metal ions
TOC	total organic carbon
toe	tonne of oil equivalent
TOMS	total ozone mapping spectrometer
TPM	total particulate matter
TSP	total suspended matter (= PM)

UT	upper troposphere
UV	ultraviolet
UNECE	United Nations economic commission for Europe
UVOC	unsaturated volatile organic compounds
VFS	vegetation fire smoke
VOC	volatile organic carbon
WCRP	world climate research program
WHO	world health organization of the United Nations
WMO	world meteorological organization of the United Nations
WSOC	water soluble organic compounds

A.2 Quantities, units and some useful numerical values

The SI (abbreviated from the French *Le Système International d'Unités*), the modern metric system of measurement was developed in 1960 from the old meter-kilogram-second (mks) system, rather than the centimeter-gram-second (cgs) system, which, in turn, had a few variants. Because the SI is not static, units are created and definitions are modified through international agreement among many nations as the technology of measurement progresses, and as the precision of measurements improve.

Long the language universally used in science, the SI has become the dominant language of international commerce and trade. The system is nearly universally employed, and most countries do not even maintain official definitions of any other units. A notable exception is the United States, which continues to use customary units in addition to SI. In the United Kingdom, conversion to metric units is government policy, but the transition is not quite complete. Those countries that still recognize non-SI units (e. g., the US and UK) have redefined their traditional non-SI units in SI units.

It is important to distinguish between the definition of a unit and its realization. The definition of each base unit of the SI is carefully drawn up so that it is unique and provides a sound theoretical basis upon which the most accurate and reproducible measurements can be made. The realization of the definition of a unit is the procedure by which the definition may be used to establish the value and associated uncertainty of a quantity of the same kind as the unit.

Some useful definitions:

- A *quantity in the general sense* is a property ascribed to phenomena, bodies, or substances that can be quantified for, or assigned to, a particular phenomenon, body, or substance. Examples are mass and electric charge.
- A *quantity in the particular sense* is a quantifiable or assignable property ascribed to a particular phenomenon, body, or substance. Examples are the mass of the moon and the electric charge of the proton.
- A *physical quantity* is a quantity that can be used in the mathematical equations of science and technology.
- A *unit* is a particular physical quantity, defined and adopted by convention, with which other particular quantities of the same kind are compared to express their value.
- The *value of a physical quantity* is the quantitative expression of a particular physical quantity as the product of a number and a unit, the number being its numerical value. Thus, the numerical value of a particular physical quantity depends on the unit in which it is expressed.

Other quantities, called *derived quantities*, are defined in terms of the seven base quantities via a system of quantity equations. The SI derived units for these derived quantities are obtained from these equations and the seven SI base units. Examples of such SI derived units are given in the Table 2, where it should be noted that the symbol 1 for quantities of dimension 1 such as mass fraction is generally omitted.

Table A.1 The seven basic units of SI.

quantity	name of unit	symbol
length	meter	m
mass	kilogram	kg
time	second	s
thermodynamic temperature	kelvin	K
amount of substance	mole	mol
electric current	ampere	A
luminous intensity	candela	cd

Table A.2 Derived units (examples only; the list can be continued, e. g. for electric and magnetic quantities).

quantity	name of unit	symbol	special definition	definition from SI
force	newton	N	–	$kg\ m\ s^{-2}$
pressure	pascal	Pa	$N\ m^{-2}$	$kg\ m^{-1}\ s^{-2}$
energy, work, quantity of heat	joule	J	$N\ m$	$kg\ m^2\ s^{-2}$
power, radiant flux	watt	W	$J\ s^{-1}$	$kg\ m^2\ s^{-3}$
mass density	–	–	–	$kg\ m^{-3}$
dynamic viscosity	–	–	$Pa\ s$	$kg\ m^{-1}\ s^{-1}$
moment of force	–	–	$N\ m$	$kg\ m^2\ s^{-2}$
surface tension	–	–	$N\ m^{-1}$	$kg\ s^{-2}$
heat flux density, irradiance	–	–	$W\ m^{-2}$	$kg\ s^{-3}$
frequency	hertz	Hz	–	s^{-1}
wave number	–	–	–	m^{-1}
area	–	–	–	m^2
volume	–	–	–	m^3
speed, velocity	–	–	–	$m\ s^{-1}$
acceleration	–	–	–	$m\ s^{-2}$

Table A.3 Units outside the SI that are accepted for use with the SI.

quantity	symbol	value in SI units
minute (time)	min	1 min $=$ 60 s
hour	h	1 h $=$ 60 min $=$ 3600 s
day	d	1 d $=$ 24 h $=$ 86400 s
degree (angle)	°	$1° = (\pi/180)$ rad
minute (angle)	′	$1′ = (1/60)° = (\pi/10\ 800)$ rad
second (angle)	″	$1″ = (1/60)′ = (\pi/648\ 000)$ rad
liter	L[a]	$1\ L = 1\ dm^3 = 10^{-3}\ m^3$
metric ton[b]	t	$1\ t = 10^3$ kg
electronvolt[c]	eV	$1\ eV = 1.602\ 18 \cdot 10^{-19}$ J, approximately
nautical mile	–	1 nautical mile $=$ 1852 m knot
knots	–	nautical mile per hour $= (1852/3600)$ m s^{-1}
hectare	ha	$1\ ha = 1\ hm^2 = 10^4\ m^2$
bar	bar	$1\ bar = 0.1\ MPa = 100\ kPa = 1000\ hPa = 10^5$ Pa
ångström	Å	$1\ Å = 0.1\ nm = 10^{-10}$ m
curie	Ci	$1\ Ci = 3.7 \cdot 10^{10}$ Bq
roentgen	R	$1\ R = 2.58 \cdot 10^{-4}$ C kg^{-1}
rad	rad	$1\ rad = 1\ cGy = 10^{-2}$ Gy
rem	rem	$1\ rem = 1\ cSv = 10^{-2}$ Sv

[a] This unit and its symbol l were adopted by the CIPM in 1879. The alternative symbol for the liter, L, was adopted by the CGPM in 1979 in order to avoid the risk of confusion between the letter l and the number 1. Thus, although both l and L are internationally accepted symbols for the liter, to avoid this risk the preferred symbol for use in the United States is L.
[b] In many countries, this unit is called "tonne".
[c] The electron volt is the kinetic energy acquired by an electron passing through a potential difference of 1 V in vacuum. The value must be obtained by experiment, and is therefore not known exactly.

Table A.4 Prefixes used to construct decimal multiples of units (SI).

Y	yotta	10^{24}
z	zetta	10^{21}
E	eta	10^{18}
P	peta	10^{15}
T	tera	10^{12}
G	giga	10^9
M	mega	10^6
k	kilo	10^3
h	hecto	10^2
da	deca	10
d	deci	10^{-1}
c	centi	10^{-2}
m	milli	10^{-3}
μ	micro	10^{-6}
n	nano	10^{-9}
p	pico	10^{-12}
f	femto	10^{-15}
a	atto	10^{-18}
z	zepto	10^{-21}
y	yocto	10^{-24}

Table A.5 Some useful numerical values.

name	symbol	value	definition
Boltzmann constant	$\mathbf{k}$	$1.3897 \cdot 10^{-23}$ J K^{-1}	$\mathbf{k} = R/N_A$
Avogadro constant	N_A	$6.0221 \cdot 10^{23}$ mol^{-1}	[a]
Loschmidt constant	n_o	$2.68678 \cdot 10^{25}$ m^{-3}	[b]
Gas constant	R	8.31451 J K^{-1} mol^{-1}	$R = \mathbf{k}\, N_A$
Planck constant	h	$6.62607 \cdot 10^{-34}$ J s^{-1}	[c]
Stefan-Boltzmann constant	σ	$5.6705 \cdot 10^{-8}$ J m^{-2} K^{-4} s^{-1}	$\sigma = 2\pi^5\mathbf{k}^4/15h^3c^2$
Faraday constant	F	96485 C mol^{-1}	$F = e\, N_A$
elementary charge	e	$1.60217649 \cdot 10^{-19}$ C	[d]
gravitational constant	G	$6.6726 \cdot 10^{-11}$ m^3 kg^{-1} s^{-2}	$G = f \cdot r^2/m_1 m_2$ [e]
standard gravity	g	9.8129 m s^{-1}	$g = G \cdot m_{\text{earth}}/r_{\text{earth}}^2$ [f]
speed of light	c	$2.99792 \cdot 10^8$ m s^{-1}	

[a] number of atoms in 0.012 kg ^{12}C
[b] number of atoms/molecules of an ideal gas in 1 m^{-3} at 0 °C and 1 atm
[c] *Planck-Einstein* relation: $E = h\nu$
[d] fundamental constant, equal to the charge of a proton and used as atomic unit of charge
[e] empirical constant (also called Newton's constant) gravitational attraction (force f) between objects with mass m_1 and m_2 and distance r
[f] nominal acceleration due to gravity at the earth's surface at sea level

A.3 The geological timescale

eon	era	period	epoch	time ago (Myr)
Phanerozoic	Cenozoic	Neocene	Holocene	0.0118
			Pleistocene	1.81
			Pliocene	5.33
			Miocene	23.0
		Paleocene	Oligocene	33.9
			Eocene	55.8
			Paleocene	65.5
	Mesozoic	Cretaceous		145.5
		Jurassic		200
		Triassic		251
	Paleozoic	Permian		299
		Carboniferous		359
		Devonian		416
		Silurian		444
		Ordovician		488
		Cambrian		542
Proterozoic	Neoproterozoic	Ediacaran		630
		Cryogenian		850
		Tonian		1000
	Mesoproterozoic	Stenian		1200
		Ectasian		1400
		Calymnian		1600
	Paleoproterozoic	Statherian		1800
		Orocirian		2050
		Rhyacian		2300
		Siderian		2500
Archean	Neoarchean			2800
	Mesoarchean			3200
	Paleoarchean			3600
	Eoarchean			3800
Hadean				4560

The Geological Time-Scale is hierarchical, consisting of (from smallest to largest units) ages, epochs, periods, eras and eons. Each era, lasting many tens or hundreds of millions of years, is characterized by completely different conditions and unique ecosystems.

It is followed the geological time scale as determined by the International Commission on Stratigraphy (ICS). The ICS has not finished its job and gaps remain, particularly in the Early Paleozoic. Where gaps occur, it is generally followed the Russian system for the Cambrian and the British system for the Silurian. Epochs are subdivided further into ages not listed here. The periods from Cretaceons and older are subdivided into epochs and ages not shown here.

A.4 Biography

Adhémar, Joseph-Alphonse (1797–1862): French mathematician; his theory on ice ages was further developed and greatly modified, first by James Croll and later by Milutin Milankovic.

Aitken, John (1839–1919): Scottish physicist and meteorologist in Edinburgh, one of the founders of cloud physics and aerosol science.

al-Khazini, Abd al-Rahman (flourished 1115–1130): Greek Muslim scientist, astronomer, physicist, biologist, alchemist, mathematician and philosopher from Merv, then in the Khorasan province of Persia but now in Turkmenistan, who made important contributions to physics and astronomy. Al-Khazini is better known for his contributions to physics in his treatise *The Book of the Balance of Wisdom*, completed in 1121, which remained an important part of Islamic physics.

al-Bīrūnī, Abū 'r-Raihān Muhammad ibn Ahmad (937–1048): Persian polymath scholar, one of the very greatest scientists of Islam. Biruni's works number 146 in total.

Anaximander (c. 610 BC–c. 546 BC): Scholar of Thales. He is known to have conducted the earliest recorded scientific experiment and can be considered the first true scientist.

Anaximenes of Miletus (about 585–528 BC): Greek Pre-Socratic philosopher, probably a younger contemporary of Anaximander and together with Thales one of the three Milesian philosophers.

Andrews, Thomas (1813–1885): Irish chemist and physicist in Belfast who did important work on phase transitions between gases and liquids.

Archimedes of Syracuse (287–212 BC): Greek mathematician, physicist, engineer, inventor, and astronomer on Sicily, at that time a colony of Magna Graecia; considered to be the greatest mathematician of antiquity and one of the greatest of all time.

Aristotle (384–322 BC): Greek philosopher, a student of Plato and teacher of Alexander the Great. Together with Plato and Socrates (Plato's teacher), Aristotle is one of the most important founding figures in Western philosophy.

Arrhenius, Svante August (1859–1927): Swedish scientist in Stockholm, originally a physicist, but often referred to as a chemist, and one of the founders of the science of physical chemistry.

Avogadro, Lorenzo Romano Amedeo Carlo di Quaregna e di Cerreto (1776–1856): Italian physicist in Turin; he is most noted for his contributions to molecular theory.

Baxendale, John H. (1916–1982): British chemist at the University of Manchester; pioneer in radiation chemistry.

Barnes, Charles Reid (1858–1910): The first professor of plant biology at the University of Chicago.

Becher, Johann Joachim (1635–1682): German physician, alchemist, precursor of chemistry, scholar and adventurer, mainly in Mainz and Vienna; best known for his development of the phlogiston theory.

Beer, August (1825–1863): German physicist and mathematician in Bonn; he published his famous book *Einleitung in die höhere Optik* [Introduction in Higher Optics].

Benedict, Francis Gano (1870–1957): American nutritionist (director of the nutrition institute in Boston): who developed a calorimeter and a spirometer used to determine oxygen consumption and measure metabolic rate; known for his precise CO_2 and O_2 measurements.

Bergmann, Torbern Olof (1735–1784): Swedish chemist and mineralogist in Uppsala; he greatly contributed to the advancement of quantitative analysis, and he developed a mineral classification scheme based on chemical characteristics and appearance; finally he is noted for his sponsorship of Carl Wilhelm Scheele.

Berthelot, Marcellin (or Marcelin) Pierre Eugène (1827–1907): French chemist and politician noted for the Thomsen-Berthelot principle of thermochemistry; he synthesized many organic compounds from inorganic substances and disproved the theory of vitalism.

Berthollet, Claude Louis (1748−1822): Savoyard-French chemist; he was one of the first chemists to recognize the characteristics of a reverse reaction, and hence, chemical equilibrium; along with Antoine Lavoisier and others, he devised a chemical nomenclature.

Berzelius, Jöns Jacob (1779−1848): Swedish chemist in Uppsala and Stockholm; discovered the elements Si, Se, Th and Ce; in discovering that atomic weights are not integer multiples of the weight of hydrogen, Berzelius also disproved Prout's hypothesis that elements are built up from atoms of hydrogen.

Black, Joseph (1728−1799): Scottish chemist and physician (born in France). He was professor of chemistry at Glasgow (1756−66) and from 1766 at Edinburgh; he is best known for his theories of latent heat and specific heat.

Blackman, Frederick Frost (1866−1947): British plant physiologist in Cambridge; he studied plant photosynthesis and proposed the Law of Limiting Factors.

Blagden, Sir Charles Brian (1748−1820): British physician and scientist in Cambridge.

Bodenstein, Max Ernst August (1871−1942): German physical chemist in Göttingen, Leipzig, Hannover and Berlin; he was first to postulate a chain reaction mechanism.

Bolin, Bert Rickard Johannes (1925−2007): Swedish meteorologist and professor at Stockholm University and involved in international climate research cooperation from the 1960s; first chairman of the IPCC, from 1988 to 1998.

Boltzmann, Ludwig Eduard (1844−1906): Austrian physicist in Graz and Vienna; famous for his founding contributions in the fields of statistical mechanics and statistical thermodynamics.

Boussingault, Jean Baptiste Joseph Dieudonne (1802−1887): French agricultural chemist; professor in Lyon and Paris who contributes much to understanding of nitrogen in nature.

Boyle, Robert (1627−1691): Irish-British natural philosopher, chemist, physicist, and inventor in Oxford and London; despite he was an alchemist and believing the transmutation, for him chemistry was the science of the composition of substances, not merely an adjunct to the arts of the alchemist.

Brønsted, Johannes Nicolaus (1879−1947): Danish physical chemist; in 1923 he introduced the protonic theory of acid-base reactions, simultaneously with the English chemist Thomas Martin Lowry.

Bunsen, Robert Wilhelm Eberhard (1811−1899): German chemist, professor in Marburg, Kassel, Breslau and Heidelberg; he investigated emission spectra of heated elements, and with Gustav Kirchhoff he discovered the elements Cs and Ru; developed several gas-analytical methods and was pioneering in photochemistry.

Calvin, Melvin (1911−1997): Born of Russian emigrant parents in Minnesota. University of California at Berkeley in 1937, Lawrence Radiation Laboratory since 1946 and full professor since 1947; Nobel Prize in Chemistry 1961.

Castelli, Benedetto, born Antonio Castelli (1578−1643): Italian mathematician in Pisa (replacing Galileo), and later in Rome; one of his students was Evangelista Torricelli.

Cauer, Hans (1899−1962): German chemist, who worked in the fields of balneology, indoor pollution and precipitation chemistry and first introduced the term *Atmosphärische Chemie*.

Cavendish, Henry (1731−1810): British chemist and physicist in London; he is considered to be one of the so-called pneumatic chemists of the eighteenth and nineteenth centuries, along with, for example, Joseph Priestley, Joseph Black, and Daniel Rutherford; by combining metals with strong acids, Cavendish made hydrogen (H_2) gas, which he isolated and studied.

Callendar, Guy Stewart (1898−1964): English steam engineer and inventor; known for propounding the theory that linked rising carbon dioxide concentrations in the atmosphere to global temperature, but he thought this warming would be beneficial.

Chamberlain, Thomas Chrowder (1843−1928): American geologist and educator in Wisconsin and from 1892 professor in Chicago; founded the *Journal of Geology*.

Charles, Jacques Alexandre César (1746–1823): French inventor, scientist, mathematician and balloonist.

Clapeyron, Benoît Paul Émile (1799–1864): French mathematician and engineer, graduated from École Polytechnique; he presented the Carnot cycle in mathematical formulation (Sadi Carnot was unknown before).

Clapham, Arthur Roy (1904–1990): Was a British botanist and worked at Rothamsted Experimental Station as a crop physiologist (1928–30), professor of botany at Sheffield University 1944–1969.

Clarke, Frank Wigglesworth (1847–1931): American chemist; he is credited with having determined composition of the earth's crust.

Clausius, Rudolf Julius Emmanuel (1822–1888): German physicist and mathematician, graduated from the university of Berlin, professor in Zürich, Würzburg and Bonn.

Cohen, Julius Berend (1859–1935): Professor for organic chemistry in Leeds.

Croll, James (1821–1890): Scottish natural philosopher; known for his several books on climate, cosmology and evolution.

Crutzen, Paul Jozef (born 1933): Dutch Nobel prize winning atmospheric chemist; worked at the Stockholm University, University of California, Georgia Institute of Technology and 1980–2000 director of the Department of Atmospheric Chemistry (MPI) in Mainz (successor of Christian Junge).

Dalton, John (1766–1844): British natural scientist and teacher (chemist, meteorologist and physicist); best known for his pioneering work in the development of modern atomic theory.

Dansgaard, Willi (1922–2011): Danish paleoclimatologist; Professor of Geophysics at the University of Copenhagen.

Descartes, René, also known as Renatus Cartesius (1596–1650): French philosopher, mathematician, physicist, and writer who spent most of his adult life in the Dutch Republic. He has been dubbed the "Father of Modern Philosophy".

Dobson, Gordon Miller Bourne (1889–1976): British physicist and meteorologist; He built the first ozone spectrophotometer and studied the results over many years (now called Dobson spectrometer or Dobsonmeter giving the Dobson unit).

Dorno, Carl Wilhelm Max (1865–1942): German physicist and founder (1906) of the Physical-Meteorological Observatory Davos (Switzerland); father of modern bioclimatology.

Drebbel, Cornelius Jacobszoon (1572–1633): Dutch inventor; since about 1604 in London; he designed and built telescopes and microscopes and he also built the first navigable submarine in 1620.

Emich, Friedrich (1860–1940): Austrian chemist at Technical University Graz; one of the founders of microchemistry.

Empedocles (495–435 BC): Greek pre-Socratic philosopher and a citizen of Agrigentum, a Greek city in Sicily; he is best known for being the origin of the cosmogenic theory of the four classical elements.

Engelmann, Theodore Wilhelm (1843–1909): German botanist. 1897 succeeded Emil du Bois-Reymond at University of Berlin (1818–1876); 1858 professor of physiology at the Physiological Institute.

Engels, Friedrich (1820–1895): Founder, along with Karl Marx (1818–1883) of the dialectical materialism, the philosophy which is not merely a philosophy of history, but a philosophy which illuminates all events whatever, from the falling of a stone to a poet's imaginings (from the preface to *Dialectics of Nature*, written by J. B. S. Haldane in 1939).

Euler, Leonhard (1707–1783): Swiss mathematician, professor in St. Petersburg and Basel.

Evelyn, John (1620–1706): English educated writer and one of the founders of the Royal Society in London (1660).

Faraday, Michael (1791–1867): English chemist and physicist; known for many discoveries and being an excellent experimentalist.

Fahrenheit, Gabriel Daniel (1686–1736): German physicist and engineer but lived most of his life in the Dutch Republic; in 1717, he settled in The Hague with the trade of glassblowing, making barometers, altimeters, and thermometers. From 1718 onwards, he lectured in chemistry in Amsterdam.

Fenton, Henry John Horstman (1854–1929): British chemist at Cambridge in 1878 and was University Lecturer in chemistry from 1904 to 1924.

Franklin, Benjamin (1706–1790): American scientist, inventor, writer and statesman; Founding Father of the United States of America.

Fourier, Jean Baptiste Joseph (1768–1830): French mathematician and physicist in Paris (1816–1822 in England), as Permanent Secretary of the French Academy of Sciences; he is also credited with the discovery in 1824 that gases in the atmosphere might increase the surface temperature of the earth.

Fourcroy, Antoine François, comte de (1755–1809): French chemist; he collaborated with Lavoisier, Guyton de Morveau, and Claude Berthollet on the *Méthode de nomenclature chimique.*

Fresenius, Carl Remigius (1818–1897): German chemist in Wiesbaden, known for his studies in analytical chemistry.

Galilei, Galileo (1564–1642): Italian physicist, mathematician, astronomer, and philosopher; his achievements include improvements to the telescope and consequent astronomical observations, and support for Copernicanism. Galileo has been called the "father of modern observational astronomy".

Gautier, Armand Émile Justin (1837–1920): Professor for Chemistry at the university in Paris

Gay-Lussac, Joseph Louis (1778–1850): French chemist and physicist in Paris (from 1808 to 1832 at Sorbonne); he is known mostly for two laws related to gases.

Gehler, Johann Samuel Traugott (1751–1795): German physicist, known for his "Physicalisches Wörterbuch" (Physical Dictionary) in five volumes (1787–1795), enlarged by Heinrich Wilhelm Brandes and others (Gmelin, Horner, Muncke, Pfaff) to 24 Vol. (1825–1845), the world largest lexicon on natural sciences and comprising the whole knowledge in that time.

Gibbs, Josiah Willard (1839–1903): Professor for theoretical physics at Yale University, New Haven (USA).

Gmelin, Leopold (1788–1853): German chemist in Heidelberg; wrote the *Handbuch der Chemie* (first edition 1817–1819).

Gold, Thomas (1920–2004): Austrian-born astrophysicist, a professor of astronomy at Cornell University: he was one of the most prominent exponents postulating a theory on the abiogenic formation of fossil fuels.

Gorup-Besanez, Eugen Freiherr von (1817–1878): Austrian-German chemist at University Erlangen.

Graham, Thomas (1805–1869): Scottish chemist in London who is best-remembered today for his pioneering work in dialysis and the diffusion of gases.

Guericke, Otto von (1602–1686): German scientist, inventor, and politician; his major scientific achievement was the establishment of the physics of vacuums; he served as the mayor of Magdeburg from 1646 to 1676.

Haldane, John Burdon Sanderson (1892–1964): British geneticist and evolutionary biologist in Cambridge and London; he was one of the founders (along with Ronald Fisher and Sewall Wright) of population genetics.

Hales, Stephen (1677–1761): English physiologist, chemist and inventor in Cambridge; he studied the role of air and water in the maintenance of both plant and animal life

Hammett, Louis Plack (1894–1987): American physical chemist, Columbia University New York.

Hann, Julius Ferdinand von (1839–1921): Austrian meteorologist and climatologist; 1877–1897 director of the *Zentralanstalt für Meteorologie* (Central Institute for Meteorology) in Vienna.

Hasselbalch, Karl Albert (1874–1962): Danish physician and chemist.

Helmholtz, Hermann Ludwig Ferdinand von (1821–1894): German physician and physicist in Königsberg, Bonn, Heidelberg and Berlin, who made significant contributions to several widely varied areas of modern science.

Helmont, Johann Baptist (Jan) van (1577–1644): Flemish chemist, physiologist, and physician; he worked during the years just after Paracelsus and iatrochemistry, and is sometimes considered to be "the founder of pneumatic chemistry".

Hempel, Walther Matthias (1851–1916): German chemist; founder of the chemical institute in Dresden Technical University and professor 1889–1912, scholar of Bunsen and pioneer in gas analysis and electrolysis.

Henderson, Lawrence Joseph (1878–1942): American biochemist: leading biochemists of the first decades of the 20th century.

Henry, William (1775–1836): English chemist in Manchester; founder and afterwards president of the Manchester Literary and Philosophical Society.

Herodotus of Halicarnassus (ca. 484 BC–ca. 425 BC): Greek historian who lived in the 5th century BC.

Hill, Robert (1899–1991): British biochemist at the Cambridge University.

Hippocrates of Cos or Hippokrates of Kos (ca. 460 BC–ca. 370 BC) Greek physician, referred as "father of medicine".

Högbom, Arvid Gustaf (1857–1940): Swedish professor of mineralogy at Uppsala University

Houzeau, Auguste (1829–1911): French agriculture chemist in Rouen who carried out the first atmospheric O_3 measurements in 1853.

Humboldt, Alexander von (1769–1859): German naturalist and explorer, Prussian minister, philosopher, and linguist; founder of the University of Berlin (1810), which has strongly influenced other European and Western universities.

Humphreys, William Jackson (1862–1949): American meteorological physicist of the U.S. Weather Bureau from 1905 to 1935.

Huntington, Ellsworth (1876–1947): American geographer (at Yale University), known for his studies on climatic determinism, economic growth and economic geography.

Ingenhousz (or Ingen-Housz as well Ingen Housz and IngenHousz) Jan (1730–1799): Dutch physician and plant physiologist working in London since 1753, often travelling and staying in Vienna.

Junge, Christian (1912–1996): German meteorologist and geophysicist in Frankfurt, 1953–1961 he worked at the Cambridge Air Force Research Center, Bedford, Mass. (USA); in 1962 he returned to Germany, first as Professor of Meteorology and Director of the Meteorological Institute at the University of Mainz (1962–1968), then as Director of the Air Chemistry Department of the *Max-Planck-Institut für Chemie* (*Otto-Hahn-Institut*) from which he retired in 1977; he is one of the founder of Atmospheric Chemistry.

Kamen, Martin David (1913–2002): American physical chemist and biochemist at Universities of Berkeley, Massachusetts and San Diego.

Keeling, Charles David (1928–2005): American chemist, most known for his CO_2 measurement at Mauna Loa, Hawaii, affiliated with Scripps Institution of Oceanography, University of California, San Diego, from 1956 until his death in 2005.

Kernbaum, Mirosław (1882–1911): Polish physicist in radiation chemistry, worked at the Marie Curie laboratory in Paris

Kirchhoff, Gustav Robert (1824–1887): German physicist in Berlin, Breslau and Heidelberg, who contributed to the fundamental understanding of electrical circuits, spectroscopy, and the emission of black-body radiation by heated objects.

Köhler, Hilding (1888–1982): Swedish theoretical meteorologist at University Uppsala (also written *Koehler* but *Kohler* is wrong)

Krogh, August (1874–1949): Danish physiologist; discovered how capillaries regulate oxygen and was awarded with the Nobel Prize in Physiology or Medicine in 1920.

Kudryavtsev, Nikolai Alexandrovich – Кудрявцев, Николай Александрович (1893–1971): Russian geologist, the founder of the modern theories of abiogenic petroleum formation. 1941 professor in Leningrad.

Lagrange, Joseph-Louis (1735–1813): French mathematician in Torino and Paris.

Lakatos, Imre; birth name: Lipschitz (1922–1974): Hungarian mathematician, physicist and philosopher at the University of Debrecen, Hungary. In 1953 he fled to Vienna and finally to London to study at the University of Cambridge for a doctorate in philosophy. In 1960 Lakatos was appointed to the London School of Economics. Many works appeared after his death in editions.

Lambert, Johann Heinrich (1728–1777): Swiss-born mathematician, physicist and astronomer in Augsburg and Berlin; he studied light intensity and also developed a theory of the generation of the universe that was similar to the nebular hypothesis that Thomas Wright and Immanuel Kant had (independently) developed.

Lampadius, Wilhelm August Eberhard (1772–1842): German chemist and professor in Freiberg (Saxonia), founder of modern metallurgy; well-known with Humboldt and Goethe.

Langevin, Paul (1872–1946): French physicist in Paris; in 1926 he became director of the *École de Physique et Chimie.*

Laplace, Pierre-Simon (Marquis) de (1749–1827): French mathematician and astronomer whose work was pivotal to the development of mathematical astronomy and statistics.

Lavoisier, Antoine-Laurent de (1743–1794): French chemist ("father of modern chemistry"); 1764 onward at the Academy of Sciences, 1775 commissioner of the Royal Gunpowder Administration and residence in the Paris Arsenal; opening of a laboratory in the Arsenal; 1785–1789 collaboration with Claude Berthollet, Antoine de Fourcroy and Guyton de Morveau; 1789 co-editor of the newly founded *Annales de chimie.* Wrote 1789 the revolutionary book „*Traité élémentaire de chimie*". 8 May 1794 executed on the guillotine.

Lawes, John Bennet, Sir (1814–1900): English entrepreneur and agricultural scientist; he founded an experimental farm at Rothamsted, where he developed a superphosphate that would mark the beginnings of the chemical fertilizer industry.

Lémery, Nicolas (1645–1715): French chemist, who was one of the first to develop theories on acid-base chemistry.

Le Roy, Édouard Louis Emmanuel Julien (1870–1954): French philosopher and mathematician; became in 1909 professor of mathematics at the Lycée Saint-Louis in Paris.

Leibniz, Gottfried Wilhelm (1646–1716): German philosopher, polymath and mathematician who wrote primarily in Latin and French; belong Humboldt the greatest German polymath.

Lessing, Gotthold Ephraim (17291781): German writer, dramatist, publicist, and art critic; one of the most prominent philosophers of the Enlightenment era.

Liebig, Justus von (1803–1873): German chemist, Professor in Gießen 1824–1852, founder of the agricultural-chemical theory and together with Wöhler the radical theory.

Loewy, Adolf (1862–1937): German physiologist; in 1921 he was a professor and in charge of the *Schweizerisches Institut für Hochgebirgsphysiologie und Tuberkuloseforschung* (Swiss Institute for Altitude Physiology and Tuberculosis) at Davos.

Loschmidt, Johann Josef (1821–1895): Austrian professor of physical chemistry at the University of Vienna in 1868.

Lovelock, James Ephraim (born 1919): English chemist; He is known for proposing the Gaia hypothesis, in which he postulates that the earth functions as a kind of superorganism.

Lyell, Sir Charles (1797–1875): Scottish geologist. He was the foremost geologist of his day, and an influence on the young Charles Darwin. Professor of Geology at King's College London in the 1830s and President of the Geological Society of London 1835–1837.

Marggraf, Andreas Sigismund (1709–1782): German chemist and pioneer of analytical chemistry from Berlin; in 1747 he announced his discovery of sugar in beets and devised a method using alcohol to extract it.

Margulis, Lynn (1938–2011): American biologist; she is best known for her theory on the origin of eukaryotic organelles, and her contributions to the endosymbiotic theory.

Marum, Martinus van (1750–1837): Dutch chemist.

Marx, Karl (1818–1883): German philosopher and economist; fundamentally, he assumed that human nature involves transforming nature. To this process of transformation he applies the term "labour", and to the capacity to transform nature the term "labour power".

Matthaei, Gabrielle Louise Caroline (1876–1930): British botanist, assistant of Blackman at the Cambridge University.

Mayer, Julius Robert von (1814–1878): German physician and physicist in Heilbronn and one of the founders of thermodynamics. He was the first person to develop the law of the conservation of energy (first law of thermodynamics).

Mayow, John (1643–1679): English chemist, physician, and physiologist who is remembered today for conducting early research into respiration and the nature of air.

Meissner, Georg (1829–1905): German anatomist and physiologist; university professor at Basel (from 1855), Freiburg (from 1857) and Göttingen (from 1860 to 1901).

Mendeleev, Dimitri Ivanovich (1834–1907): Russian chemist in Saint Petersburg; develops the modern periodic table.

Mie, Gustav Adolf Feodor Wilhelm Ludwig (1868–1957): German physicist. Studied in Rostock and Göttingen; Professor in Greifswald (1902), Halle (1917) and Freiburg (1924)

Milanković, Milutin (1879–1958): Serbian civil engineer and geophysicist, best known for his theory of ice ages, relating variations of the earth's orbit and long-term climate change, now known as Milankovitch cycles.

Miller, Stanley Lloyd (1930–2007): US biologist and chemist.

Morveau, Louis Bernard Guyton de (1737–1816): French chemist and politician; he is credited with producing the first systematic method of chemical nomenclature.

Muntz, Achille Charles (1846–1917): French agricultural chemist (student of Boussingault), 1876 professor and director of the "l'Institut National Agronomique" in Paris, 18976 member of the Academy of Sciences.

Nägeli, Carl Wilhem von (1817–1891): German botanist in Munic.

Nernst, Walther Hermann (1864–1941): German physical chemist and physicist in Leipzig, Göttingen and Berlin; Nobel Prize in 1920.

Newton, Isaac (1643–1727): English physicist, mathematician, astronomer, natural philosopher, alchemist, and theologian who is perceived and considered by a substantial number of scholars and the general public as one of the most influential men in history.

Odling, William (1829–1921): English chemist who contributed to the development of the periodic table.

Oeschger, Hans (1927–1998): Swiss geologist; founder of the Division of Climate and Environmental Physics at the Physics Institute of the University of Bern in 1963 and director until his retirement in 1992; pioneer and leader in ice core research.

Osann, Gottfried Wilhelm (1797–1866): German chemist in Dorpat, Erlangen, Jena and Würzburg.

Ostwald, Carl Wilhelm Wolfgang (1883–1943): Founder of colloid chemistry in Germany and professor in Leipzig; sun of Wilhelm Ostwald.

Oparin, Aleksander Ivanovich (1894–1980): Soviet biochemist in Moscow; notable for his contributions to the theory of the origin of life; he developed the foundations for industrial biochemistry.

Paracelsus; Philippus Aureolus Theophrastus Bombastus von Hohenheim (1493−1541): Swiss-German Renaissance physician, botanist, alchemist, astrologer and general occultist.

Parmenides of Elea (about 540−480 BC): Greek philosopher; founder of the Eleatic school, one of the most significant of the pre-Socratic philosophers.

Pascal, Blaise (1623−1662): French mathematician, physicist, and religious philosopher; he studied fluids, and clarified the concepts of pressure and vacuum by generalizing the work of Evangelista Torricelli.

Pasteur, Louis (1822−1895): French chemist and microbiologist; He is regarded as one of the three main founders of microbiology, together with Ferdinand Cohn and Robert Koch.

Pauling, Linus Carl (1901−1994): American chemist and peace activist; was among the first scientists to work in the fields of quantum chemistry, molecular biology, and orthomolecular medicine. He is one of only two people to have been awarded a Nobel Prize in two different fields (the Chemistry and Peace prizes), the other being Marie Curie (the Chemistry and Physics prizes), and the only person to have been awarded each of his prizes without sharing it with another recipient.

Pettenkofer, Max Joseph von (1818−1901): Bavarian chemist and hygienist; in 1865 he became also professor of hygiene in Munic.

Planck, Max (1858−1947): German physicist; he is considered to be the founder of the quantum theory, and thus one of the most important physicists of the twentieth century and was awarded the Nobel Prize in Physics in 1918.

Plass, Gilbert Norman (1920−2004): Canadian physicist from Harvard and Princeton University, later professor at Texas A&M University.

Pflüger, Eduard Friedrich Wilhelm (1829−1910): German physiologist in Berlin (1851) and Bonn (1859).

Priestley, Joseph (1733−1804): English scientist; due to his sympathy with the French revolution he moved to Philadelphia / USA in 1794; discovered oxygen and many gases in air (parallel to Scheele).

Prout, William (1785−1850): English chemist, physician, and natural theologian; in 1815 he hypothesized that the atomic weight of every element is an integer multiple of that of hydrogen, suggesting that the hydrogen atom is the only truly fundamental particle.

Pythagoras of Samos (about 540−500 BC): Ionian Greek mathematician. He was the first man to call himself a philosopher. Pythagoras and his students believed that everything was related to mathematics and that numbers were the ultimate reality.

Ramsay, Sir William (1852−1916): Scottish chemist who discovered the noble gases and received the Nobel Prize in Chemistry in 1904.

Raoult, François-Marie (1830−1901): French chemist in Grenoble who conducted research into the behavior of solutions, especially their physical properties.

Rayleigh, Lord (John William Strutt, 1842−1919): British mathematician and physicist. Professor in Cambridge (1879) and London, Nobel prize in 1904 (together with Ramsay).

Reslhuber, P. Augustin (1808−1875): Austrian, director of the observatory Kremsmünster (Austria)

Ruben Samuel − born Charles Rubenstein (1913−1943): US chemist at the University of California, Berkeley. Together with Kamen he discovered carbon-14.

Russell, Francis Albert Rollo (1849−1914): British meteorologist.

Rutherford, Daniel (1749−1819): Scottish chemist and physician who is most famous for the isolation of nitrogen in 1772.

Sabatier, Paul (1854−1941): French chemist; he greatly facilitated the industrial use of hydrogenation. Nobel Prize in 1912.

Sachs, Julius von (1832−1897): German botanist; 1856 Privatdozent for plant physiology in Prague, assistant to the Agricultural Academy at Tharandt in Saxony, 1861 Professor of

Botany in the University of Freiburg in Breisgau. 1868 he took over the chair in Botany at the University of Würzburg [Wursburg].

Santorio, called Sanctorius of Padu (1561−1636): Italian physiologist, physician in Padua where he performed experiments in temperature, respiration and weight.

Saussure, Horace-Bénédict de (1740−1799): Swiss aristocrat, physicist and Alpine traveller; he directed his attention to the geology and physics of that region; he made experiments with various forms of hygrometer in all climates and at all temperatures.

Saussure, Nicolas-Théodore de (1767−1845): Swiss botanist and naturalist from Geneve who made seminal advances in phytochemistry; oldest son of Horace-Bénédict de Saussure.

Scheele, Carl Wilhelm (1742−1786): German chemist, born in Stralsund/Pomerania (Sweden in that time), druggist in Gothenburg, Malmö and Stockholm, member of the Royal Academy of Sweden (1775), discovered oxygen in air.

Schloesing, Jean Jacques Théophile (1824−1919): French soil scientist who (with A. Muntz) proved (1877) that nitrification is a biological process in the soil by using chloroform vapors to inhibit the production of nitrate.

Schmaus, August (1877−1952): German meteorologist and Professor in Munich.

Schönbein, Christian Friedrich (1799−1868): Swiss chemist in Basel; best known for discovery of ozone.

Schöne, Emil Hermann (1838−1896) − Шене, Эмиль Богданович: German chemist, born in Halberstadt, studied in Halle, Berlin and Göttingen; changes to Moscow in 1863 and worked since 1864 at the Petrowskaja Agricultural and Forest Academy (Петровская земледельческая и лесная академия), founded in 1865, where he became in 1875 a professor for chemistry. He deceased in Moscow. Member of the Deutsche Chemische Gesellschaft (German Chemical Society).

Schwabe, Samuel Heinrich (1789−1875): German astronomer, discovered the solar cycle through extended observations of sunspots. Schwabe's observations were afterwards utilized in 1851 by Alexander von Humboldt in the third volume of his Kosmos.

Scoutetten, Robert Joseph Henri (1799−1871): French professor and head of medicine at the military hospital in Metz.

Senebier, Jean (1742−1809): Swiss botanist and naturalist, priest in Geneve.

Smith, Robert Angus (1817−1884): Scottish chemist, who investigated numerous environmental issues; he became in 1863 Queen Victoria's first alkali inspector (*Alkali Acts Administration*, setting limits for HCl emission by alkali plants).

Snell, Willebrord van Roijen; also called Willebrord Snelius (1580−1626): Dutch astronomer and mathematician in Leiden.

Sørensen, Søren Peter Lauritz (1868−1939): Danish biochemist in Copenhagen; suggested a convenient way of expressing acidity − the negative logarithm of hydrogen ion concentration (pH).

Sprengel, Philipp Carl (1787−1859): German agricultural chemist, first in Göttingen, in 1831 he become a college professor of agronomy and forestry (at the *Collegium Carolinum*) in Braunschweig, but in 1839 he moved and in 1842 in Regenwalde (Pommern) he founded his private academy.

Stahl, Georg Ernst (1659−1734): German chemist and physician in Weimar, Halle and Berlin; known for his obsolete phlogiston theory.

Stein, Gabriel (1920−1976): Radiation chemist, born in Budapest in 1920 and emigrated to Palestine in 1938; Professor of Physical Chemistry at the Hebrew University

Stoklasa, Julius (1857−1936): Czech agricultural chemist in Leipzig, Vienna and Prague.

Stumm, Werner (1924−1999): Swiss water chemist; he directed the Swiss Federal Institute for Water Resources and Water Pollution from 1970 to 1992.

Struve, Heinrich Wilhelm von, also Струве, Генрих Васильевич (1822−1908): German-Russian chemist in Sankt Petersburg; son of Friedrich Georg Wilhelm von Struve (In Russian, his name is often given as Vasilii Yakovlevich Struve (Василий Яковлевич

Струве); the Struve family were a dynasty of five generations of astronomers from the 18th to 20th centuries. Members of the family were also prominent in chemistry, government and diplomacy.

Suess, Eduard (1831–1914): Austrian professor of geology at the University of Vienna since1857.

Tansley, Sir Arthur George (1851–1955): English botanist who was a pioneer in the science of ecology.

Taylor, Hugh Stott (1890–1974): English chemist; worked with Arrhenius and Bodenstein, chair of the Chemistry Department at Princeton (USA) in 1926.

Teilhard de Chardin, Pierre (1881–1955): French philosopher and Jesuit priest. He teached teleological views of evolution.

Thales of Milet (624–546 BC): First known Greek philosopher, scientist and mathematician and is considered by Aristotle to be the founder of Ionic philosophy of nature. He is credited with five theorems of elementary geometry.

Thénard, Louis Jacques (1777–1857): French chemist and professor at École Polytechnique in Paris; he published a textbook, and his *Traité de chimie élémentaire, théorique et pratique* (4 vols., Paris, 1813–16), which served as a standard for a quarter of a century.

Theophrastus (about 372–287 BC): Greek native of Eressos in Lesbos, was the successor of Aristotle in the Peripatetic school. His interests were wide-ranging, extending from biology and physics to ethics and metaphysics.

Thomson, Thomas (1773–1852): Scottish chemist whose writings contributed to the early spread of Dalton's atomic theory.

Thomson, William; Lord Kelvin of Lards (1824–1907): Belfast-born mathematical physicist and engineer.

Tissandier, Gaston (1843–1899): French chemist, meteorologist, aviator (balloon rises) and editor in Paris.

Torricelli, Evangilista (1608–1647): Italian physicist and mathematician in Pisa; best known for his invention of the barometer.

Tyndall, John (1820–1893): Irish physicist known for his first explanation of atmospheric heat in terms of the capacities of various gases to absorb or transmit radiative heat.

Urey, Harold Clayton (1893–1981): US chemist, Nobel Prize in 1934 for the discovery of deuterium and known for the idea of "inorganic life formation" (Miller-Urey experiment)

van't Hoff, Jacobus Hendricus (1852–1911): Dutch physico-chemist in Amsterdam and Berlin; Nobel price in 1901.

Vernadsky, Vladimir Ivanovich / Вернадский, Владимир Иванович (1863–1945): Russian / Ukrainian geologist. His research ranged from meteorites and cosmic dust to microbiology and migration of microelements via living organisms in ecosystems.

Viviani, Vincenzo (1622–1703) Italian mathematician and scientist; he was a pupil of Torricelli and a disciple of Galileo.

Voeux, Harold (Henry) Antoine des (no birth / death data are available): French physician living in London, honourable treasurer of the Coal Smoke Abatement Society (formed in 1882) and later President of the National Smoke Abatement Society (Marsh 1947).

Wallerius, Johann Gottschalk (1709–1785): Swedish physician and professor of chemistry, mineralogy and pharmacy at the University of Uppsala, Sweden. He is regarded as the founder of agricultural chemistry.

Warburg, Otto Heinrich (1883–1970): Son of physicist Emil Warburg, was a German physiologist, medical doctor and Nobel laureate. In 1918 he was appointed professor at the Kaiser Wilhelm Institute for Biology in Berlin-Dahlem.

Wells, William Charles (1757–1817): born in South Carolina (USA) as son of Scottish immigrants, became physician, philosopher and printer.

Wöhler, Friedrich (1800–1882): German chemist in Göttingen; he is regarded as a pioneer in organic chemistry as a result of him (accidentally) synthesizing urea but also the first to isolate several chemical elements (Al, Yt, Be, Ti).

Weinschenk, Ernst (1865−1921): German pioneer of microscopy and petrography in Munich.

Wien, Wilhelm Carl Werner Otto Fritz Franz (1864−1928): German physicist in Munic; Nobel Prize in physics 1911

Wolf, Rudolf (1816−1893): Swiss astronomer, professor of astronomy in Bern (1844) and Zürich (1855), director of the Bern Observatory in 1847.

Woodward, John (1665−1728): English naturalist and geologist at Gresham College in London.

Xenophanes from Kolophon (about 570−480 BC): Greek philosopher, poet, and social and religious critic.

Zel'dovich, Yakov Borisovich / Зельдович, Яков Борисович (1914−1987): Soviet physicist at the Academy of Sciences and belong the Moscow State University.

References

Abe, T., M. Tanizawa, K. Watanabe and A. Taguchi (2009) CO$_2$ methanation property of Ru nanoparticle-loaded TiO$_2$ prepared by a polygonal barrel-sputtering method. *Energy and Environmental Science* **2**, 315−332.

Acker, K., W. Wieprecht, D. Möller, G. Mauersberger, S. Naumann and A. Oestreich (1995) Evidence of ozone destruction in clouds. *Naturwissenschaften* **82**, 86−89.

Acker, K., D. Möller, W. Wieprecht and S. Naumann (1996) Mt. Brocken, a site for a cloud chemistry programme in Central Europe. *Water, Air and Soil Pollution* **85**, 1979−1984.

Acker, K., D. Möller, W. Wieprecht, D. Kalaß und R. Auel (1998) Investigation of ground-based clouds at Mt. Brocken. *Fresenius Zeitschrift für Analytische Chemie* **361**, 59−64.

Acker, K., D. Möller, W. Wieprecht, R. Auel and D. Kalaß (1999a) The Mt. Brocken air and cloud water chemistry program and the QA/QC of its data. In: Workshop on quality assurance of EMEP measurements and data reporting. *EMEP/CCC-Report* 1/99, 161−165.

Acker, K., W. Wieprecht, D. Möller, R. Auel, D. Kalaß, A. Zwozdziak, G. Zwozdziak and G. Kmiec (1999b) Results of cloud water and air chemistry measurements at Mt. Szrenicain Poland. In: *Proc.* of EUROTRAC-2 Symposium '98, Vol 1: Transport and Chemical Transformation in the Troposphere (Eds. P. M. Borrell, P. Borrell), WITPress Comp. Mechan. Publ., Southampton, pp. 448−452.

Acker, K., W. Wieprecht, R. Auel, D. Kalaß and D. Möller (2000) Evidence for heterogeneous formation of nitrous acid on cloud droplets. *Journal of Aerosol Science* **31**, S1, S352−S353.

Acker, K., D. Möller, W. Wieprecht, R. Auel, D. Kalaß and W. Tscherwenka (2001) Nitrous and nitric acid measurements inside and outside of clouds at Mt. Brocken. *Journal of Water, Air and Soil Pollution* **130**, 331−336.

Acker, K., W Wieprecht and R. Auel R. (2001) Coupling of wet denuder and ion chromatography technique for sensitive measurements of nitrous and nitric acid in Melpitz. *Journal of Aerosol Science* **32**, S1, S523−S524.

Acker, K., D. Möller, W. Wieprecht, R. Auel, D. Kalaß and W. Tscherwenka. (2001) Nitrous and nitric acid measurements inside and outside of clouds at Mt. Brocken. *Journal of Water, Air and Soil Pollution* **130**, 331−336.

Acker, K., S. Mertes, D. Möller, W. Wieprecht, R. Auel and D. Kalaß (2002) Case study of cloud physical and chemical processes in low clouds at Mt. Brocken, Germany. *Atmospheric Research* **64**, 41−51.

Acker, K., W. Wieprecht and D. Möller (2003) Chemical and physical characterisation of low clouds: results from the ground based cloud experiment FEBUKO. *Archives of industrial hygiene and toxicology* **54**, 231−238.

Acker, K., G. Spindler and E. Brüggemann (2004) Nitrous and nitric acid measurements during the INTERCOMP2000 campaign in Melpitz. *Atmospheric Environment* **38**, 6497−6505.

Acker, K., D. Möller, R. Auel, W. Wieprecht and D. Kalaß (2005) Concentrations of nitrous acid, nitric acid, nitrite and nitrate in the gas and aerosol phase at a site in the emission zone during ESCOMPTE 2001 experiment. *Atmospheric Research* **74**, 507−524.

Acker, K., D. Möller, W. Wieprecht, F. X. Meixner, B. Bohn, S. Gilge, C. Plass-Dülmer and H. Berresheim (2005) Strong daytime production of OH from HNO$_2$ at a rural mountain site. *Geophysical Research Letters* **33**, L02809, doi:10.1029/2005GL024643.

Acker, K., A. Febo, S. Trick, C. Perrino, P. Bruno, P. Wiesen, D. Möller, W. Wieprecht, R. Auel, M. Giusto, A. Geyer, U. Platt and I. Allegrini (2006) Nitrous acid in the urban area of Rome. *Atmospheric Environment* **40**, 3123−3133.

Acker, K. and D. Möller (2007) Atmospheric variation of nitrous acid at different sites in Europe. *Environmental Chemistry* **4**, 242–255.

Acker, K., Beysens D. and D. Möller (2007) Nitrite in dew, fog, cloud and rain water – indicator for heterogeneous processes on surfaces. *Atmospheric Research* **87**, 200–212.

Ackerman, S. A. and J. A. Knox (2006) Meteorology: Understanding the atmosphere. 2nd edition. Brooks/Cole Pub Co, 467 pp.

Addamo, M., V. Augugliaro, E. García-López, V. Loddo and G. Marcì (2005) Oxidation of oxalate ion in aqueous suspensions of TiO_2 by photocatalysis and ozonation. *Catalysis Today* **107–108**, 612–618.

Adelung, J. C. (1796) Grammatisch-kritisches Wörterbuch der Hochdeutschen Mundart, Bd. 2, Leipzig (in 4 Vol., 7690 pp.).

Adewuyi, Y. G. (1989) Oxidation of biogenic sulfur compounds in aqueous media: Kinetics and environmental implications. In: Biogenic sulfur in the environment. Chapter 34, American Chemical Society, Symposium Series, Vol. 393, pp. 529–559.

Adhémar, J.-A. (1842) Révolutions de la mer – déluges périodiques. A. Lacroux, Paris, 358 pp.

Adrian, G. and F. Fiedler (1991) Simulation of unstationary wind and temperature fields over complex terrain and comparison with observations. *Beiträge zur Physik der Atmosphäre* **64**, 27–48.

Ahrens, C. D. (2007) Essentials of meteorology: An invitation to the atmosphere. 5th edition. Brooks/Cole Pub. Co (USA), 485 pp.

Aitken, J. (1880) On dust, fog and clouds. *Nature* **23**, 34–35.

Aiuppa, A., C. Federico, A. Franco, G. Giudice, S. Gurrieri, S. Inguaggiato, M. Liuzzo, A. J. S. McGonigle and M. Valenza (2005) Emission of bromine and iodine from Mount Etna volcano. *Geochemical Geophysical Geosystems* **6**, Q08008, doi:10.1029/2005GC000965.

Akimoto, H., K. Takagi and F. Sakamaki (1987) Photoenhancement of the nitrous acid formation in the surface reaction of nitrogen dioxide and water vapor: Extra radical source in smog chamber experiments. *International Journal of Chemical Kinetics* **19**, 539–551.

Allen, A. G., C. Oppenheimer, M. Ferm, P. J. Baxter, L. A. Horrocks, B. Galle, A. J. S. McGonigle and H. J. Duffell (2002) Primary sulfate aerosol and associated emissions from Masaya Volcano, Nicaragua. *Journal of Geophysical Research* **107**, 4682, doi:10.1029/2002JD002120.

Allen, A. O. (1961) The radiation chemistry of water and aqueous solutions. Van Nostrand, New York, 204 pp.

Allen, P. A. and J. L. Etienne (2008) Sedimentary challenge to Snowball Earth. *Nature Geoscience* **1**, 817–825.

Almqvist, E. (1974) An analysis of global air pollution. *Ambio* **3**, 161–167.

Alt, E. (1916) Die Physik des Klimas. In: Handbuch der Balneologie, medizinischen Klimatologie und Balneographie, Band 1 (Eds. E. Dietrich and S. Kaminer), Georg Thieme, Leipzig, pp. 423–503.

Altman, I. S., I. E. Agranovskii, M. Choi and V. A. Zagainov (2008) To the theory of homogeneous nucleation: Cluster energy. *Russian Journal of Physical Chemistry A, Focus on Chemistry* **82**, 2097–2102.

Altshuller, A. P. (1983) Review: Natural volatile organic substances and their effect on air quality in the United States. *Atmospheric Environoinment* **17**, 2131–2165.

Altshuller, A. P. (1993) Production of aldehydes as primary emissions and from secondary atmospheric reactions of alkenes and alkanes during the night and early morning hours. *Atmospheric Environment* **27**, 21–31.

Ambrose, S. H. (1998) Late Pleistocene human population bottlenecks, volcanic winter, and differentiation of modern humans. *Journal of Human Evolution* **34**, 623–651.

Amedie, S. A. (2001) Potential use of fog as an alterative water resource in the dry and semi-arid mountain chains of northern and eastern Ethiopia. Proceedings of the 2nd Conference on Fog and Fog Collection, St-John's, Canada, pp. 215–218.

Ammann, M., M. Kalberer, D. T. Jost, L. Tobler, E. Rössler, D. Piguet, H. W. Gäggeler and U. Baltensperger (1998) Heterogeneous production of nitrous acid on soot in polluted air masses. *Nature* **395**, 157–160.

Amorim, J., G. Baravian and G. Sultan (1996) Absolute density measurements of ammonia synthetized in N_2–H_2 mixture discharges. *Applied Physics Letters* 68, 1915–1917.

Amthor, J. S. and members of the Ecosystems Working Group (1998) Terrestrial Ecosystem Responses to Global Change: a research strategy. ORNL Technical Memorandum 1998/27, Oak Ridge National Laboratory, Oak Ridge, Tennessee. 37 pp. http://daac.ornl.gov/NPP/other_files/worldnpp1.txt.

Amundson, R. (2001) The soil carbon cycle. *Annual Reviews of Earth and Planetary Sciences* **29**, 535–562.

Anastas, P. T. and J. C. Warner (1998) Green chemistry: Theory and practice. Oxford University Press, New York, 30 pp.

Anastasio, C., B. C. Faust and C. J. Rao (1997) Aromatic carbonyl compounds as aqueous phase photochemical sources of hydrogen peroxide in acidic sulfate aerosols, fogs, and clouds. 1. Non-phenolic methoxybenzaldehydes and methoxyacetophenones with reductants (phenols). *Environmental Science and Technology* **31**, 218–232.

Anderson, T. R., S. A. Spall, A. Yool, P. Cipollini, P. G. Challenor and M. J. R. Fasham (2001) Global fields of sea surface dimethylsulfide predicted from chlorophyll, nutrients and light. *Journal of Marine Systems* **30**, 1–20.

Andreae, M. O. (1980) Dimethylsulfoxide in marine and freshwaters. *Limnology and Oceanography* **25**, 1054–1063.

Andreae, M. O. (1985) The emission of sulfur to the remote atmosphere: Background paper. In: The biogeochemical cycling of sulfur and nitrogen in the remote atmosphere (Eds. J. N. Galloway, R. J. Charlson, M. O. Andreae and H. Rodhe). Reidel, Dordrecht. pp. 5–25.

Andreae, M. O., R. W. Talbot, T. W. Andreae and R. C. Harriss (1988) Formic and acetic acid over the central Amazon region, Brazil 1. Dry season. *Journal of Geophysical Research* **93**, 1616–1624.

Andreae, M. O. (1990) Ocean-atmosphere interactions in the global biogeochemical sulfur cycle. *Marine Chemistry* **30**, 1–29.

Andreae, M. O. (1991) Biomass burning: Its history, use, and distribution and its impact on environmental quality and global climate. In: Levine (1991) pp. 3–27.

Andreae, M. O. and W. Jaeschke (1992) Exchange of sulphur between biosphere and atmosphere over temperate and tropical regions. In: Sulphur cycling on the continents (Eds. R. W. Howarth, J. W. B. Stewart and M. V. Ivanov). John Wiley & Sons, New York, pp. 27–61.

Andreae, M. O. (1995) Climatic effects of changing atmospheric aerosol levels. In: World survey of climatology. Vol. 16: Future climates of the world (Ed. A. Henderson-Sellers). Elsevier, Amsterdam, pp. 341–392.

Andreae, M. O. and P. J. Crutzen (1997) Atmospheric aerosols: Biogeochemical sources and role in atmospheric chemistry. *Science* **276**, 1052–1056.

Andreae, M. O. and P. Merlot (2001) Emissions of trace gases and aerosols from biomass burning. *Global Biogeochemical Cycles* **15**, 955–966.

Andreae, M. O. (2004) Assessment of global emissions from vegetation fires. *International Forest Fire News* **31**, 112–121.

Andreae, M. O. (2007) Aerosol before pollution. *Science* **315**, 50–51.

Andreae, M. O., D. A. Hegg and U. Baltensperger (2009) Sources and nature of atmospheric aerosols. In: Aerosol pollution impact on precipitation (Z. Levin, and W. R. Cotton, Eds.), Springer, Netherlands, pp. 45–89.

Andreae, M. O., B. Stevens, G. Feingold, S. Fuzzi, M. Kulmala, W. Lau, U. Lohmann, D. Rosenfeld, P. Siebesma, A. Reissell, C. O'Dowd, Th. Ackermann and G. Raga (2009) Aerosols, clouds, precipitation and climate (ACPC). Science plan and implementation strategy. www.gewex.org/ssg-22/ACPC_SciencePlan_FINAL.pdf.

Andres, R. J. and A. D. Kasgnoc (1998) A time-averaged inventory of subaerial volcanic sulfur emissions. *Journal of Geophysical Research* **103**, 25125−25261.

Andrews, D. G. (2000) An introduction to atmospheric physics. Cambridge University Press, Cambridge, U.K., 229 pp.

Andrews, T. (1867) On the identity of the body in the atmosphere which decomposes iodide of potassium with ozone. *Annalen der Physik und Chemie* **135**, 659−660; *Philosophical Magazine* **34**, 315−316.

Anklin, M. and R. C. Bales (1997) Recent increase in H_2O_2 concentration at Summit, Greenland. *Journal of Geophysical Research* **102**, 19099−19104.

ApSimon, H., Ed. (1997) Emission Abatement Strategies and the Environment (EASE), final EU project report, Imperial College, London.

Archard, G., H. D. Eva, P. Mayaux, H. J. Stibig and A. Beward (2004) Improved estimates of net carbon emission from land cover change in the tropics for the 1990s. *Global Biogeochemical Cycles* **18**, GB2008. doi:10.1029/2003GB002142.

Archer, M. D. and J. Barber (2004) Molecular to global photosynthesis. World Scientific Publ. Co., 788 pp.

Archer, S. (2007) Crucial uncertainties in predicting biological control of DMS emission. *Environmental Chemistry* **4**, 404−405.

Ardon, M. (1965) Oxygen. Elementary forms and hydrogen peroxide. W. A. Benjamin, New York, 106 pp.

Aresta, M. and G. Forti, Eds. (1987) Carbon dioxide as a source of carbon: biochemical and chemical use. NATO ASI series. Series C, Mathematical and physical sciences No. 206. D. Reidel Pub. Co, Boston and Dordrecht, 441 pp.

Aresta, M. and M. E. Aresta, Eds. (2003) Carbon dioxide recovery and utilization. Springer-Verlag, Berlin, 384 pp.

Aresta, M., Ed. (2010) Carbon dioxide as chemical feedstock. Wiley-VCH, Weinheim, 394 pp.

Aristoteles (1829) Meteorologica. Ex recensione Immanuelis Bekkeri. Typis Academicis, Berlin, 123 pp.

Aristotle (1923) Meteorologica. Translated into English by E. W. Webster. Clarendon Press, Oxford.

Armstrong, D. A., W. L. Waltz and A. Rauk (2006) Carbonate radical anion − thermochemistry. *Canadian Journal of Chemistry* **84**, 1614−1619.

Armstrong, G. F. (1879) On the variations in the amount of carbon dioxide in the air. *Proceedings of the Royal Society* **30**, 343−355.

Arneth, A., R. K. Monson, G. Schurges, Ü. Niinemets and P. I. Palmer (2009) Why are estimates of global terrestrial isoprene emissions so similar (and why is this not so for monoterpenes)? *Atmospheric Chemistry and Physics* **8**, 4605−4620.

Arneth, A., G. Schurgers, J. Lathiere, T. Duhl, D. J. Beerling, C. N. Hewitt, M. Martin and A. Guenth (2011) Global terrestrial isoprene emission models: sensitivity to variability in climate and vegetation. *Atmospheric Chemistry and Physics* **11**, 8037–8052.

Arnold, H. E., P. Kerrison and M. Steinke (2013) Interacting effects of ocean acidification and warming on growth and DMS-production in the haptophyte coccolithophore Emiliania huxleyi. *Global Change Biology* **19**, 1007−1016.

Arnts, R. R., W. B. Petersen, R. L. Seila and B. W. Gay Jr. (1982) Estimates of alpha-pinene emissions from a loblolly pine forest using an atmospheric diffusion model. *Atmospheric Environment* **16**, 2127−2137.

Arrhenius, S. (1887) Über die Dissoziation der im Wasser gelösten Stoffe. *Zeitschrift für Physikalische Chemie* **1**, 631−648.

Arrhenius, S. (1896) On the influence of carbonic acid in the air upon the temperature of the ground. *Philosophical Magazine* **41**, 237−276.

Atkins, P. (2008) Physical chemistry, Volume 1: Thermodynamics and kinetics, physical chemistry; Volume 2: Quantum chemistry and spectroscopy, student solutions manual & explorations in physical chemistry 2.0 access card. W. H. Freeman & Company, 1039 pp.

Atkinson, R., D. L. Baulch, R. A. Cox, J. N. Crowley, R. F. Hampson, R. G. Hynes, M. E. Jenkin, M. J. Rossi and J. Troe (2004) Evaluated kinetic and photochemical data for atmospheric chemistry: Volume I − gas phase reactions of O_x, HO_x, NO_x and SO_x species. *Atmospheric Chemistry and Physics* **4**, 1461−1738 .

Atkinson, R., D. L. Baulch, R. A. Cox, J. N. Crowley, R. F. Hampson, R. G. Hynes, M. E. Jenkin, M. J. Rossi, J. Troe and IUPAC Subcommittee (2006) Evaluated kinetic and photochemical data for atmospheric chemistry: Volume II − gas phase reactions of organic species. *Atmospheric Chemistry and Physics* **6**, 3625−4055.

Atkinson, R., D. L. Baulch, R. A. Cox, J. N. Crowley, R. F. Hampson, R. G. Hynes, M. E. Jenkin, M. J. Rossi and J. Troe (2007) Evaluated kinetic and photochemical data for atmospheric chemistry: Volume III − gas phase reactions of inorganic halogens. *Atmospheric Chemistry and Physics* **7**, 981−1191.

Atkinson, R., D. L. Baulch, R. A. Cox, J. N. Crowley, R. F. Hampson, R. G. Hynes, M. E. Jenkin, M. J. Rossi, J. Troe and T. J. Wallington (2008) Evaluated kinetic and photochemical data for atmospheric chemistry: Volume IV − gas phase reactions of organic halogen species. *Atmospheric Chemistry and Physics* **8**, 4141−4496.

Andrieth, L. F. (1930) Parallelism in the decomposition of NH_4^+, N_2H_4 and NH_2OH nitrites. *Journal of Physical Chemistry* **39**, 538−540.

Aumont B., S. Madronich, I. Beyand and G. S. Tyndall (2000) Contribution of secondary VOC to the composition of aqueous atmospheric particles: A modeling approach. *Journal of Atmospheric Chemistry* **35**, 59−75.

Aydin, M., W. J. De Bruyn and E. S. Saltzman (2002) Pre-industrial atmospheric carbonyl sulfide (OCS) from an Antarctic ice core. *Geophysical Research Letters* 29, doi:10.1029/2002GL014796.

Ayers, G. P. and J. P. Ivey (1988) Precipitation composition at Cape Grim, 1977−1985. *Tellus* **40B**, 297−307.

Ayers, G. P., J. P. Ivey and H. S. Goodman (1986) Sulfate and methansulfonate in the maritime aerosol at Cape Grim. *Journal Atmospheric Chemistry* **4**, 173−185.

Ayers, G. P., Penkett, S. A., Gillett, R. W., Bandy, B., Galbally, I. E., Meyer, C. P., Eisworth, C. M., Bentley, S. T. and B. W. Forgan (1996) The annual cycle of peroxides and ozone in marine air at Cape Grim. *Journal Atmospheric Chemistry* **23**, 221−252.

Bach, A. and R. Chodat (1902) Untersuchungen über die Rolle der Peroxide in der Chemie der lebenden Zelle. Erste Mitteilung: Ueber das Verhalten der lebenden Zelle gegen Hydroperoxyd. *Berichte der Deutschen Chemischen Gesellschaft* **35**, 240−262.

Bacher, C., G. S. Tyndall and J. J. Orlando (2001) The atmospheric chemistry of glycolaldehyde. *Journal of Atmospheric Chemistry* **39**, 171−189.

Baek, B. H. and V. P. Aneja (2005) Observation based analysis for the determination of equilibrium time constant between ammonia, acid gases, and fine particles. *International Journal of Environment and Pollution* **23**, 239−247.

Bahnemann, D. and E. J. Hart (1982) Rate constants of the reaction of the hydrated electron and hydroxyl radical with ozone in aqueous solution. *Journal of Physical Chemistry* **86**, 252−255.

Baillie, M. (2008) Chemical signature of the Tunguska event in Greenland ice. In: International Conference "100 years since Tunguska phenomenon: Past, present and future". June 26−28, 2008, Moscow (Russia). Abstracts Vol. p. 80.

Bakhtar, F., J. B. Young, A. K. White and D. A. Simpson (2005) Classical nucleation theory and its application to condensing steam flow calculations. *Proceedings of the Institution of Mechanical Engineers, Part C: Journal of Mechanical Engineering Science* **219**, 1315−1333.

Balashev, S. A., D. A. Varshalovich and A. V. Ivanchik (2009) Directional radiation and photodissociation regions in molecular hydrogen clouds. *Astronomy Letters* **35**, 150−166.

Baldochi, D. D. (1991) Canopy control of trace gas emissions. In: Trace gas emissions by plants (Eds. T. D. Sharkey, E. A. Holland and H. A. Mooney), Academic Press, San Diego, pp. 293−333.

Ballantyne, A. P., C. B. Alden, J. B. Miller, P. P. Tans and J. W. C. White (2012) Increase in observed net carbon dioxide uptake by land and oceans during the past 50 years. Nature, **488**, 70−72.

Ballentine, D. C., S. A. Macko, V. C. Turekian, W. P. Gilkooly and B. Martincigh (1996) Chemical and isotopic characterization of aerosols collected during sugar cane burning in South Africa. In: Biomass burning and global change (Ed. J. S. Levine), Cambridge MIT Press, Cambridge, U.K., pp. 460−465.

Balling Jr., R. C., Cerveny, R. S. and C. D. Idso (2002) Does the urban CO_2 dome of Phoenix, Arizona contribute to its heat island? *Journal of Geophysical Research Letters* **28**, 4599−4601.

Balthasar M. and M. Frenklach (2005) Detailed kinetic soot modeling of soot aggregate formation in laminar premixed flames. *Combustion and Flame* **140**, 130−145.

Bange, H. W. (2006) The importance of oceanic nitrous oxide emissions. *Atmospheric Environment* **40**, 198−199.

Barb, W. G., J. H. Baxendale, P. George and K. R. Hargrave (1949) Reactions of ferrous and ferric ions with hydrogen peroxide. *Nature* **163**, 692−694.

Barbier, B., B. Marylène, F, Boillot, A. Chabin, D. Chaput, O. Hénin and A. Brack (1998) Delivery of extraterrestrial amino acids to the primitive Earth. Exposure experiments in Earth orbit. *Biological Sciences in Space* **12**, 92−95.

Bard, A. J., R. Parsons and J. Jordan (1985) Standard potentials in aqueous solutions. Marcel Dekker, New York, 840 pp.

Barkley, M. P., P. I. Palmer, C. D. Boone, P. F. Bernath and P. Suntharalingam (2008) Global distribution of carbonyl sulphide in the upper troposphere and stratosphere. *Geophysical Research Letters* **35**, L14810, doi:10.1029/2008GL034270.

Barnes C. R. (1893) On the food of green plants. *Botanical Gazette* **18**, 403−411.

Barnes, I. and M. M. Kharytonov, Eds. (2008) Simulation and assessment of chemical processes in a multiphase environment. Springer-Verlag, Berlin, 540 pp.

Barnes, I., G. Solignac, A. Mellouki and K.-H. Becker (2010) Aspects of the atmospheric chemistry of amides. *ChemPhysChem* **11**, 3844−3857.

Barnola, J.-M., D. Raynaud, A. Neftel and H. Oeschger (1983) Comparison of CO_2 measurements by two laboratories on air from bubbles in polar ice. *Nature* **303**, 410−413.

Barnola, J.-M., D. Raynaud, Y. S. Korotkevich and C. Lorius (1987) Vostok ice core provides 160 000-year record of atmospheric CO_2. *Nature* **329**, 408−414.

Barnola, J.-M., P. Pimienta, D. Raynaud and Y.S. Korotkevich (1991) CO_2-climate relationship as deduced from the Vostok ice core: A re-examination based on new measurements and on a re-evaluation of the air dating. *Tellus* **43B**, 83−90.

Barnola, J.-M., D. Raynaud, C. Lorius and N. I. Barkov (2003) Historical CO_2 record from the Vostok ice core. In Trends: A Compendium of Data on Global Change. Carbon Dioxide Information Analysis Center, Oak Ridge National Laboratory, U.S. Department of Energy, Oak Ridge, Tenn. (USA).

Baron, P. A. and K. Willeke, Eds. (2001) Aerosol measurements. Principles, techniques and applications. John Wiley & Sons, New York, 1131 pp.

Barrett, K. (1998) Oceanic ammonia emissions in Europe and their transboundary fluxes − calculated fields and budgets 1985−1993. *Atmospheric Environment* **32**, 381−391.

Barrie, L. A. and U. Platt (1997) Arctic tropospheric chemistry: overview. *Tellus* **49B**, 450−454.

Bärring, L. (1993) Climate − change or variation? *Climate Change* **25**, 1−13.

Bartels, O. G. (1972) An estimate of volcanic contributions to the atmosphere and volcanic gases and sublimates as the source of the radio isotopes ^{10}Be, ^{35}S, and ^{22}Na. *Health Physics* **22**, 387–392.

Barth, M. C., S. Sillman, R. Hudman, M. Z. Jacobson, C.-H. Kim, A. Monod and J. Liang (2003) Summary of the cloud chemistry modeling intercomparison: Photochemical box model simulation, *Journal of Geophysical Research* **108**, 4214, doi:10.1029/2002JD002673.

Barth, T. F. W. (1952) Theoretical petrology: A textbook on the origin and the evolution of Rocks. John Wiley & Sons, New York, 387 pp.

Bates, D. R. and M. Nicolet (1950) The photochemistry of the atmospheric water vapour. *Journal of Geophysical Research* **55**, 301–327.

Bates, T. S., B. K. Lamb, A. Guenther, J. Dignon and R. E. Stoiber (1992) Sulfur emissions to atmosphere from natural sources. *Journal Atmospheric Chemistry* **14**, 315–337.

Bates, T. S., K. C. Kelly, J. E. Johnson and R. H. Gammon (1995) Regional and seasonal variations in the flux of oceanic carbon monoxide to the atmosphere. *Journal of Geophysical Research* **100**, 23093–23101.

Bates, T. S., K. C. Kelly, J. E. Johnson and R. H. Gammon (1996) A reevaluation of the open ocean source of methane to the atmosphere. *Journal of Geophysical Research* **101**, 6953–6961.

Battye, R., W. Battye, C. Overcash and S. Fudge (1994) Development and selection of ammonia emission factors. Report EPA-600/R-94-190, US-EPA, Research Triangle Park, North Carolina, USA.

Baulch, D. L., R. A. Cox, R. F. Hampson, Jr., J. A. Kerr, J. Troe and R. T. Watson (1980) Evaluated kinetic and photochemical data for atmospheric chemistry. *Journal of Physical and Chemical Reference Data* **9**, 295–471.

Baumann, A. (1893) Chemie der Atmosphäre. In: *Jahresberichte über die Fortschritte auf dem Gesamtgebiete der Agricultur-Chemie. Neue Folge*, XV. 1892, Verlag P. Parey, Berlin, pp. 3–18.

Baumgartner, A. and E. Reichel (1975) The world water balance. Elsevier Scientific Publ., New York, 179 pp.

Baur, E. (1918) Photolyse und Elektrolyse. *Helvetica Chimica Acta* **1**, 186–201.

Baur, E. and A. Perret (1924) Über die Einwirkung von Licht auf gelöste Silbersalze in Gegenwart von Zinkoxyd. *Helvetica Chimica Acta* **7**, 910–915.

Baur, E. and C. Neuweiler (1927) Über photolytische Bildung von Hydroperoxyd. *Helvetica Chimica Acta* **10**, 901–907.

Bazhanov, V. (1994) Surface ozone at Mount Areskutan: connection with ozon in the free troposphere. Report 25, Inst. of Appl. Env. Res., Stockholm University, 46 pp.

Beatty, J. T., J. Overmann, M. T. Lince, A. K. Manske, A. S. Lang, R. E. Blankenship, C. L. Van Dover, T. A. Martinson and F. G. Plumley (2005). An obligately photosynthetic bacterial anaerobe from a deep-sea hydrothermal vent. *Proceedings of the National Academy of Science* **102**, 9306–9310.

Beaumont, S. K., R. J. Gustafsson and R. M. Lambert (2008) Heterogeneous photochemistry relevant to the troposphere: H_2O_2 production during the photochemical reduction of NO_2 to HONO on UV-illuminated TiO_2 surfaces. *ChemPhysChem* **10**, 331–333.

Beck, J. P. and P. Grennfelt (1994) Estimate of ozone production and destruction over Northwestern Europe. *Atmospheric Environment* **28**, 129–140.

Becker, K.-H., J. Bechara and K. J. Brockmann (1993) Studies on the formation of H_2O_2 in the ozonolysis of alkenes. *Atmospheric Environment* **27A**, 57–61.

Becker, K.-H., I. Barnes, L. Ruppert and P. Wiesen (1999) Is the oxidising capacity of the atmosphere changing? In: Prediction of atmospheric environmental problems between technology and nature (Ed. D. Möller), Springer-Verlag, Berlin, pp. 105–132.

Beech, M. (2006) The problem of ice meteorites. *Meteorite Quarterly* **12**, 17–19.

Beekmann, M., A. Kerschbaumer, E. Reimer, R. Stern and D. Möller (2007) PM measurement campaign HOVERT in the Greater Berlin area: model evaluation with chemically specified particulate matter observations for a one year period. *Atmospheric Chemistry and Physics* **7**, 55−78.

Beerling, D. J. and R. A. Berner (2005) Feedbacks and the co-evolution of oplants and atmospheric CO_2. *Proceedings of the National Academy of Sciences* **102**, 1302−1305.

Beirle, S., U. Platt, M. Wenig and T. Wagner (2004) O_x production by lightning estimated with GOME. *Advances in Space Research* **34**, 793−797.

Bell, N., L. Hsu, D. J. Jacob, M. G. Schultz, D. R. Blake, J. H. Butler, D. B. King, J. M. Lobert and E. Maier-Reimer (2002) Methyl iodide: Atmospheric budget and use as a tracer of marine convection in global models. *Journal of Geophysical Research* **107**, 4340, doi:10.1029/2001JD001151.

Bendall, D. S., C. J. Howe, E. G. Nisbet and R. E. R. Nisbet (2008) Introduction. Photosynthetic and atmospheric evolution. *Philosophical Transactions of the Royal Society* **B 363**, 2625−2628.

Benedict, F. G. (1912) The composition of the atmosphere with special refer ence to its oxygen content. Publication by the Carnegie Institution Washington, USA, 115 pp.

Benestad, R. E. (2012) Reconciliation of global temperatures. *Environmental Research Letters* 7 011002 doi:10.1088/1748-9326/7/1/011002.

Benkovitz, C. M., M. T. Scholtz, J. Pacyna, L. Tarrason, J. Dignon, E. C. Voldner, P. A. Spiro, J. A. Logan and T. E. Graedel (1996) Global gridded inventories of anthropogenic emissions of sulfur and nitrogen. *Journal of Geophysical Research* **101**, 29,239−29,253.

Benkovitz, C. M., H. Akimoto, J. J. Corbett, J. D. Mobley, J. G. J. Olivier, T. Ohara, J. A. van Aardenne and V. Vestreng (2004) Compilation of Regional to Global Inventories of Anthropogenic Emissions. In: Emissions of atmospheric trace compounds (Eds. C. Granier, P. Artaxo and C. E. Reeves), Kluwer Academic Publishers, Dordrecht, Netherlands, pp. 17−69.

Bennett, C. L., D. Larson, J. L. Weiland, N. Jarosik, G. Hinshaw, N. Odegard, K. M. Smith, R. S. Hill, B. Gold, M. Halpern, E. Komatsu, M. R. Nolta, L. Page, D. N. Spergel, E. Wollack, J. Dunkley, A. Kogut, M. Limon, S. S. Meyer, G. S. Tucker and E. L. Wright (2013) Nine-year Wilkinson microwave anisotropy probe (WMAP) observations: Final maps and results. *The Astrophysical Journal Supplement* **208**, Issue 2, article id. 20, 54 pp. Doi:10.1088/0067-0049/208/2/20.

Benseler, G. E. and K. Schenkl (1900) Griechisch-Deutsches Schulwörterbuch. Teubner, Leipzig, 916 pp.

Benton, M. J. (1986) More than one event in the late Triassic mass extinction. *Nature* **321**, 857−861.

Benton, M. J. and D. A. T. Harper (1997) Basic palaeontology. Prentice Hall, 342 pp.

Benton (2005) When life nearly died: The greatest mass extinction of all time. Thames & Hudson, New York, 336 pp.

Bérczi, S. and B. Lukács (2001) Existence, survival and recognition of icyc meteorites on Antarctica with respect to palaeotempartures. *Acta Climatologica at Chorologica* **34−35**, 51−68.

Bérczi, S. and B. Lukács (2007) Water-ammonia ice meteorites and/or ammoniu(um)-silicates from the early Solar System: Possible sources of amino-radicals of life-molecules on Earth and Mars? 28th Lunar and Planetary Science Conference, to be held in Houston, Texas (USA), p. 97−98.

Berdowski, J., R. Guicherit and B. J. Heij, Eds. (2001) The climate system. Balkema Publishers/Swets & Zeitlinger Publishers, Lisse, The Netherlands. 186 pp.

Bernal, J. D. and R. H. Fowler (1933) A theory of water and ionic solution, with particular reference to hydrogen and hydroxyl ions. *Journal of Chemical Physics* **1**, 515−548.

Berner, E. K. and R. A. Berner (1987) The globalwater cycle: Geochemistry and environment. Prentice Hall, Englewood Cliffs, New Jersey, Prentice-Hall, 376 pp.

Berner, K. B. and R. A. Berner (1996) Global environment: Water, air, and geochemical cycles. Prentice Hall, New Jersey, 376 pp.

Berner, R. A. (2003) Overview. The long-term carbon cycle, fossil fuels and atmospheric composition. *Nature* **426**, 323−326.

Berner, R. A. (2005) A different look at biogeochemistry. *American Journal of Science* **305**, 872−873.

Berner, R. A. (2006) Geological nitrogen cycle and atmospheric N_2 over Phanerozoic time. *Geology* 34, 413−415.

Berresheim H. and W. Jaeschke (1983) The contribution of volcanoes to the global atmospheric sulfur budget. *Journal of Geophysical Research* **88**, 3732−3740.

Berresheim, H., P. H. Wine and D. D. Davis (1995) Sulfur in the atmosphere. In: Composition, chemistry, and climate of the atmosphere (Ed. H. B. Singh), Van Nostrand Reinhold, New York, pp. 251−307.

Berresheim, H. (1998) Beiträge zur Rolle des natürlichen Schwefelkreislaufes in der Atmosphäre. Berichte des Deutschen Wetterdienste 205, Offenbach, 202 pp.

Berresheim H., T. Elste, C. Plass-Dulmer, F. L. Eisele and D. J. Tanner (2000) Chemical ionization mass spectrometer for long-term measurements of atmospheric OH and H_2SO_4. *International Journal of Mass Spectrometry* **202**, 91−109.

Berresheim, H., J. McGrath, M. Adam, R. L, Mauldin III, B. Bohn and F. Röhrer (2013) seasonal measurements of OH, NO_x, and $J(O^1D)$ at Mace Head, Ireland. *Geophysical Research Letters* **40**, doi:10.1002/grl.50345.

Berry, J., A. Wolf, J. E. Campbell, I. Baker, N. Blake, D. Blake, A. S. Denning, S. R. Kawa, S. A. Montzka, U. Seibt, K. Stimler, D. Yakir and Z. Zhu (2013) A coupled model of the global cycles of carbonyl sulfide and CO_2: A possible new window on the carbon cycle. *Journal of Geophysical Research: Biogeosciences* **118**, 842−852.

Bertani, R. (2003) What is geothermal potential? *IGA News* (International Geothermal Association) 53, 1−3.

Bertie, J. E. and M. R. Shehata (1984) Ammonia dihydrate: Preparation, x-ray powder diffraction pattern and infrared spectrum of $NH_3 \cdot 2\,H_2O$ at 100 K. *Journal of Chemical Physics* **81**, doi:10.1063/1.447381.

Bettelheim, J. and A. Littler (1979) Historical trends of sulphur oxide emissions in Europe since 1865. CEGB Report PLGS/E/1/79, London, Great Britain.

Betterton, E. A. and M. R. J. Hoffmann (1988) Oxidation of aqueous SO_2 by peroxomonosulfate. *Journal of Physical Chemistry* **92**, 5962−5965.

Betz, H. D., U. Schumann and P. Laroche, Eds. (2009) Lightning: Principles, instruments and applications: Review of modern lightning research. Springer-Verlag, Berlin, 641 pp.

Beyrich, F., K. Acker, D. Kalaß, O. Klemm, D. Möller, E. Schaller, J. Werhahn and U. Weissensee (1996) Boundary layer structure and photochemical pollution in the Harz Mountains − an observational study. *Atmospheric Environment* **30**, 1271−1281.

Beysens, D., Ohayon, C, Muselli, M. and O. Clus (2006) Chemical and biological characteristics of dew and rain water in an urban coastal area (Bordeaux, France). *Atmospheric Environment* **40**, 3710−3723.

Bianchi, T. S. (2006) Biogeochemistry of estuaries. Oxford University Press, USA, 706 pp.

Bickle, M. J. (1994) The role of metamorphic decarbonation reactions in returning strontium to the silcate sediment mass. *Nature* **367**, 699−704.

Bielski, B. H. J., D. E. Cabelli, R. L. Arudi and A. B. Ross (1985) Reactivity of HO_2 / O_2^- radicals in aqueous solution. *Journal of Physical and Chemical Reference Data* **14**, 1041−1099.

Bigg, G. (2003) The oceans and climate. Cambridge University Press, Cambridge, U.K., 273 pp.

Billington, C., V. Kapos, M. Edwards, S. Blyth and S. Iremonger (1996) Estimated original forest cover map – a first attempt. WCMC (UNEP World Conservation Monitoring Centre). Cambridge, U.K.

Bischof, G. (1863) Lehrbuch der chemischen und physikalischen Geologie. Erster Band. A. Marcus, Bonn, 865 pp.

Black, J. (1754) De humore acido a cibis orto, et magnesia alba. Doctoral Thesis; published a fuller account of them in 1756 as "Experiments upon magnesia, quicklime and some other alkaline substances". On acid humor arising from foods, and on white magnesia: A translation of the Latin thesis De humore acido a cibis orto, et magnesia alba (1973) Bell Museum of Pathobiology (University of Minnesota Medical School), 40 pp.

Black, J. (1803) Lectures on elements of chemistry (ed. by J. Robinson). Mundell & Son, Edinburgh, 762 pp.

Blackall, T. D., L. J. Wilson, M.R. Theobald, C. Milford, E. Nemitz, J. Bull, P. J Bacon, K. C. Hamer, S. Wanless and M.A. Sutton (2007) Ammonia emissions from seabird colonies. *Geophysical Research Letters* **34** (L10801). doi:10.1029/2006GL028928.

Blackman, F. F. and G. L. C. Matthaei (1905) Experimental researches on vegetable assimilation and respiration. IV. – A quantitative study of carbon-dioxide assimilation and leaf-temperature in natural illumination. *Proceedings of the Royal Society London* **76**, 402–460.

Blake, N. J., et al. (2008), Carbonyl sulfide (OCS): Large-scale distributions over North America during INTEX-NA and relationship to CO_2. *Journal of Geophysical Research* **113**, D09S90, doi:10.1029/2007JD009163.

Blanchard, D. C. (1983) The production, distribution and bacterial enrichment of the sea-salt aerosol. In: Air-Sea Exchange of Gases and Particles (Eds. P. S. Liss and W. G. N. Slinn), Reidel, Boston, USA, pp. 407–454.

Blochmann (1886) Ueber den Kohlensäuregehalt der atmosphärischen Luft. *Annalen der Chemie* **237**, 39–90.

Blücher, H. (1900) Die Luft. Ihre Zusammensetzung und Untersuchung, ihr Einfluss und ihre Wirkung sowie technische Ausnutzung. Verlag von O. Wigand, Leipzig, 322 pp.

Blum, H. F. and C. R. Spealman (1933) Photochemistry of fluorescein dyes. *Journal of Physical Chemistry* **37**, 1123–1133.

Blunier, T., E. Monnin and J.-M. Barnola (2005) Atmospheric CO_2 data from ice cores: Four climatic cycles. In: A history of atmospheric CO_2 and its effects on plants, animals, and ecosystems (Eds. J. R. Ehleringer, T. E. Cerling, and M. D. Dearing), Springer, New York, pp. 62–82.

Bluth, G. J. S., C. C. Schnetzler, A. J. Krueger and L. S. Walter (1997) The contribution of explosive volcanism to global atmospheric sulphur dioxide concentrations. *Nature* **366**, 327–329.

Boden, T. A., G. Marland and R. J. Andres (2009) Global, regional, and national fossil-fuel CO_2 emissions. Carbon Dioxide Information Analysis Center, Oak Ridge National Laboratory, U.S. Department of Energy, Oak Ridge, Tenn., U.S.A. doi:10.3334/CDIAC/00001.

Boden, T. A., Marland, G. and R. J. Andres (2013) Global, Regional, and National Fossil-Fuel CO_2 Emissions, Carbon Dioxide Information Analysis Center, Oak Ridge National Laboratory, U.S. Department of Energy, Oak Ridge, Tenn., USA. doi 10.3334/CDIAC/00001_V2013.

Bodirsky, B. L., A. Popp, I. Weindl, J. P. Dietrich, S. Rolinski, L. Scheiffele, C. Schmitz and H. Lotze-Campen (2012) N_2O emissions from the global agricultural nitrogen cycle – current state and future scenarios. *Biogeosciences* **9**, 4169–4197.

Boehe, D. J., cited by J. Myers (1873) Correspondenz aus Amsterdam. *Berichte der deutschen Chemischen Gesellschaft* **6**, 439–446 [*Boehe* held a lecture at the Chemical Society but died before submitting a manuscript].

Boehm, J. (1858) Untersuchungen über das atmosphärische Ozon. In: *Sitzungsberichte der Kaiserlichen Akademie der Wissenschaften, Math.-Naturwiss. Classe*, XXIX. Bd., Nr. 11, pp. 409–440.

Boersma, K. F., H. J. Eskes, E. W. Meijer and H. M. Kelder (2005) Estimates of lightning NO_x production from GOME satellite observations. *Atmospheric Chemistry and Physics* **5**, 2311–2331.

Bohn, H. L., B. L. McNeal and G. A. O'Connor (2001) Soil chemistry. 3rd Edition. John Wiley & Sons Inc., 307 pp.

Bolin, B. and H. Rodhe (1973) A note on the concepts of age distribution and transit time in natural reservoirs. *Tellus* **25**, 58–62.

Bolin, B., G. Aspling and C. Persson (1974) Residence time of atmospheric pollutants as dependent on source characteristics, atmospheric diffusion processes and sink mechanisms. *Tellus* **27**, 281–293.

Bond, H. E., E. P. Nelan, D. A. Vandenberg, G. H. Schaefer and D. Harmer (2013) HD 140283: A star in the solar neighborhood that formed shortly after the Big Bang. *The Astrophysical Journal Letters* **765**, Issue 1, article id. L12, 5 pp., doi:10.1088/2041-8205/765/1/L12.

Bond, T. C., D. G. Streets, K. F. Yarber, S. M. Nelson, J.-H. Woo and Z. Klimont (2004) A technology-based global inventotry of blck and organic carbon emission from combustion. *Journal of Geophysical Research* **109**, D14203, doi:10.1029/2003JD003697.

Bonét-Maury, P. and M. Lefort (1948) Formation of hydrogen peroxide in water irradiated with X- and alpha-rays. *Nature* **162**, 381–382.

Boniface, J., Q. Shi, Y. Q. Li, J. L. Cheung, O. V. Rattigan, P. Davidovits, D. R. Worsnop, J. T. Jayne and C. E. Kolb (2000) Uptake of gas-phase SO_2, H_2S, and CO_2 by aqueous solutions. *Journal of Physical Chemistry* A **104**, 7502–7510.

Bonsang, B., M. Kanakidou and G. Lambert (1988) The marine source of C_2–C_6 aliphatic hydrocarbons. *Journal of Atmospheric Chemistry* **6**, 3–20.

Borch, T., R. Kretzschmar, A. Kappler, P. van Cappellen, M. Ginder-Vogelk, A. Voegelin and K. Campbell (2010) Biogeochemical redox processes and their impact on contaminant dynamics. *Environmental Science and Technology* **44**, 15–23.

Borrell, P. M. and P. Borrell, Eds. (1999) Transport and chemical transformation in the troposphere, Proceedings of EUROTRAC Symposium 2002. Vol 2: Biosphere-atmosphere exchange and anthropogenic emissions, coastal processes, acidification, mercury and POPs, modelling and model validation on regional, global and urban scales, regional and urban problems. WIT press Boston, 944 pp.

Boserup, E. (1981) Population and technology. Blackwell, Oxford, 255 pp.

Botha, C. F., J. Hahn, J. J. Pienaar and R. van Eldik (1984) Kinetics and mechanism of the oxidation of sulfur(IV) by ozone in aqueous solution. *Atmospheric Environment* **28**, 3207–3212.

Bott, A. and G. R. Carmichael (1993) Multiphase chemistry in a microphysical radiation fog model. A numerical study. *Atmospheric Environment* **27A**, 503–522.

Botta, O. and J. L. Bada (2002) Extraterrestrial organic compounds in meteorites. *Survey Geophysics* **23**, 411–467.

Bousquet, P., P. Ciais, J. B. Miller, E. J. Dlugokencky, D. A. Hauglustaine, C. Prigent, G. R. Van der Werf, P. Peylin, E. G. Brunke, C. Carouge, R. L. Langenfelds, J. Lathière, F. Papa, M. Ramonet, M. Schmidt, L. P. Steele, S. C. Tyler and J. White (2006) Contribution of anthropogenic and natural sources to atmospheric methane variability. *Nature* **443**, 439–443.

Boussingault J. B. (1856) Recherches sur la vegetation. Troisiéme mémoire. De l'action du salpêtre sur le développement des plants. *Annales de Chimie et de Physique. Serie 3* **45**, 5–41.

Boussingault, J. P. (1853) Mémoire sur la dosage de l'ammoniaque contenue dans les eaux. *Annales de Chimie et de Physique. Serie 3* **39**, 257–293.

Bouwman, A. F., J. G. J. Oliver and K. W. Van der Hoek (1995) Uncertainties in the global source distribution of nitrous oxide. *Journal of Geophysical Research* **100**, 2785−2800.

Bouwman A. F, D. S. Lee, W. A. H Asman, F. J. Dentener, K. W. van der Hoek and J. G. J. Olivier (1997) A global high-resolution emission inventory for ammonia. *Global Biogeochemical Cycles* **11**, 561−588.

Bouwman A. F., L. J. M. Boumans and N. H. Batjes (2002a) Emissions of N_2O and NO from fertilized fields. Summary of available measurement data. *Global Biogeochemical Cycles.* **16**, 1058 doi:10.1029/2001GB001811.

Bouwman A. F., L. J. M. Boumans and N. H. Batjes (2002b) Modeling global annual N_2O and NO emissions from fertilized fields. *Global Biogeochemical Cycles.* **16**, 1080 doi:10.1029/2001GB001812.

Bower, K. N., T. W. Choularton, M. W. Gallagher, R. N. Colvile, M. Wells, K. M. Beswick, A. Wiedensohler, H.-C. Hansson, B. Svenningsson, E. Swietlicki, M. Wendisch, A. Berner, C. Kruisz, P. Laj, M. C. Facchini, S. Fuzzi, M. Bizjak, G. Dollard, B. Jones, K. Acker, W. Wieprecht, M. Preiss, M. A. Sutton, K. J. Hargreaves, R. L. Storeton-West, J. N. Cape and B. G. Arends (1997) Observations and modelling of the processing of aerosol by a hill cap cloud. *Atmospheric Environment* **31**, 2527−2543.

Boyle, R. (1680): Chymista Scepticus uel Dubia et Paradoxa Chymico-Physica, circa Spagyricorum Principia, vulgo dicta Hypostatica, Prout proproni & propugnari solent a Turba Alchymistarum ... 5th Ed. (in Latin), Samuel de Tournes, Geneve, 148 pp.

Bozek, J. D., J. E. Furst, T. J. Gay, H. Gould, A. L. D. Kilcoyne, J. R. Machzacek, F. Martin, K. W. McLaughlin and J. L. Sanz-Vicario (2006) Production of exited atomic hydrogen and deuterium from H_2 and D_2 photodissociation. *Journal of Physics B: Atomic. Molecular and Optical Physics* **39**, 4871−1882.

BP (2009) Statistical review of world energy. June 2009. http://bp.com.statisticalreview.

BP (2013) Statistical review of world energy. http://www.bp.com/en/global/corporate/about-bp/statistical-review-of-world-energy-2013.

Braids, O. C. and Y. Cai, Eds. (2002) Biogeochemistry of environmentally important trace elements. ACS Symposium Series No. 835. Oxford University Press, Oxford, 448 pp.

Brasseur, G. P., J. J. Orlando and G. S. Tyndall, Ed. (1999) Atmospheric chemistry and global change. Oxford University Press, 654 pp.

Brasseur G. P., R. G. Prinn and A. P. Pszenny, Eds. (2003) Atmospheric chemistry in a changing world. An integration and synthesis of a decade of tropospheric chemistry research. Springer-Verlag, Berlin, 300 pp.

Brezonik, P. L. W. (1994) Chemical kinetic process dynamic in aquatic systems. CRC Press, Boca Raton, FL (USA), 754 pp.

Bridgham, S. D., H. Cadillo-Quiroz, J. K. Keller and Q. Zhuang (2013) Methane emissions from wetlands: biogeochemical, microbial, and modeling perspectives from local to global scales. *Global Change Biology* **19**, 1325−1346.

Briggs, M. H. and G. Mamikunian (1963) Organic constituents of the carbonaceous chondrites. *Space Science Reviews*, **1**, 647−682.

Brimblecombe, P. and I. J. Todd (1977) Sodium and potassium in dew. *Atmospheric Environment* **11**, 649−650.

Brimblecombe, P. and D. Shooter (1987) Aqueous phase chemistry of biogenic sulphur compounds In: Physico-chemical behaviour of atmospheric pollutants (Eds. G. Angeletti and G. Restelli), Proceedings of the 4th European Symposium held in Stresa, Italy Sep 23−25 1986, D. Reidel Publ. Corp., Dordrecht, pp. 249−257.

Brimblecombe, P. and S. L. Clegg (1988) The solubility and behaviour of acid gases in marine aersosol. *Journal of Atmospheric Chemistry* **7**, 1−18.

Brimblecombe, P. (1996) Air composition and chemistry. Cambridge University Press, 253 pp.

Brimblecombe, P. (2003) The global sulfur cycle. In: Treatise on geochemistry, Volume 8 (Ed. W. H. Schlesinger) Elsevier, pp. 645−682.

Broadgate, W. J., P. S. Liss and S. A. Penkett (1997) Seasonal emissions of isoprene and other reactive hydrocarbon gases from the ocean. *Geophysical Research Letters* **24**, 2675–2678.

Brocks, J. J., G. A. Logan, R. Buick and R. E. Summons (1999) Archean molecular fossils and the early rise of eukaryotes. *Science* **285**, 1033–1036.

Brønsted, J. N. (1934) Zur Theorie der Säuren und Basen und der protolytischen Lösungsmittel. *Zeitschrift für Physikalische Chemie* **169A**, 52–74.

Brosset, C. (1976) A method of measuring airborne acidity: its application for the determination of acid content on long-distance transported particles and in drainage water from spruces. *Water, Air, and Soil Pollution* **6**, 259–275.

Brown, H. T. and F. Escombe (1905a) Researches on some of the physiological processes of green leaves, with special reference to the interchange of energy between the leaf and its surroundings. *Proceedings of the Royal Society London B.* **76**, 29–111.

Brown, H. T. and F. Escombe (1905b) On the Variations in the Amount of Carbon Dioxide in the Air of Kew during the Years 1898–1901. *Proceedings of the Royal Society London B.*, **76** 118–121.

Brown, L. (1949) The story of maps. Bonanza Books, 393 pp.

Browne, C. A. (1943) A source book of agricultural chemistry. *Chronica Botanica* **8**, 1–290.

Brückner, E. (1890) Klimaschwankungen seit 1700, nebst Bemerkungen über die Klimaschwankungen der Diluvialzeit. In: *Geographische Abhandlungen*, Band 4, Heft 2 (Ed. A. Penck), 324 pp.

Bruinsma, J. (2003) World agriculture, towards 2015/2030: an FAO perspective. Earthscan, London and FAO, Rome, 440 pp.

Buckberg, G. D. (2005) Reply to Criscione et al. Nature is simple, but scientists are complicated. *European Journal of Cardio-Thoracic Surgery* **28**, 364–365.

Budyko, M. I., A. B. Ronov and A. L. Yanshin (1987) History of the earth's atmosphere. Springer-Verlag, Berlin, 139 pp.

Buijsman E., H. F. M. Mass and W. A. H. Asman (1986) Anthropogenic ammonia emissions in Europe. *Atmospheric Environment* **21**, 1009–1022.

Burford, J. R. and J. M. Bremner (1972) Is phosphate reduced to phosphine in waterlogged soils? *Soil Biology and Biochemistry* **4**, 489–495.

Buxton, G. V. (1982) Basic radiation chemistry of liquid water. In: The study of fast processes and transient species by electron pulse radiolysis (Eds. J. H. Baxendale and F. Busi), D. Reidel Publ. Comp., pp. 241–266.

Buxton, G. V., G. A. Salmon, S. Barlow, T. N. Malone, S. McGowan, S. A. Murray, J. E. Williams and N. D. Wood (1996) Photochemical sources and reactions of free radicals in the aqueous phase. In: Heterogeneous and liquid-phase processes. Vol. 2: Transport and chemical transformation of pollutants in the troposphere (Ed. P. Warneck), Springer-Verlag, Berlin, pp. 133–145.

Cachier, H. (1998) Carbonaceous combustion aerosol. In: Atmospheric particles (Eds. R. M. Harrison and R. van Grieken), John Wiley & Sons, Chichester, pp. 295–348.

Cadle, R. D. and H. F. Johnstone (1956) Chemistry of contaminated atmosphere. In: Air pollution handbook (Eds. P. L. Magill, F. R. Holden and C. Ackley), New York, Section 3.

Cadle, R. D. (1975) Volcanic emissions of halides and sulfur compounds to the troposphere and stratosphere. *Journal of Geophysical Research* **80**, 1650–1652.

Cadle, R. D. (1980) A comparison of volcanic with other fluxes of atmospheric trace gas constituents. *Reviews of Geophysics and Space Physics* **18**, 746–752.

Cai, X. and R. J. Griffin (2006) Secondary aerosol formation from the oxidation of biogenic hydrocarbons by chlorine atoms. *Journal of Geophysical Research* **111**, D14206, doi:10.1029/2005JD006857.

Caldeira, K. and R. Berner (1999) Seawater pH and atmospheric carbon dioxide (Technical comment). *Science* **286**, 2043a.

Caldeira, K. and G. H. Rau (2000) Accelerating carbonate dissolution to sequester carbon dioxide in the ocean: geochemical implications. *Geophysical Research Letters* **27**, 225–228.

Callendar, G. S. (1938) The artificial production of carbon dioxide and its influence on temperature. *Quarterly Journal of the Royal Meteorological Society* **64**, 223–240.

Callendar, G. S. (1940) Variations of the amount of carbon dioxide in different air currents. *Quarterly Journal of the Royal Meteorological Society* **66**, 395–400.

Callendar, G. S. (1949) Can carbon dioxide influence climate? *Weather* **4**, 310–314.

Callendar, G. S. (1958) On the amount of carbon dioxide in the atmosphere. *Tellus* **10**, 243–248.

Calvert, J. G., K. Theurer, G. T. Rankin and W. M. MacNevin (1954) A study of the mechanisms of the photochemical synthesis of hydrogen peroxide at zinc oxide surfaces. *Journal of the American Chemical Society* **76**, 2575–2578.

Canfield, D. E. (2004) The evolution of the earth surface sulfur reservoir. *American Journal of Science* **304**, 839–861.

Cao, L. and K. Caldeira (2010) Atmospheric carbon dioxide removal: long-term consequences and commitment. *Environmental Research Letters*. doi:10.1088/1748-9326/5/2/024011.

Cape, J. N., K. J. Hargreaves, R. L. Storeton-West, B. Jones, T. Davies, R. N. Colvile, M. W. Gallagher, T. W. Choularton, S. Pahl, A. Berner, C. Kruisz, M. Bizjak, P. Laj, M. C. Facchini, S. Fuzzi, B. G. Arends, K. Acker, W. Wieprecht, R. M. Harrison and J. D. Peak (1997) The budget of oxidised nitrogen species in orographic clouds. *Atmospheric Environment* **31**, 2625–2636.

Carmichael, R. S., Ed. (1989) CRC Practical handbook of physical properties of rocks and minerals, CRC Press, Boca Raton, FL (USA), 741 pp.

Carmichael, G. R., D. G. Streets, G. Calori, M. Amman, M. Z. Jacobson, J. Hansen and H. Ueda (2002) Changing trends in sulfur emissions in Asia: Implications for acid deposition, air pollution, and climate. *Environmental Sciences and Technology* **356**, 4707–4713.

Carroll, J. J. and A. E. Mather (1992) The system carbon dioxide-water and the Krichevsky-Kasarnovsky equation. *Journal of Solution Chemistry* **21**, 607–621.

Carlton, A. G., C. Wiedinmyer and J. H. Kroll (2009) A review of secondary organic aerosol (SOA) formation from isoprene. *Atmospheric Chemistry and Physics* **9**, 4987–5005.

Carslaw, K. S., O. Boucher, D. V. Spracklen, G. W. Mann, J. G. L. Rae, S. Woodward and M. Kulmala (2010) A review of natural aerosol interactions and feedbacks within the Earth system. *Atmospheric Chemistry and Physics* **10**, 1701–1737.

Cascade, F., G. de Patris and A. Troiani (1999) Experimental detection of hydrogen trioxide. *Science* **285**, 81–82.

Casewit, C. J. and W. A. Goddard III (1982) Energetics and mechanisms for reactions involving nitrosamide, hydroxydiazenes, and diimide N-oxides. *Journal of the American Chemical Society* **104**, 3280–3287.

Castagna, R., J. P. Eiserich, M. S. Budamagunta, P. Stipa, C. E. Cross, E. Proietti, J. C. Voss and L. Greci (2008) Hydroxyl radical from the reaction between hypochlorite and hydrogen peroxide. *Atmospheric Environment* **42**, 6551–6554.

Cauer, H. (1949a) Aktuelle Probleme der atmosphärischen Chemie. In: Meteorologie und Physik der Atmosphäre (Ed. R. Mügge). Vol. 19: Naturforschung und Medizin in Deutschland 1936–1946. Dietrich'sche Verlagsbuchhandlung, Wiesbaden, pp. 277–291.

Cauer, H. (1949b) Ergebnisse chemisch-meteorologischer Forschung. *Archiv für Meteorologie, Geophysik und Bioklimatologie. Serie B* **1**, 221–256.

Cavendish, H. (1893) Experiments on air. Alembic Club Reprint No. 3, W. F. Clay, Edinburgh, 52 pp.

CDIAC (2013) Carbon Dioxide Information Analysis Center. Global Carbon Project: Carbon budget and trends. http://cdiac.ornl.gov/GCP/carbonbudget/2013/.

Cerveny, R. S. and K. J. Coakley (2002) A weekly cycle in atmospheric carbon dioxide. *Geophysical Research Letters* **29**, doi:10.1029/2001GL013952.

Chamberlain, A. C. (1953) Aspects of the travel and deposition of aerosol clouds and vapours. Atomic Energy Research Establishment (A.E.R.E.) Report No 1271, London, 31 pp.

Chamberlain, A. C. and R. C. Chadwick (1953) Deposition of airborne radioiodine vapor. *Nucleonics* **8**, 22−25.

Chamberlain, T. C. (1916) The origin of the earth. University Chicago Press, 271 pp.

Chameides, W. L. (1979) The implication of CO production in electrical discharges. *Geophysical Research Letters* **6**, 287−290.

Chandrasekhar, S. (1960) Radiative transfer. Dover Publications Inc., New York (USA), 393 pp.

Chang, T. Y., Kuntasal, G. and W. R. Pierson (1967) Night-time N_2O_5 / NO_3 chemistry and nitrate in dew water. *Atmospheric Environment* **21**, 1345−1351.

Chapman, S. (1930) A theory of upper atmosphere ozone. *Memoirs of the Royal Meteorological Society* **3**, 103−109.

Chao, J., R. C. Wilhoit and B. J. Zwolinski (1974) Gas phase chemical equilibrium in dinitrogen trioxide and dinitrogen tetroxide. *Thermochimica Acta* **10**, 359−371.

Charlson, R. J., J. E. Lovelock, M. O. Andreae and S. G. Warren (1987) Oceanic phytoplankton, atmospheric sulfur, cloud albedo and climate: A geophysiological feedback. *Nature* **326**, 655−661.

Chen, C. T. and A. Poisson (1984) Excess carbon dioxide in the Weddell Sea. *Antarctic Journal* 1984 Rev., pp. 74−75.

Chen, K. Y. and J. C. Morris (1972) Kinetics of oxidation of aqueous sulphide by O_2. *Environmental Science and Technology* **6**, 529−537.

Chen, S. and M. Z. Hoffman (1973) Rate constants for the reaction of the carbonate radical with compounds of biochemical interest in neutral aqueous solution. *Radiation Research* **56**, 40−47.

Chen, Y.-Y. and W.-M. G. Lee (2001) The effect of surfactants on the deliquescence of sodium chloride. *Journal of Environmental Science and Health, Part A* **36**, 229−242.

Chin, M. (1992) Atmospheric studies of carbonyl sulfide and carbon disulfide and their relationship to stratospheric background sulfur aerosol. Doctoral Thesis, Georgia Institute of Technology.

Chin, M. and D. D. Davis (1993) Global sources and sinks of OCS and CS_2 and their distribution. *Global Biogeochemical Cycles* **7**, 321−337.

Chin, M. and D. D. Davis (1995) A reanalysis of carbonyl sufide as a source of stratospheric background sulfur aerosol. *Journal of Geophysical Research* **100**, 8993−9005.

Chin, M. and D. J. Jacob (1996) Anthropogenic and natural contributions to tropospheric sulfate: A global model analysis. *Journal of Geophysical Research* **101**, 18691−18699.

Chlopin, W. (1911) Formation of oxidising agents in air under the influence of ultra-violet light. *Journal of the Russian Physical-Chemical Society* **43**, 554−561.

Cho, M. and M. J. Rycroft (1997) The decomposition of CFCs in the troposphere by lightning. *Journal of Atmospheric and Solar-Terrestrial Physics* **59**, 1373−1379.

Choularton, T. W., R. N. Colvile, K. N. Bower, M. W. Gallagher, M. Wells, K. M. Beswick, B. G. Arends, J. J. Möls, G. P. A. Kos, S. Fuzzi, J. A. Lind, G. Orsi, M. C. Facchini, P. Laj, R. Gieray, P. Wiesner, T. Engelhardt, A. Berner, C. Kruisz, D. Möller, K. Acker, W. Wieprecht, J. Lüttke, K. Levsen, B. Bizjak, H.-C. Hansson, S.-I. Cederfelt, G. Frank, B. Mentes, B. Martinsson, D. Orsini, B. Svenningsson, E. Swietlicki, A. Wiedensohler, Noone, K. J., S. Pahl, P. Winkler, E. Seyffer, G. Helas, W. Jaeschke, H. W. Georgii, W. Wobrock, M. Preiss, R. Maser, D. Schell, G. Dollard, B. Jones, T. Davies, D. L. Sedlak, M. M. David, M. Wendisch, J. N. Cape, K. J. Hargreaves, M. A. Sutton, R., L. Storeton-West, D. Fowler, A. Hallberg, R. M. Harrison and J. D. Peak (1997) The Great Dun Fell Cloud Experiment 1993: an overview. *Atmospheric Environment* **31**, 2393−2405.

Chu, L. and C. Anastasio (2003) Quantum yields of hydroxyl radical and nitrogen dioxide from the photolysis of nitrate on ice. *Journal of Physical Chemistry A*, 107, 9594–9602.

Chyba, C. and C. Sagan (1992) Endogenous production, exogenous delivery and impact-shock synthesis of organic molecules: An inventory for the origins of life. *Nature* **355**, 125–132 doi:10.1038/355125a0.

Chuck A. L, S. M. Turner and P. S Liss (2002) Direct evidence for a marine source of C_1 and C_2 alkyl nitrates. *Science* **297**, 1151–1154.

Chuck, A. L., S. M. Turner and P. S. Liss (2005) Oceanic distributions and air-sea fluxes of biogenic halocarbons in the open ocean. *Journal of Geophysical Research* **110**, C10022 doi:10.1029/2004JC002741.

Clarke, F. W. (1920) The data of geochemistry. Government Printing Office, Washington D.C. (USA), 832 pp.

Clarke, A., S. R. Owens and J. Zhou (2006) An ultrafine sae-salt flux from breaking waves: Implications for cloud condensation nuclei in the remote marine atmosphere. *Journal of Geophysical Research* **111**, D06202, doi:10.1029/2005JD006565.

Claude, H., U. Köhler and W. Steinbrecht (2004) Evolution of ozone and temperature trends at Hohenpeissenberg (Germany). Proceedings. XX. Quadrennial Ozone Symposium, Kos, Greece, pp. 314–315.

Claude, H., W. Steinbrecht and U. Köhler (2008) Entwicklung der Ozonschicht. In: Klima-statusbericht des DWD 2007, Deutscher Wetterdienst, Offenbach, pp. 61–63.

Claußen, M. (1998) Von der Klimamodellierung zur Erdsystemmodellierung: Konzepte und erste Versuche. *Annalen der Meteorologie (Neue Folge)* **36**, 119–130.

Claußen, M. (2001) Earth system models. In: Understanding the earth system: Compart-ments, processes and interactions (Eds. E. Ehlers and T. Krafft), Springer-Verlag, Berlin, pp. 145–162.

Claußen, M. (2006) Klimaänderungen: Mögliche Ursachen in Vergangenheit und Zukunft. In: Klimawandel – vom Menschen verursacht? 8. Symposium Mensch – Umwelt (Ed. D. Möller). *Acta Academiae Scientiarum* (Academy of the Arts and Sciences Useful to the Public in Erfurt) **10**, 217 pp.

Clegg, S. L. and P. Brimblecombe (1989) Solubility of ammonia in pure aqueouas and multi-component solutions. *Journal of Physical Chemistry* **93**, 7237–7238.

Clifton, C. L., N. Altstein and R. E. Huie (1988) Rate constant for the reaction of nitrogen dioxide with sulfur(IV) over the pH range 5.3–13. *Environmental Science and Technology* **22**, 586–589.

Cofala, J., M. Amann, C. Heyes, F. Wagner, Z. Klimont, M. Posch, W. Schöpp, L. Tarasson, J. E. Jonson, C. Whall and A. Stavrakaki (2007) Analysis of policy measures to reduce ship emissions in the context of the revision of the National Emissions Ceilings Directive. Service Contract No 070501/2005/419589/MAR/C1, European Commission, International Institute for Applied Systems Analysis (IIASA), Laxenburg (Austria), 74 pp.

Cohen, J. B. (1895) The air of towns. In: Annual Report of the Smithsonian Institution, Washington D.C. Gov. Printing Office, pp. 349–387.

Colbourn, P., J. C. Ryden and G. J. Dollard (1987) Emission of NOx from urine treated pasture. *Environmental Pollution* **46,** 253–261.

Colvile, R. N., K. N. Bower, T. W. Choularton, M. W. Gallagher, K. M. Beswick, B. G. Arends, G. P. A. Kos, W. Wobrock, D. Schell, K. J. Hargreaves, R. L. Storeton-West, J. N. Cape, B. M. R. Jones, A. Wiedensohler, H.-C. Hansson, M. Wendisch, K. Acker, W. Wieprecht, S. Pahl, P. Winkler, A. Berner, C. Kruisza and R. Gieray (1997) Meteorology of the Great Dun Fell Cloud Experiment 1993. *Atmospheric Environment* **31**, 2407–2420.

Conrad, R. and W. Seiler (1988) Methane and hydrogen in sweater (Atlantic Ocean). *Deep Sea Research, Part A* **35**, 1903–1917.

Conrad, R. and K. Meuser (2000) Soils contain more than one activity consuming carbonyl sulphide. *Atmospheric Environment* **34**, 3635–3639.

Conway, T. J., P. P. Tans, L. S. Waterman, K. W. Thoning, D. R. Kitzis, K. A. Masarie and N. Zhang (1994) Evidence of interannual variability of the carbon cycle from the NOAA/CMDL global air sampling network. *Journal of Geophysical Research* **99**, 22831−22855.

Cook, S. F. (1932) The respiratory gas exchange in Termopsis Nevadensis. *Biological Bulletin* **63**, 246−256.

Cook, E. (1971) The flow of energy in an industrial society. *Scientific American* **224**, 135−144.

Cook, E. (1976) Man, energy, society. W. H. Freeman and Co, San Francisco, 383 pp.

Cook, W. F., C. Liousse, H. Cachier and J. Feichter (1999) Construction of a 1° × 1° degree fossil fuel emission data set for carbonaceous aerosol and implementation and radiative impact in the ECHAM4 model. *Journal of Geophysical Research* **104**, 22137−22162.

Cooper, W. S. (1957) Sir Arthur Tansley and the science of ecology. *Ecology* **38**, 658−659.

Copley, J. (2001) The story of O. *Nature* **410**, 862−864.

Corbett, J. J. and P. S. Fischbeck (1997) Emissions from ships. *Science* **278**, 823−824.

Corbett, J. J. and H. W. Koehler (2003) Updated emissions from ocean shipping. *Journal of Geophysical Research* **108**, doi:10.1029/2003JD003751.

Cossa, A. (1867) Ueber die Ozonometrie. *Zeitschrift für Analytische Chemie* **6**, 24−28.

Ćosović, B., P. Orlović Leko and Z. Kozarac (2007) Rainwater dissolved organic carbon: characterization of surface active substances by electrochemical method. *Electroanalysis* **19**, 2077−2084.

Countess, R. and J. Heicklen (1973) Kinetics of particle growth. II. Kinetics of the reaction of ammonia with hydrogen chloride and the growth of particulate ammonium chloride. *Journal of Physical Chemistry* **77**, 444−447.

Courtillot, V. (2005) New evidence for massive pollution and moratlity in Europe in 1783−1784 may have bearing on global change andmass extinctions. *Comptus Rendus Geoscience* **337**, 635−637.

Crahan, K. K., D. Hegg, D. S. David, S. Covert and H. Jonsson (2004) An exploration of aqueous oxalic acid production in the coastal marine atmosphere. *Atmospheric Environment* **38**, 3757−3764.

Cramer W. and C. B. Field, Eds. (1999) The Potsdam NPP Model Intercomparison. *Global Change Biology* **5**, Supplement 1, pp. 1−76.

Crichton, M. (1969) The Andromeda Strain. Dell, New York, 247 pp.

Croll, J. (1864) On the physical cause of the change of climate during geological epochs. *Philosophical Magazine* **28**, 121−137.

Croll, J. (1875) Climate and time in their geological relations. A theory of secular changes of the earth's climate. Edward Stanford, London, 577 pp.

Crutzen, P. J. (1971) Ozone production rates in an oxygen-hydrogen-nitrogen oxide atmosphere. *Journal of Geophysical Research* **76**, 7311−7327.

Crutzen, P. J. (1976) The possible importance of COS for the sulphur layer of the stratosphere. *Geophysical Research Letters* **3**, 73−76.

Crutzen, P. J., L. E. Heidt, J. P. Krasnec, W. H. Pollock and W. Seiler (1979) Biomass burning as a source of atmospheric gases CO, H_2, N_2O, NO, CH_3Cl and COS. *Nature* **282**, 253−256.

Crutzen, P. J., I. Aselmann and W. Seiler (1986) Methane production by domestic animals, wild ruminants, other herbivorous fauna, and humans. *Tellus* **38B**, 271−284.

Crutzen, P. J. and V. Ramanathan, Eds. (1996) Clouds, chemistry and climate. NATO ASI Series, Series 1: Global Environmental Change, Vol. 35, Springer-Verlag, Berlin, 260 pp.

Crutzen, P. (1999) Global problems of atmospheric chemistry − the story of man's impact on atmospheric ozone. In: Atmospheric environmental research − critical decisions between technological progress and preservation of nature (Ed. D. Möller), Springer-Verlag, Berlin, pp. 5−35.

Crutzen, P. J., M. L. Lawrence and U. Pöschl (1999) On the background photochemistry of tropospheric ozone. *Tellus* **51A−B**, 123−146.

Crutzen, P. J. and E. F. Stoermer (2000) The "Anthropocene". *Global Change Newsletters* **41**, 17–18.

Crutzen, P. J. (2002) A critical analysis of the Gaia hypothesis as a model for climate/biosphere interactions. *GAIA* **11**, 96–103.

Cullis, C. F. and M. M. Hirschler (1980) Atmospheric sulphur: Natural and man-made sources. *Atmospheric Environment* **14**, 1263–1278.

Curtis, P., C. P. Rinsland and J. S. Levine (1985) Free tropospheric carbon monoxide concentrations in 1950 and 1951 deduced from infrared total column amount measurements. *Nature* **318**, 250–254.

Czapski, G., S. V. Lymar and H. A. Schwarz (1999) Acidity of the Carbonate Radical. *Journal of Physical Chemistry* **A 103**, 3447–3450.

DAC (2011) Direct air capture of CO_2 with chemicals. Report for the American Physical Society (April 2011).

Dahl, E. E, E. S, Saltzman and W. J. de Bruyn (2003) The aqueous phase yield of alkyl nitrates from ROO + NO: implications for photochemical production in seawater. *Geophysical Research Letters* **30**, 1271.doi:10.1029/2002GL016811.

Damschen D. E. and L. R. Martin (1983) Aqueous aerosol oxidation of nitrous acid by O_2, O_3 and H_2O_2. *Atmospheric Environment* **17**, 2005–2011.

Danckwerts, P. V. (1970) Gas-liquid reactions. McGraw-Hill, New York, 276 pp.

D'Ans, J. and E. Lax (1943) Taschenbuch für Chemiker und Physiker. Springer-Verlag, Berlin, 1896 pp.

Darwin, C. (1839) The voyage of the Beagle. Wordsworth Classics (1997), 480 pp.

Dasgupta, S., B. Laplante, H. Wang and D. Wheeler (2002) Confronting the environmental Kuznets curve. *Journal of Economic Perspectives* **16**, 147–168.

Davidovits, P., C. E. Kolb, L. R. Williams, J. T. Jayne and D. R. Worsnop (2006) Mass accommodation and chemical reactions at gas-liquid interfaces. *Chemical Reviews* **106**, 1323–1354.

Davidson, E. A. and W. Kingerlee (1997) A global inventory of nitric oxide emissions from soils. *Nutriation and Cycling Agroecosystems* **48**, 37–50.

Davies, G., L. J. Kirschenbaum and K. Kustin (1968) Kinetics and stoichiometry of the reaction between manganese (III) and hydrogen peroxide in acid perchlorate solution. *Inorganic Chemistry* **7**, 146–151.

Dawson, G. A. (1977) Atmospheric ammonia from undisturbed land. *Journal of Geophysical Research* **82**, 3125–3133.

Day, A. L. and E. S. Shepherd (1913) Water and volcanic activity. *Bulletin Geological Society of America* **24**, 573–606.

De Bruyn, W. L., E. Swartz, J. H. Hu, J. A. Shorter, P. Davidovits, D. R. Worsnop, M. S. Zahniser and C. E. Kolb (1995) Henry's law solubility's and Śetchenow coefficients for biogenic reduced sulfur species obtained from gas-liquid uptake measurements. *Journal of Geophysical Research* **100D**, 7245–7251.

Decesari, S., M. C. Facchini, S. Fuzzi and E. Tagliavini (2000) Characterization of water soluble organic compounds in atmospheric aerosol: a new approach. *Journal of Geophysical Research* **105**, 1481–1489.

Decesari S., M. C. Facchini, E. Matta, M. Mircea, S. Fuzzi, A. R. Chughtai and D. M. Smith (2002) Water soluble organic compounds formed by oxidation of soot. *Atmospheric Environment* **36**, 1827–1832.

DeFelice, T. P. and V. K. Saxena (1991) The characterisation of extreme episodes of wet and dry deposition of pollutants on an above cloud-base forest during its growing season. *Journal of Applied Meteorology* **30**, 1548–1561.

DeFries, R. S., R. A. Houghton, M. C. Hansen, C. B. Field, D. Skole and J. Townshend (2002) Carbon emissions from tropical deforestation and regrowth based on satellite observations for the 1980s and 90s. *Proceedings of the National Academy of Sciences* **99**, 14256−14262.

Deister, U., R. Neeb, G. Helas and P. Warneck (1986) Temperature dependence of the equilibrium $CH_2(OH)_2 + HSO_3^- \rightleftharpoons CH_2(OH)SO_3^- + H_2O$ in aqueous solution. *Journal of Physical Chemistry* **90**, 3213−3217.

Delmas, R. J., J.-M. Ascencio and M. Legrand (1980) Polar ice evidence that atmospheric CO_2 20000 yr BP was 50% of present. *Nature* **284**, 155−157.

Delmas, R. and J. Servant (1987) Échanges biosphère-atmosphère d'azote et de soufre en zone intertropicale: Transferts entre les écosystème forêt et Savane en Afrique de l'Ouest. *Atmospheric Research* **21**, 53−74.

Delmas, R., D. Serça and C. Jambert (1997) Global inventory of NO_x sources. *Nutriation and Cycling Agroecosystems* **48**, 51−60.

DeMore, W. B., S. P. Sander, D. M. Golden, R. F. Hanpsoin, M. J. Kurylo, C. J. Howard, A. R. Ravishankara, C. E. Kolb and M. J. Molina (1997) Chemical kinetics and photochemicsal data for use in stratosphereic modeling. Jet Propulsion Laboratory Publication 97−4, NASA Evaluation Number 12, Pasadena, CA, USA, 278 pp.

Denlinger, R. P. (1992) A revised estimate for the temperature structure of oceanic lithosphere. *Journal of Geophysical Research* **97**, 7219−7222.

Denman, K. L., G. Brasseur, A. Chidthaisong, P. Ciais. P. M. Cox, R. E. Dickinson, D. Hauglustaine, C. Heinze, E. Holland, l. D. Jacob, U. Lohmann, S. Ramachandran, P. L. da Silva Dias, S. C. Wofsy and X. Zhang (2007) Couplings between changes in the climate system and biogeochemistry. In: Climate Change 2007: The Physical Science Basis. Contribution of Working Group I to the Fourth Assessment Report of the Intergovernmental Panel on Climate Change (Eds. S. Solomon, D. Qin, M. Manning et al.), Cambridge University Press, Cambridge, U.K., pp. 499−588.

Dentener, F. J. and P. J. Crutzen (1994) A three-dimensional model of the global ammonia cycle. *Journal of Atmospheric Chemistry* **19**, 331−369.

Deutscher, N. M., N. B. Jones, D. W. T. Griffith, S. W. Wood and F. J. Murcray (2006) Atmospheric carbonyl sulfide (OCS) variation from 1992−2004 by ground-based solar FTIR spectrometry. *Atmospheric Chemistry and Physics Dicussion* 6, 1619−1636.

Dickson, A. G., C. L. Sabine and J. R. Christian, Eds. (2007) Guide to best practices for ocean CO_2 measurements. PICES Special Publication 3, 191 pp. http://cdiac.esd.ornl.gov/oceans/Handbook_2007.html.

Diehl, T. L., M. Chin, T. C. Bond, S. A. Carn, B. N. Duncan, N. A. Krotkov and D. G.Streets (2007) Compilation of a global emission inventory from 1980 to 2000 for global model simulations of the long-term trend of tropospheric aerosols. American Geophysical Union, Fall Meeting 2007, abstract #A43D-10.

Dignon, J. and S. Hameed (1989) Global emissions of nitrogen and sulphur oxides from 1860 to 1980. *Journal Air Pollution Control Association* **39**, 180−186.

Dignon, J. (1992) NO_x and SO_x emissions from fossil fuels: a global distribution. *Atmospheric Environment* **26**, 1157−1164.

Dignon, J. (1995) Impact of biomass burning on the atmosphere. In: Ice core studies of global biogeochemical cycles (Ed. R. J. Delmas), NATO ASI Series Vol. 30, Springer-Verlag, Berlin, pp. 299−346.

Dismukes, G. C., V. V. Klimov, S. V. Baranov, Yu. N. Kozlov, J. DasGupta and A. Tyryshkin (2001) The origin of atmospheric oxygen on Earth: The innovation of oxygenic photosynthesis. *Proceedings of the National Academy of Sciences of the United States of America (PNAS)* **98**, 2170−2175.

Disselkamp, R. S., M. A. Carpenter, J. P. Cowin, C. M. Berkowitz, E. G. Chapman, R. A. Zaveri and N. S. Laulainen (2000) Ozone loss in soot aerosols. *Journal of Geophysical Research* **105**, 9767−9771.

Dlugokencky, E. J., S. Houweling, L. Bruhwiler, K. A. Masarie, P. M. Lang, J. B. Miller and P. P. Tans (2003) The growth rate and distribution of atmospheric methane. *Geophysical Research Letters* **30**, 10.1029/2003GL018126.

Doctorovich, F., D. E. Bikiel, J. Pellegrino, S. A. Suarez and M. A. Marti (2011) Anazone (HNO) interaction with hemeproteins and metalloporphyrins. *Coordination Chemistry Reviews* **255**, 2764–2784.

Dollard, G. J. and T. J. Davies (1993) Measurements of rural photochemical oxidants. In: The chemistry and deposition of nitrogen species in the troposphere (Ed. A. T. Cox), Royal Society of Chemistry, Cambridge, U.K., pp. 46–77.

Dore, J. E., R. Lukas, D. W. Sadler and D. M. Karl (2003) Climate-driven changes to the atmospheric CO_2 sink in the subtropical North Atlantic Ocean. *Nature* **424**, 754–757.

Dore, J. E., R. Lukas, D. W. Sadler, M. J. Church and D. M. Karl (2009) Physical and biogeochemical modulation of ocean acidification in the central North Pacific. *Proceedings of the National Academy of Sciences* **106**, 12235. doi:10.1073/pnas.0906044106.

Draganic, Z. D., A. Negron-Mendoza, K. Sehested, S. I. Vujosevic, R. Navarro-Gonzales, M. G. Albarran-Sacnchez and I. G. Draganic (1991) Radiolysis of aqueous solutions of ammonium bicarbonate overe a large dose range. *Radiation Physics and Chemistry* **38**, 317–321.

Drapcho C. M. and D. E. Brune (2000) The partitioned aquaculture system: impact of design and environmental parameters on algal productivity and photosynthetic oxygen production. *Aquacultural Engineering* **21**, 151–168.

Drexler, C., H. Elias, B. Fecher and K. J. Wannowius (1991) Kinetic investigation of sulfur(IV) oxidation by peroxo compounds ROOH in aqueous solution. *Fresenius Journal of Analytical Chemistry* **340**, 605–615.

Duane, W. and O. U. Scheuer (1913) Recherches sur la décomposition de l'eau par les rayons α. *Le Radium* **10**, 33–46; Décomposition de l'eau par les rayons α. *Comptes Rendus de l'Académie des Sciences* **156**, 466–467.

Dubey, M. K., H. Ziock, G. Rueff, S. Elliott and W. S. Smith (2002) Extraction of carbon dioxide from the atmosphere through engineering chemical sinkage. *Fuel Chemistry Division Preprints* **47**, 81–84.

Dubowski, Y., A. J. Colussi, C. Boxe and M. R. Hoffmann (2002) Monotonic increase of nitrite yields in the photolysis of nitrate in ice and water between 238 and 294 K. *Journal of Physical Chemistry A* **106**, 6967–6971.

Duce, R. A. (1969) On the source of gaseous chloride in the marine atmosphere. *Journal of Geophysical Research* **74**, 4597–4599.

Duce, R. A., P. S. Liss, J. T. Merril and E. L. Atlas (1991) The atmospheric input of trace species to the world ocean. *Global Biogeochemical Cycles* **5**, 193–259.

Duce, R. A. (1995) Sources, distributions, and fluxes of mineral aerosols and their relationship to climate. In: Aerosol forcing of climate (Eds. R. J. Charlson and J. Heintzenberg). John Wiley & Sons, New York, pp. 43–72.

Duce, R. A., J. LaRoche, K. Altieri, K. R. Arrigo, A. R. Baker, D. G. Capone, S. Cornell, F. Dentener, J. Galloway, R. S. Ganeshram, R. J. Geider, T. Jickells, M. M. Kuypers, R. Langlois, P. S. Liss, S. M. Liu, J. Middelburg, C. M. Moore, S. Nickovic, A. Oschlies, T. Pedersen, J. Prospero, R. Schlitzer, S. Seitzinger, L. L. Sorensen, M. Uematsu, O. Ulloa, M. Voss, B. Ward and L. Zamora (2008) Impacts of atmospheric anthropogenic nitrogen on the open ocean. *Science* **320**, 893–897.

Dufour, L. (1870) Notes sur le problème de la variation du climat. *Bulletin de la Société Vaudoise des Sciences Naturelles* **10**, 359–436.

Dumas, J.-B. (1882) Sur l'acide carbonique normal de l'air atmosphérique. *Annales de Chimie et de Physique* **26**, Ser. 5, 254–261.

Dunstan, W. R., H. A. D. Jowett and E. Goulding (1905) CLIII. – The rusting of iron. *Transaction of the Chemical Society* **87**, 1548–1574.

Durant, W. (1950) The age of faith. A history of Medieval civilization, christian, islamic and judaic, from Constantine to Dante: AD 325−1300. − The story of civilization Vol 4, Simon and Schuster, New York, 1196 pp.

Dusek, U., G. P. Frank, J. Curtius, F. Drewnick, J. Schneider, A. Kürten, D. Rose, M. O. Andreae, S. Borrmann and U. Pöschl (2010) Enhanced organic mass fraction and decreased hygroscopicity of cloud condensation nuclei (CCN) during new patricle formation events. *Geophysical Research Letters* **37**, L03804, doi:10.1019/2009GL040930.

Duyzer, J. H., A. M. H. Bouman, H. S. M. A. Diederen and R. M. van Aalst (1987) Measurement of dry deposition velocities of NH_3 and NH_4^+ over natural terrains. Report R 87/273. MT-TNO, Delft, The Netherlands.

Ebel, A., R. Friedrich and H. Rodhe, Eds. (1997) Transport and chemical transformation of pollutants in the troposphere. Vol. 7: Tropospheric modelling and emission estimation. Chemical transport and emission modelling on regional, global and urban scales. Springer-Verlag, Berlin, 440 pp.

Ebelmen, J. J. (1845) Sur les produits de la décomposition d'espèces minérales de la familie des silicates. *Annales des Mines* **7**, 3−66.

Eccleston, B. H., M. Morrison and H. M. Smith (1952) Elemental sulfur in crude oil. *Analytical Chemistry* **24**, 1745−1748.

Edenhofer, O., K. Lessmann, C. Kempert, M. Grubb and M. Köhler (2006) Induced technological change: Exploring its implications for the economics of atmospheric stabilization. Synthesis report from Innovation Modeling Comparison Project. *The Energy Journal, Special Issue* 2006: 57−107.

EDGAR (2011) Emission Database for Global Atmospheric Research. Joint project of the European Commission JRC Joint Research Centre and the Netherlands Environmental Assessment Agency (PBL). EDGAR version 4.2., http://edgar.jrc.ec.europa.eu.

Edmonds, M., T. M. Gerlach, R. A. Herd, A. J. Sutton and T. Elias (2005) The composition of volcanic gas issuing from Pu'u 'O'o, Kilauea Volcano, Hawaii, 2004−5. American Geophysical Union, Fall Meeting, abstract #V13G-08.

Edney, E. O., T. E. Kleindienst, M. Jaoui, M. Lewandowsky, J. H. Offenberg, W. Wang and M. Claeys (2005) Formation of 22-methyl tetrols and 2-methylglyceric acid in secondary organic aerosol from laboratory irradiated isoprene/NO_x/SO_2/air mixtures andf their detection in ambient PM2.5 samples collected in the eastern United States. *Atmospheric Environment* **39**, 5281−5289.

Edwards, J. H. (1995) Potential sources of CO_2 and the options for its large-scale utilisation now and in future. *Catalysis Today* **23**, 59−66.

Egli, J. (1947) Geschichte des Wortes Gas. *Das Mosaik. Kunst Kultur Natur* **2/4**, p. 125.

Egorova, T., V. Zubov, S. Jagovkina and E. Rozanov (1999) Lightning production of NO_x and ozone. *Physics and Chemistry of the Earth (C)* **24**, 473−479.

Ehhalt, D. H. (1974) The atmospheric cycle of methane. *Tellus* **26**, 58−70.

Ehhalt, D. H. (1999) Gas phase chemistry of the troposphere. In: Global aspects of atmospheric chemistry, Vol. 6 (Guest Ed. R. Zellner), Topics in physical chemistry (Eds. H. Baumgärtel, W. Grünbein, F. Hensel, Steinkopff, Darmstadt, pp. 21−109.

EIA (2009) International Energy Outlook 2009. United States Energy Information Administration.

Eibner, A. (1911) Über Lichtwirkungen auf Malerfarbstoffe. *Chemiker-Zeitung* **35**, 53−755, 774−776, 786−788.

Elderfeld, H., Ed. (2006) The oceans and marine geochemistry. Vol 6 of Treatise on geochemistry (Eds. H. D. Holland and K. K. Turekian), Elsevier, 646 pp.

Elliott, S., K. Lackner, H. Ziock, M. Dubey, H. Hanson, S. Ball, N. Ciszkowski and D. Blake (2001) Compensation of atmospheric CO_2 buildup through engineered chemical sinkage. *Geophysical Research Letters* **28**, 1235−1238.

Elshorbany, Y. F., B. Steil, C. Bruhl and J. Lelieveld (2012) Impact of HONO on global atmospheric chemistry calculated with an empirical parameterization in the EMAC model. *Atmospheric Chemistry and Physics* **12**, 9977−10000.

EMEP/CORINAIR Emission Inventory Guidebook (2006) EEA (European Environment Agency) (http://www.eea.europa.eu/publications/EMEPCORINAIR4).

Emrich, M. and P. Warneck (2000) Photodissociation of acetone in air: Dependence on pessure and wavelength. Behaviour of the excited singlet state. *Journal of Physical Chemistry* **A 104**, 9436−9442; see also for correction: **109**, 1752.

Engels, F. (1950) The part played by labor in the transition from ape to man. International Publishers, New York, 22 pp. [This is an excerpt in English translation from Friedrich Engels, "Anteil der Arbeit an der Menschwerdung des Affen", see for example: In Karl Marx/Friedrich Engels − Werke. Vol. 20. Dialektik der Natur (Dialectic of nature), (Karl) Dietz Verlag, Berlin, 1962, pp. 444−455].

Engler, C. (1879) Historisch-kritische Studien über das Ozon. *Leopoldina* **Heft 15**, Halle, 67 pp.

Ennis, C. P., J. R. Lane, H. G. Kjaergaard and A. J. McKinleyI (2009) Identification of the water amidogen radical complex. *Journal of the American Chemical Society* **131**, 1358−1359.

EPA (1972) Air Quality Data for Sulfur Dioxide, 1969, 1970, and 1971, Office of Air Programs, Publication No. APTD-1354, U.S. Environmental Protection Agency.

Erickson, D. J., III, and R. Duce (1988) On the global flux of atmospheric sea salt. *Journal of Geophysical Research* **93**, 14079−14088.

Erickson, D. J., III (1993) A stability dependent theory for air-sea gas exchange. *Journal of Geophysical Research* **98**, 8471−8488.

Erickson, D. J., III, C. Seuzaret, W. C. Keene and S. L. Gong (1999) A general circulation model based calculation of HCl and $ClNO_2$ production from sea salt dechlorination: Reactive chlorine emissions inventory. *Journal of Geophysical Research* **104**, 8347−8372.

Erickson, R. E., L. M. Yates, R. L. Clarke and D. McEwen (1977) The reaction of sulfur dioxide with ozone in water and is possible atmospheric significance. *Atmospheric Environment* **11**, 813−817.

Eriksson, E. (1959, 1960) The yearly circulation of chloride and sulfur in nature: meteorological, geochemical and pedological implications. Part 1. *Tellus* **11**, 375−403; Part 2. **12**, 63−109.

Eriksson, E. (1963) The yearly circulation of sulfur in nature. *Journal of Geophysical Research* **68**, 4001−4008.

Erisman, J. W. and A. van Pul (1997) Assessment of dry deposition and total acidifying loads in Europe. In: Transport and chemical transformation of pollutants in the troposphere. Vol. 4: Biosphere-atmosphere exchange of pollutants and trace substances (Ed. S. Slanina), Springer, The Netherlands, pp. 93−116.

Erisman, J. W. and G. P. Wyers (1993) Continuous measurements of surface exchange of SO_2 and NH_3 − implications for their possible interaction in the deposition process. *Atmospheric Environment* **27**, 1937−949.

Ermakov, A. N., G. A. Poskrebyshev and A. P. Purmal (1997) Sulfite oxidation: The state-of-the-art of the problem. *Kinetics and Catalysis* **38**, 295−308.

Ervens, B., A. G. Carlton, B. J. Turpin, K. E. Altieri, S. M. Kreidenweis and G. Feingold (2008), Secondary organic aerosol yields from cloud-processing of isoprene oxidation products. *Geophysical Research Letters* **35**, L02816, doi:10.1029/2007GL031828.

Essington, M. E. (2003) Soil and water chemistry: An integrative approach. CRC Press, Boca Raton, FL (USA), 552 pp.

Etheridge, D. M., G. I. Pearman and F. de Silva (1988) Atmospheric trace-gas variations as revealed by air trapped in an ice core from Law Dome, Antarctica. Annals of Glaciology **10**, 28−33.

Etheridge, D. M., G. I. Pearman and P. J. Fraser (1992) Changes in tropospheric methane between 1841 and 1978 from a high accumulation rate Antarctic ice core. *Tellus* **44B**, 282−294.

Etheridge, D. M., L. P. Steele, R. L. Langenfelds, R. J. Francey, J.-M. Barnola and V. I. Morgan (1996) Natural and anthropogenic changes in atmospheric CO_2 over the last 1000 years from air in Antarctic ice and firn. *Journal of Geophysical Research* **101**, 4115−4128.

Etheridge, D. M., L. P. Steele, R. J. Francey and R.L. Langenfelds (1998) Atmospheric methane between 1000 AD and present: Evidence of anthropogenic emissions and climatic variability. *Journal of Geophysical Research* **103**, 15979−15993.

Eugster, H. P. and J. Munoz (1966) Ammonium micas: Possible sources of atmospheric ammonia and nitrogen. *Science* **151**, 683−686.

Evans, M. G. and N. Uri (1949) Dissociation constant of hydrogen peroxide and the electron affinity of the HO_2 radical. *Transaction of the Faraday Society* **45**, 224−230.

Evelyn, J. (1661) Fumifugium: or, the inconvenience of the aer, and smoake of London dissipated together with some remedies humbly proposed. Reprint (original by Godbid, London) by Swan Press, Haywards Heath, 50 pp.

Exner, M., H. Herrmann and R. Zellner (1992) Laser-based studies of reactions of the nitrate radical in aqueous solution. *Berichte der Bunsen-Gesellschaft* **96**, 470−477.

Eyring, V., H. W. Köhler, J. van Aardenne and A. Lauer (2005) Emissions from international shipping; 1. The last 50 years. *Journal of Geophysical Research* **110**, D17305, doi:10.1029/2004JD005619.

Fabian, P., R. Borchers, R. Leifer, B. H. Subbaraya, S. Lal and M. Boy (1996) Global stratospheric distribution of halocarbons. *Atmospheric Environment* **30**, 1787−1796.

Facchini, M. C., S. Fuzzi, M. Kessel, W. Wobrock, W. Jaeschke, B. G. Arends, J. J. Möls, A. Berner, I. Solly, C. Kruisz, G. Reischl, S. Pahl, A. Hallberg, J. A. Ogren, H. Fierlinger-Oberlinninger, A. Marzorati and D. Schell (1992) The chemistry of sulfur and nitrogen species in a fog system. A multiphase approach. *Tellus* **44B**, 505−521.

Fachini, M. C. and S. Fuzzi (1993) Atmospheric acidity: A useful tool to describe the distribution of chemical species among the different phases in fog. In: Photo-oxidants: Precursors and products (Eds. P. M. Borrell, P. Borrell, T. Cvitas, W. Seiler), Proceedings. EUROTRAC Symp. 1992, SPB Academic Publs., Den Haag, pp. 505−509.

Facchini, M. C., S. Decesari, M. Mircea, S. Fuzzi and G. Loglio (2000) Surface tension of atmospheric wet aerosol and cloud/fog droplets in relation to their organic carbon content and chemical composition. *Atmospheric Environment* **34**, 4853−4857.

Facchini M. C. and C. D. O'Dowd (2007) Organic marine aerosol: State-of-the-art and new findings. In: Nucleation and atmospheric aerosols, Proceeding of the 17th International Conference, Galway, Ireland (Eds. C. D. O'Dowd and P. Wagner). Springer, Netherlands pp. 1045−1049.

Faghri, A. and Y. Zhang (2006) Transport phenomena in multiphase systems. Elsevier, 1030 pp.

Falkowski, P. G. and L. V. Godfrey (2008) Electrons, life and the evolution of Earth's oxygen cycle. *Philosophical Transaction of the Royal Society, Series B* **363**, 2705−2716.

FAO (2001) Global forest resources assessment 2000. UN Food and Agriculture Organization, Rome, 479 pp.

FAO (2003) State of the worlds forest 2003. UN Food and Agriculture Organization, Rome, 151 pp. (http://www.fao.org/DOCREP/005/Y7581E/Y7581E00.HTM).

FAO/IFA (2001) Global estimates of gaseous emissions of NH_3, NO and N_2O from agricultural land. Published by FAO and IFA. Rome, 106 pp.

Farman, J. C., B. G. Gardiner and J. D. Shanklin (1985) Large losses of total ozone in Antarctica reveal seasonal ClO_x/NO_x interaction. *Nature* **315**, 207−210.

Farquhar, J. and D. T. Johnston (2008) The oxygen cycle of the terrestrial planets: Insights into the processing and history of oxygen in surface environments. *Reviews in Mineralogy and Geochemistry* **68**, 463−492.

Farsky, F. (1877) Bestimmungen der atmosphärischen Kohlensäure in den Jahren 1874–1875 zu Tabor in Böhmen. *Sitzungsberichte der Kaiserlichen Akademie der Wissenschaften,* Math.-Nat. Classe. Wien, Band 74, 67–77.

Fasham, M. J. R., Ed. (2003) Ocean biogeochemistry. Springer-Verlag, Berlin, 297 pp.

Faust, B. C. and J. M. Allen (1992) Aqueous-phase photochemical sources of peroxyl radicals and singlet molecular oxygen in clouds and fog. *Journal of Geophysical Research* **97**, 12913–12926.

Fearnside, P. M. (2000) Global warming and tropical land-use change: grennhouse gas emissions from biomass burning, decomposition and soils in forest conversion, shifting cultivation and secondary vegetation. *Climatic Change* **46**, 115–158.

Feely, R. A., C. L. Sabine, J. M. Hernandez-Ayon, D. Lanson and B. Hales (2008) Evidence for upwelling of corrosive "acidified" water onto the continental shelf. *Science* **320**, 1490–1492.

Feingold, G., W. R. Cotton, S. M. Kreidenweis, and J. T. Davis (1999) Impact of giant cloud condensation nuclei on drizzle formation in marine stratocumulus: Implications for cloud radiative properties. *Journal of Atmospheric Science* **56**, 4100–4117.

Feng, J. and D. Möller (2004) Characterization of water-soluble macromolecular substances in cloud water. *Journal of Atmospheric Chemistry* **48**, 217–233.

Fenger, J. (2009) Air pollution in the last 50 years – from local to global. *Atmospheric Environment* **43**, 13–22.

Fenton, H. J. H. (1876) The oxidation of tartaric acid in presence of iron. *Chemical News* **33**, 190.

Fenton, H. J. H. (1894) Constitution of a new dibasic acid, resulting from the oxidation of tartaric acid. *Journal of the Chemical Society, Transactions* **69**, 546–562.

Fersman, A. E. (1923) Chemical elements of the earth and the cosmos (in Russian). Petersburg.

Finlayson-Pitts, B. J. and J. N. Pitts, Jr. (2000) Chemistry of the upper and lower atmosphere – theory, experiments and applications. Academic Press, San Diego (USA), 969 pp.

Finlayson-Pitts, B. J., L. M. Wingen, A. L. Sumner, D. Syomin and K. A. Ramazan (2003) The heterogeneous hydrolysis of NO_2 in laboratory systems and in outdoor and indoor atmospheres: An integrated mechanism. *Physical Chemistry and Chemical Physics* **5**, 223–242.

Firestone, M. K. and E. A. Davidson (1989) Microbiological basis of NO and N_2O production and consumption in soil. In: Exchange of trace gases between terrestrial ecosystems and the atmosphere (Eds. M. O. Andreae and D. S. Schimel). John Wiley & Sons, Chichester, pp. 7–21 Fischer, F. u. O. Ringe (1908) Die Darstellung von Argon aus Luft mit Calciumcarbid. *Berichte der Deutschen Chemischen Gesellschaft* **41**, 2017–2030.

Fischer, E. V., D. J. Jacob, D. B. Millet, M. Yantosca and J. Mao (2012) The role of the ocean in the global atmospheric budget of acetone. *Geophysical Research Letters* **39**, L01807, doi:10.1029/2011GL050086.

Fischer, H., M. Wahlen, J, Smith, D. Mastroianni and B. Deck (1999) Ice core records of atmospheric CO_2 around the last three glacial terminations. *Science* **283**, 1712–1714.

Fischlin, A, G. F. Midgley, J. T. Price, R. Leemans, B. Gopal, C. Turley, M. D. A. Rounsevell, O. P. Dube, J. Tarazona and A. A. Velichko (2007) Ecosystems, their properties, goods, and services. In: Climate change: Impacts, adaption and vulnerability. Contribution of Working Group II to the fourth Assessment Report of the Intergovernmental Panle of Climate Change (Eds. M. L. Parry, O. F. Canziani, J. P. Palutikof, P. J. van der Linden and C. E. Hanson), Cambridge University Press, Cambridge, U.K., pp. 211–272.

Fittbogen, J. and P. Hässelbarth (1879) Ueber locale Schwankungen im Kohlensäuregehalt der atmosphärischen Luft. *Jahresbericht ueber die gesammten Fortschritte auf dem Gebiet der Agricultur-Chemie* **22**, 67–71.

Fleming, J. R. (2006) James Croll in context: The encounter between climate dynamics and geology in the second half of the nineteenth century. *History of Meteorology* **3**, 43–53.

Fleming, J. R. (2010) Gilbert N. Plass: Climate science in perspective. *American Science* **98**, 58. doi:10.1511/2010.82.58.

Fodor, J. von (1879) Experimentelle Untersuchungen über Boden und Bodengase. *Vierteljahresschrift für öffentliche Gesundheitspflege* **7**, 205–237.

Fogg, P. (2003) Details and measurements of factors determining the uptake of gases by water. In: Chemicals in the atmosphere – solubility, sources and reactivity (Eds. P. G. T. Fogg and J. M. Sangster), John Wiley & Sons, Chichester, pp. 127–146.

Foken, Th., R. Dlugi and G. Kramm (1995) On the determination of dry deposition and emission of gaseous compounds at the biosphere-atmosphere interface. *Meteorologische Zeitschrift* **4**, 91–118.

Follows, M. and T. Oguz, Eds. (2007) The ocean carbon cycle and climate. Proceedings of the NATO ASI on Ocean Carbon Cylce and Climate, Ankara, Turkey, from 5 to 16 August 2002, Springer-Verlag, Berlin, 408 pp.

Foner, N. and R. L. Hudson (1953) Detection of the HO_2 radical by mass spectrometry. *Journal of Chemical Physics* **21**, 1608–1609.

Fonrobert, E. (1916) Das Ozon. Verlag von F. Enke, Stuttgardt, 282 pp.

Fonselius, S., F. Koroleff and K.-E. Wärme (1956) Carbon dioxide in the atmosphere. *Tellus* **8**, 176–183.

Fontenelle, J. (1823) Dissertation sur les eaux minérales connues sous le nom de Bains. De Renses. *Also*; Recherches historiques, chimiques et médicales sur l'air marécageux. Ouvrage couronné par l'Académie des sciences de Lyon. P., Gabon, 155 pp.

Forster, P., V. Ramaswamy, P. Artaxo, T. Berntsen, R. Betts, D. W. Fahey, J. Haywood, J. Lean, D. C. Lowe, G. Myhre, J. Nganga, R. Prinn, G. Raga, M. Schulz and R. van Dorland (2007) Changes in atmospheric constituents and in radiative forcing. In: Climate change 2007: The physical science basis. Contribution of Working Group I to the Fourth Assessment Report of the Intergovernmental Panel on Climate Change (Eds. S. Solomon, D. Quin, M. Manning, Z. Chen, M. Marquis, K. B. Averyt, M. Tignor and H. L. Miller), Cambridge University Press, Cambridge, U.K., pp. 129–234.

Forsythe, W., Ed. (2003) Smithsonian physical tables. Thermochemistry of chemical substances. 9th Edition, Knovel Corp., New York.

Foster, G. and S. Rahmstorf (2011) Global temperature evolution 1979–2010. *Environmental Research Letters* **6**, 044022.

Fowler, D., I. D. Leith, J. Binnie, A. Crossley, D. W. F. Inglis, T. W. Choularton, M. Gay, J. W. S. Longhurst and D. E. Conland (1995) Orographic enhancement of wet deposition in the United Kingdom: continuous monitoring. *Water, Air and Soil Pollution* **85**, 2107–2112.

Fox, C. B. (1873) Ozone and antozone. Their history and nature. When, where, why, how ozone is observed in the atmosphere? J. & J. Churchill, London, 329 pp.

Franklin, B. (1784) Meteorological imaginations and conjectures. *Transactions of the Manchester Philosophical Society* **1**, 373–377.

Franks, F. (2000) Water: A matrix of life. 2nd Edition. Royal Society of Chemistry, Cambridge, U.K., 192 pp.

Franzle, S. (2009) Chemical elements in plant and soil: Parameters controlling essentiality. Springer-Verlag, Berlin, 196 pp.

Frey, E. v. (1889) Der Kohlensäuregehalt der Luft in und bei Dorpat bestimmt in den Monaten September 1888 bis Januar 1889. Doctor Thesis, Medizinische Fakultät der Kaiserlichen Universität Dorpat, 49 pp.

Fricke, H. (1934) Reduction of oxygen to hydrogen peroxide by the irradiation of its aqueous solution with X-rays. *Journal of Chemical Physics* **2**, 556–557.

Friedfeld S., M. Fraser, K. Ensor, S. Tribble, D. Rehle, D. Leleux and F. Tittel (2002) Statistical analysis of primary and secondary atmospheric formaldehyde. *Atmospheric Environment* **36**, 4767–4775.

Friedheim, C., Ed. (1907) Gmelin-Kraut's Handbuch der anorganischen Chemie. Band 1, Abteilung 1. Carl Winters Universitätsbuchhandlung, Heidelberg, 888 pp.

Friedlander, S. K. (1977) Smoke, dust and haze. Fundamentals of aerosol behavior. John Wiley & Sons, New York, 317 pp.

Friedli, H., H. Lötscher, H. Oeschger, U. Siegenthaler and B. Stauffe (1986) Ice core record of 13C/12C ratio of atmospheric CO_2 in the past two centuries. *Nature* **324**, 37–38.

Friedlingstein P., R. A. Houghton, G. Marland, J. Hackler, T. A. Boden, T. J. Conway, J. G. Canadell, M. R. Raupach, P. Ciais and C. Le Quéré (2010) Update on CO_2 emissions. *Nature Geoscience* **3**, 811–812, doi:10.1038/ngeo1022.

Friedrich, R. and A. Obermeier (2000) Emissionen von Spurenstoffen. In: Handbuch der Umweltveränderungen und Ökotoxikologie, Band 1A: Atmosphäre (Ed. R. Guderian), Springer-Verlag, Berlin, pp. 168–173.

Friedrich, R. and S. Reis, Eds. (2004) Emissions and air pollutants. Measurements, calculations and uncertainties. Springer-Verlag, Berlin, 335 pp.

Friedrich, R. (2009) Natural and biogenic emissions of environmentally relevant atmospheric trace constituents in Europe. *Atmospheric Environment* **43** 1377–1379.

Friend, J. P. (1973) The global sulfur cycle. In: Chemistry of the lower atmosphere (Ed. I. E. Rasool), Plenum Press, New York, pp. 177–201.

Fröhlich C. and J. Lean (1998) The Suns total irradiance: Cycles, trends and related climate change uncertainties since 1978. *Geophysical Research Letters* **25**, 4377–4380.

From, E. and C. D. Keeling (1986) Reassessment of late 19th century atmospheric carbon dioxide variations in the air of western Europe and the British Isles based on an unpublished analysis of contemporary air masses by G. S. Callendar. *Tellus* **38B**, 87–105.

Fu, T.-M., D. J. Jacob and C. L. Head (2009) Aqueous-phase reactive uptake of dicarbonyls as a soruce of organic aerosols over eastern North America. *Atmospheric Environment* **43**, 1814–1821.

Fujishima, A. (1972) Electrochemical photolysis of water at a semiconductor electrode. *Nature* **238**, 37–38.

Fujishima, A., K. Hashimoto and T. Watanabe (1999) TiO_2 photocatalysis – fundamentals and applications. BKC Inc. Tokyo, 176 pp.

Fujita, S., Y. Ichikawa and R. K Kawaratan (1991) Preliminary inventory of sulfur dioxide emissions in East Asia. *Atmospheric Environment* **25A**, 1409–1411.

Fujita, E. (1999) Photochemical carbon dioxide reduction with metal complexes. *Coordination Chemistry Review* **185–186**, 372–384.

Fujun Du, F., B. Parise and P. Bergman (2011) Production of interstellar hydrogen peroxide (H_2O_2) on the surface of dust grains. *Astronomy and Astrophysics* 11/2011; doi:10.1051/0004–6361/201118013.

Fukuto, J. M. J. Y. Cho and S. H. Switzer (2000) The chemical properties of nitric oxide and related nitrogen oxides. In: Nitric oxide, biology and pathobiology (Ed. L. J. Ignarro), Academic Press, San Diego, pp. 23–40.

Fukuto, J. M, C. H. Switzer, K. M. Miranda and D. A. Win (2005) Nitroxyl (HNO): Chemistry, biochemistry, and pharmacology. *Annual Review of Pharmacology and Toxicology* **45**, 335–355.

Fuzzi, S., M. C. Fachini, G. Orsi, J. A. Lind, W. Wobrock, M. Kessel, R. Maser, W. Jaeschke, K. H. Enderle, B. G. Arends, A. Berner, I. Solly, C. Kruisz, G. Reischl, S. Pahl, U. Kaminski, P. Winkler, J. A. Ogren, K. J. Noone, A. Hallberg, H. Fierlinger-Oberlinninger, H. Puxbaum, A. Marzorati, H,-C. Hansson, A. Wiedensohler, B. Svenningsson, B. G. Martinsson, D. Schell und H.-W. Georgii (1992) The Po Valley Fog Experiment 1989. *Tellus* **44B**, 448–468.

Fuzzi, S., Ed. (1995) The Kleiner Feldberg Cloud Experiment 1990. Kluwer Academic Publishers, Dordrecht, The Netherlands. 264 pp.

Fuzzi, S., M. C. Facchini, G. Orsi, G.;Bonforte, W. Martinotti, G. Ziliani, P. Mazzali, P. Rossi, P. Natale, M. M. Grosa, E. Rampado, P. Vitali, R. Raffaelli, G. Azzini and S. Grotti (1996) The NEVALPA project: a regional network for fog chemical climatology over the Po Valley basin. *Atmospheric Environment* **30A**, 201−213.

Fuzzi, S. and D. Wagenbach, Eds. (1997) Cloud multi-phase processes and high alpine air and snow chemistry. Transport and chemical transformation of pollutants in the troposphere (EUROTRAC), Vol. 5, Springer-Verlag, Berlin, 286 pp.

Fuzzi, S. and A. Flossmann (1999) A first experimental approach to some key questions in cloud chemistry. *Proceedings of EUROTRAC-2 Symposium '98*, Vol. 1: Transport and Chemical Transformation in the Troposphere (Eds. P. M. Borrell and P. Borrell), WITPress Comp. Mechan. Publ., Southampton, pp. 430−434.

Gäb, S., E. Hellpointner, W.V. Turner and F. Korte (1985) Hydroxymethyl hydroperoxide and bis-(hydroxymethyl) peroxide from gas-phase ozonolysis of naturally occuring alkenes. *Nature* **316**, 535−536.

Galbally, I. E. (1989) Factors controlling NOx emission from soils. In: Exchange of trace gases between terrestrial ecosystems and the atmosphere (Eds. M. O. Andreae and D. S. Schimel). John Wiley & Sons, Chichester, pp. 23−37.

Galbally, I. E. and W. Kirstine (2002) The production of methanol by flowering plants and the global cycle of methanol. *Journal of Atmospheric Chemistry* **42**, 195−229.

Gallagher, P. H. (1923) Mechanism of oxidation in the plant: Part 1. The oxygenase of Bach and Chodat: Function of lecithins in respiration *Biochemical Journal* **17**, 515−529

Gallard, H. and U. von Gunten (2002) Chlorination of natural organic matter. Kinetics of chlorination and of THM formation. *Water Research* **36**, 65−74.

Galloway, J. N., G. E. Likens, W. C. Keene and J. M. Miller (1982) The composition of precipitation in remote areas of the world. *Journal of Geophysical Research* **87**, 8771−8786.

Galloway J. N., W. H. Schlesinger, H. Levy II, A. Michaels and J. L. Schnoor (1995) Nitrogen fixation: anthropogenic enhancement-environmental response. *Global Biogeochemical Cycles* **9**, 235−252.

Galloway, J. N., F. J. Dentener, D. G. Capone, E. W. Boyer, R. W. Howarth, S. P. Seitzinger, G. P. Asner, C. C. Cleveland, P. A. Green, E. A. Holland, D. M. Karl, A. F. Michaels, J. H. Porter, A. R. Townsend and C. J. Vöosmarty (2004) Nitrogen cycles: Past, present, and future. *Biogeochemistry* **70**, 153−226.

Galloway, M. M., P. S. Chhabra, A. W. H. Chahn, J. D. Surrat, R. C. Flagan, J. H. Seinfeld and F. N. Keutsch (2009) Glyoxal uptake on ammonium sulphate seed aerosol: Reaction products and reversibility of uptake under dark and irradiated conditions. *Atmospheric Chemistry and Physics* **9**, 3331−3345.

Galster, H. (1991) pH Measurement: Fundamentals, methods, applications, instrumentation. Wiley-VCH Verlag GmbH, 356 pp.

Ganzeveld, L., J. Lelieveld, F. J. Dentener, M. C. Krol, A. F. Bouwman and G.-J. Roelofs (2002) Global soil-biogenic emissions and the role of canopy processes. *Journal of Geophysical Research* **107**, doi:10.1029/2001JD001289.

Garland, J. A. (1978) Dry and wet removal of sulphur from the atmosphere. *Atmospheric Environment* **12**, 349−362.

Garratt, J. R. (1992) The atmospheric boundary layer. Cambridge University Press, Cambridge, U.K., 334 pp.

Garrison, W. M., H. R. Haymond, W. Bennet and S. Cole (1959) Radiation-induced oxidation of aqueous acetic acid − oxygen solutions. *Radiation Research* **10**, 273−282.

Gates, W. L. (1975) Numerical modelling of climatic change: A review of problems and prospects. In: Proceedings. WMO/IAMAP Symp. on Long-term Climatic Fluctuations. Norwich, WMO report No. 421, Geneva, pp. 343−354.

Gautier, A. (1900) Combustible gases of the atmosphere: Air of the sea. Existence of free hydrogen in the terrestrial atmosphere. *Comptes Rendus de l'Académie des Sciences* **130**, 86−90.

Gautier, A. (1900) Gaz combustibles de l'atmosphère: Air des villes. *Comptes Rendus de l'Académie des Sciences* **130**, 1677–1684.

Gautier, A. (1901) Hydrogen in air. *Nature* **63**, 478–479.

Gautier, A. (1903) A propos de la composition des gaz des fumerolles du mont Pelé. Remarques sur l'origine des phénomènes volcaniques. *Comptes Rendus de l'Académie des Sciences* **136**, 16–20.

Gautier, A. (1906) Action de l'oxyde de carbone au rouge sur la vapeur d'eau et de l'hydrogene sur l'acide carbonique. Application de l'etude de ces reactions aux phenomenes volcaniques. *Comptes Rendus de l'Académie des Sciences* **142**, 1382–1387.

Ge, X., A. S. Wexler und S. L. Clegg (2011) Atmospheric Amines – Part I. A review. *Atmospheric Environment* **45**, 524–546.

Geever, M., C. D. O'Dowd, S. van Ekeren, R. Flanagan, E. D. Nilsson, G. de Leuw and U. Rannik (2005) Submicron sea spray fluxes. *Geophysical Research Letters* **32**, 15 Art.No. L15810.

Gelencsér, A. (2004) Carbonaceous aerosol. Vol. 30: Atmospheric and oceanographic science library (Eds. L. A. Mysak and K. Hamilton), Springer, Dordrecht, 350 pp.

Gentleman, D. J. (2010) Biology meets geology through chemistry. *Environmental Science & Technology* **44**, 1–2.

Genz, H. (1994) Die Entdeckung des Nichts. Leere und Fülle im Universum. München, Hanser, 416 pp.

Gerber, H. (1991) Supersaturation and droplet spectra evolution in fog. *Journal of Atmospheric Sciences* **48**, 2569–2588.

Gerlach, T. M. (1991) Present-day CO_2 emissions from volcanoes. *Transactions of the American Geophysical Union (EOS)* **72**, pp. 249 and 254–255.

Gehler, J. S. T. (1789) Physikalisches Wörterbuch oder Versuch einer Erklärung der vornehmsten Begriffe und Kunstwörter der Naturlehre mit kurzen Nachrichten von der Geschichte der Erfindungen und Beschreibungen der Werkzeuge begleite. Vol. 2. Ed. B. Schwickert, Leipzig, pp. 762–770.

Gehler, J. S. T. (1830) Physikalisches Wörterbuch, neu bearbeitet von Brandes, Gmelin, Horner, Mucke, Pfaff. Fünfter Band. Zweite Abtheilung. I und K. E. B. Schwickert, Leipzig, pp. 856–900.

GEO 2000 (1999) Global environmental outlook 2000. UN Environmental Programme, Earthscan Publ. Ltd, UK.

George, P. (1947) Some experiments on the reactions of potassium superoxide in aqueous solution. *Discussions of the Faraday Society* **2**, 196–205.

George, C., R. S. Strekowski, J. Kleffmann, K. Stemmler and M. Ammann (2005) Photoenhanced uptake of gaseous NO_2 on solid organic compounds: a photochemical source of HONO? *Faraday Discussions* **130**, 195–210.

Georgii, H. W. (1970) Contribution to the atmosphere sulfur budget. *Journal of Geophysical Research* **75**, 2365–2371.

Georgiou, G. and L. Masip (2003) An overoxidation journey with a return ticket. *Science* **300**, 592–594.

Gest, H. (1993) Photosynthetic and quasi-photosynthetic bacteria. *FEMS Microbiology Letters* **112**, 1–6.

Gest, H. (2002) History of the word photosynthesis and evolution of its definition. *Photosynthesis Research* **73**, 7–10.

Ghosh, R. S., S. D. Ebbs, J. T. Bushey, E. F. Neuhauser and G. M. Wong-Chong (2006) Cyanide cycle in nature. In: Cyanides in water and soil (Eds. D. A. Dzombak, R. S. Ghosh and G. M. Wong-Chong), CRC Press, Boca Raton, FL (USA), pp. 226–236.

Gianguzza, A., E. H. Lieb and S. Sammartano (2002) Chemistry of marine water and sediments. Springer-Verlag, Berlin, 532 pp.

Gilbert, O. (1907) Die meteorologischen Theorien des griechischen Altertums. B. G. Teubner, Leipzig, 746 pp.

Gilge, St., D. Kley and A. Voltz-Thomas (2000) Messungen von Wasserstoffperoxid – ein Beitrag zur Charakterisierung der limitierenden Faktoren bei der Ozonproduktion. In: Troposphärisches Ozon. Proceedings VDI Symposium, Band 32 Schriftenreihe Kommission Reinhaltung der Luft, Düsseldorf, pp. 379–382.

Gill, A. E. (1982) Atmosphere-ocean dynamics. International Geophysics Series Vol. 30, Academic Press, 662 pp.

Gilm, V. H. (1857) Über die Kohlensäure-Bestimmung der atmosphärischen Luft. *Sitzungsberichte der Kaiserlichen Akademie der Wissenschaften*, Math.-Nat. Classe. Wien, Band 24, 279–284.

Glasby, G. P (2006) Abiogenic origin of hydrocarbons: An historical overview. *Resource Geology* **56**, 85–98.

Glindemann, D., M. Edwards and P. Kuschk (2003) Phosphine gas in the upper troposphere. *Atmospheric Environment* **37**, 2429–2433.

Glindemann, D., M. Edwards and O. Schrems (2004) Phosphine and methylphosphine production by simulated lightning – a study for the volatile phosphorous cycle and cloud formation in the earth atmosphere. *Atmospheric Environment* **38**, 6867–6874.

Glindemann, D., M. Edwards, J. Liu and P. Kuschk (2005a) Phosphine in soils, sludges, biogases and atmospheric implications – a review. *Ecological Engineering* **24**, 457–463.

Glindemann, D., M. Edwards and P. Morgenstern (2005b) Phosphine from rocks: Mechanically driven phosphate reduction? *Environmental Science and Technology* **39**, 8295–8299.

Glindemann, D., A. Dietrich, H.-J. Staerk and P. Kuschk (2006) The two odors of iron when touched or pickled: (skin) carbonyl compounds and organophosphines. *Angewandte Chemie International Edition* **45**, 7006–7009.

Gmelin, L. G. (1827) Handbuch der theoretischen Chemie. Erster Band. 3th ed. by F Varrentrapp, Frankfurt am Main.

Gmelin, L. G. (1848) Handbuch der Chemie. Vierter Band. Handbuch der organischen Chemie – Erster Band. Heidelberg, 936 pp.

Gmelin (1927) Handbuch der anorganischen Chemie. 8. Auflage, System-Nr. 6, Chlor. Verlag Chemie, Berlin, p. 251.

Gmelin (1936) Gmelins Handbuch der anorganischen Chemie. System-Nr. 23, Ammonium, Lieferung 1, Verlag Chemie Berlin, 242 pp.

Gmelin (1936) Gmelins Handbuch der anorganischen Chemie. 8. Auflage, System-Nr. 4, Stickstoff, Lieferung 3, p. 718.

Gmelin (1943) Gmelins Handbuch der anorganischen Chemie. 8. Auflage, System-Nr. 3: Sauerstoff, Lieferung 1–2, Verlag Chemie, Weinheim, pp. 298–300.

Gmelins Handbuch der anorganischen Chemie (1966) 8. Auflage, System-Nr. 3, Sauerstoff, Lieferung 7 (Wasserstoffperoxid). Verlag Chemie Weinheim, pp. 2097–2526.

Gmelin (1969) Gmelins Handbuch der anorganischen Chemie. 8. Auflage, System-Nr. 3, Sauerstoff, Lieferung 8 (Radikale OH und HO_2, Hydrogenoxid HO_3, höhere Wasserstoffperoxide). Verlag Chemie Weinheim, pp. 2527–2947.

Gmelin (1973) Gmelins Handbuch der anorganischen Chemie. 8. Auflage, System-Nr. 14: Kohlenstoff, Teil C 3, Verlag Chemie, Weinheim, 160 pp.

Gold, Th. (1999) The deep hot biosphere. Springer-Verlag, Berlin, 235 pp.

Goldemberg, J., Ed. (2000) World energy assessment. Energy and the challenge of sustainability. United Nations Development Programme, United Nations Department.

Goldewijk, K. K. (2001) Estimating global land use change over the past 300 years: the HYDE database. *Global Biogeochemical Cycles* **15**, 417–433.

Gombosi, T. I. (1994) Gaskinetic theory. Cambridge University Press, Cambridge, U.K., 294 pp.

Gómez-González, A. W., J. D. Surrat, F. Cuyckens, R. Szmigielsky, R. Vermeylen, M. Jaoui, M. Lewandowski, J. H. Offenberg, T. E. Kleindienst, E. O. Edney, F. Blockhuys, C. Van Alsenoy, W. Maenhout and M. Clayes (2008) Characterization of organosulfates from the photooxidation of isoprene and unsaturated fatty acids in ambient aerosl using liquid chromatography(-)electrospray ionization mass spectrometry. *Journal of Mass Spectrometry* **43**, 371−383.

Gomez-Peralta, D., S. F. Oberbauer, M. E. McClain and T. E. Philippi (2008) Rainfall and cloud-water interception in tropical montane forests in the eastern Andes of Central Peru. *Forest Ecology and Management* **255**, 1315−1325.

Gong, S. L., L. A. Barrie, J. Prospero, D. L. Savoie, G. P. Ayers, J.-P. Blanchet and L. Spacek (1997) Modeling sea-salt aerosols in the atmosphere, Part 2: Atmospheric concentrations and fluxes. *Journal of Geophysical Research* **102**, 3819−3830.

Goodeve, C. F. (1937) The absorption spectra and photo-sensitising activity of white pigments. *Transactions of the Faraday Society* **33**, 340-347.

Gore, R. (1989) Extinctions. *National Geographic* **175**, 662−699.

Gorup-Besanez, E. von (1859) Ueber die Einwirkung des Ozons auf organische Verbindungen. *Annalen der Chemie* **110**, 86−107.

Goudie, A. S. and N. J. Middleton (2006) Desert dust in the global system. Springer-Verlag, Berlin, 287 pp.

Gough, C. M. (2012) Terrestrial Primary Production: Fuel for Life. *Nature Education Knowledge* **3**(10) 28.

Goy, C. A., D. H. Shaw and H. O. Pritchard (1965) The reactions of CN radicals in the gas phase. *Journal of Physical Chemistry* **69**, 1504−1507.

Graedel, T. E. (1979) Chemical compounds in the atmosphere. Academic Press, New York, 440 pp.

Graedel, T. E. and C. J. Weschler (1981) Chemistry within aqueous atmospheric aerosol and raindrop. *Reviews of Geophysics and Space Physics* **19**, 505−539.

Graedel, T. E., M. L. Mandich and C. J. Weschler (1986) Kinetic model studies of atmospheric droplet chemistry. 2. Homogeneous transition metal chemistry in raindrops. *Journal of Geophysical Research* **91**, 5205−5221.

Graedel T. E., D. T. Hawkins and L. D. Claxton (1986) Atmospheric chemical compounds, sources, occurences and bioassays. Academic Press, New York, 732 pp.

Graedel, T. E., T. S. Bates, A. F. Bouman, D. Cunnold, J. Dignon, I. Fung, D. J. Jacob, B. K. Lamb, J. A. Logan, G. Marland, P. Middleton, J. M. Pacyna, M. Placet and C. Veldt (1993) A compilation of inventories of emissions to the atmosphere. *Global Biogeochemical Cycles* **7**, 1−26.

Graedel, T. E. and P. J. Crutzen (1993) Atmospheric change: An earth system perspective. W. H. Freeman, New York, 446 pp.

Graedel, T. E. and P. J. Crutzen (1995) Atmosphere, climate, and change. W. H. Freeman, New York, 208 pp.

Graedel, T. E. and W. C. Keene (1996) The budget and cycle of Earth's natural chlorine. *Pure and Applied Chemistry* **68**, 1689−1697.

Graedel T. E. and W. C. Keene (1999) Overview: Reactive chlorine emissions inventory. *Journal of Geophysical Research* **104**, 8331−8333.

Graf, H.-F., J. Feichter and B. Langmann (1997) Volcanic sulfur emissions: estimates of source strenght and its contribution to the global sulfate distribution. *Journal of Geophysical Research* **102**, 10727−10738.

Graf, H.-F., B. Langmann and J. Feichter (1998) The contribution of Earth degassing to the atmospheric sulphur budget. *Chemical Geology* **147**, 131−145.

Gräfenberg, L. (1902) Das Potential des Ozons. *Zeitschrift für Elektrochemie* **8**, 297−301.

Gräfenberg, L. (1903) Beiträge zur Kenntnis des Ozons. *Zeitschrift für anorganische Chemie* **36**, 355−379.

Graham, Th. (1861) Liquid diffusion applied to analysis. *Philosophical Transactions of the Royal Society London* **151**, 183−224.

Gran, G. (1950) Determination of the equivalence point in potentiometric titrations. *Acta Chemica Scandinavica* **4**, 559−577.

Granat, L. (1972) On the relation between pH and the chemical composition in atmospheric precipitation. *Tellus* **24**, 550−560.

Granat, L., H. Rodhe and R. O. Hallberg (1976) The global sulfur cycle. In: Nitrogen, phosphorous and sulphur − global cycles (Eds. B. H. Svensson and R. Söderlund), SCOPE Report 7, Ecol. Bull. (Stockholm) 22, pp. 89−134.

Granat, L. (1988) Ur Arsrapport till SNV. Institute for Meteorology, Stockholm University, Report 3475 (in Swedish).

Granier, C., B. Bessagnet, T. Bond, A. D'Angiola, H. D. van der Gon, G. J. Frost, A. Heil, J. W. Kaiser, S. Kinne, Z. Klimont, S. Kloster, J.-F. Lamarque, C. Liousse,T. Masui, F. Meleux, A. Mieville, T. Ohara, J.-C. Raut, K. Riahi, M. G. Schultz, S. J. Smith, A. Thompson, J. van Aardenne, G. R. van der Werf and D. P. van Vuuren (2011) Evolution of anthropogenic and biomass burning emissions of air pollutants at global and regional scales during the 1980–2010 period. *Climatic Change* **109**, 163−190.

Granqvist, C. G. (2002) Handbook of inorganic electrochromic materials. Elsevier, Amsterdam, 633 pp.

Gratani, L. and L. Varone (2005) Daily and seasonal variation of CO_2 in the city of Rome in relationship with the traffic volume. *Atmospheric Environment* **39**, 2619 2624.

Greenwood, K. and M. Pearce (1953) The removal of carbon dioxide from atmospheric air by scrubbing with caustic soda in packed towers. *Transactions of the Institute of Chemical Engineers* **31**, 201−207.

Gregg, W. T., M. E. Conkright, P. Ginoux, J. E. O'Reilly and N. W. Casey (2003) Oceanic primary production and climate: Global decadal change. *Geophysical Research Letters* **30**, 1809, doi:10.1029/2003GL016889.

Gribble, G. W. (2003) The diversity of naturally produced organohalogens. *Chemosphere* **52**, 289−297.

Gribble, G. W. (2004) Natural organohalogens: A new frontier for medicinal agents? *Journal Chemical Education* **81**, 1441−1449.

Griffing, G. W. (1977) Ozone and oxides of nitrogen production during thunderstorms. *Journal of Geophysical Research* **82**, 943−950.

Grimes, C., O. Varghese and S. Ranjan (2007) Light, water, hydrogen: The solar generation of hydrogen by water photoelectrolysis. Springer-Verlag, Berlin, 568 pp.

Grimm, H. (1931) Zur Benennung atmosphärischer disperser Systeme. *Kolloid-Zeitschrift* **54**, 1−2.

Grini, A., G. Myhre, J. K. Sundet and I. S. A. Isaksen (2002) Modeling the annual cycle of sea salt in the global 3D model Oslo CTM2: Concentrations, fluxes, and radiative impact. *Journal of Climate* **15**, 1717−1730.

Grönberg, L., P. Lövkvist and J. Å. Jönsson (1990) Measurement of aliphatic amines in ambient air and rainwater. *Chemosphere* **24**, 1533−1540.

Grünhage, L. H.-D. Haenel and H.-J. Jäger (2000) The exchange of ozone between vegetation and atmosphere: micrometeorological measurement techniques and models. *Environmental Pollution* **3**, 373−392.

Grünhage, L. and H.-D. Haenel (2008) Detailed documentation of the PLATIN (PLant-ATmosphere INteraction) model. *Landbauforschung*, special issue **319**, 1−85.

Gschwandtner, G., Gschwandtner, K., Eldridge, K., Mann, C. and D. Mobley (1986) Historic emissions of sulfur and nitrogen oxides in the United States from 1900 to 1980. *Journal of Air Pollution Control Association* **36**, 139−149.

Gubbins, D., T. G. Masters and J. A. Jacobs (1979) Thermal evolution of the earth's core. *Geophysical Journal of the Royal Astronomical Society* **59**, 57−99.

Guderian, R. (2000) Arten und Ursachen von Umweltbelastungen. In: Handbuch der Um-weltveränderungen und Ökotoxikologie, Band 1A: Atmosphäre (Ed. R. Guderian), Sprin-ger-Verlag, Berlin, pp. 1–21.

Guderian, R., Ed. (2000) Handbuch der Umweltveränderungen und Ökotoxikologie, Band 1B: Atmosphäre (Ed. R. Guderian), Springer-Verlag, Berlin, 516 pp.

Guenther, A., C. Nicholas Hewitt, D. Erickson, R. Fall, C. Geron, T. Graedel, P. Harley, L. Klinger, M. Lerdau, W. A. Mckay, T. Pierce, B. Scholes, R. Steinbrecher, R. Tallamraju, J. Taylor and P. Zimmerman (1995) A global model of natural volatile organic compound emissions. *Journal of Geophysical Research* **100**, 8873–8892.

Guenther, A. (1999) Modeling biogenic volatile organic compound emissions to the atmos-phere. In: Reactive hydrocarbons in the atmosphere (Ed. C. N. Hewitt), Academic Press, San Diego, pp. 41–94.

Guenther, A., T. Karl, P. Harley, C. Wiedinmyer, P. I. Palmer and C. Geron (2006) Estimates of global terrestrial isoprene emissions using MEGAN (Model of emissions of gases and aerosols from nature). *Atmospheric Chemistry and Physics* **6**, 3181–3210.

Guenther, A. (2013) Biological and chemical diversity of biogenic volatile organic emissions into the atmosphere. *ISRN Atmospheric Sciences*, Volume 2013, Article ID 786290, 27 pp.

Guericke, O. von (1672) Neue „Magdeburgische" Versuche über den leeren Raum. Ostwald's Klassiker der exakten Naturwiss. Nr. 59. Unveränderter Nachdruck (Drittes Buch). Akade-mische Verlagsgesellschaft (*ca.* 1930), Leipzig, 116 pp.

Guibet, J.-C. and E. Faure-Birchem (1999) Fuels and engines: Technology, energy, environ-ment. Technip Editions, Paris, 786 pp.

Guimbaud, C., A. M. Grannas, P. B. Shepson, J. D. Fuentes, H. Boudries, J. W. Bottenheim, F. Domine, S. Houdier, S. Perrier, T. B. Biesenthal and B. G. Splawn (2002) Snowpack processing of acetaldehyde and acetone in the Arctic atmospheric boundary layer. *Atmos-pheric Environment* **36**, 2743–2752.

Gunz, D. W. and M. R. Hoffmann (1990) Atmospheric chemistry of peroxides: A review. *Atmospheric Environment* **24**, 1601–1633.

Gupta, V., and K. S. Carroll (2013) Sulfenic acid chemistry, detection and cellular lifetime, *Biochimica et Biophysica Acta*, in press: http://dx.doi.org/10.1016/j.bbagen.2013.05.040.

Gurney, K. R., R. M. Law, A. S. Dennings, P. J. Rayner, D. Baker, P. Bousquet, L. Bruhwiler, Y.-H. Chen, P. Ciais, S. M. Fan, I.-Y. Fung, M. Gloor, M. Heimann, K. Higuchi, J. John, T. Maki, S. Maksyutov, K. Masarie. P. Peylin, M. Prather, B. C. Pak, J. T. Randerson, J. L. Sarmiento, S. Taguchi, T. Takahashi and C.-W. Yuen (2002) Towards robust regional estimates of CO_2 sources and sinks using atmospheric transport models. *Nature* **415**, 626–630.

Gustafsson R. J., A. Orlov, P. T. Griffiths, R. A. Cox and R. M. Lambert (2006) Reduction of NO_2 to nitrous acid on illuminated titanium dioxide aerosol surfaces: implications for photocatalysis and atmospheric chemistry. *Chemical Communication (Cambridge)* **37**, 3936–3938.

Haas, P. and T. G. Hill (1929) An introduction to the chemistry of plant products. Volume 1: On the nature and significance of the commoner organic compounds of plants. 4th Edition. Volume 2: Metabolic processes. 2nd Edition. Longman, London, 530 pp. and 220 pp.

Haber, F. and R. Willstädter (1931) Unpaarigkeit und Radikalketten im Reaktionsmechanis-mus organischer und enzymatischer Vorgänge. *Berichte der Deutschen Chemischen Gesell-schaft* **64**, 2844–2856.

Haber, F. and J. Weiss (1932) Über die Katalyse des Hydroperoxides. *Naturwissenschaften* **51**, 948–950.

Haber, F. and J. Weiss (1934) The catalytic decomposition of hydrogen peroxide ba iron salts. *Proceedings of the Royal Society* **147**, 332–352.

Haberl, H, K. H. Erb, F. Krausmann, V. Gaube, A. Bondeau, C. Plutzar, S. Gingrich, W. Lucht and M. Fischer-Kowalski (2007) Quantifying and mapping the human appropriation of net primary production in earth's terrestrial ecosystems. *Proceedings of the National Academy of Sciences USA* **104**, 12942−12947.

Haïssinsky, M. and M. Magat (1951) Sur les reactions primaires produites par les radiations ionisantes dans l'eau. *Comptes Rendus de l'Académie des Sciences* **233**, 954−956.

Haldane, J. S. (1918) Methods of air analysis. Ch. Griffin, London, 137 pp.

Haldane, J. B. S. (1928) Possible worlds and other essays. Chatto & Windus, London, 1927. 312 pp.

Haldane, J. B. S. (1954) The origin of life. *New Biology* **16**, 12−27.

Hall, A. (1999) Ammonium in granites and its petrogenetic significance. *Earth-Science Reviews* **45**, 145−165.

Hall, R. E. (1973) Al-Khazini. In: Dictionary of scientific biography (Ed. C. C. Gillispie), Vol. 7, New York, pp. 335−251.

Halliday, E. C. (1961) A historical review of atmospheric pollution. In: Air pollution (Ed. WHO), Columbia Univ. Press, New York, pp. 9−35.

Halliwell, B. and J. M. Gutteridge (1984) Oxygen toxicity, oxygen radicals, transition metals and disease. *Biochemical Journal* **219**, 1−14.

Hallquist, M., J. C. Wenger, U. Baltensperger, Y. Rudich, D. Simpson, M. Claeys, J. Dommen, N. M. Donahue, C. George, A. H. Goldstein, J. F. Hamilton, H. Herrmann, T. Hoffmann, Y. Iinuma, M. Jang, M. E. Jenkin, J. L. Jimenez, A. Kiendler-Scharr, W. Maenhaut, G. McFiggans, Th. F. Mentel, A. Monod, A. S. H. Prévôt, J. H. Seinfeld, J. D. Surratt, R. Szmigielski and J. Wildt (2009) The formation, properties and impact of secondary organic aerosol: Current and emerging issues. *Atmospheric Chemistry and Physics* **9**, 5155−5236.

Halmer, M. M., H.-U. Schmincke and H.-F. Graf (2002) The annual volcanic gas input into the atmosphere, in particular into the stratosphere: A global data set for the past 100 years. *Journal of Volcanology and Geothermal Research* **115**, 511−528.

Hammer, C. U. (1981) Past volcanism and climate revealed by Greenland ice cores. *Journal of Volcanology and Geothermal Research* **11**, 3−10.

Hammett, L. P. (1940) Physical-organic chemistry. McGraw-Hill, New York, 404 pp.

Handler, P. (1989) The effects of volcanic aerosols on global climate. *Journal of Volcanology and Geothermal Research* **37**, 233−249.

Hann, J. von (1883) Handbuch der Klimatologie. Bibliothek geographischer Handbücher (Ed. F. Ratzel), J. Engelhorn, Stuttgart, 764 pp. (translated into English: Handbook of Climatology: Part I. General Climatology, Macmillan Company, 1903).

Hann J. von (1897) Handbuch der Klimatologie, 3. Band: Spezielle Klimatologie, 2. Abteilung: Klima der gemässigten und der kalten Zone. Bibl. geogr. Handbücher (Hrsg. F. Ratzel), Stuttgart, J. Engelhorn, 576 pp.

Hann, J. von (1908) Handbuch der Klimatologie. I. Band: Allgemeine Klimalehre (Dritte Auflage). Engelhorn, Stuttgardt, 394 pp.

Hanna, J. V., W. DE. Johnson, R. A. Quezada, M. A. Wilson and L. Xiao-Qiao (1991) Characterization of aqueous humic substances before and after chlorination. *Environmental Science and Technology* **25**, 1160−1164.

Hansell D. A. and C. A. Carlson, Eds. (2002) Biogeochemistry of marine dissolved organic matter. Academic Press, San Diego, 774 pp.

Hansen, J., Mki. Sato, R. Ruedy, K. Lo, D. W. Lea and M. Medina-Elizade (2006) Global temperature change. *Proceedings of the National Academy of Sciences* **103**, 14288−14293, doi:10.1073/pnas.0606291103.

Hansen, J., M. Sato, P. Kharecha, G. Russel, D. W. Lea and M. Siddall (2007) Climate change and trace gases. *Philosophical Transactions of the Royal Sciety* A **365**, 1925−1954.

Hanson, D. R., J. B. Burkholder, C. L. Howard and A. R. Ravishankara (1992) Measurements of OH and HO_2 radical uptake coefficients on water and sulfuric acid. *Journal of Physical Chemistry* **96**, 4979−4985.

Hanson, P. J. and S. E. Lindberg (1991) Dry deposition of reactive nitrogen compounds: a review of leaf, canopy and non-foliar measurements. *Atmospheric Environment* **10**, 1127–1131.

Hantel, M., H. Kraus and C.-D. Schönwiese (1987) Climate definition. In: Landolt-Börnstein – Group V Geophysics. Numerical Data and Functional Relationships in Science and Technology. Subvolume C1 'Climatology. Part 1' of Volume 4 'Meteorology'. Springer-Verlag, Berlin, 5 pp.

Hantel, M. (2001) Klimatologie. In: Bergmann-Schaefer Lehrbuch der Experimentalphysik, Vol 7: Erde und Planeten (Ed. W. Raith), de Gruyter, Berlin and New York, pp. 311–426.

Hara, Y., I. Uno, I. and Z. Wang (2006) Long-term variation of Asian dust and related climate factors. *Atmospheric Environment* **40**, 6730–6740.

Hare, K. F. (1985) Climatic variability and change. In: Climate impact assessment – studies of the interaction of climate and society, SCOPE 27 (Eds. R. W. Kates, J. H. Ausubel and M. Berberian), John Wiley & Sons, Chichester, Chapter 2.

Harkins W. D. (1917) The evolution of the elements and the stability of complex atoms. *Journal of the American Chemical Society* **39**, 856–879.

Harnisch, J., R. Borchers, P. Fabian and K. Kourtidis (1995) Aluminium production as a source of atmospheric carbonyl sulfide (COS). *Environmental Science and Pollution Research* **2**, 161–162.

Harries, C. D. (1916) Untersuchungen über das Ozon und seine Einwirkung auf organische Verbindungen. Springer-Verlag, Berlin, 720 pp.

Harrison, R. M. and C. A. Pio (1983) A comparative study of the ionic composition of rainwater and atmospheric aerosols: implications for the mechanism of acidification of rainwater. *Atmospheric Environment* **17**, 2539–2543.

Harrison, R. M. and A.-M. N. Kitto (1992) Estimation of the rate constant for the reaction of acid sulphate aerosol with NH_3 gas from atmospheric measurements. *Journal of Atmospheric Chemistry* **15**, 133–143.

Harrison, R. M. and R. van Grieken, Eds. (1998) Atmospheric particles. John Wiley & Sons, Chichester, 610 pp.

Hart, E. J. (1959) Development of the radiation chemistry of aqueous solutions. *Journal of Chemical Education* 36, 266–272.

Hart, E. J. and M. Anbar (1970) The hydrated electron. John Wiley & Sons, New York, 267 pp.

Hatch, L. F. and S. Matar (1977) Refining processes and petrochemicals (Part I). *Hydrocarbon processing* **56**, 191–201.

Hauglustaine, D. A. and D. H. Ehhalt (2002) A three-dimensional model of molecular hydrogen in the troposphere. *Journal of Geophysical Research* **107**, 4330. doi:10.1029/2001JD001156.

Hauglustaine, D. F., L. Emmons, M. Newchurch, G. Brasseur, T. Takao, K. Matsubara, J. Johnson, B. Ridley, J. Stith and J. Dye (2001) On the Role of Lightning NO_x in the Formation of Tropospheric Ozone Plumes: A Global Model Perspective. *Journal of Atmospheric Chemistry* **38**, 277–294.

Havers, N., P. Burba, J. Lambert and D. Klockow (1998) Spectroscopic characterization of humic-like substances in airborne particulate matter. *Journal of Atmospheric Chemistry* **29**, 45–54.

Haywood, J. and O. Boucher (2000) Estimates of the direct and indirect radiative forcing due to tropospheric aerosols: A review. *Reviews of Geophysics* **38**, 513–443.

He, Y., X. Zhou, J. Hou, H. Gao and S. B. Bertman (2006) Importance of dew in controlling the air-surface exchange of HONO in rural forested environments. *Geophysical Research Letters* **33**, Lo2813, doi:10.1029/2005GL024348.

Heal, M. R., M. A. J. Harrison and J. N. Cape (2007) Aqueous-phase nitration of phenol by N_2O_5 and $ClNO_2$. *Atmospheric Environment* **41**, 3515–3520.

Heckner, H. N. (1977) The cathodic reduction of nitrous acid in the second reduction step in high acid solutions by the potential step method (formation of HNO). *Journal of Electroanalytical Chemistry* **83**, 51−63.

Heikes, B. G., Lazrus, A. L., Kok, G. L., Kunen, S. M., Gandrud, B. W., Gitlin, S. N. and P. D. Sperry (1982) Evidence for aqueous phase H_2O_2 synthesis in the troposphere. *Journal of Geophysical Research* **87**, 3045−3051.

Heimann, J. (1888) Der Kohlensäuregehalt der Luft in Dorpat bestimmt in den Monaten Juni bis September 1888. Doctor Thesis, Medizinische Fakultät der Kaiserlichen Universität Dorpat, 53 pp.

Heintzenberg, J., M. Wendisch, B. Yuskiewicz, D. Orsini, A. Wiedensoehler, F. Stratmann, G. Frank, B. G. Martinsson, D. Schell, S. Fuzzi and G. Orsi (1998) Characteristics of haze, mist and fog. *Beiträge zur Physik der Atmosphäre* **71**, 21−31.

Heintzenberg, J. and R. J. Charlson, Eds. (2009) Clouds in the perturbed climate system. Their relationship to energy balance, atmospheric dynamics, and precipitation. MIT Press, Cambridge, U.K., 576 pp.

Heldt, H.-W. (2004) Plant biochemistry. 3rd Edition, Academic Press, 656 pp.

Heller, M. I. and P. L. Croot (2010) Superoxide decay kinetics in the Southern Ocean. *Environmental Sciences and Technology* **44**, 191−196.

Helmig, R. and P. Schulz, P. (1997) Multiphase flow and transport processes in the subsurface − a contribution to the modeling of hydrosystems. Springer-Verlag, Berlin, 367 pp.

Hempel, W. (1913) Gasanalytische Methoden. 4th Ed. (first 1889), Fr. Vieweg & Sohn, Braunschweig, 427 pp.

Hempel, W. (1913) Gasanalytische Methoden. Vieweg & Sohn, Braunschweig, 427 pp.

Herrmann, H., A. Reese and R. Zellner (1995) Time-resolved UV/VIS Diode Array absorption Spectroscopy of SO_x − (x = 3, 4, 5) Radical Anions in Aqueous Solution. *Journal of Molecular Structure* **348**, 183−186.

Herrmann, H., R. Wolke, K. Müller, E. Brüggemann, T. Gnauk, P. Barzaghi, S. Mertes, K. Lehmann, A. Massling, W. Birmili, A. Wiedensohler, W. Wieprecht, K. Acker, W. Jaeschke, H. Kramberger, B. Svrcina, K. Bächmann, J. L. Collett Jr., D. Galgon, K. Schwirn, A. Nowak, D. van Pinxteren, A. Plewka, R. Chemnitzer, C. Rüd, D. Hofmann, A. Tilgner, K. Diehl, B. Heinold, D. Hinneburg, O. Knoth, A. M. Sehili, M. Simmel, S. Wurzler, Z. Majdik, G. Mauersberger and F. Müller (2005) FEBUKO und Modmep: Field measurements and modelling of aerosol and cloud multiphase processes. *Atmospheric Environment* **39**, 4169−4183.

Henninger, S. and W. Kuttler (2004) Mobile measurements of carbon dioxide in the urban boundary layer of Essen, Germany. In: 5th Urban Environment Symposium, Vancouver, Canada, American Meteorological Society, J 12.3. (5 pp.).

Henry, W. (1803) Experiments on the quantity of gases absorbed by water, at different temperatures, and under different pressures. *Philosophical Transaction of the Royal Society London* **93**, 29−274.

Herbst, E. and W. Klemperer (1973) The formation and depletion of molecules in dense interstellar clouds. *Astrophysical Journal* **185**, 505−533.

Herckes, P., P. Mirabel and H. Wortham (2002) Cloud water deposition at a high-elevation site in the Vosges Mountains (France). *The Science of the Total Environment* **16**, 59−75.

Herrmann, H., H. W. Jacobi, G. Raabe, A. Reese and R. Zellner (1996) Laser-spectroscopic laboratory studies of atmospheric aqueous phase free radical chemistry. *Fresenius Journal of Analytical Chemistry* **355**, 343−344.

Herrmann, H., A. Tilgner, P. Barzaghi, Z. Majdik, S. Gligorovski, L. Poulain and A. Monod (2005) Towards a more detailed description of tropospheric aqueous phase organic chemistry: CAPRAM 3.0. *Atmospheric Environment* **39**, 4351−4363.

Hewitt, C. N. and A. V. Jackson, Eds. (2003) Handbook of Atmospheric Science: Principles and Applications. Blackwell Science Ltd., 633 pp.

Heymann, M. (2009) Klimakonstruktionen. Von der klassischen Klimatologie zur Klimaforschung. *NTM Zeitschrift für Geschichte der Wissenschaften, Technik und Medizin* **17**, 171–197.

Hicks, B. B. and P. S. Liss (1976) Transfer of SO_2 and other reactive gases across the air-sea interface. *Tellus* **28**, 248–254.

Hill, R. (1937) Oxygen evolution by isolated chloroplasts. *Nature* **139**, 881–882.

Hill, R. D. (1979) On the production of nitric oxide by lightning. *Geophysical Research Letters* **6**, 945–947.

Hinds, W. C. (1982) Aerosol technology. Properties, behaviour, and measurement of airborn particles. J. Wiley & Sons, New York, 424 pp.

Hinshelwood, C. N. and A. T. Williamson (1934) The reaction between hydrogen and oxygen. Clarendon Press, Oxford, 108 pp.

Hirsch, R., R. Bezdek and R. Wendling (2005) Peaking of world oil production: Impacts, mitigation, and risk management. US Dept. Energy/National Energy Technology Lab., 91 pp.

Hitchcock, D. R., L. L. Spiller and W. E. Wilson (1980) Sulfuric acid aerosols and HCl release in costal atmosphere: Evidence of rapid formation of sulphuric acid particulates. *Atmospheric Environment* **14**, 165–182.

Hobbs, P. V., Ed. (1993) Aerosol-cloud climate interaction. Academic Press, San Diego, 233 pp.

Hobbs, P. V. (2000) Introduction to atmospheric chemistry. A companion text to basic physical chemistry for the atmospheric sciences. Cambridge University Press, Cambridge (U.K.), 206 pp.

Hodgson, P. E. (2010) Energy, the environment and climate change. World Scientific Publ. Co., 200 pp.

Hoff, H. E. (1964) Nicolaus of Cusa, Van Helmont, and Boyle: The first experiment of the renaissance in quantitative biology and medicine. *Journal of the History of Medicine and Allied Sciences* **19**, 99–117.

Hofmann, A. W. and W. M. White (1982) Mantle plumes from ancient oceanic crust. *Earth and Planetary Science Letters* **57**, 421–436.

Hofmann, D., J. Barnes, M. O'Neill, M. Trudeau and R. Neely (2009) Increase in background stratospheric aerosol observed with lidar at Mauna Loa Observatory and Boulder, Colorado. *Geophysical Research Letters* **36**, L15808, doi:10.1029/2009GL039008.

Hoffmann M. R. and J. O. Edwards (1975) Kinetics of the oxidation of sulphite by hydrogen peroxide in acidic solution. *Journal of Physical Chemistry* **79**, 2096–2098.

Hoffmann, M. R. (1977) Kinetics and mechanism of oxidation of hydrogen sulphide by hydrogen peroxide in acidic solution. *Environmental Science and Technology* **11**, 61–66.

Hoffmann, M. R. and B. C. H. Lim (1979) Kinetics and mechanism of the oxidation of sulfide by oxygen: Catalysis by homogeneous metal phthalocyanine complexes. *Environmental Science and Technology* **13**, 1406–1414.

Hoffmann, M. R. (1986) On the kinetics and mechanism of the oxidation of aquated sulfur dioxide by ozone. *Atmospheric Environment* **20**, 1145–1154.

Hoffmann, M. R. (1990) Catalysis in aquatic environment. In: Aquatic chemical kinetics (Ed. W. Stumm), John Wiley & Sons, New York, pp. 71–112.

Hoffmann, M. R., S. T. Martin, W. Choi and D. Bahnemann (1995) Environmental applications of semiconductor photocatalysis. *Chemical Reviews* **95**, 69–75.

Högbom, A. (1894) Svensk kemisk Tidskrift, Bd. vi. pp. 69.

Höglund-Isaksson, L. (2012) Global anthropogenic methane emissions 2005–2030: technical mitigation potentials and costs. *Atmospheric Chemistry and Physics Discussion* **12**, 11275–11315.

Hoigné, J. and H. Bader (1976) The role of hydroxyl radical reactions in ozonation processes in aqueous solutions. *Water Research* **10**, 377–386.

Hoigné, J., H. Bader, W. R. Haag and J. Staehelin (1985) Rate constants of reactions of ozone with organic and inorganic compounds in water-III. *Water Research* **19**, 993−1004.

Höll, K. (2002) Wasser. Nutzung im Kreislauf, Hygiene, Analyse und Bewertung. DeGruyter, Berlin, 955 pp.

Holland, G. and C. J. Ballentine (2006) Seawater subduction controls the heavy noble gas composition of the mantle. *Nature* **441**, 186−191.

Holland, G., M. Cassidy and C. J. Ballentine (2009) Meteorite Kr in earth's mantle suggests a late accretionary source for the atmosphere. *Science* **326**, 1522−1525.

Holland, H. D. (1984) The chemical evolution of the atmosphere and oceans. Princeton Univ. Press, New Jersey (USA), 582 pp.

Holland, H. D. (2002) Volcanic gases, black smokers, and the Great Oxidation Event. *Geochimica et Cosmochimica Acta* **66**, 3811−3826.

Holland, H. D. (2006) The oxygenation of the atmosphere and oceans. *Philosophical Transaction of the Royal Society, Series B*, **361**, 903−915.

Holland, H. L., M. Carey and S. Kumaresan (1993) Fungal biotransformation of organophosphines. *Xenobiotica* **23**, 519−524.

Hollemann-Wirberg (2007) Lehrbuch der Anorganischen Chemie (Eds. N. Wiberg, E. Wiberg and A. F. Holleman), 102nd Ed., DeGruyter, Berlin, 2033 pp.

Hollenbach D. F. and J. M. Herndon (2001) Deep-earth reactor: Nuclear fission, helium, and the geomagnetic field. *Proceedings of the National Academy of Sciences of the United States of America (PNAS)* **98**, 11085−11090.

Holmén, K. (1992) The global carbon cycle (pp. 239−262). In: Global biogeochemical cycles (Eds. S. S. Butcher, R. J. Charlson, G. H. Orian and G. V. Wolfe), Acad. Press London, 377 pp.

Holmes, G. and D. W. Keith (2012) An air-liquid contactor for large-scale capture of CO_2 from air. *Philosophical Transactions of the Royal Society A – Mathematical, Physical & Engineering Sciences* **370**, 4380−4403.

Holton, J. R. (2004) An introduction to dynamic meteorology, 4th edition. International Geophysics Series, Vol. 88, Elsevier, Oxford, 550 pp.

Holze, R.(2007) Electrode potentials. In: Landolt-Börnstein − Group IV Physical chemistry. Numerical data and functional relationships in science and technology. 9A: Electrochemical thermodynamics and kinetics (Ed. M. D. Lechner), Springer-Verlag, Berlin.

Holzinger, R. C. Warneke, A. Hansel, A. Jordan, W. Lindinger, D. H. Scharffe, G. Schade and P. J. Crutzen (1999) Biomass burning as a source of formaldehyde, acetaldehyde, methanol, acetone, acetonitrile, and hydrogen cyanide. *Geophysical Research Letters* **26**, 1161−1164.

Holzinger, R., J. Williams, G. Salkisbury, T. Klüpfel, M. de Reus, M. Traub, P. J. Crutzen and J. Lelieveld (2005) Oxygenated compounds in aged biomass burning plumes over the Eastern Mediterranean: evidence for strong secondary production of methanol and acetone. *Atmospheric Chemistry and Physics Discussion* **4**, 6321−6340.

Holzwarth, G., R. C. Balmer and L. Soni (1984) The fate of chlorine and chloramines in cooling towers. *Water Research* **18**, 1421−1427.

Hoover, R. B. (2006) Comets, carbonaceous meteorites, and the origin of the biosphere. *Biogeosciences Discussions* **3**, 23−70.

Hopkins, F., P. Nightingale and P. Liss (2011) Effects of ocean acidification on the marine source of atmospherically-active trace gases. In: *Ocean Acidification* (Eds. J-P. Gattuso and L. Hansson,), Oxford Univ. Press, pp. 210–229.

Hoppe P., S. Amari, E. Zinner, T. Ireland and R. S. Lewis (1994) Carbon, nitrogen, magnesium, silicon and titanium isotopic compositions of single interstellar silicon carbide grains from the Murchison carbonaceous chondrite. *Astrophysical Journal* **430**, 870−890.

Horgan, J. (1997) The end of science: Facing the limits of knowledge in the twilight of the scientific age. Broadway Books, New York 319 pp.

Horváth, L. (1983) Concentration and near surface vertical flux of ammonia in the air in Hungary. *Idöjárás* **87**, 65–70.

Horváth, L. and E. Mészáros (1984) The composition and acidity of precipitation in Hungary. *Atmospheric Environment* **18**, 1843–1847.

Horváth, L. and D. Möller (1987) On the "natural" acid deposition and the possible consequences of decreased SO_2 and NO_2 emission in Europe. *Idöjaras* **91**, 217–223.

Hosker, R. P. and S. E. Lindberg (1989) Review: Atmospheric deposition and plant assimilation of gases and particles. *Atmospheric Environment* **5**, 889–920.

Hough, A. M. and R. G. Derwent (1987) Computer modelling studies of the distribution of photochemical ozone production between different hydrocarbons. *Atmospheric Environment* **21**, 2015–2034.

Hough, A. M. (1988) The calculation of photolysis rates for use in global modelling studies. Technical Report, UK Atomic Energy Authority, Harwell, Oxon. (UK), 347.

Houghton, H. G. (1955) On the chemical composition of fog and cloud water. *Journal of Meteorology* **12**, 355–357.

Houghton, J. T., L. G. Meira Filho, J. Bruce, Hoesung Lee, B. A. Callander, E. Haites, N. Harris and K. Maskell, Eds. (1995) Climate change 1994: Radiative forcing of climate change and an evaluation of the IPCC IS92 emissions scenarios. Cambridge University Press, Cambridge, U.K., 339 pp.

Houghton, R. A. (2003) Revised estimates of the annual net flux of carbon to thre atmosphere from change in land use and land use management 1850–2000. *Tellus* **55B**, 378–390.

Houghton, R. A. (2005a) Aboveground forest biomass and the global carbon balance. *Global Change Biology* **11**, 945–958.

Houghton, R. A. (2005b) Tropical deforestation as a source of greenhouse gas emission. In: Tropical Deforestation and Climate Change (Eds. P. Mountino and S. Schwartzman), Amazon Institute for Environmental research (131 pp.), pp. 13–21.

Houghton, R. A. (2008) Carbon Flux to the Atmosphere from Land-Use Changes: 1850–2005. In TRENDS: A Compendium of Data on Global Change. Carbon Dioxide Information Analysis Center, Oak Ridge National Laboratory, U.S. Department of Energy, Oak Ridge, Tenn., U.S.A. (http://cdiac.ornl.gov/trends/landuse/houghton/houghton.html).

Houghton, R. A., J. I. House, J. Pongratz, G. R. van der Werf, R. S. DeFries, M. C. Hansen, C. Le Quéré and N. Ramankutty (2012) Carbon emissions from land use and land-cover change. *Biogeosciences* **9**, 5125–5142.

Houghton, R. A. and J. L. Hackler (2013) Annual Flux of Carbon from Land Use and Land-Cover Change 1850 to 2010. *Global Biogeochemical Cycles*, in review.

Houston, P. L. (2006) Chemical kinetics and reaction dynamics. Dover Publications, Mineola (USA), 329 pp.

Howard, P. H. (1997) Handbook of environmental fate and exposure data for organic chemicals. CRC Press, Boca Raton, FL (USA), 528 pp.

Hoyle, F. and N. C. Wickramasinghe (1977) Polysaccharides and the infrared spectra of galactic sources. *Nature* **268**, 610–612.

Huang, P. M. and G. R. Gobran, Eds. (2005) Biogeochemistry of trace elements in the rhizosphere. Elsevier Science, 480 pp.

Hudman, R. C., N. E. Moore, R. V. Martin, A. R. Russell, A. K. Mebust, L. C. Valin and R. C. Cohen (2012) A mechanistic model of global soil nitric oxide emissions: implementation and space based-constraints. *Atmospheric Chemistry and Physics Discussion* **12**, 3555–3359.

Hughes, G. and C. R Lobb (1976) Reactions of solvated electrons. In: Comprehensive chemical kinetics, Vol. 18 (Eds. C. H. Bamford and C. F. H. Tipper), Elsevier Sci. Publ. Amsterdam, pp. 429–461.

Hughes, M. N. (1999) Relationships between nitric oxide, nitroxyl ion, nitrosonium cation and peroxynitrite. *Biochimica et Biophysica Acta (BBA) – Bioenergetics* **1411**, 263–272.

Humboldt, A. v. (1799) Versuche über die chemische Zerlegung des Luftkreises und über einige andere Gegenstände der Naturlehre. Vieweg, Braunschweig, 258 pp.

Humboldt, A. v. (1845) Kosmos. Entwurf einer physischen Weltbeschreibung. J. G. Cotta'scher Verlag, Stuttgart und Tübingen, in 5 Vols., over 3600 pp. [First English Edition: Cosmos – a General Survey of the Physical Phenomena of the Universe. Hippolyte Bailliere Publisher, London 1845].

Humboldt, A. v. (1850–1852) Cosmos: A sketch of a physical description of the universe. Translated from German by E. C. Otté. Harper & Brothers, New York, Vol. 1 (1850) 275 pp., Vol 2 (1850) 367 pp., Vol. IV (1851) 219 pp., Vol. IV (181852) 234 pp.

Humphrey, W. J. (1913) Volcanic dust and other factors in the production of climate change and their relation to ice ages. *Bulletin of the Mount Weather Observatory* (Washington) **6**, 1–34.

Hunter, D. E., S. E. Schwartz, R. Wagener and C. M. Benkovitz (1993) Seasonal, latitudinal, and secular variations in temperature trend: Evidence for influence of anthropogenic sulfate. *Geophysical Research Letters* **20**, 2455–2488.

Huntington, E. (1915) Civilization and climate. Yale University Press, New Haven, Connecticut (USA), 453 pp.

Huntrieser, H, H. Schlager, C. Feigl and H. Höller (1998) Transport and production of NO$_x$ in electrified thunderstorms: Survey of previous studies and new observations at midlatitudes. *Journal of Geophysical Research* **103**, 28247–28264.

Hutchinson, D. (2003) Emissions inventories. In: Hewitt and Jackson (2003), pp. 473–502.

Hutchison, R. (2007) Meteorites: A petrologic, chemical and isotopic synthesis. Cambridge University Press, Cambridge, U.K., 524 pp.

Huthwelker, T., S. L. Clegg, T. Peter, K. Carslaw, B. P. Luo and P. Brimblecombe (1995) Solubility of HOCl in water and aqueous H$_2$SO$_4$ to stratospheric temperature. *Journal of Atmospheric Chemistry* **21**, 81–95.

Idso, S. B., Idso, C. D. and R.C. Balling Jr. (2002) Seasonal and diurnal variations of near-surface atmospheric CO$_2$ concentrations within a residential sector of the urban CO$_2$ dome of Phoenix, AZ, USA. *Atmospheric Environment* **36**, 1655–1660.

IEA (2009) Key world energy statistics. International Energy Agency, Paris, 82 pp.

Innes, J. L. (2000) Biomass burning and climate: An introduction. In: Biomass burning and its inter-relationships with the climate system, Advances in Global Change Research, Vol. 3 (Eds. J. L. Innes, M. Beniston and M. M. Verstraete), Kluwer Acad. Press, pp. 1–13.

IPCC (1995) Climate Change 1995: The Science of Climate Change (Eds. J. T. Houghton, L. G. Meira Filho, B. A. Callander, N. Harris, A. Kattenberg, and K. Maskell), Cambridge University Press, Cambridge, U.K., 572 pp.

IPCC (2001) Climate Change 2001: The Scientific Basis (Eds. J. T. Houghton, Y. Ding, D. J. Griggs, M. Noguer, P. J. van der Linden, X. Dai, K. Maskell and C. A. Johnson), Cambridge University Press, Cambridge, U.K., 881 pp.

IPCC (2007) Climate Change 2007: The Physical Science Basis. Contribution of Working Group I to the Fourth Assessment Report of the Intergovernmental Panel on Climate Change (Eds. S. Solomon, D. Qin, M. Manning, Z. Chen, M. Marquis, K. B. Averyt, M. Tignor and H. L. Miller), Cambridge University Press, Cambridge, U.K., 996 pp.

IPCC (2013) Climate Change 2013: The physical science basis. Working Group I Contribution to the IPCC 5th Assessment Report – Changes to the Underlying Scientific/Technical Assessment.

Irvine, J. C., R. H. Ritchie, J. L. Favaloro, K. L. Andrews, R. E. Widdop and B. K. Kemp-Harpe (2008) Nitroxyl (HNO): the Cinderella of the nitric oxide story. *Trends in Pharmacological Sciences* **29**, 601–608.

Irvine, W. M. (1998) Extraterrestrial organic matter: A review. *Origins of Life and the Evolution of the Biosphere* **28**, 365–383.

Isidorov, V. A., I. G. Zenkevich and B. V. Ioffe (1985) Volatile organic compounds in the atmosphere of forests. *Atmospheric Environment* **19**, 1−8.

Isidorov, V. A. (1990) Organic chemistry of the earth's atmosphere. Springer-Verlag, Berlin, 215 pp. (translation of the Russian edition from 1985: Organicheskaya Khimiya Atmosfery, Izd. Khimiya, Leningrad, 265 pp.).

Ito, A. and J. E. Penner (2004) Global estimates of biomass burning emissions based on satellite imagery for the year 2000. *Journal of Geophysical Research* **109**, D14S05, doi:10.1029/2003JD004423.

Jackman, C. H., E. I. Fleming and F. M. Vitt (2000) Influences of extremely large solar proton events in a changing atmosphere. *Journal of Geophysical Research* **105**, 11659−11670.

Jackson, A. V. (2003) Sources of air pollution. In: Hewitt and Jackson (2003), pp. 124−155.

Jacob, D. J., J. M. Waldman, J. W. Munger and M. R. Hoffmann (1985) Chemical composition of fogwater collected along the California coast. *Environmental Science and Technology* **19**, 730−736.

Jacob, D. J., B. D. Field, E. M. Jin, I. Bey, Q. Li, J. A. Logan, R. M. Yantosca and H. B. Singh (2002) Atmospheric budget of acetone. *Journal of Geophysical Research* **107**, 4100, (D10) doi:10.1029/2001JD000694.

Jacob, D. J., B. D. Field, Q. Li, D. R. Blake, J. de Gouw, C. Warneke, A. Hansel, A. Wisthaler, H. B. Singh and A. Guenther (2005) Global budget of methanol: Constraints from atmospheric observations. *Journal of Geophysical Research* **110**, D08303doi:10.1029/2004JD005172, 2005.

Jacobi, H.-W., H. Herrmann and R. Zellner (1996) Kinetic investigation of the Cl_2^- radical in the aqueous phase. In: Air pollution research report 57 − homogeneous and heterogeneous chemical processes in the troposphere (Ed. Ph. Mirabel), EU Publ., Luxembourg, pp. 172−176.

Jacobs, A. F. G., B. G. Heusinkveldand and S. M. Berkowicz (2000) Dew measurements along a longitudinal sand dune transect, Negev Desert, Israel. *International Journal of Biometeorology* **43**,184−190.

Jacobson, M. J., H.-C. Hanson, K. J. Noone and R. J. Charlson (2000) Organic atmospheric aerosols: review and state of the art. *Reviews of Geophysics* **38**, 267−294.

Jaenicke, R. (1982) Physical aspects of the atmospheric aerosol. In: Chemistry of the unpolluted and polluted troposphere (Eds. H.-W. Georgii und W. Jaeschke), D. Reidel Publ., Dordrecht, pp. 341−373.

Jaenicke, R. (1993) Tropospheric aerosols. In: Aerosol-cloud-climate interactions (Ed. P. V. Hobbs), Academic Press, San Diego, CA, pp. 1−31.

Jaenicke, R. (1999) Atmospheric aerosol size distribution. In: Atmospheric particles (Eds. R. M. Harrison and R. van Grieken), John Wiley & Sons, Chichester, pp. 1−29.

Jaeschke, W., Ed. (1986) Chemistry of multiphase atmospheric systems. Springer-Verlag, Berlin, 773 pp.

Jaffe D. A. (1992) The nitrogen cycle. In: Global biogeochemical cycles (Ed. S. S. Butcher, R. J. Charlson, G. H. Orian and G. V. Wolfe), Acad. Press, London, pp. 263−284.

Jain, A. K., P. Meiyappan, Y. Song and J. I. House (2013) CO_2 emission from land-use change affected more by nitrogen cycle, than by the choice of land-cover data. *Global Change Biology* **19**, 2893–2906.

Jang, M., Czoschke, N. M., S. Lee and K. M. Kamens (2002) Heterogeneous atmospheric aerosol production by acid-catalyzed particle-phase reactions. *Science* **298**, 814−817.

Jensen, J. N. (2003) A problem-solving approach to aquatic chemistry. John Wiley & Sons Inc., 480 pp.

Jensen, J. S. (2015) Aquatic chemistry: Chemical equilibria and rates in natural waters (Environmental Science and Technology: A Wiley-Interscience Series of Texts and Monographs), 4th Revised edition. Wiley-Blackwell, 1240 pp.

Janssens-Maenhout, G., F. Dentener, J. Van Aardenne, S. Monni, V. Pagliari, L. Orlandini, Z. Klimont, J. Kurokawa, H. Akimoto, T. Ohara, R. Wankmueller, B. Battye, D. Grano, A. Zuber and T. Keating (2012) EDGAR-HTAP: A harmonized gridded air pollution emission dataset based on national inventories. Ispra (Italy), European Commission Publications Office, JRC68434, EUR report No EUR 25.

Johnson, J. E. and I. S. A. Isaksen (1993) Tropospheric ozone chemistry: the impact of cloud chemistry. *Journal of Atmospheric Chemistry* **16**, 99–122.

Johnson, M. T. and T. G. Bell (2008) Concept: Coupling between DMS emissions and the ocean-atmosphere exchange of ammonia. *Environmental Chemistry* **5**, 259–267.

Johnston, H. S. (1971) Reduction of stratospheric ozone by nitrogen catalysts from supersonic transport exhaust. *Science* **173**, 517–522.

Jonas, P. R., R. J. Charlson and H. Rodhe (1995) Aerosols. In: Climate change 1994 (Eds. J. T. Houghton, L. G. Meira Filho, J. Bruce, H. Lee, B. A. Callander, E. Haites, N. Haris and K. Maskell), Cambridge University Press, Cambridge, U.K., pp. 127–162.

Jones, D. S. J. and P. P. Pujadó, Eds. (2003) Handbook of petroleum processing. Springer, Dordrecht, 1353 pp.

Joo, O.-S, K.-D. Jung, I. Moon, A. Y. Rozovskii, G. I. Lin, S.-H. Han, and S.-J. Uhm (1999) Carbon dioxide hydrogenation to form methanol via a reverse-water-gas-shift reaction (the CAMERE Process). *Industrial and Engineering Chemical Research* **35**, 1808–1812.

Jourdain, J. and D. A. Hauglustaine (2001) The global distribution of lightning NO_x simulated on-line in a general circulation model. *Physics and Chemistry of the Earth, Part C: Solar, Terrestrial and Planetary Science* **26**, 585–591.

Junge, C. E. (1954) The chemical composition of atmospheric aerosols. I: Measurements at Round Hill Field Station. June–July 1953. *Journal of Meteorology* **11**, 323–333.

Junge, C. E. and T. G. Ryan (1958) Study of the SO_2 oxidation in solution and its role in atmospheric chemistry. *Quarterly Journal of the Royal Meteorological Society* **84**, 46–55.

Junge, C. E. (1958) Atmospheric chemistry. In: Advances in geophysics. Vol. IV, Acad. Press. New York, (separate volume), 108 pp.

Junge, C. E. (1963) Air chemistry and radioactivity. Vol. 4 Int. Geophys. Ser. (Ed. J. van Mieghem), Academic Press, New York and London, 382 pp. Junge, C. E. (1974) Residence time and variability of tropospheric trace gases. *Tellus* **26**, 477–488.

Junker, C. and C. Liousse (2006) A global emission inventory of carbonaceous aerosol from historic records of fossil fuels and biofuel consumption for the period 1860–1997. *Atmospheric Chemistry and Physics Discussion* **6**, 4897–4927.

Junkermann, W., B. Vogel and M. Sutton (2011) The climate penalty for clean fossil fuel combustion. *Atmospheric Chemistry and Physics* **11**, 12917–12924.

Kaplan, I. R., E. T. Degens and J. H. Reuter (1963) Organic compounds in stony meteorites. *Geochimica et Cosmochimica Acta* **27**, 805–808.

Kanakidou, M., B. Bonsang, J. C. Le Roulley, G. Lambert and D. Martin (1988) Marine source of atmospheric acetylene. *Nature* **333**, 51–52.

Kanakidou, M. and P. J. Crutzen (1999) The photochemical source of carbon monoxide: Importance, uncertainties and feedbacks. *Chemosphere* **1**, 91–109.

Kanakidou, M., J. H. Seinfeld, S. N. Pandis, I. Barnes, F. J. Dentener, M. C. Facchini, R. Van Dingenen, B. Ervens, A. Nenes, C. J. Nielsen, E. Swietlicki, J. P. Putaud, Y. Balkanski, S. Fuzzi, J. Horth, G. K. Moortgat, R. Winterhalter, C. E. L. Myhre, K. Tsigaridis, E. Vignati, E. G. Stephanou and J. Wilson (2005) Organic aerosol and global climate modelling: A review. *Atmospheric Chemistry and Physics* **5**, 1053–1123.

Kargel, J. S. (1992) Ammonia-water ion icy planets: Phase relations at 1 atmosphere. *Icarus* **100**, 556–590.

Kasting, J. F., S. C. Liu and T. M. Donahue (1979) Oxygen levels in the prebiological atmosphere. *Journal of Geophysical Research* **84**, 3097–3107.

Kasting J. F. and T. M. Donahue (1980) The evolution of the atmospheric ozone. *Journal of Geophysical Research* **85**, 3255−3263.

Kasting J. F. (1993) Earth's early atmosphere. *Science* **259**, 920−926.

Kasting, J. F. (2001) The rise of atmospheric oxygen. *Science* **293**, 819−820.

Kasting, J. F. and J. L. Siefert (2002) Life and the evolution of earth's atmosphere. *Science* **296**, 1066−1068.

Kasting, J. F. and D. Catling (2003) Evolution of a habitable planet. *Annual Review of Astronomy and Astrophysics* **41**, 429−463.

Kasting, J. F. and M. T. Howard (2006) Atmospheric composition and climate on the early Earth. *Philosophical Transaction of the Royal Society, Series B* **361**, 1733−1742.

Kasting, J. F. (2006) Earth sciences − ups and downs of ancient oxygen. *Nature* **443**, 643−645.

Kato, N. and H. Akimoto (1992) Anthropogenic emissions of SO_2 and NO_x in Asia: Emission inventories. *Atmospheric Environment* **26A**, 2997−3017.

Katz, M. and S. E. Gale (1971) Mechanism of photooxidation of SO_2 in atmosphere. In: Procceedings of the 2nd Intern. Clean Air Congress (Eds. H. M. Englung and W. T. Berry), Academic Press, New York, pp. 336−343.

Kausch, O. (1937) Das Wasserstoffsuperoxyd. Eigenschaften, Herstellung und Verwendung. Verlag W. Knapp, Halle, 254 pp.

Kavanaugh, M. C. and R. R. Trussel (1980) Design of aeration towers to strip volatile contaminants from drinking water. *Journal of American Water Works Association* **72**, 684−692.

Kavarnos, G. J. and N. J. Turro (1987) Photosensitization by reversible electron transfer: Theories, experiment al evidence, and examples. *Chemical Reviews* **86**, 401−449.

Kay, M. J., M. A. Box, Th. Trautmann and J. Landgraf (2001) Actinic flux and net flux calculations in radiative transfer − A comparative study of computational efficiency. *Journal of the Atmospheric Sciences* **58**, 3752−3761.

Kazadzis, S., C. Topaloglou, A. F. Bais, M. J. Blumthaler, D. Ballis, A. Kazantzidis and B. Schalhart (2004) Actinic flux measurements and $O(^1D)$ photolysis frequencoies retrieved from spectral measurements of irradiance at Thessaloniki, Greece. *Atmospheric Chemistry and Physics Discussion* **4**, 4191−4225.

Keeling C. D. and T. P. Whorf (2004) Atmospheric carbon dioxide concentrations at 10 locations spanning latitudes 82° N to 90° S (Eds. T. J. Blasing and S. Jones), ORNL/CDIAC-147, NDP-001a. Carbon Dioxide Information Analysis Center, Oak Ridge National Laboratory, U.S. Department of Energy, Oak Ridge, Tennessee, 30 pp. doi:10.3334/CDIAC/atg.ndp001.2004.

Keeling, R. F., R. B. Bacastow, A. E. Bainbridge, C. A. Ekdahl, P. R. Guenther and L. S. Waterman (1976) Atmospheric carbon dioxide variations at Mauna Loa Observatory, Hawaii. *Tellus* **28**, 538−551.

Keeling, R. F. (1988) Development of an interferometer oxygen analyser for precise measurement of the atmospheric O_2 mole fraction. Doctoral Thesis, Harvard University, Cambridge, Mass. (USA).

Keeling, R. F. and S. R. Shertz (1992) Seasonal and interannual variations in atmospheric oxygen and implications for the global carbon cycle. *Nature* **358**, 723−727.

Keene, W. C. and J. N. Galloway (1984) Organic acidity in precipitation of North America. *Atmospheric Environment* **18**, 2491−2497.

Keene, W. C. and D. L. Savoie (1998) The pH of deliquesced sea-salt aerosol in the polluted marine boundary layer. *Geophysical Research Letters* **25**, 2181−2184.

Keene W. C., M. A. K. Khalil, D. J. Erickson, A. McCulloch, T. E. Graedel, J. M. Lobert, M. L. Aucott, S.-L. Gong, D. B. Harper, G. Kleiman, P. Midgley, R. M. Moore, C. Seuzaret, W. T. Sturges, C.M. Benkovitz, V. Koropalov, L. A. Barrie and Y.-F. Li (1999) Composite global emissions of reactive chlorine from anthropogenic and natural sources: Reactive chlorine emissions inventory. *Journal of Geophysical Research* **104**, 8429−8440.

Keene, W. C., J. M. Lobert, P. J. Crutzen, J. R. Maben, D. H. Scharffe, T. Landman, C. Hély and C. Brain (2006) Emissions of major gaseous and particulate species during experimental burns of southern African biomass. *Journal of Geophysical Research* **111**, D04301, doi:10.1029/2005JD006319.

Keidel, E. (1929) Die Beeinflussung der Lichtechtheit von Teerfarblacken durch Titanweiss. *Farben-Zeitung* **34**, 242−243.

Keil, A. D. and P. B. Shepson (2006) Chlorine and bromine atom ratios in the springtime Artic troposphere as determined from measurements of halogenated volatile organic compounds. *Journal of Geophysical Research* **111**, D17303, doi:10.1029/2006JD007119.

Keiner, L. (1934) Die Mattkunstseide. *Mellior Textilberichte* **15**, 118−120.

Keith, D. W. (2000). Geoengineering the Climate: History and Prospect. *Annual Review of Energy and the Environment* **25**, 245−284.

Keith, D. W., M. Ha-Duong and J. K. Stolaroff (2005) Climate strategy with CO_2 capture from the air. *Climate Change* **74**, 17−45.

Keith, D. W. (2009) Why capture CO_2 from the atmosphere? *Science* **325**, 1654−1655.

Keith, H., B. G. Mackey and D. B. Lindemayer (2009) Re-evaluation of forest biomass carbon stocks and lessons from the world's most carbon-dense forest. *Proceedings of the National Academy of Sciences* **106**, 11635−11640.

Keller, S. C., M. S. Bessell, A. Frebel, A. R. Casey, M. Asplund, H. R. Jacobson, K. Lind, J. E. Norris, D. Yong, A.Heger, Z.Magic, G. S. Da Costa, B. P. Schmidt and P. Tisserand (2014) A single low-energy, iron-poor supernova as the source of metals in the star SMSS J031300.36−670839.3. *Nature* arXiv:1402.1517. doi:10.1038/nature12990.

Kellog, W. W., R. D. Cadle, E. R. Allen, A. L. Lazrus and E. A. Martell (1972) The sulfur cycle. *Science* **175**, 587−595.

Keppler, F., J. T. G. Hamilton, M. Braß and Th. Röckmann (2006) Methane emissions from terrestrial plants under aerobic conditions. *Nature* **439**, 187−191.

Kerminen, V. M., K. Teinilä, R. Hillamo and T. Pakkanen (1998) Substitution of chloride in sea-salt particles by inorganic and organic anions. *Journal of Aerosol Science* **29**, 929−942.

Kern, S. (1878) *Chemical News* **87**, 35 (title unknown, cited after Schöne 1893).

Kernbaum, M. (1909a) Action chimique suv l'eau des rayons penetrants de radium. *Comptes Rendus de l'Académie des Sciences* **148**, 705−707.

Kernbaum, M. (1909b) Décomposition de l'eau par les rayons ultra-violets. *Comptes Rendus de l'Académie des Sciences* **149**, 273−275.

Kernbaum, M. (1910) Décomposition de l'eau par l'aigrette. *Comptes Rendus de l'Académie des Sciences Comptes Rendus de l'Académie des Sciences* **151**, 275−278.

Kernbaum, M. (1911) Sur la décomposition de l'eau par les métaux. *Comptes Rendus de l'Académie des Sciences Comptes Rendus de l'Académie des Sciences* **152**, 1668−1670.

Kerstiens, G., L. Schreiber and K. J. Lendzian (2006) Quantification of cuticular permeability in genetically modified plants. *Journal of Experimental Botany* **57**, 2547−2552.

Kesselmeier, J., P. Schröder and J. W. Erisman (1997) Exchange of sulfur gases between the biosphere and the atmosphere. In: Biosphere-atmosphere exchange of pollutants and trace substances. Vol. 4: Transport and chemical transformation of pollutants in the troposphere. (Ed. J. Slanina), Springer-Verlag, Berlin, pp. 167−200.

Kesselmeier, J. and M. Staudt (1999) Biogenic volatile organic compounds (VOC): an overview on emission, physiology and ecology. *Journal of Atmospheric Chemistry* **33**, 23−88.

Kesselmeier, J. (2001) Exchange of short-chain oxygenated volatile organic compounds (VOCs) between plants and the atmosphere: A compilation of field and laboratory studies. *Journal of Atmospheric Chemistry* **39**, 219−233.

Kesselmeier, J., P. Ciccioli, U. Kuhn, P. Stefani, T. Biesenthal, S. Rottenberger, A. Wolf, M. Vitullo, R. Valentini, A. Nobre, P. Kabat and M. O. Andreae (2002) Volatile organic compound emissions in relation to plant carbon fixation and the terrestrial carbon budget. *Global Biogeochemical Cycles* **16**, 1126, doi:10.1029/2001GB001813.

Kesselmeier, J. and A. Hubert (2002) Exchange of reduced volatile sulphur compounds between leaf litter and the atmosphere. *Atmospheric Environment* **36**, 4679−4686.

Kessler, C. (1984) Gasförmige salpetrige Säure (HNO₂) in der belasteten Atmosphäre. Doctoral Thesis, University of Cologne, 97 pp.

Kettle, A. J., Andreae, M. O., Amouroux, D., Andreae, T. W., Bates, T. S., Berresheim, H., Bingemer, H., Boniforti, R., Curran, M. A. J., DiTullio, G. R., Helas, G., Jones, G. B., Keller, M. D., Kiene, R. P., Leck, C., Levasseur, M., Malin, G., Maspero, M., Matrai, P., McTaggart, A. R., Mihalopoulos, N., Nguyen, B. C., Novo, A., Putaud, J. P., Rapsomanikis, S., Roberts, G., Schebeske, G., Sharma, S., Simó, R., Staubes, R., Turner, S. and G. Uher (1999) A global database of sea surface dimethyl sulfide (DMS) measurements and a procedure to predict sea surface DMS as a function of latitude, longitude, and month. *Global Biogeochemical Cycles* **13**, 399−444.

Kettle, A. J., and M. O. Andreae (2000) Flux of dimethylsulfide from the oceans: A comparison of updated data sets and flux models. *Journal of Geophysical Research* **105**, 26793−26808.

Kettle, A. J., U. Kuhn, M. von Hobe, J. Kesselmeier and M. O. Andreae (2002) The global budget of atmospheric carbonyl sulfide: Temporal and spatial modulation of the dominant sources and sinks. *Journal of Geophysical Research* **107**, 4658, doi:10.1029/2002JD002187.

Khalil, M. A. K. and R. A. Rasmussen (1989) Distribution and mass balance of molecular hydrogen in the earth's atmosphere. In: Geophysical monitoring for climate change (Eds. J. W. Elkings and R. M. Rossen), NOAA, Boulder (USA), pp. 11−113.

Khalil, M. A. K., and R. A. Rasmussen (1990a) The global cycle of carbon monoxide: Trends and mass balance. *Chemosphere* **20**, 227−242.

Khalil, M. A. K. and R. A. Rasmussen (1990b) Global increase of atmospheric molecular hydrogen. *Nature* **347**, 743−745.

Khalil, M. A. K. and R. A. Rasmussen (1992) The global sources of nitrous oxide. *Journal of Geophysical Research* **97**, 14651−14660.

Khalil, M. A. K. and R. A. Rasmussen (1995) The changing composition of the earth's atmosphere. In: Composition, chemistry, and climate of the atmosphere (Ed. H. B. Singh), Van Nostrand Reinhold, New York, pp. 50−87.

Khalil, M. A. K. (1999) Preface: Atmospheric carbon monoxide. *Chemosphere: Global Change Science*, **1**, ix−xi.

Khalil, M. A. K. and R. A. Rasmussen (1999) Atmospheric methyl chloride. *Atmospheric Environment* **33**, 1305−1321.

Khalil, M. A. K., R. M. Moore, D. B. Harper, J. M. Lobert, D. J. Erickson, V. Koropalov, W. T. Sturges and W. C. Keene (1999) Natural emissions of chlorine containing gases: Reactive chlorine emissions inventory. *Journal Geophysical Research* **104**, 8333−8346.

Khalil, M. A. K., R. A. Rasmussen and M. J. Shearer (2002) Atmospheric nitrous oxide: patterns of global change during recent decades and centuries. *Chemosphere* **47**, 807−821.

Khrgian, A. K. (1970) Meteorology. A historical survey. Vol. 1 (translation from Russian). Jerusalem, 387 pp.

Killops, S. D. and V. J. Killops (2005) Introduction to organic geochemistry. 2ⁿᵈ ed., Blackwell Publ. (USA), 393 pp.

Kindermann, G. E., I. McCallum, S. Fritz and M. Obersteiner (2008) A global forest growing stock. Biomass and carbon map based on FAO statistics. *Silva Fennica* **42**, 387−396.

Kirschke, S., P. Bousquet, P. Ciais, M. Saunois, J. G. Canadell, E. J. Dlugokencky, P. Bergamaschi, D. Bergmann, D.R. Blake, L. Bruhwiler, P. Cameron-Smith, S. Castaldi, F. Chevallier, L. Feng, A. Fraser, M. Heimann, E. L. Hodson, S. Houweling, B. Josse, P. J. Fraser, P. B. Krummel, J.-F. Lamarque, R. L. Langenfelds, C. Le Quéré, V. Naik, S. O'Doherty, P. I. Palmer, I. Pison, D. Plummer, B. Poulter, R. G. Prinn, M. Rigby, B. Ringeval, M. Santini, M. Schmidt, D. T. Shindell, I. J. Simpson, R. Spahni, L. P. Steele, S. A. Strode, K. Sudo, S. Szopa, G. R. van der Werf, A. Voulgarakis, M. van Weele, R. F. Weiss, J.

E.Williams and G. Zeng (2013) Three decades of global methane sources and sinks. *Nature Geoscience* **6**, 813−823.

Kiss, G., E. Tombácz and H.-Chr. Hansson (2005) Surface tension effects of humic-like substances in the aqueous extract of tropospheric fine aerosol. *Journal of Atmospheric Chemistry* **50**, 279−294.

Klaassen, G. (1992) Emissions of ammonia in Europe as incorporated in RAINS. In: Ammonia emissions in Europe − emissions coefficients and abatement strategies (Ed. G. Klaassen), IIASA, Report CP-92-4, Laxenburg (Austria), pp. 25−40.

Kleffmann, J., J. Heland, R. Kurtenbach, J. C. Lörzer and P. Wiesen (2002) A new instrument (LOPAP) for the detection of nitrous Acid (HONO). *Environmental Science and Technology* **9**, 48−54.

Kleffmann, J., T. Gavriloaiei, A. Hofzumahaus, F. Holland, R. Koppmann, L. Rupp, E. Schlosser, M. Siese and A. Wahner (2005) Daytime formation of nitrous acid: a major source of OH radicals in a forest. *Geophysical Research Letters* **32**, L05818. doi:10.1029/2005GL022524.

Kleffmann, J. (2007) Daytime sources of nitrous acid (HONO) in the atmospheric boundary layer. *ChemPhysChem* **8**, 1137−1144.

Klemm, O., R. S. Schemenauer, A. Lummerich, P. Cereceda, V. Marzol, D. Corell, J. van Heerden, D. Reinhard, T. Gherezghiher, J. Olivier, P. Osses, J. Sarsour, E. Frost, M. J. Estrela, J. A. Valiente and G. M. Fessehaye (2012) Fog as a fresh-water resource: overview and perspectives. *Ambio* **41**, 221−234.

Klemperer, W. (2006) Interstellar chemistry *Proceedings of the National Academy of Science* **103**, 12232−12234.

Kley, D., H. Geiss and V. A. Mohnen (1994) Tropospheric ozone at elevated sites and precursor emissions in the United States and Europe. *Atmospheric Environment* **28**, 149−158.

Klimont, Z., S. J. Smith and J. Cofala (2013) The last decade of global anthropogenic sulfur dioxide: 2000–2011 emissions. *Environmental Research Letters* **8**, 014003.

Kmiec, G., A. Zwozdziak, K. Acker and W. Wieprecht (1998) Cloud/fog water chemistry at high elevation in the Sudeten Mountains, south-western Poland. In: Proceedings of the First International Conference on Fog and Fog Collection, pp. 69−72, Vancouver, Canada, July 1998.

Klingenberg, H. (1996) Automobile exhaust emission testing. Measurements of regulated and unregulateed exhaust gas components, exhaust emission tests. Springer-Verlag, Berlin, 383 pp.

Kloster, S. (2006) DMS cycle in the ocean-atmosphere system and its response to anthropogenic perturbations. Doctoral Thesis. MPI für Meteorologie/Atmosphere in the Earth System, Hamburg (Germany), 93 pp.

Kloster, S., J. Feichter, E. Maier-Reimer, K. D. Six, P. Stier and P. Wetzel (2006) DMS cycle in the marine ocean-atmosphere system − a global model study. *Biogeosciences* **3**, 29−51.

Knop, W. (1868) Der Kreislauf des Stoffs. Lehrbuch der Agricultur-Chemie. H. Haessel, Leipzig, 925 + 309 pp.

Knorr, W. (2009) Is the airborne fraction of anthropogenic CO_2 emissions increasing? *Geophysical Research Letters* **36**, L21710, doi:10.1029/2009GL040613.

Kobozev, N. I., L. I. Nekrasov and I. I. Skorokhodov (1959) The physical chemistry of concentrated ozone: II A study of the synthesis of the higher peroxides of hydrogen H_2O_4 by interaction of concentrated ozone with atomic hydrogen. National Research Council of Canada, 36 pp.

Koeberl, C. (2006) Impact processes on the early earth. *Elements* **2**, 211−216.

Koide, M. and E. D. J. Goldberg (1971) Atmospheric sulphur and fossil fuel combustion. *Journal of Geophysical Research* **76**, 6589−6596.

Kok, G. L., K. R. Darnall, A. M. Winer, J. Pitts and B. Gay (1978) Ambient air measurements of hydrogen peroxide in the California south coast air basin. *Environmental Science and Technology* **12**, 1077−1080.

Konstantinova-Schlesinger, M. A. (1937a) Определение флуоресцентным методам содержания озона в восдухе на высоте 9620 м. (In Russian: Estimation of ozone in air at an altitude of 9620 m by a fluorescence method). *Doklady Akademii Nauk SSSR* **14**, 187−188.

Konstantinova-Schlesinger, M. A. (1937b) Резултаты определения содержания озона в восдухе флуоресцентным методам (In Russian: Results of ozone measurements in air with a fluorescence method). *Izvestiya Akademii Nauk SSSR*, 213−222.

Konstantinova-Schlesinger, M. A. (1938) Определения содержания озона в пробах восдухе с высот 13 и 14 км на уровнем моря (In Russian: Estimation of ozone in air samples from altitudes 13 and 14 km above sea level). *Doklady Akademii Nauk SSSR*, **18**, 337−338.

Konya, J. and N. M. Nagy (2009) Interfacial chemistry of rocks and soils. CRC Press, Boca Raton, FL (USA), 244 pp.

Koop, T., B. Luo, A. Tsias and T. Peter (2000) Water activity as the determinant for homogeneous ice nucleation in aqueous solution. *Nature* **406**, 611−614.

Kopp, H. (1869) Die Entdeckung der Zusammensetzung des Wassers. In: Beiträge zur Geschichte der Chemie (Hermann Kopp), Fr. Vieweg und Sohn, Braunschweig, pp. 235−310.

Kopp, H. (1931) Geschichte der Chemie. Neudruck der Originalausgabe in zwei Bänden (1843−1847), A. Lorentz Leipzig, Dritter Teil, 372 pp.

Köppen, W. (1906) Klimakunde I: Allgemeine Klimakunde. Göschen, Leipzig, 132 pp.

Köppen, W. (1923) Die Klimate der Erde. Grundriß der Klimakunde. 1. Auflage. de Gruyter, Berlin, 369 pp.

Köppen, W. (1931) Grundriß der Klimakunde. 2. verbesserte Auflage der "Klimate der Erde". de Gruyter, Berlin, 388 pp.

Koppenol, W. H. (2001) The Haber-Weiss cycle − 70 years later. *Redox Report* **6**, 229−234.

Koppmann, R., A. Khedim, J. Rudolph, D. Poppe, M.O. Andreae, G. Helas, M. Welling and T. Zenker (1997) Emissions of organic trace gases from savanna fires in southern Africa during the 1992 Southern African Fire. Atmosphere Research Initiative and their impact on the formation of tropospheric ozone. *Journal of Geophysical Research* **102**, 18879−18888.

Koppmann, R., K. von Czapiewski and J. S. Reid (2005) A review of biomass burning emisisons, part I: gaseous emisisons of carbon monoxide, methane, volatile organic compounds, and nitrogen containing compounds. *Atmospheric Chemistry and Physics Discussion* **5**, 10455−10516.

Koppmann, R., Ed. (2007) Volatile organic compounds in the atmosphere. Blackwell Pub. Professional, 512 pp.

Korcak, R. F. (1992) Early roots of the organic movement: A plant nutrition perspective. In: Hort technology Vol. 2, No. 2 From "Proceedings of the workshop: History of the organic movement," meeting held at the 88th American Society for Horticultural Science, The Pennsylvania State University, University Park, PA (USA).

Kormann, C., D. W. Bahnemann and M. R. Hoffmann (1988) Photocatalytic production of H_2O_2 and organic peroxides in aqueous suspensions of TiO_2, ZnO and desert sand. *Environmental Science and Technology* **22**, 798−806.

Kosaka, Y. and S.-P. Xie (2013) Recent global-warming hiatus tied to equatorial Pacific surface cooling. *Nature*. doi:10.1038/nature12534.

Kouimtzis, T. and C. Samara, Eds. (1995) Airborne particulate matter. Vol. 4 Part D: The handbook of environmental chemistry (Ed. O. Hutzinger), Springer-Verlag, Berlin, 339 pp.

Kozlovsky, Y. A., Ed. (1987) The Superdeep Well of the Kola Peninsula. Springer-Verlag, Berlin, 558 pp.

Kraus, C. A. (1908) Solutions of metals in non-metallic solvents. IV. Material effects accompanying the passage of an electrical current through solutions of metals in liquid ammonia. Migration experiments. *Journal of American Chemical Society* **30**, 1323−1344.

Kraus H. (2004) Die Atmosphäre der Erde. Eine Einführung in die Meteorologie. Springer-Verlag, 422 pp.

Krausmann, F., K. H. Erb, S. Gingrich, C. Laukand and H. Haberl (2008) Global patterns of socioeconomic biomass flows in the year 2000: A comprehensive assessment of supply, consumption and constraints. *Ecological Economics* **15**, 471−487.

Krikorian A. D. and F. C. Steward (1968) Water and Solutes in Plant Nutrition: With Special Reference to van Helmont and Nicholas of Cusa. *BioScience*, *18*, 286−292.

Krogh, A. (1919) The composition of the atmosphere. Det Kgl. Danske Videnskabernes Selskab. Math.-fys. Meddelser. I, 12, København, 19 pp.

Krummel, P., P. Fraser and Nada Derek (2012) The 2010 Antarctic Ozone Holeand Ozone Science Summary. A Report prepared for Refrigerant Reclaim Australia. Centre for Australian Weather and Climate Research CSIRO Marine and Atmospheric Research, 12 pp.

Kryzhanovskii, G. N. (1970) Friedrich Engels and natural science (on the 150th anniversary of his birth). *Bulletin of Experimental Biology and Medicine* **70**, 1229−1237.

Kudryavtsev, N. A. (1959) Geological proof of the deep origin of Petroleum (in Russ.). *Trudy Vsesoyuz. Neftyan. Nauch. Issledovatel Geologoraz Vedoch. Inst.* **132**, 242−262.

Kulich, D. M. and J. R. Shelton (1978) The role of certain organic sulfur compounds as preventive antioxidants. III. Reactions of tert-butyl tert-butanethiolsulfinate and hydroperoxide. *Advances in Chemistry* **169**, 226−236.

Kulich, D. M. and J. R. Shelton (1991) Organosulfur antioxidants: Mechanisms of action. *Polymer Degradation and Stability* **33,** 397−410.

Kulmala, M., T. Vesala and A. Laaksonen (2000) Physical chemistry of aerosol formation. In: *Aerosol chemical processes in the environment* (Ed. S. Spurny), Lewis Publ. Boca Raton, London, New York, Washington, pp. 23−46.

Kulmala, M., H. Vehkamäki, T. Petäjä, M. Dal Maso, A. Lauri, V.-M. Kerminen, W. Birmili and P. H. McMurry (2004) Formation and growth rates of ultrafine atmospheric particles: a review of observations. *Journal of Aerosol Science* **35**, 143−176.

Kumar, R., A. Rani, S. P. Singh, K. M. Kumari and S. S. Srivastava (2002) A long term study on chemical composition of rainwater at Dayalbagh, a suburban site of semiarid region. *Journal of Atmospheric Chemistry* **41**, 265−279.

Kump, L. R. and M. E. Barley (2007) Increased subaerial volcanism and the rise of atmospheric oxygen 2.5 billion years ago. *Nature* **448**, 1033−1036.

Kunen, S. M., A. L. Lazrus, G. L. Kopk and B. H. Heikes (1983) Aqueous oxidation of SO_2 by H_2O_2. *Journal of Geophysical Research* **88**, 3671−3674.

Kuwajima, K., Goto, Y. and F. Hirata, Eds. (2009) Water and biomolecules: Physical chemistry of life phenomena. Springer-Verlag, Berlin, 358 pp.

Kvenvolden, K. A. and T. D. Lorenson (2001) The global occurrence of natural gas hydrate. In: Natural gas hydrates: Occurrence, distribution, and dynamics (Eds. C. K. Paull and W. P. Dillon), AGU Geophysical Monograph Series 124, American Geophysical Union, Washington, D.C. USA), pp. 1−23.

Kwok, S. (2004) The synthesis of organic and inorganic compounds in evolved stars. *Nature* **439**, 985−991.

Kyle, T. G. (1991) Atmospheric transmission. Emission and scattering. Pergamon Press, Oxford (U.K.), 288 pp.

Labitzke, K. and M. P. McCormick (1992) Stratospheric temperature increases due to Pinatubo aerosols. *Geophysical Research Letters* **19**, 207−210.

Labrador, L. J., R. von Kuhlmann and M. G. Lawrence (2005) The effects of lightning-produced NO_x and its vertical distribution on atmospheric chemistry: Sensitivity simulations with MATCH-MPIC. *Atmospheric Chemistry and Physics* **5**, 1815−1834.

Lacis, A. A., J. E. Hansen and M. Sato (1992) Climate forcing by stratospheric aerosols *Geophysical Research Letters* **19**, 1607−1610.

Lack, D. A. (2003) Modelling the formation of atmospheric aerosol from gaseous organic precursors. Doctoral Thesis. School of Natural Resource Sciences, Queensland University of Technology, 341 pp.

Lackner, K. S., P. Grimes and H. J. Ziock (1999) Capturing carbon dioxide from air. In: 24th Annual Technical Conference on Coal Utilization, Clearwater, Florida (USA).

Lagrange, J., C. Pallares, G. Wenger and P. Lagrange (1993) Electrolyte effects on aqueous atmospheric oxidation of sulphur dioxide by hydrogen peroxide. *Atmospheric Environment* **27A**, 129−137.

Laj, P., S. Fuzzi, M. C. Facchini, J. A. Lind, G. Orsi, M. Preiss, R. Maser, W. Jaeschke, E. Seyffer, G. Helas, K. Acker, W. Wieprecht, D. Möller, B. G. Arends, J. J. Möls, R. N. Colvile, M. W. Gallagher, K. M. Beswick, K. J. Hargreaves, R. L. Storeton-West und M. A. Sutton (1997) Cloud processing of soluble gases. *Atmospheric Environment* **31**, 2589−2598.

Lakatos, I. (1980) History of science and its rational reconstructions. In: Scientific revolutions (Ed. J. Hacking), Oxford University Press, pp. 107−127.

Lamb, H. H. (1970) Volcanic dust in the atmosphere; with a chronology and assessment of its meteorological significance. *Philosophical Transaction of the Royal Society, Series A*, **266**, 425−533.

Lambert, G., M.-F. Le Cloarec and M. Pennisi (1988) Volcanic output of SO_2 and trace metals: a new approach. *Geochimica et Cosmochimica Acta* **52**, 39−42.

Lambert, G., and S. Schmidt (1993) Reevaluation of the oceanic flux of methane: Uncertainties and long term variations. *Chemosphere* **26**, 579−589.

Lambin, E. F. and H. J. Geist, Eds. (2006) Land use and land cover change: Local processes and global impacts (Global change − the IGBP), Springer-Verlag, Berlin, 222 pp.

Lampadius, W. A. (1806) Systematischer Grundriß der Atmosphärologie. Craz and Gerlach, Freyberg (Saxonia), 392 pp.

Langenfelds, R. L., R. Francey, B. C. Pak, L. P. Steele, J. Lloyd, C. M. Trudinger and C. E. Allison (2002) Interannual growth rate variations of atmospheric CO_2 and its delta ^{13}C, H_2, CH_4, and CO between 1992 and 1999 linked to biomass burning. *Global Biogeochemical Cycles* **16**, 21−1 to 21−22.

Langevin, P. (1905) Les ions dans l'atmosphère. *Comptes Rendus de l'Académie des Sciences* **140**, 232−234.

Langmuir, D. (1997) Aqueous environmental geochemistry. Prentice Hall, New Jersey, 600 pp.

Langner, J. and H. Rodhe (1991) A global three-dimensional model of the tropospheric sulfur cycle. *Journal Atmospheric Chemistry* **13**, 225−263.

Lathière, J., D. A. Hauglustaine, A. Friend, N. de Noblet-Ducoudré, N. Viovy and G. Folberth (2005) Impact of climate variability and land use changes on global biogenic volatile organic compound emissions. *Atmospheric Chemistry and Physics Discussions* **5**, 10613−10656.

Lavoisier, A.-L. (1789) Elements of Chemistry in a new systematic order, containing all the modern discoveries. Transl. by R. Kerr, reprint 1965, Dover Publ. New York, 511 pp.

Lawrence, M. G., W. L. Chameides, P. Kasibhatla, H. Levy II and W. Moxim (1995) Lightning and atmospheric chemistry: The rate of atmospheric NO production. In: Handbook of Atmospheric Electrodynamics, 1 (Ed. H. Volland), CRC Press, Boca Raton, FL (USA), pp. 189−202.

Lawrence, M. G., O. Hov, M. Bekmann, J. Brandt, H. Elbern, H. Eskes, H. Feichter and M. Takigawa (2005) The chemical weather. *Environmental Chemistry* **2**, 6−8.

Lazcano, A. and S. L. Miller (1994) How long did it take for life to begin and evolve to cyanobacteria? *Journal of Molecular Evolution* **39**, 546−554.

Lee, D. S., I. Köhler, E. Grobler, F. Rohrer, R. Sausen, L. Gallardo-Klenner, J. G. J. Olivier, F. J. Dentener and A. F. Bouwman (1997) Fossil fuel combustion is the largest global source of NO_x to the troposphere. *Atmospheric Environment* **12**, 1735−1749.

Lee, M., B. G. Heikes, D. J. Jacob, G. Sachse and B. Anderson (1997) Hydrogen peroxide, organic hydroperoxide and formaldehyde as primary pollutants from biomass burning. *Journal of Geophysical Research* **102**, 1301−1309.

Lee, M., B. G. Heikes and D. W. O'Sullivan (2000) Hydrogen peroxide and organic hydroperoxide in the troposphere: A review. *Atmospheric Environment* **34**, 3475−3494.

Lee, P. A. and S. J. de Mora (1999) A review of dimethylsulfoxide in aquatic environments. *Atmosphere − Ocean* **37**, 439−456.

Lee, T. J. and A. P. Rendell (1993) Ab initio characterization of peroxyhypochlorous acid: implications for atmospheric chemistry. *Journal of Physical Chemistry* **97**, 6999−7002.

Lee, Y.- N., J. Shen, P. J. Klotz, S. E. Schwartz and L. Newman (1986) Kinetics of hydrogen peroxide − sulfur(IV) to sulfur(VI) reaction in rainwater collected at a northwestern U.S. site. *Journal of Geophysical Research* **91**, 13264−13274.

Lee-Taylor, J. and K. R. Redeker (2005) Reevaluation of global emissions from rice paddies of methyl iodide and other species. *Geophysical Research Letters* **32**, L15801, doi:10.1029/2005GL022918.

Lefohn, A. S., J. D. Husar and R. B. Husar (1999) Estimating historical anthropogenic global sulfur emission patterns for the period 1850−1990. *Atmospheric Environment* **33**, 3435−3444.

Lehr, J. H. and J. Keeley, Eds. (2005) Water encyclopedia: Oceanography, meteorology, physics and chemistry, water law, and water history, art, and culture. John Wiley & Sons, New Jersey, 832 pp.

Leighton, P. A. (1961) Photochemistry of air pollution. Vol. 9 in "Physical chemistry" series of monographs (Ed. E. M. Loebl), Academic Press, New York and London, 300 pp.

Lejeune, B., F. Mahieu, P. Demoulin, C. Servais, W. Bader, B. Bovy, O. Flock, R. Zander and G. Roland (2012) Trend evolution of carbonyl sulphide aboce Jungfraujoch deduced from ground-based FTIR and ACE-FTS satellite observations. ACE (Atmospheric Chemistry Experiment) Science Team Meeting, May 23−25, 2012, Toronto, Canada.

Lelieveld, J. and P. J. Crutzen (1991) The role of clouds in tropospheric photochemistry. *Journal of Atmospheric Chemistry* **12**, 229−268.

Lelieveld, J. and R. van Dorland (1995) Ozone chemistry changes in the troposphere and consequent radiative forcing of climate. In: Atmospheric ozone as a climate gas: General circulation model simulation (Eds. W. S. Wank and I. S. A. Isaksen), NATO ASI Series Vol. 32, Springer, New York, pp. 227−258.

Lelieveld, J., P. J. Crutzen and F. J. Dentener (1998) Changing concentration, lifetime and climate forcing of atmospheric methane. *Tellus* **50B**, 128−150.

Lelieveld, J. and F. J. Dentener (2000) What controls tropospheric ozone? *Journal of Geophysical Research* **105**, 3531−3551.

Lelieveld, J., T. M. Butler, J. N. Crowley, T. J. Dillon, H. Fischer, L. Ganzeveld, H. Harder, M. G. Lawrence, M. Martinez, D. Taraborrelli and J. Williams (2008) Atmospheric oxidation capacity sustained by a tropical forest. *Nature* **452**, 737−740.

Lenard, P. (1900) Ueber Wirkungen des ultravioletten Lichtes auf gasförmige Körper. *Annalen der Physik* **306**, 486−507.

Lenhard, U. and H.-W. Georgii (1980) Der Ozean als Quelle reaktiver Stickstoffverbindungen. *Pageoph* **118**, 1145−1154.

Lenoble, J. (1985) Radiative transfer in scattering and absorbing atmospheres: Standard computational procedures. A. Deepak Publishing, Poquoson (USA), 532 pp.

Lenton, T. M. and N. E. Vaughan (2009) The radiative forcing potential of different climate geoengineering options. *Atmospheric Chemistry and Physics Discussion* **9**, 2559−2608.

Lenz, H. P. and Chr. Cozzarini (1999) Emissions and air quality. Soc. of Automotive Eng., Warrendale, 124 pp.

Lerdau, M. (2007) Ecology: a positive feedback with negative consequences. *Science* **316**, 212−213.

Le Quéré, C. (2006) The unknown and the uncertain in earth system modeling. *Eos, Transactions American Geophysical Union* **87**, 496.

Le Quéré, C., C. Rödenbeck, E. T. Buitenhuis, T. J. Conway, R. Langenfels, A. Gomez, C. Labuschagne, M. Ramonet, T. Nakazawa, N. Metzl, N. Gilet and M. Heiman (2007) Saturation of the Southern Ocean CO_2 sink due to recent cliamte change. *Science* **316**, 1735–1738.

Le Roy, E. (1927) L'exigence idéaliste et lafait de l'évolution (Idealistic exigency and the feat of evolution), Boivin, Paris, 270 pp.

Lessing, G. E. (1883) Theologische Streitschriften – Eine Duplik. In: Lessings Werke in sechs Bänden, Band 6, p. 202, Verlag von Grimme und Trömel, Leipzig.

Letts, E. A. and R. F. Blake (1900) The carbonic anhydride of the atmosphere. *Proceedings of the Royal Dublin Society* **9** (1899–1903), 107–270.

Letts, E. A. and R. F. Blake (1901) On some problems connected with atmospheric carbonic anhydride, and on a new and accurate method for determining its amount suitable for scientific expeditions. *Proceedings of the Royal Dublin Society* **9** (1899–1903), 435–453.

Leung, I., G. Wenxiang, I. Friedman and J. Gleason (1990) Natural occurrence of silicon carbide in a diamondiferous kimberlite from Fuxian. *Nature* **346**, 352–354.

Leverson, A. I. (1967) Geology of petroleum. Freemann, San Fransico, 724 pp.

Levine, J. S., Ed. (1991) Global biomass burning: Atmospheric, climatic, and biospheric implications (Ed. J. S. Levine,), MIT press, Cambridge, U.K., MA, 580 pp.

Levine, J. S., W. R. Cofer III and J. H. Pinto (1993) Biomass burning. In: Atmospheric methane – sources, sinks, and role in global change (Ed. M. A. K. Khalil), NATO ASI Series I (Global environmental change), Vol. 13, Springer-Verlag, Berlin, pp. 299–313.

Lévy, A. and P. Miquel (1891) Annuaire de l'Observatoire Municipal de Montsouros pour l'an 1891; cited from Baumann (1893).

Lewis, W. M., M. C. Grant and S. K. Hamilton (1985) Evidence that filterable phosphorus is a significant atmospheric link in the phosphorus cycle. *Oikus* **45**, 428–432.

Lewis, J. S. (2004) Physics and chemistry of the solar system. Acad. Press, San Diego, 591 pp.

Lewis, A. C., J. R. Hopkins, L. J. Carpenter, J. Stanton, K. A. Read and M. J. Pilling (2005) Sources and sinks of acetone, methanol, and acetaldehyde in North Atlantic marine air. *Atmospheric Chemistry and Physics* **5**, 1963–1974.

Liang, J. and D. J. Jacob (1997) Effect of aqueous phase cloud chemistry on tropospheric ozone. *Journal of Geophysical Research* **102**, 5993–6001.

Libes, S. (2009) Introduction to marine biogeochemistry. Academic Press, 2nd edition, 580 pp.

Lide, D. R. and H. P. R. Frederikse, Eds. (1995) CRC Handbook of Chemistry and Physics, 76th Edition. CRC Press, Boca Raton, FL (USA).

Lide, D. R., Ed. (2005) CRC Handbook of Chemistry and Physics, 85th Edition. CRC Press. Boca Raton, FL (USA).

Liebig, J. von (1827) Une note sur la nitrification. *Annales de Chimie et de Physique* **35**, 329–333.

Liebig, J. von (1840) Die organische Chemie in ihrer Anwendung auf Agricultur und Physiologie [Organic chemistry in its applications to agriculture and physiology]). Friedrich Vieweg und Sohn, Braunschweig, 344 pp.

Likens, G. E., W. C. Keene, J. M. Miller and J. N. Galloway (1987) Chemistry of precipitation from a remote, terrestrial site in Australia. *Journal of Geophysical Research.* **92**, 13299–13314.

Lilie, R., J. Hanrahan and A. Henglkein (1978) O^- transfer reactions of the carbonate radical anion. *Radiation Physics and Chemistry* **11**, 225–227.

Liljestrand, H. M. (1985) Average rainwater pH, concepts of atmospheric acidity, and buffering in open systems. *Atmospheric Environment* **19**, 487–500.

Liljequist, G. H. and K. Cehak (1984) Allgemeine Meteorologie. Fr. Vieweg und Sohn, Braunschweig, 368 pp.

Liou, K. N. (1992) Radiation and cloud processes in the atmosphere. Oxford University Press, Oxford, 487 pp.

Liss, P. S. and W. G. N. Slinn, Eds. (1983) Air-sea exchange of gases and particles. NATO Science Series C 108, Reidel Publishing Co., Dordrecht, 561 pp.

Liss, P. and L. Merlivat (1986) Air-sea gas exchange rates: Introduction and synthesis. In:The role of sea-air exchange in geochemical cycling (Ed. P. Menard) Reidel, Dordrecht, pp. 113–127.

Liss, P. S. (2007) Trace gas emissions from the marine biosphere. *Philosophical Transaction of the Royal Society* **A365**, 1697–1704.

Litter, M. I. (1999) Heterogeneous photocatalysis – transition metal ions in photocatalytic systems. *Applied Catalysis B: Environmental* **23**, 89–114.

Liu, X., G. Mauersberger and D. Möller (1997) The effects of cloud processes on the tropospheric photochemistry: an improvement of the EURAD model with a coupled gaseous and aqueous chemical mechanism. *Atmospheric Environment* **19**, 3119–3135.

Livingstone, D. A. (1963) Chemical composition of rivers and lakes. In: U.S. Geological Survey Professional Paper, 6th ed. (Ed. M. Fleischer), pp. 1–64.

Lloyd, G. E. R. (1970) Early Greek science: Thales to Aristotle. W. W. Norton & Co., New York, 156 pp.

Lobert, J. M., D. H. Scharffe, W.-M Hao, T. A. Kuhlbusch, R. Seuwen, P. Warneck and P. J. Crutzen (1991) Experimental evaluation of biomass burning: Nitrogen and carbon containing compounds. In: Levine (1991) pp. 289–304.

Lobert J M., W. C. Keene, J. A. Logan and R. Yevich (1999) Global chlorine emissions from biomass burning: Reactive Chlorine Emissions Inventory. *Journal of Geophysical Research* **104**, 8373–8390.

Lode, A. (1911) Atmosphäre. In: Handbuch der Hygiene. I. Band (Eds. M. Rubner, M., M. v. Grubner and M. Ficker), Verlag Hirzel, Leipzig, pp. 367–518.

Loewy, A. (1924) Allgemeine Klimatophysiologie. In: Klimaphysiologie und Strahlenphysiologie, Vol. 3 Handbuch der Balneologie, medizinischen Klimatologie und Balneographie (Eds. E. Dietrich and S. Kaminer), Georg Thieme, Leipzig, pp. 3–166.

Logager, T., K. Sehestad and J. Holcman (1993) Rate constants of the equilibrium reactions $SO_4^- + HNO_3 \rightleftarrows HSO_4^- + NO_3^-$ and $SO_4^- + NO_3^- \rightleftarrows SO_4^{2-} + NO_3$. *Radiation Physical Chemistry* **41**, 539–543.

Logager, T. and K. Sehestad (1993) Formation and decay of peroxynitric acid: a pulse radiolysis study. *Journal of Physical Chemistry* **97**, 10047–10052.

Logan, S. R. (1967) The solvated electron – The simplest ion and reagent. *Journal of Chemical Education* **44**, 344.

Lohmann, U. and J. Feichter (2005) Global indirect aerosol effects: A review. *Atmospheric Chemistry and Physics* **5**, 715–737 Lomborg, B. (2001) The skeptical environmentalist: Measuring the real state of the world. Cambridge University Press. Cambridge, UK, 548 pp.

Los, A., M. van Weele and P. G. Duynkerke (1997) Actinic fluxes in broken cloud fields. *Journal of Geophysical Research* **102**, 4257–4266.

Lorius, C., J. Jouzel, C. Ritz, L. Merlivat, N. I. Barkov, Y. S. Korotkevich, and V. M. Kotlyakov (1985) A 150,000-year climatic record from Antarctic ice. *Nature* 316, 591–596.

Loveland, T. R., B. C. Reed, J. F. Brown, D. O.Ohlen, Z. Zhu, L.Yang and J. W. Merchant (2000) Development of a global land cover characteristics database and IGBP DISCover from 1 km AVHRR data. *International Journal of Remote Sensing* **21**, 1303–1330.

Lovelock, J. E. (1979) Gaia: A new look at life on earth. Oxford University Press, Oxford, 157 pp.

Lovelock, J. E. and L. Margulis (1974) Atmospheric homeostasis by and for the biosphere: The Gaia hypothesis. *Tellus* **26**, 2–10.

Lovett, G. M. and J. D. Kinsman (1990) Atmospheric pollutant deposition to high-elevation ecosystems. *Atmospheric Environment* **24A**, 2767–2786.

Lueker, T. J., A. G. Dickerson and C. D. Keeling (2000) Ocean pCO_2 calculated from dissolved inorganic carbon, alkalinity, and eqiations for K_1 and K_2: validation based on laboratory measurements of CO_2 in gas and seawater equilibrium. *Marine Chemistry* **70**, 105–119.

Lundegårdh, H. (1924) Der Kreislauf der Kohlensäure in der Natur. Ein Beitrag zur Pflanzenökologie und zur landwirtschaftlichen Düngungslehre. Gustav Fischer, Jena, 308 pp.

Lunine, J. I., S. K.Croft and J. Kargel (1988) Clathrate and ammonia hydrates: Physical chemistry and applications to the solar system. NASA Technical Memorandum, NASA TM-4041, pp. 135–137.

Luria M. and Sievering H. (1991) Heterogeneous and homo- geneous oxidation of SO_2 in the remote marine atmosphere. *Atmospheric Environment* **25A**, 1489–1496.

Lutgens, F. K., E. J. Tarbuck and D. Tasa (2009) The atmosphere: An introduction to meteorology. 11th edition. Prentice Hall, Toronto, 544 pp.

Lüthi, D., M. Le Floch, B. Bereiter, T. Blunier, J.-M. Barnola, U. Siegenthaler, D. Raynaud, J. Jouzel, H. Fischer, K. Kawamura and T. F. Stocke (2008) High-resolution carbon dioxide concentration record 650 000–800 000 years before present. *Nature* **453**, 379–382.

Lüttke, J., Scheer, V., Levsen, K., Wünsch, G., Cape, J. N., Hargreaves, K. J., Storeton-West, R. L., Acker, K., Wieprecht, W. and B. Jones (1997) Occurrence and formation of nitrated phenols in and out of clouds. *Atmospheric Environment* **31**, 2637–2648.

Lüttke, J., V. Scheer, K. Levsen, G. Wünsch, J. N. Cape, K. J. Hargreaves, R. L. Storeton-West, K. Acker, W. Wieprecht and B. Jones (1997) Occurrence and formation of nitrated phenols in and out of clouds. *Atmospheric Environment* **31**, 2637–2648.

Lüttke, J., K. Levsen, K. Acker, W. Wieprecht und D. Möller (1999) Phenols and nitrated phenols in clouds at Mt. Brocken. *International Journal of Environmental Analytical Chemistry* **74**, 69–99.

Lymar, S. V., V. Shafirovich and G. A. Poskrebyshev (2005) One-electron reduction of aqueous nitric oxide: a mechanistic revision. *Inorganic Chemistry* **44**, 5212–5221.

Maahs, H. G. (1983) Measurement of the oxidation rate of sulphur(IV) by ozone in aqueous solution and their relevance to SO_2 conversion in nonurban tropospheric clouds. *Atmospheric Environment* **17**, 341–345.

Maaß, F., H. Elias and K. J. Wannowius (1999) Kinetics of the oxidation of hydrogen sulfite by hydrogen peroxide in aqueous solution: ionic strength effects and temperature dependence. *Atmospheric Environment* **33**, 4413–4419.

MacFarling Meure, C., D. Etheridge, C. Trudinger, P. Steele, R. Langenfelds, T. van Ommen, A. Smith and J. Elkins (2006) The law dome CO_2, CH_4 and N_2O ice core records extended to 2000 years BP. *Geophysical Research Letters* **33**, No. 14, L14810 10.1029/2006GL026152.

Machu, W. (1937) Das Wasserstoffperoxyd und die Perverbindungen. Springer-Verlag, Wien, 408 pp.

Mackay, D., W. Y. Shin, K.-C. Ma and S.-C. Lee, Eds. (2006) Handbook of physical-chemical properties and environmental fate for organic chemicals. Four Volumes, CRC Press, Boca Raton, FL (USA), 4182 pp.

Mackenzie, F. T. and R. A. Garrels (1966) Chemical mass balance between rivers and oceans. *American Journal of Sciences* **264**, 507–525.

Mader, P. M. (1958) Kinetics of the hydrogen peroxide-sulfite reaction in alkaline solution. *Journal of the American Chemical Society* **80**, 2634-2639.

Madronich, S. (1987) Photodissociation in the atmosphere. 1. Actinic flux and the effects of ground reflection and clouds. *Journal of Geophysical Research* **92**, 9740–9752.

Magiels, G. (2001) Dr. Jan IngenHousz or why don't we know who discovered photosynthesis. 11th Conf. Europ. Phil. Sci. Assoc., Madrid, 18 pp.

Magill, P. L., F. R. Holden and C. Ackley, Eds. (1956) Air pollution handbook. McGraw-Hill, New York, pp. 2–45.

Mahowald, N., J.-F. Lamarque, X. Tie and E. Wolff (2006) Sea salt aerosol response to climate change: last glacial maximum, pre-industrial and doubled carbon dioxide climates. *Journal of Geophysical Research* **111**, D05303, doi:10.1029/2005JD006459.

Maiss, M. and C. A. M. Brenninkmeijer (1998) Atmospheric SF$_6$, trends sources and prospects. *Environmental Science and Technology* **32**, 3077−3086.

Malingreau, J.-P. and Y.H. Zhuang (1998) Biomass burning: An ecosystem process of global significance. In: Asian change in the context of global climate change (Eds. J. Galloway and J. Melillo), Cambridge University Press, Cambridge, U.K., pp. 101−127.

Manning, A. C., R. F. Keeling and J. P. Severinghaus (1999) Precise atmospheric measurements with a paramagnetic oxygen analyzer. *Global Biogeochemical Cycles* **13**, 1107−1115.

Manning, A. C. and R. F. Keeling (2006) Global oceanic and land biotic carbon sinks from the Scripps atmospheric oxygen flask sampling network. *Tellus* **58B**, 95−116.

Manning, F. S. and R. E. Thompson (1995) Oilfield processing of petroleum: Crude oil. PennWell Publishing Co., Tulsa, 400 pp.

Mano, S. and M. O. Andreae (1994) Emission of methyl bromide from biomass burning. *Science* **263**, 1255−1257.

Manuel, O., J. T. Lee, D. E. Ragland, J. M. D. MacElroy, B. Lee and W. K. Brown (1998) Origin of the solar system and its elements. *Journal of Radioanalytical and Nuclear Chemistry* **238**, 213−225.

Marais, D. J. Des (2000) Evolution: When did photosynthesis emerge on earth? *Science* **289**, 1703−1705.

Marandino, C. S., W. J. De Bruyn, S. D. Miller, M. J. Prather and E. S. Saltzman (2005) Oceanic uptake and the global atmospheric acetone budget. *Geophysical Research Letters* **32**, L15806.1−L15806.4. doi:10−1029/2005GL023285.

Marchand, R. F. (1850) Ueber die Eudiometrie. *Jounal für praktische Chemie* **49**, 449−468.

Marchetti, C. (1977) On geoengineering and the CO$_2$ problem. *Climate Change* **1**, 59−68.

Marchi, L. de (1895) Le cause dell'era glaciale. Bernardoni e Rebeschini, Milano, 231 pp.

Marenco, A., N. Philippe and G. Hérve (1994) Ozone measurements at Pic du Midi observatory. *EUROTRAC Annual report part 9: TOR*, EUROTRAC ISS, Garmisch-Partenkirchen, pp. 121−130.

Marggraf, A. S. (1753) Examen chymique de l'eau. Histoire de l'Academie Royale des Sciences et Belles Lettres, Berlin, pp. 131−157.

Marggraf, A. S. (1786) Chemische Untersuchung des gemeinen Wassers. *Physikalische und medicinische Abhandlungen der Königlichen Academie der Wissenschaften zu Berlin* **4**, 69−95.

Mark, D. (1999) Atmospheric aerosol sampling. In: Atmospheric particles (Eds. R. M. Harrison and R. van Grieken), J. Wieley & Sons, Chichester, pp. 29−94.

Mari, C., J. P. Chaboureau, J. P. Pinty, J. Duron, P. Mascart, J. P. Cammas, F. Gheusi, T. Fehr, H. Schlager, A. Roiger, M. Lichtenstern and P. Stock (2006) Regional lightning NO$_x$ sources during the TROCCINOX experiment. *Atmospheric Chemistry and Physics* **6**, 5559−5572.

Marie-Davy, M. (1880) L'acide carbonique de l'air, dans ses rapports avec les grands mouvements de l'atmosphére. *Comptus Rendus* **90**, 32−36.

Marland, G., T. A. Boden and R. J. Andres (2008) Global, regional, and national CO$_2$ emissions. In Trends: A compendium of data on global change. Carbon Dioxide Information Analysis Center, Oak Ridge National Laboratory, U.S. Department of Energy, Oak Ridge, Tenn. (USA).

Marquardt, W. and P. Ihle (1988) Acidic and alkaline precipitation components in the mesoscale range under the aspect of meteorological factors and the emissions. *Atmospheric Environment* **22**, 2707−2716.

Marquardt, W., E. Brüggemann and P. Ihle (1996) Trends in the composition of wet deposition: effects of the atmospheric rehabilitation in East-Germany. *Tellus* **48B**, 361−371.

Marquardt, W., E. Brüggemann, R. Auel, H. Herrmann and D. Möller (2001) Trends of pollution in rain over East Germany caused by changing emissions. *Tellus* **53B**, 529−545.

Marsh, A. (1947) Smoke. The problem of coal and the atmosphere. Faber and Faber, London. 306 pp.

Marsh, G. E. (2009) Seawater pH and anthropogenic carbon. In: Climate change (Ed. S. P. Saikia), see also: arXiv:0810.3596v1 [physics.ao-ph].

Marshall, J. G. and P. V. Rutledge (1959) A higher hydrogen peroxide, H_2O_4? *Nature* **184**, 2013−2014.

Marshall, J. and R. A. Plumb (2007) Atmosphere, ocean and climate dynamics: An introductory text. International Geophysics Series Vol. 93, Academic Press, 334 pp.

Martin, L. R. and D. E. Damschen (1981) Aqueous oxidation of sulfur dioxide by hydrogen peroxide at low pH. *Atmospheric Environment* **15**, 1615−1621.

Martin, L. R. (1984) Kinetic studies of sulfite oxidation in aqueous solution. In: SO_2, NO and NO_2 oxidation mechanisms: Atmospheric considerations (Ed. J. G. Calvert), Butterworth Publishers, Boston, MA, USA, pp. 63−100.

Martin, D., M. Tsivou, B. Bonsang, C. Abonnel, T. Carsey, M. Springer-Young, A. Pszenney and K. Suhre (1997) Hydrogen peroxide in the marine atmospheric boundary layer during the Atlantic Stratocumulus Transition Experiment / Marine Aerosol and Gas Exchange experiment in the eastern subtropical North Atlantic. *Journal of Geophysical Research* **102**, 6003−6015.

Martin, D., M. Tsivou, B. Bonsang, C. Abonnel, T. Carsey, M. Springer-Young, A. Pszenney and K. Suhre (1997) Hydrogen peroxide in the marine atmospheric boundary layer during the Atlantic Stratocumulus Transition Experiment / Marine Aerosol and Gas Exchange experiment in the eastern subtropical North Atlantic. *Journal of Geophysical Research* **102**, 6003−6015.

Martin, R. V., D. J. Jacob, K. Chance, T. P. Kurosu, P. I. Palmer, and M. J. Evans (2003) Global inventory of nitrogen oxide emissions constrained by space-based observations of NO_2 columns. *Journal of Geophysical Research* **108**, 4537, doi:10.1029/2003JD003453.

Martin R. V., B. Sauvage, I. Folkins, C. E. Sioris, C. Boone, P. Bernath and J. Ziemke (2007) Space-based constraints on the production of nitric oxide by lightning. *Journal of Geophysical Research* **112** (D9), D09309.

Martins, Z., O. Botta, M. L. Fogel, M. A. Sephton, D. P. Glavin, J. S. Watson, J. P. Dworkin, A. W. Schwartz, and P. Ehrenfreund (2008) Extraterrestrial nucleobases in the Murchison meteorite. *Earth and Planetary Science Letters* **270**, 130−136.

Marufu, L., F. J. Dentener, J. Lelieveld, M. O. Andreae and G. Helas (1000) Photochemistry of the African troposphere: Influence of biomass burning emissions. *Journal of Geophysical Research* **105**, 14513−14530.

Mason, B. J. and F. H. Ludlam (1951) The microphysics of clouds. *Reports on Progress in Physics* **14**, 147−195.

Mason, B. J. and C. B. Moore (1982) Principles of geochemistry (Fourth Edition) John Wiley & Sons, New York, 344 pp.

Matar, S. and L. G. Hatch (2001) Chemistry of petrochemical processes. Gulf Publishing, 392 pp.

Mather, T. A., A. G. Allen, C. Oppenheimer, D. M. Pyle and A. J. S. McGonigle (2003) Size-resolved characterisation of soluble ions in the particles in the tropospheric plume of Masaya volcano, Nicaragua: Origins and plume processing. *Journal of Atmospheric Chemistry* **46**, 207−237.

Mather, T. A., V. I. Tsanev, D. M. Pyle,1 A. J. S. McGonigle, C. Oppenheimer and A. G. Allen (2004a) Characterization and evolution of tropospheric plumes from Lascar and Villarrica volcanoes, Chile. *Journal of Geophysical Research* **109**, D21303, doi:10.1029/2004JD004934, 2004.

Mather, T. A., D. M. Pyle and A. G. Allen (2004b) Volcanic source for fixed nitrogen in the early Earth's atmosphere. *Geology* **32**, 905–908.

Mather, T. A. (2008) Volcanism and the atmosphere: the potential role of the atmosphere in unlocking the reactivity of volcanic emissions. *Philosophical Transactions of the Royal Society* **A28**, 4581–4595.

Matson, P. A. and R. C. Harriss, Eds. (1995) Biogenic trace gases: Measuring emissions from soil and water (Ecological methods and concepts). Wiley-Blackwell, 408 pp.

Matthaei, G. L. C. (1902) The effect of temperature on carbon dioxide assimilation. *Annals of Botany* **16**, 591–592.

Matthews, E. and I. Fung (1987) Methane emission from natural wetlands: Global area, distribution and environmental characteristics of sources. *Global Biogeochemical Cycles* **1**, 61–86.

Mattews, E. (1993) Wetlands. In: Atmospheric methane: sources, sinks, and role in global change (Ed. M. A. K. Khalil), NATO ASI Series, Series I: Global Environmental Sciences, Vol. 13, Springer-Verlag, Berlin, pp. 314–361.

Matthews, E. (1997) Global litter production, pools, and turnover times: Estimates from measurement data and regression models. *Journal of Geophysical Research* **102**, 18771–18800.

Matthias-Maser, S. (1998) Primary biological aerosol particles: their significance, sources, sampling methods and size distribution in the atmosphere. In: Atmospheric aerosol particles (Eds. R. M. Harrison and R. E. van Grieken), John Wiley & Sons, Chichester, pp. 349–383.

Matthijsen, J., P. J. H. Builtjes, E. W. Meijer and G. Boersen (1997) Modelling cloud effects on ozone on a regional scale: A case study. *Atmospheric Environment* **31**, 3227–3238.

Mauldin III, R. L., S. Madronich, S. J. Focke and F. L. Eisele (1997) News insight on OH: measurements around and in clouds. *Geophysical Research Letters* **24**, 3033–3036.

Maurette, M. (2006) Micrometeorites and the mysteries of our origins. Springer-Verlag, Berlin, 200 pp.

Mayer, J. R. von (1845) Die organische Bewegung in ihrem Zusammenhange mit dem Stoffwechsel. C. Drechsler, Heilbronn, 115 pp.

Mayow, J. (1674) Tractatus quinque medico-physici. Quorum primus agit de sal-nitro, et spiriti nitro-aereo. Secundus de respiratione. Tertius de respiratione foetus in utero, et ovo. Quartus de motu musculari, et spiritius animalibus. Ultimus de rhachixite. Oxonii [Oxford], E Theatro Sheldoniano, 487 pp.

McArdle, J. V. and M. R. Hoffmann (1983) Kinetics and mechanism of the oxidation of aquated sulfur dioxide by hydrogen peroxide. *Journal of Physical Chemistry* **87**, 5425–5429.

McClelland, L., T. Simkin, M. Summers, E. A. Nielsen and T. C. Stein, Eds. (1989) Global volcanism 1975–1985: The first decade of reports from the Smithsonian Institution's Scientific Event Alert Network (SEAN). Prentice Hall and American Geophysical Union, Englewood Cliffs, 657 pp.

McCulloch, A., M. L. Aucott, C. M. Benkovitz, T. E. Graedel, G. Kleiman, P. M. Midgley and Y.-F. Li (1999) Global emissions of hydrogen chloride and chloromethane from coal combustion, incineration, and industrial activities: Reactive chlorine emissions inventory. *Journal of Geophysical Research* **104**, 8391–8404.

McElroy, W. J. (1986) Sources of hydrogen peroxide in cloudwater. *Atmospheric Environment* **20**, 427–438.

McFiggans, G., P. Artaxo, U. Baltensperger, H. Coe, M. C. Facchini, G. Feingold, S. Fuzzi, M. Gyselk, A, Laaksonen, U. Lohmann, T. F. Mentel, D. M. Murphy, C. D. O'Dowd, J. R. Snider and E. Weingartner (2005a) The effect of physical and chemical aerosol properties on warm cloud droplet activation. *Atmospheric Chemistry and Physics Diskussion* **5**, 8507–8646.

McFiggans, G., M. R. Alfarra, J. Allan, K. Bower, H. Coe, M. Cubison, D. Topping, P. Williams, S. Decesari, M. C. Facchini and S. Fuzzi (2005b) Simplification of the representation of the organic component of atmospheric particulates. *Faraday Discussion* **130**, 341–362.

McInnes, L., D. S. Covert, P. K. Quinn and M. S. Germani (1994) Measurements of chloride depletion and sulfur enrichment in individual sea-salt particles collected from the remote marine boundary layer. *Journal of Geophysicl Research* **99**, 8257–8268.

McKetta, J. (1992) Petroleum processing handbook. Marcell Dekker, New York, 792 pp.

Mechie, J., S. V. Sobolev, L. Ratschbacher, A. Y. Babeyko, G. Bock, A. G. Jones, K. D. Nelson, K. D. Solon, L. D. Brown and W. Zhao (2004) Precise temperature estimation in the Tibetan crust from seismic detection of the α-β quartz transition. *Geology* **32**, 601–604.

Medinas, D. B., G. Cerchiaro, D. F. Trindade and O. Augusto (2007) The carbonate radical and related oxidants derived from bicarbonate buffer. *IUBMB Life* **59**, 255–262.

Megonigal, J. P., M. E. Hines and P. T. Visscher (2004) Anaerobic metabolism: linkages to trace gases and aerobic processes. In: Biogeochemistry (Ed. W. H. Schlesinger), Elsevier-Pergamon, Oxford, UK, pp. 317–424.

Meiklejohn, J. (2006) The nitrifying bacteria: A review. *European Journal of Soil Science* **4**, 59–68.

Meissner, G. (1863) Ueber die Bestandtheile des Regenwassers. *Göttinger Nachrichten* **21**, 264–268.

Melián-Cabrera, I., M. López Granados, P. Terreros, and J. L. G. Fierro (1998) CO_2 hydrogenation over Pd-modified methanol synthesis catalysts. *Catalysis Today* **45**, 251–256.

Mellor, J. W. (1940) A comprehensive treatise of inorganic and theoretical chemistry. Vol. VI, Longmans, London, 1024 pp.

Melton, C. E. and A. A. Giardini (1974) The nature and significance of occluded fluids in three Indian diamonds. *American Minerologist* **66**, 746–750.

Mendeleev, D. (1877) L'origine du petrole. *Revue Scientifique, 2e Ser.* **VIII**, 409–416.

Meng L., P. G. M. Hess, N. M. Mahowald, J. B. Yavitt, W. J. Riley, Z. M. Subin, D. M. Lawrence, S. C. Swenson, J. Jauhiainen and D. R. Fuka (2012) Sensitivity of wetland methane emissions to model assumptions: application and model testing against site observations. *Biogeosciences* **9**, 2793–2819.

Merckens, W. (1906) Über strahlenartige Einwirkungen auf die photographische Bromsilbergelatine. *Annalen der Physik* **16**, 667–683.

Mészáros, E. (1981) Atmospheric chemistry – fundamental aspects. Akadémiai Kiadó, Budapest, 201 pp.

Mészáros, E. (1999) Fundamentals of atmospheric aerosol chemistry. Akadémiai Kiadó, Budapest, 308 pp. .

Meskhidze, N. and A. Nenes (2006) Phytoplankton and cloudiness in the Southern Ocean. *Science.* **314**, 1419–1423.

Metzger, R. A. and G. Benford (2001) Sequestering of atmospheric carbon through permanent disposal of crop residue. *Climatic Change* **49**, 11–19.

Mihalopoulos, N., J. P. Putand and N. C. Ngyen (1993) Seasonal variation of methane sulfonic acid in precipitation at Amsterdam Island in the southern Indian ocean. *Atmospheric Environment* **27**, 2069–2073.

Milanković, M. (1920) Théorie mathématique des phénomènes thermiques produits par la radiation solaire. Gauthier-Villars, Paris, 340 pp.

Milanković, M. (1930) Mathematische Klimalehre und astronomische Theorie der Klimaschwankungen. In: Handbuch der Klimatologie, Band 1: Allgemeine Klimalehre, Teil A (Eds. W. Köppen and R. Geiger), Gebrüder Bornträger, Berlin, pp. 1–176.

Milanković, M. (1941) Kanon der Erdbestrahlung und seine Anwendung auf das Eiszeitenproblem. Royal Serbian Academy, Special Publication Vol. 133 [first translation from German by A. Kh. Khrgian: Canon of insolation and the ice-age problem. Published by the Israel Program for Scientific Translations, Jerusalem, 1969].

Milazzo, G. and S. Caroli (1978) Tables of standard electrode potential. John Wiley & Sons, New York, 421 pp.

Miles, N. L., J. Verlinde and E. E. Clothiaux (2000) Cloud droplet size distribution in low-level stratiform clouds. Journal *Atmospheric Sciences* **57**, 295-311.

Miller, S. (1953) A production of amino acids under possible primitive earth conditions. *Science* **117**, 528–529.

Miller, S. and H. Urey (1959) Organic compound synthesis on the primitive earth. *Science* **130**, 245–251.

Millero, F. J. (1986) The thermodynamics and kinetics of the hydrogen sulfide system in natural waters. *Marine Chemistry* **18**, 121–147.

Millero, F. J., S. Hubinger, M. Fermandez and S. Garnett (1987) Oxydation of H_2S as a function of temperature, pH and ionic strength. *Environmental Science and Technology* **21**, 439–443.

Millero, F. J. (2006) Chemical oceanography. Third edition, CRC Press, Boca Raton, FL (USA), 505 pp.

Minnen, van J. G., G. K. Klein, S. Stehfest, B. Eickhout, G. van Drecht and R. Leemans (2009) The importance of three centuries of land-use change for the global and regional terrestrial carbon cycle. *Climatic Change* **97**, 123–144.

Miranda, K. M., M. G. Espey, D. Jourd'heuil, M. B. Grisham, J. M. Fukuta, M. Feelisch and D. A. Wick (2000) The chemical biology of nitric oxide. In: Nitric oxide, biology and pathobiology (Ed. L. J. Ignarro), Academic Press, San Diego, pp. 41–57.

Misselbrook, T. H., J. Webb, D. R. Chadwick, S. Ellis and B. F. Pain (2001) Gaseous emissions from outdoor concrete yards used by livestock. *Atmospheric Environment* **35**, 5331–5338.

Mitchell, P. and M. Frenklach (2003) Particle aggregation with simultaneous surface growth. *Physical Review E* **67** 061407-1-11.

Mitsch, W. J. and J. G. Gosselink (2007) Wetlands. John Wiley & Sons, Inc., New York, 582 pp.

Miyaoka, Y., H. Y. Inoue, Y. Sawa, H. Matsueda and S. Taguchi (2007) Diurnal and seasonal variations in atmospheric CO_2 in Sapporo, Japan: Anthropogenic sources and biogenic sinks. *Geochemical Journal* **41**, 429–439.

Möhler, O., P. J. DeMott, G. Vali and Z. Levin (2007) Microbiology and atmospheric processes: the role of biological particles in cloud physics. *Biogeosciences Discussion* **4**, 2559–2591.

Mohnen V. A. and J. A. Kadlecek (1989) Cloud chemistry research at Whiteface Mountain. *Tellus* **41B**, 79–91.

Mohnen, V. A. and R. J. Vong (1993) A climatology of cloud chemistry for the eastern United States derived from the cloud chemistry project. *Environmental Review* **1**, 38–54.

Mohnen, V. A. (1995) From TSP to nanaparticles: Background and status. Colloqium „Zur Geschichte der Luftchemie und Luftreinhaltung in Deutschland", Brandenburg Technical University, Cottbus, June 9, 1995.

Mojzsis, S. J., Arrhenius, G., McKeegan, K. D., Harrison, T. M., Nutman, A. P. and C. R. L. Friend (1996) Evidence for life on earth before 3800 million years ago. *Nature* **384**, 55–59.

Moldan, B., J. Kopacek and J. Kopacek (1988) Chemical composition of atmospheric precipitation in Czechoslovakia, 1978–1984 – II: Event samples. *Atmospheric Environment* **22**, 1901–1908.

Molina, M. J. and F. S. Rowland (1974) Stratospheric sink for chlorofluoromethanes. Chlorine atom catalyzed destruction of ozone. *Nature* **249**, 810–812.

Molina, M. J., A. V. Ivanov, S. Traktenberg and L. T. Molina (2004) Atmospheric evolution of organic aerosol. *Geophysical Research Letters* **31**, L22104, doi:10.1029/2004GL020910.

Möller, D. (1977) Der Schadstoff Schwefeldioxid. *Zeitschrift für die gesamte Hygiene* **23**, 310–316.

Möller, D. (1980) Kinetic model of atmospheric SO_2 oxidation based on published data. *Atmospheric Environment* **14**, 1067–1076.

Möller, D. and H. Schieferdecker (1982) Zur Rolle des atmosphärischen Ammoniaks im biogeochemischen Stickstoff-Zyklus. *Zeitschrift für die gesamte Hygiene* **28**, 797–802.

Möller, D. (1983) The global sulphur cycle. *Idöjaras* **87**, 121–143.

Möller, D. (1984a) On the global natural sulphur emission. *Atmospheric Environment* **18**, 29–39.

Möller, D. (1984b) Estimation of the global man- made sulphur emission. *Atmospheric Environment* **18**, 19–27.

Möller, D. and R. Zierath (1986) On the composition of precipitation water and its acidity. *Tellus* **38B**, 44–50.

Möller, D. (1988) Production of free radicals by an ozone-alkene reaction – a possible factor in the new-type forest decline? *Atmospheric Environment* **22**, 2607–2611.

Möller, D. (1989) The possible role of H_2O_2 in new-type forest decline. *Atmospheric Environment* **23**, 1187–1193.

Möller, D. and L. Horváth (1989) Estimation of natural acidity of precipitation water on global scale. *Idöjaras* **93**, 324–335.

Möller, D. and H. Schieferdecker (1989) Ammonia emission and deposition of NH_x in the GDR. *Atmospheric Environment* **23**, 1187–1193.

Möller, D. and G. Mauersberger (1990a) Auswaschen von Gasen und Aerosolen durch Niederschläge unter Berücksichtigung einer komplexen Flüssigphasenchemie. 2. Modellergebnisse. *Zeitschrift für Meteorologie* **40**, 330–339.

Möller, D. and G. Mauersberger (1990b) The role of clouds in removal of air pollutants. In: Air Pollution (Eds. P. Zannetti, J. E. Garcia and G. A. Millan), Elsevier Sci. Publ., London, pp. 373–380.

Möller, D. (1990) The Na/Cl ratio in rain water and the seasalt chloride cycle. *Tellus* **42B**, 254–262.

Möller, D. and G. Mauersberger (1992) Cloud chemistry effects on tropospheric photooxidants in polluted atmosphere – model results. *Journal of Atmospheric Chemistry* **14**, 153–165.

Möller, D., K. Acker and W. Wieprecht (1993) Cloud chemistry at the Brocken in the Harz Mountains, Germany. *EUROTRAC Newsletters* **12**, 24–29.

Möller, D. (1995a) Sulfate aeorosol and their atmospheric precursors. In: Aeorosol Forcing of Climate (Eds. R. J. Charlson and J. Heintzenberg). John Wiley & Sons, New York, pp. 73–90.

Möller, D. (1995b) Cloud processes in the troposphere. In: Ice core studies of global biogeochemical cycles (Ed. R. J. Delmas), NATO ASI Series Vol. I 30, Springer-Verlag, Berlin, pp. 39–63.

Möller, D. and G. Mauersberger (1995) An aqueous-phase chemical reaction mechanism. In: Clouds – models and mechanism, EUROTRAC Special Publication, Garmisch-Partenkirchen, pp. 77–93.

Möller, D., K. Acker, W. Marquardt and E. Brüggemann (1996) Precipitation and cloud chemistry in the Neue Bundesländer of Germany in the background of changing emissions. *Idöjaras* **100**, 117–133.

Möller, D., K. Acker and W. Wieprecht (1996) A relationship between liquid water content and chemical composition in clouds. *Atmospheric Research* **41**, 321–335.

Möller, D. (1996) Global sulfur and nitrogen biogeochemical cycles. In: Physics and chemistry of the atmosphere of the Earth's and other objects of the solar system (Ed. C. Boutron), ERCA Vol. 2, les éditions de physique Les Ulis (Frankreich), pp. 125–156.

Möller, D. (1999a) Explanation for the recent dramatic increase of H_2O_2 concentrations found in Greenland ice cores. *Atmospheric Environment* **33**, 2435−2437.

Möller, D. (1999b) Acid rain − gone? In: Prediction of atmospheric environmental problems between technology and nature (Ed. D. Möller), Springer-Verlag, Berlin, pp. 141−178.

Möller, D., K. Acker, D. Kalaß and W. Wieprecht (1999) Five-year record of ozone at Mt. Brocken (Germany) − implications for changing heterogeneous chemistry. In: Atmospheric environmental research − critical decisions between technological progress and preservation of nature (Ed. D. Möller), Springer-Verlag, Berlin, pp. 133−139.

Möller, D. (2002a) Rethinking the tropospheric ozone problem. In: Problems of atmospheric boundary-layer physics and air pollution − to the 80th birthday of professor M. E. Berlyand (Ed. S. S. Chicherin), Hydrometeoizdat, St. Petersburg, pp. 252−269.

Möller, D. (2002b) Hydrogen peroxide trends in Greenland glaciers. In: Encyclopedia of global environmental change (Ed. T. Munn) Vol. 1: The Earth's system: physical and chemical dimensions of global environmental change (Eds. M. C. MacCracken and J. S. Perry) John Wiley & Sons, Chichester, pp. 447−450.

Möller, D.(2003) Luft. DeGruyter, Berlin, 750 pp.

Möller, D. (2004) The tropospheric ozone problem. *Archives of Industrial Hygiene and Toxicology* (Zagreb) **55**, 11−23.

Möller D. and K. Acker (2007) Chlorine phase partitioning at Melpitz near Leipzig. In: Nucleation and Atmospheric Aerosols (Eds. C. D. O'Dowd and P. Wagner), Proceedings of 17th Int. Conference Galway, Ireland 2007, Springer, pp. 654−658.

Möller, D. (2008) On the history of the scientific exploration of fog, dew, rain and other atmospheric water. *Die Erde* **139**, 11−44.

Möller, D. (2009a) Atmospheric hydrogen peroxide: Evidence for aqueous-phase formation from a historic perspective and a one-year measurement campaign. *Atmospheric Environment* **43**, 5923−5936.

Möller (2009b) Feinstaubbelastung: Ursachen und Gesundheitsgefährdung. *Forum der Forschung* (BTU Cottbus) **22**, 117−126.

Möller, D. (2012) SONNE: solar-based man-made carbon cycle, and the carbon dioxide economy. *Ambio* **41**, 413−419.

Monahan, E. C., D. E. Spiel and K. L. Davidson (1986) A model of marine aerosol generation via whitecaps and wave disruption in oceanic whitecaps. In: Oceanic whitecaps and their role in air-sea exchange processes (Eds. E. C. Monahan and G. M. Niocaill), D. Reidel Publishing, Dordrecht, Holland, pp. 167−174.

Monnin, E., A. Indermühle, A. Dällenbach, J. Flückiger, B. Stauffer, T. F. Stocker, D. Raynaud, J.-M. Barnola (2001) Atmospheric CO_2 concentrations over the last glacial termination. *Science* **291**, 112−114.

Montreal Protocol on Substances that Deplete the Ozone Layer (2000) United Nations Environment Programme, Nairobi (Kenia), 48 pp.

Montzka, S. A., P. Calvert, B. D. Hall, J. W. Elkins, T. J. Conway, P. P. Tans and C. Sweeney (2007) On the global distribution, seasonality, and budget of atmospheric carbonyl sulfide (COS) and some similarities to CO_2. *Journal of Geophysical Research* **112**, D09302, doi:10.1029/2006JD007665.

Moore, E. B. and V. Molinero (2011) Structural transformation in supercooled water controls the crystallization rate of ice. *Nature* **479**, 506−508. doi:10.1038/nature10586.

Morgan, J. W. and E. Anders (1980) Chemical composition of Earth, Venus, and Mercury. *Proceedings of the National Academy of Science* **71**, 6973−6977.

Morgan, V. I., C. W. Wookey, J. Li, T. D. van Ommen, W. Skinner and M. F. Fitzpatrick (1997) Site information and initial results from deep ice drilling on Law Dome. *Journal of Glaciology* **43**, 3−10.

Morita, A. and B. C. Garrett (2008) Molecular theory of mass transfer kinetics and dynamics at gas−water interface. *Fluid Dynamics Research* **40,** 459−473.

Moriwaki, R., Kanda, M. and H. Nitta (2006) Carbon dioxide build-up within a suburban canopy layer in winter night. *Atmospheric Environment* **40**, 1394−1407.

Morkovnik, A. F. and O. Yu. Okhlobystin (1979) Inorganic radical-ions and their organic reactions. *Russian Chemical Review* **48**, 1055−1075.

Morse, J. W., and F. T. MacKenzie (1998) Hadean ocean carbonate chemistry. *Aquatic Geochemistry* **4**, 301−319.

Moss A. R., J.-P. Jouany and J. Newbold (2000) Methane production by ruminants: its contribution to global warming. *Annales Zootechnica* **49**, 231−253.

Mouillot, F., A. Narasimha, Y. Balkanski, J.-F. Lamarque and C. B. Field (2006) Global carbon emissions from biomass burning in the 20th century. *Geophysical Research Letters* **33**, L01801, doi:10.1029/2005GL024707.

Mozurkevich, M. (1986) Comment on "Posssible role of NO_3 in the nighttime chemistry of clouds" by W. L. Chameides. *Journal of Geophysical Research* **91D**, 14569−14570.

Mrose, H. (1966) Measurements of pH, and chemical analyses of rain-, snow, and fog water. *Tellus* **18**, 266−270.

Mudgal, P. K., S. P. Bansal and K. S. Gupta (2007) Kinetics of atmospheric oxidation of nitrous acid by oxygen in aqueous medium. *Atmospheric Emvironment* **41**, 4097−4105.

Muller, P. (1994) Glossary of terms used in physical organic chemistry (IUPAC Recommendations 1994). *Pure and Applied Chemistry* **66**, 1077−1184.

Müller, J.-F., T. Stavrakou, S. Wallens, I. De Smedt, M. Van Roozendael, M. J. Potosnak, J. Rinne, B. Munger, A. Goldstein and A. B. Guenther (2007) Global isoprene emissions estimated using MEGAN, ECMWF analyses and a detailed canopy environment model. *Atmospheric Chemistry and Physics Discussion* **7**, 15373−15407.

Munger, J. W., D. J. Jacob, and M. R. Hoffmann (1984) The occurence of bisulfite-aldehyde addition products in fog and cloudwater. *Journal of Atmospheric Chemistry* **1**, 335−350.

Munger, J. W., C. Tiller and M. R. Hoffmann (1986) Identification and quantification of hydroxymethanesulfonic acid in atmospheric water droplets. *Science* **231**, 247−249.

Müntz, A. and E. Aubin (1882) Sur la proportion d'acid carbonique dans les hautes regions de l'atzmosphére. *Annales de Chimie et de Physique*, Ser. 5, **26**, 222−254.

Murphy, D. M., J. R. Anderson, P. K. Quinn, L. M. McInnes, F. J. Brechtel, S. M. Kreidenweis, A. M. Middlebrook, M. Posfai, D. S. Thomson and P. R. Buseck (1998) Influence of sea-salt on aerosol radiative properties in the Southern Ocean marine boundary layer. *Nature* **392**, 62−65.

Muselli, M., Beysens, D. and E. Soyeux (2006) Is dew water potable? Chemical and biological analyses of dew water in Ajaccio (Corsica Island, France). *Journal of Environmental Quality* **35**, 1812−1817.

Mylona, S. (1996) Sulphur dioxide emissions in Europe 1880−1991 and their effect on sulphur concentrations and depositions. *Tellus* **48B**, 662−68.

Myriokefalitakis, S., M. Vrekoussis, K. Tsigaridis, F. Wittrock, A. Richter, C. Brühl, R. Volkamer, J. P. Burrows and M. Kanakidou (2008) The influence of natural and anthropogenic secondary sources on the glyoxal global distribution. *Atmospheric Chemistry and Physics* **8**, 4965−4981.

Nabuurs, G. J., O. Masera, K. Andrasko, P. Benitez-Ponce, R. Boer, M. Dutschke, E. Elsiddig, J. Ford-Robertson, P. Frumhoff, T. Karjalainen, O. Krankina, W. A. Kurz, M. Matsumoto, W. Oyhantcabal, N. H. Ravindranath, M. J. Sanz-Sanchez and X. Zhang (2007) Forestry. In: Climate change 2007: Mitigation. Contribution of Working Group III to the fourth assessment report of the Intergovernmental Panel on Climate Change (Eds. B. Metz, O. R. Davidson, P. R. Bosch, R. Dave and L. A. Meyer), Cambridge University Press, Cambridge, U.K., pp. 541−584.

Nahir, T. M. and G. A. Dawson (1987) Oxidation of sulfur dioxide by ozone in highly dispersed water droplets. *Journal of Atmospheric Chemistry* **5**, 373−383.

Nakamura, T. (2005) Post-hydration thermal metamorphism of carbonaceous chondrites. *Journal of Mineralogical and Petrological Sciences* **100**, 260−272.

Nakicenovic, N., J. Alcamo, G. Davis, B. de Vries, J. Fenhann, S. Gaffin, K. Gregory, A. Grübler, T. Y. Jung, T. Kram, E. L. La Rovere, L. Michaelis, S. Mori, T. Morita, W. Pepper, H. Pitcher, L. Price, K. Riahi, A. Roehrl, H.-H. Rogner, A. Sankovski, M. Schlesinger, P. Shukla, S. Smith, R. Swart, S. van Rooijen, N. Victor and Z. Dadi (2000) Special Report on Emissions Scenarios: A Special Report of Working Group III of the Intergovernmental Panel on Climate Change, Cambridge University Press, Cambridge, U.K., 599 pp.

Naqvi, S. W. A., M. Voss and J. P. Montoya (2008) Recent advances in the biogeochemistry of nitrogen in the ocean. *Biogeosciences* **5**, 1033−1041.

Naughton, J. J., V. Lewis, D. Thomas and J. B. Finlayson (1975) Fume compositions found at various stages of activity at Kilauea Volcano, Hawaii. *Journal of Geophysical Research* **80**, 2963−2966.

Neeb, P., O. Horie and G. K. Moortgat (1998) The ethene-ozone reaction in the gas-phase. *Journal of Physical Chemistry* **102**, 6778−6785.

Neftel, A., H. Oeschger, J. Schwander, B. Stauffer and R. Zumbrunn (1982) Ice core measurements give atmospheric CO_2 content during the past 40000 yr. *Nature* **295**, 220−223.

Neftel, A., E. Moor, H. Oeschger and B. Stauffer (1985) Evidence from polar ice cores for the increase in atmospheric CO_2 in the past two centuries. *Nature* **315**, 45−47.

Nelli, S, M. Hillen, K. Buyukafsar and W. Martin (2000) Oxidation of nitroxyl anion to nitric oxide by copper ions. *British Journal of Pharmacology* **131**, 356−362.

Neta, P. and R. E. Huie (1985) Free-radical chemistry of sulfite. *Environmental Health Perspectives* **64**, 209−217.

Nevison, C. D., R. F. Weiss, and D. J. Erickson III (1995) Global oceanic emissions of nitrous oxide. *Journal of Geophysical Research* **100**, 15809−15820.

Nevison, C. D., G. Esser and E. A. Holland (1996) A global model of changing N_2O emissions from natural and perturbed soils. *Climatic Change* **32,** 327−378.

Nguyen, B. C., N. Mihalopoulos, J. P. Putaud and B. Bonsang (1995) Carbonyl sulfide emissions from biomass burning in the tropics. *Journal of Atmospheric Chemistry* **22**, 55−65.

Nielsen, C. J., B. D'Anna, C. Dye, C. George, M. Graus, A. Hansel, M. Karl, S. King, M. Musabila, M. Müller, N. Schmidbauer, Y. Stengstrøm und A. Wisthaler (2010) Atmospheric degradation of Amines (ADA). Summary report: Gas phase photo-oxidation of 2-aminoethanol (MEA). CLIMIT projekt no. 193438, Norwegian Institute of Air Research.

Nienow, A. M. and J. T. Roberts (2006) Heterogeneous chemistry of carbon aerosols. *Annual Review of Physical Chemistry* **57**, 105−128.

Niethammer, J. and F. Krapp, Eds. (1986) Handbuch der Säugetiere Europas, Band 2/II, Paarhufer − Artiodactyla (Suidae, Cervidae, Bovidae). Aula, Wiesbaden, 462 pp.

Nightingale, P. D., G. Malin, C. S. Law, A. J. Watson, P. S. Liss, M. I. Liddicoat, J. Boutin, and R. C. Upstill-Goddard (2000) In situ evaluation of air-sea gas exchange parameterizations using novel conservative and volatile tracers. *Global Biogeochemical Cycles* **14**, 373−387.

Nightingale, P. D. and P. S. Liss (2003) Gases in seawater. In: The ocean and marine geochemistry, Vol. 6, Treatise in geochemistry (Eds. H. D. Holland and K. K. Turekian), Elsevier-Pergamon, Oxford, pp. 49−81.

Nikulshina, V., C. Gebald and A. Steinfeld (2009) CO_2 capture from atmospheric air via consecutive CaO-carbonation and $CaCO_3$-calcination cycles in a fluidized-bed solar reactor. *Chemical Engineering Journal* **146**, 244−248.

Nilsson, E. D., U. Rannik, E. Swietlicki, C. Leck, P. P. Aalto, J. Zhou and M. Normann (2001) Turbulent aerosol fluxes over the Arctic Ocean 2. Wind-driven sources from the sea. *Journal of Geophysical Research* **106**, 32139−32154.

Nilsson, J. and P. Grennfelt, Eds. (1988) Critical loads for sulphur and nitrogen. UNECE/ Nordic Council workshop report, Skokloster, Sweden. March 1988. Nordic Council of Ministers: Copenhagen.

Nisbet, E. G. (2006) Journal club: Methane. *Nature* **444**, 5.

Nisbet, E. G. (2007) Cinderella science. *Nature* **450**, 789–790.

Nisbet, R. E. R., R. Fisher, R. H. Nimmo, D. S. Bendall, P. M. Crill, A. Gallego-Sala, E. R. C. Hornibrook, E. López-Juez, D. Lowry, P. B. R. Nisbet, E. F. Shuckburgh, S. Sriskantharajah, C. J. Howe and E. G. Nisbet (2009) Emission of methane from plants. *Proceedings of the Royal Society B: Biological Sciences* doi:10.1098/rspb.2008.1731.

Nishigami, Y., H. Sano and T. Kojima (2000) Estimation of forest area near deserts – production of Global Bio-Methanol from solar energy. *Applied Energy* **67**, 383–393.

Norrish, R. G. W. and G. A. Oldershaw (1961) The oxidation of phosphine studied by flash photolysis and kinetic spectroscopy. *Proceedings of the Royal Society of London. Series A, Mathematical and Physical Sciences* **262**, 10–18.

Notholt, J., Z. Kuang, C. P. Rinsland, G. C. Toon, M. Rex, N. Jones, T. Albrecht, H. Deckelmann, J. Krieg, C. Weinzierl, H. Bingemer, R. Weller and O. Schrems (2003) Enhanced upper tropical tropospheric COS: Impact on the stratospheric aerosol layer. *Science* **300**, 307–310.

Novelli, P. C., K. A. Masarie and P. M. Lang (1998) Distributions and recent changes in atmospheric carbon monoxide. *Journal Geophysical Research* **103**, 19015–19033.

Novelli, P., C., P. M. Lang, K. A. Masarie, D. F. Hurst, R. Myers and J. W. Elkins (1999) Molecular hydrogen in the troposphere: Global distribution and budget. *Journal of Geophysical Research* **104**, 30427–30444.

Noxon, J. F. (1976) Atmospheric nitrogen fixation by lightning. *Geophysical Research Letters* **3**, 463–465.

Nriagu, J. O. (1979) Global inventory of natural and anthropogenic emissions of trace metals to the atmosphere. *Nature* **279**, 409–411.

Nriagu, J. O. and J. M. Pacyna (1988) Quantitative assessment of worldwide contamination of air, water and soils by trace metals. *Nature* **333**, 134–139.

Nriagu, J. O. (1989) A global assessment of natural sources of atmospheric trace metals. *Nature* **338**, 47–49.

Öberg, G. (1998) Chloride and organic chlorine in soil. *Acta hydrochimica et hydrobiolobiologica* **26**, 137–144.

Öberg, G. (2002) The natural chlorine cycle – fitting the scattered pieces. *Applied Microbiology and Biotechnology* **58**, 565–581.

Odén, S. (1976) The acidity problem – an outline of concepts. *Water, Air and Soil Pollution* **6**, 137–166.

Odling, W. (1861) A manual of chemistry, descriptive and theoretical. Longman, London, Part 1, p. 92.

O'Dowd, C. D., M. H.Smith, I. E.Consterdine and J. A. Lowe (1997) Marine aerosol, sea-salt, and the marine sulphur cycle: A short review. *Atmospheric Environment* **31**, 73–80.

O'Dowd, C. D., P. Aalto, K. Hameri, M. Kulmala and T. Hoffmann (2002) Aerosol formation – Atmospheric particles from organic vapours. *Nature* **416**, 497–498.

O'Dowd, C. D., J. L. Jimenez, R. Bahreini, R. C. Flagan J. H. Seinfeld, L. Pirjola, M. Kulmala, S. F. G. Jennings and T. Hoffmann (2005) Marine aerosols and iodine emissions. *Nature* **433**, doi:10.1038/nature03373, 2005.

O'Dowd, C. D. and G. de Leeuw (2007) Marine aerosol production: A review of the current knowledge. *Philosophical Transactions of the Royal Society* **A 365**, 1753–1774.

OECD-FAO (2009) Agricultural outlook 2009–1018. Highlights. 73 pp.

Oke, T. R. (1987) Boundary layer climates, 2nd Ed., Methuen, London, 435 pp.

Okita, T. (1968) Concentration of sulfate and other inorganic materials in fog and cloud water and in aerosol. *Journal of Meteorological Society Japan* **46**, 120–127.

Olah, G. A. (2005) Beyond oil and gas: The methanol economy. *Angewandte Chemie International Edition* **44**, 2636–2639.

Olivier, J. G. J., A. F. Bouwman, K.W. Van Der Hoek and J. J. M. Berdowski (1998) Global Air Emission Inventories for Anthropogenic Sources of NO_x, NH_3 and N_2O in 1990. *Environmental Pollution* **102**, 135−148.

Olivier, J. G. J, J. P.J. Bloos, J. J. M. Berdowski, A. J. H. Visschedijk and A. F. Bouwman (1999a) A 1990 global emission inventory of anthropogenic sources of carbon monoxide on 1 × 1 degree developed in the framework of EDGAR/GEIA. *Chemosphere: Global Change Science* **1**, 1−17.

Olivier, J. G. J., A. F. Bouwman, K. W. van der Hoek and J. J. M. Berdowski (1999b) Global air emission inventories for anthropogenic sources of NO_x, NH_3 and N_2O in 1990. *Environmental Pollution* **102**, 135−148.

Olivier, J. G. J. and J. J. M. Berdowski (2001) Global emissions sources and sinks. In: The climate system (Ed. J. Berdowski, R. Guicheritand B. J. Heij), A. A. Balkema Publishers/ Swets & Zeitlinger Publishers, Lisse, The Netherlands. pp. 33−78.

Olivier, J. G. J. (2002) Greenhouse gas emissions: 1. Shares and trends in greenhouse gas emissions; 2. Sources and methods. In: CO_2 emissions from fuel combustion 1971−2000. International Energy Agency (IEA), Paris, pp. III.1−III.31.

Olivier, J., J. Peters, C. Granier, G. Pétron, J. F. Müller and S. Wallens (2003) Present and future surface emissions of atmospheric compounds. POET Report #2, EU project EVK2-1999-00011.

Olivier, J. G. J., J. A. Van Aardenne, F. Dentener, L. Ganzefeld and J. A. H. W. Peters (2005) Recent trends in global greenhouse gas emissions: regional trends and spatial distribution of key sources. In: Non-CO_2 greenhouse gases (NCGG-4), A. van Amstaek (Coord.), Millpress, Rotterdam, pp. 325−330.

Olivier, J. G. J., G. Janssens-Maenhout, M. Muntean and J. A. H. W. Peters (2013) Trends in global CO_2 emissions: 2013 Report. PBL Netherlands Environmental Assessment Agency, The Hague, 64 pp.

Ollis, D. F., E. Pellizzetti and N. Serpone (1989) Heterogeneous photocatalysis in the environment: Application to water purification. In: Photocatalysis: Fundamentals and applications (Eds. N. Serpone and E. Pelizzetti), John Wiley & Sons, USA, pp. 603−637.

Olson, T. M. and M. R. Hoffmann (1988a) The kinetics, mechanism, and thermodynamics of glyoxal-S(IV) adduct formation. *Journal of Physical Chemistry* **92**, 533−540.

Olson, T. M. and M. R. Hoffmann (1988b) The kinetics, mechanism, and thermodynamics of glyoxylic acid-S(IV) adduct formation. *Journal of Physical Chemistry* **92**, 4246−4253.

Olson, T. M., L. A. Torry and M. R. Hoffmann (1988) Kinetics of the formation of hydroxyacetaldehyde-S(IV) adducts at low pH. *Environmental Science and Technology* **22**, 1284−1289.

Olson, T. M. and M. R. Hoffmann (1989) Hydroxyalkylsulfonate formation: Its role as a S(IV) reservoir in atmospheric water droplets. *Atmospheric Environment* **23**, 985−997.

Oltmans, S. J. and H. Levy II (1994) Surface ozone measurements from a global network. *Atmospheric Environment* **28**, 9−24.

Oltmans, S. J., A. S. Lefohn, J. M. Harris, I. Galbally, H. E. Scheel, G. Bodeker, E. Brunke, H. Claude, D. Tarasick, B. J. Johnson, P. Simmonds, D. Shadwick, K. Anlauf, K. Hayden, F. Schmidlin, F T. Fujimoto, K. Akagi, C. Meyer, S. Nichol, J. Davies, R. Aedondas and E. Cuevas (2006) Long term changes in tropospheric ozone. *Atmospheric Environment* **40**, 3156−3173.

Oparin, A. I. (1924) Происхождение жизни (in Russian: Origin of life). M. Moskowski rabotschi.

Oparin, A. I. (1938) Origin of life. Macmillan, New York, 270 pp.

Oppenheimer, C., D. M. Pyle and J. Barclay, Eds. (2003) Volcanic degassing. Geological Society, London, Special Publications 213, 420 pp.

Orlando, J. J., G. S. Tyndall, G. K. Moortgat and J. G. Calvert (1993) Quantum yields for NO_3 photolysis between 570 and 635 nm. *Journal of Physical Chemistry* **97**, 10996−11000.

Örn, G., Hansson, U. and H. Rodhe (1996) Historical worldwide emissions of anthropogenic sulfur: 1860−1985. International Meteorological Institute in Stockholm Report CM-91.

Ortiz, R., H. Hagino, K. Sekiguchi, Q. Wang and K. Sakamoto (2006) Ambient air measurements nof six bicunctional carbonyls in a suburban area. *Atmospheric Research* **82**, 709−718.

Ostwald, Wi. (1894) Die wissenschaftlichen Grundlagen der Analytischen Chemie. Engelmann, Leipzig, 187 pp.

Ostwald, Wo. (1909) Grundriss der Kolloidchemie. Theodor Steinkopff, Dresden, 525 pp.

Oswald, R., T. Behrendt, M. Ermel, D. Wu, H. Su, Y. Cheng, C. Breuninger, A. Moravek, E. Mougin, C. Delon, B. Loubet, A. Pommerening-Röser, M. Sörgel, U. Pöschl, T. Hoffmann, M. O. Andreae, F. X. Meixner and I. Trebs (2013) HONO emissions from soil bacteria as a major source of atmospheric reactive nitrogen. *Science* **341**, 1233−1235.

Pacyna, E. G., J. M. Pacyna and N. Pirrone (2001) European emissions of atmospheric mercury from anthropogenic sources in 1995. *Atmospheric Environment* **35**, 2987−2996.

Pacyna, J. and E. Pacyna (2002) Global emissions of mercury from anthropogenic sources in 1995. *Water, Air and Soil Pollution* **137**, 149−165.

Pacyna, J., E. Pacyna, F. Steenhuisen and S. Wilson (2003) Mapping 1995 global anthropogenic emissions of mercury. *Atmospheric Environment* **37** Supp. No. 1 S-109-S117.

Pacyna, J., E. Pacyna, F. Steenhuisen and S. Wilson (2006) Global anthropogenic mercury emission inventory for 2000 *Atmospheric Environment* **40**, 4048−4063.

Padro, J. (1996) Summary of ozone dry deposition velocity measurements and model estimates over vineyard, cotton, grass and deciduous forest in summer. *Atmospheric Environment* **13**, 2363−2369.

Pahl, S. (1996) Feuchte Deposition auf Nadelwälder in den Hochlagen der Mittelgebirge. Berichte des Deutschen Wetterdienstes 198, Offenbach/M., Selbstverlag, 137 pp.

Pahl, S., P. Winkler, B. G. Arends, G. P. A. Kos, D. Schell, M. C. Facchini, S. Fuzzi, M. W. Gallagher, R. N. Colvile, T. W. Choularton, A, Berner, C. Kruisz, M. Bizjak, K. Acker and W. Wieprecht (1997) Vertical gradients of dissolved chemical constituents in evaporating clouds. *Atmospheric Environment* **31**, 2577−2588.

Paik, D. H., I.-R. Lee, D.-S. Yang, J. S. Baskin and A. H. Zewail (2004)Electrons in finite-sized water cavities: Hydration dynamics observed in real time. *Science* **306**, 672−675.

Pakkanen, T. A. (1996) Study of formation of coarse particle nitrate aerosol. *Atmospheric Environment* **30**, 2475−2482.

Palenik, B. and F. M. M. Morel (1988) Dark production of H_2O_2 in the Sargossa Sea. *Limnology and Oceanography* **33**, 1606−1611.

Papoular, R. (2001) The use of kerogen data in understanding the properties and evolution of interstellar carbonaceous dust. *Astronomy and Astrophysics* **378**, 597−607.

Paracelsus, Th. (1577) Aurora thesaurusque philosophorum. Accessit Monarchia physica per Gerardum Dorneum; in defensionem Paracelsicorum principiorum ... Præterea Anatomia viva. In English: Paracelsus his Aurora, & Treasure of the Philosophers. As also The Water-Stone of The Wise Men; Describing the matter of, and manner how to attain the universal Tincture. Faithfully Englished. And Published by J. H. Oxon. London, Printed for Giles Calvert, and are to be sold at the Black Spred Eagle, at the West end of Pauls (1659) 229 pp.

Park, S.-E., J.-S. Chang, and K.-W. Lee, eds. (2004) Carbon dioxide utilization for global sustainability: Proceedings of the 7th International Conference on Carbon Dioxide Utilization, Seoul, Korea. Elsevier Science and Technology, 626 pp.

Pasteur, L. (1862) Die in der Atmosphäre vorhandenen organisierten Körperchen, Prüfung der Lehre von der Urzeugung. Ostwald's Klassiker der exakten Wiss., Nr. 39, W. Engelmann, Leipzig, 98 pp.

Payne, C. H. (1925) Stellar atmospheres. Harvard Observatory Monograph No.1, Cambridge, MA, (USA), pp. 177−189.

Pearman, G. I., D. Etheridge, F. de Silva and P. J. Fraser (1986) Evidence of changing concentrations of atmospheric CO_2, N_2O, and CH_4 from air bubbles in Antarctic ice. *Nature* **320**, 248–250.

Pearson, P. N. and M. R. Palmer (1999) Middle Eocene seawater pH and athropogenic carbon dioxide concentrations. *Science* **284**, 1824–1826.

Pédro, G. (2007) Cycles biogéochimiques et écosystèmes continentaux. Académie des sciences France, Paris, 427 pp.

Pena, R. M., S. Garcia, C. Herrero, M. Losada, A. Vazquez and T. Lucas (2002) Organic acids and aldehydes in rainwater in a northwest region of Spain. *Atmospheric Environment* **36**, 5277–5288.

Peixoto, J. P. and A. H. Oort (1992) Physics of climate. Springer, New York, 520 pp.

Penkett, S. A., B. M. R. Jones, K. A. Brice and A. E. J. Eggleton (1979) The importance of atmospheric O_3 and H_2O_2 in oxidizing SO_2 in cloud and rainwater. *Atmospheric Environment* **13**, 123–137.

Penner, J. E., H. Eddleman and T. Novakov (1993) Towards the development of a global inventory of black carbon emissions. *Atmospheric Environment* **27**, 1277–1295.

Penner, J. E. (1995) Carbonaceous aerosols influencing atmospheric radiation: Black and organic carbon. In: Aerosol forcing of climate (Eds. R. J. Charlson and J. Heintzenberg). John Wiley & Sons, Chichester, pp. 1–108.

Penner, J. E., M. Andreae, H. Annegarn, L. Barrie, J. Feichter. D. Hegg, A. Jayeraman, R. Leaitch, D. Murphy, J. Nganga and G. Pitari (2001) Aerosols, their direct and indirect effects. In: Climate change 2001: The scientific basis (Eds. J. T. Houghton, Y. Ding, D. J. Griggs, N. Noguer, P. J. van der Lingen, X. Dai, K. Maskell and C. A. Johnson). Report to Intergovernmental Panel on Climate Change from the Scientific Assessment Working Group (WGI), Cambridge University Press, Cambridge, U.K., pp. 289–348.

Perner, D. and U. Platt (1979) Detection of nitrous acid in the atmosphere by differential optical absorption. *Geophysical Research Letters* **6**, 917–920.

Petermann, A. and J. Graftiau (1891) Recherches sur la composition de l'atmosphère. Acide carbonique. Combinaisons azotées contenues dans l'air atmosphérique et dans l'eau de pluie. Première partie: Acide carbonique contenu dans l'air atmosphérique, par A. In: *Mémoires couronnés et autres mémoires publiés par l'academie royal de Belgique.* Tom XLVII (1892–1893), 79 pp.

Peters, G. P., R. M. Andrew, T. Boden, J. G. Canadell, P. Ciais, C. Le Quéré, G. Marland, M. R. Raupach and C. Wilson (2013) The challenge to keep global warming below 2 °C. *Nature Climate Change* **3**, 4–6.

Peterson, J. T. and C. E. Junge (1971) Sources of particulate matter in the atmosphere. In: Man's impact on climate (Eds. W. H. Matthews, W. W. Kellogg and G. D. Robinson), Cambridge, Massachusetts (USA), MIT Press, pp. 310–320.

Petit, J. R., I. Basile, A. Leruyuet, D. Raynaud, C. Lorius, J. Jouzel, M. Stievenard, V.Y. Lipenkov, N. I. Barkov, B. B. Kudryashov, M. Davis, E. Saltzman, and V. Kotlyakov (1997) Four climate cycles in Vostok ice core. *Nature* **387**, 359–360.

Petit, J. R., J. Jouzel, D. Raynaud, N. I. Barkov, J.-M. Barnola, I. Basile, M. Bender, J. Chappellaz, M. Davis, G. Delaygue, M. Delmotte, V. M. Kotlyakov, M. Legrand, V. Y. Lipenkov, C. Lorius, L. Pépin, C. Ritz, E. Saltzman and M. Stievenard (1999) Climate and atmospheric history of the past 420 000 years from the Vostok ice core, Antarctica. *Nature* **399**, 429–436.

Petit, J. R., D. Raynaud, C. Lorius, J. Jouzel, G. Delaygue, N. I. Barkov, and V. M. Kotlyakov (2000) Historical isotopic temperature record from the Vostok ice core. In: Trends: A compendium of data on global change. Carbon Dioxide Information Analysis Center, Oak Ridge National Laboratory, U.S. Department of Energy, Oak Ridge, Tenn., U.S.A. doi: 10.3334/CDIAC/cli.006.

Petrenchuk, O. P. and V. M. Drozdova (1966) On the chemical composition of cloud water. *Tellus* **18**, 280–286.

Petrenchuk, O. P. (1979) Экспериментальные исследования атмосферного аэрозоля (in Russian: Experimental studies of atmospheric aerosol), Gidrometeoisdat, Leningrad, 264 pp.

Petroff, A., A. Mailliat, A. Muriel and A. Fabien (2008) Aerosol dry deposition on vegetative canopies. Part I. Review of present knowledge. *Atmospheric Environment* **42**, 3625–3653.

Petsch, S. T., T. I. Eglinton and K. J. Edwards (2001) ^{14}C-dead living biomass: Evidence for microbial assimilation of ancient organic carbon during scale weathering. *Science* **292**, 1127–1131.

Pflüger, W. (1872) Ueber den Einfluss des Centralnervensystems auf die thierische Wärme. *Archiv für die Gesammte Physiologie des Menschen und der Thiere* **6**, 647–647.

Pham, M., J.-F. Müller, G. P. Brasseur, C. Granier and G. Mégie (1995) A three-dimensional study of the tropospheric sulfur cycle. *Journal of Geophysical Research* **100**, 26061–26092.

Pichat, P., J. Disdier, C. Hoang-Van, D. Mas, G. Goutailler and C. Gaysse (2000) Purification / deodorization of indoor air and gaseous effluents by TiO_2 photocatalysis. *Catalysis Today* **63**, 363–369.

Pickering, K., Y. Wang, W.-K. Tao, C. Price and J.-F. Muller (1998) Vertical distributions of lightning NO_x for use in regional and global chemical transport models. *Journal of Geophysical Research* **103**, 31203–31216.

Pickering, K., H. Huntrieser and U. Schumann (2009) Lightning and NO_x production in global models. In: Betz et al. (2009), pp. 551–572.

Pidwirny, M. (2006) The hydrologic cycle. Fundamentals of physical geography, 2nd Edition. Date Viewed. http://www.physicalgeography.net/fundamentals/8b.html.

Pidwirny, M. (2008) The carbon cycle. In: Encyclopedia of earth (Ed. C. J. Cleveland), Washington (D.C.), USA, Environmental Information Coalition, National Council for Science and the Environment (NCSE). <http://www.eoearth.org/article/Carbon_cycle>.

Pierre, J. I. (1859) Chimie agricole ou l'agriculture considéreé dans ses rapports principaux avec la chimie. 10th ed., Paris, Libr. Agricole de la Maison Rustique, 532 pp. .

Pierrehumbert, R. T. (2004) High levels of atmospheric carbon dioxide necessary for the termination of global glaciation. *Nature* **429**, 646–649.

Pilson, M. E. Q. (1998) An Introduction to the chemistry of the sea. Prentice Hall, New Jersey (USA), 431 pp.

Pimentel, D. and N. Kounang (1998) Ecology of soil erosion in ecosystems. *Ecosystems* **1**, 416–426.

Pimentel, D. and M. H. Pimentel (2007) Food, energy, and society. 3rd Edition, CRC Press, 380 pp.

Pitts, J. N. Jr. (1970) Singlet molecular oxygen and the photochemistry of urban atmospheres. *Annals of the New York Academy of Sciences* **171**, 239–272.

Pizzarello, S., Y. Hang, L. Becker, R. J. Poreda, R. A. Nieman, G. Cooper and M. Williams (2001) The organic content of the Tagish Lake meteorite. *Science* **293**, 2236–2239.

Plass, G. N. (1953) The carbon dioxide theory of climatic change. Bulletin of the American Meteorological Society **34**, 80.

Plass, G. N. (1956a) Carbon dioxide and the climate. American Scientist **44**, 302–316.

Plass, G. N. (1956b) The carbon dioxide theory of climatic change. Tellus **8**, 140–154.

Plass-Dülmer, C., R. Koppmann, M. Ratte and J. Rudolph (1995) Light nonmethane hydrocarbons in seawater. *Global Biogeochemical Cycles* **9**, 79–100.

Platt, U. and G. K. Mortgaat (1999) Heterogeneous and homogeneous chemistry of reactive halogen compounds in the lower troposphere. *Journal of Atmospheric Chemistry* **34**, 1–8.

Platt, U., W. Allan and D. Lowe (2004) Hemispheric average Cl atom concentration from ^{13}C/^{12}C ratios in atmospheric methane. *Atmospheric Chemistry and Physics* **4**, 2393–2399.

Plekhanov, G. F. (2008) Paradoxes of Tungusk meteorite. In: International Conference "100 years since Tunguska phenomenon: Past, present and future". June 26−28, 2008, Moscow (Russia), Abstracts Vol. p. 50.

Plessow, K., K. Acker, H. Heinrichs and D. Möller (2001) Time study of trace elements and major ions during two cloud events at the Mt. Brocken. *Atmospheric Environment* **35**, 367−378.

Pollack, H. N., S. J. Hunter and J. R. Johnson (1993) Heat flow from the earth's interior: Analysis of the global data set. *Reviews of Geophysics* **31**, 267−280.

Pollock, J. A. (1909) Les ions dans l'atmosphère. *Le Radium* **6**, 129−135.

Pollock, J. A. (1915a) The nature of the large ions in the air. *Philosophical Magazine Series 6* **29**, 514−526.

Pollock, J. A. (1915b) The larger Ions in the air. *Nature* **95**, 286−288.

Popovicheva, O., D. Baumgardner, K. Gierens, R. Miake-Lye, R. Niessner, M. Rossi, M. Petters, J. Suzanne and E. Villenave (2007) Atmospheric soot: Environmental fate and impact. ASEFI 2006 Meeting Summary and the Atmospheric Soot Network (ASN) definition.

Poulton, J. E. (1990) Cyanogenesis in plants. *Plant Physiology* **94**, 401−405.

Pöschl, U., Rudich, Y. and M. Ammann (2005) Kinetic model framework for aerosol and cloud surface chemistry and gas-particle interactions: Part 1 − general equations, parameters, and terminology, *Atmospheric Chemistry and Physics Discussion* **5**, 2111−2191.

Potter, C. S., J. T. Randerson, C. B. Field, P. A. Matson, P. M. Vitousek, H. A. Mooney and S. A. Klooster (1993) Terrestrial ecosystem production: A process model based on global satellite and surface data. *Global Biogeochemical Cycles* **7**, 811−841.

Potter, C. S., P. A. Matson, P. M. Vitousek and E. Davidson (1996) Process modeling of controls on nitrogen trace gas emissions from soils worldwide. *Journal of Geophysical Research* **101**, 1361−1377.

Potter, C. S., S. A. Klooster and R. B. Chatfield (1996) Consumption and production of carbon monoxide in soils: A global model analysis of spatial and seasonal variation. *Chemosphere* **33**, 1175−1193.

Power, H. and J. M. Baldasano (1998) Air pollution emissions inventory. Advances in air pollution. CMP Series Vol. 3. WIT Press, 256 pp.

Pozdnyakov, I. P., E. M. Glebov, V. F. Plyusnin, V. P. Grivin, Y. V. Ivanov, D. Yu. Vorobyev and N. M. Bahzin (2000) Mechanism of $Fe(OH)^{2+}(aq)$ photolysis in aqueous solution. *Pure and Applied Chemistry* **72**, 2187−2197.

Prasad, M. N. V., K. S. Sajwan and R. Naidu, Eds. (2005) Trace elements in the environment: Biogeochemistry, biotechnology, and bioremediation. CRC Press, Boca Raton, FL (USA), 726 pp.

Prater, M. J., C. D. Holmes and J. Hsu (2012) Reactive greenhouse gas scenarios: Systematic exploration of uncertainties and the role of atmospheric chemistry. *Geophysical Research Letters* **39**, L09803, doi:10.1029/2012GL051440, 2012.

Prather, M. J., R. Derwent, D. H. Ehhalt, P. Fraser, E. Sanhueza and X. Zhou (1995) Other trace gases and atmospheric chemistry. In: Climate change (Eds. J. T. Houghton, L. G. Meira Filho, J.Bruce, H. Lee, B. A. Callender, E. Haites, N. Harris and K. Maskell), Cambridge University Press, Cambridge, U.K., pp. 73−126.

Prather, M. J. (2007) Lifetimes and time scales in atmospheric chemistry. *Philosophical Transactions of the Royal Society* **A365**, 1705−1726.

Pratt, K. A., L. E. Hatch and K. A. Prather (2009) Seasonal volatility dependence of ambient particle phase amines. *Environmental Science and Technology* **43**, 5276−5281.

Preis, M., R. Maser, H. Franke, W. Jaeschke and J. Graf (1994) Distribution of trace substances inside and outside of clouds. *Beiträge zur Physik der Atmosphäre* **67**, 341−351.

Prentice, I. C., G. D. Farquhar, M. J. R. Fasham, M. L. Goulden, M. Heimann, V. J. Jaramillo, H. S. Kheshgi, C. Le Quéré, R. J. Scholes and D. W. R. Wallace (2001) The carbon cycle and atmospheric carbon dioxide. In: Climate change 2001: The scientific basis (Eds. J. T. Houghton, Y. Ding, D. J. Griggs, M. Noguer, P. J. van der Linden, X. Dai, K. Maskell and C. Johnson), Cambridge University Press, Cambridge, U.K., pp. 183–237.

Prestel, M. A. F. (1865) Die jährliche, periodische Änderung des atmosphärischen Ozons und die ozonoskopische Windrose als Ergebniss der Beobachtungen zu Emden von 1857–1864. Nova Acta Physico-Medica Academia Caesarea Leopoldino-Carolina, Vol. 32, E. Blochmann & Sohn, Dresden, pp. 13.

Prestel, M. A. F. (1872) Der Boden, das Klima und die Witterung von Ostfriesland. Th. Hahn, Emden, 438 pp.

Price, C., J. Penner and M. Prather (1997) NO_x from lightning. 1. Global distribution based on lightning physics. *Journal of Geophysical Research* **102**, 5929–5941.

Priestley, J. (1775) The discovery of oxygen. Part I. Experiments. Alembic Club Reprints No. 7, re-issue edition. Livingstone, Edinburgh (1947), 55 pp.

Prieto-Ballesteros, O., J. S. Kargel, M. Fernández-Sampedro, F. Selsis, E. S. Martínez and D. L. Hogenboom (2005) Evaluation of the possible presence of clathrate hydrates in Europe's icy shell or seafloor. *Icarus* **177** 491–505.

Prinn, R. G., R. F. Weiss, P. j. Fraser, P. G. Simmonds, D. M. Cunnold, F. N. Alyea, S. O'Doherty, P. Salameh, B. R. Miller, J. Huang, R. H. J. Wang, D. E. Hartley, C. Harth, L. P. Steele, G. Sturrock, P. M. Midgely and A. McCulloch (2000) A history of chemically and radiatively important gases in air deduced from ALE/GAGE/AGAGE. *Journal Geophysical Research* **105**, 17751–17792.

Prinn, R. G., J. Huang, R. F. Weiss, D. M. Cunnold, P. J. Fraser, P. G. Simmonds, A. McCulloch, C. Harth, P. Salameh, S. O'Doherty, R. H. J. Wang, L. Porter and B. R. Miller (2001) Evidence for substantial variations of atmospheric hydroxyl radicals in the past two decades. *Science* **292**, 1882–1888.

Prinn, R. G., J. Huang, R. F. Weiss, P. J. Fraser, P. G. Simmonds, A. McCulloch, C. Harth, S. Reimann, P. Salameh, S. O'Doherty, R. H. J. Wang, L. W. Porter, B. R. Miller, P. B. Krummel (2005) Evidence for variability of atmospheric hydroxyl radicals over the past quarter century. *Geophysical Research Letters* **32**, L07809, doi:10.1029/2004GL022228.

Prospero, J. M. (1999) Long-term measurements of the transport of African mineral dust to the southeastern United States: Implications for regional air quality. *Journal of Geophysical Research* **104**, 15917–15928.

Prout, W. (1834) Chemistry, meteorology and the function of digestion. W. Pickering, London, 570 pp.

Pruppacher, H. R. and J. D. Klett (1997) Microphysics of clouds and precipitation. Kluwer Acad. Publ., Dordrecht, 954 pp.

Pueschel, R. F. (1995) Atmospheric aerosols. In: Atmospheric aerosols: Composition, chemistry, and climate of the atmosphere (Ed. H. B. Singh), Van Nostrand Reinhold, pp. 120–175.

Puxbaum, H. and G. König (1997) Observation of dipropenyldisulfide and other organic sulfur compounds in the atmosphere of a beech forest with *Allium Ursinum* ground cover. *Atmospheric Environment* **31**, 291–294.

Quéré, C. Le, G. P. Peters, R. J. Andres, R. M. Andrew, T. Boden, P. Ciais, P. Friedlingstein, R. A. Houghton, G. Marland, R. Moriarty, S. Sitch, P. Tans, A. Arneth, A. Arvanitis, D. C. E. Bakker, L. Bopp, J. G. Canadell, L. P. Chini, S. C. Doney, A. Harper, I. Harris, J. I. House, A. K. Jain, S. D. Jones, E. Kato, R. F. Keeling, K. Klein Goldewijk, A. Körtzinger, C. Koven, N. Lefèvre, A. Omar, T. Ono, G.-H. Park, B. Pfeil, B. Poulter, M. R. Raupach, P. Regnier, C. Rödenbeck, S. Saito, J. Schwinger, J. Segschneider, B. D. Stocker, B. Tilbrook, S. van Heuven, N. Viovy, R. Wanninkhof, A. Wiltshire, C. Yue and S. Zaehle (2013) Global Carbon Budget 2013. Earth Syst. Sci. Data Discuss., doi:10.5194/essdd-6-689-2013, http://www.earth-syst-sci-data-discuss.net/6/689/2013/essdd-6-689-2013.html.

Quinn, E. L. and C. L. Jones (1936) Carbon dioxide. Amer. Chem. Monograph Ser., Reinhold Publ., New York, 294 pp.

Quinn, P. K., D. J. Coffman, V. N. Kapustin, T. S. Bates and D. S. Covert (1998) Aerosol optical properties in the marine boundary layer during the First Aerosol Characterization Experiment (ACE 1) and the underlying chemical and physical aerosol properties. *Journal of Geophysical Research* **103**, 16547−16563.

Radke, L. F., D. A. Hegg, P. V. Hobbs, J. D. Nance, J. H. Lyons, K. K. Laursen, R. E. Weiss, P. J. Riggan, and D. E. Ward (1991) Particulate and trace gas emissions from large biomass fires in North America. In: Levine (1991) pp. 209−224.

Raef, Y. and A. J. Swallow (1966) Reaction of the hydrated electron with carbon monoxide as studied by pulse radiolysis. *Journal of Physical Chemistry* **70**, 4072−4073.

Raga, G. B. and P. R. Jonas (1995) Vertical distribution of aerosol particles and CCN in clean air around the British Isles. *Atmospheric Environment* **29**, 673−684.

Rahn, K. A., Brosset, C., Ottar, B. and E. M. Patterson (1982) Black and white episodes, chemical evolution of Eurasian air masses, and the long-range transport of carbon to the Arctic. In Particulate carbon: atmospheric life cycle (Eds. G. T. Wolff and R. L. Klimisch). New York, Plenum Press, pp. 327−340.

Rajeshwar, K., S. Licht and R. Mcconnell, Eds. (2008) Solar hydrogen generation: Toward a renewable energy future. Springer-Verlag, Berlin, 250 pp.

Rakov, V. A. and M. A. Uman (2007) Lightning: physics and effects. Cambridge University Press, Cambridge, U.K., 697 pp.

Ramakrishna, R. N., Ch. D. Keeling, H. Hashimoto, W. M. Jolly, St. C. Piper, C. D. Tucker, R. B. Myneni and St. W. Running (2003) Climate-driven increases in global terrestrial net primary production from 1982 to 1999. *Science* **300**, 1560−1563.

Ramanathan, V., P. J. Crutzen, J. T. Kiehl and D. Rosenfeld (2001) Aerosols, climate, and the hydrological cycle. *Science* **294**, 2119−2124.

Ramanathan, V. and P. J. Crutzen (2003) Atmospheric brown clouds. *Atmospheric Environment* **37**, 4033−4035.

Ramanathan V., M. V. Ramana, G. Roberts, D. Kim, C. E. Corrigan, C. E. Chung and D. Winker (2007) Warming trends in Asia amplified by brown cloud solar absorption. *Nature* **448**, 575−578.

Rampino, M. R. and S. Self (1982) Historic eruptions of Tambora (1815) Krakatau (1883) and Agung (1963) their stratospheric aerosols and climatic impact. *Quaternary Research* **18**, 127−143.

Ramsay, W. (1907) Die Gase der Atmosphäre und die Geschichte ihrer Entdeckung. W. Knapp, Halle, 160 pp.

Randerson, J. T., G. R. van der Werf, L. Giglio, G. J. Collatz and P. S. Kasibhatla (2007) Global fire emissions database, Vesrion 2 (GFEDv2.1). Data set. Available on-line [http://daac.ornl.gov/] from Oak Ridge National Laboratory Distributed Active Archive Center, Oak Ridge, Tennessee, U.S.A.

Raschig, F. (1904) Zur Theorie des Bleikammerprozesses II. *Zeitschrift für Angewandte Chemie* **17**, 1777−1785.

Rasmussen, R. A. (1972) What do the hydrocarbons from trees contribute to air pollution? *Journal Air Pollution Control Association* **22**, 537−543.

Rasmussen, K. L., H. B. Clausen and T. Risbo (1984) Nitrate in the Greenland ice sheet in the years following Tunguska event. *Icarus* **58**, 101−108.

Rasmussen, K. L., H. J. F. Ohlsen, R. Gwozdz and E. M. Kolesnikov (1999) Evidence for a very high carbon/iridium ratio in the Tunguska impactor. *Meteorites and Planetary Science* **34**, 891−895.

Rasmussen, K. L., P. Ditlevsen and F. Praëm-Larsen (2008) A CO_2-plume over the Tunguska area in the days after the 1908 impact. In: International Conference "100 years since

Tunguska phenomenon: Past, present and future". June 26–28, 2008, Moscow (Russia). Abstracts Vol. p. 88.

Raupach, M. R. (2013) The exponential eigenmodes of the carbon-climate system, and their implications for ratios of responses to forcings. *Earth System Dynamics* **4**, 31–34.

Rayleigh, J. W. S. Lord and W. Ramsay (1896) Argon, a new constituent of the atmosphere. Smithsonian Institute, Washington D.C., 43 pp.

Rayleight, J. W. S. Lord (1901) Spectroscopic notes concerning the gases of the atmosphere. *Philosophical Magazine* **1**, 100–105.

Read, P. and J. Lermit (2005) Bio-Energy with carbon storage (BECS): A sequential decision approach to the threat of abrupt climate change. *Energy* **30**, 2654–2671.

Reay, D. S., E. A. Davidson, K. A. Smith, P. Smith, J. M. Melillo, F. Dentener and P. J. Crutzen (2012) Global agriculture and nitrous oxide emission. *Nature Climate Change* **2**, 410–416.

Reddy, K. R. and R. D. DeLaune (2008) Biogeochemistry of wetlands: Science and applications. CRC Press, Boca Raton, FL (USA), 800 pp.

Rehder, D. (2010) Chemistry in space. Wiley-VCH, Weinheim, 292 pp.

Reimann, S., J. M. Alister, P. G. Simmonds, D. M. Cunnold, R. H. J. Wang, J. Li, A. McCulloch, R. G. Prinn, J. Huang, R. F. Weiss, P. Fraser, S. O'Doherty, B. B. Greally, K. Stemmler, M. Hill and D. Folini (2004) Low European methyl chloroform emissions inferred from long-term atmospheric measurements. *Nature* **433**, 506–509.

Reiset, J. (1879) Recherches sur la proportion de l'acide carbonique dans l'air. *Comptus Rendus* **88**, 1007–1012.

Reiset, J. (1880) Recherches sur la proportion de l'acide carbonique dans l'air. *Comptus Rendus* **90**, 1144–1148.

Remy, H. (1965) Lehrbuch der anorganischen Chemie. Band. I. 12. Auflage. Akademische Verlagsgesellschaft, Leipzig, 1120 pp.

Renk, F. (1886) Die Luft. Handbuch der Hygiene und der Gewerbekrankheiten. Erster Theil. Individuelle Hygiene. 2. Abtheilung. 2. Heft (Eds. M. v. Pettenkofer and H. v. Ziemssen), F. C. W. Vogel, Leipzig, 242 pp.

Reslhuber, P. A. (1856) Untersuchungen über das atmosphärische Ozon. *Berichte der Kaiserlichen Akademie der Wissenschaften, Math.-Nat. Klasse,* Wien, Bd. XXI., pp. 351–378.

Rethinaraj, T. S. G. and C. E. Singer (2010) Historical energy statistics. Global, regional and national trends since industrialization. World Scientific Publ. Co., 400 pp.

RETRO (2008) Emission data sets and methodologies for estimating emissions. (Eds. M. Schultz and S. Rast). RETRO deliverable D1–6: Report on emission / 2 (5th European Union framework programme), 77 pp.

Reuder, J. (1999) Untersuchungen zur Variabiliät von Photolysefrequenzen. Doctoral Thesis, Brandenburg University of Technology (BTU), Cottbus (Germany).

Rhee, T. A., A. J. Kettle and M. O. Andreae (2009) Methane and nitrous oxide emissions from the ocean: A reassessment using basin-wide observations in the Atlantic. *Journal of Geophysical Research* **114**, D12304, doi:10.1029/2008JD011662.

Richards, P. G., D. G. Torr and M. R. Torr (1981) Photodissociation of N_2 – A significant source of thermospheric atomic nitrogen. *Journal of Geophysical Research* **86**, 1495–1498.

Rick, S. W. (2004) A reoptimization of the five-site water potential (TIP5P) for use with Ewald sums. *Journal of Chemical Physics* **120**, 6085–6093.

Riddick, S. N., U. Dragosits, T. D. Blackall, F. Daunt, S. Wanless and M. A. Sutton (2012) The global distribution of ammonia emissions from seabird colonies. *Atmospheric Emvironment* **55**, 319–327.

Riebsame, W. E. (1985) Research in climate-society interaction. In: Climate impact assessment – studies of the interaction of climate and aociety, SCOPE 27 (Eds. R. W. Kates, J. H. Ausubel and M. Berberian), John Wiley & Sons, Chichester, Chaper 3.

Rigby, M.,R. G. Prinn, P. J. Fraser, P. G. Simmonds, R. L. Langenfelds, JU. Huang, D. M. Cunnold, L. P. Steele, P. B. Krummel, R. F. Weis, S. O'Doherty, P. K. Salameh, H. J. Wang, C. M. Harth, J. Mühle and L. W. Porter (2008) Renewed growth of atmospheric methane. *Geophysical Research Letters* **35**, L22805, doi:10.1029/2008GL036037.

Rihko-Struckmann, L. K., A. Peschel, R. Hanke-Rauschenbach and K. Sundmacher (2010) Assessment of methanol synthesis utilizing exhaust CO_2 for chemical storage of eectrical energy. *Industrial & Engineering Chemical Research* **49**, 11073–11078.

Rinsland, C. P., A. Goldman, J. W. Elkins, L. S. Chiou, J. W. Hannigan, S. W. Wood, E. Mahieu and R. Zander (2006) Long-term trend of CH_4 at northern mid-latitudes: Comparison between ground-based infrared solar and surface sampling measurement. *Journal of Quantitative Spectroscopy and Radiative Transfer* **97**, 457–466.

Rinsland, C. P., A. Goldman, J. W. Hannigan, S.W. Wood, L. S. Chiou and E. Mahieu (2007) Long-term trends of tropospheric carbon monoxide and hydrogen cyanide from analysis of high resolution infrared solar spectra. *Journal of Quantitative Spectroscopy and Radiative Transfer* **104**, 40–51.

Rinsland, C. P., L. S. Chiou, E. Mahieu, R. Zander, C. D. Boone and P. Bernath (2008) Measurements of long-term changes in atmospheric OCS (carbonyl sulfide) from infrared solar observations. *Journal of Quantitative Spectroscopy and Radiative Transfer* **109**, 2679–2686.

Rinsland, C. P., L. Chiou, C. Boone, P. Bernath, E. Mahieu and R. Zander (2009) Trend of lower stratospheric methane (CH_4) from atmospheric chemistry experiment (ACE) and atmospheric trace molecules spectroscopy (ATMOS) measurement. *Journal of Quantitative Spectroscopy and Radiatave Transfer* **110**, 1066–1071.

Risse, O. (1929) Einige Bemerkungen zum Mechanismus chemischer Röntgenreaktionen in wässrigen Lösungen. *Strahlentherapie* **34**, 578–581.

Roberston, J. B. (1919) The chemistry of coal. Chemistry monographs (Ed. A. C. Cumming), Oliver and Boyd, Edoinbutgh, 96 pp.

Robinson E. and R. C. Robbins (1970) Gaseous sulphur pollutants from urban and natural sources. *Journal Air Pollution Control Association* **20**, 233–235.

Robock, A. (1991) The volcanic contribution to climate change of the past 100 years. In: Greenhouse-gas-induced climatic change: A critical appraisal of simulations and observations (Ed. M. E. Schlesinger), Elsevier, Amsterdam, pp. 429–444.

Robock, A. (2003) Volcanoes: Role in climate. In: *Encyclopedia of atmospheric sciences* (eds. J. Holton, J. A. Curry and J. Pyle), Academic Press, London, pp. 2494–2500.

Robock, A. and C. Oppenheimer, Eds. (2003) Volcanism and the earth's atmosphere. Geophysical monograph 139, American Geophysical Union, Washington, DC, 360 pp.

Robock, A., C. M. Ammann, L. Oman, D. Shindell, S. Levis and G. Stenchikov (2009) Did the Toba volcanic eruption of ~74k BP produce widespread glaciation? *Journal of Geophysical Research* **114**, D10107. doi:10.1029/2008JD011652.

Rodhe, H. (1978) Budgets and turn-over times of atmospheric sulfur compounds. *Atmospheric Environment* **12**, 671–680.

Rohrer, F. and H. Berresheim (2006) Strong correlation between levels of tropospheric hydroxyl radical and solar ultraviolet radiation. *Nature* **442**, 184–187.

Rolleston, T. W. (1889) Life of Gotthold Ephraim Lessing. London: Walter Scott Limited, 233 pp.

Ronov, A. B. and A. A. Yaroshevsky (1969) Earth's crust geochemistry. In: Encyclopedia of geochemistry and environmental sciences (Ed. R.W. Fairbridge), Van Nostrand, New York.

Rosenfeld, D. (2006) Aerosols, clouds, and climate. *Science* **312**, 1323–1324.

Rossi, M. (2003) Heterogeneous reactions on salts. *Chemical Review* **103**, 4823–4882.

Roth, B. and K. Okada (1998) On the modification of sea-salt particles in the coastal atmosphere. *Atmospheric Environment* **32**, 1555–1569.

Ruben, S. M. (1939): Photosynthesis and phosphorylation. *Journal of the American Chemical Society* **61**, 661–663.

Ruben, S. M., M. Randall, M. D. Kamen and J. L. Hyde (1941) Heavy oxygen (^{18}O) as a tracer in the study of photosynthesis. *Journal of the American Chemical Society* **63**, 877–879.

Rubin, M. B. (2001) The history of ozone. The Schönbein period, 1839–1868. *Bulletin for the History of Chemistry* **26**, 40–56.

Rubio, M. A., Guerrero, M. J., Guillermo, V. and E. Lissi (2006) Hydroperoxides in dew water in downtown Santiago, Chile. A comparison with gas-phase values. *Atmospheric Environment* **40**, 6165–6172.

Rubner, M. (1907) Lehrbuch der Hygiene. Systematische Darstellung der Hygiene und ihrer wichtigsten Untersuchungs-Methoden. Deuticke, Leipzig und Wien, 1029 pp.

Ruddiman, W. F. (2003) The anthropogenic greenhouse era began thousands of years ago. *Climatic Change* **61**, 261–293.

Ruddiman, W. F. (2007) The early anthropogenic hypothesis: Challenges and responses. *Reviews of Geophysics* **45**, RG40.

Rudolph, J. (1997) Biogenic sources of atmospheric alkenes and acetylene. In: Biogenic volatile organic compounds in the atmosphere (Eds. G. Helas, J. Slanina and R. Steinbrecher), SPB Academic Publishing, Amsterdam, pp. 53–66.

Ruggaber, A., R. Forkel and R. Dlugi (1993) Spectral actinic flux measurements and its ratio to spectral irradiance by radiation transfer calculations. *Journal of Geophysical Research* **98**, 1151–1162.

Ruse, M. (1999) Evolutionary ethics: Julian Huxley and George Simpson. In: Biology and the foundation of ethics (Eds. J. Maienschein and M. Ruse), Cambridge University Press, 338 pp.

Russell, E. J. (1926) Plant nutrition and crop production. University of California Press, Berkeley, 115 pp.

Russell, F. A. R. (1895) The atmosphere in relation to human and health. In: Annual Report of the Smithsonian Institution, Washington D.C. Gov. Printing Office, pp. 203–348.

Russell, H. N. (1929) On the composition of the Sun's atmosphere. *Astrophysical Journal* **70**, 11–82.

Russell M. J. and N. T. Arndt (2005) Geodynamic and metabolic cycles in the Hadean. *Biogeosciences* **2**, 97–111.

Russell, M. J. and A. J. Hall (2006) The onset and early evolution of life. In: Evolution of early earth's atmosphere, hydrosphere, and biosphere – vonstraints from ore deposits (Eds. S. E. Kesler and H. Ohmoto), *Geological Society of America, Memoir* **198**, 1–32.

Ryaboshapko, A. G. (1983) Atmospheric sulfur cycle. In: The global biogeochemical sulphur cycle and influence on it of human activity (Eds. M. V. Ivanov and J. R. Freeney), Nauka Publ., Moskau 1983, pp. 170–255.

Ryaboshapko, A. (2001) Anthropogenic ammonia emissions in the former USSR in 1990. *Water, Soil and Air Pollution* **130**, 205–210.

Ryan, S., E. J. Dlugokencky, P. P. Tans, M. E. Trudeau (2006) Mauna Loa volcano is not a methane source: Implications for Mars. *Geophysical Research Letters* **33**, L12301. doi:10.1029/2006GL026223.

Ryder, G. (2001) Mass flux during the ancient Lunar bombardment: the cataclysm. *Lunar and Planetary Science* **32**, 1326.

Ryder, G. (2003) Bombardment of the Hadean earth: Wholesome or deleterious? *Astrobiology* **3**, 3–6.

Rye, R. and H. D. Holland (1998) Paleosols and the evolution of atmospheric oxygen: A critical review. *American Journal of Science* **298**, 621–672.

Sabine, C. L., R. A. Feely, N. Gruber, R.M. Key, K. Lee, J. L. Bullister, R.Wanninkhof, C. S. Wong, D. W. R. Wallace, B. Tilbrook, F. J. Millero, T.-H. Peng, A. Kozyr, T. Ono and A. F. Rios (2004) The oceanic sink for anthropogenic CO_2. *Science* **305**, 367–371.

Saikawa, E., C. A. Schlosser and R. G. Prinn (2013) Global modeling of soil nitrous oxide emissions from natural processes. *Global Biogeochemical Cycle* **27**, 972–989.

Sammonds, P. R. and J. M. T. Thompson (2007) Advances in earth science. From earthquake to global warming. World Scientific Publ. Co., 344 pp.

Sanadze, G. A. (2004) Biogenic isoprene (a review). *Russian Journal of Plant Physiology* **51**, 729–741.

Sander, R. and P. J. Crutzen (1996) Model study indicating halogen activation and ozone destruction in polluted air masses transported to the sea. *Journal of Geophysical Research* **101D**, 9121–9138.

Sanderson, M. G. (1996) Biomass of termites and their emissions of methane and carbon dioxide: a global database. *Global Biogeochemical Cycles* **10**, 543–557.

Sanderson, M. (1999) The classicification of climates from Pythagoras to Koeppen. *Bulletin of the American Meteorological Society* **80**, 669–673.

Sauer, F., C. Schäfer, P. Neeb, O. Horie and G. K. Moortgat (1999) Formation of hydrogen peroxide in the ozonolysis of isoprene and simple alkenes under humid conditions. *Atmospheric Environment* **33**, 229–241.

Saukov, A. A. (1953) Geochemie. Verlag Technik, Berlin, 311 pp.

Saussure, De, H. B. (1796) Voyages dans les Alpes, précédés d'un essai sur l'histoire naturelle des environs de Genève. Tome IV. Fouche-Boree, Neuchâtel, pp. 199–201.

Saussure Th. de (1804) Recherches chimiques sur la végétation. Didot jeune, Paris, 327 pp.

Sawyer, D. T. (1991) Oxygen chemistry Oxford University Press, New York, 223 pp

Schade, G. W. and P. J. Crutzen (1995) Emission of aliphatic amines from animal husbandry and their reactions: Potential source of N_2O and HCN. *Journal of Atmospheric Chemistry* **22**, 319–346.

Schade, G. W. and P. J. Crutzen (1999) CO emissions from degrading plant matter (II). Estimates of a global source strength. *Tellus* **51B**, 909–918.

Scharnow, U., W. Berth and W. Keller (1965) Wetterkunde. VEB Verlag für Verkehrswesen, Berlin, 415 pp.

Scheele, C. W. (1777) Chemische Abhandlung von der Luft und dem Feuer. Ostwald's Klassiker der exakten Wissenschaften. Nr. 58. Engelmann, Leipzig (1894), 112 pp.; see also: Discovery of oxygen. Part 2. Alembic Club Reprints No. 8. Clay, Edinburgh (1980), 46 pp.

Schellnhuber, H.-J. and V. Wenzel, Eds. (1998) Earth system analysis – integrating science for sustainability. Springer-Verlag, Berlin, 530 pp.

Schellnhuber H.-J. (1999) Earth system analysis and the second Copernican Revolution. *Nature, Suppl.* **402**, C19–C23.

Schemenauer, R. S., C. M. Banic and N. Urquizo (1995) High elevation fog and precipitation chemistry in southern Quebec, Canada. *Atmospheric Environment* **29**, 2235–2252.

Schemenauer, R. S. and H. Bridgman, Eds. (1998) Proceedings of the First International Conference on Fog ans fog collection. Vancouver, Canada, 19–24 July, ISBN 0-9683887-0-1, 492 pp.

Schidlowski, M., R. Eichmann and C. E. Junge (1975) Precambrium sedimentary carbonates: carbon and oxygen isotope geochemistry and implications for the terrestrial oxygen budget. *Precambrium Research* **2**, 1–69.

Schieber, J. and H. J. Arnott (2003) Nannobacteria as a by-product of enzyme-driven tissue decay. *Geology* **31**, 717–720.

Schiefferdecker, W. (1855) Bericht über die vom Verein für wissenschaftliche Heilkunde in Königsberg in Preussen angestellten Beobachtungen über den Ozongehalt der atmosphärischen Luft und sein Verhältnis zu den herrschenden Krankheiten. In: *Berichte der Kaiserlichen Akademie der Wissenschaften, Math.-Nat. Kl.* Wien, XXII. Band. II. Heft, pp. 191–237.

Schiff, H. I. (1970) Role of singlet oxygen in the upper atmosphere. *Annals of the New York Academy of Sciences* **171**, 188–198.

Schlesinger, W. H. and A. E. Hartley (1992) A global budget for ammonia. *Biogeochemistry* **15**, 191−211.

Schlesinger, W. H. (1997) Biogeochemistry − an analysis of global change. Academic Press, San Diego, 588 pp.

Schlesinger, W. H., Eds. (2003) Biogeochemistry: Treatise on geochemistry, Vol. 8 (Eds. H. D. Holland and K. K. Turekian), Elsevier Science & Technology, Amsterdam, 702 pp.

Schloesing, T. and A. Muntz (1877) Sur la nitrification par les ferments organisés. *Comptes Rendus de l'Académie des Sciences* **84**, 301−304.

Schloesing, T. (1880) Sur la constance de proportion d'acide carbonique dans l'air. *Comptes Rendus de l'Académie des Sciences* **90**, 1410−1413.

Schmauß, A. (1920) Kolloidchemie und Meteorologie. *Meteorologische Zeitschrift* **37**, 1−8.

Schmauß, A. and A. Wigand (1929) Die Atmosphäre als Kolloid. Sammlung Vieweg, Tagesfragen aus den Gebieten der Naturwissenschaften und der Technik, Heft 96, Fr. Vieweg & Sohn, Braunschweig, 74 pp.

Schmitt, B., C. de Bergh and M. Festou, Eds. (2007) Solar system ices (Astrophysics and Space Science Library), Springer-Verlag, Berlin, 828 pp.

Schneider, St. H. (2006) Climate change: Do we know enough for policy action? *Science and Engineering Ethics* **12**, 607−636.

Schneider-Carius, K. (1961) Das Klima, seine Definition und Darstellung. Zwei Grundsatzfragen der Klimatologie. In: Veröffentlichungen der Geophysik, Institut der Universität Leipzig, 2. Serie, Bd. XVII, Heft 2, Berlin.

Scholes, M. and M. O. Andreae (2000) Biogenic and pyrogenic emissions from Africa and their impact on the global atmosphere. *Ambio* **29**, 23−29.

Schönbein, C. F. (1844) Ueber die Erzeugung des Ozons auf chemischem Wege. Basel, pp. 62, 94, 110.

Schönbein, C. F. (1861) Ueber die Bildung des Wasserstoffsuperoxydes während der langsamen Oxydation der Metalle in feuchtem gewöhnlichem Sauerstoff oder atmosphärischer Luft. *Annalen der Physik und Chemie* **188**, 445−451.

Schönbein, C. F (1864) Ueber das Vorkommen von Wasserstoffhyperoxyd im menschlichen Körper. *Zeitschrift für analytische Chemie* **3**, 245−247.

Schönbein, C. F. (1866) Mittheilungen. *Journal für Praktische Chemie* **98**, 257−283 and **99**, 11−21.

Schönbein, C. F. (1869) Letzte Arbeiten. *Journal für Praktische Chemie* **106**, 257−273.

Schöne, E. (1874) Ueber das atmosphärische Wasserstoffhyperoxyd. *Berichte der Deutschen Chemischen Gesellschaft* **7**, 1693−1708.

Schöne, E. (1878) Ueber das atmosphärische Wasserstoffhyperoxyd. *Berichte der Deutschen Chemischen Gesellschaft* **11**, 482−492, 561−566, 874−877, 1028−1031.

Schöne, E. (1893) Zur Frage über das Vorkommen des Wasserstoffhyperoxyds in der atmosphärischen Luft und den atmosphärischen Niederschlägen. *Berichte der Deutschen Chemischen Gesellschaft* **26**, 3011−3027.

Schöne, E. (1894) Zur Frage über das atmosphärische Wasserstoffhyperoxyd. *Berichte der Deutschen Chemischen Gesellschaft* **27**, 1233−1235.

Schossberger, F. (1942) Über die Umwandlung des Titandioxyds. *Zeitschrift für Kristallographie* **104**, 358−374.

Schroeter, L. C. (1963) Kinetics of air oxidation of sulfurous acid salts. *Journal of Pharmaceutical Science* **52**, 559−563.

Schultz, M. G., A. Heil, J. J. Hoelzemann, A. Spessa, K. Thonicke, J. G. Goldammer, A. C. Held, J. M. C. Pereira and M. van het Bolscher (2008) Global wildland fire emissions from 1960 to 2000. *Global Biogeochemical Cycles* **22**, GB2002, doi:10.1029/2007GB003031.

Schulze, E.-D, M. Heimann, S. Harrison, E. Holland, J. Lloyd, I. C. Prentice and D. Schimel (2001) Global biogeochemical cycles in the climate system. Academic Press, San Diego (USA), 350 pp.

Schulze, F. (1871) Tägliche Beobachtungen über den Kohlen-Säuregehalt der Atmosphäre zu Rostock vom 18. Oktober 1868 bis 31. Juli 1871. Festschrift für die 44. Versammlung Deutscher Naturforscher und Aerzte. Leopoldsche Universitätsbuchhandlung, Rostock, 34 pp.

Schumann, U. and H. Huntrieser (2007) The global lightning-induced nitrogen oxides source. *Atmospheric Chemistry and Physics* **7**, 3823−3907.

Schumb, W. C., C. N. Satterfield and R. L. Wentworth (1955) Hydrogen peroxide. Reinhold Publ. Series, New York, 759 pp.

Schwabe, K. (1959) Über Azidität smaße. *Abhandlungen der Sächs. Akademie der Wissenschaften der DDR*, Akademie-Verlag Berlin, Vol. 46, No. 2, pp. 3−24.

Schwabe, K. (1976) pH-Messtechnik. Verlag Th. Steinkopf, Dresden, 412 pp.

Schwartz, S. E. and W. H. White (1981) Solubility equilibria of the nitrogen oxides and oxyacids in dilute aqueous solution. In: Advances in environmental sciences and engineering. Vol. 4 (Eds. J. R. Pfafflin and E. N. Ziegler), Gordon and Breach Science Publishers, New York, pp. 1−45.

Schwartz, S. E. and J. E. Freiberg (1981) Mass-transport limitation to the rate of reaction of gases in liquid droplets: Application to oxidation of SO_2 in aqueous solution. *Atmospheric Environment* **15**, 1129−1144.

Schwartz, S. E. (1984) Gas- and aqueous-phase chemistry of HO_2 in liquid water clouds. *Journal of Geophysical Research* **89**, 11589−11598

Schwartz, S. E. (1986) Mass-transport considerations pertinent to aqueous phase reactions of gases in liquid-water clouds. In: Chemistry of multiphase atmospheric systems (Ed. W. Jaeschke), Springer-Verlag, Berlin, pp. 415−172.

Schweitzer, F., L. Magi, P. Mirabel and C. George (1998) Uptake rate measurements of methanesulfonic acid and glyoxal by aqueous droplets. *Journal of Physical Chemistry A* **102**, 593−600.

Sciare, J., E. Baboukas, R. Hancy, N. Mihalopoulos and B. C. Nguyen (1998) Seasonal variation of dimethylsulfoxide in rainwater at Amsterdam Island in the Southern Indian Ocean: Implications on the biogenic sulfur cycle. *Journal of Atmospheric Chemistry* **30**, 229−240.

Scoutetten, R. J. H. (1856) L'ozone, ou recherches chimiques, météorologiques, physiologiques et médicales sur l'oxygène électrisé. Victor Masson & Metz, Paris, 287 pp.

Scurlock, J. M. O. and D. O. Hall (1991) The carbon cycle. *New Scientist* **132**, No. 1793 (Inside Science suppl. No. 51) pp. S1−S4.

Scurlock J. M. and R. J. Olson (2002) Terrestrial net primary productivity − A brief history and a new worldwide database. *Environmental Reviews* **10**, 91−109.

Sedlak, D. L., J. Hoigné, M. M. David, R. N. Colvile, E. Seyffer, K. Acker, W. Wieprecht, J. A. Lind and S. Fuzzi (1997) The cloudwater chemistry of iron and copper at Great Dun Fell, U.K. *Atmospheric Environment* **31**, 2515−2526.

Segalstad, T. V. (2009) Correct timing is everything − also for CO_2 in the air. http://www.co2science.org/articles/V12/N31/EDIT.php.

Segura, A., V. S. Meadows, J. F. Kasting, A. Segura, M. Cohen and J. Scalo (2007) Abiotic formation of O_2 and O_3 in high-CO_2 terrestrial atmospheres. *Astronomy and Astrophysics* **472**, 665−679.

Sehested, K., J. Holcman and E. J. Hart (1983) Rate constants and products of the reactions e_{aq}^-, O_2^- and H with ozone in aqueous solution. *Journal of Physical Chemistry* **87**, 1951−1954.

Sehested, K., J. Holcman, E. Bjergbakke and E. J. Hart (1984) Formation of ozone in the reaction of O_3^- and the decay of the ozonide ion radical at pH 10−13. *Journal of Physical Chemistry* **88**, 269−273.

Sehmel, G. A. (1980) Particle and gas dry deposition: A review. *Atmospheric Environment* **14**, 983−1011.

Seidl, W. and G. Hänel (1983) Surface-active substances on rainwater and atmospheric particles. *Pure and Applied Geophysics* **121**, 1077–1093.

Seiler, W., and P. J. Crutzen (1980) Estimates of gross and net fluxes of carbon between the
biosphere and the atmosphere from biomass burning. *Climatic Change* **2**, 207–247.

Seiler, W., R. Conrad and D. Scharffe (1984) Field studies of methane emission from termite
nests into the atmosphere and measurements of methane uptake by tropical soils. *Journal
of Atmospheric Chemistry* **1**, 171–186.

Seiler, W. and R. Conrad (1987) Contribution of tropical ecosystems to the global budgets
of trace gases, especially CH_4, H_2, CO, and N_2O. In: The geophysiology of Amazonia (Ed.
R. E. Dickinson), John Wiley & Sons, New York, pp. 133–162.

Seinfeld, J. H. (1986) Atmospheric chemistry and physics of air pollution. John Wiley &
Sons, New York, 739 pp.

Seinfeld, J. H. and S. N. Pandis (1998) Atmospheric chemistry and physics – from air pollution to climate change. John Wiley & Sons, New York, 1326 pp.

Selander, N. E. (1888) Luftundersökningar vid Vaxholms fästning okt. 1885–juli 1886. Kung-
liga Svenska Vetenskapsakad. Stockholm, Vol. 13, Afd. II, No. 9, 38 pp.

Self, S. and A. J. King (1996) Petrology and sulfur and chlorine emissions of the 1963 eruption of Gunung Agung, Bali, Indonesia. *Bulletin Volcanology* **58**, 263–285.

Self, S. (2006) The effects and consequences of very large explosive volcanic eruptions. *Philosophical Transactions of the Royal Society* **A 364**, 2073-2097.

Sephton, M. A. (2002) Organic compounds in carbonaceous meteorites. *Natural Products
Reports* **19**, 292–311.

Seventh Field Workshop on Volcanic Gases Satsuma-Iwojima, held from 17 till 25 October
2000 in Kyushu Japan.

Seyler, C. A. (1900) The chemical classification of coal. *Proceedings of South Wales Institute
of Engineering* **21**, 483–526.

Shafirovich, V. and S. V. Lymar (2002) Nitroxyl and its anion in aqueous solutions: Spin
states, protic equilibria, and reactivities toward oxygen and nitric oxide. *Proceedings of the
National Academy of Sciences of the United States of America* **99**, 7340–7345.

Sharkey, T. D. (1991) Stomatal control of trace gas emission. In: Trace gas emissions by
plants (Eds. T. D. Sharkey, E. A. Holland and H. A. Mooney), Academic Press, San
Diego, pp. 335–339.

Sharkey, T. D., A. E. Wiberley and A. R. Donohue (2008) Isoprene Emission from Plants:
Why and How. *Annals of Botany* **101**, 5–18; doi:10.1093/aob/mcm240.

Shelef, G., G. Oron, and R. Moraine (1978) Economic aspects of microalgae production on
sewage. *Archiv für Hydrobiologie. Beihefte* **11**, 281–294.

Shim, C., Y. Wang, Y. Choi, P. I. Palmer, D. S. Abbot and K. Chance (2005) Constraining
global isoprene emissions with Global Ozone Monitoring Experiment (GOME) formaldehyde column measurements. *Journal of Geophysical Research* **110**, D24301, doi:10.1029/
2004JD005629.

Sies, H. (1986) Biochemie des oxidativen Stress. *Angewandte Chemie* **98**, 1061–1075.

Siese, M. and C. Zetzsch (1995) Addition of OH to acetylene and consecutive reactions of
the adduct with O_2. *Zeitschrift für Physikalische Chemie* **188**, 75–89.

Sigg, A. and A. Neftel (1991) Evidence for a 50% increase in H_2O_2 over the past 200 years
from a Greenland ice core. *Nature* **351**, 557–559.

Sigg, L. and W. Stumm (1996) Aquatische Chemie. Eine Einführung in die Chemie wässriger
Lösungen und natürlicher Gewässer. Vdf Hochschulverl. an der ETH Zürich, Teubner,
Stuttgart, 498 pp.

Sigurdsson, H. (1982) Volcanic pollution and climate: the 1783 Laki eruption. *Transactions
of the American Geophysical Union (EOS)* **63**, 601–602.

Sillén, L. G. (1967) The ocean as a chemical system. *Science* **156**, 1189–1197.

Simanzhenkov, V. and R. Idem (2003) Crude oil chemistry. Marcel Dekker Inc., New York, 402 pp.

Simó, R., and J. Dachs (2002), Global ocean emission of dimethylsulfide predicted from biogeophysical data. *Global Biogeochemical Cycles* **16**, 1018, doi:10.1029/2001GB001829.

Simoneit, B. R. T. (2002) Biomass burning − A review of organic tracers for smoke from incomplete combustion. *Applied Geochemistry* **17**, 129−162.

Singh, H. B., D. O'Hara, D. Herlth, W. Sachse, D. R. Blake, J. D. Bradshaw, M. Kanakidou and P. J. Crutzen (1994) Acetone in the atmosphere: Distribution, sources, and sinks. *Journal of Geophysical Research* **99**, 1805−1819.

Singh, H. B., Y. Chen, A. Tabazadeh, Y. Fukui, I. Bey, R. Yantosca, D. Jacob, F. Arnold, K. Wohlfrom, E. Atlas, F. Flocke, D. Blake, N. Blake, B. Heikes, J. Snow, R. Talbot, G. Gregory, G. Sachse, S. Vay, Y. Kondo (2000) Distribution and fate of selected oxygenated organic species in the troposphere and lower stratosphere over the Atlantic. *Journal of Geophysical Research* **105**, 3795−3806.

Singh, H. B., Y. Chen, A. Staudt, D. Jacob, D. Blake, B. Heikes and J. Snow (2001) Evidence from the Pacific troposphere for large global sources of oxygenated organic compounds. *Nature* **410**, 1078−1081.

Singh, S. K., M. Bajpai and V. K. Tyagi (2006) Amine oxides: A review. *Journal of Oleo Sciences* **55**, 99−119.

Sinha, V., J. Williams, M. Meyerhöfer, U. Riebesell, A. I. Paulino and A. Larsen (2007) Air-sea fluxes of methanol, acetone, acetaldehyde, isoprene and DMS from a Norwegian fjord following a phytoplankton bloom in a mesocosm experiment. *Atmospheric Chemistry and Physics* **7**, 739−755.

Sinnhuber, B.-M., P. von der Gathen, M. Sinnhuber, M. Rex, G. König-Langlo and S. J. Oltmans (2006) Large decadal scale changes of polar ozone suggest solar influence. *Atmospheric Chemistry and Physics* **6**, 1835−1841.

Sinreich, R., R. Volkamer, F. Filsinger, U. Frieß, C. Kern, U. Platt, O. Sebastián and T. Wagner (2007) MAX-DOAS detection of glyoxal during ICARTT 2004. *Atmospheric Chemistry and Physics* **7**, 1293−1303.

Six, K. D., S. Kloster, T. Ilyina, S. D. Archer, K. Zhang and E. Maier-Reimer (2013) Global warming amplified by reduced sulphur fluxes as a result of ocean acidification. *Nature Climate Change* **3**, 975–978.

Skorockhodov, I. I., L. I. Nekrasov, N. I. Kobozev and V. B. Evdokimov (1962) On the higher peroxide of hydrogen and frozen radicals (in Russ.). *Zhurnal Fizicheskoi Khimii* **36**, 236−240.

Sleep N. H., K. J. Zahnle J. F. Kasting and H. J. Morowitz (1989) Annihilation of ecosystems by large asteroid impacts on the early earth. *Nature* **384**, 139−142.

Slemr, F. and W. Seiler (1984) Field measurements of NO and NO_2 emissions from fertilized and unfertilized soils. *Journal of Atmospheric Chemistry* **2**, 1−24.

Slinn, W. G. N. (1977) Some approximations for the wet and dry removal of particles and gases from the atmosphere. *Water, Air, and Soil Pollution* **7**, 513−543.

Smart, D. F., M. A. Shea, G. A. M. Dreschhoff and K. G. MacCracken (2007) Solar proton fluence for 31 solar cycles derived from nitrate enhancement in polar ice. Proceedings of the 30th Int. Cosmic Ray Conference (Eds. R. Caballero, J. C. D'Oliva, G. Medina-Tanco, L. Nellen, F. A. Sánchez, J. F. Valdés-Galicia), University Mexico City, Vol. 1, pp. 725−728.

Smil, V. (1999) Nitrogen in crop production: An account of global flows. *Global Biochemical Cycles* **13**, 647–662.

Smil, V. (2002) The earth's biosphere, The MIT Press, Cambridge, Massachusetts (USA), 346 pp.

Smith, R. A. (1845) Some remarks on the air and water of towns. *Memoirs and Proceedings, Chemical Society* **3**, 311−315.

Smith, R. A. (1852) On the air and rain of Manchester. *Memoirs of the Literary and Philosophical Society of Manchester* (Sevens Series) **10**, 207–217.

Smith, R. A. (1872) Air and rain. The beginnings of a chemical climatology. Longmans, Green and Co., London, 600 pp.

Smith, K. R., R. A. Rasmussen, F. Manegdeg and M. Apte (1992) Greenhouse gases from small-scale combustion in developing countries. US EPA Report EPA-600-R-92-005, Washington DC (USA).

Smith, S. (2001) Global and regional anthropogenic sulfur dioxide emissions. *Global and Planetary Change* **29**, 99–119.

Smith, S. J., E. Conception, R. Andres and J. Lurz (2004) Historical sulphur dioxide emissions 1850–2000. Methods and results. Report PNNL-14537 (Pacific Northwest National Laboratory) for the U.S. Dep. of Energy, 11 pp.

Smith, S. J., J. van Aardenne, Z. Klimont, R. Andres, A. C. Volke and S. Delgado Arias (2011) Anthropogenic sulfur dioxide emissions: 1850–2005. *Atmospheric Chemistry and Physics* **11**, 1101–1116.

Smith, P., D. Martino, Z. Cai, D. Gwary, H. Janzen, P. Kumar, B. McCarl, S. Ongle, F. O'Mara, C. Rice, B. Scholes and O. Sirotenko (2007) Agriculture. In: Climate change 2007: Mitigation. Contribution of Working Group III to the Fourth Assessment Report of the Intergovernmental Panel on Climate Change (Eds. B. Metz, O. R. Davidson, P. R. Bosch, R. Dave and L. A. Meyer), Cambridge University Press, Cambridge, U.K., pp. 498–540.

Söderlund R. and B. H. Svensson (1976) The global nitrogen cycle. In: Nitrogen, phosphorus and sulphur – global cycles (Eds. B. H. Svensson and R. Söderlund), Swedish Natural Science Research Council, pp. 23–73.

Solomon, S., G.-K. Plattner, R. Knutti and P. Friedlingstein (2009) Irreversible climate change due to carbon dioxide emissions. *Proceedings of the National Academy of Science* **106**, 1704–1709.

Sonntag, D. and D. Heinze (1981) Sättigungsdampfdruck- und Sättigungsdampfdichtetabellen für Wasser und Eis (in German). Deutscher Verlag für Grundstoffindustrie, Leipzig, 54 pp.

Sörensen, S. P. L. (1909) Enzymstudien II. Über die Messung und die Bedeutung der Wasserstoffionen-Konzentration bei enzymatischen Prozessen. *Biochemische Zeitschrift* **21**, 131–199.

Sörensen, P. E. and V. S. Anderson (1970) The formaldehyde-hydrogen sulphite system in alkaline solution. Kinetics, mechanism and equilibria. *Acta Chemica Scandinavia* **24**, 1301–1306.

Soret, J.-L. (1864) Ueber die volumetrischen Beziehungen des Ozons. *Annalen der Chemie und Pharmacie (Lieb. Ann.)* **130**, 95–101; Ueber das volumetrische Verhalten des Ozons. *Annalen der Physik und Chemie (Pogg. Ann.)* **197**, 168–283.

Sorooshian, A., M.-L. Lu, F. J. Brechtel, H. Jonsson, G. Feingold, R. C. Flagan and J. H. Seinfeld (2007) On the source of organic acid aerosol layers above clouds. *Environmental Science and Technology* **41**, 4647–4654.

Sowers, T., R. B. Alley and J. Jubenville (2003) Ice core records of atmospheric N_2O covering the last 106,000 Years. *Science* **301**, 945–948.

Spahni, R., R. Wania, L. Neef, M. van Weele, I. Pison, P. Bousquet, C. Frankenberg, P. N. Foster, F. Joos, I. C. Prentice and P. van Velthoven (2011) Constraining global methane emissions and uptake by ecosystems. *Biogeosciences* **8**, 1643–1665.

Spector, N. A. and B. F. Dodge (1946) Removal of carbon dioxide from atmospheric air. *Transaction American Institute Chemical Engineering* **42**, 827–848.

Speight, J. G. (2006) The chemistry and technology of coal. 4th Edition. Marcel Dekker Inc., New York, 945 pp.

Spindler, G., K. Müller, E. Brüggemann, T.Gnauk and H. Herrmann (2004) Long-term size-seggregated characterization of PM_{10}, $PM_{2.5}$ and PM_1 at the IfT research station Melpitz downwind of Leipzig (Germany) using high and low-volume filter samplkers. *Atmospheric Environment* **38**, 5333−5347.

Spindler, G., E. Brüggemann, T.Gnauk, A. Grüner, K. Müller and H. Herrmann (2009) A four-year size-seggregated characterization study of particles PM_{10}, $PM_{2.5}$ and PM_1 depending on air mass origin at Melpitz. *Atmospheric Environment* **44**, 164−173.

Spindler, G., A. Grüner, K. Müller, S. Schlimper and H. Herrmann (2013) Long-term size-segregated particle (PM_{10}, $PM_{2.5}$, PM_1) characterization study at Melpitz – influence of air mass inflow, weather conditions and season. *Journal of Atmospheric Chemistry* **70**, 165−195.

Spiro, P. A., D. J. Jacob and J. A. Logan (1992) Global inventory of sulfur emissions with a $1° × 1°$ resolution. *Journal of Geophysical Research* **97**, 6023−6036.

Sposito, G. (2008) The chemistry of soils. Oxford University Press, USA, 329 pp.

Spracklen, D. V., B. Bonn and K. S Carslaw (2008) Boreal forests, aerosols and the impacts on clouds and climate. *Philosophical Transaction of the Royal Society A* **366**, 4613−4626.

Sprengel C. (1826) Ueber Pflanzenhumus, Humussäure und humussaure Salze. *Archiv für die Gesammte Naturlehre* **8**, 145−220.

Sprengel C. (1828) Von den Substanzen der Ackerkrume und des Untergrundes. *Journal für Technische und Ökonomische Chemie* **2**, 423−474 and **3**, 2−99, 313−352, and 397−421.

Spurny, K. R., Ed. (1999) Analytical chemistry of aerosols. Lewis Publishers, CRC Lewis Publishers, Boca Raton, FL (USA), 486 pp.

Spurny, K. R., Ed. (2000) Aerosol chemical processes in the environment. Lewis Publishers, Boca Raton, FL (USA), 615 pp.

Staehelin, J. and J. Hoigné (1982) Decomposition of ozone in water: rate of initiation by hydroxide ions and hydrogen peroxide. *Environmental Science and Technology* **16**, 676−681.

Staehelin, J., R. E. Bühler and J. Hoigné (1984) Ozone decomposition in water studied by pulse radiolysis − 2. OH and SO_4 as chain intermediates. *Journal of Physical Chemistry* **88**, 5999−6004.

Staehelin, J., J. Thudium, R. Buehler, A. Volz-Thomas and W. Grabers (1994) Trends in surface ozone concentrations at Arosa (Switzerland). *Atmospheric Environment* **28**, 75−101.

Staehelin, J. and C. Schnadt Poberaj (2007) Long-term tropospheric ozone trends: A critical review. In: Advances in global change research. Vol. 33: Climate variability and extremes during the past 100 years (Eds. St. Brönnimann, J. Luterbacher, T. Ewen, H. F. Diaz, R. S. Stolarski and U. Neu), Springer-Verlag, Berlin, pp. 271−282.

Staley, J. T. and G. H. Orians (2002) Evolution and the biosphere. In: Earth system science (Eds. M. K. J. Jacobson, R. J. Charlson, H. Rodhe and G. H. Orians), Vol. 72 in the International Geophysics Science (Ed. J. R. Holton), Academic Press, San Diego, pp. 29−61.

Stanbury, D. M. (1989) Reduction potentials involving inorganic free radicals in aqueous solution. *Advances in Inorganic Chemistry* **33**, 69−138.

Stanhill, G. (1982) The Montsouris series of carbon dioxide concentration measurements, 1877−1910. Climate Change 4, 221−237.

Statheropoulos, M. and J. G. Goldammer (2007) Vegetation fire smoke: Nature, impacts and policies to reduce negative consequences on humans and the environment. A contribution to the 4th Intern Wildland Fire Conference, Sevilla, Spain, 13−17 Mai 2007, 36 pp.

Stavrakou, T., J.-F. Müller, I. de Smedt, M. van Roozendael, M. Kanakidou, M. Vrekoussis, F. Wittrock, A. Richter and J. P. Burrows (2009a) The continental source of glyoxal estimated by the synergistic use of spaceborne measurements and inverse modelling. *Atmospheric Chemistry and Physics Discussions* **9**, 13593−13628.

Stavrakou, T., J.-F. Müller, I. De Smedt, M. Van Roozendael, G. R. van der Werf, L. Giglio and A. Guenther (2009b) Evaluating the performance of pyrogenic and biogenic inventories against one decade of space-based formaldehyde columns. *Atmospheric Chemistry and Physics* **9**, 1037−1060.

Steele, L. P., P. B. Krummel and R. L. Langenfelds (2007) Atmospheric methane, carbon dioxide, hydrogen, carbon monoxide and nitrous oxide from Cape Grim flask air samples analysed by gas chromatography. In: Baseline Atmospheric Program (Australia), 2005–2006 (Eds. J. M. Cainey, N. Derek and P. B. Krummel), Australian Bureau of Meteorology and CSIRO Marine and Atmospheric Research, Melbourne, Australia, pp. 62–65.

Steffen, W., P. J. Crutzen and J. R. McNeill (2007) The Anthropocene: Are humans now overwhelming the Great Forces of Nature? *Ambio* **36**, 614–621.

Stehfest, E. and L. Bouwman (2006) N_2O and NO emission from agricultural fields and soils under natural vegetation: Summarizing available measurement data and modeling of global annual emissions. *Nutrient Cycling in Agroecosystems* **74**, 207–228.

Stein, C. A. and S. Stein (1992) A model for the global variation in oceanic depth and heat flow with lithospheric age. *Nature* **359**, 123–129.

Stein, G., Ed. (1968) Radiation chemistry and aqueous systems. Proceedings of the 19th L. Farkas Memorial Sympsium held at the Hebrew University of Jerusalem, Israel, 27–29 December 1967. The Weizmann Science Press of Israel, Jerusalem in cooperation with John Wiley & Sons, 306 pp.

Steinbrecher, R. (1997) Isoprene: production by plants and ecosystem-level estimates. In: Biogenic volatile organic compounds in the atmosphere (Eds. G. Helas, J. Slanina and R. Steinbrecher), SPB Academic Publishing, Amsterdam, pp. 101–114.

Steinbrecht, W., H. Claude, F. Schönenborn, I. S. McDermid, T. Leblanc, S. Godin, T. Song, D. P. J. Swart, Y. J. Meijer, G. E. Bodeker, B. J. Connor, N. Kämpfer, K. Hocke, Y. Calisesi, N. Schneider, J. de la Noë, A. D. Parrish, I. S. Boyd, C. Brühl, B. Steil, M. A. Giorgetta, E. Manzini, L. W. Thomason, J. M. Zawodny, M. P. McCormick, J. M. Russell, P. K. Bhartia, R. S. Stolarski and S. M. Hollandsworth-Frith (2006) Long-term evolution of upper stratospheric ozone at selected stations of the Network for the Detection of Stratospheric Change (NDSC). *Journal of Geophysical Research* **111**, D10308, doi:10.1029/2005JD006454.

Steinbrecht, W., H. Claude, F. Schönenborn, I. S. McDermid, T. Leblanc, S. Godin-Beekmann, P. Keckhut, A. Hauchecorne, J. A. E. Van Gijsel, D. P. J. Swart, G. E. Bodeker, A. Parrish, I. S. Boyd, N. Kämpfer, K. Hocke, R. S. Stolarski, S. M. Frith, L. W. Thomason, E. E. Remsberg, C. von Savigny, A. Rozanov and J. P. Burrows (2009) Ozone and temperature trends in the upper stratosphere at five stations of the Network for the Detection of Atmospheric Composition Change. *International Journal of Remote Sensing* **30**, 3875–3886.

Steinle-Neumann, G., L. Stixrude, R. E. Cohen and O. Gülseren (2001) Elasticity of iron at the temperature of the earth's inner core. *Nature* **413**, 57–60.

Stemmler, K., M. Ammann, C. Donders, J. Kleffmann and C. George (2006) Photosensitized reduction of nitrogen dioxide on humic acid as a source of nitrous acid. *Nature* **440**, 195–198.

Stepanov, Yu. S. and V. A. Andrushchenko (1990) Self-ignition mechanism for coal. *Combustion, Explosion, and Shock Waves* **26**, 147–152.

Stern, D. I., and R. K. Kaufmann (1998) Annual estimates of global anthropogenic methane emissions: 1860–1994. Trends Online: A Compendium of Data on Global Change. Carbon Dioxide Information Analysis Center, Oak Ridge National Laboratory, U.S. Department of Energy, Oak Ridge, Tenn., U.S.A. doi:10.3334/CDIAC/tge.00.

Stern, D. I. (2005) Global sulfur emissions from 1850 to 2000. *Chemosphere* **58**, 163–175.

Stern, D. I. (2005) Reversal in the trend of global anthropogenic sulfur emissions. Rensselaer Working papers in Economics. Rensselaer Polytechnic Institute, New York, 50 pp.

Stewart, I. and J. Cohen (1994) Why are there simple rules in a complicated universe? *Futures* **26**, 648–664.

Stöckhardt, J. A. (1850) Über einige durch den Bergbau und Hüttenbetrieb für die Landescultur entstehende Benachteiligungen. *Zeitschrift für deutsche Landwirthe* **4**, 33–38 und 129–137.

Stöckhardt, J. A. (1871) Untersuchungen über die schädliche Einwirkung des Hütten- und Steinkohlenrauches auf das Wachstum der Pflanzen, insbesondere der Fichte und Tanne. *Tharandter Forstliches Jahrbuch* **21**, 218−254.

Stoiber, R. E. and A. Jepsen (1973) SO_2 contribution to the atmosphere by volcanoes. *Science* **182**, 577−578.

Stoiber, R. E., S. N. Williams and B. Huebert (1987) Annual contribution of sulfur dioxide to the atmosphere by volcanoes. *Journal of Volcanology and Geothermal Research* **33**, 1−8.

Stoklasa, J. (1906) Ueber die Menge und den Ursprung des Ammoniaks in den Producten der Vesuveruption im April 1906. *Berichte der Deutschen Chemischen Gesellschaft* **39**, 3530−3537.

Stolaroff, J. K., D. W. Keith and G. V. Lowry (2008) Carbon dioxide capture from atmospheric air using sodium hydroxide spray. *Environmental Science and Technology* **42**, 2728−1735.

Stolarski, R. S. and R. J. Cicerone (1974) Stratospheric chlorine: A possible sink for ozone. *Canadian Journal of Chemistry* **52**, 1610−1615.

Stothers, R. B. (1996) The great dry fog of 1783. *Climate Change* **32**, 79−89.

Strachan, R., Ed. (1866) Wells, W. C.: An essay on dew (reprint from 1814). Edited, with annotations by L. P. Castella, and an appendix by R. Strachan. Longmans, London pp. 153.

Strangeways, I. (2007) Precipitation. Theory, measurements and distribution. Cambridge University Press, Cambridge, U.K., 290 pp.

Streets, D. G. and S. T. Waldhoff (2000) Present and future emissions of air pollutants in China: SO_2, NO_x, and CO. *Atmospheric Environment* **34**, 363−374.

Stribling, R. and S. L. Miller (1987) Energy yields for hydrogen cyanide and formaldehyde synthesis: the HCN and amino acid concentrations in the primitive ocean. *Origins of Life and Evolution of Biospheres* **17**, 261−273.

Strogies, M. and D. Kallweit (1996) Nitrogen emissions in Germany and potential for their reduction. In: Atmospheric ammonia: Emission deposition and environmental impacts: Poster proceedings (Eds. M. A. Sutton, D. S. Lee, G. J. Dollard et al.), Midlothian, UK: Institute of Terrestrial Ecology, pp. 53−56.

Strunz, H. (1955) Zur Kristallchemie des wasserreichsten Zeolithes Faujasit. *Naturwissenschaften* **42**, 485.

Struve, H. (1869) Ueber die Gegenwart von Wasserstoffhyperoxyd in der Luft. *Zeitschrift für analytische Chemie* **8**, 315−321.

Struve, H. (1871) Studien über Ozon, Wasserstoffhyperoxyd und salpetrigsaures Ammoniak. *Zeitschrift für analytische Chemie* **10**, 292−298.

Stumm, W. and J. J. Morgan (1981): Aquatic chemistry. An introduction emphasizing chemical equilibria in natural waters. 2nd edition, John Wiley & Sons, New York, 780 pp.

Stumm, W., J. J. Morgan and L. D. Schnoor (1983): Saurer Regen, eine Folge der Störung hydrogeochemischer Kreisläufe. *Naturwissenschaften* **70**, 216−223.

Stumm, W., Ed. (1987) Aquatic surface chemistry: Chemical processes at the particle-water interface. John Wiley & Sons, New York, 520 pp.

Stumm, W. and J. J. Morgan (1995) Aquatic chemistry: Chemical equilibria and rates in natural waters. 3th ed., John Wiley & Sons, New York, 1040 pp.

Sturges, W. T., T. J. Wallington, M. D. Hurley, K. P. Shine, K. Sihra, A. Engel, D. E. Oram, S. A. Penkett, R. Mulvaney and C. A. M. Brenninkmeijer (2000) A potent greenhouse gas identified in the atmosphere: SF_5CF_3. *Science* **289**, 611−613.

Suess, E. (1875) Die Entstehung der Alpen. W. Braunmüller, Vienna, 168 pp.

Summer, G. (1988) Precipitation. Process and analysis. John Wiley & Sons, New York, 455 p.

Sun J. and P. A. Ariya (2006) Atmospheric organic and bio-aerosols as cloud condensation nuclei (CCN): A review. *Atmospheric Environment* **40**, 795−820.

Surrat, J. D., J. H. Kroll, T. E. Kleindienst, E. O. Edney, M. Claeys, A. Q. Sorooshian, N. L. Ng, J. H. Offenberg, M. Lewandowski, M. Jaoui, R. C. Flagan and J. H. Seinfeld

(2007) Evidence for organosulfates in secondary organic aerosol. *Environmental Sciences and Technology* **41**, 517−527.

Surrat, J. D., D., Y. Gómez-González, A. W., H. Chan, R. Vermeulen, M. Shahgholi, T. E. Kleindienst, E. O. Edney, J. H. Offenberg, M. Lewandowski, M. Jaoui, W. Maenhout, M. Clayes, R. C. Flagan and J. H. Seinfeld (2008) Organosulfate formation in biogenic secondary organic aerosol. *Journal of Physical Chemistry* A **112**, 8345−8378.

Sutton, M. A., Pitcairn, C. E. R. and D. Fowler (1993) The exchange of ammonia between the atmosphere and plant communities. *Advances in Ecological Research* **24**, 301−393.

Sutton, M. A., C. J. Place, M. Eager, D. Fowler and R. I. Smith (1995) Assessment of the magnitude of ammonia emissions in the United Kingdom. *Atmospheric Environment* **29**, 1393−1411.

Sutton, M. A., J. W. Erisman, F. Dentener and D. Möller (2008) Ammonia in the environment: From ancient times to the present. *Environmental Pollution* **156**, 583−604.

Svehla, G. (1993) Nomenclature of kinetic methods of analysis (IUPAC Recommendations 1993). *Pure and Applied Chemistry* **65**, 2291−2298.

Swietlicki, E., H.-C. Hansson, B. Martinsson, B. Mentes, D. Orsini, B. Svenningsson, A. Wiedensohler, M. Wendisch, S. Pahl, P. Winkler, R. N. Colvile, R. Gieray, J. Lüttke, J. Heintzenberg, J. N. Cape, K. J. Hargreaves, R. L. Storeton-West, K. Acker, W. Wieprecht, A. Berner, C. Kruisz, M. C. Facchini, P. Laj, S. Fuzzi, B. Jones and P. Nason (997) Source identification during the Great Dun Fell Cloud Experiment 1993. *Atmospheric Environment* **31**, 2441−2451.

Symonds, R. B., W. I. Rose and M. H. Reed (1988) Contribution of Cl- and F-bearing gases to the atmosphere by volcanoes. *Nature* **334**, 415−418.

Symonds, R. B., W. I. Rose, G. Bluth and T. M. Gerlach (1994) Volcanic gas studies: methods, results, and applications. In: Volatiles in magmas: Mineralogical Society of America Reviews in Mineralogy (Eds. M. R. Carroll and J. R. Holloway) pp. 1−66.

Szabadváry, F. (1966) History of analytical chemistry. Pergamon, Oxford. 419 pp.

Tab, K. H. (2010) Principles of soil chemistry. CRC Press, 390 pp.

Tadić, J., G. K. Moortgat and K. Wirtz (2006) Photolysis of glyoxal in air. *Journal of Photochemistry and Photobiology A: Chemistry* **177**, 116−124.

Taguchi, S., S. Murayama and K. Hihuchi (2002) Sensitivity of inter-annual variation of CO_2 seasonal cycle at Mauna Loa to atmospheric transport. *Tellus B*, **55**, 547−554.

Takemura T., H. Okamoto, Y. Maruyama, A. Numaguti, A. Higurashi, and T. Nakajima (2000) Global three-dimensional simulation of aerosol optical thickness distribution of various origins. *Journal of Geophysical Research* **105**, 17853−17873.

Takenaka, N., A. Ueda, H. Bandow, T. Daimon, T. Dohmaru and Y. Maeda (1996) Acceleration mechanism of chemical reaction by freezing: The reaction of nitrite with dissolved oxygen. *Journal of Physical Chemistry* **100**, 13874−13884.

Takenaka, N., T. Daimon, A. Ueda, K. Sato, M. Kitano, H. Bandow and Y. Maeda (1998) Fast oxidation of nitrite by dissolved oxygen in freezing process in the tropospheric aqueous phase. *Journal of Atmospheric Chemistry* **29**, 135−150.

Tanaka, K. (1925) Versuche zur Prüfung der Wielandschen Atmungstheorie. *Biochemische Zeitschrift* **157**, 425−433.

Tanaka, T. (2009) Global dust budget. In: Encyclopedia of earth (Ed. C. J. Cleveland). Washington, D.C.: Environmental Information Coalition, National Council for Science and the Environment). http://www.eoearth.org/article/Global_dust_budget.

Tansley, A. G. (1935) The use and abuse of vegetational terms and concepts. *Ecology* **16**, 284−307.

Taylor, J. A. and J. C. Orr (2000) The natural latitudinal distribution of atmospheric CO_2. In: Climate and Global Change, Series No. ANL/CGC-002-0400, 25 pp.

Tegen, I. and I. Fung (1995). Contribution to the atmospheric mineral aerosol load from land surface modification. *Journal of Geophysical Research* **100**, 18707−18726.

Tegen, I., M. Werner, S. P. Harrison and K. E. Kohfeld (2004) Relative importance of climate and land use in determining present and future global soil dust emission. *Geophysical Research Letters* **31**, L05105, doi:10.1029/2003GL019216.

Teichmüller, M. and R. Teichmüller (1968) Geological aspects of coal metaphormism. In: Coal and coal-bearing strata (Eds. D. G. Murchison and T. S. Westoll), Oliver and Boyd, Edenburgh, pp. 233–267.

Teilhard de Chadin, P. (1961) The phenomenon of Man (English translation of original 1955 French edition). Harper Torchbooks, New York, 318 pp.

Tepe J. B. and B. F. Dodge (1943) Absorption of carbon dioxide by sodium hydroxide solutions in a packed column. *Transactions of the American Institute Chemical of Engineers* **39**, 255–276.

Textor, C., H.-F. Graf, C. Timmreck and A. Robock (2004) Emission from volcanoes. In: Granier et al. (2004), pp. 269–304.

Theloke, J., Calaminus, B., Dünnebeil, F., Friedrich, R., Helms H., Kuhn, A., Lambrecht, Z., Nicklass, D., Pregger, T. Reis, S. and S. Wenzel (2007) Umweltforschungsplan des Bundesministeriums für Umwelt, Naturschutz und Reaktorsicherheit; Luftreinhaltung: Maßnahmen zur weiteren Verminderung der Emissionen an NO$_x$, SO$_2$ und NMVOC in Deutschland (in German). UBA research report No. 205 42 221, 233 pp.

Thénard, L.-J. (1819) New results on the combination of oxygen with water. Annals of Philosophy **14**, 209–210; *Annales de Chimie et de Physique* **10**, 335.

Thiele, H. (1907) Einige Reaktionen im ultravioletten Lichte. *Berichte der Deutschen Chemischen Gesellschaft* **40**, 4914–4916.

Thomas, G. E. and K. Stamnes (1999) Radiative transfer in the atmosphere and ocean. Cambridge University Press, Cambridge (U.K.), 517 pp.

Thomas, J. K. (1965) Rates of reaction of the hydroxyl radical. *Transaction of the Faraday Society* **61**, 702–707.

Thomas, K., A. Volz-Thomas, D. Mihelcic, H. G. J. Smit and D. E. Kley (1998) On the exchange of NO$_3$ radicals with aqueous solutions. Solubility and sticking coefficient. *Journal of Atmospheric Chemistry* **29**, 17–43.

Thompson, A. M. (1992) The oxidizing capacity of the earth's atmosphere: Probable past and future changes. *Science* **256**, 1157–1165.

Thoning, K. W., P. P. Tans and W. D. Komhyr (1989) Atmospheric carbon dioxide at Mauna Loa Observatory, 2. Analysis of the NOAA/GMCC data, 1974–1985. *Journal of Geophysical Research* **94**, 8549–8565.

Thordarson, T. and S. Self (1996) Sulfur, chlorine and fluorine degassing and atmospheric loading by the Roza eruption, Columbia River Basalt Group, Washington, USA. *Journal of Volcanology and Geothermal Research* **74**, 49–73.

Thordarson, T., S. Self, N. Óskarsson, and T. Hulsebosch (1996) Sulfur, chlorine, and fluorine degassing and atmospheric loading by the 1783–1784 AD Laki (Skaftár Fires) eruption in Iceland, *Bulletin Volcanology* **58**, 205–225.

Thornton, D. C., A. R. Bandy, B. W. Blomquist, D. D. Davis and R. W. Talbot (1996) Sulfur dioxide as a source of condensation nuclei in the upper troposphere of the Pacific Ocean. *Journal of Geophysical Research* **101**, 1883–1890.

Thornton, J. A, J. P. Kercher, T. P. Riedel, N. L. Wagner, J. Cozic, J. S. Holloway, W. P. Dubé, G. M. Wolfe, P. K. Quinn, A. M. Middlebrook, B. Alexander and S. S. Brown (2010) A large atomic chlorine source inferred from mid-continental reactive nitrogen chemistry. *Nature* **464**, 271–274.

Thorpe, T. E. (1867) On the amount of carbonic acid contained in sea air. *Journal of Chemical Society* **20**, 189–199.

Thorsheim, P. (2006) Inventing pollution. Coal, smoke, and culture in Britain since 1800. Ohio University Press, Athens (USA), 307 pp.

Tian, A. (1911) Radiations decomposing water and the extreme ultraviolet spectrum of the mercury arc. *Comptes Rendus de l'Académie des Sciences* **152**, 1483–1483.

Tian, F., O. B. Toon, A. A. Pavlov and H. De Sterck (2005) A hydrogen-rich early earth atmosphere. *Science* **308**, 1014–1017.

Tie, X., A. Guenther and E. Holland (2003) Biogenic methanol and its impacts on tropospheric oxidants. *Geophysical Research Letters* **30**, 1881, doi:10.1029/2003GL017167.

Tie, X., R. Zhang, G. Brasseur and W. Lei (2004) Global NO_x Production by Lightning. *Journal of Atmospheric Chemistry* **43**, 61–74.

Timonen, H., S. Saarikoski, O. Tolonen-Kivimäki, M. Aurela, K. Saarnio, T. Petäjä, P. P. Aalto, M. Kulmala, T. Pakkanen and R. Hillamo (2008) Size distributions, sources and source areas of water-soluble organic carbon in urban background air. *Atmospheric Chemistry and Physics* **8**, 5635–5647.

Tissandier, G. (1879) Observations métérologiques en ballon. Gauthier-Villars, Paris, 51 pp.

Tobie G., O. Grasset, J. Lunine, Mocquet A. and C. Sotin (2005) Titan's orbit provides evidence for a subsurface ammonia-water ocean. *Icarus* **175**, 496–502.

Todres, Z. V. (2008) Ion-radical organic chemistry. Principles and applications, 2nd Edition, CRC Press, Boca Raton, FL (USA), 476 pp.

Tohjima, Y., Machida, T., Watai, T., Akama, I., Amari, T. and Y. Moriwaki (2005) Preparation of gravimetric standards for measurements of atmospheric oxygen and reevaluation of atmospheric oxygen concentration. *Journal of Geophysical Research* **110**, D11302, doi:10.1029/2004JD005595, 2005.

Tojima, S., S. Tobita and H. Shizuka (1999) Electron transfer from triplett 1.4–dimethoxybenzen to hydronium ion in aqueous solution. *Journal of Physical Chemistry.* **A103**, 6097–6105.

Tolstikhin, I. and J. Kramers (2008) The evolution of matter: From the Big Bang to the present day. Cambridge University Press, Cambridge, U.K., 532 pp.

Tomkeiev, S. I (1944) Geochemistry in the U.S.S.R. *Nature* **154**, 814–816.

Towe, K. M., D. Catling, K. Zahnle and C. McKay (2002) The problematic rise of Archean oxygen. *Science* **295**, 1419a.

Traube, M. (1882) Über Aktivierung des Sauerstoffs. *Berichte der Deutschen Chemischen Gesellschaft* **15**, pp. 222, 659, 2421, 2423.

Traube, M. (1887) Über die elektrolytische Entstehung des Wasserstoffhyperoxyds an der Kathode. *Sitzungsberichte der Akademie der Wissenschaften zu Berlin L* **60**, 1041–1050.

Trebs, I., B. Bohn, C. Amman, U. Rummel, M. Blumthaler, R. Königsstedt, F. X. Meixner, S. Fan and M. O. Andreae (2009) Relationship between the NO_2 photolysis frequency and the solar global irradiance. *Atmospheric Measurement Technique* **2**, 725–739.

Treffeisen, R., K. Grunow, D. Möller and A. Hainsch (2002) Quantification of source region influences on the ozone burden. *Atmospheric Environment* **36**, 3565–3582.

Trenberth, K. E., L. Smith, T. Qian, A. Dai and J. Fasullo (2007) Estimates of the global water budget and its annual cycle using observational and model data. *Journal of Hydrometeorology* **8**, 758–769.

Trick, S. (2004) Formation of nitrous acid on urban surfaces-a physical-chemical perspective. Doctoral Thesis, University Heidelberg, 290 pp.

Trimmer M., N. Risgaard-Petersen, J. C. Nicholls and P. Engström (2006) Direct measurement of anaerobic ammonium oxidation (anammox) and denitrification in intact sediment cores. Marine Ecology Progress Series **326**, 37–47.

Tsonis, A. A. (2007) An introduction to atmospheric thermodynamics. 2nd Edition. Cambridge University Press, Cambridge, U.K., 181 pp.

Turchi, C. S., D. F. Ollis (1990) Photocatalytic degradation of organic water contaminants: Mechanisms involving hydroxyl radical attack. *Journal of* Catalysis **122**, 178–192.

Turco, R. P., R. C. Whitten, O. B. Toon, J. B. Pollack and P. Hamill (1980) OCS, stratospheric aerosols and climate. *Nature* **283**, 283–285.

Turcotte, D. L. (1980) On the thermal evolution of the earth. *Earth and Planetary Science Letters* **48**, 53–58.

Turekian, K. K. (1968) Oceans. Prentice-Hall, New Jersey (USA), 1173 pp.

Turner, D. R. and K. A. Hunter, Eds. (2002) The biogeochemistry of iron in seawater. John Wiley & Sons, 410 pp.

Twomey, S. (1974) Pollution and the planetary albedo. *Atmospheric Environment* **8**, 1251–1256.

Twomey, S. (1991) Aerosols, clouds and radiation. Atmospheric Environment **25**, 2435–2442.

Tyndall, J. (1861) On the absorption and radiation of heat by gases and vapours, and on the physical connection of radiation, absorption, and conduction. *Philosophical Magazine, Series 4*, **22**, 169–194, 273–285.

Tyndall, J. (1870) On haze and dust. *Nature* **1**, 339–342.

UBA – Umweltbundesamt (Federal Environmental Agency), Germany.

Uematsu M., Toratani M., Kajino M., Narita Y., Senga Y. and T. Kimoto (2004) Enhancement of primary productivity in the western North Pacific caused by the eruption of the Miyake-jima Volcano. *Geophysical Research Letters* **31**, L06106.

Umlauft, F. (1891) Das Luftmeer. Die Grundzüge der Meteorologie und Klimatologie. A. Hartleben's Verlag, Wien, 488 pp.

Umschlag, T. and H. Herrmann (1999) The carbonate radical ($HCO_3\cdot/CO_3^{-\cdot}$) as a reactive intermediate in water chemistry: Kinetics and modelling. *Acta hydrochimica et hydrobiologica* **27**, 214–222.

Urey, H. C., L. H. Dawsey and F. O. Rice (1929) The absorption spectrum and decomposition of hydrogen peroxide by light. *Journal of the American Chemical Society* **51**, 1371–83.

Uri, N. (1952) Inorganic free radicals in solution. *Chemical Reviews* **50**, 375–454.

Vali, G. (2008) Repeatability and randomness in heterogeneous freezing nucleation. *Atmospheric Chemistry and Physics* **8**, 5017–5031.

Vallis, G. K. (2006) Atmospheric and oceanic fluid dynamics: Fundamentals and large-scale circulation. Cambridge University Press, Cambridge, U.K., 745 pp.

van Aardenne, J., F. Dentener, J. Olivier, C. K. Goldewijk and J. Lelieveld (2001) A 1°·1° resolution data set of historical anthropogenic trace gas emissions for the period 1890–1990. *Global Biogeochemical Cycles* **15**, 909–928.

Varghese, O. K., M. Paulose, T. J. LaTempa and C. A. Grimes (2009) High-rate solar photocatalytic conversion of CO_2 and water vapor to hydrocarbon fuels. *Nano Letters* **9**, 731–737.

Várhelyi, G. and G. Gravenhorst (1981) An attempt to estimate biogenic sulfur emissions into the atmosphere. *Idöjáras* **85**, 126–133.

Várhelyi, G. (1985) Continental and global sulfur budgets-I. Anthropogenic SO_2 emissions. *Atmospheric Environment* **19**, 1029–1040.

Vernadski, V. I. (1926) Biosphera (*in Russian*). Nautschnyi-chim.-techn. Izdateltsvo, Leningrad 48 pp. translated into French (*La Biosphére*, 1929), German (*Biosphäre*, 1930), and English (*The Biosphere*, 1998; a reduced version in English appeared in 1986).

Vernadski, V. I. (1944) Some words on noosphere (in Russian). Успехи соврем. биол. (uspechi sovrem. biol.) 18, 113–120.

Vernadsky, V. I. (1945) The biosphere and the noosphere. *Scientific American* 33, 1–12.

Vernadsky, V. I. (1998) The biosphere. Translated by D. B. Langmuir. opernicus, Springer-Verlag New York, 192 pp.

Vernier, J.-P., T. D. Fairlie, J. J. Murray, A. Tupper, C. Trepte, D. Winker, J. Pelon, A. Garnier, J. Jumelet, M. Pavolonis, A. H. Omar and K. A. Powell (2013) An advanced system to monitor the 3D structure of diffuse volcanic ash clouds. *Journal of Applied Meteorology and Climatology* **52**, 2125–2138.

Verschueren, K., Ed. (2009) Handbook of environmental data on organic chemicals. Four Volume set, Fifth Edition, John Wiley & Sons, 4358 pp.

Villemant, B., G. Boudon, S Nougrigat, S. Poteaux and A Michel (2003) Water and halogens in volcanic clasts: Tracers of degassing processes during Plinian and dome-building eruptions. In: Oppenheimer et al. (2003), pp. 63−77.

Vione, D., V. Maurino, C. Minero and E. Pelizzeti (2003) The atmospheric chemistry of hydrogen peroxide: A review. *Annali di chimica* **93**, 477−488.

Voldner E. C., Barrie L. A. and A. Sirois (1985) A literature review *of* dry deposition of oxides of sulphur and nitrogen with emphasis on long range transport modelling in North America. *Atmospheric Environment* **20**, 2101−2123.

Volkamer, R. L. T. Molina, M. J. Molina, T, Shirley, B. Terry H. William H. (2005) DOAS measurement of glyoxal as an indicator for fast VOC chemistry in urban air. *Geophysical Research Letters* **32**, L08806. doi:10.1029/2005GL022616.

Volkamer R., P. J. Ziemann and M. J. Molina (2009) Secondary organic aerosol formation from acetylene (C_2H_2): Seed effect on SOA yields due to organic photochemistry in the aerosol aqueous phase. *Atmospheric Chemistry and Physics* **9**, 1907−1928.

Volz-Thomas, A. et al. (1997) Photochemical ozone production rates at different TOR sites. In: Transport and chemical transformation of pollutants in the troposphere (EURO-TRAC). Vol 6: Tropospheric ozone research (Ed. Ø. Hov), Springer-Verlag, Berlin, pp. 65−94.

Wagener, K. (1979) The carbonate system of the ocean. In: The global carbon cycle. Scope 13 (Eds. B. Bolin, E. T. Degens, S. Kempe and P. Ketner), John Wiley & Sons, New York, pp. 251−258.

Wagner, H. (1939) Die Körperfarben. Chemie in Einzeldarstellungen, Vol. XIII (Ed. J. Schmidt), Wissenschaftliche Verlagsgesellschaft, Stuttgart, 688 pp.

Wai, K.-M. and P. A. Tanner (2004) Wind-dependent sea salt aerosol in a Western Pacific coastal area. *Atmospheric Environment* **38**, 1167−1171.

Walden, P. (1941) Geschichte der organischen Chemie seit 1880. Zweiter Band zu C. Graebe: Geschichte der organischen Chemie. Springer-Verlag, Berlin, 946 pp.

Waldman, J. M., J. W. Munger, D. L. Jacob and M. R. Hoffmann (1985) Chemical characterization of stratus cloudwater and its role as a vector for pollutant deposition in a Los Angeles pine forest. *Tellus* **37B**, 91−108.

Walker, J. (1900) Estimation of atmospheric carbon dioxide. *Journal of Chemical Society, Transactions* **77**, 1110−1114.

Walker, J. C. C. (1977) Evolution of the atmosphere. Macmillan Publ. Co., New York, 318 pp.

Walker, S. J., R. F. Weiss and P. K. Salameh (2000) Reconstructed histories of the annual mean atmospheric mole fractions for the halocarbons CFC-11, CFC-12, CFC-113 and carbon tetrachloride. *Journal Geophysical Research* **105**, 14285−14296.

Wallace, J. M. and P. V. Hobbs (2006) Atmospheric science: An introductory survey. 2nd edition. International Geophysics Series Vol. 92, Academic Press, 484 pp.

Wallerius, J. G. (1750) Mineralogie, oder Mineralreich von Ihm eingeteilt und beschrieben. Ins Deutsche übersetzt von Johann Daniel Denso. G. Nicolai, Berlin, 600 pp. [First German edition of this important mineralogical classic first published in Swedish in 1747].

Wallerius, J. G. (1770) The natural and chemical elements of agriculture. John Bell, London, 200 pp. [First English edition of agriculturae fundamenta chemica (1761), which was published as a dissertation at Uppsala with Wallerius as praeses and Count Gustaf Adolph Gyllenborg as respondens].

Walling, C. (1957) Radicals in solution. John Wiley & Sons, New York, 631 pp.

Walter, K. M., L. C. Smith and F. S. Chapin (2007) Methane bubbling from the northern lakes: Present and future contributions to the global methane budget. *Philosophical Transaction of the Royal Society* **A365**, 1657−1676.

Wang, Y., D. J. Jacob and J. A. Logan (1998) Global simulation of tropospheric O_3-NO_x-hydrocarbon chemistry. 3. Origin of tropospheric ozone and effects of nonmethane hydrocarbons. *Journal of Geophysical Research* **156**, 148−227.

Wanninkhof, R. (1992) Relationship between wind speed and gas exchange over the ocean. *Journal of Geophysical Research* **97**, 7373−7382.

Warburg, E. (1914) Über den Energieumsatz bei photochemischen Vorgängen in Gasen. *Sitzungsberichte der Königlich Preussischen Akademie der Wissenschaften*. Phys.-math. Klasse II, pp. 872−885.

Ward, D. E., R. A. Susott, J. B. Kauffman, R. E. Babbit, D. L. Cummings,. B. Dias, B. N. Holben, Y. J. Kaufman, R. A. Rasmussen and A. W. Setzer (1992) Smoke and fire characteristics for cerrado and deforestation burns in Brazil-Base-B Experiment. *Journal of Geophysical Research* **97**, 14601−14619.

Wardman, P. (1989) Reduction potentials of one-electron couples involving free radicals in aqueous solution. *Journal of Physical Chemical Reference Data* **18**, 1637−1755.

Warneck, P. and C. Wurzinger (1988) Product yields for the 305-nm photodecomposition of NO_3^- in aqueous solution. *Journal of Physical Chemistry* **92**, 6278−6283.

Warneck, P. (1988) Chemistry of the natural atmosphere. Academic Press, San Diego, 753 pp.

Warneck, P. (1989) Sulfur dioxide in rain clouds: gas-liquid scavenging efficiencies and wet deposition rates in the presence of formaldehyde. *Journal of Atmospheric Chemistry* **8**, 99−117.

Warneck, P. (2000) Chemistry of the natural atmosphere. Academic Press, New York, 969 pp.

Warneck, P. (2003) In-cloud chemistry opens pathway to the formation of oxalic acid in the marine atmosphere. *Atmospheric Environment* **37**, 2423−2427.

Wasai, K., G. Kaptay, K. Mukai and N. Shinozaki (2007) Modified classical homogeneous nucleation theory and a new minimum in free energy change: 1. A new minimum and Kelvin equation. *Fluid Phase Equilibria* **254**, 67−74.

Waterman, L. S. (1983) Comments on "The Montsouris series of carbon dioxide concentration measurements, 1877−1910 by Stanhill". *Climate Change* **5**, 413−415.

Watson, W. W. (1924) Spectrum of water vapour. *Astrophysical Journal* **60**, 145−158.

Watts, S. F. (2000) The mass budgets of carbonyl sulfide, dimethyl sulfide, carbon disulfide and hydrogen sulfide. *Atmospheric Environment* **34**, 761−779.

Wayne, R. P., I. Barnes, J. P. Burrows, C. E. Canosa-Mas, J. Hjorth, G. Le Bras, G. K. Moortgat, D. Perner, G. Poulet, G. Restelli and H. Sidebottom (1991) The nitrate radical: Physics, chemistry, and the atmosphere. *Atmospheric Environment* **25A**, 1−203.

Wayne, R. P. (1994) Singlet oxygen in the environmental sciences. *Research on Chemical Intermediates* **20**, 395−422.

Wayne, R. P. (1996) Chemistry of the atmospheres. Clarendon Press, Oxford, 447 pp.

Wayne, R. P. (2000) Chemistry of atmospheres. An introduction to the chemistry of the atmospheres of earth, the planets, and their satellites. Third Edition. Oxford University Press, 808 pp.

WCI (2005) The coal resources: A comprehensive overview. World Coal Institute, London, 49 pp.

Weast, R. C., Ed. (1980) CRC handbook of chemistry and physics. New York, CRC Press.

Wedepohl, K. H. (1996) The composition of the continental crust. *Geochimica et Cosmochimica Acta* **59**, 1217−1232.

Weingartner, E., M. Gysel and U. Baltensperger (2002) Hygroscopicity of aerosol particles at low temperature. Part I: A new G-TDMA system: Setup and first applications. *Environmental Science and Technology* **36**, 55−62.

Weinstein-Llyod J. B., J. H. Lee, P. H. Daum, L. I. Kleinman, L. J. Nunnenmacker aund S. R. Springston (1998) Measurements of peroxides and related species during the 1995 summer intensive of the southern oxidants study in Nashville, Tennessee. *Journal of Geophysical Research* **103**, 23361−23373.

Weiss, J. (1935) Investigations on the radical HO_2 in solution. *Transactions of the Faraday Society* **31**, 668−681.

Weiss, J. (1944) Radiochemistry of aqueous solutions. *Nature* **153**, 748−750.

Weiss, J. and C. W. Humphrey (1949) reaction between hydrogen peroxide and iron salts. *Nature* **163**, 691.

Weiss, R. F. (1970) The solubility of nitrogen, oxygen, and argon in water and seawater. *Deep-Sea Research* **17**, 721–735.

Wells, M., K. N. Bower, T. W. Choularton, J. N. Cape, M. A. Sutton, R. L. Storeton-West, D. Fowler, A. Wiedensohler, H.-C. Hansson, B. Svenningsson, E. Swietlicki, M. Wendisch, B. Jones, G. Dollard, S. Fuzzi, K. Acker, W. Wieprecht, M. Preiss, B. G. Arends, S. Pahl, A. Berner, C. Kruisz, P. Laj, M. C. Facchini and S. Fuzzi (1997) The reduced nitrogen budget of an orographic cloud. *Atmospheric Environment* **31**, 2599–2614.

Wells, W. Ch. (1814) Essay on dew and several apearances connected with it. Longmans, London (reprint 1866) 120 pp. + appendix (by R. Strachan) 32 pp.

Wendisch, M. and P. Yang (2012) Theory of atmospheric radiations transfer. Wiley, 366 pp.

Went, F. W. (1955) Air pollution. *Scientific American.* **192**, 63–72.

Went, F. W. (1960) Blue hazes in the atmosphere. *Nature* **187**, 641–643.

Werf, G. R. van der, J. T. Randerson, L. Giglio, G. J. Collatz, P. S. Kasibhatla and A. F. Arellano Jr. (2006) Interannual variability of global biomass burning emissions from 1997 to 2004. *Atmospheric Chemistry and Physics Discussion* **6**, 3423–3441.

Werner, C. (2007) Compilation of a global N_2O emission inventory for tropical rainforest soils using a detailed biogeochemical model. Report from Forschungszentrum Karlsruhe FZKA 7345 (URL: http://www.freidok.uni-freiburg.de/volltexte/3350/).

Wesely, M. L. and B. B. Hicks (1977) Some factors that affect the deposition rates of sulfur dioxide and similar gases on vegetation. *Journal of Air Pollution Control Association* **27**, 1110–1116.

Weseley, M. L. (1989) Parametrizations of surface resistance to gaseous dry deposition in regional-scale, numerical models. *Atmospheric Environment* **23**, 1293–1304.

Wesely, M. L. and B. B. Hicks (2000) A review of the current status of knowledge on dry deposition. *Atmospheric Environment* **34**, 2261–2282.

Wexler, H. (1952) Volcanoes and world climate. *Scientific American* **186**, 74–80.

Whalen, S. C. (2005) Biogeochemistry of methane exchange between natural wetlands and the atmosphere. *Environmental Engineering and Science* **22**, 73–94.

Whitney, W. R. (1903) The corrosion of iron. *Journal of the American Chemical Society* **25**, 394–406.

WHO (1988) Phosphine and selected metal phosphides. IPCS, Environmental Health Criteria 73, World Health Organization, Geneva.

WHO (2006) Air quality guidelines. Global update 2005. Particulate matter, ozone, nitrogen oxides and sulphur dioxide.

Wiberg, N., A. Holleman, E. Wiberg, Eds. (2001) Hollemann-Wiberg's inorganic chemistry. Academic Press, 1924 pp.

Wicke, E., M. Eigen and T. Ackermann (1954) Über den Zustand des Protons (Hydroniumions) in wäßriger Lösung. *Zeitschrift für Physikalische Chemie (Neue Folge)* **1**, 343–364.

Wickramasinghe, N. C. (1974) Formaldehyde polymers in interstellar space. *Nature* **252**, 462–463.

Wickramasinghe, N. C. (2004) The universe: A cryogenic habitat for microbial life. *Cryobiology* **48**, 113–125.

Widory, D. and M. Javoy (2003) The carbon isotope composition of atmospheric CO_2 in Paris. *Earth and Planetary Science Letters* **215**, 289–298.

Wiedensohler, A., H.-C. Hansson, D. Orsini, M. Wendisch, F. Wagner, K. N. Bower, T. W. Choularton, M. Wells, K. Acker, W. Wieprecht, M. C. Facchini, J. A. Lind, S. Fuzzi, B. G. Arends and M. Kulmala (1997) Night-time formation and occurrence of new particles associated with orographic clouds. *Atmospheric Environment* **31**, 2545–2559.

Wieland, H. and W. Franke (1929) Über den Mechanismus der Oxydationsvorgänge. XVII. Über das Verhältnis der Oxydationsgeschwindigkeiten von molekularem Sauerstoff und Hydroperoxyd. *Annalen der Chemie* **473**, 289–300.

Wigand, A. (1913) Zur Erkenntnis der atmosphärischen Trübung. *Meteorologische Zeitschrift* **30**, 1−2.

Wigand, A. (1919) Die vertikale Verteilung der Kondensationskerne in der freien Atmosphäre. *Annalen der Physik* **364**, 689−741.

Wigand, A. (1930) Das atmosphärische Aerosol. *Die Naturwissenschaften* **18**, 31−33.

Wilks, S. S. (1959) Carbon monoxide in green plants. *Science* **129**, 964−966.

Williams, E., R. Battino and R. J. Wilcock (1977) Low-pressure solubility of gases in liquid water. *Chemical Review* **77**, 219−262.

Williams, J. E., F. J. Dentener and A. R. van den Berg (2002) The influence of cloud chemistry on HO_x and NO_x in the moderately polluted marine boundary layer: a 1-D modelling study. *Atmospheric Chemistry and Physics* **2**, 39−54.

Williams, J., V. Gros, E. Atlas, K. Maciejczyk, A.Batsaikhan, H. F. Schöler, C. Forster, B. Quack, N. Yassaa, R. Sander and R. Van Dingenen (2007) Possible evidence for a connection between methyl iodide emissions and Saharan dust. *Journal of Geophysical Research* **112**, doi:10.1029/2005JD006702.

Williams, J. and R. Koppmann (2007) Volatile organic compounds in the atmosphere: An overview. In: Koppmann (2007), pp. 1−32.

Willis, A. J. (1997) The ecosystem: An evolving concept viewed historically. *Functional Ecology* **11**, 268−271.

Wilson, E. H. and S. K. Atreya (2000) Sensitivity studies of methane photolysis and its impact on hydrocarbon chemistry in the atmosphere of Titan. *Journal of Geophysical Research* **105**, 20263−20273.

Winer, A. M., M. C. Dodd, D. R. Fitz, P. R. Miller, E. R. Stephens, K. Neisess, M. Meyers, D. E. Brown, and C. W. Johnson (1982) Assembling a vegetative hydrocarbon emission inventory for the California South Coast Air Basin: Direct measurement of emission rates, leaf biomass and vegetative distribution, In: For presentation at the 75th Annual Meeting of the Air Pollution Control Association, New Orleans, 20 pp.

Wislicenus, H., O. Schwarz, H. Sertz, F. Schröder, F. Müller and F. Bender (1916) Experimentelle Rauchschäden. Versuche über die äußeren und inneren Einwirkungen von Ruß, sauren Nebeln und stark verdünnten Gasen auf die Pflanze. In: Sammlung von Abhandlungen über Abgase von Rauchschäden, Heft 10 (Ed. H. Wislicenus), Paul Parey, Berlin.

WMO (1986) World Meteorological Organization. Atmospheric Ozone 1985. Assessment of our Understanding of the Processes Controlling its Present Distribution and Change, WMO Global Ozone Research and Monitoring Project, Report No. 16, WMO, Geneva, Switzerland.

WMO (1992) World Meteorological Organization. International meteorological vocabulary 2nd ed., WMO-No. 182, Geneve, 784 pp.

WMO (2009) World Data Centre for Greenhouse Gases, Japan Meteorological Agency, Tokyo, http://gaw.kishou.go.jp/wdcgg/products/summary/sum33/16_04ch4.pdf.

Wolf, G. M. E. and M. Hidy (1997) Aerosols and climate: anthropogenic emissions and trends. *Journal of Geophysical Research* **102**, 11113−11121.

Wolff, E. and R. Spahni (2007) Methane and nitrous oxide in the ice core record. *Philosophical Transaction of the Royal Society* **A365**, 1775−1792.

Wood, B. J. and A. N. Halliday (2005) Cooling of the earth and core formation after the giant impact. *Nature* **437**, 1345−1348.

Wood, B. J., M. J. Walter and J. Wade (2006) Accretion of the earth and segregation of its core. *Nature* **441**, 825−833.

Worldwatch Institute (2006) Biofuels for transportation. Global potential and implications for sustainable agriculture and energy in the 21st century. Washington, D.C. (USA), 38 pp.

Wu, C. Y. R., D. L. Judge and J. K. Chen (1983) Atomic nitrogen emissions from photodissociation of N_2. *Journal of Geophysical Research* **88**, 2163−2169.

Wu, G., Y. Katsumura, Y. Muroya, M. Lin and T. Morioka (2002) Temperature dependence of carbonate radical in $NaHCO_3$ and Na_2CO_3 solutions: Is the radical a single anion? *Journal of Physical Chemistry A* **106**, 2430−2437.

Wu, J. C. S. (2009) Photocatalytic reduction of greenhouse gas CO_2 to fuel. *Catalysis Surveys from Asia* **13**, 1571−1013.

Wuebbles, D. J., J. Edmonds, J. Dignon, W. Emanual, D. Fisher, R. Gammon, R. Hangebrock, R. Harriss, M. A. K. Khalil, J. Spence and T. Thompson (1997) Emissions and budgets of radiatively important atmospheric constituents: In: Engineering response to global climate change: Planning a research and development agenda (Ed. R. G. Watts), CRC Lewis Publishers, Boca Raton, FL (USA).

Xiao, Y., D. J. Jacob and S. Turquety (2007) Atmospheric acetylene and its relationship with CO as an indicator of air mass age. *Journal of Geophysical Research* **112**, D12305, doi:10.1029/2006JD008268.

Xiu, G., D. Zhang, J. Chen, X. Huang, Z. Chen, H. Guo and J. Pan (2004) Characterization of major water-soluble inorganic ions in size-fractionated particulate matters in Shanghai campus ambient air. *Atmospheric Environment* **38**, 227−236.

Yao, X., M. Fang and C. C. Chan (2003) The size dependence of chloride depletion in fine and coarse sea-salt particles. *Atmospheric Environment* **37**, 743−751.

Yienger, J. and H. Levy (1995) Global inventory of soil-biogenic NO_x emissions. *Journal of Geophysical Research* **100**, 11447−11464.

Yokelson, R. J., T. Karl, P. Artaxo, D. R. Blake, T. J. Christian, D. W. T. Griffith, A. Guenther and W. M. Hao (2007) The tropical forest and fire emissions experiment: Overview and airborne fire emission factor measurements. *Atmospheric Chemistry and Physics Discussion* **7**, 6903−6958.

Yonemura, S., L. Sandoval-Soto, J. Kesselmeier, U. Kuhn, M. Von Hobe, D. Yakir and S. Kawashima (2005) Uptake of carbonyl sulfide (COS) and emission of dimethyl sulfide (DMS) by plants. *Phyton* **45**, 17−24.

Yttri, K. E., W. Aas, A. Bjerke, J. N. Cape, F. Cavalli, D. Ceburnis, C. Dye, L. Emblico, M. C. Facchini, C. Forster, J. E. Hanssen, H. C. Hansson, S. G. Jennings, W. Maenhaut, J. P. Putaud and K. Tørseth (2007) Elemental and organic carbon in PM_{10}: a one year measurement campaign within the European Monitoring and Evaluation Programme EMEP. *Atmospheric Chemistry and Physics* **7**, 5711−5725.

Yurganov, L. (2000) Carbon monoxide inter-annual variations and trends in the Northern Hemisphere: Role of OH. *IGAC Newsletter* **21**, 15−18.

Yves, M. (2006) The hydrogen bond and the water molecule: The physics and chemistry of water, aqueous and bio-media. Elsevier Science, 332 pp.

Yvon-Lewis, S. A., D. B. King, R. Tokarczyk, K. D. Goodwin, E. S. Saltzman and J. H. Butler (2004) Methyl bromide and methyl chloride in the Southern Ocean. *Journal of Geophysical Research* **109**, C02008, doi:10.1029/2003JC001809.

Zalasiewicz, J., M. Williams, A. Smith, T. L. Barry, A. L. Coe, P. R. Bown, P. Brenchley, D. Cantrill, A. Gale, P. Gibbard, F. J. Gregory, M. W. Hounslow, A. C. Kerr, P. Pearson, R. Knox, J. Powell, C. Waters, J. Marshall, M. Oates, P. Rawson and P. Stone (2008) Are we now living in the Anthropocene? *GSA Today* **18**, 4−8.

Zdunkowski, W. and A. Bott (2004) Thermodynamics of the atmosphere: A course in theoretical meteorology. Cambridge University Press, Cambridge, U.K., 266 pp.

Zdunkowski, W., Th. Trautmann and A. Bott (2007) Radiation in the atmosphere: A course in theoretical meteorology. Cambridge University Press, Cambridge, U.K., 482 pp.

Zeebe, R. and J.-P. Gattuso (2006) Marine carbonate chemistry. In: Encyclopedia of earth (Ed. C. J. Cleveland), Washington, D.C.: Environmental Information Coalition, National Council for Science and the Environment, Published in the Encyclopedia of Earth September 21, 2006; Retrieved November 29, 2009]. http://www.eoearth.org/article/Marine_carbonate_chemistry.

Zeman, F. S. and K. S. Lackner (2004) Capturing carbon dioxide directly from the atmosphere. *World Resource Review* **16**, 62–68.

Zeman, F. S. (2007) Energy and material balance of CO_2 capture from ambient air. *Environmental Science and Technology* **41**, 7558–7563.

Zeman, F. S., and D. W. Keith (2008) Carbon neutral hydrocarbons. *Philosophical Transaction of the Royal Society* **A366**, 3901–3918.

Zhang, J. and F. Millero (1993) The products from the oxidation of H_2S in seawater. *Geochimica et Cosmochimica Acta* **57**, 1705–1718.

Zhang, L., S. Gong, J. Padro and L. Barrie (2001) A size-segregated particle dry deposition scheme for an atmospheric aerosol module. *Atmospheric Environment* **35,** 549–560.

Zhang, P., F. Liang, G. Yu, Q. Chen and W. Zhu (2003) A comparative study on decomposition of gaseous toluene by O_3/UV, TiO_2/UV and O_3/TiO_2/UV. *Journal of Photochemistry and Photobiology A: Chemistry.* **156**, 189–194.

Zhang, R., X. Tie and D. W. Bond (2003) Impacts of anthropogenic and natural NO_x sources over the U.S. on tropospheric chemistry. *Proceedings of the National Academy of Sciences of the United States of America* **100**, 1505–1509.

Zhang, R., L. Wang, A. F. Khalizov, J. Zhao, J. Zheng, R. L. McGraw and L. T. Molina (2009) Formation of nanoparticles of blue haze enhanced by anthropogenic pollution. *Proceedings of the National Academy of Sciences of the United States of America* **106**, 17650–17654.

Zhang, N., X. Zhou, P. B. Shepson, H. Gao, M. Alaghmand and B. Stirm (2009) Aircraft measurement of HONO vertical profiles over a forested region, *Geophysical Research Letters* **36**, L15820, doi:10.1029/2009GL038999.

Zhao, D., J. Xiong, Y. Yu and W. H. Chan (1988) Acid rain in Southwestern China. *Atmospheric Environement* **22**, 349–358.

Zhou, X. and K. Mopper (1997) Photochemical production of low-molecular-weight carbonyl compounds in seawater and surface microlayer and their air-sea exchange. *Marine Chemistry* **56**, 201–213.

Zhou, X., K. Civerolo, H. Dai, G. Huang, J. Schwab and K. Demerjian (2002) Summertime nitrous acid chemistry in the atmospheric boundary layer at a rural site in New York State. *Journal of Geophysical Research* **107**, 4590, doi:10.1029/2001JD001539.

Zhou, X., H. Gao, Y. He, G. Huang, S. B. Bertman, K. Civeralo and J. Schwab (2003) Nitric acid photolysis on surfaces in low-NO_x environments: Significant atmospheric implications. *Geophysical Research letters* **30**, 2217, doi:10.1029/2003GL018620.

Zhu, R., D. Glindemann, D. Kong, L. Sun, J. Geng and X. Wang (2007) Phosphine in the marine atmosphere along a hemispheric course from China to Antarctica. *Atmospheric Environment* **41**, 1567–1573.

Zhuang, H., C. K. Chan, M. Fang, and A. S. Wexler (1999) Formation of nitrate and non-seasalt sulfate on coarse particles. *Atmospheric Environment* **33**, 4223–4233.

Zhuang, Q., Y. Lu and M. Chen (2012) An inventory of global N_2O emissions from the soils of natural terrestrial ecosystems. *Atmospheric Environment* **47**, 66–75.

Zierath, R. (1981) Inhaltsstoffe atmosphärischer Niederschläge und ihr Einfluß auf die Sicker- und Grundwasserbeschaffenheit am Beispiel ausgewählter Gebiete (in German). Doctoral Thesis, Technical University Dresden.

Zimbrick, J. D. (2002) Radiation chemistry and the Radiation Research Society: A history from the beginning. *Radiation Research* **158**, 127–140.

Zimmermann, P. R. (1979) Determination of emission rates of hydrocarbons from indigenous species of vegetation in the Tampa/St. Petersburg Florida area. Appendix C, Tampa Bay Area Photochemical Oxidant Study, EPA 904/9-77, U.S. E.P.A., Research Triangle Park, NC.

Zimmermann, P. R., J. P. Greenberg, S. O. Wandiga and P. J. Crutzen (1982) Termites: A potentially large source of atmospheric methane, carbon dioxide, and molecular hydrogen. *Science* **218**, 563–565.

Zinner, E., S. Amari, B. Wopenka and R. S. Lewis (1995) Interstellar graphites in meteorites: Isotopic compositions and structural properties of single graphite grains from Murchison. *Meteoritics* **30**, 209−226.

Zobrist, B., C. Marcolli, T. Knoop, B. P. Luo, D. M. Murphy, U. Lohmann, A. A. Zardini, U. K. Krieger, T. Corti, D. J. Cziczo, S. Fueglistaler, P. K. Hudon, D. S. Thomson and T. Peter (2006) Oxalic acid as a heterogeneous ice nucleus in the upper tropopehere and ist indirect aerosl effect. *Atmospheric Chemistry and Physics* **6**, 3115−3129.

Zobrist, J. (1987) Methoden zur Bestimmung der Azidität in Niederschlagsproben. VDI Berichte 608, Düsseldorf, pp. 401−420.

Zolensky, M. E., J. R. Bodnar, E. K. Gibson Jr., L. E. Nyquist, Y. Reese, Chi-Yu Shih and H. Wiesmann (1999) Asteroidal water within fluid inclusion-bearing Halite in an H5 Chondrite, Monahans (1998). *Science* **285**, 1377−1379.

Zumdahl, St. S. (2009) Chemical principles. 7th Ed. Houghton Mifflin Company, Boston, 1073 pp.

Zuo, Y. and J. Hoigné (1992) Formation of hydrogen peroxide and depletion of oxalic acid in atmospheric water by photolysis of iron(III)-oxalato complexes. *Environmental Science and Technology* **26**, 1014−1022.

Zuo, Y. and Y. Deng (1999) Evidence for lightning induced production of hydrogen peroxide during thunderstorms. *Geochimica et Cosmochimica Acta* **63**, 3451−3455.

Author Index

Subject Index